3 Air
Coast Guard
Department
of Defense
Marine Corps

The
1998
GUIDE to the
EVALUATION of
EDUCATIONAL
EXPERIENCES in the
ARMED SERVICES

AMERICAN COUNCIL ON EDUCATION   One Dupont Circle • Washington, D. C. 20036-1193

PRODUCTION STAFF FOR THE 1998 GUIDE

Military Evaluations Program

Eugene Sullivan, Director
Judith Cangialosi, Assistant Director for Military Occupation Evaluations
Penelope West Suritz, Assistant Director for Military Course Evaluations
Gwendolyn L. Dozier, Program Coordinator
DoRita Alford, Staff Assistant

The material in the 1998 edition of the *Guide to the Evaluation of Educational Experiences in the Armed Services* is not copyrighted.

The work reported or presented herein was performed pursuant to a contract with the Defense Activity for Non-Traditional Education Support (DANTES) on behalf of the Department of Defense (Contract N00140-97-C-H168). However, the opinions expressed herein do not necessarily reflect the position or policy of the U.S. Department of Defense, and no official endorsement by the U.S. Department of Defense should be inferred.

Library of Congress Cataloging in Publication Data:

American Council on Education.
    Guide to the evaluation of educational experiences
    in the armed services.
ISBN 1-57356-104-5

Contents

Foreword	vii
How to Find and Use Course Exhibits	ix
Sample Course Exhibit	xi
How to Find and Use Coast Guard Occupation Exhibits	xiii
Sample Coast Guard Rating Exhibit	xv
How to Find and Use Coast Guard Warrant Officer Exhibits	xvii
Sample Coast Guard Warrant Officer Exhibit	xix
How to Find and Use Marine Corps MOS Exhibits	xxi
Sample Marine Corps Enlisted MOS Exhibit	xxiii
Questions and Answers	xxv
Awarding Credit for Extrainstitutional Learning	xxxiii
Transfer and Award of Credit	xxxv
Elements of a Model Policy on Awarding Credit for Extrainstitutional Learning	xxxix
Air Force Course Exhibits	1
Coast Guard Course Exhibits	81
Department of Defense Course Exhibits	127
Marine Corps Course Exhibits	153
Coast Guard Aviator Exhibits	209
Coast Guard Enlisted Ratings Exhibits	213
Coast Guard Warrant Officer Exhibits	247
Marine Corps Enlisted MOS Exhibits	253
Appendix A: The Evaluation Systems	A–1
Appendix B: Sample Military Records	B–1
Occupation Title Index	C–1
Keyword Index	D–1
Course Number Index	E–1
Request for Course Recommendation	F–1
Request for Coast Guard Rating and Warrant Officer Exhibits	G–1
Request for Marine Corps MOS Exhibits	H–1

American Council on Education

Board of Directors

Executive Committee

Lois DeFleur, *Chair*

John A. DiBiaggio, *Vice Chair/Chair Elect*

Michele Tolela Myers, *Immediate Past Chair*

Vernon O. Crawley, *Secretary*

Edward B. Fort

Freeman A. Hrabowski, III

Kenneth A. Shaw

Elisabeth Zinser

Stanley O. Ikenberry, President

Class of 1998

Raymond C. Bowen, President, LaGuardia Community College

Edward B. Fort, Chancellor, North Carolina Agricultural & Technical State University

Janet L. Holmgren, President, Mills College

Martin C. Jischke, President, Iowa State University

Steve S. Koblik, President, Reed College

Michele Tolela Myers, President, Denison University

Eduardo J. Padron, President, Miami-Dade Community College

Elisabeth Zinser, Chancellor, University of Kentucky

Class of 1999

Vernon O. Crawley, President, Moraine Valley Community College

Lois B. DeFleur, President, State University of New York at Binghamton

John V. Lombardi, President, University of Florida

Walter E. Massey, President, Morehouse College

Anne S. McNutt, President, Technical College of the Lowcountry

Miguel A. Nevarez, President, University of Texas-Pan American

Kenneth A. Shaw, President/Chancellor, Syracuse University

Julianne Still Thrift, President, Salem College

Class of 2000

Michael F. Adams, President, University of Georgia

Robert M. Berdahl, Chancellor, University of California

Philip R. Day, Jr., President, Daytona Beach Community College

John A. DeBiaggio, President, Tufts Uninversity

Vera King Farris, President, Richard Stockton College of New Jersey

Freeman A. Hrabowski, III, President, University of Maryland

Neil Rudenstine, President, Harvard University

William Segura, Chancellor, Texas State Technical College System

Elected Officers of Associations—Ex Officio For Three-Year Terms:

Howard W. Eickhoff, President, The College of New Jersey, *Association of American Colleges & Universities*

Augustine P. Gallego, Chancellor, San Diego Community College District, *American, Association of Community Colleges*

Ed M. Elliott, President, Central Missouri State University, *American Association of State College and Universities*

F. Patrick Ellis, FSC, President, The Catholic University of America, *Association of American Universities*

Karen M. Kennelly, CSJ, President, Mount St. Mary's College, *Association of Catholic Colleges & Universities*

John P. Schlegel, SJ, President, University of San Francisco, *Association of Jesuit Colleges & Universities*

Harold M. Kolenbrander, President, Mount Union College, *Council of Independent Colleges*

National Association for Equal Opportunity in Higher Education
(to be announced)

Ann H. Die, President, Hendrix College, *National Association of Independent Colleges & Universities*

Peter McPherson, President, Michigan State University, *National Association of State Universities & Land-Grant Colleges*

Elected Officers of Associations—Ex Officio For One-Year Terms:

Allen D. Glenn, Dean, University of Washington, *American Association of Colleges for Teacher Education*

Sheila Trice Bell, Executive Director/Chief Executive Officer, *National Association of College and University Attorneys*

James E. Morley, Jr., President, National Association of College and University Business Officers, *Washington Higher Education Secretariat*

Executive Secretary
Irene L. Gomberg
American Council on Education

Commission on Adult Learning and Educational Credentials of the American Council on Education

Franklin C. Ashby, Vice President for Instruction, Dale Carnegie & Associates

Luke Baldwin, Provost, Lesley College

Zerrie D. Campbell, President, Malcolm X College

Robert L. Caret, President, San Jose State University

Glenda McGaha Curry, President, Troy State University in Montgomery

Lawrence A. Davis, Jr., President, University of Arkansas-Pine Bluff

Robert G. Elkins, Manager, Training Performance Improvement Team, Union Pacific Railroad Company

Jerry Evans, Assistant Executive Director for Programs, Institute for Career Development, Inc.

Dennis M. Faber, Director, Resource Center, Essex Community College

Leon E. Flancher, Associate Vice President and Chief Operating Officer, Embry-Riddle Aeronautical University

Grace Ann Geibel, President, Carlow College

Bonnie Gordon, Vice President for College Relations, Ithaca College

Jerome Greene, Jr., Chancellor, Southern University at Shreveport

James W. Hall, Chancellor, Antioch University

Merle W. Harris, President, Charter Oak State College

Sharon Y. Hart, President, Middlesex Community-Technical College

Edward Hernandez, Chancellor, Rancho Santiago Community College

Edison O. Jackson, President, Medgar Evers College

Mary S. Knudten, Dean, University of Wisconsin-Waukesha

Margaret Lee, President, Oakton Community College

E. Timothy Lightfield, President, Prarie State College

Donald J. MacIntyre, President, Fielding Institute

Sigfredo Maestas, President, Northern New Mexico Community College

Roberto Marrero-Corletto, Chancellor, University of Puerto Rico-Humacao

Byron N. McClenney, President, Community College of Denver

John W. Moore, President, Indiana State University

Gregory S. Prince, Jr., President, Hampshire College

Leslie N. Purdy, President, Coastline Community College

Allan Quigley, Associate Professor, Adult Education Department, St. Francis Xavier University

Richard Rush, President, Mankato State University

Michael Sheeran, S.J., President, Regis University

Portia Holmes Shields, President, Albany State University

L. Dennis Smith, President, University of Nebraska System

James J. Stukel, President, University of Illinois Central Administration

Ronald Taylor, President, DeVry Institute

David Voight, Director, Small Business Center, US Department of Commerce

David J. Ward, Sr., Vice President for Academic Affairs, University of Wisconsin System

Craig D. Weidemann, Graduate Dean, College of Graduate & Extension Education, Towson State University

Larry L. Whitworth, President, Tidewater Community College

Staff Officer
Susan Robinson, Vice President and Director, The Center for Adult Learning and Educational Credentials
American Council on Education

Foreword

For more than a half century the *Guide to the Evaluation of Educational Experiences in the Armed Services* has been the standard reference work for recognizing learning acquired in military life. Since 1942, the American Council on Education has worked cooperatively with the U.S. Department of Defense, the armed services and the U.S. Coast Guard in helping hundreds of thousands of individuals earn academic credit for learning achieved while serving their country.

The ACE *Guide* evaluation system enables students to apply their experience, training, and expertise to their degree work at colleges and universities. While this is a sound educational practice, it is also an efficient use of financial and educational resources.

Special recognition must be paid to the many faculty members who have served as subject-matter experts on ACE evaluation teams and to their institutions for their wholehearted cooperation in this effort. For all this generous support and assistance, we are deeply indebted.

We are pleased to commend this *Guide* to you as you work with servicemembers and veterans in helping them to integrate ACE *Guide* recommended credit into their degree programs.

Stanley O. Ikenberry
President
American Council on Education

How to Find and Use Course Exhibits

This volume contains recommendations for formal courses offered by the Air Force, the Coast Guard, the Marine Corps, and the Department of Defense with exhibit dates of 1/90 and later.

The instructions that follow provide a step-by step procedure for finding and using the exhibits and recommendations. Readers unfamiliar with ACE evaluation procedures should read Appendix A.

Step 1

Have the applicant complete a Request for Course Recommendation form.

A *Request for Course Recommendation* form appears at the back of this volume. It is to be reproduced and should always be filled out by the applicant, using the information provided on official and personal records, as well as the applicant's own knowledge of the service course. *Applicants should not refer to the Guide while completing the form.*

Step 2

Verify course completion from military records.

It is the responsibility of the school official to verify course completion. The following military records are used to verify successful completion of course requirements:

> DD Form 295, *Application for the Evaluation of Learning Experiences During Military Service*. This is available to active-duty servicemembers, Reservists, and National Guard members from military education officers. (Form must be certified by an authorized commissioned officer or his/her designee in order to be official)

> DD Form 214, *Armed Forces of the United States Report of Transfer or Discharge*. If the veteran does not have a copy, one can be obtained, together with other in-service training records, from the National Personnel Records Center (Military Personnel Records), 9700 Page Avenue, St. Louis, MO 63132-5100. The veteran may request service records by submitting U.S. Government Standard Form 180, which is available from a state veterans affairs office, the Veteran's Administration, or the National Personnel Records Center, or reproduced from this volume (Appendix B).

Course Completion Certificates. These may be used to complement other records or when service courses are not recorded on official records.

Refer to the Sample Course Exhibit when reading the following steps.

Step 3

Find the course exhibit by identifying the American Council on Education (ACE) ID number in the Course Number Index or the Keyword Index.

- *Course Number Index.* All available military course numbers are listed in the Course Number Index in alphanumeric sequence. If the applicant's military course number cannot be located in the Course Number Index, search for the course title in the Keyword Index.
- *Keyword Index.* Identify all possible keywords within a course title. For example, the keywords in the title "Digital, Analog, and Hybrid Computer Fundamentals" are *Digital, Analog, Hybrid,* and *Computer*. Find one of those keywords in the Keyword Index, and search the listing under the keyword for the course title. If the title cannot be found under one keyword, search all other possible keywords.
- *Identify ACE ID number.* When the title or military course number has been located, note the corresponding ACE ID number. This number refers to the course exhibit's location in the *Guide*. The two-letter prefix refers to the specific service, i.e., CG = Coast Guard, DD = Department of Defense, MC = Marine Corps, AF = Air Force. Within each section, ACE ID numbers are presented in numeric sequence.

Step 4

Match the course-identifying information with the corresponding data in the course exhibit.

First determine that the dates of attendance fall within the dates in the exhibit, and select the appropriate version. Other course-identifying information includes the official military title, military course number, length of course, dates of attendance, location, etc. This information is provided by the applicant on the *Request for Course Recommendation* form. It is important to match all items.

Step 5

Read the course description.

Consideration should be given not only to the amount of credit and to the subject area, but also to the learning outcomes or course objectives, and instruction cited in the course exhibit. These portions of the exhibit outline the course content and scope and also provide essential information about the nature of the course. This information will help you determine the appropriate placement of credit for each individual student within the requirements and programs at your institution.

Step 6

Award credit as appropriate.

Users are free to modify the credit recommendations in accordance with institutional policy and the educational goals of each applicant.

Step 7

If assistance is required, contact the American Council on Education.

Whenever problems arise and assistance is needed, the school official should submit a properly completed *Request for Course Recommendation* form to:

Military Evaluations Program
American Council on Education
One Dupont Circle NW, Suite 250
Washington, DC 20036-1193
(202) 939-9470
Fax: (202) 775-8578

E-mail: mileval@ace.nche.edu

For course exhibits before 1/90, see the *1954–1989 Guide to the Evaluation of Educational Experiences in the Armed Services.*

Sample Course Exhibit

ID Number. An identification number assigned by ACE to identify each course.

Military Course Number. The number assigned to the course by the military. Listed by version, if appropriate.

Length. The length of the course in weeks, with contact hours in parentheses. Listed by version.

Learning Outcome. Competencies students acquire during the course. Older courses have *Objectives*.

Credit Recommendation. By version. Recommended in four categories: vocational certificate, lower-division baccalaureate/associate degree, upper division baccalaureate, and graduate. Expressed in semester hours. (See Appendix A for detailed explanation of credit categories.)

CG-1704-0046

HU-25A Electrical, Class C

Course Number: None.
Location: Aviation Technical Training Center, Elizabeth City, NC.
Length: 4 weeks (144-145 hours).
Exhibit Dates: 1/90–Present.
Learning Outcomes: Upon completion of the course, the student will be able to inspect, maintain, troubleshoot, and repair the HU-25A aircraft electrical system.
Instruction: Lectures and hands-on experience in the operation, troubleshooting, inspection, and routine maintenance of the HU-25A electrical power generation, distribution, indicating and control systems, autopilot, speed control, air data, heading, flight guidance, operation and interface for cockpit management, inertial sensor, and area navigation systems.
Credit Recommendation: In the lower-division baccalaureate/associate degree category, 2 semester hours in airframe/electrical familiarization (10/94).

Title. Version 1 (if applicable) is the more recent. If course has only one version, version number is omitted throughout exhibit.

Location. By version. The service school, military installation, state.

Exhibit Dates. Listed by version. Training start date on materials evaluated and, if applicable, the date the training was eliminated. "Present" denotes publication cut-off for this edition of the Guide (4/98). The earliest start date in this edition is 1/90.

Instruction. Description of instruction, including teaching methods, facilities, equipment, major subject areas covered.

Evaluation Date. Date when the credit recommendation was established. Month and year are given in parentheses following each recommendation.

How to Find and Use Coast Guard Rating Exhibits

This volume contains exhibits for Coast Guard general rates and ratings with exhibit dates of 1/90 and later.

The instructions that follow provide a step-by-step procedure for finding and using Coast Guard general rates and ratings exhibits. Readers unfamiliar with the Coast Guard enlisted system, how it is structured, and how occupational proficiency is demonstrated should read Appendix A.

Step 1

Have the applicant submit official documentation.

- DD Form 295, *Application for the Evaluation of Learning Experiences During Military Service.* This is available to active-duty servicemembers, Reservists, and National Guard members from military education officers. (Form must be certified by an authorized commissioned officer or his/her designee in order to be official)

- DD Form 214, *Armed Forces of the United States Report of Transfer or Discharge.* If the veteran does not have a copy, one can be obtained, together with other in-service training records, from the National Personnel Records Center (Military Personnel Records), 9700 Page Avenue, St. Louis, MO 63132-5100. The veteran may request service records by submitting U.S. Government Standard Form 180, which is available from a state veterans affairs office, the Veteran's Administration, or the National Personnel Records Center, or reproduced from this volume (Appendix B).

- *Achievement Sheet.* Ratings are listed in section 1.

Step 2

Verify each occupational specialty the person has successfully held.

Eligibility for the recommendation for ratings is easily determined. Advancement means that the person is automatically eligible for the rating recommendation; that is, to be advanced, the person had to demonstrate occupational proficiency by meeting all the requirements for advancement, including passing written and performance tests.

Find the information necessary for locating the correct exhibit(s): occupational designations and the date of advancement to each.

Step 3

Find the appropriate exhibit in the Guide.

The first component in a Coast Guard rating exhibit is CGR. This identifies the exhibit as one that pertains to Coast Guard enlisted ratings. The second component consists of the rating designation, e.g., AT, GM, or ASM. The third component, a three-digit sequentially assigned number, e.g., -001, uniquely identifies the exhibit. The exhibit ID numbers have either eight or nine characters, depending on the rating designation, e.g., CGR-QM-001 and CGR-MST-001. If you know the title of an occupational specialty, the exhibit ID number can be found by referring to the *Occupational Title Index.*

Step 4

Read the entire exhibit.

Before applying a given recommendation, read the entire exhibit. Each item has been prepared to help you interpret the credit recommendations. (See the sample exhibit.)

The descriptions, which are similar to learning outcome statements of postsecondary courses and programs of study, provide essential information about the learning required for proficiency in the occupation. Comparing the description with a description of the course or program of study that the student will pursue will help you

- determine how much of the recommended credit applies to the course or program of study at your institution;
- identify additional areas of possible credit;
- resolve problems with duplication of credit, when the applicant has applied for credit for more than one military learning experience; and
- place the student at the appropriate level in the course sequence or program of study.

Step 5

Award credit as appropriate.

The credit recommendations are advisory. They are intended to assist in placing active-duty Coast Guard personnel and veterans in postsecondary programs of study and jobs. The recommendations may be modified.

When an applicant has applied for credit for more than one military learning experience, you may have to reduce the total amount of credit recommended to avoid granting duplicate credit.

Step 6

If assistance is required, contact the American Council on Education.

Whenever problems arise and assistance is needed, the school official should submit a properly completed *Request for Coast Guard Rating and Warrant Officer Exhibit* form to:

> Military Evaluations Program
> American Council on Education
> One Dupont Circle
> Washington, DC 20036-1193
> (202) 939-9470
> Fax: (202) 775-8578
> E-mail: mileval@ace.nche.edu

For ratings exhibits before 1/90, see the *1954–1989 Guide to the Evaluation of Educational Experiences in the Armed Services.*

Sample Coast Guard Rating Exhibit

ID Number. An identification number assigned by ACE to identify each exhibit.

Rating Designation. The official Coast Guard system of identifying occupations and their levels; listed in ascending order according to rate. (See Question 20 for definitions of the Coast Guard identifiers.)

Occupational Field. Number and title designating a group of related ratings.

Career Pattern. Path of advancement. Shows prerequisite general rate and subsequent progression in a rating; the designation is provided for each rate, followed by the title and paygrade (in parentheses).

Recommendation. By rate. Only the recommendation for the *highest* rate held should be used; the recommendations should *not* be added.

Educational credit is expressed in semester hours and recommended in four categories: vocational certificate, lower-division baccalaureate/associate degree, upper-division baccalaureate, and graduate degree. (See Appendix A for category definitions.)

CGR-AD-002

AVIATION MACHINIST'S MATE
- AD3
- AD2
- AD1
- ADC
- ADCS
- ADCM

Exhibit Dates: 10/94–Present.

Occupational Group: V (Aviation).

Career Pattern

SN: Seaman (E-3). *AD3:* Aviation Machinist's Mate, Third Class (E-4). *AD2:* Aviation Machinist's Mate, Second Class (E-5). *AD1:* Aviation Machinist's Mate, First Class (E-6). *ADC:* Chief Aviation Machinist's Mate (E-7). *ADCS:* Senior Chief Aviation Machinist's Mate (E-8). *ADCM:* Master Chief Aviation Machinist's Mate (E-9).

Description

Summary: Services and maintains aircraft engines and their related systems, including propellers; performs.... *AD3:* Uses drawings, diagrams, blueprints, charts, publications, and work cards to perform inspections.... *AD2:* Able to perform the duties required for AD3; performs aircraft preflight, through-flight, and.... *AD1:* Able to perform the duties required for AD2; supervises aircraft ground handling.... *ADC:* Able to perform the duties required for AD1; serves as shop supervisor; trains and.... *ADCS:* Able to perform the duties required for ADC; serves as enlisted technical or specialty expert; plans.... *ADCM:* Able to perform the duties required for ADCS; serves as senior enlisted technical or specialty administrator; manages....

Recommendation, AD3

In the lower-division baccalaureate/associate degree category, 1 semester hour in propeller systems, 1 in cleaning and corrosion control, 1 in ground operation and servicing, and 3 in introduction to computers (6/95).

Recommendation, AD2

In the lower-division baccalaureate....

Recommendation, AD1

In the lower-division baccalaureate....

Recommendation, ADC

In the lower-division baccalaureate....

Recommendation, ADCS

In the lower-division baccalaureate....

Recommendation, ADCM

In the lower-division baccalaureate/associate degree category, 1 semester hour in propeller systems, 1 in cleaning and corrosion, and 1 in ground operation and servicing, 3 in introduction to computers, 3 in aircraft fuel systems, 3 in aircraft flight control systems, 3 in helicopter rotor systems, 3 in assembly and rigging, 1 in fluid lines and fittings, 3 in turbine engine inspections, 1 in cabin atmospheric control systems, 3 in personnel supervision, 3 in maintenance management, 3 in shop management, and 3 in organizational management. In the upper-division baccalaureate category, 6 semester hours for field experience in management and 3 in management problems (6/95).

Title. The official Coast Guard title of the rating during the period of the exhibit dates.

Exhibit Dates. Start and end dates by month and year. The term "Present" indicates that the exhibit is current as of April 1998. "Pending evaluation" means that the rating is scheduled to be evaluated. The earliest start date in this edition is 1/90. (See questions 1–4 in Questions and Answers.)

Description. Summary applying to all rates and a separate description of the skills, competencies, and knowledge required for proficiency in each rate.

Date of Evaluation. Month and year are given. Appears in parentheses following the recommendation for each rate.

xv

How to Find and Use Coast Guard Warrant Officer Exhibits

This volume contains exhibits for Coast Guard warrant officer specialties with exhibit dates of 1/90 and later.

The instructions that follow provide a step-by-step procedure for finding and using exhibits for Coast Guard warrant officer specialties. Readers unfamiliar with the Coast Guard warrant officer system, how it is structured, and how occupational proficiency is demonstrated should read Appendix A.

Step 1

Have the applicant submit official documentation.

- *Form CG-5311, Officer Evaluation Report (OER).* Refer to section 1-i for the title of the specialty.
- DD Form 295, *Application for the Evaluation of Learning Experiences During Military Service.* This is available to active-duty servicemembers, Reservists, and National Guard members from military education officers. (Form must be certified by an authorized commissioned officer or his/her designee in order to be official)
- DD Form 214, *Armed Forces of the United States Report of Transfer or Discharge.* If the veteran does not have a copy, one can be obtained, together with other in-service training records, from the National Personnel Records Center (Military Personnel Records), 9700 Page Avenue, St. Louis, MO 63132-5100. The veteran may request service records by submitting U.S. Government Standard Form 180, which is available from a state veterans affairs office, the Veteran's Administration, or the National Personnel Records Center, or reproduced from this volume (Appendix B).

Step 2

Verify each occupational specialty the person has successfully held.

Warrant officers have normally progressed through the enlisted ranks and are technical specialists in their fields. Determine eligibility by consulting the Officer Evaluation Report.

Find the information necessary for locating the correct exhibit(s): occupational designations and the date of advancement to each.

Step 3

Find the appropriate exhibit in the Guide.

The exhibit can be found when the designation is known. Each exhibit is assigned an ACE ID number that has three components. The first component of a warrant officer exhibit is CGW. The second component consists of the designator, e.g., ENG, MED. The third component is a three-digit number that uniquely identifies the exhibit (CGW-PERS-001, CGW-FS-001).

Step 4

Read the entire exhibit.

Before applying a given recommendation, read the entire exhibit. Each item has been prepared to help you interpret the credit recommendations. (See the sample exhibit.)

The descriptions, which are similar to learning outcome statements of postsecondary courses and programs of study, provide essential information about the learning required for proficiency in the occupation. Comparing the description with a description of the course or program of study that the student will pursue will help you

- determine how much of the recommended credit applies to the course or program of study at your institution;
- identify additional areas of possible credit;
- resolve problems with duplication of credit, when the applicant has applied for credit for more than one military learning experience; and
- place the student at the appropriate level in the course sequence or program of study.

Step 5

Award credit as appropriate.

The credit recommendations are advisory. They are intended to assist in placing active duty Coast Guard personnel and veterans in postsecondary programs of study and jobs. The recommendations may be modified.

When an applicant has applied for credit for more than one military learning experience, you may find that you will have to reduce the total amount of credit recommended to avoid granting duplicate credit.

Step 6

If assistance is required, contact the American Council on Education.

Whenever problems arise and assistance is needed, the school official should submit a properly completed *Request for Coast Guard Rating and Warrant Officer Exhibit* form to:

Military Evaluations Program
American Council on Education
One Dupont Circle
Washington, DC 20036-1193
(202) 939-9470
Fax: (202) 775-8578
E-mail: mileval@ace.nche.edu

For exhibits before 1/90, see the *1954–1989 Guide to the Evaluation of Educational Experiences in the Armed Services.*

Sample Coast Guard Warrant Officer Exhibit

ID Number. An identification number assigned by ACE to identify each exhibit. The first three digits for warrant officers are "CGW."

Exhibit Dates. Start and end dates by month and year. The term "Present" indicates that the exhibit is current as of April 1998. "Pending evaluation" means that an evaluation is scheduled. The earliest start date in this edition is 1/90. (See questions 1–4 in Questions and Answers.)

Recommendation. Educational credit is expressed in semester hours and recommended in four categories: vocational certificate, lower-division baccalaureate/associate degree, upper-division baccalaureate, and graduate degree. (See Appendix A for category definitions.)

CGW-ELC-001

ELECTRONICS

Exhibit Dates: 1/90–Present.

Career Pattern

May have progressed to Electronics Warrant Officer from Aviation Electronics Technician (AT), Electronics Technician (ET), Sonar Technician (ST), or Telephone Technician (TT).

Description

Serves as an officer technical specialist in the field of electronics; supervises and directs personnel in all aspects of electronic repair of equipment, operation of electronic repair facilities, and maintenance of operational equipment; plans and supervises instructional programs; handles personnel duties, budget matters, records, and requisition of supplies; serves as technical liaison with other services and commercial organizations; prepares technical and administrative reports.

Recommendation

In the lower-division baccalaureate/associate degree category, 3 semester hours in technical communications, 3 in personnel supervision, and 3 for field experience in electronics systems operations. In the upper-division baccalaureate category, 3 semester hours in principles of management, 3 in management problems, 3 in budget administration, and 3 for field experience in management (2/86).

Title. The official Coast Guard title during the period of the exhibit dates.

Career Pattern. Shows the various possible career paths for the warrant officer. Titles and designators are provided.

Description. The skills, competencies, and knowledge required for proficiency in the officer specialty.

Date of Evaluation. By month and year. Appears in parentheses following the recommendation.

How to Find and Use Marine Corps MOS Exhibits

This volume contains exhibits for selected Marine Corps enlisted Military Occupational Specialties (MOS's) with exhibit dates of 1/90 and later.

The instructions that follow provide a step-by-step procedure for finding and using Marine Corps MOS exhibits. Readers unfamiliar with the Marine Corps enlisted MOS system, how it is structured, and how occupational proficiency is demonstrated should read Appendix A.

Step 1

Have the applicant submit official documentation.

- *ITSS MATMEP Summary Sheet.* A copy of the ITSS (Individual Training Standards Systems) MATMEP (Maintenance Training Management and Evaluation Program) summary sheet is in Appendix B.

- DD Form 295, *Application for the Evaluation of Learning Experiences During Military Service.* This is available to active-duty servicemembers, Reservists, and National Guard members from military education officers. (Form must be certified by an authorized commissioned officer or his/her designee in order to be official)

- DD Form 214, *Armed Forces of the United States Report of Transfer or Discharge.* If the veteran does not have a copy, one can be obtained, together with other in-service training records, from the National Personnel Records Center (Military Personnel Records), 9700 Page Avenue, St. Louis, MO 63132-5100. The veteran may request service records by submitting U.S. Government Standard Form 180, which is available from a state veterans affairs office, the Veteran's Administration, or the National Personnel Records Center, or reproduced from this volume (Appendix B).

Step 2

Verify each occupational specialty the person has successfully held.

When verifying Marine Corps MOS's in the aircraft maintenance and avionics fields, it is important to identify the level (I-IV) of training the individual has completed. Refer to the ITSS MATMEP summary sheet and locate the phrase "Completed Level." When the summary sheet indicates that Level III has been completed, this indicates that all the MOS-requisite tasks have been certified. Credit is only recommended for Levels III and IV.

Step 3

Find the appropriate exhibit in the Guide.

The exhibit can be found when the designation is known. Each exhibit is assigned an ACE ID number that has three components. The first component in a Marine Corps MOS exhibit is MCE. This identifies the exhibit as one that pertains to Marine Corps enlisted MOS's. The second component consists of the four-digit MOS designation, e.g., 6052 or 6053. The third component, a three-digit sequentially assigned number, e.g., -001, uniquely identifies the exhibit. The exhibit ID numbers have ten characters.

If you know the title of an MOS, the exhibit ID number can be found by referring to the *Occupational Title Index*.

Step 4

Read the entire exhibit.

Before applying a given recommendation, read the entire exhibit. Each item in the exhibit has been prepared to help you interpret the credit recommendations. (See the sample exhibit.)

The NOTE found in the item Career Pattern explains the levels. The description, which is similar to learning outcome statements of postsecondary courses and programs of study, provides essential information about the learning required for proficiency in the MOS. Comparing the exhibit description with a description of the course or program of study that the student will pursue will help you:

- determine how much of the recommended credit applies to the course or program of study at your institution;

- identify additional areas of possible credit;

- resolve problems with duplication of credit, when the applicant has applied for credit for more than one military learning experience; and

- place the student at the appropriate level in the course sequence or program of study.

Step 5

Award credit as appropriate.

The credit recommendations are advisory. They are intended to assist in placing active-duty Marines and veterans in postsecondary programs of study and jobs. The recommendations may be modified.

When an applicant has applied for credit for more than one military learning experience, you may have to reduce the total amount of credit recommended to avoid granting duplicate credit.

Step 6

If assistance is required, contact the American Council on Education.

Whenever problems arise and assistance is needed, the school official should submit a properly completed *Request for Marine Corps MOS Exhibit* form to:

Military Evaluations Program
American Council on Education
One Dupont Circle
Washington, DC 20036-1193
(202) 939-9470
Fax: (202) 775-8578
E-mail: mileval@ace.nche.edu

For exhibits before 1/90, see the *1954–1989 Guide to the Evaluation of Educational Experiences in the Armed Services.*

Sample Marine Corps Enlisted MOS Exhibit

ID Number. An identification number assigned by ACE to identify each exhibit.

MOS Designation. The official Marine Corps system of identifying occupations and their levels.

Occupational Field. Number and title designating a group of related MOS's.

Description. Summary applying to all levels and a separate description of the skills, competencies, and knowledge required for proficiency in each level.

Recommendation. By level. Only the recommendation for the *highest* rate held should be used; the recommendations should *not* be added. Educational credit is expressed in semester hours and recommended in four categories: vocational certificate, lower-division baccalaureate/associate degree, upper-division baccalaureate, and graduate degree. (See Appendix A for category definitions.)

MCE-6084-001

AIRCRAFT SAFETY EQUIPMENT MECHANIC, F-4/RF-4

6084

Exhibit Dates: 1/90–Present.

Occupational Field: 60 (Aircraft Maintenance).

Career Pattern

PVT: Private (E-1). *PFC:* Private First Class (E-2). *LCP:* Lance Corporal (E-3). *CPL:* Corporal (E-4). *SGT:* Sergeant (E-5). *SSGT:* Staff Sergeant (E-6). *GYSGT:* Gunnery Sergeant (E-7). NOTE: MOS duties are not defined in terms of rank or paygrade, but by levels. Level I identifies tasks that are taught in the classroom; the individual is able to perform simple parts of tasks. Level II identifies those tasks that are part of on-the-job training; the trainee can do most tasks pertaining to a particular system and needs supervision on more difficult parts of tasks. Level III indicates that the individual can perform all essential tasks without supervision. Level IV indicates a high level of proficiency in job performance; the individual can perform advanced technical functions, instruction, inspection, and supervision. The MOS designator 6081 (Aircraft Safety Equipment Mechanic--Trainee), is used to identify individuals in Levels I and II. Upon completion of Level II, MOS 6084 is awarded. Credit is recommended for Levels III and IV.

Description

Summary: Inspects and maintains aircraft safety equipment. *Levels I and II:* Performs, under close supervision and training, routine duties incident to inspection or replacement of aircraft safety systems and components; uses.... *Level III:* Able to perform the duties required for Levels I and II; tests aircraft for pressure tightness; removes, checks, rigs, and adjusts ejection seats, components, and/or seat.... *Level IV:* Able to perform the duties required for Level III; plans, schedules, and directs work assignments; performs duties as quality assurance inspector; prepares and submits required reports and records.

Recommendation, Level III

In the lower-division baccalaureate/associate degree category, 3 semester hours in industrial safety (8/89).

Recommendation, Level IV

In the lower-division baccalaureate/associate degree category, 3 semester hours in industrial safety, 3 in personnel supervision, and 3 in maintenance management. In the upper-division baccalaureate category, 2 semester hours for field experience in management, if rank was Staff Sergeant (SSGT) and 3 semester hours for field experience in management, if rank was Gunnery Sergeant (GYSGT) (8/89).

Title. The official Marine Corps title during the period of the exhibit dates.

Exhibit Dates. Start and end dates by month and year. The term "Present" indicates that the exhibit is current as of April 1998. The earliest start date in this edition is 1/90. (See questions 1–4 in Questions and Answers.)

Career Pattern. Path of advancement. Indicates military rank and paygrade. Special note defines the levels which are the basis for the credit recommendations.

Date of Evaluation. Month and year are given. Appears in parentheses following the recommendation.

Questions and Answers

This section is designed to answer questions that may arise about using the Guide and awarding credit.

1. I have several editions of the *Guide to the Evaluation of Educational Experiences in the Armed Services*. Do I need to keep all of these?

The *Guide to the Evaluation of Educational Experiences in the Armed Services* has been published biennially since 1974. Each edition replaces the previous edition, which should be discarded.

ACE has now published the *1954-1989 Guide*, which contains courses and occupations with exhibit dates of 1/54 through 12/89. The 1998 *Guide* contains courses and occupations with exhibit dates of 1/90 or later.

Please retain the *1954-1989 Guide* as a permanent reference and use it with the *1996 Guide*.

2. Which *Guide* do I use if the course spans the years 1989-1990?

If the student began the course anytime in 1989, use the *1954-1989 Guide*. If records show the student began the course 1/90 or later, use the 1998 *Guide*.

3. Where do I find course and occupation exhibits dated prior to 1954?

The Military Evaluations Program began evaluating military training and experience in 1940. For credit recommendations for courses offered before 1954, send a completed *Request for Course Recommendation* form (found in the back of this volume) to ACE. The earliest occupation evaluations date back only to the 1960s and are in the *1954-1989 Guide*.

4. What is the *Handbook to the Guide to the Evaluation of Educational Experiences in the Armed Services*?

The *Handbook to the Guide* provides credit recommendations for courses and occupations evaluated since the latest *Guide* was printed. The *Handbook* also includes information on examination evaluations, ACE, forms and instruction, and helpful hints. The *Handbook* is distributed to military education offices and civilian institutions between publications of the *Guides*.

5. Which learning experiences have been evaluated?

Formal courses offered by the active Coast Guard, Marine Corps, Air Force, and selected Department of Defense schools are eligible to be evaluated. In addition, selected Marine Corps MOSs have been evaluated, as have Coast Guard enlisted and warrant officer ratings. Courses and occupations with start dates of 1/90 or later are in this volume.

6. An applicant has submitted a DD Form 214 with abbreviations that I cannot decipher. What should I do?

Military records often provide insufficient information for civilian education officials to properly identify courses. Require students to complete the *Request for Course Recommendation* form found in the back of this volume. Applicants, for credit, should interpret the information on their records and present the data in readable form. You may also use course completion certificates and other training records to supplement information on the DD Forms 214 and 295. If you need help, submit the completed request form to ACE. *No abbreviations please.*

The submission of a *Request* form does *not initiate* an ACE evaluation. Evaluations are conducted in the field using teams of subject-matter specialists. They are not evaluated at the request of a student or institution.

The purpose of these forms is to locate an evaluation that might have been conducted after the *Guide* publication date or to locate an exhibit you might have missed in your own search.

Please do not submit forms for courses or occupations already in the *Guide*.

7. Why aren't all the military courses in the *Guide*?

The course evaluations conducted by ACE do not include all courses offered by the armed services. Many cannot be evaluated. It is the responsibility of each military school to provide ACE with materials for all new and all revised courses. In general, courses evaluated and published in the *Guide* are those offered on a full-time basis. After 1981, ACE began evaluating courses that are 45 academic contact hours in length. Prior to that time, courses evaluated were at least of two weeks duration; or, if less than two weeks in length, the courses had to include a minimum total of 60 contact hours of academic instruction. Before 1973, the minimum length requirement was three weeks or 90 contact hours. Another requirement for evaluation is that a course meet servicewide training requirements and be cited in the formal school listing for the service. This requirement generally excludes locally organized and command-level training programs, as well as courses offered on a one-time basis. Classified courses cannot be evaluated unless they are sanitized.

Keep in mind that the *Guide* is constantly updated. Courses evaluated after the publication date are printed in the *Handbook to the Guide to the Evaluation of Educational Experiences in the Armed Services*. If you cannot locate the exhibit in either publication, submit a completed *Request for Course Recommendation* form.

8. Are correspondence courses evaluated?

ACE currently evaluates selected Air Force, Coast Guard,

and Marine Corps correspondence courses. Correspondence courses were not evaluated until the mid-1970s. These courses will be listed in the Keyword Index under *Correspondence* for easy identification. One criterion for reviewing correspondence courses is the establishment of an ongoing proctored end-of-course examination program.

9. What is the purpose of the *Request for Course Recommendation* form?

One purpose of the *Request for Course Recommendation* form is to obtain information about a course from the student, (verified by official records), based on personal knowledge or memory of the course. With this firsthand information, you may find the correct course exhibit in the *Guide*.

The form should be completed by the applicant and submitted to the institution official or military representative, who will then locate the appropriate course exhibit. If assistance is needed, the institution official or military counselor can send the request form to ACE.

ACE staff will use this information to search its extensive data base and files for matching information. When students attempt to identify a course taken years ago by extracting current titles or course numbers from the *Guide*, they may in fact be identifying a similar course but not the one they actually completed. A form filled out by a student who copies information from the *Guide* cannot be used for research purposes because that information only duplicates published data. Do not ask the student to find the course listings in the *Guide*. *Only accurate and verified information is to be submitted.*

While we can provide credit recommendations to applicants at their request, we encourage them to apply through their schools. We do not normally send a credit recommendation to a college at the request of a student applicant or a military counselor.

Remember that it is your responsibility to verify course completion; ACE has no access to individual records and does not verify course completions.

10. Why is so much information needed on the *Request for Course Recommendation* form?

You cannot be sure that you have identified the correct exhibit in the *Guide* unless all the information on the form matches the corresponding items in the course exhibit. The course title, course number, name and location of the service school, and length of the course shown on the form should be identical to the information in the exhibit. The date the student started the course must fall within the exhibit dates. All items must be identical if an accurate identification is to be made.

11. What do I do when the information on the *Request for Course Recommendation* form does not exactly match the information in the course exhibit?

Send the completed form to ACE. Do not give it to the student to submit. If just one item differs from that in the exhibit you have located (for example, length), circle the item to bring it to the attention of the research staff. Send *copies* of military records only if you think they will provide additional information. If the staff cannot identify the course and supply a credit recommendation, you may still grant credit to the applicant by conducting your own assessment of the applicant's learning.

12. What is the significance of the date that appears after each credit recommendation?

That date represents the month and year the credit recommendation was established. Each time an evaluation is conducted, a date is provided to indicate when the course or occupation was last considered for a credit equivalency so that you can judge the currency of the credit recommended. This information is particularly useful in subject areas where state of the art is important in determining the applicability of credit. You can also use the evaluation date when your institution has established a statute of limitations for acceptance of transfer credit. The date is provided for your information; do not confuse the date with exhibit dates.

13. An applicant completed a course in 1991, but the *Guide* exhibit dates are 5/92 to present. Should I grant credit based on the *Guide*?

First check the Keyword or Course Number index to see if another exhibit is referenced. The exhibit dates indicate the time period for which ACE has information on the course. The course may have been offered for several years prior to the exhibit start date, but since the service branch did not submit information on the course during that time period, we cannot backdate the exhibit to cover it. If you can be reasonably sure, from other information provided by the applicant (length, course content description), that the course was the same or similar to the course listed in the *Guide*, then you may want to grant credit similar to the *Guide* recommendation. You may be able to grant credit based on a comparison of the applicant's information with the descriptive information in the *Guide*. ACE encourages you to conduct your own assessment of courses for which no credit recommendation is available. For older courses, ACE may be able to provide you with military formal school catalog course descriptions for periods of time outside the exhibit dates. This should help you determine whether you might apply the credit listed in the *Guide* exhibit based on your own comparison.

14. Can a servicemember receive credit for a course that has been completed after the ACE exhibit end date?

Yes, as long as the student started the course during the time span listed in the exhibit dates, credit may be granted. For example, when a new edition of a Coast Guard correspondence course is brought on line, the Coast Guard grants a window of 26 weeks to allow the student to complete the older edition, provided the Education Services Officer has the end-of-course examination for that edition in stock. The official student record that can be obtained from the Coast Guard Institute will verify that the member completed a particular edition of the course. ACE exhibit dates show the date the new course was instituted.

15. I found the correct exhibit in my 1998 *Guide*, but under Credit Recommendation it says "Pending evaluation." What should I do?

Check the *Handbook to the Guide*. Staff members regularly conduct site evaluations. When a course or occupation is listed as "Pending evaluation," and does not appear in the *Handbook*, you can call to find out if a credit recommendation is available, or send in the *Request for Course or Occupation Recommendation* form. The *Guide* system is constantly growing as courses and occupations are added to the data base.

16. Can you offer some tips on locating courses by title?

You can generally locate an exhibit by searching in the *Keyword Index* under the main words in the title. Common terms such as *Repair* and *Officer* are not keywords.

Keywords may consist of two or three words. For example, *Electronics Technician, Aviation Electronics*, and *Air Traffic Control* are keywords.

The computer's system of alphabetizing places abbreviations and acronyms at the end of a sequence. For example:

  Nurse
  Nursing
  NBC
  NCO

Numbers are arranged in the following sequence:

  Mk 1
  Mk 1A
  Mk 10
  Mk 100
  Mk 108
  Mk 11
  Mk 111
  Mk 119
  Mk 12

As you can see, 12 appears *after* 100. Letters appear before numbers.

17. One of my students completed a formal military course but does not have a DD Form 295 or DD Form 214. Should I grant credit for the course? How can I verify that the student actually completed the course?

The requirement that school officials use the DD 295 or DD 214 to verify successful completion of a service course is not intended to exclude those individuals who do not have access to the forms. These forms are unavailable to civilians attending military courses. In such cases, alternative forms of certification, such as course completion certificates or other records of training, may be used to verify eligibility for credit. The DD 295 is available to Reservists and National Guard members through their education services officers.

Veterans may use *Request Pertaining to Military Records*, Standard Form 180, available from Veterans Administration offices and reproduced in Section B of this volume, to obtain records of their military training.

18. When credit is recommended in more than one category, what should I do?

Credit is sometimes recommended in more than one category. One reason for multiple-category recommendations is that the scope of a given rating or course may reflect learning in several subject fields at different levels of complexity.

A thorough reading of the exhibit will help you determine which category to apply. Compare learning outcomes or course objectives and content with those of your own institution.

Credit recommended in a given subject field that is applied to two or more categories should not be added. Determine how the credits apply to the student's program of study at your institution.

For example:

In the vocational certificate category, 15 semester hours in electricity or electronics. In the lower-division baccalaureate/associate degree category, 10 semester hours in electricity or electronics. In the upper-division baccalaureate category, 5 semester hours in electricity or electronics (6/75).

Compare the exhibit description with the outcomes of electricity or electronics or related courses and programs of study at your institution. Award credit based on comparison of these outcomes.

This example may occur in older exhibits found in the *1954-1989 Guide*.

Evaluators now place the credit in a given subject area in the highest, most appropriate category. If credit is placed in more than one category, specific subjects will be cited. Some military courses may reflect competencies at the lower division, such as personnel supervision, as well as more advanced competencies at the upper division, such as human resource management. Thus it would be appropriate to grant credit at both levels.

As a general rule, you should read the course or occupation description and then award credit as it best applies to the student's program of study, as determined through academic counseling.

19. Should a two-year college use only the lower-division credit recommendations? And should a four-year college use only upper-division credit recommendations?

Not necessarily. Evaluators place a course recommendation in the highest appropriate category. If your institution teaches a given course at a different level, you are encouraged to grant credit at that level. Depending on the recommendation, the programs of study at the institution, and the objectives of the student, all types of institutions can use any or all of the four credit categories. Please see Appendix A.

20. Do I have to grant credit *exactly* as it appears in the recommendation?

The use of ACE recommendations is the prerogative of education officials and employers. The recommendations are provided to assist you in assessing the applicability of military learning experiences to an educational program or occupation. You may modify the recommendations in accordance with your institution's policies and practices.

Credit may be applied to a student's program in various ways: (1) applied to the major to replace a required course, (2) applied as an optional course within the major, (3) applied as a general elective, (4) applied to meet basic degree requirements, or (5) applied to waive a prerequisite. Credit granted by a postsecondary institution will depend on institutional policies and degree requirements.

The learning of some service personnel may exceed the skills, competencies, and knowledge evaluated for a specific course or occupation. In these cases, you may wish to conduct further assessment.

For additional help in this area, see the section "Elements of a Model Policy on Awarding Credit for Extrainstitutional Learning" in this volume.

21. May I conduct my own assessment of an applicant's learning?

In a sense, you are *always* conducting your own assessment, even when you use the recommendations in this book. The *Guide* is one reference available to assess what an applicant has learned and how that learning can be applied to a specific program of study at your institution. When you cannot find a recommendation in the *Guide* or the *Handbook*, or cannot obtain one from ACE, we encourage you to use other means to assess what the person has learned.

There is a wide variety of assessment techniques that you can use: written examinations, oral examinations, faculty committee assessment, evaluation of materials supplied by the applicant, personal interviews, performance tests, and standardized examinations such as CLEP. A combination of several techniques will usually result in a reliable assessment of the person's learning.

Learn more about assessment techniques through the publications of the Council for Adult and Experiential Learning (CAEL). Publications may be purchased from CAEL, 243 South Wabash Avenue, Suite 800, Chicago, IL 60604 (312-922-5909).

The ACE *Guide* Series now comprises this *Guide*, the *1954-1989 Guide*, and companion volumes, *The National Guide to Educational Credit for Training Programs* (which lists recommendations for courses offered by business and industry, government agencies, professional and voluntary associations, and labor unions) and the *Guide to Educational Credit by Examination* (which provides descriptions and credit recommendations for nationally recognized testing programs). *The National Guide* is available from Oryx Press, (800) 279-6799. The *Guide to Educational Credit by Examination* may be purchased from the American Council on Education (202-939-9434).

22. We have received a transcript from a military service school listing semester credits and carrying the statement that the school is accredited by one of the regional accrediting associations. How do we handle this?

Although some military schools are accredited, these are the only ones with degree-granting status at publication time:

- Community College of the Air Force, Maxwell Air Force Base, AL, accredited at the two-year community college level;

- Army Command and General Staff College at Fort Leavenworth, KS, accredited to grant an MA;

- Air Force Institute of Technology at Wright-Patterson Air Force Base, OH, accredited through the doctoral level;

- Joint Military Intelligence College (formerly Defense Intelligence College), Washington, DC, accredited to grant an MA;

- Naval Postgraduate School, Monterey, CA, accredited through the doctoral level;

- Naval War College, Newport, RI, accredited to grant an MA;

- Uniformed Services University of the Health Sciences School of Medicine, Bethesda, MD, accredited through the doctoral level;

- the National Defense University's National War College, accredited to grant an MA;

- Industrial College of the Armed Forces, accredited to grant an MA;

- the service academies.

23. I am an employer. How will the *Guide* be useful to me?

You may find the exhibits helpful in identifying the skills and knowledge of veterans when hiring and placing them in jobs. The recommendations and descriptions enable you to compare a veteran's training and experience with the qualifications and requirements for the particular job. The recommendations relate learning to postsecondary courses and curricula.

A new document, *DD Form 2586, Verification of Military Experiences and Training*, has been designed specifically for your use. Although not appropriate for the granting of credit, this form is a comprehensive listing of military education and training and is ideal for an employer.

24. Why does ACE end-date some courses that are still offered?

First, check the keyword or course number index to determine whether another exhibit is referenced. If not, the exhibit may be end-dated because the military school has not provided information on the course in ten years. Other course exhibits may be end-dated when ACE receives military documentation showing that a course has changed significantly. The modified course may appear as a new exhibit or as a new version of an old exhibit.

25. Why are most Defense Language Institute (DLI) courses end-dated in 1990? What do I do for foreign language courses taken after that time?

Credit recommendations are now based on successful completion of the Defense Language Proficiency Test (DLPT) end-of-course examination.

ACE's Credit by Examination Program began evaluating DLI's end-of-course examinations in October 1990. Credit recommendations are now based on successful completion of the DLPT III and IV series. These examinations may be taken by servicemembers who complete the DLI courses, as well as by servicemembers who must demonstrate on going language proficiency without having taken formal training. Evaluation of the tests is a way to ensure that *both* groups receive appropriate recognition. See the DLPT chart in this volume for the credit recommendations.

An official DLI transcript listing the test results is available to the examinee from Commandant, Defense Language Institute, Foreign Language Center, ATTN: ATFL-DAA-AR, The Presidio of Monterey, Monterey, CA 93944-5006. Test scores are available by writing to DLI/FLC, The Presidio of Monterey, Monterey, CA 93944-5006, ATT: ATFL-EST-M (DLPT Score Reports). Or you may call (408) 242-5291 for the office of the Provost, or (408) 242-5825 or -6459 for the Registrar's office. For information about the status of a transcript, call (408) 242-5366. The Fax number is (408) 242-5146.

Additional information on these examinations may be obtained from ACE's *Guide to Educational Credit by Examination,* 4th edition.

26. What are USAFI and DANTES? Should I grant credit for the courses and tests listed on an applicant's USAFI or DANTES military test reports?

USAFI was the United States Armed Forces Institute, which offered an extensive educational program to active-duty personnel. USAFI correspondence, seminar, self-study courses, end-of-course tests, and Subject Standardized Tests (SSTs) were made available to servicemembers worldwide until 1974, when USAFI was disestablished. Credit recommendations for USAFI courses and examinations are listed in the *Handbook to the Guide.*

The Defense Activity for Non-Traditional Education Support (DANTES) was established in 1974, and that agency continues the development and administration of SSTs and other educational services. ACE continues to recommend credit for USAFI offerings and DANTES SSTs. Credit recommendations for DANTES SSTs are in the *Handbook to the Guide*. In addition, detailed descriptions are in the *Guide to Educational Credit by Examination*, which is available from the ACE Credit by Examination Program.

In verifying completion of USAFI or DANTES courses or tests, the military test report should not be considered official. That report is given to all servicemembers who have taken a course or test. To obtain official USAFI or DANTES transcripts, refer to the addresses provided in Appendix A, under "Other Resources."

27. I have looked up the exhibits for several courses and occupations for one applicant. It appears that a lot of the credit recommended is in the same subject area. How can I avoid granting duplicate credit to this person?

You may grant credit for any combination of learning experiences. In doing so, however, you must be alert to the possibility of overlapping credit recommendations. When the student is applying for credit for more than one learning experience, the credit recommendations might cover some of the same learning. In such cases, awarding a simple total of the recommended credit could result in the award of more credit than the learning merits.

To determine how much credit may be awarded without duplication, use the following steps:

- Using official military records, locate the appropriate course and occupation exhibits in the *Guide*.
- Read and compare all the descriptions, and, on the basis of the person's program of study, identify the appropriate recommendations in each exhibit.
- Read and compare all the recommendations. It may be necessary to obtain additional information from the individual through interview or further assessment.

- Make decisions on how much credit might be awarded without duplication. Credit should be awarded as appropriate to the educational goals of the individual and the policies of the institution.

If you cannot determine whether duplication exists, write to the Advisory Service.

28. A Coast Guard student I'm counseling has taken an Air Force course not in the *Guide*. What should I do?

Any servicemembers who started an Air Force course prior to March 31, 1996 can obtain a CCAF transcript. Servicemembers who take courses after this date should contact CCAF for information on transcripting availability. The transcript may be used to request transfer of credit to another institution or to otherwise document college credit. CCAF does not award credit for all Air Force courses, but only those regularly attended by Air Force enlisted personnel and taught by CCAF-affiliated schools.

To obtain a CCAF transcript, send a completed AF Form 2099 or a brief letter requesting transcript service to CCAF/RRR, Simler Hall, Suite 128, 130 West Maxwell Boulevard, Maxwell Air Force Base, AL 36112-6613. A certificate of training must accompany the request. DD Form 214 is not acceptable. Specify the type of transcript (personal or official) and the address to which the transcript is to be mailed. Provide full name and social security number, and ensure that the person whose transcript is requested signs the request form. The transcripts are free. Official transcripts will be mailed only to institutions.

29. I've found exhibits for most enlisted occupations, but not for the Air Force and Marine Corps. How come?

Air Force enlisted personnel should write to the Community College of the Air Force (CCAF) to obtain a transcript, which will include credit for the Air Force Specialty Codes (AFSCs) held. The address is listed in question 28.

Selected Marine Corps MOS recommendations are included in the *Guide*. At present, evaluations have been conducted for MOS's in the aircraft maintenance and avionics occupational fields.

30. How can I distinguish among the terms paygrade, general rate, rating, and rate?

A *paygrade* is a position from 1 to 9 on the Coast Guard's pay scale for enlisted personnel; in referring to a paygrade, the letter E (enlisted) precedes the number (E-1, E-2, E-3, through E-9). A general rate is an apprenticeship that indicates eligibility for entrance into various ratings. A rating is an occupation, e.g., Radioman. A rate is an identifying term or title associated with a given paygrade, for example, the rate petty officer third class for paygrade E-4. A rate may also be associated with a specific rating; for example, a petty officer third class (paygrade E-4) whose rating is Radioman will usually refer to his or her rate as Radioman Third Class. Coast Guard members usually refer to themselves by their rate. Refer to Appendix A for a description of the Coast Guard Enlisted Rating Structure.

31. Do all ratings provide paths of advancement and career development for paygrades E-4 through E-9?

All the Coast Guard ratings begin at paygrade E-4 and culminate at paygrade E9.

32. Should persons be granted credit for their Coast Guard general rate (paygrade E-3) as well as rating (paygrades E-4 to E-9)?

Yes. Anyone holding a rating is also eligible for the credit recommended in the general rate exhibit. The two general rates are Airman and Seaman. The item *Career Pattern* in the rating exhibit will indicate the general rate the individual held prior to progressing to the rating. The general rate exhibit should then be located and the credit recommended considered in conjunction with the credit recommendation for the rating.

33. I've checked both this *Guide* and the *1954-1989 Guide* and my student held an occupational specialty prior to the start date of the exhibit. Should I still use the recommendation?

Probably not. You need to do some additional investigating to help you decide whether to accept or modify the recommendation. The start date established by ACE is based on how far back we can verify that the occupation was the same as it was when our team evaluated it. The verification is based on official documentation provided by the military services. We do not have the means to confirm that an occupation was the same before the start date, but you might. There are two steps to follow in making your decision:

1. Ask the person to provide a copy of the military regulation pertaining to occupational qualifications or standards that were in effect when he or she held the specialty. Use the qualifications or standards to identify the learning outcomes represented by the general rate or rating.

2. Determine how much credit might be granted. Remember that advancement connotes occupational proficiency in the Coast Guard. A careful comparison of the description in the occupation exhibit and the qualifications or standards obtained in step one may reveal if there was a substantive difference. If not, the credit recommendation in the exhibit may be appropriate. If specific differences are identified, then the recommended credits need to be modified accordingly. One approach to modifying the credit recommendation is to have faculty members in appropriate subject areas assess the person's learning, identifying the discrete learning outcomes achieved by the person and relating them to the educational objectives of

courses and programs of study. Equivalent credit should be granted when the person has demonstrated achievement of the same learning outcomes for which the faculty members grant credit to their students.

34. Why is the number of credits recommended for the occupation greater than the number recommended for the course that leads to it?

It is rare for the subject matter covered in a course to perfectly coincide with the learning represented by occupational proficiency. In most cases, there is quite a difference in scope, both in the subject matter and in the depth and breadth of the learning.

Usually, the scope of a course is narrower than that of the occupation. Most military courses are designed to prepare servicemembers to function on the job or to take on additional tasks. As such, the courses normally provide entry-level occupational skills and competencies. Occupational proficiency is predicated on the additional factors of work experience and/or extensive self-instruction.

35. Several courses that I've looked up have an end date of 12/73, but the student took them in 1976. Why do so many Air Force courses carry an end date of 12/73?

The courses that are end-dated are part of the Community College of the Air Force (CCAF). ACE has no direct knowledge of the content of the courses beyond December 1973. Accredited by the Southern Association's Commission on Colleges, CCAF grants an associate in applied science degree. Lower-division credit recommendations are no longer listed for those courses. However, you may grant the upper-division credit if the student attended the course within the exhibit dates.

Anyone attending an Air Force school or course can obtain a transcript. The transcript documents training if the course is recognized by CCAF for credit. The transcript may be used to request transfer of credit to another institution or to otherwise document college credit. CCAF does not award credit for all Air Force courses, but only those regularly attended by Air Force enlisted personnel and taught by CCAF-affiliated schools.

To obtain a CCAF transcript, send a completed AF Form 2099 or a brief letter requesting transcript service to CCAF/RRR, Simler Hall, Suite 128, 130 West Maxwell Blvd, Maxwell Air Force Base, AL 36112-6613. A certificate of training must accompany the request. DD Form 214 is not acceptable. Specify the type of transcript (personal or official) and the address to which the transcript is to be mailed. Provide full name and social security number, and ensure that the person whose transcript is requested signs the request form. The transcripts are free. Official transcripts will be mailed only to institutions.

Awarding Credit for Extrainstitutional Learning

*The following statement by the ACE Commission on Educational Credit and Credentials
has been approved by the ACE Board of Directors
and endorsed by the Commission on Recognition of Postsecondary Accreditation.*

The American Council on Education recommends that postsecondary education institutions develop policies and procedures for measuring and awarding credit for learning attained outside their sponsorship.[1]

American society abounds in resources for learning at the postsecondary level. Public, private, and proprietary education institutions exercise the central but not exclusive responsibility for instruction and learning. Associations, business, government, industry, the military, and unions sponsor formal instruction. In addition, independent study and reading, work experiences, the mass media, and social interaction contribute to learning and competency.

Full and effective use of all educational resources is a worthy educational and social goal. Achieving this goal will depend to a large extent on providing equitable recognition for extrainstitutional learning:

- Educational credentials have a significant bearing on the economic, professional, and social status of the individual. Thus, social equity requires that equivalent learning, regardless of where and how it is achieved, be incorporated into the system of rewards for learning and competency.
- Recognition encourages learning and contributes to pedagogical effectiveness. Teaching students what they already know is both stultifying to them and wasteful of educational and personal resources.

Postsecondary education institutions legally authorized and accredited to award degrees and other educational credentials have a special responsibility to assess extrainstitutional learning as part of their credentialing function.

In the development of institutional policies and procedures, the American Council on Education recommends the following guidelines and resources.

Guidelines

1.

Reliable and valid evaluation of student achievement is the sine quo non in awarding credit. Experience, whether acquired at work, in social settings, in the library, at home, or in the formal classroom, is in itself an inadequate basis for awarding credit. Increased attention in choosing evaluation procedures and techniques and more thorough evaluation are necessary when learning has been attained without participation in a program of study prescribed by an educational institution and offered by its faculty.

2.

In determining whether it is appropriate to accept a student's extrainstitutional learning for credit, the governing considerations should be its applicability to the student's program of study, including graduation requirements, and the relationship of the learning to the institution's mission, curricula and standards for student achievement. Learning should be articulated, documented and measured in these terms.

3.

Institutions should evaluate extrainstitutional learning only in subject-matter fields in which they have or can arrange for faculty expertise or where they can rely on nationally validated examinations or other procedures for establishing credit equivalencies. Institutions should award credit in these areas only if the quality of learning meets their standards for student achievement. Normally, institutions should evaluate learning and award credit only in subject fields in which they offer courses or curricula. However, if the acquisition of college level learning outcomes has been demonstrated in an area not taught by the institution, but related to the student's program of study, an exception may and ought to be made.

4.

Institutions awarding credit for extrainstitutional learning should develop clearly stated policies regarding administrative responsibility, student eligibility, means of assessment, recording of results on transcripts, storage of documentation, student fees, and maximum number of credits allowable. Information on these and related institutional policies and procedures should be disseminated to students and faculty for maximum awareness and utilization.

5.

Institutional policy should include provision that the institution's policies and procedures for awarding credit for extrainstitutional learning should be subject to periodic reevaluation.

---

[1] "Extrainstitutional learning" is defined as learning that is attained outside the sponsorship of legally authorized and accredited postsecondary educational institutions. The *term* applies to learning acquired from work and life experiences, independent reading and study, the mass media, and participation in formal courses sponsored by associations, business, government, industry, the military, and unions.

Resources

1.

Guide to the Evaluation of Educational Experiences in the Armed Services, published by the American Council on Education.

2.

The National Guide to Educational Credit for Training Programs, published by the American Council on Education.

3.

Credentialing Educational Accomplishment, edited by Jerry W. Miller and Olive Mills, published by the American Council on Education.

4.

Guide to Educational Credit by Examination, published by the American Council on Education and available from the ACE Credit by Examination Program.

5.

Publications of the Council for Adult and Experiential Learning, 243 South Wabash Avenue, Suite 800, Chicago, IL, 60604.

Transfer and Award of Credit

The following set of guidelines was developed by the three national associations whose member institutions are directly involved in the transfer and award of academic credit the American Association of Collegiate Registrars and Admissions Officers, the American Council on Education, and the Commission on Recognition of Postsecondary Accreditation. The need for such a statement came from an awareness of the growing complexity of transfer policies and practices, which have been brought about, in part, by the changing contours of postsecondary education. With increasing frequency, students are pursuing their education in a variety of institutional and extrainstitutional settings. Social equity and the intelligent use of resources require that validated learning be recognized wherever it takes place.

The statement is intended to serve as a guide for institutions developing or reviewing policies dealing with transfer and award of credit. The statement is under periodic review by the three associations, and reactions to it would, of course, be welcome. Comments may be directed to Henry Spille, Vice President, ACE, and Director, The Center for Adult Learning and Educational Credentials.

Robert H. Atwell President
American Council on Education

This statement is directed to institutions of postsecondary education and others concerned with the transfer of academic credit among institutions and award of academic credit for extrainstitutional learning. Basic to this statement is the principle that each institution is responsible for determining its own policies and practices with regard to the transfer and award of credit. Institutions are encouraged to review their policies and practices periodically to assure that they accomplish the institutions' objectives and that they function in a manner that is fair and equitable to students. Any statements, this one or others referred to, should be used as guides, not as substitutes, for institutional policies and practices.

Transfer of credit is a concept that now involves transfer between dissimilar institutions and curricula and recognition of extrainstitutional learning, as well as transfer between institutions and curricula with similar characteristics. As their personal circumstances and educational objectives change, students seek to have their learning, wherever and however attained, recognized by institutions where they enroll for further study. It is important for reasons of social equity and educational effectiveness, as well as the wise use of resources, for all institutions to develop reasonable and definitive policies and procedures for acceptance of transfer credit. Such policies and procedures should provide maximum consideration for the individual student who has changed institutions or objectives. It is the receiving institution's responsibility to provide reasonable and definitive policies and procedures for determining a student's knowledge in required subject areas. All institutions have a responsibility to furnish transcripts and other documents necessary for a receiving institution to judge the quality and quantity of students' work. Institutions also have a responsibility to advise the students that the work reflected on the transcript may or may not be accepted by a receiving institution.

INTERINSTITUTIONAL TRANSFER OF CREDIT

Transfer of credit from one institution to another involves at least three considerations:

(1) the educational quality of the institution from whitey student transfers

(2) the comparability of the nature, content, and level of credit earned to that offered by the receiving institution

(3) the appropriateness and applicability of the credit earned to the programs offered by the receiving institution, in light of the student's educational goals.

Accredited Institutions

Accreditation speaks primarily to the first of these considerations, serving as the basic indicator that an institution meets certain minimum standards. Users of accreditation are urged to give careful attention to the accreditation conferred by accrediting bodies recognized by the Commission on Recognition of Postsecondary Accreditation (CORPA) (formerly the Council on Postsecondary Accreditation). CORPA has a formal process of recognition which requires that all accrediting bodies so recognized must meet the same standards. Under these standards, CORPA has recognized a number of accrediting bodies, including:

(1) regional accrediting commissions (which historically accredited the more traditional colleges and universities but which now accredit proprietary, vocational-technical, and single-purpose institutions as well)

(2) national accrediting bodies that accredit various kinds of specialized institutions

(3) certain professional organizations that accredit free-standing professional schools, in addition to programs within multipurpose institutions (CORPA) annually publishes a list of recognized accrediting bodies, as well as a directory of institutions accredited by these organizations).

Although accrediting agencies vary in the ways they are organized and in their statements of scope and mission, all accrediting bodies that meet CORPA's standards for recognition function to assure that the institutions or programs they accredit have met generally accepted minimum standards for accreditation.

Accreditation affords reason for confidence in an institution's or a program's purposes, in the appropriateness of its resources and plans for carrying out these purposes, and in its effectiveness in accomplishing its goals, insofar as these things can be judged. Accreditation speaks to the probability, but does not guarantee, that students have met acceptable standards of educational accomplishment.

Comparability and Applicability

Comparability of the nature, content, and level of transfer credit and the appropriateness and applicability of the credit earned to programs offered by the receiving institution are as important in the evaluation process as the accreditation status of the institution at which the transfer credit was awarded. Since accreditation does not address these questions, this information must be obtained from catalogues and other materials and from direct contact between knowledgeable and experienced faculty and staff at both the receiving and sending institutions. When such considerations as comparability and appropriateness of credit are satisfied, however, the receiving institution should have reasonable confidence that students from accredited institutions are qualified to undertake the receiving institution's educational program.

Admissions and Degree Purposes

At some institutions there may be differences between the acceptance of credit for admission purposes and the applicability of credit for degree purposes. A receiving institution may accept previous work, place a credit value on it, and enter it on the transcript. However, that previous work, because of its nature and not its inherent quality, may be determined to have no applicability to a specific degree to be pursued by the student. Institutions have a responsibility to make this distinction, and its implications, clear to students before they decide to enroll. This should be a matter of full disclosure, with the best interests of the student in mind. Institutions also should make every reasonable effort to reduce the gap between credits accepted and credits applied toward an educational credential.

Unaccredited Institutions

Institutions of postsecondary education that are not accredited by CORPA-recognized accrediting bodies may lack that status for reasons unrelated to questions of quality. Such institutions, however, cannot provide a reliable, third-party assurance that they meet or exceed minimum standards. That being the case, students transferring from such institutions may encounter special problems in gaining acceptance and transferring credits to accredited institutions. Institutions admitting students from unaccredited institutions should take special steps to validate credits previously earned.

Foreign Institutions

In most cases, foreign institutions are chartered and authorized by their national governments, usually through a ministry of education or similar body. Although this policy provides for a standardization within a country, it does not produce useful information about comparability from one country to another. No other nation has a system comparable to voluntary accreditation as it exists in the United States. The American Association of Collegiate Registrars and Admissions Officers (AACRAO) can assist institutions by distributing general guidelines on admission and placement of foreign students. Their World Education Series, and PIER (Projects for International Education Research) Workshop Reports are studies of foreign educational systems and contain placement recommendations as to the appropriate placement of foreign students into the US educational system. The placement recommendations are approved by the National Council on the Evaluation of Foreign Educational Credentials (CEC). The CEC membership consists of representatives from ten educational associations involved in the admission and placement of foreign students in the US. Equivalency or placement of foreign students is to be evaluated in terms of the programs and policies of the individual institution

VALIDATION OF EXTRAINSTITUTIONAL AND EXPERIENTIAL LEARNING FOR TRANSFER PURPOSES

Transfer-of-credit policies should encompass educational accomplishment attained in extrainstitutional settings as well as at accredited postsecondary institutions. In deciding on the award of credit for extrainstitutional learning, institutions will find the services of the American Council on Education's Center for Adult Learning and Educational Credentials helpful. One of The Center's functions is to operate and foster programs to determine credit equivalencies for various modes of extrainstitutional learning. The Center maintains evaluation programs for formally structured courses offered by the military, and civilian noncollegiate sponsors such as business, corporations, government agencies, and labor unions. Evaluation services are also available for examination programs, for occupations with validated job proficiency evaluation systems, and for correspondence courses offered by schools accredited by the National Home Study Council. The results are published in a Guide series. Another resource is the General Education Development (GED) Testing Program, which provides a means for assessing high school equivalency.

For learning that has not been validated through the ACE formal credit recommendations process or through credit-by-examination programs, institutions are urged to explore the Council for Adult and Experiential Learning (CAEL) procedures and processes.

USES OF THIS STATEMENT

This statement has been endorsed by the three national associations most concerned with practices in the area of transfer and award of academic credit—the American Association of Collegiate Registrars and Admissions Officers, the American Council on Education, and the Commission on Recognition of Postsecondary Accreditation.

Institutions are encouraged to use this statement as a basis for discussions in developing or reviewing institutional policies with regard to transfer. If the statement reflects an institution's policies, that institution might want to use this publication to inform faculty, staff, and students.

It is recommended that accrediting bodies reflect the essential precepts of this statement in their criteria.

Approved by the Executive Committee, American Association of Collegiate Registrars and Admissions Officers November 21, 1978 and reaffirmed in 1989.

Approved by the American Council on Education/Commission on Educational Credit and Credentials December 5, 1978 and reaffirmed in 1989.

Approved by the COPA Board
October 10, 1978

Elements of a Model Policy on Awarding Credit for Extrainstitutional Learning

American Council on Education
The Center for Adult Learning and Educational Credentials
Washington, DC

Extrainstitutional learning is defined as learning that is attained outside the sponsorship of legally authorized and appropriately accredited postsecondary education institutions. The term applies to learning acquired from work and life experiences, independent reading and study, the mass media, and participation in formal courses sponsored by associations, businesses, government, industry, the military and labor unions.

I. EXAMPLE STATEMENT OF GENERAL POLICY

Degree credit shall be awarded for postsecondary-level extrainstitutional learning. Such credit awards shall be made under protocols designed to assure that reliable and valid measures of learning outcomes are or have been applied.

Rationale

Adult students are demanding that their learning acquired in extrainstitutional settings be assessed and that appropriate degree-related credit be awarded. This is a reasonable request and also one in keeping with the national interest. Full and effective use of all educational resources is a worthy educational and social goal. Achieving this goal will depend to a large extent on providing equitable recognition for extrainstitutional learning.

American society abounds in resources for learning at the postsecondary level. Postsecondary education institutions legally authorized and appropriately accredited to award degrees and other educational credentials accordingly have a special responsibility to assess extrainstitutional learning as part of their credentialing function.

Educational credentials have a significant bearing on the economic, professional, and social status of the individual. It follows that social equity requires that equivalent learning regardless of where and how it is achieved be incorporated into the system of rewards for learning and competency. Recognition encourages learning and contributes to teaching effectiveness. Teaching students what they already know is both stultifying to them and wasteful of educational and personal resources.

Through reliable and valid assessment of extrainstitutional learning, more persons can be enabled to enter college and work successfully toward degrees and other academic goals. Many of these individuals are those who have been educationally disadvantaged in the past.

II. EXAMPLE POLICY FOR ADMINISTRATIVE RESPONSIBILITY

Responsibility for administering policies and procedures for awarding credit for extrainstitutional learning rests with the Academic Vice President (or other designated senior academic official).

Rationale

Awarding credit is an academic matter. Therefore, administrative responsibility for the policies and procedures pertaining to awarding credit for extrainstitutional learning should be assigned to the institution's senior academic official. Policy implementation and application may be delegated to other institutional officials such as deans, department chairpersons, and registrars. However, oversight of their policy implementation and application is performed by a senior academic official.

Example of Application of the Policy

The academic vice president delegates to the institutional registrar the responsibility for transcripting credit awarded on the basis of recommendations in the ACE *Guides* (The policy was approved through the institution's usual governance procedures.) and provides the registrar with guidelines for performing this responsibility. Included in the guidelines are a series of steps to follow in determining how to apply the ACE credit recommended for a course whose title is not the same as the title of any course offered by the institution.

The ACE *Guide* recommends 3 semester hours in automotive maintenance, but the institution's courses are more specific and carry titles such as Tune-Up I, Auto Brakes, and Automatic Transmission I. In accordance with the guidelines, the registrar compares course descriptions in the institution's catalog with the course description and learning outcomes in the *Guide,* finds that the learning outcomes acquired through Tune-Up I match very closely the learning outcomes acquired through the military's automotive maintenance course, and transcripts 3 semester hours in Tune-Up I.

III. EXAMPLE POLICY ON LEARNING THAT WILL BE EVALUATED AND WHO WILL CONDUCT THE EVALUATIONS

Extrainstitutional learning related to subject areas, courses and programs of study will be evaluated for registered students. Established institutional evaluation procedures will be followed, using one or more of the authorized publications and methods for effecting assessment. Students of any age may apply for credit for extrainstitutional learning at any time during a period of enrollment.

Rationale

The learning that is evaluated must be in subject areas in which faculty expertise is available or in which expertise can be arranged. This expertise is essential for conducting evaluations and making credit-award decisions. If subject-matter experts do not conduct the evaluations and make the decisions, the credibility of the activity is suspect.

While the great majority of students requesting credit for extrainstitutional learning will be of relatively mature years, it is possible that "young students" may have acquired learning in extrainstitutional settings. Providing a student can document the attainment of appropriate learning outcomes, he or she should receive credit regardless of age. While most applications for award of extrainstitutional learning credit will be received at the time of enrollment from new students, it is possible that students already enrolled may realize that they, too, have extrainstitutional learning experiences that might qualify for credit awards. It is also possible that enrolled students, especially those who are employed, will continue to learn in noncollegiate settings, and will want their learning assessed. Student initiatives with respect to extrainstitutional learning assessment shall not be limited or restricted to the initial term of enrollment.

Example of Application of the Policy

A 19-year old student has been enrolled at Local College for two years and is about to enter her junior year. At the time of enrollment, it did not occur to her that the foreign language requirement might be satisfied with a language other than one of those taught at the institution. This student's parents were refugees from Hungary who entered the United States in 1958. The Hungarian language is spoken in the home and the student speaks the language fluently. In addition, she regularly reads journal and magazine articles written in Hungarian. The student is able to document to the satisfaction of the institution's language faculty, through evaluation done by a person in the community fluent in Hungarian, that she is, indeed, proficient in Hungarian. She received 6 credits in Hungarian that satisfy the foreign language requirement for the bachelor's degree.

IV. EXAMPLE POLICY ON PUBLICATIONS AND METHODS FOR USE IN ASSESSING AND AWARDING CREDIT FOR EXTRAINSTITUTIONAL LEARNING

Nine publications and methods are permitted for use in assessing and awarding credit for extrainstitutional learning. These publications and methods are:

(1) ACE *Guide to the Evaluation of Educational Experiences in the Armed Forces,* (2) ACE *National Guide to Educational Credit for Training Programs,* (3) ACE *Guide to Educational Credit By Examination,* (4) New York Regents *Guide to Educational Programs in Noncollegiate Organizations,* (5) College Entrance Examination Board Advanced Placement Program (CEEB/AP), (6) Department challenge examinations and/or faculty end-of-course examinations, (7) Degree-relevant extrainstitutional learning credit awarded and transcripted by other accredited institutions, as well as that credit transcripted by ACE on the Army/ACE Registry Transcript System (AARTS) and the Registry of Credit Recommendations (ROCR), (8) Subject-matter experts who are not members of the institution's faculty but who evaluate extrainstitutional learning at the faculty's request and oversight, and (9) individual portfolios using Council for Adult and Experiential Learning (CAEL) or other standardized guidelines.

In addition, advice on using other publications and methods for awarding credit for extrainstitutional learning may be sought from the American Council on Education.

Rationale

For nearly 50 years, the *Guide to the Evaluation of Educational Experiences in the Armed Services* has been the standard reference work for recognizing learning acquired in the military. More recently *The National Guide to Educational Credit for Training Programs* and the *Guide to Educational Credit By Examination* have attained similar stature. ACE has worked cooperatively with the Department of Defense, the armed services, business, industry, labor unions, and other organizations providing education and training, and test publishers, as well as with appropriately accredited colleges and universities in developing the credit recommendations presented in the *Guide* series. All recommendations carried are the result of careful and periodic review by evaluation teams that consist of university and college faculty members who are subject-matter experts. All evaluation teams work under policies and procedures approved by the ACE Commission on Educational Credit and Credentials and under protocols and supervision provided by the staff of the ACE Center for Adult Learning and Educational Credentials.

The policies and procedures used to develop the exhibits in the New York Regents *Guide to Educational Programs in Noncollegiate Organizations* are similar to those used by the American Council on Education.

The CEEB/AP program permits high school students to demonstrate the attainment of college-level learning outcomes; the important factor is that learning outcomes are attained, not where the learning took place (in a high school) or who taught the course (usually a high school teacher).

The practice of administering departmental or individual faculty member challenge examinations or end-of-course examinations is long-standing and permits the faculty members to make direct judgements about the equivalence of learning for specific courses.

Credit transcripted on the basis of extrainstitutional learning should be treated the same as that transcripted for courses; in both instances the faculty of the sending institution has attested to the student's attainment of required learning outcomes. Credit transcripted on ACE's Army/ACE Registry Transcript System (AARTS) and Registry of Credit Recommendations (ROCR) for the Army component of the Military Evaluations Program and participants in the Program on Noncollegiate Sponsored Instruction (PONSI®) should also be treated as that transcripted for courses; in these instances, teams of faculty members from appropriately accredited colleges and universities have evaluated the learning and judged its equivalence to learning acquired on campuses.

Sometimes the expertise to evaluate extrainstitutional learning is not available on the faculty; in these instances, the faculty can designate an expert to conduct the evaluation. Criteria for selection of "outside" experts should be established.

The portfolio assessment guidelines promulgated by CAEL are widely accepted by faculty assessors and are considered exemplary. The portfolio provides a reliable method for evaluating learning outcomes acquired individually, as opposed to a set of learning outcomes acquired by large numbers of persons, such as through a course offered by the U.S. Army for soldiers.

Example of Application of the Policy

A student (soldier), upon entering Local College, provides an AARTS transcript that confirms completion of a course entitled "Software Analyst" at the Institute of Administration, Ft. Ben Harrison, Indiana, in October 1993. Information on the transcript and in the ACE *Guide to the Evaluation of Educational Experiences in the Armed Services* indicates that the soldier completed this course when it was taught at Ft. Ben Harrison beginning in August 1992. Its objectives were to train soldiers to use personal computer software such as DOS, Lotus, and dBase III, and to create structural programs in UNIX and ADA. The course was delivered by means of classroom lectures, practical exercises and administrations. The *Guide* recommends that a college or university award 4 semester hours in computer literacy and 2 semester hours in JCL programming at the lower division baccalaureate level; this recommendation also appears on the AARTS transcript. After review of this recommendation, Local College awards this student a total of 6 semester hours of credit in computer programs toward completion of a planned major in mathematics.

V. EXAMPLE POLICY ON MAXIMUM NUMBER OF CREDITS THAT MAY BE AWARDED FOR EXTRAINSTITUTIONAL LEARNING

The maximum number of credits that may be awarded for extrainstitutional learning is three-fourths the number required for a degree.

Rationale

Most postsecondary education institutions have a requirement that stipulates that a minimum of one-fourth of the credits required for a degree must be earned in residence. The balance of the required credits, then, could be awarded on the basis of extrainstitutional learning. However, the credits awarded for extrainstitutional learning and those earned in residence would have to satisfy all degree requirements before a degree could be awarded. In the absence of a residency requirement, all degree requirements conceivably could be met through extrainstitutional learning.

Example of Application of the Policy

John Smith is a newly registered student at Local College. He is a financial vice president of a major U.S. corporation, CPA, and licensed realtor; he is fluent in three foreign languages—German, French, and Spanish; he has published seven articles in domestic and foreign journals and has contributed chapters in two books on international economics; he is knowledgeable about foreign countries and cultures as a result of having lived abroad for eleven years; and he is proficient in computer programming.

On the basis of Mr. Smith's demonstrated knowledge, skills, and competencies, the faculty awarded him 106 credits. The college requires 124 credits for a degree and has a 24 credit residency requirement. Therefore, Mr. Smith must earn 24 credits in residence. What courses does he take to earn 24 credits? His academic advisers decided, and Mr. Smith concurred, that he lacked knowledge about the environment, a special focus and thrust of the college's general education requirement. He took courses to satisfy this requirement and was awarded a degree.

VI. EXAMPLE POLICY ON RECORDING CREDITS AWARDED FOR EXTRAINSTITUTIONAL LEARNING ON TRANSCRIPTS

The credit awarded for extrainstitutional learning shall be transcripted in the same manner as credits awarded for course completions.

Rationale

The faculty members have either evaluated the learning or approved the other means by which credit can be awarded. They have verified that learning has occurred and is equivalent to the learning acquired in their courses and programs. Therefore, special notation or differentiation of courses and credits awarded on the basis of extrainstitutional learning is unnecessary.

Example of the Application of the Policy

Examples of the way in which Mr. Smith's credit was transcripted are:

Accounting 261	Financial Information Systems—3 semester hours
Accounting 360	Financial Reporting Theory and Practice—3 semester hours
Computer Science 251	Introductory Principles of Programming—4 semester hours
Computer Science 252	Intermediate Principles of Programming—4 semester hours
German 101	Elementary German—4 semester hours
German 201	Intermediate German—4 semester hours
German 309	Advanced Conversation—3 semester hours

VII. EXAMPLE POLICY ON RETENTION OF THE DOCUMENTATION OF EXTRAISTITUTIONAL LEARNING

Documents used to verify the award of credit for extrainstitutional learning shall be retained for one calendar year.

Rationale

The length of time documentation for the credits awarded for extrainstitutional learning shall be retained should be consistent with the length of time documentation for credits awarded for classroom learning are retained. Consistency facilitates equitable handling of student documents and represents sound recordkeeping practices.

Example of Application of the Policy

Local College keeps students' final exams on file for one calendar year, after which they are shredded. Mr. Smith's documentation—copies of articles and chapters he authored, copies of his computer programs, notes and summaries of interviews faculty members had with him, etc.—were kept on file for one calendar year and were then shredded.

VIII. EXAMPLE POLICY OF STUDENT FEES FOR ASSESSING EXTRAINSTITUTIONAL LEARNING

A fee for the assessment of extrainstitutional learning shall be charged.

Rationale

Assessing extrainstitutional learning, transcripting the credits awarded, etc., are valuable services to students. These services require faculty and staff time. Students, should, therefore, pay a fee for these services just as they pay fees for other institutional services.

Example of Application of the Policy

The fee for assessing extrainstitutional learning and transcripting the results at local college is based on its average hourly salaries for various faculty ranks (assistant professor, professor, etc.) and administrative positions plus the costs for benefits and administrative services which are computed as percentages of salaries. The average hourly salaries are multiplied by the amount of time spent in performing the evaluation and administrative functions, rounded up to the half hour. The fee for assessing Mr. Smith's extrainstitutional learning and transcripting the results was $1240.

IX. EXAMPLE POLICY ON OVERSIGHT OF THE "PROGRAM"/ACTIVITY

An oversight committee, broadly representative of the institution's divisions/departments, shall be appointed (by the "program" administrator or through the institution's usual means of appointing committees). The major function of this committee is to review the credit award decisions to assure consistency across the institution.

Rationale

To protect the integrity and credibility of the activity, consistency is absolutely necessary. If either the faculty or the students perceive that credit awards are made capriciously, their confidence in the activity will be lost. When this occurs, the activity will no longer be viable.

Example Application of the Policy

The Oversight Committee detected that, when a faculty evaluation team for portfolio assessment was comprised of three specific faculty members, the credit awards seemed overly "generous." The Oversight Committee recommended to the administrator that these three faculty members not comprise a team. The administrator discussed the credit-award information with the three faculty members, and they agreed with the Oversight Committee's recommendation, which was then implemented.

X. EXAMPLE POLICY ON "PROGRAM" EVALUATION

The "Program" for assessing extrainstitutional learning shall be evaluated on an annual basis by the academic vice-president or his designee(s).

Rationale

It is sound administrative practice to evaluate "programs" on a regular basis to determine if "program" objectives are being met. It is especially important to evaluate new "programs" so that problems are detected early on and changes can be made, if necessary. Evaluation should probably be done jointly by "program" staff and the staff of the Office of Institutional Research.

Example of Application of the Policy

At Local College, responsibility for "program" evaluation is assigned to the Office of Institutional Research, but the Office involves the "program" staff. "Program" evaluation includes but is not limited to, an in-depth look at the academic performance of students who have received credit for extrainstitutional learning. Recent data showed that their overall performance, in terms of GPA, was higher than that of the student body as a whole; the average number of credits carried each semester was equivalent to the average number carried by the student body as a whole; they were enrolled in all academic departments of the college, etc. Credit awards were examined by age categories and by assessment instrument used. This information was useful to the college faculty and administrators in deciding to continue the "program."

revised 11/95

Course Exhibits

AF

AF-0101-0005

PEST MANAGEMENT SPECIALIST BY CORRESPONDENCE

Course Number: 56650.
Location: Extension Course Institute, Gunter AFS, AL.
Length: Maximum, 52 weeks.
Exhibit Dates: 1/90–8/98.
Learning Outcomes: Upon completion of the course, the student will be able to develop, implement, conduct, and evaluate management plans and programs in prevention and control of insect, animal, and plant pests. The student will be able to conduct pest management surveys to determine type and extent of pest management actions; apply control principles and prepare management plans; conduct preventive plans; operate and calibrate pesticide dispersal equipment, select proper chemical treatment, and apply pesticides; maintain tools, equipment, and ensure proper use of personal protective equipment; and supervise pest management personnel by serving as a crew leader.
Instruction: Student enrolling must have an Air Force Specialty Code assignment as a prerequisite. This provides an application of the principles learned to real requirements and constitutes an essential part of the instruction by serving as the laboratory component of the course. The student's supervisor must complete an evaluation survey of the course and its effectiveness when the student takes the final examination which measures the student's mastery of the cognitive content. Course consists of self-paced independent study with open book progress checks and satisfactory completion of a proctored final examination. Correspondence materials include instruction in identification of pests as insect, rodent, bird, reptile (venomous), or vegetative; chemicals and dispersion equipment and safety requirements; disease transmission by pests; economic factors of pests and pest management; and general military contingencies. Emphasis is placed upon identification and abatement of pest problems.
Credit Recommendation: In the lower-division baccalaureate/associate degree category, 3 semester hours in sanitation and/or pest control management (environmental/health technology) (11/91).

AF-0104-0002

DIET THERAPY SUPERVISOR BY CORRESPONDENCE

Course Number: 92670.
Location: Extension Course Institute, Gunter AFS, AL.
Length: Maximum, 52 weeks.
Exhibit Dates: 4/90–Present.
Learning Outcomes: Upon completion of the course, the student will be able to discuss the concepts of supervision and management as related to the delivery of care of therapeutic dietetics in a treatment setting.
Instruction: Student enrolling must have an Air Force Specialty Code (AFSC) assignment as a prerequisite. This provides an application of the principles learned to real requirements, and constitutes an essential part of the instruction by serving as the laboratory component of the course. The student's supervisor must complete an evaluation survey of the course and its effectiveness when the student takes the final examination which measures the student's mastery of the cognitive content. Course consists of self-paced independent study with open book progress checks and satisfactory completion of a final examination. Content includes dietetic planning and management, dietary consultation, patient relations and principles of special diets. Management concepts emphasize principles of facility sanitation, food handling, and financial, functional, and equipment management.
Credit Recommendation: In the upper-division baccalaureate category, 2 semester hours in food production management, and 2 in food safety and sanitation (11/91).

AF-0104-0003

DIET THERAPY SPECIALIST BY CORRESPONDENCE

Course Number: 92650.
Location: Extension Course Institute, Gunter AFS, AL.
Length: Maximum, 52 weeks.
Exhibit Dates: 1/90–4/98.
Learning Outcomes: Upon completion of this course, students will be able to describe the aspects of management, food production, and therapeutic nutrition appropriate to dietetic support as a diet therapy specialist. Students will be able to feed quality, nutritious food to patients and to provide dietary education in response to patient problems.
Instruction: Student enrolling must have an Air Force Specialty Code assignment as a prerequisite. This provides an application of the principles learned to real requirements and constitutes an essential part of the instruction by serving as the laboratory component of the course. The student's supervisor must complete an evaluation survey of the course and its effectiveness when the student takes the proctored final examination which measures the student's mastery of the cognitive content. Course consists of self-paced independent study with open book progress checks and satisfactory completion of a final examination. The course topics include responsibilities of a diet specialist including sanitation, safety and security, management principles of scheduling, and maintaining funds and financial reports; quality assurance; subsistence control; medical material procedures; energy conservation; medical readiness; field kitchen management; menu planning; food preparation; food serving techniques; nutrition and dietetics; therapeutic nutrition and diet modification; and community nutrition, communication, and consultation.
Credit Recommendation: In the upper-division baccalaureate category, 2 semester hours in food safety and sanitation, 1 in food preparation, and 1 in principles of nutrition (11/91).

AF-0401-0001

PUBLIC AFFAIRS SPECIALIST BY CORRESPONDENCE

Course Number: *Version 1:* 79150. *Version 2:* 79150.
Location: Extension Course Institute, Gunter AFS, AL.
Length: *Version 1:* Maximum, 52 weeks. *Version 2:* Maximum, 52 weeks.
Exhibit Dates: *Version 1:* 11/91–Present. *Version 2:* 1/90–10/91.
Learning Outcomes: *Version 1:* Upon completion of the course, the student will be able to describe the basic principles and functions of public relations activities. In addition, the student will be able to gather news through interviews and research; write news, sports, and feature stories; follow editorial guidelines; and plan and use visual communication media. *Version 2:* Upon completion of the course, the student will be able to identify and describe the basic principles and functions of public relations activities. Specific knowledge and skills include news gathering and reporting techniques, principles of newswriting, identifying and composing feature stories, the use of photography and art work, and the types of files and documents that are useful. Student will be able to describe the different requirements and activities of internal information, media relations, and community relations. With appropriate on-the-job exercises, the student will have entry-level skills for a public relations position.
Instruction: *Version 1:* This is a two-part correspondence course which provides an overview of public affairs and its administration. Course topics include how to get information, writing, editing, and visual communication, as well as the disciplines of internal information, media relations, and community relations. *Version 2:* This is a two-part correspondence course available to personnel assigned as public affairs specialists. This provides an application of the principles learned to real requirements and constitutes an essential part of the instruction by serving as the laboratory component of the course. The course materials consist of two written booklets with 98 exercises to measure student mastery. References are provided for additional sources of information but no requirements are imposed for using the references. Course materials are reviewed and revised for currency annually. The delivery method is self-instruction with a proctored examination administered upon completion of the course. Phone numbers of course authors are provided. The student's supervisor must complete an evaluation survey of the course and

its effectiveness when the student takes the final exam which measure students mastery of the cognitive content. A score of 65 is required for a satisfactory completion certificate.

Credit Recommendation: *Version 1:* In the lower-division baccalaureate/associate degree category, 1 semester hour in journalism and 2 in introduction to public relations principles and practices (4/93). *Version 2:* In the lower-division baccalaureate/associate degree category, 2 semester hours in introduction to public relations principles and practices (3/91).

AF-0401-0002

PUBLIC AFFAIRS OFFICER BY CORRESPONDENCE

Course Number: 07920.
Location: Extension Course Institute, Gunter AFS, AL.
Length: 52 weeks.
Exhibit Dates: 1/90–Present.
Learning Outcomes: Upon completion of the course, the student will be able to plan and conduct public affairs activities for an organization. Specific knowledge and skills include internal information programs, principles of newswriting, feature story composition, news editing and layout, newspaper publishing, principles and practices of dealing with the press in both print and electronic media, community relations activities, organizational principles and personnel supervision related to public affairs activities, and public affairs program administration. The student will have entry-level skills to supervise a public affairs office.
Instruction: This is a five part correspondence course with printed booklets and workbook exercises for each component. There are self-administered tests for the student to measure progress. Only students whose job positions involve public affairs are permitted to enroll and this provides an internship or laboratory component for the instruction. The student's supervisor must complete an evaluation survey of the course and its effectiveness when the student takes the proctored final exam which measures student mastery of the cognitive content. A score of 65 is required for a satisfactory completion certificate.
Credit Recommendation: In the lower-division baccalaureate/associate degree category, 3 semester hours in principles and practices of public affairs (3/91); in the upper-division baccalaureate category, 1 semester hour in public affairs management and planning (3/91).

AF-0419-0004

AIR CARGO SPECIALIST BY CORRESPONDENCE

Course Number: 60551.
Location: Extension Course Institute, Gunter AFS, AL.
Length: Maximum, 52 weeks.
Exhibit Dates: 1/90–6/92.
Objectives: This course provides a basic introduction to air cargo policies and procedures, material handling equipment, and occupational safety standards.
Instruction: Course instruction covers career opportunities, terminology, shipping and receiving documents, storage, hazardous materials, loading and unloading procedures, weight limitations, and various equipment handling systems.
Credit Recommendation: In the lower-division baccalaureate/associate degree category, 3 semester hours in basics in air cargo (10/85).

AF-0419-0021

AIR PASSENGER SPECIALIST BY CORRESPONDENCE

Course Number: 60550.
Location: Extension Course Institute, Gunter AFS, AL.
Length: Maximum, 52 weeks.
Exhibit Dates: 1/90–6/92.
Objectives: This course provides background information related to the duties and responsibilities of a military air passenger specialist. Procedures for processing passenger reservations are detailed. Supervision and employee/customer relations are also discussed.
Instruction: Course instruction covers management/supervisory functions, passenger safety, air transport regulations, documentation, aircraft identification, and automated data processing of passenger reservations.
Credit Recommendation: Credit is not recommended because of the limited, specialized nature of the course (2/87).

AF-0419-0023

VEHICLE OPERATIONS SUPERVISOR BY CORRESPONDENCE

Course Number: 60370.
Location: Extension Course Institute, Gunter AFS, AL.
Length: Maximum, 52 weeks.
Exhibit Dates: 1/90–10/96.
Objectives: This course provides initial training for vehicle operations supervisors in planning and scheduling vehicle activities, personnel management, inspection and evaluation, and maintenance operations.
Instruction: Course instruction covers care and servicing of wheeled vehicles, personnel employment standards, fuel consumption, maintenance checks, on-the-job training needs, as well as supervisory management policies and practices.
Credit Recommendation: In the lower-division baccalaureate/associate degree category, 1 semester hour in principles of supervision (10/85).

AF-0419-0026

APPRENTICE VEHICLE OPERATOR/DISPATCHER BY CORRESPONDENCE

Course Number: 60330.
Location: Extension Course Institute, Gunter AFS, AL.
Length: Maximum, 52 weeks.
Exhibit Dates: 1/90–7/97.
Objectives: To provide apprentice training in passenger/cargo vehicle selection and operation. General upkeep and maintenance training are also included.
Instruction: Course training covers vehicle operations, defensive driving, emergency repairs, passenger/cargo load factors, and dispatching procedures.
Credit Recommendation: Credit is not recommended because of the limited, specialized nature of the course (2/87).

AF-0419-0028

SERVICES SUPERVISOR BY CORRESPONDENCE

Course Number: 61170.
Location: Extension Course Institute, Gunter AFS, AL.
Length: Maximum, 52 weeks.
Exhibit Dates: 1/90–9/93.
Learning Outcomes: Upon completion of the course, the student will be able to describe how to manage equipment, supplies, furnishings; understand services readiness mission; and follow government procedures dealing with safety, operational publications, billeting, and mortuary affairs.
Instruction: Course includes the management of resources and assets, supervisory responsibilities for safety, operational publications, mortuary affairs, billeting financial management, and front desk procedures and administration of equipment, supplies, furnishing, and service readiness missions.
Credit Recommendation: Credit is not recommended because of the military-specific nature of the course (2/87).

AF-0419-0029

APPRENTICE AIR CARGO SPECIALIST BY CORRESPONDENCE

Course Number: 60531.
Location: Extension Course Institute, Gunter AFS, AL.
Length: Maximum, 52 weeks.
Exhibit Dates: 1/90–6/93.
Learning Outcomes: Upon completion of this course the student will be able to function as a beginning air cargo specialist with knowledge of cargo documentation, storage, and handling.
Instruction: Topics included in this course are military airlift services, aerial port work centers, cargo storage and handling, palletization, load planning procedures, aircraft loading and off-loading, and aircraft ground servicing.
Credit Recommendation: Credit is not recommended because of the military-specific nature of the course (2/87).

AF-0419-0031

PASSENGER AND HOUSEHOLD GOODS SPECIALIST BY CORRESPONDENCE

Course Number: 60250.
Location: Extension Course Institute, Gunter AFS, AL.
Length: Maximum, 52 weeks.
Exhibit Dates: 1/90–10/95.
Learning Outcomes: Upon completion of this course the student will be able to function as a specialist in moving people and personal property throughout the world.
Instruction: Topics covered include passenger transportation and supervision, safety, supervision and security, movement of personal property, and transportation documentation.
Credit Recommendation: In the lower-division baccalaureate/associate degree category, 3 semester hours in introduction to transportation (2/87).

AF-0419-0032

FREIGHT TRAFFIC SPECIALIST BY CORRESPONDENCE

Course Number: 60251.
Location: Extension Course Institute, Gunter AFS, AL.
Length: Maximum, 52 weeks.
Exhibit Dates: 1/90–10/95.
Learning Outcomes: At the completion of this course the student will have knowledge of the fundamentals of freight traffic management, traffic, safety, planning and packaging of ship-

ments and freight routing, documentation, and expediting.

Instruction: Topics covered include transportation planning, packaging, modes of transportation, preparation of transportation documents, scheduling and routing, freight expediting, and training.

Credit Recommendation: In the lower-division baccalaureate/associate degree category, 3 semester hours in principles of transportation/traffic management (2/87).

AF-0419-0033

BASE TRANSPORTATION OFFICER

Course Number: J30RR6051 002.
Location: 396 TCHTG/CC, Sheppard AFB, TX.
Length: 2 weeks (80 hours).
Exhibit Dates: 12/91–Present.
Learning Outcomes: Upon completion of the course, the student will be able to demonstrate knowledge and skills needed to function as a base transportation officer. This involves familiarity with the techniques and procedures of advanced traffic management; vehicle management/maintenance; transportation planning, programming, and operations, including military and commercial air, motor, rail, water, and pipeline movements; terminal management; and the management of transportation resources.
Instruction: Methods include classroom lectures and discussions, demonstration/performance, computer-aided simulations, oral and written reports, exercises, and applied case studies. Topics include transportation regulations, movement of personnel and cargo, hazardous material management, vehicle acquisition, asset management, maintenance management, and mobility planning.
Credit Recommendation: In the graduate degree category, 3 semester hours in transportation management (3/93).

AF-0419-0034

TRANSPORTATION OFFICER

Course Number: J30BR6051 000.
Location: 396 TCHTG/CC, Sheppard AFB, TX.
Length: 10 weeks (400 hours).
Exhibit Dates: 3/90–Present.
Learning Outcomes: Upon completion of the course, the student will be familiar with the knowledge and skills needed to perform the duties of a transportation officer. This includes organization of Air Force functions, financial management, base supply organization and procedures, manpower documents, cargo movement, packaging, passenger and personal property movements, airlift operations, cargo/mail operations, aircraft load planning, air terminal functions and organizations, passenger operations, commercial airlift augmentation, duty officer functions and responsibilities, vehicle maintenance and management responsibilities, transportation planning, interrelation of base level functions, and aerial port operations.
Instruction: Classroom lectures and discussions, performance, directed study, assigned readings, and applied exercises provide practice and evaluation of course objectives. Topics covered include transportation organization, freight, passenger and personal property traffic management, vehicle operations and maintenance management, airlift, aircraft load planning, transportation planning, and mobility planning.
Credit Recommendation: In the graduate degree category, 3 semester hours in operations management and 3 in transportation and logistics (5/93).

AF-0419-0035

TRANSPORTATION STAFF OFFICER

Course Number: J30AR6011 000.
Location: 396 TCHTG/CC, Sheppard AFB, TX.
Length: 5 weeks (200 hours).
Exhibit Dates: 1/90–Present.
Learning Outcomes: Upon completion of the course, the student will be familiar with the knowledge and skills needed to perform the duties of squadron commanders, equivalent chiefs of transportation, and field grade transportation officers assigned to each service or joint staff above wing level. This includes transportation and maintenance, automated transportation data system, current developments and innovations in transportation operations and management, and contracts/contracting.
Instruction: Methods include classroom lectures and discussions, directed study, field trips, and assigned readings. Topics include national transportation policy, traffic management responsibilities, operations, regulatory agencies, air transportation management, vehicle management and maintenance, joint planning, Air Force planning, base planning, automated transportation data system, financial management, and contracts/contracting.
Credit Recommendation: In the graduate degree category, 3 semester hours in transportation planning and 3 in operations management (5/93).

AF-0419-0036

AERIAL PORT OPERATION AND MANAGEMENT

Course Number: J30RR6011 000.
Location: 396 TCHTG/CC, Sheppard AFB, TX.
Length: 2 weeks (80 hours).
Exhibit Dates: 4/92–Present.
Learning Outcomes: Upon completion of the course, the student will be able to demonstrate knowledge and skills needed to to perform the duties of aerial port operations and management. This includes air lift management, hazardous materials, vehicle management, and mobility planning.
Instruction: Methods include classroom lectures and discussions, student critiques, assigned readings, exercises, computer simulations, and case studies. Topics include air lift operations, aerial port operations, passenger movements, cargo/mail movements, transportation of hazardous materials, air lift load planning, mobility planning, and vehicle management.
Credit Recommendation: In the graduate degree category, 3 semester hours in operations management (3/93).

AF-0419-0037

TRAFFIC MANAGEMENT JOURNEYMAN BY CORRESPONDENCE

Course Number: 60253.
Location: Extension Course Institute, Gunter AFS, AL.
Length: Maximum, 52 weeks.
Exhibit Dates: 3/93–Present.
Learning Outcomes: Upon completion of the course, the student will be able to plan, document, and meet requirements for the transportation of materials, equipment, and personnel, including personal belongings, to all parts of the world.
Instruction: Course introduces the fundamentals of the transportation systems which provide for the movement of materials, equipment, and personnel, including their belongings, to all parts of the world. Emphasis is given to shipment planning and packaging, safety mobility, preservation, packaging control, reusable containers, and documentation.
Credit Recommendation: In the lower-division baccalaureate/associate degree category, 1 semester hour in transportation management (6/94).

AF-0419-0038

AIR TRANSPORTATION JOURNEYMAN BY CORRESPONDENCE

Course Number: 60555.
Location: Extension Course Institute, Gunter AFS, AL.
Length: Maximum, 52 weeks.
Exhibit Dates: 5/93–7/99.
Learning Outcomes: Upon completion of the course, the student will have the skill required to function as a beginning air transportation journeyman and have knowledge of materials handling, cargo documentation, cargo storage, and cargo handling. The student will also have the skills required to determine aircraft and passenger loading for correct weight and balance.
Instruction: Topics included in this correspondence course are military airlift services, aerial port work centers, palletization, load planning procedures, aircraft loading and off-loading, and passenger manifesting.
Credit Recommendation: Credit is not recommended because of the military-specific nature of the course (6/94).

AF-0504-0003

TELECOMMUNICATIONS OPERATIONS SPECIALIST BY CORRESPONDENCE

Course Number: 29150.
Location: Extension Course Institute, Gunter AFS, AL.
Length: Maximum, 52 weeks.
Exhibit Dates: 1/90–1/97.
Objectives: The purpose of this course is to provide information about the composition, organization, and utilization of communications systems.
Instruction: This course covers communications security, publications, telephone operations, budgeting, and processing of messages by teletypewriter, magnetic tape, keypunch, and standard remote terminals. Instruction includes a basic understanding of the main components and peripheral devices of a computerized communications network.
Credit Recommendation: In the vocational certificate category, 2 semester hours in communications (10/85); in the lower-division baccalaureate/associate degree category, 2 semester in information processing theory. Credit should be awarded at the vocational certificate level or at the lower-division level (10/85).

4 COURSE EXHIBITS

AF-0505-0002

RADIO AND TELEVISION BROADCASTING SPECIALIST BY CORRESPONDENCE

Course Number: 79151.
Location: Extension Course Institute, Gunter AFS, AL.
Length: Maximum, 52 weeks.
Exhibit Dates: 1/90–Present.
Learning Outcomes: Upon completion of the course, the student will be able to plan, perform, and coordinate radio and television production operations for studio and field production.
Instruction: This is a two part correspondence course. Instruction is supplied through reading, exercises, and examinations. Topics include audio production, video production, lighting, graphics, set design, and field production. Students enrolling must have an AFSC assignment as a prerequisite. This provides an application of the principles learned to real requirements and constitutes an essential part of the instruction by serving as the laboratory component of the course. The student's supervisor must complete an evaluation survey of the course and its effectiveness when the student takes the proctored final exam which measures student mastery of the cognitive content. A score of 65 is required for a satisfactory completion certificate.
Credit Recommendation: In the upper-division baccalaureate category, 4 semester hours in radio-film-TV field production (3/91).

AF-0505-0003

VISUAL INFORMATION PRODUCTION-DOCUMENTATION SPECIALIST BY CORRESPONDENCE

Course Number: 23250.
Location: Extension Course Institute, Gunter AFS, AL.
Length: Maximum, 52 weeks.
Exhibit Dates: 1/90–1/94.
Learning Outcomes: Upon completion of the course, the student will be able to manage and perform all duties relating to an audiovisual production documentation unit. These include the operation and maintenance of motion picture and television equipment.
Instruction: This is a five-part correspondence course, taught through reading, exercises, and examinations. Topics include the basic concepts of motion picture and television methods and procedures, including composition, optics, exposure, sound recording, editing, and equipment. Students enrolling must have an AFSC assignment as a prerequisite. This provides an application of the principles learned to real requirements and constitutes an essential part of the instruction by serving as the laboratory component of the course. The student's supervisor must complete an evaluation survey of the course and its effectiveness when the student takes the proctored final exam which measures student mastery of the cognitive content. A score of 65 is required for a satisfactory completion certificate.
Credit Recommendation: In the lower-division baccalaureate/associate degree category, 6 semester hours in introduction to radio-TV-film (3/91); in the upper-division baccalaureate category, 3 semester hours in intermediate radio-TV-film (3/91).

AF-0505-0004

CIVIL AIR PATROL PUBLIC AFFAIRS OFFICER BY CORRESPONDENCE

Course Number: 02010.
Location: Extension Course Institute, Gunter AFS, AL.
Length: Maximum, 52 weeks.
Exhibit Dates: 1/90–Present.
Learning Outcomes: Upon completion of the course, the student will be able to perform all public affairs officer responsibilities for the Civil Air Patrol through the news media.
Instruction: This is a two-part correspondence course. The method of instruction is through reading, exercises and examination. Topics include writing for newsletters, magazines, newspapers, news releases, radio, and television and the preparation and distribution of these materials to internal and external consumers and media. Students enrolling must have an AFSC assignment as a prerequisite. This provides an application of the principles learned to real requirements and constitutes an essential part of the instruction by serving as the laboratory component of the course. The student's supervisor must complete and evaluation survey of the course and its effectiveness when the student takes the proctored final exam which measures student mastery of the cognitive content.
Credit Recommendation: In the lower-division baccalaureate/associate degree category, 1 semester hour in radio-TV-film journalism (3/91).

AF-0505-0005

VISUAL INFORMATION PRODUCTION-DOCUMENTATION JOURNEYMAN BY CORRESPONDENCE
(Visual Information Production-Documentation Specialist by Correspondence)

Course Number: 23153.
Location: Extension Course Institute, Gunter AFS, AL.
Length: Maximum, 52 weeks.
Exhibit Dates: 1/92–1/97.
Learning Outcomes: Upon completion of the course, the student will meet qualifications as a journeyman-level visual information production-documentation specialist and understand the fundamentals of motion picture and media production including the operation and maintenance of equipment.
Instruction: This correspondence course consists of self-paced, independent instruction. Topics covered include fundamentals of motion picture/media operation; safety and security; optics; recorded image; camera operation; frame, composition, and movement; equipment testing; the craft of constructing a motion media picture; the skills needed to support the production of the videographer; combat camera documentation; and television production.
Credit Recommendation: In the lower-division baccalaureate/associate degree category, 6 semester hours in introduction to radio-TV-film (6/94); in the upper-division baccalaureate category, 3 semester hours in intermediate radio-TV-film (6/94).

AF-0701-0003

DENTAL ASSISTANT SPECIALIST BY CORRESPONDENCE

Course Number: 98150.
Location: Extension Course Institute, Gunter AFS, AL.
Length: Maximum, 52 weeks.
Exhibit Dates: 1/90–12/96.
Learning Outcomes: Upon completion of the course, the student will be able to identify and explain the duties of the dental assistant specialist.
Instruction: Monitored home-study includes readings, written exercises, on-the-job training, and proctored examinations. Topics include role and scope of the dental assistant specialist, interpersonal communication, basic sciences, chemistry, histology, dental materials, therapeutic agents, dental instruments, preventive dentistry, nutrition, prophylaxis, dental radiography, and chairside assisting.
Credit Recommendation: In the lower-division baccalaureate/associate degree category, 3 semester hours in oral anatomy and physiology and 3 in instruments of oral prophylaxis (5/88).

AF-0701-0018

DENTAL LABORATORY SPECIALIST BY CORRESPONDENCE

Course Number: 98250.
Location: Extension Course Institute, Gunter AFS, AL.
Length: Maximum, 52 weeks.
Exhibit Dates: 1/90–Present.
Learning Outcomes: Upon completion of the course the student will be able to identify and explain the correct procedures, techniques, devices, and appliances developed in the general dental laboratory.
Instruction: Monitored home-study includes readings, written exercises, on-the-job supervision, and proctored examinations. Topics include tooth anatomy, mandibular movement, articulation, occlusion, crown and fixed dentures, waxing, spruing, investing, burnout, casting, finishing, polishing, denture construction, and removable partial dentures.
Credit Recommendation: In the lower-division baccalaureate/associate degree category, 3 semester hours in oral anatomy and physiology and 5 in dental laboratory technology (5/88).

AF-0702-0008

MEDICAL LABORATORY CRAFTSMAN (CHEMISTRY AND URINALYSIS) BY CORRESPONDENCE
(Medical Laboratory Technician (Chemistry and Urinalysis) by Correspondence)

Course Number: 92470A.
Location: Extension Course Institute, Gunter AFS, AL.
Length: Maximum, 52 weeks.
Exhibit Dates: 1/90–Present.
Learning Outcomes: Upon completion of the course, the student will be able to apply current principles of laboratory material and general laboratory safety; demonstrate knowledge of general laboratory administration with an emphasis on laboratory ethics and interpersonal relations, laboratory reports, and accreditation programs; apply knowledge and understanding of basic chemistry, basic nature and concentration of laboratory solutions, basic principles of photometry and spectrophotomy; demonstrate knowledge of the principles, techniques, and instruments used in automation; and demonstrate knowledge of the principles and techniques for collection and handling of

specimens. Student will also be able to apply quality control in the clinical laboratory and apply principles and techniques in performing chemical analysis in the clinical laboratory.

Instruction: Student enrolling must have an Air Force Specialty Code assignment as a prerequisite. This provides an application of the principles learned to real requirements and constitutes an essential part of the instruction by serving as the laboratory component of the course. The student's supervisor must complete an evaluation survey of the course and its effectiveness when the student takes the proctored final examination which measures mastery of the cognitive content. Course consists of self-paced, independent study with open book progress checks and satisfactory completion of a final examination. Topics consist of general laboratory safety, general laboratory administration, clinical chemistry, urinalysis, specimen procurement, and quality control.

Credit Recommendation: In the lower-division baccalaureate/associate degree category, 3 semester hours in clinical chemistry and 1 in urinalysis (11/91).

AF-0702-0009

MEDICAL LABORATORY CRAFTSMAN (MICROBIOLOGY) BY CORRESPONDENCE
(Medical Laboratory Technician (Microbiology) by Correspondence)

Course Number: 92470B.
Location: Extension Course Institute, Gunter AFS, AL.
Length: Maximum, 52 weeks.
Exhibit Dates: 3/91–2/99.
Learning Outcomes: Upon completion of the course, the student will be able to apply laboratory techniques and use the equipment necessary for safe handling of infectious microorganisms; explain how knowledge of bacterial morphology and physiology permits isolation and cultivation of pathogenic microbes; apply knowledge of basic antimicrobial sensitivity testing and drug assays; apply quality control to microbiology; identify gram-positive organisms of greatest significance to humans; identify gram-negative cocci and coccoid organisms; identify acid-fast bacilli and spirochetes; and apply knowledge and techniques in identifying mycological and parasitological pathogens of medical importance.
Instruction: Student enrolling must have an Air Force Specialty Code assignment as a prerequisite. This provides an application of the principles learned to real requirements and constitutes an essential part of the instruction by serving as the laboratory component of the course. The student's supervisor must complete an evaluation survey of the course and its effectiveness when the student takes the proctored final examination which measures mastery of the cognitive content. Course consists of self-paced, independent study with open book progress checks and satisfactory completion of a final examination. Topics include selection, collection, and transport of bacteriological specimens, theoretical concepts and basic techniques in clinical microbiology, theoretical concepts and basic techniques in parasitology, and theoretical concepts and basic techniques in mycology.
Credit Recommendation: In the lower-division baccalaureate/associate degree category, 1 semester hour in mycology/parasitology and 3 in microbiology (11/91).

AF-0702-0010

MEDICAL LABORATORY CRAFTSMAN (HEMATOLOGY, SEROLOGY, BLOOD BANKING AND IMMUNOHEMATOLOGY) BY CORRESPONDENCE
(Medical Laboratory Technician (Hematology, Serology, Blood Banking and Immunohematology) by Correspondence)

Course Number: 92470C.
Location: Extension Course Institute, Gunter AFS, AL.
Length: Maximum, 52 weeks.
Exhibit Dates: 6/90–Present.
Learning Outcomes: Upon completion of the course, the student will be able to procure blood samples for hematological examination; apply the knowledge of manual cell count methods; apply knowledge of leukocyte differential testing; apply knowledge of hemoglobin and hematocrit testing analysis; apply knowledge of erythrocyte studies and testing; apply knowledge of the hemostatic mechanism and the methods of testing available; apply quality assurance principles in hematology and blood banking; apply basic antigen-antibody reactions in transfusion and transfusion practices; identify donor selection and collection; apply laboratory techniques and equipment in blood storage and shipment in the Air Force program; apply basic concepts and techniques in serology testing; and apply general concepts and causes of hepatitis in hepatitis testing.
Instruction: Student enrolling must have an Air Force Specialty Code assignment as a prerequisite. This provides an application of the principles learned to real requirements, and constitutes an essential part of the instruction by serving as the laboratory component of the course. The student's supervisor must complete an evaluation survey of the course and its effectiveness when the student takes the proctored final examination which measures mastery of the cognitive content. Course consists of self-paced, independent study with open book progress checks and satisfactory completion of a final examination. Topics include theory and basic fundamentals of hematology; theory and basic fundamentals of serology; theory and basic fundamentals of blood banking; and theory and basic fundamentals of immunohematology.
Credit Recommendation: In the lower-division baccalaureate/associate degree category, 2 semester hours in hematology/immunohematology, 1 in blood banking, and 1 in serology (11/91).

AF-0703-0011

MEDICAL SERVICE TECHNICIAN BY CORRESPONDENCE

Course Number: 90270.
Location: Extension Course Institute, Gunter AFS, AL.
Length: Maximum, 52 weeks.
Exhibit Dates: 1/90–10/91.
Objectives: To prepare personnel to assist in patient treatment and to learn supervisory management skills at the unit level.
Instruction: This correspondence course includes the teaching of administrative skills, administration of medication, management functions and techniques of patient care, field medical care and sanitation, field preparation and disaster preparedness, oxygen therapy and resuscitation, therapy techniques, supervision of medical service activities, and preoperative and postoperative nursing. This course includes procedures found in licensed practical nursing programs.
Credit Recommendation: In the lower-division baccalaureate/associate degree category, 3 semester hours in nursing care (7/84).

AF-0703-0012

MEDICAL SERVICE SPECIALIST BY CORRESPONDENCE

Course Number: 90250.
Location: Extension Course Institute, Gunter AFS, AL.
Length: Maximum of, 52 weeks.
Exhibit Dates: 1/90–11/92.
Objectives: To prepare personnel to assist in nursing care and treatment of patients.
Instruction: This correspondence course teaches principles of general nursing care techniques; the student learns to assist in medical examinations and treatments; perform procedures using therapeutic equipment; peform aeromedical evacuation duties and independent duties; and learns body systems and their disorders, signs and symptoms; student also covers principles of diagnostic and therapeutic procedures; adult, geriatric, amd pediatric medicine are introduced as well as mental illness, burned patient trauma, critical care, and chronic diseases; and pathophysiology of common disorders. The course is equivalent to the basic courses commonly found in LPN programs.
Credit Recommendation: In the lower-division baccalaureate/associate degree category, 3 semester hours in basic principles of nursing assisting. Course concludes with a proctored examination (7/84).

AF-0703-0013

NURSE ANESTHETIST

Course Number: 50AY9746.
Location: Wilford Hall USAF Medical Center, Lackland AFB, TX.
Length: 104 weeks (5069-5246 hours).
Exhibit Dates: 1/90–1/97.
Objectives: To instruct qualified Air Force nurse corps officers in both the art and science of anesthesia so that they may become certified registered nurse anesthetists.
Instruction: Areas of instruction include hospital orientation; professional aspects of anesthesiology; anatomy, physiology, and pathophysiology; chemistry, physics, and biochemistry; pharmacology; principles of anesthesia practice; basic and advanced cardiac life support; EKG; research paper; educational review; clinical conference; journal club; seminars; and clinical experience.
Credit Recommendation: In the upper-division baccalaureate category, 6 semester hours in anatomy/physiology I, 6 in anatomy/physiology II, 3 in pathophysiology, and 4 in chemistry/physics/biochemistry of the digestive, neural, and respiratory systems (9/85); in the graduate degree category, 8 in pharmacology, 8 in principles of anesthesia practice, 8 in clinical conference/journal club, and 6 in clinical experience (9/85).

AF-0703-0014

NURSING SERVICE MANAGEMENT BY CORRESPONDENCE

Course Number: 9711.

Location: Extension Course Institute, Gunter AFS, AL.
Length: Maximum, 52 weeks.
Exhibit Dates: 1/90–4/97.
Learning Outcomes: Upon completion of the course the student will understand theory and principles of management of a nursing unit, personnel evaluation, and assignment and supervision of staff for patient care.
Instruction: Monitored home study course includes reading, written exercises, on-the-job supervision, and proctored examination. Topics include administration, management, personnel evaluation, and management of patient-centered nursing care. The nursing process includes patient assessment for care plans and staff assignment.
Credit Recommendation: In the upper-division baccalaureate category, 3 semester hours in nursing management (5/88).

AF-0703-0015

SURGICAL SERVICE SPECIALIST BY CORRESPONDENCE

Course Number: 90252.
Location: Extension Course Institute, Gunter AFS, AL.
Length: Maximum, 52 weeks.
Exhibit Dates: 1/90–8/98.
Learning Outcomes: Upon completion of the course the student will be able to assist in operating room aseptic techniques, manage operating room supplies and equipment, and deliver preoperative and postoperative patient care.
Instruction: Monitored home study course includes reading, written exercises, on-the-job supervision, and proctored examinations. Topics include introduction to the study of aseptic techniques, operating room procedures, operating room equipment and supplies, and preoperative and postoperative patient care. Also includes basic medical terminology and anatomy and physiology in relation to surgical procedures.
Credit Recommendation: In the lower-division baccalaureate/associate degree category, 4 semester hours in operating room techniques and procedures (5/88).

AF-0703-0016

MEDICAL SERVICE SPECIALIST BY CORRESPONDENCE

Course Number: 90250B.
Location: Extension Course Institute, Gunter AFS, AL.
Length: Maximum, 52 weeks.
Exhibit Dates: 1/90–Present.
Learning Outcomes: Upon completion of the course, the student will be able to describe the application of basic nursing techniques to the treatment of patients.
Instruction: Student enrolling must have an Air Force Specialty Code assignment as a prerequisite. This provides an application of the principles learned to real requirements and constitutes an essential part of the instruction by serving as the laboratory component of the course. The student's supervisor must complete an evaluation survey of the course and its effectiveness when the student takes the proctored final examination which measures mastery of the cognitive content. Course consists of self-paced, independent study with open book progress checks and satisfactory completion of a final examination. Topics include nursing procedures related to patient care, planning, comfort, infection control, aspects of drug administration; and specialty care for obstetrical and newborn patients. Also included is care of surgery patients, mental and systematic disorders, and basic concepts of emergency and clinical care.
Credit Recommendation: In the lower-division baccalaureate/associate degree category, 4 semester hours in nursing procedures (11/91).

AF-0703-0017

MENTAL HEALTH SERVICE SPECIALIST BY CORRESPONDENCE

Course Number: 91450.
Location: Extension Course Institute, Gunter AFS, AL.
Length: Maximum, 52 weeks.
Exhibit Dates: 11/91–11/98.
Learning Outcomes: Upon completion of the course, the student will be able to describe the basic mental health concepts appropriate to the provision of diagnostic and therapeutic mental health care for psychiatric patients.
Instruction: Student enrolling must have an Air Force Specialty Code assignment as a prerequisite. This provides an application of the principles learned to real requirements and constitutes an essential part of the instruction by serving as the laboratory component of the course. The student's supervisor must complete an evaluation survey of the course and its effectiveness when the student takes the proctored final examination which measures mastery of the cognitive content. Course consists of self-paced, independent study with open book progress checks and satisfactory completion of a final examination. Content covers basic concepts of mental health and treatment interventions and procedures including symptoms of mental disorders and techniques of intervention. Basic nursing principles are discussed as related to the care of psychiatric patients. A brief overview of anatomy and physiology is provided.
Credit Recommendation: In the lower-division baccalaureate/associate degree category, 1 semester hour in anatomy and physiology, 1 in nursing procedures, and 2 in mental health principles (11/91).

AF-0703-0018

MEDICAL SERVICE SPECIALIST BY CORRESPONDENCE

Course Number: 90250A.
Location: Extension Course Institute, Gunter AFS, AL.
Length: Maximum, 52 weeks.
Exhibit Dates: 11/90–11/97.
Learning Outcomes: Upon completion of the course, the student will be able to describe basic concepts of nursing care and basic anatomy and physiology that will apply to patient treatment.
Instruction: Student enrolling must have an Air Force Specialty Code assignment as a prerequisite. This provides an application of the principles learned to real requirements and constitutes an essential part of the instruction by serving as the laboratory component of the course. The student's supervisor must complete an evaluation survey of the course and its effectiveness when the student takes the proctored final examination which measures mastery of the cognitive content. Course consists of self-paced, independent study with open book progress checks and satisfactory completion of a final examination. Topics include nursing concepts such as ethical, legal, and safety considerations; resource management; and health maintenance. Sections relating to anatomy and physiology cover the major body systems with examples related to pathological conditions.
Credit Recommendation: In the lower-division baccalaureate/associate degree category, 2 semester hours in anatomy and physiology and 1 in principles of nursing assisting (11/91).

AF-0703-0019

OPERATING ROOM NURSING

Course Number: J50Z046S1000; J3OZP9731 003.
Location: 396 MTG/CC, Sheppard AFB, TX.
Length: 16 weeks (636 hours).
Exhibit Dates: 1/90–Present.
Learning Outcomes: Upon completion of the course, the student will assume perioperative roles and responsibilities of an operating room staff nurse, including circulating, scrubbing, and other patient-related activities and relate Air Force and American Association of Medical Instrumentalogy standards and recommended practices to perioperative care of patients.
Instruction: Course covers fundamentals of operating room practice, specialized dress of operating room practice, elements of operating room practice, directed clinical experience; and duties of scrub nurse and circulating nurse.
Credit Recommendation: In the upper-division baccalaureate category, 19 semester hours in operating room nursing electives (3/93).

AF-0703-0020

NURSING STAFF DEVELOPMENT OFFICER BASIC

Course Number: J30ZR46N3D002; J30ZR9756D002.
Location: Technical Training Center, Sheppard AFB, TX.
Length: 4 weeks (156 hours).
Exhibit Dates: 1/90–Present.
Learning Outcomes: Upon completion of the course, the student will assume the role of staff development officer as defined by the Joint Commission on Accreditation of Health Organizations, American Nursing Association, and Air Force standards.
Instruction: Lectures, discussions, and learning/performance activities cover process of staff development, teaching the adult learner, computers in staff development, research in nursing staff development, ethical and legal principles, marketing of programs, preparing learning objectives, and resources required for staff development programs.
Credit Recommendation: In the graduate degree category, 6 semester hours in a nursing education elective (8/93).

AF-0703-0021

OPERATING ROOM MANAGEMENT

Course Number: J50Z046S3000; J3OZP46S3 002; J3OZP9736 002.

Location: Training Center, Lackland AFB, TX; Training Center, Keesler AFB, MS; Technical Training Center, Sheppard AFB, TX.

Length: 4 weeks (158 hours).

Exhibit Dates: 1/90–Present.

Learning Outcomes: Upon completion of the course, the student will be able to function as an operating room supervisor according to appropriate OSHA, Joint Commission on Accreditation of Health Organizations, American Organization of Operating Room Nurses, and Air Force standards. Student will implement and evaluate quality assurance/risk management programs; manage resources; and develop job descriptions, performance standards, and employee enrichment programs.

Instruction: Lectures, discussions, and performance activities cover total quality management, change conflict, contingency planning and operations, infection control, staff development, medico-legal implications, standards of practice/standardized care, policies and procedures, medical logistics, manpower and budget, and conducting inspections.

Credit Recommendation: In the graduate degree category, 10 semester hours in perioperative nursing (8/93).

AF-0703-0022

OB/GYN NURSE PRACTITIONER

Course Number: J30ZR9756A.

Location: Technical Training Center, Sheppard AFB, TX.

Length: Phase 1, 10 weeks (394 hours); Phase 2, 8 weeks (320 hours).

Exhibit Dates: 1/90–1/96.

Learning Outcomes: Upon completion of Phases 1 and 2, the student demonstrates knowledge and skills appropriate to the expanded role of providing health care to women and is eligible to sit for the NCC Certification examination for OB/GYN nurse practitioners after having completed the six-month preceptorship.

Instruction: Lectures, discussions, performance activities, self-instruction, and clinical practicum cover reproductive anatomy, physical assessment, patient communication, diagnostic therapeutic tools, common GYN problems, pregnancy and postpartum period, family planning and human sexuality, role of the nurse practitioner, principles and practices of obstetrics and gynecology, and clinical practicum in OB/GYN.

Credit Recommendation: In the upper-division baccalaureate category, for Phase 1, 2 semester hours in obstetrical nursing (8/93); in the graduate degree category, for Phase 1, 15 semester hours in obstetrical and gynecological practitioner; for Phase 2, 6 additional semester hours in obstetrical and gynecological practitioner (8/93).

AF-0703-0023

ACCELERATED BASIC OBSTETRICAL NURSING

Course Number: J50Z046N3000; J3OZP9736 002.

Location: 396 MTG/CC, Sheppard AFB, TX.

Length: 6 weeks (238 hours).

Exhibit Dates: 1/90–Present.

Learning Outcomes: Upon completion of the course, the student will utilize the nursing process in meeting the nursing needs of the childbearing family in accordance with Joint Commission on Accreditation of Health Care Organizations, Air Force, and appropriate professional nursing organization standards.

Instruction: Preceptorship and clinical experience cover genetics, physiological changes of pregnancy, psychological aspects of pregnancy, assessment and prenatal care, high-risk pregnancy, stages of labor, complications of labor, adolescent pregnancy, maternal/fetal surveillance, newborn assessment and care, complications of the newborn, perinatal standards, puerperium, complications of the puerperium, and discharge planning.

Credit Recommendation: In the upper-division baccalaureate category, 2 semester hours in obstetrical nursing theory and 3 in obstetrical nursing laboratory (3/93).

AF-0704-0004

PHYSICAL THERAPY SPECIALIST BY CORRESPONDENCE

Course Number: 91350.

Location: Extension Course Institute, Gunter AFS, AL.

Length: Maximum, 52 weeks.

Exhibit Dates: 1/90–10/97.

Learning Outcomes: Upon completion of the course the student will be able to list, describe, and explain the general duties of the physical therapy student.

Instruction: Monitored home-study course includes readings, written exercises, on-the-job supervision, and proctored examinations. Topics include osteology and myology of all joints; joint pathology; and indications, contraindications, hazards, and goals of all major therapeutic modalities.

Credit Recommendation: In the lower-division baccalaureate/associate degree category, 3 semester hours in neuromusculoskeletal pathology, 3 in therapeutic exercise, 7 in treatment modalities, and 2 in introduction to physical therapy (5/88).

AF-0705-0003

RADIOLOGY TECHNICIAN BY CORRESPONDENCE

Course Number: 90370.

Location: Extension Course Institute, Gunter AFS, AL.

Length: Maximum, 52 weeks.

Exhibit Dates: 1/90–1/99.

Learning Outcomes: Upon completion of the course, the student will understand the basic functions of a radiology technician, including management of the radiology department, diagnostic radiology, and imaging and contrast studies procedure.

Instruction: Monitored home study includes reading, written exercises, on-the-job supervision, and proctored examination. Topics include management of the radiology department, anatomy and physiology related to positioning, physics, radiation safety, and special techniques and procedures.

Credit Recommendation: In the lower-division baccalaureate/associate degree category, 2 semester hours in radiation physics and 7 in radiography (5/88).

AF-0706-0003

OPTOMETRY SPECIALIST BY CORRESPONDENCE

Course Number: 91255.

Location: Extension Course Institute, Gunter AFS, AL.

Length: Maximum, 52 weeks.

Exhibit Dates: 1/90–12/98.

Learning Outcomes: Upon completion of the course, the student will be able to describe the procedures of the general optometric specialist.

Instruction: Monitored home-study course includes readings, written exercises, on-the-job supervision, and proctored examinations. Topics include clinic safety, scheduling, maintenance of optometric equipment, record keeping, ocular anatomy and physiology, screening tests, lenses, prescription verification, dispensing, vision training, tonometry, and the instillation of ophthalmic drugs.

Credit Recommendation: In the lower-division baccalaureate/associate degree category, 5 semester hours in optometric technology (5/88).

AF-0707-0009

ENVIRONMENTAL MEDICINE SPECIALIST BY CORRESPONDENCE

Course Number: 90850.

Location: Extension Course Institute, Gunter AFS, AL.

Length: Maximum, 52 weeks.

Exhibit Dates: 1/90–11/97.

Learning Outcomes: Upon completion of the course, the student will understand the mission and organization of the Air Force medical service; the functions of the human body; the principles of epidemiology; the different types of communicable diseases, as well as their transmission and control; the different types of work place hazards and their effects on workers; the responsibilities of the Air Force environmental health service; the management of a food facility sanitation program; the prevention of foodborne illnesses by proper food technology and food inspection procedures; field sanitation; and the responsibilities of an environmental health specialist in medical readiness for peacetime disaster and for wartime duties.

Instruction: Student enrolling must have an Air Force Specialty Code assignment as a prerequisite. This provides an application of the principles learned to real requirements and constitutes an essential part of the instruction by serving as the laboratory component of the course. The student's supervisor must complete an evaluation survey of the course and its effectiveness when the student takes the proctored final examination which measures mastery of the cognitive content. Course consists of self-paced, independent study with open book progress checks and satisfactory completion of a final examination. Topics covered include the mission and organization of the medical service, environmental health, occupational health, facility sanitation, food technology and inspections, and medical readiness.

Credit Recommendation: In the upper-division baccalaureate category, 2 semester hours in environmental and occupational health and 2 in food sanitation (11/91).

AF-0707-0010

BIOENVIRONMENTAL ENGINEERING SPECIALIST BY CORRESPONDENCE

Course Number: 90750.

Location: Extension Course Institute, Gunter AFS, AL.

Length: Maximum, 52 weeks.

COURSE EXHIBITS

Exhibit Dates: 11/91–Present.
Learning Outcomes: Upon completion of the course, the student will be able to understand the environmental effects of air pollution, water pollution, and hazardous waste; the evaluation of the occupational environment by air sampling; and the various aspects of industrial hygiene, such as ventilation, noise, illumination, and radiation protection.
Instruction: Topics covered include air and water pollution control; hazardous waste management; air sampling of the occupational environment; and other aspects of industrial hygiene, such as ventilation, noise, illumination, and radiation protection.
Credit Recommendation: In the upper-division baccalaureate category, 2 semester hours in environmental protection and 6 in industrial hygiene (11/91).

AF-0707-0011

BASIC INFECTION CONTROL SURVEILLANCE (INFECTION CONTROL AND EPIDEMIOLOGY)

Course Number: J30ZR4000 011; J3OZR9576 001.
Location: 396 MTG/CC, Sheppard AFB, TX.
Length: 4 weeks (155 hours).
Exhibit Dates: 1/90–Present.
Learning Outcomes: Upon completion of the course, the student will be able to develop, implement, and conduct on-going evaluation of an effective infection control program.
Instruction: Lectures and discussions cover basic infection control and surveillance, principles of epidemiology, surveillance systems, isolation, Center for Disease Control guidelines, patient care practices, principles of adult learning, Joint Commission on Accreditation of Health Care Organizations guidelines, and management of infection control programs.
Credit Recommendation: In the upper-division baccalaureate category, 6 semester hours as an elective in infection control (3/93).

AF-0708-0002

MENTAL HEALTH UNIT SPECIALIST BY CORRESPONDENCE

Course Number: 91451.
Location: Extension Course Institute, Gunter AFS, AL.
Length: Maximum, 52 weeks.
Exhibit Dates: 1/90–8/94.
Learning Outcomes: Upon completion of the course, the student will understand the basic mental health concepts necessary to provide diagnostic and therapeutic nursing for psychiatric patients.
Instruction: Monitored home study course includes reading, written exercises, on-the-job supervision, and proctored examination. Topics include basic concepts of mental health, diagnostic nomenclature of mental disorders, basic vital signs, diagnostic and therapeutic physical nursing procedures, patient care, and communication skills.
Credit Recommendation: In the vocational certificate category, 3 semester hours in nursing assistant (5/88); in the lower-division baccalaureate/associate degree category, 3 semester hours in mental health associate (5/88).

AF-0708-0004

AIR RESERVE FORCES SOCIAL ACTIONS TECHNICIAN (DRUG/ALCOHOL) BY CORRESPONDENCE

Course Number: 07340.
Location: Extension Course Institute, Gunter AFS, AL.
Length: Maximum, 52 weeks.
Exhibit Dates: 1/90–Present.
Learning Outcomes: Upon completion of the course, the student will be able to describe the various administrative and policy aspects of the Air Force substance abuse control program and introductory concepts of intervention and support procedures.
Instruction: Course consists of self-paced, independent study with open book progress checks and satisfactory completion of a proctored final examination. There is on-the-job supervision and an opportunity for application of learning. Content includes management concepts and information regarding the Air Force abuse control program. There is brief discussion on intervention and support procedures, such as group process development and facilitation and the maintenance of case files.
Credit Recommendation: In the lower-division baccalaureate/associate degree category, 1 semester hour in mental health assisting (11/91).

AF-0709-0027

MEDICAL ADMINISTRATIVE SPECIALIST BY CORRESPONDENCE

Course Number: 90650.
Location: Extension Course Institute, Gunter AFS, AL.
Length: Maximum, 52 weeks.
Exhibit Dates: 1/90–6/90.
Objectives: To prepare the student to serve as a medical officer or unit manager.
Instruction: This correspondence course includes organization of facilities; personnel administration; written and oral communication; medical terminology; and data collection and reporting, including biostatistical data collection, supply equipment, and bookkeeping control. Course is similar to those found in programs for medical office assistants.
Credit Recommendation: In the lower-division baccalaureate/associate degree category, 4 semester hours in principles of medical office management (9/84).

AF-0709-0030

MEDICAL LABORATORY TECHNICIAN (MICROBIOLOGY) BY CORRESPONDENCE

Course Number: 90412.
Location: Extension Course Institute, Gunter AFS, AL.
Length: Maximum, 52 weeks.
Exhibit Dates: 1/90–3/93.
Learning Outcomes: Upon completion of the course, the student will be able to describe the theory and basic fundamentals of microbiology and parasitology.
Instruction: This monitored home-study course includes reading, written exercises, and written proctored examination. Topics include the theoretical concepts of basic and clinical microbiology, parasitology, and clinical mycology.
Credit Recommendation: In the lower-division baccalaureate/associate degree category, 3 semester hours in microbiology (lecture course without lab) (5/88).

AF-0709-0031

MEDICAL LABORATORY TECHNICIAN (HEMATOLOGY, SEROLOGY, BLOOD BANKING AND IMMUNOHEMATOLOGY) BY CORRESPONDENCE

Course Number: 90413.
Location: Extension Course Institute, Gunter AFS, AL.
Length: Maximum, 52 weeks.
Exhibit Dates: 1/90–6/92.
Learning Outcomes: Upon completion of the course, the student will be able to describe the theory and basic fundamentals of hematology, serology, blood banking, and immunohematology.
Instruction: This monitored home-study course includes reading, written exercises, and proctored written examination. Introduction to the theoretical concepts of the Complete Blood Count, blood coagulation studies, serological studies, blood grouping, typing and transfusion practices, agglutination, latex-fixation, precipitin, and ASO tests are covered.
Credit Recommendation: In the lower-division baccalaureate/associate degree category, 3 semester hours for hematology, 2 for serology, 2 for blood banking, and 2 for immunohematology (5/88).

AF-0709-0032

AEROSPACE PHYSIOLOGY SPECIALIST BY CORRESPONDENCE

Course Number: 91150.
Location: Extension Course Institute, Gunter AFS, AL.
Length: Maximum, 52 weeks.
Exhibit Dates: 1/90–6/96.
Learning Outcomes: Upon completion of the course, the student will be able to describe the theory and basic fundamentals of aerospace physiology, including duties, supervision, and training; first aid procedures; medical terminology; and cardiopulmonary physiology related to altitude problems.
Instruction: This monitored home-study course includes reading, written exercises, and proctored written examinations. Topics include introduction to the theoretical concepts of aerospace physiology, fundamentals of first aid, utilization of life support equipment, hyperbaric chamber operation and maintenance, and inspection and overview of cardiopulmonary physiology as it relates to aerospace applications.
Credit Recommendation: In the lower-division baccalaureate/associate degree category, 1 semester hour in first aid (5/88).

AF-0709-0033

MEDICAL SERVICE SPECIALIST, AEROMEDICAL BY CORRESPONDENCE

Course Number: 90150.
Location: Extension Course Institute, Gunter AFS, AL.
Length: Maximum, 52 weeks.
Exhibit Dates: 7/91–Present.
Learning Outcomes: Upon completion of the course, the student will be able to describe introductory concepts of aerospace medicine as it relates to the service treatment setting.
Instruction: Student enrolling must have an Air Force Specialty Code assignment as a prerequisite. This provides an application of the principles learned to real requirements and con-

stitutes an essential part of the instruction by serving as the laboratory component of the course. The student's supervisor must complete an evaluation survey of the course and its effectiveness when the student takes the proctored final examination which measures mastery of the cognitive content. Course consists of self-paced, independent study with open book progress checks and satisfactory completion of a final examination. Content includes an overview of the Air Force medical service system and administrative and support activities for which the medical services specialist is responsible. Introductory concepts of anatomy and physiology are covered along with rudimentary pharmacological routines and basic diagnostic and treatment procedures.

Credit Recommendation: In the lower-division baccalaureate/associate degree category, 1 semester hour in medical assisting, 1 in anatomy and physiology, and 2 in principles of emergency medicine (11/91).

AF-0709-0034

HEALTH SERVICES ADMINISTRATION

Course Number: J30BR41A1 001; J30BR9021 001.
Location: 882d Training Group, Sheppard AFB, TX..
Length: 12 weeks (412-448 hours).
Exhibit Dates: 11/94–Present.
Learning Outcomes: Upon completion of the course, the student will effectively assume an administrative role in one of six specialty areas: management of patient administration, medical resource management, medical readiness, medical logistics management, personnel and administrative services and medical information systems.
Instruction: Lectures and discussions, computer-assisted instruction, group seminars, simulations, and demonstrations cover the following topics: oral and written communications, total quality management, organization of patient care administrative functions, ambulatory care services, managed care, medical resource management, financial programs, management analysis programs, microcomputers as management tools, and simulated administrative work experience.
Credit Recommendation: In the graduate degree category, 11 semester hours in health administration (8/95).

AF-0709-0035

NURSING SERVICE MANAGEMENT FOR AIR RESERVE COMPONENTS

Course Number: J30RR46A3004; J3ORR9711 004.
Location: 396 MTG/CC, Sheppard AFB, TX.
Length: 2 weeks (79 hours).
Exhibit Dates: 1/90–Present.
Learning Outcomes: Upon completion of the course, the student will demonstrate effective nursing management as related to Air Reserve components (ARC) congruent with appropriate Joint Commission on Accreditation of Health Care Organizations, Air Force, and quality assurance standards.
Instruction: Topics covered in small groups discussions include organization and mission of ARC, impact of wartime environment on nursing personnel, nursing leadership, retention of ARC personnel, nursing practice standards and appraisal statements, nurse recognition programs for ARC personnel, medical/legal issues affecting ARC personnel, staff development for ARC personnel, quality assurance with ARC, nursing personnel, medical readiness issues, effective strategies for controlling nursing service unit dysfunction, nursing service unit dysfunction, and UTA and annual tour training plans.
Credit Recommendation: Credit is not recommended because of the limited, specialized nature of the course (3/93).

AF-0709-0036

NURSING SERVICE MANAGEMENT

Course Number: J30ZR46A3005; J3OZR9711 005.
Location: 396 MTG/CC, Sheppard AFB, TX.
Length: 8 weeks (296 hours).
Exhibit Dates: 1/90–Present.
Learning Outcomes: Upon completion of the course, the student will demonstrate knowledge and skills necessary to effectively perform in a middle management nursing position, incorporating appropriate Joint Commission on Accreditation, health care organizations, health services management inspection, and nursing practice standards.
Instruction: Lectures, discussions, student group activities, performance activities cover nursing practice standards, ambulatory services, professional standards, staffing patterns, stress management, policies and procedures, negotiating, power through image, power and politics, leadership qualities, decision making delegation, marketing health care services, biomedical ethics, medical law, personnel management, quality improvement, conflict management, quality improvement/risk management, infection control, nursing research, application of computers in nursing, and publishing in nursing.
Credit Recommendation: In the graduate degree category, 18 semester hours in nursing administration (3/93).

AF-0709-0037

NURSING EXECUTIVE DEVELOPMENT

Course Number: J5OZO9716 002.
Location: 396 MTG/CC, Sheppard AFB, TX.
Length: 12 weeks (478 hours).
Exhibit Dates: 1/90–4/94.
Learning Outcomes: Upon completion of the course, the student will effectively assume a nurse executive position congruent with Joint Commission on Accreditation of Health Care Organizations (JCAHO).
Instruction: Preceptorship, discussions, interviews, self-instruction, and performance activities cover nurse executive roles and accountability, structure and interdisciplinary functions of executive administration, concepts of human resource management, health care delivery systems, role and responsibility of nurse executive in continuous quality improvement, and nurse executive's preparation for JCAHO.
Credit Recommendation: In the graduate degree category, 10 semester hours in nursing administration (3/93).

AF-0799-0003

PHARMACY SPECIALIST BY CORRESPONDENCE

Course Number: 90550.
Location: Extension Course Institute, Gunter AFS, AL.
Length: Maximum, 52 weeks.
Exhibit Dates: 1/90–10/97.
Learning Outcomes: Upon completion of the course, the student will understand the basics of pharmacy guidelines, law, quality control, procurement, principles of compounding, dispensing of unit dose, and intravenous admixtures.
Instruction: This monitored home-study course includes reading, written exercises, on-the-job supervision, and proctored examination. Topics include quality control, phamaceutical calculations, chemistry, and pharmacology.
Credit Recommendation: In the lower-division baccalaureate/associate degree category, 9 semester hours in pharmacy technology (5/88).

AF-0799-0008

HEALTH SERVICES MANAGEMENT JOURNEYMAN BY CORRESPONDENCE

Course Number: 90650A.
Location: Extension Course Institute, Gunter AFS, AL.
Length: Maximum, 52 weeks.
Exhibit Dates: 3/92–7/97.
Learning Outcomes: Upon completion of the course, students will be able to maintain and monitor the systems to support the management of inpatient and outpatient services.
Instruction: Course reviews the duties and responsibilities of patient administration and includes military policy and procedure and interpersonal relations to ensure that quality service is provided.
Credit Recommendation: In the vocational certificate category, 1 semester hour in clerical procedures (6/94).

AF-0799-0009

HEALTH SERVICES MANAGEMENT JOURNEYMAN BY CORRESPONDENCE

Course Number: 90650B.
Location: Extension Course Institute, Gunter AFS, AL.
Length: Maximum, 52 weeks.
Exhibit Dates: 8/92–Present.
Learning Outcomes: Upon completion of the course, the student will be able to maintain and monitor a management information system for health services.
Instruction: Course reviews the duties and responsibilities in the area of health services management information systems. Special emphasis is placed on written and electronic communication; establishing, maintaining, and disposing of office records; word processing; and records information management system.
Credit Recommendation: In the vocational certificate category, 1 semester hour in clerical procedures (6/94).

AF-0799-0010

HEALTH SERVICES MANAGEMENT JOURNEYMAN BY CORRESPONDENCE

Course Number: 90650C.
Location: Extension Course Institute, Gunter AFS, AL.
Length: Maximum, 52 weeks.
Exhibit Dates: 3/93–Present.

Learning Outcomes: Upon completion of the course, the student will be able to describe the duties and responsibilities involved in the management of the physical, financial, and human resources in a quality health services environment.

Instruction: Course covers medical resource management and management information systems and includes financial management, budgeting, collections, performance reporting, manpower requirements, and customer satisfaction.

Credit Recommendation: In the vocational certificate category, 1 semester hour in clerical procedures (6/94).

AF-0801-0001

AIR FORCE JOINT SERVICE SUPERVISOR SAFETY BY CORRESPONDENCE

Course Number: 1900.
Location: Extension Course Institute, Gunter AFS, AL.
Length: Maximum, 52 weeks.
Exhibit Dates: 1/90–1/97.
Learning Outcomes: At the conclusion of the course, the student will be able to describe the role of the supervisor as it relates to safety, prepare the worker for the job, supervise the job, identify and correct hazards, describe the safety considerations in the work place environment, identify the issues related to fire protection, and identify off-duty safety concerns.
Instruction: Topics include the major issues and concerns related to safety on the job. Students learn to identify unsafe conditions, develop and implement safety training programs, and describe the specific role of the supervisor in the administration of the safety function in the work place.
Credit Recommendation: In the vocational certificate category, 1 semester hour in safety (2/87).

AF-0801-0002

ARMY JOINT SERVICE SUPERVISOR SAFETY BY CORRESPONDENCE

Course Number: 1901.
Location: Extension Course Institute, Gunter AFS, AL.
Length: Maximum, 52 weeks.
Exhibit Dates: 1/90–9/94.
Learning Outcomes: At the conclusion of this course, students will be able to describe the role of the supervisor as it relates to safety, prepare the worker for the job, supervise the job, identify and correct hazards, describe safety considerations in the work place, describe fire protection issues, and identify off-duty safety concerns.
Instruction: This course covers major issues and concerns related to safety on the job. Students learn to identify unsafe conditions, develop and implement safety training programs, and identify the specific role of the supervisor in the administration of the safety function in the work place.
Credit Recommendation: In the vocational certificate category, 1 semester hour in safety (2/87).

AF-0801-0003

NAVY JOINT SERVICE SUPERVISOR SAFETY BY CORRESPONDENCE

Course Number: 1902.
Location: Extension Course Institute, Gunter AFS, AL.
Length: Maximum, 52 weeks.
Exhibit Dates: 1/90–9/92.
Learning Outcomes: At the conclusion of the course the students will be able to describe the role of the supervisor as it relates to safety, prepare the worker for the job, supervise the job, identify and correct hazards, describe the safety considerations in the work place, describe fire protection issues, and identify off-duty safety concerns.
Instruction: This course covers the major issues and concerns related to safety on the job. Students learn to identify unsafe conditions, develop and implement safety training programs, and the specific role of the supervisor in the administration of the safety function in the work place.
Credit Recommendation: In the vocational certificate category, 1 semester hour in safety (2/87).

AF-0801-0004

COAST GUARD JOINT SERVICE SUPERVISOR SAFETY BY CORRESPONDENCE

Course Number: 1903.
Location: Extension Course Institute, Gunter AFS, AL.
Length: Maximum, 52 weeks.
Exhibit Dates: 1/90–9/94.
Learning Outcomes: At the conclusion of this course, students will be able to describe the role of the supervisor as it relates to safety, prepare the worker for the job, supervise the job, identify and correct hazards, describe safety considerations in the work place, describe fire protection issues, and off-duty safety concerns.
Instruction: This course covers the major issues and concerns related to safety on the job. Students learn to identify unsafe conditions, develop and implement safety training programs, and describe the specific role of the supervisor in the administration of the safety function in the work place.
Credit Recommendation: In the vocational certificate category, 1 semester hour in safety (2/87).

AF-0801-0005

CIVIL AIR PATROL-SAFETY OFFICER (LEVEL II TECHNICAL RATING) BY CORRESPONDENCE

Course Number: 02170.
Location: Extension Course Institute, Gunter AFS, AL.
Length: Maximum, 52 weeks.
Exhibit Dates: 1/90–1/97.
Learning Outcomes: At the conclusion of this course, the student will be able to describe the general responsibilities of commanding officers at all levels with regard to the safety function, the criteria for the administration of safety programs, the principles and philosophy of accident prevention, the duties of the safety officer, and the various safety awards in the Civil Air Patrol.
Instruction: This course covers the general safety responsibilities of the various levels of commanding officers, establishing a safety program, the practice and philosophy of accident prevention, specific accident procedures, and safety awards.
Credit Recommendation: In the vocational certificate category, 1 semester hour in safety management (2/87).

AF-0802-0026

DISASTER PREPAREDNESS SPECIALIST BY CORRESPONDENCE

Course Number: 24250.
Location: Extension Course Institute, Gunter AFS, AL.
Length: Maximum, 52 weeks.
Exhibit Dates: 1/90–12/96.
Objectives: To train disaster/safety personnel in procedures for response to chemical, biological, and nuclear warfare situations.
Instruction: This correspondence course of instruction consists of on-the-job supervision and at-home study. The course of instruction includes topics in disaster preparedness as military career field, methods of managing and planning for disaster situations, and use of equipment specific to disaster created by nuclear, biological, and chemical warfare situations.
Credit Recommendation: Credit is not recommended because of the military-specific nature of the course (9/86).

AF-0804-0001

MORALE, WELFARE, RECREATION, AND SERVICES JOURNEYMAN BY CORRESPONDENCE

Course Number: 78150A.
Location: Extension Course Institute, Gunter AFS, AL.
Length: Maximum, 52 weeks.
Exhibit Dates: 6/93–10/97.
Learning Outcomes: Upon completion of the course, the student will be able to perform basic duties and responsibilities within the areas of food preparation and service, lodging, mortuary operations, morale, and recreation programs.
Instruction: Course introduces the fundamentals of food service operations; food preparation; lodging/mortuary affairs; and morale, welfare, and recreation services.
Credit Recommendation: Credit is not recommended because of the limited, specialized nature of the course (6/94).

AF-0804-0002

MORALE, WELFARE, RECREATION, AND SERVICES BY CORRESPONDENCE

Course Number: 78150B.
Location: Extension Course Institute, Gunter AFS, AL.
Length: Maximum, 52 weeks.
Exhibit Dates: 6/93–Present.
Learning Outcomes: Upon completion of the course, the student will have the knowledge to perform the basic duties of nonappropriated funds control program and activities management; student will also oversee the overall operation of physical fitness, nutrition, and sports facility programs.
Instruction: Course introduces the fundamentals of funds resource management and recreation, sports, and wellness programming.
Credit Recommendation: Credit is not recommended because of the limited, specialized nature of the course (6/94).

AF-1107-0002

MAINTENANCE DATA SYSTEMS ANALYSIS SPECIALIST BY CORRESPONDENCE

Course Number: 39170.

Location: Extension Course Institute, Gunter AFS, AL.
Length: Maximum, 52 weeks.
Exhibit Dates: 1/90–6/97.
Learning Outcomes: Upon completion of the course, the student will be able to evaluate graduates, factors, and standards; identify the various types of capabilities; analyze maintenance data, man-hour data, and data presentations; and describe the statistical methods for analyzing data and data presentations, index numbers and population estimates, correlation and time series, control charts, probability, and hypothesis testing.
Instruction: Student enrolling must have an Air Force Specialty Code assignment as a prerequisite. This provides an application of the principles learned to real requirements and constitutes an essential part of the instruction by serving as the laboratory component of the course. The student's supervisor must complete an evaluation survey of the course and its effectiveness when the student takes the proctored final examination which measures mastery of the cognitive content. Course consists of self-paced, independent study with open book progress checks and satisfactory completion of a final examination. Course includes the evaluation of graduates, factors, and standards; maintenance and man-hour data analysis; and statistical methods of summarizing data, index numbers and population estimates, correlation and time analysis, control charts, probability, and hypothesis testing.
Credit Recommendation: In the upper-division baccalaureate category, 3 semester hours in statistics for non—mathematics majors (11/91).

AF-1115-0008

POWER MEASUREMENTS BY CORRESPONDENCE

Course Number: 3037.
Location: Extension Course Institute, Gunter AFS, AL.
Length: Maximum, 52 weeks.
Exhibit Dates: 1/90–Present.
Learning Outcomes: Upon completion of the course, the student will be able to describe the mathematical concepts specific to electronic power level measurement including the logarithm and the manipulation of powers of ten. Other competencies include the use of units of measurement and specialized test equipment used to determine signal power relationships and noise at any test point.
Instruction: Course includes a review of basic mathematics including logarithms and the use of the powers of ten. Decibel and noise-weighting network characteristics and computations are covered. Principles and use of measuring devices appropriate for frequencies from the low range to the microwave range are presented.
Credit Recommendation: In the lower-division baccalaureate/associate degree category, 1 semester hour in applied physics—electronic measurement (12/89).

AF-1304-0014

WEATHER EQUIPMENT SPECIALIST BY CORRESPONDENCE

Course Number: 30250.
Location: Extension Course Institute, Gunter AFS, AL.
Length: Maximum, 52 weeks.
Exhibit Dates: 1/90–11/92.
Objectives: To provide instruction in principles of design and operational techniques of weather measuring equipment.
Instruction: This correspondence course covers general need for weather instrumentation; fundamentals and prevention of corrosion; design of instrumentation to measure temperature, dew point, wind, and ceiling heights; fundamentals of digital techniques; numbering systems; fundamentals of Boolean algebra; transistor logic; operational amplifiers; fundamentals of FPS-77 weather radar; and use of oscilloscope.
Credit Recommendation: In the lower-division baccalaureate/associate degree category, 3 semester hours in meteorological instrumentation and 3 in digital techniques (11/86).

AF-1304-0015

METEOROLOGICAL TECHNICIAN BY CORRESPONDENCE (SPECIALIZED)

Course Number: 2570.
Location: Extension Course Institute, Gunter AFS, AL.
Length: Maximum, 52 weeks.
Exhibit Dates: 1/90–10/98.
Learning Outcomes: Upon completion of the course, the student will be able to function as a forecaster or station chief in the Air Weather Service (AWS). This course is also applicable to commissioned officers newly entering the weather career field.
Instruction: Course includes study of history and administration of the AWS, weather station management, meteorology, climatology, weather data analysis, forecasting techniques, and AWS support to the mission of the Air Force.
Credit Recommendation: In the lower-division baccalaureate/associate degree category, 3 semester hours in meteorology/climatology and 3 in administrative management (12/89).

AF-1304-0016

WEATHER SPECIALIST BY CORRESPONDENCE

Course Number: 25150.
Location: Extension Course Institute, Gunter AFS, AL.
Length: Maximum, 52 weeks.
Exhibit Dates: 1/90–1/98.
Learning Outcomes: Upon completion of the course, the student will be able to make and report accurate weather observations utilizing visual and radar techniques, plot weather charts and diagrams, and assist the forecaster in preparing necessary forecasts.
Instruction: Course includes background and history of the air weather service, surface visual and radar observations, recording meteorological phenomena and measurements, weather codes, communications, weather station operations, structure and movement of the atmosphere, basic meteorology, and climatology data analysis, satellite observations, and basic mathematics and physics.
Credit Recommendation: In the lower-division baccalaureate/associate degree category, 2 semester hours in fundamentals of meteorology and 1 in general mathematics (12/89).

AF-1304-0017

METEOROLOGICAL AND NAVIGATION SYSTEMS JOURNEYMAN BY CORRESPONDENCE (Meteorological and Navigation Systems Specialist by Correspondence)

Course Number: 30452B.
Location: Extension Course Institute, Gunter AFS, AL.
Length: Maximum, 52 weeks.
Exhibit Dates: 8/93–Present.
Learning Outcomes: Upon completion of the course, the student will be able to operate and maintain airport meteorological sensors such as transmission meter and instruments measuring temperature, dew point, and cloud height.
Instruction: Course presents functional block diagrams, selected circuits, and alignment procedures for meteorological sensors. The course requires significant prerequisites in the area of electronics and atmospheric physics.
Credit Recommendation: Credit is not recommended because of the limited, specialized nature of the course (6/94).

AF-1401-0002

FINANCIAL MANAGEMENT OFFICER (FINANCIAL ANALYSIS)

Course Number: J30BR6721 003; J3OBR65F1 003.
Location: 396 TCHTG/CC, Sheppard AFB, TX.
Length: 11 weeks (437 hours).
Exhibit Dates: 5/92–Present.
Learning Outcomes: Upon completion of the course, the student will be able to describe the federal budget system, the components of automated accounting systems, and the procedures to analyze financial programs; explain the automated accounting systems interface; apply statistical techniques; perform a cost analysis of a support agreement; and analyze financial statements and a financial plan for operations and maintenance.
Instruction: Methods include classroom lectures and discussions, case studies, and directed studies. Topics cover financial analysis, cost and economic analysis, the use of statistical techniques, the use of minicomputers as an analytical tool, cost and resource analysis to assist decision making, and preparation of operating budgets and financial plans.
Credit Recommendation: In the graduate degree category, 3 semester hours in public finance, 3 in cost accounting, and 3 in applied statistical techniques (3/93).

AF-1401-0018

COMPTROLLER STAFF OFFICER, AIR RESERVE FORCES

Course Number: J30ZR6751 000.
Location: 396 TCHTG/CC, Sheppard AFB, TX.
Length: 2 weeks (80 hours).
Exhibit Dates: 1/90–Present.
Learning Outcomes: Upon completion of the course, the student will be able to identify the functions of the planning, programming, and budget system; prepare a response to a case study on accounting and finance operations; identify the techniques used in performing cost analysis; and describe the use of micro/mini-computers to support future comptroller operations.
Instruction: Classroom lectures and discussions, case study, and directed studies. Topics include principles and concepts of accounting

12 COURSE EXHIBITS

and finance, budget, cost, and comptroller plans and systems.

Credit Recommendation: In the graduate degree category, 3 semester hours in public accounting (3/93).

AF-1401-0019

COST ANALYSIS OFFICER

Course Number: J30BR6741 001.
Location: 396 TCHTG/CC, Sheppard AFB, TX.
Length: 9 weeks (334 hours).
Exhibit Dates: 7/91–Present.
Learning Outcomes: Upon completion of the course, the student will be able to identify the concepts of cost analysis, the concepts of economic analysis, the PPB system, and the resource management system; develop and validate an economic analysis; and perform an independent review of a commercial activity cost comparison study.
Instruction: Classroom lectures and discussions, demonstration/performance, and directed studies. Topics covered include financial analysis, cost and economic analysis, management services, the use of statistical and quantitative techniques, and familiarization with minicomputers as an analytical tool.
Credit Recommendation: In the upper-division baccalaureate category, 3 semester hours in cost accounting systems and 3 in applied economic analysis (3/93).

AF-1402-0069

MAINTENANCE SYSTEMS ANALYSIS SPECIALIST BY CORRESPONDENCE

Course Number: 39150.
Location: Extension Course Institute, Gunter AFS, AL.
Length: Maximum, 52 weeks.
Exhibit Dates: 1/90–7/94.
Learning Outcomes: Upon completion of this course, the student will have knowledge of concepts of maintenance management, oral and written communications, fundamentals of statistics and statistical analysis, data collection and reporting, and data processing.
Instruction: Topics covered include maintenance management, oral and written communications, fundamentals of statistics, statistical analysis, probability distributions, correlation, data collection and reporting, and data processing.
Credit Recommendation: In the vocational certificate category, 5 semester hours in maintenance management and 1 in data processing (2/87); in the lower-division baccalaureate/associate degree category, 2 semester hours in business communications or technical writing and 2 in statistics (2/87).

AF-1402-0070

INFORMATION SYSTEMS OPERATOR BY CORRESPONDENCE

Course Number: 49151.
Location: Extension Course Institute, Gunter AFS, AL.
Length: Maximum, 52 weeks.
Exhibit Dates: 1/90–5/96.
Learning Outcomes: At the conclusion of this course the student will be able to describe the safety, security, and maintenance issues in computer center installation; understand data processing terminology and have a working knowledge of computer equipment concepts and operation of a computer room; and describe the concepts of the Air Force's automated communications system.
Instruction: This course provides the student with a working knowledge of the production control and computer room functions of a computer facility. Topics include data processing terminology, storage media and equipment concepts, library management, and production control functions. Also covered is an overview of the Air Force's automated communication system.
Credit Recommendation: In the lower-division baccalaureate/associate degree category, 1 semester hour in introduction to computers or introduction to data processing (2/87).

AF-1402-0071

INFORMATION SYSTEMS PROGRAMMING SPECIALIST BY CORRESPONDENCE

Course Number: 49152.
Location: Extension Course Institute, Gunter AFS, AL.
Length: Maximum, 52 weeks.
Exhibit Dates: 1/90–2/97.
Learning Outcomes: At the conclusion of this course, the student will be able to describe the safety and security issues in computer operations; understand data processing terminology; describe how data is stored in computers; demonstrate flowcharting techniques; code elementary assembly language and COBOL and FORTRAN programs; describe systems analysis and design concepts; and describe the programming cycle.
Instruction: This course provides the student with a working knowledge of data processing concepts and terminology; flowcharting techniques; the programming function; the assembler, COBOL, and FORTRAN programming languages; and the systems analysis and design function.
Credit Recommendation: In the lower-division baccalaureate/associate degree category, 1 semester hour in introduction to computers, introduction to data processing, or introduction to computer programming (2/87).

AF-1402-0072

FUNDAMENTAL PRINCIPLES OF ELECTRONIC DATA PROCESSING EQUIPMENT BY CORRESPONDENCE

Course Number: 5123.
Location: Extension Course Institute, Gunter AFS, AL.
Length: Maximum, 52 weeks.
Exhibit Dates: 1/90–4/94.
Learning Outcomes: At the conclusion of this course, the student will be able to describe data processing terminology and concepts.
Instruction: This course covers data processing terminology and programming concepts. A survey of the COBOL and FORTRAN programming languages is presented.
Credit Recommendation: In the lower-division baccalaureate/associate degree category, 1 semester hour in introduction to data processing or introduction to computer programming (2/87).

AF-1402-0073

APPRENTICE OPERATIONS RESOURCE MANAGEMENT SPECIALIST BY CORRESPONDENCE

Course Number: *Version 1:* 27132. *Version 2:* 27132.
Location: Extension Course Institute, Gunter AFS, AL.
Length: *Version 1:* Maximum, 52 weeks. *Version 2:* Maximum, 52 weeks.
Exhibit Dates: *Version 1:* 11/91–Present. *Version 2:* 1/90–10/91.
Learning Outcomes: *Version 1:* Upon completion of the course, the student will be able to perform basic duties position in flight operations management including preparing and retaining historical documents with minimal direct supervision. *Version 2:* Upon completion of the course, the student will have a basic understanding of duties related to flight operations management including preparing and retaining historical documents with minimal direct supervision.
Instruction: *Version 1:* This correspondence course consists of a program with tests and a proctored comprehensive final examination. Subject areas include introduction to operations, resource management, publications, aviation service, ratings and badges, flight activity, aircrew training, and scheduling. *Version 2:* The course consists of self-paced, independent study with open book progress checks and satisfactory completion of a proctored comprehensive final examination. Content areas include introduction to tasks related to administrative control and accountability of flight records. Topics include aircrew schedule and training, aircraft schedule, and data processing.
Credit Recommendation: *Version 1:* In the lower-division baccalaureate/associate degree category, 1 semester hour in introduction to data processing and 1 in clerical procedures (4/93). *Version 2:* In the lower-division baccalaureate/associate degree category, 1 semester hour in introduction to data processing and 1 in clerical procedures (business) (12/89).

AF-1402-0074

AUDITOR RETRIEVAL SYSTEMS

Course Number: J30ZR6784 002.
Location: 396 TCHTG/CC, Sheppard AFB, TX.
Length: 2 weeks (80 hours).
Exhibit Dates: 10/90–Present.
Learning Outcomes: Upon completion of the course, the student will be able to identify the functions of software, define data base management systems, identify the functions of a data communication system, and identify the retrieval languages.
Instruction: Methods include classroom lectures, discussions, demonstrations, and performance. Topics cover computer systems, including data communication, systems functions/utility functions, file organization and data base management systems, and retrieval systems/languages.
Credit Recommendation: In the upper-division baccalaureate category, 3 semester hours in computer systems application (3/93).

AF-1402-0075

PERSONNEL SYSTEMS MANAGEMENT JOURNEYMAN BY CORRESPONDENCE

Course Number: 73150A.
Location: Extension Course Institute, Gunter AFS, AL.
Length: Maximum, 52 weeks.
Exhibit Dates: 3/93–5/96.

Learning Outcomes: Upon completion of the course, the student will be able to describe the personnel systems management specialist role within the military personnel flight (MPF) organization structure.

Instruction: Course offers a review of fundamentals of the personnel systems management career field, including MPF organization and its functional responsibilities, career progression, and the role of the personnel systems management specialist. Additional topics covered are office administration procedures, data base philosophy, and other system basics related to this career field.

Credit Recommendation: Credit is not recommended because of the military-specific nature of the course (6/94).

AF-1402-0076

PERSONNEL SYSTEMS MANAGEMENT JOURNEYMAN BY CORRESPONDENCE

Course Number: 73150B.
Location: Extension Course Institute, Gunter AFS, AL.
Length: Maximum, 52 weeks.
Exhibit Dates: 7/93–Present.
Learning Outcomes: Upon completion of the course, the student will be able to build, use, and maintain computer tables and interpret computer terminal messages.
Instruction: Course reviews fundamentals of computer tables with focus on their development, use, and maintenance. Course also includes the discussion of computer terminal messages, transaction identifiers, and other terminal input.
Credit Recommendation: In the lower-division baccalaureate/associate degree category, 1 semester hour in introduction to electronic data processing (6/94).

AF-1402-0077

PERSONNEL SYSTEMS MANAGEMENT JOURNEYMAN BY CORRESPONDENCE

Course Number: 73150C.
Location: Extension Course Institute, Gunter AFS, AL.
Length: Maximum, 52 weeks.
Exhibit Dates: 7/93–Present.
Learning Outcomes: Upon completion of the course, the student will demonstrate knowledge of the base-level personnel system processor and the ability to manage output products. The student will know the computer language used by the military Direct English Speaking Information Retrieval (DESIRE) system.
Instruction: Course covers the areas of personnel system processor, including job scheduling, computer run procedures, and data output management. A review of the DESIRE system is stressed.
Credit Recommendation: In the lower-division baccalaureate/associate degree category, 1 semester hour in introduction to electronic data processing (6/94).

AF-1402-0078

PERSONNEL SYSTEMS MANAGEMENT JOURNEYMAN BY CORRESPONDENCE

Course Number: 73150D.
Location: Extension Course Institute, Gunter AFS, AL.
Length: Maximum, 52 weeks.
Exhibit Dates: 9/93–6/96.

Learning Outcomes: Upon completion of the course, the student will be able to operate and troubleshoot the equipment and software required for the PC-III system.
Instruction: This correspondence course presents a review of equipment hardware, software, software maintenance, and troubleshooting for the PC-III system.
Credit Recommendation: In the lower-division baccalaureate/associate degree category, 1 semester hour in introduction to electronic data processing (6/94).

AF-1405-0013

MATERIEL FACILITIES SPECIALIST BY CORRESPONDENCE

Course Number: 64551.
Location: Extension Course Institute, Gunter AFS, AL.
Length: Maximum, 52 weeks.
Exhibit Dates: 1/90–3/92.
Objectives: After 3/92 see AF-1405-0070. This course provides the student with the basic fundamentals of materiel supply systems, inspection and equipment-handling operations, and the storing and issuing of supplies.
Instruction: Instruction cover security operations, materiel sources, storage and distribution systems, and records control. Documentation, storage procedures, and the proper selection of materials-handling equipment are also discussed.
Credit Recommendation: In the vocational certificate category, 3 semester hours in materials handling and inventory control (2/87).

AF-1405-0018

APPRENTICE MATERIEL FACILITIES SPECIALIST BY CORRESPONDENCE

Course Number: 64531.
Location: Extension Course Institute, Gunter AFS, AL.
Length: Maximum, 52 weeks.
Exhibit Dates: 1/90–8/96.
Learning Outcomes: Upon completion of this course, the student will be able to describe the standard base supply system, receiving and materials handling operations, and storage function.
Instruction: The instruction in this course provides an introduction to the supply distribution system, supply publications, and supply classifications; property receipt procedures and internal and external records; material handling principles and equipment; pickup and delivery functions; storage functions; material locater system; inventory procedures; special storage procedures; and shipments and transfers.
Credit Recommendation: In the lower-division baccalaureate/associate degree category, 2 semester hours in materials handling (10/85).

AF-1405-0033

MATERIEL FACILITIES SUPERVISOR BY CORRESPONDENCE

Course Number: 64571.
Location: Extension Course Institute, Gunter AFS, AL.
Length: Maximum, 52 weeks.
Exhibit Dates: 1/90–8/96.
Objectives: This course offers comprehensive coverage of the operational and supervisory functions of a supply distribution system. It emphasizes the importance of a computer system for purchasing, storing, controlling, and distributing materiel.
Instruction: This course covers supply sources, correspondence, requisition support services, inventory control, storage facilities and equipment, and hazardous materials. Included also are management and supervisory responsibilities, computerized control systems, and documentation.
Credit Recommendation: In the lower-division baccalaureate/associate degree category, 3 semester hours in materiel management (10/85); in the upper-division baccalaureate category, 3 semester hours in materiel control systems (10/85).

AF-1405-0051

INVENTORY MANAGEMENT SPECIALIST BY CORRESPONDENCE

Course Number: 64550.
Location: Extension Course Institute, Gunter AFS, AL.
Length: Maximum, 52 weeks.
Exhibit Dates: 1/90–8/96.
Objectives: The inventory management course provides information regarding inventory procedures, record keeping, and filing of records concerning materials and equipment in stock.
Instruction: The course content provides background information and specific procedures for completing inventory records, requisitioning materials, preparing stock control records and reports, issuing supplies and materials, handling retail sales, preparing cash receipts, and filing various records.
Credit Recommendation: In the vocational certificate category, 2 semester hours in inventory control (2/87); in the lower-division baccalaureate/associate degree category, 2 semester hours in records management, 1 in filing, 1 in data processing (2/87).

AF-1405-0059

INVENTORY MANAGEMENT SUPERVISOR BY CORRESPONDENCE

Course Number: 64570.
Location: Extension Course Institute, Gunter AFS, AL.
Length: Maximum, 52 weeks.
Exhibit Dates: 1/90–6/91.
Objectives: After 6/91 see AF-1405-0071. This courses provides a review of functions and procedures in inventory management in the Air Force; and it provides information for supervisors in supervising, training, and assigning of personnel in inventory control.
Instruction: This course includes a review of various functions involved in inventory management such as completing inventory records, requisitioning materials, preparing stock control records and reports, issuing supplies and materials, handling retail sales, and manual accounting procedures. In addition, it provides information on assigning, training, and evaluating personnel; identifying, planning, developing, and changing procedures concerning inventory management; interpreting policies; and developing inventory schedules.
Credit Recommendation: In the lower-division baccalaureate/associate degree category, 1 semester hour in records management (10/85); in the upper-division baccalaureate cat-

14 COURSE EXHIBITS

egory, 2 semester hours in principles of supervision (10/85).

AF-1405-0060

PRODUCTION CONTROL SPECIALIST BY CORRESPONDENCE

Course Number: 55530.
Location: Extension Course Institute, Gunter AFS, AL.
Length: Maximum, 52 weeks.
Exhibit Dates: 1/90–3/96.
Objectives: To provide specialty knowledge and upgrade training to apprentice production control specialists.
Instruction: This correspondence course consists of basic organization and supervision in production control, basic civil engineer automated management systems, work control and labor subsystems and maintenance, work authorization, and work planning.
Credit Recommendation: Credit is not recommended because of the limited, specialized nature of the course (9/86).

AF-1405-0064

INVENTORY MANAGEMENT SPECIALIST (MUNITIONS) BY CORRESPONDENCE

Course Number: 64550A.
Location: Extension Course Institute, Gunter AFS, AL.
Length: Maximum, 52 weeks.
Exhibit Dates: 1/90–10/92.
Learning Outcomes: Upon completion of the course, the student will be able to provide information on supply systems, procedures, inventory management, and automation procedures in munitions management.
Instruction: Instruction includes elements of data processing, records management, inventory control, requisitioning procedures, shipping and receiving, document control, and report preparation.
Credit Recommendation: In the vocational certificate category, 1 semester hour in data processing, 1 in records management, and 1 in report writing (2/87).

AF-1405-0066

SUPPLY SYSTEMS ANALYSIS SUPERVISOR BY CORRESPONDENCE

Course Number: *Version 1:* 64572. *Version 2:* 64572.
Location: Extension Course Institute, Gunter AFS, AL.
Length: *Version 1:* Maximum, 52 weeks. *Version 2:* Maximum, 52 weeks.
Exhibit Dates: *Version 1:* 7/91–12/95. *Version 2:* 1/90–6/91.
Learning Outcomes: *Version 1:* Upon completion of the course, the student will be able to describe the components of the Air Force logistics system, the automated standard base supply system, and its interfacing systems; code in Sperry job control language; understand a data processing center and the functions of its components; understand data base concepts and operations; and access, update, and maintain a complex computerized data base of supply information. *Version 2:* Upon completion of the course, the student will be able to describe the components of the Air Force Logistics system; understand the automated standard base supply system and its interfacing systems; code in Sperry job control language; understand data base concepts and operations; and access, update, and maintain a complex computerized data base of supply information.
Instruction: *Version 1:* Instruction includes two separate subject areas. One topic involves the specific components, jobs, procedures, and policies of Air Force automated standard base supply systems. The other topic presented is data processing. This topic is introduced at the conceptual level and quickly progresses into the area of data base management systems. The student learns the commands to access, update, and maintain the complex computerized data bases supporting Air Force supply systems. *Version 2:* Instruction includes two separate subject areas. One topic involves the specific components, jobs, procedures, and policies of Air Force logistics and supply. The other topic presented is data processing. This topic is introduced at the conceptual level and quickly progresses into the area of data base management systems. The student learns the commands to access, update, and maintain the complex computerized data bases supporting Air Force supply systems.
Credit Recommendation: *Version 1:* In the lower-division baccalaureate/associate degree category, 1 semester hour in data processing and 3 in data base management systems (4/93). *Version 2:* In the lower-division baccalaureate/associate degree category, 1 semester hour in data processing and 3 in data base management systems (2/87).

AF-1405-0067

INTRODUCTION TO THE QUALITY FUNCTION BY CORRESPONDENCE

Course Number: 6601.
Location: Extension Course Institute, Gunter AFS, AL.
Length: Maximum, 52 weeks.
Exhibit Dates: 1/90–1/97.
Learning Outcomes: At the conclusion of the course, the student will be able to describe the function of quality control and assurance; identify the components of a quality control program; appreciate the special problems of quality control and assurance; and evaluate and measure quality control.
Instruction: This course covers the quality control and assurance function, the various problems of quality control and assurance, developing a quality control program, and methods for evaluating and measuring quality. Although much of the orientation is on quality control in defense logistics, the instruction parallels the logistics cycle in industry.
Credit Recommendation: In the lower-division baccalaureate/associate degree category, 1 semester hour in production control (2/87).

AF-1405-0068

MANAGEMENT OF VALUE ENGINEERING BY CORRESPONDENCE

Course Number: 6603.
Location: Extension Course Institute, Gunter AFS, AL.
Length: Maximum, 52 weeks.
Exhibit Dates: 1/90–1/97.
Learning Outcomes: Upon completion of the course, the student will be able to understand the necessity of obtaining top quality products and services at the lowest prices; the need for cost reduction without sacrificing necessary functions and performance; the need for creative problem solving techniques; the need for studying problem analysis and functional cost analysis; and the need for minute planning in determining job, product, and service specifications.
Instruction: Instruction includes creative thinking and and brainstorming techniques in the acquisition of products and services and how they relate to DoD activities. Also included are cost estimating, bid functions, contractual aspects, value judgement methodology, and incentives in value engineering.
Credit Recommendation: Credit is not recommended because of the limited, specialized nature of the course (2/87).

AF-1405-0069

INTEGRATED ORGANIZATIONAL AVIONIC SYSTEMS SPECIALIST BY CORRESPONDENCE

Course Number: 45250A.
Location: Extension Course Institute, Gunter AFS, AL.
Length: Maximum, 52 weeks.
Exhibit Dates: 1/90–8/94.
Learning Outcomes: Upon completion of the course, the student will be able to describe material management procedures; identify aircraft tools and hardware; perform solderless wire connections; and describe the operating principles of a voltmeter, ohmmeter, and ammeter.
Instruction: The student receives instruction in material management procedures, various tools and hardware used on aircraft, identification and construction of solderless electrical connections, and the care and use of electrical measuring instruments.
Credit Recommendation: In the lower-division baccalaureate/associate degree category, 1 semester hour in material management (12/89).

AF-1405-0070

MATERIEL STORAGE AND DISTRIBUTION SPECIALIST BY CORRESPONDENCE

Course Number: 64551.
Location: Extension Course Institute, Gunter AFS, AL.
Length: Maximum, 52 weeks.
Exhibit Dates: 4/92–4/96.
Learning Outcomes: Before 4/92 see AF-1405-0013. Upon completion of the course, the student will be able to describe materiel supply, storage, and distribution systems and their functions.
Instruction: Instruction in this correspondence course covers career field duties; responsibilities, sources, and functions of supply systems; publications; internal records; inspection; receiving procedures; bench stock replenishment; warehouse layout; property storage; location designators; safety factors; and materiel-handling principles and equipment, as well as the disposal of hazardous material.
Credit Recommendation: In the vocational certificate category, 3 semester hours in materiel handling and inventory control (4/93).

AF-1405-0071

INVENTORY MANAGEMENT SUPERVISOR BY CORRESPONDENCE

Course Number: 64570.

Location: Extension Course Institute, Gunter AFS, AL.
Length: Maximum, 52 weeks.
Exhibit Dates: 7/91–8/96.
Learning Outcomes: Before 7/91 see AF-1405-0059. Upon completion of the course, the student will be able to manage a base automated inventory system including problem solving related to the system and systems analysis for new applications and upgrading.
Instruction: This correspondence course includes a review of various functions involved in automated inventory management, such as completing inventory records, requisitioning materials, preparing stock control records and reports, issuing supplies and materials, handling retail sales, and manual accounting procedures. In addition, it provides a system for identifying, planning, developing, and changing procedures concerning inventory management; a system for interpreting policies; and and a system for developing inventory control. There is heavy emphasis placed on the Air Force automated inventory system and its applications in asset control and management of base assets.
Credit Recommendation: In the lower-division baccalaureate/associate degree category, 1 semester hour in records management (4/93); in the upper-division baccalaureate category, 2 semester hours in principles of systems management (4/93).

AF-1405-0072

SUPPLY SYSTEMS ANALYST JOURNEYMAN BY CORRESPONDENCE

Course Number: 2S052.
Location: Extension Course Institute, Gunter AFS, AL.
Length: Maximum, 52 weeks.
Exhibit Dates: 12/93–Present.
Learning Outcomes: Upon completion of the course, the student will be able to describe a totally computerized supply system and will be able to function within this specialized area.
Instruction: Course provides the student with the basic fundamentals of a computerized supply system encompassing supply processes and procedures, computer interface with a central system, computer security and resource management, system hardware, data base retrieval, and internal records.
Credit Recommendation: In the lower-division baccalaureate/associate degree category, 1 semester hour in inventory control (6/94).

AF-1406-0033

PROFESSIONAL PERSONNEL MANAGEMENT

Course Number: LMDC 501.
Location: Leadership and Management Development Center, Air University, Maxwell AFB, AL.
Length: 6 weeks (204-248 hours).
Exhibit Dates: 1/90–12/91.
Objectives: To train senior personnel in advanced management techniques.
Instruction: Course includes lectures and practical exercises in the skills necessary to manage and handle facets of personnel management. Course includes Department of Defense management environment, management processes, human relations, management systems, quantitative methods, and personnel processes.
Credit Recommendation: In the upper-division baccalaureate category, 6 semester hours in management and 3 in labor relations (2/74).

AF-1406-0045

INFORMATION MANAGEMENT SPECIALIST BY CORRESPONDENCE
(Administration Specialist by Correspondence)

Course Number: *Version 1:* 70250. *Version 2:* 70250.
Location: Extension Course Institute, Gunter AFS, AL.
Length: *Version 1:* Maximum, 52 weeks. *Version 2:* Maximum, 52 weeks.
Exhibit Dates: *Version 1:* 8/91–7/96. *Version 2:* 1/90–7/91.
Learning Outcomes: *Version 1:* Upon completion of the course, the student will be able to describe the mission and function of administrative communications, administrative correspondence, quality control, security, and records management. *Version 2:* Upon completion of this course, the student will be able to describe the flow of communication in administration offices, administrative correspondence, documentation, word processing, and records management.
Instruction: *Version 1:* Topics in this correspondence course include the objectives of publications management, Freedom of Information act, administrative correspondence, preparation of performance reports, documentation, records and forms management, preparation of publications, and distribution of administrative communications. *Version 2:* Topics included in this course are objectives of publications management, Freedom of Information act, administrative correspondence, preparation of performance reports, documentation, word processing, records and forms management, preparation of publications, and distribution of administrative communications.
Credit Recommendation: *Version 1:* In the lower-division baccalaureate/associate degree category, 1 semester hour in records management and 1 in office administration (4/93). *Version 2:* In the lower-division baccalaureate/associate degree category, 3 semester hours in business communication (2/87).

AF-1406-0046

ADMINISTRATION TECHNICIAN BY CORRESPONDENCE

Course Number: 70270.
Location: Extension Course Institute, Gunter AFS, AL.
Length: Maximum, 52 weeks.
Exhibit Dates: 1/90–1/97.
Learning Outcomes: Upon completion of this course, the student will be able to describe the communications process, the flow of communications, administrative correspondence, documentation, word processing, and records management.
Instruction: Topics in this course include objectives of publications management, administrative correspondence, preparation of performance reports, documentation, word processing, records management, and distribution of administrative communications.
Credit Recommendation: In the lower-division baccalaureate/associate degree category, 3 semester hours in business communications (2/87).

AF-1406-0047

LEGAL SERVICES SPECIALIST BY CORRESPONDENCE

Course Number: 70550.
Location: Extension Course Institute, Gunter AFS, AL.
Length: Maximum, 52 weeks.
Exhibit Dates: 1/90–12/94.
Learning Outcomes: Upon completion of this course, the student will have knowledge of legal office administration and management, legal research, claims procedures, the tort claims act, and claims investigations.
Instruction: Topics in this course include legal office administration and management, military justice administration, automated military justice analysis and management systems, claims procedures, federal tort act, and property damage tort claims.
Credit Recommendation: In the lower-division baccalaureate/associate degree category, 1 semester hour in business management (2/87).

AF-1406-0049

PERSONAL AFFAIRS SPECIALIST BY CORRESPONDENCE

Course Number: 73251.
Location: Extension Course Institute, Gunter AFS, AL.
Length: Maximum, 52 weeks.
Exhibit Dates: 1/90–8/94.
Learning Outcomes: At the conclusion of this course, the student will be able to describe the benefits, support services, and various aid groups in place to assist and advise Air Force personnel in their personal affairs and describe the duties and responsibilities of a supervisor.
Instruction: This course covers the benefit programs, support services, and various assistance groups available to Air Force personnel concerning their personal affairs. Topics include insurance programs, hospitalization coverage, awards and decorations, legal issues, payroll, retirement, and training. Also covered are interviewing and briefing techniques and a survey of supervision and training.
Credit Recommendation: In the lower-division baccalaureate/associate degree category, 1 semester hour in personnel administration (2/87).

AF-1406-0051

CAREER ADVISORY TECHNICIAN BY CORRESPONDENCE

Course Number: 73274.
Location: Extension Course Institute, Gunter AFS, AL.
Length: Maximum, 52 weeks.
Exhibit Dates: 1/90–2/93.
Learning Outcomes: At the conclusion of the course, the student will be able to describe the duties of the Career Assistance Unit, the components of the Air Force reenlistment and bonus programs, the various Air Force publicity and advertising initiatives, the factors affecting human behavior and motivation, and various selling and interview techniques.
Instruction: Instruction covers the specialized areas of recruitment and retention of Air Force personnel. Instruction includes the specific duties of the Career Assistant Unit, the psychology of human behavior and motivation, various selling and interview techniques, and a

detailed presentation on the reenlistment options and bonus programs in the Air Force.

Credit Recommendation: In the vocational certificate category, 1 semester hour in personnel supervision (2/87).

AF-1406-0052

MANPOWER MANAGEMENT TECHNICIAN BY CORRESPONDENCE

Course Number: *Version 1:* 73371. *Version 2:* 73371.

Location: Extension Course Institute, Gunter AFS, AL.

Length: *Version 1:* Maximum, 52 weeks. *Version 2:* Maximum, 52 weeks.

Exhibit Dates: *Version 1:* 6/91–2/99. *Version 2:* 1/90–5/91.

Learning Outcomes: *Version 1:* Upon completion of the course, the student will be able to demonstrate knowledge of the organizational structure of the Air Force and will be able to apply management sciences and specific quantitative methods. *Version 2:* At the conclusion of this course, the student will be able to describe the organizational structure of the Air Force and the function of personnel administration/management.

Instruction: *Version 1:* This correspondence course includes career opportunities, planning, budgeting, organizational structures, supervision, standards development theory and practice, work measurement, time studies, data analysis, probability applications, statistics, queuing theory, layout analysis, productivity review, and interviewing techniques. *Version 2:* Subject areas include career opportunities; planning; budgeting; supervision; training; human relations; employee standards development theory and practice; work measurement; time studies; data analysis and presentation through probability, statistics, and queuing theory; interviewing techniques; and conducting workshops.

Credit Recommendation: *Version 1:* In the lower-division baccalaureate/associate degree category, 2 semester hours in introduction to statistics, 2 in management sciences, and 2 in principles of management (4/93). *Version 2:* In the lower-division baccalaureate/associate degree category, 1 semester hour in statistics and probability or quantitative methods (2/87); in the upper-division baccalaureate category, 3 semester hours in personnel administration/management (2/87).

AF-1406-0053

SOCIAL ACTIONS TECHNICIAN (EQUAL OPPORTUNITY/HUMAN RELATIONS) BY CORRESPONDENCE

Course Number: 73470A.
Location: Extension Course Institute, Gunter AFS, AL.
Length: Maximum, 52 weeks.
Exhibit Dates: 1/90–1/97.

Learning Outcomes: Upon completion of this course, the student will be able to function as a social actions technician; interpret the Freedom of Information act; prepare lesson plans and teaching materials; provide counseling on social problems; and handle human relations problems in the Air Force.

Instruction: Topics included in this course are the Freedom of Information act, teaching methods, communication skills, lesson planning, the lecture method, guided discussion group leadership, instructional aids, organizational assessment, the interview process, survey techniques organizational management tools, personal growth and development, types of counseling, recognition of social problems, social tensions, racism, sexism, minority awareness, and prevention and treatment of social problems.

Credit Recommendation: In the lower-division baccalaureate/associate degree category, 2 semester hours in methods of teaching vocational education and 1 in personnel management (2/87).

AF-1406-0054

SOCIAL ACTIONS TECHNICIAN (DRUGS/ALCOHOL) BY CORRESPONDENCE

Course Number: 73470B.
Location: Extension Course Institute, Gunter AFS, AL.
Length: Maximum, 52 weeks.
Exhibit Dates: 1/90–1/97.

Learning Outcomes: Upon completion of the course, the student will be able to function as a social action technician, prepare lesson plans and teaching materials, provide counseling in social problems, and handle human relations problems.

Instruction: Instruction includes tracking methods, communication skills, lesson planning, the lecture method, guided discussions, group leadership, instructional aids, organizational assessment, the interview process, survey techniques, organizational management tools, personal growth and development, types of counseling, recognition of social problems, social tensions, prevention and treatment of social problems, and program management.

Credit Recommendation: In the lower-division baccalaureate/associate degree category, 2 semester hours in methods of teaching vocational education and 1 in personnel administration (2/87).

AF-1406-0055

APPRENTICE FITNESS AND RECREATION SPECIALIST BY CORRESPONDENCE

Course Number: 74131.
Location: Extension Course Institute, Gunter AFS, AL.
Length: Maximum, 52 weeks.
Exhibit Dates: 1/90–1/95.

Learning Outcomes: Upon completion of the course, the student will have a basic understanding of recreational programming.

Instruction: This is a survey course which introduces students to the field of recreational and sports programs. Topics include Air Force recreation policies and procedures, recreational program bookkeeping, types of recreational activities, planning recreation programs, rules of selected games, and tournament play.

Credit Recommendation: In the lower-division baccalaureate/associate degree category, 1 semester hour in recreation/recreational programming (2/87).

AF-1406-0056

FITNESS AND RECREATION SPECIALIST BY CORRESPONDENCE

Course Number: 74151.
Location: Extension Course Institute, Gunter AFS, AL.
Length: Maximum, 52 weeks.
Exhibit Dates: 1/90–11/95.

Learning Outcomes: Upon completion of the course, the student will know the mission, policies, procedures, and practices of the Air Force's recreational programs.

Instruction: Instruction includes the concepts of recreational programs with an emphasis on specific Air Force policies, procedures, and forms in the morale, welfare, and recreation section.

Credit Recommendation: In the vocational certificate category, 1 semester hour in record keeping (2/87); in the lower-division baccalaureate/associate degree category, 1 semester hour in recreation/recreational programming (2/87).

AF-1406-0057

FITNESS AND RECREATION SUPERVISOR BY CORRESPONDENCE

Course Number: 74171.
Location: Extension Course Institute, Gunter AFS, AL.
Length: Maximum, 52 weeks.
Exhibit Dates: 1/90–1/97.

Learning Outcomes: At the conclusion of the course, the student will have a thorough understanding of the Air Force's morale, welfare, and recreation programs and the supervisor's duties and responsibilities.

Instruction: The course presents detailed coverage in the administration and operation of Air Force's morale, welfare, and recreation programs. Specific tropics covered include the record keeping, accounting, budgeting, personnel management, training, program development necessary for an effective program.

Credit Recommendation: In the lower-division baccalaureate/associate degree category, 1 semester hour in recreation/recreation programming/recreation services (2/87).

AF-1406-0058

OPEN MESS MANAGEMENT SPECIALIST BY CORRESPONDENCE

Course Number: 74250.
Location: Extension Course Institute, Gunter AFS, AL.
Length: Maximum, 52 weeks.
Exhibit Dates: 1/90–10/93.

Learning Outcomes: At the conclusion of the course the student will have a working knowledge of the administration and operation of open mess facilities.

Instruction: This is a comprehensive course in the administration and operation of an open mess (food service) facility. The broad range of materials covered includes the administrative topics of food service such as supervision, budgeting, food costing techniques, inventory control, safety, insurance, facility issues, and promotion. The operational topics include nutrition, food preparation, serving techniques, identification and classification of food products, and sanitation. Students are also introduced to special retail activities such as bar, package goods, and barber shop operations.

Credit Recommendation: In the lower-division baccalaureate/associate degree category, 3 semester hours in introduction to food service or introduction to restaurant management and 1 in retail management (2/87).

AF-1406-0059

CLUB MANAGEMENT SUPERVISOR BY CORRESPONDENCE

Course Number: 74270.
Location: Extension Course Institute, Gunter AFS, AL.
Length: Maximum, 52 weeks.
Exhibit Dates: 1/90–9/91.
Learning Outcomes: At the conclusion of the course, the student will have a working knowledge in the many areas of club administration, including principles of supervision, food service budgeting and costing, and the fundamentals of food service and special activities management.
Instruction: This course covers the duties and responsibilities of the club management supervisor. Subject matter includes fundamentals of supervision, budgeting, food costing techniques, inventory control, safety issues, insurance, facility considerations, and promotion.
Credit Recommendation: In the vocational certificate category, 1 semester hour in bookkeeping (2/87); in the lower-division baccalaureate/associate degree category, 3 semester hours in introduction to food service or introduction to restaurant management and 1 in retail management (2/87).

AF-1406-0061

FIGHTER WEAPONS INSTRUCTOR A-10

Course Number: *Version 1:* A10001DOPN. *Version 2:* A10001DOPN. *Version 3:* A10001DOPN. *Version 4:* A10001DOPN.
Location: Fighter Weapons School, Nellis AFB, NV.
Length: *Version 1:* 15-20 weeks (428-599 hours). *Version 2:* 15-20 weeks (428-599 hours). *Version 3:* 15-20 weeks (428-599 hours). *Version 4:* 14-18 weeks (439-518 hours).
Exhibit Dates: *Version 1:* 1/94–Present. *Version 2:* 1/92–12/93. *Version 3:* 4/91–12/91. *Version 4:* 1/90–3/91.
Learning Outcomes: *Version 1:* Upon completion of the course, the student will be able to recognize, interpret, and explain advanced aircraft systems and the performance environment; instruct and evaluate crewmembers in a dynamic, high-activity flight management arena; assess the needs for various instructional systems; prepare appropriate classroom and laboratory activities; and deliver and evaluate the educational event. *Version 2:* Upon completion of the course, the student will be able to recognize, interpret, and explain advanced aircraft systems and the performance environment; instruct and evaluate crewmembers in a dynamic, high-activity flight management arena; assess the needs for various instructional systems; prepare appropriate classroom and laboratory activities; and deliver and evaluate the educational event. *Version 3:* Upon completion of the course, the student will be able to recognize, interpret, and explain advanced aircraft systems and the performance environment; instruct and evaluate crewmembers in a dynamic, high-activity flight management arena; assess the needs for various instructional systems; prepare appropriate classroom and laboratory activities; and deliver and evaluate the educational event. *Version 4:* Upon completion of the course, the student will be able to design and participate in instructional programs for aircrew members on the function and capabilities of the A-10 aircraft and its systems. The student evaluates systems and advises regarding deficiencies including such areas as application and maintenance and participates in and monitors training programs. Students develop skills in communication and instructional areas.
Instruction: *Version 1:* This course consists of classroom lectures, student independent research, formal student-led briefings, and appropriate laboratory experiences in advanced aircraft systems and performance, aircrew systems management, instructional design, and teaching methodologies. *Version 2:* This course consists of classroom lectures, student independent research, formal student-led briefings, and appropriate laboratory experiences in advanced aircraft systems and performance, aircrew systems management, instructional design, and teaching methodologies. *Version 3:* This course consists of classroom lectures, student independent research, formal student-led briefings, and appropriate laboratory experiences in advanced aircraft systems and performance, aircrew systems management, instructional design, and teaching methodologies. *Version 4:* This course is specifically designed to provide the experienced Air Force pilot with a comprehensive pattern of academic study, independent research, and appropriate laboratory activity (flight) on the A-10 weapons system. Major segments of this course address the technical subjects of aerodynamics/performance and avionics/electronics. In addition, a comprehensive treatment of teaching techniques and methodology as related to the presentation of technical data is included.
Credit Recommendation: *Version 1:* In the graduate degree category, 4 semester hours in advanced aircraft systems and performance, 5 in aircrew systems management, 2 in instructional design and teaching methodologies, and 2 in advanced flight laboratory (7/94). *Version 2:* In the graduate degree category, 5 semester hours in advanced aircraft systems and performance, 4 in aircrew systems management, 2 in instructional design and teaching methodologies, and 2 in advanced flight laboratory (7/94). *Version 3:* In the graduate degree category, 4 semester hours in advanced aircraft systems and performance, 2 in aircrew systems management, 2 in instructional design and teaching methodologies, and 2 in advanced flight laboratory (7/94). *Version 4:* In the graduate degree category, 3 semester hours in applied aerodynamics in high performance aircraft, 3 in advanced avionics/electronics, and 2 in instructional methodology (1/92).

AF-1406-0062

1. WEAPONS INSTRUCTOR (AWC)
2. WEAPONS INSTRUCTOR (AWC)
3. FIGHTER WEAPONS INSTRUCTOR AIR WEAPONS CONTROLLER

Course Number: *Version 1:* 1745IDWN. *Version 2:* 1745IDWN. *Version 3:* 1745IDWN.
Location: Fighter Weapons School, Nellis AFB, NV.
Length: *Version 1:* 20 weeks (595-642 hours). *Version 2:* 20 weeks (595-642 hours). *Version 3:* 14 weeks (277 hours).
Exhibit Dates: *Version 1:* 7/93–Present. *Version 2:* 1/92–6/93. *Version 3:* 1/90–12/91.
Learning Outcomes: *Version 1:* Upon completion of the course, the student will be able to recognize, interpret, and explain advanced aircraft systems and the performance environment; instruct and evaluate crewmembers in a dynamic, high-activity flight management arena; assess the needs for various instructional systems; prepare appropriate classroom and laboratory activities; and deliver and evaluate the educational event. *Version 2:* Upon completion of the course, the student will be able to recognize, interpret, and explain advanced aircraft systems and the performance environment; instruct and evaluate crewmembers in a dynamic, high-activity flight management arena; assess the needs for various instructional systems; prepare appropriate classroom and laboratory activities; and deliver and evaluate the educational event. *Version 3:* Upon completion of the course, the student will be able to instruct in the areas of radar theory and systems for F-15 and F-16 aircraft plus AWACS, ground TACS radar, and HORAD radar systems. Student will be able to apply and integrate all radar systems specified in the course.
Instruction: *Version 1:* The course consists of classroom lectures, student independent research, formal student-led briefings, and appropriate laboratory experiences in advanced aircraft systems and performance, aircrew systems management, instructional design, and teaching methodologies. *Version 2:* The course consists of classroom lectures, student independent research, formal student-led briefings, and appropriate laboratory experiences in advanced aircraft systems and performance, aircrew systems management, instructional design, and teaching methodologies. *Version 3:* The course consists of classroom lectures, hands-on application of radar scopes in a laboratory situation, student independent research leading to a paper and presentation in appropriate radar systems, and teaching techniques and methodology. Information systems management approach is stressed throughout the course.
Credit Recommendation: *Version 1:* In the graduate degree category, 4 semester hours in advanced aircraft systems and performance, 8 in aircrew systems management, 2 in instructional design and teaching methodologies, and 1 in advanced flight laboratory (7/94). *Version 2:* In the graduate degree category, 3 semester hours in advanced aircraft systems and performance, 7 in aircrew systems management, 2 in instructional design and teaching methodologies, and 1 in advanced flight laboratory (7/94). *Version 3:* In the graduate degree category, 5 semester hours in information systems and 2 in instructional methodology (9/87).

AF-1406-0063

FIGHTER WEAPONS INSTRUCTOR F-111

Course Number: *Version 1:* F1110IDOWC; F1110IDOAC. *Version 2:* F11101DOA1; F11101DOW1.
Location: *Version 1:* Fighter Weapons School, Cannon AFB, NM. *Version 2:* Fighter Weapons School, Nellis AFB, NV.
Length: *Version 1:* 19 weeks (537-546 hours). *Version 2:* 15 weeks (454-455 hours).
Exhibit Dates: *Version 1:* 1/92–Present. *Version 2:* 1/90–12/91.
Learning Outcomes: *Version 1:* Upon completion of the course, the student will be able to recognize, interpret, and explain advanced aircraft systems and the performance environment; instruct and evaluate crewmembers in a dynamic, high-activity flight management arena; assess the needs for various instructional systems; prepare appropriate classroom and laboratory activities; and deliver and evaluate the educational event. *Version 2:* Upon completion of the course, the student will be able to

design and participate in instructional programs for aircrew members relating to the function and capabilities of the F-111 aircraft and its systems; evaluate related systems and advise regarding deficiencies including such areas as application and maintenance; participate in and monitor training programs; and develop skills in communication and instructional areas.

Instruction: *Version 1:* This course consists of classroom lectures, student independent research, formal student-led briefings, and appropriate laboratory experiences in advanced aircraft systems and performance, aircrew systems management, and instructional design and teaching methodologies. *Version 2:* This course is specifically designed to provide the experienced Air Force pilot with a comprehensive pattern of academic study, independent research, and appropriate laboratory activity (flight) on the F-111 weapon system. Major segments of this course address the technical subjects of aerodynamics/performance, avionics/electronics, and navigation systems. In addition, a comprehensive treatment of teaching techniques and methodology as related to the presentation of technical data is included.

Credit Recommendation: *Version 1:* In the graduate degree category, 4 semester hours in advanced aircraft systems and performance, 3 in aircrew systems management, 2 in instructional design and teaching methodologies, and 3 in advanced flight laboratory (7/94). *Version 2:* In the graduate degree category, 3 semester hours in applied aerodynamics in high speed aircraft, 3 in advanced avionics/electronics, 1 in navigation systems, and 2 in instructional methodology (1/87).

AF-1406-0064

FIGHTER WEAPONS INSTRUCTOR F-16

Course Number: *Version 1:* F16001DOPN. *Version 2:* F16001DOPN. *Version 3:* F16001DOPN. *Version 4:* F16001DOPN.
Location: Fighter Weapons School, Nellis AFB, NV.
Length: *Version 1:* 15-20 weeks (477-554 hours). *Version 2:* 15-20 weeks (477-554 hours). *Version 3:* 15-20 weeks (477-554 hours). *Version 4:* 15 weeks (487 hours).
Exhibit Dates: *Version 1:* 7/93–Present. *Version 2:* 1/92–6/93. *Version 3:* 1/91–12/91. *Version 4:* 1/90–12/90.
Learning Outcomes: *Version 1:* Upon completion of the course, the student will be able to recognize, interpret, and explain advanced aircraft systems and the performance environment; instruct and evaluate crewmembers in a dynamic, high-activity flight management arena; assess the needs for various instructional systems; prepare appropriate classroom and laboratory activities; and deliver and evaluate the educational event. *Version 2:* Upon completion of the course, the student will be able to recognize, interpret, and explain advanced aircraft systems and the performance environment; instruct and evaluate crewmembers in a dynamic, high-activity flight management arena; assess the needs for various instructional systems; prepare appropriate classroom and laboratory activities; and deliver and evaluate the educational event. *Version 3:* Upon completion of the course, the student will be able to recognize, interpret, and explain advanced aircraft systems and the performance environment; instruct and evaluate crewmembers in a dynamic, high-activity flight management arena; assess the needs for various instructional systems; prepare appropriate classroom and laboratory activities; and deliver and evaluate the educational event. *Version 4:* Upon completion of the course, the student will be able to instruct in the areas of aerodynamics, aircraft systems, and avionics/electronics as they apply to the F-16 aircraft.

Instruction: *Version 1:* The course consists of classroom lectures, student independent research, formal student-led briefings, and appropriate laboratory experiences in advanced aircraft systems and performance, aircrew systems management, and instructional design and teaching methodologies. *Version 2:* The course consists of classroom lectures, student independent research, formal student-led briefings, and appropriate laboratory experiences in advanced aircraft systems and performance, aircrew systems management, and instructional design and teaching methodologies. *Version 3:* The course consists of classroom lectures, student independent research, formal student-led briefings, and appropriate laboratory experiences in advanced aircraft systems and performance, aircrew systems management, and instructional design and teaching methodologies. *Version 4:* The course consists of classroom lectures, in-flight laboratories, and independent student research into all aspects of the F-16 aircraft. Particular emphasis is directed at the study of aerodynamics, and to advanced systems and avionics/electronics as they apply to the operational mission of the F-16 aircraft. Teaching techniques and methodology, as they relate to the subjects listed above, are developed throughout the course.

Credit Recommendation: *Version 1:* In the graduate degree category, 5 semester hours in advanced aircraft systems performance, 4 in aircrew systems management, 2 in instructional design and teaching methodologies, and 2 in advanced flight laboratory (7/94). *Version 2:* In the graduate degree category, 5 semester hours in advanced aircraft systems performance, 3 in aircrew systems management, 2 in instructional design and teaching methodologies, and 2 in advanced flight laboratory (7/94). *Version 3:* In the graduate degree category, 4 semester hours in advanced aircraft systems performance, 3 in aircrew systems management, 2 in instructional design and teaching methodologies, and 2 in advanced flight laboratory (7/94). *Version 4:* In the graduate degree category, 3 semester hours in applied aerodynamics in high performance aircraft, 3 in advanced aircraft systems and components, 3 in advanced avionics/electronics, and 2 in instructional methodology (3/89).

AF-1406-0065

ADVERSARY TACTICS INSTRUCTOR (AFSC K1115H)

Course Number: F50000A1AN.
Location: Fighter Weapons School, Nellis AFB, NV.
Length: 13-15 weeks (532 hours).
Exhibit Dates: 1/90–9/97.
Learning Outcomes: Upon completion of the course, the student will be able to function as a USAF aggressor pilot. This assignment requires providing academic instruction on and airborne emulation of enemy tactical fighter pilots, aircraft avionics, weapons, formations, tactics, and employment philosophy.
Instruction: This course has been developed to present an advanced program including academic and laboratory experience (flight) to a select group of Air Force pilots. The curriculum includes specialized work in applied aerodynamics/performance and aircraft systems and components that are readily equated to university topics. Class sizes are very small and the level of support, both staff and facilities, is very high. It is well to note that appreciable emphasis is placed on the comparison between US and foreign equipment, systems, and philosophy.
Credit Recommendation: In the graduate degree category, 3 semester hours in applied aerodynamics in high performance aircraft and 2 in advanced aircraft systems and components (9/87).

AF-1406-0066

FIGHTER WEAPONS INSTRUCTOR F-15

Course Number: *Version 1:* F1500IDOPN. *Version 2:* F1500IDOPN.
Location: Fighter Weapons School, Nellis AFB, NV.
Length: *Version 1:* 20 weeks (449-496 hours). *Version 2:* 14-19 weeks (409-460 hours).
Exhibit Dates: *Version 1:* 1/93–Present. *Version 2:* 1/90–12/92.
Learning Outcomes: *Version 1:* Upon completion of the course, the student will be able to recognize, interpret, and explain advanced aircraft systems and the performance environment; instruct and evaluate crewmembers in a dynamic, high-activity flight management arena; assess the need for various instructional systems; prepare appropriate classroom and laboratory activities; and deliver and evaluate the educational event. *Version 2:* Upon completion of the course, the student will be able to teach aerodynamics, aircraft systems, and avionics/electronics as they apply to the F-15 aircraft.

Instruction: *Version 1:* The course consists of classroom lectures, student independent research, formal student-led briefings, and appropriate laboratory experiences in advanced aircraft systems and performance, aircrew systems management, and instructional design and teaching methodologies. *Version 2:* The course consists of classroom lectures, in-flight laboratories, and independent student research into all aspects of the F-15 aircraft. Particular emphasis is directed at the study of aerodynamics, advanced systems and avionics/electronics as they apply to the operational mission of the F-15 aircraft. Teaching techniques and methodology, as they relate to the subjects listed above, are developed throughout the course.

Credit Recommendation: *Version 1:* In the graduate degree category, 4 semester hours in advanced aircraft systems and performance, 2 in aircrew systems management, 2 in instructional design and teaching methodologies, and 2 in advanced flight laboratory (7/94). *Version 2:* In the graduate degree category, 3 semester hours in applied aerodynamics in high performance aircraft, 3 in advanced aircraft systems and components, 3 in advanced avionics/electronics, and 2 in instructional methodology (1/92).

AF-1406-0067

ADVERSARY TACTICS INSTRUCTOR (AFSC K1745)

Course Number: 174501AA.
Location: Fighter Weapons School, Nellis AFB, NV.
Length: 10 weeks (196-206 hours).

Exhibit Dates: 1/90–Present.
Learning Outcomes: Upon completion of the course, the student will be able to provide academic instruction on radar theory and aircraft control and communications.
Instruction: The course consists of classroom lectures and student independent research leading to a paper and presentation in appropriate radar systems. Information systems management approach is stressed throughout the course.
Credit Recommendation: In the graduate degree category, 3 semester hours in information systems (6/89).

AF-1406-0068

TACTICAL AIR COMMAND AND CONTROL SPECIALIST BY CORRESPONDENCE

Course Number: *Version 1:* 27550. *Version 2:* 27550.
Location: Extension Course Institute, Gunter AFS, AL.
Length: *Version 1:* Maximum, 52 weeks. *Version 2:* Maximum, 52 weeks.
Exhibit Dates: *Version 1:* 8/91–8/96. *Version 2:* 1/90–7/91.
Learning Outcomes: *Version 1:* Upon completion of the course, the student will be able to describe personnel supervision techniques appropriate to the management of any organization. Skills are included in the use and application of radio equipment as well as in weather reporting. The graduate has an understanding of the differences between the tactical air control system and the US Army structure and can function at a entry management level. *Version 2:* Upon completion of the course, the student will be able to describe personnel supervision techniques appropriate to the management of any organization. Skills are included in the use and application of radio equipment as well as in weather reporting. The graduate will have an understanding of the differences between the tactical air control system and the US Army structure and can function at a entry management level.
Instruction: *Version 1:* The course consists of self-paced, independent study with open book progress checks and satisfactory completion of a proctored comprehensive final examination. Content areas have been constructed to provide study in the broad area of responsibility encompassed by the tactical air command and control specialist. Basic management techniques such as employee relations, conflict solving, and performance evaluation are covered. In addition, publications and communications as well as safety are included. Basic radio communication theory, field utilization, and military systems comprise a portion of the required study. Specific military topics relate to tactical air control, surface navigation (compass/map), Russian political philosophy, and specialized equipment. *Version 2:* The course consists of self-paced, independent study with open book progress checks and satisfactory completion of a proctored comprehensive final examination. Content areas have been constructed to provide study in the broad area of responsibility encompassed by the tactical air command and control specialist. Basic management techniques such as employee relations, conflict solving, and performance evaluation are covered. In addition, publications and communications as well as safety are included. Basic radio communication theory, field utilization, and military systems comprise a portion of the required study. Specific military topics relate to tactical air control, surface navigation (compass/map), Russian political philosophy, and specialized equipment.
Credit Recommendation: *Version 1:* In the lower-division baccalaureate/associate degree category, 1 semester hour in principles of supervision and 1 in radio communications (4/93). *Version 2:* In the lower-division baccalaureate/associate degree category, 1 semester hour in personnel supervision management and 1 in radio communication (12/89).

AF-1406-0069

EDUCATION SPECIALIST BY CORRESPONDENCE

Course Number: 75150.
Location: Extension Course Institute, Gunter AFS, AL.
Length: Maximum, 52 weeks.
Exhibit Dates: 1/90–12/95.
Learning Outcomes: Upon completion of the course, the student will be able to describe the various programs of education available in the Air Force and provide advising and counseling services in the area of education.
Instruction: The course consists of self-paced, independent study with open book progress checks and satisfactory completion of a proctored comprehensive final examination. Course covers the Air Force Education Services Program including education assistance programs, Veteran's Administration educational allowances, and Air Force preparatory programs. Testing and test security is addressed as well as educational advisement and counseling. Basic education office administration practices are also studied.
Credit Recommendation: In the upper-division baccalaureate category, 1 semester hour in educational counseling (12/89).

AF-1406-0070

TRAINING TECHNICIAN BY CORRESPONDENCE

Course Number: 75172.
Location: Extension Course Institute, Gunter AFS, AL.
Length: Maximum, 52 weeks.
Exhibit Dates: 1/90–Present.
Learning Outcomes: Upon completion of the course, the student will be able to develop an instructional system and operate and manage a formal training program with no supervision.
Instruction: The course consists of self-paced, independent study with open book progress checks and satisfactory completion of a proctored comprehensive final examination. Content includes fundamentals of instruction, conduct of training, training aids and equipment, and evaluation and measurement.
Credit Recommendation: In the upper-division baccalaureate category, 2 semester hours in educational system/program development (12/89).

AF-1406-0071

TRAINING SPECIALIST BY CORRESPONDENCE

Course Number: 75132.
Location: Extension Course Institute, Gunter AFS, AL.
Length: Maximum, 52 weeks.
Exhibit Dates: 1/90–Present.
Learning Outcomes: Upon completion of the course, the student will be able to support the establishment and maintenance of a training program for a work unit related to a job location or class environment.
Instruction: The course consists of self-paced, independent study with open-book progress checks and satisfactory completion of a proctored comprehensive final examination. Content includes role of on-the-job training, management and control for on-the-job training, administration and supervisory responsibilities pertaining to training programs, fundamentals of instruction, instructional techniques, training aids and materials, and tests and measurement.
Credit Recommendation: In the upper-division baccalaureate category, 1 semester hour in principles of instruction (12/89).

AF-1406-0072

TRAINING SYSTEMS TECHNICIAN BY CORRESPONDENCE

Course Number: 75171.
Location: Extension Course Institute, Gunter AFS, AL.
Length: Maximum, 52 weeks.
Exhibit Dates: 1/90–10/93.
Learning Outcomes: Upon completion of the course, the student will understand theory of instruction and instructional system development and apply that understanding to the establishment and management of an effective on-the-job training program.
Instruction: This course consists of self-paced, independent study with open book progress checks and satisfactory completion of a proctored comprehensive final examination. Content areas include instructional theory; application of instructional system development principles; establishment, control, and operation of a viable on-the-job training program; and related administrative tasks.
Credit Recommendation: In the upper-division baccalaureate category, 2 semester hours for field training program development (12/89).

AF-1406-0073

AUDIOVISUAL MEDIA SPECIALIST BY CORRESPONDENCE

Course Number: 23150.
Location: Extension Course Institute, Gunter AFS, AL.
Length: Maximum, 52 weeks.
Exhibit Dates: 1/90–1/97.
Learning Outcomes: Upon completion of the course, the student will be able to manage and operate a uniform system for cataloging, storing, distributing, maintaining, and using visual information materials and equipment.
Instruction: This is a two-part correspondence course. The method of instruction is through reading, exercises, and examinations. Topics include visual information services, products, equipment, and presentation support. Specific exercises include 16mm projector, overhead and opaque projection, slide projectors and control units, 35mm still cameras, and multimedia and video equipment. Those enrolling will have an AFSC assignment as prerequisite. This provides an application of principles learned to real requirements and constitutes an essential part of the instruction by serving as the laboratory component of the course. The student's supervisor must complete an evaluation survey of the course and its effectiveness

when the student takes the proctored final exam which measures mastery of the cognitive content. A score of 65 is required for a satisfactory completion certificate.

Credit Recommendation: In the lower-division baccalaureate/associate degree category, 3 semester hours in introduction to educational technology (3/91).

AF-1406-0074

PILOT INSTRUCTOR TRAINING
(Pilot Instructor Training (T-37))
(Pilot Instructor Training (T-38))

Course Number: F-V5A-Y; F-V5A-Z; F-V5A-A(T-37); F-V5A-B(T-38).
Location: Air Training Command, Randolph AFB, TX.
Length: 9-13 weeks (180-200 hours).
Exhibit Dates: 1/90–Present.
Learning Outcomes: Upon completion of the course, the student will be able to effectively present technical aviation data; evaluate flying performance; utilize communication skills in teaching situations; be knowledgeable in the areas of flight training, human factors, teaching methods, instructional philosophy, and technical data on specific equipment (T-37 or T-38).
Instruction: Course includes lectures, role playing, flying aircraft or simulations on a one-to-one basis, and briefing/debriefing sessions on instructor development, flight and ground safety, pertinent aircraft systems, applied aerodynamics, flight planning, and centrifuge training.
Credit Recommendation: In the graduate degree category, 3 semester hours in aviation science/instruction (10/91).

AF-1406-0075

HUMAN RESOURCE MANAGEMENT (HRMC)
(Professional Manpower (PMPMC))

Course Number: MLMDC 400.
Location: Center for Professional Development, Air University, Maxwell AFB, AL.
Length: 3-4 weeks (158-160 hours).
Exhibit Dates: 1/92–Present.
Learning Outcomes: Upon completion of the course, the student, in written and oral form, will be able to analyze the concepts, practices, procedures, and problems associated with military and nonmilitary personnel administration. In addition, the student will be able to interpret the dynamics of labor-management relations, the managerial-organizational environment, leadership values, ethics and processes, and behavioral research findings applied to the work ethic.
Instruction: Methods of instruction include lectures, seminars, case studies, research papers, and simulations. These methods are applied to administration and evaluation, manpower issues, military/civilian personnel procedures, programs, problems, the application of behavioral science to the human resource organizational climate, leadership and communication skills, and functional integration policies.
Credit Recommendation: In the upper-division baccalaureate category, 3 semester hours in public management or introduction to public administration (1/95); in the graduate degree category, 3 semester hours in public personnel administration and 3 in labor relations or collective bargaining (1/95).

AF-1406-0076

PERSONNEL SPECIALIST BY CORRESPONDENCE

Course Number: 73250.
Location: Extension Course Institute, Gunter AFS, AL.
Length: Maximum, 52 weeks.
Exhibit Dates: 1/90–2/96.
Learning Outcomes: Upon completion of the course, the student will be able to interpret personnel regulations and procedures pertaining to personnel documentation, military benefits, career planning and advising, communication, customer assistance, human resource management, classification systems, and training.
Instruction: This correspondence course details the duties and responsibilities of the personnel technician and personnel specialist positions. Emphasis is placed on detailed interpretations of rules and regulations in all areas of personnel administration and the forms and data systems used to document personnel actions. Human resource management related to staffing needs, classifications, and training is also covered.
Credit Recommendation: In the lower-division baccalaureate/associate degree category, 3 semester hours in personnel administration or in human resource management (4/93).

AF-1406-0077

PERSONNEL SYSTEMS MANAGEMENT SPECIALIST BY CORRESPONDENCE

Course Number: 73150.
Location: Extension Course Institute, Gunter AFS, AL.
Length: Maximum, 52 weeks.
Exhibit Dates: 8/91–3/95.
Learning Outcomes: Upon completion of the course, the student will be able to describe the Air Force personnel system operations and management. Student will be able to apply elementary computer operations in DOS and data base software and will be able to manage the output products. In addition, the student will be able to run the UNIX Visual Editor in office automation applications for the construction, editing, and preparation of documents. Troubleshooting equipment problems is covered in this correspondence course.
Instruction: This is a four-part correspondence course. Topics in this course include an introduction to personnel systems management and personnel system operations through applications software in office administration, data base concepts, risk management, computer systems interface, managing output products, software maintenance, and hardware troubleshooting.
Credit Recommendation: In the lower-division baccalaureate/associate degree category, 2 semester hours in computer applications (4/93).

AF-1406-0078

EDUCATION AND TRAINING MANAGER BY CORRESPONDENCE

Course Number: 3S200.
Location: Extension Course Institute, Gunter AFS, AL.
Length: Maximum, 52 weeks.
Exhibit Dates: 11/93–Present.
Learning Outcomes: Upon completion of the course, the student will be able to develop, coordinate, facilitate, and maintain a work center on-the-job training program.
Instruction: Course covers the five phases of the instructional system design process as it applies to work center on-the-job training. Course includes analyzing and defining the work center, determining duty position requirements, training objectives and measurement devices, determining method of work center training, and evaluation. Course also includes coordinating support for training and education service function.
Credit Recommendation: In the upper-division baccalaureate category, 2 semester hours in principles of instruction (6/94).

AF-1406-0079

EDUCATION AND TRAINING OFFICER BY CORRESPONDENCE

Course Number: 7504.
Location: Extension Course Institute, Gunter AFS, AL.
Length: Maximum, 52 weeks.
Exhibit Dates: 1/90–10/98.
Learning Outcomes: Upon completion of the course, the student will be able to support the establishment and maintenance of education and training programs for a work unit. The utilization of education and training in the field is emphasized.
Instruction: This correspondence course consists of self-paced, independent study. Topics covered include general management, budgeting principles, computer operation, curriculum development, and conducting and evaluating training. Emphasis is on instructional system development and how it works in curriculum development. There is a proctored final examination required.
Credit Recommendation: In the upper-division baccalaureate category, 2 semester hours in teaching methods (6/94).

AF-1406-0080

INSTRUCTIONAL SYSTEM DEVELOPMENT FOR TRAINING MANAGERS BY CORRESPONDENCE

Course Number: 7515.
Location: Extension Course Institute, Gunter AFS, AL.
Length: Maximum, 52 weeks.
Exhibit Dates: 1/90–3/96.
Learning Outcomes: Upon completion of the course, the student will be able to support the establishment and maintenance of education and training programs for a work unit. The utilization of instructional system development principles to on-the-job training is applied by the students.
Instruction: This correspondence course consists of self-paced, independent study which focuses on general instructional system development principles. Students learn how to develop duty and task inventories; analyze the task; define education and training requirements; develop learning objectives; construct tests; plan, develop, and validate instruction; conduct and evaluate instruction; and apply these principles to on-the-job training.
Credit Recommendation: In the upper-division baccalaureate category, 2 semester hours in principles of instruction (6/94).

AF-1406-0081

QUALITY MANAGEMENT BY CORRESPONDENCE

Course Number: 06613.
Location: Extension Course Institute, Gunter AFS, AL.
Length: Maximum, 52 weeks.
Exhibit Dates: 10/93–Present.
Learning Outcomes: Upon completion of the course, the student will be able to describe the concept of total quality management and the different viewpoints of the major contributors to the evolution of the quality management process.
Instruction: Course introduces the principles of quality management and provides an overview of the leading philosophies in total quality management, including those of W. Edwards Deming, Joseph M. Juran, Philip Crosby, and others. Included is the history of quality in America, statistical process control, and problem solving principles.
Credit Recommendation: In the lower-division baccalaureate/associate degree category, 1 semester hour in total quality management (6/94).

AF-1406-0094

MEDICAL MATERIEL SPECIALIST BY CORRESPONDENCE

Course Number: 91550.
Location: Extension Course Institute, Gunter AFS, AL.
Length: Maximum, 52 weeks.
Exhibit Dates: 1/90–11/97.
Learning Outcomes: Upon completion of the course, the student will be able to describe the general duties of the medical materiel management specialist.
Instruction: This monitored home study course includes readings, written exercises, on-the-job supervision, and proctored examinations. Topics include regulations, computerization, basic first aid, record keeping, control, requisitions, warehouse operations, stock rotation, excess management, financial management, and biomedical equipment repair.
Credit Recommendation: In the vocational certificate category, 6 semester hours in central supply technology (5/88).

AF-1406-0095

FIGHTER WEAPONS INSTRUCTOR F-15E

Course Number: *Version 1:* F15E01DOAN/WN. *Version 2:* F15E01DOAN/WN.
Location: Fighter Weapons School, Nellis AFB, NV.
Length: *Version 1:* 17-20 weeks (510-546 hours). *Version 2:* 17-20 weeks (510-546 hours).
Exhibit Dates: *Version 1:* 7/93–Present. *Version 2:* 6/91–6/93.
Learning Outcomes: *Version 1:* Upon completion of the course, the student will be able to recognize, interpret, and explain advanced aircraft systems and the performance environment; instruct and evaluate crewmembers in a dynamic, high-activity flight management arena; assess the needs for various instructional systems; prepare appropriate classroom and laboratory activities; and deliver and evaluate the educational event. *Version 2:* Upon completion of the course, the student will be able to recognize, interpret, and explain advanced aircraft systems and the performance environment; instruct and evaluate crewmembers in a dynamic, high-activity flight management arena; assess the needs for various instructional systems; prepare appropriate classroom and laboratory activities; and deliver and evaluate the educational event.
Instruction: *Version 1:* The course consists of classroom lectures, student independent research, formal student-led briefings, aircrew systems management, and instructional design, teaching methodologies, and appropriate laboratory experiences in advanced aircraft systems and performance. *Version 2:* The course consists of classroom lectures, student independent research, formal student-led briefing, and appropriate laboratory experiences in advanced aircraft systems and performance, aircrew systems management, and instructional design and teaching methodologies.
Credit Recommendation: *Version 1:* In the graduate degree category, 6 semester hours advanced aircraft systems and performance, 3 in aircrew systems management, 2 in instructional design and teaching methodology, and 2 in advanced flight laboratory (7/94). *Version 2:* In the graduate degree category, 6 semester hours in advanced aircraft systems and performance, 2 in aircrew systems management, 2 in instructional design and teaching methodology, and 2 in advanced flight laboratory (7/94).

AF-1406-0096

INTELLIGENCE WEAPONS INSTRUCTOR
(Fighter Weapons Instructor Intelligence)

Course Number: *Version 1:* 807501DOZN; INTELIDOZN. *Version 2:* 807501DOZN; INTELIDOZN. *Version 3:* INTELIDOZN; 807501DOZN. *Version 4:* 807501DOZN; INTELIDOZN.
Location: Fighter Weapons School, Nellis AFB, NV.
Length: *Version 1:* 15-20 weeks (491-642 hours). *Version 2:* 15-20 weeks (491-642 hours). *Version 3:* 15-20 weeks (491-642 hours). *Version 4:* 15-20 weeks (491-642 hours).
Exhibit Dates: *Version 1:* 7/94–Present. *Version 2:* 7/93–6/94. *Version 3:* 1/92–6/93. *Version 4:* 8/90–12/91.
Learning Outcomes: *Version 1:* Upon completion of the course, the student will be able to recognize, interpret, and explain advanced aircraft systems and the performance environment; instruct and evaluate crewmembers in a dynamic, high-activity flight management arena; assess the needs for various instructional systems; prepare appropriate classroom and laboratory activities; and deliver and evaluate the educational event. *Version 2:* Upon completion of the course, the student will be able to recognize, interpret, and explain advanced aircraft systems and the performance environment; instruct and evaluate crewmembers in a dynamic, high-activity flight management arena; assess the needs for various instructional systems; prepare appropriate classroom and laboratory activities; and deliver and evaluate the educational event. *Version 3:* Upon completion of the course, the student will be able to recognize, interpret, and explain advanced aircraft systems and the performance environment; instruct and evaluate crewmembers in a dynamic, high-activity flight management arena; assess the needs for various instructional systems; prepare appropriate classroom and laboratory activities; and deliver and evaluate the educational event. *Version 4:* Upon completion of the course, the student will be able to recognize, interpret, and explain advanced aircraft systems and the performance environment; instruct and evaluate crewmembers in a dynamic, high-activity flight management arena; assess the needs for various instructional systems; prepare appropriate classroom and laboratory activities; and deliver and evaluate the educational event.
Instruction: *Version 1:* The course consists of classroom lectures, student independent research, formal student-led briefings, and appropriate laboratory experiences in advanced aircraft systems and performance, aircrew systems management, instructional design, and teaching methodologies. *Version 2:* The course consists of classroom lectures, student independent research, formal student-led briefings, and appropriate laboratory experiences in advanced aircraft systems and performance, aircrew systems management, instructional design, and teaching methodologies. *Version 3:* The course consists of classroom lectures, student independent research, formal student-led briefings, and appropriate laboratory experiences in advanced aircraft systems and performance, aircrew systems management, instructional design, and teaching methodologies. *Version 4:* The course consists of classroom lectures, student independent research, formal student-led briefings, and appropriate laboratory experiences in advanced aircraft systems and performance, aircrew systems management, instructional design, and teaching methodologies.
Credit Recommendation: *Version 1:* In the graduate degree category, 6 semester hours in advanced aircrew systems and performance, 7 in aircrew systems management, and 3 in instructional design and teaching methodologies (7/94). *Version 2:* In the graduate degree category, 6 semester hours in advanced aircrew systems and performance, 5 in aircrew systems management, and 4 instructional design and teaching methodologies (7/94). *Version 3:* In the graduate degree category, 6 semester hours in advanced aircrew systems and performance, 5 in aircrew systems management, and 4 in instructional design and teaching methodologies (7/94). *Version 4:* In the graduate degree category, 5 semester hours in advance aircrew systems and performance, 4 in aircrew systems management, and 5 in instructional design and teaching methodologies (7/94).

AF-1406-0097

INSTRUCTOR QUALIFICATION TRAINING F-15E

Course Number: *Version 1:* F15EOIOOAL/WL. *Version 2:* F15EOIOOAL/WL.
Location: *Version 1:* 333 Fighter Squadron, Seymour Johnson AFB, NC; 58th Tactical Training Wing, Luke AFB, AZ. *Version 2:* 58th Tactical Training Wing, Luke AFB, AZ.
Length: *Version 1:* 11-12 weeks (88-94 hours). *Version 2:* 11-12 weeks (88-94 hours).
Exhibit Dates: *Version 1:* 11/93–Present. *Version 2:* 6/92–10/93.
Learning Outcomes: *Version 1:* Upon completion of the course, the student will be able to instruct and evaluate pilots/weapons systems operators in high performance jet aircraft. *Version 2:* Upon completion of the course, the student will be able to instruct and evaluate pilots/weapons systems operators in high performance jet aircraft.
Instruction: *Version 1:* The course consists of classroom lectures, programmed or com-

puter-based training and appropriate aircraft and flight training device laboratory. *Version 2:* The course consists of classroom lectures, programmed or computer-based training and appropriate aircraft and flight training device laboratory.

Credit Recommendation: *Version 1:* In the graduate degree category, 1 semester hour in Instructional design and teaching methodology and 4 in advanced flight laboratory (6/95). *Version 2:* In the graduate degree category, 1 semester hour in instructional design and teaching methodology and 4 in advanced flight laboratory (6/95).

AF-1406-0098

INSTRUCTOR QUALIFICATION TRAINING F-15E

Course Number: F15EOIOAL/WL.
Location: 333 Fighter Squadron, Seymour Johnson AFB, NC; 58th Tactical Training Wing, Luke AFB, AZ.
Length: 9 weeks (69-73 hours).
Exhibit Dates: 1/90–Present.
Learning Outcomes: Upon completion of the course, the student will be able to instruct and evaluate pilots/weapon systems operators in high performance jet aircraft.
Instruction: The course consists of classroom lectures, programmed or computer-based training and appropriate aircraft and flight training device laboratory.
Credit Recommendation: In the graduate degree category, 1 semester hour in instructional design and teaching methodology and 3 in advanced flight laboratory (6/95).

AF-1406-0099

T-1A PILOT INSTRUCTOR TRAINING

Course Number: F-V5A-E.
Location: 19th Air Force, Randolph AFB, TX.
Length: 13-14 weeks (201 hours).
Exhibit Dates: 12/93–Present.
Learning Outcomes: Upon completion of the course, the student will be able to instruct and evaluate pilots in high performance jet aircraft.
Instruction: The course consists of classroom lectures, programmed on computer-based training, and appropriate aircraft and flight training device laboratory exercises.
Credit Recommendation: In the graduate degree category, 3 semester hours in instructional design and teaching methodologies and 4 in advanced flight laboratory (6/95).

AF-1406-0100

IFF INSTRUCTOR TRAINING (AT-38B)

Course Number: 19AF syllabus B/F-V5A-M.
Location: 88th Flying Training Squadron, Sheppard AFB, TX.
Length: 12 weeks (111-114 hours).
Exhibit Dates: 1/95–Present.
Learning Outcomes: Upon completion of the course, the student will be able to recognize, interpret, and explain aircraft systems and performance environment; instructs and evaluate crewmembers in a dynamic and high activity flight management arena.
Instruction: The course consists of classroom lectures, formal student-led briefings and appropriate laboratory experiences in the following areas: advanced aerodynamics and aircraft systems, instructional methodology and advanced flight laboratory.

Credit Recommendation: In the graduate degree category, 1 semester hour in instructional methodology, 1 in advanced aerodynamics and systems and 2 in advanced flight laboratory (7/95).

AF-1406-0101

ENJPT PILOT INSTRUCTOR TRAINING SHEPHERD (T-37/T-38)

Course Number: F-V5N-A/B.
Location: 88th Flying Training Squadron, Sheppard AFB, TX.
Length: 17 weeks (362-376 hours).
Exhibit Dates: 1/90–Present.
Learning Outcomes: Upon completion of the course, the student will be able to recognize, interpret, and explain advanced aircraft systems and the performance environment; instruct and evaluate crewmembers in the dynamic and high-activity flight management arena.
Instruction: The course consists of classroom lectures, formal student led-briefings, and appropriate laboratory experiences in the following areas: advanced aerodynamics and aircraft systems, instructional methodology and advanced flight laboratory.
Credit Recommendation: In the graduate degree category, 1 semester hour in instructional methodology, and 2 in advanced aerodynamics and systems management, and 5 in advanced flight laboratory (6/95).

AF-1406-0102

INSTRUCTOR PILOT UPGRADE TRAINING F-16 C/D (Instructor Pilot Upgrade Training F-16 C)

Course Number: F16COIOOAL.
Location: 58th Tactical Training Wing, Luke AFB, AZ.
Length: 7-8 weeks (80-109 hours).
Exhibit Dates: 1/90–5/91.
Learning Outcomes: Upon completion of the course, the student will be able to recognize, interpret, and explain advanced aircraft systems and the performance environment; and perform advanced jet aircraft maneuvers and operations.
Instruction: The course consists of classroom lectures, programmed or computer-based training and appropriate aircraft and flight training device laboratory exercises.
Credit Recommendation: In the graduate degree category, Track 1, 1 semester hour in instructional design and teaching methodologies and 2 in advanced flight laboratory. Track 2, 2 semester hours in instructional design and teaching methodologies and 2 in advanced flight laboratory (6/95).

AF-1407-0003

PARALEGAL JOURNEYMAN BY CORRESPONDENCE

Course Number: 88150.
Location: Extension Course Institute, Gunter AFS, AL.
Length: Maximum, 52 weeks.
Exhibit Dates: 12/92–4/97.
Learning Outcomes: Upon completion of the course, the student will be knowledgeable of the Air Force's law firm, civil law military justice administration, and Air Force claims.
Instruction: Course includes review of the organizational structure and resource management of the judge advocate general department, civil law or related civil matters, nonjudicial punishment, military justice administration, the automated military justice analysis and management system, and Air Force claims.
Credit Recommendation: Credit is not recommended because of the military-specific nature of the course (6/94).

AF-1408-0004

PROFESSIONAL MILITARY COMPTROLLER

Course Number: LMDC 400.
Location: Leadership and Management Development Center, Air University, Maxwell AFB, AL.
Length: 8 weeks (233 hours).
Exhibit Dates: 1/90–9/91.
Objectives: After 9/91 see AF-1408-0099. To train officers to serve as comptrollers or resource managers.
Instruction: Course includes lectures, seminars, and case studies in economics, finance, managerial accounting, human relations, computer management including theories and concepts relating to research techniques, executive expression, quantitative methods, and resource management. Programs include individual research and application of computer techniques in the subject areas.
Credit Recommendation: In the upper-division baccalaureate category, 19 semester hours in business administration (4/76); in the graduate degree category, 3 semester hours in organizational management and 3 in financial and economic analysis (4/76).

AF-1408-0071

CONTRACTING SPECIALIST BY CORRESPONDENCE

Course Number: 65150.
Location: Extension Course Institute, Gunter AFS, AL.
Length: Maximum, 52 weeks.
Exhibit Dates: 1/90–4/93.
Learning Outcomes: Upon completion of the course, the student will be able to describe the overall job of contracting and purchasing procedures; follow a career ladder into purchasing and contracting specialties; interpret contract negotiations and agreements; gain an understanding of data base files; describe the needs of advertising and negotiations; make small purchases through the use of proper methods and procedures; and work with contracting programs of bonds, insurance, taxes, legal principles of government, and private contracts.
Instruction: Instruction includes contracting and purchasing, their history, organization, management, publications, and data base records. Methods of contracting are subdivided into technical, basic, and desirable terms, advertising, negotiation, small purchases, and simplified purchase procedures.
Credit Recommendation: In the lower-division baccalaureate/associate degree category, 1 semester hour in purchasing (2/87).

AF-1408-0072

CONTRACTING SUPERVISOR BY CORRESPONDENCE

Course Number: 65170.
Location: Extension Course Institute, Gunter AFS, AL.
Length: Maximum, 52 weeks.
Exhibit Dates: 1/90–5/95.
Learning Outcomes: Upon completion of the course, the student will be able to describe

the progression within the contracting career field; comprehend and utilize contracting procedures and laws governing contract negotiations; understand socioeconomic programs and EA/EO rules in small business and government; comprehend the legal principles of government contracts; work with contract concepts and tasks; understand and contract for formal advertising; make small purchases; understand contract administration; deal with pricing, pricing analysis, and negotiations; and deal with cost analysis and negotiation documentation.

Instruction: Instruction includes career training, contracting mission, socioeconomic programs, and legal principles of government contracts and contracting management programs. Also covered are methods of contracting, sealed bidding, negotiations, small purchases, price and cost analysis, negotiations, and purchasing analysis and control.

Credit Recommendation: In the lower-division baccalaureate/associate degree category, 3 semester hours in purchasing and 2 in business law (covering contracts) (2/87).

AF-1408-0073

FINANCIAL MANAGEMENT SPECIALIST BY CORRESPONDENCE

Course Number: 67251A.
Location: Extension Course Institute, Gunter AFS, AL.
Length: Maximum, 52 weeks.
Exhibit Dates: 1/90–7/96.
Learning Outcomes: Upon completion of the course, the student will be able to describe career opportunities and procedural steps to achieve skill levels within the financial field, identify organization and responsibilities of each skill and career level, perform basic accounting principles and practices, and describe mechanized and computerized risk management procedures.

Instruction: Instruction includes financial management orientation, career knowledge, organization, responsibilities of basic accounting systems, and computerized risk management. Accounting systems, mechanized and systems integration are interwoven with financial management principles.

Credit Recommendation: In the vocational certificate category, 1 semester hour in bookkeeping (2/87); in the lower-division baccalaureate/associate degree category, 1 semester hour in accounting or finance (2/87).

AF-1408-0074

1. COMMERCIAL SERVICES AND AUTOMATED TRAVEL RECORD ACCOUNTING SYSTEM BY CORRESPONDENCE
   (Financial Management Specialist (Commercial Services) by Correspondence)
   (Financial Management Specialist (Military Pay) by Correspondence)
   (Financial Management Specialist by Correspondence)
2. FINANCIAL MANAGEMENT SPECIALIST BY CORRESPONDENCE

Course Number: *Version 1:* 67251B. *Version 2:* 67251B.
Location: Extension Course Institute, Gunter AFS, AL.
Length: *Version 1:* Maximum, 52 weeks. *Version 2:* Maximum, 52 weeks.
Exhibit Dates: *Version 1:* 7/91–7/96. *Version 2:* 1/90–6/91.

Learning Outcomes: *Version 1:* Upon completion of the course, the student will be able to describe the criterion for commitments and how it effects annual and multiple year appropriations; use commitment documents and fund availability; use receipt documents for goods and services; and use balance sheets, accounts records, disbursement records, and procedures. *Version 2:* Upon completion of the course, the student will be able to describe the criterion for commitments and how it effects annual and multiple year appropriations; use commitment documents and fund availability; use receipt documents for goods and services; and use balance sheets, accounts records, disbursement records, and procedures.

Instruction: *Version 1:* Instruction includes source documents, specific types and uses of commitment documents, purchase records and receipts, disbursements, and transaction processes and report preparation. *Version 2:* Instruction includes source documents, specific types and uses of commitment documents, purchase records and receipts, disbursements, and transaction processes and report preparation.

Credit Recommendation: *Version 1:* In the vocational certificate category, 2 semester hours in bookkeeping (4/93); in the lower-division baccalaureate/associate degree category, 1 semester hour in accounting/financial management (4/93). *Version 2:* In the vocational certificate category, 1 semester hour in bookkeeping (2/87); in the lower-division baccalaureate/associate degree category, 1 semester hour in accounting/financial management (2/87).

AF-1408-0075

FINANCIAL MANAGEMENT SPECIALIST BY CORRESPONDENCE

Course Number: 67251C.
Location: Extension Course Institute, Gunter AFS, AL.
Length: Maximum, 52 weeks.
Exhibit Dates: 1/90–Present.
Learning Outcomes: Upon completion of the course, the student will be able to describe purpose and authority for stock funds and commissary trust revolving funds; understand the functions and responsibilities of the fuels divisions; define management and operational characteristics of support divisions; organize and utilize required accounting reports and records; originate and follow-up requirements for requisitions, receipts, and disbursements of government monies for fuel, supplies, and equipment; and maintain records and tables on the stock fund reporting data base.

Instruction: Instruction includes various types of fund accounting, inventory, and financial accounting and their relationship to general ledgers and the automated computerized accounting systems. Specifics of the Air Force stock fund trial balances, reporting systems, and report audit requirements are covered in the aviation fuel management system.

Credit Recommendation: In the vocational certificate category, 1 semester hour in bookkeeping (3/93); in the lower-division baccalaureate/associate degree category, 1 semester hour in accounting/financial management (3/93).

AF-1408-0076

FINANCIAL MANAGEMENT SPECIALIST BY CORRESPONDENCE

Course Number: *Version 1:* 67251D. *Version 2:* 67251D.
Location: Extension Course Institute, Gunter AFS, AL.
Length: *Version 1:* Maximum, 52 weeks. *Version 2:* Maximum, 52 weeks.
Exhibit Dates: *Version 1:* 11/91–Present. *Version 2:* 1/90–10/91.

Learning Outcomes: *Version 1:* Upon completion of the course, the student will be able to understand data base control summary record, establish fund summary records utilizing proper codes and indicative data, specify base variable files and their uses, set up control log procedures, comprehend foreign currency rates and their differences in accounting procedures for currency conversion, and understand and utilize a mechanized data base. *Version 2:* Upon completion of the course, the student will be able to understand data base control summary record, establish fund summary records utilizing proper codes and indicative data, specify base variable files and their uses, set up control log procedures, comprehend foreign currency rates and their differences in accounting procedures for currency conversion, and understand and utilize a mechanized data base.

Instruction: *Version 1:* Course covers the procedures on data base control, accountability and fund report processing, financial reports, and daily procedures needed for an accurate mechanized data base. *Version 2:* Course covers the procedures on data base control, accountability and fund report processing, financial reports, and daily procedures needed for an accurate mechanized data base.

Credit Recommendation: *Version 1:* In the vocational certificate category, 3 semester hours in data entry (4/93). *Version 2:* In the vocational certificate category, 1 semester hour in bookkeeping (2/87); in the lower-division baccalaureate/associate degree category, 1 semester hour in accounting/financial management (2/87).

AF-1408-0077

FINANCIAL MANAGEMENT SPECIALIST (BUDGET) BY CORRESPONDENCE

Course Number: 67251E.
Location: Extension Course Institute, Gunter AFS, AL.
Length: Maximum, 52 weeks.
Exhibit Dates: 1/90–8/94.
Learning Outcomes: Upon completion of the course, the student will be able to distinguish Air Force and Department of Defense budgets from the Federal Budget, understand the necessity for financial planning, understand and assist in developing operations budgets and financial plans, and understand and identify financial manager's responsibilities at base level within the resource management system.

Instruction: Course includes procedures of the Federal Budget System, financial planning, developing operations, operating budgets, financial plans, and funds management.

Credit Recommendation: In the vocational certificate category, 1 semester hour in bookkeeping (2/87); in the lower-division baccalaureate/associate degree category, 1 semester hour in accounting/financial management (2/87).

AF-1408-0078

FINANCIAL SERVICES SPECIALIST (INTRODUCTION) BY CORRESPONDENCE

COURSE EXHIBITS

(Financial Services Specialist by Correspondence)

Course Number: *Version 1:* 67252A. *Version 2:* 67252A.

Location: Extension Course Institute, Gunter AFS, AL.

Length: *Version 1:* Maximum, 52 weeks. *Version 2:* Maximum, 52 weeks.

Exhibit Dates: *Version 1:* 4/92–7/96. *Version 2:* 1/90–3/92.

Learning Outcomes: *Version 1:* Upon completion of the course, the student will be able to describe the basic organization of the accounting and finance office, understand the responsibilities of the various individuals in and out of the accounting and finance office, gain a working knowledge of the basic accounting systems, and understand and utilize some basic information in computer risk management. *Version 2:* Upon completion of the course, the student will follow basic career knowledge in upgrading, understand the basic organization of the accounting and finance office, understand the responsibilities of the various individuals in and out of the accounting and finance office, gain a working knowledge of the basic accounting systems, and understand and use some basic information in computer risk management.

Instruction: *Version 1:* Course includes career knowledge in financial services, organization, responsibilities, basic accounting systems, and computer risk management. It also covers disbursements and collections, public funds control, checks and bond operations, voucher control, cashier operations, accountability, and reporting. *Version 2:* Course includes career knowledge in financial services, organization, responsibilities, basic accounting systems, and computer risk management. It also covers disbursements and collections, public funds control, checks and bond operations, voucher control, cashier operations, accountability, and reporting.

Credit Recommendation: *Version 1:* In the vocational certificate category, 3 semester hours in banking operations (clerical tellering) or 3 semester hours in financial records/data management (4/93). *Version 2:* In the vocational certificate category, 1 semester hour in bookkeeping (2/87); in the lower-division baccalaureate/associate degree category, 3 semester hours in financial management, 2 in records management, and 1 in data processing (2/87).

AF-1408-0079

FINANCIAL SERVICES SPECIALIST (MILITARY PAY) BY CORRESPONDENCE

Course Number: 67252B.

Location: Extension Course Institute, Gunter AFS, AL.

Length: Maximum, 52 weeks.

Exhibit Dates: 1/90–8/96.

Learning Outcomes: Upon completion of the course, the student will be able to describe the Joint Uniform Military Pay system including pay and allowances for active military members; figure deductions, collections, and payment processes; describe status and event transactions and leave accounting; and describe the retiree/annuitant pay system.

Instruction: Course includes topics in Joint Military Pay systems and pay and allowances. Information also includes deductions and collections, status/event transactions, pay processing, leave accounting, and retiree/annuitant pay systems.

Credit Recommendation: Credit is not recommended because of the limited, specialized nature of the course (2/87).

AF-1408-0080

FINANCIAL SERVICES SPECIALIST (TRAVEL) BY CORRESPONDENCE

Course Number: *Version 1:* 67252C. *Version 2:* 67252C.

Location: Extension Course Institute, Gunter AFS, AL.

Length: *Version 1:* Maximum, 52 weeks. *Version 2:* Maximum, 52 weeks.

Exhibit Dates: *Version 1:* 1/92–Present. *Version 2:* 1/90–12/91.

Learning Outcomes: *Version 1:* Upon completion of the course, the student will be able to compute military and dependent travel entitlements for permanent and temporary change in duty status; compute civilian travel entitlements when traveling for government and military; deal with advances, claims, and recoupments; and use automated travel record/accounting systems frames processing and systems products and follow-up. *Version 2:* Upon completion of the course, the student will be able to compute military and dependent travel entitlements for permanent and temporary change in duty status; compute civilian travel entitlements when traveling for government and military; deal with advances, claims, and recoupments; and use automated travel record/accounting systems frames processing and systems products and follow-up.

Instruction: *Version 1:* Course includes travel computation and accounting in military, dependent, and civilian travel and entitlements and record keeping. Also included is a finite study of automated travel records and accounting systems, frames processing and system products, and follow-up procedures of travel. The course provides for in-depth instruction in all phases of travel-related expenses. Procedures for measuring and minimizing travel time, distances, and expenses are covered. Skills for comparative cost analysis for different modes of transportation are taught. Student will gain competence in all phases of relocation expense computation, analysis, and reimbursement. *Version 2:* Course includes travel computation and accounting in military, dependent, and civilian travel and entitlements and record keeping. Also included is a finite study of automated travel records and accounting systems, frames processing and system products, and follow-up procedures of travel.

Credit Recommendation: *Version 1:* In the lower-division baccalaureate/associate degree category, 3 semester hours in travel administration (4/93). *Version 2:* In the vocational certificate category, 1 semester hour in travel agency operation (2/87); in the lower-division baccalaureate/associate degree category, 3 semester hours in tourist travel information (2/87).

AF-1408-0081

FINANCIAL MANAGEMENT SUPERVISOR BY CORRESPONDENCE

(Financial Management/Services Supervisor by Correspondence)

Course Number: 67273A.

Location: Extension Course Institute, Gunter AFS, AL.

Length: Maximum, 52 weeks.

Exhibit Dates: 1/90–7/94.

Learning Outcomes: Upon completion of the course, the student will be able to describe pay and leave entitlements and deductions for all classifications of civilian employees; use the proper payroll system for civilian employees; follow proper procedures for making out all reports and records dealing with the civilian payroll; and use the automated payroll system, computer files, and storage banks.

Instruction: Course includes the different types and classifications of civilian employees, the rates of pay, entitlements, and deductions. A comprehensive study of the automated payroll system, computer files, storage banks, and the procedures for payroll activities and support data required for accounting procedures using the Remote Job Entry Terminal System is also included.

Credit Recommendation: In the vocational certificate category, 2 semester hours in bookkeeping and finance (2/87); in the lower-division baccalaureate/associate degree category, 1 semester hour in accounting/finance and 1 in data processing (2/87).

AF-1408-0082

FINANCIAL MANAGEMENT/SERVICES SUPERVISOR (FUNCTIONS AND RESPONSIBILITIES) BY CORRESPONDENCE

Course Number: 67273B.

Location: Extension Course Institute, Gunter AFS, AL.

Length: Maximum, 52 weeks.

Exhibit Dates: 1/90–7/94.

Learning Outcomes: Upon completion of the course, the student will be able to plan, organize, and direct accounting and finance activities of the financial office; identify liability to the government in event of loss, damage, or destruction of government property and to identify dollar amounts for activities listed; understand the objectives of the Noncommissioned Officers Professional Military Education Program; describe the Air Force Comptroller Civilian Career Management Program; and describe activities dealing with budgeting, disbursement, and reimbursement of Air Force (government) funds.

Instruction: Course includes an introduction to the duties of financial management, personnel aspects of supervising of military and civilian personnel, budgeting, customer services, and financial control.

Credit Recommendation: Credit is not recommended because of the limited, specialized nature of the course (2/87).

AF-1408-0083

COST AND MANAGEMENT ANALYSIS SPECIALIST BY CORRESPONDENCE

Course Number: 69150.

Location: Extension Course Institute, Gunter AFS, AL.

Length: Maximum, 52 weeks.

Exhibit Dates: 1/90–1/97.

Learning Outcomes: Upon completion of this course, the student will have knowledge of the role of management information systems analysis, basic mathematics, and elementary statistical concepts and methods.

Instruction: Topics included are management information systems, basic mathematics, data distribution measures of central tendency, trend analysis, probability distribution, index numbers, correlation and regression, and applications of quantitative methods.

Credit Recommendation: In the lower-division baccalaureate/associate degree category, 3 semester hours in basic statistics (2/87).

AF-1408-0084

COST AND MANAGEMENT ANALYSIS TECHNICIAN BY CORRESPONDENCE

Course Number: 69170.
Location: Extension Course Institute, Gunter AFS, AL.
Length: Maximum, 52 weeks.
Exhibit Dates: 1/90–1/97.
Learning Outcomes: Upon completion of the course, the student will have a working knowledge of management assistance services and management information systems as they apply to Air Force regulations and practices and will be able to interpret cost and economic analysis, commercial activities programs, and resource management systems.
Instruction: Course includes topics in cost and management analysis, the comptroller organization, management assistance services, and management information systems and statistical methods as they apply to the Air Force. More specific information is included in the topics of cost analysis, economic analysis, commercial activities programs, and resource management systems.
Credit Recommendation: Credit is not recommended because of the military-specific nature of the course (2/87).

AF-1408-0085

CHAPEL MANAGEMENT SPECIALIST BY CORRESPONDENCE

Course Number: 70150.
Location: Extension Course Institute, Gunter AFS, AL.
Length: Maximum, 52 weeks.
Exhibit Dates: 1/90–7/94.
Learning Outcomes: At the conclusion of the course, the student will be able to perform tasks in chapel support activities and chapel management. The student's duties cover office procedures, communication, reports, publications, chapel finances and resources, and the preparation of chapel activities.
Instruction: Instruction covers chapel management careers, office procedures and practices, equipment operation, communication, reports, and other office duties as well as preparation of chapel activities and managing chapel finances, funds, and resources.
Credit Recommendation: In the vocational certificate category, 1 semester hour in records management and correspondence (2/87).

AF-1408-0086

CHAPEL MANAGEMENT TECHNICIAN BY CORRESPONDENCE

Course Number: 70170.
Location: Extension Course Institute, Gunter AFS, AL.
Length: Maximum, 52 weeks.
Exhibit Dates: 1/90–2/95.
Learning Outcomes: At the conclusion of the course, the student will be able to identify the duties and responsibilities of chapel management, comprehend safety procedures as they relate to chapel work, prepare and explain staff study reports, develop self-inspection checklists and logistical support, and develop financial reports and manage funds and properties assigned to chapel activities.
Instruction: Course covers duties and responsibilities of the career field, occupational safety in chapel work, reports, publications, communication, position papers, logistical support, and financial management.
Credit Recommendation: In the vocational certificate category, 1 semester hour in bookkeeping (2/87).

AF-1408-0087

CONTRACT ADMINISTRATION BY CORRESPONDENCE

Course Number: 6606.
Location: Extension Course Institute, Gunter AFS, AL.
Length: Maximum, 52 weeks.
Exhibit Dates: 1/90–3/94.
Learning Outcomes: Upon completion of the course, the student will be able to describe contracting and the funding process; understand contract types and administration of contracts; understand contract pricing, evaluation, and financing; perform minor subcontract managing; evaluate contract progress; perform quality assurances and profit analysis; negotiate minor contract modifications; and act as negotiator in working with major contracts.
Instruction: Instruction covers types of contracts, accounting and cost principles, contract pricing, contractor evaluation, contractor financing, subcontract management, progress evaluation, quality assurance, profit analysis, contract modifications, and negotiations.
Credit Recommendation: In the lower-division baccalaureate/associate degree category, 3 semester hours in business law or 3 in contract purchasing or 1 in cost accounting (2/87).

AF-1408-0088

INTRODUCTION TO LABOR RELATIONS FOR AIR FORCE SUPERVISORS BY CORRESPONDENCE

Course Number: 6604.
Location: Extension Course Institute, Gunter AFS, AL.
Length: Maximum, 52 weeks.
Exhibit Dates: 1/90–4/94.
Learning Outcomes: At the conclusion of the course, the student will be able to describe the history, development, purposes, activities, and governance of organized labor unions and the Civil Service Reform act of 1978.
Instruction: Instruction includes the history and development of organized labor from the colonial era to the present, public policy toward organized labor, unionization of federal service, and the Civil Service Reform act of 1978.
Credit Recommendation: In the vocational certificate category, 1 semester hour in labor relations or orientation (2/87).

AF-1408-0089

AIR FORCE LOGISTICS COMMAND DIRECTORATE OF MATERIEL MANAGEMENT BY CORRESPONDENCE

Course Number: 6605.
Location: Extension Course Institute, Gunter AFS, AL.
Length: Maximum, 52 weeks.
Exhibit Dates: 1/90–1/97.
Learning Outcomes: Upon completion of the course, the student will be able to determine the function of each specific job within logistics and materiel management as it relates to the Air Force; describe the necessity of replenishment of weapons, materials, and supplies; understand how logisticians provide initial support; and understand the responsibilities of the Air Force Logistics Command.
Instruction: Course includes an introduction to Air Force Logistics Command and job breakdowns, financial management and planning, major weapons acquisition and support, the responsibilities of Directorate of Materiel Management, replenishment, maintenance, specialized management system, contracting, and distribution.
Credit Recommendation: Credit is not recommended because of the limited, specialized nature of the course (2/87).

AF-1408-0090

INTRODUCTION TO AIR FORCE INITIAL PROVISIONING BY CORRESPONDENCE

Course Number: 6608.
Location: Extension Course Institute, Gunter AFS, AL.
Length: Maximum, 52 weeks.
Exhibit Dates: 1/90–1/97.
Learning Outcomes: Upon completion of the course, the student will be able to describe provisioning concepts, their historical background, provisioning processes, and functions and advances in initial provisioning.
Instruction: Instruction includes an overview of the functions, responsibilities, and procedures concerning initial provisioning policy and procedures within the Air Force.
Credit Recommendation: Credit is not recommended because of the military-specific nature of the course (2/87).

AF-1408-0091

PRINCIPLES OF CONTRACT PRICING BY CORRESPONDENCE

Course Number: *Version 1:* 6610. *Version 2:* 6610.
Location: Extension Course Institute, Gunter AFS, AL.
Length: *Version 1:* Maximum, 52 weeks. *Version 2:* Maximum, 52 weeks.
Exhibit Dates: *Version 1:* 8/92–5/98. *Version 2:* 1/90–7/92.
Learning Outcomes: *Version 1:* Upon completion of the course, the student will be able to describe contract pricing, pricing analysis, and price systems; negotiate cost proposals and analysis with pricing for profit or use; and describe facilities capital investments in relation to pricing of products and services as a part of total cost. *Version 2:* Upon completion of the course, the student will be able to describe contract pricing, pricing analysis, and price systems; negotiate cost proposals and analysis with pricing for profit or use; and comprehend facilities capital investments in relation to pricing of products and services as a part of total cost.
Instruction: *Version 1:* Course covers contract pricing, mathematics of pricing, cost-volume profit analysis, price analysis, cost factors, cost proposals, comparison techniques, experience curve, direct and indirect cost analysis, capital investment considerations, profit motives and analysis, negotiating processes, and documentation. *Version 2:* Course covers

contract pricing, mathematics of pricing, cost-volume profit analysis, price analysis, cost proposals, comparison techniques, experience curve, direct and indirect cost analysis, capital investment considerations, profit motives and analysis, negotiating processes, and documentation.

Credit Recommendation: *Version 1:* In the vocational certificate category, 1 semester hour in applied mathematics and 1 in retailing (pricing) (4/93); in the lower-division baccalaureate/associate degree category, 2 semester hours in business mathematics and 2 in purchasing (4/93). *Version 2:* In the vocational certificate category, 1 semester hour in applied mathematics and 1 in retailing (pricing) (2/87); in the lower-division baccalaureate/associate degree category, 2 semester hours in business mathematics and 2 in marketing (2/87).

AF-1408-0092

BUDGET BY CORRESPONDENCE

Course Number: 6701.
Location: Extension Course Institute, Gunter AFS, AL.
Length: Maximum, 52 weeks.
Exhibit Dates: 1/90–11/92.
Learning Outcomes: Upon completion of the course, the student will know the budgetary process of the Air Force; describe the concepts of fund accounting, budgeting, and financial management; and be able to identify and complete the various Air Force forms related to the budgeting and accounting cycle.
Instruction: Instruction includes the basic concepts and principles of governmental/fund accounting, budgeting, and financial management. Also covered are the specific Air Force regulations and instructions on how to complete Air Force forms in these functions.
Credit Recommendation: In the vocational certificate category, 1 semester hour in bookkeeping (2/87); in the lower-division baccalaureate/associate degree category, 1 semester hour in fund accounting and 1 semester hour in budgeting (2/87).

AF-1408-0093

RESOURCE ADVISOR BY CORRESPONDENCE

Course Number: 6702.
Location: Extension Course Institute, Gunter AFS, AL.
Length: Maximum, 52 weeks.
Exhibit Dates: 1/90–Present.
Learning Outcomes: At the conclusion of this course, the student will be able to describe the budgeting and fund control processes within the Air Force, have a working knowledge of financial planning concepts, and identify various forms related to the functions of budgeting and fund control.
Instruction: This course offers an overview of fund accounting, budgeting, and financial planning in the Air Force. Also covered are some of the specific forms and procedures related to these functions.
Credit Recommendation: In the lower-division baccalaureate/associate degree category, 1 semester hour in fund accounting, 1 in financial planning, and 1 in budgeting (4/93).

AF-1408-0095

OPERATIONS RESOURCE MANAGEMENT BY CORRESPONDENCE

Course Number: 27152.
Location: Extension Course Institute, Gunter AFS, AL.
Length: Maximum, 52 weeks.
Exhibit Dates: 1/90–4/98.
Learning Outcomes: Upon completion of the course, the student will be able to monitor and audit aircrew data with little supervision; describe flight management policies and programs; and retrieve, summarize, and analyze data to facilitate decision making on the part of supervisory staff.
Instruction: The course consists of self-paced, independent study with open book progress checks and satisfactory completion of a proctored comprehensive final examination. Content includes a review of basic math with introduction to statistical methods. Data processing is covered to a base level of understanding and application. Emphasis is on administrative record keeping pertaining to flight operations and training.
Credit Recommendation: In the lower-division baccalaureate/associate degree category, 2 semester hours in records and information management (4/93).

AF-1408-0096

ELECTRICAL POWER PRODUCTION TECHNICIAN BY CORRESPONDENCE

Course Number: 54272A.
Location: Extension Course Institute, Gunter AFS, AL.
Length: Maximum, 52 weeks.
Exhibit Dates: 1/90–5/98.
Learning Outcomes: Upon completion of the course, the student will be able to supervise an electrical power production organization shop.
Instruction: Course covers workshop management; electrical and electronic principles; operation of generation switching gear and electronic test equipment; and troubleshooting of electrical switching gear, portable generators, and automatic transfer panels.
Credit Recommendation: In the lower-division baccalaureate/associate degree category, 1 semester hour in maintenance management, 1 in principles of supervision, and 1 in electrical/electronic fundamentals (12/89).

AF-1408-0097

B-1B AVIONICS TEST STATION AND COMPONENT SPECIALIST BY CORRESPONDENCE

Course Number: 45157.
Location: Extension Course Institute, Gunter AFS, AL.
Length: Maximum, 52 weeks.
Exhibit Dates: 1/90–6/97.
Learning Outcomes: Upon completion of the course, the student will be able to use, troubleshoot, and replace or repair components for B-1B using automated test equipment.
Instruction: This is a correspondence course with practical exercises in the use, repair, and maintenance of components. Topics include maintenance material management and data collection systems, maintenance and inspection system, aircraft systems, avionics concepts, intermediate maintenance, and automated test equipment. Also included are core test station power and cooling stimulus measurement and radio frequency and video pneumatic modules as well as support equipment. Students enrolling must have an AFSC assignment as a prerequisite. This provides an application of the principles learned to real requirements and constitutes an essential part of the instruction by serving as the laboratory component of the course. The student's supervisor must complete an evaluation survey of the course and its effectiveness when the student takes the final exam which measures mastery of the cognitive content. A score of 65 is required for a satisfactory completion certificate.
Credit Recommendation: In the lower-division baccalaureate/associate degree category, 1 semester hour in maintenance and material management and 2 in test equipment and measurement (3/91).

AF-1408-0098

REPROGRAPHIC SPECIALIST BY CORRESPONDENCE

Course Number: 70350.
Location: Extension Course Institute, Gunter AFS, AL.
Length: Maximum, 52 weeks.
Exhibit Dates: 1/90–Present.
Learning Outcomes: Upon completion of the course, the student will be able to perform administrative and management duties associated with directing, controlling, and supervising a printing plant, duplicating center, microfilm production facility, and related reprographic activities.
Instruction: Topics covered include management within the Air Force organizational ladder, health and safety in the work place, procurement procedures, record keeping, operational costs, and production compilation. The course also covers management reports and budgeting as well as policies, procedures, and issues to consider in determining reprographic equipment requirements, including securing approval for acquisition, customer support, copier selection, and operation control. Students enrolling must have an AFSC assignment as a prerequisite. This provides an application of the principles learned to real requirements and constitutes an essential part of the instruction by serving as the laboratory component of the course. The student's supervisor must complete an evaluation survey of the course and its effectiveness when the student takes the final exam which measures mastery of the cognitive content. A score of 65 is required for a satisfactory completion certificate.
Credit Recommendation: In the upper-division baccalaureate category, 1 semester hour in the principles of management and supervision (3/91).

AF-1408-0099

PROFESSIONAL MILITARY COMPTROLLER (PMCC)

Course Number: MLMDC 501.
Location: Center for Professional Development, Air University, Maxwell AFB, AL.
Length: 6 weeks (256 hours).
Exhibit Dates: 10/91–Present.
Learning Outcomes: Before 10/91 see AF-1408-0004. Upon completion of the course, the student will have an in-depth knowledge of the theoretical foundation and practical applications required to successfully function as a comptroller in a public agency/organization.
Instruction: Emphasis is placed on the interrelationship of business disciplines that complement a sophisticated/advanced level of finance, economics, and accounting at the managerial level. Methodology includes case analy-

sis, game theory/playing, seminars, guest speakers, and a research project.

Credit Recommendation: In the graduate degree category, 3 semester hours in managerial accounting or public finance and 3 in organizational management or managerial economics (7/95).

AF-1408-0100
BUDGET OFFICER

Course Number: J30BR6731 000.
Location: 396 TCHTG/CC, Sheppard AFB, TX.
Length: 8 weeks (304 hours).
Exhibit Dates: 1/90–Present.
Learning Outcomes: Upon completion of the course, the student will be able to identify the functions and responsibilities of the budget officer, the resource management system, the purpose of financial plans, and the procedures used to analyze financial programs; describe the federal budget system, the components of the automated accounting system, and their relationships; prepare and present a recommended budget; and verify approved fund targets.
Instruction: Course includes classroom lectures and discussions and demonstration/performance directed studies. Topics cover fundamentals of Air Force budgeting, Air Force financial planning system, developing estimates of requirements, preparing operating budgets and financial plans, and Air Force financial management.
Credit Recommendation: In the graduate degree category, 3 semester hours in public accounting and 3 in public budgeting (3/93).

AF-1408-0101
ACCOUNTING AND FINANCE OFFICER

Course Number: J30BR6721 001.
Location: 396 TCHTG/CC, Sheppard AFB, TX.
Length: 8 weeks (294 hours).
Exhibit Dates: 1/90–Present.
Learning Outcomes: Upon completion of the course, the student will be able to describe the Air Force accounting system, the civilian pay system, the Joint Uniform Pay System, governmental accountability functions, and the organization, functions, and responsibilities of accounting and finance; identify the document flow and accounting stages used to process transactions, methods used to control agent operations, the procedures for operating specialized equipment; and analyze selected case problems.
Instruction: Course includes classroom lectures and discussions, case studies, directed studies, and field trip. Topics include organizing, managing, and coordinating activities for disbursing, collecting, and accounting of funds. These subjects are enhanced by a field trip to the Air Force Accounting and Finance Center.
Credit Recommendation: In the graduate degree category, 3 semester hours in public finance and 3 in government accounting systems (3/93).

AF-1408-0102
FINANCIAL MANAGEMENT OFFICER (FINANCIAL SERVICES)

Course Number: J30BR6721 002.
Location: 396 TCHTG/CC, Sheppard AFB, TX.
Length: 8 weeks (293 hours).
Exhibit Dates: 5/92–Present.
Learning Outcomes: Upon completion of the course, the student will be able to describe the federal budget system, the components of the automated accounting system, the principles of the automated supply system, and procedures for effective management; identify methods used to control agent operations, computer accounts control, and solutions to management problems; and prepare cash accountability statements and pay record accessibility audits.
Instruction: Methods include classroom lectures, discussions, case study, and directed studies. Topics cover organizing, managing, and coordinating activities for disbursing, collecting, and accounting of funds.
Credit Recommendation: In the graduate degree category, 3 semester hours in public finance and 3 in government accounting systems (3/93).

AF-1408-0103
FINANCIAL MANAGEMENT STAFF OFFICER

Course Number: J30AR6711 001.
Location: 396 TCHTG/CC, Sheppard AFB, TX.
Length: 6 weeks (221 hours).
Exhibit Dates: 2/93–Present.
Learning Outcomes: Upon completion of the course, the student will be able to describe the techniques used in conducting cost and economic analysis, the federal budget system, the process for development of financial plans; prepare a response to case studies on financial services operations; and describe the five-year plan for financial management and comptroller.
Instruction: Methods include classroom lectures, discussions, case studies, and directed studies. Topics include financial management and comptroller disciplines, financial analysis, financial services, plans, and systems.
Credit Recommendation: In the graduate degree category, 3 semester hours in public financial management and 3 in public financial systems (3/93).

AF-1408-0104
COMPTROLLER STAFF OFFICER

Course Number: J30AR6751 000.
Location: 396 TCHTG/CC, Sheppard AFB, TX.
Length: 6 weeks (230 hours).
Exhibit Dates: 1/90–Present.
Learning Outcomes: Upon completion of the course, the student will be able to identify the functions of the planning, programming, and budget system; prepare a response to a case study on accounting and finance operations;, identify the techniques used in performing cost analysis; and describe the use of micro/minicomputers to support future comptroller operations.
Instruction: Methods include classroom lectures and discussions, case study, and directed studies. Topics cover principles and concepts of accounting and finance, budget, cost, and comptroller plans and systems.
Credit Recommendation: In the graduate degree category, 3 semester hours in public finance and 3 in public accounting (3/93).

AF-1408-0105
FINANCIAL MANAGEMENT STAFF OFFICER, AIR RESERVE FORCES

Course Number: J3OZR6711 000.
Location: 396 TCHTG/CC, Sheppard AFB, TX.
Length: 2 weeks (80 hours).
Exhibit Dates: 1/90–Present.
Learning Outcomes: Upon completion of the course, the student will be able to describe the techniques used in conducting cost and economic analysis, the federal budget system, and the process for development of financial plans; prepare a response to case studies on financial services operations; and describe the five-year plan for financial management and comptroller.
Instruction: Methods include classroom lectures, discussions, case studies, and directed studies. Topics include financial management and comptroller disciplines, financial analysis, financial services, plans, and systems.
Credit Recommendation: In the graduate degree category, 3 semester hours in public financial management (5/93).

AF-1408-0106
INTRODUCTION TO ACQUISITION MANAGEMENT BY CORRESPONDENCE

Course Number: 6611.
Location: Extension Course Institute, Gunter AFS, AL.
Length: Maximum, 52 weeks.
Exhibit Dates: 9/90–Present.
Learning Outcomes: Upon completion of the course, the student will be able to define the acquisition process as it relates to research, development, test and evaluation, and procurement of weapons systems and security assistance programs; describe an acquisition program and its components; describe the procurement process; describe planning, programming, and budgeting as related to procurement of weapons and weapons life cycle; describe the systems engineering and logistics supply required to support weapons systems through their life cycle; describe the data management systems and how the data are used in the test and evaluation phase of weapons systems; and describe software support systems required to support the weapons systems.
Instruction: This correspondence course covers the acquisition program used by the Air Force related to the research, development, test and evaluation, and procurement of weapons systems and security assistance programs providing a new or improved capacity. Also covered are planning, programming, and budgeting; contract solicitation and management; systems engineering, configuration management, and logistical support required by the systems; and computer and software required to support weapons both in the testing and evaluation phases and the weapons life cycle.
Credit Recommendation: In the lower-division baccalaureate/associate degree category, 3 semester hours in purchasing and 1 in systems management (4/93).

AF-1408-0107
SERVICES SUPERVISOR BY CORRESPONDENCE

Course Number: 62370.
Location: Extension Course Institute, Gunter AFS, AL.
Length: Maximum, 52 weeks.
Exhibit Dates: 9/91–4/96.
Learning Outcomes: Upon completion of the course, the student will be able to describe the various responsibilities of the the service

supervisor regarding manpower management and funds management; describe the process of developing service contracts and the survey process used before the contract is awarded; describe the process of facility upgrade and the role of base-level advisory boards; use the various accounting reports and forms; interpret military pay entitlements and allowances; understand the management of funds for and from base facilities; describe the mortuary affairs program, and define the prime readiness services program.

Instruction: This correspondence program covers the responsibilities of the services supervisor regarding manpower management and various funds uses in the services. The responsibilities are applied in the areas of contracting for goods and services, facilities upgrades, food management, military pay entitlements and allowances, mortuary affairs, and the use of prime readiness in base services programs.

Credit Recommendation: In the lower-division baccalaureate/associate degree category, 2 semester hours in supervision, 1 in budgeting, and 1 in management (4/93).

AF-1408-0108

CHAPEL MANAGEMENT SPECIALIST BY CORRESPONDENCE

Course Number: 89350.
Location: Extension Course Institute, Gunter AFS, AL.
Length: Maximum, 52 weeks.
Exhibit Dates: 7/91–Present.
Learning Outcomes: Upon completion of the course, the student will be able to describe the practices, policies, and history of the four major religions. Additionally, a functional knowledge of administration will be gained in the areas of personnel, accounting, finance, data base management, and emergency training procedures/policies that relate to the management of chapel operations.

Instruction: Topics covered are from a variety of disciplines. Student is introduced to comparative religion (history, practices, and policies); principles of management (personnel, organization behavior, facilities design, construction, repair and maintenance); accounting (journalizing, financial report preparation); finance (funds management, asset acquisition/management); internal control concepts; budgeting; business report writing; database management; and coordination of functions to facilitate effective chapel and office administration.

Credit Recommendation: In the lower-division baccalaureate/associate degree category, 1 semester hour in religion, 1 in principles of accounting, 1 in principles of finance, and 3 in office administration (4/93).

AF-1408-0109

FINANCIAL ANALYSIS SPECIALIST BY CORRESPONDENCE
(Cost Analysis Specialist by Correspondence)

Course Number: 67450.
Location: Extension Course Institute, Gunter AFS, AL.
Length: Maximum, 52 weeks.
Exhibit Dates: 8/92–Present.
Learning Outcomes: Upon completion of the course, the student will be able to describe the theory and techniques required to successfully function as a financial analyst specialist with a specialized emphasis in the areas of cost controls, computations, and strategic financial/economic planning and analysis.

Instruction: This is a comprehensive correspondence review of quantitative methods utilized in business applications. Additional areas addressed include cost/benefit analysis; lease versus buy decisions in capital budgeting; review of cost terminology; and budgeting and financial packaging.

Credit Recommendation: In the lower-division baccalaureate/associate degree category, 2 semester hours in introduction to algebra and 2 in principles of finance (4/93).

AF-1408-0110

INTRODUCTION TO LOGISTICS BY CORRESPONDENCE

Course Number: 6612.
Location: Extension Course Institute, Gunter AFS, AL.
Length: Maximum, 52 weeks.
Exhibit Dates: 10/91–9/97.
Learning Outcomes: Upon completion of the course, the student will be able to define and describe logistics as it relates to planning and carrying out the movement and maintenance of forces; define and describe the Air Force logistical process and organization; describe the planning, programming, and budgeting system as a part of logistics; describe logistics as applied to the design, development, and acquisition of major weapons systems; describe total quality management and its applications to quality, reliability, and maintenance of systems; describe the elements of an integrated weapons system during the system's life cycle; describe provisioning management as it relates to weapons systems; describe the management of contracts; and describe supply systems, management maintenance, and information systems needed to support movement of troops.

Instruction: Course covers logistics as the science of planning and carrying out movement and maintenance of forces. Content covers design and development, acquisition, storage, movement, distribution maintenance, and disposition of materiel. Management of services, planning and managing facilities, and providing for the needs of troops is heavily emphasized.

Credit Recommendation: In the lower-division baccalaureate/associate degree category, 3 semester hours in maintenance management (4/93).

AF-1408-0111

FINANCIAL ANALYSIS JOURNEYMAN BY CORRESPONDENCE

Course Number: 67450.
Location: Extension Course Institute, Gunter AFS, AL.
Length: Maximum, 52 weeks.
Exhibit Dates: 8/92–Present.
Learning Outcomes: Upon completion of the course, the student will be able to describe the basic principles of financial analysis and planning.

Instruction: This course discusses a number of basic financial areas beginning with the overall concepts of resource management, internal management control, and the federal budget system. A review of basic mathematical techniques with a brief introduction to statistical concepts is provided along with techniques and principles of cost/financial analysis. Course culminates with the basic principles of financial planning. There is a proctored final examination.

Credit Recommendation: Credit is not recommended because of the limited, specialized nature of the course (6/94).

AF-1408-0112

CHAPLAIN SERVICE SUPPORT CRAFTSMAN BY CORRESPONDENCE

Course Number: 89370.
Location: Extension Course Institute, Gunter AFS, AL.
Length: Maximum, 52 weeks.
Exhibit Dates: 2/93–9/97.
Learning Outcomes: Upon completion of the course, the student will be able to plan, organize, supervise, and manage appropriated funds budgets, supply and equipment accounts, and readiness operations for the support of chapel service activities.

Instruction: Course covers planning, organizing, communicating, training, and supervising the resources necessary to support the activities for chaplain services.

Credit Recommendation: In the lower-division baccalaureate/associate degree category, 1 semester hour in supervisory management (6/94).

AF-1408-0113

CHAPEL SERVICE SUPPORT JOURNEYMAN BY CORRESPONDENCE

Course Number: 89350.
Location: Extension Course Institute, Gunter AFS, AL.
Length: Maximum, 52 weeks.
Exhibit Dates: 7/92–9/96.
Learning Outcomes: Upon completion of the course, the student will be able to describe the requirements for maintenance of facilities, utilization of volunteers, management of correspondence, bookkeeping responsibilities, budget appropriation and submission, and beliefs and worship practices of major religious denominations.

Instruction: Course covers mission and organizational structure of the USAF chaplain service, background information on the beliefs and worship practices of the major religious denominations, maintenance of religious facilities, utilization of lay volunteers, logistical support of chapel programs, written communication including preparation of reports, and financial and logistical resources used to support the chaplain program.

Credit Recommendation: Credit is not recommended because of the military-specific nature of the course (6/94).

AF-1511-0001

AIR COMMAND AND STAFF COLLEGE

Course Number: None.
Location: Air University, Maxwell AFB, AL.
Length: 40 weeks (746-898 hours).
Exhibit Dates: 1/90–8/90.
Objectives: To train officers in the factors affecting national behavior and policy formation, in military management processes, and in military subjects including employment of aerospace power.

Instruction: Lectures, seminars, readings, experiential exercises, and student research in national security studies; strategic theory; the history and theory of various kinds of war; international and regional studies; and management, including organizational behavior, leadership, communication, human resource management, and management systems.

Credit Recommendation: In the upper-division baccalaureate category, 15 semester hours in international relations, national security studies, political science, and economics; 12 in management, including management systems, leadership, communication, and human resource management; and 3 in research methods (3/83); in the graduate degree category, 9 semester hours in international relations and national security studies; 6 in management systems, leadership, communication, and human resource management; and 3 in organizational behavior (3/83).

AF-1511-0007

CIVIL LAW BY CORRESPONDENCE

Course Number: 7802A.
Location: Extension Course Institute, Gunter AFS, AL.
Length: Maximum, 52 weeks.
Exhibit Dates: 1/90–5/92.
Learning Outcomes: Upon completion of the course, the student will be able to describe the basic principles of administrative, martial, and civil law as applicable to the Air Force.
Instruction: This monitored home-study course includes readings, written exercises, and a proctored examination in the application of civil law within the Air Force. Topics include Air Force administrative law; interrelationships between the Department of Defense, the armed services, and other governmental agencies; status of officers, warrant officers and enlisted personnel; legal aspects of military pay and allowances; martial law; and international law and treaties as they apply to the armed forces.
Credit Recommendation: In the lower-division baccalaureate/associate degree category, 1 semester hour in political science or government (5/88).

AF-1511-0009

AIR COMMAND AND STAFF NONRESIDENT SEMINAR PROGRAM
(Air Command and Staff Nonresident Seminar Associate Program)

Course Number: None.
Location: Air University, Maxwell AFB, AL.
Length: 52 weeks (369 hours).
Exhibit Dates: 1/90–Present.
Learning Outcomes: Upon completion of the course, the student, in either written or oral form, will be able to analyze those specific domestic and international conditions which affect the national security policy-making process; critically interpret the historical evolution of military thought and strategic, conventional, and low-intensity warfare; delineate those crucial changes occurring in the regional areas of the globe which collectively impact upon the present and future deployment of aerospace power; and apply those managerial and communication skills essential to effective Air Force command organizations.
Instruction: The seminar program covers patterns and processes implicit in international dynamics, the national security policy-making process, the evolution of military history and warfare, military administration and leadership, and the deployment of aerospace power. The program includes group discussion, briefings, guest experts, videotapes, microcomputer-assisted instruction, and faculty teaching visits to supplement individual study.
Credit Recommendation: In the graduate degree category, 9 semester hours in military history and evolution of strategic thought, 6 in regional studies, 6 in defense resource management, and 6 in defense policy/national and international security (9/92).

AF-1511-0010

AIR COMMAND AND STAFF COLLEGE RESIDENT PROGRAM

Course Number: None.
Location: Air University, Maxwell AFB, AL.
Length: 40 weeks (883 hours).
Exhibit Dates: 1/90–Present.
Learning Outcomes: Upon completion of the course, the student, in either written or oral form, will be able to: analyze those specific domestic and international conditions which affect the national security policy-making process; critically interpret the historical evolution of military thought and strategic, conventional, and low-intensity warfare; delineate those crucial changes occurring in the regional areas of the globe which collectively impact upon the present and future deployment of aerospace power; apply those managerial and communication skills essential to effective Air Force command organization.
Instruction: Course includes seminars and reading involving the patterns and processes implicit in interstate dynamics, the national security policy-making process, military history and weapons evolution, defense resource allocation, military administration and leadership, and the deployment of aerospace power.
Credit Recommendation: In the graduate degree category, 9 semester hours in military history and evolution of strategic thought, 6 in regional studies, 6 in defense resource management, and 6 in defense policy/national and international security (9/92).

AF-1511-0011

AIR WAR COLLEGE

Course Number: None.
Location: Air University, Maxwell AFB, AL.
Length: 44 weeks (846-864 hours).
Exhibit Dates: 1/90–Present.
Learning Outcomes: Upon completion of the course, graduates will be able to analyze and provide sound advice on national and international security policy, in particular concerning the development and employment of aerospace power.
Instruction: This is the senior professional school in the Air Force. The program offered includes lectures, seminars, gaming and crisis simulations, case studies, field trips, and individualized student reading and research on military strategy, national, and international security and resource allocations for the purpose of developing and employing aerospace power. The curriculum, in particular, focuses on the historical evolution of US and foreign military strategy and doctrine; the nature of power and its role in the international system; changing threats to national and international security; regional studies with an assessment of the effects of regional developments on US and global security; the impact of science and technology on weapons systems and on national and international security; the formulation and implementation of national defense and security policy; and the allocation and management of defense resources.
Credit Recommendation: In the graduate degree category, 6 semester hours in military history and evolution of strategic thought, 3 in regional studies, 6 in defense resource management, and 12 in defense policy/national and international security (9/92).

AF-1511-0012

AIR WAR COLLEGE NONRESIDENT SEMINAR PROGRAM
(Air War College Correspondence Program)
(Air War College Associate Programs Nonresident and Correspondence)

Course Number: None.
Location: Air University, Maxwell AFB, AL.
Length: 44 weeks (420-480 hours).
Exhibit Dates: 1/90–Present.
Learning Outcomes: Upon completion of the course, graduates will be able to analyze and provide sound advice on national and international security policy, in particular concerning the development and employment of aerospace power.
Instruction: This is the senior professional school in the US Air Force. The program is offered through correspondence, providing for individualized student reading and research including case studies and supplementary student-organized seminars on military strategy, national and international security and aerospace allocations for the purpose of developing and employing aerospace power. The curriculum, in particular, focuses on the historical evolution of US and foreign military strategy and doctrine; the nature of power and its role in the international system; changing threats to national and international security; regional studies with an assessment of the effects of regional developments on US and global security; the impact of science and technology on weapons systems and on national and international security; the formulation and implementation of national defense and security policy; and the allocation and management of defense resources.
Credit Recommendation: In the graduate degree category, 6 semester hours in military history and evolution of strategic thought, 3 in regional studies, 6 in defense resource management, and 12 in defense policy/national and international security (9/92).

AF-1511-0013

AIR COMMAND AND STAFF CORRESPONDENCE PROGRAM
(Air Command and Staff Correspondence Associate Program)

Course Number: None.
Location: Air University, Maxwell AFB, AL.
Length: Maximum, 52 weeks.
Exhibit Dates: 1/90–Present.
Learning Outcomes: Upon completion of the course, the student, in either written or oral form, will be able to analyze those specific domestic and international conditions which

affect the national security policy making process; critically interpret the historical evolution of military thought and strategic, conventional, and low-intensity warfare; delineate those crucial changes occurring in the regional areas of the globe which collectively impact upon the present and future deployment of aerospace power; and apply those managerial and communication skills essential to effective Air Force command organization.

Instruction: The correspondence program covers patterns and processes implicit in international dynamics, the national security policymaking process, the evolution of military history and warfare, military administration and leadership, and the deployment of aerospace power. The program involves correspondence study and students may supplement the learning process with student-led seminars.

Credit Recommendation: In the graduate degree category, 9 semester hours in military history and evolution of strategic thought, 6 in regional studies, 6 in defense resource management, and 6 in defense policy/national and international security (9/92).

AF-1601-0030

LIQUID FUEL SYSTEMS MAINTENANCE SPECIALIST BY CORRESPONDENCE

Course Number: 54551.
Location: Extension Course Institute, Gunter AFS, AL.
Length: Maximum, 52 weeks.
Exhibit Dates: 1/90–Present.
Objectives: To train individuals in the maintenance of liquid fuel distribution systems.
Instruction: Course consists of on-the-job supervision and at-home study. The at-home study provides knowledge regarding fundamentals of fuel tank construction and tank entry procedures; operation and maintenance of mechanical systems components, including valves, pumps, meters, and filtration equipment; and operation of hydrant systems and components. Automotive gasoline dispensing units, fuel characteristics, hand tools, and safety procedures are also covered.
Credit Recommendation: In the vocational certificate category, 4 semester hours in liquid fuel systems, distribution and maintenance, and tank farm operations (9/86).

AF-1601-0046

ENGINEERING ASSISTANT SPECIALIST BY CORRESPONDENCE

Course Number: 55350.
Location: Extension Course Institute, Gunter AFS, AL.
Length: Maximum, 52 weeks.
Exhibit Dates: 1/90–8/99.
Objectives: This course is intended to train individuals to be civil/construction engineering assistants.
Instruction: This correspondence course consists of on-the-job supervision and at-home studies. The home study provides basic knowledge of applied mathematics, surveying, soils analysis and testing, pavement analysis and testing, drafting, and construction management. The course provides knowledge related to real estate property design and construction, maintenance, and repair of civil engineering and construction projects and functions. Field inspection and management techniques are also covered. The course has a proctored final examination.
Credit Recommendation: In the vocational certificate category, 3 semester hours in applied technical mathematics (9/86); in the lower-division baccalaureate/associate degree category, 6 semester hours in basic surveying and/or construction surveying, 3 in soil analysis and testing, 3 in asphalt and concrete testing, 6 in engineering drafting and/or architectural drafting, and 3 in construction management (9/86).

AF-1601-0047

PAVEMENTS MAINTENANCE SPECIALIST BY CORRESPONDENCE

Course Number: 55150.
Location: Extension Course Institute, Gunter AFS, AL.
Length: Maximum, 52 weeks.
Exhibit Dates: 1/90–10/94.
Objectives: To train individuals to design, construct, and maintain concrete and bituminous pavements.
Instruction: Course consists of on-the-job supervision and at-home study. The home study course provides basic knowledge in the design, construction, and maintenance of concrete and bituminous pavements. Subgrade preparation, materials and design mixtures, placement of pavement, maintenance of functions, drainage structures, fencing, and prefabrication surface mats are covered in this course. Also covered are safety, management techniques, and general military contingency plans.
Credit Recommendation: In the vocational certificate category, 4 semester hours in highway design and maintenance (9/86); in the lower-division baccalaureate/associate degree category, 3 semester hours in highway design and construction (9/86).

AF-1601-0048

APPRENTICE PAVEMENTS MAINTENANCE SPECIALIST BY CORRESPONDENCE

Course Number: 55130.
Location: Extension Course Institute, Gunter AFS, AL.
Length: Maximum, 52 weeks.
Exhibit Dates: 1/90–10/94.
Objectives: To train individuals to maintain rigid and flexible pavements.
Instruction: Course consists of on-the-job supervision and at-home study. The home study course provides basic knowledge in job safety, construction, and maintenance. Flexible pavement, prefabricated surface mats, painting, fencing, and explosives are taught. Primary emphasis is on maintenance functions. Military contingency plans and operation are also covered.
Credit Recommendation: In the vocational certificate category, 4 semester hours in highway maintenance (9/86).

AF-1601-0049

STRUCTURAL TECHNICIAN BY CORRESPONDENCE

Course Number: 55271.
Location: Extension Course Institute, Gunter AFS, AL.
Length: Maximum, 52 weeks.
Exhibit Dates: 1/90–1/97.
Objectives: To train personnel in the materials and methods used in light building construction.
Instruction: This course consists of on-the-job training and at-home studies. The home study provides training in the materials and methods of building construction, and the on-the-job training provides reinforcement for the home studies. The course contains materials used for light construction.
Credit Recommendation: In the vocational certificate category, 3 semester hours in materials of building construction and 3 in methods of light construction (9/86); in the lower-division baccalaureate/associate degree category, 3 semester hours in materials and methods of building construction (light construction) (9/86).

AF-1601-0050

FUELS MANAGEMENT OFFICER

Course Number: J3OZR6421 000; C3OZR6421 000.
Location: 396 TCHTG/CC, Sheppard AFB, TX.
Length: 5 weeks (200 hours).
Exhibit Dates: 2/90–Present.
Learning Outcomes: Upon completion of the course, the student will be able to identify responsibilities of the fuels management branch, relationships between base organization and fuels management, and DoD organizations and fuels management; identify functions of fuels quality control and inspection, cryogenics production and maintenance section, fuels support section, fuels operations section, and accounting and administrative sections; audit fuels inventory management report; evaluate daily logs for trends; define the responsibilities of the fuels management officer for contract operations and support agreements; develop support planning document based upon requirements; calculate defueler allowances using published tables of allowances; review requirements of resource management for compliance; develop support planning documents; calculate refueler allowances; and review requirements of resource management.
Instruction: Course includes classroom lectures and discussions using text, audiovisuals, and simulations to effect planning, organizing, directing, and coordinating fuel operation activities, including control of fuels, lubricants, demineralized water, and liquid oxygen. Topics also include Air Force stock fund, audit trails, trend analysis, environmental protection, energy conservation, fire hazards, and fuel safety precautions.
Credit Recommendation: In the graduate degree category, 3 semester hours in chemical engineering and 3 in operations management (5/93).

AF-1601-0051

PAVEMENT AND CONSTRUCTION EQUIPMENT JOURNEYMAN BY CORRESPONDENCE
(Pavement and Construction Equipment Specialist by Correspondence)

Course Number: 55151A.
Location: Extension Course Institute, Gunter AFS, AL.
Length: Maximum, 52 weeks.
Exhibit Dates: 1/93–1/99.
Learning Outcomes: Upon completion of the course, the student will be able to perform duties as a pavement and construction equipment specialist.

Instruction: This correspondence course consists of on-the-job training and at-home study. The topics covered include the operation and maintenance of pavement and construction equipment, with emphasis on work place safety, drainage structures, general pavement functions, rigid and flexible pavement construction/maintenance, flexible pavement maintenance, and fencing.

Credit Recommendation: In the vocational certificate category, 3 semester hours in basic construction equipment operation and maintenance (6/94).

AF-1601-0052

PAVEMENT AND CONSTRUCTION EQUIPMENT JOURNEYMAN BY CORRESPONDENCE
(Pavement and Construction Equipment Specialist by Correspondence)

Course Number: 55151B.
Location: Extension Course Institute, Gunter AFS, AL.
Length: Maximum, 52 weeks.
Exhibit Dates: 2/93–Present.
Learning Outcomes: Upon completion of the course, the student will be able to perform duties as a pavement and construction equipment specialist.
Instruction: This correspondence course consists of on-the-job training and at-home study. The topics covered include hauling equipment and materials; pavement sweeping and inspection; snow removal and ice control; rigging, lifting, and crane operation; operating wheeled loading and compaction equipment; operating motorized graders, crawler tractors, and excavation equipment; expedient repair and construction; oxyacetylene welding; and rapid runway repair.
Credit Recommendation: In the vocational certificate category, 6 semester hours in advanced construction equipment operation and maintenance (6/94).

AF-1606-0020

SPECIAL OPERATIONS TRAINING, AC-130E PILOT

Course Number: 104140Z.
Location: 415th Special Operations Training Squadron, Hurlburt Field, FL.
Length: 4 weeks (94 hours).
Exhibit Dates: 1/90–3/90.
Objectives: To train pilots in the operation of multi-engine aircraft.
Instruction: Course includes ground egress training and multi-engine flight experience and lectures on electrical systems, hydraulics, landing gear and brakes, flight controls, fuel systems, instrumentation, communication, navigation, performance and mission planning, and flight characteristics and procedures.
Credit Recommendation: In the lower-division baccalaureate/associate degree category, 3 semester hours in flight experience (2/74).

AF-1606-0023

C-5 PILOT

Course Number: None.
Location: 443d Military Airlift Wing, Altus AFB, OK.
Length: 4 weeks (127 hours).
Exhibit Dates: 1/90–3/90.
Objectives: To train pilots in the operation of the C-5 aircraft.
Instruction: Course covers light and academic training, including an introduction to the C-5 aircraft and its systems, component parts, and normal and emergency functions.
Credit Recommendation: In the lower-division baccalaureate/associate degree category, 3 semester hours in flight experience (2/74).

AF-1606-0119

INTELLIGENCE FUNDAMENTALS BY CORRESPONDENCE

Course Number: 8000.
Location: Extension Course Institute, Gunter AFS, AL.
Length: Maximum, 52 weeks.
Exhibit Dates: 1/90–Present.
Learning Outcomes: Upon completion of the course, the student will be able to present intelligence briefings, set standards for intelligence performance, appreciate the complexity and scope of intelligence, and control utilization of classified materials.
Instruction: This monitored home-study course includes readings, written exercises, use of maps and charts, and proctored examinations. Topics include history and nature of intelligence, handling and presenting intelligence data, and the nature of counterintelligence and psychological operations.
Credit Recommendation: In the lower-division baccalaureate/associate degree category, 3 semester hours in political science and 1 in geography (5/88).

AF-1606-0120

INTELLIGENCE OFFICER BY CORRESPONDENCE

Course Number: 8001.
Location: Extension Course Institute, Gunter AFS, AL.
Length: Maximum, 52 weeks.
Exhibit Dates: 1/90–Present.
Learning Outcomes: Upon completion of the course, the student will be able to collect, process, and disseminate intelligence and conduct briefings about prisoner survival and enemy interrogation.
Instruction: This monitored home-study course includes readings, written exercises, and proctored examination. Topics include collection and dissemination of intelligence, preparing for missions, and comparison of US and Soviet capabilities.
Credit Recommendation: In the upper-division baccalaureate category, 2 semester hours in political science or international relations (5/88).

AF-1606-0121

INTELLIGENCE OPERATIONS SPECIALIST BY CORRESPONDENCE

Course Number: 20150.
Location: Extension Course Institute, Gunter AFS, AL.
Length: Maximum, 52 weeks.
Exhibit Dates: 1/90–10/96.
Learning Outcomes: Upon completion of the course, the student will be able to demonstrate a basic understanding of the military intelligence system including doctrine, strategy, and tactics.
Instruction: This monitored home-study course includes readings, exercises, and proctored examinations. Topics include a review of basic military intelligence fundamentals, fundamentals of maps and charts, target planning and selection, weapon characteristics and effects, and combat mission activities.
Credit Recommendation: In the lower-division baccalaureate/associate degree category, 1 semester hour in political science or government (5/88).

AF-1606-0122

TARGET INTELLIGENCE SPECIALIST BY CORRESPONDENCE

Course Number: 20151.
Location: Extension Course Institute, Gunter AFS, AL.
Length: Maximum, 52 weeks.
Exhibit Dates: 1/90–10/98.
Learning Outcomes: Upon completion of the course, the student will demonstrate a basic understanding of the military intelligence system including doctrine, strategy, and tactics with a focus on target selection.
Instruction: This monitored home-study course includes readings, exercises, and proctored examinations. Topics include a review of basic military intelligence fundamentals, fundamentals of maps and charts, target planning and selection, weapon characteristics and effects, combat mission activities, and the basics of imagery and radar intelligence.
Credit Recommendation: In the lower-division baccalaureate/associate degree category, 1 semester hour in political science or government (5/88).

AF-1606-0123

INTELLIGENCE OPERATIONS TECHNICIAN BY CORRESPONDENCE

Course Number: 20170.
Location: Extension Course Institute, Gunter AFS, AL.
Length: Maximum, 52 weeks.
Exhibit Dates: 1/90–2/93.
Learning Outcomes: Upon completion of the course, the student will be able to apply general principles of intelligence to specific US and Soviet cases.
Instruction: This monitored home-study course offers readings, written exercises, and proctored final examinations. Topics include the general intelligence mission and information about Soviet military forces, Warsaw Pact forces, and China's forces.
Credit Recommendation: In the upper-division baccalaureate category, 1 semester hour in international relations (5/88).

AF-1606-0150

EXPERIMENTAL TEST PILOT

Course Number: TPS 2865.
Location: Test Pilot School, Edwards AFB, CA.
Length: 46 weeks (585 hours).
Exhibit Dates: 1/90–Present.
Learning Outcomes: Upon completion of the course, the student will be able to plan, fly, evaluate, and report on projects relating to the testing and analysis of aircraft and their related systems. Prerequisite undergraduate degree is required in engineering, physics, mathematics, or related physical sciences provide the foundation for advanced study in integrated academic and flying laboratory exercises. The graduate is qualified to function as a team member in the

evaluation of advanced aircraft/systems. Competency is developed in the area of avionic systems, advanced aerodynamics, appropriate control and flight mechanics theory, and well as aircraft performance. Project proposal planning and writing as well as overall management skills and responsibilities are emphasized.

Instruction: The curriculum is divided into three phases: performance, flying qualities, and systems test. All three phases have integrated academic and flying programs. Examinations in calculus and mechanics are given at the start of the performance phase. This phase develops the theory and methods of evaluating aircraft performance. A performance test program is conducted and a formal written report is generated. Evaluation of flying qualities examines the stability parameters and derivations found in the aircraft equations of motion. A flight controls project is conducted and the results are presented in an oral report. The systems test phase evaluates installed systems and associated equipment. Evaluations are conducted and the results are reported in a series of oral and written reports. As a final student test project, the student test teams conduct a test management project on an aircraft just as they will in flight test projects after graduation.

Credit Recommendation: In the graduate degree category, 3 semester hours in avionics system integration, 2 in aerodynamics, 2 in linear control theory, 4 in flight mechanics, 4 in performance, and 2 in systems management (7/88).

AF-1606-0151

AIR NATIONAL GUARD FIGHTER WEAPONS INSTRUCTOR

Course Number: F-400 FWS.
Location: Air National Guard Reconnaissance Weapons School, McConnell AFB, KS.
Length: 14-15 weeks (166-167 hours).
Exhibit Dates: 1/90–Present.
Learning Outcomes: Upon completion of the course, the student will be able to design and participate in instructional programs aimed at teaching aircrew the skills necessary to provide weapons, weapons-related systems and tactical expertise at the squadron, wing, and headquarters level. In order to perform these functions, the student must possess communication and instructional skills in both academic and flying areas. Specific requirements include a competency in an array of weapons and weapons-related systems and avionics devices and their systems to support delivery, as well as the tactics to employ the various systems. In addition, it is necessary to have the skills necessary to coordinate and implement modifications to both tactics and the weapons systems.

Instruction: This course is designed for experienced F-4 pilot instructors and is of an advanced nature. Instructional methods involved in the delivery of this curriculum involve a wide variety of techniques. The primary teaching methods are those of classroom lecture supported by very specialized laboratory assignments, namely flight problems. This core of work is supported by 50 hours of self-instruction, 4 hours of seminars, 9 hours of examinations, and 4 hours of critique integrated into the 78 days of the program as appropriate. The lecture phase treats such topics as aerodynamics and performance instructor preparation and communication, special weapons systems and tactics, aggressor military and social information, and flight mission planning. The flying portion of the program provides a unique opportunity for planning, executing, and evaluating the material covered in the classroom. Qualified instructors monitor all phases of the program.

Credit Recommendation: In the graduate degree category, 2 semester hours in computer applications, 3 in avionics/electronic/systems, 3 in instructional methods, and 3 in applied aerodynamics (9/89).

AF-1606-0152

RF-4 FIGHTER WEAPONS INSTRUCTOR

Course Number: RF-400 RWS.
Location: Air National Guard Reconnaissance Weapons School, Gowan AFB, ID.
Length: 14 weeks (198 hours).
Exhibit Dates: 1/90–Present.
Learning Outcomes: Upon completion of the course, the graduate will be able to teach instructors all phases of the RF-4 fighter weapons system. The graduate will also be qualified to function as the weapons officer in an operational unit. The student will possess communication skills and effective instructional techniques in both the academic and flight areas of the RF-4 weapons system.

Instruction: The course is restricted to pilots who have an minimum of 300 hours in RF-4; are instructor pilots or instructor weapons system officers; are current mission-ready crew members, ready training unit instructor pilots, or instructor weapons systems officers; are qualified in very low altitude training to 300 feet above ground level;, and are defense-maneuvering qualified. Pilots only must be air combat instructor pilot qualified. Further, recent experience requirements are 16 sorties in preceding 60 days, 10 sorties in preceding 30 days, at least 3 of which are devoted to air combat training and all physical and flight checks accomplished prior to arrival with none due during the course. Pilots must possess a top secret clearance. The curriculum consists of 152 hours of academic work supported by 42.5 hours of related flight activities and 2.9 hours of device (simulator) training. Each flight sortie requires approximately 6 hours of additional time for planning, briefing, and critique. Classes are small (approximately 4 students), permitting a high instructor-student ratio and providing concentrated instruction. Highly experienced and academically qualified faculty are involved in all phases of the program. Adequate equipment and facilities are available and maintained in a high state of readiness.

Credit Recommendation: In the graduate degree category, 1 semester hours in computer applications, 4 in avionics/electronics, systems, and photography, 3 in instructional methodology, and 3 in applied aerodynamics (9/89).

AF-1606-0153

FLIGHT TEST ENGINEER/NAVIGATOR

Course Number: TPS 28XX; TPS 2875.
Location: Test Pilot School, Edwards AFB, CA.
Length: 46 weeks (585 hours).
Exhibit Dates: 1/90–Present.
Learning Outcomes: Upon completion of the course, the student will be able to plan, fly, evaluate, and report on projects relating to the testing and analysis of aircraft and their related systems. Prerequisite undergraduate degree in engineering, physics, mathematics, or related physical science provides the foundation for advanced study in integrated academic and flying laboratory exercises. The graduate is qualified to function as a team member in the evaluation of advanced aircraft/systems. Competency is developed in the area of avionic systems, advanced aerodynamics, appropriate control and flight mechanics theory, and well as aircraft performance. Project proposal planning and writing as well as overall management skills and responsibilities are emphasized.

Instruction: The curriculum is divided into three phases: performance, flying qualities, and systems test. All three phases have integrated academic and flying programs. Examinations in calculus and mechanics are given at the start of the performance phase. This phase develops the theory and methods of evaluating aircraft performance. A performance test program is conducted and a formal written report is generated. Evaluation of flying qualities examines the stability parameters and derivations found in the aircraft equations of motion. A flight controls project is conducted and the results are presented in an oral report. The systems test phase evaluates installed systems and associated equipment. Evaluations are conducted and the results are reported in a series of oral and written reports. As a final student test project, the student test teams conduct a test management project on an aircraft just as they will in flight test projects after graduation.

Credit Recommendation: In the graduate degree category, 3 semester hours in avionics system integration, 2 in aerodynamics, 2 in linear control theory, 4 in flight mechanics, 4 in performance, and 2 in systems management (7/88).

AF-1606-0154

UNDERGRADUATE PILOT TRAINING

Course Number: P-V4A-B.
Location: Air Training Command, Columbus AFB, MS; Air Training Command, Reese AFB, TX; Air Training Command, Laughlin AFB, TX; Air Training Command, Williams AFB, AZ; Air Training Command, Vance AFB, OK.
Length: 52 weeks (1000-1200 hours).
Exhibit Dates: 2/90–Present.
Learning Outcomes: Upon completion of the course, the student will be able to perform the duties of an Air Force pilot and function in the national airspace system.

Instruction: Course includes classroom lectures, computer-aided instruction, flight simulation, individualized instruction, and actual flight training in aerodynamics, meteorology, aerospace physiology, aviation systems and components, flight safety, basic/instrument/advanced flight training, and aircraft performance.

Credit Recommendation: In the lower-division baccalaureate/associate degree category, 3 semester hours in basic flight training, 3 semester hours in instrument pilot, and 2 in aerodynamics (10/91); in the upper-division baccalaureate category, 3 semester hours in aerospace physiology, 3 in aviation systems and components, 3 in advanced pilot training, 1 in flight safety, and 3 semester hours in navigation (10/91).

AF-1606-0157

INSTRUCTOR PILOT TRAINING (F-15)

Course Number: F15AC100AT; F1500100AL/T.
Location: 325 Training Squadron, Tyndall AFB, FL.
Length: 11 weeks (107 hours).
Exhibit Dates: 6/90–Present.
Learning Outcomes: Upon completion of the course, the student will be able to assume the responsibilities of the instructor in the F-15 aircraft. Specifically, the student will be able to develop, deliver, manage, and maintain the associated instructional system for that airplane.
Instruction: The graduate of this program will have experience in recognizing and correcting student error, student briefing and debriefing, the use of training aids, employing differing instructional techniques, appropriate grading and evaluation techniques, and preparation of instructional goals and objectives. In addition, the student will have experience in the use of training devices, syllabus preparation, and instructional responsibilities. The inclusion of extensive teaching practice is an integral part of this course.
Credit Recommendation: In the graduate degree category, 4 semester hours in advanced instructional techniques and 2 in teaching practicum (8/93).

AF-1606-0158

1. BASIC OPERATIONAL TRAINING F-15 (F-15 Basic Qualification Training)
2. F-15 BASIC QUALIFICATION TRAINING
3. F-15 BASIC QUALIFICATION TRAINING

Course Number: *Version 1:* F15ACB00AT. *Version 2:* F1500B00 AL/T. *Version 3:* F1500B00 AL/T.
Location: 325 Training Squadron, Tyndall AFB, FL.
Length: *Version 1:* 22 weeks (540-586 hours). *Version 2:* 16 weeks (455 hours). *Version 3:* 18 weeks (427 hours).
Exhibit Dates: *Version 1:* 1/93–Present. *Version 2:* 9/90–12/92. *Version 3:* 1/90–8/90.
Objectives: *Version 1:* Upon completion of the course, the student will be able to employ various advanced aircraft systems such as propulsion, electrical, fuel, hydraulic, inertial navigation, heads-up-display, and air-to-air attack radar. The student will also be able to employ the F-15 in a variety of air-to-air combat maneuvers, select and use appropriate weapons, and respond to threats. *Version 2:* Upon completion of the course, the student will be able to operate and employ the various advanced aircraft systems such as propulsion, electrical, fuels, hydraulics, inertial navigation, heads-up-display, and air-to-air attack radar systems. The student will also be able to employ the F-15 in a variety of air-to-air combat maneuvers, select and use appropriate weapons, and respond to extend threats. *Version 3:* Upon completion of the course, the student will be able to employ various advanced aircraft systems such as propulsion, electrical, fuel, hydraulic, inertial navigation, heads-up-display, and air-to-air attack radar. The student will also be able to employ the F-15 in a variety of air-to-air combat maneuvers, select and use appropriate weapons, and respond to threats.
Instruction: *Version 1:* This course of instruction is designed to prepare the undergraduate pilot training graduate to become mission capable in the F-15 aircraft. Instructional methodologies include lecture, laboratory (including simulator and self-paced instruction), flying, and mission support (flight brief/debrief). Major segments of the course include aircraft systems, radar systems, aerial attack concepts, and specialized (weapons) systems. *Version 2:* This course of instruction is designed to prepare the undergraduate pilot training graduate to become mission capable in the F-15 aircraft. Instructional methodologies include lecture, laboratory (including simulator and self-paced instruction), flying, and mission support (flight brief/debrief). Major segments of the course include aircraft systems, radar systems, aerial attack concepts, and specialized (weapons) systems. *Version 3:* This course of instruction is designed to prepare the undergraduate pilot training graduate to become mission capable in the F-15 aircraft. Instructional methodologies include lecture, laboratory (including simulator and self-paced instruction), flying, and mission support (flight brief/debrief). Major segments of the course include aircraft systems, radar systems, aerial attack concepts, and specialized (weapons) systems.
Credit Recommendation: *Version 1:* In the graduate degree category, 1 semester hour in advanced aircraft systems, 2 in airspace positioning radar systems, 3 in advanced aircraft maneuvers, 2 in specialized systems, and 3 in advanced flight (1/98). *Version 2:* In the graduate degree category, 1 semester hour in advanced aircraft systems, 2 in airspace positioning radar systems, 3 in advanced aircraft maneuvers, 2 in specialized systems, and 2 in advanced flight (8/93). *Version 3:* In the graduate degree category, 1 semester hour in advanced aircraft systems, 2 in airspace positioning radar systems, 2 in specialized systems, and 2 in advanced flight (8/93).

AF-1606-0159

1. TRANSITION/REQUALIFICATION F-15 (Conversion Training F-15 Track 1A)
2. CONVERSION TRAINING F-15 TRACK 1A
3. CONVERSION TRAINING F-15 TRACK 1A

Course Number: *Version 1:* F15ACTX00AT; F15ACTX0AT. *Version 2:* F1500TX0AL/T. *Version 3:* F1500TX0AL/T.
Location: 325 Training Squadron, Tyndall AFB, FL.
Length: *Version 1:* 16 weeks (395-586 hours). *Version 2:* 11 weeks (326 hours). *Version 3:* 11 weeks (326 hours).
Exhibit Dates: *Version 1:* 1/93–Present. *Version 2:* 9/90–12/92. *Version 3:* 1/90–8/90.
Learning Outcomes: *Version 1:* Upon completion of the course, the student will be able to manage and utilize advanced electronics/radar systems as well as be thoroughly familiar with the aerodynamics and operational systems of the F15 high-performance aircraft. The flight portion of this program is designed to produce a pilot proficient in F-15 air-to-air mission assignments. This requires a high degree of flying skill coupled with the technical knowledge necessary to select and utilize sophisticated navigation, radar, armament, and flight systems. *Version 2:* Upon completion of the course, the student will be able to manage and utilize advanced electronics/radar systems as well as be thoroughly familiar with the aerodynamics and operational systems of the F15 high-performance aircraft. The flight portion of this program is designed to produce a pilot proficient in F-15 air-to-air mission assignments. This requires a high degree of flying skill coupled with the technical knowledge necessary to select and utilize sophisticated navigation, radar, armament, and flight systems. *Version 3:* Upon completion of the course, the student will be able to manage and utilize advanced electronics/radar systems as well as be thoroughly familiar with the aerodynamics and operational systems of the F15 high-performance aircraft. The flight portion of this program is designed to produce a pilot proficient in F-15 air-to-air mission assignments. This requires a high degree of flying skill coupled with the technical knowledge necessary to select and utilize sophisticated navigation, radar, armament, and flight systems.
Instruction: *Version 1:* Lecture and discussion methods and demonstration/performance (laboratory) activities are effectively utilized. Computers, simulators, and high-performance aircraft are an integral component of the teaching program. *Version 2:* Lecture and discussion methods and demonstration/performance (laboratory) activities are effectively utilized. Computers, simulators, and high-performance aircraft are an integral component of the teaching program. *Version 3:* Lecture and discussion methods and demonstration/performance (laboratory) activities are effectively utilized. Computers, simulators, and high-performance aircraft are an integral component of the teaching program.
Credit Recommendation: *Version 1:* In the graduate degree category, 2 semester hours in advanced aircraft systems, 2 in airspace positioning radar systems, 3 in advanced aircraft maneuvers, 1 in specialized systems, and 3 in advanced flight (1/98). *Version 2:* In the graduate degree category, 2 semester hours in advanced aircraft systems, 2 in airspace positioning radar systems, 3 in advanced aircraft maneuvers, 1 in specialized systems, and 2 in advanced flight (8/93). *Version 3:* In the graduate degree category, 2 semester hours in advanced aircraft systems, 1 in airspace positioning radar systems, 3 in advanced aircraft maneuvers, 1 in specialized systems, and 2 in advanced flight (8/93).

AF-1701-0009

HEATING SYSTEMS TECHNICIAN BY CORRESPONDENCE

Course Number: 54572.
Location: Extension Course Institute, Gunter AFS, AL.
Length: Maximum, 52 weeks.
Exhibit Dates: 1/90–1/97.
Objectives: To train personnel to be specialists in the installation and maintenance of heating systems.
Instruction: Course consists of on-the-job training and at-home studies in the installation and maintenance of various types of heating systems. The home-study course provides the principles and practice of heating systems, and on-the-job training provides the application. The course includes heating principles, fuels, electrical and pneumatic control systems, warm air and water heating systems, and central heating systems.
Credit Recommendation: In the vocational certificate category, 3 semester hours in advanced plumbing (9/86); in the lower-division baccalaureate/associate degree category, 2 semester hours in mechanical and electrical equipment in buildings (heating) (9/86).

COURSE EXHIBITS

AF-1701-0010

HEATING SYSTEMS SPECIALIST BY CORRESPONDENCE

Course Number: 54552A.
Location: Extension Course Institute, Gunter AFS, AL.
Length: Maximum, 52 weeks.
Exhibit Dates: 1/90–10/94.
Learning Outcomes: Upon completion of the course, the student will be able to work as a civil engineering heating systems specialist and demonstrate knowledge in the areas of basic heating concepts, electricity, and control systems. The student will operate and maintain various fuel systems and fuel burning equipment.
Instruction: Student enrolling must have an Air Force Specialty Code assignment as a prerequisite. This provides an application of the principles learned to real requirements and constitutes an essential part of the instruction by serving as the laboratory component of the course. The student's supervisor must complete an evaluation survey of the course and its effectiveness when the student takes the final examination which measures mastery of the cognitive content. Course consists of self-paced, independent study with open book progress checks and satisfactory completion of a proctored final examination. Topics include base civil engineering organization, readiness programs, resources and work force management, safety, oxyacetylene welding, heating principles, electrical wiring, circuit protective devices, circuit control devices, electrical meters and troubleshooting, electromagnetic operating devices, flame safeguard, pneumatic controls, and operation and maintenance of equipment burning various fuels. The course content does not duplicate that listed in AF-1701-0011.
Credit Recommendation: In the lower-division baccalaureate/associate degree category, 2 semester hours in mechanical and electrical equipment for buildings (heating systems) (11/91).

AF-1701-0011

HEATING SYSTEMS SPECIALIST BY CORRESPONDENCE

Course Number: 54552B.
Location: Extension Course Institute, Gunter AFS, AL.
Length: Maximum, 52 weeks.
Exhibit Dates: 1/90–11/96.
Learning Outcomes: Upon completion of the course, the student will be able to work as a civil engineering heating systems specialist and demonstrate knowledge of warm-air heating and distribution systems, low and high temperature water systems, steam heating equipment, water treatment, and external corrosion.
Instruction: Student enrolling must have an Air Force Specialty Code assignment as a prerequisite. This provides an application of the principles learned to real requirements and constitutes an essential part of the instruction by serving as the laboratory component of the course. The student's supervisor must complete an evaluation survey of the course and its effectiveness when the student takes the final examination which measures mastery of the cognitive content. Course consists of self-paced, independent study with open book progress checks and satisfactory completion of a proctored final examination. Topics include: warm-air heating equipment, maintaining warm-air heating equipment, types and operating principles of warm-air distribution systems, low and high temperature water systems, steam heating equipment, water testing and treatment, and external corrosion. This does not duplicate course listed as AF-1701-0010.
Credit Recommendation: In the lower-division baccalaureate/associate degree category, 2 semester hours in mechanical and electrical equipment for buildings (heating systems) (11/91).

AF-1701-0012

HEATING, VENTILATION, AIR CONDITIONING, AND REFRIGERATION JOURNEYMAN BY CORRESPONDENCE
(Heating, Ventilation, Air Conditioning, and Refrigeration Specialist by Correspondence)

Course Number: 54550A.
Location: Extension Course Institute, Gunter AFS, AL.
Length: Maximum, 52 weeks.
Exhibit Dates: 1/93–4/99.
Learning Outcomes: Upon completion of the course, the student will be able to install, maintain, and operate various heating, ventilating, air conditioning, and refrigeration equipment.
Instruction: This correspondence course provides the principles and practices of heating, ventilating, air conditioning, and refrigeration equipment. Principles covered include electrical fundamentals and controls, safety and service practices, control fundamentals, psychometrics, pneumatic fundamentals and auxiliaries, energy management systems, fuels and fuel burning equipment, and heating equipment.
Credit Recommendation: In the vocational certificate category, 3 semester hours in refrigeration and air conditioning installation and maintenance (6/94); in the lower-division baccalaureate/associate degree category, 2 semester hours in refrigeration, ventilation, and air conditioning fundamentals (6/94).

AF-1701-0013

HEATING, VENTILATION, AIR CONDITIONING, AND REFRIGERATION JOURNEYMAN BY CORRESPONDENCE
(Heating, Ventilation, Air Conditioning, and Refrigeration Specialist by Correspondence)

Course Number: 54550B.
Location: Extension Course Institute, Gunter AFS, AL.
Length: Maximum, 52 weeks.
Exhibit Dates: 6/93–Present.
Learning Outcomes: Upon completion of the course, the student will be able to install, maintain, and operate various heating, ventilating, air conditioning, and refrigeration equipment.
Instruction: This correspondence course includes refrigeration service and control, ventilating equipment and systems, air and hydronic distribution systems, water treatment and corrosion control, and special auxiliary heating equipment.
Credit Recommendation: In the lower-division baccalaureate/associate degree category, 3 semester hours in refrigeration, heating, ventilating, and air conditioning (6/94).

AF-1703-0018

GENERAL PURPOSE VEHICLE MECHANIC BY CORRESPONDENCE

Course Number: 47252.
Location: Extension Course Institute, Gunter AFS, AL.
Length: Maximum, 52 weeks.
Exhibit Dates: 1/90–3/99.
Objectives: To train enlisted personnel to repair general purpose vehicles.
Instruction: This correspondence course offers individualized reading and study of the basic theory of operation and the repair of buses, trailers, semitrailers, light and medium trucks, and staff cars. The course includes shop organization, publications, supply, vehicle inspection, engines, fuel systems, exhaust and emission systems, electrical systems, hydraulics, power trains, heating, air conditioning, and front end alignment.
Credit Recommendation: In the lower-division baccalaureate/associate degree category, 3 semester hours in general automotive repair, 1 in shop practices, and 1 in preventive maintenance systems (9/85).

AF-1703-0019

FUEL SPECIALIST BY CORRESPONDENCE

Course Number: 63150.
Location: Extension Course Institute, Gunter AFS, AL.
Length: Maximum, 52 weeks.
Exhibit Dates: 1/90–5/98.
Objectives: To train selected enlisted personnel to become fuel specialists.
Instruction: This course presents individualized readings dealing with airmen fuel career fields, publications and operations security, safety, fuel control centers, fuel quality assurance and product accountability, fueling systems, air transportable fuel systems, cryotainers, and mobile fueling equipment.
Credit Recommendation: In the lower-division baccalaureate/associate degree category, 3 semester hours in aviation technology (12/84).

AF-1703-0020

SPECIAL PURPOSE VEHICLE AND EQUIPMENT MAINTENANCE APPRENTICE BY CORRESPONDENCE
(Special Purpose Vehicle and Equipment Mechanic by Correspondence)

Course Number: 47230.
Location: Extension Course Institute, Gunter AFS, AL.
Length: Maximum, 52 weeks.
Exhibit Dates: 1/90–3/94.
Learning Outcomes: Upon completion of the course, the student will demonstrate the general knowledge necessary to perform maintenance on a variety of special purpose vehicles, including cranes, crawler tractors and attachments, sweepers, loaders, towing and servicing vehicles, and auxiliary personnel heaters.
Instruction: This correspondence course presents information about the operation and maintenance of special purpose vehicles and equipment. Topics covered include vehicle maintenance management, operating principles of gasoline and diesel engines, basic automotive electrical principles, hydraulics, power trains, suspension, steering, brakes, heating and

air conditioning, and material-handling equipment.

Credit Recommendation: In the lower-division baccalaureate/associate degree category, 2 semester hours in general introduction to heavy vehicle and equipment mechanics (6/94).

AF-1703-0021

SPECIAL PURPOSE VEHICLE AND EQUIPMENT CRAFTSMAN BY CORRESPONDENCE
(Special Purpose Vehicle and Equipment Supervisor by Correspondence)

Course Number: 47271.
Location: Extension Course Institute, Gunter AFS, AL.
Length: Maximum, 52 weeks.
Exhibit Dates: 4/91–7/99.
Learning Outcomes: Upon completion of the course, the student will be able to perform maintenance on a variety of special purpose vehicles, including fire truck systems, aircraft refueling vehicles, mobile cranes, crawler tractors, sweepers, forklifts, aircraft cargo loaders, towing and servicing vehicles, and auxiliary personnel heaters.
Instruction: This correspondence course presents information about the operation and maintenance of special purpose vehicles and equipment, including troubleshooting, maintenance, characteristics, and operation of auxiliary systems. The course does not have laboratory exercises but does include reading assignments and written lessons consisting of objective questions.
Credit Recommendation: In the lower-division baccalaureate/associate degree category, 2 semester hours in heavy equipment operation/maintenance (6/94).

AF-1703-0022

GENERAL PURPOSE VEHICLE AND BODY MAINTENANCE BY CORRESPONDENCE
(General Purpose Vehicle and Body Maintenance Supervisor by Correspondence)

Course Number: 47275.
Location: Extension Course Institute, Gunter AFS, AL.
Length: Maximum, 52 weeks.
Exhibit Dates: 7/91–7/99.
Learning Outcomes: Upon completion of the course, the student will be able to perform and supervise general vehicle repair and auto body maintenance on passenger cars and light and medium duty trucks.
Instruction: This correspondence course presents the mechanics and repair of passenger cars and light and medium duty trucks. Course includes vehicle maintenance management, diesel and gasoline engines, power trains, hydraulic systems, chassis units, alignment, brakes, heating and air conditioning, and auto body and repair.
Credit Recommendation: In the lower-division baccalaureate/associate degree category, 2 semester hours in automotive technology (6/94).

AF-1703-0023

VEHICLE OPERATOR/DISPATCHER JOURNEYMAN BY CORRESPONDENCE

Course Number: 60350.
Location: Extension Course Institute, Gunter AFS, AL.
Length: Maximum, 52 weeks.
Exhibit Dates: 9/93–Present.
Learning Outcomes: Upon completion of the course, the student will apply the principles of vehicle operations, radio communications, and proper handling procedures for normal and hazardous cargo.
Instruction: This correspondence course covers the organization and operation of vehicles including dispatching procedures, the handling of hazardous materials, general maintenance, and defensive driving.
Credit Recommendation: In the vocational certificate category, 2 semester hours in vehicle operations (6/94).

AF-1704-0031

C-141 FLIGHT ENGINEER TECHNICIAN

Course Number: A435X0C-1.
Location: 443d Military Airlift Wing, Altus AFB, OK.
Length: 9 weeks (131 hours).
Exhibit Dates: 1/90–3/90.
Objectives: To provide flight engineers with knowledge of the normal and emergency functions of the C-141 aircraft.
Instruction: Course includes lectures and practical exercises on the C-141 aircraft, its systems and components, and troubleshooting procedures; flight simulator training;' and flight training.
Credit Recommendation: In the upper-division baccalaureate category, 2 semester hours in flight engineer technician (2/74).

AF-1704-0033

C-5 FLIGHT ENGINEER TECHNICIAN
(Flight Engineer School, C-5)

Course Number: A43570C-4.
Location: 443d Military Airlift Wing, Altus AFB, OK.
Length: 9 weeks (130 hours).
Exhibit Dates: 1/90–3/90.
Objectives: To provide enlisted personnel with a practical knowledge of the C-5 aircraft, its subsystems, and troubleshooting procedures.
Instruction: Course includes lectures and practical exercises in troubleshooting and solution of performance problems; simulator training designed to enable students to perform preflight checks without assistance, to understand both normal and emergency procedures, and to manage subsystems with and without related component failures; flight training in which students demonstrate the ability to accomplish normal and emergency operation procedures; preflight checks and scanner preflight checks; and management and/or operation of pertinent aircraft subsystems.
Credit Recommendation: In the upper-division baccalaureate category, 2 semester hours in flight engineering (2/74).

AF-1704-0070

MASTER CREW CHIEF

Course Number: None.
Location: 443d Military Airlift Wing, Altus AFB, OK.
Length: 2 weeks (68 hours).
Exhibit Dates: 1/90–3/90.
Objectives: To provide enlisted personnel with a basic understanding of maintenance management.
Instruction: Course includes lectures and practical experience in systems management and the fundamentals of organization, maintenance data collection, supply procedures, and standards and inspection.
Credit Recommendation: In the upper-division baccalaureate category, 3 semester hours in aviation maintenance management (2/74).

AF-1704-0094

FLIGHT ENGINEER SPECIALIST BY CORRESPONDENCE

Course Number: 11350C.
Location: Extension Course Institute, Gunter AFS, AL.
Length: Maximum, 52 weeks.
Exhibit Dates: 1/90–8/96.
Objectives: To teach individuals to perform as flight engineers without direct supervision.
Instruction: The course consists of a combination of self-instruction supplemented by close supervision. Introduction to flight management procedures and aircraft record keeping are presented. The course also covers ground handling and service of aircraft before and after flight. There is an in-depth study of factors which influence aircraft cargo distribution for safe loading and calculations of aircraft performance from take-off to landing. The course includes climb, cruise, and descent performance. Other topics involve the understanding of electrical, hydraulic, instrument, fuel, and related aircraft systems; power plants, auxiliary power units, bleed air, air conditioning, and pressurization systems; adverse weather; fire emergency; and oxygen, communications, and navigation systems. Systems study leads to qualification to operate systems in normal and emergency situations.
Credit Recommendation: In the upper-division baccalaureate category, 2 semester hours in flight engineer (3/85).

AF-1704-0102

AIRCRAFT MAINTENANCE SPECIALIST, TACTICAL AIRCRAFT BY CORRESPONDENCE

Course Number: 43151.
Location: Extension Course Institute, Gunter AFS, AL.
Length: Maximum, 52 weeks.
Exhibit Dates: 1/90–1/97.
Objectives: To train students to maintain tactical aircraft.
Instruction: This correspondence course consists of on-the job supervision and at-home study. The course provides general knowledge about maintenance management, maintenance and inspection concepts, and aircraft hardware and tools. The course provides information about aircraft cleaning and corrosion control; ground equipment; ground handling; egress systems; and maintenance tasks related to electrical, pneudralic, flight controls, landing gears, oxygen, and utility systems. The course provides general information about turbojet, and reciprocating engines.
Credit Recommendation: In the lower-division baccalaureate/associate degree category, 1 semester hour in aircraft cleaning and corrosion control and 2 in aircraft ground servicing and handling (3/85).

AF-1704-0137

JET ENGINE MECHANIC BY CORRESPONDENCE

Course Number: 42652.
Location: Extension Course Institute, Gunter AFS, AL.
Length: Maximum, 52 weeks.
Exhibit Dates: 1/90–Present.
Learning Outcomes: Upon completion of the course, the student will be able to perform effective maintenance in the area of jet propulsion, including management of jet mechanics, maintenance forms and records, jet theory and maintenance information, practices and procedures, jet engine systems, and operation and adjustment of jet engines and small gas turbine engines.
Instruction: Course covers maintenance forms and records; jet engine operation and construction; tools and hardware; repair and maintenance of the jet engine; corrosion control; support equipment; fuel system; oil system; starter; ignition and electrical system; anti-icing; thrust reversers; constant speed drive; test equipment; jet operation and adjustment; engine removal; troubleshooting; and small turbine engine disassembly, inspection, reassembly, and testing.
Credit Recommendation: In the lower-division baccalaureate/associate degree category, 4 semester hours in gas turbine theory and maintenance (10/88).

AF-1704-0181

STRATEGIC AIRCRAFT MAINTENANCE SPECIALIST BY CORRESPONDENCE
(Aircraft Maintenance Specialist, Airlift and Bombardment Aircraft by Correspondence)

Course Number: 43152C.
Location: Extension Course Institute, Gunter AFS, AL.
Length: Maximum, 52 weeks.
Exhibit Dates: 1/90–1/97.
Objectives: To train the student in maintenance of strategic aircraft.
Instruction: This correspondence course consists of on-the-job supervision and at-home study. The course provides general knowledge of aircraft maintenance and job-related information about airframe systems, ground handling, and servicing; fuel; turboprop and electrical systems; hydraulics; landing gear; flight controls; and utility systems.
Credit Recommendation: In the lower-division baccalaureate/associate degree category, 2 semester hours in aircraft ground servicing and handling (3/85).

AF-1704-0184

AIRLIFT AIRCRAFT MAINTENANCE SPECIALIST BY CORRESPONDENCE
(Aircraft Maintenance Specialist and Bombardment Aircraft by Correspondence)

Course Number: 43152G.
Location: Extension Course Institute, Gunter AFS, AL.
Length: Maximum, 52 weeks.
Exhibit Dates: 1/90–1/97.
Objectives: To train the student in maintenance of airlift and bombardment aircraft.
Instruction: This correspondence course consists of at-home study and on-the-job supervision on the C-141 and C-5A jet aircraft in the area of technical and material deficiency reporting. Inspection concepts, tools and material supply, ground handling equipment, airframe systems, fuel systems, power plant and auxiliary power unit, electrical system, hydraulic and cargo door system, landing gear, flight controls, and utility systems are all covered.
Credit Recommendation: In the lower-division baccalaureate/associate degree category, 2 semester hours in aircraft ground servicing and handling (3/85).

AF-1704-0187

AEROSPACE GROUND EQUIPMENT MECHANIC BY CORRESPONDENCE

Course Number: 42153.
Location: Extension Course Institute, Gunter AFS, AL.
Length: Maximum, 52 weeks.
Exhibit Dates: 1/90–10/91.
Objectives: To advance the technical skills of ground equipment mechanics in aerospace ground equipment.
Instruction: This correspondence course and practical exercises focus on operation and maintenance of aerospace ground equipment. Topics include maintenance policies, practices, and procedures; electrical circuits and components; engines (gas and turbine); generators; air compressors; heating; air conditioning; bomb lifts; diesels; hydraulics; and test equipment.
Credit Recommendation: In the lower-division baccalaureate/associate degree category, 3 semester hours in ground equipment repair (3/85).

AF-1704-0188

STRATEGIC AIRCRAFT MAINTENANCE SPECIALIST BY CORRESPONDENCE

Course Number: 43152E.
Location: Extension Course Institute, Gunter AFS, AL.
Length: Maximum, 52 weeks.
Exhibit Dates: 1/90–1/97.
Objectives: To provide education in aircraft maintenance.
Instruction: This correspondence course consists of on-the-job supervised application and at-home study in the areas of maintenance inspection concepts, tools, maintenance material, supply, ground handling requirements, airframe systems, pneudralic systems, landing gears, flight control, and ground support equipment of the KC 135 jet aircraft.
Credit Recommendation: In the lower-division baccalaureate/associate degree category, 2 semester hours in aircraft ground servicing and handling (3/85).

AF-1704-0191

AIRCRAFT FUEL SYSTEMS MECHANIC BY CORRESPONDENCE

Course Number: 42353.
Location: Extension Course Institute, Gunter AFS, AL.
Length: Maximum, 52 weeks.
Exhibit Dates: 1/90–1/94.
Objectives: To train students in aircraft fuel system maintenance.
Instruction: This correspondence course of instruction consists of on-the-job supervision and at-home study in the areas of maintenance management; shop and flight line safety; hand tools and hardware; operational checks and troubleshooting; fuel systems, fuel system safety, tools, and equipment; protective clothing; and leak detection and repair of integral tanks and bladder tanks/cells.
Credit Recommendation: In the lower-division baccalaureate/associate degree category, 1 semester hour in aircraft fuel systems (3/85).

AF-1704-0192

AEROSPACE GROUND EQUIPMENT TECHNICIAN BY CORRESPONDENCE

Course Number: 42173.
Location: Extension Course Institute, Gunter AFS, AL.
Length: Maximum, 52 weeks.
Exhibit Dates: 1/90–10/91.
Objectives: To advance the skills of the ground equipment technician in the area of aerospace ground equipment.
Instruction: This is a correspondence course with practical exercises in the operation and maintenance of aerospace ground equipment. Topics include maintenance policies, practices, and procedures; electrical and electronic circuits and components; and aircraft servicing equipment, including generators, air compressors, air conditioning, heating, ventilation, and hydraulic equipment.
Credit Recommendation: In the lower-division baccalaureate/associate degree category, 3 semester hours in ground equipment repair (3/85).

AF-1704-0193

FIGHTER WEAPONS INSTRUCTOR

Course Number: A7000FW.
Location: Air National Guard, A-7 Fighter Weapons School, Tucson, AZ.
Length: 16 weeks (292 hours).
Exhibit Dates: 1/90–Present.
Learning Outcomes: Upon completion of the course, the student will be able to train instructors in all phases of the A-7 fighter weapons system. The graduate will also be qualified to function as the weapons officer in an operational unit and will possess communication skills and effective instructional techniques in both the academic and flight areas of the A-7 weapons system.
Instruction: This course is restricted to pilots who have an average of ten years of military flight experience. They have been selected by their commanding officers as individuals who possess the capacity to assimilate advanced training and to function as the weapons expert in their unit. The curriculum consists of 232.5 hours of academic work supported by 53.4 hours of related flight activity (each sortie requires 6 hours of additional time for such activity as planning, briefing, and critique) and 6 hours of device (simulator) training. Classes are very small (approximately 6 students). Adequate instructional equipment in the form of training devices as well as operational aircraft are available. Experienced academically qualified faculty on long term assignment are involved in the presentation of the educational program. Emphasis is placed on avionics and systems that support the weapons delivery, the procedures for initiating equipment and tactical changes, the potential engine threat capability, and the tactics available for effective weapons employment.
Credit Recommendation: In the graduate degree category, 3 semester hours in applied aerodynamics (high performance aircraft), 3 in advanced aircraft systems and components, 3 in

computer applications, and 3 in instructional methodology (3/88).

AF-1704-0194

CIVIL AIR PATROL SCANNER BY CORRESPONDENCE

Course Number: 01230A.
Location: Extension Course Institute, Gunter AFS, AL.
Length: Maximum, 52 weeks.
Exhibit Dates: 1/90–Present.
Learning Outcomes: Upon completion of the course, the student will be able to systematically scan a specified area for prolonged periods of time while maintaining concentration and accurately identifying specific objects.
Instruction: This is a supervised, self-paced extension course covering scanning techniques, aircraft identification, and aircraft familiarization.
Credit Recommendation: Credit is not recommended due to the limited and military-specific nature of the content (10/88).

AF-1704-0195

CIVIL AIR PATROL MISSION OBSERVER LEVEL II BY CORRESPONDENCE

Course Number: 02130B.
Location: Extension Course Institute, Gunter AFS, AL.
Length: Maximum, 52 weeks.
Exhibit Dates: 1/90–Present.
Learning Outcomes: Upon completion of the course, the student will be able to assume the duties of a mission observer coordinating rescue operations under the leadership of the pilot.
Instruction: This is a supervised, self-paced extension course covering search planning, electronic search, and navigation assistance; navigation elements; the grid systems for locating position; navigational aids, primarily automatic direction-finding devices and very high frequency omnirange devices; communication techniques; nonverbal signals; weather conditions; and aircraft familiarity.
Credit Recommendation: In the lower-division baccalaureate/associate degree category, 1 semester hour in visual navigation (10/88).

AF-1704-0196

INTRODUCTION TO CIVIL AIR PATROL EMERGENCY SERVICES BY CORRESPONDENCE

Course Number: 02130D.
Location: Extension Course Institute, Gunter AFS, AL.
Length: Maximum, 52 weeks.
Exhibit Dates: 1/90–Present.
Learning Outcomes: Upon completion of the course, the student will be able to initiate, operate, and terminate an emergency service mission.
Instruction: This is a supervised, self-paced extension course covering the operating concepts, policies, and procedures which govern supervisory, ground, and flight personnel when performing civil air patrol emergency services.
Credit Recommendation: Credit is not recommended because of the limited, specialized nature of the course (10/88).

AF-1704-0197

FLIGHT ENGINEER SPECIALIST (HELICOPTER QUALIFIED) BY CORRESPONDENCE

Course Number: 11350B.
Location: Extension Course Institute, Gunter AFS, AL.
Length: Maximum, 52 weeks.
Exhibit Dates: 1/90–5/99.
Learning Outcomes: Upon completion of the course, the student will be able to perform, with little or no supervision and with the aid of technical publications, helicopter maintenance procedures. The course includes use of hand tools. The planning and organization of tasks will be directed by a supervisor who coordinates and monitors efforts of the specialist.
Instruction: The student works under individual supervision and a trainer or training manager in areas of maintenance safety, maintenance records and publications, care and use of hand tools, aircraft maintenance troubleshooting, and aircraft engines and systems. Emphasis is placed on landing gear, flight controls, transmissions, hydraulics, and fuel and electrical systems. Planning considerations are given for computation of weight and balance and performance factors (e.g., takeoff and landing computations).
Credit Recommendation: In the lower-division baccalaureate/associate degree category, 3 semester hours in flight engineering principles (9/93).

AF-1704-0198

AIRBORNE COMMUNICATIONS SYSTEMS OPERATOR BY CORRESPONDENCE

Course Number: 11650.
Location: Extension Course Institute, Gunter AFS, AL.
Length: Maximum, 52 weeks.
Exhibit Dates: 1/90–1/99.
Learning Outcomes: Upon completion of the course, the student will be able to provide an accurate, efficient exchange of information between headquarters and field units or between aircraft and their home stations.
Instruction: This is a supervised, self-paced extension course covering communication agencies, command control and communications, various airborne communication systems, security, radio wave propagation and collection, voice operations and procedures, aviation weather, and record communication operation and procedures. Other topics covered are equipment/system, inflight and flight line safety, and inspection.
Credit Recommendation: In the vocational certificate category, 2 semester hours in basic public communications (10/88).

AF-1704-0199

AIRCREW LIFE SUPPORT SPECIALIST BY CORRESPONDENCE

Course Number: 12250.
Location: Extension Course Institute, Gunter AFS, AL.
Length: Maximum, 52 weeks.
Exhibit Dates: 1/90–3/97.
Learning Outcomes: Upon completion of the course, the student will be able to inspect and maintain aircrew life support equipment; issue, fit, and adjust aircrew life support and chemical defense equipment; provide life support continuation training to aircrew; and supervise aircrew life support personnel.
Instruction: The student works under individual supervision and a trainer or training manager in the areas of industrial shop safety; aircrew survival emergency procedures and techniques; inspection, maintenance, and use of aircrew life support and survival equipment; fabrics and composite equipment maintenance and testing; and physiology of flight.
Credit Recommendation: In the lower-division baccalaureate/associate degree category, 1 semester hour in industrial shop management and 1 in speech/teaching techniques (10/88).

AF-1704-0200

AIRCRAFT ENVIRONMENTAL SYSTEMS MECHANIC BY CORRESPONDENCE

Course Number: 42351.
Location: Extension Course Institute, Gunter AFS, AL.
Length: Maximum, 52 weeks.
Exhibit Dates: 1/90–Present.
Learning Outcomes: Upon completion of the course, the student will be able to apply specialty knowledge required to perform as aircraft environmental systems mechanics. The student will be capable in field fundamentals, maintenance fundamentals, aircraft familiarization, environmental systems, and aircraft oxygen and utility systems.
Instruction: This is a supervised, self-paced extension course covering maintenance management, maintenance system, publications, forms and records, safety, maintenance of tools and safety devices; general maintenance; elements of physics, AC and DC motor circuits; electrical measuring devices; aircraft familiarization; engine bleed air and aircraft ducting; aircraft air conditioning system and air conditioning pressurization; gaseous oxygen system, liquid oxygen system and servicing, and storage units; and utility system.
Credit Recommendation: In the lower-division baccalaureate/associate degree category, 3 semester hours in basic electricity and 3 in aircraft systems (10/88).

AF-1704-0201

AIRCRAFT PNEUDRAULIC SYSTEMS MECHANIC BY CORRESPONDENCE

Course Number: 42354.
Location: Extension Course Institute, Gunter AFS, AL.
Length: Maximum, 52 weeks.
Exhibit Dates: 1/90–Present.
Learning Outcomes: Upon completion of the course, the student will be able to function as a maintenance management technician and demonstrate knowledge of pneumatics fundamentals, material and equipment, and components and systems.
Instruction: Instruction is self-paced, extension study in maintenance management, supply, and records; fundamentals of pneudraulics and electricity; pneudraulic system materials, tools, and test instruments; ground and shop equipment; hydraulic power supply components; selector valves, actuators, and flow valves; landing gear components; pneudraulic power systems; landing gears and related systems; nose wheel steering and brake system; aircraft flight control; and pneumatic systems.
Credit Recommendation: In the lower-division baccalaureate/associate degree category, 1 semester hour in maintenance management, 4 in hydraulics and pneumatics, and 3 in aircraft air conditioning material and processes (10/88).

AF-1704-0202

AIRCRAFT FUEL SYSTEMS TECHNICIAN BY CORRESPONDENCE

Course Number: 42373.

Location: Extension Course Institute, Gunter AFS, AL.

Length: Maximum, 52 weeks.

Exhibit Dates: 1/90–6/97.

Learning Outcomes: Upon completion of the course, the student will be able to perform duties in the personnel management of specialized fuel systems and perform supervisory functions for fuel systems technicians.

Instruction: Instruction is presented through a self-paced, extension course in personnel management and supervisory functions, safety and performance evaluation, maintenance data publications, maintenance supply and inspection, refueling and defueling, vent pressurization, troubleshooting, corrosion control, and basic electricity.

Credit Recommendation: In the lower-division baccalaureate/associate degree category, 2 semester hours in maintenance management; 1 in records, forms, and publications; and 3 in basic electricity (10/88).

AF-1704-0204

TURBOPROP PROPULSION MECHANIC BY CORRESPONDENCE

Course Number: 42653.

Location: Extension Course Institute, Gunter AFS, AL.

Length: Maximum, 52 weeks.

Exhibit Dates: 1/90–11/91.

Learning Outcomes: Upon completion of the course, the student will be able to perform effective maintenance in turboprop propulsion, including general turboprop engine operation, turboprop equipment and engine system, turboprop engine maintenance, turboprop operation, turbopropeller maintenance, and helicopter and OV-10 propulsion systems.

Instruction: Topics include turboprop engine principles; engine reduction gear box and compressor section construction; turboprop engine combustion and turbine section construction; small gas turbine engines, inspection, maintenance forms and records, tools, and equipment; ground equipment, fuels and fuel systems, and oils and oil systems; engine instruments and indicating systems; general engine maintenance, assembly, disassembly, removal, troubleshooting of engine systems, and engine testing, turbopropeller construction and operation; electrical principles; maintenance of turbopropeller and propeller control systems; and turboshaft engine construction on various types of turboshaft engines including turboprop propulsion system.

Credit Recommendation: In the lower-division baccalaureate/associate degree category, 4 semester hours in turbopropeller and 4 in gas turbine engines and operation (10/88).

AF-1704-0205

F-100 JET ENGINE MECHANIC BY CORRESPONDENCE

Course Number: 42654.

Location: Extension Course Institute, Gunter AFS, AL.

Length: Maximum, 52 weeks.

Exhibit Dates: 1/90–Present.

Learning Outcomes: Upon completion of the course, the student will be able to perform effective maintenance in the area of jet propulsion, including management, jet engine mechanic operating principles, maintenance practices and procedures, associated jet engine systems, and operation and adjustment of jet engine.

Instruction: This is a supervised, self-paced extension study course in responsibilities and supply authorization; publications; jet operation, engine construction, tools, and hardware; maintenance and inspection systems and forms; corrosion control; support equipment; engine and start system troubleshooting and test equipment; engine removal and installation; engine operating; and testing of the jet fuel starter and gearbox.

Credit Recommendation: In the lower-division baccalaureate/associate degree category, 2 semester hours in gas turbine theory and operation (10/88).

AF-1704-0206

HELICOPTER MECHANIC BY CORRESPONDENCE

Course Number: *Version 1:* 43150. *Version 2:* 43150.

Location: Extension Course Institute, Gunter AFS, AL.

Length: *Version 1:* Maximum, 52 weeks. *Version 2:* Maximum, 52 weeks.

Exhibit Dates: *Version 1:* 11/92–2/97. *Version 2:* 1/90–10/92.

Learning Outcomes: *Version 1:* Upon completion of the course, the student will be able to provide technical assistance and supervise subordinates in ground handling, inspection, repair, and maintenance of the H-1, H-3, and H-53 helicopters including power plants and related equipment. *Version 2:* Upon completion of the course, the student will be able to provide technical assistance and supervise subordinates in ground handling, inspection, repair, and maintenance of the H-1, H-3, and H-53 helicopters including power plants and related equipment.

Instruction: *Version 1:* This is a supervised, self-paced extension course covering topics of job-related information on Air Force publications, technical order deficiency reporting, maintenance concepts and forms, inventory records, special tools, airframe construction, helicopter markings, helicopter aerodynamics, and hydraulics and utility systems. Rotor systems, electrical power and distribution, fuel systems, ground handling, corrosion control, and cleaning and specific information on inspection, repair, alteration, and maintenance of the H-3 and H-53 helicopters, systems, engine operation, and construction features are also included. *Version 2:* This is a supervised, self-paced extension course covering topics of job-related information on Air Force publications, technical order deficiency reporting, maintenance concepts, maintenance forms, inventory records, special tools, airframe construction, helicopter markings, helicopter aerodynamics, ground handling, corrosion control, and cleaning and specific information on testing, repair, alteration, and maintenance of the H-3 and H-53 helicopters, systems, engine operation, and construction features.

Credit Recommendation: *Version 1:* In the lower-division baccalaureate/associate degree category, 3 semester hours in maintenance administration, 3 in basic electricity, 3 in helicopter power plants, and 3 in general helicopter mechanics (4/93). *Version 2:* In the lower-division baccalaureate/associate degree category, 3 semester hours in general helicopter mechanics, 3 in advanced helicopter systems, and 3 in helicopter power plants (10/88).

AF-1704-0209

STRATEGIC AIRCRAFT MAINTENANCE SPECIALIST BY CORRESPONDENCE

Course Number: 43152A.

Location: Extension Course Institute, Gunter AFS, AL.

Length: Maximum, 52 weeks.

Exhibit Dates: 1/90–Present.

Learning Outcomes: Upon completion of the course, the student will be able to describe the principles of operation of major aircraft systems (e.g. pneumatic, hydraulic, fuel, and electrical) and jet engines. The student will also have the foundation to plan and organize maintenance efforts with the use of technical publications, basic troubleshooting techniques, and supervisory techniques.

Instruction: The student works under individual supervision and a trainer or training manager in areas of technical publications, maintenance and inspection concepts, basic tools and maintenance materials, aircraft maintenance supply system, and ground support equipment. Major airframe systems and aircraft and ground handling are also covered. The course includes, in greater detail, the principles of operation and servicing of the pneumatic, hydraulic, landing gear, flight control, fuel, and electrical systems.

Credit Recommendation: In the lower-division baccalaureate/associate degree category, 1 semester hour in aircraft maintenance management and 1 in aircraft systems (10/88).

AF-1704-0210

STRATEGIC AIRCRAFT MAINTENANCE SPECIALIST BY CORRESPONDENCE

Course Number: 43152J.

Location: Extension Course Institute, Gunter AFS, AL.

Length: Maximum, 52 weeks.

Exhibit Dates: 1/90–Present.

Learning Outcomes: Upon completion of the course, the student will to able to describe the principles of major aircraft systems (e.g. pneumatic, hydraulic, fuel, and electrical) and jet engines. The student will also have the foundation to plan and organize maintenance efforts with the use of technical publications, basic troubleshooting techniques, and supervisory techniques.

Instruction: The student works under individual supervision and a trainer or training manager in the areas of technical publications, maintenance and inspection concepts, basic tools, and maintenance materials. Aircraft maintenance supply systems and ground support equipment are covered as are major airframe systems and aircraft and ground handling. The principles of operation and service of the pneumatic, hydraulic, landing gear, flight control, fuel, and electrical systems are also covered.

Credit Recommendation: In the lower-division baccalaureate/associate degree category, 1 semester hour in aircraft maintenance management and 1 in aircraft systems (10/88).

Air Force

AF-1704-0211

STRATEGIC AIRCRAFT MAINTENANCE SPECIALIST BY CORRESPONDENCE

Course Number: 43152Z.
Location: Extension Course Institute, Gunter AFS, AL.
Length: Maximum, 52 weeks.
Exhibit Dates: 1/90–Present.
Learning Outcomes: Upon completion of the course, the student will be able to describe principles of operation of The student will also have the foundation to plan and organize maintenance efforts with the use of technical publications, basic troubleshooting techniques, and supervisory techniques.
Instruction: The student works under individual supervision and a trainer or training manager in areas of technical publications, maintenance and inspection concepts, basic tools and maintenance materials, aircraft maintenance supply system, ground support equipment, major airframe systems, and aircraft and ground handling. In greater detail, the course covers principles of operation and service of pneumatic, hydraulic, landing gear, flight control, fuel, and electrical systems.
Credit Recommendation: In the lower-division baccalaureate/associate degree category, 1 semester hour in aircraft maintenance management and 1 in aircraft systems (10/88).

AF-1704-0214

APPRENTICE AIRFIELD MANAGEMENT SPECIALIST BY CORRESPONDENCE

Course Number: 27131.
Location: Extension Course Institute, Gunter AFS, AL.
Length: Maximum, 52 weeks.
Exhibit Dates: 1/90–Present.
Learning Outcomes: Upon completion of the course, the student will have a basic level of understanding of duties related to an entry-level position in the National Airspace System. The individual will be able to log information on flights (departing/arriving) and maintain flight information and planning data with little direct supervision.
Instruction: Course is self-paced, independent study. Intermediate open book exams are utilized for individualized progress checks. A comprehensive final exam is administered under a controlled testing environment and must be satisfactorily completed. Material is studied in conjunction with supervised on-the-job training. Course includes introduction of material pertaining to basic air traffic control and flight service station duties, as well as specific and general tasks related to the individual's entry-level assignment.
Credit Recommendation: In the vocational certificate category, 2 semester hours in air traffic control (terminology and procedures) (4/93).

AF-1704-0215

AIRFIELD MANAGEMENT SPECIALIST BY CORRESPONDENCE

Course Number: 27151.
Location: Extension Course Institute, Gunter AFS, AL.
Length: Maximum, 52 weeks.
Exhibit Dates: 1/90–1/97.
Learning Outcomes: Upon completion of the course, the student will be able to assist flight crew members with preparation of documents required for flight operations. An understanding of aeronautical charts and publications is needed to monitor and document aircraft movement (departing /arrival). A basic understanding of supervisory practices is achieved to assist with the on-the-job training of subordinates.
Instruction: Course covers an introduction to duties related to aircraft dispatching. Individual tasks and subject areas pertain to basic aviation weather and airport operations.
Credit Recommendation: In the lower-division baccalaureate/associate degree category, 1 semester hour in introduction to aircraft dispatching (4/93).

AF-1704-0216

COMBAT CONTROL OPERATOR BY CORRESPONDENCE

Course Number: 27350.
Location: Extension Course Institute, Gunter AFS, AL.
Length: Maximum, 52 weeks.
Exhibit Dates: 1/90–10/96.
Learning Outcomes: Upon completion of the course, the student who as a prerequisite must have completed basic air traffic control training prior to entering this course, will be qualified for upgrading to a skill level five (skilled) controller. In addition, student will have competency in the tactical aspects of the material covered in the Combat Control School Program.
Instruction: Instruction is accomplished through correspondence course methods. The course includes topics related to FAA, ICAO, and military aircraft navigation and control. Included are such areas as basic flight regulations, weather, navigation, airway/route, and terminal nonradar approaches. A portion of the program is devoted to communication techniques and equipment common to ATC utilization. Approximately one-half of the study hours cover tactical equipment specific to the tasks assigned to this military specialty.
Credit Recommendation: In the lower-division baccalaureate/associate degree category, 2 semester hours in air traffic control (4/93).

AF-1704-0217

COMMAND AND CONTROL SPECIALIST BY CORRESPONDENCE

Course Number: 27450.
Location: Extension Course Institute, Gunter AFS, AL.
Length: Maximum, 52 weeks.
Exhibit Dates: 1/90–3/96.
Learning Outcomes: Upon completion of the course, the student will be able to function as a command post controller. Completion of this course culminates the resident air traffic controller training required by the airman in the control center who is then qualified to assume the responsibilities of the controller.
Instruction: The course consists of self-paced, independent study with open-book progress checks and satisfactory completion of a proctored comprehensive final examination. This consists of a review of related technician publications, filing and record keeping, cryptographic and physical security methods, flight operations, basic weather, and computer applications.
Credit Recommendation: In the lower-division baccalaureate/associate degree category, 1 semester hour in air traffic control (4/93).

AF-1704-0218

INTEGRATED AVIONICS ATTACK CONTROL SYSTEMS SPECIALIST (F-15) BY CORRESPONDENCE

Course Number: 32656B.
Location: Extension Course Institute, Gunter AFS, AL.
Length: Maximum, 52 weeks.
Exhibit Dates: 1/90–5/92.
Learning Outcomes: Upon completion of the course, the student will be able to describe the fundamentals and operation of avionics combat control and navigation systems.
Instruction: The student receives instruction concerning aircraft familiarization, basic navigation, inertial navigation systems, and radar principles and operation.
Credit Recommendation: Credit is not recommended because of the military-specific nature of the course (12/89).

AF-1704-0220

INTEGRATED AVIONICS COMMUNICATION, NAVIGATION, AND PENETRATION AIDS SYSTEMS SPECIALIST (F-16) BY CORRESPONDENCE

Course Number: 32658C.
Location: Extension Course Institute, Gunter AFS, AL.
Length: Maximum, 52 weeks.
Exhibit Dates: 1/90–3/92.
Learning Outcomes: Upon completion of the course, the student will be able to describe the fundamentals and operation of basic electronic and avionic communication and navigation systems.
Instruction: The student receives instruction in aircraft familiarization, aircraft communication systems, navigation systems, and electronic identification systems.
Credit Recommendation: Credit is not recommended because of the military-specific nature of the course (12/89).

AF-1704-0221

AIRCRAFT ELECTRO-ENVIRONMENTAL SYSTEM TECHNICIAN BY CORRESPONDENCE

Course Number: 45070.
Location: Extension Course Institute, Gunter AFS, AL.
Length: Maximum, 52 weeks.
Exhibit Dates: 1/90–11/95.
Learning Outcomes: Upon completion of the course, the student will be able to conduct maintenance on aircraft using correct ground safety practices, maintenance publications, and hand tools and will be able to service and maintain the battery and electrical systems using the correct test equipment.
Instruction: Through correspondence, the student receives information on aircraft ground safety, maintenance publications, inspection fundamentals, hand tools and soldering, basic electricity and batteries, and the use of the volt-ohm meter.
Credit Recommendation: In the lower-division baccalaureate/associate degree category, 1 semester hour in in aircraft maintenance and 1 in aircraft maintenance management (12/89).

AF-1704-0222

AIRCRAFT ELECTRO-ENVIRONMENTAL SYSTEMS TECHNICIAN BY CORRESPONDENCE

Course Number: 45070A.
Location: Extension Course Institute, Gunter AFS, AL.
Length: Maximum, 52 weeks.
Exhibit Dates: 1/90–4/97.
Learning Outcomes: Upon completion of the course, the student will be able to identify the basic operation of the AC and DC power systems found in the aircraft and the basic layout of the trim system, flap system, fuel system, power plant electrical system, warning systems, and steering system.
Instruction: Through correspondence, the student receives basic knowledge of the operation of the aircraft electrical generation system, fuel system, utility system, warning systems, and steering system.
Credit Recommendation: Credit is not recommended because of the limited, specialized nature of the course (12/89).

AF-1704-0223

AIRCRAFT ELECTRO-ENVIRONMENTAL SYSTEMS TECHNICIAN BY CORRESPONDENCE

Course Number: 45070B.
Location: Extension Course Institute, Gunter AFS, AL.
Length: Maximum, 52 weeks.
Exhibit Dates: 1/90–4/97.
Learning Outcomes: Upon completion of the course, the student will be able to troubleshoot and correct faults in the engine bleed air system, aircraft air conditioning systems, cabin pressurization systems, fire detection and extinguishing systems, oxygen systems, and utility systems (life raft, liquid cycle refrigeration) and will be able to supervise others in the maintenance and repair of the above systems.
Instruction: Through correspondence, the student will receive information on engine bleed air systems, aircraft air conditioning, cabin pressurization and oxygen systems, fire protection systems, and utility systems (life raft, liquid cycle refrigeration).
Credit Recommendation: In the lower-division baccalaureate/associate degree category, 2 semester hours in aircraft maintenance (12/89).

AF-1704-0224

F/FB-111 AVIONICS MANUAL/ELECTRONIC COUNTERMEASURES (ECM) TEST STATION AND COMPONENT SPECIALIST BY CORRESPONDENCE

Course Number: 45156B.
Location: Extension Course Institute, Gunter AFS, AL.
Length: Maximum, 52 weeks.
Exhibit Dates: 1/90–8/96.
Learning Outcomes: Upon completion of the course, the student will be able to describe fundamentals of electronics and the operation of electronic countermeasure test equipment and components.
Instruction: The student receives instruction in electronic principles, microcomputer operation, maintenance management, inspection procedures, and electronic testing equipment and procedures.
Credit Recommendation: Credit is not recommended because of the limited, specialized nature of the course (12/89).

AF-1704-0225

INTEGRATED AVIONICS COMMUNICATIONS, NAVIGATION, AND PENETRATION AIDS SYSTEMS SPECIALIST (F-15) BY CORRESPONDENCE

Course Number: 45251C.
Location: Extension Course Institute, Gunter AFS, AL.
Length: Maximum, 52 weeks.
Exhibit Dates: 1/90–6/97.
Learning Outcomes: Upon completion of the course, the student will be able to describe the operating principles of avionics communications and navigation systems and the electronic principles involved.
Instruction: The student receives instruction in basic electronic principles, communications, navigation, and electronic identification principles.
Credit Recommendation: Credit is not recommended because of the military-specific nature of the course (12/89).

AF-1704-0226

INTEGRATED AVIONIC INSTRUMENT AND FLIGHT CONTROL SYSTEMS SPECIALIST BY CORRESPONDENCE

Course Number: 45252B.
Location: Extension Course Institute, Gunter AFS, AL.
Length: Maximum, 52 weeks.
Exhibit Dates: 1/90–9/93.
Learning Outcomes: Upon completion of the course, the student will be able to perform maintenance on integrated avionic instrument and flight control systems incorporated in the F-16 aircraft.
Instruction: The course consists of self-paced, independent study with open-book progress checks and satisfactory completion of a proctored comprehensive final examination. Content areas include a treatment of the integrated aerodynamic control system for the F-16 aircraft. Individual systems are reviewed including electrical and hydraulic power, air conditioning and pressurization, engine instrumentation, fuel quantity, navigation aids, the flight environment system and inflight recorder, and system diagrams. Checkout techniques and troubleshooting are covered. Safety measures relative to fuels, engine, radiation, and egress systems, as well as work pertinent to these and their attendant maintenance hazards are stressed.
Credit Recommendation: In the lower-division baccalaureate/associate degree category, 2 semester hours in aircraft instrument and flight control maintenance (12/89).

AF-1704-0227

INTEGRATED AVIONIC INSTRUMENT AND FLIGHT CONTROL SYSTEMS SPECIALIST BY CORRESPONDENCE

Course Number: 45253B.
Location: Extension Course Institute, Gunter AFS, AL.
Length: Maximum, 52 weeks.
Exhibit Dates: 1/90–1/96.
Learning Outcomes: Upon completion of the course, the student will be able to perform maintenance and troubleshooting on the engine, airframe, instruments, and their related systems. In addition, the student will have the ability to perform maintenance on the integrated flight control system and its components.
Instruction: The course consists of self-paced, independent study with open-book progress checks and satisfactory completion of a proctored comprehensive final examination. Pressure, temperature, and quantity instruments relative to the F-111 aircraft are presented as are their systems, including navigation instruments, sensors, and supporting testing devices. Airborne signal data recording and engine air induction as applied to this aircraft are analyzed. The integrated flight control system and its related components including the autopilot are covered. Both troubleshooting and maintenance aspects are emphasized.
Credit Recommendation: In the lower-division baccalaureate/associate degree category, 2 semester hours in aircraft instrument and flight control maintenance (12/89).

AF-1704-0228

TACTICAL AIRCRAFT MAINTENANCE TECHNICIAN BY CORRESPONDENCE

Course Number: 45274.
Location: Extension Course Institute, Gunter AFS, AL.
Length: Maximum, 52 weeks.
Exhibit Dates: 1/90–3/97.
Learning Outcomes: Upon completion of the course, the student will be able to describe the basic construction and maintenance requirements for a tactical aircraft and will understand ground safety, aircraft corrosion control, battle damage repair, and ground handling of aircraft. The student will have a basic knowledge of aircraft pneudraulics and landing gear systems. Student will have a basic understanding of aircraft engines, fuel systems, electrical systems, and environmental control systems.
Instruction: Through correspondence, the student receives basic information on aircraft ground handling, aircraft electrical, pneudraulic, and landing gear systems, engine construction and theory of operation, engine electrical systems, fuel systems, and utility systems (environmental control, fire protection, and ice and rain removal).
Credit Recommendation: In the lower-division baccalaureate/associate degree category, 3 semester hours in aircraft maintenance (12/89).

AF-1704-0229

AEROSPACE PROPULSION TECHNICIAN (JET ENGINE) BY CORRESPONDENCE

Course Number: 45470A.
Location: Extension Course Institute, Gunter AFS, AL.
Length: Maximum, 52 weeks.
Exhibit Dates: 1/90–3/94.
Learning Outcomes: Upon the completion of the course, the student will be able to describe the construction of jet engines of conventional and modular design and be familiar with the tools and hardware required for engine overhaul. Student gains the knowledge necessary for engine evaluation during overhaul in the areas of inspection, measurement, and corrosion control. Student will be able to describe the engine fuel system, oil system, start and electrical systems, bleed air system, afterburner,

thrust reverse, and constant speed drive. Student will be able to perform engine test runs using calibration instruments to adjust the engine and will be able to remove and install the engine in the airframe. Student gains knowledge on engine noise suppression, is able to troubleshoot the engine before overhaul, and is familiar with the construction and maintenance of the auxiliary power unit.

Instruction: Through correspondence, the student receives information on engine operation, construction, tools and hardware, and support equipment for overhaul and repair. Student receives information on the fuel, oil, start and electrical, bleed air, afterburner, thrust reverse, constant speed drive, and relationships of these systems to one another. Also covered is information on engine testing, test instruments, adjustments to engine systems, noise suppression, and the removal and installation of the engine in the airframe. Student learns about engine troubleshooting and the auxiliary power unit.

Credit Recommendation: In the lower-division baccalaureate/associate degree category, 3 semester hours in jet engine maintenance (12/89).

AF-1704-0230

AIRCRAFT PNEUDRAULIC SYSTEMS TECHNICIAN BY CORRESPONDENCE

Course Number: 45474.
Location: Extension Course Institute, Gunter AFS, AL.
Length: Maximum, 52 weeks.
Exhibit Dates: 1/90–12/96.
Learning Outcomes: Upon completion of the course, the student will be able to describe the theory of hydraulic systems, electric motors and generators, and related components and will gain basic knowledge of the construction of fluid lines using hose and tubing. Student will be able to operate hydraulic test equipment and will learn the basic operation of hydraulic power systems, landing gear systems, flight control systems, pneumatic systems, inflight refueling systems, and cargo door and ramp operating systems.
Instruction: Through correspondence, the student receives information on pneudraulic system theory and maintenance, troubleshooting, repair, and testing. Course covers hydraulic power, landing gear, flight control, pneumatic, inflight refueling, and cargo door and ramp systems.
Credit Recommendation: In the lower-division baccalaureate/associate degree category, 1 semester hour in aircraft maintenance (12/89).

AF-1704-0231

AIRCRAFT PNEUDRAULIC SYSTEMS TECHNICIAN BY CORRESPONDENCE

Course Number: 45474A.
Location: Extension Course Institute, Gunter AFS, AL.
Length: Maximum, 52 weeks.
Exhibit Dates: 1/90–3/95.
Learning Outcomes: Upon completion of the course, the student will be able to perform maintenance on aircraft landing gear, flight control, cargo and ramp door systems; jack the aircraft for landing gear maintenance; and inspect aircraft wheels and tires for defects.

Instruction: Through correspondence, the student learns about aircraft landing gear, flight control, cargo and ramp door, jacking, and wheel and tire systems.
Credit Recommendation: Credit is not recommended because of the limited, specialized nature of the course (12/89).

AF-1704-0232

AVIONIC GUIDANCE AND CONTROL SYSTEMS TECHNICIAN BY CORRESPONDENCE

Course Number: 45571A.
Location: Extension Course Institute, Gunter AFS, AL.
Length: Maximum, 52 weeks.
Exhibit Dates: 1/90–5/99.
Learning Outcomes: Upon completion of the course, the student will be able to troubleshoot or supervise the troubleshooting of avionics guidance and control systems of Air Force aircraft, including compass, attitude heading reference, stability augmentation, and automatic flight control and digital automatic flight control systems and utilize the MC-1 compass calibrator set.
Instruction: Course includes system operation, inspections, operational checks, troubleshooting, and bench check of compass system, attitude heading reference system, stability augmentation system, and components; repair of avionics system malfunctions; and operation of the MC-1 autopilot system.
Credit Recommendation: Credit is not recommended because of the limited, specialized nature of the course (12/89).

AF-1704-0233

GUIDANCE AND CONTROL SYSTEMS TECHNICIAN BY CORRESPONDENCE

Course Number: 45571B.
Location: Extension Course Institute, Gunter AFS, AL.
Length: Maximum, 52 weeks.
Exhibit Dates: 1/90–5/99.
Learning Outcomes: Upon completion of the course, the student will be able to troubleshoot or supervise the troubleshooting of avionics guidance and control systems of Air Force aircraft compass, attitude heading reference, stability augmentation, and automatic flight control and digital automatic flight control systems and utilize the MC-1 compass calibrator set.
Instruction: Course includes system operation, inspections, operational checks, troubleshooting and bench check of compass system, attitude heading reference system, stability augmentation system, automatic flight control system, and components; repair of avionics system malfunctions; and operation of the MC-1 autopilot system.
Credit Recommendation: Credit is not recommended because of the limited, specialized nature of the course (12/89).

AF-1704-0234

AVIONICS GUIDANCE AND CONTROL SYSTEMS TECHNICIAN BY CORRESPONDENCE

Course Number: 45571C.
Location: Extension Course Institute, Gunter AFS, AL.
Length: Maximum, 52 weeks.
Exhibit Dates: 1/90–5/99.

Learning Outcomes: Upon completion of the course, the student will be able to troubleshoot or supervise the troubleshooting of avionics guidance and control systems of Air Force aircraft compass, attitude heading reference, stability augmentation, and automatic flight control systems and utilize the MC-1 compass calibrator.
Instruction: Course covers the basics of aircraft compass systems, including system operation, inspections, operational checks and bench check of compass, attitude heading reference, stability augmentation and automatic flight control system components and the MC-1 compass calibrator including component operation, setup, measuring compass errors, magnetic survey, and compass surveying procedures.
Credit Recommendation: Credit is not recommended because of the limited, specialized nature of the course (12/89).

AF-1704-0235

AVIONICS GUIDANCE AND CONTROL SYSTEMS TECHNICIAN BY CORRESPONDENCE

Course Number: 45571D.
Location: Extension Course Institute, Gunter AFS, AL.
Length: Maximum, 52 weeks.
Exhibit Dates: 1/90–5/99.
Learning Outcomes: Upon completion of the course, the student will display an updated understanding of avionics instrument systems including position, engine, fuel quantity indicating, flight data recording, central air data and flight director systems, vertical scale flight instrument and flight and navigational aids.
Instruction: Course includes a review of the theory and operation of the above cited systems in order to enhance the technician's ability to maintain the systems.
Credit Recommendation: Credit is not recommended because of the limited, specialized nature of the course (12/89).

AF-1704-0236

AVIONICS GUIDANCE AND CONTROL SYSTEMS TECHNICIAN BY CORRESPONDENCE

Course Number: 45517E.
Location: Extension Course Institute, Gunter AFS, AL.
Length: Maximum, 52 weeks.
Exhibit Dates: 1/90–5/99.
Learning Outcomes: Upon completion of the course, the student will display an updated understanding of avionics instrument systems, including position, engine, fuel quantity, flight loads data recording, basic flight, central air data, flight director systems, and flight and navigational aids.
Instruction: Course includes a review of the theory and operation of the above systems in order to enhance the technician's ability to maintain the system.
Credit Recommendation: Credit is not recommended because of the limited, specialized nature of the course (12/89).

AF-1704-0237

AVIONICS GUIDANCE AND CONTROL SYSTEMS TECHNICIAN BY CORRESPONDENCE

Course Number: 45571F.
Location: Extension Course Institute, Gunter AFS, AL.
Length: Maximum, 52 weeks.

Exhibit Dates: 1/90–5/99.

Learning Outcomes: Upon completion of the course, the student will better understand avionics instrument systems previously learned through training and experience.

Instruction: The course consists of self-paced, independent study with open-book progress checks and satisfactory completion of a proctored comprehensive final examination. Content includes a review of elementary aspects of solenoid, selsyn, and synchro system operation; engine indicating systems (tachometer, temperature, fuel flow, pressure); fuel quantity indicators (resistance, capacitance, maintenance); flight data records; pilot-static systems; altitude encoders; and Central SINS Data Computers including angle-of-attack and flight director systems.

Credit Recommendation: Credit is not recommended because of the limited, specialized nature of the course (12/89).

AF-1704-0238

AVIONICS GUIDANCE AND CONTROL SYSTEMS TECHNICIAN BY CORRESPONDENCE

Course Number: 45571G.
Location: Extension Course Institute, Gunter AFS, AL.
Length: Maximum, 52 weeks.
Exhibit Dates: 1/90–5/99.
Learning Outcomes: Upon completion of the course, the student will demonstrate a fundamental knowledge of concepts related to avionic navigation systems. This course is an extension to training and experience previously acquired.
Instruction: Course includes fundamentals of computer programming and hardware, navigation and computer elements, inertial operation principles, and fuel savings advisory systems.
Credit Recommendation: Credit is not recommended because of the limited, specialized nature of the course (12/89).

AF-1704-0239

AVIONICS GUIDANCE AND CONTROL SYSTEMS TECHNICIAN BY CORRESPONDENCE

Course Number: 45571H.
Location: Extension Course Institute, Gunter AFS, AL.
Length: Maximum, 52 weeks.
Exhibit Dates: 1/90–5/99.
Learning Outcomes: Upon completion of the course, the student will be able to describe the basic principles of navigation and computer elements and basic concepts of computer function and data storage. Student will learn the basic operation of inertial navigation and Doppler navigation systems and the role these systems play in the fuel savings advisory system.
Instruction: Through correspondence, the student gains knowledge of navigation, computer function and data storage, inertial and Doppler navigation systems, and fuel savings advisory systems.
Credit Recommendation: In the lower-division baccalaureate/associate degree category, 1 semester hour in avionics (12/89).

AF-1704-0240

AVIONICS GUIDANCE AND CONTROL SYSTEMS TECHNICIAN BY CORRESPONDENCE

Course Number: 45571J.
Location: Extension Course Institute, Gunter AFS, AL.
Length: Maximum, 52 weeks.
Exhibit Dates: 1/90–5/99.
Learning Outcomes: Upon completion of the course, the student will comprehend basic navigation, operate gyroscopes, solve navigation problems, and understand computer navigation operations.
Instruction: The student receives instructions concerning basic, trigonometric, and computer navigation and the operations of gyroscope and accelerometer systems.
Credit Recommendation: In the lower-division baccalaureate/associate degree category, 1 semester hour in basic navigation principles (12/89).

AF-1704-0241

AVIONICS GUIDANCE AND CONTROL SYSTEMS TECHNICIAN BY CORRESPONDENCE

Course Number: 45571K.
Location: Extension Course Institute, Gunter AFS, AL.
Length: Maximum, 52 weeks.
Exhibit Dates: 1/90–5/99.
Learning Outcomes: Upon completion of the course, the student will be able to describe weapon control and monitoring systems, television transmission, and heads-up displays.
Instruction: The student receives instruction on weapons control and monitoring through review, television transmission, and heads-up display systems in the A-10.
Credit Recommendation: Credit is not recommended because of the military-specific nature of the course (12/89).

AF-1704-0242

COMMUNICATION/NAVIGATION SYSTEMS TECHNICIAN BY CORRESPONDENCE

Course Number: 45572A.
Location: Extension Course Institute, Gunter AFS, AL.
Length: Maximum, 52 weeks.
Exhibit Dates: 1/90–Present.
Learning Outcomes: Upon completion of the course, the student will be able to repair, maintain, and supervise the repair and maintenance of aerial communications/navigation systems.
Instruction: Course is an introduction to avionics maintenance, repair procedures, safety practices, and administration of aviation electronics maintenance program; a basic grounding in electronics; and use of electronics test equipment.
Credit Recommendation: In the lower-division baccalaureate/associate degree category, 2 semester hours in avionics maintenance (12/89).

AF-1704-0243

COMMUNICATION/NAVIGATION SYSTEMS TECHNICIAN BY CORRESPONDENCE

Course Number: 45572B.
Location: Extension Course Institute, Gunter AFS, AL.
Length: Maximum, 52 weeks.
Exhibit Dates: 1/90–Present.
Learning Outcomes: Upon completion of the course, the student will be able to maintain and repair aerial communications, intercommunications, and direction-finding equipment.
Instruction: Course is an introduction to intercommunication equipment, communication systems, modulation, detection, mixing, receivers, high-frequency receivers, VHF and UHF transceivers, UHF DF, and emergency communications.
Credit Recommendation: In the lower-division baccalaureate/associate degree category, 1 semester hour in avionics maintenance (12/89).

AF-1704-0244

COMMUNICATION/NAVIGATION SYSTEMS TECHNICIAN BY CORRESPONDENCE

Course Number: 45572C.
Location: Extension Course Institute, Gunter AFS, AL.
Length: Maximum, 52 weeks.
Exhibit Dates: 1/90–Present.
Learning Outcomes: Upon completion of the course, the student will be able to repair and maintain aerial electronic navigation equipment, precision landing equipment, and electronic aircraft terrain clearance and identification equipment.
Instruction: Course is an introduction to electronic aerial direction, distance-finding and landing equipment, and theory and principles of radar in aircraft height-finding and identification.
Credit Recommendation: Credit is not recommended because of the limited, specialized nature of the course (12/89).

AF-1704-0245

COMMUNICATION/NAVIGATION SYSTEMS TECHNICIAN BY CORRESPONDENCE

Course Number: 45572D.
Location: Extension Course Institute, Gunter AFS, AL.
Length: Maximum, 52 weeks.
Exhibit Dates: 1/90–Present.
Learning Outcomes: Upon completion of the course, the student will be able to describe the functions of airborne radars and the three special functions of the AN/APN-59E search and weather radar.
Instruction: Course is an introduction to basic radar theory, nomenclature, phraseology and the functional capabilities of the AN/APN-59E airborne radar in its three modes of operation.
Credit Recommendation: Credit is not recommended because of the limited, specialized nature of the course (12/89).

AF-1704-0246

COMMUNICATION/NAVIGATION SYSTEMS TECHNICIAN BY CORRESPONDENCE

Course Number: 45572E.
Location: Extension Course Institute, Gunter AFS, AL.
Length: Maximum, 52 weeks.
Exhibit Dates: 1/90–Present.
Learning Outcomes: Upon completion of the course, the student will be able to describe the theory, operation, capabilities, limitations, and the necessity of Doppler navigation systems in high-speed aircraft.
Instruction: Course is an introduction to traditional basic navigation, analog computer elements, Doppler principles and techniques, Doppler receiver transmitter theory, frequency tracker and components, antenna, indicator,

tracker, Doppler navigation computer, and the common strategic Doppler.

Credit Recommendation: In the lower-division baccalaureate/associate degree category, 1 semester hour in avionics maintenance (12/89).

AF-1704-0247

STRATEGIC AIRCRAFT MAINTENANCE SPECIALIST BY CORRESPONDENCE

Course Number: 45750.
Location: Extension Course Institute, Gunter AFS, AL.
Length: Maximum, 52 weeks.
Exhibit Dates: 1/90–9/96.
Learning Outcomes: Upon completion of the course, the student will be able to describe the basic construction and maintenance requirements for strategic aircraft, ground safety, aircraft corrosion control, battle damage repair, and ground handling of aircraft. The student will have a basic knowledge of aircraft pneudraulics, and landing gear systems and will also have a basic understanding of engines, fuel systems, electrical systems, and environmental control systems for aircraft.
Instruction: Through correspondence the student receives basic information on aircraft ground handling, aircraft electrical, pneudraulic, and landing gear systems; engine construction and theory of operation; and engine electrical, fuel, and utility systems (environment control, fire protection, and ice and rain removal).
Credit Recommendation: In the lower-division baccalaureate/associate degree category, 2 semester hours in aircraft maintenance (12/89).

AF-1704-0248

OFFENSIVE AVIONICS SYSTEMS SPECIALIST (B-1B) BY CORRESPONDENCE

Course Number: 45753A.
Location: Extension Course Institute, Gunter AFS, AL.
Length: Maximum, 52 weeks.
Exhibit Dates: 1/90–3/97.
Learning Outcomes: Upon completion of the course, the student will be able to describe the fundamental operation of offensive avionic systems, maintenance and material management systems, and basic aircraft familiarization.
Instruction: The student receives instruction on avionic system maintenance, organizational maintenance, electronic multiplexing and integrated testing systems, and navigation display systems.
Credit Recommendation: In the lower-division baccalaureate/associate degree category, 1 semester hour in aviation maintenance management (12/89).

AF-1704-0249

AIRLIFT AIRCRAFT MAINTENANCE TECHNICIAN BY CORRESPONDENCE

Course Number: 45772.
Location: Extension Course Institute, Gunter AFS, AL.
Length: Maximum, 52 weeks.
Exhibit Dates: 1/90–Present.
Learning Outcomes: Upon completion of the course, the student will be able to describe the basic construction and maintenance requirements for airlift aircraft, ground safety, aircraft corrosion control, and ground handling of aircraft. The student will gain a basic knowledge of aircraft pneudraulics and landing gear systems, engines, fuel systems, electrical systems, and environmental control systems.
Instruction: Through correspondence, the student receives basic information on aircraft ground handling; aircraft electrical, pneudraulic, and landing gear systems; engine construction and theory of operation; and engine electrical, fuel, and utility systems (environmental control, fire protection, and ice and rain removal).
Credit Recommendation: In the lower-division baccalaureate/associate degree category, 3 semester hours in aircraft maintenance (12/89).

AF-1704-0250

1. AIRCRAFT STRUCTURAL MAINTENANCE TECHNICIAN (AIRFRAME REPAIR) BY CORRESPONDENCE
2. STRUCTURAL MAINTENANCE TECHNICIAN BY CORRESPONDENCE

Course Number: *Version 1:* 45872B. *Version 2:* 45872B.
Location: Extension Course Institute, Gunter AFS, AL.
Length: *Version 1:* Maximum, 52 weeks. *Version 2:* Maximum, 52 weeks.
Exhibit Dates: *Version 1:* 7/92–7/94. *Version 2:* 1/90–6/92.
Learning Outcomes: *Version 1:* Upon completion of the course, the student will be able to identify types of corrosion; select correct removal methods; replace protective coatings to aircraft components; and select correct cleaning methods to prevent further corrosive attack on aircraft components. *Version 2:* Upon completion of the course, the student will be able to identify types of corrosion; select correct removal methods; replace protective coatings to aircraft components; and select correct cleaning methods to prevent further corrosive attack on aircraft components.
Instruction: *Version 1:* Through correspondence, the student receives information on identification of corrosion, methods of treatment to remove corrosion, preparation of surfaces for organic coating to prevent corrosion, and cleaning aircraft to prevent the formation of corrosion. *Version 2:* Through correspondence, the student receives information on identification of corrosion, methods of treatment to remove corrosion, preparation of surfaces for organic coating to prevent corrosion, and cleaning aircraft to prevent the formation of corrosion.
Credit Recommendation: *Version 1:* In the lower-division baccalaureate/associate degree category, 1 semester hour in aircraft corrosion control (4/93). *Version 2:* In the lower-division baccalaureate/associate degree category, 1 semester hour in aircraft maintenance (12/89).

AF-1704-0251

1. AIRCRAFT MAINTENANCE OFFICER
2. AIRCRAFT MAINTENANCE/MUNITIONS OFFICER

Course Number: *Version 1:* J3OBR21A1 006. *Version 2:* J30BR4021; C30BR4021 004; C3OBR4001 001/002.
Location: *Version 1:* Technical Training Center, Sheppard AFB, TX. *Version 2:* Technical Training Center, Chanute AFB, IL; Technical Training Center, Sheppard AFB, TX.
Length: *Version 1:* 12 weeks (480 hours). *Version 2:* 18 weeks (704 hours).
Exhibit Dates: *Version 1:* 6/95–Present. *Version 2:* 8/90–5/95.
Learning Outcomes: *Version 1:* Upon completion of the course, the student will be able to identify appropriate aviation maintenance strategies, coordinate responsibilities of maintenance control and other maintenance staff agencies, identify necessary maintenance training activities, apply appropriate supply and other logistics principles, identify and employ both analog and digital data collection methodologies, apply propulsion systems management concepts, use appropriate inspection policies and procedures, identify preventive maintenance strategies, confront health and safety issues, and perform various operations management and human resource management activities. *Version 2:* Upon completion of the course, the student will be able to identify and manage Air Force maintenance publications; identify basic aerodynamic effects of aircraft weight and balance; describe principles of operation, components, malfunctions, and foreign-object damage on aircraft; describe system analysis and propulsion management; identify fuels and fuel systems; describe operating principles of pneudraulic, A/C environmental, fire control, and ice prevention systems; describe structural inspection and repair, corrosion control, and nuclear weapons maintenance/inspection; describe electricity, power supply, and avionic systems; work with maintenance forms, microcomputers, and munitions, including their maintenance, storage, transportation, handling, and control; describe preventive maintenance, resource management, personnel management, occupational safety, security, and inspection systems; and use effective communication techniques.
Instruction: *Version 1:* Classroom and on-the-job lectures and laboratory activities use audiovisuals, mock-ups, serviceable equipment, various publications, and hands-on laboratory experiences in the maintenance of aircraft and aircraft systems and equipment. Case studies are used to reinforce concepts in operations, maintenance, and human resource management. *Version 2:* Classroom instruction includes lectures, audiovisuals, mock-ups, research using various publications and documents, and hands-on experience in performing maintenance of aircraft, munitions and support equipment. The course also includes case studies of maintenance and munitions, including personnel performance and evaluations, production, handling, storage, transportation, and communication techniques.
Credit Recommendation: *Version 1:* In the graduate degree category, 3 semester hours in operations and human resource management, 3 in aviation maintenance operations and management, and 1 in industrial hygiene and safety (12/96). *Version 2:* In the graduate degree category, 3 semester hours in operations management and human resources, 3 in mechanical engineering, and 3 in industrial hygiene and safety (3/93).

AF-1704-0252

1. AIRCRAFT MAINTENANCE OFFICER (ACCELERATED/AIR RESERVE FORCES)
2. AIRCRAFT MAINTENANCE OFFICER, AIR RESERVE

Course Number: *Version 1:* J3OBR21A1 009. *Version 2:* C3OBR4021 003.
Location: *Version 1:* Technical Training Center, Sheppard AFB, TX. *Version 2:* Technical Training Center, Chanute AFB, IL.
Length: *Version 1:* 4 weeks (152 hours). *Version 2:* 4 weeks (150 hours).
Exhibit Dates: *Version 1:* 1/96–Present. *Version 2:* 1/90–12/95.
Learning Outcomes: *Version 1:* Upon completion of the course, the student will be able to identify appropriate aviation maintenance strategies, coordinate responsibilites of maintenance control and other maintenance staff agencies identify necessary maintenance training activities; apply appropriate supply and other logistic principles, identify and employ both analog and digital data collection methods, apply propulsion systems management concepts, use appropriate inspection policies and procedures, identify preventive maintenance strategies; confront health and safety issues, and perform various operations management and human resource management activities. *Version 2:* Upon completion of the course, the student will be able to identify selected concepts employed in Air Force maintenance policy; coordinate maintenance control staff agencies and maintenance operations division staff agencies, dual-channel training, and depot responsibilities; describe supply principles, munitions operations requirements, maintenance data collection systems, the Core Automated Maintenance System, propulsion management modification program, Air Force inspection system, preventive maintenance program, supervisor responsibility for Air Force occupational safety and health program, principles of non-nuclear explosives, and operations security vulnerabilities unique to AFSC 40XX.
Instruction: *Version 1:* Classroom and on-the-job lectures and laboratory activities use audiovisuals, mock-ups, serviceable equipment, various publications, and hands-on laboratory experiences in the maintenance of aircraft and aircraft systems and equipment. Case studies are used to reinforce concepts in operations, maintenance, and human resource management. *Version 2:* Classroom and on-the-job lectures use audiovisuals, case studies, mock-ups, and real equipment. Topics include regulations and publications in aircraft maintenance and munitions, personnel and equipment evaluation, production, handling, storage, transportation, use of resources, and communication techniques.
Credit Recommendation: *Version 1:* In the graduate degree category, 1 semester hour in operations and human resource management, 1 in aviation maintenance operations and management, and 1 in industrial hygiene and safety (12/96). *Version 2:* In the graduate degree category, 1 semester hour in operations management and human resources, 1 in mechanical engineering, and 1 in industrial hygiene and safety (3/93).

AF-1704-0253

1. AIRCRAFT MAINTENANCE OFFICER
    (ACCELERATED/AIR RESERVE FORCES)
2. AIRCRAFT MAINTENANCE OFFICER
    (ACCELERATED)

Course Number: *Version 1:* J3OBR21A1 008. *Version 2:* J3OBR4021 001; C3OBR4021 004.
Location: *Version 1:* Technical Training Center, Sheppard AFB, TX. *Version 2:* Technical Training Center, Sheppard AFB, TX; Technical Training Center, Chanute AFB, IL.
Length: *Version 1:* 4 weeks (152 hours). *Version 2:* 3 weeks (103 hours).
Exhibit Dates: *Version 1:* 1/96–Present. *Version 2:* 4/90–12/95.
Learning Outcomes: *Version 1:* Upon completion of the course, the student will be able to identify appropriate aviation maintenance strategies, coordinate responsibilities of maintenance control and other maintenance staff agencies, identify necessary maintenance training activities, apply appropriate supply and other logistics principles, identify and employ both analog and digital data collection methodologies, apply propulsion systems management concepts, use appropriate inspection policies and procedures, identify preventive maintenance strategies, confront health and safety issues, and perform various operations management and human resource management activities. *Version 2:* Upon completion of the course, the student will be able to identify responsibilities of deputy commander for maintenance, including functions of maintenance management agencies, maintenance control, job control/munitions control, maintenance operations division, planning and scheduling, documentation section, quality assurance and maintenance standardization, and evaluation; identify responsibilities of maintenance squadron commander; describe the manpower authorization and control process of centralized and decentralized units; and identify personnel classification and career management fields, principles of the Quality Force Management Program, responsibilities of unit management and on-the-job training, maintenance management information system, Air Force reliability program, unit status reporting, supply, records, publication system, planned inspection concepts, maintenance security, occupational safety, and conventional and nuclear munitions.
Instruction: *Version 1:* Classroom and on-the-job lectures and laboratory activities use audiovisuals, mock-ups, serviceable equipment, various publications, and hands-on laboratory experience in the maintenance of aircraft and aircraft systems and equipment. Case studies are used to reinforce concepts in operations, maintenance, and human resource management. *Version 2:* Classroom and on-the-job lectures use audiovisuals, mock-ups, real equipment, various publications, and hands-on experience in maintenance of aircraft, equipment, and munitions. Case studies include maintenance and munitions matters, personnel evaluation, production, handling, storage, transportation and use of resources, and communication techniques.
Credit Recommendation: *Version 1:* In the graduate degree category, 1 semester hour in operations and human resource management, 1 in aviation maintenance operations and management, and 1 in industrial hygiene and safety (12/96). *Version 2:* In the graduate degree category, 1 semester hour in operations management and human resources, 1 in mechanical engineering, and 1 in industrial hygiene and safety (3/93).

AF-1704-0254

F-15 INTEGRATED ORGANIZATIONAL AVIONICS
    SYSTEMS SPECIALIST BY CORRESPONDENCE

Course Number: 45251.
Location: Extension Course Institute, Gunter AFS, AL.
Length: Maximum, 52 weeks.
Exhibit Dates: 4/92–Present.
Learning Outcomes: Upon completion of the course, the student will be able to perform as an F-15 avionics maintenance specialist in the areas of specific safety, maintenance fundamentals, procedures, and check-out methodology.
Instruction: This correspondence course is required for advancement in the avionics maintenance discipline. Contents include a general description of aircraft with an emphasis on safety, maintenance fundamentals, and supply discipline.
Credit Recommendation: In the vocational certificate category, 2 semester hours in introduction to aircraft maintenance (avionics) (4/93).

AF-1704-0255

INTEGRATED AVIONIC INSTRUMENT AND FLIGHT
    CONTROL SYSTEMS BY CORRESPONDENCE

Course Number: 45252B.
Location: Extension Course Institute, Gunter AFS, AL.
Length: Maximum, 52 weeks.
Exhibit Dates: 1/90–Present.
Learning Outcomes: Upon completion of the course, students are able to diagnose faults by using test equipment to check out and adjust avionic instruments, aircraft computers, displays, and flight control systems.
Instruction: This correspondence course is designed to prepare advanced maintenance technicians to maintain sophisticated F-16 instrument systems. Instrument maintenance procedures are covered such as auto pilot compass swing, fuel system operation, and operation of flight controls.
Credit Recommendation: In the vocational certificate category, 2 semester hours in autopilot maintenance, fuel system maintenance, and operation of flight controls (4/93); in the lower-division baccalaureate/associate degree category, 2 semester hours in introduction to instruments and flight control systems (4/93).

AF-1704-0256

AEROSPACE PROPULSION SPECIALIST (JET ENGINE)
    BY CORRESPONDENCE

Course Number: 45450A.
Location: Extension Course Institute, Gunter AFS, AL.
Length: Maximum, 52 weeks.
Exhibit Dates: 8/92–6/97.
Learning Outcomes: Upon completion of the course, the student will be able to maintain, repair, remove and install, troubleshoot, and test jet engines.
Instruction: Through correspondence, the student receives information on general maintenance, theory of jet engines, construction differences, and small gas turbine engine operation principles. General maintenance procedures, repair, corrosion control, storage and shipment, and nonpowered support equipment are covered. A review of jet engine fuels and contaminants, oil systems, starter ignition, electrical systems, typical anti-icing bleed air system, thrust augmentation systems, and thrust reversers are all presented. The student studies test cell operation, instrumentation, checks and adjustments, noise supressers, removal and

installation, and systematic troubleshooting of various engine systems.

Credit Recommendation: In the lower-division baccalaureate/associate degree category, 2 semester hours in jet engine maintenance (4/93).

AF-1704-0257

AEROSPACE PROPULSION SPECIALIST (TURBOPROP) BY CORRESPONDENCE

Course Number: 45450B.
Location: Extension Course Institute, Gunter AFS, AL.
Length: Maximum, 52 weeks.
Exhibit Dates: 8/92–7/97.
Learning Outcomes: Upon completion of the course, the student will be able to maintain, troubleshoot, and repair turbopropeller systems and equipment.
Instruction: This correspondence course involves home study and practical exercises in the maintenance of turboprops. Topics include maintenance management policies, engine management, career management, reliability and maintainability, static electricity, fuels, foreign-object damage prevention, and safety. Also included are supply systems, inspections, and material deficiency reporting system. A review of the theory of jet propulsion with information on the construction and operating principles of the T-56-A-7 and the T-56-A-15 and small gas turbine compressor engines is presented. A review of the different tools and safety equipment required to maintain propulsion systems is provided. Engine fuel and oil systems are covered. Various engine operating systems, instruments, and indicating systems are reviewed. Engine rigging, operation, and adjustments and the T58-GE-5 are all reviewed. Technical information on the Hamilton Standard Turbo propeller model 54H60-91 is presented in depth.
Credit Recommendation: In the lower-division baccalaureate/associate degree category, 2 semester hours in gas turbine engines and 2 in turboprop engines and systems (4/93).

AF-1704-0258

AIRCRAFT FUEL SYSTEMS MECHANIC BY CORRESPONDENCE

Course Number: 45453.
Location: Extension Course Institute, Gunter AFS, AL.
Length: 52 weeks.
Exhibit Dates: 1/92–11/97.
Learning Outcomes: Upon completion of the course, the student will be able to describe the fundamentals of aircraft fuel systems, components, operating principles, and repair.
Instruction: Subject matter includes maintenance management, shop and flight-line safety, hand tools and hardware, troubleshooting and operational checks, fuel systems, fuel system safety, leak detection, and repair of integral and bladder tanks/cells.
Credit Recommendation: In the lower-division baccalaureate/associate degree category, 2 semester hours in aircraft fuel systems (4/93).

AF-1704-0259

AVIONICS GUIDANCE AND CONTROL SYSTEMS TECHNICIAN BY CORRESPONDENCE

Course Number: 45571D.
Location: Extension Course Institute, Gunter AFS, AL.
Length: Maximum, 52 weeks.
Exhibit Dates: 10/91–9/96.
Learning Outcomes: Upon completion of the course, the student will be able to apply a general knowledge of advanced navigation systems (essential to maintenance technicians but also applicable in flight operations) and conduct maintenance checkout procedures at the technician level.
Instruction: This correspondence course is a review of avionics instrument systems for avionic guidance and control systems technicians. Instrument systems covered include position, engine, fuel quality, flight data recording, standard flight, control air data computer, vertical seal flight, flight director, and navigational aids.
Credit Recommendation: In the lower-division baccalaureate/associate degree category, 1 semester hour in introduction to advanced navigation systems (4/93); in the upper-division baccalaureate category, 1 semester hour in advanced navigation systems (4/93).

AF-1704-0260

COMMUNICATION/NAVIGATION SYSTEMS TECHNICIAN (DOPPLER SYSTEMS) BY CORRESPONDENCE

Course Number: 45572E.
Location: Extension Course Institute, Gunter AFS, AL.
Length: Maximum, 52 weeks.
Exhibit Dates: 3/92–Present.
Learning Outcomes: Upon completion of the course, the student will demonstrate knowledge of air navigation terms, physics of the Doppler effect and Doppler beam patterns, and perform specific maintenance checkout procedures.
Instruction: This correspondence course provides technical information of the Doppler system needed to become a communications/navigational systems technician. The units include explanation of Doppler principles and beam patterns, description of basic flight computer fundamentals, and characteristics of Doppler systems.
Credit Recommendation: In the lower-division baccalaureate/associate degree category, 1 semester hour in introduction to communication/navigation systems (4/93); in the upper-division baccalaureate category, 1 semester hour in advanced communication/navigation systems (4/93).

AF-1704-0261

AIRCRAFT GUIDANCE AND CONTROL SYSTEMS TECHNICIAN BY CORRESPONDENCE

Course Number: 45371.
Location: Extension Course Institute, Gunter AFS, AL.
Length: Maximum, 52 weeks.
Exhibit Dates: 3/93–2/98.
Learning Outcomes: Upon completion of the course, the student will be able to apply a basic knowledge of electronics principles, aircraft guidance and control systems, specific maintenance checkouts and avionics instruments and control at a level appropriate to avionics technicians.
Instruction: This correspondence study course with on-the-job training offers general career field principles and information on compass and automatic flight control systems, avionic instrument systems, and inertial navigation and fuel advisory systems.
Credit Recommendation: In the vocational certificate category, 3 semester hours in introduction to aircraft maintenance and avionics equipment (4/93); in the lower-division baccalaureate/associate degree category, 3 semester hours in introduction to avionics and flight control systems (4/93).

AF-1704-0263

INTEGRATED AVIONICS COMMUNICATIONS, NAVIGATION, AND PENETRATION AIDS SYSTEMS SPECIALIST (F/FB-111) BY CORRESPONDENCE

Course Number: 45253C.
Location: Extension Course Institute, Gunter AFS, AL.
Length: Maximum, 52 weeks.
Exhibit Dates: 1/90–6/97.
Learning Outcomes: Upon completion of the course, the student will be able to use a general overview of F/FB-111 communications systems and avionics to perform specific operational checkouts for aircraft integrated avionic systems.
Instruction: This correspondence course presents material to the 5-level maintenance specialist supporting the F/FB-111 and includes aircraft familiarization and safety; communications systems; intercommunications; HF, UHF radio, and secure voice communications systems; direction finder systems; and the FB-111 satellite communications system.
Credit Recommendation: In the lower-division baccalaureate/associate degree category, 3 semester hours in introduction to communications systems (4/93).

AF-1704-0264

MISSILE FACILITIES SPECIALIST BY CORRESPONDENCE

Course Number: 41152A.
Location: Extension Course Institute, Gunter AFS, AL.
Length: Maximum, 52 weeks.
Exhibit Dates: 9/91–7/98.
Learning Outcomes: Upon completion of this course, the student will be able to provide technical assistance and supervise subordinates in weapons systems characteristics, hardness assurance, maintenance management, publications, tools, hardware, and launch facility entry and exit. Also, the student will perform similar duties on power generation and distribution; stationary inertial combustion engine maintenance; waste disposal; ground support equipment; and heating, ventilation, and air conditioning systems.
Instruction: This is a correspondence course with home study, practical exercises, and a proctored final examination in the general maintenance and administration of a missile facility. Topics include fundamentals of missile maintenance, hardness assurance, maintenance management, publications, hardware, hand tools, and entry and exit to the launch facility. Power generation and distribution are included along with stationary internal combustion engines; power systems; waste disposal systems; air conditioning, heating, and ventilation systems; and ground support systems.
Credit Recommendation: In the lower-division baccalaureate/associate degree category, 3 semester hours in internal combustion

engine fundamentals, 3 in electrical fundamentals, 3 in industrial electricity, 3 in air conditioning principles, and 3 in maintenance administration (4/93).

AF-1704-0265

MISSILE AND SPACE SYSTEMS MAINTENANCE JOURNEYMAN BY CORRESPONDENCE
(Missile Maintenance Specialist by Correspondence)

Course Number: 41151A.
Location: Extension Course Institute, Gunter AFS, AL.
Length: Maximum, 52 weeks.
Exhibit Dates: 8/93–4/98.
Learning Outcomes: Upon completion of the course, the student will be able to describe the functional design of a hardened missile site. Student will have a working knowledge of maintenance management, hydraulic/pneumatic systems, and vehicles.
Instruction: This is a correspondence course where the student learns material by home study, review exercises, and proctored end-of-course examination. Topics include pneumatics, hydraulics, vehicle maintenance, and missile complex design.
Credit Recommendation: Credit is not recommended because of the military-specific nature of the course (6/94).

AF-1704-0266

TRANSITION/REQUALIFICATION TRAINING F-15E

Course Number: *Version 1:* F15EOTXOAL/WL. *Version 2:* F15EOTXOAL/WL.
Location: 56th Training Squadron, Luke AFB, AZ.
Length: *Version 1:* 14-18 weeks (268-314 hours). *Version 2:* 13-16 weeks (222-280 hours).
Exhibit Dates: *Version 1:* 10/93–Present. *Version 2:* 7/92–9/93.
Learning Outcomes: *Version 1:* Upon completion of the course, the student will be able to recognize, interpret, and explain advanced aircraft systems and the performance environment and perform advanced jet aircraft flight maneuvers and operations. *Version 2:* Upon completion of the course, the student will be able to recognize, interpret, and explain advanced aircraft systems and the performance environment and perform advanced jet aircraft flight maneuvers and operations.
Instruction: *Version 1:* This course consists of classroom lectures, computer-based training, and appropriate aircraft and flight training device laboratory experiences. *Version 2:* This course consists of classroom lectures, computer-based training, and appropriate aircraft and flight training device laboratory experiences.
Credit Recommendation: *Version 1:* In the graduate degree category, for Track 1A, 4 semester hours in advanced aircraft systems and performance and 2 in advanced flight laboratory. For Track 1B/1C, 5 semester hours in advanced aircraft systems and performance and 3 in advanced flight laboratory (9/94). *Version 2:* In the graduate degree category, for Track 1A, 3 semester hours in advanced aircraft systems and performance and 2 in advanced flight laboratory. For Track 1B/1C, 4 semester hours in advanced aircraft systems and performance, and 3 in advanced flight laboratory (9/94).

AF-1704-0267

CONVERSION TRAINING F-16C/D

Course Number: *Version 1:* F16C0CX0PL. *Version 2:* F16COCXOPL/M.
Location: *Version 1:* 56th Training Squadron, Luke AFB, AZ. *Version 2:* 56th Training Squadron, MacDill AFB, FL; 56th Training Squadron, Luke AFB, AZ.
Length: *Version 1:* 1-3 weeks (48 hours). *Version 2:* 1-3 weeks (48 hours).
Exhibit Dates: *Version 1:* 2/94–Present. *Version 2:* 10/92–1/94.
Learning Outcomes: *Version 1:* Upon completion of the course, the student will be able to recognize, interpret, and explain advanced aircraft systems and the performance environment and perform advanced jet aircraft flight maneuvers and operations. *Version 2:* Upon completion of the course, the student will be able to recognize, interpret, and explain advanced aircraft systems and the performance environment and perform advanced jet aircraft flight maneuvers and operations.
Instruction: *Version 1:* The course consists of classroom lectures, computer-based training, and appropriate aircraft and flight training device laboratory experiences. *Version 2:* The course consists of classroom lectures, computer-based training, and appropriate aircraft and flight training device laboratory experiences.
Credit Recommendation: *Version 1:* In the graduate degree category, for Track 1, 1 semester hour in advanced aircraft systems and performance. For Track 2, 1 semester hour in advanced aircraft systems and performance. For Track 3, credit is not recommended (9/94). *Version 2:* In the graduate degree category, for Track 1, 1 semester hour in advanced aircraft systems and performance. For Track 2, 1 semester hour in advanced aircraft systems and performance. For Track 3, credit is not recommended (9/94).

AF-1704-0268

CONVERSION TRAINING F-16C/D

Course Number: F16CGCXOPL.
Location: 56th Training Squadron, Luke AFB, AZ.
Length: 2-5 weeks (41-77 hours).
Exhibit Dates: 10/92–Present.
Learning Outcomes: Upon completion of the course, the student will be able to recognize, interpret, and explain advanced aircraft systems and the performance environment and perform advanced jet aircraft flight maneuvers and operations.
Instruction: The course consists of classroom lectures, computer-based training, and appropriate aircraft and flight training device laboratory experiences.
Credit Recommendation: In the graduate degree category, for Track 1, 1 semester hour in advanced aircraft systems and performance. For Track 2, 1 semester hour in advanced aircraft systems and performance. For Track 3, 2 semester hours in advanced aircraft systems and performance (9/94).

AF-1704-0269

BASIC QUALIFICATION TRAINING F-15E
(Basic Operational Training F-15E)

Course Number: *Version 1:* F15EOBOOAL/WL. *Version 2:* F15E0BOOAL/WL.
Location: 56th Training Squadron, Luke AFB, AZ.
Length: *Version 1:* 25-26 weeks (343-374 hours). *Version 2:* 25-26 weeks (343-374 hours).
Exhibit Dates: *Version 1:* 9/93–Present. *Version 2:* 6/92–8/93.
Learning Outcomes: *Version 1:* Upon completion of the course, the student will be able to recognize, interpret, and explain advanced aircraft and the performance environment and perform advanced jet aircraft flight maneuvers and operations. *Version 2:* Upon completion of the course, the student will be able to recognize, interpret, and explain advanced aircraft and the performance environment and perform advanced jet aircraft flight maneuvers and operations.
Instruction: *Version 1:* The course consists of classroom lectures, computer-based training, and appropriate aircraft and flight training device laboratory experiences. *Version 2:* The course consists of classroom lectures, computer-based training, and appropriate aircraft and flight training device laboratory experiences.
Credit Recommendation: *Version 1:* In the graduate degree category, 6 semester hours in advanced aircraft systems and performance and 4 in advanced flight laboratory (9/94). *Version 2:* In the graduate degree category, 5 semester hours in advanced aircraft systems and performance and 4 in advanced flight laboratory (9/94).

AF-1704-0270

BASIC OPERATIONAL TRAINING F-16C/D

Course Number: *Version 1:* F16COBOOPL/M. *Version 2:* F16COBOOPL/M.
Location: 56th Training Squadron, Luke AFB, AZ; 56th Training Squadron, MacDill AFB, FL.
Length: *Version 1:* 22-26 weeks (337-407 hours). *Version 2:* 22-26 weeks (337-407 hours).
Exhibit Dates: *Version 1:* 1/93–Present. *Version 2:* 7/90–12/92.
Learning Outcomes: *Version 1:* Upon completion of the course, the student will be able to recognize, interpret, and explain advanced aircraft and the performance environment and perform advanced jet aircraft flight maneuvers and operations. *Version 2:* Upon completion of the course, the student will be able to recognize, interpret, and explain advanced aircraft and the performance environment and perform advanced jet aircraft flight maneuvers and operations.
Instruction: *Version 1:* The course consists of classroom lectures, computer-based training, and appropriate aircraft and flight training device laboratory experiences. *Version 2:* The course consists of classroom lectures, computer-based training, and appropriate aircraft and flight training device laboratory experiences.
Credit Recommendation: *Version 1:* In the graduate degree category, 6 semester hours in advanced aircraft systems and performance and 4 in advanced flight laboratory (9/94). *Version 2:* In the graduate degree category, 5 semester hours in advanced aircraft systems and perfor-

mance and 3 in advanced flight laboratory (9/94).

AF-1704-0271

INSTRUCTOR PILOT UPGRADE TRAINING F-16C/D

Course Number: F-16 COIOOPL/M; F-16COIOOPL/M.
Location: 56th Training Squadron, MacDill AFB, FL; 56th Training Squadron, Luke AFB, AZ.
Length: 8 weeks (59 hours).
Exhibit Dates: 3/91–4/94.
Learning Outcomes: Upon completion of the course, the student will be able to instruct and evaluate pilots in high performance jet aircraft.
Instruction: The course consists of classroom lectures, computer-based training, and appropriate aircraft and flight training device laboratory.
Credit Recommendation: In the graduate degree category, for Track 1, 2 semester hours in instructional design and teaching methodologies. For Track 2, 1 semester hour in instructional design and teaching methodologies (9/94).

AF-1704-0272

TRANSITION/REQUALIFICATION TRAINING F-16C/D

Course Number: *Version 1:* F-16COTXOPL. *Version 2:* F-16COTXOPL/M. *Version 3:* F-16COTXOPL/M.
Location: 56th Training Squadron, Luke AFB, AZ; 56th Training Squadron, MacDill AFB, FL.
Length: *Version 1:* 1-16 weeks (48-345 hours). *Version 2:* 1-13 weeks (46-344 hours). *Version 3:* 4-13 weeks (136-328 hours).
Exhibit Dates: *Version 1:* 6/94–Present. *Version 2:* 1/93–5/94. *Version 3:* 10/92–12/92.
Learning Outcomes: *Version 1:* Upon completion of the course, the student will be able to recognize, interpret, and explain advanced aircraft systems and the performance environment and perform advanced jet aircraft maneuvers and operations. *Version 2:* Upon completion of the course, the student will be able to recognize, interpret, and explain advanced aircraft systems and the performance environment and perform advanced jet aircraft maneuvers and operations. *Version 3:* Upon completion of the course, the student will be able to recognize, interpret, and explain advanced aircraft systems and the performance environment and perform advanced jet aircraft maneuvers and operations.
Instruction: *Version 1:* The course consists of classroom lectures, computer-based training, and appropriate aircraft and flight training device laboratory exercises. *Version 2:* The course consists of classroom lectures, computer-based training, and appropriate aircraft and flight training device laboratory exercises. *Version 3:* The course consists of classroom lectures, computer-based training, and appropriate aircraft and flight training device laboratory exercises.
Credit Recommendation: *Version 1:* In the graduate degree category, for Track 1, 5 semester hours in advanced aircraft systems and performance and 3 in advanced flight laboratory. Track 2, 2 semester hours in advanced aircraft systems and performance and 1 in advanced flight laboratory. Track 3, 2 semester hours in advanced aircraft systems and performance and 1 in advanced flight laboratory. Track 4, 1 semester hour in advanced aircraft systems and performance (9/94). *Version 2:* In the graduate degree category, for Track 1, 5 semester hours in advanced aircraft systems and performance and 3 in advanced flight laboratory. Track 2, 3 semester hours in advanced aircraft systems and performance and 1 in advanced flight laboratory. Track 3, 2 semester hours in advanced aircraft systems and performance and 1 in advanced flight laboratory. Track 4, 1 semester hour in advanced aircraft systems and performance (9/94). *Version 3:* In the graduate degree category, for Track 1, 5 semester hours in advanced aircraft systems and performance and 2 in advanced flight laboratory. Track 2A, 2 semester hours in advanced aircraft systems and performance and 1 in advanced flight laboratory. Track 2B, 2 semester hours in advanced aircraft systems and performance and 1 in advanced flight laboratory. Track 3, 2 semester hours in advanced aircraft systems and performance and 1 in advanced flight laboratory (9/94).

AF-1704-0273

B-52 BOMBER WEAPONS INSTRUCTOR
(Bomber Weapons Instructor B-52)

Course Number: *Version 1:* B52001DOAB/JB/EB; B52001DOAE/WE/EE. *Version 2:* B52BWIC. *Version 3:* B52BWIC.
Location: *Version 1:* Detachment 5, 57th Wing, Barksdale AFB, LA; Weapons School, Ellsworth AFB, SD. *Version 2:* Weapons School, Ellsworth AFB, SD. *Version 3:* Weapons School, Ellsworth AFB, SD.
Length: *Version 1:* 11-18 weeks (219-431 hours). *Version 2:* 11-18 weeks (219-431 hours). *Version 3:* 11-18 weeks.
Exhibit Dates: *Version 1:* 7/94–Present. *Version 2:* 2/93–6/94. *Version 3:* 1/90–1/93.
Learning Outcomes: *Version 1:* Upon completion of the course, the student will be able to recognize, interpret, and explain advanced aircraft systems and the performance environment; instruct and evaluate crewmembers in a dynamic, high-activity flight management arena; provide integrated systems management expertise to key decision makers; perform needs assessments; and design instructional programs to meet educational goals and objectives. *Version 2:* Upon completion of the course, the student will be able to recognize, interpret, and explain advanced aircraft systems and the performance environment; instruct and evaluate crewmembers in a dynamic, high-activity flight management arena; provide integrated systems management expertise to key decision makers; perform needs assessment; and design instructional programs to meet educational goals and objectives. *Version 3:* Upon completion of the course, the student will be able to recognize, interpret, and explain advanced aircraft systems and the performance environment; instruct and evaluate crewmembers in a dynamic, high-activity flight management arena; provide integrated systems management expertise to key decision makers; perform needs assessments; and design instructional programs to meet educational goals and objectives.
Instruction: *Version 1:* The course consists of classroom lectures, student independent research, and formal student-led briefings covering integrated systems management, research paper, and advanced flight laboratory. Course includes appropriate laboratory experiences in advanced aerodynamics and aircraft systems. *Version 2:* The course consists of classroom lectures, student independent research, formal student-led briefings, and appropriate laboratory experiences in advanced aerodynamics and aircraft systems, integrated systems management, research paper, and advanced flight laboratory. *Version 3:* The course consists of classroom lectures, student independent research, formal student-led briefings, and appropriate laboratory experiences in advanced aerodynamics and aircraft systems, integrated systems management, research paper, and advanced flight laboratory.
Credit Recommendation: *Version 1:* In the graduate degree category, 1 semester hour in advanced aerodynamics and aircraft systems, 2 in integrated systems management, 1 in research paper, and 5 in advanced flight laboratory (2/97). *Version 2:* In the graduate degree category, 1 semester hour in advanced aerodynamics and aircraft systems, 2 in integrated systems management, 1 in research paper, and 4 in advanced flight laboratory (2/95). *Version 3:* In the graduate degree category, 1 semester hour in advanced aerodynamics and aircraft systems, 1 in integrated systems management, 1 in research paper, and 5 in advanced flight laboratory (2/95).

AF-1704-0274

BOMBER WEAPONS INSTRUCTOR B1

Course Number: *Version 1:* B1000IDOAE/WE. *Version 2:* B1BWIC.
Location: Weapons School, Ellsworth AFB, SD.
Length: *Version 1:* 17-19 weeks (288-409 hours). *Version 2:* 17-18 weeks (396-409 hours).
Exhibit Dates: *Version 1:* 8/94–Present. *Version 2:* 2/93–7/94.
Learning Outcomes: *Version 1:* Upon completion of the course, the student will be able to recognize, interpret, and explain advanced aircraft systems and the performance environment; instruct and evaluate crewmembers in a dynamic and high-activity flight management areas; provide integrated systems management expertise to key decision makers; perform needs assessment; and design instructional programs to meet educational goals and objections. *Version 2:* Upon completion of the course, the student will be able to recognize, interpret, and explain advanced aircraft systems and the performance environment; instruct and evaluate crewmembers in a dynamic and high-activity flight management areas; provide integrated systems management expertise to key decision makers; perform needs assessment; and design instructional programs to meet educational goals and objections.
Instruction: *Version 1:* The course consists of classroom lectures, student independent research, formal student-led briefings, and appropriate laboratory experiences in advanced aerodynamics and aircraft systems, integrated systems management, research paper, instructional methodology, and advanced flight laboratory. *Version 2:* The course consists of classroom lectures, student independent research, formal student-led briefings, and appropriate laboratory experiences in advanced aerodynamics and aircraft systems, integrated systems management, research paper, instructional methodology, and advanced flight laboratory.

Credit Recommendation: *Version 1:* In the graduate degree category, 1 semester hours in advanced aerodynamics and systems management, 2 in integrated systems management, 1 in research paper, 1 in instructional methodology, and 3 in advanced flight laboratory (5/96). *Version 2:* In the graduate degree category, 1 semester hour in advanced aerodynamics and systems management, 2 in integrated systems management, 1 in research paper, 1 in instructional methodology, and 2 in advanced flight laboratory (2/95).

AF-1704-0275

TRANSITION/REQUALIFICATION TRAINING F-16 A/B

Course Number: F16AOTXOPL/M.
Location: 56th Tactical Training Wing, MacDill AFB, FL; 58th Tactical Training Wing, Luke AFB, AZ.
Length: 4-13 weeks (83-273 hours).
Exhibit Dates: 1/90–Present.
Learning Outcomes: Upon completion of the course, the student will be able to recognize, interpret, and explain advanced aircraft systems and the performance environment; and perform advanced jet aircraft maneuvers and operations.
Instruction: The course consists of classroom lectures, programmed or computer-based training, and appropriate aircraft and flight training device laboratory exercises.
Credit Recommendation: In the graduate degree category, Track 1, 3 semester hours in advanced aircraft systems and performance and 3 in advanced flight laboratory. Track 2, 2 semester hours in advanced aircraft systems and performance and 2 in advanced flight laboratory. Track 3, 1 semester hour in advanced aircraft systems and performance. Track 4, 1 semester hour in advanced aircraft systems performance (6/95).

AF-1704-0276

TRANSITION/REQUALIFICATION TRAINING F-16 C/D (BLOCK 40/42)

Course Number: F16CGTXOPL.
Location: 58th Tactical Training Wing, Luke AFB, AZ.
Length: 4-13 weeks (124-295 hours).
Exhibit Dates: 1/90–Present.
Learning Outcomes: Upon completion of the course, the student will be able to recognize, interpret and explain advanced aircraft systems and the performance environment; and perform advanced jet aircraft maneuvers and operations.
Instruction: The course consists of classroom lectures, programmed or computer-based training, and appropriate aircraft and flight training device laboratory exercises.
Credit Recommendation: In the graduate degree category, Track 1, 2 semester hours in advanced aircraft systems and performance and 3 in advanced flight laboratory. Track 2, 1 semester hour in advanced aircraft systems and performance and 1 in advanced flight laboratory. Track 3, 1 semester hour in advanced aircraft systems and performance and 1 in advanced flight laboratory (6/95).

AF-1704-0277

BASIC OPERATIONAL TRAINING

Course Number: F16COB00AL.
Location: 58th Tactical Training Wing, Luke AFB, AZ.
Length: 23 weeks (388 hours).
Exhibit Dates: 1/90–9/90.
Learning Outcomes: Upon completion of the course, the student will be able to recognize, interpret, and explain advanced aircraft systems and the performance environment and perform advanced jet aircraft flight maneuvers and operations.
Instruction: The course consists of classroom lectures, programmed and computer based training, and appropriate aircraft and flight training device laboratory experiences.
Credit Recommendation: In the graduate degree category, 3 semester hours in advanced aircraft systems performance and 5 in advanced flight laboratory (6/95).

AF-1704-0278

LANTIRN TRAINING F-16C/D BLOCK 40/42
(Conversion and Lantirn Training F16 C/D Block 40/42)

Course Number: F16CGLOOPL; F16CGCLOPL.
Location: 58th Tactical Training Wing, Luke AFB, AZ.
Length: 3-6 weeks (51-107 hours).
Exhibit Dates: 1/90–Present.
Learning Outcomes: Upon completion of the course, the student should be able to recognize, interpret, and explain advanced aircraft systems and the performance environment and perform advanced jet aircraft maneuvers and operations.
Instruction: This course consists of classroom lecture, programmed or computer-based training, and appropriate aircraft and flight training device laboratory experiences.
Credit Recommendation: In the graduate degree category, Track 1, 1 semester hour in advanced aircraft systems and performance. Track 2, 1 semester hour in advanced aircraft systems and performance. Track 3, 1 semester hour in advanced aircraft systems and performance and 1 in advanced flight laboratory. Track 4, 1 semester hour in advanced aircraft systems and performance (6/95).

AF-1704-0279

BASIC OPERATIONAL TRAINING F16 C/D BLOCK 40/42

Course Number: F16CGBOOPL.
Location: 58th Tactical Training Wing, Luke AFB, AZ.
Length: 22-23 weeks (367-373 hours).
Exhibit Dates: 7/90–6/94.
Learning Outcomes: Upon completion of the course, the student will be able to recognize, interpret, and explain advanced aircraft systems and the performance environment and perform advanced jet aircraft flight maneuvers and operations.
Instruction: This course consists of classroom lectures, programmed or computer-based training, and appropriate aircraft and flight training device laboratory experiences.
Credit Recommendation: In the graduate degree category, 3 semester hours in advanced aircraft system and performance and 4 in advanced flight laboratory (6/95).

AF-1704-0280

BASIC OPERATIONAL TRAINING F-16
(Basic Operational Training F-16 A/B)

Course Number: F16A0B00AL/M.
Location: 56th Tactical Training Wing, MacDill AFB, FL.
Length: 23 weeks (379 hours).
Exhibit Dates: 1/90–4/90.
Learning Outcomes: Upon completion of the course, the student will be able to recognize, interpret, and explain advanced aircraft systems and the performance environment, and perform advanced jet aircraft flight maneuvers and operations.
Instruction: The course consists of classroom lectures, programmed and computer based training, and appropriate aircraft and flight training device laboratory experiences.
Credit Recommendation: In the graduate degree category, 3 semester hours in advanced aircraft systems and performance and 5 in advanced flight laboratory (6/95).

AF-1704-0281

BASIC OPERATIONAL TRAINING A/OA-10

Course Number: *Version 1:* A1000BOAPD. *Version 2:* A1000BOAPD.
Location: 355th Operations Group, Davis-Montham AFB, Tucson, AZ.
Length: *Version 1:* 17-20 weeks (254-326 hours). *Version 2:* 17-20 weeks (254-326 hours).
Exhibit Dates: *Version 1:* 12/95–Present. *Version 2:* 4/93–11/95.
Learning Outcomes: *Version 1:* Upon completion of the course, the student will be able to recognize, interpret, and explain advanced aircraft systems and the performance environment and perform advanced jet aircraft flight maneuvers and operations. *Version 2:* Upon completion of the course, the student will be able to recognize, interpret, and explain advanced aircraft systems and the performance environment and perform advanced jet aircraft flight maneuvers and operations.
Instruction: *Version 1:* This course consists of classroom lectures, programed or computer-based training, and appropriate aircraft and flight training device laboratory experience. *Version 2:* This course consists of classroom lectures, programed or computer-based training, and appropriate aircraft and flight training device laboratory experience.
Credit Recommendation: *Version 1:* In the graduate degree category, 2 semester hours in advanced aircraft systems and performance and 4 in advanced flight laboratory (3/96). *Version 2:* In the graduate degree category, 3 semester hours in advanced aircraft systems and performance and 4 in advanced flight laboratory (3/96).

AF-1704-0282

UPGRADE TRAINING A-10
(Operation Training, A10A)
(Transition Training, A10A)
(Requalification Training, A10A)

Course Number: A1000B; A1000TXC; A1000TXB; A1000TXA; A1000B/TXA/TXB/TXC.
Location: 355th Operations Group, Davis-Montham AFB, Tucson, AZ.
Length: 4-15 weeks (90-226 hours).
Exhibit Dates: 1/90–Present.
Learning Outcomes: Upon completion of the course, the student will be able to recognize, interpret, and explain advanced aircraft systems and the performance environment and perform

advanced jet aircraft flight maneuvers and operations.

Instruction: This course consists of classroom lectures, programed or computer-based training, and appropriate aircraft and flight training device laboratory experiences.

Credit Recommendation: In the graduate degree category, Track B: 2 semester hours in advanced aerodynamics and aircraft systems and 3 in advanced flight laboratory. Track TXA: 2 semester hours in advanced aerodynamics and aircraft systems and 2 in advanced flight laboratory. Track TXB: 1 semester hour in advanced aerodynamics and aircraft systems and 2 in advanced flight laboratory. Track TXC: 1 semester hour in advanced aerodynamics and aircraft systems and 1 in advanced flight laboratory (3/96).

AF-1704-0283

TRANSITION/REQUALIFICATION TRAINING A/OA-10

Course Number: *Version 1:* A1000TXAPD. *Version 2:* A1000TXAPD.

Location: 355th Operations Group, Davis-Montham AFB, Tucson, AZ.

Length: *Version 1:* 4-17 weeks (58-281 hours). *Version 2:* 4-17 weeks (58-281 hours).

Exhibit Dates: *Version 1:* 4/93–Present. *Version 2:* 12/92–3/93.

Learning Outcomes: *Version 1:* Upon completion of the course, the student will be able to recognize, interpret, and explain advanced aircraft systems and the performance environment and perform advanced jet aircraft flight maneuvers and operations. *Version 2:* Upon completion of the course, the student will be able to recognize, interpret, and explain advanced aircraft systems and the performance environment and perform advanced jet aircraft flight maneuvers and operations.

Instruction: *Version 1:* This course consists of classroom lectures, programed or computer-based training, and appropriate aircraft and flight training device laboratory experiences. *Version 2:* This course consists of classroom lectures, programed or computer-based training, and appropriate aircraft and flight training device laboratory experiences.

Credit Recommendation: *Version 1:* In the graduate degree category, Track TX1: 2 semester hours in advanced aerodynamics and aircraft systems and 4 in advanced flight laboratory. Track TX2A: 2 semester hours in advanced aerodynamics and aircraft systems and 3 in advanced flight laboratory. Track TX2B: 2 semester hours in advanced aerodynamics and aircraft systems and 3 in advanced flight laboratory. Track TX3: 1 semester hour in advanced aerodynamics and aircraft systems and 1 in advanced flight laboratory. Track TX4: 1 semester hour in advanced flight laboratory (3/96). *Version 2:* In the graduate degree category, Track TX1: 2 semester hours in advanced aerodynamics and aircraft systems and 3 in advanced flight laboratory. Track TX2A: 2 semester hours in advanced aerodynamics and aircraft systems and 2 in advanced flight laboratory. Track TX2B: 2 semester hours in advanced aerodynamics and aircraft systems and 2 in advanced flight laboratory. Track TX2C: 1 semester hour in advanced aerodynamics and aircraft systems and 1 in advanced flight laboratory. Track TX4: 1 semester hour in advanced flight laboratory (3/96).

AF-1704-0284

INSTRUCTOR PILOT UPGRADE

Course Number: *Version 1:* A10001A0PD. *Version 2:* A1000IOAPD.

Location: 355th Operations Group, Davis-Montham AFB, Tucson, AZ.

Length: *Version 1:* 4-8 weeks (54-89 hours). *Version 2:* 4-8 weeks (54-89 hours).

Exhibit Dates: *Version 1:* 1/93–Present. *Version 2:* 11/90–12/92.

Learning Outcomes: *Version 1:* Upon completion of the course, the student will be able to instruct and evaluate crewmembers in a dynamic high-activity flight management area. *Version 2:* Upon completion of the course, the student will be able to instruct and evaluate crewmembers in a dynamic high-activity flight management area.

Instruction: *Version 1:* This course consists of classroom lectures, programmed or computer-based training and appropriate aircraft and flight training laboratory experiences. *Version 2:* This course consists of classroom lectures, programed or computer-based training and appropriate aircraft and flight training laboratory experiences.

Credit Recommendation: *Version 1:* In the graduate degree category, Track IPA: 1 semester hour in instructional methodology and 2 in advanced flight laboratory. Track IPB: 1 semester hour in instructional methodology and 2 in advanced flight laboratory. IPC: 1 semester hour in instructional methodology and 2 in advanced flight laboratory (3/96). *Version 2:* In the graduate degree category, Track IPA: 1 semester hour in instructional methodology and 2 in advanced flight laboratory. Track IPB: 1 semester hour in instructional methodology and 2 in advanced flight laboratory. Track IPC: 1 semester hour in instructional methodology and 1 in advanced flight laboratory (3/96).

AF-1704-0285

BASIC OPERATIONAL AND TRANSITION/
REQUALIFICATION TRAINING A-10

Course Number: A1000TRAAD; A1000BOOAD.

Location: 355th Operations Group, Davis-Montham AFB, Tucson, AZ.

Length: 8-13 weeks (118-158 hours).

Exhibit Dates: 3/91–Present.

Learning Outcomes: Upon completion of the course, the student will be able to recognize, interpret, and explain advanced aircraft systems and the performance environment and perform advanced jet aircraft flight maneuvers and operations.

Instruction: This course consist of classroom lectures, programed or computer-based training, and appropriate aircraft and flight training device laboratory experiences.

Credit Recommendation: In the graduate degree category, Track B: 2 semester hours in advanced aircraft systems and performance and 3 in advanced flight laboratory. Track TX1: 2 semester hours in advanced aircraft systems and performance and 2 in advanced flight laboratory. Track TX2A: 1 semester hour in advanced aircraft systems and performance and 1 in advanced flight laboratory. Track TX2B: 2 semester hours in advanced aircraft systems and performance and 1 in advanced flight laboratory (3/96).

Air Force   49

AF-1704-0286

AIRCRAFT MAINTENANCE OFFICER (BRIDGE)

Course Number: J30LR21A1 008.

Location: Technical Training Center, Sheppard AFB, TX.

Length: 4 weeks (152 hours).

Exhibit Dates: 1/96–Present.

Learning Outcomes: Upon completion of the course, the student will be able to identify appropriate aviation maintenance strategies, coordinate responsibilities of maintenance control and other maintenance staff agencies, identify necessary maintenance training activities, apply appropriate supply and other logistics principles, identify and employ both analog and digital data collection methodologies, apply propulsion systems management concepts, use appropriate inspection policies and procedures, identify preventive maintenance strategies, confront health and safety issues, and perform various operations management and human resource management activities.

Instruction: Classroom and on-the-job lectures and laboratory activities use audiovisuals, mock-ups, serviceable equipment, various publications, and hands-on laboratory experiences in the maintenance of aircraft and aircraft systems and equipment. Case studies are used to reinforce concepts in operations, maintenance, and human resource management.

Credit Recommendation: In the graduate degree category, 1 semester hour in operations and human resource management, 1 in aviation maintenance operations and management, and 1 in industrial hygiene and safety (12/96).

AF-1709-0027

IMAGERY INTERPRETER SPECIALIST BY
  CORRESPONDENCE

Course Number: 20650.

Location: Extension Course Institute, Gunter AFS, AL.

Length: Maximum, 52 weeks.

Exhibit Dates: 1/90–8/98.

Learning Outcomes: Upon completion of the course, the student will be able to exploit aerial imagery, assist in the planning of reconnaissance missions, photogrammetrically measure the imagery, and prepare intelligence interpretation reports.

Instruction: Course method is monitored home study and includes readings, exercises, and proctored examinations. Topics include a review of basic military intelligence fundamentals; fundamentals and uses of maps and charts; coordinate systems; multisensor aerial reconnaissance techniques; imagery interpretation fundamentals, equipment, and facilities; photogrammetry; plotting; titling; forwarding and reporting aerial imagery; and tactical and strategic imagery interpretation including stereograms.

Credit Recommendation: In the lower-division baccalaureate/associate degree category, 2 semester hours in photographic interpretation or photogrammetry (5/88).

AF-1709-0028

STILL PHOTOGRAPHIC SPECIALIST BY
  CORRESPONDENCE

Course Number: 23152.

Location: Extension Course Institute, Gunter AFS, AL.

Length: Maximum, 52 weeks.

Exhibit Dates: 1/90–1/97.

Learning Outcomes: Upon completion of the course, the student will be able to take photographs in daylight, existing light and studio conditions in both black and white and color and process them in a darkroom. The student will also be able to use a variety of cameras, lenses, and other photographic accessories to enhance the subject matter as appropriate to the assignment. There is some emphasis in this course on record keeping, picture/negative storage, darkroom supervision, and quality control, as well as picture composition and artistic expression. The student will have an excellent background in technical aspects of optics, film composition, film processing, and filter control.

Instruction: Course content duplicates that in AF-1709-0029 but has more depth in the areas of darkroom procedures, quality control, and record keeping. Students enrolling must have an AFSC assignment as a prerequisite. This provides an application of the principles learned to real requirements and constitutes an essential part of the instruction by serving as the laboratory component of the course. The student's supervisor must complete an evaluation of the course and its effectiveness when the student takes the final exam which measures mastery of the cognitive content. A score of 65 is required for a satisfactory completion certificate.

Credit Recommendation: In the lower-division baccalaureate/associate degree category, 3 semester hours in principles of still photography (3/91); in the upper-division baccalaureate category, 3 semester hours in advanced photographic applications (3/91).

AF-1709-0029

APPRENTICE STILL PHOTOGRAPHIC SPECIALIST BY CORRESPONDENCE

Course Number: 23132.
Location: Extension Course Institute, Gunter AFS, AL.
Length: Maximum, 52 weeks.
Exhibit Dates: 1/90–3/96.
Learning Outcomes: Upon completion of the course, the student will be able to take photographs in daylight, existing light, and studio conditions in both black and white and in color and will be able to process and print them in a darkroom. The student will also be able to use a variety of cameras, lenses, and other photographic accessories to enhance the subject matter as appropriate to the assignment. There is some emphasis in this course in record keeping, picture/negative storage, darkroom supervision and quality control, as well as picture composition and artistic expression. The student will have an excellent background in the technical aspects of optics, film composition, film processing, and filter control.

Instruction: This is a four-part correspondence course. Method of instruction is through manuals which thoroughly discuss each segment of the course followed by exercises and examinations. Topics include photographic safety, photo lab supervision, light sources, exposure control, black and white film, optics, filters, documentary photography, photojournalism, studio photo, black and white and color film processing, and black and white and color printing. Students enrolling must have an AFSC assignment as a prerequisite. This provides an application of the principles learned to real requirements and constitutes an essential part of the instruction by serving as the laboratory components of the course. The student's supervisor must complete an evaluation of the course and its effectiveness when the student takes the final exam which measures mastery of the cognitive content. A score of 65 is required for a satisfactory completion certificate.

Credit Recommendation: In the lower-division baccalaureate/associate degree category, 3 semester hours in principles of still photography (3/91); in the upper-division baccalaureate category, 3 semester hours in advanced photographic applications (3/91).

AF-1709-0030

IMAGERY PRODUCTION SPECIALIST BY CORRESPONDENCE

Course Number: 23350.
Location: Extension Course Institute, Gunter AFS, AL.
Length: Maximum, 52 weeks.
Exhibit Dates: 1/90–1/97.
Learning Outcomes: Upon completion of the course, the student will be able to perform the duties of a laboratory technician in a still and motion picture processing laboratory. Knowledge and skills include laboratory management, safety, and operations; black and white and color film processing and reproductions; motion picture processing, printing, and editing; film chemical analysis and controls, calibration of test equipment; sensitometric controls in photographic printing; and statistical methods for quality control.

Instruction: Method of instruction consists of a six-part correspondence program with reading materials, exercises, and self-instruction examinations. Students enrolling must have an AFSC assignment as prerequisite. This provides an application of the principles learned to real requirements and constitutes an essential part of the instruction by serving as the laboratory component of the course. The student's supervisor must complete an evaluation of the course and its effectiveness when the student takes the final exam which measures mastery of the cognitive content. A score of 65 is required for a satisfactory completion certificate.

Credit Recommendation: In the vocational certificate category, 1 semester hour in calibration and testing of film processing equipment (3/91); in the lower-division baccalaureate/associate degree category, 3 semester hours in principles of still photography (3/91); in the upper-division baccalaureate category, 1 semester hour in photo laboratory management, 3 in motion picture film processing and editing, 3 in film chemistry, 1 in densitometry and sensitometry measurement, and 2 in statistical methods of quality control (3/91).

AF-1710-0021

PLUMBING SPECIALIST BY CORRESPONDENCE

Course Number: 55255.
Location: Extension Course Institute, Gunter AFS, AL.
Length: Maximum, 52 weeks.
Exhibit Dates: 1/90–10/94.
Objectives: This course trains individuals to be plumbing specialists in installation and maintenance of waste systems and water supply systems.

Instruction: This correspondence course consists of on-the-job supervision and at-home studies. The home study provides basic knowledge of installation and maintenance of waste and water supply systems. Also covered are basic plan reading, job safety, tool and equipment use, and project planning. Fire prevention and military contingency training are also covered.

Credit Recommendation: In the vocational certificate category, 3 semester hours in plumbing trades (9/86); in the lower-division baccalaureate/associate degree category, 2 semester hours in mechanical systems for buildings (9/86).

AF-1710-0023

PLUMBING TECHNICIAN BY CORRESPONDENCE

Course Number: 55275.
Location: Extension Course Institute, Gunter AFS, AL.
Length: Maximum, 52 weeks.
Exhibit Dates: 1/90–9/91.
Objectives: To train personnel to perform as plumbing technicians.

Instruction: This course of instruction consists of on-the-job training and at-home studies. The home study provides training to become a plumbing supervisor and the on-the-job training provides reinforcement for the home study. The course contains plumbing supervision and construction and operation of waste and water systems.

Credit Recommendation: In the vocational certificate category, 3 semester hours in advanced plumbing (9/86); in the lower-division baccalaureate/associate degree category, 2 in mechanical and electrical equipment in buildings (plumbing) (9/86).

AF-1710-0027

APPRENTICE PLUMBER BY CORRESPONDENCE

Course Number: 55235.
Location: Extension Course Institute, Gunter AFS, AL.
Length: Maximum, 52 weeks.
Exhibit Dates: 1/90–10/94.
Objectives: To train personnel to perform as apprentice plumbers.

Instruction: This course of instruction consists of on-the-job training and at-home studies. The home study provides training to perform as a plumber, and the on-the-job training provides reinforcement for the home study. The course includes water supply systems, building distribution systems, and interior and exterior plumbing systems.

Credit Recommendation: In the vocational certificate category, 3 semester hours in plumbing (9/86).

AF-1710-0028

CONSTRUCTION EQUIPMENT OPERATOR BY CORRESPONDENCE

Course Number: 55151.
Location: Extension Course Institute, Gunter AFS, AL.
Length: Maximum, 52 weeks.
Exhibit Dates: 1/90–10/94.
Objectives: To train individuals in the operation and maintenance of construction equipment with the primary emphasis being placed on earth-moving and compaction equipment.

Instruction: Course consists of on-the-job supervision and and at-home study. The at-home study provides knowledge regarding the operation and maintenance of construction equipment, and primary emphasis is placed upon earth moving and compaction. Topics covered include operating wheeled earth-loading equipment, crawler tractors, motorized graders, and excavation equipment and soil and surface preparation. Maintenance of equipment is covered as well as snow removal, ice control, and crane operation and rigging. The individual will also study basic welding principles and the use of explosives for clearing and quarrying operations.

Credit Recommendation: In the vocational certificate category, 6 semester hours in advanced construction equipment operation and maintenance (9/86).

AF-1710-0029

APPRENTICE CONSTRUCTION EQUIPMENT OPERATOR BY CORRESPONDENCE

Course Number: 55131.
Location: Extension Course Institute, Gunter AFS, AL.
Length: Maximum, 52 weeks.
Exhibit Dates: 1/90–10/95.
Objectives: To train individuals in the basic operation and maintenance of construction equipment.
Instruction: Course consists of on-the-job training and at-home study. The at-home study provides knowledge in the operation and maintenance of basic construction equipment. Topics covered include the operation of wheeled loading equipment, hauling equipment, motorized graders, and bulldozers. Operation and maintenance of auxiliary construction equipment such as pavement sweepers and snow plows are also discussed. Accident prevention, job safety, and first aid techniques are also covered.
Credit Recommendation: In the vocational certificate category, 3 semester hours in basic construction equipment operation and maintenance (9/86).

AF-1710-0030

APPRENTICE MASON BY CORRESPONDENCE

Course Number: 55231.
Location: Extension Course Institute, Gunter AFS, AL.
Length: Maximum of, 52 weeks.
Exhibit Dates: 1/90–1/97.
Objectives: To train personnel to perform as a mason's helper.
Instruction: This correspondence course consists of on-the-job training and at-home studies. The home study provides the background principles for the apprentice to perform masonry construction using concrete block, structural tile, surface tile, plaster, and stucco. The on-the-job training provides the apprentice with the practical application of these principles.
Credit Recommendation: In the vocational certificate category, 3 semester hours in masonry construction (9/86).

AF-1710-0031

MISSILE MAINTENANCE SPECIALIST (WS-133) BY CORRESPONDENCE

Course Number: 44350G.
Location: Extension Course Institute, Gunter AFS, AL.
Length: Maximum, 52 weeks.
Exhibit Dates: 1/90–5/92.

Learning Outcomes: Upon completion of the course, the student will be able to supervise subordinates and provide technical assistance in basic Minuteman assembly, repair, maintenance, modification, configuration, inspection, and servicing of missile subsystems and related support equipment.
Instruction: This is a supervised self-paced extension course covering topics of missile maintenance, technical orders, maintenance data collection systems, security and safety, corrosion, hand tools, aerospace hardware, principles of basic electricity, pneudraulics, layout of Minuteman facilities, operation and maintenance of equipment and special-purpose vehicles, equipment operations, and proofload testing equipment.
Credit Recommendation: In the vocational certificate category, 3 semester hours in heavy equipment operation (10/88).

AF-1710-0032

SPECIAL PURPOSE VEHICLE AND EQUIPMENT MECHANIC BY CORRESPONDENCE

Course Number: 47250.
Location: Extension Course Institute, Gunter AFS, AL.
Length: Maximum, 52 weeks.
Exhibit Dates: 1/90–3/99.
Learning Outcomes: Upon completion of the course, the student will be able to perform duties in maintenance and maintenance management (vehicles), including diesel and gasoline engine theory, power trains, hydraulic systems, alignment, wheel balancing, heating and air conditioning, vehicle equipment, forklifts and loaders, and towing and servicing.
Instruction: Course covers publications; maintenance management and procedures; tools and equipment management; internal combustion engines; cooling and lubricating systems; engine fuel system; air, exhaust, and emission control; automotive electrical system; electronic ignition; power trains; hydraulic system; chassis unit; front end alignment and wheel balancing; heating and air conditioning; heavy earth-moving equipment; sweepers; snow-removal equipment and miscellaneous equipment; fork lifts; aircraft cargo loaders; heaters; spray de-icers; and towing tractors.
Credit Recommendation: In the vocational certificate category, 3 semester hours in heavy equipment maintenance and 3 in heavy equipment operation (10/88).

AF-1710-0033

SPECIAL VEHICLE MECHANIC (FIRETRUCKS) BY CORRESPONDENCE

Course Number: 47251A.
Location: Extension Course Institute, Gunter AFS, AL.
Length: Maximum, 52 weeks.
Exhibit Dates: 1/90–3/99.
Learning Outcomes: Upon completion of the course, the student will be able to describe vehicle maintenance management; apply the theory of operation of diesel and gasoline engines and their power trains and hydraulic systems; perform front end alignment and wheel balancing, heating and air conditioning maintenance, and fire truck system maintenance.
Instruction: This is a supervised, self-paced extension course covering publications, maintenance forms, and records; vehicle management

Air Force 51

procedures; tools and equipment management; internal combustion engines; cooling and lubrication systems; air, exhaust, and emission control systems; automotive electrical system; electronic ignition; power train; hydraulic systems; chassis unit; front end alignment and wheel balancing; heating and air conditioning; and fire truck drive trains, dispensing systems, and winterization systems.
Credit Recommendation: In the vocational certificate category, 3 semester hours in heavy equipment mechanics (10/88); in the lower-division baccalaureate/associate degree category, 1 semester hour in fire science (10/88).

AF-1710-0034

SPECIAL VEHICLE MECHANIC (REFUELING VEHICLES) BY CORRESPONDENCE

Course Number: 47251B.
Location: Extension Course Institute, Gunter AFS, AL.
Length: Maximum, 52 weeks.
Exhibit Dates: 1/90–3/99.
Learning Outcomes: Upon completion of the course, the student will be able to perform duties as a special purpose vehicle (refueling vehicle) mechanic, including maintenance management; theory of operation of diesel and gasoline engines and their systems, power trains, hydraulic system, and chassis unit; front end alignment and wheel balancing; heating and air conditioning; and refueling vehicle maintenance.
Instruction: This is a supervised, self-paced extension course which includes publications, forms, and records; vehicle management and procedures; tools and equipment management; internal combustion engines and cooling and lubricating systems; air, exhaust, and emission control systems; electronic ignition; power trains; automotive electrical system; chassis unit; front end alignment and wheel balancing; heating and air conditioning; power take-off and throttle interlock system; dispensing system; hoses and reels; meters and static ground reels; bottom loading system; and winterization system.
Credit Recommendation: In the vocational certificate category, 3 semester hours in heavy equipment mechanic (10/88).

AF-1710-0035

VEHICLE BODY MECHANIC BY CORRESPONDENCE

Course Number: 47253.
Location: Extension Course Institute, Gunter AFS, AL.
Length: Maximum, 52 weeks.
Exhibit Dates: 1/90–Present.
Learning Outcomes: Upon completion of the course, the student will be able to perform as a vehicle mechanic and work with technical orders and supply publications, policies and material deficiency reports, vehicle operation and safety, and the allied trades.
Instruction: Instruction covers maintenance policies, publications, vehicle maintenance management, maintenance procedures, documents and reports, equipment management and use, automotive body repair, vehicle frame and hardware, sewing machine and upholstery, automotive gas arc welding, gas-shielded and oxyacetylene welding, heat exchanger, and fixed tanks.

Credit Recommendation: In the vocational certificate category, 3 semester hours in automotive body mechanics (10/88).

AF-1710-0036

VEHICLE MAINTENANCE CONTROL AND ANALYSIS SPECIALIST BY CORRESPONDENCE

Course Number: 47254.
Location: Extension Course Institute, Gunter AFS, AL.
Length: Maximum, 52 weeks.
Exhibit Dates: 1/90–Present.
Learning Outcomes: Upon completion of the course, the student will be able to qualify as a vehicle maintenance manager and perform maintenance data collection analysis.
Instruction: Instruction covers publications, maintenance management, maintenance procedures, documents and reports, tool and equipment management, data processing, data collection system, data inquiry system, standard analysis, and visual media and written presentations.
Credit Recommendation: In the lower-division baccalaureate/associate degree category, 1 semester hour in maintenance control (10/88).

AF-1710-0037

AEROSPACE PROPULSION TECHNICIAN (TURBOPROP) BY CORRESPONDENCE

Course Number: 45470B.
Location: Extension Course Institute, Gunter AFS, AL.
Length: Maximum, 52 weeks.
Exhibit Dates: 1/90–3/94.
Learning Outcomes: Upon completion of the course, the student will be able to maintain, troubleshoot, and repair turbopropeller systems and equipment.
Instruction: This is a correspondence course with home study and practical exercises in maintenance of turboprops. Topics include maintenance site orientation and supply functions, inspection systems and forms, gas turbine engines and systems. Inspection, maintenance and repair topics include prevention, troubleshooting, and testing of gas turbines, turboshaft, and turbopropellers including Hamilton Standard propellers. Students enrolling must have an AFSC assignment as a prerequisite. This provides an application of principles learned to real requirements and constitutes an essential part of the instruction by serving as a laboratory component to the course. Student's supervisor must complete an evaluation of the course and its effectiveness when the student takes the final examination which measures mastery of the cognitive content. A score of 65 is required for a satisfactory completion certificate.
Credit Recommendation: In the lower-division baccalaureate/associate degree category, 3 semester hours in gas turbine engines and 3 in turboprop engines and systems (3/91).

AF-1710-0038

AEROSPACE GROUND EQUIPMENT MECHANIC BY CORRESPONDENCE

Course Number: 45451.
Location: Extension Course Institute, Gunter AFS, AL.
Length: Maximum, 52 weeks.
Exhibit Dates: 1/90–6/97.
Learning Outcomes: Upon completion of the course, the student will be able to operate, maintain, troubleshoot, and repair aerospace ground equipment.
Instruction: This correspondence course combines home study and practical exercises in operation and maintenance of aerospace ground equipment. Topics include maintenance policies, practices, and procedures; electrical and electronic circuits and components; aircraft servicing equipment, including principles and functions of electrical motors and motor controls; gasoline, diesel, and gas turbine engines, generators, and controls; hydraulic test stands; bomb lifts; air compressors; and air conditioning, heating and ventilation systems. Course concludes with a proctored examination. Students enrolling must have an AFSC assignment as prerequisite. This provides an application of the principles learned to real requirements and constitutes an essential part of the instruction by serving as the laboratory component of the course. The student's supervisor must complete an evaluation of the course and its effectiveness when the student takes the final exam which measures mastery of the cognitive content. A score of 65 is requires for a satisfactory completion certificate.
Credit Recommendation: In the lower-division baccalaureate/associate degree category, 3 semester hours in aircraft ground equipment servicing and repair (3/91).

AF-1710-0039

CIVIL ENGINEERING CONTROL SYSTEM SPECIALIST BY CORRESPONDENCE

Course Number: 54533A.
Location: Extension Course Institute, Gunter AFS, AL.
Length: Maximum, 52 weeks.
Exhibit Dates: 1/90–10/94.
Learning Outcomes: Upon completion of this course, the student will be able to demonstrate general contingency responsibilities for survival, including first aid, moving and transporting injured personnel, and extreme weather protection; personal hygiene and sanitation in the field; camouflage, movement, fire control, and field fortification; planning, coordinating, and controlling convoys; terrorism; contingency airfield surfaces and expedient field construction methods; career progression and operations management; publications; security and safety; blueprints, scales, and diagrams; testing equipment; soldering; principles of electricity and alternating current; motors and motor controls; electronic principles; and solid state circuits.
Instruction: Course consists of self-paced independent study with open book progress checks and satisfactory completion of a proctored final examination. There is on-the-job supervision and an opportunity for the application of learning. Course includes first aid, maintenance and repair of electrical systems, and basic electrical and electronics theory.
Credit Recommendation: In the vocational certificate category, 1 semester hour in basic electrical repair (11/91); in the lower-division baccalaureate/associate degree category, 1 semester hour in first aid and 1 in basic electricity and electronics (11/91).

AF-1710-0040

CIVIL ENGINEERING CONTROL SYSTEMS SPECIALIST BY CORRESPONDENCE

Course Number: 54533B.
Location: Extension Course Institute, Gunter AFS, AL.
Length: Maximum, 52 weeks.
Exhibit Dates: 1/90–4/96.
Learning Outcomes: Upon completion of the course, the student will be able to describe heating, ventilation, and air conditioning (HVAC) fundamentals and theory and will be familiar with pneumatic control, electric control, and electronic hardware including complete control systems.
Instruction: Student enrolling must have an Air Force Specialty Code assignment as a prerequisite. This provides an application of the principles learned to real requirements and constitutes an essential part of the instruction by serving as the laboratory component of the course. The student's supervisor must complete an evaluation of the course and its effectiveness when the student takes the final examination which measures the mastery of the cognitive content. Course consists of self-paced, independent study with open book progress checks and satisfactory completion of a proctored final examination. Course includes HVAC fundamentals and theory and HVAC equipment and systems.
Credit Recommendation: In the lower-division baccalaureate/associate degree category, 2 semester hours in mechanical and electrical equipment of buildings (HVAC) (11/91).

AF-1710-0041

ENVIRONMENTAL SUPPORT SPECIALIST BY CORRESPONDENCE

Course Number: 56651.
Location: Extension Course Institute, Gunter AFS, AL.
Length: Maximum, 52 weeks.
Exhibit Dates: 1/90–10/94.
Learning Outcomes: Upon completion of the course, the student will be able to describe the management, organization, and general contingency responsibilities of civil engineering sanitation careers; water treatment processes; operation and maintenance requirements of water treatment plant units; analysis and processing of waste water; water distribution and waste water collection systems; and fundamentals as well as safety principles of electrical motors.
Instruction: Student enrolling must have an Air Force Specialty Code assignment as a prerequisite. This provides an application of the principles learned to real requirements and constitutes an essential part of the instruction by serving as the laboratory component of the course. The student's supervisor must complete an evaluation of the course and its effectiveness when the student takes the final examination which measures mastery of the cognitive content. Course consists of self-paced, independent study with open book progress checks and satisfactory completion of a proctored final examination. Topics covered include general knowledge of civil engineering; environmental support fundamentals; operation and maintenance of water plants; waste water analysis, treatment, equipment maintenance, and contingency operations; water distribution and waste water collection systems; and electrical motor fundamentals.
Credit Recommendation: In the lower-division baccalaureate/associate degree category, 1 semester hour in civil engineering and

electrical motor fundamentals and 3 in water analysis, treatment, processing, distribution, and collection (11/91).

AF-1714-0003

AIRCRAFT ELECTRICAL SYSTEMS SPECIALIST BY CORRESPONDENCE

Course Number: 42350.
Location: Extension Course Institute, Gunter AFS, AL.
Length: Maximum, 52 weeks.
Exhibit Dates: 1/90–1/97.
Objectives: To teach enlisted personnel maintenance management and documentation and the basic electrical theory necessary to maintain and repair aircraft electrical systems.
Instruction: This is a self-instruction course presenting materials pertaining to maintenance management and documentation, test equipment, basic AC and DC electrical theory, and the aircraft control and warning system. Aircraft electrical systems discussed include landing gear and related systems, flight control systems, starter systems, fuel systems, aircraft warning systems, lighting systems, and nesa glass systems.
Credit Recommendation: In the vocational certificate category, 3 semester hours in basic electrical theory (3/85).

AF-1714-0023

APPRENTICE ELECTRICIAN BY CORRESPONDENCE

Course Number: 54230.
Location: Extension Course Institute, Gunter AFS, AL.
Length: Maximum, 52 weeks.
Exhibit Dates: 1/90–10/94.
Objectives: To provide knowledge of electrical fundamentals to enable the apprentice electrician to increase skills.
Instruction: Course includes installation and maintenance of interior electrical power distribution systems for lighting and motor controls. Electrical codes, troubleshooting, estimating, and distribution are also presented. Three-phase power systems, motor controls, and emergency power systems are covered. The course concludes with a proctored examination.
Credit Recommendation: In the vocational certificate category, 2 semester hours in basic electric fundamentals or electrical wiring (3/85).

AF-1714-0024

ELECTRICAL POWER PRODUCTION TECHNICIAN BY CORRESPONDENCE

Course Number: 54370.
Location: Extension Course Institute, Gunter AFS, AL.
Length: Maximum, 52 weeks.
Exhibit Dates: 1/90–1/97.
Objectives: To train personnel in electrical power production operation and maintenance.
Instruction: Course consists of on-the-job training and at-home studies in the operation and maintenance of electrical power production. The course includes engine maintenance, engine systems, power generation and control, electrical devices and power plant switch gear, power plant equipment, and gas turbines.
Credit Recommendation: In the vocational certificate category, 3 semester hours in advanced electrical power plant operation and maintenance (9/86).

AF-1714-0028

ELECTRICIAN BY CORRESPONDENCE

Course Number: 54250.
Location: Extension Course Institute, Gunter AFS, AL.
Length: Maximum, 52 weeks.
Exhibit Dates: 1/90–10/94.
Objectives: Upon completion of the course, the student will be trained in the installation of electrical systems and the installation and maintenance of motors, controls, and special equipment.
Instruction: This course consists of on-the-job training and at-home studies in the installation and maintenance of various types of electrical systems, motors, controls, and special equipment. The course also includes basic electricity.
Credit Recommendation: In the vocational certificate category, 3 semester hours in electrical systems and 3 in electric motors and equipment (9/86); in the lower-division baccalaureate/associate degree category, 2 semester hours in electrical and mechanical equipment in buildings (electrical) (9/86).

AF-1714-0029

FUNDAMENTALS OF ELECTRICITY BY CORRESPONDENCE

Course Number: 3030.
Location: Extension Course Institute, Gunter AFS, AL.
Length: Maximum, 52 weeks.
Exhibit Dates: 1/90–7/96.
Learning Outcomes: Upon completion of the course, the student will be able to accomplish basic algebraic and trigonometric functions; construct graphs; and describe basic elements of electricity and direct current circuits as well as magnetism and electromagnetism. Student will be able to describe the principles of direct current generators and motors as well as operation of meters, theory of capacitors, physical factors affecting inductance, and transient phenomena in electronics. Students will have a mastery of alternating current and circuits and will be able to perform graphical analysis of AC circuits.
Instruction: This is a self-paced extension course designed to cover review of algebra, graphs, and simultaneous linear equations; elements of electricity; DC circuits and magnetism; direct current, including generators, motors, meters, inductors, transients, and alternating current; elementary trigonometry; AC circuit principles; and analysis of AC circuits.
Credit Recommendation: In the lower-division baccalaureate/associate degree category, 1 semester hour in introduction to electricity (10/88); in the upper-division baccalaureate category, 1 semester hour in advanced electricity (AC) and 1 in advanced electricity (DC) (10/88).

AF-1714-0030

ELECTRICAL POWER PRODUCTION SPECIALIST BY CORRESPONDENCE

Course Number: 54252A.
Location: Extension Course Institute, Gunter AFS, AL.
Length: Maximum, 52 weeks.
Exhibit Dates: 1/90–5/93.
Learning Outcomes: After 5/93 see AF-1714-0043. Upon completion of the course, the student will be able to maintain electrical generators, electronic-controlled equipment, and switch gear components and associated circuits.
Instruction: Course topics include magnetism, fundamentals of AC/DC circuits and generators, exciters, and alternators; types and characteristics of solid state devices; power supplies; logic functions; truth tables; Boolean equations; logic applications and gates; principles of voltage regulators, battery chargers, governors, rectifiers, and brushless exciters; and switch gear components.
Credit Recommendation: In the vocational certificate category, 3 semester hours in fundamentals of electricity (AC/DC circuits) (12/89); in the lower-division baccalaureate/associate degree category, 3 semester hours in electronics (12/89).

AF-1714-0031

ELECTRICAL POWER PRODUCTION TECHNICIAN BY CORRESPONDENCE

Course Number: 54272B.
Location: Extension Course Institute, Gunter AFS, AL.
Length: Maximum, 52 weeks.
Exhibit Dates: 1/90–Present.
Learning Outcomes: Upon completion of the course, the student will be able to maintain gasoline and diesel engines as well as power production auxiliary equipment and aircraft arresting systems.
Instruction: Course covers maintenance of gasoline and diesel engines, including two-stroke and four-stroke-cycle engine operation, fuel and lubricating system maintenance, cooling systems, governing systems, induction and exhaust systems, internal engine analysis, engine protective devices, maintenance of power production auxiliary equipment, and aircraft arresting systems.
Credit Recommendation: In the vocational certificate category, 2 semester hours in diesel engine system maintenance, 2 in electrical power plant maintenance, and 2 in electrical power production engine maintenance (12/89).

AF-1714-0032

RELAYS, GENERATORS, MOTORS, AND ELECTROMECHANICAL DEVICES BY CORRESPONDENCE

Course Number: 3035.
Location: Extension Course Institute, Gunter AFS, AL.
Length: Maximum, 52 weeks.
Exhibit Dates: 1/90–Present.
Learning Outcomes: Upon completion of the course, the student will be able to describe the principles of electromagnetism and relays, time-delay, AC and DC motors and generators, synchro devices, control transformers, and servomechanisms.
Instruction: Through correspondence, the student receives information on electromagnetism and relays, protection, time-delay and voltage control circuits, AC and DC motors and generators, synchro devices, control transformers, and servomechanisms.
Credit Recommendation: Credit is not recommended because of the limited, specialized nature of the course (12/89).

AF-1714-0033

SOLDERING AND ELECTRICAL CONNECTORS BY CORRESPONDENCE

Course Number: 3036.
Location: Extension Course Institute, Gunter AFS, AL.
Length: Maximum, 52 weeks.
Exhibit Dates: 1/90–Present.
Learning Outcomes: Upon completion of the course, the student will be able to prepare and solder electrical wiring and components to terminal ends and circuit boards and install wire terminals and connector pins using the crimp process. Student will know the safety aspects and correct tools and materials necessary to form aerospace-quality solder connections.
Instruction: Through correspondence, the student receives information on soldering safety and tools, types of wires and terminals, correct procedures for soldering, PC board preparation for soldered leads and terminals, and pin installation in electrical connectors.
Credit Recommendation: Credit is not recommended because of the limited, specialized nature of the course (12/89).

AF-1714-0034

POWER MEASUREMENTS BY CORRESPONDENCE

Course Number: 03037.
Location: Extension Course Institute, Gunter AFS, AL.
Length: Maximum, 52 weeks.
Exhibit Dates: 1/90–Present.
Learning Outcomes: Upon completion of the course, the student will be able to measure power from the lowest line frequencies to those in the microwave range.
Instruction: This is a correspondence course with home study and practical exercises in the operation of power-measuring instruments. Topics include noise and signal power level units, power measurements and basic measuring techniques, and power measurements at high RF and microwave frequencies.
Credit Recommendation: Credit is not recommended because of the limited technical nature of the course content (3/91).

AF-1714-0036

COMMUNICATIONS CABLE SYSTEMS INSTALLATION/MAINTENANCE SPECIALIST BY CORRESPONDENCE

Course Number: 36151A.
Location: Extension Course Institute, Gunter AFS, AL.
Length: Maximum, 52 weeks.
Exhibit Dates: 11/91–10/97.
Learning Outcomes: Upon completion of the course, the student will be able to demonstrate knowledge of cable systems and their characteristics, installation, and tools and techniques for cable splicing.
Instruction: Instruction covers basic cables including their electrical characteristics, aerial and underground installation, and cable splicing, terminating, protecting, and tagging.
Credit Recommendation: Credit is not recommended because of the limited, specialized nature of the course (4/93).

AF-1714-0037

COMMUNICATIONS CABLE SYSTEMS INSTALLATION/MAINTENANCE SPECIALIST BY CORRESPONDENCE

Course Number: 36151B.
Location: Extension Course Institute, Gunter AFS, AL.
Length: Maximum, 52 weeks.
Exhibit Dates: 11/91–Present.
Learning Outcomes: Upon completion of the course, the student will be able to apply principles, techniques, and procedures for installing, testing, and splicing various types of cable.
Instruction: This is a continuation of course 36151A and presents a more advanced study of cables, including installation, testing, fault isolation, pressurization, leak location, conductor splicing, system maintenance, and fiber optic cable installation, testing, and troubleshooting.
Credit Recommendation: Credit is not recommended because of the limited, specialized nature of the course (4/93).

AF-1714-0038

MISSILE CONTROL COMMUNICATIONS SYSTEMS SPECIALIST BY CORRESPONDENCE

Course Number: 36253.
Location: Extension Course Institute, Gunter AFS, AL.
Length: Maximum, 52 weeks.
Exhibit Dates: 8/92–Present.
Learning Outcomes: Upon completion of the course, the student will be able to repair and maintain the Minuteman I and Minuteman II missile control communications systems and the peripheral equipment associated with these systems.
Instruction: This is a correspondence course presenting a review of basic electricity and electronics, test equipment, troubleshooting and repair which lays the foundation for the repair and maintenance of the Minuteman I and II missile control communications systems and their peripheral equipment.
Credit Recommendation: Credit is not recommended because of the limited, specialized nature of the course (4/93).

AF-1714-0039

INTEGRATED AVIONIC COMMUNICATION, NAVIGATION AND PENETRATION AIDS SYSTEMS SPECIALIST BY CORRESPONDENCE

Course Number: 45252C.
Location: Extension Course Institute, Gunter AFS, AL.
Length: Maximum, 52 weeks.
Exhibit Dates: 3/90–3/98.
Learning Outcomes: Upon completion of the course, the student will be able to perform fault diagnosis, troubleshooting, and repair/adjustment/replacement of integrated avionics systems. The student will be able to maintain sophisticated computer, radar, navigation, and RF energy systems.
Instruction: This correspondence course presents job-related knowledge required by a 5-level specialist in integrated avionic communication, navigation, and penetration aids systems. The course includes aircraft safety and maintenance, radio selection circuits, and operational checkouts; provides for an overview of radio communications concepts and principles, UHF, VHF, air navigation principles, and instrument landing systems; and describes identification systems and radar signal interception.
Credit Recommendation: In the lower-division baccalaureate/associate degree category, 3 semester hours in introduction to communications systems (3/93).

AF-1714-0040

ELECTRICAL POWER PRODUCTION JOURNEYMAN BY CORRESPONDENCE
(Electrical Power Production Specialist)

Course Number: 54252B.
Location: Extension Course Institute, Gunter AFS, AL.
Length: Maximum, 52 weeks.
Exhibit Dates: 8/93–Present.
Learning Outcomes: Upon completion of the course, the student will be able to maintain and troubleshoot gasoline and diesel engines, electrical power equipment, auxiliary systems, and aircraft arresting systems.
Instruction: This correspondence course presents information on engine fundamentals, intake and exhaust systems, cooling and lubricating systems, fuel and governor systems, pneumatics and motors, and power generators.
Credit Recommendation: In the lower-division baccalaureate/associate degree category, 2 semester hours in diesel engine systems maintenance, 2 in electrical power plant maintenance, and 2 in electrical power production engine maintenance (6/94).

AF-1714-0041

MISSILE AND SPACE SYSTEMS ELECTRONIC MAINTENANCE JOURNEYMAN BY CORRESPONDENCE
(Air Launched Missile Systems Specialist by Correspondence)

Course Number: 46650A.
Location: Extension Course Institute, Gunter AFS, AL.
Length: Maximum, 52 weeks.
Exhibit Dates: 6/90–1/98.
Learning Outcomes: Upon completion of the course, the student will be able to describe the fundamentals of DC and AC electrical/magnetism, transformers, generator synchros, electrical circuits, and solid state devices and circuits.
Instruction: This is a correspondence course where, through written exercises, the student learns motor theory, Ohm's law, Watt's law, and Kirchhoff's law. Other topics include transistor theory and test equipment.
Credit Recommendation: In the lower-division baccalaureate/associate degree category, 1 semester hour in basic electricity and 1 in solid state circuits (6/94).

AF-1714-0042

ELECTRICAL SYSTEMS JOURNEYMAN BY CORRESPONDENCE
(Electrical Systems Technician by Correspondence)

Course Number: 54250A.
Location: Extension Course Institute, Gunter AFS, AL.
Length: Maximum, 52 weeks.
Exhibit Dates: 2/93–3/99.
Learning Outcomes: Upon completion of the course, the student will be able to install electrical systems.
Instruction: This course consists of on-the-job training and at-home studies in the installation and maintenance of various types of electrical systems, motors, controls, and special equipment. The course includes basic electricity.
Credit Recommendation: In the vocational certificate category, 2 semester hours in electri-

cal equipment installation and maintenance (6/94).

AF-1714-0043

ELECTRICAL POWER PRODUCTION JOURNEYMAN BY CORRESPONDENCE

(Electrical Power Production Specialist by Correspondence)

Course Number: 54252A.
Location: Extension Course Institute, Gunter AFS, AL.
Length: Maximum, 52 weeks.
Exhibit Dates: 6/93–10/98.
Learning Outcomes: Before 6/93 see AF-1714-0030. Upon completion of the course, the student will be able to maintain electrical generators, electronic-controlled equipment, and switch gear components and associated circuits.
Instruction: Course includes magnetism; fundamentals of AC/DC circuits and generators, exciters and alternators; types and characteristics of solid state devices; power supplies; logic functions; truth tables; Boolean equations; logic application and gates; principles of voltage regulators, battery chargers, governors, rectifiers, and brushes exciters; and switch gear components.
Credit Recommendation: In the lower-division baccalaureate/associate degree category, 1 semester hour in AC/DC circuits and 1 in solid state electronics (6/94).

AF-1714-0044

ELECTRICAL SYSTEMS JOURNEYMAN BY CORRESPONDENCE

Course Number: 54250B.
Location: Extension Course Institute, Gunter AFS, AL.
Length: Maximum, 52 weeks.
Exhibit Dates: 7/93–Present.
Learning Outcomes: Upon completion of the course, the student will be able to install and maintain motors, controls, and special equipment.
Instruction: This course of instruction consists of on-the-job training and at-home studies in the installation and maintenance of various types of electrical systems, motors, controls, and special equipment.
Credit Recommendation: In the vocational certificate category, 2 semester hours in electrical equipment installation and maintenance (6/94).

AF-1714-0045

AIRCRAFT ELECTRICAL AND ENVIRONMENTAL SYSTEMS JOURNEYMAN BY CORRESPONDENCE

Course Number: 45455.
Location: Extension Course Institute, Gunter AFS, AL.
Length: Maximum, 52 weeks.
Exhibit Dates: 11/93–Present.
Learning Outcomes: Upon completion of the course, the student will be able to troubleshoot and correct faults in the engine bleed air system, aircraft air conditioning systems, cabin pressurization systems, fire detection and extinguishing systems, oxygen systems, utility systems (life raft, liquid cycle refrigeration), and aircraft electrical control and warning systems.
Instruction: Through correspondence and on-the-job training, the student receives information on engine bleed air systems; aircraft air conditioning, cabin pressurization, and oxygen systems; fire protection systems; utility systems (life raft, liquid cycle refrigeration); and aircraft electrical control and warning systems.
Credit Recommendation: Credit is not recommended because content is general and not subject-specific in sufficient depth (6/94).

AF-1715-0003

AIR TRAFFIC CONTROL RADAR REPAIRMAN BY CORRESPONDENCE

(Air Traffic Control Radar Specialist by Correspondence)

Course Number: *Version 1:* 30351. *Version 2:* 30351.
Location: Extension Course Institute, Gunter AFS, AL.
Length: *Version 1:* Maximum, 52 weeks. *Version 2:* Maximum, 52 weeks.
Exhibit Dates: *Version 1:* 9/91–7/96. *Version 2:* 1/90–8/91.
Learning Outcomes: *Version 1:* Upon completion of the course, the student will be able to identify specialized hand tools used in soldering, cable boring, and removal and replacement of electronic devices. The student will be able to describe the theory of the operation of TWTs, magnetrons, klystrons, wave guides, wave guide devices, and the theory of operation and maintenance of a radar set. Types of radar sets include air traffic control, precision approach, and AIMS. *Version 2:* Upon completion of the course, the student will have expertise in troubleshooting and repair of air traffic radar systems, including air traffic control transmitters and receiving systems, moving target indicators, precision approach radar, and the AIMS system.
Instruction: *Version 1:* This course covers Air Force procedures for training, general maintenance procedures, soldering techniques, fundamentals and operation of electronic test equipment, description and limits of various air traffic control radars, fundamentals of wave guides, antennas and scanning systems, moving radar target detection, remote data transmission, and principles of precision approach radars and aircraft transponder systems. *Version 2:* This course covers Air Force procedures for training, general maintenance procedures, soldering techniques, fundamentals and operation of electronic test equipment, description and limits of various air traffic control radars, fundamentals of wave guides, antennas and scanning systems, moving radar target detection, remote data transmission, and principles of precision approach radars and aircraft transponder systems.
Credit Recommendation: *Version 1:* In the lower-division baccalaureate/associate degree category, 3 semester hours in basic microwave systems (4/93). *Version 2:* In the vocational certificate category, 3 semester hours in electronic troubleshooting (11/86); in the lower-division baccalaureate/associate degree category, 3 semester hours in fundamentals of radar (11/86).

AF-1715-0005

AIRCRAFT CONTROL AND WARNING RADAR SPECIALIST BY CORRESPONDENCE

Course Number: 30352.
Location: Extension Course Institute, Gunter AFS, AL.
Length: Maximum, 52 weeks.
Exhibit Dates: 1/90–7/96.
Objectives: Provide a basic knowledge of digital electronics as well as a basic overview of high-power search radars.
Instruction: This correspondence course covers Air Force training procedures; digital techniques; various numbering systems; digital logic and Boolean algebra; clock and pulse generators; sequential logic; combination logic and converters; peripherals and storage devices; semiconductors; corrosion control; description of high-power radar systems; radar receivers and calibration devices; and support equipment such as generators, aircraft transponders, and portable search radars.
Credit Recommendation: In the lower-division baccalaureate/associate degree category, 3 semester hours in fundamentals of digital techniques and 3 in radar fundamentals (11/86).

AF-1715-0007

AUTOMATIC TRACKING RADAR SPECIALIST BY CORRESPONDENCE

Course Number: 30353.
Location: Extension Course Institute, Gunter AFS, AL.
Length: Maximum, 52 weeks.
Exhibit Dates: 1/90–9/93.
Objectives: To enable the electronic technician to troubleshoot automatic tracking radar systems.
Instruction: This course covers Air Force training procedures; basic electronic maintenance techniques; soldering and cabling; basic electronic test equipment; fundamentals of transistors and vacuum tube amplifiers; synchro devices; operational amplifiers; fundamentals of tracking radars; and analog computation and its use in tracking radars, radar receivers, and radar control systems.
Credit Recommendation: In the vocational certificate category, 3 semester hours in electronic maintenance procedures and 2 in radar fundamentals (11/86).

AF-1715-0009

AIR TRAFFIC CONTROL RADAR TECHNICIAN BY CORRESPONDENCE

Course Number: 30371.
Location: Extension Course Institute, Gunter AFS, AL.
Length: Maximum, 52 weeks.
Exhibit Dates: 1/90–5/95.
Learning Outcomes: Upon completion of the course, the student will be able to troubleshoot and repair air traffic control radar and electronic circuits to the component level and will have an in-depth knowledge of digital electronics.
Instruction: The course covers material management, maintenance organization, electronic fundamentals, DC and AC circuits, basics of electron tubes and semiconductors, transistor and associated circuits, amplifiers, oscillators, multivibrators, power supplies, numbering systems, digital logic, storage devices, timing, sequential logic, combination logic, and converters.
Credit Recommendation: In the vocational certificate category, 3 semester hours in maintenance techniques (11/91); in the lower-division baccalaureate/associate degree category, 1 semester hour in basic electronics and 3 in digital techniques (11/91).

AF-1715-0011

AUTOMATIC TRACKING RADAR TECHNICIAN BY CORRESPONDENCE

Course Number: 30373.
Location: Extension Course Institute, Gunter AFS, AL.
Length: Maximum, 52 weeks.
Exhibit Dates: 1/90–1/97.
Objectives: To enhance the radar technician's understanding of the principles of tracking radars and overall maintenance management.
Instruction: Course provides instruction in maintenance organization and record keeping, basic television scan techniques, basic principles of radar target acquisition and tracking, and analog computation.
Credit Recommendation: Credit is not recommended because of the limited, specialized nature of the course (11/86).

AF-1715-0012

NAVIGATIONAL AIDS EQUIPMENT SPECIALIST BY CORRESPONDENCE

Course Number: 30451.
Location: Extension Course Institute, Gunter AFS, AL.
Length: Maximum, 52 weeks.
Exhibit Dates: 1/90–11/92.
Objectives: To produce an individual who is knowledgeable in the general principles of air navigational ground equipment.
Instruction: Course provides a very limited discussion of electronic principles, a fundamental discussion of aircraft landing systems (localizer and glide slope), very high frequency and low frequency marker and beacon transmitters, and an excellent discussion of the principles of the omnirange, tactical air navigation (TACAN) systems as well as digital monitoring equipment.
Credit Recommendation: In the vocational certificate category, 3 semester hours in air navigational systems (11/86).

AF-1715-0013

TELEVISION EQUIPMENT REPAIRMAN BY CORRESPONDENCE

Course Number: 30455.
Location: Extension Course Institute, Gunter AFS, AL.
Length: Maximum, 52 weeks.
Exhibit Dates: 1/90–10/93.
Learning Outcomes: Upon completion of the course, enlisted personnel will be able to repair and maintain television equipment.
Instruction: Course offers correspondence study with on-the-job supervision and proctored testing for training television equipment repairmen. The student is introduced to such topics as electric circuits, magnetism, and semiconductor and transistor circuits. Also covered are numbering systems, Boolean algebra, digital circuits, conversion circuits, and storage devices. The television portion covers television test equipment, television system theory, repair, and maintenance.
Credit Recommendation: In the vocational certificate category, 2 semester hours in electronics, 2 in digital circuits, and 6 in television equipment specialist (11/86).

AF-1715-0014

SPACE COMMUNICATIONS SYSTEMS EQUIPMENT BY CORRESPONDENCE
(Satellite Communications System Equipment Specialist by Correspondence)

Course Number: 30456.
Location: Extension Course Institute, Gunter AFS, AL.
Length: Maximum, 52 weeks.
Exhibit Dates: 1/90–11/95.
Objectives: Upon completion of the course, the student will be able to troubleshoot and repair satellite communications systems including super high frequency and ultra high frequency systems.
Instruction: Course is offered as correspondence study with on-the-job supervision and proctored testing to train enlisted personnel to be space communications systems equipment operators/specialists. The student is introduced to waveform measurement using an oscilloscope, electronic circuits, digital systems, semiconductor devices, and sequential logic. Multiplexing techniques, digital communications, and installation and maintenance techniques for satellite systems are also covered.
Credit Recommendation: In the vocational certificate category, 3 semester hours in electronics (11/91); in the lower-division baccalaureate/associate degree category, 3 semester hours in digital circuits (11/91); in the upper-division baccalaureate category, 3 semester hours in digital communications (11/91).

AF-1715-0015

WIDEBAND COMMUNICATIONS EQUIPMENT TECHNICIAN BY CORRESPONDENCE

Course Number: 30470.
Location: Extension Course Institute, Gunter AFS, AL.
Length: Maximum, 52 weeks.
Exhibit Dates: 1/90–5/92.
Objectives: Upon completion of the course, enlisted personnel will be trained as wideband communication equipment technicians.
Instruction: This is correspondence study with on-the-job supervision and proctored testing to train enlisted personnel as wideband communication equipment technicians. Topics such as system performance monitoring, wideband system siting, topographic maps, microwave propagation, waveguides and antennas are covered.
Credit Recommendation: In the upper-division baccalaureate category, 1 semester hour in communications theory (11/86).

AF-1715-0017

NAVIGATIONAL AIDS EQUIPMENT TECHNICIAN BY CORRESPONDENCE

Course Number: 30471.
Location: Extension Course Institute, Gunter AFS, AL.
Length: Maximum, 52 weeks.
Exhibit Dates: 1/90–2/92.
Objectives: The technician is instructed in fundamentals of electronics, digital operations, and installation of instrument landing systems.
Instruction: The course presents a general discussion of maintenance management, technical publications, test equipment, use of the oscilloscope, and basics of DC and AC circuits. Higher level discussion of numbering systems, digital logic, Boolean algebra, clock and pulse generators, sequential logic, combination logic and converters, computer peripherals and storage devices, as well as installation of instrument landing systems are included.
Credit Recommendation: In the vocational certificate category, 1 semester hour in fundamentals of electronic circuits (11/86); in the lower-division baccalaureate/associate degree category, 3 semester hours in fundamentals of digital techniques (11/86).

AF-1715-0021

SPACE COMMUNICATIONS SYSTEMS EQUIPMENT OPERATOR/TECHNICIAN BY CORRESPONDENCE

Course Number: 30476.
Location: Extension Course Institute, Gunter AFS, AL.
Length: Maximum, 52 weeks.
Exhibit Dates: 1/90–6/90.
Objectives: To upgrade space communications systems equipment specialists to the technician level.
Instruction: This correspondence study course covers security of space communications systems, maintenance management, processing and controlling materiel, and a limited review of defense satellite communications systems and satellite access and tracking.
Credit Recommendation: Credit is not recommended because of the limited, specialized nature of the course (11/86).

AF-1715-0024

ELECTRONIC CRYPTOGRAPHIC COMMUNICATIONS EQUIPMENT SPECIALIST BY CORRESPONDENCE

Course Number: 30650.
Location: Extension Course Institute, Gunter AFS, AL.
Length: Maximum, 52 weeks.
Exhibit Dates: 1/90–9/92.
Objectives: The student will be provided the training required to upgrade enlisted personnel from the apprentice level to electronic communication equipment specialist.
Instruction: This is offered through correspondence study with on-the-job supervision and proctored testing to train electronic cryptographic communications equipment specialists. The student is introduced to such topics as electronics, communication systems, modems, modulation, filters, multiplexers, interface devices, data conversion, and data and line conditioning. Maintenance practices and test equipment are also covered.
Credit Recommendation: In the vocational certificate category, 2 semester hours in electronics and 1 in maintenance practices (11/86); in the upper-division baccalaureate category, 2 semester hours in communications theory (11/86).

AF-1715-0031

TELECOMMUNICATIONS SYSTEM MAINTENANCE SPECIALIST BY CORRESPONDENCE

Course Number: 30653.
Location: Extension Course Institute, Gunter AFS, AL.
Length: Maximum, 52 weeks.
Exhibit Dates: 1/90–9/92.
Objectives: To train enlisted personnel in telecommunications system maintenance.

Instruction: This correspondence course with proctored examinations includes topics on test equipment, soldering, hand tools, basic circuit theory, transistors, and a little logic theory. Also covered are computers, peripherals, and storage media. Extensive coverage is given to the mechanical operation of three teletypewriter models.

Credit Recommendation: In the vocational certificate category, 3 semester hours in basic electronics and 3 in computer maintenance (11/86).

AF-1715-0034

TELECOMMUNICATIONS SYSTEMS CONTROL TECHNICIAN BY CORRESPONDENCE

Course Number: 30770.
Location: Extension Course Institute, Gunter AFS, AL.
Length: Maximum, 52 weeks.
Exhibit Dates: 1/90–1/97.
Objectives: To upgrade technical personnel in the field of telecommunications system control and supervision.
Instruction: This correspondence course with proctored examinations includes topics in decibels, noise, voice channel multiplexing, signal compression, modems, and data transmission systems. Telecommunication system supervision and quality assurance procedures are also covered.
Credit Recommendation: In the vocational certificate category, 3 semester hours in management technology (11/86); in the upper-division baccalaureate category, 2 semester hours in telecommunication theory (11/86).

AF-1715-0035

AIRCRAFT ARMAMENT SYSTEMS SPECIALIST BY CORRESPONDENCE

Course Number: 46250.
Location: Extension Course Institute, Gunter AFS, AL.
Length: Maximum, 52 weeks.
Exhibit Dates: 1/90–4/97.
Objectives: To provide upgrade training in aircraft munitions and armament maintenance systems.
Instruction: This correspondence course consists of on-the-job supervision and home study. The home study provides technical knowledge in the field of aircraft armament systems, and the on-the-job supervision provides training to reinforce and apply the technical knowledge. Course includes security and hardware/hand tool control; fundamentals of electricity; armament maintenance; loading support equipment; and weapons launch and release systems for attack, interceptor, fighter, and bomber aircraft.
Credit Recommendation: In the vocational certificate category, 3 semester hours in fundamentals of electricity, 3 in electronics, and 2 in shop management and property control (9/86).

AF-1715-0036

SPACE SYSTEMS EQUIPMENT MAINTENANCE SPECIALIST BY CORRESPONDENCE

Course Number: 30950.
Location: Extension Course Institute, Gunter AFS, AL.
Length: Maximum, 52 weeks.
Exhibit Dates: 1/90–8/99.
Objectives: The student will be trained in space systems equipment maintenance.
Instruction: This correspondence course includes topics on test equipment, circuits, solid state devices, digital logic circuits, computers, peripherals, and storage. Also included are transmitters, receivers, and antennas.
Credit Recommendation: In the vocational certificate category, 2 semester hours in electronics (11/86); in the lower-division baccalaureate/associate degree category, 3 semester hours in fundamentals of digital techniques (11/86).

AF-1715-0038

MISSILE SYSTEMS MAINTENANCE SPECIALIST BY CORRESPONDENCE

Course Number: 31650G.
Location: Extension Course Institute, Gunter AFS, AL.
Length: Maximum, 52 weeks.
Exhibit Dates: 1/90–1/97.
Objectives: Personnel are trained as missile systems maintenance specialists.
Instruction: This correspondence course provides instruction in missile systems maintenance including maintenance management and maintenance and repair of the security and access systems, the command and control systems, and the power and environmental systems in the specialized missile systems.
Credit Recommendation: In the vocational certificate category, 3 semester hours in industrial maintenance (11/86).

AF-1715-0043

MISSILE SYSTEMS MAINTENANCE SPECIALIST BY CORRESPONDENCE

Course Number: 31650T.
Location: Extension Course Institute, Gunter AFS, AL.
Length: Maximum, 52 weeks.
Exhibit Dates: 1/90–5/92.
Objectives: Personnel are trained as missile maintenance specialists.
Instruction: This correspondence course provides instruction in the maintenance of military missile systems. Specific content includes maintenance management, corrosion control, electronic principles, special circuits, digital principles, digital logic, Boolean algebra, sequential logic, maintenance of specific missile power and control systems, electronic systems testing, and analysis of missile control system.
Credit Recommendation: In the vocational certificate category, 3 semester hours in industrial maintenance and 3 in instrumentation (11/86); in the lower-division baccalaureate/associate degree category, 2 semester hours in fundamentals of digital techniques and 1 in electronics (11/86).

AF-1715-0045

INSTRUMENTATION MECHANIC BY CORRESPONDENCE

Course Number: 31653.
Location: Extension Course Institute, Gunter AFS, AL.
Length: Maximum, 52 weeks.
Exhibit Dates: 1/90–8/97.
Objectives: Personnel are trained as instrumentation mechanics.
Instruction: This correspondence course provides instruction in instrumentation mechanics, including maintenance management, corrosion control, instrumentation principles and systems, instrumentation test equipment and tools, transducers, lasers, digital logic, Boolean algebra, analog signal conditioners, modulation and multiplexing systems, radio frequency components, and instrumentation ground stations.
Credit Recommendation: In the vocational certificate category, 3 semester hours in instrumentation (11/86); in the lower-division baccalaureate/associate degree category, 3 semester hours in electronic instrumentation and 1 in fundamentals of digital techniques (11/86).

AF-1715-0046

BOMB NAVIGATION SYSTEMS SPECIALIST (B-52G/H: ASQ-176, ASQ-151 SYSTEMS) BY CORRESPONDENCE

Course Number: 32150.
Location: Extension Course Institute, Gunter AFS, AL.
Length: Maximum, 52 weeks.
Exhibit Dates: 1/90–10/91.
Objectives: A review of electronic principles as applied to maintenance of B-52G/H systems, ASQ-176, and ASQ-151.
Instruction: This is a correspondence course which offers a review of electronic principles, sequential logic, semiconductor devices, electro-optical viewing systems, and radar.
Credit Recommendation: Credit is not recommended because of the military-specific nature of the course (11/86).

AF-1715-0047

DEFENSIVE FIRE CONTROL SYSTEMS MECHANIC (B-52D/G: MD-9, ASG-15 TURRETS) BY CORRESPONDENCE

Course Number: 32151G.
Location: Extension Course Institute, Gunter AFS, AL.
Length: Maximum, 52 weeks.
Exhibit Dates: 1/90–10/94.
Objectives: The student will be trained in the maintenance and alignment of fire control systems, ASG-15 turret.
Instruction: Course covers basics of radar, modes of operation of ASG-15 radar, and operation of 0.50 caliber machine gun and specific test equipment such as AN/UPM-33 Radar Test Set and AN/UPM-141 radar test set.
Credit Recommendation: In the vocational certificate category, 1 semester hour in basics of radar (11/86).

AF-1715-0050

WEAPON CONTROL SYSTEMS MECHANIC (A-7D, A-10, AC-130, F-5) BY CORRESPONDENCE

Course Number: 32152.
Location: Extension Course Institute, Gunter AFS, AL.
Length: Maximum, 52 weeks.
Exhibit Dates: 1/90–6/92.
Objectives: The student will be trained in maintenance of fire control systems, including the APQ-153, AN/ASG-29, and tactical computer set AN/ASN-91.
Instruction: This is a correspondence course that offers principles of electronics (diodes and power supplies, transistor circuits, oscillators, logic circuits) as well as description of specific military equipment.

COURSE EXHIBITS

Credit Recommendation: In the vocational certificate category, 1 semester hour in basics of electronics (11/86).

AF-1715-0052

WEAPON CONTROL SYSTEMS MECHANIC (F-4C/D: APQ-109/APA-165) BY CORRESPONDENCE

Course Number: 32152P.
Location: Extension Course Institute, Gunter AFS, AL.
Length: Maximum, 52 weeks.
Exhibit Dates: 1/90–7/90.
Objectives: To provide training in maintenance of fire control systems including the APQ-109/APA-165.
Instruction: This correspondence course includes electric circuit analysis and basics of vacuum tube circuits, along with theory and operation of radio equipment APQ-109/APA-165.
Credit Recommendation: In the vocational certificate category, 1 semester hour in basic DC/AC circuit analysis (11/86).

AF-1715-0053

WEAPON CONTROL SYSTEMS MECHANIC (F-4E: APQ-120) BY CORRESPONDENCE

Course Number: 32152Q.
Location: Extension Course Institute, Gunter AFS, AL.
Length: Maximum, 52 weeks.
Exhibit Dates: 1/90–7/94.
Objectives: To provide training for maintenance of fire control systems including the APQ-120.
Instruction: Contents include basic electronic circuits using semiconductors, along with theory and maintenance of radar equipment APQ-120.
Credit Recommendation: In the vocational certificate category, 1 semester hour in semiconductor circuits (11/86).

AF-1715-0054

WEAPON CONTROL SYSTEMS MECHANIC (A-7D: AN/APQ-126) BY CORRESPONDENCE

Course Number: 32152S.
Location: Extension Course Institute, Gunter AFS, AL.
Length: Maximum, 52 weeks.
Exhibit Dates: 1/90–10/92.
Objectives: To provide training for maintenance of fire control systems including the AN/APQ-126.
Instruction: Materials are presented by correspondence. Topics include basic electronic circuitry, along with specific maintenance instructions for the AN/APQ-126 radar system.
Credit Recommendation: In the vocational certificate category, 1 semester hour in transistor circuitry and 1 in introduction to number systems and digital logic (11/86).

AF-1715-0056

WEAPON CONTROL SYSTEMS TECHNICIAN (F-4C/D: APQ-109/APA-165) BY CORRESPONDENCE

Course Number: 32172P.
Location: Extension Course Institute, Gunter AFS, AL.
Length: Maximum, 52 weeks.
Exhibit Dates: 1/90–8/92.
Objectives: The student will be trained in the maintenance of radar sets APQ-109 and APA-165 as required for upgrade to Level 7 AFSC.
Instruction: This is a correspondence course with topics including specific information on operation and maintenance of radar sets APQ-109 and APA-165.
Credit Recommendation: Credit is not recommended because of the military-specific nature of the course (11/86).

AF-1715-0063

WEAPON CONTROL SYSTEMS TECHNICIAN (F-4E: APQ-120) BY CORRESPONDENCE

Course Number: 32172Q.
Location: Extension Course Institute, Gunter AFS, AL.
Length: Maximum, 52 weeks.
Exhibit Dates: 1/90–7/94.
Objectives: The student is provided a review of digital logic principles and devices as required for upgrade training to Level 7 AFSC.
Instruction: This is a correspondence course presenting a brief review of number systems, digital components, and standard test equipment.
Credit Recommendation: Credit is not recommended because of the limited, specialized nature of the course (11/86).

AF-1715-0068

AVIONIC SENSOR SYSTEMS SPECIALIST (RECONNAISSANCE ELECTRONIC SENSORS) BY CORRESPONDENCE

Course Number: 32252A.
Location: Extension Course Institute, Gunter AFS, AL.
Length: Maximum, 52 weeks.
Exhibit Dates: 1/90–1/97.
Objectives: Upon completion of the course, the student will be upgraded from the apprentice level to the specialist level in avionic sensor systems specializing in reconnaissance electronic sensors.
Instruction: Correspondence materials for this course include shop practices and test equipment, electronics principles, infrared sensors, and radar sensors.
Credit Recommendation: In the vocational certificate category, 2 semester hours in electronics principles and 1 in radar principles (11/86); in the lower-division baccalaureate/associate degree category, 2 semester hours in digital principles (11/86).

AF-1715-0070

AVIONIC SENSOR SYSTEMS SPECIALIST (TACTICAL/REAL TIME DISPLAY ELECTRONIC SENSORS) BY CORRESPONDENCE

Course Number: 32252B.
Location: Extension Course Institute, Gunter AFS, AL.
Length: Maximum, 52 weeks.
Exhibit Dates: 1/90–2/92.
Objectives: The student will be provided the training required to upgrade from the apprentice level to the specialist level in avionics sensor systems specializing in display electronic sensors.
Instruction: Course provides correspondence materials covering shop practices and test equipment, basic electricity and electronics, and a number of specific military display systems.
Credit Recommendation: In the vocational certificate category, 3 semester hours in basic electricity and electronics (11/86).

AF-1715-0071

AVIONIC SENSOR SYSTEMS SPECIALIST (ELECTRO-OPTICAL SENSORS) BY CORRESPONDENCE

Course Number: 32252C.
Location: Extension Course Institute, Gunter AFS, AL.
Length: Maximum, 52 weeks.
Exhibit Dates: 1/90–1/97.
Objectives: The student will be provided the training required to upgrade from the apprentice level to the specialist level in avionics sensor systems specializing in electro-optical sensors.
Instruction: Course is presented through correspondence materials including shop practices and test equipment, electronics, and the maintenance of specific military electro-optical systems.
Credit Recommendation: In the vocational certificate category, 3 semester hours in basic electronics and troubleshooting (11/86).

AF-1715-0073

PRECISION MEASURING EQUIPMENT SPECIALIST BY CORRESPONDENCE

Course Number: 32450.
Location: Extension Course Institute, Gunter AFS, AL.
Length: Maximum, 52 weeks.
Exhibit Dates: 1/90–8/98.
Objectives: Training is provided to upgrade enlisted personnel from the apprentice level to the specialist level in precision measurement equipment.
Instruction: These correspondence materials includes specific military equipment used for voltage and power measurement, waveform analysis, microwave measurements, and time domain data and standards.
Credit Recommendation: Credit is not recommended because of the limited, specialized nature of the course (11/86).

AF-1715-0074

PRECISION MEASURING EQUIPMENT TECHNICIAN BY CORRESPONDENCE

Course Number: 32470.
Location: Extension Course Institute, Gunter AFS, AL.
Length: Maximum, 52 weeks.
Exhibit Dates: 1/90–1/98.
Objectives: Training is provided to upgrade enlisted personnel from the specialist level to technician level in precision measuring equipment.
Instruction: These correspondence materials include survey treatment of typical precision laboratory equipment. Brief descriptions of the theory of operation of equipment for dimensional measurements, electrical measurements, and nuclear radiation measurements are included.
Credit Recommendation: Credit is not recommended because of the limited, specialized nature of the course (11/86).

AF-1715-0075

AUTOMATIC FLIGHT CONTROL SYSTEMS SPECIALIST BY CORRESPONDENCE

Course Number: 32550.

Location: Extension Course Institute, Gunter AFS, AL.
Length: Maximum, 52 weeks.
Exhibit Dates: 1/90–6/90.
Objectives: To provide the training required to upgrade enlisted personnel from apprentice level to specialist level in automatic flight control systems.
Instruction: These correspondence materials include a survey of basic electrical and electronic theory and more in-depth treatments of specific aircraft flight control systems including those for the B52 G/H and the C141.
Credit Recommendation: In the vocational certificate category, 3 semester hours in basic electrical theory (11/86).

AF-1715-0076

AVIONICS AEROSPACE GROUND EQUIPMENT SPECIALIST BY CORRESPONDENCE

Course Number: 32650.
Location: Extension Course Institute, Gunter AFS, AL.
Length: Maximum, 52 weeks.
Exhibit Dates: 1/90–4/93.
Objectives: The student will be knowledgeable in the basics of semiconductor theory, amplifiers, and computer systems, as well as in test stations and avionics.
Instruction: This course consists of correspondence study, with on-the-job supervision, and proctored testing. Topics include transistor theory, amplifiers, computer systems, and maintenance.
Credit Recommendation: In the vocational certificate category, 2 semester hours in semiconductor circuits and 1 in digital fundamentals (11/86).

AF-1715-0077

AVIONICS AEROSPACE GROUND EQUIPMENT SPECIALIST (F/RF-4 PECULIAR AVIONICS AGE) BY CORRESPONDENCE

Course Number: 32650C.
Location: Extension Course Institute, Gunter AFS, AL.
Length: Maximum, 52 weeks.
Exhibit Dates: 1/90–3/93.
Objectives: The student will be knowledgeable in the basics of amplifiers, logic devices, computers, and avionics maintenance.
Instruction: Materials are presented through correspondence study with on-the-job supervision and proctored testing. Topics include transistor amplifiers, logic devices, computers, and maintenance.
Credit Recommendation: In the vocational certificate category, 1 semester hour in transistor amplifiers and 1 in computers (11/86).

AF-1715-0078

A-7D AVIONICS AEROSPACE GROUND EQUIPMENT SPECIALIST BY CORRESPONDENCE

Course Number: 32650D.
Location: Extension Course Institute, Gunter AFS, AL.
Length: Maximum, 52 weeks.
Exhibit Dates: 1/90–1/97.
Objectives: To provide the basics of semiconductors, digital circuits, and test equipment.
Instruction: This a correspondence course with on-the-job supervision and proctored testing. Topics includes semiconductor devices, digital logic, Boolean algebra, sequential and combination logic, and test equipment.
Credit Recommendation: In the lower-division baccalaureate/associate degree category, 1 semester hour in digital circuits (11/86).

AF-1715-0080

AVIONICS INSTRUMENT SYSTEMS SPECIALIST BY CORRESPONDENCE

Course Number: 32551.
Location: Extension Course Institute, Gunter AFS, AL.
Length: Maximum, 52 weeks.
Exhibit Dates: 1/90–6/90.
Objectives: To provide the training required to upgrade enlisted personnel from apprentice level to specialist level in avionic instrument systems.
Instruction: Correspondence materials include a survey of basic electrical and electronic theory and the theory of operation of typical flight instruments. Specific flight instruments are used as examples and to illustrate the theory.
Credit Recommendation: In the vocational certificate category, 3 semester hours in basic electricity (11/86).

AF-1715-0081

INTEGRATED AVIONICS ELECTRONIC WARFARE EQUIPMENT AND COMPONENT SPECIALIST (F/FB-111) BY CORRESPONDENCE

Course Number: 32653A.
Location: Extension Course Institute, Gunter AFS, AL.
Length: Maximum, 52 weeks.
Exhibit Dates: 1/90–1/97.
Objectives: The student will be knowledgeable in the basics of electronic principles and electronic countermeasure (ECM) principles, as well as in ECM systems and ECM test equipment for F-111 aircraft.
Instruction: This is a correspondence course with on-the-job supervision and proctored testing. Topics include solid state regulators, basic logic principles, digital integrated circuits, digital troubleshooting techniques, receiver fundamentals, and ECM principles. F-111 ECM systems and test equipment are also covered.
Credit Recommendation: In the lower-division baccalaureate/associate degree category, 2 semester hours in digital fundamentals (11/86).

AF-1715-0082

INTEGRATED AVIONICS ELECTRONIC WARFARE EQUIPMENT AND COMPONENT SPECIALIST (F-15) BY CORRESPONDENCE

Course Number: 32653B.
Location: Extension Course Institute, Gunter AFS, AL.
Length: Maximum, 52 weeks.
Exhibit Dates: 1/90–10/91.
Objectives: To provide the basics of solid state technology, an introduction to the TEWS intermediate test station, and maintenance management.
Instruction: This is a correspondence course with on-the-job supervision and proctored testing. Topics include basic semiconductor theory, basic logic principles, digital integrated circuits, and TEWS test station control and management.
Credit Recommendation: In the lower-division baccalaureate/associate degree category, 2 semester hours in digital fundamentals (11/86).

AF-1715-0086

INTEGRATED AVIONICS COMPUTERIZED TEST STATION AND COMPONENT SPECIALIST (F-15) BY CORRESPONDENCE

Course Number: 32654B.
Location: Extension Course Institute, Gunter AFS, AL.
Length: Maximum, 52 weeks.
Exhibit Dates: 1/90–10/91.
Objectives: To provide the basics of digital computer fundamentals, test station computers, and automatic test equipment for the F-15.
Instruction: Materials are presented through correspondence study with on-the-job supervision and proctored testing. Topics include digital fundamentals, troubleshooting, automatic test equipment, and test stations.
Credit Recommendation: In the lower-division baccalaureate/associate degree category, 2 semester hours in digital fundamentals (11/86).

AF-1715-0087

INTEGRATED AVIONICS COMPUTERIZED TEST STATION AND COMPONENT SPECIALIST (F/FB-111) BY CORRESPONDENCE

Course Number: 32654A.
Location: Extension Course Institute, Gunter AFS, AL.
Length: Maximum, 52 weeks.
Exhibit Dates: 1/90–1/97.
Objectives: To provide the basics of digital computers, automatic test equipment, and test stations for the F-111.
Instruction: Course is offered as correspondence study with on-the-job supervision and proctored testing. Topics include fundamentals of digital computers and circuits, troubleshooting techniques, automatic test equipment, and test stations.
Credit Recommendation: In the lower-division baccalaureate/associate degree category, 2 semester hours in digital computer fundamentals (11/86).

AF-1715-0090

INTEGRATED AVIONICS ATTACK CONTROL SYSTEMS SPECIALIST (F/FB-111) BY CORRESPONDENCE

Course Number: 32656A.
Location: Extension Course Institute, Gunter AFS, AL.
Length: Maximum, 52 weeks.
Exhibit Dates: 1/90–1/97.
Objectives: The basics of the F-111 radar and navigation systems are provided.
Instruction: This course is presented through correspondence study with on-the-job supervision and proctored testing. Topics include F-111 familiarization, display systems, and radar and navigation systems.
Credit Recommendation: Credit is not recommended because of the military-specific nature of the course (11/86).

AF-1715-0091

INTEGRATED AVIONICS ATTACK CONTROL SYSTEMS SPECIALIST BY CORRESPONDENCE

Course Number: 32656.

Location: Extension Course Institute, Gunter AFS, AL.
Length: Maximum, 52 weeks.
Exhibit Dates: 1/90–1/97.
Objectives: An overview of electronics, computer systems, and communications is provided.
Instruction: This is a correspondence course with on-the-job supervision and proctored testing. Topics includes basic tools, wiring, solid state, integrated circuits, logic gates, computer systems, communications, and maintenance management.
Credit Recommendation: In the vocational certificate category, 2 semester hours in mechanical tools and wiring and 2 in computer fundamentals and systems (11/86); in the lower-division baccalaureate/associate degree category, 2 semester hours in communications (11/86).

AF-1715-0092

INTEGRATED AVIONICS MANUAL TEST STATION AND COMPONENT SPECIALIST (F-15) BY CORRESPONDENCE

Course Number: 32655B.
Location: Extension Course Institute, Gunter AFS, AL.
Length: Maximum, 52 weeks.
Exhibit Dates: 1/90–10/91.
Objectives: The student will be knowledgeable in the basic test equipment, airborne navigational systems, and test stations.
Instruction: This is a correspondence course with on-the-job supervision and proctored testing. Equipment covered includes circuit, voltage, and resistance measurements; waveform analysis; and UHF communications testing.
Credit Recommendation: In the vocational certificate category, 3 semester hours in test equipment and measurements (11/86).

AF-1715-0093

INTEGRATED AVIONICS MANUAL TEST STATION AND COMPONENT SPECIALIST (F/FB-111) BY CORRESPONDENCE

Course Number: 32655A.
Location: Extension Course Institute, Gunter AFS, AL.
Length: Maximum, 52 weeks.
Exhibit Dates: 1/90–1/97.
Objectives: The student will become knowledgeable in basic test equipment, airborne navigation systems, and test stations.
Instruction: This is a correspondence course with on-the-job supervision and proctored testing. Equipment covered includes current, voltage, and resistance measurements; waveform analysis, and UHF communications testing.
Credit Recommendation: In the vocational certificate category, 3 semester hours in test equipment and measurements (11/86).

AF-1715-0094

INTEGRATED AVIONICS INSTRUMENT AND FLIGHT CONTROL SYSTEMS SPECIALIST (F/FB/EF-111) BY CORRESPONDENCE

Course Number: 32657A.
Location: Extension Course Institute, Gunter AFS, AL.
Length: Maximum, 52 weeks.
Exhibit Dates: 1/90–1/97.
Objectives: Training is provided to upgrade to specialist level in integrated avionics instrument and flight controls for F/FB/EF-111 aircraft.
Instruction: Course consists of correspondence materials including aircraft familiarization and aircraft (F/FB/EF-111) instruments and flight controls.
Credit Recommendation: Credit is not recommended because of the limited, specialized nature of the course (11/86).

AF-1715-0095

INTEGRATED AVIONIC INSTRUMENT AND FLIGHT CONTROL SYSTEMS SPECIALIST BY CORRESPONDENCE

Course Number: 32657.
Location: Extension Course Institute, Gunter AFS, AL.
Length: Maximum, 52 weeks.
Exhibit Dates: 1/90–1/97.
Objectives: The required training is provided to upgrade enlisted personnel from apprentice to specialist level in integrated instrument and flight controls.
Instruction: Correspondence materials include aircraft maintenance management, survey materials in digital and computer fundamentals, and common laboratory hand tools.
Credit Recommendation: In the vocational certificate category, 1 semester hour in tool usage and 2 in digital and computer fundamentals (11/86).

AF-1715-0096

INTEGRATED AVIONIC ATTACK CONTROL SYSTEMS SPECIALIST (F-16) BY CORRESPONDENCE

Course Number: 32656C.
Location: Extension Course Institute, Gunter AFS, AL.
Length: Maximum, 52 weeks.
Exhibit Dates: 1/90–8/91.
Objectives: To instruct students in the basics of the F-16 navigation and fire control systems.
Instruction: Curriculum materials for this correspondence course offer on-the-job supervision and proctored testing. Topics include F-16 familiarization, display systems, airborne video systems, and fire control systems.
Credit Recommendation: Credit is not recommended because of the military-specific nature of the course (11/86).

AF-1715-0097

INTEGRATED AVIONICS ATTACK CONTROL SYSTEMS SPECIALIST (F-15) BY CORRESPONDENCE

Course Number: 32656B.
Location: Extension Course Institute, Gunter AFS, AL.
Length: Maximum, 52 weeks.
Exhibit Dates: 1/90–1/97.
Learning Outcomes: Upon completion of the course, the student will be able to describe the fundamentals and operation of avionic combat control and navigation systems.
Instruction: Topics include F-15 familiarization, display and control system, radar, and test systems. Student will receive instruction concerning aircraft familiarization, basic navigation, inertial navigation systems, radar principles, and operation.
Credit Recommendation: Credit is not recommended because of the military-specific nature of the course (11/86).

AF-1715-0098

INTEGRATED AVIONICS INSTRUMENT AND FLIGHT CONTROL SYSTEMS SPECIALIST (F-15) BY CORRESPONDENCE

Course Number: 32657B.
Location: Extension Course Institute, Gunter AFS, AL.
Length: Maximum, 52 weeks.
Exhibit Dates: 1/90–2/92.
Objectives: Training is provided to upgrade enlisted personnel from apprentice level to specialist level in integrated avionic instrument and flight controls for F-15 aircraft.
Instruction: Correspondence materials including fundamentals of F-15 aircraft maintenance and F-15 instruments and flight control systems comprise this course.
Credit Recommendation: Credit is not recommended because of the limited, specialized nature of the course (11/86).

AF-1715-0099

INTEGRATED AVIONICS INSTRUMENT AND FLIGHT CONTROL SYSTEMS SPECIALIST (F-16) BY CORRESPONDENCE

Course Number: 32657C.
Location: Extension Course Institute, Gunter AFS, AL.
Length: Maximum, 52 weeks.
Exhibit Dates: 1/90–1/97.
Objectives: Training is provided to upgrade enlisted personnel from apprentice level to specialist level in integrated avionics instruments and flight control systems for F-16 aircraft.
Instruction: The correspondence materials for this course include basic F-16 aircraft maintenance and F-16 instruments and flight control system maintenance.
Credit Recommendation: Credit is not recommended because of the limited, specialized nature of the course (11/86).

AF-1715-0100

INTEGRATED AVIONICS COMMUNICATION, NAVIGATION, AND PENETRATION AIDS SYSTEMS SPECIALIST BY CORRESPONDENCE

Course Number: 32658.
Location: Extension Course Institute, Gunter AFS, AL.
Length: Maximum, 52 weeks.
Exhibit Dates: 1/90–1/97.
Objectives: Training is provided to upgrade enlisted personnel from apprentice level to specialist level in integrated avionics communications, navigation, and penetration aids systems.
Instruction: Correspondence materials include aircraft maintenance management, survey materials in hand tool usage, and digital and computer fundamentals.
Credit Recommendation: In the vocational certificate category, 1 semester hour in tool usage and 2 in digital and computer fundamentals (11/86).

AF-1715-0101

INTEGRATED AVIONICS COMMUNICATION, NAVIGATION, AND PENETRATION AIDS SYSTEMS SPECIALIST (F/FB-111) BY CORRESPONDENCE

Course Number: 32658A.
Location: Extension Course Institute, Gunter AFS, AL.
Length: Maximum, 52 weeks.

Exhibit Dates: 1/90–1/97.

Objectives: Training is provided to upgrade enlisted personnel from apprentice level to specialist level in integrated avionics communication and navigation and penetration aids systems for F/FB-111 aircraft.

Instruction: Offered are correspondence materials including F/FB-111 aircraft familiarization and F/FB-111 communication, navigation, and penetration aids system maintenance.

Credit Recommendation: Credit is not recommended because of the limited, specialized nature of the course (11/86).

AF-1715-0106

INTEGRATED AVIONICS COMMUNICATIONS, NAVIGATION, AND PENETRATION AIDS SYSTEMS SPECIALIST (F-15) BY CORRESPONDENCE

Course Number: 32658B.
Location: Extension Course Institute, Gunter AFS, AL.
Length: Maximum, 52 weeks.
Exhibit Dates: 1/90–7/91.

Objectives: Training is provided to upgrade enlisted personnel from apprentice level to specialist level in integrated avionics communications, navigation, and penetration aids systems for F-15 aircraft.

Instruction: This program comprises correspondence materials including F-15 aircraft familiarization and F-15 communication, navigation, and penetration aids system maintenance.

Credit Recommendation: Credit is not recommended because of the limited, specialized nature of the course (11/86).

AF-1715-0116

COMMUNICATIONS/NAVIGATION SYSTEMS TECHNICIAN BY CORRESPONDENCE

Course Number: 32870.
Location: Extension Course Institute, Gunter AFS, AL.
Length: Maximum, 52 weeks.
Exhibit Dates: 1/90–8/92.

Objectives: To present avionics supervision and management concepts and fundamentals for upgrading technical personnel into supervisory positions.

Instruction: The course presents avionics supervision and management, inspections, and maintenance systems. Also presented are specific Air Force publications and Air Force supply procedures including maintenance management.

Credit Recommendation: Credit is not recommended because of the limited, specialized nature of the course (11/86).

AF-1715-0122

AIRBORNE COMMAND POST COMMUNICATIONS EQUIPMENT SPECIALIST BY CORRESPONDENCE

Course Number: 32855.
Location: Extension Course Institute, Gunter AFS, AL.
Length: Maximum, 52 weeks.
Exhibit Dates: 1/90–3/97.

Objectives: To provide specialty knowledge for upgrade training.

Instruction: This correspondence course offers a review of tube and solid state circuitry, number systems, digital logic, and maintenance techniques.

Credit Recommendation: In the vocational certificate category, 1 semester hour in electronic circuits (11/86).

AF-1715-0123

AIRBORNE WARNING AND CONTROL RADAR SPECIALIST BY CORRESPONDENCE

Course Number: 32852.
Location: Extension Course Institute, Gunter AFS, AL.
Length: Maximum, 52 weeks.
Exhibit Dates: 1/90–10/91.

Objectives: To provide training to upgrade enlisted personnel from apprentice level to specialist level in airborne warning and control radar.

Instruction: The correspondence course includes maintenance supervision, hand tools, general test equipment operation, and a survey of digital and computer fundamentals and radar principles. Also included are maintenance training in typical airborne radar equipment.

Credit Recommendation: In the vocational certificate category, 1 semester hour in tool usage and general test equipment operation (11/86); in the lower-division baccalaureate/associate degree category, 1 semester hour in digital and computer fundamentals and 1 in radar principles (11/86).

AF-1715-0126

CABLE SPLICING INSTALLATION AND MAINTENANCE SPECIALIST BY CORRESPONDENCE

Course Number: 36151.
Location: Extension Course Institute, Gunter AFS, AL.
Length: Maximum, 52 weeks.
Exhibit Dates: 1/90–11/93.

Objectives: Personnel are trained as cable splicing installation and maintenance specialists.

Instruction: This correspondence course provides instruction in cable splicing, installation, and maintenance. The course includes cable system fundamentals, cable splicing and sealing, cable system maintenance, splice case installation, cable maintenance, and hardened intersite cable system's pressure systems.

Credit Recommendation: In the vocational certificate category, 4 semester hours in communications systems (cable installation/maintenance) (11/86).

AF-1715-0127

TELEPHONE CENTRAL OFFICE SWITCHING EQUIPMENT SPECIALIST, ELECTRONIC/ ELECTROMECHANICAL BY CORRESPONDENCE

Course Number: 36251.
Location: Extension Course Institute, Gunter AFS, AL.
Length: Maximum, 52 weeks.
Exhibit Dates: 1/90–7/91.

Objectives: Personnel are trained as telephone central office switching equipment specialists (electronic/electromechanical).

Instruction: This correspondence course provides instruction in the electronic/electromechanical telephone central office switching equipment. The course includes electronic principles; telephone fundamentals; manual telephone systems; specific telephone systems; and fundamentals of digital techniques, including digital logic, Boolean algebra, clock/pulse generators, sequential logic, combination logic and converters, and computers, peripherals and storage devices.

Credit Recommendation: In the vocational certificate category, 3 semester hours in communications systems (electromechanical telephone systems) (11/86); in the lower-division baccalaureate/associate degree category, 3 semester hours in the fundamentals of digital techniques (11/86).

AF-1715-0130

MISSILE CONTROL COMMUNICATIONS SYSTEMS REPAIRMAN BY CORRESPONDENCE

Course Number: 3625B.
Location: Extension Course Institute, Gunter AFS, AL.
Length: Maximum, 52 weeks.
Exhibit Dates: 1/90–1/97.

Objectives: Personnel are taught to repair missile control communications systems.

Instruction: This correspondence course provides instruction in missile control communications systems. The course includes electrical fundamentals; basic electronic circuits; fundamentals of missile control communications systems; specific missile communications systems; and maintenance, inspection, and testing management of missile control systems.

Credit Recommendation: In the vocational certificate category, 3 semester hours in communications systems (11/86).

AF-1715-0131

TELEPHONE EQUIPMENT INSTALLATION AND REPAIR SPECIALIST BY CORRESPONDENCE

Course Number: 36254.
Location: Extension Course Institute, Gunter AFS, AL.
Length: Maximum, 52 weeks.
Exhibit Dates: 1/90–9/98.

Objectives: The student will be trained as a telephone equipment installation and repair specialist.

Instruction: This correspondence course provides instruction in telephone equipment installation and repair. The course includes basic electricity, electronic circuits, telephone construction and repair, substation installation, and key telephone systems.

Credit Recommendation: In the vocational certificate category, 3 semester hours in communications systems (11/86).

AF-1715-0136

PRECISION IMAGERY AND AUDIOVISUAL MEDIA MAINTENANCE SPECIALIST BY CORRESPONDENCE

Course Number: 40450.
Location: Extension Course Institute, Gunter AFS, AL.
Length: Maximum, 52 weeks.
Exhibit Dates: 1/90–11/94.

Objectives: To provide training in precision imagery and audiovisual media maintenance.

Instruction: This correspondence course provides instruction in precision imagery and audiovisual media maintenance. The course includes electronics fundamentals; advanced electronics and test equipment; maintenance of photographic equipment; maintenance of processors; quality control equipment; relocatable laboratories; audiovisual production systems; and the fundamentals of digital techniques, including digital logic, Boolean algebra, clock/pulse generators, and sequential logic. It also includes semiconductor devices and waveform measurement using an oscilloscope.

Credit Recommendation: In the lower-division baccalaureate/associate degree category, 2 semester hours in fundamentals of digital techniques, 2 in basic electricity/electronics, and 3 in photography (11/86).

AF-1715-0139

AEROSPACE PHOTOGRAPHIC SYSTEMS SPECIALIST BY CORRESPONDENCE

Course Number: 40451.
Location: Extension Course Institute, Gunter AFS, AL.
Length: Maximum, 52 weeks.
Exhibit Dates: 1/90–1/97.
Learning Outcomes: Upon completion of the course, students will be able to function as aerospace photographic systems specialists.
Instruction: This correspondence course provides instruction in aerospace photographic systems. The course includes electronic fundamentals; advanced electronics; test equipment; photographic principles;, aerospace photographic systems repair; airborne video systems and related equipment; and the fundamentals of digital techniques, including digital logic, Boolean algebra, clocks/pulse generators, and sequential logic. It also includes corrosion control, semiconductor devices, and measurement using an oscilloscope.
Credit Recommendation: In the lower-division baccalaureate/associate degree category, 2 semester hours in the fundamentals of digital techniques, 2 in basic electricity/electronics, and 3 in photography (11/86).

AF-1715-0208

WIDEBAND COMMUNICATIONS EQUIPMENT SPECIALIST BY CORRESPONDENCE

Course Number: 30450.
Location: Extension Course Institute, Gunter AFS, AL.
Length: Maximum, 52 weeks.
Exhibit Dates: 1/90–11/95.
Objectives: To upgrade the technical skills of communications technicians in analog and digital data communications, wideband radio transmission methods, and equipment maintenance.
Instruction: The course provides an introduction to analog and digital signals and systems for data communications. It includes continuous and sampled-data waveform characteristics and modulation and multiplexing methods with hardware implementations. Microwave principles, propagation, wideband communication systems including noise, and performance measurement are included.
Credit Recommendation: In the lower-division baccalaureate/associate degree category, 2 semester hours in introduction to communications systems (3/85).

AF-1715-0227

GROUND RADIO COMMUNICATIONS SPECIALIST BY CORRESPONDENCE

Course Number: 30454.
Location: Extension Course Institute, Gunter AFS, AL.
Length: Maximum, 52 weeks.
Exhibit Dates: 1/90–3/92.
Objectives: To upgrade the technical skills of communications technicians in electronic devices and circuits, antennas, communications systems, and maintenance techniques.
Instruction: This program provides an introduction to electronics fundamentals, devices, and circuits for both linear and digital applications and includes electronic measurements. Also covered are a description of communications circuits, modulation and demodulation methods, and radio transmission systems. The course also includes mobile radio communications, AM, FM, SSB, antenna theory, and propagation. A final proctored examination is required.
Credit Recommendation: In the lower-division baccalaureate/associate degree category, 2 semester hours in introduction to communications systems (3/85).

AF-1715-0390

TELECOMMUNICATIONS SYSTEM CONTROL SPECIALIST BY CORRESPONDENCE

Course Number: 30750.
Location: Extension Course Institute, Gunter AFS, AL.
Length: Maximum, 52 weeks.
Exhibit Dates: 1/90–12/91.
Objectives: To teach personnel to monitor and analyze performance of radio and wire telecommunications circuits and equipment and to maintain and repair that equipment.
Instruction: The course is presented through self-instruction materials in microwave circuits and theory, antennas and transmission theory, and data transmission and telecommunications. Also included are review materials in semiconductor devices and digital electronics.
Credit Recommendation: In the lower-division baccalaureate/associate degree category, 5 semester hours in communications systems and telecommunications (3/85); in the upper-division baccalaureate category, 3 semester hours in data communications for non—engineering degree program students (3/85).

AF-1715-0465

GROUND RADIO COMMUNICATIONS TECHNICIAN BY CORRESPONDENCE

Course Number: 30474.
Location: Extension Course Institute, Gunter AFS, AL.
Length: Maximum, 52 weeks.
Exhibit Dates: 1/90–2/93.
Objectives: To provide management training necessary to raise the ground radio communications specialist to the supervisory level.
Instruction: This is a self-instruction program supplemented by on-the-job experience in supervision, evaluating and rating personnel, safety, security, publications, and training as they apply to a maintenance supervisor.
Credit Recommendation: In the upper-division baccalaureate category, 1 semester hour in personnel supervision (3/85).

AF-1715-0466

ELECTRONIC WARFARE SYSTEMS SPECIALIST BY CORRESPONDENCE

Course Number: 32853.
Location: Extension Course Institute, Gunter AFS, AL.
Length: Maximum, 52 weeks.
Exhibit Dates: 1/90–12/91.
Objectives: To advance the electronic warfare systems specialist into leadership (middle management) positions involving personnel supervision and equipment and shop management.
Instruction: This is a correspondence course with practical exercises in personnel supervision, management, and maintenance of electronic warfare equipment and electronic warfare systems. Topics include introduction to supervision and management, electronic equipment maintenance, computer principles, basic electricity and electronics, and accident prevention.
Credit Recommendation: In the lower-division baccalaureate/associate degree category, 3 semester hours in communications systems and 3 in introduction to computers (3/85).

AF-1715-0493

DCLF REFERENCE MEASUREMENT AND CALIBRATION

Course Number: G3AZR32470 017.
Location: Technical Training Center, Lowry AFB, CO.
Length: 6 weeks (230 hours).
Exhibit Dates: 1/90–1/97.
Objectives: To provide training on measurement and calibration methods.
Instruction: Course trains personnel to repair; calibrate; and use electronic test and measuring equipment, including meters, scopes, bridges, current and voltage standards, sampling devices, and signal generators.
Credit Recommendation: In the lower-division baccalaureate/associate degree category, 2 semester hours in electronic tests and measurements and 2 in electronic systems troubleshooting and maintenance (11/86).

AF-1715-0740

MICROWAVE MEASUREMENT AND CALIBRATION

Course Number: G3AZR32470 018.
Location: Technical Training Center, Lowry AFB, CO.
Length: 6 weeks (256 hours).
Exhibit Dates: 1/90–2/97.
Learning Outcomes: Upon completion of the course, the student will be able to maintain and calibrate frequency counters, phase locking sychronizers, attenuator calibrators, power-measuring equipment, and sweep oscillators.
Instruction: Course includes lectures and practical exercises on mathematics, transmission and line theory, and the maintenance and calibration of microwave test equipment.
Credit Recommendation: In the lower-division baccalaureate/associate degree category, 2 semester hours in transmission line theory and 3 in microwave test equipment (2/87).

AF-1715-0741

AUDIOVISUAL EQUIPMENT REPAIRER

Course Number: G3ABR40430 005.
Location: Technical Training Center, Lowry AFB, CO.
Length: 6-7 weeks (233 hours).
Exhibit Dates: 1/90–2/97.
Learning Outcomes: Upon completion of the course, the student will be able to repair precision imagery, audiovisual, darkroom, and film developing equipment.

Instruction: Course includes lectures and practical exercises on the repair and maintenance of darkroom equipment, including timers, dryers, enlargers, densitometers, and sensitometers. Course also includes repair of slide and motion picture projectors and still and motion picture cameras.

Credit Recommendation: In the vocational certificate category, 2 semester hours in still camera repair, 2 in motion picture camera repair, 2 in slide and motion picture projector repair, and 2 in darkroom equipment repair (2/87).

AF-1715-0743

PHYSICAL MEASUREMENT AND CALIBRATION

Course Number: G3AZR32470 023.
Location: Technical Training Center, Lowry AFB, CO.
Length: 4 weeks (159 hours).
Exhibit Dates: 1/90–2/97.
Learning Outcomes: Upon completion of the course, the student will be able to perform duties in physical measurement and calibration of instruments.
Instruction: Personnel are trained in linear, torque, mass, weight, temperature, pressure, vacuum, laser, and intercomparison measurements. Emphasis is placed on error analysis, applied mathematics, use of test equipment, and calibration standards.
Credit Recommendation: In the lower-division baccalaureate/associate degree category, 3 semester hours in fundamentals of metrology (2/87).

AF-1715-0744

ENEMY DEFENSE PENETRATION AIDS

Course Number: T1115005.
Location: Air National Guard, A-7 Fighter Weapons School, Tucson, AZ.
Length: 2 weeks (69-70 hours).
Exhibit Dates: 1/90–7/96.
Learning Outcomes: Upon completion of the course, the student will be able to assist in the establishment, development, and execution of a unit electronic combat (EC) program. This course is designed for mission-ready A-7, A-10, and F-16 pilots.
Instruction: Methods include classroom lectures and working group assignments. The academics include a detailed study of history and future trends in the use of the electromagnetic spectrum as a medium of warfare. Particular emphasis is directed at radar principles, tracking techniques, countermeasures, electro-optical techniques, and EM tactics. Students will develop a practical application problem with real world target environments.
Credit Recommendation: In the upper-division baccalaureate category, 2 semester hours in advanced electronics/radar systems (3/88).

AF-1715-0745

BIOMEDICAL EQUIPMENT MAINTENANCE SPECIALIST BY CORRESPONDENCE

Course Number: 91850.
Location: Extension Course Institute, Gunter AFS, AL.
Length: Maximum, 52 weeks.
Exhibit Dates: 1/90–1/96.
Learning Outcomes: Upon completion of the course, the student will be able to describe the theory and basic fundamentals of biomedical electronics and repair and evaluate clinical laboratory equipment and computers.
Instruction: This monitored home-study course includes readings, written exercises, and a written proctored examination. Topics include theoretical concepts of repair and evaluation of clinical and medical equipment, the basic and advanced principles of electronics and troubleshooting procedures, and introductory and intermediate concepts in computer technology.
Credit Recommendation: In the lower-division baccalaureate/associate degree category, 3 semester hours in basic electronic principles, 3 in advanced electronics, 3 in theory of maintenance of therapeutic equipment, 3 in fundamentals of X-ray and imaging systems, 3 in theory of cardiac equipment and computer applications of biomedical equipment (5/88).

AF-1715-0746

ELECTRONICS FUNDAMENTALS BY CORRESPONDENCE

Course Number: 3025.
Location: Extension Course Institute, Gunter AFS, AL.
Length: Maximum, 52 weeks.
Exhibit Dates: 1/90–12/94.
Learning Outcomes: Upon completion of the course, the student, with a minimum of supervision, will be able to perform maintenance and repair of basic navigation communications and pulse equipment including antenna installation and repair.
Instruction: This is a supervised, self-paced extension course covering direct and alternating current theory; electronic device theory, motion conversion, and behavior of electrons; communication principles dealing with characteristics and functional operations of AM transmitters, single-sideband transmitters and receivers, frequency and phase modulated transmitters, and FM receivers; and transmission lines, antennas, and wave propagation fundamentals.
Credit Recommendation: In the lower-division baccalaureate/associate degree category, 3 semester hours in electronics fundamentals (10/88).

AF-1715-0747

COMMUNICATIONS-ELECTRONICS SYSTEMS TECHNOLOGY BY CORRESPONDENCE

Course Number: 3026.
Location: Extension Course Institute, Gunter AFS, AL.
Length: Maximum, 52 weeks.
Exhibit Dates: 1/90–4/97.
Learning Outcomes: Upon completion of the course, the student will be able to describe radio and radar theory and applications necessary for intermediate management of a communications and electronics organization.
Instruction: This is a self-paced follow-on course which goes into radio and radar theory and application. Radio bands from extremely low (ELF) to extremely high (EHF) are examined. Emphasis includes line of sight considerations, tropospheric scatter, and safety. Probability (distribution), reliability, maintainability, applications, and multiplex operations are also included. The course is structured for those with limited previous experience.
Credit Recommendation: In the lower-division baccalaureate/associate degree category, 3 semester hours in radio wave theory and applications (10/88).

AF-1715-0748

COMMUNICATIONS-ELECTRONICS EMPLOYMENT BY CORRESPONDENCE

Course Number: 3028.
Location: Extension Course Institute, Gunter AFS, AL.
Length: Maximum, 52 weeks.
Exhibit Dates: 1/90–4/97.
Learning Outcomes: Upon completion of the course, the student will be able to describe telephone principles, signaling and inside plant support equipment, telephone switching system, outside plant and traffic engineering, air traffic control systems, and traffic control and landing systems.
Instruction: This is a self-paced, supervised extension course including automatic telephone systems, supervision and signaling, basic switching techniques and terminology, inside-plant support, telephone and wire maintenance documents, automatic telephone—direct control, automatic telephone—common control system, tactical switching systems, multiline subscriber equipment, outside plant, outside-plant in-place records, traffic engineering, control of air traffic, navigational aids, air traffic control, tower, radar principles, facilities and equipment, and omnirange instrument landing system. The course is geared mostly to telephone systems with very little on air traffic control other than principles of air traffic control and their related systems.
Credit Recommendation: In the vocational certificate category, 1 semester hour in basic telephone principles (10/88).

AF-1715-0749

ELECTRONIC TUBES AND SPECIAL PURPOSE TUBES BY CORRESPONDENCE

Course Number: 3031.
Location: Extension Course Institute, Gunter AFS, AL.
Length: Maximum, 52 weeks.
Exhibit Dates: 1/90–Present.
Learning Outcomes: Upon completion of the course, the student will be able to describe vacuum tube principles and operation. This course is designed as a review for a person who has previously completed more extensive training in the area or for a brief overview of this equipment.
Instruction: This is a self-paced extension course which serves as an overview of electron tubes (diodes, triodes, and multigrid), tube checking and circuits, amplifiers, and cathode ray tubes (cold and hot). Also covered are oscillators (lighthouse, reflex klystron, and magnetron).
Credit Recommendation: Credit is not recommended as the course is a review of limited material (10/88).

AF-1715-0750

FUNDAMENTALS OF SOLID STATE DEVICES BY CORRESPONDENCE

Course Number: 3032.
Location: Extension Course Institute, Gunter AFS, AL.
Length: Maximum, 52 weeks.
Exhibit Dates: 1/90–11/97.
Learning Outcomes: Upon completion of the course, the student will be able to describe principles and concepts of semiconductors and fundamental circuitry.

Instruction: This self-paced, extension course begins with an overview of solid state devices from fundamental physics and mathematics to theory of simple rectifying and special solid state devices and transistors. Material goes through functional principles of electronic circuits and applications for solid state equipment.

Credit Recommendation: In the lower-division baccalaureate/associate degree category, 1 semester hour in semiconductor theory and application (10/88).

AF-1715-0751

SINEWAVE OSCILLATORS-MODULATION/ DEMODULATION BY CORRESPONDENCE

Course Number: 3034.
Location: Extension Course Institute, Gunter AFS, AL.
Length: Maximum, 52 weeks.
Exhibit Dates: 1/90–11/92.
Learning Outcomes: Upon completion of the course, the student will be able to describe the principles of sine wave oscillators and modulation/demodulation techniques and concepts of amplitude modulation, single-sideband, frequency modulation, phase modulation, and pulse modulation.
Instruction: This self-paced extension course covers principles of oscillation, operation, and troubleshooting of common oscillators. The course discusses frequency synthesizers and AM modulation and demodulation as well as principles of single-sideband and FM. The material is constructed to include a block diagram and brief review in each stage.
Credit Recommendation: Credit is not recommended due to the limited application and coverage of content (10/88).

AF-1715-0752

POWER SUPPLIES BY CORRESPONDENCE

Course Number: 3038.
Location: Extension Course Institute, Gunter AFS, AL.
Length: Maximum, 52 weeks.
Exhibit Dates: 1/90–Present.
Learning Outcomes: Upon completion of the course, the student will be able to describe details of the power supply's relationship to circuits in order to more efficiently troubleshoot power supply circuits.
Instruction: This self-paced extension course is designed for the person with prior theoretical and practical training and experience. The course discusses operation of transformers, rectifiers, and filters and principles and uses of voltage dividers and regulators. The course concludes with polyphase power supply concept review.
Credit Recommendation: Credit is not recommended as the course contains a very limited review of material (10/88).

AF-1715-0753

COMMUNICATION-ELECTRONICS SYSTEMS ACQUISITION AND MANAGEMENT BY CORRESPONDENCE

Course Number: 3027.
Location: Extension Course Institute, Gunter AFS, AL.
Length: Maximum, 52 weeks.
Exhibit Dates: 1/90–6/97.
Learning Outcomes: Upon completion of the course, the student will be able to describe the steps required to plan, program, and budget for major government systems acquisition. This is a foundation course for middle managers.
Instruction: This self-paced extension course for managers and prospective managers contains an overview of planning on the national level with specific applications to the Department of Defense. The course offers a more detailed view of communications and electronics planning, programming, and documentation. Standard and nonstandard programming involving program management, action agencies, integration of logistics support, operational test and evaluation, resource management, and the budgeting process are all included.
Credit Recommendation: In the upper-division baccalaureate category, 2 semester hours in acquisition and management of government systems (10/88).

AF-1715-0754

AEROSPACE CONTROL AND WARNING SYSTEMS OPERATOR BY CORRESPONDENCE

Course Number: 27650.
Location: Extension Course Institute, Gunter AFS, AL.
Length: Maximum, 52 weeks.
Exhibit Dates: 1/90–11/95.
Learning Outcomes: Upon completion of the course, the student will be able to describe the principles of operation, system limitations, and the equipment applicable to both atmospheric and space radar detection systems.
Instruction: The three volumes that comprise the study materials for this course provide a broad survey of radar applications in the aerospace control and warning field. The basic theory of radar, weather effects, related communication systems, and computer integration are presented. An overview of major military radar as applied to space and space-related systems is covered. Atmospheric radar systems, both airborne and ground based, are described, and their performance is evaluated.
Credit Recommendation: In the lower-division baccalaureate/associate degree category, 2 semester hours in introduction to radar systems (12/89).

AF-1715-0755

SPACE SYSTEMS OPERATIONS SPECIALIST BY CORRESPONDENCE

Course Number: 27750.
Location: Extension Course Institute, Gunter AFS, AL.
Length: Maximum, 52 weeks.
Exhibit Dates: 1/90–7/97.
Learning Outcomes: Upon completion of the course, the student will be able to operate and maintain digital computers and microprocessors under medium supervision in support of space systems.
Instruction: The student will receive instruction on basic electricity and electronics, basic communication systems, digital computer operation, and fundamental space mechanics.
Credit Recommendation: In the lower-division baccalaureate/associate degree category, 2 semester hours in computer system troubleshooting and maintenance (12/89).

AF-1715-0756

AVIONICS TEST STATION AND COMPONENT SPECIALIST (F-16/A-10) BY CORRESPONDENCE

Course Number: 45155.
Location: Extension Course Institute, Gunter AFS, AL.
Length: Maximum, 52 weeks.
Exhibit Dates: 1/90–5/96.
Learning Outcomes: Upon completion of the course, the student will be able to maintain, troubleshoot, repair, and replace electronic equipment using computerized testing and maintenance equipment and test and repair flight control and instrument systems.
Instruction: The student receives instruction on electronic theory and application, test equipment operation and maintenance, and test and maintenance of flight control systems.
Credit Recommendation: In the lower-division baccalaureate/associate degree category, 2 semester hours in electronic systems troubleshooting and 2 semester hours in electronic test station operation (12/89).

AF-1715-0758

INTEGRATED ORGANIZATIONAL AVIONIC SYSTEM SPECIALIST BY CORRESPONDENCE

Course Number: 45250B.
Location: Extension Course Institute, Gunter AFS, AL.
Length: Maximum, 52 weeks.
Exhibit Dates: 1/90–8/94.
Learning Outcomes: Upon completion of the course, the student will be able to describe the fundamentals of DC and AC electricity, magnetism, transformers, generators, motors and electrical circuit interpretation and work with solid state electronics systems and circuits, numbering systems, digital logic and computation systems, and digital computer theory and operations.
Instruction: The student receives instruction on the use of Ohm's and Kirchhoff's laws to analyze electrical circuits, the operating principles and uses of transformers, and theory of operation of generators and motors. Course also covers solid state electronics and associated systems, digital computer theory and operation, and material management.
Credit Recommendation: In the lower-division baccalaureate/associate degree category, 2 semester hours in basic electricity and 2 in solid state electronics and digital computer fundamentals (12/89).

AF-1715-0759

INTEGRATED AVIONIC ATTACK CONTROL SYSTEMS SPECIALIST BY CORRESPONDENCE

Course Number: 45252A.
Location: Extension Course Institute, Gunter AFS, AL.
Length: Maximum, 52 weeks.
Exhibit Dates: 1/90–8/98.
Learning Outcomes: Upon completion of the course, the specialist who is at the skilled or advanced level will be able to maintain the integrated avionic attack control system installed on the F-16 aircraft.
Instruction: The course consists of self-paced, independent study with open-book progress checks and satisfactory completion of a proctored comprehensive final examination. Content areas cover an orientation to the F-16

series aircraft with particular reference to the integrated avionic attack control system. Specific topics include head-up and head-down systems, fire control sensing, pulse Doppler radar, inertial navigation, synchronizer, transmitter and modulator circuits, radio waves and receivers, transmission lines, waveguides, antennas, and radar displays.

Credit Recommendation: In the lower-division baccalaureate/associate degree category, 2 semester hours in electronic radar systems (2/89).

AF-1715-0760

INTEGRATED AVIONICS ATTACK CONTROL SYSTEMS SPECIALIST (F/FB-111) BY CORRESPONDENCE

Course Number: 45253A.
Location: Extension Course Institute, Gunter AFS, AL.
Length: Maximum, 52 weeks.
Exhibit Dates: 1/90–6/97.
Learning Outcomes: Upon completion of the course, the student who is at the skilled or advanced level will be able to maintain the integrated avionic attack control system installed on the F-111 Aircraft.
Instruction: The course consists of self-paced, independent study with open-book progress checks and satisfactory completion of a proctored comprehensive final examination. Content area covers an orientation to the F-111 series aircraft with particular reference to the integrated avionic attack control system. Specific topics include navigation computers, including computer complexes and multiplex busses, control and display systems (optical and integrated), airborne video tape recording, terrain following, navigation and altimeter radar, and inertial navigation systems.
Credit Recommendation: In the upper-division baccalaureate category, 2 semester hours in electronic radar systems (12/89).

AF-1715-0761

AVIONICS GUIDANCE AND CONTROL SYSTEMS TECHNICIAN BY CORRESPONDENCE

Course Number: *Version 1:* 45571; 45571X. *Version 2:* 45571; 45571X.
Location: Extension Course Institute, Gunter AFS, AL.
Length: *Version 1:* Maximum, 52 weeks. *Version 2:* Maximum, 52 weeks.
Exhibit Dates: *Version 1:* 3/93–2/98. *Version 2:* 1/90–2/93.
Learning Outcomes: *Version 1:* Upon completion of the course, the student will be able to describe personnel and shop safety around aircraft; describe maintenance and material complexes and tracking; use basic electrical theory, synchro-indicating systems, and electronic systems including solid state; apply avionics fundamentals for flight and navigation; apply data processing methods; and use test equipment. *Version 2:* Upon completion of the course, the student will be able to describe personnel and shop safety around aircraft; describe maintenance and material complexes and tracking; use basic electrical theory, synchro-indicating systems, and electronic systems including solid state; use avionics fundamentals for flight and navigation; apply data processing methods; and use test equipment.
Instruction: *Version 1:* The student receives information on personnel and shop safety, maintenance and material complexes and tracking, basic electricity, electronics, solid state electronics, and avionics fundamentals. Basic theory of flight, navigation principles, principles of data processing, and test equipment use are also covered. *Version 2:* The student receives information on personnel and shop safety, maintenance and material complexes and tracking, basic electricity, electronics, solid state electronics, and avionics fundamentals. Basic theory of flight, navigation principles, principles of data processing and test equipment use are also covered.
Credit Recommendation: *Version 1:* In the lower-division baccalaureate/associate degree category, 1 semester hour in electronic fundamentals and 1 in introduction to avionics systems (6/94). *Version 2:* In the lower-division baccalaureate/associate degree category, 1 semester hour in basic electricity and 2 in electronics and avionics (12/89).

AF-1715-0762

SCIENTIFIC MEASUREMENTS TECHNICIAN BY CORRESPONDENCE

Course Number: 99105-5.
Location: Extension Course Institute, Gunter AFS, AL.
Length: Maximum, 52 weeks.
Exhibit Dates: 1/90–5/91.
Learning Outcomes: Upon completion of the course, the student will be able to operate seismic, hydroacoustic, electromagnetic pulse, and satellite data acquisition and processing equipment and program and operate computer systems which process digital and analog inputs. Student will also process, compile, analyze, prepare, and transmit scientific information.
Instruction: This correspondence course covers topics in AFTAC history, career progression, security, basic mathematics, basic electronics, basic computer principles, introduction to data processing, and basic principles of physics.
Credit Recommendation: In the vocational certificate category, 2 semester hours in basic electronics and 1 in introduction to data processing (3/91).

AF-1715-0763

SYSTEMS REPAIR TECHNICIAN BY CORRESPONDENCE

Course Number: 99104-5.
Location: Extension Course Institute, Gunter AFS, AL.
Length: Maximum, 52 weeks.
Exhibit Dates: 1/90–5/91.
Learning Outcomes: Upon completion of the course, the student will be able to inspect, troubleshoot, repair, modify, install, and calibrate seismographic, hydroacoustic, electromagnetic pulse, and associated equipment.
Instruction: This correspondence course deals with fundamentals of electricity, electronics, and computer systems as applied to monitoring equipment. Topics include security and supply maintenance management, DC/AC circuits, relays and transformers, generators and motors, semiconductors, power supplies, amplifiers and coupling, sensors and detectors, basic and advanced logic circuits, magnetic data storage, communication and computer operations and systems, and microprocessors. Also seismic, hydroacoustic, ground-base sampling and laboratory techniques are included.

Credit Recommendation: In the vocational certificate category, 3 semester hours in basic electronics and 1 in introduction to computer systems (3/91).

AF-1715-0764

PHOTO-SENSORS MAINTENANCE SPECIALIST BY CORRESPONDENCE

Course Number: 45550B.
Location: Extension Course Institute, Gunter AFS, AL.
Length: Maximum, 52 weeks.
Exhibit Dates: 1/90–8/95.
Learning Outcomes: Upon completion of the course, the student will be able to identify the various types and operating fundamentals of photographic equipment and video systems; describe electronic principles; and maintain electro-optical and photo-sensing systems.
Instruction: This correspondence course covers fundamentals of photography, photographic equipment maintenance, camera types, cockpit television, VCRs, systems displays, integrated video systems, electro-optical systems, and infrared and photo-sensing systems. Also covered are the principles of electricity and electronics and electronic circuitry troubleshooting. Troubleshooting photographic, video, electro-optical, and photo-sensing systems and conducting function analysis are part of the course.
Credit Recommendation: In the lower-division baccalaureate/associate degree category, 3 semester hours in solid state electronics and 1 in electronic systems troubleshooting and maintenance (3/91).

AF-1715-0765

ELECTRONIC COMPUTER AND SWITCHING SYSTEMS SPECIALIST BY CORRESPONDENCE

Course Number: 30554.
Location: Extension Course Institute, Gunter AFS, AL.
Length: Maximum, 52 weeks.
Exhibit Dates: 1/90–2/98.
Learning Outcomes: Upon completion of the course, the student will be able to install, troubleshoot, repair, and maintain computer systems equipment.
Instruction: Student enrolling must have an Air Force Specialty Code assignment as a prerequisite. This provides an application of the principles learned to real requirements and serves as the laboratory component of the course. The student's supervisor must complete an evaluation of the course and its effectiveness when the student takes the final examination which measures mastery of the cognitive content. Course consists of self-paced, independent study with open-book progress checks and satisfactory completion of a final examination. Topics include basic digital electronics used to supplement study in semiconductor devises and test equipment operations. Includes computer systems maintenance and operations.
Credit Recommendation: In the lower-division baccalaureate/associate degree category, 3 semester hours in digital electronics and 1 in semiconductor circuit analysis (11/91).

AF-1715-0766

WEAPON CONTROL SYSTEMS TECHNICIAN (F-4D/E) BY CORRESPONDENCE

Course Number: 45573C.

Location: Extension Course Institute, Gunter AFS, AL.

Length: Maximum, 52 weeks.

Exhibit Dates: 7/90–7/94.

Learning Outcomes: Upon completion of the course, the student will be able to troubleshoot and repair weapon control systems of F-4D/E.

Instruction: This is a correspondence course with practical exercises in maintenance and material management. Topics include DC and AC fundamentals, semiconductor devices, active devices and circuits, microprocessors, AM/FM transmitter principles, radar and optical sight systems, and specialized test equipment.

Credit Recommendation: In the lower-division baccalaureate/associate degree category, 1 semester hour in maintenance and material management, 2 in DC/AC fundamentals, 2 in semiconductor devices, 1 in AM/FM transmitter principles, and 2 in radar systems (3/91).

AF-1715-0767

F-15 INTEGRATED AVIONIC ATTACK CONTROL SYSTEMS SPECIALIST BY CORRESPONDENCE

Course Number: 45251A.

Location: Extension Course Institute, Gunter AFS, AL.

Length: Maximum, 52 weeks.

Exhibit Dates: 5/90–6/97.

Learning Outcomes: Upon completion of the course, the student will be able to troubleshoot attack control systems on the F-15.

Instruction: This correspondence course contains practical exercises in troubleshooting, maintenance, and repair of avionic attack control systems and components. Topics include aircraft familiarization, computer systems, navigation radar, and display systems.

Credit Recommendation: In the lower-division baccalaureate/associate degree category, 1 semester hour in radar principles (3/91).

AF-1715-0768

RELAYS, GENERATORS, MOTORS, AND ELECTROMECHANICAL DEVICES BY CORRESPONDENCE

Course Number: 03035.

Location: Extension Course Institute, Gunter AFS, AL.

Length: Maximum, 52 weeks.

Exhibit Dates: 1/90–Present.

Learning Outcomes: Upon completion of the course, the student will be able to describe the fundamentals and operating theory of relay, generators, motors, and electromechanical devices.

Instruction: This correspondence course covers magnetism, electromagnetism, electromechanical devices, switching, protection alarm, time delay, and voltage control relay devices. Principles of AC and DC electricity generation, alternating current and direct current generation devices, AC and DC motor operating principles, operating principles of remote indicating systems, synchros, differential synchros, synchronous motor operation servo mechanisms, and control transformers are also included.

Credit Recommendation: Credit is not recommended because of the limited, specialized nature of the course (3/91).

AF-1715-0769

ELECTRONIC WARFARE SYSTEMS SPECIALIST BY CORRESPONDENCE

Course Number: 45651.

Location: Extension Course Institute, Gunter AFS, AL.

Length: Maximum, 52 weeks.

Exhibit Dates: 1/90–10/95.

Learning Outcomes: Upon completion of the course, the student will be able to troubleshoot and repair electronic warfare receivers, transmitters, pods, and chaff/flare dispensers.

Instruction: This is a correspondence course with practical exercises in maintenance and material management. Topics include electronic circuits lab, active devices and circuits, digital logic and fundamentals, and electronic systems troubleshooting and repair of electronic warfare receivers, transmitters pods, and chaff/flare dispensers.

Credit Recommendation: In the lower-division baccalaureate/associate degree category, 1 semester hour in maintenance and material management, 1 in active devices and circuits, 2 in digital logic and fundamentals, and 3 in communications systems (3/91).

AF-1715-0770

INSTRUMENTS/FLIGHT CONTROL SYSTEMS SPECIALIST (B-1B) BY CORRESPONDENCE

Course Number: 45753B.

Location: Extension Course Institute, Gunter AFS, AL.

Length: Maximum, 52 weeks.

Exhibit Dates: 1/90–7/97.

Learning Outcomes: Upon completion of the course, the student will be able to troubleshoot and repair flight environmental, hydraulic attitude, direction, engine indicating, and fuel management and primary/secondary/auto/flight control systems.

Instruction: This is a correspondence course with practical exercises in maintenance and material management. Topics include flight controls (primary, secondary, and auto), environmental, hydraulic, indicating flight instruments, engine indicators, and fuel management systems.

Credit Recommendation: In the lower-division baccalaureate/associate degree category, 1 semester hour in maintenance and material management, 4 in instrument systems, and 2 in flight control systems (3/91).

AF-1715-0771

F-15 AVIONICS MANUAL/ECM TEST STATION AND COMPONENT SPECIALIST BY CORRESPONDENCE

Course Number: 45154B.

Location: Extension Course Institute, Gunter AFS, AL.

Length: Maximum, 52 weeks.

Exhibit Dates: 1/90–8/96.

Learning Outcomes: Upon completion of the course, the student will be able to use the electronic warfare test station to troubleshoot and repair electronic warfare line replaceable units on the F-15 aircraft.

Instruction: This is a correspondence course with practical exercises in maintenance management. Topics include radar principles, test station controls, stimulus, switching, and measurements. Operation and use of both the communication-navigation-identification test station and the antenna test station are covered.

Credit Recommendation: In the lower-division baccalaureate/associate degree category, 1 semester hour in maintenance management, 1 in radar principles, and 4 in test equipment and measurement (3/91).

AF-1715-0772

F-15 AVIONICS AUTOMATIC TEST STATION AND COMPONENT SPECIALIST BY CORRESPONDENCE

Course Number: 45154A.

Location: Extension Course Institute, Gunter AFS, AL.

Length: Maximum, 52 weeks.

Exhibit Dates: 1/90–8/96.

Learning Outcomes: Upon completion of the course, the student will be able to troubleshoot and repair the avionics systems installed on the F-15 aircraft.

Instruction: This is a correspondence course with practical exercises in maintenance management. Topics include airborne radar, navigation, flight control, instruments, automatic flight control and specialized avionic test equipment, microprocessor applications, digital interface adapters, and computer control and display panel systems.

Credit Recommendation: In the lower-division baccalaureate/associate degree category, 1 semester hour in maintenance management, 2 in avionic systems, and 2 in microprocessor applications (3/91).

AF-1715-0773

F/FB-111 AUTOMATIC TEST STATION AND COMPONENT SPECIALIST BY CORRESPONDENCE

Course Number: 45156A.

Location: Extension Course Institute, Gunter AFS, AL.

Length: Maximum, 52 weeks.

Exhibit Dates: 1/90–8/96.

Learning Outcomes: Upon completion of the course, the student will be able to utilize, troubleshoot, and replace or repair components for F/FB-111 using an automatic test station.

Instruction: This correspondence course contains practical exercises in use, repair, and maintenance of components. Topics include electronic principles, solid state devices, DC power supplies, waveguides, microcomputers, principles of aerial navigation, radar and components, central processor and controller, and automatic test equipment. Avionics intermediate shop replacement, test equipment, computers, interface systems, and system software are included. In addition, line replaceable units, power distribution and interface, stimulus generating and measurement, and avionics test calibrator with ancillary equipment are covered.

Credit Recommendation: In the lower-division baccalaureate/associate degree category, 3 semester hours in test equipment measurement and 2 in avionics fundamentals (3/91).

AF-1715-0774

BOMB NAVIGATION SYSTEMS SPECIALIST BY CORRESPONDENCE

Course Number: 45650.

Location: Extension Course Institute, Gunter AFS, AL.

Length: Maximum, 52 weeks.
Exhibit Dates: 1/90–5/97.
Learning Outcomes: Upon completion of the course, the student will be able to troubleshoot and repair bomb navigation and offensive avionics systems on the B-52.
Instruction: This is a correspondence course with practical exercises in maintenance and material management. Topics include semiconductor devices, digital fundamentals, display processing, steerable television, forward-looking infrared, video recording, and camera systems. Theory and application of basic radar systems and components are included. Topics in transmitter, receiver, timing and built-in test circuits, antenna controls, and strategic radar are covered along with topics in operation of offensive avionic systems and interface with navigational systems.
Credit Recommendation: In the lower-division baccalaureate/associate degree category, 1 semester hour in maintenance and material management, 1 in digital fundamentals, 2 in electro-optical systems, and 4 in radar systems (3/91).

AF-1715-0775

APPLIED SCIENCES TECHNICIAN BY CORRESPONDENCE

Course Number: 99106-5.
Location: Extension Course Institute, Gunter AFS, AL.
Length: Maximum, 52 weeks.
Exhibit Dates: 1/90–5/91.
Learning Outcomes: Upon completion of the course, the student will be able to describe the history and mission of AFTAC, organizational structure, security, supply and records management, data processing, mathematics, physics, chemistry, and earth sciences.
Instruction: This correspondence course covers the history and missions of AFTAC, organizational structure, security, and supply systems. Records management systems; laboratory procedures and safety; sample handling; and data processing systems, including development, files, data bases, and numbering systems are included in the course. Also covered are general mathematics; physics; chemistry; and an introduction to meteorology, navigation, and mineralogy.
Credit Recommendation: In the vocational certificate category, 1 semester hour in introduction to computer systems and 1 in introduction to facilities management (3/91).

AF-1715-0776

AUTOMATIC TRACKING RADAR SPECIALIST BY CORRESPONDENCE

Course Number: 30353A.
Location: Extension Course Institute, Gunter AFS, AL.
Length: Maximum, 52 weeks.
Exhibit Dates: 11/91–Present.
Learning Outcomes: Upon completion of the course, the student will be able to apply principles of RF and electrical and chemical safety while maintaining electronic equipment; interpret oscilloscope and spectrum analyzer displays; operate multimeters, differential voltmeters, pulse and signal generators, tube testers, frequency counters, and power meters; explain operation of amplifier circuits employing vacuum tubes, bipolar function transistors, and field effect transistors; analyze rectifiers, filter circuits, and voltage regulators; troubleshoot systems which employ linear and digital integrated circuits; verify proper operation of microwave amplifier and oscillator circuits using lighthouse tubes, klystrons, magnetrons, and TWTs; and demonstrate the proper operation of analog and digital servomechanisms.
Instruction: Students enrolling must have an Air Force Specialty Code assignment as a prerequisite. This serves as the laboratory component of the course. The student's supervisor must complete an evaluation of the course and its effectiveness when the student takes the proctored final examination which measures mastery of the cognitive content. Course consists of self-paced, independent study with open-book progress checks and satisfactory completion of a final examination. Topics include shop safety, semiconductors and electron tubes, power supply circuits, linear and digital circuits, waveform generators, microwave amplifiers and oscillators, synchros and servos, and operation of electronic test equipment.
Credit Recommendation: In the lower-division baccalaureate/associate degree category, 1 semester hour in analog circuits, 1 in digital circuits, and 1 in microwave systems (11/91).

AF-1715-0777

PHOTO-SENSORS MAINTENANCE SPECIALIST (TACTICAL/RECONNAISSANCE ELECTRONIC SENSORS) BY CORRESPONDENCE

Course Number: 45550A.
Location: Extension Course Institute, Gunter AFS, AL.
Length: Maximum, 52 weeks.
Exhibit Dates: 2/90–8/95.
Learning Outcomes: Upon completion of the course, the student will be able to apply shop safety principles; use and maintain test equipment; and describe principles of AC and DC circuits, magnetism, solid state electronics, rectifiers, oscillators, number systems, logic circuits, and A/D and D/A converters. Student will be able to describe transmission and reception of television communications, infrared and low light sensors, and lasers and the principles of microwave transmissions.
Instruction: Course consists of self-paced, independent study with open-book progress checks and satisfactory completion of a final examination. There is on-the-job supervision and an opportunity for application of learning. Subjects include shop safety, shop maintenance, use of test equipment, DC, AC, solid state devices and rectifiers, oscillators, number systems, logic circuits, A/D and D/A converters, television fundamentals, infrared and low light sensors, lasers, and microwave fundamentals.
Credit Recommendation: In the lower-division baccalaureate/associate degree category, 1 semester hour in electronics fundamentals, 1 in introduction to television, and 1 in microwave fundamentals (11/91).

AF-1715-0778

COMMUNICATIONS COMPUTER SYSTEMS CONTROL SPECIALIST BY CORRESPONDENCE

Course Number: 49350A.
Location: Extension Course Institute, Gunter AFS, AL.
Length: Maximum, 52 weeks.
Exhibit Dates: 1/90–5/99.
Learning Outcomes: Upon completion of the correspondence course, the student will be able to apply basic algebra and number systems; solder wires, components, and connectors; discuss the characteristics of amplitude modulation, frequency modulation, phase, pulse, delta, frequency division, time division modulation, and multiplexing techniques; and understand computer architecture, digital communications, protocols, networking, digital interfacing, and conditioning devices and facsimile.
Instruction: This course consists of self-paced, independent study with open-book progress checks and satisfactory completion of a proctored final examination. There is on-the-job supervision and an opportunity for the application of the learning. Subjects covered include a review of electrical and electronic principles, algebra, number systems, soldering techniques, modulating and multiplexing techniques, digital communication, and computer fundamentals.
Credit Recommendation: In the lower-division baccalaureate/associate degree category, 1 semester hour in review of electronic and digital principles (11/91).

AF-1715-0779

INTEGRATED AVIONIC COMMUNICATIONS, NAVIGATION, AND PENETRATION AIDS SYSTEMS SPECIALIST (F/FB-111) BY CORRESPONDENCE

Course Number: 45243C.
Location: Extension Course Institute, Gunter AFS, AL.
Length: Maximum, 52 weeks.
Exhibit Dates: 1/90–Present.
Learning Outcomes: Upon completion of this correspondence course, the student will be able to identify differences in operational capabilities of UHF, HF, and satellite communications systems; explain proper operation of ADF, ILS, TACAN, and transponder systems; correlate pulse repetition time, pulse repetition frequency, pulse width, resting time, duty cycle, peak power, and average power for a radar pulse train; and compare the significant operating properties of the klystron, magnetron, and the traveline wave tube.
Instruction: This course consists of self-paced, independent study with open-book progress checks and satisfactory completion of a proctored final examination. There is on-the-job supervision and an opportunity for the application of learning. Topics include fundamentals of radio communications, fundamentals of navigation aids, basic radar principles, automatic direction finders, satellite communications, communications by HF and UHF radio, TACAN, ILS, and transponder systems.
Credit Recommendation: In the lower-division baccalaureate/associate degree category, 1 semester hour in aviation electronics systems (11/91).

AF-1715-0780

F-16 A/B AVIONICS TEST STATION AND COMPONENT SPECIALIST BY CORRESPONDENCE

Course Number: 45155A.
Location: Extension Course Institute, Gunter AFS, AL.
Length: Maximum, 52 weeks.
Exhibit Dates: 4/91–6/97.

Learning Outcomes: Upon completion of the course, the student will recognize computer system architecture as applied to avionics testing, signal tracing, and troubleshooting. Student will be able to program, set up, and debug operations and identify and troubleshoot interface circuitry. Student will recognize and troubleshoot digital-to-analog conversion circuitry, pneumatic and light transducers, and power systems.

Instruction: Topics include CPU operations, input/output operations, electrical interface, registers, disk drive and controller, and computer memory organization (byte, bit, word). Operating characteristics of the test compiler, editor functions, and test software applications are included as are topics in electronic systems of video processing, transducer interface circuitry, and digital elements. Course concludes with a proctored examination.

Credit Recommendation: In the lower-division baccalaureate/associate degree category, 1 semester hour in computer system architecture (11/91).

AF-1715-0781

COMMUNICATIONS SYSTEMS RADIO OPERATOR BY CORRESPONDENCE

Course Number: 49251.
Location: Extension Course Institute, Gunter AFS, AL.
Length: Maximum, 52 weeks.
Exhibit Dates: 1/90–11/93.
Learning Outcomes: Upon completion of the course, the student will be able to describe the general principles of radio wave creation and propagation, electromagnetic frequency (EMF) characteristics, the uses and limitations of high frequency (HF) wave propagation, the characteristics line-of-sight communications, antenna fundamentals, general principles of transmitters and receivers, digital communications, communications satellite systems, and operating procedures.

Instruction: Course consists of self-paced, independent study with open-book progress checks and satisfactory completion of a proctored final examination. There is on-the-job supervision and an opportunity for the application of learning. Subjects covered are general principles of radio wave creation and propagation, EMF characteristics, the use and limitations of HF wave propagation, line-of-sight communications, antenna fundamentals, transmitters, receivers, digital and satellite communications, and operating procedures.

Credit Recommendation: In the lower-division baccalaureate/associate degree category, 1 semester hour in survey of communications theory and equipment (11/91).

AF-1715-0782

GROUND RADIO COMMUNICATIONS SPECIALIST (UNIT) BY CORRESPONDENCE

Course Number: 30454A.
Location: Extension Course Institute, Gunter AFS, AL.
Length: Maximum, 52 weeks.
Exhibit Dates: 5/90–Present.
Learning Outcomes: Upon completion of the course, the student will demonstrate proper operation of electronic test equipment including multimeters; megohmmeters; wattmeters and dummy loads; frequency counters; AF and RF signal generators; impedance and RCL bridges; oscilloscopes; spectrum and distortion analyzers; and tube, semiconductor, and IC testers. The student will be able to comply with test equipment maintenance and calibration requirements.

Instruction: Course consists of self-paced, independent study with open-book progress checks and satisfactory completion of a proctored final examination. There is on-the-job supervision and an opportunity for the application of learning. Topics covered include a review of AC/DC principles, introduction to semiconductors and electron tubes, power supply and regulator circuits, amplifier fundamentals, and test equipment theory and operation.

Credit Recommendation: In the lower-division baccalaureate/associate degree category, 1 semester hour in electronic test and measurement (11/91).

AF-1715-0783

ELECTRONIC WARFARE SYSTEMS TECHNICIAN (UNIT) BY CORRESPONDENCE

Course Number: 45671.
Location: Extension Course Institute, Gunter AFS, AL.
Length: Maximum, 52 weeks.
Exhibit Dates: 8/90–3/97.
Learning Outcomes: Upon completion of the course, the student will be able to maintain a safe work place; solder; handle chemicals and compressed gases; and protect against electrical shock and electrostatic discharge. Student will be able to describe centralized and decentralized management concepts, management support services, maintenance control and core automated data collection system (CAMS), and quality assurance.

Instruction: Course consists of self-paced, independent study with open-book progress checks and satisfactory completion of a proctored final examination. There is on-the-job supervision and an opportunity for the application of learning. Topics covered include safety in soldering, handling of chemicals and compressed gases, electrical shock, and prevention of electrostatic discharge problems. The principles of centralized and decentralized management concepts, management support principles, maintenance control and core automated data collection system are covered, as is quality assurance.

Credit Recommendation: In the lower-division baccalaureate/associate degree category, 1 semester hour in safety training and 2 in maintenance management (11/91).

AF-1715-0784

BASIC TECHNIQUES OF WAVE FORM MEASUREMENT USING AN OSCILLOSCOPE BY CORRESPONDENCE

Course Number: 03039.
Location: Extension Course Institute, Gunter AFS, AL.
Length: Maximum, 52 weeks.
Exhibit Dates: 2/90–Present.
Learning Outcomes: Upon completion of the course, the student will be able to connect the oscilloscope as a test instrument; adjust the oscilloscope so that correct measurements can be made; interpret the data shown on the oscilloscope; select the correct probes for particular applications; and understand the purpose of vertical, horizontal, trigger, dual trace, and delay controls.

Instruction: Course consists of self-paced, independent study with open-book progress checks and satisfactory completion of a final examination. There is on-the-job supervision and an opportunity for the application of learning. Topics covered include purpose of vertical and horizontal controls, trigger, dual trace and delay controls, probe selection, and measured data.

Credit Recommendation: In the lower-division baccalaureate/associate degree category, 1 semester hour in use of the oscilloscope (11/91).

AF-1715-0785

F-16 C/D AVIONICS TEST STATION AND COMPONENT SPECIALIST BY CORRESPONDENCE

Course Number: 45155C.
Location: Extension Course Institute, Gunter AFS, AL.
Length: Maximum, 52 weeks.
Exhibit Dates: 5/91–Present.
Learning Outcomes: Upon completion of the course, the student will be able to describe the proper operation of the F-16 avionics test station (military specific).

Instruction: Course consists of self-paced, independent study with open-book progress checks and satisfactory completion of a final examination. There is on-the-job supervision and an opportunity for the application of learning. Topics covered include hardware and software for testing F-16 avionics equipment.

Credit Recommendation: Credit is not recommended because of the military-specific nature of the course (11/91).

AF-1715-0786

SECURE COMMUNICATION SYSTEMS MAINTENANCE SPECIALIST BY CORRESPONDENCE

Course Number: 30656.
Location: Extension Course Institute, Gunter AFS, AL.
Length: Maximum, 52 weeks.
Exhibit Dates: 10/90–11/93.
Learning Outcomes: After 11/93 see AF-1715-0813. Upon completion of the course, the student will be able to apply principles of electrical safety in the work place; perform basic wiring, casing, and soldering operations; use basic electronic hand tools; identify common electronic components; analyze series, parallel, and series/parallel AC and DC circuits; explain forward and reverse bias of rectifier and zener diodes; bias bipolar junction and field-effect transistors; troubleshoot power supply rectifiers, filters, and regulators; distinguish between Colpitts, Hartley, Butler, blocking, phase-shift, and crystal oscillators; convert between decimal, binary, hexadecimal, octal, and BCD numbering systems; produce truth tables for inverters, and, or, nand, nor, and exclusive or gates, latches, flip-flops, and shift registers; explain voltage, current, phase, and torque relationships for AC and DC motors; differentiate between Bandot and ASCII codes; convert between baud rate, Bit time, and characters-per-minute; identify characteristics of various digital transmission media; explain the effects of signal distortion on a digital waveform; identify the functions and elements of a telecommunications terminal; and analyze multiplexer/demultiplexer and modem input and output signals.

Instruction: Course consists of self-paced, independent study with open-book progress checks and satisfactory completion of a proctored final examination. There is on-the-job supervision and an opportunity for the application of learning. Correspondence topics include electrical safety; soldering, wiring, and interconnect; basic hand tools; shop practices; AC/DC fundamentals; semiconductors; power supplies; oscillator circuits; digital logic and numbering systems; servomechanisms; basic electronic troubleshooting; telecommunications terminals; principles of digital communications; troubleshooting of telecommunications equipment; multiplexing; and modems.

Credit Recommendation: In the lower-division baccalaureate/associate degree category, 1 semester hour in AC/DC circuits, 1 in telecommunications, 1 in digital principles, and 1 in solid state electronics (11/91).

AF-1715-0787

COMMUNICATIONS-COMPUTER SYSTEMS PROGRAM MANAGEMENT SPECIALIST BY CORRESPONDENCE

Course Number: 49650.
Location: Extension Course Institute, Gunter AFS, AL.
Length: Maximum, 52 weeks.
Exhibit Dates: 1/90–10/96.
Learning Outcomes: Upon completion of the course, the student will be familiar with the communications-computer systems program manager's job.
Instruction: Course consists of self-paced, independent study with open-book progress checks and satisfactory completion of a proctored final examination. There is on-the-job supervision and an opportunity for the application of learning. This military-specific course covers Air Force-specific job training.
Credit Recommendation: Credit is not recommended because of the military-specific nature of the course (11/91).

AF-1715-0788

COMMUNICATIONS COMPUTERS SYSTEMS CONTROL SPECIALIST BY CORRESPONDENCE

Course Number: 49350B.
Location: Extension Course Institute, Gunter AFS, AL.
Length: Maximum, 52 weeks.
Exhibit Dates: 1/90–Present.
Learning Outcomes: Upon completion of the course, the student will be able to discuss radio communications propagation, microwave communications systems, high frequency propagation, antennas, satellite communications, conducting media such as optical fibers and local area network configurations (LANs).
Instruction: Course consists of self-paced, independent study with open-book progress checks and satisfactory completion of a proctored final examination. There is on-the-job supervision and an opportunity for the application of learning. Topics include radio communications, propagation characteristics, microwave systems, high frequency propagations, antennas, satellite communications, transmission media (cables, atmosphere, optical fibers), and local area network configurations (LANs).
Credit Recommendation: In the vocational certificate category, 1 semester hour in survey of transmission media and 1 in introduction to telecommunications systems (11/91).

AF-1715-0789

DEFENSIVE AVIONICS/COMMUNICATIONS/ NAVIGATION SYSTEMS BY CORRESPONDENCE

Course Number: 45753C.
Location: Extension Course Institute, Gunter AFS, AL.
Length: Maximum, 52 weeks.
Exhibit Dates: 1/90–7/97.
Learning Outcomes: Upon completion of the course, the student will be able to draw a block diagram of a pulsed radar system; discuss several different types of radar; differentiate between mezical, spiral, monopulse, and raster scans; compute range of a target from a radar display; identify the standard radar bands within the electromagnetic spectrum; distinguish between electronic support measures, electronic countermeasures, and electronic warfare; explain the functions of each block of a superheterodyne receiver; name and describe the functions of the four components of an instrument landing system.
Instruction: Course consists of self-paced, independent study with open-book progress checks and satisfactory completion of a proctored final examination. There is on-the-job supervision and an opportunity for the application of learning. Correspondence topics include history of electronic warfare, radar fundamentals, directional antennas, transmitters and receivers, modulation and demodulation techniques, HF wave propagation, and aircraft communications and navigation systems.
Credit Recommendation: In the lower-division baccalaureate/associate degree category, 1 semester hour in microwave systems and 1 in aviation electronics (11/91).

AF-1715-0790

COMMUNICATIONS-COMPUTER SYSTEMS CONTROL TECHNICIAN BY CORRESPONDENCE

Course Number: 49370.
Location: Extension Course Institute, Gunter AFS, AL.
Length: Maximum, 52 weeks.
Exhibit Dates: 1/90–Present.
Learning Outcomes: Upon completion of the course, the student will be able to recognize component elements of a communications system and discuss digital systems architecture, packet switching, Arpanet, DDN, telecommunications, satellite uses, protocol, concentrators, multiplexers, and antenna configurations.
Instruction: Course consists of self-paced, independent study with open-book progress checks and satisfactory completion of a proctored final examination. There is on-the-job supervision and an opportunity for the application of learning. Topics include packet switching, Arpanet, DDN, telecommunications, satellite communications, digital communications protocol, SDLC, BSC, HDLC, concentrators, multiplexers, and antenna configurations.
Credit Recommendation: In the lower-division baccalaureate/associate degree category, 1 semester hour in introduction to telecommunications systems (11/91).

AF-1715-0791

AVIONICS TEST STATION AND COMPONENT SPECIALIST BY CORRESPONDENCE

Course Number: 45155B.
Location: Extension Course Institute, Gunter AFS, AL.
Length: Maximum, 52 weeks.
Exhibit Dates: 8/90–6/97.
Learning Outcomes: Upon completion of the course, the student will be able to operate the computer system application software of the avionics systems of the A-10 and discuss computer system operation, aircraft systems operation, and test conditions for systems integration.
Instruction: Course consists of self-paced, independent study with open-book progress checks and satisfactory completion of a proctored final examination. There is on-the-job supervision and an opportunity for the application of learning. Topics include computer operating system, editor, compiler and application test software, computer system components and I/O elements, system power sources and test instruments, and test applications for aircraft systems.
Credit Recommendation: In the lower-division baccalaureate/associate degree category, 1 semester hour in computer systems architecture or 1 in aircraft instrumentation (11/91).

AF-1715-0792

COMMUNICATIONS ELECTRONICS EQUIPMENT, CIRCUITS AND SYSTEMS BY CORRESPONDENCE

Course Number: 30454B.
Location: Extension Course Institute, Gunter AFS, AL.
Length: Maximum, 52 weeks.
Exhibit Dates: 7/90–Present.
Learning Outcomes: Upon completion of the course, the student will be able to apply Boolean algebra to digital circuit analysis; convert between decimal, binary, octal, hexadecimal, and binary-coded decimal numbering systems; distinguish between ECL, DTL, RTL, and TTL digital circuit implementations; analyze circuits employing inverters, and, or, nand, and nor gates, flip-flops, multivibrators, and shift registers; demonstrate knowledge of amplitude and frequency modulation and demodulation principles; troubleshoot superheterodyne receivers and exciters at the block-diagram level; and describe the propagation of MF, VMF, and UHF signals, including antenna and transmission line characteristics.
Instruction: Course consists of self-paced, independent study with open-book progress checks and satisfactory completion of a proctored final examination. There is on-the-job supervision and an opportunity for the application of learning. Correspondence topics covered include review of basic algebra, numbering systems, Boolean algebra, and digital logic; AM and FM fundamentals; modulator and demodulator circuits; receivers, exciters, and power amplifiers; HF, VHF, and UHF propagation; antenna fundamentals; and data communications and facsimile.
Credit Recommendation: In the lower-division baccalaureate/associate degree category, 1 semester hour in digital circuits and 1 in electronic communications fundamentals (11/91).

AF-1715-0793

B-1B AVIONICS TEST STATION AND COMPONENT TECHNICIAN BY CORRESPONDENCE

Course Number: 45177.
Location: Extension Course Institute, Gunter AFS, AL.
Length: Maximum, 52 weeks.
Exhibit Dates: 12/90–12/97.
Learning Outcomes: Upon completion of the course, the student will be able to distinguish between operational characteristics of pulsed and Doppler radar systems; correlate pulse repetition time, pulse repetition frequency, duty cycle, pulse width, resting time, peak power, and average power for a radar pulse train; correctly interpret type A, B, C, E, and PPI radar displays; explain proper operation of inertial navigation, instrument landing system, TACAN, and IFF transponder systems; and identify differences in operation and capabilities of UHF and HF airborne communications systems.
Instruction: Course consists of self-paced, independent study with open-book progress checks and satisfactory completion of a proctored final examination. There is on-the-job supervision and an opportunity for the application of learning. Correspondence topics include pulsed and Doppler radar systems; airborne navigation basics; radar altimeters; IFF transponders; instrument landing systems; inertial navigation; and TACAN, HF, and UHF communications equipment.
Credit Recommendation: In the lower-division baccalaureate/associate degree category, 1 semester hour in aviation electronics systems (11/91).

AF-1715-0794

AUTOMATIC TRACKING RADAR SPECIALIST BY CORRESPONDENCE

Course Number: 30353B.
Location: Extension Course Institute, Gunter AFS, AL.
Length: Maximum, 52 weeks.
Exhibit Dates: 2/92–Present.
Learning Outcomes: Upon completion of the course, the student will be able to demonstrate a general knowledge of the applications of radar system used in Air Defense Networks, the operation and maintenance of automatic tracking radar, and the operation of the IFF decoders equipment. The student will also be able to troubleshoot and repair the equipment.
Instruction: This correspondence course consists of an estimated 111 clock hours of self-instruction on automatic tracking radar systems, including precision tracking radar, electronic bomb scoring, and aircraft identification. The operation and maintenance of electronic warfare equipment is included.
Credit Recommendation: Credit is not recommended because of the military-specific content of the course (4/93).

AF-1715-0795

AUTOMATIC TRACKING RADAR SPECIALIST BY CORRESPONDENCE

Course Number: 30353C.
Location: Extension Course Institute, Gunter AFS, AL.
Length: Maximum, 52 weeks.
Exhibit Dates: 7/92–Present.
Learning Outcomes: Upon completion of the course, the student will be able to maintain and troubleshoot automatic tracking radar systems, including transmitters, receivers, indicators, and monitors.
Instruction: This is a correspondence course providing instruction in the primary components and circuits that comprise automatic tracking radar systems, including modulators, transmitters, receivers, CRT deflection methods, and special purpose computer monitors.
Credit Recommendation: Credit is not recommended because of the limited, specialized nature of the course (4/93).

AF-1715-0796

TELEVISION SYSTEMS SPECIALIST BY CORRESPONDENCE

Course Number: 30455A.
Location: Extension Course Institute, Gunter AFS, AL.
Length: Maximum, 52 weeks.
Exhibit Dates: 8/92–8/98.
Learning Outcomes: Upon completion of the course, the student will be able to maintain, troubleshoot, and repair basic television equipment.
Instruction: This is a correspondence course covering electronic devices and circuits, digital circuits and systems, television fundamentals, and the use of common laboratory test equipment.
Credit Recommendation: In the vocational certificate category, 1 semester hour in fundamentals of electronics, 3 in digital logic circuits and systems, and 3 in electronics laboratory (4/93).

AF-1715-0797

SATELLITE AND WIDEBAND COMMUNICATIONS EQUIPMENT SPECIALIST BY CORRESPONDENCE

Course Number: 30457A.
Location: Extension Course Institute, Gunter AFS, AL.
Length: Maximum, 52 weeks.
Exhibit Dates: 4/92–Present.
Learning Outcomes: Upon completion of the course, the student will have reviewed AC/DC circuit analysis and be able to make complex Z calculations; apply basic semiconductor theory (basic junctions and field effect devices) in troubleshooting and repair of electronic equipment; apply the theory of analog and digital system modulators to multiplex equipment; and make a limited application of RF transmission theory to wideband communications in the microwave spectrum.
Instruction: This course consists of an estimated 69 clock hours of self-instruction. A review of AC and DC circuit analysis is included. Course also covers solid state devices including basic junction and field effect devices, a limited coverage of digital logic including application of integrated circuits, the theory of analog and digital system modulation as applied to wideband system modulators, and basic RF transmission theory as it applies to the microwave spectrum.
Credit Recommendation: In the lower-division baccalaureate/associate degree category, 1 semester hour in wideband communications and 2 in digital logic theory (4/93).

AF-1715-0798

TELEVISION SYSTEMS SPECIALIST BY CORRESPONDENCE

Course Number: 30455B.
Location: Extension Course Institute, Gunter AFS, AL.
Length: Maximum, 52 weeks.
Exhibit Dates: 8/92–Present.
Learning Outcomes: Upon completion of the course, the student will be able to operate and maintain television cameras, monitors, and receivers; perform limited troubleshooting and repair of videotape recorders; and demonstrate a general knowledge of television studio equipment, operational controls, and satellite microwave systems.
Instruction: This correspondence course provides for an estimated 108 hours of self-instruction on television studio equipment, including cameras, monitors, VTRs, and audio systems. The theory of operation of the television transmitter, receiver, and associated equipment is included. The course also includes the makeup of the composite TV signal and analysis of waveforms. The operation and maintenance of the total TV system is included throughout the course.
Credit Recommendation: In the lower-division baccalaureate/associate degree category, 4 semester hours in introduction to television system operation and maintenance (4/93).

AF-1715-0799

SATELLITE AND WIDEBAND COMMUNICATIONS EQUIPMENT SPECIALIST BY CORRESPONDENCE

Course Number: 30457B.
Location: Extension Course Institute, Gunter AFS, AL.
Length: Maximum, 52 weeks.
Exhibit Dates: 9/92–Present.
Learning Outcomes: Upon completion of the course, the student will be able to apply basic multiplex principles to wideband communications; apply basic theory of transmission paths for line-of-site, tropospheric scatter, and satellite communications; perform functional analysis of line-of-site, tropospheric scatter, and satellite microwave communication equipment in performance assessment and troubleshooting equipment; and apply theory of microwave amplifiers, including klystrons, TWTs, and low-noise solid state amplifiers.
Instruction: Course consists of an estimated 96 clock hours of self-instruction on basic multiplex principles (includes both frequency division and time division mux), microwave transmitters and receivers as they apply to wideband communications, and microwave devices as appropriate for wideband communications. The correspondence course includes use of test equipment in performance testing and in troubleshooting.
Credit Recommendation: In the lower-division baccalaureate/associate degree category, 3 semester hours in wideband communications (4/93).

AF-1715-0800

WEAPONS CONTROL SYSTEMS MECHANIC (A-7D, A-10, AC-130, F-5) BY CORRESPONDENCE

Course Number: 32152.
Location: Extension Course Institute, Gunter AFS, AL.
Length: Maximum, 52 weeks.
Exhibit Dates: 7/92–12/95.
Learning Outcomes: Upon completion of the course, the student will be able to maintain

weapons control systems, including displays, radar systems, and tactical computers.

Instruction: Course includes instruction in the principles of electronics (diodes and power supplies, transistor circuits, oscillators, logic circuits) in addition to the description of specific military equipment.

Credit Recommendation: In the vocational certificate category, 3 semester hours in basic electronics (4/93).

AF-1715-0801

AIRBORNE WARNING AND CONTROL RADAR SPECIALIST BY CORRESPONDENCE

Course Number: 32852.
Location: Extension Course Institute, Gunter AFS, AL.
Length: Maximum, 52 weeks.
Exhibit Dates: 11/91–1/98.
Learning Outcomes: Upon completion of the course, the student will be able to operate and maintain major types of surveillance radar systems and their related equipment.
Instruction: Course presents equipment maintenance principles, hand tools, general test equipment operation, digital circuit fundamentals, computer systems, and principles of radar. Also included are maintenance procedures for typical airborne radar equipment.
Credit Recommendation: In the vocational certificate category, 1 semester hour in electronic shop principles and 1 in basic electronic laboratory (4/93); in the lower-division baccalaureate/associate degree category, 3 semester hours in digital and computer fundamentals and 3 in principles of radar (4/93).

AF-1715-0802

ANTENNA AND CABLE SYSTEMS PROJECTS/ MAINTENANCE SPECIALIST BY CORRESPONDENCE

(Antenna and Cable Systems Installation/ Maintenance Specialist by Correspondence)

Course Number: 36150.
Location: Extension Course Institute, Gunter AFS, AL.
Length: Maximum, 52 weeks.
Exhibit Dates: 1/90–Present.
Learning Outcomes: Upon completion of the course, the student will be able to install pole lines and aerial and underground cable systems and install and maintain antennas.
Instruction: This correspondence course presents the tools and techniques necessary for the installation and maintenance of cables and antennas, including safety procedures, pole construction and climbing, aerial and underground installations, cable storage, elementary surveying, antenna system supports, and antenna and cable system maintenance.
Credit Recommendation: Credit is not recommended because of the limited, specialized content of the course (4/93).

AF-1715-0803

TELEPHONE SWITCHING SPECIALIST BY CORRESPONDENCE

Course Number: 36251.
Location: Extension Course Institute, Gunter AFS, AL.
Length: Maximum, 52 weeks.
Exhibit Dates: 8/91–9/98.
Learning Outcomes: Upon completion of the course, the student will be able to apply the theory and normal function of telephone central office switching equipment (electrical/electromechanical) to troubleshoot telephone equipment.
Instruction: This correspondence course provides instruction in electronic/electromechanical telephone central office switching equipment. The course includes electronic principles; telephone fundamentals; manual telephone systems; specific telephone systems; and fundamentals of digital electronics, including digital logic, Boolean algebra, clock/pulse generators, sequential logic, combination logic and converters, and computers, peripherals and storage devices.
Credit Recommendation: In the vocational certificate category, 3 semester hours in electronic/electromechanical communications systems (4/93); in the lower-division baccalaureate/associate degree category, 3 semester hours in digital fundamentals and 3 in communications systems (4/93).

AF-1715-0804

GENERAL SUBJECTS FOR F-111 AVIONICS SYSTEM SPECIALIST BY CORRESPONDENCE

Course Number: 45253.
Location: Extension Course Institute, Gunter AFS, AL.
Length: Maximum, 52 weeks.
Exhibit Dates: 8/92–6/95.
Learning Outcomes: Upon completing the course, the student will be able to use specific electronic test equipment for specific maintenance checkout procedures on military aircraft systems.
Instruction: This correspondence course contain the job-related knowledge needed to upgrade to the 5-level specialist level. Information pertains to the F-111 aircraft characteristics, flight-line safety, technical orders, general supply information, the automated maintenance system, maintenance forms, tools and test equipment, wiring, basic electronics, basic computer theory, and digital logic gates.
Credit Recommendation: In the vocational certificate category, 2 semester hours in introduction to electronics or electronic test equipment (4/93).

AF-1715-0805

F-4 AND AC-130 WEAPONS CONTROL SYSTEMS TECHNICIAN BY CORRESPONDENCE

Course Number: 45573A.
Location: Extension Course Institute, Gunter AFS, AL.
Length: Maximum, 52 weeks.
Exhibit Dates: 7/92–5/98.
Learning Outcomes: Upon completion of the course, the student will be able to operate and maintain the F-4 and AC-130 weapon control systems, including operational checks, fault isolation, and system test equipment.
Instruction: This is a correspondence course providing instruction in basic electric circuits, electronics, microwave devices, and weapons system testing and maintenance.
Credit Recommendation: In the vocational certificate category, 3 semester hours in introduction to electronics (4/93).

AF-1715-0806

IMAGERY SYSTEMS MAINTENANCE SPECIALIST BY CORRESPONDENCE

Course Number: 40450A.
Location: Extension Course Institute, Gunter AFS, AL.
Length: Maximum, 52 weeks.
Exhibit Dates: 11/92–Present.
Learning Outcomes: Upon completion of the course, the student will be able to perform tests and troubleshoot faults found in electronic imagery equipment.
Instruction: This is a self-instruction course containing career knowledge information in imagery systems maintenance. It requires on-the-job training and is a prerequisite to the 5-level classification. Content covers OSHA standards, job safety hazards, general shop practices, tools and measuring devices, electronics fundamentals, basic AC/DC circuits, electromechanical devices, relays and power supplies, amplifier circuits, wave generator devices, and an introduction to computers.
Credit Recommendation: In the vocational certificate category, 1 semester hour in electronics shop operations, 2 in digital logic, and 2 in introduction to electronics (4/93); in the lower-division baccalaureate/associate degree category, 3 semester hours in introduction to electronics or introduction to principles of electrical circuits (4/93).

AF-1715-0807

IMAGERY SYSTEMS MAINTENANCE SPECIALIST BY CORRESPONDENCE

Course Number: 40450B.
Location: Extension Course Institute, Gunter AFS, AL.
Length: Maximum, 52 weeks.
Exhibit Dates: 11/92–Present.
Learning Outcomes: Upon completion of the course, the student will be able to troubleshoot and test equipment, apply photographic fundamentals, use cameras, process and print film, and perform maintenance on video display and audiovisual production equipment.
Instruction: Content covers troubleshooting, test equipment, photographic fundamentals, processing and printing equipment, support and quality control equipment, and video display and audiovisual production equipment.
Credit Recommendation: In the vocational certificate category, 3 semester hours in introduction to test equipment and 3 in introduction to audiovisual equipment (4/93); in the lower-division baccalaureate/associate degree category, 3 semester hours in introduction to photography (4/93).

AF-1715-0808

F-16 INTEGRATED ORGANIZATIONAL AVIONICS SYSTEMS SPECIALIST BY CORRESPONDENCE

Course Number: 45252.
Location: Extension Course Institute, Gunter AFS, AL.
Length: Maximum, 52 weeks.
Exhibit Dates: 8/92–3/98.
Learning Outcomes: Upon completion of the course, the student will be able to apply a knowledge of avionics systems to Air Force maintenance procedures and records with emphasis on aircraft systems safety and flight-line safety.

Instruction: This course presents job-related knowledge on F-16 avionics and maintenance concepts. The course also includes basic electronics necessary for flight-line maintenance.

Credit Recommendation: In the vocational certificate category, 2 semester hours in introduction to basic electronics (4/93); in the lower-division baccalaureate/associate degree category, 2 semester hours in industrial or electronics safety (4/93).

AF-1715-0810

ELECTRONIC WARFARE (EW) SYSTEM
JOURNEYMAN BY CORRESPONDENCE

Course Number: 2A252B.
Location: Extension Course Institute, Gunter AFS, AL.
Length: Maximum, 52 weeks.
Exhibit Dates: 12/93–Present.
Learning Outcomes: Upon completion of the course, the student will be able to troubleshoot and repair electronic warfare receivers, transmitters, pods, and chaff/flare dispensers.
Instruction: This is a correspondence course with topics in electronic circuits; active devices and circuits; digital logic fundamentals; and electronic troubleshooting and repair of electronic warfare receivers, transmitters, pods, and chaff and flare dispensers.
Credit Recommendation: In the lower-division baccalaureate/associate degree category, 1 semester hour in solid state principles and 2 in digital circuits (6/94).

AF-1715-0811

AIRBORNE COMMAND AND CONTROL
    COMMUNICATIONS EQUIPMENT
    JOURNEYMAN BY CORRESPONDENCE
    (Airborne Command and Control Equipment Specialist by Correspondence)

Course Number: 11851B.
Location: Extension Course Institute, Gunter AFS, AL.
Length: Maximum, 52 weeks.
Exhibit Dates: 1/94–Present.
Learning Outcomes: Upon completion of the course, the student will be able to operate, troubleshoot, and maintain electronic communications systems.
Instruction: This course consists of on-the-job training and at-home studies on radio communications circuits and systems.
Credit Recommendation: In the lower-division baccalaureate/associate degree category, 3 semester hours in analog communications systems (6/94).

AF-1715-0812

AUTOMATIC TRACKING RADAR JOURNEYMAN BY
    CORRESPONDENCE
    (Automatic Tracking Radar Specialist by Correspondence)

Course Number: 30353C.
Location: Extension Course Institute, Gunter AFS, AL.
Length: Maximum, 52 weeks.
Exhibit Dates: 7/92–Present.
Learning Outcomes: Upon completion of this correspondence course, the student will be able to troubleshoot automatic tracking radar systems.
Instruction: This course covers Air Force training procedures; basic electronic maintenance techniques; soldering and cabling; basic electronic test equipment; fundamentals of transistor and vacuum tube amplifiers, synchro devices, and operational amplifiers; fundamentals of tracking radars; and analog computation and its use in tracking radars, radar receivers, and radar control systems.
Credit Recommendation: In the lower-division baccalaureate/associate degree category, 1 semester hour in radar systems (6/94).

AF-1715-0813

SECURE COMMUNICATIONS SYSTEMS
    MAINTENANCE BY CORRESPONDENCE

Course Number: 30656B.
Location: Extension Course Institute, Gunter AFS, AL.
Length: Maximum, 52 weeks.
Exhibit Dates: 12/93–3/98.
Learning Outcomes: Before 12/93 see AF-1715-0786. Upon completion of the course, the student will be able to differentiate between Bandot and ASCII codes; convert between baud rate, bit time, and characters-per minute; identify characteristics of various digital transmission media; explain the effects of signal distortion on a digital waveform; identify the functions and elements of a telecommunications terminal; and analyze multiplexer/demultiplexer and modem input and output signals.
Instruction: Course consists of self-paced, independent study with open-book progress checks and satisfactory completion of a proctored final examination. There is on-the-job supervision and an opportunity for the application of learning. Correspondence topics include basic electronic troubleshooting, telecommunications terminals, principles of digital communications, troubleshooting telecommunications equipment, multiplexing, and modems.
Credit Recommendation: In the lower-division baccalaureate/associate degree category, 1 semester hour in telecommunications (6/94).

AF-1715-0814

MISSILE AND SPACE SYSTEMS ELECTRONIC
    MAINTENANCE JOURNEYMAN BY
    CORRESPONDENCE
    (Air Launched Missile Systems Specialist by Correspondence)

Course Number: 46650B.
Location: Extension Course Institute, Gunter AFS, AL.
Length: Maximum, 52 weeks.
Exhibit Dates: 12/92–Present.
Learning Outcomes: Upon completion of the course, the student will have obtained the understanding of the basic principles of air launched missile systems necessary to become proficient as a maintenance specialist. Proficiency requires practical experience.
Instruction: This course presents material covering career progression, Air Force publications, maintenance management procedures, munitions maintenance squadron, and automated maintenance.
Credit Recommendation: Credit is not recommended because of the military-specific nature of the course (6/94).

AF-1715-0815

F-15/F-111 AVIONIC SYSTEMS JOURNEYMAN,
    INSTRUMENT, BY CORRESPONDENCE

Course Number: 2A351B.
Location: Extension Course Institute, Gunter AFS, AL.
Length: Maximum, 52 weeks.
Exhibit Dates: 1/94–Present.
Learning Outcomes: Upon completion of the course, the student will be able to troubleshoot avionics systems on the F-15 and F-111 aircraft.
Instruction: This correspondence course consists of practical exercises in troubleshooting, maintenance, and repair of avionic systems and components. Topics include aircraft indicating systems and instrumentation, flight instruments, navigational instruments, hydraulic systems, signal data recorders, aircraft built-in test systems, automatic terrain-following systems, and automatic flight control systems.
Credit Recommendation: In the vocational certificate category, 2 semester hours in aircraft navigation and control systems (6/94).

AF-1715-0816

AIRBORNE COMMAND AND CONTROL
    COMMUNICATIONS EQUIPMENT
    JOURNEYMAN BY CORRESPONDENCE
    (Airborne Command and Control Communications Equipment Specialist by Correspondence)

Course Number: 11851A.
Location: Extension Course Institute, Gunter AFS, AL.
Length: Maximum, 52 weeks.
Exhibit Dates: 11/92–Present.
Learning Outcomes: Upon completion of the course, the student will be able to demonstrate a knowledge of general aircrew skills and procedures, safety related items, and radiotelephone operating procedures.
Instruction: Course includes a presentation of information on security programs, security procedures, and security vulnerabilities. Instructional material pertains to operations management, safety related items, and first aid procedures.
Credit Recommendation: Credit is not recommended because of the limited, technical nature of the course (6/94).

AF-1715-0817

AIRCRAFT COMMUNICATION/NAVIGATION SYSTEMS
    JOURNEYMAN BY CORRESPONDENCE

Course Number: 45352.
Location: Extension Course Institute, Gunter AFS, AL.
Length: Maximum, 52 weeks.
Exhibit Dates: 12/93–2/97.
Learning Outcomes: Upon completion of the course, the student will be able to describe a maintenance organization; AC, DC, electronic, and digital circuits; and the operation of navigational systems such as AOF, VOR/ILS, TACAN, and IFF.
Instruction: This is a correspondence course where, through reading and exercises, the student learns the concepts and applications of electronic and digital circuits and navigational systems.
Credit Recommendation: In the lower-division baccalaureate/associate degree category, 1 semester hour in basic electricity, 1 in basic electronics, and 1 in digital fundamentals (6/94).

AF-1715-0818

AVIONICS SENSORS MAINTENANCE JOURNEYMAN BY CORRESPONDENCE

Course Number: 45550.
Location: Extension Course Institute, Gunter AFS, AL.
Length: Maximum, 52 weeks.
Exhibit Dates: 8/93–1/98.
Learning Outcomes: Upon completion of the course, the student will be able to apply shop safety principles; maintain and use test equipment; apply elementary AC and DC and magnetic and solid state principles; and describe rectifiers, oscillators, number systems, logic circuits, and A/D and D/A converters. Student will be able to describe transmission and reception of television communications and infrared and low-light sensors and lasers. Student will also apply the principles of microwave transmissions.
Instruction: This course consists of self-paced independent study with open-book progress checks and satisfactory completion of a proctored final examination. There is on-the-job supervision and an opportunity for the application of learning. Subjects include shop safety, shop maintenance, use of test equipment, DC, AC, solid state devices and rectifiers, oscillators, number systems, logic circuits, A/D and D/A converters, television fundamentals, IF and low-light sensors, lasers, and microwave fundamentals.
Credit Recommendation: In the lower-division baccalaureate/associate degree category, 1 semester hour in electronics fundamentals, 1 in introduction to television, and 1 in microwave fundamentals (6/94).

AF-1715-0819

ELECTRONIC WARFARE SYSTEMS JOURNEYMAN BY CORRESPONDENCE

Course Number: 45651A.
Location: Extension Course Institute, Gunter AFS, AL.
Length: Maximum, 52 weeks.
Exhibit Dates: 10/93–8/97.
Learning Outcomes: Upon completion of the course, the student will be able to troubleshoot and repair electronic warfare receivers, transmitters, pods, and chaff/flare dispensers.
Instruction: This is a correspondence course with practical exercises in maintenance and material management. Topics include electronic circuits lab; active devices and circuits; digital logic and fundamentals; and troubleshooting and repair of electronic warfare receivers, transmitters pods, and chaff/flare dispensers.
Credit Recommendation: In the lower-division baccalaureate/associate degree category, 3 semester hours in avionics systems (6/94).

AF-1715-0820

SECURE COMMUNICATIONS SYSTEMS MAINTENANCE BY CORRESPONDENCE

Course Number: 30656A.
Location: Extension Course Institute, Gunter AFS, AL.
Length: Maximum, 52 weeks.
Exhibit Dates: 11/93–Present.
Learning Outcomes: Upon completion of the course, the student will be able to apply principles of electrical safety in the work place; perform basic wiring, casing, and soldering operations; properly use basic electronic hand tools; identify common electronic components; analyze series, parallel, and series/parallel AC and DC circuits; explain forward and reverse bias of rectifier and zener diodes; properly bias bipolar junction and field effect transistors; troubleshoot power supply rectifiers, filters, and regulators; distinguish between Colpitts, Hartley, Butler, blocking, phase-shift, and crystal oscillators; convert between decimal, binary, hexadecimal, octal, and BCD numbering systems; produce truth tables for inverters, and, or, nand, nor, and exclusive or gates, latches, flip-flops, and shift registers; explain voltage, current, phase, and torque relationships for AC and DC motors.
Instruction: Course consists of self-paced, independent study with open-book progress checks and satisfactory completion of a proctored final examination. There is on-the-job supervision and an opportunity for the application of learning. Correspondence topics include electrical safety; soldering, wiring, and interconnect; basic hand tools; shop practices; AC/DC fundamentals; semiconductors; power supplies; oscillator circuits; digital logic and numbering systems; and servomechanisms.
Credit Recommendation: In the lower-division baccalaureate/associate degree category, 1 semester hour in AC/DC circuits, 1 in digital principles, and 1 in solid state electronics (6/94).

AF-1715-0821

METEOROLOGICAL AND NAVIGATION SYSTEMS JOURNEYMAN BY CORRESPONDENCE
(Meteorological and Navigation Systems Specialist by Correspondence)

Course Number: 30452A.
Location: Extension Course Institute, Gunter AFS, AL.
Length: Maximum, 52 weeks.
Exhibit Dates: 1/94–3/98.
Learning Outcomes: Upon completion of the course, the student will be capable of performing maintenance of meteorological and radio navigation equipment with emphasis on VOR/ILS and TACAN.
Instruction: This course includes subject matter on work center administration, electronic fundamentals, VOR/ILS, and TACAN.
Credit Recommendation: In the lower-division baccalaureate/associate degree category, 1 semester hour in radio navigation (6/94).

AF-1717-0005

AIRCREW EGRESS SYSTEMS MECHANIC BY CORRESPONDENCE

Course Number: 42352.
Location: Extension Course Institute, Gunter AFS, AL.
Length: Maximum, 52 weeks.
Exhibit Dates: 1/90–10/97.
Learning Outcomes: Upon completion of the course, the student will be able to perform general maintenance duties; follow appropriate technical orders and publications; and applying good maintenance management concepts. Emphasis is placed on proper handling of explosives and on demonstrating intermediate knowledge of electrical and pneumatic principles.
Instruction: The student works under individual supervision and with a trainer or training manager to control progress in the areas of aircrew egress systems mechanical fundamentals, management and supply fundamentals, and fundamentals and specific knowledge of egress systems (B52,F-4, F015, F-111, and F-38). Test equipment such as multimeters and common hand tools are used.
Credit Recommendation: In the lower-division baccalaureate/associate degree category, 2 semester hours in maintenance principles and 1 in industrial safety (10/88).

AF-1717-0013

CORROSION CONTROL SPECIALIST BY CORRESPONDENCE

Course Number: 42751.
Location: Extension Course Institute, Gunter AFS, AL.
Length: Maximum, 52 weeks.
Exhibit Dates: 1/90–Present.
Learning Outcomes: Upon completion of the course, the student will be able to provide technical assistance and supervise subordinates in missiles, aircraft, and support systems and in corrosion control procedures.
Instruction: This is a supervised, self-paced extension course covering security training publications; maintenance equipment and supply; supervision, management, and inspection systems; preparation for corrosion control and corrosion removal; and the application of organic coatings.
Credit Recommendation: In the lower-division baccalaureate/associate degree category, 3 semester hours in corrosion control (10/88).

AF-1717-0026

MISSILE FACILITIES SPECIALIST BY CORRESPONDENCE

Course Number: 44550G.
Location: Extension Course Institute, Gunter AFS, AL.
Length: Maximum, 52 weeks.
Exhibit Dates: 1/90–5/93.
Learning Outcomes: Upon completion of the course, the student will be able to provide technical assistance and supervise subordinates in inspecting, monitoring, troubleshooting, operating, maintaining, and repairing missile weapons system support facilities and equipment.
Instruction: This is a supervised, self-paced extension course covering missile maintenance; technical orders; maintenance data collections systems; security and safety; corrosion; hand tools; aerospace hardware; principles of electricity and pneudraulics; layout of Minuteman facilities; hardness assurance and construction and maintenance of missile; maintaining Minuteman facilities; operation and maintenance special purpose, missile handling, and transporting vehicles and equipment; and operation and proofload of testing equipment.
Credit Recommendation: In the lower-division baccalaureate/associate degree category, 3 semester hours in plant maintenance (10/88).

AF-1717-0027

AIRCRAFT METALS TECHNOLOGY (MACHINIST) BY CORRESPONDENCE

Course Number: 45870A.

Location: Extension Course Institute, Gunter AFS, AL.
Length: Maximum, 52 weeks.
Exhibit Dates: 1/90–8/98.
Learning Outcomes: Upon completion of the course, the student will be able to identify and utilize various machining techniques and equipment.
Instruction: This is a correspondence course with home study and practical exercises in the operation, maintenance, and use of machining equipment. Topics include drafting practices, drawings, hand and special tools, lubricants and coolants, contour machine, drill press and lathe work, milling machine, gear cutters, tool and cutter grinders, and tool design and fabrication of jigs and fixtures. Students enrolling must have an AFSC assignment as a prerequisite. This serves as the laboratory component of the course. The student's supervisor must complete an evaluation of the course and its effectiveness when the student takes the final exam which measures mastery of the cognitive content. A score of 65 is required for a satisfactory completion certificate.
Credit Recommendation: In the vocational certificate category, 3 semester hours in machine shop (3/91); in the lower-division baccalaureate/associate degree category, 3 semester hours in machine/mechanical technology (3/91).

AF-1717-0028

AIRCRAFT STRUCTURAL MAINTENANCE TECHNICIAN (AIRFRAME REPAIR) BY CORRESPONDENCE

Course Number: 45872A.
Location: Extension Course Institute, Gunter AFS, AL.
Length: Maximum, 52 weeks.
Exhibit Dates: 1/90–9/99.
Learning Outcomes: Upon completion of the course, the student will be able to troubleshoot and repair airframe structures.
Instruction: This correspondence course consists of home study and practical exercises in the maintenance of airframes and structures. Topics include aircraft metals, basic hand tools, layout techniques, metal forming and cutting, drilling and riveting, repair procedures, damage removal, fastener layout fabrication, installation of repair parts, and special repair situations. Specialized repairs include use of blind rivets, high-strength fasteners, common hardware, aircraft cables, tubing, and balancing of components. In addition, plastic and fiberglass structure, honeycomb, and advanced composite structures are covered. Course concludes with a proctored examination.
Credit Recommendation: In the lower-division baccalaureate/associate degree category, 3 semester hours in materials and methods, 3 in aircraft structural repair, and 3 in aircraft rigging and balance (3/91).

AF-1719-0008

APPRENTICE REPROGRAPHIC SPECIALIST BY CORRESPONDENCE

Course Number: 70330.
Location: Extension Course Institute, Gunter AFS, AL.
Length: Maximum, 52 weeks.
Exhibit Dates: 1/90–3/96.
Learning Outcomes: Upon completion of the course, the student will be able to perform entry level duties, functions, and operations associated with electrostatic platemaking, photo lithography, offset duplication, microfilming; use the collator, sorter, cutter, and paper drill; and troubleshoot and maintain equipment.
Instruction: Subject matter is introduced in manuals in small sections or modules for ease of learning and is followed by self-test questions and unit review exams. Topics covered include those above, as well as safety and health in the work place, general administration duties, historical developments in printing, and ink and paper characteristics. Students enrolling must have an AFSC assignment as a prerequisite. This serves as the laboratory component of the course. The student's supervisor must complete an evaluation of the course and its effectiveness when the student takes the final exam which measures mastery of the cognitive content. A score of 65 is required for a satisfactory completion certificate.
Credit Recommendation: In the lower-division baccalaureate/associate degree category, 3 semester hours in printing (3/91).

AF-1719-0009

APPRENTICE GRAPHICS SPECIALIZED BY CORRESPONDENCE

Course Number: 23131.
Location: Extension Course Institute, Gunter AFS, AL.
Length: Maximum, 52 weeks.
Exhibit Dates: 1/90–3/96.
Learning Outcomes: Upon completion of the course, the student will be able to identify the basic principles of graphic design and describe their application to specific visual communication requirements. Specifically the student will have knowledge and skills in identification, use, and care of basic graphics and illustration tools and equipment; the fundamentals of lettering styles, their uses and characteristics; the fundamental characteristics and principles of drawing in terms of shapes, colors, and textures; representation of the human form; cartoons and caricature techniques; terms and principles of geometric construction, composition, and layout; use of graphics for visual communication; basic design requirements for projected materials; and the preparation of materials for printing. With appropriate drawing practice, the student will have entry level skills for an illustration or graphic design position.
Instruction: This is a two-part correspondence course consisting of printed documents with some eighty-four exercises to measure the cognitive components of the course. The exercises may be either self-scored or scored by the student's supervisor. Students enrolling must have an AFSC assignment as prerequisite. This serves as the laboratory component of the course. The students supervisor must complete an evaluation of the course and its effectiveness when the student takes the final exam which measures mastery of the cognitive content. A score of 65 is required for a satisfactory completion certificate.
Credit Recommendation: In the lower-division baccalaureate/associate degree category, 3 semester hours in principles of graphic design (3/91).

AF-1719-0010

GRAPHICS SPECIALIST BY CORRESPONDENCE

Course Number: 23151.
Location: Extension Course Institute, Gunter AFS, AL.
Length: 52 weeks.
Exhibit Dates: 1/90–2/98.
Learning Outcomes: Upon completion of the course, the student will be able to perform advanced drawing and visual communication duties as a graphic artist.
Instruction: This duplicates the instruction of AF-1719-0009 (23131) but has greater depth in visualization concepts, use of air brush techniques, silk screen, and other special effects.
Credit Recommendation: In the upper-division baccalaureate category, 3 semester hours in advanced graphics illustration and design (3/91).

AF-1723-0005

NONDESTRUCTIVE INSPECTION SPECIALIST BY CORRESPONDENCE

Course Number: 42752.
Location: Extension Course Institute, Gunter AFS, AL.
Length: Maximum, 52 weeks.
Exhibit Dates: 1/90–Present.
Learning Outcomes: Upon completion of the course, the student will be able to provide technical assistance and supervise subordinates in the methods and techniques of nondestructive testing.
Instruction: This is a supervised, self-paced extension course covering topics of terminology, technique, Air Force safety, metallurgy, optical, liquid penetrant, magnetic particle inspection, ultrasonics, eddy current, oil analysis, radiography, technique development, and bonding test.
Credit Recommendation: In the lower-division baccalaureate/associate degree category, 4 semester hours in nondestructive inspection (10/88).

AF-1723-0007

METALS PROCESSING SPECIALIST BY CORRESPONDENCE

Course Number: 53151.
Location: Extension Course Institute, Gunter AFS, AL.
Length: Maximum, 52 weeks.
Exhibit Dates: 1/90–1/97.
Objectives: This course trains individuals in the fabrication and repair of metal parts and components.
Instruction: This correspondence course consists of on-the-job training and at-home study. The at-home study provides basic knowledge of metal repair and fabrication operations. Oxyacetylene welding, electric welding, metallic arc equipment, gas-shielded welding, TIG welding, and resistance welding are covered. Also taught are metal processing, heat treatment, testing and inspection, and metal treatment.
Credit Recommendation: In the vocational certificate category, 6 semester hours in welding (9/86).

AF-1723-0008

MACHINIST BY CORRESPONDENCE

Course Number: 42750.
Location: Extension Course Institute, Gunter AFS, AL.
Length: Maximum, 52 weeks.

Exhibit Dates: 1/90–11/91.

Learning Outcomes: Upon completion of the course, the student will be able to provide technical assistance and supervise subordinates and operate metalworking machines in fabricating, reworking, and repairing metal parts.

Instruction: This is a supervised, self-paced extension course covering manufacturing and reworking mechanical parts, assembling and fitting machine parts, maintaining hand and machine tools, and supervising machine shop personnel. Course does not require proficiency in computer-assisted machinery.

Credit Recommendation: In the vocational certificate category, 3 semester hours in basic machine shop and 3 in advanced machine tools (10/88).

AF-1723-0009

METALS PROCESSING SPECIALIST BY CORRESPONDENCE

Course Number: 42754.
Location: Extension Course Institute, Gunter AFS, AL.
Length: Maximum, 52 weeks.
Exhibit Dates: 1/90–Present.
Learning Outcomes: Upon completion of the course, the student will be able to test, inspect, and treat metals; identify metals to determine appropriate process (e.g. lead on silver, solder, oxyacetylene, or electric welding); work with assorted hand tools from pliers, nippers, wrenches, files, and chisels to measuring devices such as micrometers and calipers; work with some power equipment such as grinders and buffers; and have basic knowledge of mechanical drawing, maintenance documentation and management, and maintenance safety.

Instruction: The student in this course works under supervision and with a trainer or training manager to oversee progress in the use of assorted hand tools and limited power tools. Extensive coverage of metalwork from the most basic grinding to more sophisticated efforts with gas-shielded, tungsten-inert gas, and resistance welding are covered. Carbon steel, ferrous alloys, pipes, castings, and forgings are all included.

Credit Recommendation: In the lower-division baccalaureate/associate degree category, 3 semester hours in welding and 3 in metal fabrication (10/88).

AF-1723-0010

AIRFRAME REPAIR SPECIALIST BY CORRESPONDENCE

Course Number: 42755.
Location: Extension Course Institute, Gunter AFS, AL.
Length: Maximum, 52 weeks.
Exhibit Dates: 1/90–11/91.
Learning Outcomes: Upon completion of the course, the student will be able to provide technical assistance and supervise subordinates in the repair and maintenance of sheet metal and composite aircraft structures.

Instruction: This is a supervised, self-paced extension course covering operation of an airframe repair shop; composition and processing of aluminum, titanium, steel, and magnesium to obtain a desired end product; general repair techniques and procedures; methods of fastening including installation and removal of many types of blind fasteners; and general repair techniques and procedures for fiberglass, advanced composites, and metal-bonded components.

Credit Recommendation: In the lower-division baccalaureate/associate degree category, 3 semester hours in sheet metal repair, 3 in production processes, and 3 in composite structures repair (10/88).

AF-1723-0011

AIRCRAFT METALS TECHNOLOGY (WELDING) BY CORRESPONDENCE

Course Number: 45870B.
Location: Extension Course Institute, Gunter AFS, AL.
Length: Maximum, 52 weeks.
Exhibit Dates: 1/90–7/99.
Learning Outcomes: Upon completion of the course, the student will be able to identify various welding techniques and their uses and equipment, including oxyacetylene and electric welding, ferrous and non-ferrous metals, heat-treatment, inspection, and metal surface treatments and perform basic welding, brazing, and cutting techniques under direct supervision.

Instruction: Correspondence methods cover study of oxyacetylene, electronic arc, tungsten inert gas, metals inert gas welding, oxyacetylene, and electric methods of cutting; the assembly, operation, shutdown, troubleshooting, and maintenance of various equipment; ferrous and nonferrous alloy designations and heat-treatment processes and equipment; use and interpretation of the standard Rockwell, superficial Rockwell, Riechle portable, and Barcol portable testers; visual inspection; and metal surface treatments for corrosion. Course of study also includes resistance welding, pipe welding, engine repair, and brazing.

Credit Recommendation: In the vocational certificate category, 3 semester hours in welding and cutting practices (3/91); in the lower-division baccalaureate/associate degree category, 3 semester hours in metallurgy, heat-treatment, and testing (3/91).

AF-1723-0012

SOLDERING AND ELECTRICAL CONNECTORS BY CORRESPONDENCE

Course Number: 03036.
Location: Extension Course Institute, Gunter AFS, AL.
Length: Maximum, 52 weeks.
Exhibit Dates: 1/90–Present.
Learning Outcomes: Upon completion of the course, the student will be able to identify the fundamentals of solder materials, equipment, and techniques. Student will solder, under direct supervision.

Instruction: This correspondence course covers solder, flux, work area safety, soldering tools and equipment, soldering techniques, wire types, wire terminals, and ends. Also included are components, accessories, and contacts of electrical connectors including assembly, installation and maintenance. Introduction to the preparation, handling, and repair of printed circuit boards, board termination devices, lead bends, and terminations is included.

Credit Recommendation: In the vocational certificate category, 2 semester hours in soldering techniques and applications (3/91).

AF-1723-0013

METAL FABRICATION SPECIALIST BY CORRESPONDENCE

Course Number: 55252.
Location: Extension Course Institute, Gunter AFS, AL.
Length: Maximum, 52 weeks.
Exhibit Dates: 1/90–8/93.
Learning Outcomes: Upon completion of the course, the student will be able to use and maintain tools and equipment used in sheet metal fabrication; layout and install duct systems; install doors and other metal equipment; and perform basic oxyacetylene and electric arc welding operations.

Instruction: Student enrolling must have an Air Force Specialty Code assignment as a prerequisite. This serves as the laboratory component of the course. The student's supervisor must complete an evaluation of the course and its effectiveness when the student takes the final examination which measures the mastery of the cognitive content. Course consists of self-paced, independent study with open-book progress checks and satisfactory completion of a final examination. Students receive instruction in types, uses, and composition of metals; use, care, and maintenance of tools and equipment; seam and joint connections; installation of ducts, vents, doors, hoists, and other metal equipment; and oxyacetylene and electric arc welding procedures.

Credit Recommendation: In the vocational certificate category, 4 semester hours in sheet metal fabrication and installation (11/91).

AF-1723-0014

METAL FABRICATING SPECIALIST BY CORRESPONDENCE

Course Number: 55252A.
Location: Extension Course Institute, Gunter AFS, AL.
Length: Maximum, 52 weeks.
Exhibit Dates: 6/91–10/94.
Learning Outcomes: Upon completion of the course, the student will be able to use and maintain tools and equipment used in sheet metal fabrication; perform layout of sheet metal systems; and perform sheet metal seam and connection operations.

Instruction: Student enrolling must have an Air Force Specialty Code assignment as a prerequisite. This serves as the laboratory component of the course. The student's supervisor must complete an evaluation of the course and its effectiveness when the student takes the proctored final examination which measures the mastery of the cognitive content. Course consists of self-paced, independent study with open-book progress checks and satisfactory completion of a final examination. Students receive instruction in types, uses, and compositions of metals; use, care, and maintenance of metals; use, care, and maintenance of tools and equipment; layout of sheet metal systems; and seam and connection construction. The credits listed for this course do not duplicate those in AF-1723-0015.

Credit Recommendation: In the vocational certificate category, 2 semester hours in introductory sheet metal fabrication and installation (11/91).

AF-1723-0015

METAL FABRICATING SPECIALIST BY CORRESPONDENCE

Course Number: 55252B.
Location: Extension Course Institute, Gunter AFS, AL.

COURSE EXHIBITS

Length: Maximum, 52 weeks.
Exhibit Dates: 7/91–Present.
Learning Outcomes: Upon completion of the course, the student will be able to install duct work, doors, metal roofing; repair fixed metal equipment; perform oxyacetylene welding and cutting operations; and perform electric arc, TIG, and MIG welding operations.
Instruction: Student enrolling must have an Air Force Specialty Code assignment as a prerequisite. This serves as the laboratory component of the course. The student's supervisor must complete an evaluation of the course and its effectiveness when the student takes the final examination which measures mastery of the cognitive content. Course consists of self-paced, independent study with open-book progress checks and satisfactory completion of a proctored final examination. Students will receive instruction in installation of ducts, doors, awnings, canopies, hoists, and other metal equipment; practices and procedures of oxyacetylene welding and cutting operations; practices and procedures of electric arc, TIG, and MIG welding operations. Credits listed here do not duplicate those listed in AF-1723-0014.
Credit Recommendation: In the vocational certificate category, 2 semester hours in sheet metal fabrication and installation (11/91).

AF-1724-0006

APPRENTICE NONDESTRUCTIVE INSPECTION SPECIALIST BY CORRESPONDENCE

Course Number: 45851.
Location: Extension Course Institute, Gunter AFS, AL.
Length: Maximum, 52 weeks.
Exhibit Dates: 1/90–2/98.
Learning Outcomes: Upon completion of the course, the student will be able to perform nondestructive testing procedures using magnification, penetrant, ultrasonic, eddy current, and X-ray techniques and to evaluate the results under supervision.
Instruction: Student receives instructions on the use of visual, dye penetrant, ultrasonic, eddy current, and X-ray equipment and methods and development and evaluation of the results.
Credit Recommendation: In the lower-division baccalaureate/associate degree category, 3 semester hours in nondestructive inspection (12/89).

AF-1724-0007

AIRCRAFT METALS TECHNOLOGY (WELDING) BY CORRESPONDENCE

Course Number: 45870B.
Location: Extension Course Institute, Gunter AFS, AL.
Length: Maximum, 52 weeks.
Exhibit Dates: 1/90–Present.
Learning Outcomes: Upon completion of the course, the student will be able to construct components using oxyacetylene, electrical arc, or resistance welding methods; cut metals using flame cutting equipment; identify properties of both steel and aluminum; heat treat metals and test for hardness; treat metal surfaces to resist corrosion; and supervise others in metal processing using welding for joining.
Instruction: Through correspondence, the student receives information on types of metals and welding using oxyacetylene, electrical arc and resistance methods. Also covered are heat treatment to develop strength, plating to prevent corrosion, and testing to provide quality control of welding.
Credit Recommendation: In the lower-division baccalaureate/associate degree category, 3 semester hours in metals joining and processes (12/89).

AF-1728-0038

FIRE PROTECTION SPECIALIST BY CORRESPONDENCE

Course Number: 57150.
Location: Extension Course Institute, Gunter AFS, AL.
Length: Maximum, 52 weeks.
Exhibit Dates: 1/90–7/95.
Learning Outcomes: Upon completion of the course, the student will be able to fight fires in aircraft, structures, grass or forests, vehicles, and electronic components; operate and maintain various pieces of fire fighting equipment; drive fire fighting vehicles; perform fire investigations and fire prevention inspections; supervise fire crews; and assist in rescue operations.
Instruction: The student receives instruction in the combustion process; principles of fire suppression, control, and extinguishing; equipment inspection and maintenance; use of personal protective clothing; driving, specifications, and operation of various types of fire fighting vehicles; principles of crew supervision; command and control of fire sites; and fire prevention and investigation techniques. This correspondence course consists of on-the-job supervision and at-home studies to provide basic knowledge in fire protection, equipment, and fire fighting.
Credit Recommendation: In the lower-division baccalaureate/associate degree category, 3 semester hours in introduction to fire science (11/91).

AF-1728-0039

SPECIAL INVESTIGATIONS AND COUNTERINTELLIGENCE TECHNICIAN BY CORRESPONDENCE

Course Number: 82170.
Location: Extension Course Institute, Gunter AFS, AL.
Length: Maximum, 52 weeks.
Exhibit Dates: 1/90–2/97.
Objectives: Officers and enlisted personnel will be trained in the techniques of conducting and managing an investigation.
Instruction: This correspondence course consists of on-the-job supervision and at-home studies. Study includes an examination of the military justice system and the importance of ethics in agent conduct. Rules of evidence as applied to the military setting and the collection and preservation of physical evidence are studied. Investigations of crimes against persons and property and incidents of fraud are examined. The course includes an in-depth study of interviews and interrogation techniques and emphasizes the importance of formulating an investigative plan. Also included is a section on counterintelligence operations and personnel security.
Credit Recommendation: In the lower-division baccalaureate/associate degree category, 3 semester hours in criminal investigation and 2 in interview and interrogation (9/86).

AF-1728-0042

CRIME PREVENTION BY CORRESPONDENCE

Course Number: 8100.
Location: Extension Course Institute, Gunter AFS, AL.
Length: Maximum, 52 weeks.
Exhibit Dates: 1/90–Present.
Learning Outcomes: Upon completion of the course, the student will be able to describe the basics of the Air Force Crime Prevention program and a number of basic security measures.
Instruction: This course consists of monitored home-study with readings, exercises, and a proctored final examination. Topics include the history and philosophy of crime prevention, definitions of problems and priorities, risk management, prevention programs and management, evaluations of crime prevention programs, and security measures.
Credit Recommendation: Credit is not recommended because of limited, specialized nature of the course (5/88).

AF-1728-0043

MILITARY JUSTICE BY CORRESPONDENCE

Course Number: 8800.
Location: Extension Course Institute, Gunter AFS, AL.
Length: Maximum, 52 weeks.
Exhibit Dates: 1/90–5/92.
Learning Outcomes: Upon completion of the course, the student will be able to describe the scope of the military justice program as administered by the Air Force and identify the fundamental differences between the military and civilian justice systems, the need for those differences, and the ramifications of those differences on the individual service member, the military community, and American society.
Instruction: Course consists of monitored home study with written exercises, on-the-job training, and proctored examinations. Topics include the need for a military justice system, the sources of military law, the punitive provisions of the Uniform Code of Military Justice, the concept of nonjudicial punishment as a corollary to the civilian concept of misdemeanor, the rulings of the USSC that impact on the Manual for Courts Martial, and the imposition of military justice. Instruction also includes reviews of specific areas of nonjudicial punishment, the rights of the accused, pretrial activities, the various types of judicial proceedings, the Military Appellate review process, and the procedure (writ of certiorari) to petition review of a military finding by the Supreme Court. Instruction also includes addenda that allow for an appreciation of the scope and breadth of the Constitution and the Uniform Code of Military Justice.
Credit Recommendation: Credit is not recommended because of the military-specific nature of the course (5/88).

AF-1728-0044

LAW ENFORCEMENT SPECIALIST BY CORRESPONDENCE

Course Number: 81152.
Location: Extension Course Institute, Gunter AFS, AL.
Length: Maximum, 52 weeks.
Exhibit Dates: 1/90–7/96.

Learning Outcomes: Upon completion of the course, the student will be able to describe the general functions and responsibilities of an Air Force Security Police member including the various tactics and techniques inherent in enforcement.

Instruction: This monitored home-study course includes reading, written exercises, drills, and proctored examinations. Topics include general police functions, individual and team movement, target detection and engagement, challenges and identification, search procedures, evidence protection, resource and correction programs, traffic control and operations, systems security, emergency response, defense management, antiterrorism, and threat identification.

Credit Recommendation: In the lower-division baccalaureate/associate degree category, 2 semester hours in criminal justice (5/88).

AF-1728-0045

SECURITY SPECIALIST BY CORRESPONDENCE

Course Number: 81150.
Location: Extension Course Institute, Gunter AFS, AL.
Length: Maximum, 52 weeks.
Exhibit Dates: 1/90–7/96.
Learning Outcomes: Upon completion of the course, the student will be able to carry out security police functions; observe; investigate; and deal with hostage situations, hijacking operations, terrorism, and nuclear security accidents.

Instruction: This monitored home-study program includes readings, written exercises, and proctored examinations. Topics include requirements of integrity and ethics, legality, conduct of searches, maintaining systems security, how and when to use dogs, methods of riot control, and terrorism.

Credit Recommendation: In the lower-division baccalaureate/associate degree category, 2 semester hours in criminal justice (5/88).

AF-1728-0046

FIRE PROTECTION SUPERVISOR BY CORRESPONDENCE

Course Number: 57170.
Location: Extension Course Institute, Gunter AFS, AL.
Length: Maximum, 52 weeks.
Exhibit Dates: 1/90–5/96.
Objectives: To provide upgrade training in fire protection supervision.

Instruction: This correspondence course consists of on-the-job supervision and at-home studies. The home study provides general knowledge in fire protection operations and technical services related to supervision of fire protection activities. This course consists of general instruction in fire protection safety and security programs, fire protection operations command, structural and crash fire fighting, technical services related to building design and construction, and fire protection supervision activities.

Credit Recommendation: In the vocational certificate category, 2 semester hours in fire ground supervision (9/86).

AF-1729-0010

APPRENTICE FOOD SERVICE SPECIALIST BY CORRESPONDENCE

Course Number: 62230.
Location: Extension Course Institute, Gunter AFS, AL.
Length: Maximum, 52 weeks.
Exhibit Dates: 1/90–2/92.
Objectives: This course provides an introduction to the food service industry and emphasizes sanitation procedures, food service equipment, food selections, and preparation. Management techniques relating to the industry are also discussed.

Instruction: Instruction covers food service careers, hygiene standards, communicable diseases, nutrition, and baking fundamentals. The course also examines cooking equipment, safety precautions, energy conservation, storeroom operations, stock control, and quality assurance.

Credit Recommendation: In the lower-division baccalaureate/associate degree category, 2 semester hours in food service fundamentals (10/85).

AF-1729-0012

FOOD SERVICE SUPERVISOR BY CORRESPONDENCE

Course Number: 62270.
Location: Extension Course Institute, Gunter AFS, AL.
Length: Maximum, 52 weeks.
Exhibit Dates: 1/90–9/93.
Objectives: This is an introductory course for food service supervisors and training is provided in the planning and scheduling of food service activities, accounting procedures, quality control, and manpower requirements.

Instruction: Course covers work load scheduling, performance standards, dining hall operations, menu preparation, accounting fundamentals, sanitation, and performance evaluation.

Credit Recommendation: In the lower-division baccalaureate/associate degree category, 2 semester hours in food service management (10/85).

AF-1729-0013

APPRENTICE SERVICES SPECIALIST BY CORRESPONDENCE

Course Number: 61130.
Location: Extension Course Institute, Gunter AFS, AL.
Length: Maximum, 52 weeks.
Exhibit Dates: 1/90–2/92.
Objectives: The purpose of this course is to provide basic information concerning the handling of linen exchanges and housing facilities in the Air Force.

Instruction: This course provides basic information about inventory procedures, requisition, and stock control. Also included are front desk operations, including making reservations, maintaining occupancy rate, handling sundry sales, operating cash register, maintaining equipment, and handling customer complaints.

Credit Recommendation: In the vocational certificate category, 1 semester hour in introduction to hotel/motel operations (10/85).

AF-1729-0014

SERVICES SPECIALIST BY CORRESPONDENCE

Course Number: 61150.
Location: Extension Course Institute, Gunter AFS, AL.
Length: Maximum, 52 weeks.
Exhibit Dates: 1/90–1/97.
Objectives: This course provides specific information on handling linen, housing arrangements, and mortuary services in the military.

Instruction: This course includes basic procedures for the requisition, inventory, and stock control of linens. Also covered are front desk operations for housing, including reservations, handling sundry sales, operating cash register, maintaining equipment, and handling customer complaints. The mortuary services content describes the military honors and ceremonies procedures; search, recovery and identification of remains; and transportation and shipment procedures for the remains.

Credit Recommendation: In the vocational certificate category, 1 semester hour in introduction to hotel/motel operations or front office procedures and 1 in institutional housekeeping (10/85).

AF-1729-0015

APPRENTICE SUBSISTENCE OPERATIONS SPECIALIST BY CORRESPONDENCE

Course Number: 61231.
Location: Extension Course Institute, Gunter AFS, AL.
Length: Maximum, 52 weeks.
Exhibit Dates: 1/90–12/92.
Learning Outcomes: Upon completion of the course, the student will be able to describe the organization and operations of the subsistence program of the Air Force.

Instruction: This course includes specialized operations of the troop support section, commissary resale operations, control, and Air Force Commissary Service Mission.

Credit Recommendation: Credit is not recommended because of the military-specific nature of the course (2/87).

AF-1729-0016

SUBSISTENCE OPERATIONS SPECIALIST BY CORRESPONDENCE

Course Number: 61251.
Location: Extension Course Institute, Gunter AFS, AL.
Length: Maximum, 52 weeks.
Exhibit Dates: 1/90–10/94.
Learning Outcomes: Upon completion of the course, the student will be able to describe the Air Force's subsistence operations and assist in commissary operation and troop support subsistence.

Instruction: Instruction includes the career field of subsistence operations, commissary organization and operations, control and resale functions, troop support, and front-end functions.

Credit Recommendation: Credit is not recommended because of the military-specific nature of the course (2/87).

AF-1729-0017

SUBSISTENCE OPERATIONS TECHNICIAN BY CORRESPONDENCE

Course Number: 61271.
Location: Extension Course Institute, Gunter AFS, AL.
Length: Maximum, 52 weeks.

Exhibit Dates: 1/90–3/93.
Learning Outcomes: Upon completion of the course, the student will be able to perform supervisory tasks in commissaries and resale stores and assist in subsistence procurement for troop support and resource protection.
Instruction: Instruction includes career training and supervision in the subsistence operations career field. Topics covered are complex and control office operations, resale store operations, troops support, and resource protection.
Credit Recommendation: Credit is not recommended because of the military-specific nature of the course (2/87).

AF-1730-0016
REFRIGERATION AND CRYOGENICS SPECIALIST BY CORRESPONDENCE

Course Number: 54550.
Location: Extension Course Institute, Gunter AFS, AL.
Length: Maximum, 52 weeks.
Exhibit Dates: 1/90–10/94.
Objectives: To train personnel in refrigeration and air conditioning installation and maintenance.
Instruction: This course consists of on-the-job training and at-home study in the installation of various types of refrigeration and air conditioning systems. The course provides training in the principles of refrigeration and air conditioning, and the on-the-job training provides the application. The course includes electrical principles, refrigeration principles, air conditioning principles, and cryogenic operations.
Credit Recommendation: In the vocational certificate category, 6 semester hours in refrigeration and air conditioning installation and maintenance (9/86); in the lower-division baccalaureate/associate degree category, 2 in mechanical and electrical equipment of buildings (air conditioning) (9/86).

AF-1732-0004
UTILITIES SYSTEM JOURNEYMAN BY CORRESPONDENCE

Course Number: 56651A.
Location: Extension Course Institute, Gunter AFS, AL.
Length: Maximum, 52 weeks.
Exhibit Dates: 2/93–2/99.
Learning Outcomes: Upon completion of the course, the student will be able to install and maintain plumbing fixtures. The student will also be trained as a wastewater treatment specialist.
Instruction: This correspondence course covers topics in safety, plumbing fundamentals, water systems, wastewater systems, and operation and maintenance of water plants. This course ends with a proctored final exam.
Credit Recommendation: In the vocational certificate category, 1 semester hour in basic plumbing fundamentals (6/94); in the lower-division baccalaureate/associate degree category, 2 semester hours in wastewater management (6/94).

AF-1732-0005
UTILITIES SYSTEM JOURNEYMAN BY CORRESPONDENCE

Course Number: 56651B.
Location: Extension Course Institute, Gunter AFS, AL.
Length: Maximum, 52 weeks.
Exhibit Dates: 6/93–Present.
Learning Outcomes: Upon completion of the course, the student will be able to install and maintain waste systems.
Instruction: This correspondence course covers sewage systems and their construction; sewage fixtures and auxiliaries; fire protection systems; basic electricity; and troubleshooting of pumps, motors, and appurtenances.
Credit Recommendation: In the vocational certificate category, 1 semester hour in electrical motors and pumps (6/94); in the lower-division baccalaureate/associate degree category, 2 semester hours in wastewater management (6/94).

AF-1732-0015
STRUCTURAL JOURNEYMAN BY CORRESPONDENCE

Course Number: 55250A.
Location: Extension Course Institute, Gunter AFS, AL.
Length: Maximum, 52 weeks.
Exhibit Dates: 12/92–10/98.
Learning Outcomes: Upon completion of the course, the student will be able to work in the area of sheet metal layout and construction.
Instruction: This is a fundamental course in the fabrication and installation of sheet metal duct work. Other topics include the application and use of oxyacetylene and electric arc welding equipment. This is a self-paced, independent study correspondence course with a proctored final exam.
Credit Recommendation: In the vocational certificate category, 3 semester hours in sheet metal fabrication and installation (6/94).

AF-1732-0016
STRUCTURAL JOURNEYMAN BY CORRESPONDENCE

Course Number: 55250B.
Location: Extension Course Institute, Gunter AFS, AL.
Length: Maximum, 52 weeks.
Exhibit Dates: 1/93–Present.
Learning Outcomes: Upon completion of the course, the student will have skills in basic building construction, materials, and methods. The structural journeyman combines the skills of a carpenter, painter, and a metal fabricator.
Instruction: Students receive instruction in the field of built-up roofing, metal roofing, building parts, awnings and canopies, ceilings and wall coverings, and other building materials. This is a self-paced, independent study correspondence course with a proctored final exam. Additional topics cover masonry materials and construction practices.
Credit Recommendation: In the vocational certificate category, 2 semester hours in carpentry and 1 in masonry construction (6/94).

AF-1733-0002
FABRICATION AND PARACHUTE SPECIALIST BY CORRESPONDENCE

Course Number: 42753.
Location: Extension Course Institute, Gunter AFS, AL.
Length: Maximum, 52 weeks.
Exhibit Dates: 1/90–10/97.
Learning Outcomes: Upon completion of the course, the student will be able to troubleshoot and repair or adjust industrial sewing machines. With very limited supervision, the individual will inspect, repair, and pack aircrew survival kits, flotation equipment, parachutes, and related devices.
Instruction: Students work under supervision of a trainer or a training manager to monitor progress in the use of hand tools and nomenclature and the troubleshooting of sewing machines. The course also covers hand and machine repair of fabrics, safety considerations, and management practices.
Credit Recommendation: In the vocational certificate category, 1 semester hour in parachute rigging, inspection, and repair; 1 in fabric repair (hand and machine); and 1 in sewing machine service and repair (10/88).

AF-2203-0051
AIR NATIONAL GUARD ACADEMY OF MILITARY SCIENCE

Course Number: YAMS-000.
Location: Air National Guard Support Center, Knoxville, TN.
Length: 6 weeks (296 hours).
Exhibit Dates: 1/90–8/96.
Objectives: This course provides the student with an understanding of the interaction of economic and political determinants of power and military strategy and decisions. It also provides knowledge on the processes and functions of management and the role of communication in management and leadership.
Instruction: Course includes lectures, seminars, laboratory experiences, and applied experiences in management, leadership, and in communication. There are also lectures and seminars on defense studies.
Credit Recommendation: In the lower-division baccalaureate/associate degree category, 1 semester hour in defense studies and 3 in management (leadership and communications) (8/86).

AF-2203-0052
FIRST SERGEANT BY CORRESPONDENCE

Course Number: 1090.
Location: Extension Course Institute, Gunter AFS, AL.
Length: Maximum, 52 weeks.
Exhibit Dates: 1/90–8/92.
Learning Outcomes: At the conclusion of the course, the student will be able to describe the duties and responsibilities of a first sergeant, understand the policies and procedures involving the duties and responsibilities of the first sergeant, and have a working knowledge of military justice.
Instruction: This course is primarily a summary of an Air Force first sergeant's duties and responsibilities and relevant policies, procedures, and military justice regulations.
Credit Recommendation: Credit is not recommended because of the limited, specialized nature of the course (2/87).

AF-2203-0053
AIR FORCE TECHNICAL ORDER SYSTEM BY CORRESPONDENCE

Course Number: 1200.
Location: Extension Course Institute, Gunter AFS, AL.
Length: Maximum, 52 weeks.
Exhibit Dates: 1/90–1/97.

Learning Outcomes: Upon completion of the course, the student will be able to describe the administration, the necessity, and the acquisition of technical orders and use technical orders in job assignments.

Instruction: This course includes information about technical orders for the Air Force and covers the administration, types, issuance, numbering system, filing, use, and improvement. It also includes the distribution system of technical orders, establishing requirements, documentation, and the checking of technical order files.

Credit Recommendation: Credit is not recommended because of the military-specific nature of the course (2/87).

AF-2203-0054

COMBAT ARMS TRAINING AND MAINTENANCE SPECIALIST/TECHNICIAN BY CORRESPONDENCE

Course Number: 75350.

Location: Extension Course Institute, Gunter AFS, AL.

Length: Maximum, 52 weeks.

Exhibit Dates: 1/90–1/94.

Learning Outcomes: Upon completion of the course, the student will be able to establish and maintain a small arms training program with minimal supervision. Program could be appropriate at the community college level or in a law enforcement unit.

Instruction: This course consists of self-paced, independent study with open-book progress checks and satisfactory completion of a proctored comprehensive final examination. Content areas include combat arms operations and facility management, marksmanship fundamentals, fundamentals of instruction, conduct of training, training aids, evaluation and measurement, and maintenance and repair of small arms.

Credit Recommendation: In the lower-division baccalaureate/associate degree category, 2 semester hours in small arms instruction (12/89).

AF-2203-0055

SQUADRON OFFICERS SCHOOL RESIDENT

Course Number: None.

Location: Air University, Maxwell AFB, AL.

Length: 7 weeks (229 hours).

Exhibit Dates: 1/90–Present.

Learning Outcomes: Upon completion of the course, the student will demonstrate understanding of the historical importance of the profession of arms; interpersonal and communication skills useful for directing military units of increasing complexity and size; mastery of key leadership concepts, including team building, situational leadership, leadership styles, and an awareness of when various styles are appropriate; and proficiency in communication theory and skill, including briefing, researching, writing, and listening skills.

Instruction: This school is the first of three major schools for Air Force officers as they pursue their professional military careers. The school stresses military organization, leadership, management, and communication. Approaches to subject mastery include lectures, seminars, simulations, and group problem solving sessions. Emphasis is placed on knowledge acquisition and the development of operational skills in four major areas officership, force employment, leadership, and communication skills.

Credit Recommendation: In the upper-division baccalaureate category, 3 semester hours in managerial communication and 3 in national security studies and military strategy (9/92); in the graduate degree category, 3 semester hours in leadership/managerial human relations (9/92).

AF-2203-0056

SQUADRON OFFICERS SCHOOL NONRESIDENT

Course Number: None.

Location: Air University, Maxwell AFB, AL.

Length: Maximum, 118 weeks.

Exhibit Dates: 1/90–Present.

Learning Outcomes: Upon completion of the course, the student will demonstrate understanding of the historical importance of the profession of arms; interpersonal and communication skills useful for directing military units of increasing complexity and size; mastery of key leadership concepts, including team building, situational leadership, leadership styles, and an awareness of when various styles are appropriate; and proficiency in communication theory and skill, including briefing, researching, writing, and listening skills.

Instruction: Nonresident students complete their work and their assignments through correspondence. This school is the first of three major schools for Air Force officers as they pursue their professional military careers. The school stresses military organization, leadership, management, and communication. Approaches to subject mastery include readings and computer simulations. Emphasis is placed on knowledge acquisition and the development of operational skills in four major areas officership, force employment, leadership, and communication skills.

Credit Recommendation: In the upper-division baccalaureate category, 3 semester hours in managerial communications and 3 in national security studies and military strategy (9/92).

AF-2203-0057

GENERAL CONTINGENCY RESPONSIBILITIES BY CORRESPONDENCE

Course Number: 3E050.

Location: Extension Course Institute, Gunter AFS, AL.

Length: Maximum, 52 weeks.

Exhibit Dates: 12/93–Present.

Learning Outcomes: Upon completion of the course, the student will be able to respond during readiness and wartime deployment/employment, and to apply various principles learned to saving the life of others.

Instruction: This correspondence course consists of self-paced, independent instruction. The topics covered include readiness and wartime deployment/employment; first aid; field sanitation, personal hygiene, and pestborne disease; self-protection from extreme weather; personal and work party security; airbase ground defense, including camouflage, concealment, and deception; rapid runway repair; and basic fire fighting.

Credit Recommendation: Credit is not recommended because of the military-specific nature of the course (6/94).

Course Exhibits

CG

CG-0419-0002

1. HC-130 LOADMASTER BY CORRESPONDENCE
2. HC-130 LOADMASTER (AC130L) BY CORRESPONDENCE

Course Number: *Version 1:* 0447-2. *Version 2:* 447-1.
Location: Coast Guard Institute, Oklahoma City, OK.
Length: *Version 1:* Maximum, 156 weeks. *Version 2:* Maximum, 104 weeks.
Exhibit Dates: *Version 1:* 6/97–Present. *Version 2:* 1/90–5/97.
Learning Outcomes: *Version 1:* This version is pending evaluation. *Version 2:* Provides specialized training as loadmaster on HC-130 aircraft.
Instruction: *Version 1:* This version is pending evaluation. *Version 2:* Presents familiarization on HC-130 aircraft, aircraft engines, props and prop systems, ground handling and servicing, Loadmaster duties, and aircraft loading equipment and furnishings. Students are required to successfully complete a proctored end-of-course examination.
Credit Recommendation: *Version 1:* Pending evaluation. *Version 2:* In the vocational certificate category, 1 semester hour in weight and balance and aircraft loading (3/97).

CG-0419-0003

COASTAL DEFENSE COMMAND AND STAFF, CLASS C

Course Number: MS 733R.
Location: Reserve Training Center, Yorktown, VA.
Length: 2 weeks (79-80 hours).
Exhibit Dates: 9/90–Present.
Learning Outcomes: Upon completion of the course, the student will be able to develop and plan coastal support operations during national emergencies.
Instruction: Methods of instruction include lectures, practical exercises, examinations, and reviews. Topics include planning and preparation, base case planning, and crisis/contingency planning.
Credit Recommendation: In the upper-division baccalaureate category, 2 semester hours in strategic planning (5/96).

CG-0701-0002

DENTAL TECHNICIAN, CLASS C

Course Number: None.
Location: Training Center, Cape May, NJ.
Length: 5 weeks (177 hours).
Exhibit Dates: 1/90–11/96.
Objectives: To train personnel in chairside dental assisting, dental radiography, and related topics.
Instruction: Lectures, laboratory, and supervised clinical experience in dental anatomy, chairside assisting, oral pathology, and radiography.
Credit Recommendation: In the lower-division baccalaureate/associate degree category, 1 semester hour in dental anatomy, 2 in chairside assisting, 1 in radiology, and 2 in clinical experience (6/83).

CG-0709-0004

1. EMERGENCY MEDICAL TECHNICIAN BASIC
2. EMERGENCY MEDICAL TECHNICIAN, CLASS C
3. EMERGENCY MEDICAL TECHNICIAN, CLASS C

Course Number: *Version 1:* None. *Version 2:* None. *Version 3:* None.
Location: Training Center, Petaluma, CA.
Length: *Version 1:* 3 weeks. *Version 2:* 3 weeks (158-159 hours). *Version 3:* 2-3 weeks (117 hours).
Exhibit Dates: *Version 1:* 1/94–Present. *Version 2:* 1/93–12/93. *Version 3:* 1/90–12/92.
Learning Outcomes: *Version 1:* Upon completion of the course, the student will be able to provide emergency medical care at a basic life support level with an ambulance service or other specialized service. The student will be capable of performing a series of functions at the minimum entry level. Student will recognize the nature and seriousness of the patient's condition or extent of injuries to assess requirements for emergency medical care; administer appropriate emergency medical care based on assessment lift, move, position, and otherwise handle the patient to minimize discomfort and prevent further injury; and safely and effectively the expectations of the job. *Version 2:* Upon completion of the course, the student will be able to assess the seriousness of the patient's condition; provide appropriate emergency care to stabilize the patient's condition; and lift, move, position, or otherwise handle the patient in such a way as to minimize discomfort and additional injury. *Version 3:* Upon completion of the course, the student will be able to perform as an entry-level emergency medical technician with the capacity for assessing requirements for emergency care, administering emergency care, stabilizing a patient's condition, and acting to minimize further injury and discomfort.
Instruction: *Version 1:* This course is designed to prepare the student to be an entry-level emergency medical technician. This includes all knowledge, skills, and practice necessary for unassisted emergency medical care at a basic life support level. Topics include patient assessment; airway obstruction and respiratory arrest; basic cardiac life support; shock and bleeding; soft tissue injuries; musculoskeletal injuries; head, neck, skull, and spine injuries; poisonous, bites, and stings; respiratory illnesses; acute abdominal distress; seizures, strokes, and diabetic emergencies; pediatric emergencies; infectious diseases; emergency childbirth; environmental injuries; burns and hazardous materials; and water rescue, near drowning, and diving accidents. Please note that the student will provide a service in an environment requiring special skills and knowledge in such areas of communications, transportation, and keeping records. *Version 2:* This course is designed to prepare the student to be an entry-level emergency medical technician. This includes all knowledge, skills, and practice necessary for unassisted emergency medical care at a basic life support level. Topics include patient assessment; airway obstruction and respiratory arrest; basic cardiac life support; shock and bleeding; soft tissue injuries; musculoskeletal injuries; head, neck, skull, and spine injuries; poisonous, bites, and stings; respiratory illnesses; acute abdominal distress; seizures, strokes, and diabetic emergencies; pediatric emergencies; infectious diseases; emergency childbirth; environmental injuries; burns and hazardous materials; and water rescue, near drowning, and diving accidents. *Version 3:* This course, utilizing lectures, simulation, and laboratory experience, covers legal aspects of emergency care; anatomy and physiology of the cardiac and respiratory systems; use of oropharyngeal airways, suction equipment, oxygen delivery systems and other ventilation devices; the signs and symptoms of patients suffering from cardiac problems, dyspnea, acute abdominal problems, poisoning, stings, bites, diabetic emergencies and seizure disorders, heat cramps and heat exhaustion, alcohol and drug abuse, emotional disturbances, rape, and burns. Student will learn proper treatment and response to the above. Total proficiency will be required through examinations in CPR, dressing and bandaging techniques, splinting techniques, and the measurement of vital signs.
Credit Recommendation: *Version 1:* In the lower-division baccalaureate/associate degree category, 6 semester hours in emergency medical technology (12/96). *Version 2:* In the lower-division baccalaureate/associate degree category, 6 semester hours in emergency medical technology (5/93). *Version 3:* In the lower-division baccalaureate/associate degree category, 5 semester hours in emergency technology (7/87).

CG-0709-0005

HEALTH SERVICES TECHNICIAN SECOND CLASS BY CORRESPONDENCE

Course Number: *Version 1:* 0230-2. *Version 2:* 0230-1; 0230.
Location: Coast Guard Institute, Oklahoma City, OK.
Length: *Version 1:* Maximum, 156 weeks. *Version 2:* Maximum, 52 weeks.
Exhibit Dates: *Version 1:* 4/93–Present. *Version 2:* 1/90–3/93.
Learning Outcomes: *Version 1:* Upon completion of the course, the student will be able to manage the materials, records, and supplies of a medical clinic; assess patients and patient care with prescribed limits; and administer limited

public health/prevention services. *Version 2:* Upon completion of the course, the student will be able to demonstrate the knowledge and skills to perform as a second class health services technician.

Instruction: *Version 1:* This is a self-paced independent study course that includes physical assessment; inventory management; hazardous medical waste; dental radiography; alcohol abuse; minor surgery; pest control; food service sanitation; emergency medical procedures; sexually transmitted diseases; and specific common illnesses and their management. The course includes a summative written final examination. Many performance evaluations are conducted by the student's supervisor. *Version 2:* Topics include physical diagnosis and treatment, emergency medicine, occupational health, preventive medicine, basic care techniques, pharmacy, medical laboratory, X-ray, and health services organization and administration.

Credit Recommendation: *Version 1:* In the lower-division baccalaureate/associate degree category, 2 semester hours in physical assessment, 2 in emergency medical procedures, and 2 in medical clinic management (10/95). *Version 2:* In the lower-division baccalaureate/associate degree category, 3 semester hours in health care procedures and techniques (10/87).

CG-0709-0006

HEALTH SERVICES TECHNICIAN FIRST CLASS BY CORRESPONDENCE

Course Number: *Version 1:* 130-2. *Version 2:* 0130-1; 0110.
Location: Coast Guard Institute, Oklahoma City, OK.
Length: *Version 1:* Maximum, 156 weeks. *Version 2:* Maximum, 52 weeks.
Exhibit Dates: *Version 1:* 1/96–Present. *Version 2:* 1/90–12/95.
Learning Outcomes: *Version 1:* Upon completion of the course, the student will be able to counsel military personnel regarding health benefits, including federal and nonfederal services, completion of appropriate forms, and medical and dental benefits; prepare requests for health care contracts; process out-of-system bills and arrange for transfer of patients between federal and nonfederal institutions. *Version 2:* Upon completion of the course, the student will have the knowledge and skills to provide advanced services as a health care technician.

Instruction: *Version 1:* This is a correspondence course followed by a timed and proctored open book examination. *Version 2:* Topics include physical examinations, microbiology and parasitology, occupational health, splints and casting, and medical administration.

Credit Recommendation: *Version 1:* In the lower-division baccalaureate/associate degree category, 1 semester hour in health care insurance administration (5/97). *Version 2:* In the lower-division baccalaureate/associate degree category, 3 semester hours in health care procedures and administration (10/87).

CG-0709-0009

HEALTH SERVICES TECHNICIAN, CLASS A

Course Number: *Version 1:* None. *Version 2:* None.
Location: *Version 1:* Training Center, Petaluma, CA. *Version 2:* Coast Guard Academy, New London, CT.
Length: *Version 1:* 17 weeks (500 hours). *Version 2:* Multi-phased, 24 weeks (886 hours).
Exhibit Dates: *Version 1:* 10/90–Present. *Version 2:* 1/90–9/90.
Learning Outcomes: *Version 1:* Upon completion of the course, the student will be able to provide direct patient care, including treatment of disease and injuries, basic dental care, and administrative assignments as appropriate. *Version 2:* Upon completion of the course, the student will be able to provide direct patient care, including prevention and treatment of disease and injuries, basic dental care, and to carry out administrative assignments as appropriate. The 24-week long course had a 15-week long medical phase, 5-week long dental phase, and 4-week long clinical experience phase.

Instruction: *Version 1:* Topics include general anatomy and physiology, medical and surgical conditions, principles and techniques of patient care, medical administration, pharmacology, laboratory techniques, military and administrative regulations, introductory aspects of dental assisting, pathology, radiography, and supervised clinical experiences. *Version 2:* Topics include general anatomy and physiology, medical and surgical conditions, preventive medicine, principles and techniques of patient care, medical administration, medical mathematics and pharmacology, laboratory techniques, first aid, military and administrative regulations, introductory aspects of dental assisting, pathology, radiography, and supervised clinical experience. NOTE: The Health Services Technician rating was formerly a combination of the Hospital Corpsman (HM) and Dental Technician (DT) ratings. In order to qualify for the HS rating, new recruits complete all 24 weeks of this course; those who held the DT rating take the 19-week medical course (15 weeks classroom instruction and 4 weeks clinical experience); those who held the HM rating take the 9-week dental course (5 weeks classroom instruction and 4 weeks clinical experience). Partial credit may be awarded as appropriate.

Credit Recommendation: *Version 1:* In the lower-division baccalaureate/associate degree category, 3 semester hours in anatomy and physiology, 1 in infection control procedures, 2 in health facility management, 2 in pharmacology, 3 in medical laboratory procedures, 3 in medical/surgical procedures, 1 in dental anatomy, 2 in chairside dental assisting, 1 in dental radiography, 1 in health care clinical practice, and 1 in dental care clinical practice (5/93). *Version 2:* In the lower-division baccalaureate/associate degree category, 4 semester hours in anatomy and physiology, 1 in introduction to health services, 5 in emergency medicine, 2 in preventive medicine, 2 in medical administration, 2 in pharmacology, 3 in laboratory practices, 3 in medical-surgical procedures, 2 in general health-oriented clinical experience, 1 in dental anatomy, 2 in chairside assisting, 1 in radiology, and 2 in dental clinical experience. For the 19-week medical course: in the lower-division baccalaureate/associate degree category, 4 semester hours in anatomy and physiology, 1 in introduction to health services, 5 in emergency medicine, 2 in preventive medicine, 2 in medical administration, 2 in pharmacology, 3 in laboratory practices, 3 in medical-surgical procedures, and 4 in general health-oriented clinical experience (10/87). For the 9-week dental course: In the lower-division baccalaureate/associate degree category, 1 semester hour in dental anatomy, 2 in chairside assisting, 1 in radiology, and 4 in dental clinical experience (10/87).

CG-0801-0001

SMALL CUTTER DAMAGE CONTROL

Course Number: DC-3.
Location: Reserve Training Center, Yorktown, VA.
Length: 2 weeks (75-76 hours).
Exhibit Dates: 1/90–2/92.
Learning Outcomes: Upon completion of the course, the student will be able to perform tests to detect the presence of chemical and biological agents and radiation; perform decontamination duties; calculate displacement and moment to trim; plot an uncorrected stability curve; calculate sine and cosine corrections; correct the stability curve; log angle of final list; compute affects of partial flooding; compute affects of longitudinal weight movements; perform tests to detect oxygen, explosive gases, toxic vapors; and state the methods of ventilation.

Instruction: Lectures, demonstrations, and practical exercises cover chemical, biological, radiological detection; CBR decontamination; mathematical calculations relating to ship stability; oxygen, explosive, and toxic vapor gas detection; and fire extinguishing.

Credit Recommendation: In the lower-division baccalaureate/associate degree category, 2 semester hours in detection of toxic materials and other gases, decontamination, and ventilation and 2 in fundamentals of mathematics (8/90).

CG-0802-0007

HU-25A AVIONICSMAN BY CORRESPONDENCE

Course Number: 0512-1.
Location: Coast Guard Institute, Oklahoma City, OK.
Length: Maximum, 52 weeks.
Exhibit Dates: 1/90–Present.
Objectives: To give students a working knowledge of the aircraft, selected systems, and emergency procedures.

Instruction: Military personnel must be supervised by qualified HU-25A (Falcon aircraft) avionicsman instructor to complete the course. Course includes communication equipment location, internal sensor system, radar and Loran familiarization, and indicator and annunciator location. Students are required to successfully complete a proctored final examination.

Credit Recommendation: Credit is not recommended because of the limited, specialized nature of the course (10/91).

CG-0802-0008

SEARCH AND RESCUE HU-25A BASIC AIRCREW BY CORRESPONDENCE

Course Number: 0513-1.
Location: Coast Guard Institute, Oklahoma City, OK.
Length: Maximum, 104 weeks.
Exhibit Dates: 1/90–Present.
Objectives: To provide student with aircraft familiarization and description of crew duties of the Falcon aircraft (HU-25A).

Instruction: Instruction in aircraft familiarization, crew duties, types of flight and maintenance inspections, emergency procedures, communications and ground handling procedures. Students are required to complete a proctored end-of-course examination.

Credit Recommendation: In the vocational certificate category, 1 semester hour in aircraft familiarization, ground handling, inspection, and emergency procedures (5/85).

CG-0802-0009

BASIC SEARCH AND RESCUE BY CORRESPONDENCE

Course Number: 422-1.
Location: Coast Guard Institute, Oklahoma City, OK.
Length: Maximum, 104 weeks.
Exhibit Dates: 1/90–11/96.
Objectives: To provide personnel with a basic knowledge of search and rescue procedures.
Instruction: Course introduces the student to fundamental search and rescue (SAR) procedures and terminology, and basic SAR patterns and techniques used by ships and aircraft in conducting a search. Students are required to successfully complete a proctored end-of-course examination.
Credit Recommendation: In the lower-division baccalaureate/associate degree category, 1 semester hour in basic navigation and 1 in community safety (5/85).

CG-0802-0011

SAFETY AND SECURITY OF THE PORT BY CORRESPONDENCE

Course Number: 429-1.
Location: Coast Guard Institute, Oklahoma City, OK.
Length: Maximum, 104 weeks.
Exhibit Dates: 1/90–11/96.
Objectives: To provide training in the safety of port and pier areas regulated by the Coast Guard.
Instruction: Course covers the topics of recognition and identification of safety hazards, including fire and electrical hazards, cargo loading, vehicle use and storage, and waste material controls. Students are required to successfully complete a proctored end-of-course examination.
Credit Recommendation: In the vocational certificate category, 1 semester hour in industrial safety and health (5/85).

CG-0802-0012

BASIC SEARCH AND RESCUE AIRCREWMAN BY CORRESPONDENCE

Course Number: 440-2.
Location: Coast Guard Institute, Oklahoma City, OK.
Length: Maximum, 104 weeks.
Exhibit Dates: 1/90–3/90.
Objectives: After 2/90 see CG-1704-0040. To provide training in aircraft familiarization, ground handling, and safety procedures.
Instruction: Topics include ground handling, fire fighting, crew survival, rescue equipment, pyrotechnics, basic communications, and navigation. Students are required to successfully complete a proctored end-of-course examination.
Credit Recommendation: In the vocational certificate category, 2 semester hours in aircraft familiarization, ground handling, and safety (5/85).

CG-0802-0013

EXPLOSIVE HANDLING SUPERVISOR

Course Number: MS 496R; MS 426R.
Location: Marine Safety School, Yorktown, VA.
Length: 2 weeks (87-88 hours).
Exhibit Dates: 1/90–11/96.
Objectives: To train Coast Guard personnel to supervise explosive cargo handling operations at military and commercial ports.
Instruction: Lectures and practical experience in explosive properties, occupational safety, permits, and cargo handling operations.
Credit Recommendation: Credit is not recommended because of the limited, specialized nature of the course (4/86).

CG-0802-0014

NATIONAL BOATING SAFETY

Course Number: NBSC.
Location: Reserve Training Center, Yorktown, VA.
Length: 2 weeks (69 hours).
Exhibit Dates: 1/90–Present.
Learning Outcomes: Upon completion of the course, the student will be able to perform the duties of a law enforcement officer in the areas of boarding safety procedures and regulations.
Instruction: Includes lectures in federal laws and regulations, vessel equipment requirements, boating accident reporting and investigation, stolen boat detection, intoxication, identification, and boating safety instructor training.
Credit Recommendation: In the vocational certificate category, 3 semester hours in federal and state boating regulations (2/96); in the lower-division baccalaureate/associate degree category, 1 semester hour in instructor methods or teaching methods (2/96).

CG-0802-0015

MARINE SAFETY EXPLOSIVE HANDLING SUPERVISOR

Course Number: MS 496R.
Location: Reserve Training Center, Yorktown, VA.
Length: 2 weeks (60-80 hours).
Exhibit Dates: 1/90–9/92.
Learning Outcomes: Upon completion of the course, the student will be able to perform the duties of an explosive handling supervisor in a port operations department, a marine safety unit, or a large mobilization billet.
Instruction: Includes lectures, demonstrations, simulations, and practical exercises in verification of compliance with federal regulations for the safe handling of commercial and military explosives. Topics include packaged hazardous material regulations, military regulations regarding explosives, dangerous goods code of international maritime organization, permit processing, vessel preload examinations, supervision of cargo operations, and the requirements for proper blocking and bracing of cargo.
Credit Recommendation: In the lower-division baccalaureate/associate degree category, 3 semester hours in industrial security, transportation safety, or hazardous materials (4/89).

CG-0802-0016

SEARCH AND RESCUE FUNDAMENTALS BY CORRESPONDENCE

Course Number: 0431-1.
Location: Coast Guard Institute, Oklahoma City, OK.
Length: Maximum, 26 weeks.
Exhibit Dates: 1/90–6/97.
Learning Outcomes: Upon completion of the course, the student will be able to demonstrate basic knowledge of terminology, concepts, organization, and principles used in the conduct of search and rescue missions.
Instruction: Instruction is provided through a nonresident, self-paced study program using Coast Guard study material. Subject matter covers National Search and Rescue (SAR) policies, the National SAR System, SAR Awareness and initial action stages, planning and operation stages, and the conclusion stages. The purpose of the course is to provide prerequisite knowledge necessary to attend resident course at the National SAR School. The course includes a proctored final examination.
Credit Recommendation: In the lower-division baccalaureate/associate degree category, 1 semester hour in fundamentals of ocean search and rescue (9/90).

CG-0802-0017

COMPUTER AIDED SEARCH PLANNING BY CORRESPONDENCE

Course Number: 0414-1.
Location: Coast Guard Institute, Oklahoma City, OK.
Length: Maximum, 26 weeks.
Exhibit Dates: 1/90–Present.
Learning Outcomes: Upon completion of the course, the student will be able to demonstrate proficiency in the use of an Operations Information System Computer to assist in a search and rescue mission.
Instruction: Instruction is provided through a nonresident, self-paced study program using Coast Guard study material. Subject matter covers that Coast Guard Computer Aided Search Planning (CASP) System, law enforcement, ice operations, Automated Mutual Assistance Rescue System, and computer-generated search areas. Emphasis is on menu selection and file entries. The course includes a proctored final examination.
Credit Recommendation: Credit is not recommended because of the limited, specialized nature of the course (9/90).

CG-0802-0018

SEARCH AND RESCUE BY CORRESPONDENCE

Course Number: 0409-5.
Location: Coast Guard Institute, Oklahoma City, OK.
Length: Maximum, 78 weeks.
Exhibit Dates: 1/90–Present.
Learning Outcomes: Upon completion of the course, the student will be able to demonstrate proficiency in the organization and conduct of a search and rescue mission. This includes knowledge of proper search patterns, navigation, aircraft emergency procedures over water, and international search and rescue (SAR) obligations under the International Civil Aviation Organization. Volumes I and II of the National Search and Rescue Manual are the primary texts.

Instruction: Instruction is provided through a nonresident, self-paced study program using Coast Guard study material. Subject matter covers navigational procedures and practices to be followed for search and rescue mission. The course includes a proctored final examination.

Credit Recommendation: In the lower-division baccalaureate/associate degree category, 2 semester hours in advanced navigation and 2 in community safety (9/90).

CG-1104-0001

ELEMENTARY ALGEBRA BY CORRESPONDENCE

Course Number: 0486-1.
Location: Coast Guard Institute, Oklahoma City, OK.
Length: Maximum, 156 weeks.
Exhibit Dates: 1/94–Present.
Learning Outcomes: Upon completion of the course, the student will be able to perform algebraic operations, including the ability to solve linear equations and inequalities; solve problems involving exponents; factor trinomial, quadratic equations, and sum/difference of three cubes; solve equations containing rational expressions, graph linear equations, and inequalities; solve systems of linear equations using graphing, addition, and substitution; solve equations containing radical expressions; and solve quadratic equations using quadratic formulae.
Instruction: The student is provided a standard elementary algebra text book with worked examples and exercises. Students study the text, work exercises, and check answers against the answers provided in the text. A proctored final examination is given, in which the student must work at least 80 percent of the problems correctly.
Credit Recommendation: In the lower-division baccalaureate/associate degree category, 3 semester hours in elementary algebra (10/94).

CG-1107-0001

BASIC MATHEMATICS BY CORRESPONDENCE

Course Number: 0485-1.
Location: Coast Guard Institute, Oklahoma City, OK.
Length: Maximum, 52 weeks.
Exhibit Dates: 8/90–Present.
Learning Outcomes: Upon completion of the course, the student will be able to perform basic mathematical functions.
Instruction: Course includes self-instruction in addition, subtraction, multiplication, division, fractions, and decimals. Includes a proctored end-of-course examination.
Credit Recommendation: Credit is not recommended because of the limited, specialized nature of the course (11/91).

CG-1303-0001

MARINE ENVIRONMENTAL PROTECTION (MEP) BY CORRESPONDENCE

Course Number: 611-1.
Location: Coast Guard Institute, Oklahoma City, OK.
Length: Maximum, 104 weeks.
Exhibit Dates: 1/90–6/96.
Objectives: To provide training in marine environmental protection and investigation of oil pollution.
Instruction: Topics include contingency planning, containment and recovery equipment, oil pollution investigations, identification of chemical pollutants, on-scene coordination and pollution cleanup monitoring. Students are required to successfully complete a proctored end-of-course examination.
Credit Recommendation: In the lower-division baccalaureate/associate degree category, 2 semester hours in oil and chemical spill control (5/85).

CG-1304-0006

ICE OBSERVER BY CORRESPONDENCE

Course Number: 476-2; 476-1.
Location: Coast Guard Institute, Oklahoma City, OK.
Length: Maximum, 52 weeks.
Exhibit Dates: 1/90–Present.
Objectives: To train Coast Guard personnel to observe land and sea ice concentrations and topographical conditions and to report these observations to superiors.
Instruction: This is a brief nonresident course covering responsibilities in ice breaking and describing basic ice breaking and ice reconnaissance procedures. Course teaches stages of water freezing and thawing and provides analysis of ice charts.
Credit Recommendation: Credit is not recommended because of the limited, technical nature of the course (10/91).

CG-1304-0014

WEATHER FORECASTING AND FLIGHT BRIEFING BY CORRESPONDENCE

Course Number: 420-1.
Location: Coast Guard Institute, Oklahoma City, OK.
Length: Maximum, 104 weeks.
Exhibit Dates: 1/90–11/96.
Objectives: To train Coast Guard personnel in weather briefing skills and in the identification and analysis of meteorological conditions as these relate to marine and flight weather.
Instruction: Course includes training in climatology, atmospheric physics, world weather, atmospheric conditions, and the effect of these conditions on the intensity and movement of weather systems. Students are required to successfully complete a proctored end-of-course examination.
Credit Recommendation: In the lower-division baccalaureate/associate degree category, 3 semester hours in meteorology (5/85).

CG-1304-0016

MARINE SCIENCE TECHNICIAN SECOND CLASS BY CORRESPONDENCE

Course Number: *Version 1:* 0234-5; 0234-4. *Version 2:* 0234-3.
Location: Coast Guard Institute, Oklahoma City, OK.
Length: *Version 1:* Maximum, 52 weeks. *Version 2:* Maximum, 52 weeks.
Exhibit Dates: *Version 1:* 2/91–Present. *Version 2:* 1/90–1/91.
Learning Outcomes: *Version 1:* Upon completion of the course, the student will be able to perform advanced word processing applications as they apply to a Coast Guard standard computer work station; identify and observe appropriate marine safety measures relating to harbor patrols, cargo vessels, and tank vessels; identify environmental hazards and determine appropriate response measures; identify Coast Guard responsibilities relative to all facilities which require physical security measures; and identify types of port security equipment and systems. *Version 2:* Upon completion of the course, the student will be able to perform advanced word processing applications as they apply to a Coast Guard standard computer work station; identify and observe appropriate marine safety measures relating to harbor patrols, cargo vessels, and tank vessels; and identify environmental hazards and determine appropriate response measures.
Instruction: *Version 1:* Topics include word processing, marine safety, marine environmental response, and physical security survey. Training is provided in a nonresident, self-instruction format. Students are required to successfully complete a proctored end-of-course examination. *Version 2:* Topics include word processing, marine safety, and marine environmental response. Training is provided in a nonresident, self-instruction format. Students are required to successfully complete a proctored end-of-course examination.
Credit Recommendation: *Version 1:* In the lower-division baccalaureate/associate degree category, 2 semester hours in advanced word processing, 1 in marine safety, 1 in environmental management, and 1 in port security (10/92). *Version 2:* In the vocational certificate category, 1 semester hour in marine safety (6/90); in the lower-division baccalaureate/associate degree category, 3 semester hours in advanced word processing and 1 in environmental management (6/90).

CG-1304-0017

MARINE SCIENCE TECHNICIAN, CLASS A

Course Number: MS 729R.
Location: Reserve Training Center, Yorktown, VA.
Length: 6 weeks (120-145 hours).
Exhibit Dates: 1/90–Present.
Learning Outcomes: Upon completion of this course, the student will be able to utilize software, and, using a Coast Guard terminal, perform common word processing operations and use spread sheets and databases; use computer terminals to transmit and receive messages; operate facsimile and teletype machines; and demonstrate understanding of basic concepts of physical oceanography; and, to a greater extent, the basic concepts of meteorology.
Instruction: Course includes lectures and practice in word processing and general lectures in meteorology and oceanography.
Credit Recommendation: In the vocational certificate category, 1 semester hour in word processing (6/90); in the lower-division baccalaureate/associate degree category, 3 semester hours in computer applications and 3 in introduction to meteorology (6/90).

CG-1304-0018

MARINE SCIENCE TECHNICIAN FIRST CLASS BY CORRESPONDENCE

Course Number: 0134-2.
Location: Coast Guard Institute, Oklahoma City, OK.
Length: Maximum, 52 weeks.
Exhibit Dates: 1/90–12/90.

Learning Outcomes: After 12/90 see CG-1304-0022. Upon completion of the course, the student will be able to respond to hazardous chemical releases according to Comprehensive Environmental Response, Compensation and Liability Act guidelines and establish guidelines for training personnel as Coast Guard standard terminal operators.

Instruction: Course is provided by correspondence using examples and problems in weather forecasting, hazardous chemical release response situations, marine safety, and word processing applications. Training is provided in a nonresident, self-instruction format. Students are required to successfully complete a proctored end-of-course examination.

Credit Recommendation: In the vocational certificate category, 1 semester hour in marine safety and 1 in word processing (6/90); in the lower-division baccalaureate/associate degree category, 3 semester hours in meteorology and 1 in environmental management (6/90).

CG-1304-0020

WEATHER BRIEFER ANALYST, CLASS C
(Weather Briefer, Class C)

Course Number: MS 722R.
Location: Reserve Training Center, Yorktown, VA.
Length: 6 weeks.
Exhibit Dates: 1/90–Present.
Learning Outcomes: Upon completion of the course, the student will be able to procure and utilize applicable meteorological data for preparation of forecasts of sea states, winds and temperatures at various altitudes; and convective outlook and probability of turbulence, fog, icing conditions, air mass thunderstorms, and hail that are of importance to Coast Guard surface and aviation operations. In addition, the student shall have the necessary expertise to properly utilize Skew-T and Log P diagrams to derive condensation levels; Showalter Index; freezing levels and air mass stability; standard facsimile charts, radar plots and satellite visible and infrared imagery to designate areas of convergence and divergence as well as surface and upper level frontal positions and arcas of precipitation. The student will have the additional ability to analyze twelve and twenty-four hour constant pressure charts for continuity with the current constant pressure chart set; analyze for isotachs, isotherms, height centers, troughs, ridges, and moisture; analyze vorticity charts for maxima and minima, vorticity lobes, and advection. In addition, the student will be able to identify mission-specific parameters; integrate available information with sound reasoning; prepare a complete written report; and conduct weather briefings.

Instruction: The course includes lectures and programmed instruction on properties of air masses, structure of the atmosphere, frontal systems, formation of various types of severe weather, practical experience in methods of meteorological data acquisition, techniques of weather briefing, analysis of satellite and radar imagery, and analysis of weather charts. In addition, practical experience provides in-depth analysis of weather charts, preparation of written reports, and conduct of complete weather briefing.

Credit Recommendation: In the lower-division baccalaureate/associate degree category, 3 semester hours in meteorological observation, 3 in basic meteorology, and 3 in practical meteorological forecasting (6/90); in the upper-division baccalaureate category, 3 semester hours in public speaking (6/90).

CG-1304-0021

CELESTIAL NAVIGATION BY CORRESPONDENCE

Course Number: 0463-5.
Location: Coast Guard Institute, Oklahoma City, OK.
Length: Maximum, 52 weeks.
Exhibit Dates: 1/90–Present.
Learning Outcomes: Upon completion of the course, the student will be able to demonstrate knowledge of the principles of celestial navigation.

Instruction: Instruction is provided through a nonresident, self-paced study program using Coast Guard study material. Subject matter includes coordinate systems, timing accuracy, description of the marine sextant, use of the nautical almanac and HO tables 229, the celestial fix and running fix, the Rude starfinder, and how to determine gyrocompass error at sea. There is a proctored final examination.

Credit Recommendation: In the lower-division baccalaureate/associate degree category, 3 semester hours in celestial navigation (9/90).

CG-1304-0022

MARINE SCIENCE TECHNICIAN FIRST CLASS BY CORRESPONDENCE

Course Number: 0134-4; 0134-3.
Location: Coast Guard Institute, Oklahoma City, OK.
Length: Maximum, 156 weeks.
Exhibit Dates: 1/91–Present.
Learning Outcomes: Before 1/91 see CG-1304-0018. Upon completion of the course, the student will be able to identify and analyze atmospheric physical data; read meteorological charts; relate upper atmosphere phenomena to surface features; identify responsibilities on harbor patrols; identify all regulations relating to noxious liquid substances; identify specific performance tasks during chemical/oil spills; identify various oil pollution containment and removal concerns; identify standard workstation steps and procedures and user file modifications; identify system manager's responsibility; identify command file customizing features; identify methods used to maintain system security and to prevent system overload; and identify system formatting/initializing procedures for volume storage disks.

Instruction: Topics include weather analysis, marine safety and response, and system management. Course includes a proctored end-of-course examination.

Credit Recommendation: In the lower-division baccalaureate/associate degree category, 1 semester hour in weather analysis, 2 in computer applications, and 1 in environmental management (9/95).

CG-1402-0001

ELECTRONICS TECHNICIAN CHIEF BY CORRESPONDENCE

Course Number: 021-9; 0021-1.
Location: Coast Guard Institute, Oklahoma City, OK.
Length: Maximum, 78 weeks.
Exhibit Dates: 1/90–Present.
Learning Outcomes: Upon completion of the course, the student will be able to describe the principles of operation of communication systems, including PBX, private line telephone, TWX/TELEX, Federal Telecommunications system; list the requirements for handling classified information; define terms such as data processing, input/output, multiplexing, simplex, duplex, accumulator, register; describe how magnetic tape and hard disk storage devices are constructed; identify programming concepts, including a description of design materials, flowchart definitions, and digital computer concepts influencing programming.

Instruction: The major topics covered are communication systems, handling of classified information, computer and computer operator terms, data storage, flow charts, and computer concepts. Training is provided in a nonresident, self-instruction format. Students are required to successfully complete a proctored end-of-course examination.

Credit Recommendation: In the vocational certificate category, 3 semester hours in introduction to computer operations (6/90); in the lower-division baccalaureate/associate degree category, 1 semester hour in project administration (6/90).

CG-1402-0002

AN/UYK-7 COMPUTER AND DATA AUXILIARY CONSOLE COURSE DEVELOPMENT
(AN/UYK-7 Computer and OJ-172 Data Exchange Auxiliary Console)

Course Number: FT-3.
Location: Reserve Training Center, Yorktown, VA.
Length: 8 weeks (297-298 hours).
Exhibit Dates: 1/90–1/95.
Learning Outcomes: Upon completion of the course, the student, without references and given an illustration of the OJ-172 DEAC, will be able to label each unit with its correct designation without error. When given the AN/UYK-7 computer with a fault inserted, technical manuals, diagnostic program tape, and test equipment, the student will be able to troubleshoot the AN/UYK-7 to the lowest replaceable component. The faulty component(s) must be correctly identified without error in accordance with the performance checklist criteria. The student will be able to label each unit with its correct designation without error, record specific information four out of five items correctly, energize the computer in accordance with performance checklist, measure DC-DC converter voltages in accordance with performance checklist, replace a 90 volt switching regulator, and state in writing the principles pertaining to the theory of operation of the power supply unit. Student will be able to replace a core memory, perform the memory port test, state in writing the principles pertaining to the theory of operation of the core memory unit, state in writing the principles pertaining to the theory of operation of the IOA/IOC unit, troubleshoot IOA/IOC unit as a team member, perform supplemental diagnostics, perform the automated confidence test, perform wire wrapping and unwrapping without error, and perform the NORO test procedure in accordance with the performance check list criteria. The student will be able to state in writing principles of the theory of operation of the TTY, paper tape punch, and paper tape reader; perform maintenance panel checkout procedures; state in writing

principles pertaining to the theory of operation of the control logic; and duplicate a magnetic tape as a team member in accordance with performance checklist criteria. The student will be able to maintain MYUs, solar cells, lamp and EOT/BOT currents, capstan speed, read/write amplifier outputs, start and stop times, read/write deskew delays in accordance with the performance checklist criteria; measure time delays; perform POFA tests; and measure tape reader power supply voltages in accordance with performance checklist criteria.

Instruction: Course is presented through lectures, practical exercises, and examination review of topics, including diagnostics, operation, maintenance, troubleshooting, repair, testing of electrical systems, duplication of magnetic tapes, maintenance checkout procedures, use of specific hand tools, and predictive maintenance.

Credit Recommendation: In the lower-division baccalaureate/associate degree category, 6 semester hours in computer operations, 2 in computer maintenance, and 1 in computer repair (9/90).

CG-1402-0003

SYSTEM MANAGER
(Standard Workstation Regional System Manager)

Course Number: None.
Location: Training Center, Petaluma, CA.
Length: 2 weeks (63 hours).
Exhibit Dates: 6/91–Present.
Learning Outcomes: Upon completion of the course, the student will be able to manage a local area network of computer workstations, including hardware and software installation, protocol implementation, system audit, and error tracing and fault isolation.

Instruction: Instruction consists of team-taught demonstrations and hands-on workshops with a series of structured exercises in LAN management, and operation, including software installation, communication, electronic mail, and file server. Instruction is system specific in terms of hardware and software applications for X.25 and Ethernet LANs.

Credit Recommendation: In the lower-division baccalaureate/associate degree category, 2 semester hours in computer local area network management (9/93).

CG-1402-0004

TELECOMMUNICATIONS SPECIALIST SECOND CLASS BY CORRESPONDENCE

Course Number: 0241-9.
Location: Coast Guard Institute, Oklahoma City, OK.
Length: Maximum, 156 weeks.
Exhibit Dates: 1/95–Present.
Learning Outcomes: Upon completion of the course, the student will be able to prepare broadcast violation and interference reports; maintain a communications tactical doctrine (COMTAC) library; prepare inventory and security investigation reports; prepare and monitor special broadcast messages; identify distress and emergency frequencies; use a narrowband direct printing telegraphy system; and use a standard workstation to create, install, and manage a local area network (LAN) for user environment applications and all aspects of electronic mail.

Instruction: Course covers preparation and monitoring of special broadcast messages, security investigation reports, international distress and emergency frequencies, narrow-band direct-printing telegraphy systems (NAVTEX), workstation managed local area network system for spooled printing, protected user environment applications to install and establish GPS access information and all aspects of electronic mail. Training is provided in a nonresident, self-study format using five publications, broadcast and network related manuals, texts include reading assignments with objectives to be mastered and appropriate review exercises. Students' supervisor evaluates their performance with a checklist provided for each section and an end-of-course test (EOCT) is administered. The student must receive a score of 80 percent or more to successfully complete the course.

Credit Recommendation: In the lower-division baccalaureate/associate degree category, 3 semester hours in local area network management (3/96).

CG-1403-0001

SMALL UNIT PAPERWORK BY CORRESPONDENCE

Course Number: 470-1.
Location: Coast Guard Institute, Oklahoma City, OK.
Length: Maximum, 104 weeks.
Exhibit Dates: 1/90–11/96.
Objectives: To train persons in the field of small unit paperwork management.
Instruction: Topics include the Coast Guard directives system, publications and reports index, letters, security and ordnance administration, procurement and property accountability, and commissary administration. Students are required to successfully complete a proctored end-of-course examination.
Credit Recommendation: Credit is not recommended because of the military-specific nature of the course (5/85).

CG-1403-0002

TELECOMMUNICATIONS SPECIALIST FIRST CLASS BY CORRESPONDENCE

Course Number: 0141-8.
Location: Coast Guard Institute, Oklahoma City, OK.
Length: Maximum, 156 weeks.
Exhibit Dates: 10/94–Present.
Learning Outcomes: Upon completion of the course, the student will be able to prepare communications watch lists, update standard operating procedures, perform follow-up procedures for security clearances, review communication logs, prepare communication shift and spot messages, update emergency action plan, set up an effective training program for CMS users, initiate actions for loss or compromise, and maintain accountability of classified material.
Instruction: Course material provided directs the student through reference material and practical exercise completion. Supervisors are responsible for reviewing practical exercises and guiding students in corrective measures to develop proficiency. Course requires completion of a closed book examination at the 80 percent level.
Credit Recommendation: In the lower-division baccalaureate/associate degree category, 1 semester hour in record keeping and 1 in planning and scheduling (3/96).

CG-1404-0002

RADIOMAN SECOND CLASS BY CORRESPONDENCE

Course Number: 241-8; 241-7; 241-6.
Location: Coast Guard Institute, Oklahoma City, OK.
Length: Maximum, 52 weeks.
Exhibit Dates: 1/90–Present.
Learning Outcomes: Upon completion of the course, the student will be able to provide information and carry out proper security procedures for radio communications; and select proper frequencies, antennas, and operating procedures.
Instruction: The technical portion of this correspondence course provides a brief coverage of the basic operating characteristics of communication receivers and transmitters used in CW, voice, and telephone modes, and antenna and wave propagation. Includes proctored end-of-course examination.
Credit Recommendation: Credit is not recommended because of the military-specific content of the course (11/91).

CG-1404-0004

COMMUNICATIONS OFFICER BY CORRESPONDENCE

Course Number: 0401-6; 0401-7.
Location: Coast Guard Institute, Oklahoma City, OK.
Length: Maximum, 52 weeks.
Exhibit Dates: 1/90–Present.
Learning Outcomes: Upon completion of the course, the student will be able to draft and format messages, select means of transmission, provide security and administration, use radio telephone procedures, and provide basic electromagnetic radiation and radio frequency management. Also includes a brief overview of communication equipment.
Instruction: Method is by self-instruction and covers the administration of communication systems. Includes a proctored end-of-course examination.
Credit Recommendation: In the lower-division baccalaureate/associate degree category, 3 semester hours in telecommunications (6/94).

CG-1404-0005

1. TELECOMMUNICATIONS SPECIALIST, CLASS A
2. RADIOMAN, CLASS A

Course Number: *Version 1:* None. *Version 2:* None.
Location: Training Center, Petaluma, CA.
Length: *Version 1:* 9 weeks (315 hours). *Version 2:* 18 weeks.
Exhibit Dates: *Version 1:* 5/94–Present. *Version 2:* 6/90–4/94.
Learning Outcomes: *Version 1:* This version is pending evaluation. Upon completion of the course, the student will be able to operate both voice and teletype communication systems for message transmission, reception, and processing. Knowledge includes concepts of RF propagation and signal circuitry related to frequency selection and atmospheric conditions for establishing and maintaining radio communication. Student will be able to maintain appropriate administration and security procedures for message traffic and control. *Version 2:* Upon completion of the course, the student

will be able to operate both voice and teletype communication systems for message transmission, reception, and processing. Knowledge includes concepts of RF propagation and signal circuitry related to frequency selection and atmospheric conditions for establishing and maintaining radio communication. Student will be able to maintain appropriate administration and security procedures for message traffic and control.

Instruction: *Version 1:* This version is pending evaluation. Instruction includes lectures, demonstrations, laboratory exercises, and extensive computer-assisted instruction for simulation and drill, and practice in voice and teletype communication procedures. Instruction includes radio teletype and voice systems and circuits, Morse code, software applications, message format, network transmission and receiving procedures, and message traffic and control. *Version 2:* Instruction includes lectures, demonstrations, laboratory exercises, and extensive computer-assisted instruction for simulation and drill, and practice in voice and teletype communication procedures. Instruction includes radio teletype and voice systems and circuits, Morse code, software applications, message format, network transmission and receiving procedures, and message traffic and control.

Credit Recommendation: *Version 1:* Pending evaluation. *Version 2:* In the vocational certificate category, 3 semester hours in radio voice and teletype applications for network telecommunications (9/93); in the lower-division baccalaureate/associate degree category, 3 semester hours in theory and principles of RF propagation and radio/teletype circuit operations (9/93).

CG-1405-0002

STOREKEEPER THIRD CLASS BY CORRESPONDENCE

Course Number: 0350-1; 349-9.
Location: Coast Guard Institute, Oklahoma City, OK.
Length: Maximum, 104 weeks.
Exhibit Dates: 1/90–Present.
Objectives: To provide potential storekeepers with an understanding of selected topics in administration and clerical procedures.
Instruction: Course includes administration, clerical procedures, fiscal matters, transportation, and supply. Students are required to complete a proctored end-of-course examination.
Credit Recommendation: In the vocational certificate category, 1 semester hour in record keeping and 1 in office procedures (7/93).

CG-1405-0005

AVIATION ENGINEERING ADMINISTRATION
(Aircraft Logs and Records)

Course Number: None.
Location: Aviation Technical Training Center, Elizabeth City, NC.
Length: 1-2 weeks (51 hours).
Exhibit Dates: 1/90–11/96.
Objectives: Provides training in the proper procedures used in correspondence, aeronautical engineering forms and reports, and the procurement and maintenance of technical publications.
Instruction: Topics include an introduction to work as a technical librarian or computerized maintenance program manager.
Credit Recommendation: In the lower-division baccalaureate/associate degree category, 1 semester hour in technical records maintenance (6/85).

CG-1405-0006

STOREKEEPER, CLASS A

Course Number: *Version 1:* None. *Version 2:* None.
Location: Training Center, Petaluma, CA.
Length: *Version 1:* 9 weeks (294 hours). *Version 2:* 10 weeks (320 hours).
Exhibit Dates: *Version 1:* 12/91–Present. *Version 2:* 1/90–11/91.
Learning Outcomes: *Version 1:* Upon completion of the course, the student will be able to use a computer workstation for word processing, document filing, property management, and tracking of materials for inspection, packaging handling, storage, and transportation. *Version 2:* Upon completion of the course, the student will be able to use keyboard, word processing software, file documents and procure transport and manage the supplies.
Instruction: *Version 1:* Instruction includes computer workstation experiences using word processing and document preparation. Classroom instruction includes lecture, practical experience, video, examination, and review. *Version 2:* Instruction includes typing, word processing, correspondence, procurement, requisitions, publications, inventory control, transportation, and property management. Methodology includes lecture, discussion, and practical exercises.
Credit Recommendation: *Version 1:* In the lower-division baccalaureate/associate degree category, 3 semester hours in business correspondence and 2 in word processing (9/93); in the upper-division baccalaureate category, 2 semester hours in property management and inventory control (9/93). *Version 2:* In the lower-division baccalaureate/associate degree category, 2 semester hours in supply management, 1 in merchandise procurement, 1 in word processing, and 1 in office procedures (8/87).

CG-1405-0008

STOREKEEPER FIRST CLASS BY CORRESPONDENCE

Course Number: 0149-8; 0150-1.
Location: Coast Guard Institute, Oklahoma City, OK.
Length: Maximum, 156 weeks.
Exhibit Dates: 1/90–Present.
Learning Outcomes: Upon completion of the course, the student will be able to manage travel and transportation in accordance with the travel manual, Joint Federal Travel regulations, and the transportation manual; manage fiscal procedures in accordance with claims manual and the Comptroller manual; and manage personal property.
Instruction: Instruction is provided through a nonresident, self-paced study program containing three workbooks and a proctored end-of-course test. Course consists of selected subjects in travel and transportation, fiscal procedures, and supply support.
Credit Recommendation: In the lower-division baccalaureate/associate degree category, 3 semester hours in military science (travel, transportation, fiscal procedures, and supply support) (9/94).

CG-1405-0009

STOREKEEPER SECOND CLASS BY CORRESPONDENCE

Course Number: *Version 1:* 250-1. *Version 2:* 0249-8.
Location: Coast Guard Institute, Oklahoma City, OK.
Length: *Version 1:* Maximum, 156 weeks. *Version 2:* Maximum, 52 weeks.
Exhibit Dates: *Version 1:* 6/94–Present. *Version 2:* 1/90–5/94.
Learning Outcomes: *Version 1:* Upon completion of the course, the student will be able to follow standard office procedures required of a Second Class Storekeeper. Course includes receiving stores and equipment, maintaining inventory documents, and storing supplies and equipment. The student must also have a working knowledge of accounting procedures that are used in receiving, shipping, and transportation of stores and equipment. *Version 2:* Upon completion of the course, the student will be able to use a computer terminal, select proper software, and follow standard office procedures required of a Second Class Storekeeper. Course includes receiving stores and equipment, maintaining inventory documents, and storing supplies and equipment. The student must also have a working knowledge of accounting procedures that are used in receiving, shipping, and transportation of stores and equipment.
Instruction: *Version 1:* Instruction in office administration, fiscal procedures, supply, and logistics maintenance. *Version 2:* Instruction in office administration, fiscal procedures, supply, and logistics maintenance.
Credit Recommendation: *Version 1:* In the lower-division baccalaureate/associate degree category, 1 semester hour in office procedures and 1 in supply management (9/94). *Version 2:* In the lower-division baccalaureate/associate degree category, 2 semester hours in office procedures, 1 in computer applications, and 2 in supply management (11/91).

CG-1405-0010

STOREKEEPER BASIC RESERVE

Course Number: None.
Location: Training Center, Petaluma, CA.
Length: 2 weeks (73 hours).
Exhibit Dates: 5/92–Present.
Learning Outcomes: Upon completion of the course, the student will be able to use a computer workstation to update publications, process materials for handling and storage, update inventory through requisition or change orders, utilize E-mail, process supply and equipment orders, and provide accountability reports for damaged, lost, or stolen inventory.
Instruction: Methods include lectures, discussions, and practical exercises utilizing the computer workstation.
Credit Recommendation: In the lower-division baccalaureate/associate degree category, 1 semester hour in computer word processing and 1 in supply inventory and management (9/93).

CG-1406-0004

INSTRUCTOR/COURSE DEVELOPER, CLASS C

Course Number: None.
Location: Aviation Technical Training Center, Elizabeth City, NC.
Length: 2-3 weeks (74 hours).
Exhibit Dates: 1/90–11/96.

Objectives: Provides skills and knowledge necessary to analyze, design, review, revise, develop, and instruct in accordance with accepted principles of instructional technology.

Instruction: Course includes lectures and laboratory experience in performing job analyses; defining target populations; preparing training analyses; writing instructional objectives and criterion-referenced test items; organizing and preparing lesson plans; selecting, designing, and developing training aids; delivering instruction/instructional packages; performing test item analysis; designing and administering instructional feedback devices. Methods consist of lecture, testing, and laboratory.

Credit Recommendation: In the upper-division baccalaureate category, 1 semester hour in instructional methods (6/85).

CG-1406-0005

BASIC INSTRUCTOR

Course Number: None.
Location: Training Center, Petaluma, CA.
Length: 2 weeks (73 hours).
Exhibit Dates: 1/90–12/92.
Learning Outcomes: Upon completion of the course, the student will be able to prepare course materials, deliver instruction utilizing multiple methods, evaluate instruction, and resolve or manage instructor/student and student/student relationships.

Instruction: Instruction focuses on development of instructor skills and course materials, enhancement and management of the learning environment, instruction ethics, and individual skill development. The course also focuses on design and effective use of print and audiovisual support materials. Methods of instruction include proctored exercise, lecture, discussion, and case study.

Credit Recommendation: In the lower-division baccalaureate/associate degree category, 1 semester hour in educational media and 2 in educational methods (8/87).

CG-1406-0006

COURSE DESIGNER

Course Number: None.
Location: Training Center, Petaluma, CA.
Length: 2 weeks (58 hours).
Exhibit Dates: 1/90–Present.
Learning Outcomes: Upon completion of the course, the student will be able to prepare a new course or revise an existing course utilizing a systems approach to training.

Instruction: Instruction focuses on design of instructional plans for training a specific group and includes analysis of training needs, strategies and methods for development of training materials, implementation of training, and evaluation of instruction/training. Methodology includes lecture, individual instruction, conferences, and practical exercises.

Credit Recommendation: In the lower-division baccalaureate/associate degree category, 2 semester hours in instructional systems development (8/87).

CG-1406-0007

RECRUITER TRAINER

Course Number: None.
Location: Reserve Training Center, Yorktown, VA.
Length: 3-4 weeks (90-170 hours).
Exhibit Dates: 1/90–Present.
Learning Outcomes: Upon completion of the course, the student will be able to recruit, process, and enlist qualified officers, officer candidates, and enlisted personnel in the United States Coast Guard.

Instruction: Course includes lectures, demonstrations, practical exercises, and independent study in general policies and objectives of the Coast Guard recruiting program, effective sales techniques, sales training, product knowledge, and applicant processing.

Credit Recommendation: In the upper-division baccalaureate category, 3 semester hours in human resource management (4/89).

CG-1406-0008

LEADERSHIP AND MANAGEMENT

Course Number: None.
Location: Leadership and Management Training Centers, USA; Leadership and Management Training Center, Petaluma, CA.
Length: 1 week (46 hours).
Exhibit Dates: 10/95–Present.
Learning Outcomes: Upon completion of the course, the student will be able to recognize and apply styles of leadership; analyze situations and select appropriate leadership techniques; practice constructive communication skills; and recognize and apply motivation methods for subordinates' personal and professional performance improvement.

Instruction: This course is taught in a series of case studies, situational exercises, and models of leadership styles. Methods include seminar discussion, self-critique, and workshop sessions in small groups.

Credit Recommendation: In the upper-division baccalaureate category, 2 semester hours in leadership (3/96).

CG-1406-0009

OFFICER IN CHARGE, AIDS TO NAVIGATION TEAM

Course Number: ANC-ANT.
Location: Reserve Training Center, Yorktown, VA.
Length: 2 weeks (66 hours).
Exhibit Dates: 1/94–Present.
Learning Outcomes: Upon completion of the course, the student will be able to supervise the performance of the Aids To Navigation System (ATON) of an operational buoy tender unit, identify specific problems and issues in ATON, and establish a basis for passing information along to successors.

Instruction: Methods of instruction include lecture, practical exercise, and laboratory. Topics include US aids to navigation system, minor ATON servicing structures, 12V DC, lighthouse maintenance, minor ATON servicing buoys, ATON visual signals, ATON legal issues, aids to navigation positioning, ATON record administration, servicing equipment, minor ATON team operations, waterway analysis and management system review, and hazardous waste management objectives.

Credit Recommendation: In the lower-division baccalaureate/associate degree category, 1 semester hour in personnel supervision (5/96).

CG-1406-0010

CHIEF, INSPECTION DEPARTMENT

Course Number: MS 457R.
Location: Reserve Training Center, Yorktown, VA.
Length: 2 weeks (70-71 hours).
Exhibit Dates: 5/92–Present.
Learning Outcomes: Upon completion of the course, the student will be able to perform in a mid-level management position involving the organization, training, and evaluation of personnel assigned to specific tasks and missions.

Instruction: Methods of instructions include lectures, videos, practical exercises, and role-playing. The focus of this course is management methodologies. Some topics covered are policy issues, time management, correspondence management, budget preparation and execution, and personnel training and management.

Credit Recommendation: In the upper-division baccalaureate category, 2 semester hours in human resources management (5/96).

CG-1408-0003

AVIATION MAINTENANCE ADMINISTRATION BY CORRESPONDENCE

Course Number: 448-2; 448-1.
Location: Coast Guard Institute, Oklahoma City, OK.
Length: Maximum, 104 weeks.
Exhibit Dates: 1/90–2/90.
Objectives: After 2/90 see CG-1704-0038. To train enlisted persons and officers in administrative practice of aviation units.

Instruction: Course presents information describing the organization of a typical air station engineering section, the purpose and use of the aircraft inspection system, standard aircraft record forms and proper entries into the forms (aircraft logs) and reports, use of the technical order system, identification of major categories or types of aviation material, and procedures for requisitioning material. Students are required to successfully complete a proctored end-of-course examination.

Credit Recommendation: In the vocational certificate category, credit in aviation maintenance administration and management on the basis of institutional evaluation (10/86).

CG-1408-0004

YEOMAN SECOND CLASS BY CORRESPONDENCE

Course Number: *Version 1:* 0275-2; 0255-1; 255-7; 255-8. *Version 2:* 0255-1; 255-7; 255-8.
Location: Coast Guard Institute, Oklahoma City, OK.
Length: *Version 1:* Maximum, 36 weeks. *Version 2:* Maximum, 104 weeks.
Exhibit Dates: *Version 1:* 1/93–Present. *Version 2:* 1/90–12/92.
Learning Outcomes: *Version 1:* Upon completion of the course, the student will be able prepare and distribute forms and reports; follow policies and procedures related to personnel pay, retirement benefits, appointments, and travel regulations; and update files. *Version 2:* Upon completion of the course, the student will be able to perform basic paperwork management.

Instruction: *Version 1:* Training in the completion and use of specialized documents includes desertion and unauthorized absences forms, confinement orders, transfers, travel authorized absence forms, confinement orders, transfers, travel authorization and claims, pay related issues, and retirement applications. Instruction is also provided in the proper counseling of personnel regarding the significance of each document. *Version 2:* Topics include the administration of the paperwork management program, requirements for promotion within the Coast Guard, security clearances, and other specialized training related to Coast Guard operating procedures. Students are required to successfully complete a proctored end-of-course examination.

Credit Recommendation: *Version 1:* In the lower-division baccalaureate/associate degree category, 1 semester hour in clerical procedures (3/94). *Version 2:* In the lower-division baccalaureate/associate degree category, 1 semester hour in office administration (2/92).

CG-1408-0005

YEOMAN THIRD CLASS BY CORRESPONDENCE

Course Number: *Version 1:* 0375-1. *Version 2:* 355-3; 355-2; 355-1.
Location: Coast Guard Institute, Oklahoma City, OK.
Length: *Version 1:* Maximum, 156 weeks. *Version 2:* Maximum, 104 weeks.
Exhibit Dates: *Version 1:* 9/95–Present. *Version 2:* 1/90–8/95.
Learning Outcomes: *Version 1:* This version is pending evaluation. To provide enlisted personnel with an understanding of communication procedures and personnel policies. *Version 2:* To provide enlisted personnel with an understanding of communication procedures and personnel policies.
Instruction: *Version 1:* This version is pending evaluation. Topics include office communications procedures, service record maintenance, paperwork processing, and military pay entitlements and procedures. *Version 2:* Topics include office communications procedures, service record maintenance, paperwork processing, and military pay entitlements and procedures.
Credit Recommendation: *Version 1:* Pending evaluation. *Version 2:* Credit is not recommended because of the military-specific nature of the course (1/87).

CG-1408-0006

YEOMAN FIRST CLASS BY CORRESPONDENCE

Course Number: *Version 1:* 0175-1; 0155-1; 155-8. *Version 2:* 0155-1; 155-8.
Location: Coast Guard Institute, Oklahoma City, OK.
Length: *Version 1:* Maximum, 36 weeks. *Version 2:* Maximum, 104 weeks.
Exhibit Dates: *Version 1:* 1/93–Present. *Version 2:* 1/90–12/92.
Objectives: *Version 1:* Upon completion of the course, the student will be able to draft, prepare, and proofread correspondence relating to personnel procedures and Coast Guard organization and administration. *Version 2:* To provide enlisted personnel with an understanding of the Coast Guard's administration and legal system.
Instruction: *Version 1:* Training is provided in basic letter and memorandum structure with emphasis on conciseness. The use of short paragraphs, key points, and various letter and memorandum formats are emphasized. *Version 2:* Training is provided in forms management, personnel procedures, the security of classified material, military justice, Coast Guard organization and administration, and the Coast Guard morale and education programs. Students are required to successfully complete a proctored end-of-course examination.

Credit Recommendation: *Version 1:* In the lower-division baccalaureate/associate degree category, 3 semester hours in business communications (3/94). *Version 2:* Credit is not recommended due to the military-specific nature of the course (2/92).

CG-1408-0007

WEAPONS ADMINISTRATION BY CORRESPONDENCE

Course Number: 410-1.
Location: Coast Guard Institute, Oklahoma City, OK.
Length: Maximum, 104 weeks.
Exhibit Dates: 1/90–1/98.
Objectives: To teach ordnance personnel the administrative skills to function on board ship.
Instruction: Course covers ordnance records, casualty reports, property management, and ordnance safety. Students must successfully complete a proctored end-of-course examination.
Credit Recommendation: In the vocational certificate category, 1 semester hour in office administration (3/97).

CG-1408-0035

1. ENGINEERING PETTY OFFICER INDOCTRINATION
2. ENGINEERING ADMINISTRATION

Course Number: *Version 1:* MK-1. *Version 2:* MK-1(R).
Location: Reserve Training Center, Yorktown, VA.
Length: *Version 1:* 2 weeks. *Version 2:* 1-2 weeks (58-62 hours).
Exhibit Dates: *Version 1:* 10/92–Present. *Version 2:* 1/90–4/92.
Learning Outcomes: *Version 1:* Upon completion of the course, the student will be able to manage and administer engineering department operations at group and shore stations, as well as on Coast Guard vessels. *Version 2:* Upon completion of the course, the student will be able to manage engineering department operations at group and shore stations, as well as on Coast Guard vessels.
Instruction: *Version 1:* Topics include preparation of department forms, records, reports, and publications; administration of Coast Guard maintenance programs; procurement systems for obtaining supplies; and procedures for receiving, issuing, and accounting for engineering supplies. *Version 2:* Topics include performance evaluation and training programs; preparation of department forms, records, reports, and publications; administration of Coast Guard safety program and motor vehicle program; procurement systems for obtaining supplies; and procedures for receiving, issuing, and accounting for engineering supplies.
Credit Recommendation: *Version 1:* In the lower-division baccalaureate/associate degree category, 1 semester hour in maintenance management (5/98). *Version 2:* In the upper-division baccalaureate category, 2 semester hours in field experience in management, 3 when this course is combined with an assignment as a senior engineer (Engineering Petty Officer—EPO) (10/88).

CG-1408-0036

RADARMAN FIRST CLASS BY CORRESPONDENCE

Course Number: *Version 1:* 0139-6. *Version 2:* 139-5.
Location: Coast Guard Institute, Oklahoma City, OK.
Length: *Version 1:* Maximum, 156 weeks. *Version 2:* Maximum, 52 weeks.
Exhibit Dates: *Version 1:* 9/94–Present. *Version 2:* 1/90–8/94.
Learning Outcomes: *Version 1:* Upon completion of the course, the student will be able to identify procedures to plan, organize, schedule, and supervise training programs including identification of appropriate training styles and qualification of instructors; identify duties and responsibilities of search and rescue (SAR) mission coordinators and on scene commander; identify procedures involved in the prosecution and search patterns of a SAR case; prepare reports and recommendations pertaining to the prosecution of a SAR case; and identify technical parameters and proper procedures for constructing intelligence and vulnerability reports for an electronic warfare equipment. *Version 2:* Upon completion of the course, the student will be able to identify procedures used to plan, organize, schedule, and supervise training programs; identify procedures used to prepare reports and communication plans, and implement operational plans; identify procedures to implement preventive maintenance plans and coordinate antiaircraft operations; and identify procedures for the reporting of equipment status.
Instruction: *Version 1:* The major topics covered are administration and supply, operational and personnel procedures, operations, navigational plotting procedures, safety and training, electronic warfare and communications. Training is provided in a nonresident, self-instruction format. Students are required to successfully complete a proctored end-of-course examination. *Version 2:* The major topics covered are administration and supply, operational and personnel procedures, operations, navigational plotting procedures, safety and training, electronic warfare and communications. Training is provided in a nonresident, self-instruction format. Students are required to successfully complete a proctored end-of-course examination.
Credit Recommendation: *Version 1:* In the lower-division baccalaureate/associate degree category, 1 semester hour in project administration (3/96). *Version 2:* In the lower-division baccalaureate/associate degree category, 1 semester hour in project administration (6/90).

CG-1408-0037

DAMAGE CONTROLMAN CHIEF BY CORRESPONDENCE

Course Number: 0015-1.
Location: Coast Guard Institute, Oklahoma City, OK.
Length: Maximum, 78 weeks.
Exhibit Dates: 1/90–8/92.
Learning Outcomes: Upon completion of the course, the student will be able to perform carpentry maintenance inspections, determine design requirements and material for all types of piping and plumbing systems, train and supervise personnel in all aspects of CBR (chemical, biological, radiological) defense, administer a shop on a ship or base, and inspect shipyard work.
Instruction: Topics include carpentry, plumbing, CBR warfare defense, and adminis-

tration. Training is provided in a nonresident, self-instruction format. Students are required to successfully complete a proctored end-of-course examination.

Credit Recommendation: In the lower-division baccalaureate/associate degree category, 1 semester hour in project administration (6/90).

CG-1408-0039

SONAR TECHNICIAN CHIEF BY CORRESPONDENCE

Course Number: 0043-1; 043-9.
Location: Coast Guard Institute, Oklahoma City, OK.
Length: Maximum, 78 weeks.
Exhibit Dates: 1/90–2/92.
Learning Outcomes: Upon completion of the course, the student will be able to identify the proper installation of new or overhauled assemblies, subassemblies, or components of electronic equipment in accordance with Coast Guard standards; select the correct interface cables for assemblies; verify the performance of assemblies; use WQM-series sonar test set to test transducers; describe the procedures used to analyze operational and maintenance logs; and requisition replacement parts and tools.
Instruction: The major topics covered are test equipment, test procedures, preventive maintenance, and equipment installation. Training is provided in a nonresident, self-instruction format. Students are required to successfully complete a proctored end-of-course examination.
Credit Recommendation: In the lower-division baccalaureate/associate degree category, 1 semester hour in project administration (9/92).

CG-1408-0041

SMALL ARMS INSTRUCTOR, CLASS C
(Small Arms Instructor)

Course Number: SAI.
Location: Reserve Training Center, Yorktown, VA.
Length: 3 weeks (103-112 hours).
Exhibit Dates: 1/90–Present.
Learning Outcomes: Upon completion of the course, the student will be able to present lectures and demonstrations relating to weapon and range safety and operator maintenance on .45 caliber and 9mm service pistols; qualify on the .45 service pistol, M16 service rifle, 12 gauge riot service shotgun; fire the 9mm pistol for familiarization; manage weapons range; and perform duties of the line instructor.
Instruction: Lectures, practical exercises, and videos cover topics, such as preparation of lesson plans, first aid emergencies, line instruction, range preparation and management, operator weapon maintenance, and weapon qualification.
Credit Recommendation: In the lower-division baccalaureate/associate degree category, 2 semester hours in small arms range management (1/96).

CG-1408-0042

AIDS TO NAVIGATION OPERATIONS MANAGEMENT

Course Number: ANC-OPS.
Location: Reserve Training Center, Yorktown, VA.
Length: 2 weeks (70 hours).
Exhibit Dates: 10/94–Present.

Learning Outcomes: Upon completion of the course, the student will be able to administer the aids to navigation system (ATON) of an operational buoy tender unit.
Instruction: Methods of instruction include classroom instruction case studies, and practical training. Topics covered include US aids to navigation system, minor ATON servicing structures, 12V and 120V lighthouse maintenance, ATON record administration, ATON visual signals, ATON legal issues, aids to navigation positioning (calculator, computer generated), buoy deck operations, discrepancy response, and short range ATON systems design and evaluation.
Credit Recommendation: In the lower-division baccalaureate/associate degree category, 1 semester hour in materiel management (5/96).

CG-1409-0003

YEOMAN SCHOOL, CLASS A

Course Number: *Version 1:* None. *Version 2:* None.
Location: Training Center, Petaluma, CA.
Length: *Version 1:* 9 weeks (302 hours). *Version 2:* 12 weeks (412 hours).
Exhibit Dates: *Version 1:* 1/92–Present. *Version 2:* 1/90–12/91.
Learning Outcomes: *Version 1:* Upon completion of the course, the student will be able to compose business correspondence and process personnel actions for pay, travel, transportation, and other personnel administrative requirements in accordance with Coast Guard procedures, using a computer workstation. *Version 2:* Upon completion of the course, the student will be able to utilize keyboard, word processing, and software to effectively communicate with internal and external agencies; and maintain records and files associated with personnel and office reports.
Instruction: *Version 1:* Instruction includes basic word processing followed by external application of word processing to real experiences within an office. Record maintenance, office communications, personnel issues, and legal procedures are also covered. Methodology includes lectures, proctored exercises, demonstrations, and case studies. *Version 2:* Instruction includes basic keyboard and word processing followed by external application of word processing to real experience within an office. Record maintenance, office communication, personnel issues, and legal procedures are also covered. Methodology includes lecture, proctored exercise, demonstration, and case study.
Credit Recommendation: *Version 1:* In the lower-division baccalaureate/associate degree category, 3 semester hours in word processing and office procedures, 2 in business communication, and 2 in personnel administration procedures (9/93). *Version 2:* In the lower-division baccalaureate/associate degree category, 3 semester hours in word processing and 3 in office procedures (8/87).

CG-1409-0004

YEOMAN RESERVE
(Reserve Yeoman Basic)

Course Number: None.
Location: Training Center, Petaluma, CA.
Length: 2 weeks (70-74 hours).
Exhibit Dates: 1/90–Present.

Learning Outcomes: Upon completion of the course, the student will be able to effectively utilize publications, maintain office records, and maintain personnel records and internal and external correspondence for the unit.
Instruction: Instruction includes office communication procedures, forms and procedural publications, record maintenance, and legal procedures for correspondence and records. Methods of instruction include lecture, proctored exercise, and discussion.
Credit Recommendation: In the lower-division baccalaureate/associate degree category, 2 semester hours in office procedures (2/92).

CG-1511-0002

CHIEF PETTY OFFICER ACADEMY

Course Number: *Version 1:* None. *Version 2:* None.
Location: Training Center, Petaluma, CA.
Length: *Version 1:* 6 weeks (264 hours). *Version 2:* 6 weeks (259 hours).
Exhibit Dates: *Version 1:* 1/93–Present. *Version 2:* 1/90–12/92.
Learning Outcomes: *Version 1:* Upon completion of the course, the student will be able to perform duties and responsibilities in planning, directing, and coordinating organizational programs. *Version 2:* Upon completion of the course, the student will be able to identify leadership and management functions related to the Coast Guard. Student will have a basic knowledge of management and organizational systems, organizational behavior, roles, responsibilities, and policies of the agencies associated with the Coast Guard. Student will be able to plan personnel programs including retirement, health and educational benefits, and other programs.
Instruction: *Version 1:* Instruction is presented through lectures, seminars, case studies, panel discussions, and visiting lecturers prominent in the field of organizational management and leadership. Curriculum includes individual and organizational behavior and business communication, personnel administration, policies, and career counseling. Case studies and laboratory exercises are specific to Coast Guard policies and programs. *Version 2:* Course is presented via lecture, demonstration, case study, panel discussion, supervised assignments, video presentations, practical exercises, and examinations. Technical writing and communication skills are an integral part of the course.
Credit Recommendation: *Version 1:* In the lower-division baccalaureate/associate degree category, 2 semester hours in business communication and 2 in principles of organizational management and personnel administration (9/93); in the upper-division baccalaureate category, 3 semester hours in advanced personnel policies and management (9/93). *Version 2:* In the lower-division baccalaureate/associate degree category, 3 semester hours in personnel management, 2 in business communication, and 3 in oral communications (9/93).

CG-1512-0001

MILITARY CIVIL RIGHTS BY CORRESPONDENCE

Course Number: 0800-1.
Location: Coast Guard Institute, Oklahoma City, OK.

Length: Maximum, 104 weeks.
Exhibit Dates: 1/90–11/96.
Objectives: To present an overview of military civil rights and affirmative action.
Instruction: Course provides instruction in the military civil rights program and its organization within the Coast Guard. Students are required to successfully complete a proctored end-of-course examination.
Credit Recommendation: Credit is not recommended because of the military-specific nature of the course (5/85).

CG-1701-0002

AIR CONDITIONING AND REFRIGERATION, CLASS C
(Refrigeration/Air Conditioning Operation and Maintenance, Class C)

Course Number: MK-22.
Location: Reserve Training Center, Yorktown, VA.
Length: 6 weeks (219-223 hours).
Exhibit Dates: 1/90–Present.
Learning Outcomes: Upon completion of the course, the student will be able to operate, test, and maintain refrigeration and air conditioning equipment.
Instruction: Lectures and practical exercises cover gas laws, the properties of heat, pressure-temperature relationship, the refrigeration cycle, refrigerants, safety leak detectors, and temperatures as they pertain to refrigeration, air conditioning, and heat pumps. Instruction includes the use of service equipment; evacuating and charging procedures; testing electrical circuits; and troubleshooting heating, ventilating, and air conditioning systems.
Credit Recommendation: In the lower-division baccalaureate/associate degree category, 3 semester hours in refrigeration, 1 in refrigeration laboratory, and 1 in air conditioning laboratory (5/96).

CG-1704-0001

AVIATION MACHINIST'S MATE, CLASS A
(Aviation Machinist's Mate (AD), Class A)
(Aviation Machinist's Mate)

Course Number: *Version 1:* None. *Version 2:* None.
Location: *Version 1:* Aviation Technical Training Center, Elizabeth City, NC. *Version 2:* Aviation Technical Training Center, Elizabeth City, NC; Aircraft Repair and Supply Center, Elizabeth City, NC.
Length: *Version 1:* 11 weeks (440-550 hours). *Version 2:* 8-9 weeks (198 hours).
Exhibit Dates: *Version 1:* 12/95–Present. *Version 2:* 1/90–11/95.
Learning Outcomes: *Version 1:* This version is pending evaluation. To provide fundamental principles of mathematics, physics, and aerodynamics for further study on specific aircraft systems; and to provide training at the entry level in aircraft maintenance on several specific aircraft. *Version 2:* To provide fundamental principles of mathematics, physics, and aerodynamics for further study on specific aircraft systems; and to provide training at the entry level in aircraft maintenance on several specific aircraft.
Instruction: *Version 1:* This version is pending evaluation. Course trains students to service and maintain ground support equipment, select and use maintenance forms and publications, select and use hand tools and aircraft hardware, taxi aircraft, perform basic lubrication operations, and perform the basic skills of a search and rescue (SAR) aircrewman. NOTE: After completing the basic phase students normally progress to either the Fixed Wing track (CG-1704-0022) or Rotary Wing track (CG-1704-0023). *Version 2:* Course trains students to service and maintain ground support equipment, select and use maintenance forms and publications, select and use hand tools and aircraft hardware, taxi aircraft, perform basic lubrication operations, and perform the basic skills of a search and rescue (SAR) aircrewman. NOTE: After completing the basic phase students normally progress to either the Fixed Wing track (CG-1704-0022) or Rotary Wing track (CG-1704-0023).
Credit Recommendation: *Version 1:* Pending evaluation. *Version 2:* In the lower-division baccalaureate/associate degree category, 1 semester hour in introduction to aviation maintenance, 1 in ground operation and servicing procedures, 1 in aircraft technical publications, 1 in basic shop practices, and 2 in materials and processes (6/85).

CG-1704-0004

T58-GE-8B ENGINE MAINTENANCE, CLASS C

Course Number: None.
Location: Aviation Technical Training Center, Elizabeth City, NC; Aircraft Repair and Supply Center, Elizabeth City, NC.
Length: 4 weeks (147 hours).
Exhibit Dates: 1/90–11/96.
Objectives: To train students in the maintenance and adjustment of T58-GE-8B turboshaft engines.
Instruction: Topics include engine and systems operation, maintenance publications, corrosion control, service, maintenance, inspection, and repair of the T58-8 engine.
Credit Recommendation: In the lower-division baccalaureate/associate degree category, 3 semester hours in aircraft gas turbine engine inspection and 1 in aircraft gas turbine engine maintenance (6/85).

CG-1704-0006

HC-131A AIRCRAFT MAINTENANCE, CLASS C

Course Number: None.
Location: Aircraft Repair and Supply Center, Elizabeth City, NC.
Length: 6 weeks (210 hours).
Exhibit Dates: 1/90–1/90.
Objectives: Provides training in the repair, troubleshooting, and general maintenance of HC-131A aircraft.
Instruction: Introduction to HC-131A aircraft and ground handling equipment; instruction in the operation and maintenance of major engine subsystems, including fuel instrumentation and propellers, and in other aircraft subsystems, including electrical, safety, hydraulic, landing gear, air conditioning, pressurization, fire detection, and automatic pilot.
Credit Recommendation: In the vocational certificate category, 4 semester hours in aircraft maintenance (9/77).

CG-1704-0007

AVIATION MACHINIST'S MATE SECOND CLASS BY CORRESPONDENCE

Course Number: 205-2; 205-1; 205-9.
Location: Coast Guard Institute, Oklahoma City, OK.
Length: Maximum, 104 weeks.
Exhibit Dates: 1/90–Present.
Objectives: To provide the second class machinist's mate with those basic aircraft subjects needed for the mission.
Instruction: This correspondence course outlines aviation safety precautions, states the purpose and scope of the Coast Guard aircraft maintenance management system, lists the kinds of aviation directives and publications, explains technical orders, and identifies and explains the purpose of the Coast Guard aviation forms and reports. Aircraft hardware selection and identification is presented along with various safety methods. Basic corrosion control and preventive measures are discussed. Jet propulsion engines, turbojet engines, and their systems are presented. A proctored end-of-course examination is required.
Credit Recommendation: In the lower-division baccalaureate/associate degree category, 2 semester hours in power plant maintenance (11/91).

CG-1704-0008

AVIATION MACHINIST'S MATE FIRST CLASS BY CORRESPONDENCE
(Aviation Machinist Mate First Class by Correspondence)

Course Number: 105-6.
Location: Coast Guard Institute, Oklahoma City, OK.
Length: Maximum, 52 weeks.
Exhibit Dates: 1/90–7/90.
Objectives: After 7/90 see CG-1704-0052. To provide the student with aircraft maintenance management training, including records and forms, and to present general propeller principles.
Instruction: This correspondence course describes the duties of each subsection of an air station engineering section. The technical order system and related records and maintenance publications are discussed. The operating principles of the 43D50 and 54H60 propellers are covered including their components and electrical systems. The course includes a proctored end-of-course examination.
Credit Recommendation: In the lower-division baccalaureate/associate degree category, 1 semester hour in propeller maintenance (1/79).

CG-1704-0009

HC-130 FLIGHT ENGINEER BY CORRESPONDENCE

Course Number: *Version 1:* 0445-3. *Version 2:* 445-1; 445-2; 0445-2.
Location: Coast Guard Institute, Oklahoma City, OK.
Length: *Version 1:* Maximum, 156 weeks. *Version 2:* Maximum, 52 weeks.
Exhibit Dates: *Version 1:* 6/97–Present. *Version 2:* 1/90–5/97.
Learning Outcomes: *Version 1:* This version is pending evaluation. *Version 2:* To provide standardized special training and introductory information to flight engineer candidates.
Instruction: *Version 1:* This version is pending evaluation. *Version 2:* This correspondence course includes descriptions of the mechanical, communication, and navigation systems of the HC-130 aircraft. It also includes descriptions of preflight, inflight, and postflight duties of the flight engineer on the HC-130 air-

CG-1704-0011
HH-3F FLIGHT MECHANIC BY CORRESPONDENCE

Course Number: 0443-2.
Location: Coast Guard Institute, Oklahoma City, OK.
Length: Maximum, 52 weeks.
Exhibit Dates: 1/90–Present.
Objectives: To introduce the HH-3F helicopter to flight mechanic candidates.
Instruction: Flight mechanics are introduced to the HH-3F aircraft, aircraft systems, power train, aircraft equipment and furnishings, rescue equipment and recovery procedures, flight mechanic duties, helicopter aerodynamics, and theory of flight. Students must take a proctored final exam.
Credit Recommendation: In the lower-division baccalaureate/associate degree category, 2 semester hours in helicopter aircraft systems (9/90).

CG-1704-0018
AVIATION SURVIVALMAN SECOND CLASS BY CORRESPONDENCE

Course Number: 208-5; 208-4; 208-2.
Location: Coast Guard Institute, Oklahoma City, OK.
Length: Maximum, 52 weeks.
Exhibit Dates: 1/90–12/93.
Objectives: After 12/93 see CG-1704-0043. To provide the student with the fundamentals of survival equipment.
Instruction: This correspondence course contains a review of physics and electricity and describes pyrotechnics and parachute lofts. Identification of basic survival equipment, protective clothing, small arms and ammunition, the function of the oxygen system, and use of related hand tools are covered, as well as the inspection and repair of parachutes.
Credit Recommendation: Credit is not recommended due to the military-specific nature of the training involved (3/87).

CG-1704-0021
T58-GE-8B AND T58-GE-5 TURBOSHAFT ENGINE FAMILIARIZATION BY CORRESPONDENCE

Course Number: 0551-1.
Location: Coast Guard Institute, Oklahoma City, OK.
Length: Maximum, 104 weeks.
Exhibit Dates: 1/90–11/96.
Objectives: To provide the aviation machinist's mate with basic information on construction and operation of the T58-GE-8B and T58-GE-5 turboshaft engines.
Instruction: Instruction in engine components, operation of engine and components, and basic operational troubleshooting. Students are required to successfully complete a proctored end-of-course examination.
Credit Recommendation: In the vocational certificate category, 1 semester hour in power plant theory of operation and troubleshooting procedures (5/85).

craft. The course includes ground handling, servicing, and safety precautions. Course includes proctored end-of-course examination.
Credit Recommendation: *Version 1:* Pending evaluation. *Version 2:* Credit is not recommended because of the military-specific nature of the course (9/90).

CG-1704-0022
AVIATION MACHINIST'S MATE FIXED WING, CLASS A

Course Number: None.
Location: Aviation Technical Training Center, Elizabeth City, NC.
Length: 4-5 weeks (136 hours).
Exhibit Dates: 1/90–11/96.
Objectives: To provide training necessary to perform the maintenance, configuration, and inspection of fixed-wing aircraft.
Instruction: Topics include airframe, airframe electrical, hydraulic, landing gear, wheel, brake, tire, air, environmental, oxygen, and fuel systems, fuselage components, and flight controls.
Credit Recommendation: In the lower-division baccalaureate/associate degree category, 2 semester hours in airframe systems and 1 in aircraft turbine engine theory and maintenance (6/85).

CG-1704-0023
AVIATION MACHINIST'S MATE ROTARY WING, CLASS A

Course Number: None.
Location: Aviation Technical Training Center, Elizabeth City, NC.
Length: 6-7 weeks (226 hours).
Exhibit Dates: 1/90–11/96.
Objectives: To provide the knowledge and skill necessary to perform the maintenance, configuration, operation, and inspection of rotary-wing aircraft.
Instruction: Topics include airframe systems, aircraft turbine engine theory and maintenance, ground handling procedures, engine gearbox, main and tail rotor blades, landing gear, and technical manuals and publications.
Credit Recommendation: In the lower-division baccalaureate/associate degree category, 3 semester hours in airframe systems and 1 in aircraft turbine engine theory and maintenance (6/85).

CG-1704-0025
AVIATION ELECTRICIAN'S MATE, CLASS A ROTARY WING TRAINING

Course Number: None.
Location: Aviation Technical Training Center, Elizabeth City, NC.
Length: 4-5 weeks (157 hours).
Exhibit Dates: 1/90–11/96.
Objectives: Provides the knowledge and skill necessary to perform the configuration, application, operation, inspection, and repair of rotary wing aircraft electrical systems.
Instruction: Instruction and hands-on experience in the preparation, inspection, operation, modification, repair, and maintenance of selected electrical systems on the HH3F and the HH52A aircraft.
Credit Recommendation: In the lower-division baccalaureate/associate degree category, 2 semester hours in airframe systems (6/85).

CG-1704-0026
T58-5 AND T62 ENGINE MAINTENANCE, CLASS C

Course Number: None.
Location: Aviation Technical Training Center, Elizabeth City, NC.
Length: 4-5 weeks (161 hours).
Exhibit Dates: 1/90–11/96.
Objectives: To provide training in servicing, maintaining, inspecting, and repairing the T58-5 engine in accordance with the computerized maintenance system.
Instruction: Lectures on engine and systems operation; practical exercises in engine removal from aircraft, disassembly, inspection, repair, reassembly, and reinstallation along with required rigging, T58-5 engine run, and adjustments.
Credit Recommendation: In the lower-division baccalaureate/associate degree category, 2 semester hours in aircraft gas turbine engine inspection and 3 in aircraft gas turbine engine maintenance (6/85).

CG-1704-0027
AVIATION STRUCTURAL MECHANIC (AM), CLASS A

Course Number: None.
Location: Aviation Technical Training Center, Elizabeth City, NC.
Length: 18-19 weeks (581 hours).
Exhibit Dates: 1/90–1/90.
Objectives: After 1/90 see CG-1704-0047. To provide the knowledge and skills necessary to perform the configuration, application, operation, inspection, and repair of aircraft structures and aircraft components.
Instruction: Provides hands-on experience to handle, service, inspect, and maintain aircraft fuselages, wings, and control surfaces; fabricate and assemble metal parts and make repairs to aircraft skins; spray paint and maintain paint equipment; perform nondestructive testing inspections using visual dye and florescent penetrant methods; maintain hydraulic systems, landing gear, wheels, and tires; repair integral fuel tanks; perform oxyacetylene welding, bronze, and silver solder metal; perform cable swaging operations; identify and remove corrosion from aircraft and components; make repairs to thermosetting plastics.
Credit Recommendation: In the lower-division baccalaureate/associate degree category, 7 semester hours in aircraft structural repair, 2 in principles of hydraulics, 1 in aircraft cleaning and corrosion control, 1 in aircraft ground handling and servicing (6/85).

CG-1704-0031
HU-25A AVIONICS, CLASS C

Course Number: None.
Location: Aviation Technical Training Center, Elizabeth City, NC.
Length: 3-4 weeks (136 hours).
Exhibit Dates: 1/90–3/93.
Objectives: After 3/93 see CG-1704-0045. To provide knowledge and skills necessary to inspect, maintain, troubleshoot, and repair the HU-25A aircraft.
Instruction: Lectures and hands-on experience in the operation, troubleshooting, inspection, and routine maintenance of HU-25A avionics systems, including cockpit management system, audio control, radio altimeter, direction finder, LF-ADF receiver, VHF-AM transceiver, UHF transmitter/receiver, inertial sensor system, radar, area navigation system, and LORAN.
Credit Recommendation: In the lower-division baccalaureate/associate degree category, 3 semester hours in avionic systems maintenance (6/85).

CG-1704-0032

HYDRAULIC SYSTEMS AND EQUIPMENT, CLASS C (Hydraulic Systems and Equipment Operation and Maintenance)

Course Number: MK-6.
Location: Reserve Training Center, Yorktown, VA.
Length: 2 weeks (68-75 hours).
Exhibit Dates: 1/90–Present.
Learning Outcomes: Upon completion of the course, the student will apply basic principles involved in the operation, maintenance, troubleshooting, and repair industrial fluid power systems.
Instruction: Lectures and practical exercises cover fundamentals of hydraulics and associated subjects, including operation and repair of hydraulics systems and components; hydraulic circuit construction and operation; and troubleshooting and maintenance procedures.
Credit Recommendation: In the lower-division baccalaureate/associate degree category, 2 semester hours in fluid power systems (5/96).

CG-1704-0033

AVIATION STRUCTURAL MECHANIC FIRST CLASS BY CORRESPONDENCE

Course Number: 0107-8; 0107-7.
Location: Coast Guard Institute, Oklahoma City, OK.
Length: Maximum, 78 weeks.
Exhibit Dates: 1/90–Present.
Learning Outcomes: Upon completion of the course, the student will be familiar with aircraft maintenance management systems, standard aircraft maintenance procedures, including records, reports, publications, technical orders and inspections, aircraft drawings, blueprints, riveting, jacking and weighing, rigging, heat treatment, nondestructive inspections, arc welding, hardness testing, balancing of flight controls, and decontamination of hydraulic systems.
Instruction: Instruction is provided through a nonresident, sclf-paccd study program consisting of pamphlets covering the areas of aircraft maintenance management systems, sheet metal layout and forming, aircraft damage repair, and aircraft maintenance practices. Course includes proctored, comprehensive end-of-course examination. After 12/92 this course no longer covered heat treatment or electric arc welding, balancing of flight control surfaces, metal hardness test, or plating and anodizing.
Credit Recommendation: In the lower-division baccalaureate/associate degree category, 1 semester hour in aircraft maintenance (9/90).

CG-1704-0034

AIRCRAFT ENGINE FAMILIARIZATION BY CORRESPONDENCE

Course Number: 0552-1; 0552-2.
Location: Coast Guard Institute, Oklahoma City, OK.
Length: Maximum, 52 weeks.
Exhibit Dates: 1/90–Present.
Learning Outcomes: Upon completion of the course, the student will be able to demonstrate proficiency in the knowledge of the jet, fanjet, turboprop, and turboshaft aircraft engines.
Instruction: Instruction is provided in familiarization of the basic theory and operation of the gas turbine engine. Power plants covered include those of the T-56 turboprop, T-58 turboshaft, ATF3 turbofan, and the LTS turboshaft engines and their related systems.
Credit Recommendation: In the vocational certificate category, 1 semester hour in basic aircraft gas turbines (5/95).

CG-1704-0035

HH-65A FLIGHT MECHANIC BY CORRESPONDENCE

Course Number: *Version 1:* 0512-2. *Version 2:* 0515-1.
Location: Coast Guard Institute, Oklahoma City, OK.
Length: *Version 1:* Maximum, 156 weeks. *Version 2:* Maximum, 52 weeks.
Exhibit Dates: *Version 1:* 5/97–Present. *Version 2:* 1/90–4/97.
Learning Outcomes: *Version 1:* This version is pending evaluation. *Version 2:* Upon completion of the course, the student will be able to demonstrate knowledge of HU-65A nomenclature, systems, avionics, and flight preparation.
Instruction: *Version 1:* This version is pending evaluation. *Version 2:* Instruction is provided in familiarization of the helicopter, including helicopter aerodynamics, main rotor, transmission, flight controls, hydraulics, brakes, landing gear, electrical, fire protection, fuel, and environmental and avionics systems. Also included are flight preparations, engine familiarization, energencies, rescue, and handling procedures.
Credit Recommendation: *Version 1:* Pending evaluation. *Version 2:* In the vocational certificate category, 2 semester hours in aircraft systems, power train, aircraft equipment, and helicopter aerodynamics (9/90).

CG-1704-0036

HH-52A FLIGHT MECHANIC BY CORRESPONDENCE

Course Number: 0441-3.
Location: Coast Guard Institute, Oklahoma City, OK.
Length: Maximum, 52 weeks.
Exhibit Dates: 1/90–5/90.
Learning Outcomes: Upon completion of the course, the student will be familiar with HH-52A nomenclature systems, flight preparation, helicopter aerodynamics, brakes, landing gear, electrical, lighting, fire protection, fuel, air conditioning, avionic systems, engine familiarization, emergencies, rescue, and handling procedures.
Instruction: Instruction is provided through a non-resident, self-paced study program consisting of pamphlets, text material, quizzes, and a proctored, comprehensive end-of-course examination.
Credit Recommendation: In the vocational certificate category, 2 semester hours in aircraft systems, powertrains, aircraft equipment, and helicopter aerodynamics (9/90).

CG-1704-0037

HU-25A DROPMASTER BY CORRESPONDENCE

Course Number: 0514-1.
Location: Coast Guard Institute, Oklahoma City, OK.
Length: Maximum, 52 weeks.
Exhibit Dates: 1/90–Present.
Learning Outcomes: Upon completion of the course, the student will be able to demonstrate proficiency in the knowledge of proper procedures and safety precautions for rescue equipment and airdrop on the HU-25A aircraft.
Instruction: Instruction is provided in familiarization of the aircraft including general arrangement, publications and forms, weight and balance, aerial delivery, drag chutes, engines and auxiliary power units, hydraulic, landing gear, brake, flight control, fuel, environmental, electrical and fire proctection systems.
Credit Recommendation: Credit is not recommended because of the limited, specialized nature of the course (9/90).

CG-1704-0038

AVIATION ADMINISTRATION BY CORRESPONDENCE

Course Number: 0448-4; 0448-3.
Location: Coast Guard Institute, Oklahoma City, OK.
Length: Maximum, 26 weeks.
Exhibit Dates: 3/90–Present.
Learning Outcomes: Before 3/90 see CG-1408-0003. Upon completion of the course, the student will be familiar with aircraft maintenance functions, aircraft records and reports, aircraft publications and directives, aircraft inspections, use of Air Force and Coast Guard technical orders, Navy manuals, methods of aviation supply management, and procedures for management of support programs.
Instruction: Instruction includes reading assignments with objectives to be mastered and review exercises in aviation administration policies and procedures. Includes a proctored end-of-course examination.
Credit Recommendation: Credit is not recommended because of the limited, specialized nature of the course (1/93).

CG-1704-0039

AVIATION STRUCTURAL MECHANIC SECOND CLASS BY CORRESPONDENCE

Course Number: 207-7.
Location: Coast Guard Institute, Oklahoma City, OK.
Length: Maximum, 156 weeks.
Exhibit Dates: 6/93–Present.
Learning Outcomes: Upon completion of the course, the students will have basic knowledge of aircraft support equipment, servicing, tools, publications, and inspection requirements. They will have basic knowledge of aircraft system and structure defects and repair.
Instruction: Instruction is through correspondence work books. Section self-quizzes indicate student performance. A proctored final test is submitted for evaluating student success.
Credit Recommendation: In the lower-division baccalaureate/associate degree category, 1 semester hour in general aviation maintenance technology (10/94).

CG-1704-0040

AIRCREW BASIC BY CORRESPONDENCE

Course Number: 0440-4; 0440-3.
Location: Coast Guard Institute, Oklahoma City, OK.
Length: Maximum, 156 weeks.
Exhibit Dates: 2/90–Present.
Learning Outcomes: Before 3/90 see CG-0802-0012. Upon completion of the course, the

student will have basic knowledge of aircraft construction and operation, navigation, radio operation procedures, and survival.

Instruction: Instruction consists of three workbooks, section self-quizzes, and a proctored end-of-course examination.

Credit Recommendation: Credit is not recommended because of military-specific nature of the course (9/97).

CG-1704-0041

AVIATION SURVIVALMAN FIRST CLASS BY CORRESPONDENCE

Course Number: 0108-4.
Location: Coast Guard Institute, Oklahoma City, OK.
Length: Maximum, 156 weeks.
Exhibit Dates: 2/93–Present.
Learning Outcomes: Upon completion of the course, the student will be able to prepare, operate, inspect, and maintain aerial delivery systems, parachute systems, survival equipment, oxygen masks, safety belts, fire extinguishers, small arms, and aviation ordnance.
Instruction: This correspondence course includes topics related to aircraft safety procedures.
Credit Recommendation: Credit is not recommended because of the limited, specialized nature of the course (10/94).

CG-1704-0042

HH-60 FLIGHT MECHANIC BY CORRESPONDENCE

Course Number: 0483-2; 0483-1.
Location: Coast Guard Institute, Oklahoma City, OK.
Length: Maximum, 156 weeks.
Exhibit Dates: 7/91–Present.
Learning Outcomes: Students will have a basic knowledge of the HH-60 helicopter systems and ground operation. They will know how to check the helicopter prior to flight and assist the crew during rescue operations.
Instruction: Through correspondence, students learn basic principles of HH-60 helicopter systems, ground handling, engine operation, and duties during rescue hoist operation.
Credit Recommendation: Credit is not recommended because of the military-specific nature of the course (10/94).

CG-1704-0043

AVIATION SURVIVALMAN SECOND CLASS BY CORRESPONDENCE

Course Number: 208-5.
Location: Coast Guard Institute, Oklahoma City, OK.
Length: Maximum, 156 weeks.
Exhibit Dates: 1/94–Present.
Learning Outcomes: Before 1/94 see CG-1704-0018. Upon completion of the course, the student will be able to perform safety procedures related to aircraft servicing equipment, fire extinguisher systems, flight clothing, and rescue and emergency briefings. Students will also be able to identify the types of information contained in aircraft logbooks and to identify how to procure and stock certain equipment.
Instruction: This correspondence course includes topics related to aircraft safety procedures and aircraft record keeping.
Credit Recommendation: Credit is not recommended because of the limited, specialized nature of the course (10/94).

CG-1704-0044

HU-25A AIRFRAME, CLASS C

Course Number: None.
Location: Aviation Technical Training Center, Elizabeth City, NC.
Length: 2-3 weeks (86 hours).
Exhibit Dates: 1/90–Present.
Learning Outcomes: Upon completion of the course, the student will be familiar with the HU-25A airframe, subsystems, and engines in order to perform operation troubleshooting, inspection, and routine maintenance of the aircraft.
Instruction: Instruction provided in familiarization of the HU-25A, utilizing applicable inspection documents, drawings, and publications.
Credit Recommendation: In the lower-division baccalaureate/associate degree category, 1 semester hour airframe familiarization (10/94).

CG-1704-0045

HU-25A AVIONICS MAINTENANCE, CLASS C

Course Number: None.
Location: Aviation Technical Training Center, Elizabeth City, NC.
Length: 4 weeks (145-146 hours).
Exhibit Dates: 4/93–Present.
Learning Outcomes: Before 4/93 see CG-1704-0031. Upon completion of the course, the student will be able to operate, troubleshoot, inspect, and provide routine maintenance of the HU-25A avionic systems.
Instruction: Lectures, demonstrations, self-paced instruction, and laboratory exercises in avionic systems, including audio control, direction finder, VHF-AM transceiver, UHF transceiver, LORAN, and radar.
Credit Recommendation: In the lower-division baccalaureate/associate degree category, 3 semester hours in avionic systems maintenance (10/94).

CG-1704-0046

HU-25A ELECTRICAL, CLASS C

Course Number: None.
Location: Aviation Technical Training Center, Elizabeth City, NC.
Length: 4 weeks (144-145 hours).
Exhibit Dates: 1/90–Present.
Learning Outcomes: Upon completion of the course, the student will be able to inspect, maintain, troubleshoot, and repair the HU-25A aircraft electrical system.
Instruction: Lectures and hands-on experience in the operation, troubleshooting, inspection, and routine maintenance of the HU-25A electrical power generation, distribution, indicating and control systems, autopilot, speed control, air data, heading, flight guidance, operation and interface for cockpit management, inertial sensor, and area navigation systems.
Credit Recommendation: In the lower-division baccalaureate/associate degree category, 2 semester hours in airframe/electrical familiarization (10/94).

CG-1704-0047

AVIATION STRUCTURAL MECHANIC, CLASS A

Course Number: None.
Location: Aviation Technical Training Center, Elizabeth City, NC.
Length: 19 weeks (631 hours).
Exhibit Dates: 2/90–Present.
Learning Outcomes: Before 2/90 see CG-1704-0027. Upon completion of the course, the student will be able to perform basic tasks pertaining to inspection, service, and maintenance of aircraft and related systems. An introduction to aircraft handling, signaling, and aerodynamics is provided.
Instruction: Course is conducted through a combination of lectures and hands-on experience. The student will handle, service, inspect, and maintain aircraft fuselages, wings, and control surfaces; fabricate and assemble metal parts; make repairs to aircraft skins; spray paint and maintain paint equipment; perform nondestructive testing inspections using visual dye and fluorescent penetrant methods; maintain hydraulic systems, landing gear wheels, and tires; perform cable swaging operations; identify and remove corrosion from aircraft and components; and make repairs to composite structures.
Credit Recommendation: In the lower-division baccalaureate/associate degree category, 3 semester hours in aircraft structural repair, 3 in aircraft corrosion control, 2 in aircraft ground handling and servicing, 1 in principles of hydraulics, 1 in shop safety, and 1 in maintenance record keeping practices and procedures (10/94).

CG-1704-0048

HH-60J AIRFRAME AND POWER TRAIN MAINTENANCE, CLASS C

Course Number: None.
Location: Aviation Technical Training Center, Elizabeth City, NC.
Length: 3 weeks (105-106 hours).
Exhibit Dates: 8/93–Present.
Learning Outcomes: Upon completion of the course, the student will be able to locate, maintain, and troubleshoot airframe and power train components on the HH-60J aircraft.
Instruction: Instruction is provided through practical instruction and includes component location and function, systems operation, airframe inspections, maintenance procedures, flight control rigging and adjustment, and troubleshooting of system drive train malfunctions.
Credit Recommendation: In the lower-division baccalaureate/associate degree category, 1 semester hour in airframe familiarization (10/94).

CG-1704-0049

HH-65A ELECTRICAL/AUTOMATIC FLIGHT CONTROL SYSTEMS MAINTENANCE, CLASS C

Course Number: None.
Location: Aviation Technical Training Center, Elizabeth City, NC.
Length: 5 weeks (184 hours).
Exhibit Dates: 8/93–Present.
Learning Outcomes: Upon completion of the course, the student will be able to operate, troubleshoot, inspect, and provide routine maintenance on the HH-65A electrical and automatic flight control systems through the use of inspection documents, drawings, publications, and test equipment.
Instruction: Course includes lectures, demonstrations, and computer-assisted instruction on the electrical systems of the HH-65A air-

craft. Included are the fuel, hydraulic, landing gear, and flight control electrical systems.

Credit Recommendation: In the lower-division baccalaureate/associate degree category, 3 semester hours in aircraft electrical systems (10/94).

CG-1704-0050

HH-65A AIRFRAME/POWER TRAIN MAINTENANCE, CLASS C

Course Number: None.
Location: Aviation Technical Training Center, Elizabeth City, NC.
Length: 3-4 weeks.
Exhibit Dates: 4/93–Present.
Learning Outcomes: Upon completion of the course, the student will be able to locate, maintain, and troubleshoot airframe and power train components on the HH-65A aircraft.
Instruction: Instruction is provided through lecture, demonstration, and exercises and includes component and system operation, inspections, removal and installation, and troubleshooting of system malfunctions.
Credit Recommendation: In the lower-division baccalaureate/associate degree category, 1 semester hour in assembly and rigging (10/94).

CG-1704-0051

HH-60J ELECTRICAL/AUTOMATIC FLIGHT CONTROL SYSTEMS MAINTENANCE, CLASS C

Course Number: None.
Location: Aviation Technical Training Center, Elizabeth City, NC.
Length: 5-6 weeks (196 hours).
Exhibit Dates: 4/93–Present.
Learning Outcomes: Upon completion of the course, the student will be able to operate, troubleshoot, inspect, and provide routine maintenance on the HH-60J electrical systems through the use of inspection documents, drawings, publications, and test equipment.
Instruction: Instruction includes lectures, demonstrations, and computer-assisted instruction on the electrical systems of the HH-60J aircraft. Included are electrical systems of the engine, starting, fuel, hydraulic, landing gear, blade de-ice, engine, and lighting systems.
Credit Recommendation: In the lower-division baccalaureate/associate degree category, 3 semester hours in aircraft electrical systems (10/94).

CG-1704-0052

AVIATION MACHINIST'S MATE FIRST CLASS BY CORRESPONDENCE

Course Number: 0105-7; 0105-8.
Location: Coast Guard Institute, Oklahoma City, OK.
Length: Maximum, 52 weeks.
Exhibit Dates: 8/90–Present.
Learning Outcomes: Before 8/90 see CG-1704-0008. Upon completion of the course, the student will be able to properly utilize related maintenance forms and records necessary to apply general propeller maintenance procedures.
Instruction: Topics include the Technical Order (T.O.) and related forms; records; maintenance publications; and the operating principles, components, and electrical systems of the 43D50 and 54H60 propellers.

Credit Recommendation: In the lower-division baccalaureate/associate degree category, 1 semester hour in maintenance record-keeping (10/94).

CG-1708-0004

BOATSWAIN'S MATE FIRST CLASS BY CORRESPONDENCE

Course Number: Version 1: 0109-8; 0109-7. Version 2: 109-6.
Location: Coast Guard Institute, Oklahoma City, OK.
Length: Version 1: Maximum, 156 weeks. Version 2: Maximum, 104 weeks.
Exhibit Dates: Version 1: 8/94–Present. Version 2: 1/90–7/94.
Learning Outcomes: Version 1: Upon completion of the course, the student will, through reading and studying textual materials, be capable of solving sample problems, answering lesson questions, and passing the proctored end-of-course examination. Version 2: To provide training to boatswain's mates at the second class level through reading and studying textual material, solving sample problems, answering lesson questions, and passing the proctored end-of-course examination.
Instruction: Version 1: Topics include basic search and rescue, shipboard damage control and stability, the duties of members of a 5/38 gun mount team, electronic navigation and maneuvering board navigation with emphasis on the use of relative motion radar, and small-unit administrative paperwork management. Students are required to successfully complete a proctored end-of-course examination. Version 2: Topics include basic search and rescue, shipboard damage control and stability, the duties of members of a 5/38 gun mount team, electronic navigation and maneuvering board navigation with emphasis on the use of relative motion radar, and small unit administrative paperwork management. Students are required to successfully complete a proctored end-of-course examination.
Credit Recommendation: Version 1: In the vocational certificate category, 1 semester hour in technical communications (4/98); in the lower-division baccalaureate/associate degree category, 2 in advanced navigation and 1 in advanced seamanship (4/98). Version 2: In the vocational certificate category, 1 semester hour in technical communications (1/88); in the lower-division baccalaureate/associate degree category, 2 semester hours in advanced navigation and 1 in advanced seamanship (1/88).

CG-1708-0005

NAVIGATION RULES BY CORRESPONDENCE

Course Number: 469-2.
Location: Coast Guard Institute, Oklahoma City, OK.
Length: Maximum, 104 weeks.
Exhibit Dates: 1/90–11/96.
Objectives: To provide training on the international regulations for preventing collisions at sea dated 1972 (72 COLREGS) and the United States Inland Rules of the Road.
Instruction: A nonresident training course that includes steering and sailing rules, lights and shapes, sound and light signals, inland rules and pilot rules. There is a proctored end-of-course examination.

Credit Recommendation: In the lower-division baccalaureate/associate degree category, 2 semester hours in navigation (5/85).

CG-1708-0006

PILOTING NAVIGATION BY CORRESPONDENCE

Course Number: 416-4.
Location: Coast Guard Institute, Oklahoma City, OK.
Length: Maximum, 104 weeks.
Exhibit Dates: 1/90–Present.
Objectives: To train personnel in basic navigation practices under piloting and dead reckoning conditions.
Instruction: Course introduces the student to navigation terminology and the use of applicable sections of the Defense Mapping Agency publication American Practical Navigator. The course sets forth the basic principles of determining a vessel's position by the use of bearings, soundings, and applicable tables. The use of sextant angles, marine radar, and compass bearings to determine position are covered. Course describes the basic principles and steps involved in setting up a piloting team in order to safely navigate a vessel. Students must pass proctored final exam.
Credit Recommendation: In the lower-division baccalaureate/associate degree category, 2 semester hours in basic navigation (10/91).

CG-1708-0008

COMPASS SYSTEMS BY CORRESPONDENCE

Course Number: 418-1.
Location: Coast Guard Institute, Oklahoma City, OK.
Length: Maximum, 104 weeks.
Exhibit Dates: 1/90–6/97.
Objectives: To provide basic knowledge of gyrocompasses and magnetic compasses and degaussing ranges and systems and their operation.
Instruction: Course provides a basic introduction to gyrocompass use, magnetic compass adjustment to reduce deviation, basic description of magnetism, causes of deviation, and descriptions of degaussing systems including ranges and settings.
Credit Recommendation: Credit is not recommended because of the military-specific nature of the course (5/85).

CG-1708-0009

SEAMAN BY CORRESPONDENCE

Course Number: 451-9.
Location: Coast Guard Institute, Oklahoma City, OK.
Length: Maximum, 104 weeks.
Exhibit Dates: 1/90–Present.
Objectives: To introduce personnel to the duties of seaman.
Instruction: Course includes the operation and maintenance of deck equipment, rigging, wire and line size determination, proper watchstanding procedures, boat lowering operations, life saving equipment, small boat mooring, and rating identification. Students are required to successfully complete a proctored end-of-course examination.
Credit Recommendation: In the vocational certificate category, 1 semester hour in basic seamanship (3/97).

CG-1708-0010

DECK SEAMANSHIP BY CORRESPONDENCE

Course Number: 433-1.
Location: Coast Guard Institute, Oklahoma City, OK.
Length: Maximum, 104 weeks.
Exhibit Dates: 1/90–6/97.
Objectives: To train Coast Guard personnel in deck seamanship.
Instruction: Course includes deck maintenance, including the use and maintenance of painting/coating application equipment and the planning, supervision, and control of shipboard painting operations; seamanship, including the use of wire and synthetic fiber rope, rope maintenance, and ground tackle maintenance; shipboard use of accommodation ladders, cargo nets, and scaffolding; cold weather maintenance of deck equipment and de-icing procedures. Also includes the operation and maintenance of deck machinery and the adjustment of standard rigging. Students are required to successfully complete a proctored end-of-course examination.
Credit Recommendation: In the vocational certificate category, 1 semester hour in seamanship and 1 in nautical science (5/85).

CG-1708-0011

NAVIGATION SYSTEMS BY CORRESPONDENCE

Course Number: 424-1.
Location: Coast Guard Institute, Oklahoma City, OK.
Length: Maximum, 104 weeks.
Exhibit Dates: 1/90–11/96.
Objectives: To provide an introduction to electronic navigation systems aboard ship.
Instruction: Course includes description and use of basic shipboard electronic navigation systems, including LORAN-C, NAVSAT, radio direction finding, OMEGA, DECCA, console, inertial navigation, acoustic Doppler, and bathymetric navigation. Students are required to successfully complete a proctored end-of-course examination.
Credit Recommendation: In the vocational certificate category, 1 semester hour in nautical science (5/85).

CG-1708-0012

WATCHSTANDING: THE CONNING OFFICER BY CORRESPONDENCE

Course Number: 430-1.
Location: Coast Guard Institute, Oklahoma City, OK.
Length: Maximum, 104 weeks.
Exhibit Dates: 1/90–Present.
Objectives: To introduce the fundamental duties, responsibilities, and procedures of the conning officer.
Instruction: Course includes theory and basic knowledge in the fundamentals of ship handling and general duties and related responsibilities of conning officers in restricted waters and in the open sea. Students are required to successfully complete a proctored end-of-course examination.
Credit Recommendation: Credit is not recommended because of the limited, specialized nature of the course (3/97).

CG-1708-0013

SHIPHANDLING BY CORRESPONDENCE

Course Number: 426-1.
Location: Coast Guard Institute, Oklahoma City, OK.
Length: Maximum, 104 weeks.
Exhibit Dates: 1/90–6/97.
Objectives: To provide basic knowledge of ship handling for single screw or multiple screw vessels in heavy weather, restricted, and unrestricted waters.
Instruction: Course is designed to develop basic knowledge and skills in physical characteristics of ships, maneuvering characteristics, ship handling fundamentals, heavy weather handling and survival, small boat operation in surf, and anchoring, mooring, and towing procedures. Students are required to successfully complete a proctored end-of-course examination.
Credit Recommendation: In the vocational certificate category, 1 semester hour in basic seamanship (5/85).

CG-1708-0014

DECK WATCH OFFICER NAVIGATION RULES BY CORRESPONDENCE

Course Number: 631-1.
Location: Coast Guard Institute, Oklahoma City, OK.
Length: Maximum, 104 weeks.
Exhibit Dates: 1/90–6/97.
Objectives: To provide training and familiarization for students in prevention of collisions at sea.
Instruction: Topics include rules of the road encompassing all vessels, both power and sail, lights and shapes, and sound and light signals in inland and international waters. Students are required to successfully complete a proctored end-of-course examination.
Credit Recommendation: In the lower-division baccalaureate/associate degree category, 3 semester hours in nautical science (5/85).

CG-1708-0015

MANEUVERING BOARDS BY CORRESPONDENCE

Course Number: 0464-2; 464-1.
Location: Coast Guard Institute, Oklahoma City, OK.
Length: Maximum, 104 weeks.
Exhibit Dates: 1/90–Present.
Objectives: To provide training in solving maneuvering board problems.
Instruction: This course includes reading assignments and solving of maneuvering board problems, including relative motion, course, speed, closest point of approach, danger of collision, station taking, contact interception, true/relative wind vectors, and direct plotting methods.
Credit Recommendation: In the lower-division baccalaureate/associate degree category, 1 semester hour in nautical science (7/92).

CG-1708-0016

BOATSWAIN'S MATE, CLASS A

Course Number: *Version 1:* BMA. *Version 2:* BMA.
Location: Reserve Training Center, Yorktown, VA.
Length: *Version 1:* 7 weeks (255-260 hours). *Version 2:* 6-7 weeks (262-264 hours).
Exhibit Dates: *Version 1:* 3/96–Present. *Version 2:* 1/90–2/96.
Learning Outcomes: *Version 1:* Upon completion of the course, the student will be able to demonstrate personal physical fitness; use equipment found in a boat crew personal survival kit; deploy and board rescue rafts using a boatswain's chair; handle lines; splice lines; demonstrate anchoring and weighing anchor; identify equipment; perform the duties of a lookout; moor and unmoor; and stand a helm watch onboard a 41 foot UTB. Student will apply knowledge of rules of the road using a fathometer, radar, Polaris, GPS, and Loran-C; recover a person overboard; participate in helicopter operations; use a nautical chart; tie standard knots and know their uses; throw a heaving line; operate a CG-PL pump and an eductor; identify the classes of fire and the extinguishing agent for each; operate a carbon dioxide fire extinguisher and a dry chemical extinguisher; demonstrate CPR; repair and paint various types of surfaces; maintain records and logs in accordance with the latest Coast Guard instructions; and use a boatswain's pipe for standard calls. *Version 2:* Upon completion of the course, the student will be able to demonstrate personal physical fitness; demonstrate correct method to enter water wearing anti-exposure clothing; identify and use equipment found in a boat crew personal survival kit; deploy and board a four person rescue raft; set up and hoist off the deck using stages and a boatswain's chair; properly handle lines while acting as a member of a boat crew; splice double braid and three-strand lines; anchor weighing; state the duties of the boatswain's mate of the watch onboard a vessel underway; identify the meanings of various nautical terms; identify equipment found onboard a 41 foot UTB; perform the duties of a lookout onboard a 41 foot UTB; perform as part of the crew of a 41 foot UTB while towing another vessel astern; moor and unmoor a 41 foot UTB at a pier; stand a helm watch on board a 41 foot UTB; identify required lights and shapes in accordance with the rules of the road; determine the depth of water using a fathometer; use information from the Raynav 66 radar; use a VHF-FM Polaris to locate another vessel; use Loran-C to fix a vessel's position; recover a person overboard using a 41 foot UTB; participate in helicopter operations; use a nautical chart; tie standard knots and know their uses; throw a heaving line; operate a CG-PL pump and an eductor; identify the classes of fire and the extinguishing agent for each; operate a carbon dioxide fire extinguisher and a dry chemical extinguisher; demonstrate CPR; repair and paint various types of surfaces; maintain records and logs in accordance with the latest Coast Guard instructions; and use a boatswain's pipe for standard calls.
Instruction: *Version 1:* Lectures and practical exercises cover physical fitness and survival, line handling and ground tackle, underway operations and watches, rescue and evacuation procedures, navigation and piloting, boat communications, salvage operations, fire fighting, first aid, and leadership. *Version 2:* Lectures and practical exercises cover physical fitness and survival, line handling and ground tackle, underway operations and watches, rescue and evacuation procedures, navigation and piloting, boat communications, salvage operations, fire fighting, first aid, and leadership.
Credit Recommendation: *Version 1:* In the vocational certificate category, 4 semester hours in seamanship (5/96); in the lower-division baccalaureate/associate degree category, 2 semes-

ter hours in navigation (5/96). *Version 2:* In the vocational certificate category, 10 semester hours in seamanship (2/96); in the lower-division baccalaureate/associate degree category, 8 semester hours in marine technology (2/96).

CG-1708-0017

COXSWAIN, CLASS C

Course Number: *Version 1:* None. *Version 2:* COX C.
Location: Reserve Training Center, Yorktown, VA.
Length: *Version 1:* 4 weeks (154 hours). *Version 2:* 4-5 weeks (154 hours).
Exhibit Dates: *Version 1:* 6/94–Present. *Version 2:* 1/90–5/94.
Learning Outcomes: *Version 1:* Upon completion of the course, the student will be able to perform as a coxswain aboard standard and nonstandard and Coast Guard utility boats. Duties include navigation and piloting, search and rescue, boat handling, and towing. *Version 2:* Upon completion of the course, the student will be able to qualify as a competent coxswain on the 41-ft. UTB; this includes a thorough knowledge of small boat handling, navigation and piloting, search and rescue, construction and design, stability, rules of the road, and towing and engineering casualty drills.
Instruction: *Version 1:* Methods of instruction include lectures, case studies, evaluations, reviews, self-instruction, written exams, and demonstrations. Topics include coxswain's authority and responsibilities, boat familiarization, basic piloting and navigation, rescue and assistance, search patterns, and engineering casualties. *Version 2:* Instruction is provided by lectures, practical demonstrations, and on-the-water experience in small boat handling, chart navigation, rules of road, safety and rescue techniques, and piloting and coastwise navigation.
Credit Recommendation: *Version 1:* In the lower-division baccalaureate/associate degree category, 3 semester hours in small boat handling and 1 in navigation (5/96). *Version 2:* In the lower-division baccalaureate/associate degree category, 3 semester hours in boat handling and 1 in navigation (4/89).

CG-1708-0018

QUARTERMASTER, CLASS A

Course Number: *Version 1:* None. *Version 2:* None. *Version 3:* None.
Location: Reserve Training Center, Yorktown, VA.
Length: *Version 1:* 12 weeks (392 hours). *Version 2:* 8 weeks (280 hours). *Version 3:* 12 weeks (420 hours).
Exhibit Dates: *Version 1:* 2/94–Present. *Version 2:* 3/93–1/94. *Version 3:* 1/90–2/93.
Learning Outcomes: *Version 1:* Upon completion of the course, the student will be able to apply knowledge of international Morse code, rules of the road, navigation, and plotting. *Version 2:* Upon completion of the course, the student will be able to navigate, operate signaling equipment, send and receive Morse code at eight words per minute, apply rules of the road, plot, and apply communications and navigational information while afloat. *Version 3:* Upon completion of the course, the student will be able to perform tasks in computation of navigational data, signaling and navigation equipment, and the application of communications and navigational information in organizational and intermediate levels of navigation afloat.
Instruction: *Version 1:* Instruction is provided by lectures, role play, demonstrations, practical exercises, evaluations, and review. Topics included are petty officer duties, communications, voyage preparation, vessel movement, record keeping, helmsmanship, and quartermaster of the watch. *Version 2:* Methods of instruction include lectures, practical exercises, and laboratory. Topics included are administration; communication; communications procedures; and navigation systems, including tides and currents, plotting, sun abstractions and time, navigation rules, compasses, and watchstanding. *Version 3:* Instruction is provided by lectures and practical exercises in administration, communications, communications procedures, navigation systems and instruments, navigation rules, compasses, and watchstanding.
Credit Recommendation: *Version 1:* In the lower-division baccalaureate/associate degree category, 6 semester hours in navigation and plotting (5/96). *Version 2:* In the lower-division baccalaureate/associate degree category, 4 semester hours in navigation and plotting (5/96). *Version 3:* In the vocational certificate category, 1 semester hour in navigation rules (rules of road) (4/89); in the lower-division baccalaureate/associate degree category, 6 semester hours in navigation (4/89).

CG-1708-0019

MARINE SAFETY INITIAL INDOCTRINATION LESSON PLAN SERIES BY CORRESPONDENCE
(Marine Safety Initial Indoctrination by Correspondence)

Course Number: 0575-1; 0575.
Location: Coast Guard Institute, Oklahoma City, OK.
Length: Maximum, 52 weeks.
Exhibit Dates: 1/90–Present.
Learning Outcomes: Upon completion of the course, the student will be able to demonstrate knowledge of laws, regulations, and directives, introduction to marine industry, terminology, ship design and construction, and safety.
Instruction: Instruction is provided through a nonresident, self-paced study program. Topics include marine safety applications in a variety of situations, including commercial vessel safety, port safety and security, and marine environmental response. Includes a proctored, comprehensive end-of-course examination.
Credit Recommendation: In the lower-division baccalaureate/associate degree category, 1 semester hour in fundamentals of marine safety and 1 in military science (security and safety) (9/90).

CG-1708-0020

NAVIGATION RULES BY CORRESPONDENCE

Course Number: *Version 1:* 0469-4. *Version 2:* 0469-3.
Location: Coast Guard Institute, Oklahoma City, OK.
Length: *Version 1:* Maximum, 156 weeks. *Version 2:* Maximum, 52 weeks.
Exhibit Dates: *Version 1:* 6/94–Present. *Version 2:* 1/90–5/94.
Learning Outcomes: *Version 1:* Upon completion of the course, the student will be able to demonstrate knowledge of inland and international rules of the road related to safe navigation in various situations, including potential collisions and limited visibility, sailing vessels, lights and shapes used on vessels, and sound signals. *Version 2:* Upon completion of the course, the student will be able to demonstrate knowledge of inland and international rules of the road related to safe navigation in various situations, including potential collisions and limited visibility, sailing vessels, lights and shapes used on vessels, and sound signals.
Instruction: *Version 1:* This course includes reading and studying two pamphlets covering introduction to navigation rules, steering and sailing rules, sound and light signals, and annexes to the rules. Includes a proctored, comprehensive end-of-course examination. *Version 2:* Includes reading and studying two pamphlets covering introduction to navigation rules, steering and sailing rules, sound and light signals, and annexes to the rules. Includes a proctored, comprehensive end-of-course examination.
Credit Recommendation: *Version 1:* In the lower-division baccalaureate/associate degree category, 2 semester hours in basic navigation (6/94). *Version 2:* In the lower-division baccalaureate/associate degree category, 2 semester hours in basic navigation (9/90).

CG-1708-0021

BOATSWAIN'S MATE SECOND CLASS BY CORRESPONDENCE

Course Number: *Version 1:* 0209-9. *Version 2:* 209-8.
Location: Coast Guard Institute, Oklahoma City, OK.
Length: *Version 1:* Maximum, 156 weeks. *Version 2:* Maximum, 52 weeks.
Exhibit Dates: *Version 1:* 4/94–Present. *Version 2:* 1/90–3/94.
Learning Outcomes: *Version 1:* Upon completion of the course, the student will demonstrate proficiency in administration, communication, and ordnance. *Version 2:* Upon completion of the course, the student will be able to demonstrate proficiency in deck seamanship, particularly as it relates to selecting surface coating and splicing wire cable; LORAN operation; maneuvering small craft in search and rescue operations; maintaining fire fighting equipment; identifying aids to navigation; and selecting electrical systems.
Instruction: *Version 1:* The course consists of a correspondence course booklet and end of course test, provided in an non-resident self-study format. Subject material is specifically designed for BM3s as a required component for advancement to BM2. Topics covered include requisition forms; inventory at personal property; report of survey; abstract of operations report; charts and publication corrections; chart ordering; unit directives system; SITREP; small arms maintenance; small arms ammunition; pyrotechnics; instructor for a boat crewman; and performance evaluation. *Version 2:* Methodology includes reading, studying, and responding to questions from a pamphlet covering four subject areas. Areas of instruction include deck seamanship, navigation and maneuvering, damage control, and search and rescue operations. A proctored exam is administered at the end of the course.
Credit Recommendation: *Version 1:* Credit is not recommended because of the limited, specialized nature of the course (5/95). *Version 2:* In the lower-division baccalaureate/associate

degree category, 2 semester hours in advanced seamanship (9/90).

CG-1708-0022
BOATSWAIN'S MATE THIRD CLASS BY CORRESPONDENCE

Course Number: *Version 1:* 0309-4. *Version 2:* 0309-3; 0309-2.
Location: Coast Guard Institute, Oklahoma City, OK.
Length: *Version 1:* Maximum, 156 weeks. *Version 2:* Maximum, 78 weeks.
Exhibit Dates: *Version 1:* 2/97–Present. *Version 2:* 1/90–1/97.
Learning Outcomes: *Version 1:* This version is pending evaluation. *Version 2:* Upon completion of the course, the student will be familiar with boat seamanship, navigation and piloting, boat operations, deck seamanship, law enforcement, search and rescue, boat/helicopter operations, damage control, administration, watchstanding, and communications.
Instruction: *Version 1:* This version is pending evaluation. *Version 2:* Methodology includes reading and studying 14 pamphlets covering navigation, seamanship, administration, piloting and boat operations, watchstanding and communications, law enforcement, pyrotechnics, and damage control. Includes a proctored, comprehensive end-of-course examination. After 9/92 this course includes no instruction in helicopter operations, underway operations, seaman's eye navigation, anchoring, or fire fighting.
Credit Recommendation: *Version 1:* Pending evaluation. *Version 2:* In the lower-division baccalaureate/associate degree category, 2 semester hours in seamanship and 2 in navigation (9/92).

CG-1708-0023
HC-130H NAVIGATOR BASIC, CLASS C

Course Number: None.
Location: Aviation Technical Training Center, Elizabeth City, NC.
Length: 4 weeks (143 hours).
Exhibit Dates: 3/93–Present.
Learning Outcomes: Upon completion of the course, the student will be familiar with the equipment and tools used in navigation of aircraft and will be able to perform simple navigation steps required prior to flight and during flight.
Instruction: Instruction includes lectures, demonstrations, and practical exercises in determination of course and position. Dead reckoning under usual and instrument rules is addressed with reference to appropriate flight publications and use of electronic aids to navigation. The student is introduced to time and distance conversions and principles involved in the determination of work angle and velocity and the affects of wind on aircraft during flight.
Credit Recommendation: In the lower-division baccalaureate/associate degree category, 3 semester hours in navigation principles (10/94).

CG-1708-0024
MARINE SAFETY INITIAL INDOCTRINATION MARINE INSPECTION BY CORRESPONDENCE

Course Number: 0580-1.
Location: Coast Guard Institute, Oklahoma City, OK.
Length: Maximum, 156 weeks.
Exhibit Dates: 10/94–Present.
Learning Outcomes: Upon completion of the course, the student will be able to attend marine safety inspectors entry-level resident training and will have basic knowledge of marine safety programs and organization; marine safety authorities, references, and standards; and marine inspection.
Instruction: This is a correspondence course covering marine safety programs, marine safety authorities, and marine inspection, with reading and writing assignments. A proctored final examination is required.
Credit Recommendation: In the upper-division baccalaureate category, 2 semester hours in marine transportation: vessels (5/95).

CG-1708-0025
MARINE SAFETY INITIAL INDOCTRINATION PORT OPERATIONS BY CORRESPONDENCE

Course Number: 0585-1.
Location: Coast Guard Institute, Oklahoma City, OK.
Length: Maximum, 156 weeks.
Exhibit Dates: 10/94–Present.
Learning Outcomes: Upon completion of the course, the student will be able to attend all marine safety entry-level resident training courses and will have a basic knowledge of marine safety programs and organization; marine safety authorities, references, and standards; port safety and security; and marine environmental problems.
Instruction: This is a correspondence course covering marine safety programs; marine safety authorities; port safety; and marine protection. It is conducted through reading and writing assignments. A proctored final examination is required.
Credit Recommendation: In the upper-division baccalaureate category, 2 semester hours marine transportation: ports (5/95).

CG-1709-0001
PUBLIC AFFAIRS SPECIALIST SECOND CLASS BY CORRESPONDENCE

Course Number: 236-4.
Location: Coast Guard Institute, Oklahoma City, OK.
Length: Maximum, 52 weeks.
Exhibit Dates: 1/90–Present.
Learning Outcomes: Upon completion of the course, the student will be able to produce print materials for internal publications and public media.
Instruction: This course covers introductory knowledge in the areas of newswriting, photography, editing, and layout and public affairs operations. The training is conducted through self-instruction format with a journalism text, public affairs manual, and student workbook as resource materials. There are seven self-administered tests for students to measure progress and an end-of-course proctored examination for certifying subject mastery.
Credit Recommendation: In the lower-division baccalaureate/associate degree category, 1 semester hour in introduction to public relations (6/91).

CG-1709-0002
PUBLIC AFFAIRS SPECIALIST FIRST CLASS BY CORRESPONDENCE

Course Number: 0136-2.
Location: Coast Guard Institute, Oklahoma City, OK.
Length: Maximum, 78 weeks.
Exhibit Dates: 1/90–Present.
Learning Outcomes: Upon completion of the course, the student will be able to use the skills, principles, and techniques of community, internal, and public relations, and journalism required by the rating.
Instruction: Training is provided in a non-resident, self-instruction format, with reading assignments, review exercises, and a proctored end-of-course examination. Topics include theory of public relations and public affairs management.
Credit Recommendation: In the lower-division baccalaureate/associate degree category, 2 semester hours in introduction to public relations (6/91).

CG-1710-0008
FIRE TUBE AND BOILER/FLASH TYPE EVAPORATOR, CLASS C

Course Number: MK-5.
Location: Reserve Training Center, Yorktown, VA.
Length: 2 weeks (63 hours).
Exhibit Dates: 1/90–11/96.
Objectives: To train engineers in the operation of fire tube boilers and flash evaporators.
Instruction: Lectures and laboratory experience in the theory, maintenance, and operation of boilers and evaporators.
Credit Recommendation: In the vocational certificate category, 3 semester hours in boiler operation, maintenance, and repair (7/84).

CG-1710-0011
DAMAGE CONTROL DECK GROUP RATINGS BY CORRESPONDENCE

Course Number: 477-1.
Location: Coast Guard Institute, Oklahoma City, OK.
Length: Maximum, 104 weeks.
Exhibit Dates: 1/90–11/96.
Objectives: To provide training in damage control including organization, fire fighting, and watertight integrity.
Instruction: Course provides limited coverage of damage control organization, shipboard fires, battle damage repair procedures, compartment testing and inspection, and water tight integrity determinations. Students are required to successfully complete a proctored end-of-course examination.
Credit Recommendation: Credit is not recommended because of the limited, specialized nature of the course (5/85).

CG-1710-0012
HIGH RELIABILITY SOLDERING, CLASS C

Course Number: None.
Location: Aviation Technical Training Center, Elizabeth City, NC.
Length: 2 weeks (71 hours).
Exhibit Dates: 1/90–11/96.
Objectives: To provide specialized training in soldering using the PRC-151 soldering unit.
Instruction: Practical training in the use and maintenance of the PRC-151 power soldering system.
Credit Recommendation: In the lower-division baccalaureate/associate degree cate-

gory, 1 semester hour in high reliability soldering laboratory practices (6/85).

CG-1710-0013

1. 270' WMEC Machinery Plant Control and Monitoring System, Class C
2. 270' WMEC Machinery Plant Control and Monitoring System
3. 270' WMEC Main Propulsion Control and Monitoring System (Electrical) Operation and Maintenance, Class C

Course Number: *Version 1:* EM-25. *Version 2:* EM-25. *Version 3:* EM-25.

Location: Reserve Training Center, Yorktown, VA.

Length: *Version 1:* 4 weeks (144 hours). *Version 2:* 4 weeks (140-148 hours). *Version 3:* 3 weeks (110-112 hours).

Exhibit Dates: *Version 1:* 5/94–Present. *Version 2:* 10/90–4/94. *Version 3:* 1/90–9/90.

Learning Outcomes: *Version 1:* Upon completion of the course, the student will be trained on propulsion system control, control console, and control software; and will be able to operate, maintain, troubleshoot, and calibrate instrumentation used as input for the control system data processor. *Version 2:* Upon completion of the course, the student will be trained on propulsion system control, control console, and control software and will be able to make necessary repairs and calibrate instrumentation used as input for the control system data processor. *Version 3:* Upon completion of the course, the student will be trained on propulsion system control, control console and control software, and will be able to make necessary repairs.

Instruction: *Version 1:* Lectures and laboratory experiences cover electrical engine control systems, propeller pitch control, monitoring systems, and systems troubleshooting. In addition, the course covers calibration of transducers, thermistors, and running of diagnostic tapes to troubleshoot the data processor. *Version 2:* Lectures and laboratory experiences cover electrical diesel engine systems, propeller pitch control, monitoring systems, and systems troubleshooting. In addition, the course covers calibration of transducers, thermistors, and running of diagnostic tapes to troubleshoot the data processor. *Version 3:* Lectures and laboratory experiences cover electrical diesel engine systems, propeller pitch control, monitoring systems, and systems troubleshooting.

Credit Recommendation: *Version 1:* In the lower-division baccalaureate/associate degree category, 3 semester hours in marine propulsion electrical control maintenance (5/96). *Version 2:* In the lower-division baccalaureate/associate degree category, 3 semester hours in marine propulsion electrical control maintenance (8/90). *Version 3:* In the lower-division baccalaureate/associate degree category, 2 semester hours in marine propulsion electrical control maintenance (10/88).

CG-1710-0014

Damage Controlman, Class A

Course Number: *Version 1:* DCA. *Version 2:* DCA.

Location: Reserve Training Center, Yorktown, VA.

Length: *Version 1:* 13 weeks (446 hours). *Version 2:* 13 weeks (450 hours).

Exhibit Dates: *Version 1:* 1/95–Present. *Version 2:* 1/90–12/94.

Learning Outcomes: *Version 1:* Upon completion of the course, the student will be able to demonstrate the use of hand and power woodworking and carpentry tools, perform rough construction tasks, and describe and use pipefitting and plumbing tools. Student will be able to use oxyacetylene equipment for cutting and welding, observing all safety procedures, and will be able to explain various types of fire fighting equipment and breathing apparatus. *Version 2:* Upon completion of the course, the student will be able to demonstrate the use of hand and power woodworking tools, perform rough carpentry tasks, and describe and use pipefitting and sheet metal tools. Student will be able to use oxyacetylene equipment for cutting and welding, observing all safety procedures, and will be able to explain various types of fire fighting equipment and breathing apparatus.

Instruction: *Version 1:* Methodology includes lectures, demonstrations, and laboratories in the use of woodworking tools and equipment; rough carpentry; concrete forming; pipefitting; damage control; and oxyacetylene and electric arc cutting, brazing, and welding. Shipboard fire fighting and hazardous material handling are covered. *Version 2:* Methodology includes lectures, demonstrations, and laboratories in the use of woodworking tools and equipment; rough carpentry; concrete forming; pipefitting and sheet metal work; oxyacetylene and electric arc cutting, brazing, and welding. Shipboard fire fighting and hazardous material handling are covered.

Credit Recommendation: *Version 1:* In the lower-division baccalaureate/associate degree category, 2 semester hours in building construction technology; 2 in oxyacetylene welding; 2 in electric arc welding; 2 in plumbing pipefitting; and 2 in shipboard fire fighting, damage control, and hazardous material handling (5/96). *Version 2:* In the lower-division baccalaureate/associate degree category, 2 semester hours in building construction technology, 2 in oxyacetylene welding, 3 in electric arc welding, and 1 in shipboard fire fighting and hazardous material handling (10/88).

CG-1710-0015

1. Damage Controlman Refresher Reserve Active Duty for Training, Class C
2. Damage Controlman Advanced Reserve Active Duty for Training, Class C

Course Number: *Version 1:* DC ADV(R). *Version 2:* DC ADV(R).

Location: Reserve Training Center, Yorktown, VA.

Length: *Version 1:* 1-2 weeks (69-72 hours). *Version 2:* 1-2 weeks (69-72 hours).

Exhibit Dates: *Version 1:* 4/92–1/95. *Version 2:* 1/90–3/92.

Learning Outcomes: *Version 1:* Upon completion of the course, the student will be able to erect a light-frame structure, determine the building materials required, show the proper use of plumbing tools, perform fire fighting exercises, and show damage control skills. The student will be able to describe the testing and handling of various hazardous materials. *Version 2:* Upon completion of the course, the student will be able to erect a light-frame structure, determine the building materials required, show the proper use of plumbing tools, and perform visual and dye-penetrant weld tests. The student will be able to describe the testing and handling of various hazardous materials.

Instruction: *Version 1:* Lectures, demonstrations, and practical exercises cover light-frame structures, including construction and estimating; firefighting and damage control; and testing and handling of hazardous materials. *Version 2:* Lectures, demonstrations, and practical exercises cover light-frame structures, including construction and estimating; visual and dye-penetrant weld tests; testing and handling of hazardous materials.

Credit Recommendation: *Version 1:* Credit is not recommended because of limited content in any one subject area (5/96). *Version 2:* In the lower-division baccalaureate/associate degree category, 1 semester hour in building construction estimating (10/88).

CG-1710-0016

Aluminum Welding, Class C

Course Number: DC-2.

Location: Reserve Training Center, Yorktown, VA.

Length: 4 weeks (144-148 hours).

Exhibit Dates: 1/90–Present.

Learning Outcomes: Upon completion of the course, the student will be able to describe, set up, and complete various welding projects using both the gas and metal arc and the gas tungsten arc processes and equipment, observing all safety procedures.

Instruction: Lectures, demonstrations, and practical exercises in the use and operation of the gas metal arc and gas tungsten arc welding processes.

Credit Recommendation: In the lower-division baccalaureate/associate degree category, 2 semester hours in gas metal arc welding and 2 in gas tungsten arc welding (1/96).

CG-1710-0017

Steel Welding, Class C

Course Number: DC-1.

Location: Reserve Training Center, Yorktown, VA.

Length: 3 weeks (84 hours).

Exhibit Dates: 1/90–Present.

Learning Outcomes: Upon completion of the course, the student will be able to describe, set up, and complete various welding projects using the shielded metal arc process and equipment, observing all safety procedure; and demonstrate the procedures to perform nondestructive testing and air carbon arc cutting.

Instruction: Course includes lectures, demonstrations, and practical exercises in shielded metal arc welding, nondestructive testing, and air carbon arc cutting.

Credit Recommendation: In the lower-division baccalaureate/associate degree category, 2 semester hours in shielded metal arc welding (1/96).

CG-1710-0018

Pratt and Whitney FT4A Gas Turbine

Course Number: MK-4.

Location: Reserve Training Center, Yorktown, VA.

Length: 2 weeks (70-74 hours).

Exhibit Dates: 1/90–Present.

Objectives: Upon completion of the course, the student will be able to operate, maintain, and troubleshoot malfunctions in a Pratt and Whitney FT4A gas turbine engine of 18,000 hp; and will demonstrate inspection techniques and the use of inspection instruments.

Instruction: Course includes lectures concentrating on gas turbine lubrication, speed control, and burner and component inspection. The manufacturer's maintenance procedures are studied and practiced in a laboratory setting. There is some instruction in troubleshooting and some in malfunction analysis.

Credit Recommendation: In the lower-division baccalaureate/associate degree category, 1 semester hour in gas turbine laboratory (2/96).

CG-1710-0019

DAMAGE CONTROLMAN THIRD CLASS BY CORRESPONDENCE

Course Number: *Version 1:* 0315-8. *Version 2:* 315-7.
Location: Coast Guard Institute, Oklahoma City, OK.
Length: *Version 1:* Maximum, 156 weeks. *Version 2:* Maximum, 78 weeks.
Exhibit Dates: *Version 1:* 7/95–Present. *Version 2:* 1/90–6/95.
Learning Outcomes: *Version 1:* Upon completion of the course, the student will be able to read basic construction blueprints and identify components of fire fighting, flood control, and emergency repair equipment; and to understand principles of basic carpentry, basic plumbing, and basic masonry techniques. Course also included ships' handling and stowage of DC hazardous material; repair of roofing; repair of cabinets, interior wall sheathing, and ceilings; fastener; floor, wall, roof, and stair framing; compartment checkoff sitss; damage control diagrams; and joining of plastic pipes. *Version 2:* Upon completion of the course, the student will be able to read basic construction blueprints and identify components of fire fighting, flood control, and emergency repair equipment; and to understand principles of basic carpentry, basic plumbing, and basic masonry techniques.
Instruction: *Version 1:* Topics include general safety, measuring devices, safety, anchoring, blueprints and diagrams, carpentry, masonry, plumbing, sewage, welding, cutting, allied processes, damage control, fire fighting, and CBR (Chemical, Biological, Radiological) warfare, shielded arc welding and oxyacetylene cutting and brazing, and administration. Training is provided in a nonresident, self-instruction format. Students are required to successfully complete a proctored end-of-course examination. *Version 2.* Topics include general safety, measuring devices, blueprints and diagrams, carpentry, masonry, plumbing, sewage, welding, cutting, allied processes, damage control, fire fighting, and CBR (Chemical, Biological, Radiological) defense, and administration. Training is provided in a nonresident, self-instruction format. Students are required to successfully complete a proctored end-of-course examination.
Credit Recommendation: *Version 1:* In the vocational certificate category, 2 semester hours in building maintenance and 1 in welding technology (3/96). *Version 2:* In the vocational certificate category, 2 semester hours in building maintenance and 1 in welding technology (6/90).

CG-1710-0020

DAMAGE CONTROLMAN FIRST CLASS BY CORRESPONDENCE

Course Number: 0115-5; 0115-6.
Location: Coast Guard Institute, Oklahoma City, OK.
Length: Maximum, 52 weeks.
Exhibit Dates: 1/90–Present.
Learning Outcomes: Upon completion of the course, the student will be able to identify methods of light wood frame construction, masonry construction procedures, aluminum welding GTA and MTA procedures, safety precautions, duties of damage control repair party leaders, components of Chemical Hazard Response Initiation System, and procedures used for setup and operation of a chemical/biological/radiological decontamination station.
Instruction: Topics include carpentry, masonry, plumbing, welding, cutting, allied processes, damage control, fire fighting, CBR defense, and administration. Training is provided in a nonresident, self-instruction format. Students are required to successfully complete a proctored end-of-course examination.
Credit Recommendation: In the lower-division baccalaureate/associate degree category, 1 semester hour in construction technology and 1 in aluminum welding technology (6/90).

CG-1710-0021

DAMAGE CONTROLMAN SECOND CLASS BY CORRESPONDENCE

Course Number: 0215-7; 0215-8.
Location: Coast Guard Institute, Oklahoma City, OK.
Length: Maximum, 52 weeks.
Exhibit Dates: 1/90–Present.
Learning Outcomes: Upon completion of the course, the student will be able to identify wood framing systems and structural members; identify roofing materials and installation procedures, various types of pipe and pipe fittings, copper tube and tube fittings; identify various welding, architectural, piping, and damage control symbols and diagrams; and identify actions taken by scene leader of a damage control party at a fire.
Instruction: Topics include framing and roofing buildings; finishing exterior walls; installing acoustical tile ceilings; plumbing pipe, valves, and fixtures; welding castings; and reading blueprint and diagrams. Training is provided in a nonresident, self-instruction format. Students are required to successfully complete a proctored end-of-course examination.
Credit Recommendation: In the vocational certificate category, 1 semester hour in basic welding and 2 in carpentry and plumbing or 3 in basic construction trades (6/90).

CG-1710-0022

LTS-101 ENGINE MAINTENANCE, CLASS C

Course Number: None.
Location: Aviation Technical Training Center, Elizabeth City, NC.
Length: 1-2 weeks (59-60 hours).
Exhibit Dates: 1/94–Present.
Learning Outcomes: Upon completion of the course, the student will be able to maintain the LTS-101 engine.
Instruction: Instruction consists of lectures and practical exercises on engine and systems operation, maintenance, disassembly, inspection, repair, and reassembly of the LTS-101 engine.
Credit Recommendation: In the lower-division baccalaureate/associate degree category, 1 semester hour aircraft gas turbine engine maintenance (10/94).

CG-1712-0004

CATERPILLAR DIESEL ENGINE MAINTENANCE, CLASS C

Course Number: MK-24.
Location: Reserve Training Center, Yorktown, VA.
Length: 2 weeks (68-72 hours).
Exhibit Dates: 1/90–9/93.
Learning Outcomes: Upon completion of the course, the student will be able to perform routine maintenance, troubleshooting, and repair on small diesel engines. Caterpillar model D379 and similar are used as example engines.
Instruction: Students learn maintenance and repair procedures for small Caterpillar diesel engines by a combination of lectures and laboratories. Engines are studied, disassembled, repaired, reassembled, and tested. Also, engine condition is assessed by performance tests.
Credit Recommendation: In the lower-division baccalaureate/associate degree category, 2 semester hours in diesel engine maintenance/repairs (10/88).

CG-1712-0005

270' WMEC S/S GENERATOR, WASTE HEAT RECOVERY SYSTEM AND EVAPORATOR OPERATION AND MAINTENANCE, CLASS C

Course Number: MK-27.
Location: Reserve Training Center, Yorktown, VA.
Length: 2-3 weeks (105-106 hours).
Exhibit Dates: 1/90–2/97.
Learning Outcomes: Upon completion of the course, the student will be able to start up, operate, and service the Caterpillar D398 diesel engine; and service the engine systems related to lubrication, cooling, fuel, electrical, and air starting. The student will understand the operation and servicing of the waste heat recovery system and the distilling plant.
Instruction: Approximately twenty percent of this course is devoted to Caterpillar 398 diesel engine component operation and another twenty percent to servicing its components. The remaining sixty percent of the course concentrates on the waste heat recovery system and the operation of the distilling plant. Students are instructed in its component troubleshooting and repair. Instruction is by combination of one-third lecture and two-thirds laboratory using actual equipment.
Credit Recommendation: In the lower-division baccalaureate/associate degree category, 2 semester hours in diesel engine with heat recovery system (10/88).

CG-1712-0006

VT903M CUMMINS DIESEL ENGINE DISASSEMBLY-ASSEMBLY AND TUNE-UP

Course Number: CUMM (R).
Location: Reserve Training Center, Yorktown, VA.
Length: 2 weeks (70-72 hours).
Exhibit Dates: 1/90–11/96.

Learning Outcomes: Upon completion of the course, the student will be able to disassemble, inspect components, and reassemble a Cummins VT903M diesel engine. Engine tune-up procedures are covered.

Instruction: Lectures and practical laboratory exercises are used to teach the basic principles of diesel engines along with disassembly/assembly and tune-up of an actual engine.

Credit Recommendation: In the vocational certificate category, 1 semester hour in diesel engine laboratory (10/88).

CG-1712-0007

LISTER DIESEL OVERHAUL AND POWER SYSTEM EQUIPMENT MAINTENANCE
(Lister Diesel Overhaul and Engine Power System Equipment Maintenance)

Course Number: ANC-M.
Location: Reserve Training Center, Yorktown, VA.
Length: 1-2 weeks (44-45 hours).
Exhibit Dates: 1/90–10/94.
Learning Outcomes: Upon completion of the course, the student will be able to inspect and repair a Lister Diesel engine, perform inspection and preventive maintenance on a standard fuel oil day tank and associated environmental control unit and power controller.

Instruction: Lectures and practice cover the inspection, complete disassembly, repair, reassembly, and testing of the Lister Diesel engine and associated power system controller.

Credit Recommendation: In the vocational certificate category, 1 semester hour in diesel engine repair (5/96).

CG-1712-0008

COXSWAIN 41

Course Number: None.
Location: Reserve Training Center, Yorktown, VA.
Length: 2 weeks (67 hours).
Exhibit Dates: 4/92–Present.
Learning Outcomes: Upon completion of the course, the student will be able to apply knowledge of small boats and small boat handling; coxswain responsibilities and authority; and basic navigation, piloting skills, and helicopter rescue operations as these relate to the 41 foot utility boat.

Instruction: Classroom and field instruction cover small boat construction and design, small boat handling, basic navigation, and piloting.

Credit Recommendation: In the lower-division baccalaureate/associate degree category, 1 semester hour in small boat handling (5/96).

CG-1714-0005

ELECTRICIAN'S MATE THIRD CLASS BY CORRESPONDENCE
(Electrician's Mate Third Class by Correspondence)

Course Number: 319-7.
Location: Coast Guard Institute, Oklahoma City, OK.
Length: Maximum, 52 weeks.
Exhibit Dates: 1/90–4/90.
Objectives: To present basic electricity through semiconductor devices and the theory of operation and maintenance on AC and DC motors and generators.

Instruction: This correspondence course offers basic AC and DC electricity including series and parallel circuits, inductance, capacitance, and impedance. Theory and maintenance of AC and DC motors and generators, transformers, and batteries are covered. There is a brief introduction to the theory and testing procedures on transistor and semiconductor devices. Course provides no laboratory experience. There is a proctored end-of-course examination.

Credit Recommendation: In the lower-division baccalaureate/associate degree category, 1 semester hour in electricity/electromechanical technology (1/79).

CG-1714-0009

GUNNER'S MATE SECOND CLASS BY CORRESPONDENCE

Course Number: 229-4.
Location: Coast Guard Institute, Oklahoma City, OK.
Length: Maximum, 104 weeks.
Exhibit Dates: 1/90–4/91.
Objectives: After 4/91 see CG-1714-0026. To provide personnel with the knowledge and skills to understand the operation and perform maintenance procedures on specific weapons and their components.

Instruction: Course provides instruction in DC electricity, electrical supply systems, fire control system electrical circuits, gun fire control system, safety, torpedo launching fire control equipment, torpedo loading and unloading, projectile hoist, Browning .50 caliber machine gun, and the M60 machine gun. Students are required to successfully complete a proctored end-of-course examination.

Credit Recommendation: In the vocational certificate category, 2 semester hours in DC circuits (5/85).

CG-1714-0011

ELECTRICIAN'S MATE FIRST CLASS BY CORRESPONDENCE

Course Number: Version 1: 0119-6. Version 2: 119-3; 119-4; 119-5.
Location: Coast Guard Institute, Oklahoma City, OK.
Length: Version 1: Maximum, 156 weeks. Version 2: Maximum, 52 weeks.
Exhibit Dates: Version 1: 11/94–Present. Version 2: 1/90–10/94.
Learning Outcomes: Version 1: Upon completion of the course, the student will be able to repair and maintain motors and generators; conduct operational tests on motors and generators in DC propulsion; maintain and repair components of switchboards; and repair, maintain, and operate various shipboard communications systems and degaussing systems. Version 2: To present the procedures for maintenance and repair of motors and generators, and an introduction to digital computer concepts, terminology, and circuits.

Instruction: Version 1: This correspondence course presents the procedures for the maintenance and repair of motors and generators including cleaning, removal and repair of bearings, commutators, and slip rings, etc. Included in the course is an overview explanation of diodes, transistors, and some thyristors as they relate to electric power distribution. The concept of power distribution is expanded to include distribution panels, relays, wiring, and conduit placement. The course includes sections on repair of serial shipboard communications systems (such as the engine order telegraph) and associated IC equipment. Finally the course covers the use of several Coast Guard reporting systems (CASREPT/CASCOR, yard reports, budget submission reports, and maintenance procedures reports) as they relate to engineering administration. Version 2: This correspondence course presents the procedures for the maintenance and repair of motors and generators including cleaning, removing and replacing bearings, commutators and slip rings, and testing. Also covered is an introduction to voltage and frequency regulators, power protective devices, and types of controllers, as well as an introduction to digital computer principles, including number systems, Boolean algebra, logic gates, flip-flops, counters, registers, decoders and adders. The course requires a proctored end-of-course examination.

Credit Recommendation: Version 1: In the lower-division baccalaureate/associate degree category, 1 semester hour in electric motors and generators and 2 in electric power distribution (3/96). Version 2: In the vocational certificate category, 2 semester hours in electrical power theory (1/79); in the lower-division baccalaureate/associate degree category, 1 semester hour in digital electronics theory (1/79).

CG-1714-0013

GUNNER'S MATE, CLASS A, PHASE 2 (ADVANCED ELECTRICITY, 3"/50 CALIBER GUN, 5"/38 CALIBER GUN)

Course Number: None.
Location: Training Center, Governors Island, NY.
Length: 3-8 weeks (90-220 hours).
Exhibit Dates: 1/90–11/96.
Objectives: To train personnel in advanced electricity as it pertains to two specific gun mounts.

Instruction: Lectures and practical exercises in troubleshooting and preventive maintenance of a particular gun mount including firing, lighting, heating, AC/DC circuits, magnetism, transformers, motors, generators and synchros. The eight-week version of this course includes electrohydraulic systems and drives.

Credit Recommendation: In the lower-division baccalaureate/associate degree category, for those students completing the three-week version of this course, 2 semester hours in basic electricity; for those completing the eight-week version, 2 semester hours in basic electricity and 1 in basic hydraulics (2/83).

CG-1714-0015

HH-3F AUTOMATIC FLIGHT CONTROL SYSTEM AND SELECTED ELECTRICAL MAINTENANCE, CLASS C

Course Number: None.
Location: Aviation Technical Training Center, Elizabeth City, NC.
Length: 3 weeks (110 hours).
Exhibit Dates: 1/90–11/96.
Objectives: To provide training necessary to service, inspect, maintain, and repair the automatic flight control system, a heading attitude reference system, a flight director system, and associated aircraft subsystems.

Instruction: Topics include operating individual systems, subsystems, and modules; testing, using specified test equipment; calibrating;

making necessary adjustments; and troubleshooting.

Credit Recommendation: In the lower-division baccalaureate/associate degree category, 3 semester hours in aircraft electrical systems and automatic flight control system troubleshooting (6/85).

CG-1714-0017

GUNNER'S MATE, CLASS A

Course Number: *Version 1:* GMA. *Version 2:* GMA.
Location: Reserve Training Center, Yorktown, VA.
Length: *Version 1:* 10 weeks (343-350 hours). *Version 2:* 9 weeks (311 hours).
Exhibit Dates: *Version 1:* 2/93–Present. *Version 2:* 1/90–1/93.
Learning Outcomes: *Version 1:* Upon completion of the course, the student will be able to solve basic electrical circuit problems, understand circuit symbols, read circuit diagrams, use a multimeter/DVM and megger. The student will be able to read and interpret fluid power circuit diagrams, understand how fluid power circuits are constructed, and understand the operating characteristics of fluid power components. The student will be able to practice safety precautions when working on electrical and hydraulic circuits. *Version 2:* Upon completion of the course, the student will be able to solve basic electrical circuit problems, understand circuit symbols, read circuit diagrams, use a multimeter/DVM and megger. The student will be able to read and interpret fluid power circuit diagrams, understand how fluid power circuits are constructed, and understand the operating characteristics of fluid power components, The student will be able to practice safety precautions when working on electrical and hydraulic circuits.
Instruction: *Version 1:* The course is conducted using a combination of lectures, demonstrations, and video laboratories. The first part of the course covers the basics of electricity and fluid power. The second part is entirely oriented toward operation of weapons. *Version 2:* The course is conducted using a combination of lectures, demonstrations, and video laboratories. The first part of the course covers the basics of electricity and fluid power. The second part is entirely oriented toward operation of weapons.
Credit Recommendation: *Version 1:* In the lower-division baccalaureate/associate degree category, 1 semester hour in basic electricity and 1 in safety engineering (5/96). *Version 2:* In the lower-division baccalaureate/associate degree category, 2 semester hours in electricity and 1 in fluid power (9/90).

CG-1714-0018

ELECTRICIAN MATE, CLASS A
(Electrician's Mate, Class A)

Course Number: *Version 1:* EMA. *Version 2:* EMA.
Location: Reserve Training Center, Yorktown, VA.
Length: *Version 1:* 14-15 weeks (480-513 hours). *Version 2:* 13 weeks (466 hours).
Exhibit Dates: *Version 1:* 7/91–Present. *Version 2:* 1/90–6/91.
Learning Outcomes: *Version 1:* Upon completion of the course, the student will apply the theory of AC and DC circuits, to problem solving. The student will be able to use standard electrical test equipment and electrical troubleshooting procedures; follow standard safety procedures; and explain the use and function of resistors, inductors, capacitors, generating batteries, voltage regulators, motors, circuit protectors, and electrical lighting and power distribution. The student will be able to use altimeters, voltmeters, Ohm meters, watt meters, multimeters, basic hand tools, machine tools, and wiring tools. *Version 2:* Upon completion of the course, the student will be able to apply the theory of AC and DC circuits to problem solving. The student will be able to use standard electrical test equipment and electrical troubleshooting procedures; follow standard safety procedures; and explain the use and function of resistors, inductors, capacitors, generating batteries, voltage regulators, motors, circuit protectors, and electrical lighting and power distribution. The student will be able to use electrical devices such as amplifier circuits and gyrocompasses and use basic hand, machine, and wiring tools.
Instruction: *Version 1:* Lectures and practical exercises cover AC/DC circuits and applying Ohm's law and Kirchhoff's laws to simple series and parallel circuits with one source. Topics include properties of resistors, inductors, and capacitors; RL time constant and waveform response; simple RL, RF, and RLC circuit analysis using trig functions, phaser diagrams, and power computations; transformer construction, characteristics, and operating types including 30 transformers; an in-depth analysis of AC/DC generators and motors, including AC/DC controllers and starters; maintenance of shipboard power distribution panels and cable repair; characteristics of vacuum tubes, and semiconductor diodes; gyroscope and synchro construction, operation, and maintenance; and maintenance procedures for shipboard motors, generators, and power distribution systems. *Version 2:* Lectures and practical exercises in AC/DC circuits and applying Ohm's law and Kirchhoff's laws to simple series and parallel circuits with one source. Topics include properties of resistors, inductors and capacitors; RL time constant and waveform response; simple RL, RF, and RLC circuit analysis using trig functions, phaser diagrams, and power computations; transformer construction, characteristics, and operating types including 30 transformers; an in-depth analysis of AC/DC generators and motors including AC/DC controllers and starters; maintenance of shipboard power distribution panels and cable repair; characteristics of vacuum tubes, and semiconductor diodes; gyroscope and synchro construction, operation, and maintenance; and maintenance procedures for shipboard motors, generators, and power distribution systems.
Credit Recommendation: *Version 1:* In the lower-division baccalaureate/associate degree category, 3 semester hours in AC and DC theory and 3 in AC and DC motors and generators (5/96). *Version 2:* In the lower-division baccalaureate/associate degree category, 3 semester hours in AC and DC theory and 6 in AC and DC motors and generators (9/90).

CG-1714-0019

1. SMALL BOAT ENGINEER, CLASS C
2. SMALL BOAT ENGINEERING

Course Number: *Version 1:* SBE. *Version 2:* SBE-(R).
Location: Reserve Training Center, Yorktown, VA.
Length: *Version 1:* 2 weeks (76-77 hours). *Version 2:* 2 weeks (70-75 hours).
Exhibit Dates: *Version 1:* 4/95–Present. *Version 2:* 1/90–3/95.
Learning Outcomes: *Version 1:* Upon completion of the course, the student will apply the principles of operation of electrical components to small engines (starters, generators, spark distributors, and alarm systems), as well as motors and controls for dewatering pumps, lubrication, cooling, and fuel oil systems, fire fighting equipment, and mooring equipment. Students will have mastered the use of electrical volt, current, and resistance meters and troubleshoot and repair the above equipment. Student will also be able to operate and maintain small gasoline and diesel engines, steering systems, and propulsion systems. *Version 2:* Upon completion of the course, the student will apply the principles of operation of electrical components to small engines (starters, generators, spark distributors, and alarm systems), as well as motors and controls for dewatering pumps. Students will have mastered the use of electrical volt, current, and resistance meters and troubleshoot and repair the above equipment. Student will also be able to operate and maintain small gasoline and diesel engines.
Instruction: *Version 1:* The course is devoted to teaching the principles of operation of engine components, starting with the most basic theory. Also included in the course is laboratory work on servicing and repairing engines and other auxiliary equipment. *Version 2:* Half of the course is devoted to teaching the principles of operation of engine electrical components, starting with the most basic theory of electricity. The other half of the course is laboratory work on servicing and repairing this type of equipment.
Credit Recommendation: *Version 1:* In the lower-division baccalaureate/associate degree category, 2 semester hours in small vessel engine operation and maintenance (5/96). *Version 2:* In the vocational certificate category, 1 semester hour in electrical components for engines and ship alarms (10/88).

CG-1714-0020

ELECTRICIAN MATE ADVANCED RESERVE ACTIVE DUTY FOR TRAINING

Course Number: EM-ADV-(R).
Location: Reserve Training Center, Yorktown, VA.
Length: 1-2 weeks (70-73 hours).
Exhibit Dates: 1/90–1/95.
Learning Outcomes: Upon completion of the course, the student will be able to use basic electrical meters to measure voltage current insulation resistance, frequency, and rms values. The student will also be able to do basic AC/DC circuit analysis, decimal to binary conversions, and mathematical logic gates to truth tables.
Instruction: This course is taught by a combination of lecture and laboratory. Topics covered include use of meters, Ohm's law, electrical power, series/parallel circuits, troubleshooting for faulty electrical components, binary numbers, truth tables, and logic gates.
Credit Recommendation: In the lower-division baccalaureate/associate degree category, 1 semester hour in electrical laboratory (10/88).

CG-1714-0021

378 Class WHEC Control Systems Operation and Maintenance, Class C

Course Number: EM-18.
Location: Reserve Training Center, Yorktown, VA.
Length: 2 weeks (70 hours).
Exhibit Dates: 1/90–9/94.
Learning Outcomes: Upon completion of the course, the student will be able to troubleshoot, repair, and perform preventive maintenance on console cards for the main propulsion control systems for the 378 foot Coast Guard cutter.
Instruction: Lectures and practical exercises in component and logic symbols, console operating conditions, and component troubleshooting of the main propulsion systems of the 378 foot class of Coast Guard cutter.
Credit Recommendation: In the lowerdivision baccalaureate/associate degree category, 1 semester hour in console card troubleshooting and repair (10/88).

CG-1714-0022

Aviation Electrician's Mate First Class by Correspondence

Course Number: 0101-5; 101-4; 101-5.
Location: Coast Guard Institute, Oklahoma City, OK.
Length: Maximum, 52 weeks.
Exhibit Dates: 1/90–Present.
Learning Outcomes: Upon completion of the course, the student will be able to identify the general procedures for aircraft receipt and transfer in accordance with engineering maintenance manual, follow correct procedures for requesting manufacturing and/or open purchase of aeronautical equipment, select the correct type of quality assurance inspector, identify the limitations and capabilities of local repair, and list the procedures for salvage and damage control of electronic components in accordance with accepted Coast Guard standards.
Instruction: The major topics covered are equipment purchase, quality control, salvage and damage control. Training is provided in a nonresident, self-instruction format. Students are required to successfully complete a proctored end-of-course examination.
Credit Recommendation: Credit is not recommended because of the military-specific nature of the course (12/92).

CG-1714-0023

Electrician's Mate Second Class by Correspondence

Course Number: 219-1.
Location: Coast Guard Institute, Oklahoma City, OK.
Length: Maximum, 78 weeks.
Exhibit Dates: 1/90–Present.
Learning Outcomes: Upon completion of the course, the student will be able to troubleshoot shipboard power and lighting distributions systems and AC/DC motors and generators; explain the theory, operation, and application of synchro systems, rotary amplifiers, magnetic amplifiers, and voltage regulators; use the oscilloscope to troubleshoot solid state and logic circuits; and describe the procedures for testing electrical safety devices and control circuits.
Instruction: Topics include power and lighting distribution systems, theory and application of synchro systems, amplifiers, and regulators. Training is provided in a nonresident, self-instruction format. Students are required to successfully complete a proctored end-of-course examination.
Credit Recommendation: In the lowerdivision baccalaureate/associate degree category, 3 semester hours in electrical/electronic theory (6/90).

CG-1714-0024

Electrician's Mate Third Class by Correspondence

Course Number: 319-8.
Location: Coast Guard Institute, Oklahoma City, OK.
Length: Maximum, 104 weeks.
Exhibit Dates: 5/90–Present.
Learning Outcomes: Upon completion of the course, the student will be able to maintain small boat electrical systems, troubleshoot and perform preventive maintenance on motors and generators, explain the theory and application of transformers, use the volt ohm meter in circuit troubleshooting, select and install electrical cable for a particular application, and describe common dangers encountered while working with electrical equipment.
Instruction: This correspondence course offers basic AC and DC electricity, including series and parallel circuits, inductance, capacitance, and impedance; and theory and maintenance of AC and DC motors and generators, transformers, and batteries. Training is provided in a nonresident, self-instruction format. Students are required to successfully complete a proctored end-of-course examination.
Credit Recommendation: In the vocational certificate category, 3 semester hours in electromechanical technology (6/90).

CG-1714-0025

Aviation Electrician's Mate Second Class by Correspondence

Course Number: 201-6; 201-5.
Location: Coast Guard Institute, Oklahoma City, OK.
Length: Maximum, 78 weeks.
Exhibit Dates: 1/90–Present.
Learning Outcomes: Upon completion of the course, the student will be able to describe operation of test equipment provided for fieldlevel maintenance; procure test equipment, tools and materials using aircraft parts catalogs and supply forms; maintain shop files of technical publications, directives, and manuals; list possible methods of reducing noise interference; replace aircraft wiring and associated hardware; identify the basic operating principles of aircraft heating and air conditioning systems; perform bench test procedures for aircraft flight control and stabilization systems; and describe the basic principles and application of modulator, demodulator, signal discriminator circuits, and power amplifier circuits.
Instruction: The major topics covered are field-level maintenance, document maintenance, aircraft heating and air conditioning systems, bench test flight control, and stabilization systems. Training is provided in a nonresident, self-instruction format. Students are required to successfully complete a proctored end-of-course examination.
Credit Recommendation: In the lowerdivision baccalaureate/associate degree category, 3 semester hours in avionics instruments (1/94).

CG-1714-0026

Gunner's Mate Second Class by Correspondence

Course Number: Version 1: 0229-8; 0229-7. Version 2: 0229-6; 0229-5.
Location: Coast Guard Institute, Oklahoma City, OK.
Length: Version 1: Maximum, 156 weeks. Version 2: Maximum, 104 weeks.
Exhibit Dates: Version 1: 7/95–Present. Version 2: 5/91–6/95.
Learning Outcomes: Version 1: Upon completion of the course, the student will be able to identify, inspect, test, and troubleshoot gun systems consisting of magazine sprinkler systems and gun mounts of weapon systems. This enables the student to understand operation and maintenance procedures for pyrotechnic, electrical, hydraulic, and pneumatic systems specific to gun mounts, weapon systems, and ordnance administration. Version 2: Before 5/91 see CG-1714-0009. Upon completion of the course, the student will be able to identify, inspect, test, and troubleshoot gun systems consisting of magazine sprinkler systems and gun mounts of weapon systems. This enables the student to understand operation and maintenance procedures for pyrotechnic, electrical, hydraulic, and pneumatic systems specific to gun mounts, weapon systems, and ordnance administration.
Instruction: Version 1: This course provides instruction in basic hydraulics; system troubleshooting and repair of hydraulic and pneumatic systems; safety precautions; use of ammeter and megger for electrical testing and troubleshooting; solderless connections; identification, removal, and installation of electrical and electronic modular components; use of reference publications to identify and dispose of unserviceable ammunition and pyrotechnics; and use of maintenance data systems for ordnance administration as related to specific gun mount, machine guns, and weapon systems. Version 2: This course provides instruction in basic hydraulics; system troubleshooting and repair of hydraulic and pneumatic systems; safety precautions; use of ammeter and megger for electrical testing and troubleshooting; solderless connections; identification, removal, and installation of electrical and electronic modular components; use of reference publications to identify and dispose of unserviceable ammunition and pyrotechnics; and use of maintenance data systems for ordnance administration as related to specific gun mount, machine guns, and weapon systems.
Credit Recommendation: Version 1: In the lower-division baccalaureate/associate degree category, 2 semester hours in basic electricity and 2 in basic hydraulics and pneumatics (9/97). Version 2: In the lower-division baccalaureate/associate degree category, 2 semester hours in basic electricity and 2 in basic hydraulics and pneumatics (3/94).

CG-1714-0027

Gunner's Mate First Class by Correspondence

Course Number: 0129-7; 129-6.

Location: Coast Guard Institute, Oklahoma City, OK.
Length: Maximum, 156 weeks.
Exhibit Dates: 2/93–Present.
Learning Outcomes: Upon completion of the course, the student will be able to perform as a gunner's mate first class and will function as the primary assistant to the gun chief. Version 7 includes hazardous materiel management and deleted filling high pressure pneumatic cylinders and flasks.
Instruction: Course is offered using self-instruction and self-quizzes in the areas of pyrotechnics, ammunition, hydraulics, weapons, and systems. A proctored written exam concludes the course.
Credit Recommendation: In the lower-division baccalaureate/associate degree category, 1 semester hour in hydraulics and pneumatics (5/97).

CG-1715-0001

AVIATION ELECTRONICS TECHNICIAN, CLASS A

Course Number: *Version 1:* None. *Version 2:* None.
Location: Aviation Technical Training Center, Elizabeth City, NC.
Length: *Version 1:* 22 weeks (880 hours). *Version 2:* 27 weeks (878 hours).
Exhibit Dates: *Version 1:* 12/95–Present. *Version 2:* 1/90–11/95.
Learning Outcomes: *Version 1:* This version is pending evaluation. To provide selected enlisted personnel with the understanding and knowledge necessary to fulfill the requirements for Aviation Electronics Technician, Third Class. *Version 2:* To provide selected enlisted personnel with the understanding and knowledge necessary to fulfill the requirements for Aviation Electronics Technician, Third Class.
Instruction: *Version 1:* This version is pending evaluation. Topics include electronics fundamentals, theory of operation of airborne electronics systems, troubleshooting and testing of airborne electronics systems, and operation of associated test equipment. *Version 2:* Topics include electronics fundamentals, theory of operation of airborne electronics systems, troubleshooting and testing of airborne electronics systems, and operation of associated test equipment.
Credit Recommendation: *Version 1:* Pending evaluation. *Version 2:* In the vocational certificate category, 15 semester hours toward a certificate in electronics technology (6/85); in the lower-division baccalaureate/associate degree category, 1 semester hour in theory of flight/aircraft performance and 9 in electronic fundamentals/introduction to electrical engineering (6/85); in the upper-division baccalaureate category, 2 semester hours in electricity and electronics (6/85).

CG-1715-0016

FIRE CONTROL MK 56 SYSTEM, CLASS C
(Fire Control Technician, Class C)
(Gun Fire Control Systems Mk 52 and Mk 56)

Course Number: None.
Location: Training Center, Governors Island, NY.
Length: 14 weeks (548 hours).
Exhibit Dates: 1/90–11/96.
Objectives: To train personnel to operate and maintain a gun fire control system.
Instruction: The course has a heavy emphasis on techniques to troubleshoot fixed and mixed trigger circuits, gate circuits, pulse and trigger circuits, receiver and AFC circuits, range error and sense circuits, and tracking and angle error circuits. The course also covers radar operation, train amplifiers, amplidyne generators, filter, and switch smoothing circuits.
Credit Recommendation: In the vocational certificate category, 3 semester hours in electronic circuit maintenance and troubleshooting (2/83); in the lower-division baccalaureate/associate degree category, 1 semester hour in electrical laboratory (2/83).

CG-1715-0020

OFFICER IN CHARGE AND EXECUTIVE PETTY OFFICER

Course Number: None.
Location: Training Center, Petaluma, CA.
Length: 2 weeks (67 hours).
Exhibit Dates: 3/93–Present.
Learning Outcomes: Upon completion of the course, the student will be able to perform the duties and responsibilities of leading and managing small organizational units.
Instruction: Instruction includes lectures, case study, and laboratory exercises in administration, personnel policies, organizational behavior and communication, property management, and performance evaluation.
Credit Recommendation: In the lower-division baccalaureate/associate degree category, 3 semester hours in principles of organizational behavior and personnel administration (9/93).

CG-1715-0031

AVIATION ELECTRONICS TECHNICIAN SECOND CLASS BY CORRESPONDENCE

Course Number: *Version 1:* 203-6. *Version 2:* 203-5.
Location: Coast Guard Institute, Oklahoma City, OK.
Length: *Version 1:* Maximum, 156 weeks. *Version 2:* Maximum, 104 weeks.
Exhibit Dates: *Version 1:* 1/94–Present. *Version 2:* 1/90–12/93.
Learning Outcomes: *Version 1:* Upon completion of the course, the student will be able to identify procedures, policy, and methods associated with aircraft handling equipment, fuels, safety, and repair and calibration of special tools. Student will demonstrate knowledge of logs and records; supply, salvage, and damage control; integrated logistics supply systems; principles and functions of a basic computer, including computer data busses, input/output ports, and microprocessors; the detailed operation of solid state devices; and aircraft navigation, communication, and microwave systems. *Version 2:* Upon completion of the course, the student is provided with an overview of safety, administration, corrosion, and hardware in addition to basic electronics theory.
Instruction: *Version 1:* Course is made up of self-instruction of textual material covering clean-up of salt water immersed electronic equipment, basic corrosion control and preservation, safety precautions for the technician, general aircraft safety and administration, avionic administration, operation of the oscilloscope, and solid state components and logic troubleshooting. A proctored final examination is required. *Version 2:* Training is provided in basic oscilloscope and octopus, solid state components, corrosion control, troubleshooting, safety, first aid, general aircraft hardware, publications, management, and administration. Students are required to successfully complete a proctored end-of-course examination.
Credit Recommendation: *Version 1:* In the lower-division baccalaureate/associate degree category, 3 semester hours in solid state electronic theory and troubleshooting methods and 2 in corrosion control (9/94). *Version 2:* In the lower-division baccalaureate/associate degree category, 2 semester hours in basic electronics and specialized corrosion control as related to electronic equipment (5/85).

CG-1715-0034

SONAR TECHNICIAN FIRST CLASS BY CORRESPONDENCE

Course Number: 143-2.
Location: Coast Guard Institute, Oklahoma City, OK.
Length: Maximum, 104 weeks.
Exhibit Dates: 1/90–1/93.
Objectives: To provide experienced sonar technicians with supervisory training required for advancement to first class technician status.
Instruction: Course provides training in the supervision of sonar operators, antisubmarine warfare organization and operations, sonar contact classification, evasive devices and casualty procedures, torpedo systems, sonar domes, and transducers. Students are required to successfully complete a proctored end-of-course exam.
Credit Recommendation: Credit is not recommended due to the military-specific nature of the course (5/85).

CG-1715-0037

ELECTRONICS TECHNICIAN SECOND CLASS BY CORRESPONDENCE

Course Number: 0221-7.
Location: Coast Guard Institute, Oklahoma City, OK.
Length: 104 weeks.
Exhibit Dates: 1/90–3/90.
Learning Outcomes: After 3/90 see CG-1715-0146. Upon completion of the course, the student will be able to teach fundamentals of electronics.
Instruction: Course includes topics in the general operation of solid state diodes, transistors, unifunction transistors, field effect transistors, silicon controlled rectifiers, and special semiconductor devices. An introduction to computer systems includes number systems, Boolean algebra, logic gates, counters, registers, decoders, and adders. The basic theory of operation of an oscilloscope and the interpretation of cathode ray tube oscillograms are presented. The study of cable installation and termination includes cable routing and securing coaxial cable, military specification connectors, and connector soldering procedures. Students may supplement course study with hands-on training on the job. Students are required to successfully complete a proctored end-of-course examination.
Credit Recommendation: In the lower-division baccalaureate/associate degree category, 3 semester hours in introduction to electronics (5/85).

CG-1715-0038

ELECTRONICS TECHNICIAN FIRST CLASS BY CORRESPONDENCE

Course Number: *Version 1:* 0121-7. *Version 2:* 121-6; 121-7.
Location: Coast Guard Institute, Oklahoma City, OK.
Length: *Version 1:* Maximum, 104 weeks. *Version 2:* Maximum, 104 weeks.
Exhibit Dates: *Version 1:* 1/97–Present. *Version 2:* 1/90–12/96.
Learning Outcomes: *Version 1:* This version is pending evaluation. *Version 2:* To present the theory of electromagnetic interference effects, digital electronics, digital troubleshooting techniques, and the Coast Guard paperwork system.
Instruction: *Version 1:* This version is pending evaluation. *Version 2:* Course includes electromagnetic interference as related to radio and radar receivers and transmitters; digital logic elements such as gates, flip-flops, registers, counters, adders, and decoders; digital troubleshooting methods; and Coast Guard formats for memoranda, messages, and letters. Students are required to successfully complete a proctored end-of-course examination. Students may supplement course study with hands-on experience on the job.
Credit Recommendation: *Version 1:* Pending evaluation. *Version 2:* In the lower-division baccalaureate/associate degree category, 2 semester hours in digital logic (5/85).

CG-1715-0049

AN/SPS-29D RADAR SYSTEMS, CLASS C
(AN/SPS-29 Maintenance and Repair, Class C)

Course Number: NAV-05.
Location: Training Center, Governors Island, NY.
Length: 3 weeks (109-111 hours).
Exhibit Dates: 1/90–1/90.
Objectives: To train electronics technicians in the maintenance and repair of a specific air-search radar system in accordance with its technical manuals.
Instruction: Lectures on equipment components and their function and practical exercises in the operation, alignment and troubleshooting of equipment with instructor inserted faults. Electronics theory is offered on an ""as needed" basis.
Credit Recommendation: In the vocational certificate category, 1 semester hour in radar service and maintenance (2/83).

CG-1715-0063

TELEPHONE TECHNICIAN SECOND CLASS BY CORRESPONDENCE

Course Number: *Version 1:* 0245-7. *Version 2:* 245-6.
Location: Coast Guard Institute, Oklahoma City, OK.
Length: *Version 1:* Maximum, 156 weeks. *Version 2:* Maximum, 104 weeks.
Exhibit Dates: *Version 1:* 10/95–Present. *Version 2:* 1/90–9/95.
Learning Outcomes: *Version 1:* This version is pending evaluation. To teach telephone technicians to install, operate, and repair communications and terminal equipment. *Version 2:* To teach telephone technicians to install, operate, and repair communications and terminal equipment.
Instruction: *Version 1:* This version is pending evaluation. Course contains volumes 2 and 6 of Lee's ABC of the Telephone covering station wiring, station carrier equipment, signaling, power requirements, noise, installation and maintenance, and troubleshooting. Additional topics include Model 28 and Model 40 teletypes, oscilloscopes, as well as cable installation and termination. Students are required to successfully complete a proctored end-of-course examination. Students may supplement course study with hands-on experience on the job. *Version 2:* Course contains volumes 2 and 6 of Lee's ABC of the Telephone covering station wiring, station carrier equipment, signaling, power requirements, noise, installation and maintenance, and troubleshooting. Additional topics include Model 28 and Model 40 teletypes, oscilloscopes, as well as cable installation and termination. Students are required to successfully complete a proctored end-of-course examination. Students may supplement course study with hands-on experience on the job.
Credit Recommendation: *Version 1:* Pending evaluation. *Version 2:* In the vocational certificate category, 3 semester hours in telephone and teletype theory and installation (5/85).

CG-1715-0068

TEMPEST MODEL 40 TELETYPE, CLASS C

Course Number: None.
Location: Training Center, Governors Island, NY.
Length: 3 weeks (100 hours).
Exhibit Dates: 1/90–11/96.
Objectives: To provide telephone and electronics technicians with the necessary training to install, maintain, and overhaul the Tempest model 40 teletype data terminal.
Instruction: Lectures and laboratory exercises at the functional level on the electromechanical operation of the video display monitor, printer, controller logic, and power supply of the Tempest data terminal. The main emphasis of the course is on the installation, maintenance, and repair to the defective card, module, or mechanical component level.
Credit Recommendation: In the vocational certificate category, 1 semester hour in electromechanical laboratory (2/83).

CG-1715-0070

AN/SPS-64(V)4 RADAR

Course Number: NAV-14.
Location: Training Center, Governors Island, NY.
Length: 2 weeks (71 hours).
Exhibit Dates: 1/90–6/92.
Objectives: To train electronics technicians in the maintenance and repair of a specialized radar system.
Instruction: Course provides lectures and discussions on equipment components and functions, including practical experience in the operation, alignment, and troubleshooting of instructor-inserted faults. Electronics theory is offered on an as needed basis.
Credit Recommendation: In the vocational certificate category, 1 semester hour in radar servicing and maintenance (2/83).

CG-1715-0072

COL-URG-II HF TRANSMITTING SYSTEM
(COL-URG-II High Frequency Transmitting System)

Course Number: COM-12.
Location: Training Center, Governors Island, NY.
Length: 2 weeks (71 hours).
Exhibit Dates: 1/90–11/96.
Objectives: To train experienced electronic technicians to operate and perform preventive and corrective maintenance on a specific transmitter and its associated control equipment.
Instruction: Course provides lectures and practical exercises in the operation of a specific computerized transmitter, including the use of diagnostic test equipment and programs, fault isolation, and repair to the component level.
Credit Recommendation: Credit is not recommended because of the limited, specialized nature of the course (2/83).

CG-1715-0074

RADAR OPERATOR BY CORRESPONDENCE

Course Number: 432-1.
Location: Coast Guard Institute, Oklahoma City, OK.
Length: Maximum, 104 weeks.
Exhibit Dates: 1/90–6/97.
Objectives: To teach the principles of radar systems, navigation, and piloting.
Instruction: Course introduces the principles of radar, including carrier frequency, pulse width, time-range relationship, antenna systems, repeaters, scope interpretation identification equipment, navigation, and radar piloting. Student may supplement course study with hands-on experience on the job. Students are required to successfully complete a proctored end-of-course examination.
Credit Recommendation: In the vocational certificate category, 1 semester hour in radar operation (5/85).

CG-1715-0076

BASIC RADAR USE AND OPERATION BY CORRESPONDENCE

Course Number: 651-1.
Location: Coast Guard Institute, Oklahoma City, OK.
Length: Maximum, 104 weeks.
Exhibit Dates: 1/90–Present.
Objectives: To provide training in the basic use and operation of radar.
Instruction: Course includes the history of radar development, the legalities of radar use, radar fundamentals, radar propagation, operation of a specific radar, factors affecting detection and determination of position of contacts, radar navigation, and ice detection. Also included are maneuvering board problems involving relative motion, course, speed, collision, station taking, contact interception, true wind, relative wind, desired wind, and direct plotting procedures. Students may supplement course study with hands-on training on the job. Students are required to successfully complete a proctored end-of-course examination.
Credit Recommendation: In the vocational certificate category, 1 semester hour in introduction to radar system operation (3/97).

CG-1715-0078

AVIATION ELECTRONICS TECHNICIAN AN/ARC-160, CLASS C

Course Number: None.
Location: Aviation Technical Training Center, Elizabeth City, NC.
Length: 2 weeks (71 hours).
Exhibit Dates: 1/90–11/96.
Objectives: Provides specialized training in maintenance of VHF-FM transceivers.
Instruction: Course provides practical training in the use of specialized test equipment, minimum performance testing, and troubleshooting to the module and chassis-mounted component level of a specific VHF-FM transceiver.
Credit Recommendation: In the lower-division baccalaureate/associate degree category, 1 semester hour in electronic communication repair (6/85).

CG-1715-0079

AVIATION ELECTRONICS TECHNICIAN 618M-3, CLASS C

Course Number: None.
Location: Aviation Technical Training Center, Elizabeth City, NC.
Length: 3 weeks (109 hours).
Exhibit Dates: 1/90–11/96.
Objectives: To provide specialized training in the maintenance of a specific VHF AM transceiver.
Instruction: Course provides practical training in the use of specialized test equipment, minimum performance testing, and troubleshooting of the VHF AM transceiver.
Credit Recommendation: In the lower-division baccalaureate/associate degree category, 1 semester hour in electronic communication repair (6/85).

CG-1715-0080

1. AVIATION ELECTRONICS TECHNICIAN AN/APS-127 RADAR, CLASS C
2. AVIATION ELECTRONICS TECHNICIAN AN/APS-127

Course Number: *Version 1:* None. *Version 2:* None.
Location: Aviation Technical Training Center, Elizabeth City, NC.
Length: *Version 1:* 4 weeks (148 hours). *Version 2:* 4 weeks (148 hours).
Exhibit Dates: *Version 1:* 3/97–Present. *Version 2:* 1/90–2/97.
Learning Outcomes: *Version 1:* This version is pending evaluation. *Version 2:* To provide specialized training in the theory and maintenance of a specific radar system.
Instruction: *Version 1:* This version is pending evaluation. *Version 2:* Course provides lectures and demonstrations on the electronic theory and operation of the equipment's units and modules and practical training in the use of specialized test equipment. Course includes minimum performance testing, alignment, and troubleshooting.
Credit Recommendation: *Version 1:* Pending evaluation. *Version 2:* In the lower-division baccalaureate/associate degree category, 2 semester hours in AN/APS-127 radar system maintenance, 2 in radar system maintenance, and 2 in radar theory (6/85).

CG-1715-0081

DIGITAL MICROPROCESSOR, CLASS C

Course Number: None.
Location: Aviation Technical Training Center, Elizabeth City, NC.
Length: 2-3 weeks (93 hours).
Exhibit Dates: 1/90–11/96.
Objectives: To provide the knowledge and skill training necessary to perform the configuration, operation, inspection, and repair of aircraft microprocessor-based systems at the organizational and intermediate levels of maintenance.
Instruction: Course provides hands-on experience in the preparation, operation, repair, and maintenance of digital and microprocessor-based systems using applicable test equipment, schematic diagrams, and publications.
Credit Recommendation: In the upper-division baccalaureate category, 2 semester hours in electricity and electronics (6/85).

CG-1715-0082

ADL-81 LORAN-C RECEIVER, CLASS C

Course Number: None.
Location: Aviation Technical Training Center, Elizabeth City, NC.
Length: 4-5 weeks (149-154 hours).
Exhibit Dates: 1/90–7/93.
Objectives: After 7/93 see CG-1715-0141. To provide specialized training in the theory of operation and maintenance of long range navigation equipment.
Instruction: Topics include electronic theory, Loran theory, theory of operation of the ADL-81 and associated test equipment, fault location to the component level, and the relationship between receiver and associated equipment such as the active antenna (UPS-95) system.
Credit Recommendation: In the lower-division baccalaureate/associate degree category, 3 semester hours in electronic systems maintenance laboratory (6/85).

CG-1715-0083

AVIATION ELECTRONICS TECHNICIAN AN/ARN-133V2, CLASS C

Course Number: None.
Location: Aviation Technical Training Center, Elizabeth City, NC.
Length: 2 weeks (71 hours).
Exhibit Dates: 1/90–11/96.
Objectives: To provide specialized training in the maintenance of Loran navigation equipment.
Instruction: Topics include electronic theory and theory of operation for the Loran navigator, practical training in the use of specialized test equipment, minimum performance testing, and troubleshooting.
Credit Recommendation: In the lower-division baccalaureate/associate degree category, 1 semester hour in basic electronic maintenance and repair laboratory (6/85).

CG-1715-0084

AVIATION ELECTRONICS TECHNICIAN AN/ARN-118(V), CLASS C

Course Number: None.
Location: Aviation Technical Training Center, Elizabeth City, NC.
Length: 4 weeks (146 hours).
Exhibit Dates: 1/90–11/96.
Objectives: To provide advanced specialized training in the maintenance of tactical air navigation equipment.
Instruction: Course provides training on the maintenance and operation of the equipment's units and modules, including performance testing, troubleshooting, and alignment.
Credit Recommendation: In the lower-division baccalaureate/associate degree category, 3 semester hours in electronic systems maintenance (6/85).

CG-1715-0085

1. HC-130 AVIONICS BY CORRESPONDENCE
2. HC-130 AVIONICSMAN BY CORRESPONDENCE

Course Number: *Version 1:* 0444-4. *Version 2:* 0444-3.
Location: Coast Guard Institute, Oklahoma City, OK.
Length: *Version 1:* Maximum, 156 weeks. *Version 2:* Maximum, 52 weeks.
Exhibit Dates: *Version 1:* 6/97–Present. *Version 2:* 1/90–5/97.
Learning Outcomes: *Version 1:* This version is pending evaluation. *Version 2:* Upon completion of the course, the student will be able to demonstrate proficiency in the HC-130 airframe and power plant systems, emergency and rescue procedures, and avionic systems.
Instruction: *Version 1:* This version is pending evaluation. *Version 2:* Instruction is provided in familiarization of the aircraft, including power plants; propellers; ground handling and servicing; hydraulic-powered flight controls and landing gear systems; brake, fuel, and electrical systems; and emergency and rescue procedures. Avionic topics include general communications; null field static discharge wicks; intercommunications; P.A. systems; HF, VHF, and UHF systems; and CVR, CMS, IFF, Loran, OMEGA, radar, TACAN, INS, FDS, and auto pilot systems.
Credit Recommendation: *Version 1:* Pending evaluation. *Version 2:* In the lower-division baccalaureate/associate degree category, 1 semester hour in aircraft systems and ground handling and 2 in communication and navigation equipment and electronic flight aids (9/90).

CG-1715-0086

HH-3F AVIONICSMAN BY CORRESPONDENCE

Course Number: 0442-2.
Location: Coast Guard Institute, Oklahoma City, OK.
Length: Maximum, 52 weeks.
Exhibit Dates: 1/90–Present.
Learning Outcomes: Upon completion of the course, the student will be able to demonstrate knowledge of the HH-3F airframe and power plant systems, emergency and rescue procedures, and avionic systems.
Instruction: Instruction is provided in familiarization of the helicopter airframe, power plant, transmission and rotor systems; ground handling and servicing; hydraulic, landing gear, brake, fuel, and electrical systems; and emergency and rescue procedures. Avionic topics include general and intercommunications, DF, HF, VHF, and UHF systems; and IFF, Loran, radar, TACAN, and FDS systems.
Credit Recommendation: In the lower-division baccalaureate/associate degree category, 1 semester hour in aircraft systems and ground handling and 1 in communication and

navigation equipment and electronic flight aids (9/90).

CG-1715-0090

ELECTRONICS FUNDAMENTALS COMMUNICATIONS TRACK

Course Number: None.
Location: Training Center, Governors Island, NY.
Length: 27 weeks (840 hours).
Exhibit Dates: 1/90–1/93.
Objectives: To provide enlisted personnel with the basic electronic fundamentals necessary to maintain, troubleshoot, and repair high frequency and facsimile receivers and radar units.
Instruction: The program comprises the basic ELF sequence, a common core covering transistors, power supplies, amplifiers, oscillators, logic, and microcomputer and special circuits; ELF 08, 12, and 13 sequence, including communication theory, AM/FM transceivers, antennas, and radar fundamentals; COM 03, 05, 06, and 08 sequence, including military radio single sideband transmitter, high frequency and facsimile receivers, and shipboard communications systems. The subject matter is qualitative in nature and presented at the block-diagram level with schematic signal tracing to the component level.
Credit Recommendation: In the vocational certificate category, 8 semester hours in basic electronic fundamentals (2/83); in the lower-division baccalaureate/associate degree category, 3 semester hours in logic circuits and microprocessors in a computer science curriculum, 3 in communications systems, and 1 in communications laboratory (2/83).

CG-1715-0091

ELECTRONICS FUNDAMENTALS V/S TRACK #1

Course Number: None.
Location: Training Center, Governors Island, NY.
Length: 29 weeks (900 hours).
Exhibit Dates: 1/90–1/91.
Objectives: To provide enlisted personnel with the basic electronic fundamentals necessary to maintain, troubleshoot, and repair radar systems, transceivers, and direction finders.
Instruction: The program comprises the basic ELF sequence (ELF 00, 02, 03, 04, 05, 09, 10, 11), a common core covering transistors, power supplies, amplifiers, oscillators, logic, microcomputers, and special circuits. Lectures and practical exercises in basic electrical concepts (Ohm's law, series and parallel DC circuits, use of meters), power supply construction and regulation (primarily tube circuitry, introductory AC and THREE-phase materials and filtering concepts), amplifier fundamentals (vacuum tube amplifiers relating to distortion, frequency response, degeneration, signal generator usage), transistor fundamentals (testing, biasing, operating points, temperature compensation), oscillators, logic circuits (brief overview of gates, counters, shift registers), microprocessors (overview of terminology, addressing, memory devices, I/O requirements), and specialized circuits (superficial coverage of couplers, inverters, multivibrators and timing circuits, and other pulsed circuits). ELF 08, 12, and 13 is also covered and include communications theory (operation, adjustment, and troubleshooting single sideband transceivers, AM/FM transmitters and receivers at the block diagram level), antenna operations (radio wave propagation, frequency limitations), transmission lines (wave ratio, reflections, TDR operation), as well as radar fundamentals. Also included are COM 02, 09, and NAV 04 and 13 (homers, direction finders, depth sounders, and radar systems). Each of these courses covers a specific piece of equipment with the student learning to operate, troubleshoot, and repair each piece to the component level.
Credit Recommendation: In the vocational certificate category, 8 semester hours in basic electronic fundamentals and 2 in marine electronic repair (2/83); in the lower-division baccalaureate/associate degree category, 3 semester hours in logic circuits and microprocessors in a computer science curriculum, 3 in communications systems, and 1 in electronics communications laboratory (2/83).

CG-1715-0092

LORAN TRACK #1 FOR LORSTAS, MALONE, SENCA

Course Number: None.
Location: Training Center, Governors Island, NY.
Length: 28 weeks (840 hours).
Exhibit Dates: 1/90–1/93.
Objectives: To provide enlisted personnel with the basic electronic fundamentals necessary to maintain, troubleshoot, and repair specific Loran equipment.
Instruction: The program comprises the basic ELF sequence, a common core curriculum covering transistors, power supplies, amplifiers, oscillators, logic, microcomputers, and special circuits; ELF 08 and 12 comprising communication theory and AM/FM and SSB transceivers and antennas; and LOR 06, 15, 18 and 19 comprising Loran-C timing and control, remote Austron 5000, and calculation-assisted Loran-C. The subject matter covers the theoretical and operational concepts of a navigation system, is qualitative in nature, and includes alignment procedures, block diagram analysis, and schematic signal tracing to the component level.
Credit Recommendation: In the vocational certificate category, 8 semester hours in basic electronic fundamentals and 1 in Loran service and repair (2/83); in the lower-division baccalaureate/associate degree category, 3 semester hours in logic circuits and microcomputers in a computer science curriculum, 2 in communications systems, and 1 in electronics communications laboratory (2/83).

CG-1715-0093

ELECTRONICS FUNDAMENTALS V/S TRACK #2

Course Number: None.
Location: Training Center, Governors Island, NY.
Length: 26 weeks (780 hours).
Exhibit Dates: 1/90–1/93.
Objectives: To provide enlisted personnel with the basic electronic fundamentals necessary to maintain, troubleshoot, and repair FM transceivers, direction finders, and surface-search radar.
Instruction: The program comprises the basic ELF sequence, a common core covering transistors, power supplies, amplifiers, oscillators, logic, microcomputers, and special circuits; the ELF 08, 12, and 13 sequence covering communication theory and AM/FM and SSB transceivers, antennas, and radar fundamentals; and the COM 02, 10, and NAV 04 and 12 sequences covering FM transceivers, homers, direction finders, depth sounders, SSB transceivers, and surface-search radar. The subject matter is qualitative in nature and presented at the block diagram level.
Credit Recommendation: In the vocational certificate category, 8 semester hours in electronic fundamentals and 2 in marine electronic repair (2/83); in the lower-division baccalaureate/associate degree category, 3 semester hours in logic circuits and microprocessors in a computer science curriculum, 3 in communications systems, and 1 in electronic communications laboratory (2/83).

CG-1715-0094

LORAN TRACK #5 FOR LORSTAS, ATTU, BAUDETTE, CARIBOU, CAROLINA BEACH, ESTARTIT, JUPITER, KARGABARUN, KURE, NANTUCKET, PORT CLARENCE, SHOAL COVE, SYLT, UPOLU POINT

Course Number: None.
Location: Training Center, Governors Island, NY.
Length: 26 weeks (720 hours).
Exhibit Dates: 1/90–1/93.
Objectives: To provide enlisted personnel with the basic electronics fundamentals needed to maintain, troubleshoot, and repair specific Loran equipment.
Instruction: The program is comprised of the basic ELF sequence, a core curriculum covering transistors, power supplies, amplifiers, oscillators, logic, microcomputers, and special circuits; ELF 08 and 12 comprising communications theory and AM/FM and SSB transceivers and antennas; and LOR 06 and 16 comprising Loran-C timing and control equipment and Loran-C low power vacuum tube transmitter. The subject matter covers the theoretical and operational concepts of a navigation system, is qualitative in nature, and includes block diagram analysis and schematic signal tracing to the component level.
Credit Recommendation: In the vocational certificate category, 8 semester hours in basic electronics fundamentals and 1 in Loran service and repair (2/83); in the lower-division baccalaureate/associate degree category, 3 semester hours in logic circuits and microprocessors in a computer science curriculum, 2 in communications systems, and 1 in electronics communications laboratory (2/83).

CG-1715-0095

LORAN TRACK #2 FOR LORSTAS, KODIAC, YAKOTA

Course Number: None.
Location: Training Center, Governors Island, NY.
Length: 26 weeks (720 hours).
Exhibit Dates: 1/90–1/93.
Objectives: To provide enlisted personnel with the basic electronics fundamentals needed to maintain, troubleshoot, and repair specific Loran equipment.
Instruction: The program comprises the basic ELF sequence, a common core curriculum covering transistors, power supplies, amplifiers, oscillators, logic, microcomputers, and special circuits; ELF 08 and 12 comprising communication theory and AM/FM and SSB transceivers and antennas; and LOR 13 and 19 comprising Austron 5000, Loran-C, control/

monitor equipment, and calculator-assisted Loran-C. The subject matter covers the theoretical and operational concepts of a navigation system, is qualitative in nature, and includes block diagram analysis and schematic signal tracing to the component level.

Credit Recommendation: In the vocational certificate category, 8 semester hours in electronic fundamentals and 1 in Loran service and repair (2/83); in the lower-division baccalaureate/associate degree category, 3 semester hours in circuits and microcomputers in a computer science program, 2 in communications systems, and 1 in electronic communications laboratory (2/83).

CG-1715-0096

LORAN TRACK #4 FOR LORSTAS, GRANGEVILLE, RAYMONDVILLE

Course Number: None.
Location: Training Center, Governors Island, NY.
Length: 26 weeks (780 hours).
Exhibit Dates: 1/90–1/93.
Objectives: To provide enlisted personnel with the basic electronics fundamentals necessary to maintain, troubleshoot, and repair specific Loran equipment.
Instruction: The program comprises the basic ELF sequence, a core curriculum covering transistors, power supply, amplifiers, oscillators, logic, microcomputers, and special circuits; ELF 08 and 12 comprising communication theory and AM/FM and SSB transceivers and antennas; and LOR 06 and 15 comprising Loran-C timing and control equipment and a solid state transmitter. The subject matter covers the theoretical and operational concepts of a navigation system, is qualitative in nature, and includes block diagram analysis and schematic signal tracing to the component level.
Credit Recommendation: In the vocational certificate category, 8 semester hours in electronic fundamentals and 1 in Loran service and repair (2/83); in the lower-division baccalaureate/associate degree category, 3 semester hours in circuits and microcomputers in a computer science curriculum, 2 in communications systems, and 1 in electronics communication laboratory (2/83).

CG-1715-0097

LORAN TRACK #3 FOR LORSTA, HONOLULU

Course Number: None.
Location: Training Center, Governors Island, NY.
Length: 24 weeks (720 hours).
Exhibit Dates: 1/90–1/93.
Objectives: To provide enlisted personnel with the basic electronic fundamentals necessary to maintain, troubleshoot, and repair specific Loran equipment.
Instruction: The program comprises the basic ELF sequence, a common core curriculum covering transistors, power supplies, amplifiers, oscillators, logic, microcomputers, and special circuits; ELF 08 and 12 comprising communication theory and AM/FM and SSB transceivers and antennas; LOR 18 and 19 comprising remote Austron 5000 and calculator-assisted Loran-C. The subject matter covers the theoretical and operational concepts of a navigation system, is qualitative in nature, and includes block diagram analysis and schematic signal tracing to the component level.

Credit Recommendation: In the vocational certificate category, 8 semester hours in basic electronic fundamentals and 1 in Loran service and repair (2/83); in the lower-division baccalaureate/associate degree category, 3 semester hours in logic circuits and microcomputers in a computer science curriculum, 2 in communications systems, and 1 in electronics communications laboratory (2/83).

CG-1715-0098

LORAN TRACK #6 FOR LORSTAS, DANA, FALLON, GEORGE, GESASHI, HOKKAIDO, IWO JIMA, LAMPEDUSKA, MARCUS, NARROWCAPE SEARCHLIGHT, TOK, YAP

Course Number: None.
Location: Training Center, Governors Island, NY.
Length: 26 weeks (780 hours).
Exhibit Dates: 1/90–1/93.
Objectives: To provide enlisted personnel with the basic electronics fundamentals necessary to maintain, troubleshoot, and repair a specific Loran equipment.
Instruction: The program comprises the basic ELF sequence, a common core curriculum covering transistors, power supplies, amplifiers, oscillators, logic, microcomputers and special circuits; ELF 08 and 12 comprising communication theory and AM/FM and SSB transceivers and antennas; and LOR 06 and 17 comprising Loran-C timing and control equipment and Loran-C high power vacuum tube transmitters. The subject matter covers the theoretical and operational concepts of a navigation system, is qualitative in nature, and includes block diagram analysis and schematic signal tracing to the component level.
Credit Recommendation: In the vocational certificate category, 8 semester hours in electronics fundamentals and 1 in Loran service and repair (2/83); in the lower-division baccalaureate/associate degree category, 3 semester hours in circuits and microcomputers in a computer science curriculum, 2 in communications systems, and 1 in electronics communications laboratory (2/83).

CG-1715-0099

LORAN TRACK #7 FOR LORSTA, MIDDLETOWN

Course Number: None.
Location: Training Center, Governors Island, NY.
Length: 28 weeks (820 hours).
Exhibit Dates: 1/90–1/93.
Objectives: To provide enlisted personnel with the basic electronics fundamentals necessary to maintain, troubleshoot, and repair specific Loran equipment.
Instruction: The program comprises the basic ELF sequence, a common core curriculum covering transistors, power supplies, amplifiers, oscillators, logic, microcomputers, and special circuits; ELF 08 and 12 comprising communication theory and AM/FM and SSB transceivers and antennas; and LOR 06,17,18, and 19 comprising Loran-C timing and control equipment, high power vacuum tube transmitters, remote Austron 5000, and calculator-assisted Loran-C. The subject matter covers the theoretical and operational concepts of a navigation system, is qualitative in nature, and includes block diagram analysis and schematic signal tracing to the component level.

Credit Recommendation: In the vocational certificate category, 8 semester hours in electronics fundamentals and 1 in Loran service and repair (2/83); in the lower-division baccalaureate/associate degree category, 3 semester hours in circuits and microcomputers in a computer science curriculum, 2 in communications systems, and 1 in electronic communications laboratory (2/83).

CG-1715-0100

LORAN TRACK #8 FOR LORSTA, SELLIA MARINA

Course Number: None.
Location: Training Center, Governors Island, NY.
Length: 34 weeks (1020 hours).
Exhibit Dates: 1/90–1/93.
Objectives: To provide enlisted personnel with the basic electronic fundamentals necessary to maintain, troubleshoot, and repair specific Loran equipment.
Instruction: The program is comprised of the basic ELF sequence, a common core curriculum covering transistors, power supplies, amplifiers, oscillators, logic, microcomputers, and special circuits; ELF 08 and 12 comprising communication theory and AM/FM and SSB transceivers and antennas; LOR 06,13,16, and 19 comprising Loran-C timer and control equipment, Austron 5000 control monitor equipment, low-powered vacuum tube transmitter, and calculator-assisted Loran-C. The subject matter covers the theoretical and operational concepts of a navigation system, is qualitative in nature, and includes block diagram analysis and schematic signal tracing to the component level.
Credit Recommendation: In the vocational certificate category, 8 semester hours in basic electronics fundamentals and 1 in Loran repair and service (2/83); in the lower-division baccalaureate/associate degree category, 3 semester hours in logic circuits and microcomputers in a computer science curriculum, 2 in communications systems, and 1 in electronic communications laboratory (2/83).

CG-1715-0101

LORAN TRACK #9 FOR LORSTA, JOHNSTON ISLAND, ST. PAUL ISLAND

Course Number: None.
Location: Training Center, Governors Island, NY.
Length: 26 weeks (780 hours).
Exhibit Dates: 1/90–1/93.
Objectives: To provide enlisted personnel with the basic electronic fundamentals necessary to maintain, troubleshoot, and repair specific Loran equipment.
Instruction: The program is comprised of the basic ELF sequence, a common core curriculum covering transistors, power supplies, amplifiers, oscillators, logic, microcomputers, and special circuits; ELF 09 and 12 comprising communications theory and AM/FM and SSB transceivers and antennas; LOR 06,13, and 16 comprising Loran-C timers and control equipment, Austron 5000 control monitor equipment, and low-powered vacuum tube transmitters. The subject matter covers the theoretical and operational concepts of a navigation system, is qualitative in nature, and includes block diagram analysis and schematic signal tracing to the component level.

Credit Recommendation: In the vocational certificate category, 8 semester hours in basic electronic fundamentals and 1 in Loran service and repair (2/83); in the lower-division baccalaureate/associate degree category, 3 semester hours in logic circuits and microcomputers in a computer science curriculum, 2 in communications systems, and 1 in electronics communications laboratory (2/83).

CG-1715-0102

LORAN TRACK #10 LORSTA, SHETLAND ISLAND

Course Number: None.
Location: Training Center, Governors Island, NY.
Length: 26 weeks (780 hours).
Exhibit Dates: 1/90–1/93.
Objectives: To provide enlisted personnel with the basic electronics fundamentals necessary to maintain, troubleshoot, and repair specific Loran equipment.
Instruction: The program comprises the basic ELF sequence, a common core curriculum covering transistors, power supplies, amplifiers, oscillators, logic, microcomputers, and special circuits; ELF 08 and 12 comprising communication theory and AM/FM and SSB transceivers and antennas; and LOR 10 covering the AN/FPN-46 timer. The subject matter covers the theoretical and operational concepts of a navigation system, is qualitative in nature, and includes block diagram analysis and schematic signal tracing to the component level.
Credit Recommendation: In the vocational certificate category, 8 semester hours in basic electronic fundamentals and 1 in Loran service and repair (2/83); in the lower-division baccalaureate/associate degree category, 3 semester hours in circuits and microcomputers in a computer science program, 2 in communications systems, and 1 in electronics communications laboratory (2/83).

CG-1715-0103

1. FIRE CONTROL SYSTEM MK 92 MOD 1 OPERATION AND MAINTENANCE
2. MK 92 MOD 1 FIRE CONTROL SYSTEM

Course Number: Version 1: FT-5. Version 2: FT-2.
Location: Reserve Training Center, Yorktown, VA.
Length: Version 1: 19 weeks (718 hours). Version 2: 18 weeks (672 hours).
Exhibit Dates: Version 1: 1/95–Present. Version 2: 1/90–12/94.
Learning Outcomes: Version 1: Upon completion of the course, the student will be familiar with the initialization, operation, and documentation of a specific fire control system. The student uses basic tools and instruments to detect electrical failures of components in pulse transmitters, power supplies, receivers, video equipment, and servo systems. Version 2: Upon completion of the course, the student will be familiar with the initialization, operation, and documentation of a specific fire control system. The student uses basic tools and instruments to detect electrical failures of components in pulse transmitters, power supplies, receivers, video equipment, and servo systems.
Instruction: Version 1: Well documented procedures are used to maintain, troubleshoot, and replace faulty components on the fire control system. Fundamentals theory is not covered. Version 2: Well documented procedures are used to maintain, troubleshoot, and replace faulty components on the fire control system. Fundamentals theory is not covered.
Credit Recommendation: Version 1: In the lower-division baccalaureate/associate degree category, 1 semester hour in electrical control systems maintenance and repair (5/96). Version 2: In the lower-division baccalaureate/associate degree category, 1 semester hour in electrical control systems maintenance and repair (9/90).

CG-1715-0104

1. MK 27 GYROCOMPASS SYSTEM
2. MK 27 GYROCOMPASS SYSTEM OPERATION AND MAINTENANCE

Course Number: Version 1: EM-20. Version 2: EM-20.
Location: Reserve Training Center, Yorktown, VA.
Length: Version 1: 2 weeks (60 hours). Version 2: 2 weeks (67-70 hours).
Exhibit Dates: Version 1: 8/92–Present. Version 2: 1/90–7/92.
Learning Outcomes: Version 1: Upon completion of the course, students using the manual and appropriate test equipment, the student will be able to maintain, troubleshoot, repair, and calibrate gyrocompass equipment, including synchro signal amplifier, power transfer unit, and power converter to manufacturer's standards. Version 2: Upon completion of the course, students using the manual and appropriate test equipment, the student will be able to maintain, troubleshoot, repair, and calibrate gyrocompass equipment, including synchro signal amplifier, power transfer unit, and power converter to manufacturer's standards.
Instruction: Version 1: Instruction is very specific and detailed and geared specifically to the Mk 27 Mod 1 gyrocompass. With the aid of a synchro trainer, digital meter, and job sheet, the student performs zeroing, cleaning, troubleshooting, and maintenance. Version 2: Instruction is very specific and detailed and geared specifically to the Mk 27 Mod 1 gyrocompass. With the aid of a synchro trainer, digital meter, and job sheet, the student performs zeroing, cleaning, and maintenance.
Credit Recommendation: Version 1: In the lower-division baccalaureate/associate degree category, 1 semester hour in electronics laboratory (5/96). Version 2: In the upper-division baccalaureate category, 1 semester hour in electronics laboratory (10/88).

CG-1715-0105

ADVANCED ELECTRICAL/ELECTRONICS

Course Number: EM-17.
Location: Reserve Training Center, Yorktown, VA.
Length: 11 weeks (400-409 hours).
Exhibit Dates: 1/90–12/92.
Learning Outcomes: Upon completion of the course, the student will be able to demonstrate mathematical concepts, DC and AC circuit analysis, the use of test equipment, the concepts of discrete devices, integrated circuits, logic circuits, analog converters, flip-flop circuits, binary shift registers, amplifier circuits, digital display circuits, and other electronic circuits.
Instruction: Course provides lectures, demonstrations, and practical exercises in concepts, fundamentals, and circuit analysis in electricity and electronics. It also includes the use of training devices and test equipment.
Credit Recommendation: In the lower-division baccalaureate/associate degree category, 2 semester hours in AC and DC circuits, 4 in solid state circuits, 3 in logic circuits, 1 in electronics laboratory, and 1 in technical mathematics (10/88).

CG-1715-0106

MK 29 MOD 1 GYROCOMPASS OPERATION AND MAINTENANCE

Course Number: EM-26.
Location: Reserve Training Center, Yorktown, VA.
Length: 1-2 weeks (55-58 hours).
Exhibit Dates: 1/90–Present.
Learning Outcomes: Upon completion of the course, the student will be able to operate specialized tools required to to check gyrocompass performance during simulated roll, pitch, and azimuth movements; further, the student is qualified to exchange components and lubricate the bearings.
Instruction: The course content emphasizes the preventive maintenance and troubleshooting on the gyrocompass.
Credit Recommendation: In the lower-division baccalaureate/associate degree category, 1 semester hour in gyrocompass maintenance (1/96).

CG-1715-0107

ELECTRICAL/ELECTRONIC CONTROL FOR FIRE TUBE BOILERS/OILY WATER SEPARATOR AND ELECTRICAL GENERATOR

(Electrical/Electronic Control for Fire Tube Boilers/Oily Water Separator/Electrical Generators Operation and Maintenance, Class C)

Course Number: EM-21.
Location: Reserve Training Center, Yorktown, VA.
Length: 2 weeks (66-70 hours).
Exhibit Dates: 1/90–Present.
Learning Outcomes: Upon completion of the course, the student will be able to operate and maintain the electrical and electronic control systems for auxiliary boilers, oily water separators, and electrical generators.
Instruction: Lectures and practical exercises in the testing and troubleshooting of control circuits of boilers, separators, and generators as applied to specific Coast Guard equipment.
Credit Recommendation: In the lower-division baccalaureate/associate degree category, 1 semester hour in electrical/electronic laboratory (1/96).

CG-1715-0108

ELECTRONICS TECHNICIAN, CLASS A

Course Number: Version 1: None. Version 2: None.
Location: Version 1: Training Center, Petaluma, CA. Version 2: Training Center, Governors Island, NY.
Length: Version 1: 17-18 weeks (612-636 hours). Version 2: 14-15 weeks (465-515 hours).
Exhibit Dates: Version 1: 2/91–Present. Version 2: 1/90–1/91.
Learning Outcomes: Version 1: Upon completion of the course, the student will be able to

demonstrate knowledge of electronic safety procedures; use common laboratory instruments and test equipment; apply Ohm's law to circuits and networks; analyze AC circuits; demonstrate an understanding of digital principles; demonstrate an understanding of solid state theory and circuits; use troubleshooting techniques to repair electronic equipment; demonstrate an understanding of communications theory and equipment, such as single sideband receivers; and demonstrate an understanding of radar fundamentals. *Version 2:* Upon completion of the course, the student will be able to demonstrate knowledge of electronic safety procedures; use common laboratory instruments and test equipment; apply Ohm's law to circuits and networks; analyze AC circuits; demonstrate an understanding of digital principles; demonstrate an understanding of solid state theory and circuits; use troubleshooting techniques to repair electronic equipment; demonstrate an understanding of communications theory and equipment, such as single sideband receivers; and demonstrate an understanding of radar fundamentals.

Instruction: *Version 1:* Lectures and practical exercises in electronic theory, electronic safety, corrective and preventive maintenance techniques, and electronic administrative procedures. *Version 2:* Lectures and practical exercises in electronic theory, electronic safety, corrective and preventive maintenance techniques, and electronic administrative procedures.

Credit Recommendation: *Version 1:* In the lower-division baccalaureate/associate degree category, 4 semester hours in communications diagnosis and repair, 4 in solid state electronics, 2 in DC electronics, 2 in AC electronics, 2 in digital circuits, and 2 in electronic instrumentation laboratory (5/97). *Version 2:* In the lower-division baccalaureate/associate degree category, 3 semester hours in basic electronics laboratory, 1 in DC circuits, 1 in AC circuits, 3 in digital principles, 4 in solid state electronics, and 4 in electronic communications (11/88).

CG-1715-0109

TELEPHONE TECHNICIAN, CLASS A

Course Number: *Version 1:* TT A. *Version 2:* TT A. *Version 3:* TT A.
Location: Training Center, Petaluma, CA; Training Center, Governors Island, NY.
Length: *Version 1:* 23-24 weeks (772-812 hours). *Version 2:* 19-20 weeks (812 hours). *Version 3:* 22 weeks (748 hours).
Exhibit Dates: *Version 1:* 3/93–Present. *Version 2:* 1/92–2/93. *Version 3:* 1/90–12/91.
Learning Outcomes: *Version 1:* Upon completion of the course, the student will be able to use common laboratory instruments and test equipment to troubleshoot and maintain telephone and telecommunications analog and digital equipment; demonstrate telecommunications theory; and practice safety, installation, operation, corrective and preventive maintenance techniques, pole and tower erection, pole climbing, cable assembly, and installation. *Version 2:* Upon completion of the course, the student will be able to use common laboratory instruments and test equipment to troubleshoot and perform maintenance on telephone and telecommunications equipment; demonstrate telecommunications theory; and practice safety, installation, operation, corrective and preventive maintenance techniques, pole and tower erection, pole climbing, cable assembly, and installation. *Version 3:* Upon completion of the course the student will be able to use common laboratory instruments and test equipment; apply Ohm's law to circuits and networks; analyze DC and AC circuits; demonstrate an understanding of digital principles and circuits including multiplexers and demultiplexers; demonstrate an understanding of solid state theory and circuits; use troubleshooting techniques to repair telephone and carrier circuits; and demonstrate a basic knowledge of special telephone network systems and teletypewriter systems.

Instruction: *Version 1:* Lectures and practical laboratory in electricity, electronic circuits, components, solid state theory and practice, printed circuit techniques, digital logic circuits, modulation techniques, communication cable troubleshooting and repair, telephone equipment and circuits, troubleshooting techniques, and electronic switch equipment repair. *Version 2:* Lectures and practical laboratory in electricity, electronic circuits, components, communication cable troubleshooting and repair, telephone equipment and circuits, troubleshooting techniques, and electronic switch equipment repair. *Version 3:* Lectures and practical laboratory in electricity, electronic circuits, components, solid state theory and practice, printed circuit techniques, digital logic circuits, modulation techniques, communication cable troubleshooting and repair, telephone equipment and circuits, troubleshooting techniques, and teletypewriter equipment repair.

Credit Recommendation: *Version 1:* In the vocational certificate category, 1 semester hour in consumer electronics, 2 in electronic assembly, and 3 in telephone communications (9/97); in the lower-division baccalaureate/associate degree category, 2 semester hours in basic electronic laboratory, 1 in DC circuits, 1 in AC circuits, 10 in analog and digital electronic systems troubleshooting and maintenance, and 2 in digital electronic communications (9/97). *Version 2:* In the vocational certificate category, 1 semester hour in consumer electronics, 2 in electronic assembly, and 3 in telephone communications (9/93); in the lower-division baccalaureate/associate degree category, 2 semester hours in basic electronics laboratory, 1 in DC circuits, 1 in AC circuits, 10 in electronic systems troubleshooting and maintenance, and 2 in electronic communications (9/93). *Version 3:* In the lower-division baccalaureate/associate degree category, 3 semester hours in basic electronics laboratory, 1 in DC circuits, 1 in AC circuits, 3 in digital principles, 4 in solid state electronics, 3 in electronic communications, and 3 in telephone communications (11/88).

CG-1715-0110

AN/FPN-44A TRANSMITTER MAINTENANCE, CLASS C
(AN/FPN-44A Loran Transmitter, Class C)

Course Number: *Version 1:* LOR-05. *Version 2:* LOR-05.
Location: Training Center, Petaluma, CA; Training Center, Governors Island, NY.
Length: *Version 1:* 3 weeks (89 hours). *Version 2:* 1-2 weeks (53 hours).
Exhibit Dates: *Version 1:* 10/91–Present. *Version 2:* 1/90–9/91.
Learning Outcomes: *Version 1:* Upon completion of the course, the student will be able to operate, repair, troubleshoot, and maintain a Loran-C electronic location transmitter. *Version 2:* Upon completion of the course, the student will be able to operate, repair, troubleshoot, and maintain a Loran-C electronic location transmitter.

Instruction: *Version 1:* Lectures and practical exercises in the operation, logical troubleshooting, repair, and preventive maintenance of the equipment. *Version 2:* Lectures and practical exercises in the operation, logical troubleshooting, repair, and preventive maintenance of the equipment.

Credit Recommendation: *Version 1:* In the lower-division baccalaureate/associate degree category, 1 semester hour in electronic systems troubleshooting and maintenance and 1 in electronic communications (9/93). *Version 2:* In the lower-division baccalaureate/associate degree category, 1 semester hour in electronic systems troubleshooting and maintenance (11/88).

CG-1715-0111

1. LORAN-C TIMING AND CONTROL OPERATIONS AND MAINTENANCE, CLASS C
2. TIMING AND CONTROL EQUIPMENT, CLASS C

Course Number: *Version 1:* LOR-02. *Version 2:* LOR-02.
Location: Training Center, Petaluma, CA.
Length: *Version 1:* 3 weeks (97 hours). *Version 2:* 2-3 weeks (110 hours).
Exhibit Dates: *Version 1:* 10/91–Present. *Version 2:* 1/90–9/91.
Learning Outcomes: *Version 1:* Upon completion of the course, the student will be able to operate, repair, troubleshoot, and maintain Loran-C electronic location equipment. *Version 2:* Upon completion of the course, the student will be able to operate, repair, troubleshoot, and maintain Loran-C electronic location equipment.

Instruction: *Version 1:* Lectures and practical exercises in the operation, logical troubleshooting, repair, and preventive maintenance of the specific equipment. *Version 2:* Lectures and practical exercises in the operations, logical troubleshooting, repair, and preventive maintenance of the specific equipment.

Credit Recommendation: *Version 1:* In the lower-division baccalaureate/associate degree category, 2 semester hours in electronic systems troubleshooting and maintenance and 1 in electronic communications (9/93). *Version 2:* In the lower-division baccalaureate/associate degree category, 2 semester hours in electronic systems troubleshooting and maintenance (11/88).

CG-1715-0112

1. AN/URT-41(V)2 10 KW TRANSMITTER SYSTEM MAINTENANCE
2. AN/URT-41(V) TRANSMITTING SYSTEM, CLASS C

Course Number: *Version 1:* COM-02. *Version 2:* COM-02.
Location: *Version 1:* Training Center, Petaluma, CA. *Version 2:* Training Center, Governors Island, NY.
Length: *Version 1:* 2-3 weeks (86 hours). *Version 2:* 2-3 weeks (94 hours).
Exhibit Dates: *Version 1:* 1/92–Present. *Version 2:* 1/90–12/91.
Learning Outcomes: *Version 1:* Upon completion of the course, the student will be able to operate, align, troubleshoot, and maintain a voice/teletype RF high power transmitter using appropriate safety procedures. *Version 2:* Upon completion of the course, the student will be

able to operate, align, troubleshoot, and maintain a high power radio transmitter.

Instruction: *Version 1:* Instruction is presented using lectures, demonstrations, and laboratory exercises in operation, fault isolation diagnostics, component replacement, and system maintenance. Classroom and laboratory exercises are conducted for a specific radio transmitter that will be used by graduates on the job. *Version 2:* Lectures and laboratory exercises in the operation, logical troubleshooting, repair, and preventive maintenance on a specific radio transmitter.

Credit Recommendation: *Version 1:* In the vocational certificate category, 1 semester hour in electronic troubleshooting and RF transmitter operation and maintenance (9/93); in the lower-division baccalaureate/associate degree category, 2 semester hours in high power RF transmission theory and principles including system circuitry and fault isolation (9/93). *Version 2:* In the lower-division baccalaureate/associate degree category, 2 semester hours in electronic systems troubleshooting and maintenance (11/88).

CG-1715-0113

CEJT-SX-100/200 MITEL ELECTRONIC PRIVATE AUTOMATIC BRANCH EXCHANGE GENERIC 217, CLASS C

Course Number: TEL-10.
Location: Training Center, Governors Island, NY.
Length: 2 weeks (75 hours).
Exhibit Dates: 1/90–9/93.
Learning Outcomes: Upon completion of the course, the student will be able to operate, repair, troubleshoot, and maintain an automatic telephone exchange.
Instruction: Course provides lectures and practical exercises on the operation, logical troubleshooting, repair, and preventive maintenance of the automatic telephone exchange.
Credit Recommendation: In the lower-division baccalaureate/associate degree category, 2 semester hours in electronic systems troubleshooting and maintenance (11/88).

CG-1715-0114

1. LORAN-C CONTROL STATION OPERATIONS, CLASS C
2. LORAN-C CONTROL STATION OPERATIONS

Course Number: *Version 1:* LOR-08. *Version 2:* LOR-08.
Location: *Version 1:* Training Center, Petaluma, CA. *Version 2:* Training Center, Governors Island, NY.
Length: *Version 1:* 2 weeks (65-66 hours). *Version 2:* 2 weeks (78 hours).
Exhibit Dates: *Version 1:* 10/91–Present. *Version 2:* 1/90–9/91.
Learning Outcomes: *Version 1:* Upon completion of the course, the student will be able to operate, perform basic line monitoring and calculation, troubleshoot, and maintain the calculator-assisted Loran-C electronic location system. The student will be able to rescue an electric shock victim and to extinguish an electrical fire. *Version 2:* Upon completion of the course, the student will be able to operate, perform base line monitoring, troubleshoot, and maintain the calculator-assisted Loran-C electronic location system.
Instruction: *Version 1:* Course provides lectures, demonstrations, and practical exercises in equipment functions, operation, base line control, troubleshooting, and maintenance of the Loran-C system. *Version 2:* Lectures, demonstrations, and practical exercises in equipment function, operation, base line control, troubleshooting, and maintenance of the Loran-C system.
Credit Recommendation: *Version 1:* In the lower-division baccalaureate/associate degree category, 2 semester hours in electronic systems troubleshooting and maintenance (9/93). *Version 2:* In the lower-division baccalaureate/associate degree category, 1 semester hour in electronic systems troubleshooting and maintenance (11/88).

CG-1715-0115

TEL-14 CDXC-SG-1/1A PULSE 120 ELECTRONIC PRIVATE AUTOMATIC BRANCH EXCHANGE TELEPHONE SYSTEM, CLASS C

Course Number: TEL-14.
Location: Training Center, Governors Island, NY.
Length: 1-2 weeks (56 hours).
Exhibit Dates: 1/90–11/97.
Learning Outcomes: Upon completion of the course, the student will be able to install, maintain, and troubleshoot the electronic private automatic branch exchange equipment.
Instruction: Course provides lectures and laboratory exercises in safety precautions, equipment operation, programming exercises, preventive maintenance, and troubleshooting techniques.
Credit Recommendation: In the lower-division baccalaureate/associate degree category, 1 semester hour in electronic systems troubleshooting and maintenance (11/88).

CG-1715-0116

1. MITEL SX-200D INSTALLATION AND MAINTENANCE, CLASS C
2. CEJT-SX-200 MITEL ELECTRONIC PRIVATE AUTOMATIC BRANCH EXCHANGE GENERIC 1000 OR 1001 TELEPHONE SYSTEM

Course Number: *Version 1:* TEL-13. *Version 2:* TEL-13.
Location: *Version 1:* Training Center, Petaluma, CA. *Version 2:* Training Center, Governors Island, NY.
Length: *Version 1:* 2 weeks (76 hours). *Version 2:* 2 weeks (76 hours).
Exhibit Dates: *Version 1:* 10/93–Present. *Version 2:* 1/90–9/93.
Learning Outcomes: *Version 1:* Upon completion of the course, the student will be able to install, troubleshoot, and maintain a private automated branch exchange. *Version 2:* Upon completion of the course, the student will be able to install and maintain the electronic private automatic branch exchange.
Instruction: *Version 1:* Instruction includes lecture, demonstrations, and hands-on exercises with a digital telephone switch. Topics include installation, checkout, operation, diagnostics, board replacement, and maintenance of a PABX system. Course includes analog to digital conversion and T1 telephone connect. *Version 2:* Course provides lectures and laboratory exercises in safety precautions, programming exercises, installation procedures, and troubleshooting techniques.
Credit Recommendation: *Version 1:* In the lower-division baccalaureate/associate degree category, 3 semester hours in digital electronic switching systems troubleshooting and maintenance (9/93). *Version 2:* In the lower-division baccalaureate/associate degree category, 2 semester hours in electronic systems troubleshooting and maintenance (11/88).

CG-1715-0117

AN/FPN-42 LORAN TRANSMITTER, CLASS C

Course Number: *Version 1:* LOR-04. *Version 2:* LOR-04.
Location: Training Center, Petaluma, CA; Training Center, Governors Island, NY.
Length: *Version 1:* 2 weeks (66 hours). *Version 2:* 1-2 weeks (45 hours).
Exhibit Dates: *Version 1:* 10/91–Present. *Version 2:* 1/90–9/91.
Learning Outcomes: *Version 1:* Upon completion of the course, the student will be able to operate, align, troubleshoot, and maintain the Loran electronic location transmitter. *Version 2:* Upon completion of the course, the student will be able to operate, align, troubleshoot, and maintain the Loran electronic location transmitter.
Instruction: *Version 1:* Course provides lectures and laboratory exercises in troubleshooting techniques and maintenance of specific Loran transmitter. *Version 2:* Lectures and laboratory exercises in troubleshooting techniques and maintenance of a specific Loran transmitter.
Credit Recommendation: *Version 1:* In the lower-division baccalaureate/associate degree category, 1 semester hour in electronic system troubleshooting and maintenance and 1 in electronic communications (9/93). *Version 2:* In the lower-division baccalaureate/associate degree category, 1 semester hour in electronic systems troubleshooting and maintenance (11/88).

CG-1715-0118

AN/FPN-39 TRANSMITTER MAINTENANCE, CLASS C
(AN/FPN-39 Loran Transmitter, Class C)

Course Number: *Version 1:* LOR-03. *Version 2:* LOR-03.
Location: Training Center, Petaluma, CA; Training Center, Governors Island, NY.
Length: *Version 1:* 2 weeks. *Version 2:* 1-2 weeks (45 hours).
Exhibit Dates: *Version 1:* 10/91–Present. *Version 2:* 1/90–9/91.
Learning Outcomes: *Version 1:* Upon completion of the course, the student will be able to operate, repair, troubleshoot, and maintain a Loran-C electronic location transmitter. *Version 2:* Upon completion of the course, the student will be able to operate, repair, troubleshoot, and maintain a Loran-C electronic location transmitter.
Instruction: *Version 1:* Lectures and practical exercises in the operation, logical troubleshooting, repair, and preventive maintenance of the Loran-C transmitter. *Version 2:* Lectures and practical exercises in the operation, logical troubleshooting, repair, and preventive maintenance of the Loran-C transmitter.
Credit Recommendation: *Version 1:* In the lower-division baccalaureate/associate degree category, 1 semester hour in electronic systems troubleshooting and maintenance and 1 in electronic communication (9/93). *Version 2:* In the lower-division baccalaureate/associate degree category, 1 semester hour in electronic systems troubleshooting and maintenance (11/88).

COURSE EXHIBITS

CG-1715-0119

AN/SPS-64(V)1,2,3 RADARS

Course Number: NAV-02.
Location: Training Center, Governors Island, NY.
Length: 3 weeks (117 hours).
Exhibit Dates: 1/90–Present.
Learning Outcomes: Upon completion of the course, the student will be able to operate, align, troubleshoot, and maintain the specific radar.
Instruction: Lectures and laboratory exercises in troubleshooting techniques and maintenance of specific radar systems.
Credit Recommendation: In the lower-division baccalaureate/associate degree category, 3 semester hours in electronic systems troubleshooting and maintenance (11/88).

CG-1715-0120

AN/SPS-64(V)4 RADAR

Course Number: NAV-03.
Location: Training Center, Governors Island, NY.
Length: 1-2 weeks (61 hours).
Exhibit Dates: 1/90–12/91.
Learning Outcomes: Upon completion of the course, the student will be able to operate, align, troubleshoot, and maintain the specific radar.
Instruction: Lectures and laboratory exercises in the operation, logical troubleshooting, repair, and preventive maintenance on a specific type of radar system.
Credit Recommendation: In the lower-division baccalaureate/associate degree category, 2 semester hours in electronic systems troubleshooting and maintenance (11/88).

CG-1715-0121

AN/URC-9 RADIO SET, CLASS C

Course Number: COM-04.
Location: Training Center, Governors Island, NY.
Length: 2 weeks (78 hours).
Exhibit Dates: 1/90–Present.
Learning Outcomes: Upon completion of the course, the student will be able to perform corrective and preventive maintenance on a specific transceiver.
Instruction: Course provides lectures and laboratory exercises in equipment operation, parts location, troubleshooting, and preventive maintenance.
Credit Recommendation: In the lower-division baccalaureate/associate degree category, 2 semester hours in electronic systems troubleshooting and maintenance (11/88).

CG-1715-0122

AN/SPS-66 AND AN/SPS-66A RADARS, CLASS C

Course Number: NAV-01.
Location: Training Center, Governors Island, NY.
Length: 2 weeks (69 hours).
Exhibit Dates: 1/90–Present.
Learning Outcomes: Upon completion of the course, the student will be able to operate, align, troubleshoot, and maintain the specific radars.
Instruction: Lectures and laboratory exercises in the operation, logical troubleshooting, repair, and preventive maintenance on a specific type of radar system.
Credit Recommendation: In the lower-division baccalaureate/associate degree category, 2 semester hours in electronic systems troubleshooting and maintenance (11/88).

CG-1715-0123

AN/FPN-64(V) TRANSMITTER MAINTENANCE (AN/FPN-64(V) Loran Solid State Transmitter, Class C)

Course Number: *Version 1:* LOR-06. *Version 2:* LOR-06.
Location: Training Center, Petaluma, CA; Training Center, Governors Island, NY.
Length: *Version 1:* 2 weeks (72-73 hours). *Version 2:* 2 weeks (70 hours).
Exhibit Dates: *Version 1:* 10/91–Present. *Version 2:* 1/90–9/91.
Learning Outcomes: *Version 1:* Upon completion of the course, the student will be able to operate, repair, troubleshoot, and maintain a Loran-C electronic location transmitter. *Version 2:* Upon completion of the course, the student will be able to operate, repair, troubleshoot, and maintain a Loran-C electronic location transmitter.
Instruction: *Version 1:* Lectures and practical exercises in the operation, logical troubleshooting, repair, and preventive maintenance of the transmitter. *Version 2:* Lectures and practical exercises in the operation, logical troubleshooting, repair, and preventive maintenance of the transmitter.
Credit Recommendation: *Version 1:* In the lower-division baccalaureate/associate degree category, 1 semester hour in electronic control circuits and 1 in electronic communications (9/93). *Version 2:* In the lower-division baccalaureate/associate degree category, 2 semester hours in electronic systems troubleshooting and maintenance (11/88).

CG-1715-0124

1. AN/URC-114(V) 1 KW TRANSCEIVER SYSTEM MAINTENANCE
2. AN/URC-114(V) LOW POWER (1KW) COMMUNICATIONS SYSTEM, CLASS C

Course Number: *Version 1:* COM-03. *Version 2:* COM-03.
Location: *Version 1:* Training Center, Petaluma, CA. *Version 2:* Training Center, Governors Island, NY.
Length: *Version 1:* 3 weeks (105 hours). *Version 2:* 3 weeks (118 hours).
Exhibit Dates: *Version 1:* 1/92–Present. *Version 2:* 1/90–12/91.
Learning Outcomes: *Version 1:* Upon completion of the course, the student will be able to operate, troubleshoot, and maintain a low-power voice/teletype RF transceiver system in a safe environment. *Version 2:* Upon completion of the course, the student will be able to troubleshoot, repair, and maintain a low-power communications system.
Instruction: *Version 1:* Instruction includes lectures, demonstrations, and laboratory exercises in operation, logical diagnosis for fault isolation, component replacement, and system maintenance. Classroom and practical exercises are conducted for a specific transceiver system that will be used by graduates on the job. *Version 2:* Lectures, demonstrations, and practical exercises in the operation, logical troubleshooting, repair, and maintenance of the equipment.
Credit Recommendation: *Version 1:* In the vocational certificate category, 1 semester hour in electronic logical fault isolation, low-power transceiver operation and maintenance (9/93); in the lower-division baccalaureate/associate degree category, 2 semester hours in low-power RF transmission theory and principles, including system circuits and diagnosis (9/93). *Version 2:* In the lower-division baccalaureate/associate degree category, 3 semester hour in electronic systems troubleshooting and maintenance (11/88).

CG-1715-0125

PRIMARY CHAIN MONITOR SET MAINTENANCE, CLASS C

Course Number: *Version 1:* LOR-07. *Version 2:* LOR-07.
Location: Training Center, Petaluma, CA; Training Center, Governors Island, NY.
Length: *Version 1:* 2 weeks (53 hours). *Version 2:* 2-3 weeks (85 hours).
Exhibit Dates: *Version 1:* 10/91–Present. *Version 2:* 1/90–9/91.
Objectives: *Version 1:* Upon completion of the course, the student will be able to safely maintain, repair, and operate basic AC power supply circuits, digital communication circuits, antenna coupler assemblies, notch filter assemblies, A-D converter circuits, AGC circuits, training control circuits, serial data links, and digitally controlled circuits related to the Loran-C primary chain monitor set. *Version 2:* Upon completion of the course, the student will be able to operate, repair, troubleshoot, and maintain monitor set equipment.
Instruction: *Version 1:* Course provides lectures, demonstrations, and practical exercises in the maintenance, troubleshooting, repair, and operations of specific electronic equipment. *Version 2:* Course provides lectures and practical exercises in the operation, logical troubleshooting, repair, and preventive maintenance of the specific equipment.
Credit Recommendation: *Version 1:* In the lower-division baccalaureate/associate degree category, 2 semester hours in electronic applications for troubleshooting, repair, and maintenance of digital and communication circuits (9/93). *Version 2:* In the lower-division baccalaureate/associate degree category, 2 semester hours in electronic systems troubleshooting and maintenance (11/88).

CG-1715-0127

NAUTEL RADIOBEACON MAINTENANCE (Videograph B Fog Detector Maintenance)

Course Number: ANC-RB; ANC-FD.
Location: Reserve Training Center, Yorktown, VA.
Length: 2 weeks (41 hours).
Exhibit Dates: 1/90–6/94.
Learning Outcomes: Upon completion of the course, the student will be able to install, operate, align and repair circuits boards and chassis components for the fog detector.
Instruction: Classroom lectures and laboratory exercises involving installation, operation, and preventive and corrective maintenance of the fog detector.
Credit Recommendation: In the lower-division baccalaureate/associate degree category, 2 semester hours in electromechanical technician for the last segment of the two-week program. This is a course which consisted of

two one-week courses run consecutively. Credit is recommended for course number ANC-FD only (11/88).

CG-1715-0128

1. LIGHTHOUSE TECHNICIAN, CLASS C
2. AUTOMATED AIDS TO NAVIGATION LIGHTHOUSE TECHNICIAN
(Lighthouse Technician)

Course Number: *Version 1:* ANC-LT. *Version 2:* ANC-LT.
Location: Reserve Training Center, Yorktown, VA.
Length: *Version 1:* 4 weeks (146 hours). *Version 2:* 3-4 weeks (115 hours).
Exhibit Dates: *Version 1:* 3/93–Present. *Version 2:* 1/90–2/93.
Learning Outcomes: *Version 1:* Upon completion of the course, the student will be able to operate, maintain, and repair, to component level category one through four, automated lighthouse equipment. *Version 2:* Upon completion of the course, the student will be able to operate, maintain, repair to component level category 1 through 4 automated lighthouse equipment.
Instruction: *Version 1:* Instruction includes lectures and practical exercises, that provide hands-on experience. Topics include environmental control systems, fire suppression systems, sound and light systems, and AC/DC power systems. Aids control monitor systems are also covered. Data obtained is used to upgrade/update Coast Guard directives and publications. *Version 2:* Instruction includes lectures and practical exercises, that provide hands-on experience. Topics include environmental control systems, fire suppression systems, sound and light systems, and AC/DC power systems. Aids control monitor systems are also covered. Data obtained is used to upgrade/update Coast Guard directives and publications.
Credit Recommendation: *Version 1:* In the lower-division baccalaureate/associate degree category, 2 semester hours in material management (5/96). *Version 2:* In the lower-division baccalaureate/associate degree category, 3 semester hours in AC/DC power systems and 2 in record keeping (4/89).

CG-1715-0130

RADARMAN, CLASS A

Course Number: *Version 1:* RD. *Version 2:* None.
Location: Reserve Training Center, Yorktown, VA.
Length: *Version 1:* 14 weeks (492 hours). *Version 2:* 12 weeks (470-474 hours).
Exhibit Dates: *Version 1:* 9/92–Present. *Version 2:* 1/90–8/92.
Learning Outcomes: *Version 1:* Upon completion of the course, the student will be able to gather, process, display, evaluate, and disseminate operational radar data. *Version 2:* Upon completion of the course, the student will be able to gather, process, display, evaluate, and disseminate operational data.
Instruction: *Version 1:* Course provides lectures and practical exercises in administration, communications, radar, navigation systems, rules of the road, safety, search and rescue, operational and preventative maintenance, combat information center operations, detection equipment, and data interpretation. *Version 2:* Course provides lectures and practical exercises in administration, communications, navigation systems, rules of the road, safety, search and rescue, operational and preventative maintenance, combat information center operations, detection equipment, and data interpretation.
Credit Recommendation: *Version 1:* In the lower-division baccalaureate/associate degree category, 3 semester hours in radar navigation (5/96). *Version 2:* In the vocational certificate category, 1 semester hour in basic use and operation of a radar system (4/89).

CG-1715-0131

ADVANCED MINOR AIDS TO NAVIGATION MAINTENANCE

Course Number: ANC-AM.
Location: Reserve Training Center, Yorktown, VA.
Length: 2 weeks (76 hours).
Exhibit Dates: 1/90–10/91.
Learning Outcomes: Upon completion of the course, the student will be able to install, troubleshoot, operate, and maintain minor navigation equipment.
Instruction: Instruction is given in operation, troubleshooting, and maintenance of minor aids, beacons, buoys, and sound signals; buoy tender operations encompassing buoy deck procedures, buoy positioning techniques, and piloting and safety; and minor-aid construction. Emphasis is placed upon technical report writing and oral presentations.
Credit Recommendation: In the lower-division baccalaureate/associate degree category, 3 semester hours in analysis of AC/DC circuits (4/89).

CG-1715-0132

AVIATION ELECTRONICS TECHNICIAN FIRST CLASS BY CORRESPONDENCE

Course Number: 103-7.
Location: Coast Guard Institute, Oklahoma City, OK.
Length: Maximum, 104 weeks.
Exhibit Dates: 1/90–Present.
Learning Outcomes: Upon completion of the course, the student will be able to identify potential safety hazards in the work area/flight line; repair current, voltage, and resistance measuring devices; use test equipment to measure frequency, waveforms, modulation, and standing waves; describe the operating characteristics of electromagnetic wave measuring devices including wavemeters and bridges; repair printed circuit boards; and list the advantages and disadvantages of various types of wave guides and the operating principles of microwave antennas and feed systems.
Instruction: The major topics covered are safety, voltage, current, and resistance measurement; test equipment to measure frequency, waveforms, modulation, and standing waves; and repair of printed circuit boards, wave guides, and microwave antennas. Training is provided in a nonresident, self-instruction format. Students are required to successfully complete a proctored end-of-course examination.
Credit Recommendation: In the lower-division baccalaureate/associate degree category, 3 semester hours in electronic communications (12/92).

CG-1715-0133

SONAR TECHNICIAN SECOND CLASS BY CORRESPONDENCE

Course Number: 243-2.
Location: Coast Guard Institute, Oklahoma City, OK.
Length: Maximum, 104 weeks.
Exhibit Dates: 1/90–6/97.
Learning Outcomes: Upon completion of the course, the student will be able to identify sounds produced by surface ships, submarines, and marine life; perform preventive maintenance on sonar equipment, X-Y recorders, signal generators, and megohmeters; prepare weekly and monthly preventive maintenance schedules; list calibration procedures for hydrophone; and replace sonar transducers.
Instruction: Major topics covered are sound identification, preventative maintenance for sonar equipment, acoustical test equipment operation, preparation of monthly preventive maintenance schedules, and calibration. Training is provided in a nonresident, self-instruction format. Students are required to successfully complete a proctored end-of-course examination.
Credit Recommendation: Credit is not recommended because of the military-specific nature of the course (6/90).

CG-1715-0134

TELEPHONE TECHNICIAN FIRST CLASS BY CORRESPONDENCE

Course Number: *Version 1:* 145-7. *Version 2:* 145-6.
Location: Coast Guard Institute, Oklahoma City, OK.
Length: *Version 1:* Maximum, 156 weeks. *Version 2:* Maximum, 78 weeks.
Exhibit Dates: *Version 1:* 10/96–Present. *Version 2:* 1/90–9/96.
Learning Outcomes: *Version 1:* This version is pending evaluation. *Version 2:* Upon completion of the course, the student will be able to install and maintain telephone cable, operate data communication equipment, employ cable fault locating methodology and test equipment methods, identify interior and exterior power distribution systems, construct letters and messages in proper format and meeting security requirements, use procedures for commercial procurement, and employ the principles of property management and control.
Instruction: *Version 1:* This version is pending evaluation. Training is provided in a nonresident, self-instruction format. Students are required to successfully complete a proctored end-of-course examination. *Version 2:* Topics include troubleshooting techniques and equipment for telephone cable; troubleshooting and construction techniques for power lines; transformer load balancing; power line safety procedures; residential service drops and wiring; theory of cable construction; operation and function of data communication equipment; data communication codes; use of Coast Guard directives, letters, messages, and memoranda; and use of property management and control techniques. Training is provided in a nonresident, self-instruction format. Students are required to successfully complete a proctored end-of-course examination.
Credit Recommendation: *Version 1:* Pending evaluation. *Version 2:* In the vocational certificate category, 3 semester hours in outside

telephone cable installation and maintenance (6/90).

CG-1715-0135

FIRE CONTROL TECHNICIAN FIRST CLASS BY CORRESPONDENCE

Course Number: *Version 1:* 0127-8. *Version 2:* 127-7; 127-6.
Location: Coast Guard Institute, Oklahoma City, OK.
Length: *Version 1:* Maximum, 156 weeks. *Version 2:* Maximum, 78 weeks.
Exhibit Dates: *Version 1:* 2/94–Present. *Version 2:* 1/90–1/94.
Learning Outcomes: *Version 1:* This version is pending evaluation. Upon completion of the course, the student will be able to employ safety procedures in the operation and maintenance of fire control equipment; explain the theory of operation of microwave components, antennas, and associated servo equipment; explain ballistic characteristics of major projectiles; explain the operation of fire control equipment in an operational environment; explain battery alignment of a weapons control system; identify proper sources for equipment maintenance; and identify administrative procedures and reports. *Version 2:* Upon completion of the course, the student will be able to employ safety procedures in the operation and maintenance of fire control equipment; explain the theory of operation of microwave components, antennas, and associated servo equipment; explain ballistic characteristics of major projectiles; explain the operation of fire control equipment in an operational environment; explain battery alignment of a weapons control system; identify proper sources for equipment maintenance; and identify administrative procedures and reports.
Instruction: *Version 1:* This version is pending evaluation. The major topics covered are safety, electronics, fire control fundamentals, maintenance, and administration. Training is provided in a nonresident, self-instruction format. Students are required to successfully complete a proctored end-of-course examination. *Version 2:* The major topics covered are safety, electronics, fire control fundamentals, maintenance, and administration. Training is provided in a nonresident, self-instruction format. Students are required to successfully complete a proctored end-of-course examination.
Credit Recommendation: *Version 1:* Pending evaluation. *Version 2:* In the lower-division baccalaureate/associate degree category, 3 semester hours in electronic theory (6/90).

CG-1715-0136

RADIOMAN FIRST CLASS BY CORRESPONDENCE

Course Number: 0141-7; 0141-8.
Location: Coast Guard Institute, Oklahoma City, OK.
Length: Maximum, 52 weeks.
Exhibit Dates: 1/90–Present.
Learning Outcomes: Upon completion of the course, the student will be able to prepare telecommunication reports; identify the various forms and procedures used in preparing telecommunication summaries and reports; and define terms and basic communications system components such as amplifiers, oscillators, exciters, multiplexers, receivers and transmitters.
Instruction: Topics include the construction and use of operation orders, operation plans, and pro forma messages, telecommunication systems and components, security clearances and investigations, and communication officer duties. Training is provided in a nonresident, self-instruction format. Students are required to successfully complete a proctored end-of-course examination.
Credit Recommendation: Credit is not recommended because of the limited, specialized nature of the course (6/90).

CG-1715-0137

FIRE CONTROL TECHNICIAN SECOND CLASS BY CORRESPONDENCE

Course Number: 0227-1; 227-9; 227-8.
Location: Coast Guard Institute, Oklahoma City, OK.
Length: Maximum, 156 weeks.
Exhibit Dates: 1/90–7/96.
Learning Outcomes: Upon completion of the course, the student will be able to observe safety precautions in dealing with electronic equipment, radioactive devices, toxic compounds, ordnance, and electronic radiation; explain basic electrical and electronic theory; explain the theory of operation of AC devices, vacuum tube circuitry, digital circuitry, and weapons control equipment; employ various base numbering systems and Boolean algebra; operate fire control equipment and associated computers and terminals; identify equipment alignment procedures and sources of maintenance; and identify administrative procedures and reports.
Instruction: The major topics are safety, electronic fundamentals, fire control fundamentals, maintenance, and administration. Training is provided in a nonresident, self-instruction format. Students are required to successfully complete a proctored end-of-course examination.
Credit Recommendation: In the vocational certificate category, 2 semester hours in electronic fundamentals (9/95).

CG-1715-0138

NAVMACS/SATCOM SYSTEMS MAINTENANCE

Course Number: COM-05.
Location: Training Center, Petaluma, CA.
Length: 3 weeks (106 hours).
Exhibit Dates: 4/92–Present.
Learning Outcomes: Upon completion of the course, the student will be able to operate, troubleshoot, and maintain a satellite telecommunications system with analog-digital conversion.
Instruction: Instruction includes lecture, demonstrations, and laboratory exercises on operating, logical fault isolation and diagnostics, component replacement, and system maintenance for an automated satellite telecommunication link including analog-digital conversion technology.
Credit Recommendation: In the vocational certificate category, 1 semester hour in operating, troubleshooting, and maintaining an automated satellite telecommunication system (9/93); in the lower-division baccalaureate/associate degree category, 3 semester hours in theory and diagnostics of digital, audio, and analog-digital signal conversion and satellite telecommunications (9/93).

CG-1715-0139

AN/SPS-64(V) LARGE CUTTER RADAR SYSTEMS MAINTENANCE

Course Number: NAV-06.
Location: Training Center, Petaluma, CA.
Length: 5 weeks (187 hours).
Exhibit Dates: 3/92–Present.
Learning Outcomes: Upon completion of the course, the student will be able to operate, repair, troubleshoot, and maintain a surface search radar system operating in the X band frequency with computer enhancement of displays. This course includes block-diagram analysis of faults and systematic troubleshooting techniques for component-level repair of transmitter, receiver, power supply, digital control, and display sections. This course is a combination of NAV-2, NAV-3, and NAV-4.
Instruction: Course provides small lecture/demonstration groups with laboratory practice on actual equipment. Emphasis is on using standard laboratory instruments to locate and replace defective components.
Credit Recommendation: In the lower-division baccalaureate/associate degree category, 3 semester hours in basic electronics laboratory, 2 in electronic systems troubleshooting and maintenance, 1 in electronic communications, 1 in solid state electronics, and 1 in RF power amplifiers (9/93).

CG-1715-0140

AB/SPS-64(V) SMALL CUTTER RADAR MAINTENANCE

Course Number: NAV-07.
Location: Training Center, Petaluma, CA.
Length: 4 weeks (138 hours).
Exhibit Dates: 3/92–Present.
Learning Outcomes: Upon completion of the course, the student will be able to operate, repair, troubleshoot, and maintain a surface search radar system operating in the X band frequency with computer-enhanced displays. This course includes block-diagram analysis of faults and systematic troubleshooting techniques for component-level repair of transmitter, receiver, power supply, digital control, and display sections. This course is a combination of NAV-02 and NAV-05.
Instruction: Course provides small lecture/demonstration, groups with laboratory practice on actual equipment. Emphasis is on using standard laboratory instruments to locate and replace defective components. Instruction is specific to the radar system.
Credit Recommendation: In the lower-division baccalaureate/associate degree category, 3 semester hours in basic electronics laboratory, 2 in electronic systems troubleshooting and maintenance, 1 in electronic communication, and 1 in solid state electronics (9/93).

CG-1715-0141

ADL-81/82 LORAN-C RECEIVER MAINTENANCE, CLASS C

Course Number: None.
Location: Aviation Technical Training Center, Elizabeth City, NC.
Length: 2 weeks (62 hours).
Exhibit Dates: 8/93–Present.
Learning Outcomes: Before 8/93 see CG-1715-0082. Upon completion of the course, the student will be able to operate and maintain long range navigation equipment.

Instruction: Instruction includes lectures, demonstrations, and practical exercises in the use of test equipment and troubleshooting of units of the navigation system.

Credit Recommendation: In the lower-division baccalaureate/associate degree category, 1 semester hour electronic component troubleshooting techniques and procedures laboratory (10/94).

CG-1715-0142

AVIATION ELECTRONICS TECHNICIAN 618M-3, CLASS C

Course Number: None.
Location: Aviation Technical Training Center, Elizabeth City, NC.
Length: 2 weeks (73 hours).
Exhibit Dates: 3/93–Present.
Learning Outcomes: Upon completion of the course, the student will be able to test, align, and troubleshoot the 618M-3 transceiver to the component level.
Instruction: Course provides lectures, demonstration, and laboratory exercises in the use of special test equipment to test and troubleshoot the subsystems of the transceiver, including the synthesizer, receiver, transmitter, and power supply.
Credit Recommendation: In the lower-division baccalaureate/associate degree category, 1 semester hour in electronic communication equipment repair (10/94).

CG-1715-0143

APS-127 RADAR SYSTEM MAINTENANCE, CLASS C

Course Number: None.
Location: Aviation Technical Training Center, Elizabeth City, NC.
Length: 3-4 weeks (141-142 hours).
Exhibit Dates: 8/93–Present.
Learning Outcomes: Upon completion of the course, the student will understand the function and operating characteristics of the radar system, performance tests, and user associated test equipment; and will be able to maintain and troubleshoot the system to the component level.
Instruction: Instruction includes lectures, demonstrations, and laboratory exercises in which performance tests are conducted on malfunctioning subsystems.
Credit Recommendation: In the lower-division baccalaureate/associate degree category, 3 semester hours in electronic equipment maintenance and troubleshooting (10/94).

CG-1715-0144

HH-65A AVIONICS MAINTENANCE, CLASS C

Course Number: None.
Location: Aviation Technical Training Center, Elizabeth City, NC.
Length: 3-4 weeks (136-137 hours).
Exhibit Dates: 11/93–Present.
Learning Outcomes: Upon completion of the course, the student will be able to operate, troubleshoot, and provide routine maintenance on avionics equipment as specified on the HH-65A avionics integrated logistics support plan.
Instruction: Instruction is provided through extensive hands-on experience in the operation, troubleshooting, inspection, and routine maintenance of the HH-65A avionics systems. Instruction includes audio control, V/VHF-AM communications, VHF-FM communications, high frequency communications, radar identification system, radar altimeter, UHF or VHF AM/FM direction finder, weather search radar, horizontal situation, radio display, underwater acoustic transmitter, signal interface, and global positioning.
Credit Recommendation: In the lower-division baccalaureate/associate degree category, 3 semester hours in avionic systems maintenance (10/94).

CG-1715-0145

HH-60J AVIONICS MAINTENANCE, CLASS C

Course Number: None.
Location: Aviation Technical Training Center, Elizabeth City, NC.
Length: 4 weeks (139-140 hours).
Exhibit Dates: 2/94–Present.
Learning Outcomes: Upon completion of the course, the student will be able to operate, troubleshoot, and provide routine maintenance on avionics equipment as specified on the HH-60J avionics integrated logistics support plan.
Instruction: Course provides lectures, demonstrations, self-paced instruction, and laboratory exercises in avionics systems, including HF secure voice system, HF communication system, radar identification system, Doppler radar system, and global positioning system.
Credit Recommendation: In the lower-division baccalaureate/associate degree category, 3 semester hours in avionics systems maintenance (10/94).

CG-1715-0146

ELECTRONICS TECHNICIAN SECOND CLASS BY CORRESPONDENCE (ET2)

Course Number: 0221-7.
Location: Coast Guard Institute, Oklahoma City, OK.
Length: Maximum, 156 weeks.
Exhibit Dates: 4/90–Present.
Learning Outcomes: Before 4/90 see CG-1715-0037. Upon completion of the course, the student will have learned the operation of semiconductor devices including diodes, transistors, UJTs, FETs, SCRs, and other special devices; digital operations including Boolean algebra, logistics, counters, registers, decoders and adders; cable installation and termination, routing, and securing; and military specifications of connectors and connections.
Instruction: Course includes topics in the general operation of solid state diodes, transistors, unijunction transistors, field effect transistors, silicon controlled rectifiers, and special semiconductor devices; digital circuit operations including Boolean algebra, logic gates, counters, registers, decoders and adders; theory of oscilloscope operation and interpretation of CRT oscillograms; cable termination and installation including routing and securing coaxial cable, military specification connectors and connection soldering procedures. Students are required to complete an end of course exam at an 80 percent level.
Credit Recommendation: In the lower-division baccalaureate/associate degree category, 3 semester hours in electronic circuit applications (3/96).

CG-1715-0147

ADVANCED ELECTRICITY, ELECTRONICS, AND HYDRAULICS

Course Number: GM-10.
Location: Reserve Training Center, Yorktown, VA.
Length: 4-5 weeks (182 hours).
Exhibit Dates: 10/95–Present.
Learning Outcomes: Upon completion of the course, the student will be able to take a schematic drawing of a hydraulic, AC electric, DC electric, and electronic system and construct a functioning system.
Instruction: Hydraulic, electric, and electronic systems covered using lectures, demonstrations, exercises, and practical exercises. Components are identified, operating parameters calculated, and systems constructed and tested until they meet performance standards.
Credit Recommendation: In the lower-division baccalaureate/associate degree category, 1 semester hour in hydraulics, 1 in AC circuits, 1 in DC circuits, and 1 in basic electronics laboratory (5/96).

CG-1715-0148

ADVANCED DIGITAL ELECTRONICS TECHNOLOGY, CLASS C

Course Number: EM-2.
Location: Reserve Training Center, Yorktown, VA.
Length: 5 weeks (172-182 hours).
Exhibit Dates: 11/92–Present.
Learning Outcomes: Upon completion of the course, the student will have a basic familiarity with single digital circuits, binary and hexademical number systems, troubleshoot and write code for microprocessors, and operate analog-to-digital converters.
Instruction: The course is taught by lectures and practical laboratory experiments combined with troubleshooting sessions or tutorials. Laboratory test equipment is also covered.
Credit Recommendation: In the lower-division baccalaureate/associate degree category, 2 semester hours in digital principles (5/96).

CG-1715-0149

ADVANCED ANALOG ELECTRONICS TECHNOLOGY, CLASS C

Course Number: EM-1.
Location: Reserve Training Center, Yorktown, VA.
Length: 7-8 weeks (280-295 hours).
Exhibit Dates: 11/92–Present.
Learning Outcomes: Upon completion of the course, the student will apply a basic understanding of analog electronic circuits to practical operations and troubleshooting.
Instruction: This course is taught using lectures, laboratory experiments, and laboratory problem solving activities.
Credit Recommendation: In the lower-division baccalaureate/associate degree category, 3 semester hours in electronic control circuits (5/96).

CG-1717-0007

RESERVE SENIOR PETTY OFFICER LEADERSHIP AND MANAGEMENT

Course Number: None.
Location: Training Center, Petaluma, CA.
Length: 2 weeks (73 hours).
Exhibit Dates: 1/90–8/90.
Learning Outcomes: Upon completion of the course, the student will be able to exercise

management skills in communication and leadership in assigned supervisory positions.

Instruction: Instruction includes situational leadership, power, motivation, change, group dynamics, and decision making, team building, communications, performance management, and problem solving. Instructional methods include lecture and discussion with emphasis on role playing, case studies, structured experiences, and group activities.

Credit Recommendation: In the lower-division baccalaureate/associate degree category, 3 semester hours in leadership (8/87).

CG-1717-0008

RESERVE OFFICER LEADERSHIP AND MANAGEMENT

Course Number: None.
Location: Training Center, Petaluma, CA.
Length: 2 weeks (73 hours).
Exhibit Dates: 1/90–8/90.
Learning Outcomes: Upon completion of the course, the student will be able to exercise management skills in communication, team building, and leadership in assigned supervisory positions.
Instruction: Instruction includes communications, leadership development, team building, conflict management, change management, and problem solving.
Credit Recommendation: In the lower-division baccalaureate/associate degree category, 3 semester hours in leadership (8/87).

CG-1717-0009

OFFICER LEADERSHIP AND MANAGEMENT

Course Number: None.
Location: Training Center, Petaluma, CA.
Length: 2 weeks (71 hours).
Exhibit Dates: 1/90–8/90.
Learning Outcomes: Upon completion of the course, the student will be able to exercise management skills in communication, team building, and leadership in assigned supervisory positions.
Instruction: Instruction includes communications, leadership development, team building, conflict management, change management, and problem solving. Instructional methods include lecture and discussion, practical exercises, case studies, and role playing.
Credit Recommendation: In the lower-division baccalaureate/associate degree category, 3 semester hours in leadership (8/87).

CG-1717-0010

SENIOR PETTY OFFICER LEADERSHIP AND MANAGEMENT

Course Number: None.
Location: Training Center, Petaluma, CA.
Length: 2 weeks (71 hours).
Exhibit Dates: 1/90–8/90.
Learning Outcomes: Upon completion of the course, the student will be able to exercise management skills in communication and leadership in assigned supervisory positions.
Instruction: Instruction includes situational leadership, power, motivation, change, group dynamics and decision making, team building, communications, performance management, and problem solving. Instructional methods include lecture and discussion with emphasis on role playing, case studies, structured experiences, and group activities.
Credit Recommendation: In the lower-division baccalaureate/associate degree category, 3 semester hours in leadership (8/87).

CG-1722-0006

SMALL BOAT MAGNETIC COMPASS CALIBRATION BY CORRESPONDENCE

Course Number: 0641.
Location: Coast Guard Institute, Oklahoma City, OK.
Length: Maximum, 104 weeks.
Exhibit Dates: 1/90–8/95.
Objectives: To introduce the theory and basic and practical application of magnetic compass calibration.
Instruction: Topics include the basic theory of magnetism, causes of variation and deviation and their practical application, methods of determining deviation, and procedures for adjusting small boat magnetic compasses to reduce deviation. Students are required to successfully complete a proctored end-of-course examination.
Credit Recommendation: In the lower-division baccalaureate/associate degree category, 1 semester hour in nautical science (5/85).

CG-1722-0010

COASTAL SEARCH PLANNING, CLASS C

Course Number: None.
Location: Reserve Training Center, Yorktown, VA; Training Center, Governors Island, NY.
Length: 2 weeks (76 hours).
Exhibit Dates: 1/90–10/92.
Learning Outcomes: Upon completion of the course, the student will be able to plan and coordinate, in accordance with the National Search and Rescue Manual, a search and rescue mission with emphasis on coastal waters.
Instruction: Course provides lectures in coastal search and rescue planning, including the use of private and public facilities, communications, emergency care, position plotting, water search patterns, documentation, legal aspects, and public relations.
Credit Recommendation: In the lower-division baccalaureate/associate degree category, 1 semester hour in basic navigation (11/88).

CG-1722-0011

MARITIME RESERVE COORDINATION CENTER CONTROLLER
(Comprehensive Search and Rescue Planning, Class C)

Course Number: None.
Location: Reserve Training Center, Yorktown, VA; Training Center, Governors Island, NY.
Length: 2 weeks (85 hours).
Exhibit Dates: 1/90–10/92.
Learning Outcomes: Upon completion of the course, the student will be able to plan and coordinate a search and rescue mission at sea in accordance with the National Search and Rescue Manual.
Instruction: Course provides instruction on the organization, documentation, public information procedures, international rescue facilities, navigation, and, with consideration of wind- and ocean-induced drift, determination of most probable position of emergency. Thirty-five hours are devoted to practical exercises in a laboratory environment.
Credit Recommendation: In the lower-division baccalaureate/associate degree category, 1 semester hour in basic navigation and 1 in search and rescue procedures (11/88).

CG-1722-0012

COASTAL DEFENSE PLANNER, PORT LEVEL

Course Number: MS 732.
Location: Reserve Training Center, Yorktown, VA.
Length: 2 weeks (46 hours).
Exhibit Dates: 1/90–Present.
Learning Outcomes: Upon completion of the course, the student will be able to prepare port-level operation plans to support missions of the Coast Guard and the Maritime Defense Zone in national emergencies.
Instruction: The method of instruction includes some lecture and self-paced study, but is mostly practical experiences working in groups under supervision.
Credit Recommendation: In the lower-division baccalaureate/associate degree category, 3 semester hours in management problems (4/89).

CG-1722-0013

COASTAL DEFENSE EXERCISE PLANNER, PORT LEVEL

Course Number: MS 735R.
Location: Reserve Training Center, Yorktown, VA.
Length: 2 weeks (56 hours).
Exhibit Dates: 1/90–Present.
Learning Outcomes: Upon completion of the course, the student will be able to plan, write reports, and evaluate results of field training exercises in coastal defense; write operation orders; execute planning, budgeting, billeting, and controlling of the field training exercise; and evaluate in order to identify problem areas and recommend corrective actions.
Instruction: The course contains lectures and practical exercises providing hands-on experience in planning, executing, and evaluating port-level requirements for field training exercises in coastal defense.
Credit Recommendation: In the lower-division baccalaureate/associate degree category, 3 semester hours in technical and report writing (4/89).

CG-1722-0014

1. QUARTERMASTER THIRD CLASS, BY CORRESPONDENCE
2. QUARTERMASTER THIRD CLASS BY CORRESPONDENCE

Course Number: Version 1: 0337-2; 0337-1. Version 2: 0337-1; 0337-8; 0337-9.
Location: Coast Guard Institute, Oklahoma City, OK.
Length: Version 1: Maximum, 156 weeks. Version 2: Maximum, 78 weeks.
Exhibit Dates: Version 1: 8/94–Present. Version 2: 1/90–7/94.
Learning Outcomes: Version 1: Upon completion of the course, the student will be able to demonstrate a working knowledge of basic navigation, aids to navigation, bridge communication, interpretation and use of nautical charts and publications, use of tide and current tables, maritime calculations, rules of the road, and

ceremonial procedures. *Version 2:* Upon completion of the course, the student will be able to demonstrate a working knowledge of basic navigation, bridge communication, interpretation and use of nautical charts and publications, use of tide and current tables, rules of the road and aids to navigation, and ceremonial procedures.

Instruction: *Version 1:* Documents read and studied covering navigation, bridge communication, chart projections, navigation aids, maritime calculations, watchstanding, tides and currents, and weather. A proctored test will be administered at the end of the course. *Version 2:* Eight pamphlets will be read and studied covering four subject areas. Questions can be answered at the end of each subject. Areas of instruction include navigation, bridge communication, tides and currents, and weather. A proctored test will be administered at the end of the course.

Credit Recommendation: *Version 1:* In the lower-division baccalaureate/associate degree category, 1 semester hour basic navigation (7/95). *Version 2:* In the lower-division baccalaureate/associate degree category, 1 semester hour in basic navigation (4/92).

CG-1722-0015

QUARTERMASTER SECOND CLASS BY CORRESPONDENCE

Course Number: *Version 1:* 0237-8. *Version 2:* 0237-6; 0237-7.
Location: Coast Guard Institute, Oklahoma City, OK.
Length: *Version 1:* Maximum, 156 weeks. *Version 2:* Maximum, 78 weeks.
Exhibit Dates: *Version 1:* 6/94–Present. *Version 2:* 1/90–5/94.
Learning Outcomes: *Version 1:* Upon completion of the course, the student will be able to demonstrate proficiency in procurement; search and rescue procedures; maneuvering boards; fundamentals of surface navigation; navigation of charts, and proper procedures for honor and ceremonial events. *Version 2:* Upon completion of the course, the student will be able to demonstrate proficiency in the fundamentals of meteorology; communication procedures, message designators, and formats; fundamentals of surface navigation; control of ship's magnetic field; navigational charts; and proper procedures for ceremonies.

Instruction: *Version 1:* Instruction is provided through a nonresident, self-paced study program using Coast Guard supplied study material. The course is not task-oriented, but is comprehensive in content and objectives. Subject matter covers procurement, search and rescue procedures, maneuvering boards, fundamentals of surface navigation, navigational charts, and proper procedures for ceremonial occasions. A proctored final examination is required. *Version 2:* Instruction is provided through a nonresident, self-paced study program. Subject matter covers fundamentals of surface navigation, communication procedures, fundamentals of meteorology, control of vessel's magnetic field, and proper procedures for ceremonial occasions. The course includes a proctored final examination.

Credit Recommendation: *Version 1:* In the lower-division baccalaureate/associate degree category, 2 semester hours in coastwise navigation and piloting (5/95). *Version 2:* In the lower-division baccalaureate/associate degree category, 1 semester hours in fundamentals of meteorology and 2 in coastwise navigation and piloting (6/94).

CG-1722-0016

FIREMAN BY CORRESPONDENCE

Course Number: 0450-1.
Location: Coast Guard Institute, Oklahoma City, OK.
Length: Maximum, 78 weeks.
Exhibit Dates: 1/90–Present.
Learning Outcomes: Upon completion of the course, the student will be able to perform as a fireman aboard Coast Guard vessels and shore units, maintain and repair engines and auxiliary machinery, use hand tools, and practice damage control.

Instruction: Instruction is provided through a nonresident, self-paced study program consisting of 16 reading and study assignments covering auxiliary machinery, main propulsion machinery, damage control, and hand tools. Includes proctored end-of-course examination.

Credit Recommendation: In the vocational certificate category, 1 semester hour in survey of hand tools (9/90); in the lower-division baccalaureate/associate degree category, 3 semester hours in marine engineering (9/90).

CG-1723-0005

MACHINERY TECHNICIAN, CLASS A

Course Number: *Version 1:* None. *Version 2:* None. *Version 3:* MKA.
Location: Reserve Training Center, Yorktown, VA.
Length: *Version 1:* 11 weeks (380 hours). *Version 2:* 11-12 weeks (399 hours). *Version 3:* 11 weeks (334 hours).
Exhibit Dates: *Version 1:* 11/94–Present. *Version 2:* 7/91–10/94. *Version 3:* 1/90–6/91.
Learning Outcomes: *Version 1:* Upon completion of the course, the student will be able to describe the operation and parts of internal combustion engines; explain the use and operation of marine engine cooling systems; perform tests on engine fuels and lubricants; explain the operation of engine governors, diesel injectors, gearing, clutch assemblies, and drive shafts; describe the basic principles of electricity engine ignition and starting, and battery charging; describe the operation of auxiliary boilers and related components; explain the operation of a refrigeration system; and perform the duties and tasks of a small craft engineer. *Version 2:* Upon completion of the course, the student will be able to describe the operation of internal combustion engines; explain the use and operation of marine engine cooling systems; perform tests on engine fuels and lubricants; explain the operation of engine governors, diesel injectors, gearing, clutch assemblies, and drive shafts; describe the basic principles of electricity, engine ignition and starting, and battery charging; describe the operation of auxiliary boilers and related components; explain the operation of a refrigeration system; and perform the duties and tasks of a small craft engineer. *Version 3:* Upon completion of the course, the student will be able to describe the operation and parts of two-stroke and four-stroke gasoline and diesel engines; explain the components and functions in an axial flow gas turbine engine; explain the use and operation of marine engine cooling systems; perform tests on engine fuels and lubricants; explain the operation of engine governors, diesel injectors, gearing, clutch assemblies, and drive shafts; describe the principles of electricity, engine ignition and starting, and charging; describe the operation of auxiliary boilers and related components; explain the operation of a refrigeration system; and perform the duties and tasks of a small craft engineer.

Instruction: *Version 1:* Course provides lectures and laboratories covering basic internal combustion engines, starting systems, cooling systems, lubricants, reclaiming lube oil, governors, fuel systems, diesel engine tune-up, clutches, gearing, air compressors, electricity, ignition, generators/alternators, small boat engineering, blower/turbochargers, auxiliary boilers, water distillation, hydraulics, and refrigeration. *Version 2:* Course provides lectures and laboratories covering basic internal combustion engines, starting systems, cooling systems, lubricants, reclaiming lube oil, governors, fuel systems, diesel engine tune-up, clutches, gearing, air compressors, electricity, ignition, generators/alternators, small boat engineering, blower/turbochargers, auxiliary boilers, water distillation, hydraulics, and refrigeration. *Version 3:* Course provides lectures and laboratories covering basic internal combustion engines, axial flow gas turbine engines, starting systems, cooling systems, lubricants, reclaiming lube oil, governors, fuel systems, diesel engine tune-up, clutches, gearing, air compressors, electricity, ignition, generators/alternators, small boat engineering, blower/turbochargers, auxiliary boilers, water distillation, and refrigeration.

Credit Recommendation: *Version 1:* In the lower-division baccalaureate/associate degree category, 2 semester hours in internal combustion engines, 3 in marine engine auxiliary systems, 1 in engine electrical systems, and 1 in basic refrigeration (5/96). *Version 2:* In the lower-division baccalaureate/associate degree category, 3 semester hours in internal combustion engines, 3 in marine engine auxiliary systems, 1 in engine electrical systems, and 1 in basic refrigeration (5/96). *Version 3:* In the lower-division baccalaureate/associate degree category, 3 semester hours in internal combustion engines, 1 in marine drive train, 2 in engine electrical systems, 1 in basic refrigeration, and 1 in small craft engineering (9/90).

CG-1723-0006

MACHINERY TECHNICIAN FIRST CLASS BY CORRESPONDENCE

Course Number: 0132-3.
Location: Coast Guard Institute, Oklahoma City, OK.
Length: Maximum, 52 weeks.
Exhibit Dates: 1/90–Present.
Learning Outcomes: Upon completion of the course, the student will be able to diagnose and repair internal combustion engines, both gasoline and diesel. This includes fuel injection, engine controls, power transmission, and basic repairs.

Instruction: Topics include troubleshooting, diagnosis, and repair of marine internal combustion engines (gasoline and diesel). The student will understand engine components, efficiency, power transmission, fuel systems, and related components. Instruction is self-instruction with a proctored end-of-course examination.

CG-1728-0005

FIREFIGHTING ASHORE BY CORRESPONDENCE

Course Number: 478-2.
Location: Coast Guard Institute, Oklahoma City, OK.
Length: Maximum, 104 weeks.
Exhibit Dates: 1/90–8/92.
Objectives: To provide training in the principles and techniques of fire fighting ashore.
Instruction: Topics include Coast Guard port safety authority and policy, roles of the fire fighter, chemistry and behavior of fire, personnel safety and protective breathing apparatus, ropes, knots, and hitches, portable extinguishers, fire hose, fire streams and water supply, ladders, fire apparatus, rescue, forcible entry, ventilation, salvage and overhaul, sizing up and tactics/strategy used in fire fighting, and hazardous material incident analysis. The principle textbook for this course is Fire Service Practices for Volunteer Fire Departments published by the International Fire Service Training Association. Students are required to successfully complete a proctored final examination.
Credit Recommendation: In the lower-division baccalaureate/associate degree category, 2 semester hours in fire science technology (5/85).

CG-1728-0007

FIREFIGHTING ON VESSELS BY CORRESPONDENCE

Course Number: 427; 427-3.
Location: Coast Guard Institute, Oklahoma City, OK.
Length: Maximum, 104 weeks.
Exhibit Dates: 1/90–8/92.
Objectives: To provide training in the basic principles of shipboard fire fighting.
Instruction: Course describes pumping equipment, vapor detection devices, personal protective equipment, shipboard fire fighting systems, appliances and portable fire equipment. Students are required to successfully complete a proctored end-of-course examination.
Credit Recommendation: In the vocational certificate category, 1 semester hour in fire science technology and 1 in nautical science (5/85).

CG-1728-0009

PORT SECURITYMAN THIRD CLASS BY CORRESPONDENCE

Course Number: 365-1.
Location: Coast Guard Institute, Oklahoma City, OK.
Length: Maximum, 104 weeks.
Exhibit Dates: 1/90–6/97.
Objectives: To present basic principles of port security, law enforcement, fire protection and hazardous materials control.
Instruction: Course offers basic instruction in the enforcement of federal laws governing the transportation, storage and containment of vessels and their cargos in an industrial setting. The elements of arrest and search and seizure are covered including the concepts of jurisdiction, warrant and warrantless searches. This course also covers basic firefighting information, techniques, equipment and appliances, placarding and hazardous materials control. There is also coverage of environmental protection as related to the federal Water Pollution Control Act and regulations governing vessel cargo packaging. Students are required to successfully complete a proctored end-of-course examination.
Credit Recommendation: In the lower-division baccalaureate/associate degree category, 1 semester hour in industrial security (5/85).

CG-1728-0012

WATERFRONT PROTECTION BY CORRESPONDENCE

Course Number: 423-1.
Location: Coast Guard Institute, Oklahoma City, OK.
Length: Maximum, 104 weeks.
Exhibit Dates: 1/90–6/97.
Objectives: Covers the law enforcement duties of the captain of the port in relation to waterfront protection.
Instruction: Provides an overview of arrest and search procedures, fingerprinting, and self-defense.
Credit Recommendation: Credit is not recommended because of the military-specific nature of the course (5/85).

CG-1728-0013

LAW ENFORCEMENT BY CORRESPONDENCE

Course Number: 403-3.
Location: Coast Guard Institute, Oklahoma City, OK.
Length: Maximum, 104 weeks.
Exhibit Dates: 1/90–6/97.
Objectives: To provide training in the development and application of law enforcement duties and responsibilities of the Coast Guard in marine activities.
Instruction: Course covers the topics of constitutional law, rules of evidence, statutory authority, and principles of marine law enforcement. Course also includes navigation rules and regulations, jurisdiction, customs, boating safety and inspections, and investigative techniques. Students are required to successfully complete a proctored end-of-course examination.
Credit Recommendation: In the vocational certificate category, 1 semester hour in marine patrol and law enforcement (5/85).

CG-1728-0014

INVESTIGATOR 1 BY CORRESPONDENCE

Course Number: 0166-1.
Location: Coast Guard Institute, Oklahoma City, OK.
Length: Maximum, 104 weeks.
Exhibit Dates: 1/90–6/97.
Objectives: To provide criminal investigation training.
Instruction: Topics include the theory and application of criminal investigation techniques, including rules of evidence and court procedure, search and seizure, use of force, interrogation, use of informants, evidence collection and preservation, and coordination with other agencies at the state and federal level. Also includes coverage of physical plant security techniques and devices, security measures, background checks, and counterintelligence.

Credit Recommendation: In the lower-division baccalaureate/associate degree category, 3 semester hours in criminal investigation (5/85).

CG-1728-0017

INVESTIGATION DEPARTMENT

Course Number: MS 472R.
Location: Marine Safety School, Yorktown, VA.
Length: 3 weeks (112-113 hours).
Exhibit Dates: 1/90–12/93.
Objectives: To train student in investigative research techniques.
Instruction: Lectures and practical exercises regarding violations of laws, regulations, and acts of violence in the marine industry. Includes analysis of laws, collection and preservation of evidence, court procedures, and case preparation.
Credit Recommendation: In the lower-division baccalaureate/associate degree category, 3 semester hours in technical report writing, 3 in evidence collection and preservation, and 3 in introductory risk management (4/86).

CG-1728-0021

1. MARITIME LAW ENFORCEMENT BOARDING OFFICER, CLASS C
2. MARITIME LAW ENFORCEMENT BOARDING OFFICER

Course Number: *Version 1:* None. *Version 2:* None.
Location: Reserve Training Center, Yorktown, VA.
Length: *Version 1:* 4-5 weeks (161-165 hours). *Version 2:* 3-4 weeks (144 hours).
Exhibit Dates: *Version 1:* 2/91–Present. *Version 2:* 1/90–1/91.
Learning Outcomes: *Version 1:* Upon completion of the course, the student will be able to supervise a group of investigators involved in the detection, investigation, and processing of criminal and safety violations. *Version 2:* Upon completion of the course, the student will be able to perform all duties of maritime lawyers in enforcement positions in the areas of administrative inspections, search, seizure, arrest, use of force, crime scene processing, and case file preparation.
Instruction: *Version 1:* Methods of instruction include lectures, practical exercises, and demonstrations. This course focuses on the techniques and procedures for the detection and investigation of criminal violations and safety violations. Included are use of force, concept of suspicion, pat down searches, enforcement of criminal law, searches incident to arrest, protection of crime scene, crime scene processing, interrogation, collection and preservation of evidence, and report writing. *Version 2:* Course includes lectures, group discussions, and practical exercises in topic areas such as authority and jurisdiction, use of force, applied concepts of law enforcement, enforcement of law and treaties, US flag, fisheries, and criminal law procedures.
Credit Recommendation: *Version 1:* In the lower-division baccalaureate/associate degree category, 3 semester hours in basic criminal investigation and 2 in basic law enforcement organization and administration (5/96). *Version 2:* In the lower-division baccalaureate/associate degree category, 3 semester hours in law enforcement or police science and 1 in

(Credit Recommendation continued from previous: In the lower-division baccalaureate/associate degree category, 3 semester hours in internal combustion engines and 3 in automotive fuel systems (11/91).)

political science (4/89); in the upper-division baccalaureate category, 3 semester hours in criminal justice (4/89).

CG-1728-0024

MARITIME LAW ENFORCEMENT INSTRUCTOR

Course Number: MLE IC.
Location: Reserve Training Center, Yorktown, VA.
Length: 2 weeks (75 hours).
Exhibit Dates: 1/90–2/94.
Learning Outcomes: Upon completion of the course, the student will be able to perform the duties of a maritime law enforcement training officer at a field unit; and demonstrate an understanding of such topics as defensive tactics, field sobriety, and Coast Guard training techniques.
Instruction: Course includes lectures, group discussions, and practical exercises in use of force, sobriety enforcement, and instructor methodology.
Credit Recommendation: In the lower-division baccalaureate/associate degree category, 3 semester hours in instructional methods or teaching methods and 3 in police science or law enforcement (2/96).

CG-1728-0025

PORT PHYSICAL SECURITY PRACTICAL

Course Number: MS 423R.
Location: Reserve Training Center, Yorktown, VA.
Length: 2 weeks (85 hours).
Exhibit Dates: 1/90–12/91.
Learning Outcomes: Upon completion of the course, the student will be able to carry out all assigned duties relating to security and law enforcement of waterfront facilities, vessels, and waterways during peacetime and wartime.
Instruction: Course provides lectures, Socratic and practical exercises with emphasis on hands-on training demonstrations. Subjects include facility security, physical security system capabilities, cargo/personnel facility access control, patrol techniques, vessel security, application of cargo security principles to port security, risk, threat and vulnerability assessments, including OPSEC, COMSEC, and intelligence coordination, maritime counterterrorism, and law enforcement policies and techniques.
Credit Recommendation: Credit is not recommended because of the military-specific nature of the course (4/89).

CG-1728-0026

PORT PHYSICAL SECURITY MANAGEMENT

Course Number: MS 425R.
Location: Reserve Training Center, Yorktown, VA.
Length: 2 weeks (60 hours).
Exhibit Dates: 1/90–12/91.
Learning Outcomes: Upon completion of the course, the student will demonstrate management techniques as applicable to a port environment including Coast Guard responsibilities and security procedures and methods.
Instruction: Course material will be covered in a lecture format with some practical and written exams to assure understanding and comprehension.
Credit Recommendation: In the lower-division baccalaureate/associate degree category, 3 semester hours in administrative management (4/89).

CG-1728-0027

PORT SECURITY SAFETY ENLISTED

Course Number: MS 421R.
Location: Marine Safety School, Yorktown, VA.
Length: 2 weeks (67 hours).
Exhibit Dates: 1/90–4/90.
Learning Outcomes: Upon completion of the course, the student will be able to research and apply laws, regulations, and the objectives of marine safety as related to their applications in the handling of hazardous materials and enforce the laws and regulations as they apply to safety and security functions.
Instruction: Lectures and practical exercises in techniques of inspection and identification of violations of hazardous cargo storage and handling procedures and pollution prevention regulations. Course also includes law enforcement responsibilities.
Credit Recommendation: In the lower-division baccalaureate/associate degree category, 3 semester hours in a law enforcement practicum (4/89).

CG-1728-0029

1. MARINE SAFETY OFFICER ADT
2. PORT SECURITY AND SAFETY OFFICER

Course Number: *Version 1:* MS 402R. *Version 2:* MS 402R.
Location: Reserve Training Center, Yorktown, VA.
Length: *Version 1:* 2 weeks (70 hours). *Version 2:* 2 weeks (67 hours).
Exhibit Dates: *Version 1:* 2/94–Present. *Version 2:* 1/90–1/94.
Learning Outcomes: *Version 1:* Upon completion of the course, the student will be able to research and apply laws, regulations, and the objectives of marine safety to the handling of hazardous material and enforce laws and regulations as they apply to environmental safety and security. *Version 2:* Upon completion of the course, the student will be able to research and apply laws, regulations, and the objectives of marine safety to the handling of hazardous material and enforce laws and regulations as they apply to safety and security.
Instruction: *Version 1:* Methods of instruction include lectures, practical exercises, case study, and simulations. Topics include vessel boarding, waterfront facility inspections, enforcement options, incident response efforts, vessel examinations, and waterway management. *Version 2:* Course consists of lecture overview of the marine safety program, including vessel orientation, port safety, jurisdiction, laws and statutes, Coast Guard policies, and pollution control.
Credit Recommendation: *Version 1:* In the lower-division baccalaureate/associate degree category, 2 semester hours in marine and fresh water environmental safety (5/96). *Version 2:* In the lower-division baccalaureate/associate degree category, 3 semester hours in law enforcement practicum (4/89).

CG-1728-0030

MARINE SAFETY PETTY OFFICER

Course Number: *Version 1:* MS 400R. *Version 2:* MS 400R.
Location: Reserve Training Center, Yorktown, VA.
Length: *Version 1:* 5 weeks (242 hours). *Version 2:* 6 weeks (199-201 hours).
Exhibit Dates: *Version 1:* 9/93–Present. *Version 2:* 1/90–8/93.
Learning Outcomes: *Version 1:* Upon completion of the course, the student will be able to demonstrate competence in the areas of security and pollution, facility compliance, waterways management, pollution response authority, investigations, and pollution case files. *Version 2:* Upon completion of the course, the student will be able to demonstrate competence in security and pollution concerns, cargo and vessel compliance, facility compliance, waterways management, pollution response authority, investigations, and pollution case files.
Instruction: *Version 1:* Methods of instruction include lectures, demonstrations, practical exercises, laboratory exercises, and role playing. Topics include facility inspection, harbor control, vessel boarding, pollution response authority, pollution investigations, response organizations, and marine safety administration. *Version 2:* Course includes lectures, practical exercises, and exercise reviews in law and authority; handling of dangerous cargo manifests, such as commercial explosives, radioactive materials, flammable or combustible liquids, and pressurized materials; monitoring and boarding procedures; and monitoring investigations, post security functions, oil spill and hazardous chemical response, toxicological problems, flammable and vapor detection and measurement, and hazardous chemical transportation.
Credit Recommendation: *Version 1:* In the lower-division baccalaureate/associate degree category, 2 semester hours in industrial security technician, 2 in environmental management, and 2 in report writing (5/96). *Version 2:* In the lower-division baccalaureate/associate degree category, 3 semester hours in industrial security techniques, 3 in environmental management, 3 in industrial security practicum, and 3 in report writing (4/89).

CG-1728-0032

PORT SECURITYMAN SECOND CLASS BY CORRESPONDENCE

Course Number: 0267-1; 0265-2.
Location: Coast Guard Institute, Oklahoma City, OK.
Length: Maximum, 104 weeks.
Exhibit Dates: 1/90–Present.
Learning Outcomes: Upon completion of the course, the student will be able to demonstrate proficiency in physical plant protection through use of locks, fences, electronic security systems, and access controls; safety and security surveys of storage facilities for cargo/inventory including explosives and incendiary materials; and basic security enforcement procedures. Also includes military techniques as related to physical security.
Instruction: Instruction is provided through a nonresident, self-paced study program using five workbook pamphlets with charts, diagrams, illustrations, and review quizzes. Course is comprehensive in content and translatable to civilian physical plant security activities.
Credit Recommendation: In the lower-division baccalaureate/associate degree category, 3 semester hours in introduction to physi-

cal security and 1 in military science (combat tactics) (1/93).

CG-1728-0033

SAFETY AND SECURITY OF THE PORT: MILITARY EXPLOSIVES BY CORRESPONDENCE

Course Number: 0425-1.
Location: Coast Guard Institute, Oklahoma City, OK.
Length: Maximum, 52 weeks.
Exhibit Dates: 1/90–1/92.
Learning Outcomes: Upon completion of the course, the student will be able to use tables and charts for identification and segregation of military explosives; describe the procedures for inspecting dangerous cargo manifests; list the responsibilities and duties of a supervisory detail for loading explosive materials; and cite the safety precautions that must be observed when a cargo of military explosive is handled.
Instruction: Instruction is provided through a nonresident training course consisting of one pamphlet covering three topics. It is comprehensive in content and requires reading of text materials, solving problems, answering lesson quiz questions, and taking a proctored end-of-course test.
Credit Recommendation: In the lower-division baccalaureate/associate degree category, 1 semester hour in military science (ordnance safety and security) (9/90).

CG-1728-0034

FIRE AND SAFETY TECHNICIAN SECOND CLASS BY CORRESPONDENCE

Course Number: *Version 1:* 0225-3. *Version 2:* 225-2.
Location: Coast Guard Institute, Oklahoma City, OK.
Length: *Version 1:* Maximum, 156 weeks. *Version 2:* Maximum, 52 weeks.
Exhibit Dates: *Version 1:* 4/91–1/93. *Version 2:* 1/90–3/91.
Learning Outcomes: *Version 1:* Upon completion of the course, the student will demonstrate proficiency in marine fire protection, ship board fire suppression equipment, fire safety inspections, hazardous materials placarding, haz-mat and explosive storage and transfers, safety and regulations governing hazardous materials and explosives, EPA regulations regarding waterway pollution, incident monitoring, and initial response. *Version 2:* Upon completion of the course, the student will be able to demonstrate proficiency in basic marine fire protection, prevention, and suppression; hazardous material storage and transfer; pollution incident monitoring and supervision; and facility safety.
Instruction: *Version 1:* Instruction is provided through a nonresident, self-paced study program using four publications and a end-of-course test. This course is comprehensive in content and includes reading assignments, solving problems, and answering lesson quiz questions. *Version 2:* Instruction is provided through a nonresident, self-paced study program using five pamphlets and a proctored end-of-course test. The texts include reading assignments with objectives to be mastered and appropriate review exercises designed to test mastery of the objectives. This course is comprehensive in content.
Credit Recommendation: *Version 1:* In the lower-division baccalaureate/associate degree category, 3 semester hours in marine fire safety (3/96). *Version 2:* In the lower-division baccalaureate/associate degree category, 3 semester hours in shipboard fire safety and suppression (9/90).

CG-1728-0035

SECURITY OF CLASSIFIED INFORMATION BY CORRESPONDENCE

Course Number: 0402-4; 0402-3; 0402-2.
Location: Coast Guard Institute, Oklahoma City, OK.
Length: Maximum, 52 weeks.
Exhibit Dates: 1/90–6/97.
Learning Outcomes: Upon completion of the course, the student will be able to properly handle classified material by using security regulations, including knowing who is responsible for classification, how to obtain security clearances, and how to protect and store classified materials.
Instruction: Instruction is provided through a nonresident, self-paced study program using one pamphlet and a proctored end-of-course test. The text includes reading assignments with objectives to be mastered and appropriate review exercises.
Credit Recommendation: In the lower-division baccalaureate/associate degree category, 1 semester hour in military science (security of classified information) (9/90).

CG-1728-0036

SAFETY AND SECURITY OF THE PORT: MARINE ENVIRONMENTAL PROTECTION BY CORRESPONDENCE

Course Number: 0766-1.
Location: Coast Guard Institute, Oklahoma City, OK.
Length: Maximum, 52 weeks.
Exhibit Dates: 1/90–9/90.
Learning Outcomes: Upon completion of the course, the student will be able to describe contingency planning and the operational response phases for marine environmental protection; describe the national, regional, and local response organizations; and describe the role and duties of the on-scene coordinator and the monitor.
Instruction: Instruction is provided through a nonresident, self-paced study program consisting of one pamphlet covering four topics. This course consists of text material, quizzes, and a proctored, comprehensive end-of-course test.
Credit Recommendation: In the lower-division baccalaureate/associate degree category, 1 semester hour in military science (marine environmental protection) (9/90).

CG-1728-0038

PORT SECURITYMAN FIRST CLASS BY CORRESPONDENCE

Course Number: *Version 1:* 0167-1. *Version 2:* 0165-1.
Location: Coast Guard Institute, Oklahoma City, OK.
Length: *Version 1:* Maximum, 52 weeks. *Version 2:* Maximum, 52 weeks.
Exhibit Dates: *Version 1:* 6/93–Present. *Version 2:* 9/90–5/93.
Learning Outcomes: *Version 1:* Upon completion of the course, the student will be able to identify priorities for vessel boarding and boarding personnel scheduling, identify elements for notice of violation letters and other related correspondence, oil spill sampling handling, recovery and clean up techniques, and appropriate procedures for supervising a federal clean up operation. *Version 2:* Upon completion of the course, the student will be familiar with rules of legal evidence, oil pollution investigation, bulk facilities and inspection, and tank vessel regulations.
Instruction: *Version 1:* Instruction is in a nonresident self study, with outside reading materials. Successful course completion requires an 80 percent score or a comprehensive end-of-course examination. *Version 2:* Instruction is provided in a nonresident, self-instruction format consisting of one pamphlet and a proctored, comprehensive end-of-course examination.
Credit Recommendation: *Version 1:* In the lower-division baccalaureate/associate degree category, 3 semester hours in marine environmental safety (3/96). *Version 2:* In the lower-division baccalaureate/associate degree category, 3 semester hours in military science (9/90).

CG-1728-0039

1. MARINE INSPECTOR
2. MARINE SAFETY INSPECTION

Course Number: *Version 1:* MS 452R. *Version 2:* MS 452R.
Location: Reserve Training Center, Yorktown, VA.
Length: *Version 1:* 8 weeks (297 hours). *Version 2:* 8 weeks (225 hours).
Exhibit Dates: *Version 1:* 1/94–Present. *Version 2:* 1/90–12/93.
Learning Outcomes: *Version 1:* Upon completion of the course, the student will be able to inspect shipboard systems and supervise occupational safety according to established marine safety guidelines. *Version 2:* Upon completion of the course, the student will be able to inspect shipboard systems and supervise occupational safety according to marine safety office guidelines.
Instruction: *Version 1:* Methods of instruction include lectures, demonstrations, exercises, practical exercises, exam, and exam review. Topics include administration; vessel manning; ship's papers and documents; and measurement; factory inspection; lifesaving systems; fire fighting systems; navigation and communications systems; anchoring, mooring, and boarding equipment; human factors engineering; electrical systems; nondestructive testing; stability, damage stability, and subdivision; hull inspection; accommodations and structural fire protection; occupational safety; cargo systems; main propulsion; pressure piping and unfired pressure vessels; auxiliary systems; and technical review. *Version 2:* Course provides combined lectures/labs and shipboard on-the-job training during which students are taught to classify, identify, demonstrate, evaluate, and list standards and deficiencies of marine safety inspection. Areas of instruction include life saving and fire fighting systems, hull inspection, machinery/boiler inspection, stability, technical review, and occupational safety.
Credit Recommendation: *Version 1:* In the lower-division baccalaureate/associate degree category, 1 semester hour in marine safety, 3 in marine inspection methods (deck and engine), and 3 in inspection management (5/96). *Version*

2: In the lower-division baccalaureate/associate degree category, 1 semester hour in marine safety, 4 in deck inspection methods, and 4 in engine inspection methods (9/90); in the upper-division baccalaureate category, 3 semester hours in inspection management (9/90).

CG-1728-0040

PORT SECURITYMAN, CLASS A
 (Port Security, Class A)

Course Number: MS 420R.
Location: Reserve Training Center, Yorktown, VA.
Length: 7-10 weeks (229-366 hours).
Exhibit Dates: 1/90–Present.
Learning Outcomes: Upon completion of the course, the student will be able to enforce regulations related to protection and security of vessels, harbors, ports, and waterfront facilities.
Instruction: Course provides lectures and practical exercises in port and vessel safety laws and regulations, practical application of enforcement procedures, discussion of occupational safety, and introduction to fire safety. Instruction is also provided in basic knowledge of port and vessel security laws, regulations, and enforcement procedures. Emphasis is placed on application of physical security measures, weapons training, personal defense, arrests, apprehensions, and security team tactics.
Credit Recommendation: In the lower-division baccalaureate/associate degree category, 3 semester hours in law enforcement, 3 in port and maritime security, and 3 in small arms training (5/96).

CG-1728-0041

1. PORT OPERATIONS DEPARTMENT
2. MARINE SAFETY PORT OPERATIONS

Course Number: *Version 1:* None. *Version 2:* MS 422R.
Location: Reserve Training Center, Yorktown, VA.
Length: *Version 1:* 4-5 weeks (161 hours). *Version 2:* 7 weeks (220-225 hours).
Exhibit Dates: *Version 1:* 12/93–Present. *Version 2:* 1/90–11/93.
Learning Outcomes: *Version 1:* Upon completion of the course, the student will be able to describe and write appropriate response/requests dealing with port operations, including vessel/port facilities, hazardous cargo, port security, port/vessel inspections, pollution investigations, enforcement actions, and verification/disbursement of federal funds. *Version 2:* Upon completion of the course, the student will be able to carry out advanced technician and supervision duties, including loading and transferring hazardous cargos, port/ship inspections, security surveys, pollution investigations, enforcement actions, and verification of disbursement of federal funds.
Instruction: *Version 1:* Instruction consists of lectures, practical exercises, case studies, role playing, videos, and discussions. Topics in the areas of prevention include listing resources, agency contact, hazard indication, inspection procedures, and permit approval. Topics in the areas of response include writing reports, prioritizing actions, identifying sources, writing safety plans, writing pollution reports, completing funding reports, media review, and writing final rules. Personal protective equipment requirements are also presented. *Version 2:* Instruction consists of lectures, demonstrations, and practical exercises and includes review of laws and regulations.
Credit Recommendation: *Version 1:* In the lower-division baccalaureate/associate degree category, 2 semester hours in hazardous cargo operations, 1 in harbor control and security operations, and 2 in environmental protection (5/96). *Version 2:* In the lower-division baccalaureate/associate degree category, 3 semester hours in hazardous cargo operations, 3 in harbor control and security operations, and 3 in environmental protection (9/90).

CG-1728-0042

INVESTIGATOR SECOND CLASS BY CORRESPONDENCE

Course Number: 0266-1.
Location: Coast Guard Institute, Oklahoma City, OK.
Length: Maximum, 156 weeks.
Exhibit Dates: 4/92–Present.
Learning Outcomes: Upon completion of the course, the student will demonstrate knowledge of criminal investigation, forensic equipment, report writing, photography, US intelligence community, counterintelligence, and threat counteraction operational plans. The student will learn the techniques of arrest, search, seizure, handling persons in custody, fingerprinting, court testimony, interviewing, suspect identification, and protective services. Student will gather, assemble, and report intelligence information, and conduct security clearance investigations.
Instruction: This is a correspondence course consisting of topics in criminal investigation, protective services, counterintelligence, and forensic science.
Credit Recommendation: In the lower-division baccalaureate/associate degree category, 2 semester hours in police operations (10/94).

CG-1728-0043

MARITIME SEARCH AND RESCUE PLANNING, CLASS C

Course Number: None.
Location: Reserve Training Center, Yorktown, VA.
Length: 3 weeks (134 hours).
Exhibit Dates: 8/94–Present.
Learning Outcomes: Upon completion of the course, the student will be able to plan and coordinate search and rescue missions according to standard procedures adopted by the Coast Guard.
Instruction: Methods of instruction are lecture, laboratory, search/rescue practice, and review. Topics include awareness stages (simulations, seminars), initial action stages, planning stages, operations stages, mission conclusion stages, search and rescue information, and emergency care.
Credit Recommendation: In the lower-division baccalaureate/associate degree category, 3 semester hours in marine safety (5/96).

CG-1728-0044

PORT SECURITYMAN DIRECT ENTRY

Course Number: MS 421R.
Location: Reserve Training Center, Yorktown, VA.
Length: 2 weeks (71 hours).
Exhibit Dates: 4/94–Present.
Learning Outcomes: Upon completion of the course, the student will be able to apply laws and regulations as they pertain to safety, security, occupational health, and the safety of vessels and terminals.
Instruction: The focus of this course is vessel and terminal safety and security. Topics include types of vessel boarding, waterfront facility inspections, responses to port incidents, and vessel compliance with freight and tariff regulations. Also included is study covering compliance with pollution and security standards and enforcement actions to detect and process safety and security violations including occupational health and safety of vessels and terminals.
Credit Recommendation: In the lower-division baccalaureate/associate degree category, 2 semester hours in physical security (5/96).

CG-1728-0045

MARITIME LAW ENFORCEMENT BOARDING TEAM MEMBER, CLASS C

Course Number: None.
Location: Training Center, Petaluma, CA; Reserve Training Center, Yorktown, VA.
Length: 2 weeks (74 hours).
Exhibit Dates: 6/94–Present.
Learning Outcomes: Upon completion of the course, the student will be able to perform effectively as a member of a team investigating criminal law and safety violations.
Instruction: This course focuses primarily on a team approach to the protection, investigation, and processing of criminal and safety violations. Topics include use of force, searches incident to arrest, pat down (frisk) searches, firearms, and investigation procedures.
Credit Recommendation: In the lower-division baccalaureate/associate degree category, 1 semester hour in basic criminal investigation and 1 in basic law enforcement organization and administration (5/96).

CG-1729-0004

FOOD SERVICE SPECIALIST SECOND CLASS BY CORRESPONDENCE
 (Subsistence Specialist Second Class by Correspondence)

Course Number: 251-3; 0251-4.
Location: Coast Guard Institute, Oklahoma City, OK.
Length: Maximum, 52 weeks.
Exhibit Dates: 1/90–Present.
Learning Outcomes: Upon completion of the course, the student will be able to demonstrate proficiency in menu planning and nutrition, sanitation and safety, policy and organization, procurement, receiving and storage, reporting, and record keeping.
Instruction: This course consists of reading assignments, review exercises, and a proctored final examination in a self-instruction format. Topics covered include menu planning and costing, nutrition, sanitation and safety, policy and organization, procurement, inventory control reports, and record keeping procedures.
Credit Recommendation: In the lower-division baccalaureate/associate degree category, 1 semester hour in introduction to food service supervision (5/91).

COURSE EXHIBITS

CG-1729-0005

1. FOOD SERVICE SPECIALIST FIRST CLASS BY CORRESPONDENCE
   (Subsistence Specialist First Class by Correspondence)
2. SUBSISTENCE SPECIALIST FIRST CLASS BY CORRESPONDENCE

Course Number: *Version 1:* 0151-5. *Version 2:* 151-3; 0151-4.
Location: Coast Guard Institute, Oklahoma City, OK.
Length: *Version 1:* Maximum, 156 weeks. *Version 2:* Maximum, 52 weeks.
Exhibit Dates: *Version 1:* 7/94–Present. *Version 2:* 1/90–6/94.
Learning Outcomes: *Version 1:* Upon completion of the course, the student will be able to organize and administer a food service training program; set realistic performance standards; train employees in specific food service procedures; and provide leadership and direction. *Version 2:* Upon completion of the course, the student will be able to organize and administer a food service training program; set realistic performance standards; train employees in specific food service procedures; and provide leadership and direction.
Instruction: *Version 1:* A self-instruction course containing reading assignments and review exercises. Topics include sanitation, care and cleaning procedures, safety procedures, and inventory management. *Version 2:* A self-instruction course containing reading assignments and review exercises. Topics include sanitation, care and cleaning procedures, safety procedures, and inventory management.
Credit Recommendation: *Version 1:* Pending evaluation. *Version 2:* In the lower-division baccalaureate/associate degree category, 1 semester hour in food service supervision (5/91).

CG-1729-0006

FOOD SERVICE SPECIALIST THIRD CLASS BY CORRESPONDENCE
(Subsistence Specialist Third Class by Correspondence)

Course Number: 351-3; 0351-4.
Location: Coast Guard Institute, Oklahoma City, OK.
Length: Maximum, 52 weeks.
Exhibit Dates: 1/90–Present.
Learning Outcomes: Upon completion of the course, the student will be able to identify sanitation, safety, and inspection procedures relative to preparing and serving food; identify food production and service equipment and supplies; state cleaning and maintenance procedures; identify characteristics and/or grades of meat, fish, and poultry; explain the appropriate cooking, cutting, storing, thawing, and serving procedure for each; state methods of preparation for all other major food categories; and explain organization, structure, policies, and duties of personnel.
Instruction: This course is in self-instruction format, consisting of reading assignments, review exercises, and a proctored end-of-course examination.
Credit Recommendation: In the lower-division baccalaureate/associate degree category, 1 semester hour in food preparation (5/91).

CG-1729-0007

1. FOOD SERVICE SPECIALIST, CLASS A
   (Subsistence Specialist, Class A)
2. SUBSISTENCE SPECIALIST, CLASS A

Course Number: *Version 1:* None. *Version 2:* None.
Location: Training Center, Petaluma, CA.
Length: *Version 1:* 12 weeks (420 hours). *Version 2:* 13 weeks (481 hours).
Exhibit Dates: *Version 1:* 6/92–Present. *Version 2:* 1/90–5/92.
Learning Outcomes: *Version 1:* Upon completion of the course, the student will be able to adjust recipes and prepare, serve, and store food in a safe, sanitary manner using commercial food service equipment, under the direction of supervisory personnel. *Version 2:* Upon completion of the course, the student will be able to prepare and store food under safe and sanitary conditions using modern facilities and equipment under supervision of a senior staff officer.
Instruction: *Version 1:* Instruction covers recipe expansion/adjustment, food preparation, storage, service, sanitation, safety, and equipment use. Methodology utilizes laboratory and practical exercises with related instruction and demonstration. *Version 2:* Instruction includes food preparation, sanitation, safety, storage, and equipment. Methodology includes lecture, discussion, and laboratory experiences.
Credit Recommendation: *Version 1:* In the lower-division baccalaureate/associate degree category, 8 semester hours in quantity food preparation/culinary arts (9/93). *Version 2:* In the lower-division baccalaureate/associate degree category, 4 semester hours in quantity food preparation and 4 in food sanitation and safety (8/87).

CG-1729-0008

FOOD SERVICE SPECIALIST, LARGE MESS MANAGEMENT
(Subsistence Specialist, Large Mess Management)

Course Number: None.
Location: Training Center, Petaluma, CA.
Length: 2 weeks (80 hours).
Exhibit Dates: 1/90–9/92.
Learning Outcomes: Upon completion of the course, the student will be able to manage and supervise food service in all phases of Coast Guard operations.
Instruction: Instruction includes menu planning, cost control, inventory control, record keeping, cash handling, space requirements and utilization, fast food, and manpower management. Methodology includes lectures, discussion, and practical exercises.
Credit Recommendation: In the lower-division baccalaureate/associate degree category, 3 semester hours in food service management (8/87).

CG-1729-0009

FOOD SERVICE SPECIALIST, INDEPENDENT DUTY
(Subsistence Specialist, Independent Duty)

Course Number: None.
Location: Training Center, Petaluma, CA.
Length: 2 weeks (77 hours).
Exhibit Dates: 1/90–9/92.
Learning Outcomes: Upon completion of the course, the student will be able to manage and supervise all phases of Coast Guard dining facility operations at independent duty stations.
Instruction: Instruction includes menu planning, record keeping, cost control, inventory control, cash handling, space requirements and utilization, and business correspondence. Methodology includes lecture, discussion, and practical exercises.
Credit Recommendation: In the lower-division baccalaureate/associate degree category, 3 semester hours in food service management (8/87).

CG-1729-0010

FOOD SERVICE SPECIALIST ADMINISTRATION AND MANAGEMENT
(Subsistence Specialist Administration and Management)

Course Number: None.
Location: Training Center, Petaluma, CA.
Length: 2-3 weeks (145 hours).
Exhibit Dates: 9/92–Present.
Learning Outcomes: Upon completion of the course, the student will be able to administer a food service operation.
Instruction: Instruction covers the administrative aspects of food service management including inventory systems; cash handling procedures; purchasing, receiving, and storing; record keeping; and planning menus. Methodology uses practical exercises to review pertinent administrative procedures.
Credit Recommendation: In the lower-division baccalaureate/associate degree category, 3 semester hours in food service record keeping (9/93).

CG-1730-0001

REFRIGERATION AND AIR CONDITIONING

Course Number: RAC-ADV.
Location: Reserve Training Center, Yorktown, VA.
Length: 2 weeks (70-71 hours).
Exhibit Dates: 1/90–6/95.
Learning Outcomes: Upon completion of the course, the student will be able to list the types of heat and state of matter found in pure water at various temperatures; explain refrigerant pressure temperature relationships; draw the basic refrigeration cycle; identify, state the functions, and locate refrigeration system components; list the safety precautions necessary to service refrigeration equipment; operate, read gages, defrost, and secure a water-cooled refrigeration system; select and use appropriate lead detectors and find refrigerant leaks; silver solder, flare, and swage copper and brass components; add and remove oil; evacuate and charge a system; adjust low and high pressure switches; and perform preventive maintenance.
Instruction: Course is presented through lectures, demonstrations, videos, and laboratory exercises. Topics include types of heat, state of matter, pressure temperature relationships, refrigeration components, safety, system operation, leak detection and repair, adding and removing oil, evacuation, TXV adjustment, low and high pressure switch adjustment, and preventive maintenance.
Credit Recommendation: In the lower-division baccalaureate/associate degree category, 2 semester hours in refrigeration theory and service principles (8/90).

CG-2202-0003

DIRECT COMMISSION OFFICER

Course Number: DCO.
Location: Reserve Training Center, Yorktown, VA.
Length: 2 weeks (60-80 hours).
Exhibit Dates: 1/90–Present.
Learning Outcomes: Upon completion of the course, the student will be able to serve effectively as a commissioned officer in the Coast Guard.
Instruction: Course includes lectures, discussions, demonstrations, simulations, and practical exercises on the Coast Guard's organization, missions, and duties, as well as operational, administrative, leadership, and management duties.
Credit Recommendation: Credit is not recommended because of the limited, specialized nature of the course (4/89).

CG-2202-0004

RESERVE OFFICER CANDIDATE INDOCTRINATION

Course Number: ROCI.
Location: Reserve Training Center, Yorktown, VA.
Length: 2 weeks (60-80 hours).
Exhibit Dates: 1/90–Present.
Learning Outcomes: Upon completion of the course, the student will be able to serve effectively as an officer in the Coast Guard Selected Reserve.
Instruction: Course includes lectures, discussions, and practical exercises in the duties and responsibilities required of officers in the Coast Guard. Topics cover rules, customs, and traditions. The requirement for self-discipline demands an understanding of the importance of honor and integrity in the conduct of an officer and an understanding of the operations of the Coast Guard.
Credit Recommendation: Credit is not recommended because of the military-specific nature of the course (4/89).

CG-2202-0005

OFFICER CANDIDATE

Course Number: *Version 1:* None. *Version 2:* OCS.
Location: Reserve Training Center, Yorktown, VA.
Length: *Version 1:* 17 weeks (663 hours). *Version 2:* 17 weeks (811 hours).
Exhibit Dates: *Version 1:* 8/94–Present. *Version 2:* 1/90–7/94.
Learning Outcomes: *Version 1:* Upon completion of the course, the student will be able to supervise personnel working in nautical science, perform as a public administrator, perform as a problem solver in management, and have a working knowledge of shipboard navigation. *Version 2:* Upon completion of the course, the student will serve effectively as a commissioned officer in the United States Coast Guard.
Instruction: *Version 1:* Course provides lectures, laboratories, films, discussions, role playing, and practical exercises covering Coast Guard orientation, career information, administrative procedures, leadership, seamanship, celestial and electronic navigation, physical education, small arms, military law, wellness, inventory, media relations, financial management, and safety. *Version 2:* Course provides lectures and practical exercises covering Coast Guard orientation, career information, administrative procedures, leadership, seamanship, celestial and electronic navigation, physical education, small arms, military law, and safety.
Credit Recommendation: *Version 1:* In the lower-division baccalaureate/associate degree category, 3 semester hours in personnel supervision (5/96); in the upper-division baccalaureate category, 2 semester hours in nautical science and 2 in management problems (5/96). *Version 2:* In the lower-division baccalaureate/associate degree category, 3 semester hours in human resource management (4/89); in the upper-division baccalaureate category, 3 semester hours in problem solving in management and 3 in naval science (4/89).

CG-2204-0001

20MM MK 16 MOD 5 MACHINE GUN AND MAGAZINE SPRINKLERS OPERATION AND MAINTENANCE, CLASS C

Course Number: GM-1.
Location: Reserve Training Center, Yorktown, VA.
Length: 1-2 weeks (56-59 hours).
Exhibit Dates: 1/90–Present.
Learning Outcomes: Upon completion of the course, the student will be able to load, operate, and perform maintenance on the 20mm machine gun and operate, test, and maintain the magazine sprinkler system.
Instruction: Course provides lectures and practical exercises on the 20mm machine gun and the magazine sprinkler system.
Credit Recommendation: Credit is not recommended because of the military-specific nature of the course (10/88).

CG-2204-0002

3"/50 GUN MOUNT MK 22 OPERATION AND MAINTENANCE, CLASS C

Course Number: GM-3.
Location: Reserve Training Center, Yorktown, VA.
Length: 1-2 weeks (50-55 hours).
Exhibit Dates: 1/90–11/96.
Learning Outcomes: Upon completion of the course, the student will be able to identify components, operate all subsystems, maintain, and inspect 3"/50 caliber Mk 22 guns and mounts.
Instruction: This course is taught by a combination of lectures, demonstrations, and laboratories. Topics covered include operating, maintaining, and testing the 3"/50 caliber gun.
Credit Recommendation: Credit is not recommended because of the military-specific nature of the course (10/88).

CG-2205-0008

OFFICERS ADVANCED AIDS TO NAVIGATION (Aids to Navigation Officer Advanced, Class C)

Course Number: ANC-5.
Location: Aids to Navigation School, Governors Island, NY.
Length: 2 weeks (60-64 hours).
Exhibit Dates: 1/90–11/96.
Objectives: To provide personnel with advanced training in signals engineering and administration.
Instruction: Course provides lectures and practical exercises in signal arrangements, system evaluation, administration, and legal aspects of aids to navigation.
Credit Recommendation: In the lower-division baccalaureate/associate degree category, 1 semester hour in basic seamanship (2/83).

CG-2205-0013

COAST GUARD ORIENTATION BY CORRESPONDENCE

Course Number: 404-1.
Location: Coast Guard Institute, Oklahoma City, OK.
Length: Maximum, 104 weeks.
Exhibit Dates: 1/90–8/94.
Objectives: To provide indoctrination to reserve officers and enlisted personnel in lieu of basic training.
Instruction: Course covers orientation to the Coast Guard. Topics include history and mission, organization, discipline, uniforms and awards, watches and routines, ship construction, seamanship, rules of the road, general drills, damage control and fire fighting, survival, first aid, safety, maintenance, security, ships and aircraft, customs and ceremonies, rates and ratings, career information, boating safety, law enforcement, and administration. Students are required to successfully complete a proctored end-of-course examination.
Credit Recommendation: Credit is not recommended because of the military-specific nature of the course (5/85).

CG-2205-0017

RADARMAN THIRD CLASS BY CORRESPONDENCE

Course Number: 339-7.
Location: Coast Guard Institute, Oklahoma City, OK.
Length: Maximum, 104 weeks.
Exhibit Dates: 1/90–10/90.
Objectives: To provide training to radarmen in the use of detection equipment, communications, search and rescue procedures, rules of the road, and electronic warfare.
Instruction: Course includes radar fundamentals and radar scope interpretation, principles and operation of radio direction finders, dead reckoning techniques, antisubmarine warfare operations, use of charts, use of maneuvering board, search and rescue procedures, and navigation rules. Also includes communication procedures, use of publications, and electronic warfare equipment and operations.
Credit Recommendation: In the lower-division baccalaureate/associate degree category, 3 semester hours in basic navigation (5/85).

CG-2205-0018

WEAPONS OFFICER BY CORRESPONDENCE

Course Number: 406-3.
Location: Coast Guard Institute, Oklahoma City, OK.
Length: Maximum, 104 weeks.
Exhibit Dates: 1/90–12/93.
Objectives: To provide officers with the basic ordnance knowledge to effectively administer the weapons department on board a Coast Guard cutter.
Instruction: Training includes weapons administration, weapon systems, weapons training, gunnery, and administration. Students are required to successfully complete a proctored end-of-course examination.

Credit Recommendation: In the vocational certificate category, 1 semester hour in office administration (5/85).

CG-2205-0019

COAST GUARD ORIENTATION OFFICERS BY CORRESPONDENCE

Course Number: 0407-1.
Location: Coast Guard Institute, Oklahoma City, OK.
Length: Maximum, 104 weeks.
Exhibit Dates: 1/90–11/96.
Objectives: To present a general overview of basic Coast Guard knowledge required by selected reserve officer candidates.
Instruction: Course consists of information on Coast Guard grooming standards, mission of the Coast Guard and reserve, shore units, universal and zulu time, personnel records, the Coast Guard filing system, correspondence course ordering, and Coast Guard reserve terminology. Students are required to successfully complete a proctored end-of-course examination.
Credit Recommendation: Credit is not recommended because of the military-specific nature of the course (5/85).

CG-2205-0020

OFFICER IN CHARGE/EXECUTIVE PETTY OFFICER

Course Number: *Version 1:* None. *Version 2:* None.
Location: Training Center, Petaluma, CA.
Length: *Version 1:* 2 weeks (67 hours). *Version 2:* 2 weeks (77 hours).
Exhibit Dates: *Version 1:* 3/93–Present. *Version 2:* 1/90–2/93.
Learning Outcomes: *Version 1:* Upon completion of the course, the student will be able to of lead and manage small organizational units. *Version 2:* Upon completion of the course, the student will be able to manage and administer a small operating station or unit ashore or afloat.
Instruction: *Version 1:* Instruction includes lectures, case study, and laboratory exercises in administration, personnel policies, organizational behavior and communication, property management, and performance evaluation. *Version 2:* Instruction includes small unit operation and management with emphasis on logistical and physical plant administration, budget planning, written and oral communication to individuals and groups, and human relations management. Instructional methods include lecture, discussion, practical exercises, role playing, and case studies.
Credit Recommendation: *Version 1:* In the lower-division baccalaureate/associate degree category, 3 semester hours in principles of organizational behavioral and personnel administration (9/93). *Version 2:* In the lower-division baccalaureate/associate degree category, 3 semester hours in small business management (8/87).

CG-2205-0023

RADARMAN SECOND CLASS BY CORRESPONDENCE

Course Number: 0239-2; 239-9.
Location: Coast Guard Institute, Oklahoma City, OK.
Length: Maximum, 78 weeks.
Exhibit Dates: 1/90–Present.
Learning Outcomes: Upon completion of the course, the student will be able to identify procedures for training personnel for plotting, standing watch, and ordering supplies; explain a communications plan; identify procedures to prepare navigation charts, search and rescue plans, and preventive maintenance plans; identify procedures to solve advanced maneuvering board problems, tactical control problems, electronic warfare problems, and air control problems; explain the operation of electronic warfare equipment; and explain the procedures used in radar interpretation.
Instruction: Major topics covered are administration, tactical communications, radar navigation, search and rescue techniques, preventive maintenance, operational techniques, detection equipment characteristics, and radar interpretation. An emphasis is placed on radar interpretation and navigation skills to support combat information center operations. Training is provided in a nonresident, self-instruction format. Students are required to successfully complete a proctored end-of-course examination.
Credit Recommendation: In the lower-division baccalaureate/associate degree category, 2 semester hours in radar interpretation techniques (5/93).

CG-2205-0024

DAMAGE CONTROL AND STABILITY BY CORRESPONDENCE

Course Number: 0480-2.
Location: Coast Guard Institute, Oklahoma City, OK.
Length: Maximum, 52 weeks.
Exhibit Dates: 1/90–Present.
Learning Outcomes: Upon completion of the course, the student will be able to apply the basic principles of naval architecture to calculate a vessel's stability under a variety of conditions and identify the duties and responsibilities of shipboard personnel in the damage control organization both in port and underway.
Instruction: Topics include vessel stability and damage control. Training is provided in a nonresident, self-instruction format. Students are required to successfully complete a proctored end-of-course examination.
Credit Recommendation: In the upper-division baccalaureate category, 2 semester hours in technical naval architecture or 2 in nautical science (ship stability) (6/90).

CG-2205-0025

AIDS TO NAVIGATION POSITIONING

Course Number: ANC-AP.
Location: Reserve Training Center, Yorktown, VA.
Length: 1 week (52–57 hours).
Exhibit Dates: 1/90–1/93.
Learning Outcomes: Upon completion of the course, the student will be able to perform standard Coast Guard aids to navigation positioning operations. The student will be able to adjust and use survey sextants; determine the position of aids to navigation using Loran-C, radar ranges, gyro bearings, and horizontal sextant angles; complete a gradient diagram work form using hand calculations; use a portable computer and a Cap-II program tape to establish aid and object positioning files. The student will be able to transfer programs and data files between the standard work station (with Cap-II program installed) and portable computer; determine the accuracy classification for an aid to navigation and record this data in accordance with current instructions; construct a positioning grid and a positioning template following Coast Guard procedures; troubleshoot faulty positioning grids following instructions in the Naton School positioning grid; evaluate an aid's position when at short stay and when not at short stay; and complete an aid positioning record (CG-5216) in accordance with the latest Coast Guard instruction.
Instruction: Course is conducted through classroom lectures and practical exercises include adjusting the sextant; measuring sextant angles; determining aid's position using Loran-C, radar, gyro bearings, and horizontal sextant angles; using the Cap-II portable computer; classifying aids accurately; constructing a position grid and template; identifying errors in templates; and completing an aid position record.
Credit Recommendation: In the vocational certificate category, 1 semester hour in seamanship/navigation (6/90).

CG-2205-0026

QUARTERMASTER FIRST CLASS BY CORRESPONDENCE

Course Number: 0137-5.
Location: Coast Guard Institute, Oklahoma City, OK.
Length: Maximum, 78 weeks.
Exhibit Dates: 1/90–Present.
Learning Outcomes: Upon completion of the course, the student will be able to define the responsibilities and duties of the officer of the deck, executive officer, and navigator; define the various effects and principles involved in conning a vessel in either restricted or open waters; distinguish between the magnetic and gyro compasses; define the principles of and procedures for degaussing; define the difference between permanent and induced magnetism; define the effect and corrective actions for deviation; determine a vessel's position using bearings, charts, instruments, and tables; anchor a vessel in an assigned berth; prepare a chart for use in piloting; calibrate and operate a radio direction finder; operate and use information obtained from the Loran-C system; operate a radar system; maintain and use a marine sextant; complete a solution for a celestial observation; define the responsibilities and duties for search and rescue (SAR) operations; define the responsibilities and duties of the on-scene commander of an SAR operation.
Instruction: Topics include ship conning, compasses and compass systems, piloting and dead reckoning, navigation by electronic equipment, celestial navigation, and basic search and rescue. Proctored final examination is required.
Credit Recommendation: In the lower-division baccalaureate/associate degree category, 1 semester hour in celestial navigation and 2 in coastwise navigation (6/90).

CG-2205-0027

SMALL BOAT CREWMEMBER
(Reserve Small Boat Crewmember)

Course Number: SBC.
Location: Reserve Training Center, Yorktown, VA.
Length: 2 weeks.
Exhibit Dates: 1/90–9/92.
Learning Outcomes: After 9/92 see CG-2205-0040. Upon completion of the course, the

student will be able to operate small boats according to the procedures outlined in rules of the road; identify navigational aids; moor and tow small boats; and practice basic seamanship, communication, fire drills, and survival skills.

Instruction: Lectures, including audio visual aids, and practical exercises both dockside and underway are methods of presentation. Students will use fathometers, assist coxswain with pump, stand towing watch, observe fire fighting equipment, and observe use of life raft.

Credit Recommendation: In the vocational certificate category, 2 semester hours in seamanship and 2 in small boat handling (6/90).

CG-2205-0028

MILITARY REQUIREMENTS FOR E-3 BY CORRESPONDENCE

Course Number: 0455-2; 0455-3.
Location: Coast Guard Institute, Oklahoma City, OK.
Length: Maximum, 52 weeks.
Exhibit Dates: 1/90–6/96.
Learning Outcomes: Upon completion of the course, the student will be able to demonstrate the duties and responsibilities of an enlisted person assigned the rank of E-3.
Instruction: Instruction is provided in a nonresident, self-instruction format and consists of one pamphlet and a proctored end-of-course test. The course contains topics such as customs and courtesies; conduct and justice; course information; first aid; survival; safety and occupational health; chemical, biological, and radiological defense; damage control; organization; signals and communications; administration; and clinical procedures.
Credit Recommendation: In the lower-division baccalaureate/associate degree category, 1 semester hour in military science (9/90).

CG-2205-0029

MILITARY REQUIREMENTS FOR E-4 BY CORRESPONDENCE

Course Number: 0452-4; 0452-3.
Location: Coast Guard Institute, Oklahoma City, OK.
Length: Maximum, 52 weeks.
Exhibit Dates: 1/90–6/96.
Learning Outcomes: Upon completion of the course, the student will be able to demonstrate knowledge of the duties and responsibilities of an enlisted person assigned the rank of E-4.
Instruction: Instruction is provided in a nonresident, self-instruction format and consists of one pamphlet and a proctored end-of-course test. The course contains topics such as organization; pay and allowances; supervision; conduct; security regulations; coating and color manual; ordnance; survival; chemical, biological, and radiological warfare defense; boating safety; and radio/telephone procedures.
Credit Recommendation: In the lower-division baccalaureate/associate degree category, 1 semester hour in military science (9/95).

CG-2205-0030

MILITARY REQUIREMENTS FOR E-5 BY CORRESPONDENCE

Course Number: 0453-4; 0453-3.
Location: Coast Guard Institute, Oklahoma City, OK.
Length: Maximum, 52 weeks.
Exhibit Dates: 1/90–Present.
Learning Outcomes: Upon completion of the course, the student will be able to demonstrate knowledge of the duties and responsibilities of enlisted persons assigned the rank of E-5.
Instruction: Instruction is provided in a nonresident, self-instruction format and consists of one pamphlet and a proctored end-of-course test. The course contains topics such as career information; nuclear, biological, and chemical warfare; numbering, identification, and certification of boats; and captain of the port.
Credit Recommendation: In the lower-division baccalaureate/associate degree category, 1 semester hour in military science (9/90).

CG-2205-0031

MILITARY REQUIREMENTS FOR E-6 BY CORRESPONDENCE

Course Number: 0454-3; 0454-2.
Location: Coast Guard Institute, Oklahoma City, OK.
Length: Maximum, 52 weeks.
Exhibit Dates: 1/90–6/96.
Learning Outcomes: Upon completion of the course, the student will be able to perform the duties of a first-line supervisor in a military organization.
Instruction: Instruction is provided through a nonresident, self-paced study program using one pamphlet and a proctored end-of-course test. Course gives a general overview of knowledge factors for military requirements for E-6. Such topics as law enforcement; correspondence; message organization; personnel evaluation; nuclear, biological, and chemical decontamination; and training are included.
Credit Recommendation: In the lower-division baccalaureate/associate degree category, 2 semester hours in military science (1/93).

CG-2205-0032

HONORS AND CEREMONIES BY CORRESPONDENCE

Course Number: 421-1.
Location: Coast Guard Institute, Oklahoma City, OK.
Length: Maximum, 52 weeks.
Exhibit Dates: 1/90–Present.
Learning Outcomes: Upon completion of the course, the student will be able to render international, naval, and military honors and ceremonials.
Instruction: Instruction is provided in a nonresident, self-instruction format. This course consists of one pamphlet and a proctored end-of-course test. The course contains topics such as honors and ceremonies, flags and pennants, and deaths and funerals.
Credit Recommendation: Credit is not recommended because of the military-specific nature of the course (9/90).

CG-2205-0033

COAST GUARD DIVISION OFFICER BY CORRESPONDENCE

Course Number: 0467-1.
Location: Coast Guard Institute, Oklahoma City, OK.
Length: Maximum, 78 weeks.
Exhibit Dates: 1/90–5/93.
Learning Outcomes: Upon completion of the course, the student will be able to demonstrate a knowledge of first-line supervisory skills, office administration, budget management, general personnel (human resources) practices and procedures, and records information management; and demonstrate a fundamental knowledge of ship handling and damage control.
Instruction: Instruction is provided through a nonresident self-paced study program consisting of seven comprehensive workbooks and study guides.
Credit Recommendation: In the lower-division baccalaureate/associate degree category, 2 semester hours in office management, 1 in business communication, and 1 in principles of vessel stability (9/90).

CG-2205-0034

SENIOR PETTY OFFICERS BY CORRESPONDENCE

Course Number: 0700-1.
Location: Coast Guard Institute, Oklahoma City, OK.
Length: Maximum, 52 weeks.
Exhibit Dates: 1/90–6/92.
Learning Outcomes: Upon completion of the course, the student will be able to demonstrate a knowledge of first-line supervisory skills, office administration and budget management, general personnel (human resources) practices and procedures, and records and information management.
Instruction: Instruction is provided through a nonresident self-paced study program consisting of five comprehensive workbook and study guides.
Credit Recommendation: In the lower-division baccalaureate/associate degree category, 2 semester hours in office management and 1 in business communication (9/90).

CG-2205-0035

RECRUIT TRAINING
(Basic Training)

Course Number: None.
Location: Training Center, Cape May, NJ.
Length: 8-9 weeks (321 hours).
Exhibit Dates: 1/90–Present.
Learning Outcomes: Upon completion of the course, the graduated recruit will demonstrate knowledge of general military and Coast Guard protocol, seamanship, fire fighting, basic engineering, safety, first aid, and personal health; demonstrate basic swimming and water survival skills; and meet prescribed standards for physical fitness.
Instruction: Course includes lectures, demonstration, and performance exercises in military protocol and drill, seamanship, basic engineering, fire fighting, safety, first aid, personal health, and physical conditioning.
Credit Recommendation: In the lower-division baccalaureate/associate degree category, 1 semester hour in personal fitness/conditioning, 1 in beginning swimming, 1 in boating/seamanship, and 1 in personal health and first aid (5/96).

CG-2205-0036

MILITARY REQUIREMENTS FOR CHIEF PETTY OFFICER BY CORRESPONDENCE

Course Number: *Version 1:* 0456-3. *Version 2:* 0456-1; 0456-2.

Location: Coast Guard Institute, Oklahoma City, OK.

Length: *Version 1:* Maximum, 156 weeks. *Version 2:* Maximum, 24 weeks.

Exhibit Dates: *Version 1:* 1/93–Present. *Version 2:* 1/91–12/92.

Learning Outcomes: *Version 1:* Upon completion of the course, the student will be able to provide leadership, advice, counseling, and standards for enlisted personnel. *Version 2:* Upon completion of the course, the student will be able to serve in the role of a Chief Petty Officer in the Coast Guard, providing leadership, advice, counseling, and standards for enlisted personnel.

Instruction: *Version 1:* This nonresident, self study course consists of topics including the role of the Chief Petty Officer, civil rights program, military justice, effective communications, work life, training and instruction, performance counseling and evaluation, accountability and management of federal property, safety and environmental health, nutrition and fitness, stress management, suicide prevention and substance abuse (including alcohol, drugs, and tobacco). Mastery of the content enables the Chief Petty Officer to function effectively in personnel matters. *Version 2:* This nonresident, self study course consists of 11 major topics, including the role of the Chief Petty Officer, civil rights program, military justice, effective writing, training and instruction, performance counseling and evaluation, accountability and management of federal property, safety and environmental health, nutrition and fitness, stress management, and substance abuse (including alcohol, drugs, and tobacco). Mastery of the content enables the Chief Petty Officer to function effectively in personnel matters.

Credit Recommendation: *Version 1:* In the lower-division baccalaureate/associate degree category, 2 semester hours in personnel supervision (7/95); in the upper-division baccalaureate category, 2 semester hours in management problems and 2 in communications techniques for managers (7/95). *Version 2:* In the lower-division baccalaureate/associate degree category, 2 semester hours in personnel supervision (11/92); in the upper-division baccalaureate category, 2 semester hours in management problems and 2 in communication techniques for managers (11/92).

CG-2205-0037

SONOBUOY BY CORRESPONDENCE

Course Number: 0419-1.

Location: Coast Guard Institute, Oklahoma City, OK.

Length: Maximum, 52 weeks.

Exhibit Dates: 9/90–5/93.

Learning Outcomes: Upon completion of the course, the student will be able to order, handle, preset, and deploy all types of sonobuoys used by the Coast Guard.

Instruction: Training is provided in a self-instruction format. Topics include safety precautions, inspections, loading and deployment, supply, general and specific sonobuoy characteristics, prelaunch selections, operational functions, and sonobuoy patterns.

Credit Recommendation: Credit is not recommended because of the military-specific nature of the course (10/92).

CG-2205-0038

MILITARY REQUIREMENTS FOR SECOND CLASS PETTY OFFICER BY CORRESPONDENCE

Course Number: 0453-5.

Location: Coast Guard Institute, Oklahoma City, OK.

Length: Maximum, 156 weeks.

Exhibit Dates: 12/93–6/96.

Learning Outcomes: Upon completion of the course, the student will understand how to efficiently conduct travel required by the military and the requirements placed on the family relative to finance, leave, general relocation problems, and assistance available through the Coast Guard. The course also provides guidance to the petty officer in conducting training sessions in conducting counseling sessions, and in methods of assisting in suicide prevention.

Instruction: Instruction is provided in a nonresident, self-study format and requires an end-of-course test. The course contains information relative to the military duty of the E-5 with particular emphasis on counseling and the family impact of military life.

Credit Recommendation: In the lower-division baccalaureate/associate degree category, 2 semester hours in military science (3/96).

CG-2205-0039

ALLIED VISUAL COMMUNICATIONS, CLASS C

Course Number: AVCC.

Location: Reserve Training Center, Yorktown, VA.

Length: 4 weeks (196 hours).

Exhibit Dates: 3/93–Present.

Learning Outcomes: Upon completion of the course, the quartermaster will be able to perform as a signal bridge watch stander who operates, trains, and manages allied visual communications (AVC) systems aboard Coast Guard cutters.

Instruction: Methods of instruction include lectures, exercises, practical exercises, role playing, evaluations, and review. Topics include drafting messages; flashing lights; readiness check; challenge and reply using flashing lights, semaphore, and flag hoist; and processing messages.

Credit Recommendation: Credit is not recommended because of the limited, specialized nature of the course (5/96).

CG-2205-0040

CREWMAN 41, CLASS C

Course Number: None.

Location: Reserve Training Center, Yorktown, VA.

Length: 2 weeks (50-80 hours).

Exhibit Dates: 10/92–Present.

Learning Outcomes: Before 10/92 see CG-2205-0027. Upon completion of the course, the student will be able to operate small boats according to the procedures outlined in rules of the road; identify navigational aids; moor and tow small boats; and practice basic seamanship, radio telephone communications, fire drills, and survival and rescue skills.

Instruction: Lectures, including audio visual aids, and practical exercises both dockside and underway are methods of presentation. Students will use fathometers, assist coxswain with pump, stand towing watch, observe fire fighting equipment, and observe use of life raft.

Credit Recommendation: In the lower-division baccalaureate/associate degree category, 2 semester hours in seamanship and small boat handling (5/96).

Course Exhibits

DD

DD-0326-0001

ARMED FORCES STAFF COLLEGE

Course Number: None.
Location: Armed Forces Staff College, Norfolk, VA; National Defense University, Ft. Lesley J. McNair, Washington, DC.
Length: 21-22 weeks (633-735 hours).
Exhibit Dates: 1/90–8/94.
Objectives: To train officers in joint and combined military organization, planning, and operations and in related aspects of national and international security.
Instruction: Course includes lectures, readings, student papers, and discussions in joint and combined military organization, planning, and operations and in related aspects of national and international security. Materials are divided into four major topic areas. First, US national security studies includes US national strategy formulation and execution, domestic and external factors influencing and constraining US national security policy making, geopolitics, and introduction to nuclear strategic theory and to arms control and disarmament. A political-military gaming exercise is required to help the student apply theory to the task of formulating and implementing security policy in a crisis situation. Second, principles of managerial problem solving and leadership includes managerial planning techniques, computerized information system utilization, problem solving skills, and general leadership/professional skills. Third, communicative skills include practical experience in oral communication for both large and small groups and written communication in the form of short briefing and summary papers as well as a major research paper. Fourth, defense organization includes Army, Navy, Air Force, Marine, and Department of Defense organization and operation.
Credit Recommendation: In the upper-division baccalaureate category, 3 semester hours in principles of managerial problem solving, 3 in communicative skills, 6 in US national security studies, and 3 in defense organization (7/87).

DD-0326-0002

JOINT AND COMBINED STAFF OFFICER

Course Number: None.
Location: Armed Forces Staff College, Norfolk, VA.
Length: 12 weeks (335-336 hours).
Exhibit Dates: 12/94–Present.
Learning Outcomes: Upon completion of the course, the student will be prepared for joint and combined military organization planning, intelligence, and operations and in related aspects of national and international security.
Instruction: Course includes lectures, readings, student papers, and discussions in joint and combined military organization, planning, intelligence, and operations and in related aspects of national and international security. Materials are divided into five major topic areas: US national security studies including US national strategy formulation and execution, domestic and external factors influencing and constraining US national security policy making, geopolitics, and a political-military gaming exercise to help the student apply theory to the task of formulating and implementing security policy in a crisis situation; principles of managerial problem solving and leadership including managerial planning techniques, computerized information system utilization, problem solving skills, and general leadership/professional skills; communication skills including practical experience in oral communication for both large and small groups and written communication in the form of short briefing and summary papers as well as a major research paper; defense organization including Army, Navy, Air Force, Marine, and Department of Defense organization and operation; and operational planning including collaborative campaign analysis and campaign planning exercises.
Credit Recommendation: In the upper-division baccalaureate category, 3 semester hours in joint military operations and strategic planning, 3 in national security studies, 3 in defense organization and structure, 3 in planning and management, and 3 in communication skills (11/95).

DD-0326-0003

JOINT AND COMBINED WARFIGHTING

Course Number: None.
Location: Armed Forces Staff College, Norfolk, VA.
Length: 12 weeks (350-351 hours).
Exhibit Dates: 9/94–Present.
Learning Outcomes: Upon completion of the course, the student will be prepared for joint and combined military organization planning, intelligence, and operations and in related aspects of national and international security.
Instruction: Course includes lectures, readings, student papers, and discussions in joint and combined military organization, planning, intelligence, and operations and in related aspects of national and international security. Materials are divided into five major topic areas: US national security studies including US national strategy formulation and execution, domestic and external factors influencing and constraining US national security policy making, geopolitics, and a political-military gaming exercise to help the student apply theory to the task of formulating and implementing security policy in a crisis situation; principles of managerial problem solving and leadership including managerial planning techniques, computerized information system utilization, problem solving skills, and general leadership/professional skills; communication skills including practical experience in oral communication for both large and small groups and written communication in the form of short briefing and summary papers as well as a major research paper; defense organization including Army, Navy, Air Force, Marine, and Department of Defense organization and operation; and operational planning including collaborative campaign analysis and campaign planning exercises.
Credit Recommendation: In the upper-division baccalaureate category, 3 semester hours in national security studies, 3 in defense organization and structure, 3 in planning and management, and 3 in communication skills (11/95); in the graduate degree category, 3 semester hours in joint military operations and strategic planning (11/95).

DD-0326-0004

EXECUTIVE ACQUISITION LOGISTICS MANAGEMENT

Course Number: LOG 304.
Location: Defense Acquisition University, Various locations in the Continental US.
Length: 2 weeks (80 hours).
Exhibit Dates: 8/96–Present.
Learning Outcomes: Upon completion of the course, the student will demonstrate a comprehensive understanding of the DOD systems acquisition management, technical, and business processes including the important role played by logisticians in the process.
Instruction: The course immerses students in the role the logistician plays in the acquisition process, acquaints them with specialized terminology, and current DOD policies, demonstrates how they can affect the acquisition process. Lectures, expository discussion, case studies, group case analysis, and presentations are used in this course. Principle areas covered by the course include integrated product and process development; role of the acquisition logistician in the systems engineering; reliability, maintainability, and availability; specifications and standards reform; contract solicitation process; test and evaluation; environment, safety, and health; source selection; human systems integration, ethics in negotiations, incentive performance, and enhanced support in the future through automated tools; planning for flexible sustainment; financial management for the acquisition logistician; and current topics in acquisition logistics.
Credit Recommendation: In the upper-division baccalaureate category, 3 semester hours in channel management and logistics (3/97).

DD-0326-0005

ADVANCED INFORMATION SYSTEMS ACQUISITION

Course Number: IRM 303.
Location: Defense Acquisition University, Various locations in the Continental US.
Length: 3 weeks (102-128 hours).
Exhibit Dates: 10/95–Present.

Learning Outcomes: Upon completion of the course, the student will be able to apply management and decision making skills as well as information systems/information technology knowledge to the activities associated with the planning, organizing, directing and controlling information systems acquisition programs at the senior (or executive) level. In addition, the student should have effective team development and management skills.

Instruction: This course is designed as "paperless", i.e. most of the course materials are on CD-ROM, and each student is issued a laptop computer using MS Windows and Lotus Notes. Supplementary readings are provided and the students interact with the instructor and each other using Lotus notes. Students perform both individual and group exercises using a range of DOD and non-DOD cases supported by extensive readings. Students are required to write an essay at the end of each week and to perform a major group project in class. A non-DOD text on project management is supplied. Principal areas covered in the course include integrated product teams and team building including the human dimension (resistance to change); strategic and information systems planning (information security, information systems planning, architecture, infrastructure, and interoperability); telecommunications technology planning; market research; process improvement methodologies (modeling and functional economic analysis); electronic commerce (information economics, industry perspective); system acquisition (strategy and planning, acquisition process and reform, risk and stakeholder management, requirements and configuration management, trade off of cost-schedule performance, software acquisition issues, and information systems test and evaluation); contract administration (integrated solicitation protests and alternative disputes resolution); and life cycle management (operation requirements definition, acquisition resource planning, program management metrics and deployment, and emerging issues).

Credit Recommendation: In the graduate degree category, 3 semester hours in management of information systems acquisition as an elective in a management information systems program (3/97).

DD-0326-0006

INTERMEDIATE INFORMATION SYSTEMS ACQUISITION

Course Number: IRM 201.
Location: Defense Acquisition University, Various locations in the Continental US.
Length: 3 weeks (93-117 hours).
Exhibit Dates: 10/95–Present.
Learning Outcomes: Upon completion of the course, the student will be able to apply information systems/information technology skills in planning, organizing, directing, and controlling information systems acquisition programs. This includes conceptual stages through fielding, post-development operations and support, and product improvement of information systems. The student should have improved judgment, initiative, and common sense in these areas.

Instruction: This course is designed as "paperless", i.e. most of the course materials are on CD-ROM and each student is issued a laptop computer using MS Window and Lotus notes. Supplementary readings are provided and the students interact with the instructor and each other using Lotus notes. Students perform both individual and group case exercises. Students are examined at the end of each week by essay exams which are live on the laptop (Lotus notes) with the instructor. Principle areas covered in the course include information systems technology (product development, systems modeling, architectures and infrastructure, telecommunications and networks, and security); life cycle management (configuration management, functional and commercial specifications, software metrics, and reuse); risk management, organization change, role of stakeholders; system acquisition (planning and strategy, resource management, testing and evaluation, management reviews, and reform legislation); and contract administration (source selection and solication preparation).

Credit Recommendation: In the upper-division baccalaureate category, 3 semester hours in acquisition of information systems as an elective in a management information systems program (3/97).

DD-0326-0007

GOVERNMENT CONTRACT LAW

Course Number: CON 201.
Location: Defense Acquisition University, Various locations in the Continental US.
Length: 2 weeks (54 hours).
Exhibit Dates: 1/97–Present.
Learning Outcomes: Upon completion of the course, the student will be able to apply basic statutory principles and regulatory provisions to the government acquisition process with particular emphasis on discriminating between statutory, regulatory and ethical considerations applicable to government contracts; determining the appropriate amount of delegation of a contracting officer's authority; applying the impact of social and economic concerns in contracting situations; analyzing the impact of legislation and judicial decisions on the formation of government contracts; responding to the procedures available to an adversely affected bidder; assessing the government and contractor's rights regarding property disputes; identifying the appropriate obligation of federal monies in contracting situations; making a determination on government acceptance of contractor-provided goods and services; and applying the elements of equitable adjustment in situations where the contractor has performed additional work beyond the stated scope of the contract.

Instruction: Methods of instruction include lecture, student exercises, and case analysis involving decisions of both the Comptroller General and the Board of Contract Appeals. Principle areas covered include business ethics; historical changes in the government contracting process; extent of contracting officer's authority and ability to delegate that authority; impact of legislation and judicial decisions on the formation of government contracts; protest procedures and remedies; impact of property rights on the government contracting process; obligation of federal funds; influence of social and economic issues in government contracting; identification of procurement fraud; product/service delivery, acceptance, and warranties; equitable adjustment in situations involving additional vendor effort not stipulated in the contract agreement; dispute litigation; and contract termination.

Credit Recommendation: In the upper-division baccalaureate category, 2 semester hours in public sector contract law (3/97).

DD-0419-0001

DEFENSE PRESERVATION AND INTERMEDIATE PROTECTION

Course Number: 822-F1 (JT); 8B-F1 (JT).
Location: Selected on-site locations, Continental US; Joint Military Packaging Training Center, Aberdeen Proving Ground, MD; School of Military Packaging Technology, Aberdeen Proving Ground, MD.
Length: 2 weeks (70-76 hours).
Exhibit Dates: 1/90–2/92.
Objectives: To train commissioned officers, enlisted personnel, and civilian personnel in the techniques of cleaning, drying, preserving, and packing military supplies and equipment.

Instruction: Lectures and practical exercises cover preservation and intermediate protection of military supplies and equipment, including packing specifications, cleaning, preservatives, sprayable and strippable films, controlled humidity, packaging inspection, packaging costs, and documentation.

Credit Recommendation: In the lower-division baccalaureate/associate degree category, 1 semester hour in the preservation of equipment and supplies (11/92).

DD-0419-0002

DEFENSE PACKING AND UNITIZATION

Course Number: 822-F2 (JT); 8B-F2 (JT).
Location: On-site locations, US and overseas; Joint Military Packaging Training Center, Aberdeen Proving Ground, MD; School of Military Packaging Technology, Aberdeen Proving Ground, MD.
Length: 2 weeks (70-76 hours).
Exhibit Dates: 1/90–8/95.
Objectives: To train personnel in the packing, marking, and loading of military supplies and equipment for storage and shipment.

Instruction: Course includes lectures and practical exercises on the packing, marking, and loading of military supplies and equipment for storage and shipment, including introduction to packing and containerization, crate design, cargo unitization and containerization, cushioning and blocking, bracing and anchoring, loading for shipment, utilization of packing lines, parcel shipments, and packing cost reduction.

Credit Recommendation: In the lower-division baccalaureate/associate degree category, 1 semester hour in transportation, regulations, packing, and unitization (1/92).

DD-0419-0003

DEFENSE PACKAGING OF HAZARDOUS MATERIALS FOR TRANSPORTATION

Course Number: 8B-F7 (JT); 822-F7 (JT).
Location: Joint Military Packaging Training Center, Aberdeen Proving Ground, MD; On-site locations, US; School of Military Packaging Technology, Aberdeen Proving Ground, MD.
Length: 1-2 weeks (50-78 hours).
Exhibit Dates: 1/90–9/95.
Objectives: To train military personnel, civilian government employees, and qualified members of industry in the current requirements and procedures for the preparation of hazardous materials for transportation. Course

includes marking, certifying, handling, and storing using approved government and civilian methods and techniques.

Instruction: Course makes use of Department of Defense (DoD) documents; Department of Transportation (DoT) documents; Air Force Regulation (71-4); International Air Transport Association Restricted Articles Regulations (IATA); Intergovernmental Maritime Consultative Organization (IMCO); and Official Air Transport Restricted Articles Tariff and Circular No. 6-D for the transportation of hazardous materials by rail, motor, air, and water. Course also includes comparison of DoD, DoT, IATA, and IMCO containers authorized for use in packaging hazardous materials, as well as classifications and labels; IMCO classifications and labeling requirements; and MIL-STD-129 requirements for hazardous and other materials requiring special handling data/certification on DD Form 1387-2 label.

Credit Recommendation: In the lower-division baccalaureate/associate degree category, 1 semester hour in packaging of hazardous materials for transportation (7/93).

DD-0419-0004

DEFENSE BASIC PRESERVATION AND PACKING

Course Number: 822-F13 (JT).
Location: Joint Military Packaging Training Center, Aberdeen Proving Ground, MD; School of Military Packaging Technology, Aberdeen Proving Ground, MD.
Length: 2 weeks (62-70 hours).
Exhibit Dates: 1/90–1/97.
Objectives: To train enlisted personnel and civilian employees of the Army, Navy, Air Force, Marine Corps, and the Defense Logistics Agency in basic preservation and packing principles for new and repairable material for storage or shipment.
Instruction: This course, designed for operational enlisted and civilian personnel, consists of introduction to military and industrial preservation, corrosion control, cleaning and drying, preservatives, preservation materials and equipment, preservation methods, introduction to military and industrial packing, special and general purpose shipping container, cushioning, blocking, bracing, weatherproofing, cargo utilization, marking and labeling, and packing for shipment. Forty-two percent of the course is devoted to conference time, 45 percent to demonstrations and student practice in the areas of preservation and packing, and 13 percent to examinations and nonacademic areas.
Credit Recommendation: In the vocational certificate category, 1 semester hour in basic preservation, packing, and protection of new and repairable equipment (7/93).

DD-0419-0005

DEFENSE PACKAGING MANAGEMENT TRAINING

Course Number: 8B-F26 (JT).
Location: Joint Military Packaging Training Center, Aberdeen Proving Ground, MD; School of Military Packaging Technology, Aberdeen Proving Ground, MD.
Length: 9 weeks (318-347 hours).
Exhibit Dates: 1/90–3/92.
Objectives: To train military and civilian personnel in the techniques and skills required to become packaging specialists and to manage packaging programs for the military services and Defense Logistic Agency.

Instruction: Course includes philosophies, concepts, and practices of military packaging and packaging management. Lecture/conferences and practical instruction are provided in preservation and intermediate protection, packing and unitization, marking for shipment and storage, packaging of hazardous materials for transportation, foam-in-place packaging, and packaging design. Guest speakers and field training trips are part of regular classroom instruction. Students prepare a research paper. Course includes packaging design.

Credit Recommendation: In the lower-division baccalaureate/associate degree category, 1 semester hour in preservation of equipment and supplies; 1 in transportation regulations, packing, and unitization, including marking for shipment, storage, and and foam-in-place packaging; 2 in packaging of hazardous materials for transportation; 1 in packaging design; and 1 in packaging management including field training (6/86).

DD-0419-0006

DEFENSE PACKAGING DESIGN

Course Number: 8B-F16 (JT).
Location: Joint Military Packaging Training Center, Aberdeen Proving Ground, MD; School of Military Packaging Technology, Aberdeen Proving Ground, MD.
Length: 2 weeks (70-76 hours).
Exhibit Dates: 1/90–11/92.
Objectives: To train military and civilian personnel of the Department of Defense in approved policies, methods, and techniques of packaging design, handling, and shipment. Emphasis is placed on selecting packing and cushioning materials to provide adequate protection at a minimum cost.
Instruction: Course includes identifying item characteristics, transportability, deterioration of materials, natural and transportation environment, container design and selection, packaging documentation, testing, and safety.
Credit Recommendation: In the lower-division baccalaureate/associate degree category, 1 semester hour in packaging design (1/90).

DD-0504-0001

BASIC JOURNALIST
(Information Specialist (Journalist))
(Basic Military Journalist)

Course Number: 28-R-701.1; A-570-0011 (USN); 570-71Q20; 570-71Q10; ABA79130-1 (USAF).
Location: Defense Information School, Ft. Slocum, NY; Defense Information School, Ft. Benjamin Harrison, IN.
Length: 10 weeks (337-396 hours).
Exhibit Dates: 1/90–1/91.
Objectives: To teach selected enlisted personnel the principles, techniques, and skills required for public information, service information, and community relations.
Instruction: Course includes lectures and practical experiences in print journalism, including interviewing techniques, news and feature writing, editing, newspaper layout and make-up; photojournalism, including the taking, processing, and printing of photographs; radio and television writing; speech; international relations and government; public affairs. Print media, public affairs, and photojournalism are emphasized.

Credit Recommendation: In the lower-division baccalaureate/associate degree category, 5 semester hours in news writing and reporting (print), 2 in news writing and reporting (electronic), 2 in photojournalism, and 2 in layout and design (8/87).

DD-0504-0009

PUBLIC AFFAIRS OFFICER

Course Number: 7G-46A (USCG); 7G (USMC); 5OBA7921 002 (USAF); A-7G-0010 (USN); 7G-46A.
Location: Defense Information School, Ft. Benjamin Harrison, IN.
Length: 9-10 weeks (331-367 hours).
Exhibit Dates: 1/90–10/92.
Objectives: After 10/92 see DD-0504-0018. To provide entry level public affairs training for officers and civilians from all services.
Instruction: Lectures and practical exercises in the duties of an information specialist. Course includes instruction in government and public affairs, media management, print and electronic journalism, and basic public speaking.
Credit Recommendation: In the lower-division baccalaureate/associate degree category, 3 semester hours in print journalism (news writing and editing), 2 in electronic journalism, 3 in public relations, 2 in introduction to photography, and 1 in basic public speaking (11/87); in the upper-division baccalaureate category, 2 semester hours in government and public affairs and 1 in media management (11/87).

DD-0504-0012

SHIPBOARD INFORMATION, TRAINING, AND ENTERTAINMENT (SITE) SYSTEM

Course Number: A-570-0010.
Location: Defense Information School, Ft. Benjamin Harrison, IN.
Length: 5 weeks (208 hours).
Exhibit Dates: 1/90–1/96.
Learning Outcomes: Upon completion of the course, the student will be able to administer programs and operate a shipboard information training and entertainment system. Student will be able to deliver a newscast, write and produce an audio sport announcement, prepare a news release, and produce a video or audio training program. Student will operate and maintain basic SITE audio and video equipment.
Instruction: The course is limited to ten students and includes 96 hours of conference and exam work and 120 hours of laboratory practical exercises. The topics include public affairs management and procedures; operation of SITE equipment for broadcast of armed forces programming; and production of print, audio, and video productions for information, training, and public affairs.
Credit Recommendation: In the lower-division baccalaureate/associate degree category, 3 semester hours in instructional television or 1 in public relations, 1 in broadcasting, and 1 in video production (5/90).

DD-0504-0013

1. PUBLIC AFFAIRS OFFICER RESERVE COMPONENT, PHASE 2
2. PUBLIC AFFAIRS OFFICER COURSE RESERVE COMPONENT

Course Number: *Version 1:* AFIS-PAOC-RC. *Version 2:* DCJ; G502A7921 001; A-7G0013; 7G-F3.
Location: Defense Information School, Ft. Benjamin Harrison, IN.
Length: *Version 1:* 2 weeks (74 hours). *Version 2:* 2 weeks (66 hours).
Exhibit Dates: *Version 1:* 12/94–1/96. *Version 2:* 1/90–11/94.
Learning Outcomes: *Version 1:* Upon completion of phase 2 of the course, the student will be able to function as an information/public affairs officer with training in public relations, community relations, and oral communication with specific skills in writing for print and broadcast media plus photo editing and composition. Phase 1 is taken through correspondence and has not been evaluated. *Version 2:* Upon completion of the course, the student will be able to function as an information/public affairs officer with training in public relations, community relations, and oral communication with specific skills in writing for print and broadcast media plus photo editing and composition.
Instruction: *Version 1:* Instruction includes lectures and exercises on the duties of an information specialist, including public relations principles, public opinion and mass communication, media relations, communication law, oral training, electronic newsgathering, commercial station operations, visual communication, and writing news and feature stories. Students will have completed a correspondence course as phase 1 of this course which has not been evaluated. Practical exercises are critiqued but no grades are awarded. *Version 2:* Instruction includes lectures and exercises on the duties of an information specialist, including public relations principles, public opinion and mass communication, media relations, communication law, oral training, electronic newsgathering, commercial station operations, visual communication, and writing news and feature stories.
Credit Recommendation: *Version 1:* In the lower-division baccalaureate/associate degree category, 3 semester hours in public relations (7/94). *Version 2:* In the upper-division baccalaureate category, 1 semester hour in public relations (communication) (5/90).

DD-0504-0014

PUBLIC AFFAIRS SUPERVISOR

Course Number: 570-46Q/R30.
Location: Defense Information School, Ft. Benjamin Harrison, IN.
Length: 2 weeks (70-75 hours).
Exhibit Dates: 1/90–1/96.
Learning Outcomes: Upon completion of the course, the student will be able to plan, supervise, and coordinate public affairs activities, usually in a public affairs office or in an Armed Forces Radio and Television network overseas.
Instruction: Course provides training in preparing and conducting news briefings, media and community relations, public affairs policy, newsgathering, leadership and supervisory skills in a public affairs environment, public affairs in contingency operations, electronic journalism, radio and television production, and computer literacy.
Credit Recommendation: In the upper-division baccalaureate category, 2 semester hours in media management (5/90).

DD-0504-0015

EDITORS

Course Number: A1143D1 (43D); GA2A79150; A-570-0013; 7G-F11/570-F2.
Location: Defense Information School, Ft. Benjamin Harrison, IN.
Length: 4 weeks (157 hours).
Exhibit Dates: 1/90–1/96.
Learning Outcomes: Upon completion of the course, students will be able to apply editorial principles and techniques to their base, post, ship, or station publication; employ newspaper layout, makeup, design, and readability tests to a publication; edit advanced news and feature articles; and write headlines and photo captions and outlines. Students will be able to use desktop publishing software in editing and newspaper production.
Instruction: The course is highly individualized and self-paced. Diagnostic tests are used to designate editorial strengths and weaknesses at the beginning of the course. Instructors/advisors monitor student progress carefully to ensure weaknesses receive special attention. Student are required to complete a publication improvement project requiring evaluation and redesign. Student may complete all requirements in less than four weeks.
Credit Recommendation: In the upper-division baccalaureate category, 4 semester hours of editing including desktop publishing (5/90).

DD-0504-0016

INTRODUCTION TO BROADCASTING RESERVE, PHASE 2

Course Number: AFIS-IB-RC.
Location: Defense Information School, Ft. Benjamin Harrison, IN.
Length: 2 weeks (95 hours).
Exhibit Dates: 7/94–1/96.
Learning Outcomes: Upon completion of the course, the student will be able to apply general broadcast journalism principles, radio newswriting techniques, broadcast announcing skills, interview techniques, and radio production techniques.
Instruction: Methods of instruction include lectures, demonstrations, practical exercises, and hands-on exercises. Topics include instruction on military relations with the civilian press, briefings, broadcast law, radio news, feature and spot writing, broadcast announcing, interviewing, and radio production/control room procedures.
Credit Recommendation: In the lower-division baccalaureate/associate degree category, 2 semester hours in introduction to broadcasting (7/94).

DD-0504-0017

INTRODUCTION TO JOURNALISM RESERVE, PHASE 2

Course Number: AFIS-IJ-RC.
Location: Defense Information School, Ft. Benjamin Harrison, IN.
Length: 2 weeks (86 hours).
Exhibit Dates: 7/94–1/96.
Learning Outcomes: Upon completion of the course, the student will be able to write newspaper stories including hard news and features; shoot and develop photographs; and perform the basics of newspaper layout and design.
Instruction: Lectures and practical experiences cover print journalism, including news and feature writing, interviewing, editing, newspaper layout and makeup and photojournalism, including taking, processing, and printing photographs.
Credit Recommendation: In the lower-division baccalaureate/associate degree category, 2 semester hours in print journalism (7/94).

DD-0504-0018

PUBLIC AFFAIRS OFFICER

Course Number: A-7G-0010; 7G; 5OB7921 002; 7G-46A.
Location: Defense Information School, Ft. Benjamin Harrison, IN.
Length: 9-10 weeks (353 hours).
Exhibit Dates: 11/92–1/96.
Learning Outcomes: Before 11/92 see DD-0504-0009. Upon completion of the course, the student will be able to manage a public affairs office for a military command or agency, including planning and supervising public information internal communications, media relations, and community relations. Specific skills include advising decision makers on public relations, crisis management, and information news releases to print media and electronic media.
Instruction: This course is taught with a mixture of classroom activities, lectures, and discussions and is reinforced by practical exercises. Evaluation is conducted using both paper and pencil examination and performance evaluation with an emphasis on the latter. Content includes the principles of public relations, policy and planning, legal and ethical considerations, internal communication, community relations, and media relations. International press and cultural sensitivities are also included.
Credit Recommendation: In the lower-division baccalaureate/associate degree category, 3 semester hours in news writing/editing, 1 in broadcast journalism, 3 in principles of public relations, 2 in basic photography, and 1 in public speaking (7/94); in the upper-division baccalaureate category, 3 semester hours in advanced public relations and 1 in media management (7/94).

DD-0505-0003

ELECTRONIC JOURNALISM

Course Number: 4391; 4321; 4313; 570-F3.
Location: Defense Information School, Ft. Benjamin Harrison, IN.
Length: 2 weeks (75-77 hours).
Exhibit Dates: 1/90 3/90.
Objectives: After 3/90 see DD-0505-0007. To provide training in the planning, production, and editing of news and information features for television, using electronic news gathering equipment and techniques.
Instruction: Course includes training and practical experience in videotape editing techniques, Portapak operations, and news gathering.
Credit Recommendation: In the lower-division baccalaureate/associate degree category, 2 semester hours in electronic news gathering and editing techniques (4/90).

DD-0505-0004

BASIC BROADCASTER

Course Number: *Version 1:* AFIS-BBC. *Version 2:* 570; G5ABA79131 000; A-570-0010; 570-46R10.
Location: Defense Information School, Ft. Benjamin Harrison, IN.
Length: *Version 1:* 12 weeks (461 hours). *Version 2:* 12 weeks (446 hours).
Exhibit Dates: *Version 1:* 1/95–1/96. *Version 2:* 1/90–12/94.
Learning Outcomes: *Version 1:* Upon completion of the course, the student will be able to enter the field of military radio or television station operations as either on-air talent or as a production specialist. The student will have broadcast writing, delivery, and equipment operation skills. *Version 2:* Upon completion of the course, the student will be able to perform the duties and functions of a broadcaster at a military radio or television station, including on-air announcing, radio/television production, and broadcast journalism assignments.
Instruction: *Version 1:* Methods of instruction include lecture, seminar, and demonstration combined with extensive hands-on training. The setting is well equipped with state of the art broadcast equipment. Course content includes a minimum of fundamentals dealing with service-specific public affairs procedures. Emphasis is placed upon broadcast journalism in general, broadcast announcing, radio production, and television production. *Version 2:* Methods of instruction include lecture, seminar, demonstration, and practical exercise. The Defense Information School is very well equipped for teaching these classes with radio and television school and all allied equipment. Instruction includes worldwide geopolitical studies, research and library methods, radio/television news and feature writing, intensive announcing and radio/television production exercises, and training in electronic journalism.
Credit Recommendation: *Version 1:* In the lower-division baccalaureate/associate degree category, 3 semester hours in broadcast newswriting and 3 in performance (announcing) (7/94); in the upper-division baccalaureate category, 3 semester hours in radio production/operation and 3 in television production (7/94). *Version 2:* In the lower-division baccalaureate/associate degree category, 3 semester hours in broadcast newswriting, 3 in radio production/operations, 3 in television production, and 3 in performance (announcing) (5/90).

DD-0505-0005

ARMED FORCES RADIO TELEVISION SYSTEM BROADCAST MANAGER

Course Number: DCJ; A-7G-0011/CDP 0313; G502A7921 004; FG-46B/570-F4.
Location: Defense Information School, Ft. Benjamin Harrison, IN.
Length: 4 weeks (150 hours).
Exhibit Dates: 1/90–1/96.
Learning Outcomes: Upon completion of the course, the student will be able to manage Armed Forces Radio and Television Stations. The training provides skills in broadcast operations, broadcast management, broadcast computer operations, and the public affairs aspects of broadcasting.
Instruction: The course includes lectures, seminars, demonstrations, and practical exercises. Topics include station operations, communication technology, conflicts of interest, executive duties, radio programming, television programming, news and sports programming, music selection, broadcast law, audience surveys, budgeting, engineering, and other managerial responsibilities.
Credit Recommendation: In the upper-division baccalaureate category, 3 semester hours in broadcast management (5/90).

DD-0505-0006

ADVANCED ELECTRONIC JOURNALISM

Course Number: AFIS-AEJC.
Location: Defense Information School, Ft. Benjamin Harrison, IN.
Length: 2 weeks (89 hours).
Exhibit Dates: 1/95–1/96.
Learning Outcomes: Upon completion of the course, the student will be able to shoot and edit high quality news and informational features under normal and adverse conditions.
Instruction: Course includes training and practical experience in the finer points of electronic field production, including lighting and collecting sound, as well as shooting, editing, newsgathering, and interviewing.
Credit Recommendation: In the upper-division baccalaureate category, 2 semester hours in advanced video production (7/94).

DD-0505-0007

ELECTRONIC JOURNALISM

Course Number: 4391; 4321; 4313; G5AZA79151; 570-F3.
Location: Defense Information School, Ft. Benjamin Harrison, IN.
Length: 2 weeks (77 hours).
Exhibit Dates: 4/90–1/96.
Learning Outcomes: Before 4/90 see DD-0505-0003. Upon completion of the course, the student will be able to plan, produce, and edit news and informational features for television, using electronic news gathering equipment and techniques.
Instruction: The course provides training and practical experience in videotape editing techniques, Portapak operations, and newsgathering.
Credit Recommendation: In the lower-division baccalaureate/associate degree category, 2 in electronic newsgathering and editing techniques (7/94).

DD-0602-0012

MALAY BASIC
(Indonesian-Malay Basic)

Course Number: None.
Location: Defense Language Institute, West Coast Branch, Presidio of Monterey, CA.
Length: 30-34 weeks (870 hours).
Exhibit Dates: 1/90–1/90.
Objectives: To enable students to develop operational oral and written communicative skills in the Malay language.
Instruction: The course utilizes 12 volumes of a specially prepared Indonesian Malay text series as well as additional Malay components which include oral and written materials.
Credit Recommendation: In the lower-division baccalaureate/associate degree category, 9 semester hours for beginning Malay and 6 for intermediate Malay (3/81).

DD-0602-0016

GERMAN AURAL COMPREHENSION

Course Number: None.
Location: Defense Language Institute, Presidio of Monterey, CA.
Length: 32 weeks (960 hours).
Exhibit Dates: 1/90–1/97.
Objectives: To be conversant in everyday German, with special emphasis given to listening comprehension and reading translation.
Instruction: Class activities focus on dialog adaptation, memorization, and pronunciation practice. Grammar rules are presented in conjunction with oral drill and pattern exercises. The course puts great emphasis on listening comprehension using the language laboratory to develop this aural skill.
Credit Recommendation: In the lower-division baccalaureate/associate degree category, 10 semester hours in elements of first year German and 6 in second year German for the 32-week German course (3/81).

DD-0602-0018

ALBANIAN BASIC

Course Number: None.
Location: Defense Language Institute, West Coast Branch, Presidio of Monterey, CA.
Length: 47 weeks (690-1380 hours).
Exhibit Dates: 1/90–1/90.
Objectives: Intensive basic instruction in the four skills: elements of aural comprehension, speaking, reading and writing. Introduction of the most frequent structural features through cultural/situational exercises.
Instruction: Audio-lingual small-group instruction incorporating structural and vocabulary/pronunciation drills. Grammar and vocabulary are introduced through dialogues, reading/translation exercises and controlled conversation with area background knowledge covered. Throughout the course functional oral/aural communicative skills are emphasized. NOTE: While these courses are listed as Basic, it should be understood that this is the terminology used by the Armed Forces to indicate that the courses are their regular programs in the various languages. They are not limited to what most civilian institutions would term beginning or basic courses in a language.
Credit Recommendation: In the lower-division baccalaureate/associate degree category, 7 semester hours in oral/aural, 7 in writing/translation (3/81); in the upper-division baccalaureate category, 7 semester hours in intermediate conversation and writing (3/81).

DD-0602-0021

CHINESE CANTONESE BASIC

Course Number: 0ICC47.
Location: Defense Language Institute, West Coast Branch, Presidio of Monterey, CA.
Length: 47 weeks (1068-1380 hours).
Exhibit Dates: 1/90–9/90.
Objectives: To provide an intensive course of study with the focus on speaking/comprehension, reading, and writing in descending order of emphasis. The course enables the student to develop operational oral and written communicative skills.
Instruction: The course includes oral/aural drills in both the classroom and the laboratory. Reading instruction emphasizes text comprehension and translation. In this course students are exposed to approximately 1200 characters for recognition and are expected to write approximately 600 characters. Writing exercises emphasize character writing with mini-

mum training in expressive writing. Question/answer, drill, classroom discussion, and supplementary workbooks provide drill material on pronunciation. NOTE: While these courses are listed as Basic it should be understood that this is the terminology used by the armed forces to indicate that the courses are their regular programs in the various languages. They are not limited to what most civilian institutions would term beginning or basic courses in a language.

Credit Recommendation: In the lower-division baccalaureate/associate degree category, extending into the upper division baccalaureate category 12 semester hours at the beginning level and 12 ranging from lower intermediate to upper intermediate level (3/81).

DD-0602-0022

BULGARIAN INTERMEDIATE

Course Number: None.
Location: Defense Language Institute, West Coast Branch, Presidio of Monterey, CA.
Length: 37-38 weeks (690-1080 hours).
Exhibit Dates: 1/90–1/90.
Objectives: An intensive course designed to extend the student's competency in all four skill areas beyond that of the basic language course.
Instruction: The course begins with a review of the content of the basic course. Through selected readings drawn from contemporary materials, and discussion in the designated language on a variety of topics the student achieves greater language proficiency and increased awareness of area studies. Emphasis is given to the reading component focusing on newpapers, periodicals, technical materials and special texts and the aural component using radio broadcasts, tape transcription and class discussion.
Credit Recommendation: In the upper-division baccalaureate category, 12 semester hours in advanced Bulgarian (conversation) and 6 in advanced Bulgarian (reading) (3/81).

DD-0602-0023

ARABIC EGYPTIAN/SYRIAN INTERMEDIATE

Course Number: None.
Location: Defense Language Institute, West Coast Branch, Presidio of Monterey, CA.
Length: 32 weeks (930 hours).
Exhibit Dates: 1/90–9/90.
Objectives: An intensive course designed to extend the student's competency in all four skill areas beyond that of the basic course.
Instruction: The course reviews the basic elements of Modern Standard Arabic (MSA) and offers extensive oral-aural practice in Egyptian and Syrian dialects with limited continued training in MSA. The course also exposes the student to other dialect variations such as Libyan, Palestinian, Jordanian, and Iraqi dialects. The course uses an oral-aural approach with extensive situational dialogues and texts relevant to the historical, cultural, and geopolitical aspects of the dialects taught.
Credit Recommendation: In the lower-division baccalaureate/associate degree category, 8 semester hours in Egyptian dialect, 8 in Syrian dialect, and 2 in exposure to other Arabic dialects (3/81).

DD-0602-0024

CZECH INTERMEDIATE

Course Number: None.
Location: Defense Language Institute, West Coast Branch, Presidio of Monterey, CA.
Length: 37 weeks (1080 hours).
Exhibit Dates: 1/90–9/90.
Objectives: An intensive course which is designed to develop a higher level of proficiency in all four language skills than provided in the basic course.
Instruction: There is first a review of the content of the basic course. The course then proceeds to have the student achieve greater language proficiency through the use of selected readings drawn from contemporary materials and authentic radio broadcasts and discussions in Czech on a variety of topics including culture, economics, and politics.
Credit Recommendation: In the upper-division baccalaureate category, 12 semester hours in advanced Czech (conversation) and 6 in advanced Czech (reading) NOTE: The credit recommended reflects senior year college study and does not overlap with the advanced credit awarded for the Basic Czech course (3/81).

DD-0602-0029

GREEK BASIC

Course Number: None.
Location: Defense Language Institute, West Coast Branch, Presidio of Monterey, CA.
Length: 47 weeks (690-1380 hours).
Exhibit Dates: 1/90–3/91.
Objectives: To train selected individuals in the comprehension, speaking, reading, and writing of the Greek language. The student gains the ability to communicate effectively in everyday vocabulary and language.
Instruction: The fundamentals of comprehension, speaking, reading and writing are developed with the help of integrated course materials. The approach is essentially audio-lingual with questions and answers combined with textual interpretation, grammatical exercises, and pattern drills. Students come close to fluency. This course teaches both demotic and Katharevusa languages. The students active vocabulary consists of approximately 2000 lexical items. NOTE: While this course is listed as Basic, it should be understood that this is the terminology used by the armed forces to indicate that the courses are their regular programs in the various languages. They are not limited to what most civilian institutions would term beginning or basic courses in a language.
Credit Recommendation: In the lower-division baccalaureate/associate degree category, 10 semester hours in first year Greek, 6 in second year Greek (3/81); in the upper-division baccalaureate category, 8 semester hours in third year Greek (3/81).

DD-0602-0032

INDONESIAN BASIC

Course Number: None.
Location: Defense Language Institute, West Coast Branch, Presidio of Monterey, CA.
Length: 30-34 weeks (810-870 hours).
Exhibit Dates: 1/90–1/90.
Objectives: Intensive instruction in the elements of oral comprehension, speaking and reading with functional oral/aural skills emphasized. Students develop operational communicative skills.
Instruction: The course takes an audio-lingual approach which embraces intensive small-group interaction with multimedia support materials. Grammar and vocabulary are introduced through dialogues, structural drills and controlled conversation. NOTE: While the course is listed as Basic, it should be understood that this is the terminology used by the Armed Forces to indicate that the courses are their regular programs in the various languages. They are not limited to what most civilian institutions would term beginning or basic courses in a language.
Credit Recommendation: In the lower-division baccalaureate/associate degree category, 9 semester hours in beginning Indonesian and 6 in intermediate Indonesian (3/81).

DD-0602-0042

SHORT BASIC TURKISH

Course Number: None.
Location: Defense Language Institute, West Coast Branch, Presidio of Monterey, CA.
Length: 12 weeks (360 hours).
Exhibit Dates: 1/90–9/90.
Objectives: To provide an introduction to modern standard Turkish phonology and the most frequently used grammatical structures.
Instruction: The course starts with a formal study of modern Turkish phonology and grammar utilizing the aural-oral and written techniques. A variety of situational dialogues is used to develop a working, practical knowledge of the language.
Credit Recommendation: In the lower-division baccalaureate/associate degree category, 7 semester hours in introduction to basic Turkish elements (3/81).

DD-0602-0105

UKRAINIAN BASIC

Course Number: None.
Location: Defense Language Institute, Presidio of Monterey, CA.
Length: 47 weeks (1760 hours).
Exhibit Dates: 1/90–1/97.
Objectives: To train military personnel in the interpretation and translation of Ukrainian and to provide basic military, geographic, economic, historical, and political information about the area in which the language is spoken. (These area studies are taught in the foreign language.).
Instruction: Lectures, discussions, and oral drills in the interpretation and translation of Ukrainian and additional training in basic military, geographic, economic, historical, and political information about Ukraine. NOTE: While this course is listed as Basic, it should be understood that this is the terminology used by the armed forces to indicate that the courses are their regular programs in the various languages. They are not limited to what most civilian institutions would term beginning or basic courses in a language.
Credit Recommendation: In the lower-division baccalaureate/associate degree category, extending into the upper-division baccalaureate category, 21 semester hours in Ukrainian (3/81).

DD-0602-0108

ITALIAN BASIC

Course Number: None.
Location: Defense Language Institute, West Coast Branch, Presidio of Monterey, CA.
Length: 24 weeks (690-1380 hours).

Exhibit Dates: 1/90–9/90.
Objectives: Intensive basic instruction in the four skills: elements of aural comprehension, speaking, reading, and writing. Introduction to the most frequent structural features through cultural/situational exercises.
Instruction: Course includes audio-lingual small group instruction incorporating structural and vocabulary/pronunciation drills. Grammar and vocabulary are introduced through dialogues, reading/translation exercises, and controlled conversation with area background knowledge covered. Throughout the course functional oral/aural communicative skills are emphasized. NOTE: While these courses are listed as Basic, it should be understood that this is the terminology used by the armed forces to indicate that the courses are their regular programs in the various languages. They are not limited to what most civilian institutions would term beginning or basic courses in a language.
Credit Recommendation: In the lower-division baccalaureate/associate degree category, 8 semester hours in oral/aural Italian and 4 in reading/writing Italian (8/74).

DD-0602-0112
POLISH BASIC

Course Number: 01PL47.
Location: Defense Language Institute, Presidio of Monterey, CA.
Length: 47 weeks (1380 hours).
Exhibit Dates: 1/90–9/90.
Objectives: An intensive language course which aims to develop basic proficiency in all four language skills at a level providing functional communicative competence.
Instruction: Through the use of dialog manipulation, oral practice, grammar presentation, structural drills, reading, and translation exercises, the student develops language proficiency and receives an introduction to selected areas of the culture, history, geography, socio-economic structure, and military affairs of Poland. NOTE: While this course is listed as Basic, it should be understood that this is the terminology used by the armed forces to indicate that the courses are their regular programs in the various languages. They are not limited to what most civilian institutions would term beginning or basic courses in language.
Credit Recommendation: In the lower-division baccalaureate/associate degree category, 6 semester hours in elementary Polish and 6 in intermediate Polish (3/81); in the upper-division baccalaureate category, 9 semester hours in advanced (reading, conversation) Polish (3/81).

DD-0602-0114
SERBO-CROATIAN BASIC

Course Number: None.
Location: Defense Language Institute, West Coast Branch, Presidio of Monterey, CA.
Length: 47 weeks (1380 hours).
Exhibit Dates: 1/90–2/90.
Objectives: An intensive language course which aims to develop basic proficiency in all four language skills at a level providing functional communicative competence.
Instruction: Through the use of dialog manipulation, oral practice, grammar presentation, structural drills, reading and translation exercises, the student develops language proficiency and receives an introduction to selected areas of the culture, history, geography, socio-economic structure and military affairs of the country of the language studied. NOTE: While this course is listed as Basic, it should be understood that this is the terminology used by the Armed Forces to indicate that the courses are their regular programs in the various languages. They are not limited to what most civilian institutions would term beginning or basic courses in language.
Credit Recommendation: In the lower-division baccalaureate/associate degree category, 6 semester hours in elementary and 6 in intermediate Serbo-Croatian (3/81); in the upper-division baccalaureate category, 9 semester hours in advanced (conversation, reading) Serbo-Croatian (3/81).

DD-0602-0115
RUSSIAN EXTENDED

Course Number: None.
Location: Defense Language Institute, West Coast Branch, Presidio of Monterey, CA.
Length: 27 weeks (690-780 hours).
Exhibit Dates: 1/90–9/90.
Objectives: An intensive course designed to provide a higher level of proficiency in all four language skills for those students directly out of the basic language course.
Instruction: The course draws upon selected readings from contemporary materials and Russian radio broadcasts in order to provide the student with discussion materials in a variety of topic areas. The student achieves further language proficiency and increased awareness of Russian/Soviet area studies.
Credit Recommendation: In the upper-division baccalaureate category, 10 semester hours in advanced Russian (conversation) and 3 in advanced Russian (reading) (3/81).

DD-0602-0116
CHINESE MANDARIN BASIC

Course Number: None.
Location: Defense Language Institute, West Coast Branch, Presidio of Monterey, CA.
Length: 47 weeks (1068-1380 hours).
Exhibit Dates: 1/90–9/90.
Objectives: To provide an intensive course of study with the focus on speaking/comprehension, reading, and writing in descending order of emphasis. The course enables the student to develop operational oral and written communicative skills.
Instruction: Course includes oral/aural drills in both the classroom and the laboratory. Reading instruction emphasizes text comprehension and translation. Students are exposed to approximately 1200 characters for recognition and are expected to self-write approximately 600 characters. Writing exercises emphasize character writing with a minimum training in expressive writing. There is a fair amount of military terminology interspersed with vocabulary of a more general nature, NOTE: While these courses are listed as Basic it should be understood that this is the terminology used by the armed forces to indicate that the courses are their regular programs in the various languages. They are not limited to what most civilian institutions would term beginning or basic courses in a language.
Credit Recommendation: In the lower-division baccalaureate/associate degree category, extending into the upper division baccalaureate category 12 semester hours at the beginning level and 12 ranging from lower intermediate to upper intermediate level (3/81).

DD-0602-0117
JAPANESE BASIC

Course Number: None.
Location: Defense Language Institute, West Coast Branch, Presidio of Monterey, CA.
Length: 47 weeks (1068-1380 hours).
Exhibit Dates: 1/90–3/91.
Objectives: To provide an intensive course of study with the focus on speaking/comprehension, reading, and writing in descending order of emphasis. The course enables the student to develop operational oral and written communicative skills.
Instruction: Course includes oral/aural drills in both the classroom and the laboratory. Reading instruction emphasizes text comprehension and translation. Japanese textual materials introduce the student to 500 kanji with supplementary materials providing exposure to a limited number of additional characters. Question/answer, drill, classroom discussion, and supplementary workbooks provide drill material on pronunciation. NOTE: While these courses are listed as Basic it should be understood that this is the terminology used by the armed forces to indicate that the courses are their regular programs in the various languages. They are not limited to what most civilian institutions would term beginning or basic courses in a language.
Credit Recommendation: In the lower-division baccalaureate/associate degree category, extending into the upper division baccalaureate category 12 semester hours in Japanese at the beginning level and 8 in Japanese ranging from lower intermediate to upper intermediate level (3/81).

DD-0602-0118
KOREAN BASIC

Course Number: None.
Location: Defense Language Institute, West Coast Branch, Presidio of Monterey, CA.
Length: 47 weeks (1068-1380 hours).
Exhibit Dates: 1/90–10/90.
Objectives: To provide an intensive course of study with the focus on speaking/comprehension, reading, and writing in descending order of emphasis. The course enables the student to develop operational oral and written communicative skills.
Instruction: Course includes oral/aural drills in both the classroom and the laboratory. Reading instruction emphasizes text comprehension and translation. The student is introduced to about 300 Chinese characters in addition to Hangul. There is a fair amount of military terminology interspersed with vocabulary of a more general nature. Question/answer, drill, classroom discussion, and supplementary workbooks provide drill material on pronunciation. NOTE: While these courses are listed as Basic it should be understood that this is the terminology used by the armed forces to indicate that the courses are their regular programs in the various languages. They are not limited to what most civilian institutions would term beginning or basic courses in a language.
Credit Recommendation: In the lower-division baccalaureate/associate degree category, extending into the upper division bacca-

laureate category 12 semester hours in Korean at the beginning level and 10 in Korean ranging from lower intermediate to upper intermediate level (3/81).

DD-0602-0119

THAI BASIC

Course Number: None.
Location: Defense Language Institute, West Coast Branch, Presidio of Monterey, CA.
Length: 36 weeks (1068-1380 hours).
Exhibit Dates: 1/90–3/91.
Objectives: To provide an intensive course of study with the focus on speaking/comprehension, reading, and writing in descending order of emphasis. The course enables the student to develop operational oral and written communicative skills.
Instruction: Course includes oral/aural drills in both the classroom and the laboratory. Reading instruction emphasizes text comprehension and translation. There is a fair amount of military terminology interspersed with vocabulary of a more general nature. Question/answer, drill, classroom discussion, and supplementary workbooks provide drill material on pronunciation. NOTE: While these courses are listed as Basic it should be understood that this is the terminology used by the armed forces to indicate that the courses are their regular programs in the various languages. They are not limited to what most civilian institutions would term beginning or basic courses in a language.
Credit Recommendation: In the lower-division baccalaureate/associate degree category, extending into the upper division baccalaureate category 12 semester hours in Thai at the beginning level and 6 in Thai at the lower intermediate level (3/81).

DD-0602-0121

GERMAN INTERMEDIATE

Course Number: None.
Location: Defense Language Institute, Presidio of Monterey, CA.
Length: 24 weeks (690-1080 hours).
Exhibit Dates: 1/90–9/90.
Objectives: An intensive course designed to extend the student's competency in all four skill areas beyond that of the basic language course.
Instruction: The course begins with a review of the content of the basic course. Through selected readings drawn from contemporary materials and discussion in the designated language on a variety of topics, the student achieves greater language proficiency and increased awareness of area studies. Emphasis is given to the reading component focusing on newspapers, periodicals, technical materials, and special texts and to the the aural component using radio broadcasts, tape transcription, and class discussion.
Credit Recommendation: In the upper-division baccalaureate category, 9 semester hours in advanced German (3/81).

DD-0602-0122

POLISH INTERMEDIATE

Course Number: None.
Location: Defense Language Institute, West Coast Branch, Presidio of Monterey, CA.
Length: 37 weeks (690-1080 hours).
Exhibit Dates: 1/90–9/90.
Objectives: An intensive course designed to extend the student's competency in all four skill areas beyond that of the basic language course.
Instruction: The course begins with a review of the content of the basic course. Through selected readings drawn from contemporary materials and discussion in the designated language on a variety of topics, the student achieves greater language proficiency and increased awareness of area studies. Emphasis is given to the reading component focusing on newspapers, periodicals, technical materials, and special texts and to the the aural component using radio broadcasts, tape transcription, and class discussion.
Credit Recommendation: In the upper-division baccalaureate category, 12 semester hours in advanced Polish (conversation) and 6 in advanced Polish (reading) (3/81).

DD-0602-0123

RUSSIAN INTERMEDIATE

Course Number: None.
Location: Defense Language Institute, West Coast Branch, Presidio of Monterey, CA.
Length: 37 weeks (690-1080 hours).
Exhibit Dates: 1/90–9/90.
Objectives: An intensive course designed to extend the student's competency in all four skill areas beyond that of the basic language course.
Instruction: The course begins with a review of the content of the basic course. Through selected readings drawn from contemporary materials and discussion in the designated language on a variety of topics, the student achieves greater language proficiency and increased awareness of area studies. Emphasis is given to the reading component focusing on newspapers, periodicals, technical materials, and special texts and to the the aural component using radio broadcasts, tape transcription, and class discussion.
Credit Recommendation: In the upper-division baccalaureate category, 12 semester hours in advanced Russian (conversation) and 6 in advanced Russian (reading) (3/81).

DD-0602-0125

VIETNAMESE INTERMEDIATE

Course Number: None.
Location: Defense Language Institute, West Coast Branch, Presidio of Monterey, CA.
Length: 36-37 weeks (690-1080 hours).
Exhibit Dates: 1/90–9/90.
Objectives: An intensive course designed to extend the student's competency in all four skill areas beyond that of the basic language course.
Instruction: The course begins with a review of the content of the basic course. Through selected readings drawn from contemporary materials and discussion in the designated language on a variety of topics, the student achieves greater language proficiency and increased awareness of area studies. Emphasis is given to the reading component focusing on newspapers, periodicals, technical materials, and special texts and to the the aural component using radio broadcasts, tape transcription, and class discussion.
Credit Recommendation: In the upper-division baccalaureate category, 15 semester hours in intermediate Vietnamese beyond the basic course (3/81).

DD-0602-0128

RUSSIAN ADVANCED

Course Number: None.
Location: Defense Language Institute, Presidio of Monterey, CA.
Length: 37 weeks (1080-1100 hours).
Exhibit Dates: 1/90–9/90.
Objectives: An intensive course which aims to develop maximum proficiency in all four language skills at a level higher than that provided by the intermediate course.
Instruction: The course begins with a review of the intermediate course materials and continues on with selected readings of contemporary materials and the use of radio broadcasts. The course time is also devoted to advanced composition and conversation, translation, and stylistics. Students produce oral and written reports on area background subjects.
Credit Recommendation: In the upper-division baccalaureate category, 6 semester hours in advanced Russian (conversation and composition), 6 in advanced Russian (reading), 3 in advanced Russian (syntax), and 3 in advanced Russian (stylistics) (3/81).

DD-0602-0129

KOREAN GATEWAY

Course Number: None.
Location: Defense Language Institute, West Coast Branch, Presidio of Monterey, CA.
Length: Self-paced, 8 weeks (310 hours).
Exhibit Dates: 1/90–9/90.
Objectives: To provide an orientation to the Korean language and culture on a very basic level.
Instruction: Students listen to audio recordings of various speakers and work with accompanying print materials. The materials cover such everyday topics as travel, shopping, leisure time activities, and survival needs. The student then interacts with the instructor on a one-to-one basis.
Credit Recommendation: In the lower-division baccalaureate/associate degree category, 5 semester hours in elementary Korean (3/81).

DD-0602-0131

SWEDISH BASIC

Course Number: None.
Location: Defense Language Institute, Presidio of Monterey, CA.
Length: 24-25 weeks (690-1380 hours).
Exhibit Dates: 1/90–1/90.
Objectives: To train selected individuals in the comprehension, speaking, reading and writing of a foreign language. The student gains the ability of communicating effectively in everyday vocabulary and language.
Instruction: The fundamentals of comprehension, speaking, reading and writing are developed with the held of integrated course materials. The approach is essentially audio-lingual with questions and answers combined with textual interpretation, grammatical exercises and pattern drills. Students come close to fluency. The course uses course materials identical to those presently used in comparable courses in colleges and universities. NOTE: While these courses are listed as Basic, it should be understood that this is the terminology used by the Armed Forces to indicate that the courses are their regular programs in the various lan-

guages. They are not limited to what most civilian institutions would term beginning or basic courses in a language.

Credit Recommendation: In the lower-division baccalaureate/associate degree category, 10 semester hours in first year Swedish and 6 in second year Swedish (3/81).

DD-0602-0132

HUNGARIAN BASIC

Course Number: None.
Location: Defense Language Institute, Presidio of Monterey, CA.
Length: 47 weeks (930-1380 hours).
Exhibit Dates: 1/90–1/90.
Objectives: To understand, speak, read, and write the language with an emphasis on audio lingual skills.
Instruction: Lectures, practical exercises and laboratory work in basic grammatical structures in conjunction with oral drills, pattern exercises and written dictation. The course covers over 4000 vocabulary items, 500 of which are military terms. Students draft written reports on general and technical materials and cover a fair amount of cultural information on modern Hungary. NOTE: While these courses are listed as Basic, it should be understood that this is the terminology used by the Armed Forces to indicate that the courses are their regular programs in various languages. They are not limited to what most civilian institutions would term beginning or basic courses in a language.
Credit Recommendation: In the lower-division baccalaureate/associate degree category, 10 semester hours in introductory Hungarian 6 in in advanced Hungarian (3/81); in the upper-division baccalaureate category, 5 semester hours in reading and translation and 3 in culture and civilization (3/81).

DD-0602-0180

RUSSIAN ADVANCED

Course Number: None.
Location: Defense Language Institute, Presidio of Monterey, CA.
Length: 37 weeks.
Exhibit Dates: 1/90–9/90.
Objectives: To train selected Department of Defense personnel in Russian at a more advanced level of proficiency than is provided in Defense Language Institute extended or intermediate courses and to provide a wide knowledge of cultural, geographical, economic, historical, and political information on areas in which the language is spoken.
Instruction: The advanced course places equal emphasis upon the development of all four language skills. There is no specialized or technical terminology in the course. It includes a total vocabulary of approximately 4,000 terms over and above that covered in previous courses. These terms cover all general, nontechnical communication situations that one would normally encounter in the country of the target language. The cultural complex within which the language is spoken is covered extensively: history, economics, geography, politics, military, ethnic groups, languages, attitudes, customs, and mores of the people.
Credit Recommendation: In the lower-division baccalaureate/associate degree category, extending into the upper-division baccalaureate category, 18 semester hours in Russian (8/74).

DD-0602-0205

GERMAN GATEWAY

Course Number: 03GM06; 03GM.
Location: Defense Language Institute, West Coast Branch, Presidio of Monterey, CA.
Length: 8 weeks (173 hours).
Exhibit Dates: 1/90–1/97.
Learning Outcomes: Upon completion of the course, the student will be able to function in common everyday situations, using common phrases; understand similar phrases in everyday situations when spoken slowly by native speakers; and understand cultural issues.
Instruction: The course is conducted through small-group classroom instruction focusing on communication in everyday situations. Presentation of cultural background is included.
Credit Recommendation: In the lower-division baccalaureate/associate degree category, 4 semester hours in German language and culture (8/87).

DD-0602-0206

ARABIC-SYRIAN SPECIAL

Course Number: 09AP24; 09AP.
Location: Defense Language Institute, Presidio of Monterey, CA.
Length: 24 weeks (678 hours).
Exhibit Dates: 1/90–9/90.
Learning Outcomes: Upon completion of the course, the student will have achieved listening and speaking proficiency at the ILR level 1-plus which is equivalent to the ACTFL intermediate high level.
Instruction: Small-group classroom instruction stresses communicative competence in oral/aural reading skills, taught concurrently. An integral learning format directed toward functional language use incorporates a cultural component.
Credit Recommendation: In the lower-division baccalaureate/associate degree category, 10 semester hours in Syrian dialect (8/87).

DD-0602-0207

ARABIC-IRAQI SPECIAL

Course Number: 09DG24; 09DG.
Location: Defense Language Institute, Presidio of Monterey, CA.
Length: 24 weeks (678 hours).
Exhibit Dates: 1/90–9/90.
Learning Outcomes: Upon completion of the course, the student will have achieved listening and speaking proficiency at the ILR level 1-plus which is equivalent to the ACTFL intermediate high level.
Instruction: Small-group classroom instruction stresses communicative competence in oral/aural reading skills, taught concurrently. An integral learning format directed toward functional language use incorporates a cultural component.
Credit Recommendation: In the lower-division baccalaureate/associate degree category, 10 semester hours in Iraqi dialect (8/87).

DD-0602-0208

ARABIC-EGYPTIAN SPECIAL

Course Number: 09AE24; 09AE.
Location: Defense Language Institute, Presidio of Monterey, CA.
Length: 24 weeks (678 hours).
Exhibit Dates: 1/90–9/90.
Learning Outcomes: Upon completion of the course, the student will have achieved listening and speaking proficiency at the ILR level 1-plus which is equivalent to the ACTFL intermediate high level.
Instruction: Small-group classroom instruction stresses communicative competence in oral/aural reading skills, taught concurrently. An integral learning format directed toward functional language use incorporates a cultural component.
Credit Recommendation: In the lower-division baccalaureate/associate degree category, 10 semester hours in Egyptian dialect (8/87).

DD-0602-0209

ARABIC-IRAQI EXTENDED

Course Number: 05DG16; 05DG.
Location: Defense Language Institute, Presidio of Monterey, CA.
Length: 16 weeks (456 hours).
Exhibit Dates: 1/90–9/90.
Learning Outcomes: Upon completion of the course, the student will have achieved listening and speaking proficiency at the ILR level 1-plus which is equivalent to the ACTFL intermediate high level.
Instruction: Small-group classroom instruction stresses communicative competence in oral/aural reading skills, taught concurrently. An integral learning format directed toward functional language use incorporates a cultural component.
Credit Recommendation: In the lower-division baccalaureate/associate degree category, 10 semester hours in Iraqi dialect (8/87).

DD-0602-0210

ARABIC-SYRIAN EXTENDED

Course Number: 05AP16; 05AP.
Location: Defense Language Institute, Presidio of Monterey, CA.
Length: 16 weeks (456 hours).
Exhibit Dates: 1/90–9/90.
Learning Outcomes: Upon completion of the course, the student will have achieved listening and speaking proficiency at the ILR level 1-plus which is equivalent to the ACTFL intermediate high level.
Instruction: Small-group classroom instruction stresses communicative competence in oral/aural reading skills, taught concurrently. An integral learning format directed toward functional language use incorporates a cultural component.
Credit Recommendation: In the lower-division baccalaureate/associate degree category, 10 semester hours in Syrian dialect (8/87).

DD-0602-0211

ARABIC-EGYPTIAN EXTENDED

Course Number: 05AE16; 05AE.
Location: Defense Language Institute, Presidio of Monterey, CA.
Length: 16 weeks (454 hours).
Exhibit Dates: 1/90–9/90.
Learning Outcomes: Upon completion of the course, the student will have achieved listening and speaking proficiency at the ILR level

1-plus which is equivalent to the ACTFL intermediate high level.

Instruction: Small-group classroom instruction stresses communicative competence in oral/aural reading skills, taught concurrently. An integral learning format directed toward functional language use incorporates a cultural component.

Credit Recommendation: In the lower-division baccalaureate/associate degree category, 10 semester hours in Egyptian dialect (8/87).

DD-0602-0212
PORTUGUESE BASIC

Course Number: 01PQ25; 01PQ.
Location: Defense Language Institute, Presidio of Monterey, CA.
Length: 25 weeks (728 hours).
Exhibit Dates: 1/90–9/90.
Learning Outcomes: Upon completion of the course, the student will have achieved listening, reading, and speaking proficiency at the following ILR levels: listening: 2/2+; speaking: 2; reading: 2; writing is not evaluated in the school but is approximately ILR 1+/2.
Instruction: Small-group classroom instruction stresses communicative competence in oral/aural and reading skills, taught concurrently. An integral learning format, directed toward functional language use, incorporates cultural and writing components which are not formally tested.
Credit Recommendation: In the lower-division baccalaureate/associate degree category, 3 semester hours in Portuguese elements I, 3 in Portuguese elements II, 3 in Portuguese composition and culture, and 3 in Portuguese readings (8/87).

DD-0602-0213
FRENCH BASIC

Course Number: 01FR25; 01FR.
Location: Defense Language Institute, Presidio of Monterey, CA.
Length: 25 weeks (748 hours).
Exhibit Dates: 1/90–9/90.
Learning Outcomes: Upon completion of the course, the student will have achieved listening, reading, and speaking proficiency at the following ILR levels: listening: 2/2+; speaking: 2; reading: 2; writing is not evaluated by the school but is approximately ILR 1+/2.
Instruction: Small-group classroom instruction stresses communicative competence in oral/aural and reading skills, taught concurrently. An integral learning format, directed toward functional language use, incorporates cultural and writing components which are not formally tested.
Credit Recommendation: In the lower-division baccalaureate/associate degree category, 3 semester hours in French elements I, 3 in French elements II, 3 in French composition and culture, and 3 in French readings (8/87).

DD-0602-0214
GERMAN SHORT

Course Number: 04GM12; 04GM.
Location: Defense Language Institute, Presidio of Monterey, CA.
Length: 12 weeks (356 hours).
Exhibit Dates: 1/90–9/90.
Learning Outcomes: Upon completion of the course, the student will have achieved listening, reading, and speaking proficiency at the ILR level 1-plus which is equivalent to the ACTFL intermediate/mid-level guidelines.
Instruction: Small-group classroom instruction stresses communicative competence in oral/aural and reading skills, taught concurrently. An integral learning format directed toward functional language use incorporates a cultural component.
Credit Recommendation: In the lower-division baccalaureate/associate degree category, 8 semester hours in German language skills (8/87).

DD-0602-0215
GERMAN BASIC

Course Number: 01GM34; 01GM.
Location: Defense Language Institute, Presidio of San Francisco, CA; Defense Language Institute, Presidio of Monterey, CA.
Length: 34 weeks (1018 hours).
Exhibit Dates: 1/90–9/90.
Learning Outcomes: Upon completion of the course, the student will have achieved listening, reading, and speaking proficiency at the ILR level 1-plus which is equivalent to the ACTFL intermediate-high level.
Instruction: Small-group classroom instruction stresses communicative competence in oral/aural and reading skills, taught concurrently. An integral learning format directed toward functional language use incorporates a cultural component.
Credit Recommendation: In the lower-division baccalaureate/associate degree category, 16 semester hours in German language skills (8/87).

DD-0602-0216
TURKISH GATEWAY

Course Number: 03TU12; 03TU.
Location: Defense Language Institute, Presidio of Monterey, CA.
Length: 8-12 weeks (336 hours).
Exhibit Dates: 1/90–9/90.
Learning Outcomes: Upon completion of the course, the student will be able to function in common everyday situations, using common phrases and understand phrases in everyday situations when spoken at a slow pace.
Instruction: Small-group classroom instruction stresses communicative competence in oral/aural and reading skills, taught concurrently. An integral learning format directed toward functional language use incorporates a cultural component.
Credit Recommendation: In the lower-division baccalaureate/associate degree category, 10 semester hours in Turkish language skills (8/87).

DD-0602-0217
GERMAN EXTENDED

Course Number: 10GM24; 10GM.
Location: Defense Language Institute, Presidio of Monterey, CA.
Length: 24 weeks (692 hours).
Exhibit Dates: 1/90–9/90.
Learning Outcomes: Upon completion of the course, the student will have achieved listening, reading, and speaking proficiency at the ILR level 2-plus which is equivalent to the ACTFL advanced-plus level. Students will have limited writing competence.
Instruction: Small-group classroom instruction stresses communicative competence in oral/aural and reading skills (with limited writing practice), taught concurrently. An integral learning format directed toward functional language use incorporates a cultural component.
Credit Recommendation: In the upper-division baccalaureate category, 8 semester hours in German reading, speaking, and textual analysis (8/87).

DD-0602-0218
SPANISH INTERMEDIATE

Course Number: 06LA24; 06LA.
Location: Defense Language Institute, Presidio of Monterey, CA.
Length: 24 weeks (718 hours).
Exhibit Dates: 1/90–9/90.
Learning Outcomes: Upon completion of the course, the student will have achieved listening, reading, and speaking proficiency at the ILR level 2-plus which is equivalent to the ACTFL advanced-plus level. Students will have limited writing competence.
Instruction: Small-group classroom instruction stresses communicative competence in oral/aural and reading skills (with limited writing practice), taught concurrently. An integral learning format directed toward functional language use incorporates a cultural component.
Credit Recommendation: In the lower-division baccalaureate/associate degree category, 4 semester hours in oral/aural Spanish language skills (8/87); in the upper-division baccalaureate category, 8 semester hours in Spanish reading/writing/culture (8/87).

DD-0602-0219
GERMAN INTERMEDIATE

Course Number: 06GM24.
Location: Defense Language Institute, Presidio of Monterey, CA.
Length: 24 weeks (692 hours).
Exhibit Dates: 1/90–9/90.
Learning Outcomes: Upon completion of the course, the student will have achieved listening, reading, and speaking proficiency at the ILR level 2-plus which is equivalent to the ACTFL advanced-plus level. Students will have limited writing competence.
Instruction: Small-group classroom instruction stresses communicative competence in oral/aural and reading skills (with limited writing practice), taught concurrently. An integral learning format directed toward functional language use incorporates a cultural component.
Credit Recommendation: In the upper-division baccalaureate category, 8 semester hours in German reading, speaking, and textual analysis (8/87).

DD-0602-0220
ARABIC INTERMEDIATE

Course Number: 06AD32; 06AD.
Location: Defense Language Institute, Presidio of Monterey, CA.
Length: 32 weeks (958 hours).
Exhibit Dates: 1/90–9/90.
Learning Outcomes: Upon completion of the course, the student will have achieved listening, reading, and speaking proficiency at the ILR level 2-plus which is equivalent to the

ACTFL advanced-plus level. Students will have limited writing competence.

Instruction: Small-group classroom instruction stresses communicative competence in oral/aural and reading skills (with limited writing practice), taught concurrently. An integral learning format directed toward functional language use incorporates a cultural component.

Credit Recommendation: In the upper-division baccalaureate category, 6 semester hours in Arabic speaking and writing skills (8/87); in the graduate degree category, 6 semester hours in Arabic listening and reading skills (8/87).

DD-0602-0221

FRENCH INTERMEDIATE

Course Number: 06FR24; 06FR.
Location: Defense Language Institute, Presidio of Monterey, CA.
Length: 24 weeks (738 hours).
Exhibit Dates: 1/90–9/90.
Learning Outcomes: Upon completion of the course, the student have achieved listening, reading, and speaking proficiency at the ILR level 2-plus which is equivalent to the ACTFL advanced-plus level. Students will have limited writing competence.

Instruction: Small-group classroom instruction stresses communicative competence in oral/aural and reading skills (with limited writing practice), taught concurrently. An integral learning format directed toward functional language use incorporates a cultural component.

Credit Recommendation: In the lower-division baccalaureate/associate degree category, 4 semester hours in oral/aural French language skills (8/87); in the upper-division baccalaureate category, 8 semester hours in French reading/writing/culture (8/87).

DD-0602-0222

SPANISH BASIC

Course Number: 01LA25; 01LA.
Location: Defense Language Institute, Presidio of San Francisco, CA; Defense Language Institute, Presidio of Monterey, CA.
Length: 25 weeks (748 hours).
Exhibit Dates: 1/90–9/90.
Learning Outcomes: Upon completion of the course, the student will have achieved listening, reading, and speaking proficiency at the following ILR levels: listening: 2/2+; speaking: 2; reading: 2. Writing is not evaluated by the school but is approximately ILR 1+/2.

Instruction: Small-group classroom instruction stresses communicative competence in oral/aural and reading skills, taught concurrently. An integral learning format, directed toward functional language use, incorporates cultural and writing components which are not formally tested.

Credit Recommendation: In the lower-division baccalaureate/associate degree category, 3 semester hours in Spanish elements I, 3 in Spanish elements II, 3 in Spanish composition and culture, and 3 in Spanish readings (8/87).

DD-0602-0223

DUTCH BASIC

Course Number: 01DU25; 01DU.
Location: Defense Language Institute, Presidio of Monterey, CA.
Length: 24 weeks (744 hours).
Exhibit Dates: 1/90–1/97.
Learning Outcomes: Upon completion of the course, the student will have achieved listening, reading, and speaking proficiency at the ILR level 1-plus which is equivalent to the ACTFL intermediate-high level. Students will have limited writing competence.

Instruction: Small-group classroom instruction stresses communicative competence in oral/aural and reading skills (with limited writing practice), taught concurrently. An integral learning format directed toward functional language use incorporates a cultural component.

Credit Recommendation: In the lower-division baccalaureate/associate degree category, 12 semester hours in Dutch (8/87).

DD-0602-0224

NORWEGIAN BASIC

Course Number: 01NR25; 01NR.
Location: Defense Language Institute, Presidio of Monterey, CA.
Length: 25 weeks (728 hours).
Exhibit Dates: 1/90–10/90.
Learning Outcomes: Upon completion of the course, the student will have achieved listening, reading, and speaking proficiency at the ILR level 1-plus which is equivalent to the ACTFL intermediate-high level. Students will have limited writing competence.

Instruction: Small-group classroom instruction stresses communicative competence in oral/aural and reading skills (with limited writing practice), taught concurrently. An integral learning format directed toward functional language use incorporates a cultural component.

Credit Recommendation: In the lower-division baccalaureate/associate degree category, 12 semester hours in Norwegian (8/87).

DD-0602-0225

ROMANIAN BASIC

Course Number: 01RQ34; 01RQ.
Location: Defense Language Institute, Presidio of Monterey, CA.
Length: 34 weeks (998 hours).
Exhibit Dates: 1/90–9/90.
Learning Outcomes: Upon completion of the course, the student will have achieved listening, speaking, and reading proficiency at the ILR level 2. Writing is not evaluated during the course.

Instruction: Small-group classroom instruction stresses communicative competence in oral/aural and reading skills, taught concurrently. Course work is presented through an integral learning format directed toward functional language use, incorporating cultural and writing components which are not formally tested.

Credit Recommendation: In the lower-division baccalaureate/associate degree category, 3 semester hours in Romanian elements I, 3 in Romanian elements II, 3 in Romanian composition and culture, and 3 in Romanian readings (2/88); in the upper-division baccalaureate category, 4 semester hours in Romanian intermediate (oral/aural and reading) (2/88).

DD-0602-0226

ITALIAN BASIC

Course Number: 01JT25; 01JT.
Location: Defense Language Institute, Presidio of Monterey, CA.
Length: 25 weeks (750 hours).
Exhibit Dates: 1/90–9/90.
Learning Outcomes: Upon completion of the course, the student will have achieved listening, reading, and speaking proficiency at the following ILR levels: Listening 2/2+; Speaking 2; Reading 2; Writing is not evaluated during the course.

Instruction: Small-group classroom instruction stresses communicative competence in oral/aural and reading skills, taught concurrently. Course is presented through an integrated learning format directed toward functional language use, incorporating cultural and writing components which are not formally tested.

Credit Recommendation: In the lower-division baccalaureate/associate degree category, 3 semester hours in Italian I, 3 in Italian II, 3 in Italian composition and culture, and 3 in Italian reading (2/88).

DD-0602-0227

RUSSIAN BASIC

Course Number: 01RU47; 01RU.
Location: Defense Language Institute, Presidio of Monterey, CA.
Length: 47 weeks (1415 hours).
Exhibit Dates: 1/90–9/90.
Learning Outcomes: Upon completion of the course, the student will have achieved listening, reading, and speaking proficiency at following ILR levels: Listening 2+; Speaking 2+; Reading 2; Writing is not evaluated.

Instruction: Group classroom instruction with elaborated techniques and materials emphasizing active skills (oral/aural) has somewhat less emphasis on reading skills. There is very little true translation, as befits de-emphasis on passive skills.

Credit Recommendation: In the lower-division baccalaureate/associate degree category, 7 semester hours in elementary Russian and 7 in lower-intermediate Russian (2/88); in the upper-division baccalaureate category, 10 semester hours in intermediate Russian (2/88).

DD-0602-0229

CZECH BASIC

Course Number: 01CX47; 01CX.
Location: Defense Language Institute, Presidio of Monterey, CA.
Length: 47 weeks (1382 hours).
Exhibit Dates: 1/90–9/90.
Learning Outcomes: Upon completion of the course, the student will have achieved listening, reading, and speaking proficiency at following ILR levels: Listening 1+/2: Speaking 1+/2; Reading 1+/2; Writing is not evaluated, but is approximately 1+.

Instruction: Group classroom instruction has emphasis on active skills (oral/aural) and somewhat less emphasis on reading skills. There is only very little translation, as befits de-emphasis on passive skills.

Credit Recommendation: In the lower-division baccalaureate/associate degree category, 7 semester hours in elementary Czech and 7 in lower-intermediate Czech (2/88); in the upper-division baccalaureate category, 8 semester hours in intermediate Czech (2/88).

DD-0602-0230

PERSIAN BASIC

Course Number: 01PF47; 01PF.
Location: Defense Language Institute, Presidio of Monterey, CA.
Length: (1382), 47 weeks.
Exhibit Dates: 1/90–9/90.
Learning Outcomes: Upon completion of the course, the student will have achieved listening, speaking, and reading proficiency at the following ILR levels: Listening 1+/2; Speaking 1+; Reading 1+/2; Writing 0+/1.
Instruction: The course is presented through an integrated skill method directed toward functional language use with limited writing competence. Small-group classroom instructions stresses oral/aural practice. Persian newspapers and Voice of America Persian programs are used during the last ten weeks.
Credit Recommendation: In the lower-division baccalaureate/associate degree category, 12 semester hours in elementary Persian (2/88); in the upper-division baccalaureate category, 6 semester hours in intermediate Persian and 4 in Persian news media (2/88).

DD-0602-0231

TURKISH BASIC

Course Number: 01TU47; 01TU.
Location: Defense Language Institute, Presidio of Monterey, CA.
Length: 47 weeks (1382 hours).
Exhibit Dates: 1/90–9/90.
Learning Outcomes: Upon completion of the course, the student will be able to achieve listening, speaking, and reading proficiency at the following ILR levels: Listening 1+; Speaking 1+/2; Reading 1+; and Writing 0+/1.
Instruction: This course offers an integrated skill method directed toward functional language use with limited writing competence. Instruction is accomplished through small-group classroom instruction, stressing oral/aural practice. Turkish newspapers and Voice of America Turkish programs are used during the last ten weeks.
Credit Recommendation: In the lower-division baccalaureate/associate degree category, 12 semester hours in elementary Turkish (2/88); in the upper-division baccalaureate category, 6 semester hours in intermediate Turkish and 4 in Turkish news media (2/88).

DD-0602-0232

ARABIC BASIC (MODERN STANDARD ARABIC)

Course Number: 01AD47; 01AD.
Location: Defense Language Institute, Presidio of Monterey, CA.
Length: 47 weeks (1412 hours).
Exhibit Dates: 1/90–9/90.
Learning Outcomes: Upon completion of the course, the student will have achieved listening, speaking, and reading proficiency at the following ILR levels: Listening 1+/2; Speaking 1+/2; Reading 1+; Writing 0+/1.
Instruction: The course is presented through an integrated skill method directed toward functional language use with limited writing competence. Small-group classroom instructions stresses oral/aural practice. Arabic newspapers and radio programs are used during the last ten weeks.
Credit Recommendation: In the lower-division baccalaureate/associate degree category, 12 semester hours in elementary Modern Standard Arabic (MSA) (2/88); in the upper-division baccalaureate category, 6 semester hours in intermediate Modern Standard Arabic and 4 in Arabic news media (2/88).

DD-0602-0233

CHINESE INTERMEDIATE

Course Number: 06CM37.
Location: Defense Language Institute, Presidio of Monterey, CA.
Length: 37 weeks (1084 hours).
Exhibit Dates: 1/90–9/90.
Learning Outcomes: Upon completion of the course, the student will have achieved an ILR 2/2R (ACTFL advanced/advanced high) in speaking, listening, and reading. Writing as an independent skill is not stressed in the program.
Instruction: Students meet in small groups in which listening, reading, and speaking skills are worked on through activities ranging from oral reading to free discussion. The course material includes both standard textbooks and those designed specifically for military courses, as well as selected readings from a variety of sources including newspapers, magazines, middle school textbooks. Radio broadcasts are also used.
Credit Recommendation: In the upper-division baccalaureate category, 10 semester hours in spoken Chinese, 10 in reading, and 4 in listening comprehension/radio broadcasts (2/91).

DD-0602-0234

CHINESE ADVANCED

Course Number: 07CM; 07CM37.
Location: Defense Language Institute, Presidio of Monterey, CA.
Length: 37 weeks (1082 hours).
Exhibit Dates: 1/90–9/90.
Learning Outcomes: Upon completion of the course, the student will have achieved an ILR 2+/3 (ACTFL advanced high/superior) in the three skills: reading, speaking, and listening. The course does not stress writing as an independent skill.
Instruction: Students meet in small group classes in which listening, reading, and speaking skills are worked on through activities ranging from oral reading and free discussion to lectures. The course material includes readings from newspaper, magazines, military manuals, articles on technical subjects written for the educated reader, and radio broadcasts. The readings include extensive information about Chinese society, its political and economic system, and its culture.
Credit Recommendation: In the upper-division baccalaureate category, 8 semester hours in spoken Chinese, 8 in reading, and 2 in listening/radio broadcast (2/91); in the graduate degree category, 6 semester hours in Chinese for special purposes with a focus on military and technical uses (2/91).

DD-0602-0235

HEBREW BASIC

Course Number: 01HE47.
Location: Defense Language Institute, Presidio of Monterey, CA.
Length: 47 weeks (1410 hours).
Exhibit Dates: 1/90–4/91.
Learning Outcomes: Upon completion of the course, the student will have achieved proficiency at the ILR level 2/2+ (ACTFL advanced/advanced high) in at least two of the following skills and the 1/1+ (ACTFL intermediate low intermediate high) level in the third skill: reading, listening, and speaking. Writing proficiency is stressed to a lesser degree, and students can be expected to achieve proficiency at the ILR 1/1+ (ACTFL intermediate low/high) level.
Instruction: Small group classroom instruction stresses reading and listening comprehension. The student develops language proficiency through a variety of activities including reading text books and authentic materials taken from Israeli newspapers, discussions of these texts, and grammar exercises. The cultural components include an introduction to regional geography, historical events, current events, and various aspects of daily life in Israel.
Credit Recommendation: In the lower-division baccalaureate/associate degree category, 8 semester hours in elementary Hebrew (2/91); in the upper-division baccalaureate category, 8 semester hours in intermediate Hebrew and 4 in Israeli news media (2/91).

DD-0602-0236

KOREAN INTERMEDIATE

Course Number: 06KP; 06KP47.
Location: Defense Language Institute, Presidio of Monterey, CA.
Length: 47 weeks (1362 hours).
Exhibit Dates: 1/90–9/90.
Learning Outcomes: Upon completion of the course, the student will have achieved listening and reading proficiency at the ILR level 2+ (ACTFL advanced high) and speaking proficiency at the ILR level 1+/2 (ACTFL intermediate high to advanced). Writing proficiency is stressed to a lesser degree, and in this skill students can be expected to achieve proficiency at the ILR level 2 (ACTFL advanced).
Instruction: Small group classroom instruction gives equal emphasis to listening, speaking, and reading skills. The student develops language proficiency through a variety of activities, including listening to radio broadcasts; reading newspaper articles of political, cultural, and social content; as well as reading other general publications. Hanja (Chinese characters) as well as Hangul are taught in order to give students full access to the range of Korean publications. The cultural component of the course introduces the student to Korean geography, history, current events, and aspects of daily life. The courses meets for 6 hours per day, five days a week for 47 weeks. The 6 hours includes one hour of study and one hour in the language laboratory. It is recommended that students spend at least an additional 4 hours each day in preparation for class.
Credit Recommendation: In the upper-division baccalaureate category, 10 semester hours in spoken Korean and 10 in written Korean (2/91).

DD-0602-0237

TAGALOG BASIC
(Pilipino/Tagalog Basic)

Course Number: 017A47.
Location: Defense Language Institute, Presidio of Monterey, CA.

Length: 47 weeks (1382 hours).
Exhibit Dates: 1/90–4/91.
Learning Outcomes: Upon completion of the course, the student will have achieved, reading and speaking at the following ILR levels: listening 2/2+ (ACTFL advanced to high advanced); speaking 2 (ACTFL advanced); reading 2/2+ (ACTFL advanced to high advanced); and writing 2 (ACTFL advanced).
Instruction: The course arrives at an integrated skills approach using a variety of written and oral materials. Over the course of 47 weeks, six hours a day, the students receive an average per day of the following: 1 hour of reading skill development, 2 hours a day of drill, grammar exercises, and translation exercises, 1 hour a day of group classroom or individualized instruction emphasizing communicative skills, 2 hours of laboratory work, and 1 hour of quizzes. Although English is used in the classroom, the emphasis is on the employment of Tagalog.
Credit Recommendation: In the lower-division baccalaureate/associate degree category, 8 semester hours in elementary Tagalog (2/91).

DD-0602-0238

VIETNAMESE BASIC

Course Number: 01VN; 01VN47.
Location: Defense Language Institute, Presidio of Monterey, CA.
Length: 47 weeks (1410 hours).
Exhibit Dates: 1/90–9/90.
Learning Outcomes: Upon completion of the course, the student will have achieved an ILR 2 (ACTFL advanced) in the 3 skills: reading, speaking, and listening. The course does not stress writing as an independent skill.
Instruction: Students meet in small group classes in which listening, reading, and speaking are worked on through a range of activities including drills, transcription exercises, and discussions based on readings and assigned topics. The course material includes language textbooks, tapes of radio broadcasts, and articles from Vietnamese newspapers and magazines. The course also includes lectures on Vietnamese society, its political and economic system, and its culture.
Credit Recommendation: In the lower-division baccalaureate/associate degree category, 10 semester hours in basic Vietnamese and 10 in intermediate Vietnamese (2/91).

DD-0602-0239

THAI INTERMEDIATE

Course Number: 06TH37; 06TH.
Location: Defense Language Institute, Presidio of Monterey, CA.
Length: 37 weeks (1082 hours).
Exhibit Dates: 1/90–Present.
Learning Outcomes: Upon completion of the course, the student will be able to comprehend, read, translate, and speak (in descending order of emphasis) Thai at a FSI proficiency skill level of 2/3.5. Writing proficiency will be at level 1. The student will be able to comprehend, interpret, and translate speeches and radio broadcasts on current events, politics, economics, and general military subjects based on recorded Thai radio broadcasts at a proficiency level of 3.5. Spoken proficiency dealing with concrete situations and responses in the form of connected statements that provide descriptions, explanations, and opinions aim to bring the student to a FSI level of 2. Reading proficiency covers the ability to read newspaper articles and editorials in order to extract critical information on current events, politics, and economics. Writing proficiency is the least developed skill and is limited to the ability to take dictation slowly and to relay simple messages.
Instruction: All instruction includes practice of oral/aural skills in both the classroom and laboratory with major emphasis on comprehension of connected speech in the form of brief paragraphs recorded previously or spoken by a native speaker in the classroom and segments of authentic radio and TV broadcasts. Reading of authentic materials begins with newspaper advertisements and notices and culminates with exercises on reading editorials and opinion essays. The course begins with an extensive 10 week review of the structures covered in the Basic course and continues for 30 more weeks with a variety of exercises that include recorded conversations written for developing advanced comprehension and translation and transcription abilities.
Credit Recommendation: In the upper-division baccalaureate category, for the first 10 weeks review of grammar: 3 semester hours in Thai grammar and, for the following 20-30 weeks, 6-12 semester hours in reading and translation (2/91).

DD-1115-0001

FUNDAMENTALS OF COST ANALYSIS

Course Number: *Version 1:* BCF 101. *Version 2:* BCE 101.
Location: Defense Acquisition University, Various locations in the Continental US.
Length: *Version 1:* 3 weeks (101 hours). *Version 2:* 3 weeks (88-89 hours).
Exhibit Dates: *Version 1:* 5/97–Present. *Version 2:* 6/96–4/97.
Learning Outcomes: *Version 1:* This version is pending evaluation. *Version 2:* Upon completion of the course, the student will be able to participate as a member of an estimating team employing basic techniques and policies for determining a major defense acquisition program life cycle cost.
Instruction: *Version 1:* This version is pending evaluation. *Version 2:* The focus of this course is the quantitative methods used to analyze and project contract costs. After extensive classroom instruction and practice in these quantitative methods, the students work on a week-long cost analysis case requiring the application of most of these techniques. There are two exams and a final. All are multiple choice but require some calculation. Students use PC-based software to solve the exercises and the case. Principle topics covered by this course include quantitative methods for cost estimating (simple regression analysis, learning curves, multiple regression, statistics for cost estimating, uncertainty analysis, data collection and exploratory data analysis, inflation theory and application, software cost estimating, intrinsically linear regression, and forecasting); cost estimating policies (cost analysis and design to cost); economic analysis (analysis of alternatives and life cycle cost analysis); and automated tools for cost analysis.
Credit Recommendation: *Version 1:* Pending evaluation. *Version 2:* In the lower-division baccalaureate/associate degree category, 3 semester hours in introduction to applied statistics, cost estimating, and analysis (3/97).

DD-1402-0003

AUTOMATED INFORMATION SYSTEMS MANAGEMENT FOR INTERMEDIATE MANAGERS
(Automated Information Systems Management for Intermediate Executives)
(Computer Orientation for Intermediate Executives)

Course Number: None.
Location: Department of Defense Computer Institute, Washington, DC; National Defense University, Ft. Lesley J. McNair, Washington, DC.
Length: 2 weeks (65 hours).
Exhibit Dates: 1/90–1/90.
Objectives: Course is designed to provide an educational background for high-level management personnel who are general-purpose digital computer systems users and have had little or no previous introduction to data processing principles.
Instruction: Course covers computer capabilities, limitations, and applications; the basics of computer hardware and software; systems development management considerations; planning and design; introduction to quantitative techniques. Also includes hardware and software acquisition, data base management systems, small computer systems, teleprocessing and distributed processing, and computer security. Student is provided hands-on programming experience with a remote, time-sharing computer terminal using the BASIC programming language.
Credit Recommendation: In the lower-division baccalaureate/associate degree category, 3 semester hours in automated information systems (5/84).

DD-1402-0004

SOFTWARE ACQUISITION MANAGEMENT

Course Number: SAM 301.
Location: Defense Acquisition University, Various locations in the Continental US.
Length: 3 weeks (98 hours).
Exhibit Dates: 11/95–Present.
Learning Outcomes: Upon completion of the course, the student should be able to: analyze the causes of cost, schedule, and performance problems in large software efforts and explore strategies for avoiding or correcting such problems; examine salient differences in strategy, methods, and tolls between commercial software acquisition efforts and DOD efforts; develop an ability to recognize and selectively adopt commercial practices for use in a DOD software program; understand the organizational and cultural dynamics of program offices and software development teams; evaluate the suitability of alternative organizational structures including integrated product teams; examine and select software metrics that will provide insight into program status and facilitate early detection of potential problems; and assess the current state of the Federal and DOD acquisition reform movements and incorporate new policies into current and future software acquisition programs.
Instruction: Methods of instruction include lectures, seminars, case studies, and class exercises. Principle areas covered include planning (addresses issues related to acquisition policy,

planning, and change management); requirements (explores issues related to the identification, familiarization, and implementation of requirements for a large complex software effort); integration (examines the strategies for developing software systems that are portable across different hardware platforms, that can inter-operate with other related systems, and that will foster current and future reuse of existing software products); risk (analyzes methods for identifying, avoiding, and controlling various types of risk that are likely to arise in major software programs); and quality management (examines techniques that a manager can use to measure and improvement the quality of software products).

Credit Recommendation: In the upper-division baccalaureate category, 3 semester hours in management information systems (3/97).

DD-1402-0005

SOFTWARE COST ESTIMATING

Course Number: *Version 1:* BCF 208. *Version 2:* BCE 208.
Location: Defense Acquisition University, Various locations in the Continental US.
Length: *Version 1:* 1-2 weeks (65 hours). *Version 2:* 2 weeks (57 hours).
Exhibit Dates: *Version 1:* 5/97–Present. *Version 2:* 6/96–4/97.
Learning Outcomes: *Version 1:* This version is pending evaluation. *Version 2:* Upon completion of the course, the student should be familiar with the quantitatively-based methodologies for estimating the cost of software development contracts and be able to describe the software acquisition process, determine the most appropriate cost estimating methodology and apply it, and formulate and test models for life-cycle cost estimating.
Instruction: *Version 1:* This version is pending evaluation. *Version 2:* Students prepare a major case study and are graded on their individual presentations. Principles subject areas taught in this course are software life-cycle management; contracting issues (MIL STD 498); technical fundamentals (architecture and interoperability); software development paradigms (software design approaches and methodologies); capability evaluation of software developers; software cost estimating (COCOMO (REVIC), survey of software cost estimating models, risk analysis in software cost estimating, REVIC calibration, and case study); and open systems (software reuse).
Credit Recommendation: *Version 1:* Pending evaluation. *Version 2:* Credit is not recommended because of the limited, specialized nature of the course (3/97).

DD-1402-0006

AUTOMATED INFORMATION SYSTEMS CONTRACTING

Course Number: CON 241.
Location: Defense Acquisition University, Various locations in the Continental US.
Length: 2 weeks (63 hours).
Exhibit Dates: 4/93–Present.
Learning Outcomes: Upon completion of the course, the student will be able to understand the impact of legislation and other congressional action on the acquisition of information resources and to apply correct decision making techniques to making such acquisitions.
Instruction: This course is designed for intermediate level personnel involved in the acquisition of information resources. It provides a working familiarity with pertinent government regulations, policies, and procedures. Lectures and practical exercises provide detailed instruction in all the significant aspects of contracting for information resources. Major areas covered in this course include policies and procedures, pre-acquisition planning, acquisition plan development, presolicitation phase, solicitation development phase, and award and performance phase.
Credit Recommendation: In the upper-division baccalaureate category, 3 semester hours in management information systems (3/97).

DD-1402-0007

INTERMEDIATE FACILITIES CONTRACTING

Course Number: CON 223.
Location: Defense Acquisition University, Various locations in the Continental US.
Length: 2 weeks (67 hours).
Exhibit Dates: 11/96–Present.
Learning Outcomes: Upon completion of the course, the student should be able to determine requirements, administer the contract award process, apply the provisions of the Brooks act, monitor contract performance, accept contract work, evaluate compliance in cost reimbursement contracts, and determine the appropriate resolution of contractor claims.
Instruction: Methods of instruction include lecture, discussion, workshop, and case study. Principal areas covered include acquisition planning, source selection, architect-engineering contracts, contract options, quality management, labor relations, payment, contract compliance clauses, contract delays and extensions, contract modifications, cost reimbursement contracts, and claims resolution.
Credit Recommendation: In the upper-division baccalaureate category, 2 semester hours in procurement/supply management (3/97).

DD-1402-0008

FACILITIES CONTRACTS PRICING

Course Number: CON 106.
Location: Defense Acquisition University, Various locations in the Continental US.
Length: 3 weeks (84 hours).
Exhibit Dates: 10/96–Present.
Learning Outcomes: Upon completion of the course, the student should apply knowledge and perspective of contract price and cost analysis to evaluate price reasonableness of proposed facilities contracts. Also, students should become aware of the importance of facilities contracts pricing in terms of construction, services, and environmental contacting. Finally, student should be able to evaluate contract pricing strategies in the formulation of government cost/pricing objectives.
Instruction: This course focuses on pricing principles as applied to government construction, services, and environmental contracting areas. It combines lecture material, workshops, exercises, and a week of intensive team-based negotiation simulation. Major areas covered in this course include contract pricing, market research for price analysis, price competition, award criteria for pricing, costs and cost analysis, work design and analysis, regression analysis, improvement curve, direct and indirect cost, facilities capital (cost of money); negotiations, and post-award negotiations.
Credit Recommendation: In the upper-division baccalaureate category, 3 semester hours in procurement/supply management (3/97).

DD-1402-0009

FACILITIES CONTRACTING FUNDAMENTALS

Course Number: CON 103.
Location: Defense Acquisition University, Various locations in the Continental US.
Length: 4 weeks (137 hours).
Exhibit Dates: 10/96–Present.
Learning Outcomes: Upon completion of the course, the student should be able to administer the basic contracting process, including preparation of solicitation requests with focus on the types of contracts, statement of work, and competition considerations; execution of contract awards with emphasis on sealed bid procedures, source selection, negotiation of proposals, and cost and pricing review; management of on-going contracts with emphasis on contract modifications, quality review, vendor reimbursement, and accounting; and close-out of completed contracts with emphasis on final administrative procedures, disposition of excess funds, and final invoicing.
Instruction: Methods of instruction include lectures, discussions, class exercises, and problems. Principal areas covered include contracting fundamentals, purchase requests, competition considerations, ethical standards, cost and pricing, architect and engineering contracting, negotiation skills, contract modifications and options, protests and claims, small purchase acquisitions, payment and accounting, contract modifications, and commercial contracting.
Credit Recommendation: In the upper-division baccalaureate category, 3 semester hours in procurement/supply management (3/97).

DD-1402-0011

INTERMEDIATE SYSTEMS PLANNING, RESEARCH DEVELOPMENT, AND ENGINEERING

Course Number: SYS 201.
Location: Defense Acquisition University, Various locations in the Continental US.
Length: 2 weeks (55-56 hours).
Exhibit Dates: 10/94–Present.
Learning Outcomes: Upon completion of the course, the student will be able to apply the Systems Engineering process to the acquisition life cycle within the Integrated Process/Product development methodology.
Instruction: The focus of this course is the systems engineering process. The student attends lectures and performs in-class practical exercises to provide closure on the lesson. There are three multiple choice examinations. The main topics covered by this course include: systems engineering overview - systems acquisition, integrated product and process development, systems engineering, acquisition ethics, technology development and insertion; systems engineering process - systems engineering planning, requirements analysis, functional analysis/allocation, synthesis verification, systems engineering process outputs; and systems analysis and control tools - work breakdown structure, solicitations/source selections, cost and an

independent variable, life cycle cost, trade studies, configuration management, risk management, technical performance measurement, technical reviews, product improvement, environmental safety and health planning.

Credit Recommendation: In the upper-division baccalaureate category, 3 semester hours in systems engineering (3/97).

DD-1404-0001

SENIOR MILITARY CRYPTOLOGIC SUPERVISORS

Course Number: CY-200.
Location: National Cryptologic School, Ft. Meade, MD.
Length: 7 weeks (240-250 hours).
Exhibit Dates: 1/90–7/91.
Objectives: To train mid-career noncommissioned and warrant officers to supervise communication, intelligence, and other cryptologic resources vital to US national security.
Instruction: Lectures, problem solving seminars, tours, and student presentations cover future roles and technical requirements for supervisory positions in the security community. Subject areas include seminars in contemporary management concepts, problem solving, and decision making techniques; roles in the intelligence community; resource allocation decisions; operational management philosophies; and processes involved with collection and reporting of intelligence and communications security.
Credit Recommendation: In the lower-division baccalaureate/associate degree category, 1 semester hour in communication including written and oral skills (3/82); in the upper-division baccalaureate category, 2 semester hours in field experience in applied electronic systems management, including intelligence report preparation and analysis, electronics systems operations, and communications security; 1 in general management, including facilities management, logistics management, personnel supervision, and office practices; and 1 in political science including political geography and U.S. government (3/82).

DD-1404-0004

NATIONAL COMMUNICATIONS SECURITY (COMSEC)

Course Number: CY-300.
Location: National Cryptologic School, Ft. Meade, MD.
Length: 2 weeks (80 hours).
Exhibit Dates: 1/90–12/90.
Objectives: To prepare senior-level intelligence personnel in policy and technical requirements necessary for planning communications system security.
Instruction: Course includes lectures, question and answer sessions, tours, and demonstrations. Subject areas include communications security operational processes, security policy, and doctrine; the threat to US communications; and the relation of communications security to other public and private sectors.
Credit Recommendation: In the graduate degree category, 1 semester hour in communications security management at the executive level plus 1 additional hour based on departmental interview (3/82).

DD-1405-0001

OPERATIONAL LEVEL CONTRACT PRICING

Course Number: CON 105.
Location: Defense Acquisition University, Various locations in the Continental US.
Length: 3 weeks (112 hours).
Exhibit Dates: 1/95–Present.
Learning Outcomes: Upon completion of the course, the students will be able to apply pricing theory; analysis techniques of price, cost, and profit; and contract negotiations. Students should be able to apply contract pricing tools to maximize the pricing objectives.
Instruction: This course is designed for contracting personnel in operational level contract pricing/cost analysis and contractor proposal evaluation. Lectures, exercises, demonstrations, case studies, problem solving, and a small group negotiation simulation are used to expose the students to the knowledge and skills of operational level cost estimating. Topics covered include general environment of contract pricing, cost and price analysis, methods of performing price, direct cost and indirect costs, profit analysis, methods of developing government price objectives, and negotiation.
Credit Recommendation: In the upper-division baccalaureate category, 3 semester hours in procurement/supply management (3/97).

DD-1405-0002

CONTRACT PRICING

Course Number: L40ST64P3 016; L302R64P3 016.
Location: Defense Acquisition University, Various locations in the Continental US; Defense Systems Management College, Ft. McNair, DC.
Length: 3 weeks (120 hours).
Exhibit Dates: 1/95–Present.
Learning Outcomes: Upon completion of the course, the student will be able to conduct market research for price analysis, gather and analyze data for cost estimating relationships, apply price-relative factors in contract decisions, and prepare for an engage in a competitive contract negotiation.
Instruction: This course focuses on the principles of contract pricing including the process of conducting cost analyses to maximize the pricing objectives. Lectures, case studies, group exercises, student review materials, and a team-based negotiation simulation are used. Topics covered include maximizing price competition, developing award criteria by pricing, cost-volume-profit analysis, price-related decisions in sealed bidding, costs and cost analysis, work design and analysis, direct cost of labor and materials, indirect costs; facilities capital (cost of money), negotiations, and post-award negotiation.
Credit Recommendation: In the upper-division baccalaureate category, 3 semester hours in procurement/supply management (3/97).

DD-1405-0003

OPERATIONAL LEVEL CONTRACTING FUNDAMENTALS

Course Number: CON 102.
Location: Defense Acquisition University, Various locations in the Continental US.
Length: 160 hours.
Exhibit Dates: 3/95–Present.
Learning Outcomes: Upon completion of the course, the student will demonstrate knowledge of the procurement cycle covering solicitation through sealed bid and competitive proposals, and including negotiation, pricing policies, the quality assurance evaluation (QAE) program, and contract management.
Instruction: This course focuses on the application of knowledge and skills acquired in CON 101 (Contracting Fundamentals) to the management of the contracting process. The instructional design for this course is primary group paced and involves the use of a study guide and workbook, supervised study, and lectures. Principal areas covered in the course include presolicitation (initiating the acquisition, specifications, competitive consideration, and contact types); solicitation (planning, ethics, contractor responsibility, and sealed bidding); evaluation and award (cost and pricing, contract management); and post-award functions (quality management, protest and claims, and contact close-out).
Credit Recommendation: In the upper-division baccalaureate category, 3 semester hours in procurement/supply management (3/97).

DD-1405-0004

CONTRACT PRICING

Course Number: CON 104.
Location: Defense Acquisition University, Various locations in the Continental US.
Length: 3 weeks (120 hours).
Exhibit Dates: 1/93–Present.
Learning Outcomes: Upon completion of the course, the student will demonstrate knowledge of the general environment of contract pricing; determine the sources and means of acquiring data for cost and price analysis; analyze direct and indirect costs by applying quantitative techniques; perform a cost analysis and develop a cost objective based on an actual case exercise; and apply bargaining techniques, strategies, and tactics of the negotiation process.
Instruction: This course is designed for entry-level contracting personnel. It lays the foundation for the study and practice of cost and price analysis. This is a group paced course. In addition a student's workbook, expository discussion, group exercises, case studies, and problem solving techniques are also used to expose students to contract pricing. Principal areas covered in this course include market research for price analysis, contracting environment's sources of data, price analysis, quantitative techniques for pricing, cost analysis, profit analysis, and government contract negotiations.
Credit Recommendation: In the upper-division baccalaureate category, 3 semester hours in procurement/supply management (3/97).

DD-1405-0005

INTERMEDIATE CONTRACT ADMINISTRATION

Course Number: CON 221.
Location: Defense Acquisition University, Various locations in the Continental US.
Length: 2 weeks (70 hours).
Exhibit Dates: 6/96–Present.
Learning Outcomes: Upon completion of the course, the student will be able to apply cost principles to a contractors request for contract pricing adjustments evaluate requests for contract modifications determine the legal relation-

ship between the government and the prime contractor and subcontractor, determine how and when price adjustments are accomplished under fixed-price contracts with price adjustment provisions, assess a contractor's financial status, apply relevant policies to post-award requests for government property, identify environmental issues, and determine the appropriate course of action in dispute situations.

Instruction: Methods of instruction include lectures, case studies, and exercises. Principal areas covered include cost and pricing considerations; modifications and equitable adjustments; subcontract management; costs and profit/fee adjustments; cost accounting; government property; financial management; technical support; technical data, copyrights and patents; contract terminations; claims, disputes, and remedies; environmental contract management; contract quality assurance; contract labor relations; contract close-out; and ethics.

Credit Recommendation: In the upper-division baccalaureate category, 2 semester hours in procurement/supply management (3/97).

DD-1405-0006
INTERMEDIATE CONTRACT PRICING

Course Number: CON 231.
Location: Defense Acquisition University, Various locations in the Continental US.
Length: 2 weeks (85-86 hours).
Exhibit Dates: 4/95–Present.
Learning Outcomes: Upon completion of the course, the student will be able to perform more advanced pricing duties including using quantitative methods for cost and price analysis to make pre- and post-award decisions about general contract pricing issues.

Instruction: This course is designed to provide the necessary knowledge and skills needed by contracting personnel who successfully completed CON 104. Lectures, in-class group exercises, computer and manual exercises, and computer applications are used to develop skills in performing more advanced pricing functions. Principal areas covered include selection and application of estimating techniques in cost/price analysis, using index numbers to determine price reasonableness, cost-volume-profit analysis, net present value analysis, regression analysis, moving averages, improvement curve analysis, work measurement, evaluating indirect costs, forecasting cost overruns; pricing equitable adjustments and settlements, reviewing the contractor's pricing and accounting practices, recognizing and adjusting for defective pricing; conducting cost realism analysis, and post-award negotiation.

Credit Recommendation: In the upper-division baccalaureate category, 3 semester hours in procurement/supply management (3/97).

DD-1408-0002
PROGRAM MANAGEMENT

Course Number: None.
Location: Defense Systems Management College, Ft. Belvoir, VA.
Length: 20 weeks (393-499 hours).
Exhibit Dates: 1/90–1/90.
Objectives: To offer managers from the government, the military and industry the skills and knowledge to successfully manage defense systems acquisition programs.

Instruction: Lectures, discussions and cases cover the following areas: defense program and project management including program cost management, contract management, procurement policy and logistics; production and operations management including quantitive methods; general management and organizational behavior; and managerial finance and accounting.

Credit Recommendation: In the upper-division baccalaureate category, 3 semester hours in production and operations management, 2 in managerial finance, and 1 in general management (4/80); in the graduate degree category, 9 semester hours in defense program and project management (If the student has already completed the course, Program Management for Functional Managers (Program Management for Contract Administration), only 6 hours additional credit is recommended) (4/80).

DD-1408-0006
DEFENSE RESOURCES MANAGEMENT
(Defense Management Systems)
(Navy Management Systems)

Course Number: P-00-3306; L5OZN6916 000; S-00-3306.
Location: Naval Postgraduate School, Defense Resources Management Institute, Monterey, CA.
Length: 4 weeks (105 hours).
Exhibit Dates: 1/90–Present.
Objectives: To provide officers with an introduction to the basic principles of management.

Instruction: Lectures cover resource management, program budgeting, management accounting, systems analysis, marginal analysis, and cost effectiveness. Emphasis is placed on the analytical aspects of management, including requirement studies, systems analyses, cost effectiveness, and marginal analysis.

Credit Recommendation: In the upper-division baccalaureate category, 3 semester hours in management (12/93).

DD-1408-0007
PROGRAM MANAGEMENT

Course Number: None.
Location: Defense Systems Management College, Ft. Belvoir, VA.
Length: 20 weeks (460-499 hours).
Exhibit Dates: 2/90–Present.
Learning Outcomes: Upon completion of the course, the student shall have acquired competencies in defense systems acquisition management, Department of Defense acquisition regulations, procurement processes and decision techniques, human relations and program management, and the utilization of financial data and analysis in procurement.

Instruction: Methods of instruction include lectures, demonstrations, simulations, readings, role playing, distinguished guest speakers, case studies, scenarios, and individualized learning programs. Topics include contract management, systems engineering, managerial development, funds management, manufacturing management, program principles, contractor financial management, software management, logistics support, cost scheduling, and individualized learning.

Credit Recommendation: In the upper-division baccalaureate category, 2 semester hours in financial planning and analysis (11/91); in the graduate degree category, 3 semester hours in leadership and group decision process, 3 in systems management, and 3 in survey of program/operations/manufacturing management (11/91).

DD-1408-0008
ADVANCED MANAGEMENT PROGRAM

Course Number: None.
Location: National Defense University, Ft. Lesley J. McNair, Washington, DC.
Length: 16 weeks.
Exhibit Dates: 10/90–Present.
Learning Outcomes: This program prepares selected individuals responsible for information resource management (IRM) decisions for senior leadership and staff positions in the Department of Defense. It stresses senior-level management requirements given IRM policies and contemporary issues. Graduates will comprehend the necessary integration of tasks to ensure effective allocation and application of information resources to national requirements in compliance with regulatory, policy and ethical standards.

Instruction: Methods include seminars, case studies, simulations, advanced study options, field trips, and guest lectures. Topics include leadership, managing relationships, managing organizational change, financial management, foundations in information resource management, IRM in the DoD, IRM policy, IRM planning, IRM requirements, IRM oversight, functional information management, total quality information management, principles of program management, programming planning and control, managing automated information systems (AIS) acquisitions, managing information technology, strategic decision support systems, artificial intelligence and expert systems, decision making, information engineering, and AIS security strategy. The curriculum consists of graduate-level study in managerial, financial, operational, and technical areas of IRM with an emphasis on policy issues. The course emphasis is on making and influencing critical IRM decisions. Students will select at least two of five advanced studies courses.

Credit Recommendation: In the graduate degree category, 2 semester hours in organizational behavior, 2 organizational theory, 1 in public budgeting, 2 in defense information resource management, 2 in IRM policy, 1 in IRM planning, 2 in system analysis and design, 1 in system design and implementation, 1 in information systems planning, 3 in acquisition of information systems, and 2 in management of information technology. Students will select at least two of five advanced studies courses, for which the following credit applies: for ASP519, Strategic Decision Support Systems, 2 semester hours in decision support systems; for ASP525, Artificial Intelligence and Expert Systems, 2 semester hours in knowledge-based systems; for ASP516, Decision Making: Understanding New Applications, 1 semester hour in group processes and dynamics; for ASP530, Information Engineering; 2 semester hours in information engineering; for ASP542, AIS Security Strategies, 2 semester hours in data security management (3/92).

DD-1408-0009
SYSTEMS ACQUISITION FOR CONTRACTING PERSONNEL (EXECUTIVE)

Department of Defense

Course Number: PMT 341.
Location: Defense Systems Management College, Ft. Belvoir, VA.
Length: 2 weeks (80 hours).
Exhibit Dates: 1/90–Present.
Learning Outcomes: Upon completion of the course, the student will be able to describe the concepts, principles, language, and functions required to manage major systems acquisitions; explain the relationship between the program manager and contracting officer in the acquisition of major systems; and understand the roles, activities, and relationships between government and industry in the acquisition process.
Instruction: Course covers organizational dynamics; acquisition policy in major systems acquisition; principles of program, financial, software, and manufacturing management; systems engineering; test and evaluation; and integrated logistics support. Methodology includes lectures, class discussions, group exercises, and case studies.
Credit Recommendation: In the graduate degree category, 3 semester hours in procurement management (12/93).

DD-1408-0010

ADVANCED PRODUCTION AND QUALITY MANAGEMENT
(Defense Acquisition Engineering, Manufacturing, and Quality Control)

Course Number: PQM 301; PRD 301.
Location: Defense Systems Management College, Ft. Belvoir, VA.
Length: 2 weeks (69-70 hours).
Exhibit Dates: 10/93–Present.
Learning Outcomes: Upon completion of the course, the student will be able to identify and analyze policy issues related to selection, development, and evaluation of industrial suppliers; review and define quality requirements in the design and production of manufactured products; and evaluate the effectiveness of a manufacturing organization.
Instruction: Instruction covers quality policy and deployment, quality ethics, systems engineering, integrated product/process development (concurrent engineering), manufacturing management, design of experiments, product costing, ISO 9000, theory of constraints, factory design and simulation, competitiveness, environmental issues, military acquisition policy, contract management (hardware and software), and commercial practices. Methods of instruction include lectures by regular faculty and guest lecturers, readings and texts, class exercises, and cases. Students are required to do a factory simulation on a computer and write a position paper on a topic related to the course.
Credit Recommendation: In the graduate degree category, 3 semester hours in business administration or technical management (4/97).

DD-1408-0011

INTERMEDIATE SYSTEMS ACQUISITION
(Acquisitions Basics)

Course Number: PMT 201; ACQ 201; DSMC-37.
Location: Defense Systems Management College, Ft. Belvoir, VA.
Length: 4 weeks (130 hours).
Exhibit Dates: 10/90–5/96.
Learning Outcomes: After 5/96 see DD-1408-0020. Upon completion of the course, the student will be able to describe the concepts and terminology of acquisition management processes; organize, direct, and control acquisition programs from the conceptual stages through contracting, manufacturing, delivery, and field support including future product improvement; and manage people, money, facilities, information, and time in the accomplishment of program objectives.
Instruction: Methodology includes lectures, discussions, audiovisuals with PC support, group discussions, and testing and assessment. Topics covered include managerial development; TQM concepts; acquisition strategy and policy implementation; supplier financial consideration and the management of the contractual, cost, and schedule criteria; systems and software engineering; manufacturing, delivery, and logistics support, and performance testing and evaluation results.
Credit Recommendation: In the upper-division baccalaureate category, 4 semester hours in systems management (12/93).

DD-1408-0012

FUNDAMENTALS OF SYSTEMS ACQUISITION MANAGEMENT

Course Number: ACQ 101.
Location: Defense Acquisition University, Various locations in the Continental US.
Length: 2 weeks (60 hours).
Exhibit Dates: 9/94–5/97.
Learning Outcomes: Upon completion of the course, the student will be able to define key terms used by business and the technical industry pertaining to acquisitions and identify key players and organizations who must work together to create a successful acquisition. The student should be able to describe the relationship between the government and the prime contractor, the prime contractor and the sub contractor, and the government and the prime subcontractor; how the government modifies contracts; why different types of contracts are used in the contract process; and the difference between sealed bids and the competitive process.
Instruction: Methods include lectures, discussions, case studies and exercises, videotapes, and smallgroup decisionmaking projects relate to funds management, contracting, logistics, systems engineering, software management, tests and evaluation, manufacturing, and the acquisition life cycle.
Credit Recommendation: In the lower-division baccalaureate/associate degree category, 3 semester hours in acquisition management (10/96).

DD-1408-0013

INTERMEDIATE SOFTWARE ACQUISITION MANAGEMENT

Course Number: *Version 1:* SAM 201. *Version 2:* SAM 201.
Location: Defense Acquisition University, Various locations in the Continental US.
Length: *Version 1:* 2 weeks (69 hours). *Version 2:* 3 weeks (95 hours).
Exhibit Dates: *Version 1:* 1/98–Present. *Version 2:* 6/96–12/97.
Learning Outcomes: *Version 1:* This version is pending evaluation. *Version 2:* Upon completion of the course, the student will be able to perform a wide range of tasks required during the acquisition of a software-intensive systems; including automated information; command control, communications, and intelligence systems. Students will demonstrate a level of knowledge and ability commensurate with that of a level two software acquisition manager.
Instruction: *Version 1:* This version is pending evaluation. *Version 2:* Lectures, discussions, exercises, and case studies are course methods. The cases are analyzed individually and in groups. Topics covered include life-cycle planning, risk management, software life-cycle cost factors, software development cost estimation, software capability evaluation, source selection, and acquisition strategy.
Credit Recommendation: *Version 1:* Pending evaluation. *Version 2:* In the upper-division baccalaureate category, 3 semester hours in acquisition management (10/96).

DD-1408-0014

CONTRACT PERFORMANCE MANAGEMENT FUNDAMENTALS

Course Number: BFM 102.
Location: Defense Acquisition University, Various locations in the Continental US.
Length: 2 weeks (62 hours).
Exhibit Dates: 7/95–10/97.
Learning Outcomes: Upon completion of the course, the student will be able to describe the principles of cost performance management (CPM), including cost/schedule control systems criteria, contractor performance measurement base line, contractor review and surveillance process, earned value analysis, and role of earned value analysis in the acquisition process.
Instruction: Lectures, group exercises, and presentations. Topics covered include acquisition life cycle, CPM principles, CPM in the request for proposal process, CPM in the contractor selection process, and CPM in the program execution and surveillance process.
Credit Recommendation: In the upper-division baccalaureate category, Credit is recommended only upon completion of Intermediate Contract Performance Management, BFM 203. (See DD-1408-0015) (10/96).

DD-1408-0015

INTERMEDIATE CONTRACT PERFORMANCE MANAGEMENT

Course Number: BFM 203.
Location: Defense Acquisition University, Various locations in the Continental US.
Length: 2 weeks (72 hours).
Exhibit Dates: 3/96–10/97.
Learning Outcomes: Upon completion of the course, the student will be able to apply cost/schedule control systems criteria, determine the adequacy of contractor base line analysis, establish contract surveillance program, perform earned value analysis and make appropriate recommendations, and ensure application of earned value throughout the acquisition process.
Instruction: Topics covered include contract performance management (CPM) principles, CPM in the acquisition management process, CPM in the request for proposal process, CPM in proposal evaluation cost schedule control systems criteria implementation, and CPM surveillance.

COURSE EXHIBITS

Credit Recommendation: In the upper-division baccalaureate category, 3 semester hours in management, to be awarded upon completion of this course and Contract Performance Management Fundamentals, BFM 102. (DD-1408-0014) (10/96).

DD-1408-0016

ADVANCED SYSTEMS PLANNING, RESEARCH, DEVELOPMENT, AND ENGINEERING

Course Number: SYS 301.
Location: Defense Acquisition University, Various locations in the Continental US.
Length: 2 weeks (77 hours).
Exhibit Dates: 6/96–Present.
Learning Outcomes: Upon completion of the course, the student will demonstrate competence in the technical management of engineering processes, procedures, and tools used during the acquisition cycle. Students will be able to apply appropriate tools from the areas of risk management, modeling and simulation, trade off analyses, and configuration management to evaluate and control the evaluation of a new system.
Instruction: Lectures, discussions, videos, case studies, exercises, site visitation, and demonstrations. Topics covered include risk management, modeling and simulation, trade off analysis, configuration management, engineering and manufacturing development, production, program definition, risk reduction, and concept exploration.
Credit Recommendation: In the graduate degree category, 3 semester hours in technical management (10/96).

DD-1408-0017

BUDGET, COST ESTIMATING, AND FINANCIAL MANAGEMENT WORKSHOP

Course Number: BCF 301.
Location: Defense Acquisition University, Various locations in the Continental US.
Length: 2 weeks (60 hours).
Exhibit Dates: 6/96–Present.
Learning Outcomes: Upon completion of the course, the student will be able to define the duties of budget, cost estimating, and financial management functions; define current budget, cost estimating, and financial management related laws, regulations, policies and procedures; identify and describe the interrelationship among the budget, cost estimating, and financial management functions; and select and analyze appropriate decision making information based on the integrated nature of a budget, cost estimating, and financial management task.
Instruction: Lectures, discussions, case studies, and critical incident analyses are course methods. Case studies are analyzed by teams. Topics covered include cost estimation, earned value fundamentals, earned value analysis, budget execution, budget policy, financial analysis, and contract types.
Credit Recommendation: In the upper-division baccalaureate category, 2 semester hours in financial management (10/96).

DD-1408-0018

ADVANCED PROGRAM MANAGEMENT

Course Number: PMT 302.
Location: Defense Acquisition University, Various locations in the Continental US.
Length: 14 weeks (398 hours).
Exhibit Dates: 3/95–Present.
Learning Outcomes: Upon completion of the course, the student will demonstrate, from the program management perspective, the integration of functional disciplines into the dynamic processes used to manage systems; perform the financial, operations, and technical management functions related to the system acquisition process; and develop and manage effective integrated acquisition management teams.
Instruction: Lectures, discussions, exercises, case analysis, and development of an individual learning reports are course methods. Topics covered include acquisition policy, contractor financial management, contract management, cost/schedule control, funds management, logistic support, managerial development, manufacturing management, principles of program management, systems engineering, software management, and testing and evaluation.
Credit Recommendation: In the graduate degree category, 3 semester hours in financial management, 3 in operations management, and 3 in technical management (10/96).

DD-1408-0019

EXECUTIVE PROGRAM MANAGER

Course Number: PMT 303.
Location: Defense Acquisition University, Various locations in the Continental US.
Length: 4 weeks (76-145 hours).
Exhibit Dates: 8/94–Present.
Learning Outcomes: Upon completion of the course, the student be able to coordinate among program functional areas; integrate cost, schedule, and performance; perform risk analysis and trade off analysis; develop the long-range, innovative acquisition strategy essential to the development of a technology or weapon and its manufacturing and fielding; conduct an assessment of an ACAT 1 or 2 level program; analyze problems; and distinguish between relevant and irrelevant information to make logical judgements.
Instruction: Lectures, speakers, panel discussions, and exercises are course methods. Topics covered include program management, acquisition environment, business management, and technical management.
Credit Recommendation: In the graduate degree category, 3 semester hours in program management (10/96).

DD-1408-0020

INTERMEDIATE SYSTEMS ACQUISITION

Course Number: ACQ 201.
Location: Defense Acquisition University, Various locations in the Continental US.
Length: 3 weeks (137 hours).
Exhibit Dates: 6/96–8/97.
Learning Outcomes: Before 6/96 see DD-1408-0011. Upon completion of the course, the student will be able to manage an acquisition program, at the intermediate level, using planning, organizing, directing, and contracting from the conceptual stages through fielding, post-production support, and product improvement of systems. The student should recognize internal and external factors which influence and constrain the acquisition process and deal with those factors in light of risk, uncertainty, and change.
Instruction: Lectures, discussions, case studies and group projects relate to acquisition policy and environment, contractor financial management, contract management, cost/schedule control systems criteria, principles of program management, systems engineering, software management, and test and evaluation.
Credit Recommendation: In the upper-division baccalaureate category, 4 semester hours in acquisition management (10/96).

DD-1408-0021

PROGRAM MANAGER'S SURVIVAL

Course Number: PMT 305.
Location: Defense Acquisition University, Various locations in the Continental US.
Length: 2 weeks (48 hours).
Exhibit Dates: 6/96–Present.
Learning Outcomes: Upon completion of the course, the student will be able to manage an acquisition program using current funding policies and procedures and tested acquisition strategies that provide effective management of cost and schedule performance risk while effectively using competition in contracting.
Instruction: Lectures, discussions, individual research projects, team work, and presentations. Cover funds management, acquisition reform, acquisition risk management, acquisition strategy, organization and staff concepts, software acquisition, contract management, systems engineering update, specifications and standards update, acquisition logistics, management, manufacturing management issues and costs, scheduling, and performance management using earned value principles.
Credit Recommendation: In the graduate degree category, 1 semester hour in program management (10/96).

DD-1408-0022

INTERMEDIATE TEST AND EVALUATION (T & E)

Course Number: TST 202.
Location: Defense Acquisition University, Various locations in the Continental US.
Length: 2 weeks (56 hours).
Exhibit Dates: 8/96–Present.
Learning Outcomes: Upon completion of the course, the student will demonstrate knowledge of test planning, design, and conduct; human systems integration; analysis/evaluation; reporting; software; modeling; and simulation. In addition, students should recognize and avoid the pitfalls of the past. Working as members of a team, students should be able to apply knowledge in a detailed integrative exercise which addresses the major issues in developing a complete test plan for a major weapons system.
Instruction: This course focuses on the role of testing and evaluation (T & E) as a weapons system advances through the acquisition cycle. Knowledge and skills are acquired through the use of lectures/discussions, tutorials, expository discussions, problem solving scenarios, simulations, and practical exercises (including group activities). Principal areas covered in this course include the role of T & E in the acquisition process, requirements traceability and analysis, planning process, operational suitability, reliability testing, validity, and environmental security.
Credit Recommendation: In the upper-division baccalaureate category, 3 semester hours in data analysis and modeling (3/97).

DD-1408-0023

PRODUCTION AND QUALITY MANAGEMENT FUNDAMENTALS

Course Number: PQM 101.
Location: Defense Acquisition University, Various locations in the Continental US.
Length: 2 weeks (51 hours).
Exhibit Dates: 10/94–Present.
Learning Outcomes: Upon completion of the course, the students will have some understanding of production management and quality assurance processes as they relate to in the oversight of contractor operations.
Instruction: This course consists of a series of 16 modules ranging from one to six hours each that include several in-class exercises (layout, SPC, flow charts, C/SCSC) and a facilities tour. There is a multiple choice exam at the end of each week. The focus of this course is procurement and contracting procedures. Principal areas covered in this course include systems acquisition (life cycle, integrated product and process development and integrated product teams, systems engineering, performance specifications, source selection, contract administration functions, current DOD initiatives, Federal Acquisition Regulations, and ethics); manufacturing/design technology (electronic tools, industrial capabilities, and manufacturing planning); and quality assurance (contracting incentives and warranties, progress payments and progress review, and analytical tools).
Credit Recommendation: Credit is not recommended because of the limited, specialized nature of the course (3/97).

DD-1408-0024

INTERMEDIATE PRODUCTION AND QUALITY MANAGEMENT FUNDAMENTALS

Course Number: PQM 201.
Location: Defense Acquisition University, Various locations in the Continental US.
Length: 3 weeks (86-91 hours).
Exhibit Dates: 10/94–Present.
Learning Outcomes: Upon completion of the course, the students will have some knowledge of production, manufacturing and, quality assurance processes as they relate the oversight of contractor operations.
Instruction: The course consists of a series of 20 modules ranging from 2 to 14 hours each that include some in-class exercises using a student provided case study (example). There is a multiple choice exam at the end of the first two weeks and a major graded exercise involving an instructor provided case study during the third week. Principle areas covered in this course include systems acquisition (acquisition reform, life cycle, integrated product and process development and integrated product teams, systems engineering, performance specifications, source selection, contract administration functions, current DOD initiatives, federal acquisition regulations, and ethics); manufacturing/design technology (electronic tools, industrial capabilities); manufacturing planning; and quality assurance (contracting incentives and warranties; progress payments and progress review; analytical tools; environment, safety, and health planning; and project planning and control).
Credit Recommendation: In the upper-division baccalaureate category, 3 semester hours in production and operations management procurement (supply) management (3/97).

DD-1511-0001

NATIONAL SECURITY MANAGEMENT (CORRESPONDENCE/NONRESIDENT/ SEMINAR COURSE OF THE NATIONAL DEFENSE UNIVERSITY)
(Correspondence Course of the Industrial College of the Armed Forces)

Course Number: None.
Location: Industrial College of the Armed Forces, Ft. Lesley J. McNair, Washington, DC; National Defense University, Ft. Lesley J. McNair, Washington, DC.
Length: 52-104 weeks.
Exhibit Dates: 1/90–8/90.
Objectives: To train officers in economic and industrial aspects of national security and the management of resources under all conditions and in the context of national and world affairs. The course is offered in correspondence, seminar, and nonresident modes.
Instruction: Course is designed to promote an understanding of the strategic, economic, and industrial aspects of national security and management of the nation's resources under varying circumstances and conditions. Unit I covers the environment of national security. This unit examines fundamental concepts of national security relating to the domestic and international scene. It provides a framework for understanding the problems confronting the United States as an international actor. It discusses economic theories, policies, and issues in the context of resource adequacy for defense and assesses the position of the United States in the world economy. It also explains how the political system influences the nation's conduct in world affairs. Unit II, resources of defense, analyzes major resources such as energy, transportation, and technology and assesses their importance to the economy and national security. It discusses increased dependence on foreign sources for resources critical to national security. This unit also examines the development and role of industry and the problem of foreign competition in markets once dominated by the United States. Unit III addresses defense decision making. It studies the main instruments and processes of the federal government involved in the formulation of national security policies and in decision making. Also this unit explains the organization and management of the Defense Department, the systems and processes involved in determining requirements, and the decision making procedures on resource allocation. Unit IV concerns executive management. It provides a synthesis of the traditional and newer approaches to management with an emphasis on human resource management. The unit assesses quantitative and qualitative factors relating to the work force with particular emphasis on manpower for defense. It examines the acquisition process of major defense systems, the management of defense systems, and the management of defense logistics. This unit analyzes preparedness issues and mobilization planning. The program's four units each conclude with a proctored closed-book objective examination covering approximately five books. The books, texts, and anthologies are vehicles for instruction. Unit III also requires a written report of 3,000 to 3,500 words.
Credit Recommendation: In the upper-division baccalaureate category, 3 semester hours in US foreign policy studies, 3 in US national security policies and processes, 3 in introduction to economic principles and problems, 2 in 20th century American economic history, 2 in management, and 1 in American national political institutions (1/85); in the graduate degree category, for those students in the seminar program achieving a rating of outstanding, 3 semester hours in policies and procedures of the national security establishment, and 3 in US public policy (1/85).

DD-1511-0002

NATIONAL WAR COLLEGE

Course Number: None.
Location: National War College, Ft. Lesley J. McNair, Washington, DC; National Defense University, Ft. Lesley J. McNair, Washington, DC.
Length: 40-45 weeks.
Exhibit Dates: 1/90–8/90.
Objectives: After 8/90 see DD-1511-0009. The National War College provides an educational experience intended to improve the knowledge and expertise of a practitioner in the field of US foreign affairs and national security affairs.
Instruction: This course is presented through lectures, seminars, readings, and independent student research in international relations and United States foreign policy and national security studies. The core curriculum is divided into six units: executive decision making and security challenges; the art of war; the international security environment; United States policy making studies, including politics, policy, and resource allocations and policy planning and national security decision making; major powers and regions, including the Soviet Union, Asia, Western Europe, the Middle East, Sub-Saharan Africa, and Latin America; and United States defense policy and military strategy. A two-week overseas field trip to one of the regions covered in the fifth unit is part of the course. A variety of electives related to the core curriculum is offered.
Credit Recommendation: In the upper-division baccalaureate category, 30 semester hours to be apportioned as follows: 6 semester hours in international relations, 12 in United States foreign policy studies, and 12 in United States national security policies and processes (1/85); in the graduate degree category, 1-9 semester hours in international relations, United States foreign policy studies, and United States national security policies and processes and 1-3 for research completed during the course of the program, all to be granted based on the receiving institution's review of the student's written materials. It is recommended that the receiving institution request an evaluation from the Dean of Faculty and Academic Programs, National Defense University, National War College, Ft. Lesley J. McNair, Washington, DC 20319 (1/85).

DD-1511-0003

INDUSTRIAL COLLEGE OF THE ARMED FORCES (RESIDENT PROGRAM)

Course Number: None.
Location: Industrial College of the Armed Forces, Ft. Lesley J. McNair, Washington, DC; National Defense University, Ft. Lesley J. McNair, Washington, DC.
Length: 40-45 weeks.
Exhibit Dates: 1/90–7/90.
Objectives: To teach the political, military, social, economic, and industrial aspects of

national security, resource management and the command, staff, and policy-making functions of the national and international security structure.

Instruction: This course consists of lectures, simulations, seminars, field studies, and student research in aspects of United States national security and resource management. The core curriculum is divided into five phases: public executive perspectives; national security and mobilization management; manpower resources management; materiel resources management; and joint training exercises. Field trips to domestic and foreign industrial sites are included in Phase 4. A variety of electives relating to the core curriculum is offered.

Credit Recommendation: In the lower-division baccalaureate/associate degree category, 2 semester hours in introduction to economics (5/85); in the upper-division baccalaureate category, 12 semester hours in United States foreign policy and national security studies, 6 in international relations, 4 in labor and manpower resource planning, 2 in transportation planning, 4 in industrial structure and process, and 0-6 additional credit in the appropriate field for the electives course based on the receiving institution's review of the transcript and additional information from the Dean, Industrial College of the Armed Forces, National Defense University, Ft. Lesley J. McNair, Washington, DC 20319 (5/85); in the graduate degree category, 1-9 semester hours in international relations, manpower materiel planning, United States national security studies; and 1-3 for research completed during the course, all to be granted based on the receiving institution's review of the student's written materials. It is recommended that the receiving institution request a transcript and additional information from the Academic Dean, Industrial College of the Armed Forces, National Defense University, Ft. Lesley J. McNair, Washington, DC 20319 (5/85).

DD-1511-0006

RESERVE COMPONENTS NATIONAL SECURITY SEMINAR

Course Number: None.
Location: National Defense University, Ft. Lesley J. McNair, Washington, DC; National Defense University, Ft. Bragg, NC; National Defense University, Naval Air Station, Pensacola, FL; National Defense University, Vandenberg AFB, CA.
Length: 2 weeks (60 hours).
Exhibit Dates: 1/90–12/96.
Objectives: To present and analyze issues of national security and foreign policy.
Instruction: This seminar includes global intelligence assessment (US policy perspective); US security policy (East Asia, Western Europe, the Soviet Union, the Middle East, and the developing countries); arms trade and arms acquisitions; defense analysis including a case study; DoD Management; ocean policy; economic issues; human and industrial resources; energy problems; and a simulation exercise.
Credit Recommendation: Credit is not recommended for the course (7/87).

DD-1511-0008

NATIONAL SECURITY MANAGEMENT BY CORRESPONDENCE

Course Number: None.
Location: National Defense University, Ft. Lesley J. McNair, Washington, DC.
Length: Maximum, 104 weeks.
Exhibit Dates: 9/90–Present.
Learning Outcomes: Upon completion of the course, the student will possess a general understanding of the national security issues confronting the United States and the processes by which strategies are developed to deal with these issues. A graduate will be able to analyze the strategic, economic, and industrial aspects of national security.
Instruction: The course is conducted by correspondence. Students have the option of studying individually or in small groups. Students are required to read 19 short volumes organized into four units: the environment of national security; resources for defense; defense decision making; and executive management. Four multiple choice examinations are given, and a paper of 12-14 pages is required.
Credit Recommendation: In the lower-division baccalaureate/associate degree category, 3 semester hours in introduction to economics (5/91); in the upper-division baccalaureate category, 3 semester hours in international relations, 3 in political science, 3 in management, and 3 in national security studies (5/91).

DD-1511-0009

NATIONAL WAR COLLEGE

Course Number: None.
Location: National Defense University, Ft. Lesley J. McNair, Washington, DC.
Length: 42 weeks.
Exhibit Dates: 9/90–11/94.
Learning Outcomes: Before 9/90 see DD-1511-0002. Upon completion of the course, the student will be able to deal with the complex issues pertaining to decision making and policy implementation within the national security arena. Students will be able to analyze the broad range of considerations which strategic thinking entails, including geographic, cultural, political economy, diplomatic, political, and economic considerations.
Instruction: Instruction in this course involves formal lectures, directed seminar discussion, and extensive reading in the professional literature in the areas of national security studies, international relations, and strategy. Oral and written reports are required, as well as a two-week foreign study trip. The course is divided into four units consisting of: foundations of national security strategy; national security policy process; geostrategic context; and military strategy and operations. A final component entails strategy exercises in global regions.
Credit Recommendation: In the graduate degree category, 3 semester hours in political science or public policy, 6 in American foreign policy, 6 in military studies, and 6 in international relations (5/91).

DD-1511-0010

INDUSTRIAL COLLEGE OF THE ARMED FORCES

Course Number: None.
Location: National Defense University, Ft. Lesley J. McNair, Washington, DC.
Length: 40-45 weeks.
Exhibit Dates: 9/90–11/94.
Learning Outcomes: Upon completion of the program, the student will be able to describe the interdisciplinary nature of the national security decision-making process with emphasis on the military, historical, political, economic, managerial, and ethical dimensions. Graduates will be able to make effective decisions and formulate and implement strategies within the national security environment, apply strategic thinking to security problems, and develop expertise in the resource component of national security. Emphasis is placed on the economic, logistical, acquisition, and mobilization aspects of resource management.
Instruction: Course involves formal lectures; directed seminar discussions; and reading on the military, historical, political, economic, managerial, and ethical dimensions of national security decision making with emphasis on resource management. Written assignments, including a major research paper and group exercises, are integrated throughout. The first half of the program concentrates on national security strategy which emphasizes military strategy, history, political science, economics, and decision making. The second half focuses on resource management for national security with emphasis on acquisition, logistics, mobilization, and defense industry studies. Integrated throughout the curriculum are specialized courses. These include a mandatory course in regional security studies and various electives drawn from all areas of the program. Students conduct field research on defense industries which includes two weeks in a foreign region and one week in the United States.
Credit Recommendation: In the lower-division baccalaureate/associate degree category, 3 semester hours in introduction to economics (9/91); in the upper-division baccalaureate category, 3 semester hours in international relations, 3 in history, 3 in political science, 3 in military studies, and 3 in management (9/91); in the graduate degree category, 3 semester hours in military studies, 1 in history, 6 in management (administrative decision making, logistics, defense acquisition), 3 in international relations, and 3 in the political economy of defense (9/91).

DD-1511-0011

INTER-AMERICAN DEFENSE COLLEGE

Course Number: None.
Location: Inter-American Defense College, Ft. Lesley J. McNair, Washington, DC.
Length: 43 weeks (830 hours).
Exhibit Dates: 1/90–Present.
Learning Outcomes: Upon completion of the course, the student will be able to serve as a senior leader and decision maker dealing with the geopolitical, economic, sociocultural, and security factors facing the American states—both global and regional aspects.
Instruction: This curriculum includes a review of basic theoretical topics in the area of power and general studies of the current world situation. It thus provides a framework for extensive analysis of the hemisphere's situation in light of political, social, economic, and military factors. The students learn and practice, in group discussions, the methodology of international cooperation in basic aspects of continental security planning at higher levels. The modes of instruction include numerous lectures by outside experts, seminars, symposia, study committees, and the preparation of individual research papers.

Credit Recommendation: In the graduate degree category, 3 semester hours in international relations, 3 in comparative politics, and 3 in cross-cultural communication (7/91).

DD-1512-0003

DEFENSE EQUAL OPPORTUNITY MANAGEMENT INSTITUTE

Course Number: None.
Location: Defense Equal Opportunity Management Institute, Patrick AFB, FL.
Length: 14-16 weeks (331-702 hours).
Exhibit Dates: 1/90–Present.
Learning Outcomes: Upon completion of the course the student will be able to effectively communicate in small groups; identify barriers to communication; mediate conflict situations; recognize the influences of stereotypes on perceptions; use feedback skills; use persuasive speech; speak extemporaneously; demonstrate effective writing skills using short report writing; prepare background papers, position papers, and official policy letters; identify the nature of diversity; recognize individual differences and similarities in various environments; establish norms for group dynamics; apply theories of motivation to groups; describe task functions in small group interaction and differentiate between formal and informal groups; identify interpersonal and intrapersonal conflict; negotiate; apply the concept of organization to systems; identify concepts and types of power; recognize stereotypes and the effect of perceptions on intergroup dynamics; communicate across differences of culture, gender, and race; recognize differences between racism, sexism, discrimination, and prejudice with particular attention paid to institutional discrimination; differentiate between concepts of culture and race; be cognizant of the history of racism in the military and identify contemporary racism where it exists; be knowledgeable about African American, Asian American, Jewish American, and Arab American history, sociology, and ethnicity and apply this information to contemporary issues; identify the majority white experience in the United States and consider the dynamics of majority-minority relations; possess administrative skills including action planning, intervention techniques, managing EO programs, and interviewing techniques; process assessment data, develop surveys and questionnaires; perform a unit climate assessment and interpret survey data; present briefings as an equal opportunity staff advisor; identify the impact of sexual harassment on the individual, society, and unit readiness; define affirmative action goals and processes; and use the demographic and social issues of Work Force 2000.
Instruction: Course methods include lectures and guest lectures, interactive large group discussions, student-led and trainer-led seminars, case studies, video tapes, role playing, and psychodramas. Extensive use is made of small group discussions led by trainers/facilitators. Except for the preparation of short reports using library files, no outside references/readings/research are required. Note: Recommended readings, bibliographies, and videos are very dated.
Credit Recommendation: In the lower-division baccalaureate/associate degree category, 3 semester hours in basic communication skills; 2 in public speaking, rhetoric, argument and debate, or persuasive techniques; 3 in introduction to ethnic and gender differences, pluralism and diversity, or race, ethnicity, and gender; 1 in Jewish studies; 3 in public administration or policy implementation; 1 in introduction to survey methods and analysis; and 3 in introduction to social science or introduction to behavioral science for a total of 16 semester hours (5/92); in the upper-division baccalaureate category, 1 semester hour in communication and presentation skills, 1 in field study or practicum in social science, 1 in racism/sexism in the military, 1 in human resource management, 1 in administrative practices, 1 in applied field methods in social science, and 1 in instructional methods (5/92).

DD-1601-0019

BASIC PHOTOLITHOGRAPHIC PROCESSES

Course Number: A024621; ESABD70330 000; 740-83E10; 740-306.
Location: Defense Mapping School, Ft. Belvoir, VA.
Length: 15 weeks (497-516 hours).
Exhibit Dates: 1/90–9/95.
Learning Outcomes: Upon completion of this course, the student will be able to perform all process camera and darkroom operations necessary to produce a line-copy negative, halftone negative, and line negative, using filters, continuous-tone negative, and contact paper prints from continuous tone aerial photo negatives and produce an offset plate, a bookwork dummy and flats, a multi-image combination form dummy and flat, and a multicolor proof.
Instruction: Course includes some lecture/demonstration but is predominantly hand-on activities in process camera operations, making contact paper prints and contact negatives, lithographic offset plates, bookwork flats, simple flats, multi-image combination form flats, and a multicolor proof.
Credit Recommendation: In the lower-division baccalaureate/associate degree category, 6 semester hours in process camera operations and 5 in stripping and platemaking operations (2/95).
Related Occupations: 82E.

DD-1601-0020

BASIC TERRAIN ANALYSIS

Course Number: *Version 1:* 491-400/491-81Q10; AO214J1; 491-400. *Version 2:* 491-400/491-81Q10; AO214J1; 491-400.
Location: Defense Mapping School, Ft. Belvoir, VA.
Length: *Version 1:* 14 weeks (459 hours). *Version 2:* 11-12 weeks (379 hours).
Exhibit Dates: *Version 1:* 10/90–Present. *Version 2:* 1/90–9/90.
Learning Outcomes: *Version 1:* Upon completion of the course, the student will be able to collect, map, analyze, and synthesize basic physical and cultural geographic features and apply findings to real environmental problems. *Version 2:* Upon completion of the course, the student will be able to collect, observe, map, analyze, and synthesize basic physical and cultural geographic features and apply findings to real problems.
Instruction: *Version 1:* This course includes lecture and laboratory experience in terrain analysis, basic map and air photo interpretation, review of basic mathematics through trigonometric functions, production of manual and computer-assisted data base overlays of terrain analysis, and preparation of terrain synthesis. *Version 2:* This course includes lecture, laboratory, and field experience in terrain analysis, basic map and air photo interpretation, review of basic mathematics through trigonometric functions, production of terrain analysis data base overlays, and preparation of terrain synthesis.
Credit Recommendation: *Version 1:* In the lower-division baccalaureate/associate degree category, 3 semester hours in displays of geographic information, 5 in fundamentals of physical geography, and 3 in geology and geomorphology (2/95); in the upper-division baccalaureate category, 3 semester hours in air photo interpretation, 3 in geography of land use, and 3 in problems of physical geography (2/95). *Version 2:* In the lower-division baccalaureate/associate degree category, 3 semester hours in basic map and air photo interpretation and 5 in physical geography (7/87); in the upper-division baccalaureate category, 3 semester hours in physical geography field techniques, 3 in land use analysis, and 3 in physical geography (7/87).

DD-1601-0021

MAPPING, CHARTING, AND GEODESY OFFICER

Course Number: AO214S1; 4M-21C; 4M-701; 409-701/4M-701.
Location: Defense Mapping School, Ft. Belvoir, VA.
Length: 11 weeks (299-468 hours).
Exhibit Dates: 1/90–12/93.
Learning Outcomes: Upon completion of the course, the student will have developed an overview at the manager's level of cartography, map reading, geodesy, remote sensing, surveying techniques and processes, terrain analysis, geographical information systems (GIS) technology, and current management practices at a very large map or charting facility. Student will be able to manage a large production operation. For Army and Marine Corps students there is an added emphasis on cartography, reproduction of maps and charts, and operations management of map production.
Instruction: The course involves lecture, hands-on applications, field trips, and demonstration projects including geodesy, map reading, remote sensing, surveying, GIS technology, and terrain analysis. Instruction includes lectures, laboratory, and practical experiences in geodesy and error theory, photogrammetry and remote sensing, basic surveying, terrain analysis, cartography, map reproduction, topographic operation, and digital data processing.
Credit Recommendation: In the lower-division baccalaureate/associate degree category, 3 semester hours in any one of the following: physical geography, map reading and interpretation, civil engineering, remote sensing, or cartography (2/92); in the upper-division baccalaureate category, 3 semester hours in general cartography, management, or geographic techniques (2/92). In addition, for Marine and Army 6 semester hours in management or 3 in management and 3 in cartography (2/92).
Related Occupations: 21C.

DD-1601-0022

ADVANCED GEODETIC SURVEY

Course Number: *Version 1:* 412-104. *Version 2:* 412-104.

Location: Defense Mapping School, Ft. Belvoir, VA.

Length: *Version 1:* 12-13 weeks (448 hours). *Version 2:* 13 weeks (453 hours).

Exhibit Dates: *Version 1:* 10/96–Present. *Version 2:* 1/90–9/96.

Learning Outcomes: *Version 1:* This version is pending evaluation. *Version 2:* Upon completion of the course, the student will be able to apply advanced geodetic survey techniques covering geodesy and management for survey operations. Student will have skills and knowledge to both compute and adjust geodetic surveys and will be able to perform field adjustments and calibration of survey equipment. Student will be able to observe, record, and compute azimuths and astronomical positions.

Instruction: *Version 1:* This version is pending evaluation. *Version 2:* Instruction includes lectures, demonstrations, and practical problems as well as hands-on use of equipment. Topics are designed to introduce students to electronic calculators, microcomputers, and surveying software packages. An introduction to geodesy and management for survey operations is given. There is instruction on checking field data and computations. Student will learn to compute and adjust geodetic surveys and will be able to perform field adjustments and calibration of survey equipment. Course concentrates on observing, recording, and computing astronomical azimuths and astronomical positions. Course concludes with examination of special types of surveys such as gravity, magnetic, and hydrographic surveys.

Credit Recommendation: *Version 1:* Pending evaluation. *Version 2:* In the lower-division baccalaureate/associate degree category, 3 semester hours in general engineering or 3 in introductory astronomy (2/95); in the upper-division baccalaureate category, 3 semester hours in civil engineering (surveying) (2/95).

Related Occupations: 821A; 82D.

DD-1601-0023

BASIC GEODETIC SURVEY

Course Number: *Version 1:* A021A1; 412-82D10; 412-101. *Version 2:* A0214A1; G5ABD22230 000; 412-82D10; 412-101.

Location: Defense Mapping School, Ft. Belvoir, VA.

Length: *Version 1:* 12-14 weeks (427-487 hours). *Version 2:* 19 weeks (506 hours).

Exhibit Dates: *Version 1:* 1/93–Present. *Version 2:* 1/90–12/92.

Learning Outcomes: *Version 1:* This version is pending evaluation. *Version 2:* Upon completion of this course, the student will be able to establish ground survey control under supervision for mapping, charting, and geodetic surveys.

Instruction: *Version 1:* This version is pending evaluation. *Version 2:* Instruction includes lecture, demonstration, and field exercises in the operation of precise levels, electronic distance measuring devices, theodolites, satellite receivers, analytical photogrammetric positioning systems, basic map reading, recovery and installation of control points, familiarization of gravity surveys, and operation and use of a programmable calculator.

Credit Recommendation: *Version 1:* Pending evaluation. *Version 2:* In the lower-division baccalaureate/associate degree category, 3 semester hours in map reading and interpretation (2/92); in the upper-division baccalaureate category, 3 semester hours in basic surveying or civil engineering. An additional 2 hours in planetable surveying may be awarded to Army and Marine students who have completed Annex 1: Topographic Techniques (2/92).

Related Occupations: 82D.

DD-1601-0025

BASIC CARTOGRAPHY

Course Number: *Version 1:* 411-200. *Version 2:* A0214R1; 411-81C10; 411-200; 411-200/411-81C10.

Location: Defense Mapping School, Ft. Belvoir, VA.

Length: *Version 1:* 10 weeks (11-352 hours). *Version 2:* 10-11 weeks (363 hours).

Exhibit Dates: *Version 1:* 4/94–Present. *Version 2:* 1/90–3/94.

Learning Outcomes: *Version 1:* This version is pending evaluation. *Version 2:* Upon completion of the course, the student will be able to perform basic cartographic tasks in several areas, including map interpretation and tracing of map overlays, base map preparation in UTM and graduate grids, geodetic control, construction of photo mosaics and use of the zoom-transfer scopes, map compilation focusing on hyposmetry and planimetric features, recompilation of one map from several (at a new scale), and color separation.

Instruction: *Version 1:* This version is pending evaluation. *Version 2:* This course is taught through lectures and laboratory exercises with some practical application. A solid approach to cartography includes drafting, type stick-up, registration, flop separation, scribing line work, area masks, and map proofing. Course also includes heavy reliance on aerial photography and zoom-transfer scope maintenance.

Credit Recommendation: *Version 1:* Pending evaluation. *Version 2:* In the lower-division baccalaureate/associate degree category, 3 semester hours in map interpretation (2/92); in the upper-division baccalaureate category, 3 semester hours in introductory cartography and 2 in interpretation of aerial photography (2/92).

Related Occupations: 81C.

DD-1601-0026

1. POINT POSITIONING SYSTEMS
2. ANALYTICAL PHOTOGRAMMETRIC POSITIONING SYSTEM

Course Number: *Version 1:* G5AZD22150 000; A0247B1; 411-APPS; 411-207. *Version 2:* G5AZD22150 000; A0247B1; 411-APPS; 411-207.

Location: Defense Mapping School, Ft. Belvoir, VA.

Length: *Version 1:* 2 weeks (58-60 hours). *Version 2:* 2 weeks (58-68 hours).

Exhibit Dates: *Version 1:* 10/91–3/95. *Version 2:* 1/90–9/91.

Learning Outcomes: *Version 1:* Upon completion of the course, the student will be able to operate the Analytical Photogrammetric Positioning System (APPS) for point positioning and will be able to determine UTM and geographic coordinate positions. *Version 2:* Upon completion of the course, the student will be able to operate the Analytical Photogrammetric Positioning System I (APPS) for point positioning and will be able to determine UTM and geographic coordinate positions.

Instruction: *Version 1:* Instruction includes lectures, demonstrations, and practical exercises in general operating procedures and assembly, disassembly, and performance of diagnostic procedures of the APPS. Photo interpretation is also covered. *Version 2:* Instruction includes lectures, demonstrations, and practical exercises in general operating procedures and assembly, disassembly, and performance of diagnostic procedures of the APPS. Photo interpretation is also covered.

Credit Recommendation: *Version 1:* In the upper-division baccalaureate category, 2 semester hours in photogrammetry or geodetic survey, surveying, or engineering (2/95). *Version 2:* In the upper-division baccalaureate category, 3 semester hours in photogrammetry, geodetic science, surveying, civil engineering, or cartography. Credit is to be granted in either the upper-division or the graduate category, as appropriate, but not both (4/90); in the graduate degree category, 3 semester hours in photogrammetry, geodetic science, surveying, civil engineering, or cartography. Credit is to be granted in either the upper-division or the graduate category, as appropriate, but not both (4/90).

Related Occupations: 21C; 811A; 81C; 81Q; 821A; 82D; 841A.

DD-1601-0028

ADVANCED TERRAIN ANALYSIS

Course Number: 491-81Q30; 491-402.

Location: Defense Mapping School, Ft. Belvoir, VA.

Length: 15 weeks (525-549 hours).

Exhibit Dates: 1/90–Present.

Learning Outcomes: Upon completion of the course, the student will be able to analyze and produce physical and human geographic information through observation, remote sensing, and field techniques.

Instruction: The course is presented through lectures, laboratories, and field exercises in terrain analysis, remote sensing system, production of terrain analysis date base overlays, application of computer-based digital terrain analysis, project planning and coordination, and quality control.

Credit Recommendation: In the lower-division baccalaureate/associate degree category, 3 semester hours in basic map and air photo interpretation, 5 in geology or geomorphology, and 3 in introduction to management principles (2/95); in the upper-division baccalaureate category, 3 semester hours in land use analysis, 3 in remote sensing, 3 in geographic information systems, and 3 in physical geography field techniques (2/95).

Related Occupations: 81Q.

DD-1601-0029

TERRAIN ANALYSIS WARRANT OFFICER CERTIFICATION

Course Number: *Version 1:* 4N-215D. *Version 2:* 4M-841A.

Location: Defense Mapping School, Ft. Belvoir, VA.

Length: *Version 1:* 12 weeks (386 hours). *Version 2:* 9-10 weeks (392 hours).

Exhibit Dates: *Version 1:* 2/90–Present. *Version 2:* 1/90–1/90.

Learning Outcomes: *Version 1:* This version is pending evaluation. *Version 2:* Upon completion of the course the student will understand the theory and principles necessary to coordinate, plan, manage, direct, and advise acquisition, reproduction, and and distribution of terrain analysis information and its synthesis.

Instruction: *Version 1:* This version is pending evaluation. *Version 2:* The course includes lecture, laboratory, and practical experience in geodesy, multispectral imagery, computer use, review of terrain analysis, cartography, reproduction, operation, and digital imagery.

Credit Recommendation: *Version 1:* Pending evaluation. *Version 2:* In the lower-division baccalaureate/associate degree category, 3 semester hours in map and map interpretation (7/87); in the upper-division baccalaureate category, 3 semester hours in geographic laboratory techniques including cartography, 3 in field experience in operation management (7/87).

Related Occupations: 841A.

DD-1601-0030

REMOTELY SENSED IMAGERY FOR MAPPING, CHARTING, AND GEODESY

(Remotely Sensed Imagery for MC&G)

Course Number: X502D8016 011; 3A/24-710.

Location: Defense Mapping School, Ft. Belvoir, VA.

Length: 2 weeks (73-83 hours).

Exhibit Dates: 8/91–Present.

Learning Outcomes: Upon completion of the course, the student will have a working knowledge of remote sensing of the environment. Student will understand the electromagnetic spectrum and will relate it to remote sensing. Attention is paid to spectral characteristics of earth surface materials which includes image interpretation and image analysis.

Instruction: This course is a well balanced blend of lecture and hands-on laboratory exercises. It provides a basic introduction to remote sensing of the environment. Attention is given to spectral characteristics of earth surface materials, which leads to hard copy image interpretation and image analysis. This includes hard copy image analysis problems. Elements of photographic, thermal, radar, and satellite systems are covered. The second part of the course is designed to give the student an overview and hands-on experience in digital image processing of remote sensing imagery. Students are assigned to a work station to work on ERADS. Training on this system includes basic image display, image enhancement, positional accuracy, and geocoding. Students are taught to merge MSI data with other data, including unsupervised and supervised image classification. GIS functions and applications are introduced, as is change detection. The course concludes with a comprehensive problem in which the student must call up on the computer and image and perform techniques of digital image processing without the aid of an instructor. The student must give a debriefing or critique including analysis of the image produced.

Credit Recommendation: In the upper-division baccalaureate category, 3 semester hours in remote sensing of the environment (2/95).

Related Occupations: 215D; 21C; 81Q.

DD-1601-0031

GEOGRAPHIC INFORMATION SYSTEMS (GIS)

Course Number: 4M/41-708.

Location: Defense Mapping School, Ft. Belvoir, VA.

Length: 2 weeks (80 hours).

Exhibit Dates: 5/91–9/94.

Learning Outcomes: Upon completion of the course, the student will be literate in most basic areas of geographic information systems (GIS) from a user's standpoint. The student will be grounded in history, applications to military maneuvers, analysis functions, new technology, data base trends, accuracy issues, data models and their applicability to particular projects, geographic information system management, and digital data with concentration on the digital chart of the world. Student will also have learned vector GIS (ARC/INFO), Raster GIS (ERDAS), remote sensing data (clarification and rectification), and use of GPS in GIS data collection.

Instruction: Through lectures, demonstrations, and carefully constructed practical exercises, the course covers many issues basic to geographic information systems. Many projects are rote (typing commands from a script) but not entirely. Each section allows for some independent modelling. Course covers a wide range of topics. Though in-depth study may be lacking in most areas, the course provides a comprehensive introduction to GIS.

Credit Recommendation: In the lower-division baccalaureate/associate degree category, 3 semester hours in computer literacy (2/95); in the upper-division baccalaureate category, 3 semester hours in introduction to geographic information systems (2/95).

Related Occupations: 215D; 21C; 81Q.

DD-1706-0003

REPRODUCTION EQUIPMENT REPAIR

Course Number: AO215F1; 690-ASIJ6; 690-620.

Location: Defense Mapping School, Ft. Belvoir, VA.

Length: 18 weeks (489-632 hours).

Exhibit Dates: 1/90–Present.

Learning Outcomes: Upon completion of this course, the student will be able to safely use and care for applicable hand and measuring tools for repair of reproduction equipment; adjust, repair, and replace parts on the offset duplicator; trammel and calibrate the copy camera to within prescribed specifications; troubleshoot and repair medium-sized offset presses, including troubleshooting and adjusting timing of paper train, cylinder assembly, dampening system, inking system, timing of main press; and verify timing through troubleshooting with register.

Instruction: The course includes demonstrations and lectures but is conducted as predominantly hands-on instruction in offset duplicator repair, copy camera repair, and maintenance and repair of medium-sized offset presses.

Credit Recommendation: In the upper-division baccalaureate category, 3 semester hours in offset duplicator repair, 3 in process camera repair, and 5 in medium-sized offset press repair (2/95).

Related Occupations: 83F.

DD-1709-0002

INTERMEDIATE PHOTOJOURNALISM

Course Number: *Version 1:* AFIS-IPC. *Version 2:* 570-ASIJ8; 43E; A-580-0017; GSAZA79150 002.

Location: Defense Information School, Ft. Benjamin Harrison, IN.

Length: *Version 1:* 8 weeks (314 hours). *Version 2:* 7 weeks (274-276 hours).

Exhibit Dates: *Version 1:* 1/94–1/96. *Version 2:* 1/90–12/93.

Learning Outcomes: *Version 1:* Upon completion of the course, the student will be able to operate a still photography camera to record events in a variety of situations and lighting conditions; develop and print black-and-white film in a chemical darkroom; and use E-6 color processing for transparencies. Student will be able to edit and layout photographs for news purposes, information, instruction, and promotional materials and will be able to edit and layout feature story illustrations, including photos, texts, captions, and headlines. Student will be able to capture, edit, and print photographs with digital imaging equipment. *Version 2:* Upon completion of the course, the student will be able to practice basic camera techniques, basic and advanced black-and-white darkroom procedures, E-6 slide processing, photo editing and layout, and basic audiovisual slide show production. Students prepare simple feature packages that include photos, text, captions, and headlines.

Instruction: *Version 1:* This is a two-track program that provides basic photography, writing, and editing training for military photojournalists. During the first three weeks, students with writing and reporting experience study the fundamentals of photography while those with photography experience concentrate on writing and copy editing skills. The final four weeks involve both groups in the production of word and picture packages that include newspaper and magazine format layouts and an audiovisual presentation. The course includes classroom instruction and laboratory exercises in digital imaging using a variety of workstations and software applications. *Version 2:* This is a two-track program that provides basic photography, writing, and editing training for military photojournalists. During the first three weeks, students with writing and reporting experience study the fundamentals of photography, while those with photography experience concentrate on writing and copy editing skills. The final four weeks involve both groups in the production of word and picture packages that include newspaper and magazine format layouts and an audiovisual presentation.

Credit Recommendation: *Version 1:* In the lower-division baccalaureate/associate degree category, 6 semester hours in photojournalism and 2 in digital imaging (7/94). *Version 2:* In the lower-division baccalaureate/associate degree category, 6 semester hours of photojournalism (6/92).

DD-1709-0003

ADVANCED PUBLIC AFFAIRS SUPERVISOR

Course Number: 570-46Q/R40.

Location: Defense Information School, Ft. Benjamin Harrison, IN.

Length: 2 weeks (71-73 hours).

Exhibit Dates: 10/90–11/95.

Learning Outcomes: Upon completion of the course, the student will be able to supervise the operation of public affairs offices located on posts, bases, ships, or stations of the military services and perform duties as the noncommissioned officer in charge or the public affairs officer responsible for the planning, supervising, and managing operations, personnel, and resources.

Instruction: Instruction is presented as lectures, seminars, and practical exercises in problem solving. Topics include management functions of media production, media relations, and internal information requirements of public affairs operation. Emphasis is placed on supervision, planning, staffing, training, resource management, budgeting, and staffing functions in policy decisions and contingency operations. Media operations include both print and electronic products.

Credit Recommendation: In the upper-division baccalaureate category, 3 semester hours in public relations management (10/92).

Related Occupations: 46Q; 46R.

DD-1713-0006

CARTOGRAPHIC/GEODETIC OFFICER

Course Number: *Version 1:* G50BD5731 000; 4M-701AF. *Version 2:* G50BD5731 000; 4M-701AF.

Location: Defense Mapping School, Ft. Belvoir, VA.

Length: *Version 1:* 7 weeks (249-250 hours). *Version 2:* 10-11 weeks (351 hours).

Exhibit Dates: *Version 1:* 6/90–7/91. *Version 2:* 1/90–5/90.

Learning Outcomes: *Version 1:* Upon completion of the course, the student will have developed an overview, at the manager's level, of cartography, map reading, geodesy, remote sensing, surveying techniques and processes, terrain analysis, GIS technology, and current management practices as they combine in map production at a very large map or charting facility. The student will not only have learned about cartography and map production but also will have achieved a technical understanding of the management skills necessary in large production operations. *Version 2:* Upon completion of the course, the student will have a general overview of cartography, geodesy, surveying, terrain analysis, computer graphics, and digital processing as well as operations, products, and management policies of the Defense Mapping Agency. Upon completion, the student will be trained in the overall cartographic geodetic process, and will have the technical skills necessary to manage a production facility or oversee the production of products, data, or reports from field operations.

Instruction: *Version 1:* The course involves lectures, hand-on applications, field trips, and demonstration projects in geodesy, map reading, remote sensing, surveying, GIS technology, and terrain analysis. *Version 2:* The course presents an explanation of the map-making process. Students learn computer technology as it applies to mapping and charting. Course includes the management of mapping, charting, and geodesy process. The course offers a strong hands-on component on surveying techniques, remote sensing, and terrain analysis.

Credit Recommendation: *Version 1:* In the lower-division baccalaureate/associate degree category, 3 semester hours in any one of the following: management, physical geography, map reading and interpretation, civil engineering, remote sensing, or cartography (2/92); in the upper-division baccalaureate category, 3 semester hours in general cartography, management, or geographic techniques (2/92). *Version 2:* In the lower-division baccalaureate/associate degree category, 6 semester hours in blocks of 3 hours each in: map reading and interpretation, cartography, remote sensing, surveying, geodesy, or management. In addition, 3 semester hours in introduction to physical geography. Credit may be awarded in the lower or the upper-division in any combination not to exceed 6 semester hours (7/87); in the upper-division baccalaureate category, 3 semester hours in geodesy, cartography, or management. Credit may be awarded in the lower or the upper-division in any combination not to exceed 6 semester hours (7/87).

DD-1713-0007

ADVANCED CARTOGRAPHY

Course Number: *Version 1:* 411-81C30; 411-208. *Version 2:* 411-81C30; 411-208. *Version 3:* 411-81C30; 411-208.

Location: Defense Mapping School, Ft. Belvoir, VA.

Length: *Version 1:* 8 weeks (279 hours). *Version 2:* 8 weeks (272-283 hours). *Version 3:* 8-9 weeks (272-283 hours).

Exhibit Dates: *Version 1:* 10/92–Present. *Version 2:* 2/91–9/92. *Version 3:* 1/90–1/91.

Learning Outcomes: *Version 1:* This version is pending evaluation. *Version 2:* Upon completion of the course, the student will understand the roles of Army noncommissioned officer topographic teams; understand and be familiar with map products; be able to do some compilation on graphics with ARC/INFO; construct grids; work with the analytical photogrammetric positioning system in a wide variety of settings (including setup and breakdown) while connected to a PC; determine project suitability and requirements, color separation, and use of aerial photography; and apply good program management skills. *Version 3:* Student will have completed a review of new cartographic equipment and developments and mathematics needed in the course; will have reviewed geodesy military projections, grids, and mapping controls; and will be able to describe the construction of a base map. Student will have developed the basic skills and knowledge required to set up and operate an analytical photogrammetic positioning system including troubleshooting and assembly of the system. The student will be provided with the advanced technical skills and knowledge required to supervise planimetric map revision including color separations and image based projects. There is an emphasis on the development of the management skills necessary for the collection and evaluation of sound material, quality control, and determination of technical supply requirements for cartographic projects.

Instruction: *Version 1:* This version is pending evaluation. *Version 2:* This course is taught as a combination of lectures and closely monitored hands-on laboratory exercises which develop an understanding of advanced skills in cartography and photogrammetry. Course includes cartographic photomapping applications, production management methods, equipment assembly and troubleshooting, and supervision techniques. *Version 3:* This course is taught as a combination of lectures and closely monitored hands-on laboratory exercises which develop an understanding of advanced skills in cartography and photogrammetry. Course includes cartographic photomapping applications, production management methods, equipment assembly and troubleshooting, and supervision techniques.

Credit Recommendation: *Version 1:* Pending evaluation. *Version 2:* In the lower-division baccalaureate/associate degree category, 3 semester hours in aerial photo interpretation in geography or civil engineering. Credit is to be granted in the lower or the upper-division but not in both (2/92); in the upper-division baccalaureate category, 3 semester hours in photogrammetry and 3 in cartography. Receiving school should not grant upper division credit in photogrammetry if student has already received credit in that field. Credit in cartography requires that student has completed course work in which reference mapping focus is used. Credit is to be granted in the lower or the upper division but not in both (2/92). *Version 3:* In the vocational certificate category, 6 semester hours in photogrammetry, cartography, civil engineering or 3 hours in management in combination with 3 hours in any one of the above listed subjects (7/87); in the lower-division baccalaureate/associate degree category, 3 semester hours in basic photogrammetry, map reading and interpretation, air photo interpretation, architectural planning, or management and 3 in geodesy. Credit is to be granted in the upper-division or the lower-division but not both (7/87); in the upper-division baccalaureate category, 3 semester hours in photogrammtery or advanced cartography. Credit is to be granted in the upper-division or the lower-division but not both (7/87).

Related Occupations: 81C.

DD-1715-0008

RP11/RP03 DISK PACK MAINTENANCE

Course Number: ES-403.

Location: National Cryptologic School, Ft. Meade, MD.

Length: 2 weeks (80 hours).

Exhibit Dates: 1/90–3/92.

Objectives: This course enables the student to maintain the RP11/RP03 disk system.

Instruction: Extensive practical exercises include troubleshooting and alignment techniques. Lectures discuss signal flow and logic analysis. Students will have completed a PDP-11 Processor Maintenance course.

Credit Recommendation: In the vocational certificate category, 2 semester hours in disk pack system maintenance (3/82).

DD-1715-0011

PDP-11 INSTRUCTION SET

Course Number: ES-411.

Location: National Cryptologic School, Ft. Meade, MD.

Length: 2 weeks (80 hours).

Exhibit Dates: 1/90–4/91.

Objectives: To teach machine and assembly language programming on the DEC PDP-11.

Instruction: Lecture and hands-on experience cover machine and assembly language programming on the DEC PDP-11, including full word addressing, stack operations, and system architecture including UNIBUS. Many practical examples are covered in the programming exercises.

Credit Recommendation: In the upper-division baccalaureate category, 3 semester hours in machine and assembly language programming on the PDP-11 (DEC) (3/82).

DD-1715-0016

INTERMEDIATE ELINT COLLECTION AND ANALYSIS

Course Number: EA-280.
Location: National Cryptologic School, Ft. Meade, MD.
Length: 8 weeks (280 hours).
Exhibit Dates: 1/90–3/91.
Objectives: This course is designed to prepare SCA personnel to perform the duties of collector/analyst at ELINT collection sites or analysis centers with minimum supervision.
Instruction: The course combines lecture and hands-on experience. Topics include scientific notation, basic algebra and trigonometry, unit conversions, oscilloscope fundamentals, use of counters and spectrum analyzers, and fundamentals of modulation and signal analysis with emphasis on radar signals.
Credit Recommendation: In the lower-division baccalaureate/associate degree category, 3 semester hours in technical mathematics (if credit has not been given for prerequisite), 1 in instrument familiarization laboratory, and 2 in fundamental signal analysis (3/82).

DD-1719-0006

BASIC OFFSET PRINTING

Course Number: A02PAA1; 740-83F10; 740-303.
Location: Defense Mapping School, Ft. Belvoir, VA.
Length: 8-9 weeks (286-297 hours).
Exhibit Dates: 1/90–Present.
Learning Outcomes: Upon completion of the course, the student will be able to set up and operate both sheet-fed offset lithographic duplicators and presses. The student will be able to make normal operating adjustments of sheet feed, transport and delivery systems, dampening and inking systems; produce a lithographic plate; and operate common binder equipment.
Instruction: The course covers press operating procedures, including pressure adjustments for inking and dampening systems, packing calculations, register adjustments and requirements, platemaking techniques, and proper operation of bindery equipment. Distinctions are made between duplicating and press limitations and capabilities. Instruction in general operating maintenance and lubrication is provided in both process and bindery areas.
Credit Recommendation: In the vocational certificate category, 6 semester hours in offset lithographic printing (2/95); in the lower-division baccalaureate/associate degree category, 3 semester hours in offset printing laboratory (2/95).
Related Occupations: 83F.

DD-1719-0007

NAVY/AIR FORCE BASIC LITHOGRAPHER

Course Number: A-740-0025; E5ABD70330 000; 740-309.
Location: Defense Mapping School, Ft. Belvoir, VA.
Length: 18 weeks (546-594 hours).
Exhibit Dates: 1/90–11/94.
Learning Outcomes: Upon completion of the course, the student will be proficient in offset lithographic press operation, general operating maintenance, reproduction photography, general copy preparation and image assembly, common bindery equipment operation and maintenance. Student will have some knowledge of microfilm equipment operation/metal photo process and basic knowledge of print shop administration.
Instruction: Course includes lectures, demonstrations, and practical experience in phototypesetting procedures; paste-up; line and halftone negative generation, including use of filters, image assembly techniques, and platemaking control; operating adjustments and mechanics of lithographic process/duplicators; and bindery systems operations. Navy personnel receive instruction in print shop administration and metal photo plates. Air Force personnel receive microfilm equipment operation and reduced copy preparation training.
Credit Recommendation: In the vocational certificate category, 8 semester hours in offset lithographic press work and 1 in finishing operations (2/95); in the lower-division baccalaureate/associate degree category, 5 semester hours in lithographic press work, 3 in reproduction photography, and 1 in layout and design (2/95).

DD-1719-0011

ADVANCED LITHOGRAPHY

Course Number: 740-310; 740-310 (83F30).
Location: Defense Mapping School, Ft. Belvoir, VA.
Length: 8 weeks (289 hours).
Exhibit Dates: 1/90–Present.
Learning Outcomes: Upon completion of the course, the student will be able to plan, estimate, produce, and supervise as well as make quality judgments on photolithographic production. Student will be skilled in camera operation, stripping techniques, offset platemaking procedures, and offset press operation and maintenance. Student will be versed in bindery functions, including operation and maintenance of a power paper cutter and drill, buckle folders, paper gatherers and collators, and stitching. Student will have knowledge of safety requirements and of the hazards involved.
Instruction: Course is conducted through lectures, demonstrations, and many hands-on projects that produce actual printed products.
Credit Recommendation: In the lower-division baccalaureate/associate degree category, 3 semester hours in basic graphic arts (2/95); in the upper-division baccalaureate category, 3 semester hours in camera, stripping, and platemaking, 2 in equipment operation, and 1 in supervision and operation management (2/95).

DD-1721-0003

SURVEY INSTRUMENT MAINTENANCE

Course Number: 670-601; 670-41B10; 670-602.
Location: Defense Mapping School, Ft. Belvoir, VA.
Length: 4 weeks (128-131 hours).
Exhibit Dates: 1/90–5/95.
Learning Outcomes: Upon completion of the course, the student will be able to use proper equipment, tools, and chemicals needed to clean, adjust, and repair basic surveying equipment such as tripods, leveling rods, and dumpy level.
Instruction: In-depth training is given in performing organizational and operator maintenance on the wild T-16 and the wild T-16 with compensator, i.o. and minute theodolite, and the wild T-2 and the T-2 with compensator, i.o. second theodolite. Instruction includes lectures, demonstrations, and hands-on practical exercises in the cleaning, maintenance, adjustment, and repair of optical instruments and basic survey support equipment such as tripods and tapes.
Credit Recommendation: In the lower-division baccalaureate/associate degree category, 3 semester hours in mechanical/optical instrument testing, repair, and calibration (2/92).
Related Occupations: 41B.

DD-1722-0001

HYDROGRAPHIC SURVEY

Course Number: AO2DC1; 412-103.
Location: Defense Mapping School, Ft. Belvoir, VA.
Length: 7 weeks (232-233 hours).
Exhibit Dates: 1/90–3/90.
Learning Outcomes: Upon completion of the course, the student will have a basic understanding of geodesy, including visual survey position, resection, and intersection and electronic surveying position. The course is concerned with beach signals, error theory, currents, and current measurement; the student will be able to scale and interpret echograms, annotate sounding data, and construct a grid and smooth sheet. The student will be familiar with hydrographic equipment and organizations.
Instruction: The overall purpose of the course is to provide basic skills and knowledge necessary to conduct manual hydrographic surveys from the planning stages through the implementation of a smooth sheet. There is hands-on experience in hydro-survey equipment and theoretical application of electronic hydro-survey techniques. Students is expected to convert raw hydrographic data into usable format.
Credit Recommendation: In the vocational certificate category, 3 semester hours in surveying or hydrographic surveying (7/87); in the lower-division baccalaureate/associate degree category, 3 semester hours in basic surveying, 3 in hydrographic surveying, 3 in basic marine science, 3 in coastal geomorphology, or 3 in basic geodesy for a total of not more than 6 semester hours (7/87); in the upper-division baccalaureate category, 3 semester hours in surveying, hydrographic surveying, geodesy, or marine science. Recommendation is to grant up to 6 semester hours of which 3 may be in the upper-division (7/87).

DD-1728-0003

POLYGRAPH EXAMINER TRAINING

Course Number: 7H-F11.
Location: Polygraph Institute, Ft. McClellan, AL.
Length: 12-14 weeks (560 hours).
Exhibit Dates: 10/90–Present.
Learning Outcomes: Before 10/90 see AR-1728-0008. Completion of the course will qualify military and federal civilian investigative/intelligence personnel as polygraph examiners.
Instruction: Topics include polygraph theory and maintenance management; mental and physical evaluation of examinee; polygraph instrumentation and examination; post-test procedures; and practical exercises in zone com-

parison, peak of tension, and general question techniques.

Credit Recommendation: In the upper-division baccalaureate category, 3 semester hours in interviews and interrogations and 3 in applied psychology (12/96); in the graduate degree category, 6 semester hours in a forensic science or criminal justice elective and 1-3 in research (based upon institutional evaluation of individual research project) (12/96).

Course Exhibits

MC

MC-0327-0001

1. BASIC RECRUITER
2. MARINE CORPS RECRUITER

Course Number: *Version 1:* 81C. *Version 2:* 81C.

Location: *Version 1:* Recruit Depot, San Diego, CA. *Version 2:* Recruiters School, San Diego, CA.

Length: *Version 1:* 8 weeks (280 hours). *Version 2:* 7 weeks (214-218 hours).

Exhibit Dates: *Version 1:* 4/95–Present. *Version 2:* 1/90–3/95.

Learning Outcomes: *Version 1:* Upon completion of the course, the student will be able to determine if recruit applicants meet the requirements of the Marine Corps by using proper screening and processing procedures, use sales techniques in the recruiting process, use interpersonal communication and public speaking skills use office management skills in recruiting substation operation, determine advertising and community relations necessary to meet recruiting needs, and use leadership skills to present proper image of the Marine Corps. *Version 2:* Upon completion of the course, the student will be able to determine if recruit applicants meet the requirements of the Marine Corps by using proper screening and processing procedures, use sales techniques in the recruiting process, use interpersonal communications and public speaking skills, use office management skills in recruiting substation operation, determine advertising and community relations necessary to meet recruiting needs, and use leadership skills to present proper image of the Marine Corps.

Instruction: *Version 1:* Lectures and practical exercises cover all aspects of personnel recruitment, processing procedures, public speaking, operation of recruiting substations, community relations, advertising, leadership training, and office skills. *Version 2:* Lectures and practical exercises in recruitment of personnel. Course includes screening and processing procedures, sales techniques, public speaking, operation of recruiting substations, officer selection, community relations, advertising, leadership training, and typing.

Credit Recommendation: *Version 1:* In the lower-division baccalaureate/associate degree category, 1 semester hour in salesmanship, 1 in business communication, 2 in office procedures, and 1 in leadership (12/96). *Version 2:* In the lower-division baccalaureate/associate degree category, 3 semester hours in salesmanship, 3 in business communication, 3 in clerical procedures, and 1 in leadership (6/91).

MC-0406-0001

INTRODUCTION TO RETAILING BY CORRESPONDENCE

Course Number: 41.7.

Location: Marine Corps Institute, Washington, DC.

Length: Maximum, 52 weeks.

Exhibit Dates: 1/90–Present.

Learning Outcomes: Upon completion of the course, the student will be able to identify primary elements of retailing institutions and specific examples of their influences on the economy. Student will be able to identify the marketing process, primary types of retail operations, and basic retail functions. Student will also be familiar with armed services exchange regulations and the specific operations of an exchange.

Instruction: This is a correspondence course covering the topics of retailing systems, the organizational structure of the retailing firm, retailing in the military, the Marine Corps exchange system, and field exchange operations.

Credit Recommendation: In the lower-division baccalaureate/associate degree category, 1 semester hour in introduction to retailing (5/91).

MC-0419-0001

BASIC FREIGHT OPERATION

Course Number: None.

Location: Service Support School, Cp. Lejeune, NC.

Length: 4 weeks (140 hours).

Exhibit Dates: 1/90–1/92.

Objectives: To train enlisted personnel in all aspects of freight operation.

Instruction: Lectures and practical experience in the fundamentals of shipping and receiving; the capabilities of the transportation system; rules and regulations governing transportation; freight classification, regulation, and storage; use of material-handling equipment including the forklift; and use of the manual typewriter.

Credit Recommendation: In the vocational certificate category, 3 semester hours in freight handling (1/74).

MC-0419-0005

MOTOR TRANSPORT OFFICER

Course Number: *Version 1:* CEJ. *Version 2:* CEJ.

Location: *Version 1:* Engineer School, Cp. Lejeune, NC. *Version 2:* Service Support School, Cp. Lejeune, NC.

Length: *Version 1:* 6 weeks (225 hours). *Version 2:* 6-8 weeks (208-229 hours).

Exhibit Dates: *Version 1:* 10/94–Present. *Version 2:* 1/90–9/94.

Learning Outcomes: *Version 1:* Upon completion of the course, the student will be able to manage tactical vehicle maintenance operations, vehicle recovery operations, tactical and convoy operations. *Version 2:* Upon completion of the course, the student will be able to describe all major assemblies used in current motor transport tactical vehicles; discuss the Marine Corps integrated maintenance management system; implement record keeping functions inherent to motor transport organizations; establish a motor pool; describe special vehicle operations, including vehicle recovery, deep water fording, vehicle camouflage, loading procedures, and night driving operations; and defend a tactical convoy.

Instruction: *Version 1:* Instruction will be provided by lecture, demonstration and laboratory activities on introduction to component testing, motor transportation maintenance management, maintenance administration, operations maintenance management and management of tactical convoy operations; emphasis is on lecture and demonstration work outside of class is required. *Version 2:* Course covers an introduction to automotive components and test equipment, motor transport maintenance management, administration, motor transport vehicle operations, and tactical convoy operations. Methodology includes lectures and laboratory exercises.

Credit Recommendation: *Version 1:* In the upper-division baccalaureate category, 4 semester hours in vehicle maintenance systems management and 3 in vehicle fleet operations management (8/95). *Version 2:* In the lower-division baccalaureate/associate degree category, 1 semester hour in vehicle maintenance and 2 in management (10/94).

MC-0419-0007

AIR MOVEMENT PLANNING

Course Number: H-8A-3558.

Location: Landing Force Training Command, Pacific, San Diego, CA; Mobile training units, US.

Length: 2-3 weeks (99-102 hours).

Exhibit Dates: 10/93–Present.

Learning Outcomes: Upon completion of the course, the student will be able to develop an air movement plan including aircraft loading for transport aircraft.

Instruction: Lectures and practical experience on cargo balance computation, loading and restraint, cargo placement, completion of cargo and passenger manifests, and transport aircraft characteristics. Exams include the completion of an aircraft loading and weight distribution plan.

Credit Recommendation: In the lower-division baccalaureate/associate degree category, 2 semester hours in aircraft loading and weight distribution (11/93).

MC-0419-0008

SEABEE TACTICAL SHIPBOARD PLANNING

Course Number: H-551-3553.

Location: Naval Construction Regiment, Port Hueneme, CA.

Length: 2 weeks (72 hours).

Exhibit Dates: 8/92–9/93.

Learning Outcomes: Upon completion of the course, the student will be able to load naval amphibious ships and prepare naval construction forces for deployment by sea.

Instruction: Presented by lectures, audiovisuals, and practical hands-on learning experiences, topics covered include amphibious operations, amphibious ships and landing craft, embarkation programs, ship to shore movement, cargo analysis and ship loading characteristics, and containerization.

Credit Recommendation: Credit is not recommended because of the military-specific nature of the course (11/93).

MC-0501-0001

SPELLING BY CORRESPONDENCE

Course Number: 01.18j.
Location: Marine Corps Institute, Washington, DC.
Length: Maximum, 52 weeks.
Exhibit Dates: 1/90–Present.
Learning Outcomes: Upon completion of the course, the student will be able to use a dictionary and thesaurus and understand the use of vowels, consonants, and syllables.
Instruction: This is a self-instruction course covering spelling, use of dictionary and thesaurus, suffixes, plurals, contractions, and the possessive case of nouns and pronouns. Students will learn how to distinguish between words often confused.
Credit Recommendation: Credit is not recommended because of the limited, specialized nature of the course (5/91).

MC-0501-0002

PUNCTUATION BY CORRESPONDENCE

Course Number: 01.19f.
Location: Marine Corps Institute, Washington, DC.
Length: Maximum, 52 weeks.
Exhibit Dates: 1/90–Present.
Learning Outcomes: Upon completion of the course, the student will be able to capitalize and punctuate all types of sentences. Specifically the student will understand under what circumstances words are capitalized and when to use terminal and internal punctuation.
Instruction: This is a self-instruction course covering capitalization, periods, exclamation marks, question marks, commas, semicolons, apostrophes, dashes, quotation marks, italics, parentheses, brackets, hyphens, colons, abbreviations, and numbers.
Credit Recommendation: Credit is not recommended because of the limited, specialized nature of the course (5/91).

MC-0705-0001

NUCLEAR WARFARE DEFENSE BY CORRESPONDENCE

Course Number: 57.7i.
Location: Marine Corps Institute, Washington, DC.
Length: Maximum, 52 weeks.
Exhibit Dates: 1/90–Present.
Learning Outcomes: Upon completion of the course, the student will be able to function as a nuclear defense specialist and determine the nature of the effects of a nuclear blast. Student will be able to use radiac instruments, conduct a radiological survey for exposure guidance, and provide protection and decontamination measures.

Instruction: Through the use of a study guide text, the individual will develop knowledge in five major themes: nuclear weapons effects, radiac instruments and their use, evaluating and reporting procedures, radiation exposure guidance, and protection/decontamination procedures. Each major theme has workbook exercises, and successful course completion is achieved through a two-hour proctored written examination.

Credit Recommendation: In the lower-division baccalaureate/associate degree category, 1 semester hour in nuclear safety measures (9/94).

MC-0707-0001

COLD WEATHER MEDICINE

Course Number: WAC.
Location: Mountain Warfare Training Center, Bridgeport, CA.
Length: 2-3 weeks (165 hours).
Exhibit Dates: 1/90–Present.
Learning Outcomes: Upon completion of the course, the student will be prepared to provide medical support in a cold weather mountainous environment.
Instruction: Course includes lectures, demonstrations, and performance exercises in mountain safety, weather, nutrition, survival skills, bivouac procedures, preventive medicine, evacuation procedures, and treatment of cold injuries and high altitude illness.
Credit Recommendation: In the lower-division baccalaureate/associate degree category, 2 semester hours in cold weather first aid and safety (3/92).

MC-0707-0002

MEDICAL DEPARTMENT OFFICERS ORIENTATION

Course Number: *Version 1:* M6F. *Version 2:* M6F.
Location: Field Medical Service School, Cp. Pendleton, CA; Field Medical Service School, Cp. Lejeune, NC.
Length: *Version 1:* 2-3 weeks (109 hours). *Version 2:* 1-2 weeks (128 hours).
Exhibit Dates: *Version 1:* 9/94–Present. *Version 2:* 1/90–8/94.
Learning Outcomes: *Version 1:* Upon completion of the course, the student will be able to develop a medical plan for evacuation; implement and supervise programs dealing with common environmental problems; implement and supervise preventive health care programs; and treat common emergency medical problems. *Version 2:* Upon completion of the course, the student will be able to write the medical appendix, given a scenario of an amphibious operation, to include a medical estimate, a casualty estimate, a casualty evacuation plan, and a preventive medicine tab; assemble, clean, and use various I.C. E. components including a 9mm pistol and clothing; navigate using land maps, coordinate grids, terrain features, compasses, and azimuths; and use appropriate nuclear, biological, and chemical warfare defense clothing, mask, and personal protective equipment.
Instruction: *Version 1:* Lectures, practical exercises, and field exercises in medical planning, field medical care, casualty triage, and health aspects of specialized warfare operations. *Version 2:* The course employs lecture, discussion, demonstration, practice, and repetition of the knowledge and skills needed in planning medical support of an amphibious assault operation as well as personal land navigation and individual combat equipment.
Credit Recommendation: *Version 1:* In the graduate degree category, 3 semester hours in an overview of environmental, preventive and emergency medical care (10/95). *Version 2:* In the graduate degree category, 2 semester hours in medical support planning, amphibious (6/92).

MC-0709-0002

FIELD MEDICAL SERVICE TECHNIC ENLISTED
(Field Service Medical Service Technic)

Course Number: M6D; B-300-0013; B-300-0053.
Location: Field Medical Service School, Cp. Lejeune, NC; Field Medical Service School, Cp. Pendleton, CA.
Length: 7 weeks (241-298 hours).
Exhibit Dates: 1/90–Present.
Learning Outcomes: Upon completion of the course, the student will be able to identify medical conditions resulting from combat wounds or injuries; provide treatment for specific medical conditions; triage casualties for treatment or evacuation; evaluate field sanitation related to water and waste disposal and correct deficiencies; assist in the chain of evacuation for casualties; introduce practices for preventive medicine; apply appropriate resources for decontamination of hazardous materials; treat specified dental emergencies; and respond to specified psychological problems.
Instruction: Lectures and practical exercises in combat survival and field medical practices. Course includes physical conditioning, the corpsman in the field, field medical emergency procedures, medical supply in the field, preventive medicine, and dental practices.
Credit Recommendation: In the lower-division baccalaureate/associate degree category, 5 semester hours in basic emergency care and 5 in advanced emergency care (9/94).

MC-0801-0002

AFTERCARE PROGRAM MANAGEMENT

Course Number: S-501-0001.
Location: Consolidated Drug and Alcohol Center, Cp. Lejeune, NC.
Length: 2 weeks (70 hours).
Exhibit Dates: 1/90–7/97.
Learning Outcomes: Upon completion of the course, the student will be able to assist recovering substance abusers in the process of conforming to aftercare prescription protocols, participate in conducting substance abuse education programs, and provide minimal support services to family members.
Instruction: Course introduces student to beginning skills in substance abuse, communication, data management, interviewing, and case management.
Credit Recommendation: In the lower-division baccalaureate/associate degree category, 3 semester hours in introduction to substance abuse counseling (7/87).

MC-0801-0012

MEDICAL PERSONNEL AUGMENTATION SYSTEM/
MOBILE MEDICAL AUGMENTATION
READINESS TEAM
(Mobile Medical Augmentation Readiness Team)

Course Number: M6E.
Location: Field Medical Service School, Cp. Lejeune, NC; Field Medical Service School, Cp. Pendleton, CA.
Length: 1 week (46-73 hours).
Exhibit Dates: 1/90–Present.
Learning Outcomes: Upon completion of the course, the student will be able to perform specific defensive survival procedures in a field situation and demonstrate application of casualty triage and evacuation procedures in a field situation.
Instruction: Lectures, demonstrations, and field exercises are the methodologies. Course includes knowledge and use of support equipment needed in the field, emergency defensive techniques, and applications of medical skills in a field hospital environment.
Credit Recommendation: In the lower-division baccalaureate/associate degree category, 1 semester hour in basic military science (9/94).

MC-0801-0013

CHEMICAL WARFARE DEFENSE BY CORRESPONDENCE

Course Number: 57.6d.
Location: Marine Corps Institute, Washington, DC.
Length: Maximum, 52 weeks.
Exhibit Dates: 1/90–Present.
Learning Outcomes: Upon completion of the course, the student will be able to identify the systems and effects of chemical agents on the body, perform first aid procedures, and demonstrate chemical weapons protective clothing and decontamination procedures.
Instruction: Course uses self-instruction to cover chemical warfare defense procedures, including classification of weapons, first aid, protective clothing and equipment, and site decontamination. A proctored written exam concludes the course.
Credit Recommendation: In the lower-division baccalaureate/associate degree category, 1 semester hour in toxic chemical safety (9/94).

MC-0801-0014

BASIC NUCLEAR, BIOLOGICAL, AND CHEMICAL (NBC) DEFENSE

Course Number: T3B.
Location: NBC Defense School, Ft. McClellan, AL.
Length: 9 weeks (341 hours).
Exhibit Dates: 1/95–Present.
Learning Outcomes: Upon completion of the course, the student will have skills in basic nuclear, biological, and chemical (NBC) warfare; NBC warning and reporting systems; chemical and biological hazard prediction; operation and maintenance of radiation detection equipment; nuclear fallout prediction; and decontamination of personnel and equipment.
Instruction: Instruction includes lectures, group discussions, and practical training in nuclear, biological, and chemical warfare skills.
Credit Recommendation: In the upper-division baccalaureate category, 3 semester hours in environmental science, environmental studies, or hazardous materials management (9/95).

MC-0803-0007

DESERT OPERATIONS BY CORRESPONDENCE

Course Number: 03.54.
Location: Marine Corps Institute, Washington, DC.
Length: Maximum, 52 weeks.
Exhibit Dates: 1/90–1/97.
Objectives: To provide military personnel with principles, procedures, and techniques for survival in the desert and its environment.
Instruction: This correspondence course covers topics such as characteristics of the desert, desert regions of the world, health and hygiene, and military operations and training in the desert. A proctored final examination is administered.
Credit Recommendation: Credit is not recommended because of the limited, specialized nature of the course (3/85).

MC-0804-0002

COLD WEATHER SURVIVAL

Course Number: WAB.
Location: Mountain Warfare Training Center, Bridgeport, CA.
Length: 2 weeks (115-132 hours).
Exhibit Dates: 1/90–Present.
Learning Outcomes: Upon completion of the course, the student will demonstrate mountain safety and basic preventive and rehabilitative medicine, implement a survival diet and procure water, and demonstrate survival and navigation skills in a cold weather environment.
Instruction: Lectures, demonstration, field experiences and individual and group application are all used to present mountain survival skills in a cold weather environment.
Credit Recommendation: In the lower-division baccalaureate/associate degree category, 2 semester hours in cold weather survival (3/92).

MC-0804-0003

WINTER MOUNTAIN LEADERS (A)

Course Number: M7B.
Location: Mountain Warfare Training Center, Bridgeport, CA.
Length: 4 weeks (313 hours).
Exhibit Dates: 1/90–Present.
Learning Outcomes: Upon completion of the course, the student will be able to serve as a unit advisor, planner, and instructor during cold weather mountain operations.
Instruction: This course includes lectures, demonstrations, and performance exercises in cold weather/mountain safety, first aid, survival skills, bivouac procedures, and mobility on mountain terrain.
Credit Recommendation: In the lower-division baccalaureate/associate degree category, 3 semester hours in winter mountaineering (3/92).

MC-0804-0004

WINTER MOUNTAIN LEADERS (B)

Course Number: 03Q.
Location: Mountain Warfare Training Center, Bridgeport, CA.
Length: 4 weeks (343 hours).
Exhibit Dates: 1/90–Present.
Learning Outcomes: Upon completion of the course, the student will be able to serve as a unit advisor, planner, and instructor during cold weather mountain operations.
Instruction: This course includes lectures, demonstrations, and performance exercises in cold weather/mountain safety, first aid, survival skills, bivouac procedures, and mobility on mountain terrain.
Credit Recommendation: In the lower-division baccalaureate/associate degree category, 3 semester hours in winter mountaineering (3/92).

MC-0804-0005

MOUNTAIN SURVIVAL

Course Number: M5C.
Location: Mountain Warfare Training Center, Bridgeport, CA.
Length: 2-3 weeks (127-151 hours).
Exhibit Dates: 1/90–Present.
Learning Outcomes: Upon completion of the program, the student will be able to demonstrate employment of mountain safety principles, identification of incoming weather conditions, field management for medical problems, maintenance of high level hygiene and sanitation, and survival and navigation skills in a mountainous environment.
Instruction: Instruction includes lectures, demonstrations, field experiences, individual and group applications, and experiences in mountain survival skills.
Credit Recommendation: In the lower-division baccalaureate/associate degree category, 2 semester hours in mountain survival (3/92).

MC-1107-0001

MATHEMATICS FOR MARINES BY CORRESPONDENCE

Course Number: 13.34g; 13.34h.
Location: Marine Corps Institute, Washington, DC.
Length: Maximum, 52 weeks.
Exhibit Dates: 1/90–Present.
Objectives: To provide introductory training in algebra and geometric forms.
Instruction: This individual, self-instruction course in mathematics includes number systems, fractions and percents, algebra, and geometric forms.
Credit Recommendation: In the vocational certificate category, 3 semester hours in fundamentals of mathematics (technical mathematics) (6/92).

MC-1401-0004

INTRODUCTION TO MARINE CORPS ACCOUNTING BY CORRESPONDENCE

Course Number: 34.10a.
Location: Marine Corps Institute, Washington, DC.
Length: Maximum, 52 weeks.
Exhibit Dates: 1/90–1/97.
Objectives: To teach the basic terminology and methods of accounting and to permit the student to obtain an overview of the Marine Corps financial management organizational structure.
Instruction: The course presents journalizing, posting, and preparation of financial statements. The organizational structure of financial management of the Marine Corps is examined including comptroller organization at posts and stations. Other topics include the system of accounting, directives and files, types and status of appropriated funds, the general ledger account structure, and financial structure data elements.

Credit Recommendation: In the vocational certificate category, 1 semester hour in bookkeeping (8/84).

MC-1401-0006

ADVANCED DISBURSING

Course Number: *Version 1:* 34B. *Version 2:* 34B.
Location: *Version 1:* Service Support School, Cp. Lejeune, NC. *Version 2:* Financial Management and Personnel Administration School, Cp. Lejeune, NC.
Length: *Version 1:* 7 weeks (245 hours). *Version 2:* 11-12 weeks (351-421 hours).
Exhibit Dates: *Version 1:* 10/94–Present. *Version 2:* 1/90–9/94.
Learning Outcomes: *Version 1:* Upon completion of the course, the student will be able identify, interpret, and validate travel entitlements and orders including distance and accounting for travel time; process travel vouchers; use control logs and procedures to verify accounting data; effectively use internal control measures to minimize travel-related fraud; identify the responsibilities of a disbursing officer; prepare letters of appointment; maintain US treasury checking accounts, cash funds, collections, disbursements and deposits; and compute and validate pay, allowances, and deductions utilizing automated equipment. *Version 2:* Upon completion of the course, the student will be able identify, interpret, and validate travel entitlements and orders include distance and accounting for travel time; process travel vouchers; use control logs and procedures to verify accounting data; effectively use internal control measures to minimize travel-related fraud; identify the responsibilities of a disbursing officer; prepare letters of appointment; maintain US treasury checking accounts, cash funds, collections, disbursements, and deposits; compute and validate pay, allowances, and deductions utilizing automated equipment.
Instruction: *Version 1:* Course includes lectures, computer-assisted instruction, and practical exercises in the duties of a disbursing office and includes demonstrations and an extensive laboratory/simulation phase in field disbursing office environments. *Version 2:* Course covers lectures and practical exercises in the duties of a disbursing office and includes demonstrations and an extensive laboratory/simulation phase in field disbursing office environments.
Credit Recommendation: *Version 1:* In the lower-division baccalaureate/associate degree category, 1 semester hour in computer software applications and 3 semester hours in clerical procedures (7/95). *Version 2:* In the vocational certificate category, 3 semester hours in clerical procedures (9/87).

MC-1401-0007

SUBSISTENCE SUPPLY

Course Number: 303.
Location: Service Support School, Cp. Lejeune, NC.
Length: 5 weeks (126-127 hours).
Exhibit Dates: 1/90–7/95.
Learning Outcomes: Upon completion of the course, the student will be able to perform basic mathematics functions, operate electronic calculators, use food service manuals and forms, maintain inventory control cards, and type letters and messages for management use.
Instruction: Course covers basic mathematics, calculators, dining and storeroom accounting, inventory procedures, subsistence receipts, and forms used in support of dining room facilities for large organizations. Methodology includes lectures, discussion, and use of practical applications.
Credit Recommendation: In the lower-division baccalaureate/associate degree category, 3 semester hours in office procedures (7/95).

MC-1401-0008

BASIC PAY ENTITLEMENT BY CORRESPONDENCE

Course Number: 34.21a.
Location: Marine Corps Institute, Washington, DC.
Length: Maximum, 52 weeks.
Exhibit Dates: 1/90–Present.
Learning Outcomes: Upon completion of the course, the student will be able to record, update, validate, credit, and compute basic pay entitlements for members of the Marine Corps.
Instruction: This is a self-instruction course and includes determination of creditable and noncreditable service for pay purposes, the various entitlements and types of basic pay, conditions for starting and stopping pay, computation of leave, maintenance of pay records during periods of absence, and the procedures for reporting absences.
Credit Recommendation: In the vocational certificate category, 1 semester hour in payroll accounting (6/89).

MC-1401-0009

1. PERSONAL FINANCIAL MANAGEMENT BY CORRESPONDENCE
2. PERSONAL FINANCE BY CORRESPONDENCE

Course Number: *Version 1:* 34.20d. *Version 2:* 34.20c.
Location: Marine Corps Institute, Washington, DC.
Length: *Version 1:* Maximum, 52 weeks. *Version 2:* Maximum, 52 weeks.
Exhibit Dates: *Version 1:* 8/93–Present. *Version 2:* 1/90–7/93.
Learning Outcomes: *Version 1:* Upon completion of the course, the students will be able to assist in managing their own personal financial matters and military benefits. *Version 2:* Upon completion of the course, the student will be able to prepare a personal budget, identify the services afforded by banks and other financial institutions, use proven concepts in making personal major asset purchases, identify information on leave and earnings statements, and assess the process used in pay computation.
Instruction: *Version 1:* This correspondence course has workbook, self-scoring examination, and Marine Corps Institute graded lessons and final examination. Topics include personal financial planning, use of personal checking accounts, and credit decisions. *Version 2:* Course covers budget formulation and analysis, personal financial planning, use of personal checking accounts, use of services afforded by financial institutions, major asset purchases and credit decision, and gross and net pay computation.
Credit Recommendation: *Version 1:* In the lower-division baccalaureate/associate degree category, 1 semester hour in personal finance (12/96). *Version 2:* In the lower-division baccalaureate/associate degree category, 1 semester hour in personal finance (6/89).

MC-1401-0010

FISCAL ACCOUNTING FOR SUPPLY CLERKS BY CORRESPONDENCE

Course Number: 34.11.
Location: Marine Corps Institute, Washington, DC.
Length: Maximum, 52 weeks.
Exhibit Dates: 1/90–Present.
Learning Outcomes: Upon completion of the course, the student will understand operating budgets and maintenance of financial records.
Instruction: This is a self-instruction course that covers budgeting, financial responsibilities, appropriations, requisitions, financial document transmittals, financial reports, and purchasing records.
Credit Recommendation: In the lower-division baccalaureate/associate degree category, 1 semester hour in record keeping (5/91).

MC-1401-0011

FIELD BUDGET FORMULATION BY CORRESPONDENCE

Course Number: 34.12.
Location: Marine Corps Institute, Washington, DC.
Length: Maximum, 52 weeks.
Exhibit Dates: 1/90–Present.
Learning Outcomes: Upon completion of the course, the student will understand the budgetary process of field budgeting and budget formulation, identify various types of costs and techniques of costing requirements, and identify and complete the various Marine Corps forms related to the budgeting and accounting cycle.
Instruction: Instruction includes the basic concepts and principles of field budgeting and Marine Corps regulations and instructions on how to complete the various forms in this activity.
Credit Recommendation: In the lower-division baccalaureate/associate degree category, 1 semester hour in bookkeeping (5/91).

MC-1401-0012

FINANCIAL MANAGEMENT BY CORRESPONDENCE

Course Number: 34.14.
Location: Marine Corps Institute, Washington, DC.
Length: Maximum, 52 weeks.
Exhibit Dates: 1/90–Present.
Learning Outcomes: Upon completion of the course, the student will understand accounting theory, financial statements, and internal audits.
Instruction: This nonresident course covers financial management responsibilities and training, accounting concepts, appropriations, trial balances, reconciliation reports, expense reports, performance statements, audits, internal controls, and vulnerability assessment.
Credit Recommendation: In the lower-division baccalaureate/associate degree category, 1 semester hour in principles of accounting (5/91).

MC-1401-0013

BASIC DISBURSING CLERK

Course Number: 34D.

Location: Service Support School, Cp. Lejeune, NC.
Length: 10 weeks (428-429 hours).
Exhibit Dates: 2/93–Present.
Learning Outcomes: Upon completion of this course, the student will be able to operate a microcomputer, calculator, and a customer information system; locate information in individual records and input information to report transactions; compute base pay, allowances, deductions, and travel entitlements.
Instruction: Methods include lectures, computer-assisted instruction, and practical applications.
Credit Recommendation: In the lower-division baccalaureate/associate degree category, 3 semester hours in clerical procedures (7/95).

MC-1402-0013

COBOL PROGRAMMING
(IBM System 360 (OS) COBOL Programming)
(IBM System/360 (OS) COBOL Programming Entry-level)

Course Number: 395; A-532-0015.
Location: Computer Sciences School, Quantico, VA.
Length: Self-paced, 8 weeks (286 hours).
Exhibit Dates: 1/90–Present.
Objectives: To provide technical education to entry-level personnel in the basic concepts of data processing and the COBOL language.
Instruction: Course includes lectures and practical exercises on ANSI COBOL, including the sort verb, computer programming introduction, IBM S/360 computer concepts, programming techniques, COBOL coding, documentation conventions, IBM Operating System programming, job control language, and testing and debugging techniques.
Credit Recommendation: In the lower-division baccalaureate/associate degree category, 1 semester hour in introduction to data processing, 4 in introduction to COBOL (COBOL I), and 4 in advanced COBOL (COBOL II) (3/88).

MC-1402-0023

COMPUTER OPERATOR
(IBM S/360 OS Computer Operator)
(IBM System 360 Operating System (OS) Operations)

Course Number: 40L.
Location: Computer Sciences School, Quantico, VA.
Length: 4 weeks (122 hours).
Exhibit Dates: 1/90–Present.
Objectives: To train enlisted personnel to operate the IBM system 360 computer running under the IBM operating system (OS).
Instruction: This course consists of lectures and practical exercises in basic computer fundamentals, fundamentals of virtual storage and paging, MUS commands, JESZ commands, MVS IPL operations, and the understanding and operation of the System 4300 CPU, 1052 master console, 2540 card reader/punch, 2420 tape drive, 4245 printer, 3880 storage controller, and the 3380 disk drive.
Credit Recommendation: In the lower-division baccalaureate/associate degree category, 3 semester hours in computer operations (10/86).

MC-1402-0034

FORTRAN PROGRAM SPECIALIST
(IBM System 360 OS FORTRAN Programming)

Course Number: T2H; E5OZX4924 000/RRM; A-532-0031.
Location: Computer Sciences School, Quantico, VA.
Length: Self-paced, 2 weeks (70-90 hours).
Exhibit Dates: 1/90–Present.
Objectives: To train students with a programming background in the uses of the VS FORTRAN 77 programming languages.
Instruction: This is a self-paced course requiring practical exercises in terminal entry coding and debugging and executing application programs using the FORTRAN programming language. Course includes the use of arrays, input/output operations, subprogram linkage, and processing of sequential and direct-access data sets.
Credit Recommendation: In the lower-division baccalaureate/associate degree category, 2 semester hours in computer programming (2/90).

MC-1402-0042

OS/VS2 MVS DATA CONTROL TECHNIQUES
(Data Control Techniques)
(IBM System 360 (OS) Data Control Techniques)

Course Number: 39G.
Location: Computer Sciences School, Quantico, VA.
Length: Self-paced, 6-7 weeks (212-244 hours).
Exhibit Dates: 1/90–Present.
Objectives: To train enlisted personnel in data control techniques for the IBM System 360 computer system utilizing HASP.
Instruction: Lectures and practical exercises provide the student with in-depth knowledge of the IBM OS/VS2 MVS JCL utilities and production job procedures, including catalog procedures, disk space calculation, library maintenance, buffer and blocking optimization, and memory dump readings.
Credit Recommendation: In the lower-division baccalaureate/associate degree category, 3 semester hours in computer operations/programming/systems programming (2/90).

MC-1402-0050

AUTOMATED DATA PROCESSING EQUIPMENT FOR THE FLEET MARINE
(Automated Data Processing Equipment Fleet Marine Force Programmer)

Course Number: 38U.
Location: Computer Sciences School, Quantico, VA.
Length: 4 weeks (127-128 hours).
Exhibit Dates: 1/90–1/97.
Learning Outcomes: Upon completion of the course, COBOL programmers will be able to perform duties as applications programmers for the IBM 4110 system.
Instruction: The course is designed as a basic introduction to the operation and programming of the IBM 4110 computer. Instruction is primarily lecture with hands-on lab experience. Topics covered include 4110 operating characteristics, software installation, utilities, event-driven language and COBOL programming.
Credit Recommendation: In the vocational certificate category, 3 semester hours in computer operations/programming (10/86).

MC-1402-0051

IBM SYSTEM 360 ASSEMBLER LANGUAGE CODING (ENTRY)
(Assembler Language Code (ALC) Programming (Combination))

Course Number: 38R; A-532-0030; E5OXZ4924 002.
Location: Computer Sciences School, Quantico, VA.
Length: 8 weeks (285-320 hours).
Exhibit Dates: 1/90–5/94.
Objectives: To provide technical training for experienced programmers in using assembler language coding as a second programming language or to provide background skills for entry-level systems programming.
Instruction: This course provides a detailed introduction in ALC for IBM-specific equipment via terminal data entry for coding, testing, and debugging and includes, but is not limited to, input and output processing, logical instructions, floating point, macros and subprogramming. Heavy emphasis is placed on JCL, OS, utilities and linkage editor usage, as well as various access methods and file organization techniques including QSAM, BSAM, QISAM, and BDAM.
Credit Recommendation: In the lower-division baccalaureate/associate degree category, 3 semester hours in assembly language coding (10/86); in the upper-division baccalaureate category, 3 semester hours in advanced assembly language coding (10/86).

MC-1402-0052

MULTIPLE VIRTUAL STORAGE (MVS) FUNDAMENTALS AND LOGIC

Course Number: D2C.
Location: Computer Sciences School, Quantico, VA.
Length: 6 weeks (196 hours).
Exhibit Dates: 1/90–1/97.
Objectives: To provide the experienced data processor with knowledge of the Multiple Virtual Storage (MVS) operating system.
Instruction: Lectures include system resource management, job entry subsystem, storage management, data management, diagnostics, supervisor services, service aids and utilities, system generation and maintenance, and system modification program.
Credit Recommendation: In the lower-division baccalaureate/associate degree category, 6 semester hours in introduction to system programming (10/86); in the upper-division baccalaureate category, 3 semester hours in advanced system programming (10/86).

MC-1402-0053

MULTIPLE VIRTUAL STORAGE (MVS) DIAGNOSTICS

Course Number: D2H.
Location: Computer Sciences School, Quantico, VA.
Length: 3 weeks (78 hours).
Exhibit Dates: 1/90–9/93.
Objectives: To provide the experienced system programmer with the basic knowledge and skills required to isolate system problems to the component level.

Instruction: Lectures cover job entry system control blocks and dumps, control block chains, recovery management support, MVS application dumps, SVC dumps, stand-alone dumps, traces/slipper, SMF and LOGREC, and interactive problem control system.

Credit Recommendation: In the upper-division baccalaureate category, 3 semester hours in advanced system programming—diagnostics (10/86).

MC-1402-0054

MULTIPLE VIRTUAL STORAGE (MVS) PERFORMANCE AND TRAINING

Course Number: D2J.
Location: Computer Sciences School, Quantico, VA.
Length: 3 weeks (94 hours).
Exhibit Dates: 1/90–9/93.
Objectives: To provide the experienced system programmer with the basic knowledge and skills required to fine tune an MVS system through performance-oriented reports.
Instruction: Lectures and case studies cover I/O tuning, auxiliary storage manager, system resource manager parameters, SRM guidelines, resource measurement facility, and statistical analysis of RMF data using SAS.
Credit Recommendation: In the upper-division baccalaureate category, 3 semester hours in advanced system programming—performance and tuning (10/86).

MC-1402-0055

1. FISCAL/BUDGET TECHNICIAN
   (Fiscal Accounting)
2. FISCAL ACCOUNTING

Course Number: *Version 1:* 34N. *Version 2:* 34N.
Location: *Version 1:* Service Support School, Cp. Lejeune, NC. *Version 2:* Financial Management and Personnel Administration School, Cp. Lejeune, NC.
Length: *Version 1:* 6 weeks (212-213 hours). *Version 2:* 9 weeks (285 hours).
Exhibit Dates: *Version 1:* 5/94–Present. *Version 2:* 1/90–4/94.
Learning Outcomes: *Version 1:* Upon completion of the course, the student will be able to record basic accounting transactions in accordance with generally accepted accounting principles, post to the ledger and summarize accounting data to final balance, prepare work sheets and end of accounting period tasks, prepare accounting reports, and use a mechanized (computerized) data entry device in accordance with the financial management structure of the Marine Corps. *Version 2:* Upon completion of the course, the student will be able to record basic accounting transactions in accordance with generally accepted accounting principles, post to the ledger and summarize accounting data to final balance, prepare work sheets and end of accounting period tasks, prepare accounting reports, and use a mechanized (computerized) data entry device in accordance with the financial management structure of the Marine Corps.
Instruction: *Version 1:* Course includes lectures, computer-assisted instruction, demonstrations, and illustrative problems in basic accounting procedures, providing students with a working knowledge of formal mechanized accounting techniques and their application to a computer system. *Version 2:* Course includes lectures, demonstrations, and illustrative problems in basic accounting procedures, providing students with a working knowledge of formal mechanized accounting techniques and their application to a computer system.
Credit Recommendation: *Version 1:* In the lower-division baccalaureate/associate degree category, 3 semester hours in applied accounting and 3 semester hours in computer software applications (7/95). *Version 2:* In the vocational certificate category, 3 semester hours in bookkeeping (9/87); in the lower-division baccalaureate/associate degree category, 3 semester hours in data processing and 3 in accounting survey (9/87).

MC-1402-0056

1. COMPUTER TECHNICIAN
2. GROUND COMPUTER TECHNICIAN
   (Tactical General Purpose Computer Technician)

Course Number: *Version 1:* DA9. *Version 2:* DA9.
Location: *Version 1:* Air Ground Combat Center, Twentynine Palms, CA. *Version 2:* Air Ground Combat Center, Twentynine Palms, CA; Communication-Electronics School, Twentynine Palms, CA.
Length: *Version 1:* 10-12 weeks (420 hours). *Version 2:* 17-32 weeks (635-1100 hours).
Exhibit Dates: *Version 1:* 5/93–Present. *Version 2:* 1/90–4/93.
Learning Outcomes: *Version 1:* Upon completion of the course, the student will be able to apply the principles of computer organization and architecture, basic software concepts, computer systems, and computer peripherals to the troubleshooting and repair of two different computer systems and associated peripherals. *Version 2:* Upon completion of the course, the student will be able to apply the principles of computer organization and architecture, basic software concepts, computer systems, and computer peripherals to the troubleshooting and repair of two different computer systems and associated peripherals.
Instruction: *Version 1:* Lectures, practical exercises, and hands-on laboratory work, develop the knowledge and skills necessary to operate, diagnose, and repair a computer system and peripherals. *Version 2:* Lectures and practical exercises on topics related to computer organization, troubleshooting, and repair.
Credit Recommendation: *Version 1:* In the lower-division baccalaureate/associate degree category, 5 semester hours in computer systems/microprocessors and 3 in computer troubleshooting (3/96). *Version 2:* In the lower-division baccalaureate/associate degree category, 5 semester hours in computer systems and organization and 3 in computer systems troubleshooting and maintenance (1/93).

MC-1402-0057

MARINE CORPS INTEGRATED MAINTENANCE MANAGEMENT SYSTEMS (MIMMS)

Course Number: None.
Location: Logistics Base, Albany, GA.
Length: 4 weeks (126 hours).
Exhibit Dates: 1/90–Present.
Learning Outcomes: Upon completion of the course, the student will be able to supervise the maintenance and upkeep of materials and machinery; establish systems and control for periodic maintenance, repair, and inventory; and provide management leadership for the unit.
Instruction: Lectures and practical exercises in maintenance management information systems, maintenance records, information systems, maintenance records, production and resources, supply support, leadership and management, and Marine Corps procedures.
Credit Recommendation: In the lower-division baccalaureate/associate degree category, 1 semester hour in computer applications and 3 in material management (11/91).

MC-1402-0058

BASIC LOGISTICS/EMBARKATION SPECIALIST

Course Number: 04H; G-551-4407; H-551-3555.
Location: Amphibious Base, Little Creek, VA; Landing Force Training Command, Pacific, San Diego, CA.
Length: 3 weeks (112-116 hours).
Exhibit Dates: 8/92–Present.
Learning Outcomes: Upon completion of the course, the student will possess basic administrative, clerical embarkation, and logistics knowledge and skills; use computers to determine data, calculate loading, develop and maintain data base, and load software programs for logistical tasks; use MS-DOS and military-specific software systems including one support system and two management systems; prepare correspondence and messages; maintain publication libraries and correspondence files; and prepare templates and diagrams for loading of cargo in ship or in vehicles.
Instruction: Course includes lectures, conferences, practical exercises, and computer-assisted laboratory covering computer-assisted logistics, developing loading and embarkation data, transportation planning, administrative and clerical functions, radio use, and procedures in computer operations.
Credit Recommendation: In the lower-division baccalaureate/associate degree category, 2 semester hours in computer operations and 1 in clerical procedures (11/93).

MC-1402-0059

MARINE AIR-GROUND TASK FORCE DEPLOYMENT SUPPORT SYSTEM/COMPUTER-AIDED EMBARKATION MANAGEMENT SYSTEM (Introduction to MDSS II/CAEMS)

Course Number: H-551-3554.
Location: Landing Force Training Command, Pacific, San Diego, CA.
Length: 2 weeks (75 hours).
Exhibit Dates: 5/93–Present.
Learning Outcomes: Upon completion of the course, the student will be able to perform basic tasks in the loading of amphibious ships by using computer-aided programs and demonstrate basic computer operations.
Instruction: Course includes lectures, audiovisuals, computer-aided instruction, and hands-on learning experiences. Topics covered are basic computer operations including introduction to MS-DOS, preparation of computerized ship loading plans, and use of Paradox and Auto CAD computer systems.
Credit Recommendation: In the lower-division baccalaureate/associate degree category, 2 semester hours in basic computers and 1 in basic Auto CAD (11/93).

MC-1402-0060

TEAM EMBARKATION OFFICER/ASSISTANT

Course Number: OD1; H-8A-3551.
Location: Landing Force Training Command, Pacific, San Diego, CA.
Length: 3-4 weeks (126 hours).
Exhibit Dates: 8/92–Present.
Learning Outcomes: Upon completion of the course, the student will be able to load amphibious ships; use computers to determine data, calculate loading, plan the loading of ships, and perform logistical tasks; use MS-DOS and three military-specific software systems to prepare automated documentation required for ship loading plan and build and maintain data base of equipment and personnel for embarkation; operate data collection devices and collect data for inventory and deployment purposes; and create embarkation reports for the loading plan.
Instruction: Course includes lectures, conferences, computer-assisted laboratory tasks, and practical exercises to develop and refine the skills necessary for effective functioning as an embarkation officer (or assistant). Major topics include preliminary planning, load planning, computer-assisted logistics calculation and embarkation management system, and deployment support system.
Credit Recommendation: In the lower-division baccalaureate/associate degree category, 2 semester hours in computer operations, 1 in record keeping, and 1 in report writing (11/93).

MC-1402-0061

TEAM EMBARKATION OFFICER/ASSISTANT
(Reserve Embarkation)

Course Number: H-8A-3550.
Location: Landing Force Training Command, Pacific, San Diego, CA.
Length: 2 weeks (71 hours).
Exhibit Dates: 5/94–Present.
Learning Outcomes: Upon completion of the course, the student will be able to load amphibious ships; use computer to determine data, calculate loading, plan loading of ships, and perform logistical tasks; use MS-DOS and three military-specific software systems to prepare automated documentation required for ship loading plan, build and maintain data base of equipment and personnel for embarkation, operate data collection devices, and collect data for inventory and deployment purposes; and create embarkation reports for the loading plan.
Instruction: Course includes lectures, conferences, computer-assisted laboratory task, and practical exercises to develop and refine the skills necessary for effective functioning as an embarkation officer (or assistant). Major topics include preliminary planning, load planning, computer-assisted logistics calculation and embarkation management system, and deployment support system.
Credit Recommendation: In the lower-division baccalaureate/associate degree category, 2 semester hours in computer operations and 1 in report writing (11/93).

MC-1402-0062

RESERVE FIELD RADIO OPERATOR REFRESHER

Course Number: 2EK.
Location: Air Ground Combat Center, Twentynine Palms, CA.
Length: 2 weeks (63 hours).
Exhibit Dates: 10/95–Present.
Learning Outcomes: Upon completion of the course, the student will be able to set up and operate mobile radio communication equipment. NOTE: This course is a refresher course for reserve personnel. No student evaluations are given during the course.
Instruction: Lecture, computer aided instruction, and practical hands-on instruction covers set up and utilization of radio communication equipment.
Credit Recommendation: Credit is not recommended because of the limited, specialized nature of the course (3/96).

MC-1402-0063

MOBILE TECHNICAL CONTROL

Course Number: DRJ.
Location: Air Ground Combat Center, Twentynine Palms, CA.
Length: 9-10 weeks (350 hours).
Exhibit Dates: 4/95–Present.
Learning Outcomes: Upon completion of the course, the student will be able to plan and conduct communication circuit analysis and restoration, and utilize tools and test equipment for communication circuit/component/systems troubleshooting. Emphasis is placed upon system engineering and troubleshooting, equipment, isolation and restoration.
Instruction: Course includes lectures, demonstrations and practical exercises in communication concepts and structure, basic digital theory, and multiplexing.
Credit Recommendation: In the lower-division baccalaureate/associate degree category, 3 semester hours in communications circuits and 2 in communications circuits laboratory (3/96).

MC-1402-0064

RESERVE RADIO CHIEF REFRESHER

Course Number: R2E.
Location: Air Ground Combat Center, Twentynine Palms, CA.
Length: 2 weeks (67-68 hours).
Exhibit Dates: 7/95–Present.
Learning Outcomes: Upon completion of the course, the student will be able to demonstrate the operation of radio and automated planning systems, and be able to run and operate basic computer application packages.
Instruction: Course includes lecture and practical application using single channel radio and computer based automated planning systems.
Credit Recommendation: Credit is not recommended because of the limited, specialized nature of the course (3/96).

MC-1402-0065

SYSTEM COMPUTER TECHNICIAN

Course Number: DRG.
Location: Air Ground Combat Center, Twentynine Palms, CA.
Length: 10-11 weeks (406 hours).
Exhibit Dates: 5/93–Present.
Learning Outcomes: Upon completion of the course, the student will be able to demonstrate knowledge of basic computer systems and additional skills include troubleshooting, operation, and machine level programming of this same computer system.
Instruction: Lectures, practical exercise, and hands-on laboratory are used to develop the knowledge and skills necessary to operate, diagnose, and repair computer systems.
Credit Recommendation: In the lower-division baccalaureate/associate degree category, 5 semester hours in computer systems/microprocessors and 3 in computer troubleshooting (3/96).

MC-1402-0067

RESERVE OPERATIONAL COMMUNICATION CHIEF REFRESHER

Course Number: R2D.
Location: Air Ground Combat Center, Twentynine Palms, CA.
Length: 2 weeks (67-68 hours).
Exhibit Dates: 12/95–Present.
Learning Outcomes: Upon completion of the course, the student will be able to supervise the operation of Marine Corps communication equipment, facilities, and agencies.
Instruction: Lecture, practical application, and demonstrated proficiency is used to develop additional student skills in supervising communication equipment usage.
Credit Recommendation: Credit is not recommended because of the limited, specialized nature of the course (3/96).

MC-1402-0068

AN/MSC-63A SYSTEM MANAGER

Course Number: L3Z.
Location: Air Ground Combat Center, Twentynine Palms, CA.
Length: 3 weeks (105 hours).
Exhibit Dates: 8/95–Present.
Learning Outcomes: Upon completion of the course, the student will be able to format, send, receive, reformat, and dispose of encoded messages.
Instruction: Lecture and operation of automated messaging equipment, message development, modification, and disposal are used.
Credit Recommendation: Credit is not recommended because of the limited, specialized nature of the course (3/96).

MC-1403-0009

BASIC TRAVEL CLERK

Course Number: 34A.
Location: Financial Management and Personnel Administration School, Cp. Lejeune, NC.
Length: 7-8 weeks (224-252 hours).
Exhibit Dates: 1/90–2/90.
Learning Outcomes: Upon completion of the course, the student will be able to perform the various duties involved in travel procedures and to perform the clerical duties necessary to ensure conformity with travel regulations.
Instruction: Course includes preparing travel vouchers, including verifying travel status and computing distance, per diem expenses, and travel allowances; a review of basic mathematics; a brief orientation to the electronic calculator; and the real FAMMIS travel system equipment.
Credit Recommendation: In the vocational certificate category, 2 semester hours in clerical procedures (9/87).

MC-1403-0010

PERSONAL FINANCIAL RECORDS CLERK

Course Number: 34M.
Location: Financial Management and Personnel Administration School, Cp. Lejeune, NC.
Length: 8-9 weeks (272-289 hours).
Exhibit Dates: 1/90–3/95.
Learning Outcomes: Upon completion of the course, the student will be able to set up and maintain a payroll accounting system that includes the basic concepts, fundamentals, and principles of payroll disbursement; operate a ten-key electronic calculator; and operate computerized information and video inquiry systems.
Instruction: Course covers use of computer and software applications such as DOS, spread sheets, word processing, data bases, and electronic calculator; payroll disbursement procedures such as basic pay, various allowances, leave entitlements, increases and decreases in pay brought about by extra military duties, court martial penalties, and separation payments. Methodology includes lectures, demonstrations, and practical exercises.
Credit Recommendation: In the vocational certificate category, 2 semester hours in clerical procedures (4/89); in the lower-division baccalaureate/associate degree category, 2 semester hours in payroll accounting and 2 in computer concepts and software applications (4/89).

MC-1403-0011

INDEPENDENT DUTY ADMINISTRATION

Course Number: *Version 1:* AAE. *Version 2:* AAE.
Location: *Version 1:* Service Support School, Cp. Lejeune, NC. *Version 2:* Financial Management and Personnel Administration School, Cp. Lejeune, NC.
Length: *Version 1:* 2 weeks (70 hours). *Version 2:* 3-4 weeks (125-131 hours).
Exhibit Dates: *Version 1:* 12/94–Present. *Version 2:* 1/90–11/94.
Learning Outcomes: *Version 1:* Upon completion of the course, the student will be able to perform general administrative tasks, prepare drill accounting reports, perform pay related matters, and maintain individual records. *Version 2:* Upon completion of the course, the student will be able to perform tasks general administrative tasks, including reports on Marine Corps reserve officers allowances, composite scores, fitness reports, accounting reports, training reports, reserve manpower management pay system, service records, support requirements, and correspondence and directives.
Instruction: *Version 1:* Course includes lecture, demonstration, practical application, and critique. *Version 2:* Lectures and practice applications are used to train students in the performance of general administrative tasks, including preparation of reports, correspondence, and directives.
Credit Recommendation: *Version 1:* In the lower-division baccalaureate/associate degree category, 1 semester hour in clerical procedures (7/95). *Version 2:* In the vocational certificate category, 3 semester hours in clerical procedures (9/87).

MC-1403-0013

UNIT DIARY CLERK

Course Number: *Version 1:* 01S. *Version 2:* 01S.
Location: *Version 1:* Service Support School, Cp. Lejeune, NC. *Version 2:* Schools Battalion, Cp. Pendleton, CA; Financial Management and Personnel Administration School, Cp. Lejeune, NC.
Length: *Version 1:* 5 weeks (214 hours). *Version 2:* 8-9 weeks (269-305 hours).
Exhibit Dates: *Version 1:* 10/94–Present. *Version 2:* 1/90–9/94.
Learning Outcomes: *Version 1:* Upon completion of the course, the student will be able to prepare and maintain unit diaries, write letters and reports, audit unit diary management reports, and enter and access all such records on the computer. *Version 2:* Upon completion of the course, the student will be able to prepare and maintain unit diaries, write letters and reports, and audit unit diary management reports.
Instruction: *Version 1:* Methodologies include lectures, computer-assisted instruction, practical applications, and critiques. *Version 2:* Lectures and extensive practical workshops cover special assignments (typing), unit diary preparation, audit of unit diary management documents, joint uniform pay system-manpower management system, allotments, and systems feedback.
Credit Recommendation: *Version 1:* In the lower-division baccalaureate/associate degree category, 2 semester hours in basic keyboarding and 1 in clerical procedures (7/95). *Version 2:* In the vocational certificate category, 1 semester hour in record keeping and 3 in clerical procedures (11/87); in the lower-division baccalaureate/associate degree category, 2 semester hours in typing (11/87).

MC-1403-0014

ADMINISTRATIVE CLERK

Course Number: *Version 1:* 01T. *Version 2:* 01T.
Location: *Version 1:* Service Support School, Cp. Lejeune, NC. *Version 2:* Financial Management and Personnel Administration School, Cp. Lejeune, NC; Financial Management and Personnel Administration School, Cp. Pendleton, CA.
Length: *Version 1:* 8 weeks (305-306 hours). *Version 2:* 7-8 weeks (254 hours).
Exhibit Dates: *Version 1:* 10/94–Present. *Version 2:* 1/90–9/94.
Learning Outcomes: *Version 1:* Upon completion of the course, the student will be able to perform entry-level duties in general administration, enter information into the computer on various kinds of correspondence using correct military format, maintain a correspondence filing system, and demonstrate a keyboarding speed of 25 words per minute. *Version 2:* Upon completion of the course, the student will be able to perform entry level duties in general administration, type various kinds of correspondence using correct military format, maintain a correspondence filing system, and demonstrate a typing speed of 25 words per minute.
Instruction: *Version 1:* Methods include lecture aided by transparencies and white board, computer-assisted instruction, and practical computer application. *Version 2:* This is a course of specialized instruction in typing naval correspondence, maintaining correspondence files, and preparing military directives. In addition, instruction is presented on general administrative matters, including the preparation of fitness reports, leave authorization, identification cards, and unit punishment books.
Credit Recommendation: *Version 1:* In the lower-division baccalaureate/associate degree category, 2 semester hours in basic keyboarding, 1 in record keeping, and 1 in clerical procedures (7/95). *Version 2:* In the vocational certificate category, 1 semester hour in record keeping and 2 in clerical procedures (9/87); in the lower-division baccalaureate/associate degree category, 2 semester hours in typing (9/87).

MC-1403-0015

SENIOR CLERK

Course Number: *Version 1:* AAC. *Version 2:* AAC.
Location: Service Support School, Cp. Lejeune, NC.
Length: *Version 1:* 7-8 weeks (258 hours). *Version 2:* 9-10 weeks (352 hours).
Exhibit Dates: *Version 1:* 9/94–Present. *Version 2:* 1/90–8/94.
Learning Outcomes: *Version 1:* Upon completion of the course, the student will be able to edit correspondence; supervise the administrative procedures required for promotions, reductions, fitness reports, and casualty assistance calls; supervise a records management program; produce a unit diary; supervise administration of military pay and allowances; verify the administration of such legal issues as absenteeism, punishments, courts-martial, suspensions and confinements; process separations; audit and update directives; and use computerized information and video inquiry systems. *Version 2:* Upon completion of the course, the student will be able to edit correspondence; supervise the administrative procedures required for promotions, reductions, fitness reports, and casualty assistance calls; supervise a records management program; produce a unit diary; supervise administration of military pay and allowances; verify the administration of legal issues such as absenteeism, punishments, courts-martial, suspensions and confinements; process separations; audit and update directives; and use computerized information and video inquiry systems.
Instruction: *Version 1:* Methodology includes lectures and practical applications involving computer data entry, retrieval, and manipulation; computer-assisted instruction; and videotapes. *Version 2:* Course covers correspondence and directives, individual records, general administration, records management, personnel reporting, military pay and allowances, legal administration, separations, directive maintenance, and computer and video inquiry systems. Methodology includes lectures, practical applications, computer-assisted instruction, and videotapes.
Credit Recommendation: *Version 1:* In the lower-division baccalaureate/associate degree category, 3 semester hours in office administration and 1 in computer software applications (7/95). *Version 2:* In the lower-division baccalaureate/associate degree category, 3 semester hours in office administration and 2 in computer concepts and software applications (4/89).

MC-1403-0016

INDIVIDUAL PERSONNEL RECORDS BY CORRESPONDENCE

Course Number: 01.35g.

Location: Marine Corps Institute, Washington, DC.
Length: Maximum, 52 weeks.
Exhibit Dates: 1/90–Present.
Learning Outcomes: Upon completion of the course, the student will be able to use available legal references; perform the basic duties of a legal clerk; and identify reference material pertaining to nonjudicial punishment, as well as what constitutes absentee and desertion.
Instruction: This correspondence course covers use of LEGADMINMAN documentation of police, procedure guidelines for legal administration; use of manual for Judge Advocate General, use of manual for court-martial, uniform code of military justice, and absentee and deserters.
Credit Recommendation: In the vocational certificate category, 1 semester hour in legal record keeping (6/89).

MC-1403-0017

VALIDATION AND RECONCILIATION CLERK BY CORRESPONDENCE

(Validation and Reconciliation Clerk by Correspondence)

Course Number: 04.15-2; 04.15-1.
Location: Marine Corps Institute, Washington, DC.
Length: Maximum, 52 weeks.
Exhibit Dates: 2/90–Present.
Learning Outcomes: Upon completion of the course, the student will be able to maintain Marine Corps supply and maintenance records.
Instruction: This is a self-instruction course that covers daily and biweekly validation of maintenance records and analyzing supply status through biweekly reconciliation of maintenance forms.
Credit Recommendation: Credit is not recommended because of the military-specific nature of the course (5/91).

MC-1403-0018

FIELD RADIO OPERATOR

Course Number: *Version 1:* 25U. *Version 2:* 25U.
Location: Air Ground Combat Center, Twentynine Palms, CA.
Length: *Version 1:* 8 weeks (350 hours). *Version 2:* 8 weeks (280 hours).
Exhibit Dates: *Version 1:* 10/95–Present. *Version 2:* 2/91–9/95.
Learning Outcomes: *Version 1:* Upon completion of the course, the student will be able to install, operate, perform minor maintenance on radar communication equipment, determine appropriate radio telephone procedures, publications directives, security regulations, and application testing. *Version 2:* Upon completion of the course, the student will be able to install, operate, and perform minor maintenance on a variety of radio sets. Student will be able to determine types of antennas to be used, calculate their dimensions, and build and install them using common electrical safety practices.
Instruction: *Version 1:* Course includes lectures, demonstrations, and performance exercises in application, installation, and adjustment of radio sets; basic radio telephone procedures; and determining antenna lengths, fabricating them, and installing them. *Version 2:* Course includes lectures, demonstrations, and performance exercises in application, installation, and adjustment of radio sets; basic radio telephone procedures; and determining antenna lengths, fabricating them, and installing them.
Credit Recommendation: *Version 1:* In the lower-division baccalaureate/associate degree category, 1 semester hour in radio communications technology and 1 in radio communications technology laboratory (3/96). *Version 2:* In the lower-division baccalaureate/associate degree category, 1 semester hour in radio communications technology (1/93).

MC-1405-0021

WAREHOUSING OPERATIONS BY CORRESPONDENCE

Course Number: 30.3h.
Location: Marine Corps Institute, Washington. DC.
Length: Maximum, 52 weeks.
Exhibit Dates: 1/90–1/97.
Objectives: To give added depth to materials presented in MCI 30.1k, especially in the area of storage space and packaging.
Instruction: The material presented in the course includes storage, material handling, and packaging techniques. Also included are specific storage techniques for lumber and ammunition. Course concludes with a proctored examination.
Credit Recommendation: In the vocational certificate category, 2 semester hours in warehousing (8/84); in the lower-division baccalaureate/associate degree category, 2 semester hours in physical distribution or transportation management or, if taken in conjunction with 30.1K, Basic Warehousing by Correspondence (MC-1405-0020), 3 semester hours in physical distribution or transportation management (8/84).

MC-1405-0025

ACCOUNTING FOR PLANT PROPERTY BY CORRESPONDENCE

Course Number: 34.5c.
Location: Marine Corps Institute, Washington, DC.
Length: Maximum, 52 weeks.
Exhibit Dates: 1/90–1/97.
Objectives: To enable the student to establish and maintain plant property ledgers and to prepare plant property reports required in the Marine Corps plant account system.
Instruction: Topics include the plant property accounting cycle, classes of plant property, responsibilities for plant property accounting, cost criteria, and determination of costs. A major area is that of maintaining plant property records and preparing the plant account report.
Credit Recommendation: In the lower-division baccalaureate/associate degree category, 1 semester hour in property accounting (8/84).

MC-1405-0027

SUPPLY MANAGEMENT BY CORRESPONDENCE

Course Number: 30.20.
Location: Marine Corps Institute, Washington, DC.
Length: Maximum, 52 weeks.
Exhibit Dates: 1/90–1/97.
Objectives: To provide the supply chief with an understanding of the functions necessary to organize and manage a supply office.
Instruction: This correspondence course familiarizes the supply chief with the managerial functions of planning, organizing, directing, and controlling. In addition, it defines methods for identifying the proper problem solving techniques through steps ranging from problem identification through the use of feedback in measuring goal achievement.
Credit Recommendation: In the lower-division baccalaureate/associate degree category, 1 semester hour in principles of management (8/84).

MC-1405-0031

ADJUTANT

Course Number: *Version 1:* AAF. *Version 2:* AAF.
Location: *Version 1:* Service Support School, Cp. Lejeune, NC. *Version 2:* Financial Management and Personnel Administration School, Cp. Lejeune, NC.
Length: *Version 1:* 6 weeks (204 hours). *Version 2:* 6 weeks (202-219 hours).
Exhibit Dates: *Version 1:* 10/94–Present. *Version 2:* 1/90–9/94.
Learning Outcomes: *Version 1:* Upon completion of the course, the student will be able to perform legal tasks in general administration, management of correspondence and files related to personnel pay and allowance, and personal records. *Version 2:* Upon completion of the course, the student will be able to perform functions of an adjutant including general administration, management of correspondence and directives, supervision of military pay and allowances, personal records, and legal administration of a command.
Instruction: *Version 1:* Methods include lecture, computer-assisted instruction, demonstration, practical applications, and critique. *Version 2:* The lecture method and practical applications are used as teaching methods. Topics include correspondence and directives, military pay and allowances, general administration, individual records, separations, personnel assets, and legal administration.
Credit Recommendation: *Version 1:* In the lower-division baccalaureate/associate degree category, 3 semester hours in office administration and 1 in business communication (7/95). *Version 2:* In the lower-division baccalaureate/associate degree category, 1 semester hour in business communications and 3 in office administration (3/88).

MC-1405-0032

ADVANCED PERSONNEL ADMINISTRATION

Course Number: *Version 1:* AAD. *Version 2:* AAD.
Location: *Version 1:* Service Support School, Cp. Lejeune, NC. *Version 2:* Financial Management and Personnel Administration School, Cp. Lejeune, NC.
Length: *Version 1:* 6 weeks (230 hours). *Version 2:* 9 weeks (338 hours).
Exhibit Dates: *Version 1:* 4/94–Present. *Version 2:* 1/90–3/94.
Learning Outcomes: *Version 1:* Upon completion of the course, the student will be able to prepare, process, and edit all types of correspondence; verify the maintenance of a directives control point and completeness of correspondence files; process command-issued directives; maintain personnel records accountability; verify the audit of personnel records; verify personnel reports, process pay-related documents, and advise personnel on pay matters; verify the processing of personnel forms;

and generate and access computerized personnel records. *Version 2:* Upon completion of the course, the student will be able to produce a unit diary; supervise the administration of military pay and allowances; serve as the unit administrative chief; edit correspondence and directives; verify the maintenance of correspondence files and directive systems; verify assignment eligibility and audit individual records; review processing requirements for absentees, nonjudicial punishments, courts-martial, vacations, suspensions, or confinements; and verify separation requirements and retirement and resignation eligibility.

Instruction: *Version 1:* Methods include lectures, practical applications, computer-assisted instruction, and critiques. *Version 2:* Course includes lectures, practical applications, computer-assisted instruction, critiques, videotapes, and audio cassette materials in personnel reporting, military pay and allowances, general administration, correspondence, directives, file maintenance, personnel assets, individual records, legal administration, and separation.

Credit Recommendation: *Version 1:* In the lower-division baccalaureate/associate degree category, 3 semester hours in office procedures (7/95). *Version 2:* In the lower-division baccalaureate/associate degree category, 3 semester hours in office administration (9/87).

MC-1405-0033

PERSONNEL OFFICER

Course Number: *Version 1:* 011. *Version 2:* 011.

Location: *Version 1:* Service Support School, Cp. Lejeune, NC. *Version 2:* Financial Management and Personnel Administration School, Cp. Lejeune, NC.

Length: *Version 1:* 7-8 weeks (261 hours). *Version 2:* 7 weeks (249-250 hours).

Exhibit Dates: *Version 1:* 10/94–Present. *Version 2:* 1/90–9/94.

Learning Outcomes: *Version 1:* Upon completion of the course, the student will be able to write and edit military correspondence, business letters, and directives; supervise the administration of military pay systems; supervise the administration of promotions, fitness reports, casualty reports, and casualty assistance programs; supervise the maintenance of personnel records, separations, personal assets, and processing requirements for nonjudicial punishments. *Version 2:* Upon completion of the course, the student will be able to write and edit military correspondence, business letters, and directives; supervise the administration of military pay systems; supervise the administration of promotions, fitness reports, casualty reports, and casualty assistance programs; supervise the maintenance of personnel records, separations, personal assets, and the processing of nonjudicial punishments.

Instruction: *Version 1:* Methods used include lectures, practical applications, critiques, and computer-assisted instruction. *Version 2:* Methods used include lectures, practical applications, critiques, audio cassette tapes, videotapes (films), and computer-assisted instruction. Topics include drafting of correspondence, procedures for military pay and allowances, administrative techniques, maintenance of records and personnel reports, casualty assistance programs, resignations, retirements, personnel assistance programs, and legal administration of nonjudicial punishments.

Credit Recommendation: *Version 1:* In the lower-division baccalaureate/associate degree category, 3 semester hours in office administration and 2 in principles of supervision (7/95). *Version 2:* In the lower-division baccalaureate/associate degree category, 3 semester hours in office administration, 1 in business communication, and 2 in principles of supervision (9/87).

MC-1405-0034

PERSONNEL CLERK

Course Number: *Version 1:* 01C. *Version 2:* 0IC.

Location: *Version 1:* Service Support School, Cp. Lejeune, NC. *Version 2:* Financial Management and Personnel Administration School, Cp. Lejeune, NC; Schools Battalion, Cp. Pendleton, CA.

Length: *Version 1:* 7 weeks (261 hours). *Version 2:* 7-8 weeks (229-282 hours).

Exhibit Dates: *Version 1:* 10/94–Present. *Version 2:* 1/90–9/94.

Learning Outcomes: *Version 1:* Upon completion of the course, the student will be able to keyboard at a speed of at least 25 wpm, type correspondence and military reports, maintain computerized service records, and prepare and process personnel separation/retention documents on the computer. *Version 2:* Upon completion of the course, the student will be able to type at a speed of at least 25 wpm, type correspondence and necessary military reports, maintain service records, and prepare and process personnel separation/retention documents.

Instruction: *Version 1:* Methods of instruction include computer-assisted keyboarding instruction, critiques, videotapes, lectures, practical applications, and self-paced instruction. *Version 2:* Methods of instruction include automated typing instruction, critiques, videotapes, lectures, practical applications, and self-paced instruction. Topics covered are typing power drills, typing correspondence and reports, the maintenance of service records, audit of computer-generated documents, and the processing of separation/retention documents.

Credit Recommendation: *Version 1:* In the lower-division baccalaureate/associate degree category, 3 semester hours in basic keyboarding (7/95). *Version 2:* In the lower-division baccalaureate/associate degree category, 2 semester hours in basic typing (10/87).

MC-1405-0035

ENLISTED SUPPLY BASIC
(Basic Supply Stock Control)

Course Number: *Version 1:* 30V. *Version 2:* 30V.

Location: Service Support School, Cp. Lejeune, NC.

Length: *Version 1:* 7 weeks (240 hours). *Version 2:* 7 weeks (232 hours).

Exhibit Dates: *Version 1:* 1/94–Present. *Version 2:* 1/90–12/93.

Learning Outcomes: *Version 1:* Upon completion of the course, the student will be able to perform basic inventory and supply control procedures, operate personal computers using elementary word processing and spread sheet applications, file and maintain supply records, maintain fiscal ledgers, and complete financial reports. *Version 2:* Upon completion of the course, the student will be able to perform basic inventory and supply control procedures, operate personal computers using elementary word processing and spread sheet applications, and file and maintain supply records.

Instruction: *Version 1:* Course covers technical skills for supply clerks utilizing computer software and hardware within a basic record keeping process. Methodology includes lecture, practical applications, and computer-assisted instruction. *Version 2:* Course covers technical skills for supply clerks utilizing computer software and hardware within a basic record keeping process. Methodology includes lecture and practical applications.

Credit Recommendation: *Version 1:* In the lower-division baccalaureate/associate degree category, 1 semester hour in computer software applications and 3 in records management (7/95). *Version 2:* In the lower-division baccalaureate/associate degree category, 1 semester hour in computer literacy, 3 in filing and records control, and 1 in keyboarding (4/89).

MC-1405-0036

ENLISTED WAREHOUSING INTERMEDIATE

Course Number: *Version 1:* 30H. *Version 2:* 30H.

Location: Service Support School, Cp. Lejeune, NC.

Length: *Version 1:* 5 weeks (172-173 hours). *Version 2:* 5 weeks (164 hours).

Exhibit Dates: *Version 1:* 12/94–Present. *Version 2:* 1/90–11/94.

Learning Outcomes: *Version 1:* Upon completion of the course, the student will be able to perform general warehousing functions, use associated data processing hardware, and monitor an inventory control system. *Version 2:* Upon completion of the course, the student will be able to perform general warehousing functions, use specific publications, operate automated material handling equipment, use associated data processing hardware, and monitor an inventory control system.

Instruction: *Version 1:* Course covers warehousing operations, including use of equipment, personnel facilities, space and methods and systems of receiving, storing, and shipping. Also includes the use of data processing equipment to facilitate control procedures. Methodology includes lectures, practical exercises, guest speakers, and field trips. *Version 2:* Course covers warehousing operations, including use of equipment, facilities, and space and methods and systems of receiving, storing, and shipping equipment. Also includes the use of data processing equipment to facilitate control procedures. Methodology includes lecture, practical exercises, guest speakers, and field trips.

Credit Recommendation: *Version 1:* In the lower-division baccalaureate/associate degree category, 3 semester hours in supply management and 2 in automated inventory control (7/95). *Version 2:* In the lower-division baccalaureate/associate degree category, 3 semester hours in warehousing operations and 1 in computer operations (4/89).

MC-1405-0037

ENLISTED SUPPLY REORIENTATION
(Enlisted Supply Advanced)

Course Number: *Version 1:* 30G. *Version 2:* 30G.

Location: Service Support School, Cp. Lejeune, NC.

Length: *Version 1:* 4 weeks (150 hours). *Version 2:* 4 weeks (142 hours).

Exhibit Dates: *Version 1:* 1/95–Present. *Version 2:* 1/90–12/94.

Learning Outcomes: *Version 1:* Upon completion of the course, the student will be able to use supply publications and directives, perform manual and automated accounting procedures, use automated data processing equipment, maintain supply files, and reconcile supply management reports. *Version 2:* Upon completion of the course, the student will be able to use supply publications and directives, perform manual accounting procedures, use automated data processing equipment, maintain supply files, and reconcile supply management reports.

Instruction: *Version 1:* Methods include lecture, class discussions, field trips, practical application, computer-assisted instruction, and critiques. *Version 2:* Course covers manual accounting procedures, personal effects, automated data processing functions, reporting functions, inventory control, and shipping and receiving responsibilities. Methodology includes lectures, class discussions, field trips, and practical applications.

Credit Recommendation: *Version 1:* In the lower-division baccalaureate/associate degree category, 3 semester hours office administration and 1 in computer software applications (7/95). *Version 2:* In the lower-division baccalaureate/associate degree category, 3 semester hours in supply management and 3 in introduction to data processing (4/89).

MC-1405-0038

GROUND SUPPLY OFFICER

Course Number: COG.
Location: Service Support School, Cp. Lejeune, NC.
Length: 11-12 weeks (386-387 hours).
Exhibit Dates: 1/90–Present.
Learning Outcomes: Upon completion of the course, the student will be able to manage all the personnel and assets necessary for a Marine Corps operating unit.
Instruction: Course covers supply procedures, including supply management, office management, property control, financial management, accounting, systems analysis, purchasing, transportation, distribution, warehousing, and computerized supply operations. Methodology includes lectures, demonstrations, practical applications, and performance exercises.
Credit Recommendation: In the lower-division baccalaureate/associate degree category, 1 semester hour in computer concepts and software application (4/89); in the upper-division baccalaureate category, 4 semester hours in supply management (4/89).

MC-1405-0039

ENLISTED SUPPLY INTERMEDIATE

Course Number: *Version 1:* 30A. *Version 2:* 30A.
Location: Service Support School, Cp. Lejeune, NC.
Length: *Version 1:* 9 weeks (357-358 hours). *Version 2:* 10 weeks (318-319 hours).
Exhibit Dates: *Version 1:* 12/94–Present. *Version 2:* 1/90–11/94.
Learning Outcomes: *Version 1:* Upon completion of the course, the student will be able to plan, organize, and control the purchasing function of an organization; train and direct department personnel; and use computer methods to solve procurement problems. *Version 2:* Upon completion of the course, the student will be able to plan, organize, and control the purchasing function of an organization; supervise, and direct department personnel; and use computer methods to solve procurement problems.

Instruction: *Version 1:* Course covers supply procurement, administration, and operations with emphasis on management controls. Also includes training in the use of computers related to purchasing practices, contract principles, and budgeting procedures. Methodology includes lectures, class discussions, and practical applications in procurement operations and procedures. *Version 2:* Course covers supply procurement, administration, and operations with emphasis on management controls. Also includes training in the use of computers related to purchasing practices, contract principles, and budgeting procedures. Methodology includes lectures, class discussions, and practical applications in procurement operations and procedures.

Credit Recommendation: *Version 1:* In the lower-division baccalaureate/associate degree category, 3 semester hours in office administration and 3 in supply management (7/95). *Version 2:* In the lower-division baccalaureate/associate degree category, 3 semester hours in supply management, 2 in supervisory principles, and 3 in office procedures (4/89).

MC-1405-0040

ENLISTED SUPPLY INDEPENDENT DUTY

Course Number: *Version 1:* SAF. *Version 2:* SAF.
Location: Service Support School, Cp. Lejeune, NC.
Length: *Version 1:* 2 weeks (74-75 hours). *Version 2:* 2 weeks (73-74 hours).
Exhibit Dates: *Version 1:* 12/94–Present. *Version 2:* 1/90–11/94.
Learning Outcomes: *Version 1:* Upon completion of the course, the student will be able to prepare documents related to procurement of materials and supplies, complete purchase transaction and maintain related inventory records, monitor and process bid requests and purchase documents involving imprest funds, and complete annual accounting and summary reports for management reviews. *Version 2:* Upon completion of the course, the student will be able to prepare course documents related to procurement of materials and supplies, complete purchase transaction and maintain related inventory records, monitor and process bid requests and determine purchases which involve imprest funds, and complete annual accounting and summary reports for management reviews.
Instruction: *Version 1:* Course covers procurement, supply, and requisitioning procedures; purchasing; accounting; management; and reporting of source document. Methodology includes lectures and practical applications in procurement with emphasis on contract procedures and process. *Version 2:* Course covers procurement, supply, and requisitioning procedures; purchasing; accounting; management; and reporting of source document. Methodology includes lectures and practical applications in procurement with emphasis on contract procedures and process.
Credit Recommendation: *Version 1:* In the lower-division baccalaureate/associate degree category, 3 semester hours in purchasing procedures or procurement (7/95). *Version 2:* In the lower-division baccalaureate/associate degree category, 3 semester hours in purchasing or procurement (4/89).

MC-1405-0043

RESERVE UNIT SUPPLY REFRESHER

Course Number: *Version 1:* CEX. *Version 2:* SCA; CEX.
Location: Service Support School, Cp. Lejeune, NC.
Length: *Version 1:* 2 weeks (80 hours). *Version 2:* 2 weeks (73-74 hours).
Exhibit Dates: *Version 1:* 11/94–Present. *Version 2:* 1/90–10/94.
Learning Outcomes: *Version 1:* Upon completion of the course, the student will be able to perform supply operations at a unit level using computer and manual supply accounting procedures, requisitioning, warehousing, and distribution of ground supply accounts. *Version 2:* Upon completion of the course, the student will be able to perform unit level supply operations using computer and manual supply accounting procedures, requisitioning, warehousing, and distribution of ground supply accounts.
Instruction: *Version 1:* Course covers the basic supply administration skills required in the daily operation of a unit supply account. Includes instruction in property and management and use of the computerized system to requisition, account for, monitor, distribute, and dispose of supply assets. Methodology includes lectures, demonstrations, and practical applications. *Version 2:* Covers the basic supply administration skills required in the daily operation of a unit supply account. Includes instruction in technical research, property and financial management, and use of the computerized system to requisition, account for, monitor, distribute, and dispose of supply assets. Methodology includes lectures, demonstrations, and practical applications.
Credit Recommendation: *Version 1:* In the lower-division baccalaureate/associate degree category, 1 semester hour in computer software applications and 2 in supply management (7/95). *Version 2:* In the lower-division baccalaureate/associate degree category, 2 semester hours in supply management (4/89).

MC-1405-0044

SASSY: ORGANIC PROCEDURES BY CORRESPONDENCE

Course Number: 30.9g-1.
Location: Marine Corps Institute, Washington, DC.
Length: Maximum, 52 weeks.
Exhibit Dates: 1/90–Present.
Learning Outcomes: Upon completion of the course, the student will be able to provide supply personnel with a basic knowledge of techniques for recording periodic and miscellaneous transactions between user units and a centralized management unit.
Instruction: This correspondence course trains supply clerks in the proper procedure for processing forms for warehousing operations, maintenance of equipment, and deployment procedures necessary in the operation of a supply unit. The course has a proctored final examination.
Credit Recommendation: In the lower-division baccalaureate/associate degree cate-

gory, 1 semester hour in supply procedures (6/89).

MC-1405-0045

1. BASIC COMMUNICATION OFFICER
2. BASIC COMMUNICATION OFFICERS

Course Number: *Version 1:* 28V. *Version 2:* 28V.
Location: Communication Officers School, Quantico, VA.
Length: *Version 1:* 18 weeks (817 hours). *Version 2:* 13-18 weeks (491-817 hours).
Exhibit Dates: *Version 1:* 6/94–Present. *Version 2:* 1/90–5/94.
Learning Outcomes: *Version 1:* Upon completion of the course, the student will be able to design and install electronic and radio communications systems; use and apply the fundamentals of communications theory; perform analysis of system malfunction; and maintain the system. *Version 2:* Upon completion of the course, the student will be able to design and install a wire (telephone) or a radio communications system, utilize fundamentals of communications theory and equipment capabilities in communications system design, perform analysis of system malfunction, and use standard procedures for restoring system.
Instruction: *Version 1:* Course includes lectures, practical exercises, and activities in communications center design, organization, and operation. Course also includes lectures and demonstrations in radio antenna theory and design electronic and computer communications systems, introductory lectures in AC/DC circuits, electromagnetic wave propagation, cryptographic equipment, and signal characteristics and experience in system installation, malfunction, restoration, and maintenance. *Version 2:* Course includes lectures, practical exercises, and activities in communications center design, organization, and operation. Course also includes lectures and demonstrations in radio and antenna theory and design, introductory lectures in AC/DC circuits, electromagnetic wave propagation, cryptographic equipment, and signal characteristics and experience in system installation, malfunction, restoration, and maintenance.
Credit Recommendation: *Version 1:* In the lower-division baccalaureate/associate degree category, 3 semester hours in introduction to computing, 3 in principles of communication technology, and 2 in introduction to AC/DC circuits (4/96). *Version 2:* In the vocational certificate category, 3 semester hours in installation and maintenance of communications systems (6/94); in the lower-division baccalaureate/associate degree category, 2 semester hours in introduction to AC/DC circuits and 1 in communications techniques (6/94).

MC-1405-0046

BASIC WAREHOUSING BY CORRESPONDENCE

Course Number: 30.1m; 30.1n.
Location: Marine Corps Institute, Washington, DC.
Length: Maximum, 52 weeks.
Exhibit Dates: 1/90–Present.
Learning Outcomes: Upon completion of the course, the student will understand the duties of a warehouseman, including the basics of preservation, packaging, packing, and material handling equipment.
Instruction: The course, a combination of self-instruction and on-the-job training, presents principles of physical storage, material handling, packaging, and preparation of logistical movement. The material covered by the course includes safety, space planning, records and record keeping, material handling, and packing requirements. Also included in the course is a study of elementary mathematical computations and camouflage techniques. Course concludes with a proctored examination.
Credit Recommendation: In the lower-division baccalaureate/associate degree category, 1 semester hour in physical distribution (7/92).

MC-1405-0047

AVIATION QUALITY ASSURANCE SUPERVISOR BY CORRESPONDENCE

Course Number: 60.6a.
Location: Marine Corps Institute, Washington, DC.
Length: Maximum, 52 weeks.
Exhibit Dates: 1/90–Present.
Learning Outcomes: Upon completion of the course, the student will understand the function of quality control and assurance, identify the components of a quality control program, demonstrate an appreciation of the special problems of quality control and assurance, and be able to evaluate and measure quality control.
Instruction: This course covers the quality control and assurance function, the various problems of quality control and assurance, and methods for evaluating and measuring quality.
Credit Recommendation: In the lower-division baccalaureate/associate degree category, 1 semester hour in operations management (5/91).

MC-1405-0048

COMMUNICATION NONCOMMISSIONED OFFICER (NCO)

Course Number: H-201-3334.
Location: Landing Force Training Command, Pacific, San Diego, CA.
Length: 2 weeks (52-53 hours).
Exhibit Dates: 6/92–Present.
Learning Outcomes: Upon completion of the course, the student will be able to supervise the installation, interconnection, and operation of electrical and electronic communications systems.
Instruction: Course includes lectures and demonstrations on communication systems, techniques, and management systems.
Credit Recommendation: Credit is not recommended because of the military specific nature of the course (11/93).

MC-1406-0021

BASIC TYPING
(Basic Typing and Personnel Administration)

Course Number: AAB.
Location: Schools Battalion, Cp. Pendleton, CA; Financial Management and Personnel Administration School, Cp. Lejeune, NC.
Length: 2 weeks (62-64 hours).
Exhibit Dates: 1/90–2/90.
Learning Outcomes: Upon completion of the course, the student will be able to type at least 15 words per minute by using the touch method; describe the duties of a personnel clerk, a unit diary clerk, an administrative clerk, and a personnel/administration chief; and demonstrate proper office telephone techniques.
Instruction: Basic instruction is from an instructional television system and an automated typewriting system which enables students to learn touch typing and to develop speed in an individualized environment. In addition, there are lectures, discussions, and a series of slides and tapes that address the responsibilities of the personnel clerk, unit diary clerk, administrative clerk, and the personnel/administrative chief as well as proper office telephone techniques.
Credit Recommendation: In the lower-division baccalaureate/associate degree category, 1 semester hour in basic typing or keyboarding (11/87).

MC-1406-0022

RESERVE ADMINISTRATION

Course Number: *Version 1:* 01J. *Version 2:* 01J.
Location: *Version 1:* Service Support School, Cp. Lejeune, NC. *Version 2:* Financial Management and Personnel Administration School, Cp. Lejeune, NC.
Length: *Version 1:* 2 weeks (70 hours). *Version 2:* 2 weeks (68-69 hours).
Exhibit Dates: *Version 1:* 12/94–Present. *Version 2:* 1/90–11/94.
Learning Outcomes: *Version 1:* Upon completion of the course, the student will be able to perform general administrative tasks, including drill accounting reports, pay related matters, and maintain individual records. *Version 2:* Upon completion of the course, the student will be able to perform general administrative tasks, including preparation of drill accounting reports, preparation of management pay system, and maintenance of individual records.
Instruction: *Version 1:* Methods include lecture, demonstration, practical application, and critique. *Version 2:* Lectures and practical applications are used to train students to perform tasks of general administration, drill accounting, reserve manpower management pay system, and individual records.
Credit Recommendation: *Version 1:* In the lower-division baccalaureate/associate degree category, 1 semester hour in clerical procedures (7/95). *Version 2:* In the vocational certificate category, 1 semester hour in clerical procedures (9/87).

MC-1406-0023

FUNDAMENTALS OF MARINE CORPS LEADERSHIP BY CORRESPONDENCE
(The Noncommissioned Officer (NCO) by Correspondence)
(Noncommissioned Officer (NCO) by Correspondence)

Course Number: *Version 1:* 03.3m; 03.3k. *Version 2:* 03.3K; 03.3M.
Location: Marine Corps Institute, Washington, DC.
Length: *Version 1:* Maximum, 52 weeks. *Version 2:* Maximum, 52 weeks.
Exhibit Dates: *Version 1:* 2/90–Present. *Version 2:* 1/90–1/90.
Learning Outcomes: *Version 1:* Upon completion of the course, the student will be able to demonstrate principles and techniques of leadership and their application to problems in areas such as race relations and substance abuse. *Ver-*

sion 2: Upon completion of the course, the student will be familiar with various principles and techniques of leadership and their applications to problems in areas such as race relations and substance abuse.

Instruction: *Version 1:* This is a self-instruction course covering topics of leadership traits and principles, measurements and techniques of leadership, race relations, women's issues, drug and alcohol abuse, and leader/subordinate relations. *Version 2:* This is a self-instruction course covering topics of leadership traits and principles, measurements and techniques of leadership, race relations, women's issues, drug and alcohol abuse and leader/subordinate relations.

Credit Recommendation: *Version 1:* In the lower-division baccalaureate/associate degree category, 1 semester hour in supervisory management (5/91). *Version 2:* In the lower-division baccalaureate/associate degree category, 1 semester hour in employee relations (6/89).

MC-1406-0024

1. LEGAL ADMINISTRATION CLERK BY CORRESPONDENCE
2. LEGAL ADMINISTRATION FOR THE REPORTING UNIT BY CORRESPONDENCE

Course Number: *Version 1:* 01.43a. *Version 2:* 01.43.
Location: Marine Corps Institute, Washington, DC.
Length: *Version 1:* Maximum, 52 weeks. *Version 2:* Maximum, 52 weeks.
Exhibit Dates: *Version 1:* 6/93–Present. *Version 2:* 1/90–5/93.
Learning Outcomes: *Version 1:* Upon completion of the course, the student will be able to complete the tasks of legal administration, including investigation, handling absentees/deserters, and nonjudicial punishment. *Version 2:* Upon completion of the course, the student will be able to identify the context of personnel records, describe procedures for proper entry and maintenance of records, identify documents used in personnel audits, and report casualty and emergency data.

Instruction: *Version 1:* This is a correspondence course with workbook, self-scored examinations, and Marine Corps Institute graded lessons and final examinations. Topics include handling absentees and deserters, auditing requirements, and maintenance of records. *Version 2:* Course includes maintenance of records, performance evaluations, time lost procedures, handling absentees and deserters, auditing requirements, medical and health documentation, and identification cards. Methodology includes use of scenarios. Test questions on how personnel records should be completed are based on given scenarios.

Credit Recommendation: *Version 1:* In the lower-division baccalaureate/associate degree category, 1 semester hour in record keeping (12/96). *Version 2:* In the vocational certificate category, 1 semester hour in personnel record keeping (6/89).

MC-1406-0025

GENERAL PERSONNEL ADMINISTRATION FOR RESERVE BY CORRESPONDENCE

Course Number: *Version 1:* 01.28a. *Version 2:* 01.28.
Location: Marine Corps Institute, Washington, DC.
Length: *Version 1:* Maximum, 52 weeks. *Version 2:* Maximum, 52 weeks.
Exhibit Dates: *Version 1:* 2/90–Present. *Version 2:* 1/90–1/90.
Learning Outcomes: *Version 1:* Upon completion of the course, the student will be able to prepare documents for new personnel entering the unit; continually process and audit such documents as personnel are transferred, promoted, discharged, etc.; and maintain a system of reports and files to account for all personnel movements both by individuals and the unit collectively. *Version 2:* Upon completion of the course, the student will be able to prepare documents for new personnel entering the unit, continually process and audit such documents as personnel are transferred, promoted, discharged, etc., and maintain a system of reports and files to account for all personnel movements both by individuals and the unit collectively.

Instruction: *Version 1:* This is a self-instructed course covering topics of general personnel administration, assignment of personnel, record keeping and auditing, and general office procedures. *Version 2:* This is a self-instructed course covering topics of general personnel administration, assignment of personnel, record keeping and auditing, and general office procedures.

Credit Recommendation: *Version 1:* In the lower-division baccalaureate/associate degree category, 1 semester hour in office procedures (5/91). *Version 2:* In the lower-division baccalaureate/associate degree category, 1 semester hour in personnel administration (6/89).

MC-1406-0026

PERSONNEL ADMINISTRATION FOR THE REPORTING UNIT BY CORRESPONDENCE

Course Number: 01.36f; 01.36e.
Location: Marine Corps Institute, Washington, DC.
Length: Maximum, 52 weeks.
Exhibit Dates: 1/90–Present.
Learning Outcomes: Upon completion of the course, the student will be able to perform basic personnel administrative functions at the reporting unit level and be able to use and complete various forms and book entries related to assignment, transfer, promotion, performance evaluation, and dismissal.

Instruction: This is a self-instructed course that includes general personnel record keeping, document processing, and file maintenance and retrieval.

Credit Recommendation: In the lower-division baccalaureate/associate degree category, 1 semester hour in personnel administration (6/89).

MC-1406-0027

PERSONNEL REPORTING FOR REMMPS BY CORRESPONDENCE

Course Number: 01.39.
Location: Marine Corps Institute, Washington, DC.
Length: Maximum, 52 weeks.
Exhibit Dates: 1/90–Present.
Learning Outcomes: Upon completion of the course, the student will demonstrate familiarity with the military reserve pay system and be able to use a standardized, automated system to collect, sort, and classify data, audit entries, correct errors, update and change files, store records on disks or tape, and retrieve data from existing files.

Instruction: This is a self-instructed course covering topics of data entry and retrieval, auditing procedures, and record keeping within an automated pay system.

Credit Recommendation: In the lower-division baccalaureate/associate degree category, 1 semester hour in payroll administration (6/89).

MC-1406-0028

PRINCIPLES OF INSTRUCTION FOR THE MARINE NONCOMMISSIONED OFFICER (NCO) BY CORRESPONDENCE

Course Number: 00.1a.
Location: Marine Corps Institute, Washington, DC.
Length: Maximum, 52 weeks.
Exhibit Dates: 1/90–Present.
Learning Outcomes: Upon completion of the course, the student will demonstrate an understanding of general concepts relating to teaching and learning processes designed to suit needs and apply a systems approach to training covering analysis, design, development, implementation and evaluation.

Instruction: This correspondence course is designed to provide marine NCOs with necessary background data regarding educational theory, techniques, and procedures which will assist them in becoming effective instructors.

Credit Recommendation: In the lower-division baccalaureate/associate degree category, 3 semester hours in instructional methodology (9/89).

MC-1406-0029

FORMAL INSTRUCTOR
(Formal School Instructor)

Course Number: T4B.
Location: Instructional Management School, Quantico, VA; Instructional Management School, Cp. Lejeune, NC; Instructional Management School, Cp. Pendleton, CA.
Length: 3-4 weeks (134-156 hours).
Exhibit Dates: 1/90–5/94.
Learning Outcomes: Upon completion of the course, the student will be able to develop a task list and select tasks for training, write and sequence learning objectives, write objective-referenced test items, select methods and media for lesson design, develop lectures and student handouts, conduct lectures, apply concepts of motivation and mastery learning, and evaluate and recommend revisions to a lesson. Students will be able to conduct learning analysis of a task, develop training aids, and write an instructor's guide.

Instruction: Students are oriented to the Instructional Systems Development (ISD) approach to training design and delivery. Course concepts and skills are taught within the ISD framework. Methods include lectures, guided discussions, and instructional design projects and both written and performance testing. Topics include the systems approach to instructional design, target population analysis, performance-based training, and effective verbal and nonverbal behavior for classroom teaching.

Credit Recommendation: In the upper-division baccalaureate category, 3 semester hours in instructional technology, educational technology, instructional design, development,

MC-1406-0030

1. INSTRUCTOR TRAINING
2. INSTRUCTOR ORIENTATION

Course Number: *Version 1:* XRG. *Version 2:* XRG.
Location: *Version 1:* Instructional Management School, Cp. Pendleton, CA; Service Support School, Cp. Lejeune, NC. *Version 2:* Service Support School, Cp. Lejeune, NC; Instructional Management School, Cp. Pendleton, CA; Instructional Management School, Quantico, VA.
Length: *Version 1:* 2 weeks (75-76 hours). *Version 2:* 2 weeks (66-70 hours).
Exhibit Dates: *Version 1:* 6/94–Present. *Version 2:* 1/90–5/94.
Learning Outcomes: *Version 1:* This version is pending evaluation. Upon completion of the course, the student will be able to outline and deliver a lecture using appropriate visual aids and question/answer techniques; construct test items formatted as multiple choice, short answers, essay, matching, listing, and performance; demonstrate understanding of learning objectives, construction, learning motivation, communication barriers, and counseling. The sequencing of instruction is added to content, and there is not emphasis on outlining. Presentation is emphasized. *Version 2:* Upon completion of the course, the student will be able to outline and deliver a lecture using appropriate visual aids and question/answer techniques; construct test items formatted as multiple choice, short answers, essay, matching, listing, and performance; demonstrate understanding of learning objectives, construction, learning motivation, communication barriers, and counseling. The sequencing of instruction is added to content, and there is not emphasis on outlining. Presentation is emphasized.
Instruction: *Version 1:* This version is pending evaluation. Lecture is used to cover the topics of learning objectives, learning motivation, instructional system development, evaluation, communication, and counseling students. A major portion of the course is devoted to students making and critiquing presentations and employing detailed checklists. The checklists reference voice, gestures, facial expression and eye contact, listening skills, questioning techniques, motivational strategies, and transitions. Students formulate written plans for improvement based on the feedback from instructors and peer critique. *Version 2:* Lecture is used to cover the topics of learning objectives, learning motivation, instructional system development, evaluation, communication, and counseling students. A major portion of the course is devoted to students making and critiquing presentations and employing detailed checklists. The checklists reference voice, gestures, facial expression and eye contact, listening skills, questioning techniques, motivational strategies, and transitions. Students formulate written plans for improvement based on the feedback from instructors and peer critique.
Credit Recommendation: *Version 1:* Pending evaluation. *Version 2:* In the upper-division baccalaureate category, 2 semester hours in training and development (instructional methods) (9/89); in the graduate degree category, credit is to be granted based on institution's review of student's course project (9/89).

MC-1406-0031

INSTRUCTIONAL MANAGEMENT

Course Number: RF9.
Location: Instructional Management School, Quantico, VA.
Length: 2-3 weeks (69-77 hours).
Exhibit Dates: 1/90–7/91.
Learning Outcomes: Upon completion of the course, the student will be able to train personnel in the skills required to supervise instruction and contribute to the planning and decision making process. The student will be able to apply the five phases of instructional systems development, mastery learning, criterion referenced learning objectives, and selection of instructional methods as well as enhance their verbal communication skills and management abilities.
Instruction: The instruction covers course requirements analysis, including developing objectives and tests, supervising preparation of instruction, evaluating the instructional process, developing budget proposals, supervising academic functions, and counseling students and staff. The methods used are lecture, self-paced learning, performance, group techniques, and role playing. Course was well organized. Phase 1 was self-paced individualized, and phase 2 was group-paced instruction. Training equipment includes multimedia approach, including films, slides, and tapes as well as television shows, manuals, and several texts. The methods used are comprehensive and give each student a hands-on approach to the teaching/learning process. Topics covered are the systems approach to training, instructional development, communications skills, and management skills.
Credit Recommendation: In the upper-division baccalaureate category, 3 semester hours in educational administration including instructional strategies and educational supervision training and development (9/89); in the graduate degree category, credit is to be granted based on institution's review of student's course project (9/89).

MC-1406-0032

COUNSELING FOR MARINES BY CORRESPONDENCE

Course Number: 01.12.
Location: Marine Corps Institute, Washington, DC.
Length: Maximum, 52 weeks.
Exhibit Dates: 1/90–Present.
Learning Outcomes: Upon completion of the course, the student will be able to plan and conduct a counseling session.
Instruction: This is a self-instruction course that covers an overview of counseling, the frequency of counseling, the counseling process, planning and conducting a counseling session, counseling techniques and activities, and countering problems in counseling.
Credit Recommendation: Credit is not recommended because of the limited, specialized nature of the course (5/91).

MC-1406-0033

ADVANCED LEGAL SERVICES

Course Number: 442.
Location: Service Support School, Cp. Lejeune, NC.
Length: 2 weeks (64 hours).
Exhibit Dates: 3/95–Present.
Learning Outcomes: Upon completion of this course, the student will be able to draft correspondence and legal documents and conduct legal inspections; prepare legal documents and reports of court-martial orders; and process investigations, claims, and administrative separations.
Instruction: Lectures and practical applications are used in the administrative report and legal components of this course.
Credit Recommendation: In the lower-division baccalaureate/associate degree category, 3 semester hours in legal writing (7/95).

MC-1406-0034

LEGAL SERVICES

Course Number: 58X.
Location: Service Support School, Cp. Lejeune, NC.
Length: 7-8 weeks (265 hours).
Exhibit Dates: 10/94–Present.
Learning Outcomes: Upon completion of the course, the student will be able to input drafted correspondence, legal documents, and reports; maintain case files; process requests for legal services; process administrative separations; and prepare legal assistance documentation, powers of attorney, and wills.
Instruction: Methods include lectures, practical applications, computer based training, and keyboarding.
Credit Recommendation: In the lower-division baccalaureate/associate degree category, 2 semester hours in basic keyboarding and 3 in introduction to law and legal assisting (7/95).

MC-1406-0035

MARINE CADRE TRAINER

Course Number: M4T.
Location: Marine Corps Security Force, Chesapeake, VA; Marine Corps Security Force, Mare Island Naval Shipyard, Vallejo, CA.
Length: 6 weeks (161 hours).
Exhibit Dates: 8/91–Present.
Learning Outcomes: Upon completion of the course, the student will be able to provide small arms instruction, defensive tactics, antiterrorism measures, arrest and restraint techniques, and antiterrorism instruction techniques.
Instruction: Instruction is provided through lecture, demonstration, and practical exercises covering topics in security, defensive tactics, small arms marksmans@pect restraint, antiterrorism and instructor training.
Credit Recommendation: In the lower-division baccalaureate/associate degree category, 1 semester hour in military science (leadership) and 1 in education practicum (3/96).

MC-1408-0015

COMMAND AND STAFF COLLEGE RESERVE PHASE 1, PHASE 2

Course Number: None.
Location: Command and Staff College, Quantico, VA.
Length: Phase 1, 2 weeks (65 hours); Phase 2, 2 weeks (62 hours).
Exhibit Dates: 1/90–5/90.
Objectives: To train selected reserve officers of the US Marine Corps and officers from foreign countries for higher reserve and mobilization activities.

Instruction: Workshops and lectures with emphasis on problem solving and planning and conducting amphibious warfare, rounded out by briefings and seminars relative to current policies, plans, and procedures.

Credit Recommendation: In the upper-division baccalaureate category, Phase 1, 1 semester hour in geopolitics; Phase 2, 1 additional hour in geopolitics (7/82).

MC-1408-0016

DISBURSING OFFICER

Course Number: CDA.
Location: Financial Management and Personnel Administration School, Cp. Lejeune, NC.
Length: 11-12 weeks (412-422 hours).
Exhibit Dates: 1/90–1/91.
Learning Outcomes: Upon completion of the course, the student will be able to identify directives; interpret publications related to disbursement functions; verify travel status; certify balance sheets, reports, distance computations, allotments, housing allowances, and related military expenditures for personnel; supervise fiscal internal control records; audit cash funds, bank deposits, subsistence, clothing, and other allowances; manage a video inquiry system; and operate a calculator.
Instruction: Course includes lectures and practical exercises in military pay and accounts and military travel accounting for funds and general disbursement procedures. A simulated field disbursing workshop includes problem solving and customer service aspects of disbursing.
Credit Recommendation: In the vocational certificate category, 2 semester hours in clerical procedures (9/87); in the lower-division baccalaureate/associate degree category, 3 semester hours in principles of financial statement analysis (9/87).

MC-1408-0017

MARINE CORPS COMMAND AND STAFF COLLEGE

Course Number: *Version 1:* RHA. *Version 2:* RHA.
Location: Marine Corps University, Quantico, VA.
Length: *Version 1:* 43 weeks (1578 hours). *Version 2:* 43 weeks (1108-1306 hours).
Exhibit Dates: *Version 1:* 8/93–Present. *Version 2:* 1/90–7/93.
Learning Outcomes: *Version 1:* Upon completion of the course, the student will be able to prepare and communicate ideas in written and oral presentations involving complex and diverse military concepts, tactics, strategies, and theories. Through formal research and analysis, the student will understand how military history and strategy are integrated and how the fundamental political-military relationship shapes the planning, organizing, and implementation of theater-level operations. The student will be able to employ research which draws on a wide range of government and public sources and prepare clear, specific, and unambiguous plans, directives, and orders for project and program management. The student will also be able to communicate clearly with superiors, subordinates, and peers. The student will be able to prepare logistics and operational plans to achieve the objectives and goals of the theater of operations. Students electing to take the master of military studies option will demonstrate historical and social science research and writing skills appropriate to the professional military history and military policies studies fields. Beginning in 1993-94, students may choose a professional paper involving in-depth research under faculty supervision. *Version 2:* Upon completion of this course, the student will be able to prepare and communicate ideas in writing on a variety of complex military subjects using expository research and persuasive format. The student will be able to prepare clear, specific, and unambiguous plans, directives, and orders for project and program management. Students will be able to verbally defend their position on a variety of military scenarios and will be able to demonstrate competency in historical and social science research and writing skills.

Instruction: *Version 1:* Information in this course encompasses the theory and nature of war and strategic thought, translated to the operational level of war. Topics include critical thinking, geopolitical and current international relations, executive leadership, influences on national policy, the national security policy process, strategic geography and case studies illustrating the principles and applications of the theory and strategy of war. Instruction includes lectures, multimedia presentations, role playing, seminars, small group seminars, field trips, and independent study. *Version 2:* Information in this course encompasses the theory and nature of war and strategic thought, translated to the operational level of war. Topics include critical thinking, geopolitical and current international relations, executive leadership, influences on national policy, the national security policy process, strategic geography, and case studies illustrating the principles and applications of the theory and strategy of war. Instruction includes lectures, multimedia presentations, role playing, seminars, small group seminars, field trips, and independent studies.

Credit Recommendation: *Version 1:* In the upper-division baccalaureate category, 3 semester hours in written/oral communication, 3 in organizational management, 3 in international relations, 3 in military history, 2 in field study of organizational management, and credit in the electives program based on the receiving institution's review of papers and materials provided by the student (2/94); in the graduate degree category, 3 semester hours in US national security or 3 in US foreign policy, 3 in evolution of strategic thought, and 1-3 in directed research according to the receiving institution's review of the student's major professional paper (2/94). *Version 2:* In the lower-division baccalaureate/associate degree category, 3 semester hours in written and oral communication (7/91); in the upper-division baccalaureate category, 3 semester hours in world history, 3 in principles of management, 3 in survey of world affairs, and 3 in military history (7/91); in the graduate degree category, 3 semester hours in military history/military policy or 3 semester hours in international relations (7/91).

MC-1408-0018

MOTOR TRANSPORT OPERATIONS NONCOMMISSIONED OFFICER (NCO)

Course Number: *Version 1:* CDG. *Version 2:* CDG.
Location: *Version 1:* Engineer School, Cp. Lejeune, NC. *Version 2:* Service Support School, Cp. Lejeune, NC.
Length: *Version 1:* 4 weeks (137 hours). *Version 2:* 4-5 weeks (137-157 hours).
Exhibit Dates: *Version 1:* 1/94–Present. *Version 2:* 1/90–12/93.
Learning Outcomes: *Version 1:* Upon completion of the course, the student will be able to understand and describe the maintenance and administration and management procedures; organizational procedures for motor pool operations; proper loading and unloading procedures; tactical convoy operations and the management of a preventative maintenance system. *Version 2:* Upon completion of the course, the student will be able to describe the Marine Corps Integrated Maintenance Management System, operate a motor pool, load and unload vehicles, perform vehicle inspections, prepare vehicles for tactical convoys, and perform field expedient repairs.
Instruction: *Version 1:* Course covers maintenance administration and management, transportation procedures, loading cargoes, first echelon maintenance, and tactical operations. Methodology includes lectures, practical applications, and performance and written examinations. *Version 2:* Course covers maintenance administration and management, transportation procedures, loading cargoes, first echelon maintenance, and tactical operations. Methodology includes lectures, practical applications, and performance and written examinations.
Credit Recommendation: *Version 1:* In the lower-division baccalaureate/associate degree category, 2 semester hours in vehicle maintenance administration and management (8/95). *Version 2:* In the lower-division baccalaureate/associate degree category, 2 semester hours in vehicle maintenance administration and management (1/94).

MC-1408-0019

1. LOGISTICS VEHICLE SYSTEM (LVS) MAINTENANCE
2. LOGISTICS VEHICLE SYSTEM MAINTENANCE

Course Number: *Version 1:* MBC. *Version 2:* MBC.
Location: *Version 1:* Motor Transport School, Cp. Lejeune, NC. *Version 2:* Service Support School, Cp. Lejeune, NC.
Length: *Version 1:* 4 weeks (134 hours). *Version 2:* 3-4 weeks (117-118 hours).
Exhibit Dates: *Version 1:* 1/95–Present. *Version 2:* 1/90–12/94.
Learning Outcomes: *Version 1:* Upon completion of the course, the student will be able to perform maintenance operations on LVS cooling systems, electrical systems, fuel systems, transmission systems, brake and axle mechanizing hydraulic system, and steering systems. The student will also be able to operate the LVS in conjunction with the performance of the maintenance procedures. *Version 2:* Upon completion of the course, the student will be able to maintain logistics vehicles used in emergency recovery.
Instruction: *Version 1:* Instruction covers maintenance some repair and operation of the logistic and vehicle system. Methodology includes lecture, demonstration and laboratory exercises. *Version 2:* Course covers the repair and maintenance of automotive vehicles and includes the techniques of recovering disabled

vehicles. Methodology includes lectures, laboratory exercises, and demonstrations.

Credit Recommendation: *Version 1:* In the lower-division baccalaureate/associate degree category, 3 semester hours in advanced maintenance maintenance practices and 2 in basic hydraulics (8/95). *Version 2:* In the lower-division baccalaureate/associate degree category, 3 semester hours in advanced maintenance practices and 3 in hydraulic troubleshooting and maintenance (2/92).

MC-1408-0020

RESERVE AUTOMOTIVE MECHANIC

Course Number: *Version 1:* CEU. *Version 2:* CEU.
Location: *Version 1:* Engineer School, Cp. Lejeune, NC. *Version 2:* Service Support School, Cp. Lejeune, NC.
Length: *Version 1:* 2 weeks (70 hours). *Version 2:* 2 weeks (68-74 hours).
Exhibit Dates: *Version 1:* 6/94–Present. *Version 2:* 1/90–5/94.
Learning Outcomes: *Version 1:* Upon completion of the course, the student will be able to inspect, diagnose, service, and repair systems on the High Mobility Multipurpose Wheeled Vehicle (HMMWV) or the Logistics Vehicle System (LVS); depending on course option taken. *Version 2:* Upon completion of the course, the student will be able to inspect, diagnose, service, and repair the M998 and LVS automotive vehicles.
Instruction: *Version 1:* Instruction will be provided by lecture, demonstration and practical laboratory exercises on maintenance and repair of diesel engines, fuel systems, electrical systems, drive lines, suspension systems, brake systems, cooling systems and hydraulic systems on either the HMMWV or the LVS. *Version 2:* Course includes lectures and practical exercises in the inspection, service, and repair of specified tactical automotive vehicles, including shop procedures, power plants, fuel and electrical trouble diagnosis and tune-up, power trains, chassis, brakes, suspension systems, diesel and multifuel engines. Course also includes motor vehicle operation and preventative maintenance techniques.
Credit Recommendation: *Version 1:* In the lower-division baccalaureate/associate degree category, 2 semester hours in automotive maintenance and repair (8/95). *Version 2:* In the lower-division baccalaureate/associate degree category, 3 semester hours in automotive maintenance and repair (6/94).

MC-1408-0021

MARINE CORPS TECHNICAL PUBLICATIONS SYSTEM BY CORRESPONDENCE

Course Number: 30.23a.
Location: Marine Corps Institute, Washington, DC.
Length: Maximum, 52 weeks.
Exhibit Dates: 1/90–Present.
Learning Outcomes: Upon completion of the course, the student will be able to establish a technical publications library and identify and retrieve stored information.
Instruction: This is a self-instructed course which includes topics on the establishment and maintenance of a technical publications library, indexing and cross-referencing entries, locating stock numbers and sources of supply, and identifying component and repair parts information.
Credit Recommendation: In the vocational certificate category, 2 semester hours in material management (6/89).

MC-1408-0022

ADDITIONAL DEMANDS LISTING CLERK BY CORRESPONDENCE

Course Number: 30.24.
Location: Marine Corps Institute, Washington, DC.
Length: Maximum, 52 weeks.
Exhibit Dates: 1/90–Present.
Learning Outcomes: Upon completion of the course, the student will be able to prepare, input, and submit requisitions; maintain requisitions on a data base; and understand the responsibilities of supply personnel.
Instruction: This nonresident course covers scope and responsibilities of supply units, supply publications, organization of supply units, preparing a request, inputting daily transactions, overview of computer equipment, troubleshooting faulty computer entries, and edit error codes.
Credit Recommendation: In the lower-division baccalaureate/associate degree category, 1 semester hour in automated record keeping (5/91).

MC-1408-0023

COMMUNICATION SYSTEMS CHIEF

Course Number: CHK.
Location: Air Ground Combat Center, Twentynine Palms, CA.
Length: 12 weeks (446-447 hours).
Exhibit Dates: 12/93–Present.
Learning Outcomes: Upon completion of the course, the student will be able to operate, supervise, and employ Marine Corps communication facilities.
Instruction: The course includes the fundamentals of communications planning, radio communications, data communications, satellite communications, electronic warfare, power sources, cryptographic equipment, computer literacy, and war games.
Credit Recommendation: In the vocational certificate category, 2 semester hours in communications systems administration (1/93).

MC-1408-0024

FINANCIAL MANAGEMENT OFFICER

Course Number: CDA.
Location: Financial Management and Personnel Administration School, Cp. Lejeune, NC.
Length: 8 weeks (383-384 hours).
Exhibit Dates: 4/95–Present.
Learning Outcomes: Upon completion of the course, the student will be able to operate microcomputers using financial management software, certify adjustments to pay and allowances, verify travel status and payments, prepare and monitor an operating budget, execute fiscal year close-out, monitor unit accounting functions, supervise internal management and fiscal control, conduct a resource evaluation and analysis, and conduct an internal control evaluation.
Instruction: Course includes lectures, applications, and computer-assisted instruction in the administrative functions of a financial management officer and demonstrations in the accounting, military pay, travel, and fiscal functions.
Credit Recommendation: In the lower-division baccalaureate/associate degree category, 2 semester hours in software applications, 2 in preparation of financial statement analysis, and 2 in clerical procedures (7/95).

MC-1408-0025

COMMAND AND STAFF NONRESIDENT PROGRAM BY CORRESPONDENCE

Course Number: 8700.
Location: Marine Corps Institute, Washington, DC.
Length: Maximum, 104-130 weeks.
Exhibit Dates: 2/94–Present.
Learning Outcomes: Upon completion of the course, the student will demonstrate an advanced knowledge of military history, theories of war, strategic consequences of war, and national security policy; understand national, theater, and tactical military operations; prepare, at advanced level, joint and combined operations and logistic plans for Marine Air Ground Task Force operations; be aware of and understand peace keeping, low-intensity conflict, and humanitarian efforts; and demonstrate leadership fundamentals, decision making, basic organization, administration planning, and problem solving skills.
Instruction: Methods include nonresident assignments, reading, correspondence, books, and anthologies; limited primary sources and extensive secondary source material; practical exercises, problem solving exercises, case studies, historical examples, multiple choice examinations, and some essays are all components of this course.
Credit Recommendation: In the upper-division baccalaureate category, 3 semester hours in international relations, 3 in military studies, 3 in organizational management and planning, and 3 in theory and nature of war (9/97); in the graduate degree category, 3 semester hours in military history and the evolution of strategic thought and 3 in national security studies (9/97).

MC-1408-0027

SERGEANTS NONRESIDENT PROGRAM BY CORRESPONDENCE

Course Number: 8000.
Location: Marine Corps Institute, Washington, DC.
Length: Maximum, 104 weeks.
Exhibit Dates: 9/96–Present.
Learning Outcomes: Upon completion of the course, the student will possess necessary leadership, communication, and analytic skills to become an effective noncommissioned officer in the Marine Corps.
Instruction: Instruction involves assigned readings and exercises in communication skills (basic rules of grammar and effective writing skills); leadership (fundamentals of leadership skills); military studies (procedures for drill, uniforms, inspections, nonjudicial punishment, physical fitness training, overview of Marine Corps history and customs); training management (unit training management and techniques for oral instruction); battle skills (fundamentals and techniques for combat operations at the squad level); and small arms weapons (techni-

cal instruction in characteristics and operations of Marine Corps small arms).

Credit Recommendation: Credit is not recommended because of the military-specific nature of the course (9/97).

MC-1408-0028

STAFF NONCOMMISSIONED OFFICERS ADVANCED NONRESIDENT PROGRAM (SNCOANP) BY CORRESPONDENCE

Course Number: 7200.
Location: Marine Corps Institute, Washington, DC.
Length: Maximum, 104-130 weeks.
Exhibit Dates: 1/90–Present.
Learning Outcomes: Upon completion of the course, the student will be able to develop and maintain a physical fitness program for company size units; be familiar with drill, ceremonies, inspections, and awards up to battalion level; will have an introductory knowledge of marine military justice; and will have a basic background in administration and military forms.
Instruction: Manuals with exercises and training suggestions are used. Multiple choice exams which are machine graded.
Credit Recommendation: Credit is not recommended because of the military-specific nature of the course (9/97).

MC-1409-0005

COMMUNICATION CENTER OPERATOR

Course Number: *Version 1:* 254. *Version 2:* 254.
Location: Communication-Electronics School, Twentynine Palms, CA.
Length: *Version 1:* 8-11 weeks (280-425 hours). *Version 2:* 8-11 weeks (280-425 hours).
Exhibit Dates: *Version 1:* 10/91–Present. *Version 2:* 1/90–9/91.
Learning Outcomes: *Version 1:* Upon completion of the course, the student will be able to perform the duties of a communication center operator. *Version 2:* Upon completion of the course, the student will be able to operate teletype equipment, prepare Autodin messages, type 35 words per minute.
Instruction: *Version 1:* The course includes an introduction to the operation and employment of communications equipment and systems used by the Marine Corps, including Autodin communications center procedures, computer editing, and tactical communications center equipment. *Version 2:* Lectures and equipment operation training to operate the communications center. F.
Credit Recommendation: *Version 1:* Credit is not recommended because of the military-specific nature of the course (7/93). *Version 2:* In the lower-division baccalaureate/associate degree category, 3 semester hours in typing (2/89).

MC-1511-0001

INTELLIGENCE BRIEF: SOUTHWEST ASIA BY CORRESPONDENCE

Course Number: 02.01.
Location: Marine Corps Institute, Washington, DC.
Length: Maximum, 52 weeks.
Exhibit Dates: 1/91–Present.
Learning Outcomes: Upon completion of the student will demonstrate knowledge of the major physical and climatic features of the Middle East. Students will also be familiar with the main cultural and religious characteristics of groups in the Gulf region and their implications for US national security.
Instruction: This is a correspondence course presenting the geography, military forces, security threats, culture, history, and politics of the Gulf region. The evaluation requirements consist of an objective proctored final examination.
Credit Recommendation: In the lower-division baccalaureate/associate degree category, 2 semester hours in international studies (4/91).

MC-1511-0002

MARINE CORPS WAR COLLEGE

Course Number: M59.
Location: Marine Corps Combat Development Command, Quantico, VA; Marine Corps University, Quantico, VA.
Length: 45 weeks.
Exhibit Dates: 1/92–Present.
Learning Outcomes: Upon completion of the course, the student will be able to analyze the causes, nature and history of both conventional and revolutionary/unconventional warfare; appraise the relationship among the political, strategic, and tactical levels of war in an historical context; assess the military and political strengths and weaknesses of the United States Armed Forces as well as those of other nations; assess and demonstrate knowledge on current joint and combined military operations doctrine as well as an understanding of interservice rivalry within the context the democratic system of the U.S.; understand the Marine Corp commitment to cultural diversity, gender, and racial sensitivity, and to the process of constitutional government in the United States; understand the interdisciplinary nature of decision making and policy implementation within the national security arena with emphasis on military, historical, political, and regional dimensions; and understand the macro-economic policy as an element of national power, especially relationships among fiscal, monetary, and international trade policies in promoting U.S. national interests and objectives.
Instruction: The curriculum consists of seminar participation, short papers, war games, written examinations, field trips, guest lectures, research papers, and practical applications and exercises focused on U.S. national policy and strategy and the employment of the Nation's Armed Forces in the joint and multi-national environment. Courses stress critical analysis, critical thinking, and sound military judgment by developing the abilities to analyze, assess, and formulate strategies while applying understanding, skills, and knowledge developed through the study of historical and contemporary cases. Key topics are war, policy, and strategy; Marine Air Ground Task (MAGT) operations; national security and joint warfare; regional studies; and general studies.
Credit Recommendation: In the upper-division baccalaureate category, 3 semester hours in U.S. Foreign Policy, 3 in international relations, 3 in international economics, and 3 in organizational development (4/96); in the graduate degree category, 3 semester hours in military history, 3 in national security studies, 3 in international regional security studies, and 3 in combined military operations (4/96).

Marine Corps 169

MC-1601-0012

COMBAT ENGINEER CHIEF, CONSTRUCTION SUPPORT BY CORRESPONDENCE

Course Number: 13.38c.
Location: Marine Corps Institute, Washington, DC.
Length: Maximum, 52 weeks.
Exhibit Dates: 1/90–1/97.
Objectives: To provide the skills necessary to perform as platoon sergeant, construction foreman, or engineer operations chief and as supervisor for building, road, and airfield construction.
Instruction: This is a self-instruction course with emphasis placed on job planning and management and estimation of equipment, materials, personnel, and time required for the various engineering projects that are accomplished in general support of a tactical situation. Emphasis is placed on supply, engineering equipment, and soil engineering and their relationship to construction projects.
Credit Recommendation: In the lower-division baccalaureate/associate degree category, 6 semester hours in engineering construction management (8/83).

MC-1601-0015

ENGINEER FORMS AND RECORDS BY CORRESPONDENCE

Course Number: 14.42b.
Location: Marine Corps Institute, Washington, DC.
Length: Maximum, 52 weeks.
Exhibit Dates: 1/90–6/90.
Objectives: To provide basic training in the theory of equipment maintenance and in the use of forms and records.
Instruction: This is a self-paced, self-instruction course consisting of reading assignments and written lessons with objective questions. Materials include the identification of various service forms for scheduling and recording equipment use and maintenance. The student will demonstrate achievement of objectives by answering correctly 65 percent of the questions in the final, proctored examination.
Credit Recommendation: Credit is not recommended because of the limited, specialized nature of the course (8/83).

MC-1601-0016

ENGINEER EQUIPMENT CHIEF BY CORRESPONDENCE

Course Number: 13.28d.
Location: Marine Corps Institute, Washington, DC.
Length: Maximum, 52 weeks.
Exhibit Dates: 1/90–Present.
Objectives: To provide advanced training in the theory of engineer equipment operation, maintenance, and scheduling.
Instruction: This is an self-paced, self-instruction course containing reading assignments and written lessons with objective questions. Materials include the identification of types of work projects and proper equipment selection for the task. Students will demonstrate achievement of objectives by answering correctly 65 percent of the questions on the proctored final examination.
Credit Recommendation: In the lower-division baccalaureate/associate degree category, 3 semester hours in construction planning

(construction management or civil technologies) (8/83).

MC-1601-0024

BASIC COMBAT ENGINEER

Course Number: *Version 1:* 130. *Version 2:* 130.

Location: *Version 1:* Service Support School, Cp. Lejeune, NC. *Version 2:* Engineer School, Cp. Lejeune, NC.

Length: *Version 1:* 5 weeks (179 hours). *Version 2:* 5-7 weeks (180-213 hours).

Exhibit Dates: *Version 1:* 9/92–Present. *Version 2:* 1/90–8/92.

Learning Outcomes: *Version 1:* Upon completion of the course, the student will be able to perform demolition of wood, concrete, and steel structures; engage in mine warfare techniques; erect simple wood and frame structures; place concrete slabs; and erect prefabricated bridges. *Version 2:* Upon completion of the course, the student will be able to use camouflage, erect prefabricated bridges, perform demolition, and construct field fortifications.

Instruction: *Version 1:* Instruction includes lectures, demonstrations, and practical exercises in demolition of wood, steel, and concrete structures with explosives; land mine warfare; erection of prefabricated bridges; erection of simple wood frame structures; placement of simple concrete slabs; and use of appropriate hand and power tools. *Version 2:* Instruction includes lectures, demonstrations, and practical exercises in construction of bunkers, trenches, and shelters; demolition of wood and steel with explosives; land mine warfare; erection of prefabricated bridges; construction of a simple concrete foundation; rigging; and culvert placement.

Credit Recommendation: *Version 1:* In the vocational certificate category, 1 semester hour in basic wood frame construction and 1 in basic concrete placement (7/96). *Version 2:* Credit is not recommended because of the military-specific nature of the course (3/92).

MC-1601-0025

BASIC HYGIENE EQUIPMENT OPERATOR

Course Number: *Version 1:* 110. *Version 2:* 110.

Location: Engineer School, Cp. Lejeune, NC.

Length: *Version 1:* 10 weeks (333 hours). *Version 2:* 11-12 weeks (476 hours).

Exhibit Dates: *Version 1:* 9/93–Present. *Version 2:* 1/90–8/93.

Learning Outcomes: *Version 1:* Upon completion of the course, the student will be able to install plumbing, including pipe, plumbing fixtures, water supply lines, stacks, vents, drains, and waste disposal line; perform pH determination tests, chlorine residual tests, and coagulation for tests; develop a water point and a water source; operate and maintain a water purification unit; operate and maintain hygiene equipment, including field bath unit; decontamination apparatus, delousing unit, and laundry unit; and install and maintain latrines, urinals, and leaching fields. *Version 2:* Upon completion of the course, the student will be able to install plumbing, including pipe, plumbing fixtures, water supply lines, stacks, vents, drains, and waste disposal lines; perform a pH determination test, a chlorine residual test, a coagulation for test; develop a water point and a water source; operate and maintain a water purification unit; operate and maintain hygiene equipment including field bath unit, decontamination apparatus, delousing unit, and laundry unit; and install and maintain latrines, urinals, and leaching fields.

Instruction: *Version 1:* Lectures, demonstrations and applications in basic plumbing installation; reconnaissance and development of a water point and a water source including routine control tests and operation and maintenance of water purification equipment; operation and maintenance of hygiene equipment, including field bath units, decontamination apparatus, laundry units, and delousing units; and installation and maintenance of latrines, urinals, and leaching fields. The course concludes with a field exercise where the student establishes a water point and sets up and operates hygiene equipment in a field environment. *Version 2:* Lectures, demonstrations and applications in basic plumbing installation; reconnaissance and development of a water point and a water source, including routine control tests operation and maintenance of water purification equipment; operation and maintenance of hygiene equipment, and including field bath units, decontamination apparatuses, laundry units, and delousing units; and installation and maintenance of latrines, urinals, and leaching fields. The course concludes with a field exercise where the student establishes a water point and sets up and operates hygiene equipment in a field environment.

Credit Recommendation: *Version 1:* In the lower-division baccalaureate/associate degree category, 2 semester hours in plumbing, 2 in water purification/sewage treatment, 1 in hygiene equipment operation, and 2 in water testing and purification (7/96). *Version 2:* In the vocational certificate category, 2 semester hours in basic plumbing, 3 in water/sewerage treatment, 2 in hygiene equipment operator, and 3 in water testing and purification (12/88).

MC-1601-0026

COMBAT ENGINEER OFFICER
(Combat Engineer)

Course Number: *Version 1:* ACC. *Version 2:* ACC.

Location: Engineer School, Cp. Lejeune, NC.

Length: *Version 1:* 13-447 weeks (455 hours). *Version 2:* 11-12 weeks (361-455 hours).

Exhibit Dates: *Version 1:* 5/94–Present. *Version 2:* 1/90–4/94.

Learning Outcomes: *Version 1:* Upon completion of the course, the student will be able to provide leadership in the areas of equipment capabilities, organization, and maintenance; soil stabilization; design and placement of simple drainage systems; design and layout of simple wood structures and concrete slabs; demolitions with explosives; prefabricated bridge placement; fortification and minefield design and placement; and construction management. *Version 2:* Upon completion of the course, the student will be able to provide technical assistance and leadership in engineering supervision with emphasis in equipment organization, maintenance, and transportation; fortification and camouflage; demolitions with explosives; reconnaissance for road selection; vertical construction; and prefabricated bridge erection.

Instruction: *Version 1:* Instruction includes lectures, demonstrations, and practical exercises in construction equipment maintenance and operations; use of explosives for demolition; soil testing and classification by type; basic design and layout of simple wood structures and concrete slabs; classification and erection of prefabricated bridges; construction management using GANT and CPM schedules; and fortification and minefield placement and clearing. *Version 2:* Instruction includes lectures, demonstrations, and practical exercises on management functions of equipment operation and maintenance; site clearing with explosives; wood frame structure construction; estimation techniques for cut, fill, and building; minefield clearing and installation; and job planning and management.

Credit Recommendation: *Version 1:* In the lower-division baccalaureate/associate degree category, 3 semester hours in basic construction materials and methods and 3 in construction management (7/96). *Version 2:* In the lower-division baccalaureate/associate degree category, 3 semester hours in construction management/supervision (5/94).

MC-1601-0027

ENGINEER OPERATIONS CHIEF

Course Number: *Version 1:* 13G. *Version 2:* 13G.

Location: Engineer School, Cp. Lejeune, NC.

Length: *Version 1:* 7-8 weeks (252 hours). *Version 2:* 7-8 weeks (252 hours).

Exhibit Dates: *Version 1:* 8/93–Present. *Version 2:* 1/90–7/93.

Learning Outcomes: *Version 1:* Upon completion of the course, the student will be able to perform management functions in the areas of equipment organization and maintenance, basic wood and concrete construction, demolitions with explosives, bridging techniques, basic road and drainage planning, base camp layout, and construction management. *Version 2:* Upon completion of the course, the student will be able to provide technical assistance and engineering supervision with primary emphasis on equipment organization, maintenance, and transportation, in the areas of demolitions, camouflage of fortified position, vertical construction, military road planning and construction, airfield construction, and base camp layout and construction.

Instruction: *Version 1:* Instruction includes lectures, demonstrations, and practical exercises in basic wood frame and concrete construction; construction equipment use, selection, organization, and maintenance; uses of explosives in demolitions; bridging techniques; road and drainage design and layout; base camp planning and layout; and use of GANT and CPM for construction management. *Version 2:* Instruction includes lectures, demonstrations, and practical exercises on management functions of equipment operation and maintenance; demolition of wood and steel barriers with explosives; mine warfare; camouflage techniques; construction of a wood frame structure; haul road location; cut and fill estimation; and base camp layout, including traffic flow, drainage, utilities, and billeting.

Credit Recommendation: *Version 1:* In the lower-division baccalaureate/associate degree category, 3 semester hours in basic methods and materials of construction and 3 in construction management (7/96). *Version 2:* In the lower-division baccalaureate/associate degree category, 4 semester hours in construction management/supervision (9/87).

MC-1601-0028

1. COMBAT ENGINEER NONCOMMISSIONED OFFICER (NCO)
2. JOURNEYMAN COMBAT ENGINEER
3. JOURNEYMAN COMBAT ENGINEER

Course Number: *Version 1:* ACS. *Version 2:* ACS. *Version 3:* ACS.
Location: Engineer School, Cp. Lejeune, NC.
Length: *Version 1:* 10 weeks (363 hours). *Version 2:* 9 weeks (323 hours). *Version 3:* 13-14 weeks (460-493 hours).
Exhibit Dates: *Version 1:* 3/95–Present. *Version 2:* 2/91–2/95. *Version 3:* 1/90–1/91.
Learning Outcomes: *Version 1:* Upon completion of the course, the student will be able to perform basic shop maintenance and management, classify soil types, size and install simple drainage systems, design and lay out simple wood frame structures and concrete slabs, assist in installment of fixed and floating bridges, perform demolitions, and construct fortifications. *Version 2:* Upon completion of the course, the student will be able to provide technical assistance in engineering tasks on the battlefield, including maintenance management, horizontal and vertical construction, and road and bridge classification. *Version 3:* Upon completion of the course, the student will be able to provide technical assistance for the mobilization of troops and equipment on the battlefield, for the counter mobilization of enemy forces on the battlefield, and for survival on the battlefield.
Instruction: *Version 1:* Lectures include classroom and field demonstrations in shop maintenance, soil testing and classification as to type; simple culvert sizing and placement, and drainage area determination; design of simple wood and concrete structures; types and placement of fixed and floating bridges; use of explosives for demolition; and construction of basic fortifications including trenches, bunkers, and minefields. *Version 2:* Lectures include classroom and field demonstrations in shop maintenance, soils and earthwork estimating, design of bunkers and shelters, and demolition and obstacle breaching. *Version 3:* Course includes lectures, demonstrations, application, and program instruction in engineer organization, the use of engineering tools and equipment, leadership, map reading, reconnaissance bridging, rafting, communications, road construction, and obstacle breaching. Course also covers obstacle construction, land mine warfare, fire support, obstacle planning, and camouflage and fortified positions, as well as utilities equipment, welding, blueprint reading, project management, vertical construction, airfield construction, and maintenance management.
Credit Recommendation: *Version 1:* In the lower-division baccalaureate/associate degree category, 3 semester hours in basic methods and materials of construction, specifically wood, concrete, and soil (7/96). *Version 2:* In the vocational certificate category, 3 semester hours in methods of construction (3/92). *Version 3:* In the lower-division baccalaureate/associate degree category, 2 semester hours in engineering construction (4/89).

MC-1601-0029

RESERVE COMBAT ENGINEER OFFICER

Course Number: CEZ.
Location: Engineer School, Cp. Lejeune, NC.
Length: 2 weeks (61-80 hours).
Exhibit Dates: 1/90–Present.
Learning Outcomes: Upon completion of the course, the student will be able to demonstrate knowledge of current doctrine and trends in combat engineering.
Instruction: Lectures, demonstrations, and practical exercises cover new technology in the engineering field, amphibious operations, and Soviet organization and equipment. Course also includes employment of heavy engineering equipment and utilities, bridging, US mine detectors, land mine warfare, engineer intelligence, reconnaissance, counter mobility through the use of obstacles, priming and firing demolitions, and field fortification and camouflage.
Credit Recommendation: Credit is not recommended because of the limited, specialized nature of the course (7/93).

MC-1601-0030

RESERVE COMBAT ENGINEER NONCOMMISSIONED OFFICER (NCO)

Course Number: *Version 1:* 43F. *Version 2:* 43F.
Location: Engineer School, Cp. Lejeune, NC.
Length: *Version 1:* 2 weeks (90 hours). *Version 2:* 2 weeks (78 hours).
Exhibit Dates: *Version 1:* 7/93–Present. *Version 2:* 1/90–6/93.
Learning Outcomes: *Version 1:* Upon completion of the course, the student will be able to perform basic reconnaissance functions, use explosives for demolition, construct basic fortifications, design and clear minefields, and design and lay out simple wood structures and concrete slabs. *Version 2:* Upon completion of the course, the student will demonstrate knowledge of current doctrine, equipment, and trends in combat engineering.
Instruction: *Version 1:* Course includes lectures, demonstrations, and practical exercises in placing and using explosives for demolition; designing, placing, and clearing minefields; designing and placing fighting positions; designing simple wood structures; and designing and placing simple concrete slabs. *Version 2:* Lectures, demonstrations, and practical applications in field fortifications and camouflage, including camouflage of equipment, construction of bunkers and shelters, and construction of barriers and obstacles. Course also includes demolitions, including demolition safety, tools and equipment, military explosives, torpedoes and shaped charges, priming and firing systems, explosive created obstacles, land clearing with explosives, steel cutting, and bridge demolition with explosives. Also covered is information on land mine warfare and bridging including the erection of a double-story medium girder bridge using link reinforcement.
Credit Recommendation: *Version 1:* Credit is not recommended because of the military-specific nature of the course (7/96). *Version 2:* Credit is not recommended because of the limited, specialized nature of the course (9/87).

MC-1601-0031

1. RESERVE ENGINEER EQUIPMENT SUPERVISOR PHASE 2
2. RESERVE ENGINEER EQUIPMENT SUPERVISOR

Course Number: *Version 1:* CEY. *Version 2:* CEY.
Location: *Version 1:* Phase 1, Local Reserve Training Centers, Various locations; Phase 2, Engineer School, Cp. Lejeune, NC. *Version 2:* Engineer School, Cp. Lejeune, NC.
Length: *Version 1:* 2 weeks (63-92 hours). *Version 2:* 2 weeks (63-92 hours).
Exhibit Dates: *Version 1:* 7/93–Present. *Version 2:* 1/90–6/93.
Learning Outcomes: *Version 1:* Upon completion of the course, the student will be able to perform basic equipment operations and maintenance shop operations for heavy equipment. *Version 2:* Upon completion of the course, the student will be able to perform basic equipment operations and maintenance shop operations for heavy equipment.
Instruction: *Version 1:* Instruction includes lectures, demonstrations, and practical exercises in the Marine Corps maintenance management system; equipment records and forms; inspection of a maintenance facility; operation of equipment, including backhoes, tractor/scrapers, road graders, forklifts, and cranes; preventive maintenance; logistic support estimation; and severe weather operation. *Version 2:* Instruction includes lectures, demonstrations, and practical exercises in the Marine Corps maintenance management system; equipment records and forms; inspection of a maintenance facility; operation of equipment, including backhoes, tractor/scrapers, road graders, forklifts, and cranes; preventive maintenance; logistic support estimation; and severe weather operation.
Credit Recommendation: *Version 1:* Credit is not recommended because of the military-specific nature of the course (7/96). *Version 2:* Credit is not recommended because of the limited, specialized nature of the course (6/89).

MC-1601-0032

ARTILLERY SURVEY INSTRUMENTS BY CORRESPONDENCE

Course Number: 08.12.
Location: Marine Corps Institute, Washington, DC.
Length: Maximum, 52 weeks.
Exhibit Dates: 1/90–Present.
Learning Outcomes: Upon completion of the course, the student will be able to set up, align, and operate either an M2 or M2A2 aiming circle and will be able to measure angles and declinate either instrument. The student will also be able to set up a T-16 theodolite and measure both horizontal and vertical angles. The student will be able to set up and operate Distomat/DI4L electronic distance measuring devices and be able to correct readings from slope to horizontal and for environmental changes. The student will be able to set up and align the reflection prism assembly for use with the Distomat/DI4L.
Instruction: Materials in this self-instruction course include reading assignments, diagrams, and objective questions. A proctored final exam is taken when the student is ready, and a score of 70 percent or better is needed to pass the course. Topics include setup and opera-

tion of an aiming circle, theodolite, and electronic distance measuring device.

Credit Recommendation: In the lower-division baccalaureate/associate degree category, 1 semester hour in basic surveying (6/89).

MC-1601-0033

INTERIOR WIRING BY CORRESPONDENCE

Course Number: 11.43.
Location: Marine Corps Institute, Washington, DC.
Length: Maximum, 52 weeks.
Exhibit Dates: 1/90–Present.
Learning Outcomes: Upon completion of the course, the student will be able to identify and know the purpose for conductors, boxes and covers, conduit and conduit fittings, switches and receptacles, and lighting fixtures; install, service and distribute panels; perform circuit installation with nonmetallic cable and conduit; and troubleshoot, repair, and maintain distribution systems.
Instruction: This correspondence course contains reading assignments and study questions with answers to study questions at the end of each unit. Materials include planning, installing, and maintaining interior wiring for buildings. The student will demonstrate achievement of objectives by answering 70 percent of the questions correctly on a proctored final examination.
Credit Recommendation: In the vocational certificate category, 6 semester hours in basic electrical installation (6/89); in the lower-division baccalaureate/associate degree category, 2 semester hours (mechanical and electrical equipment of buildings) in an architectural or construction management technology program (6/89).

MC-1601-0034

ENGINEER EQUIPMENT OPERATOR BY CORRESPONDENCE

Course Number: 13.31i.
Location: Marine Corps Institute, Washington, DC.
Length: Maximum, 52 weeks.
Exhibit Dates: 1/90–Present.
Learning Outcomes: Upon completion of the course, the student will be able to identify the duties of an equipment operator and determine proper procedures for equipment maintenance; understand basic engine principles and systems; identify factors for efficient use of earth moving machinery; plan order of operation; read surveyors' grade stakes; identify principal soil types; identify and operate tractors and tractor-drawn equipment; identify and operate material handling (forklifts) equipment; identify decontamination methods, fording procedures, and uses of wire rope.
Instruction: Materials for this correspondence course include reading assignments, diagrams, and objective questions. A proctored final exam is taken when the student is ready, and a score of 70 percent or better is needed to pass the course.
Credit Recommendation: In the lower-division baccalaureate/associate degree category, 1 semester hour in construction planning and 2 in heavy equipment operation and/or maintenance (6/89).

MC-1601-0035

AIMING CIRCLE OPERATOR BY CORRESPONDENCE

Course Number: 08.11.
Location: Marine Corps Institute, Washington, DC.
Length: Maximum, 52 weeks.
Exhibit Dates: 1/90–Present.
Learning Outcomes: Upon completion of the course, the student will be able to set up and operate either an M2 or an M2A2 aiming circle. The student will be able to match components with functions, declinate and orient the instrument, and pass along basic commands for the artillery firing battery. The student will also be able to determine basic direction by simultaneous observation, polaris, or directional traverse.
Instruction: Materials for this correspondence course include reading assignments, maps, diagrams, and objective questions. A proctored final exam is taken when the student is ready, and a score of 70 percent or better is needed to pass the course. Topics covered include setup, operation, and orientation of an aiming circle.
Credit Recommendation: Credit is not recommended because of the military-specific nature of the course (6/89).

MC-1601-0036

COMBAT ENGINEER NONCOMMISSIONED OFFICER (NCO) BY CORRESPONDENCE

Course Number: 13.19h; 13.19i.
Location: Marine Corps Institute, Washington, DC.
Length: Maximum, 52 weeks.
Exhibit Dates: 1/90–Present.
Learning Outcomes: Upon completion of the course, the student will be able to identify basic construction techniques, including building layout, concrete, forms and foundation, wood framing, and construction safety; perform standard rigging procedures with such devices as a gin pole, tripod, and shears; work with wire rope; layout, plan, and construct simple bridging and drainage structures, including simple stringer bridges, rope footbridges, open channels, culverts, and storm drains.
Instruction: Materials for this correspondence course include reading assignments, diagrams, charts, and objective questions. A proctored final exam is taken when the student is ready, and a score of 70 percent or better is needed to pass the course.
Credit Recommendation: In the lower-division baccalaureate/associate degree category, 2 semester hours in methods of basic and building construction in an architectural, civil, or construction management technology program (4/89).

MC-1601-0037

ENGINEER FORMS AND RECORDS BY CORRESPONDENCE

Course Number: 13.42c.
Location: Marine Corps Institute, Washington, DC.
Length: Maximum, 52 weeks.
Exhibit Dates: 1/90–Present.
Learning Outcomes: Upon completion of the course, the student will be able to complete standardized forms used in operating and maintaining engineering equipment.
Instruction: This self-paced, self-instruction course contains reading assignments and study questions with answers to study questions at the end of each unit. Materials include maintenance systems and publications, administrative control and operational records, and maintenance records. The student will demonstrate achievement of objectives by answering correctly 70 percent of the questions on a proctored final examination.
Credit Recommendation: Credit is not recommended because of the limited, specialized nature of the course (6/89).

MC-1601-0038

MAINTENANCE OF BULK FUEL EQUIPMENT BY CORRESPONDENCE

Course Number: 13.39.
Location: Marine Corps Institute, Washington, DC.
Length: Maximum, 52 weeks.
Exhibit Dates: 1/90–6/90.
Learning Outcomes: Upon completion of the course, the student will be familiar with Marine Corps maintenance system procedures and troubleshooting procedures for the care and handling of bulk fuel systems and equipment.
Instruction: This course content includes diagnosis, troubleshooting, and repair of bulk fuel systems, including pumps, tanks, lubrication requirements, safety procedures, and preventive maintenance.
Credit Recommendation: In the vocational certificate category, 1 semester hour in maintenance of bulk fuel systems (6/89).

MC-1601-0039

INTRODUCTION TO FIELD ARTILLERY SURVEY BY CORRESPONDENCE

Course Number: 08.9.
Location: Marine Corps Institute, Washington, DC.
Length: Maximum, 52 weeks.
Exhibit Dates: 1/90–Present.
Learning Outcomes: Upon completion of the course, the student will be able to conduct field surveys in support of field artillery applications, demonstrate understanding of elevations and measurement tools, and measure survey accuracy.
Instruction: This course is descriptive and informative but with limited skill transfer. Topics include conducting the field artillery survey, survey equipment and accessories, survey taking, measuring dissimilarities, and comparative taping accuracy. The course also includes practical exercises which reinforce the material.
Credit Recommendation: In the lower-division baccalaureate/associate degree category, 1 semester hour in surveying (9/89).

MC-1601-0040

INFANTRY STAFF NONCOMMISSIONED OFFICER (NCO): MILITARY SYMBOLS AND OVERLAYS BY CORRESPONDENCE

Course Number: 03.97.
Location: Marine Corps Institute, Washington, DC.
Length: Maximum, 52 weeks.
Exhibit Dates: 1/90–Present.
Learning Outcomes: Upon completion of the course, the student will be able to use one-color/multicolor map symbols, understand geometric forms used in map symboling for military units, and prepare and use overlays.
Instruction: This is a descriptive course which provides the student with the skill to use and recognize military symbols on maps and

overlays, information on the development of the military symbol, location and content of fields, marine air symbols, and tactical symbols. Two application exercises effectively reinforce all subject material.

Credit Recommendation: Credit is not recommended because of the military-specific nature of the course (9/89).

MC-1601-0041

CONSTRUCTION PRINT READING BY
  CORRESPONDENCE

Course Number: 13.44b; 13.44c.
Location: Marine Corps Institute, Washington, DC.
Length: Maximum, 52 weeks.
Exhibit Dates: 1/90–Present.
Learning Outcomes: Upon completion of the course, the student will be able to recognize terms and symbols used in construction prints; use an architect's scale; and read and intercept plot, structural, and utility plans.
Instruction: Students are required to pass a proctored end-of-course examination to successfully complete this correspondence course.
Credit Recommendation: In the lower-division baccalaureate/associate degree category, 3 semester hours in architectural blueprint reading (11/90).

MC-1601-0042

RESERVE COMBAT ENGINEER, PHASE 2

Course Number: EAT.
Location: Engineer School, Cp. Lejeune, NC.
Length: 2 weeks (92 hours).
Exhibit Dates: 1/90–Present.
Learning Outcomes: Upon completion of the course, the students will be able to estimate charge required, install explosives, and execute demolition of wood, concrete, and steel structures and fortifications, as well as describe the construction of single story bridges, military roads, and temporary airfields.
Instruction: Course includes lectures with some demonstration and limited practical experience. Instruction focuses on land mine warfare, demolitions, obstacles and obstacle breaching, engineer reconnaissance, bridging, field fortifications, and horizontal construction.
Credit Recommendation: Credit is not recommended because of the military-specific nature of the course (3/92).

MC-1601-0043

ENGINEER EQUIPMENT OFFICER

Course Number: *Version 1:* ACN. *Version 2:* ACN.
Location: Engineer School, Cp. Lejeune, NC.
Length: *Version 1:* 10 weeks (338 hours). *Version 2:* 9-10 weeks (291-331 hours).
Exhibit Dates: *Version 1:* 9/93–Present. *Version 2:* 1/90–8/93.
Learning Outcomes: *Version 1:* Upon completion of the course, the student will be able to supervise and manage the planning and construction of military roads by reviewing soil characteristics, estimating logistical requirements, and monitoring project progress. *Version 2:* Upon completion of the course, the student will be able to construct a military road and compute logistical requirements, including equipment and maintenance.

Instruction: *Version 1:* Course includes lectures, demonstrations, and practical exercises in analysis, road layout, earthwork estimating, equipment operation and maintenance, and project management. *Version 2:* Course includes lectures, demonstrations, and practical exercises in engine and power train construction and operation, road layout, earthwork, estimating, equipment operations, and maintenance management.
Credit Recommendation: *Version 1:* In the lower-division baccalaureate/associate degree category, 3 semester hours in horizontal construction and 2 in heavy equipment maintenance/management (7/96). *Version 2:* In the lower-division baccalaureate/associate degree category, 3 semester hours in civil technology and 3 in engine maintenance management (3/92).

MC-1601-0044

1. HYGIENE EQUIPMENT OPERATOR
    NONCOMMISSIONED OFFICER (NCO)
2. JOURNEYMAN HYGIENE EQUIPMENT OPERATOR

Course Number: *Version 1:* UAC. *Version 2:* UAC.
Location: Engineer School, Cp. Lejeune, NC.
Length: *Version 1:* 6-7 weeks (210 hours). *Version 2:* 15 weeks (493 hours).
Exhibit Dates: *Version 1:* 1/94–Present. *Version 2:* 1/90–12/93.
Learning Outcomes: *Version 1:* Upon completion of the course, the student will be able to maintain and repair plumbing systems, water purification and distribution equipment, and hygiene equipment. *Version 2:* Upon completion of the course, the student will be able to function as a journeyman hygiene operator.
Instruction: *Version 1:* Through lectures, demonstrations, and practical learning, students acquire the skills of a hygiene equipment operator, including maintenance management and troubleshooting of plumbing systems, water purification systems and GPM pumps, water distribution equipment, laundry and shower facilities, and water heater. *Version 2:* Course includes lectures and practical exercises in the skills of a journeyman hygiene equipment operator, including records management, plumbing, water treatment, basic electrical principles, equipment operation and maintenance (hygiene, water purification, deep well maintenance), and the development of field sanitation.
Credit Recommendation: *Version 1:* In the lower-division baccalaureate/associate degree category, 2 semester hours in plumbing and 2 in hygiene equipment operation (mechanical systems) (7/96). *Version 2:* In the vocational certificate category, 3 semester hours in plumbing, water supply, and water treatment operation; 3 in principles of electricity; and 3 in well maintenance (3/92).

MC-1601-0045

BASIC RECONNAISSANCE

Course Number: AHK; H-010-3923.
Location: Amphibious Base, Little Creek, VA; Landing Force Training Command, Pacific, San Diego, CA.
Length: 9 weeks (532 hours).
Exhibit Dates: 10/92–Present.
Learning Outcomes: Upon completion of the course, the student will be able to perform basic radio communication operations, apply land navigational techniques and prepare basic beach and hydrographic surveys, read and use topographic maps, understand basic global positioning systems (GPS), perform reconnaissance skills such as field photography and sketching, and apply principles and techniques of nonelectrical explosives.
Instruction: Methods include lectures, audiovisuals, and hands-on learning activities. The topics covered include radio communication skills, land navigation, reading and understanding of topographic maps, global positioning systems, demolitions, and physical skills.
Credit Recommendation: In the lower-division baccalaureate/associate degree category, 3 semester hours in topographic map reading and land navigation, 3 in physical education—swimming, and 2 in report writing (11/93).

MC-1601-0046

LANDING FORCE GROUND COMBAT
  COMMUNICATION STAFF PLANNING

Course Number: 0D5; H-4C-3331.
Location: Mobile training units, US; Landing Force Training Command, Pacific, San Diego, CA.
Length: 2 weeks (58 hours).
Exhibit Dates: 12/92–Present.
Learning Outcomes: Upon completion of the course, the student will be able to determine communications requirements and plan the allocation of communication means in support of the marine air-ground task force.
Instruction: A combination of prescreening tests, course orientation, lecture, practical application, and student performance is utilized.
Credit Recommendation: In the lower-division baccalaureate/associate degree category, 1 semester hour in map reading and navigation and 1 in radio and satellite communications (11/93).

MC-1601-0047

GROUND OPERATIONS SPECIALIST

Course Number: RCC; H-250-3166.
Location: Landing Force Training Command, Pacific, San Diego, CA.
Length: 2 weeks (59-66 hours).
Exhibit Dates: 11/92–Present.
Learning Outcomes: Upon completion of the course, the student will be able to describe staff organization and management, provide administrative assistance, read and prepare operations and situation maps; read and understand a topographic map, and understand fire support systems.
Instruction: Through the use of lecture, slides, transparencies and hands-on laboratory experiences topics covered include operations, supporting arms, logistics, intelligence, radio communications, embarkation, and maritime navigation.
Credit Recommendation: In the lower-division baccalaureate/associate degree category, 1 semester hour in map reading and navigation and 1 in physical education—orienteering (11/93).

MC-1601-0048

1. LAND NAVIGATION BY CORRESPONDENCE
2. INFANTRY SQUAD LEADER: LAND NAVIGATION
    BY CORRESPONDENCE

Course Number: *Version 1:* 03.81a. *Version 2:* 03.81.
Location: Marine Corps Institute, Washington, DC.
Length: *Version 1:* Maximum of, 52 weeks. *Version 2:* Maximum of, 52 weeks.
Exhibit Dates: *Version 1:* 2/92–Present. *Version 2:* 1/90–1/92.
Learning Outcomes: *Version 1:* Upon completion of the course, the student will be able to navigate over unfamiliar terrain with the use of a map, compass, and global positioning system receiver; plan routes; and prepare for travel. *Version 2:* Upon completion of the course, the student will be able to navigate over unfamiliar terrain with the use of a map and compass, plan routes, and prepare for travel.
Instruction: *Version 1:* The primary method of instruction is correspondence and includes reading, map interpretation, and practical exercises employing maps and the compass. *Version 2:* The primary method of instruction is correspondence and includes readings, map interpretations, and practical exercises employing maps and the compass.
Credit Recommendation: *Version 1:* In the lower-division baccalaureate/associate degree category, 1 semester hour in map reading and 1 in physical education—orienteering (3/94). *Version 2:* In the lower-division baccalaureate/associate degree category, 1 semester hour in map reading and 2 in physical education—orienteering (3/94).

MC-1606-0002

AERIAL NAVIGATION
(Air Navigation)

Course Number: None.
Location: Marine Aerial Navigation School, Mather AFB, CA.
Length: 24 weeks (834 hours).
Exhibit Dates: 1/90–1/97.
Objectives: To train enlisted personnel to be qualified aerial navigators.
Instruction: Course includes lectures and practical exercises in basic aerial navigation, meteorology, and radar principles and operation.
Credit Recommendation: In the lower-division baccalaureate/associate degree category, 2 semester hours in meteorology and 8 in navigation (11/86); in the upper-division baccalaureate category, 9 semester hours in advanced navigation (11/86).

MC-1606-0005

INTELLIGENCE FOR THE MARINE AIR-GROUND TASK FORCE BY CORRESPONDENCE

Course Number: 02.9a.
Location: Marine Corps Institute, Washington, DC.
Length: Maximum, 52 weeks.
Exhibit Dates: 1/90–Present.
Learning Outcomes: Upon completion of the course, the student will be able to describe the organization of a Marine intelligence section and functions of intelligence personnel at different levels; identify and understand various phases of the intelligence cycle; identify capabilities of remote sensors in developing intelligence; understand techniques and procedures in recording, processing, evaluating, and disseminating intelligence; and identify components of Marine air ground intelligence system.
Instruction: This correspondence course is designed to cover basics of intelligence collection and dissemination methodology utilized by marine air ground intelligence system.
Credit Recommendation: In the lower-division baccalaureate/associate degree category, 1 semester hour in intelligence (9/89).

MC-1606-0006

MILITARY OPERATIONS ON URBAN TERRAIN BY CORRESPONDENCE

Course Number: 03.66a.
Location: Marine Corps Institute, Washington, DC.
Length: Maximum, 52 weeks.
Exhibit Dates: 1/90–Present.
Learning Outcomes: Upon completion of the course, the student will be able to survive in an urban environment and use combat skills to dislodge an enemy from urban areas.
Instruction: This correspondence course teaches urban warfare characteristics, offensive and defensive urban warfare, and combat tactics in cities.
Credit Recommendation: Credit is not recommended because of the military-specific nature of the course (9/89).

MC-1606-0008

POSITION LOCATION REPORTING SYSTEM (PLRS) CAPABILITIES AND EMPLOYMENT BY CORRESPONDENCE

Course Number: 00.4.
Location: Marine Corps Institute, Washington, DC.
Length: Maximum, 52 weeks.
Exhibit Dates: 1/90–6/90.
Learning Outcomes: Upon completion of the course, the student will understand the ingredients of the PLRS, a technological system of tracking air and ground units.
Instruction: This descriptive course is aimed at providing information on history and mission, system description, capabilities, ground and air application, and application to artillery "hip shoots".
Credit Recommendation: Credit is not recommended because of the military-specific nature of the course (9/89).

MC-1606-0009

INTRODUCTION TO COMBAT INTELLIGENCE BY CORRESPONDENCE

Course Number: 02.8a.
Location: Marine Corps Institute, Washington, DC.
Length: Maximum, 52 weeks.
Exhibit Dates: 1/90–Present.
Learning Outcomes: Upon completion of the course, the student will be able to identify the principles of intelligence and describe the operations of a marine intelligence section.
Instruction: Course is descriptive and informative with no application or skill reinforcement. Topics include principles of organization of a battalion-level intelligence unit, methods of collection, source of collection, processing and dissemination of collected information, protecting classified information, and escape and evasion for intelligence personnel.
Credit Recommendation: Credit is not recommended because of the military-specific nature of the course (7/94).

MC-1606-0010

TERRORISM COUNTERACTION FOR MARINES BY CORRESPONDENCE

Course Number: 02.10.
Location: Marine Corps Institute, Washington, DC.
Length: Maximum, 52 weeks.
Exhibit Dates: 1/90–Present.
Learning Outcomes: Upon completion of the course, the student will be able to describe international terrorism, methods of targeting, the nature of the threat, prevention of acts of terrorism, and survival as a hostage.
Instruction: This course is descriptive and informative. Topics include background history of terrorism, protective measures against terrorism, and hostage survival.
Credit Recommendation: In the lower-division baccalaureate/associate degree category, 1 semester hour in introduction to terrorism (9/89).

MC-1606-0011

INTELLIGENCE BRIEF: THE OPPOSING FORCES NUCLEAR, BIOLOGICAL, CHEMICAL (NBC) THREAT BY CORRESPONDENCE

Course Number: 02.13.
Location: Marine Corps Institute, Washington, DC.
Length: Maximum, 52 weeks.
Exhibit Dates: 1/90–Present.
Learning Outcomes: Upon completion of the course, the student will be able to recognize and cope with chemical warfare directed at Marine units.
Instruction: This correspondence course covers the threat of chemical warfare originating from the Warsaw pact and includes equipment, personnel, and tactics.
Credit Recommendation: Credit is not recommended because of the military-specific nature of the course (9/89).

MC-1606-0012

INTELLIGENCE BRIEF: THE MILITARY THREAT BY CORRESPONDENCE

Course Number: 02.12.
Location: Marine Corps Institute, Washington, DC.
Length: Maximum, 52 weeks.
Exhibit Dates: 1/90–Present.
Learning Outcomes: Upon completion of the course, the student will be able to integrate knowledge of Soviet history, tactics, equipment, and organization in order to function as an NCO in a Marine unit.
Instruction: This correspondence course includes brief history of the Russian army, organizational structure of the army, study of the Soviet soldier, Soviet tactics, and specialized units of the Soviet military.
Credit Recommendation: Credit is not recommended because of the military-specific nature of the course (9/89).

MC-1606-0013

TACTICAL AIR CONTROL PARTY

Course Number: 674; G-2G-4318; H-2G-3615.
Location: Amphibious Base, Little Creek, VA; Landing Force Training Command, Pacific, San Diego, CA.
Length: 3 weeks (91-92 hours).

Exhibit Dates: 12/92–Present.

Learning Outcomes: Upon completion of the course, the student will be able to advise on effective deployment of aviation assets, air support, reconnaissance, and communications.

Instruction: Course includes lecture, demonstration, and practical exercises on air support requirements, planning, mission control, and fire support coordination.

Credit Recommendation: In the lower-division baccalaureate/associate degree category, 1 semester hour in map reading and 1 in radio communications (11/93).

MC-1703-0022

AUTOMOTIVE POWER TRAINS BY CORRESPONDENCE

Course Number: 35.9f.

Location: Marine Corps Institute, Washington, DC.

Length: Maximum, 52 weeks.

Exhibit Dates: 1/90–1/97.

Objectives: To introduce the Marine to automotive power trains by presenting, through correspondence course material, the principles and fundamentals of automotive power train components.

Instruction: This correspondence course in the study of automotive power trains presents clutches, clutch principles and elements, and operation and maintenance. The course also introduces transmissions, including manual shift transmission functions and types, and constant mesh and synchromesh transmission. Also included are components, bearings and lubrication, transfer assembly, spray unit construction and operation, service and adjustment, universal joints, propeller shafts and slip joints, differential and axle assemblies, and service and repair. This correspondence course does not include laboratory exercises. There is a proctored final examination.

Credit Recommendation: In the lower-division baccalaureate/associate degree category, 2 semester hours in theory of automotive power trains (8/83).

MC-1703-0024

AUTOMOTIVE ENGINE MAINTENANCE AND REPAIR BY CORRESPONDENCE

Course Number: 35.8c; 35.8b.

Location: Marine Corps Institute, Washington, DC.

Length: Maximum, 52 weeks.

Exhibit Dates: 1/90–1/97.

Objectives: To introduce the Marine candidate, through correspondence, to the field of automotive engine maintenance and repair and to increase knowledge of construction, operation, malfunction diagnosis, maintenance, repair, and overhaul of the internal combustion engine.

Instruction: This correspondence course in automotive engine maintenance and repair covers construction and operation, engines and engine component design, engine blocks, crankshaft, fly wheels, piston connecting rods, cylinder heads, valve mechanisms, engine malfunctions, diagnosis and remedy, engine disassembly, and engine repair, assembly, and testing (basic engine). This correspondence course does not include laboratory exercises. A proctored final examination is required.

Credit Recommendation: In the lower-division baccalaureate/associate degree category, 2 semester hours in maintenance and repair of automotive engines (8/83).

MC-1703-0025

MOTOR TRANSPORT STAFF NONCOMMISSIONED OFFICER (NCO)

Course Number: Version 1: 35F. Version 2: 35F. Version 3: 35F.

Location: Version 1: Motor Transport School, Cp. Lejeune, NC. Version 2: Engineer School, Cp. Lejeune, NC. Version 3: Service Support School, Cp. Lejeune, NC.

Length: Version 1: 6 weeks (281 hours). Version 2: 9-11 weeks (312 hours). Version 3: 9-11 weeks (276-348 hours).

Exhibit Dates: Version 1: 2/96–Present. Version 2: 8/92–1/96. Version 3: 1/90–7/92.

Learning Outcomes: Version 1: Upon completion of the course, the student will be able to establish and administer a motor vehicle fleet maintenance program; monitor safety programs and driver licensing procedures; perform vehicle inspections; recover disabled vehicles; perform special vehicle operations such as deep water fording, vehicle loading, and night operations; and establish a vehicle convoy system to establish routes and defend a convoy. Version 2: Upon completion of the course, the student will be able to describe the Marine Corps integrated maintenance system; perform record keeping and reporting functions related to motor transport maintenance; establish a tactical motor pool; perform vehicle inspections; supervise vehicle preventive, corrective maintenance, parts procurement, and inventory; establish and maintain a vehicle publications library; investigate accidents; describe special vehicle operation such as vehicle recovery, deep water fording, vehicle camouflage, vehicle loading, and night driving; and defend a tactical convoy. Version 3: Upon completion of the course, the student will be able to describe the Marine Corps integrated maintenance management system; perform record keeping functions related to motor transport maintenance; establish a tactical motor pool; describe special vehicle operation, including vehicle recovery, deep water fording, vehicle camouflage, vehicle loading, and night driving operations; defend a tactical convoy; describe transmissions, transfer cases, axle assemblies, brake systems, steering systems, and suspension systems employed in current motor transport tactical vehicles; and perform preventive maintenance on these vehicles.

Instruction: Version 1: Instruction includes lectures, demonstrations, and practical exercises in the operation and administration of maintenance programs; record keeping; vehicle inspection and safety procedures; driver licensing procedures; vehicle recovery; deep water fording techniques; and route selection, vehicle type, communications, and defensive tactics for motor transport convoy operations. Version 2: Course covers motor transport maintenance management/administration; operations administration; special vehicle operations; tactical vehicle operations; tactical convoy operations; supervision of preventive maintenance, correction, and inspection; and diagnosing equipment problems. Version 3: Course covers motor transport maintenance management/administration; operations administration; special motor transport vehicle operations; tactical convoy operations; introduction to automotive components; test, measurement, and diagnostic equipment; preventive maintenance; repair of diesel vehicle engines and (MOS 3529) first echelon maintenance.

Credit Recommendation: Version 1: In the lower-division baccalaureate/associate degree category, 3 semester hours in vehicle maintenance management, 2 in vehicle and driver safety, and 2 in principles of supervision (7/96). Version 2: In the lower-division baccalaureate/associate degree category, 4 semester hours in maintenance management and 3 in principles of supervision (8/95). Version 3: In the lower-division baccalaureate/associate degree category, 2 semester hours in diesel engine tune-up, 2 in diesel accessory maintenance, and 6 in motor vehicle service and maintenance management (8/92).

MC-1703-0027

MOTOR VEHICLE OPERATOR

Course Number: Version 1: 35X. Version 2: 35X.

Location: Version 1: Engineer School, Cp. Lejeune, NC. Version 2: Service Support School, Cp. Lejeune, NC; Schools Battalion, Cp. Pendleton, CA.

Length: Version 1: 4 weeks (224 hours). Version 2: 7 weeks (224-253 hours).

Exhibit Dates: Version 1: 5/92–Present. Version 2: 1/90–4/92.

Learning Outcomes: Version 1: Upon completion of the course, the student will be able to demonstrate safe and proper operation of heavy wheeled motor vehicles; perform the duties of a military driver; perform required preventive maintenance; and maintain vehicle operational records. Version 2: Upon completion of the course, the student will be able to demonstrate safe and proper operation of motor vehicles, perform the duties of a military driver, and demonstrate acquired preventive maintenance responsibilities.

Instruction: Version 1: Instruction includes lectures, laboratory exercises, demonstration, and practical application of driving ability. Also included are vehicle preventive maintenance; traffic rules and regulations; use of required trip reports; clearance forms, operation reports; weight limitations; and proper driving techniques. Version 2: Methodology includes lectures, laboratory exercises, demonstrations, and practical applications of driving ability. Content covers vehicle preventive maintenance; traffic rules and regulations; required military forms, including trip reports, clearance forms, and operations reports; weight limitations; and proper driving techniques.

Credit Recommendation: Version 1: In the vocational certificate category, 3 semester hours in truck driving (8/95); in the lower-division baccalaureate/associate degree category, 3 semester hours in vehicle preventive maintenance (8/95). Version 2: In the vocational certificate category, 2 semester hours in driver education (5/92); in the lower-division baccalaureate/associate degree category, 1 semester hour in automotive engine repair fundamentals and 2 in driveability diagnosis (5/92).

MC-1703-0028

AUTOMOTIVE ORGANIZATIONAL MAINTENANCE

Course Number: Version 1: 35H. Version 2: 35H.

Location: Service Support School, Cp. Lejeune, NC.

Length: *Version 1:* 12 weeks (459 hours). *Version 2:* 12 weeks (465 hours).

Exhibit Dates: *Version 1:* 8/92–Present. *Version 2:* 1/90–7/92.

Learning Outcomes: *Version 1:* Upon completion of the course, the student will be able to inspect, maintain, operate, service and repair heavy duty wheeled transport vehicles. *Version 2:* Upon successful completion of the course, the student will be able to inspect, service, and repair automotive vehicles.

Instruction: *Version 1:* Lectures and practical exercises in the inspection, maintenance, operation and repair of heavy duty wheeled transport vehicles, including shop procedures; introduction to tools and test equipment; heavy duty drive trains; heavy duty brakes; heavy duty steering and suspension; electrical system trouble diagnosis and component replacement. *Version 2:* Course includes lectures and practical exercises in the inspection, service, and repair of tactical automotive vehicles, including shop procedures, power plants, and fuel and electrical systems; fuel and electrical systems trouble diagnosis and tune-up; power trains; chassis, brakes, and suspension systems; diesel and multifuel engines; motor vehicle operation; and preventive maintenance techniques.

Credit Recommendation: *Version 1:* In the vocational certificate category, 2 semester hours in truck driving (8/96); in the lower-division baccalaureate/associate degree category, 3 semester hours in preventive maintenance/vehicle inspection, 2 in basic automotive electrical systems, 2 in heavy duty brakes, 2 in heavy duty steering and suspension, and 2 in heavy duty drive trains (8/96). *Version 2:* In the lower-division baccalaureate/associate degree category, 17 semester hours in automotive maintenance and repair (4/89).

MC-1703-0029

RESERVE MOTOR TRANSPORT OFFICER REFRESHER

Course Number: CEL.

Location: Service Support School, Cp. Lejeune, NC.

Length: 2 weeks (74-75 hours).

Exhibit Dates: 1/90–Present.

Learning Outcomes: Upon completion of the course, the student will be able to prepare maintenance schedules, describe Army and Marine Corps motor transport publications, discuss the Marine Corps integrated maintenance management system, use dispatch and operation forms and records, inspect vehicles to identify first echelon maintenance discrepancies, load and secure cargo, and organize elements of a convoy.

Instruction: Course covers technical publications, Marine Corps integrated maintenance management system, maintenance management, motor transport equipment, dispatch forms and records, vehicle inspection, preparation of vehicles for embarkation, vehicle loads and loading, and tactical convoy operation. Methodology includes lecture, demonstration, and practical application.

Credit Recommendation: In the lower-division baccalaureate/associate degree category, 2 semester hours in vehicle maintenance administration and management (4/89).

MC-1703-0030

LOGISTICS VEHICLE SYSTEM OPERATOR

Course Number: *Version 1:* 35Z. *Version 2:* 35Z.

Location: Service Support School, Cp. Lejeune, NC; Schools Battalion, Cp. Pendleton, CA.

Length: *Version 1:* 4-5 weeks (136-168 hours). *Version 2:* 4-5 weeks (169 hours).

Exhibit Dates: *Version 1:* 5/92–Present. *Version 2:* 1/90–4/92.

Learning Outcomes: *Version 1:* Upon completion of the course, the student will be able to perform basic vehicle preventive maintenance and operation and operate a vehicle in a tandem towing configuration. *Version 2:* Upon completion of the course, the student will be able to inspect and prepare a vehicle for safe and efficient operation for the mission and operate the vehicle in accordance with proper procedures including safety to occupants and surrounding vehicles.

Instruction: *Version 1:* Course covers proper vehicle operation techniques and emergency operating procedures, including tire changing, coupling and uncoupling, basic maneuvers, and tandem towing operations. Methodology includes lectures, demonstrations, practical applications, vehicle hands-on operation, and written and driving tests. *Version 2:* Course covers proper vehicle operation techniques, minor maintenance, emergency procedures, coupling and uncoupling procedures, and basic maneuvers. Methodology includes lab demonstration, troubleshooting using mock-ups and simulators, laboratory vehicle hands-on operation, and a written and driving examination.

Credit Recommendation: *Version 1:* In the vocational certificate category, 2 semester hours in driveability diagnosis and 1 in automotive technology (7/96). *Version 2:* In the lower-division baccalaureate/associate degree category, 6 semester hours in automotive technology (4/89); in the upper-division baccalaureate category, 2 semester hours in advanced automotive technology (4/89).

MC-1703-0031

AUTOMOTIVE FUEL AND EXHAUST SYSTEMS BY CORRESPONDENCE

Course Number: 35.25b.

Location: Marine Corps Institute, Washington, DC.

Length: Maximum, 52 weeks.

Exhibit Dates: 1/90–Present.

Learning Outcomes: Upon completion of the course, the student will be able to describe the characteristics of gasoline and diesel fuels, including volatility, detonation, and additives; describe the principles of carburetion and its related components; and describe basic diesel fuel and exhaust systems.

Instruction: This correspondence course presents information about the fundamentals of gasoline engine fuel systems including the tank lines, fuel pump, carburetor, filters, intake manifolds and governors. The maintenance of fuel tanks, fuel pumps, air cleaner carburetor, and fuel gauge are included. In addition, the fundamentals, maintenance, and repair of diesel engine pumps, injectors, governors, and superchargers are presented. Material is also presented on the purpose, type, and maintenance of exhaust systems. The course does not include laboratory exercises. There is a proctored final examination required.

Credit Recommendation: In the lower-division baccalaureate/associate degree category, 2 semester hours in carburetion and diesel fuel systems with introduction to fuel characteristics and exhaust system construction (6/89).

MC-1703-0032

1. COOLING AND LUBRICATION SYSTEM MAINTENANCE BY CORRESPONDENCE
2. AUTOMOTIVE COOLING AND LUBRICATING SYSTEMS BY CORRESPONDENCE

Course Number: *Version 1:* 35.13b. *Version 2:* 35.13a.

Location: Marine Corps Institute, Washington, DC.

Length: *Version 1:* Maximum, 52 weeks. *Version 2:* Maximum, 52 weeks.

Exhibit Dates: *Version 1:* 8/92–Present. *Version 2:* 1/90–7/92.

Learning Outcomes: *Version 1:* Upon completion of the course, the student will be able to identify the construction and main component parts of automotive cooling systems and lubricating systems and describe the correct maintenance procedures performed on these systems. *Version 2:* Upon completion of the course, the student will be able to identify the construction and main component parts of automotive cooling and lubricating systems.

Instruction: *Version 1:* This correspondence course consists of four study units which contain statements of unit objectives, text materials, illustrations of applicable equipment to explain the course objectives, and items to teach and fully demonstrate the inspection, correction, and proper maintenance of cooling systems. Course includes maintenance of cooling system parts, including radiator, water pump, belts and belt adjustments, thermostat, radiator, hoses, and causes of cooling system malfunctions, methods of flushing, and procedures for checking temperature sending units and gauges including removal and replacement. Course also covers lubrication system including sampling for oil analysis, selection and handling of lubricants, and disposal of used lubricants, filters, rags and other lubricant contaminated materials. Also included are system component operation; maintenance procedures; and oil pump removal, repair, and replacement procedures. Crank case ventilating system is included. Course does not include laboratory exercises. There is a proctored final examination required. *Version 2:* This correspondence course consists of four study units which contain statements of unit objectives, text materials, illustrations of applicable equipment to explain the course objectives, and items to teach and fully demonstrate the inspection, correction, and proper maintenance of cooling system. Course includes maintenance of cooling system parts, including radiator, water pump, belts and belt adjustments, thermostat, radiator, hoses, and causes of cooling system malfunctions, methods of flushing, and procedures for checking temperature sending units and gauges including removal and replacement. Course also covers lubrication system; system component operation; maintenance procedures; and oil pump removal, repair, and replacement procedures. Course also includes crank case ventilating system. Course does not include laboratory exercises. There is a proctored final examination required.

Credit Recommendation: *Version 1:* In the lower-division baccalaureate/associate degree category, 2 semester hours in vehicle lubrication and cooling system maintenance (8/95).

Version 2: In the lower-division baccalaureate/associate degree category, 1 semester hour in cooling and lubricating systems (6/89).

MC-1703-0033

AUTOMOTIVE BRAKE SYSTEM BY CORRESPONDENCE

Course Number: 35.15b.
Location: Marine Corps Institute, Washington, DC.
Length: Maximum, 52 weeks.
Exhibit Dates: 1/90–Present.
Learning Outcomes: Upon completion of the course, the student will be able to identify the components used in hydraulic and air brake systems used on military wheeled vehicles, describe the function of the components, and use the troubleshooting guides for each brake system. Maintenance and troubleshooting are not covered.
Instruction: This correspondence course in automotive brake systems includes introduction to the types of brake systems, brake drums, brake shoes, mechanical brake systems, hydraulic brake systems, construction, and adjustment. An introduction to air-hydraulic and vacuum assisted braking systems is included, as is introduction to air brake systems and foundation brakes. Troubleshooting charts for brake systems are introduced to the students. Course does not include laboratory exercises.
Credit Recommendation: In the lower-division baccalaureate/associate degree category, 1 semester hour in introduction to automotive hydraulic brakes and 1 in introduction to air brake systems (11/92).

MC-1703-0034

PREVENTIVE MAINTENANCE AND OPERATING TECHNIQUES FOR HEAVY VEHICLES BY CORRESPONDENCE

Course Number: 35.24a.
Location: Marine Corps Institute, Washington, DC.
Length: Maximum, 52 weeks.
Exhibit Dates: 2/90–Present.
Learning Outcomes: Upon completion of the course, individuals will be able to identify component systems and describe their function on heavy duty military wheeled vehicles, perform the necessary maintenance and operational checks on heavy duty wheeled vehicles, and assist the maintenance mechanics in the lubrication and maintenance of heavy duty wheeled vehicles.
Instruction: This is a correspondence course that presents information about the operation and maintenance of heavy duty tactical vehicles and includes characteristics and operation of power plant, power train, and chassis systems. The operator's role in preventive maintenance scheduling, record keeping, and inspection are included. The course does not have laboratory exercises, but does conclude with a proctored final examination.
Credit Recommendation: In the lower-division baccalaureate/associate degree category, 1 semester hour in motor vehicle maintenance (11/92).

MC-1703-0035

METALWORKER
(Basic Metal Worker)

Course Number: 704-1316.
Location: Marine Corps Detachment, Ordnance Center and School, Aberdeen Proving Ground, MD.
Length: 13 weeks (592 hours).
Exhibit Dates: 7/95–Present.
Learning Outcomes: Upon completion of the course, the student will be able to set up and use oxyacetylene welding, shielded metal arc welding (SMAW), gas metal arc welding (GMAW), gas tungsten arc welding (GTAW, TIG) equipment in all positions on ferrous and nonferrous metals.
Instruction: Lectures and laboratory exercises cover safety and equipment setup and operation for oxyacetylene welding and cutting; plasma cutting; and inert gas welding (both MIG and TIG) on ferrous, nonferrous, and armor plate. Also covered is care and use of tools and equipment as well as respirator use.
Credit Recommendation: In the lower-division baccalaureate/associate degree category, 2 semester hours in oxyacetylene and electric arc welding and 3 in MIG and TIG welding (6/97).

MC-1704-0004

TACTICAL AIR DEFENSE CONTROLLER
(Tactical Air Defense Controller Enlisted)

Course Number: FG5.
Location: Communication-Electronics School, Twentynine Palms, CA; Air Command and Control School, Twentynine Palms, CA.
Length: 5 weeks (171-199 hours).
Exhibit Dates: 1/90–7/91.
Objectives: After 7/91 see MC-1704-0009. To provide selected enlisted personnel with a thorough knowledge of the functioning and operations of ground controlled intercepts and air traffic control procedures in a clear environment.
Instruction: The course provides the formal training necessary for further qualification as a tactical air controller. Topics include air control communications, air intercept control, weapons systems characteristics, flight rules and regulations, and electronic warfare.
Credit Recommendation: Credit is not recommended because of the military-specific nature of the course (3/91).

MC-1704-0007

HYDRAULIC PRINCIPLES AND TROUBLESHOOTING BY CORRESPONDENCE

Course Number: 13.45.
Location: Marine Corps Institute, Washington, DC.
Length: Maximum, 52 weeks.
Exhibit Dates: 1/90–Present.
Learning Outcomes: Upon completion of the course, the student will be able to discuss the principles of force and pressure and develop introductory knowledge of hydraulic system components and troubleshooting procedures.
Instruction: This course covers hydraulic principles, system component identification, and troubleshooting procedures for hydraulic systems.
Credit Recommendation: In the vocational certificate category, 2 semester hours in hydraulic principles and introduction to hydraulic system components including troubleshooting procedures (6/89).

MC-1704-0008

AIRCRAFT MAINTENANCE NONCOMMISSIONED OFFICER (NCO) BY CORRESPONDENCE

Course Number: 60.1g.
Location: Marine Corps Institute, Washington, DC.
Length: Maximum, 52 weeks.
Exhibit Dates: 1/90–Present.
Learning Outcomes: Upon completion of the course, the student will be able to schedule required inspections as stated in maintenance publications and will describe the differences in inspection requirements. Student will apply correct methods of aviation ground handling, towing, securing, and servicing all fluid and oxygen systems. Student will identify causes and types of aircraft corrosion and methods and hazards of corrosion control.
Instruction: This is a correspondence course consisting of six units with workbooks and with test materials. The units include aircraft ground handling, aircraft inspection requirements, and aircraft corrosion control.
Credit Recommendation: In the lower-division baccalaureate/associate degree category, 2 semester hours in a general aircraft maintenance technician program (10/92).

MC-1704-0009

TACTICAL AIR DEFENSE CONTROLLER (AN/TYQ-23)

Course Number: 72N; FG5.
Location: Air Ground Combat Center, Twentynine Palms, CA.
Length: 6-7 weeks (218-229 hours).
Exhibit Dates: 8/92–Present.
Learning Outcomes: Before 8/91 see MC-1704-0004. Upon completion of the course, the student will have a thorough knowledge of the function and operation of ground controlled intercepts and air traffic control procedures.
Instruction: The course provides the formal training necessary for qualification as a tactical air controller. Topics include air control communications, air intercept control, weapons systems characteristics, flight rules and regulations, and electronic warfare.
Credit Recommendation: Credit is not recommended because of the military-specific nature of the course (3/96).

MC-1708-0002

LONG RANGE (OTH) MARITIME NAVIGATION

Course Number: H-2E-3743.
Location: Landing Force Training Command, Pacific, San Diego, CA.
Length: 3 weeks (107-129 hours).
Exhibit Dates: 1/93–Present.
Learning Outcomes: Upon completion of the course, the navigator will be able to plan and execute long range transit in small boats to specific land sites.
Instruction: Course includes lectures, demonstrations, and practical exercises on navigation, charts, piloting, electronic navigation, and mission planning.
Credit Recommendation: In the lower-division baccalaureate/associate degree category, 2 semester hours in maritime navigation techniques (11/93).

MC-1708-0003

COXSWAIN SKILLS

Course Number: 81K; H-2E-3741.
Location: Landing Force Training Command, Pacific, San Diego, CA.
Length: 6 weeks (236 hours).
Exhibit Dates: 7/93–Present.
Learning Outcomes: Upon completion of the course, the student will be able to perform duties required of a small boat coxswain in support of surface operations.
Instruction: Course includes lecture, practical application, demonstration, and testing in general navigation skills, coxswain skills, small boat operation, and launch and recovery from amphibious shipping.
Credit Recommendation: In the lower-division baccalaureate/associate degree category, 3 semester hours in small boat operations and navigation (11/93).

MC-1710-0001

BASIC ENGINEER EQUIPMENT MECHANIC

Course Number: 13B.
Location: Engineer School, Cp. Lejeune, NC.
Length: 8-11 weeks (280-412 hours).
Exhibit Dates: 1/90–2/91.
Objectives: To train enlisted personnel to operate and maintain heavy construction equipment.
Instruction: Course includes lectures, demonstrations and practical exercises in the operation, maintenance, troubleshooting, and repair of diesel and gasoline engines and basic engineering heavy construction equipment and systems.
Credit Recommendation: In the vocational certificate category, 8 semester hours in heavy construction equipment operation, maintenance, and repair (4/89).

MC-1710-0004

THEORY AND CONSTRUCTION OF TURBINE ENGINES BY CORRESPONDENCE

Course Number: 60.2.
Location: Marine Corps Institute, Washington, DC.
Length: Maximum, 52 weeks.
Exhibit Dates: 1/90–1/97.
Learning Outcomes: Upon completion of the course, the student will able to describe the basic physics of gas turbine engines and turbine engine construction and operation.
Instruction: Topics include physics of gas turbine engines, turbine engine construction, maintenance, and operation. A detailed study of turbine engine systems includes induction, exhaust, fuel metering, ignition, starting, fire protection, lubrication, and cooling.
Credit Recommendation: In the vocational certificate category, 2 semester hours in turbine engine theory, maintenance, and operation (7/87).

MC-1710-0005

PLUMBING AND SEWAGE DISPOSAL BY CORRESPONDENCE

Course Number: 11.21b.
Location: Marine Corps Institute, Washington, DC.
Length: Maximum, 52 weeks.
Exhibit Dates: 1/90–Present.
Learning Outcomes: Upon completion of the course, the student will be able to perform the functions of plumber in Marine Corps garrison and field facilities and describe the operation of home plumbing equipment.
Instruction: This is a self-instruction course containing reading assignments and study questions at the end of each unit. Students are required to pass a proctored end-of-course exam.
Credit Recommendation: In the vocational certificate category, 2 semester hours in plumbing (6/90).

MC-1710-0010

ASSAULT AMPHIBIOUS VEHICLE UNIT LEADER
(Amphibious Vehicle Unit Leader)
(Amphibious Vehicle Unit Leaders (Enlisted))

Course Number: None.
Location: Tracked Vehicle Operations School, Cp. Pendleton, CA; Schools Battalion, Cp. Pendleton, CA; Amphibious School, Cp. Pendleton, CA.
Length: 5-12 weeks (197-528 hours).
Exhibit Dates: 1/90–1/97.
Objectives: To train selected personnel to supervise amphibious vehicle crews.
Instruction: Course includes lectures and practical exercises in amphibious vehicle operations, including operation, inspection, and maintenance of power train, engines, and turrets; hull, track, and suspension systems; electrical system; communications equipment; indirect and direct fire gunnery procedures; tactics; logistics; and leadership training.
Credit Recommendation: Credit is not recommended because of the military-specific nature of the course (4/74).

MC-1710-0033

1. ENGINEER EQUIPMENT MECHANIC NONCOMMISSIONED OFFICER (NCO)
2. JOURNEYMAN ENGINEER EQUIPMENT MECHANIC

Course Number: *Version 1:* ACU. *Version 2:* ACU.
Location: Engineer School, Cp. Lejeune, NC.
Length: *Version 1:* 12 weeks (420 hours). *Version 2:* 16-18 weeks (567-613 hours).
Exhibit Dates: *Version 1:* 1/94–Present. *Version 2:* 1/90–12/93.
Learning Outcomes: *Version 1:* Upon completion of the course, the student will be able to repair gasoline and diesel engines and power trains. *Version 2:* Upon completion of the course, the student will be able to repair diesel and gasoline engines and power trains.
Instruction: *Version 1:* Course includes lectures, demonstrations, and practical exercises in the complete rebuilding of gasoline and diesel engines and their components and troubleshooting and repairing power trains, including hydraulic systems, hydraulic components, brake systems, steering systems, transmission, and track and final drives. *Version 2:* Course includes lectures, demonstrations, and practical exercises in the complete rebuilding of gasoline and diesel engines and their components and troubleshooting and repairing power trains, including hydraulic systems, hydraulic components, brake systems, steering systems, transmission, and track and final drives.
Credit Recommendation: *Version 1:* In the lower-division baccalaureate/associate degree category, 3 semester hours in engine rebuilding and 3 in engine and power trains (7/96). *Version 2:* In the lower-division baccalaureate/associate degree category, 4 semester hours in engine rebuilding and 6 in engine power trains (3/92).

MC-1710-0034

1. ENGINEER EQUIPMENT OPERATOR NONCOMMISSIONED OFFICER (NCO)
2. JOURNEYMAN ENGINEER EQUIPMENT OPERATOR
3. JOURNEYMAN ENGINEER EQUIPMENT OPERATOR

Course Number: *Version 1:* ACX. *Version 2:* ACX. *Version 3:* ACX.
Location: Engineer School, Cp. Lejeune, NC.
Length: *Version 1:* 8 weeks (269 hours). *Version 2:* 15-16 weeks (514-523 hours). *Version 3:* 15-16 weeks (514-523 hours).
Exhibit Dates: *Version 1:* 5/94–Present. *Version 2:* 7/90–4/94. *Version 3:* 1/90–6/90.
Learning Outcomes: *Version 1:* Upon completion of the course, the student will be able to operate engineer equipment and manage and supervise maintenance and supply functions including container handling; basic projects; planning and construction of roads; and the operation of earth-moving equipment, cranes, and excavating equipment. *Version 2:* Upon completion of the course, the student will be able to operate engineer equipment and supervise and manage maintenance and supply including container handling; basic projects planning and construction of roads; and the operation of earth-moving equipment, cranes, and excavating equipment. *Version 3:* Upon completion of the course, the student will be able to operate and maintain crawler tractors, tractor/scraper combination, road graders, vibratory compactors, forklifts, cranes, backhoes, and excavators; plan out hauling routes and equipment required; and compute the amount of soil to be removed or fill required to complete a job.
Instruction: *Version 1:* Course includes lectures on integrated maintenance management system, use of technical publications, stock lists, equipment records, equipment licensing procedures, transactions, and reports. Lectures also include project management, specifically production and logistical estimates; site supervision; briefing; and construction operations. Demonstrations and practical exercises are directed at expanding the student's knowledge and skills in equipment recovery, container handling, and construction operations. Safety control, maneuvering, and operation of engineer equipment are emphasized. Focus is on road construction and load testing. *Version 2:* Course includes lectures on integrated maintenance management system, use of technical publications, stock lists, equipment records, equipment licensing procedures, transactions, and reports. Lectures also include project management, specifically production and logistical estimates; site supervision; briefing; and construction operations. Demonstrations and practical exercises are directed at expanding the student's knowledge and skills in equipment recovery, container handling, and construction operations. Safety control, maneuvering, and operation of engineer equipment are emphasized. Focus is on road construction and load testing. *Version 3:* Course includes lectures on maintenance management of crawlers, tractor/scrapers, road graders, compactors, forklifts, cranes, backhoes, and excavators. Demonstrations and practical exercises cover the operation of crawlers, tractor/scrapers, road graders, com-

pactors, forklifts, cranes, backhoes, and excavators. Lectures, demonstrations, and applications in logistical estimation include planning hauling route and equipment required and computing the amount of soil to be removed or fill required.

Credit Recommendation: *Version 1:* In the vocational certificate category, 3 semester hours in heavy equipment operation and maintenance (7/96); in the lower-division baccalaureate/associate degree category, 3 semester hours in construction field operations (7/96). *Version 2:* In the vocational certificate category, 6 semester hours in operation and maintenance of heavy construction and materials handling equipment (3/92); in the lower-division baccalaureate/associate degree category, 3 semester hours in project management and 3 in construction management (3/92). *Version 3:* In the vocational certificate category, 6 semester hours in heavy equipment operation (9/87).

MC-1710-0035

UTILITIES CHIEF

Course Number: *Version 1:* 11E. *Version 2:* 11E.
Location: Engineer School, Cp. Lejeune, NC.
Length: *Version 1:* 10 weeks (344 hours). *Version 2:* 15-16 weeks (516 hours).
Exhibit Dates: *Version 1:* 9/93–Present. *Version 2:* 1/90–8/93.
Learning Outcomes: *Version 1:* Upon completion of the course, the student will be able to provide technical assistance in the supervision of the installation of utility systems, including electrical power sources, refrigeration and air conditioning, plumbing and water supply, and hygiene systems. *Version 2:* Upon completion of the course, the student will be able to provide technical assistance in the supervision of the installation of utility systems, including electrical power sources, refrigeration and air conditioning, plumbing and water supply, and hygiene systems.
Instruction: *Version 1:* Instruction includes lectures, demonstrations, and practical exercises in basic electrical concepts for AC and DC circuits, mobile electrical power sources, management of refrigeration and air conditioning systems, plumbing project management, water supply systems and purification, hygiene management for camp cleaning and sanitation, and basic maintenance procedures. *Version 2:* Instruction includes lectures, demonstrations, and practical exercises in basic electrical concepts for AC and DC circuits, mobile electrical power sources, management of refrigeration and air conditioning systems, plumbing project management, water supply systems and purification, hygiene management for camp cleaning and sanitation, and basic maintenance procedures.
Credit Recommendation: *Version 1:* In the lower-division baccalaureate/associate degree category, 2 semester hours in electrical systems, 2 in plumbing systems, and 2 in refrigeration/air conditioning for a total of 6 semester hours or 3 in mechanical and electrical systems (7/96). *Version 2:* In the vocational certificate category, 3 semester hours in refrigeration/air conditioning systems management, 3 in water supply treatment and management, and 4 in electrical systems management (9/88); in the lower-division baccalaureate/associate degree category, 1 semester hour in electrical blueprint reading, 3 in DC circuits, and 1 in AC circuits (9/88).

MC-1710-0036

UTILITIES OFFICER

Course Number: *Version 1:* ACE. *Version 2:* ACE.
Location: Engineer School, Cp. Lejeune, NC.
Length: *Version 1:* 6-7 weeks (218 hours). *Version 2:* 14 weeks (476 hours).
Exhibit Dates: *Version 1:* 1/94–Present. *Version 2:* 1/90–12/93.
Learning Outcomes: *Version 1:* Upon completion of the course, the student will be able to manage utility systems and supervise maintenance of refrigeration/air conditioning systems, water supply systems, hygiene, site sanitation, electrical systems, and field maintenance. *Version 2:* Upon completion of the course, the student will be able to manage utility systems and supervise maintenance of refrigeration/air conditioning systems, water supply systems, hygiene, site sanitation, electrical systems, and field maintenance.
Instruction: *Version 1:* Course includes lectures, demonstrations, and hands-on activities in AC/DC circuits, mobile electrical power sources, management of refrigeration and air conditioning systems, plumbing project management, water supply systems and purification, hygiene management for camp cleaning and sanitation, and basic maintenance procedures. *Version 2:* Lectures, demonstrations, and practical exercises cover basic electrical concepts for AC/DC circuits, mobile electrical power sources, management of refrigeration and air conditioning systems, plumbing project management, water supply systems and purification, hygiene management for camp cleaning and sanitation, and basic maintenance procedures.
Credit Recommendation: *Version 1:* In the lower-division baccalaureate/associate degree category, 1 semester hour in construction management (electrical), 1 in construction management (water supply/purification), and 1 in construction management (refrigeration/air conditioning) (7/96). *Version 2:* In the lower-division baccalaureate/associate degree category, 4 semester hours in construction management (electrical), 4 in construction management (refrigeration/air conditioning), and 4 in construction management (water supply and purification) (3/92).

MC-1710-0038

SEMITRAILER REFUELER OPERATOR

Course Number: *Version 1:* CO9. *Version 2:* CO9.
Location: Service Support School, Cp. Lejeune, NC.
Length: *Version 1:* 4 weeks (152-153 hours). *Version 2:* 4 weeks (152-153 hours).
Exhibit Dates: *Version 1:* 8/95–Present. *Version 2:* 1/90–7/95.
Learning Outcomes: *Version 1:* Upon completion of the course, the student will be able to perform basic preventive maintenance of a tractor semitrailer refueler, operate a tractor semitrailer refueler in the fueling services of fixed-wing aircraft, describe the safety operations and precautions in handling of fuel and hazardous materials, understand different fuel characteristics, demonstrate the loading and unloading of fuel of fixed- and rotary-wing aircraft, and diagnose troubles with the semitrailer refueler. *Version 2:* Upon completion of the course, the student will be able to perform preventive maintenance and operate a tractor semitrailer refueler in the fueling services of fixed-wing aircraft.
Instruction: *Version 1:* Lectures, demonstrations, and practical exercises cover understanding the operator's manuals, troubleshooting, and preventive maintenance; coupling and uncoupling of tractors and semitrailers; operation of tractors and semitractors; fuel characteristics and safety precautions in fuel handling and transport of hazardous materials; and fuel sampling and fueling and refueling operations for fixed- and rotary-wing aircraft. *Version 2:* Course covers the characteristics of fuel, loading and transporting of hazardous materials, fuel sampling, and fueling services to fixed-wing aircraft. Included is the operation and preventive maintenance of five-ton tactical tractor and semitrailer refueler. Methodology includes lecture, demonstration, and practical application.
Credit Recommendation: *Version 1:* In the vocational certificate category, 3 semester hours in tractor trailer operation (8/96); in the lower-division baccalaureate/associate degree category, 1 semester hour in hazardous material safety (aircraft fueling operations) (8/96). *Version 2:* In the vocational certificate category, 3 semester hours in tractor trailer operations (5/92); in the lower-division baccalaureate/associate degree category, 1 semester hour in heavy equipment maintenance and repair (5/92).

MC-1710-0039

CRANE OPERATOR BY CORRESPONDENCE

Course Number: 13.52a.
Location: Marine Corps Institute, Washington, DC.
Length: Maximum, 52 weeks.
Exhibit Dates: 1/90–Present.
Learning Outcomes: Upon completion of the course, the student will be able to identify the primary purpose of a crane and excavator and understand their lifting and fording capabilities; identify the basic attachments to cranes and excavators and understand their purpose; use and care for wire rope; determine basic safety factors in the operation of cranes and excavators and understand hand signals; and perform troubleshooting and lubrication procedures.
Instruction: This correspondence course includes reading assignments, diagrams, and objective questions. A proctored final exam is taken when the student is ready, and a score of 70 percent or better is needed to pass the course.
Credit Recommendation: In the lower-division baccalaureate/associate degree category, 1 semester hour in equipment maintenance and 1 in safety procedures (6/89).

MC-1710-0040

MATERIALS HANDLING EQUIPMENT BY CORRESPONDENCE

Course Number: 13.40.
Location: Marine Corps Institute, Washington, DC.
Length: Maximum, 52 weeks.
Exhibit Dates: 1/90–Present.
Learning Outcomes: Upon completion of the course, the student will be able to demon-

strate safe operation of the forklift, 7.5-ton crane, Lightweight Amphibious Container Handler, and Rough Terrain Container Handler.

Instruction: The topics presented include MC 4000 and MC 6000 forklift operation, 7.5-ton cranes, Lightweight Amphibious Container, and Rough Terrain Container Handler. Safety is stressed.

Credit Recommendation: In the vocational certificate category, 1 semester hour in heavy equipment operation (6/89).

MC-1710-0041

TRACTOR-TRAILER OPERATOR BY CORRESPONDENCE

Course Number: 35.33.
Location: Marine Corps Institute, Washington, DC.
Length: Maximum, 52 weeks.
Exhibit Dates: 1/90–Present.
Learning Outcomes: Upon completion of the course, the student will be able to describe the techniques necessary for safe tractor-trailer operation and gain an introductory understanding for driving, loading, safety inspection maintenance, and overall operation.
Instruction: The course of study includes vehicle inspection, control, loading, and safe operation of tractor-trailers. Emphasis is placed upon basic operation and safety procedures.
Credit Recommendation: In the vocational certificate category, 1 semester hour in introductory tractor-trailer operation and maintenance (6/89).

MC-1710-0043

ENGINEER EQUIPMENT CHIEF

Course Number: Version 1: 13E. Version 2: 13E.
Location: Engineer School, Cp. Lejeune, NC.
Length: Version 1: 10 weeks (345 hours). Version 2: 12 weeks (388 hours).
Exhibit Dates: Version 1: 9/93–Present. Version 2: 1/90–8/93.
Learning Outcomes: Version 1: Upon completion of the course, the student will be able to operate and supervise the maintenance of container-handling, earth-moving, and excavating equipment. Version 2: Upon completion of the course, the student will be able to operate and supervise the maintenance of container-handling, earth-moving, and excavating equipment.
Instruction: Version 1: Course includes lectures, demonstrations, and practical exercises in methods of power train production and transfer used in engineer equipment and in operation of engineer equipment, including crawler tractors, scrapers, rubber-tired tractors, and graders. Version 2: Course includes lectures, demonstrations, and practical exercises in methods of power train production and transfer used in engineer equipment and in operation of engineer equipment, including crawler tractors, scrapers, rubber-tired tractors, and graders.
Credit Recommendation: Version 1: In the lower-division baccalaureate/associate degree category, 2 semester hours in earth-moving equipment operation and maintenance (7/96). Version 2: In the vocational certificate category, 4 semester hours in earth-moving equipment operation and maintenance and 2 in engine repair and maintenance (3/92).

MC-1710-0044

1. BASIC EQUIPMENT OPERATOR
2. BASIC ENGINEER EQUIPMENT OPERATOR

Course Number: Version 1: 13F. Version 2: 13F.
Location: Engineer School, Cp. Lejeune, NC.
Length: Version 1: 10-11 weeks (350 hours). Version 2: 10-11 weeks (335-336 hours).
Exhibit Dates: Version 1: 1/95–Present. Version 2: 1/90–12/94.
Learning Outcomes: Version 1: Upon completion of the course, the student will be able to operate rubber-tired tractors, crawlers, road graders, backhoes, front-end loaders, excavators, and forklifts; and perform pre and post operational maintenance checks. Version 2: Upon completion of the course, the student will be able to maintain and operate rubber-tired tractors, crawlers, road graders, backhoes, front-end loaders, excavators, and forklifts.
Instruction: Version 1: Lecture, demonstration of engineering (earth moving and materials handling) equipment with practical experience in operation of rubber tired tractors, crawlers, road graders, backhoes, front-end loaders, excavators, and forklifts. Included are practical experiences in pre and post operational maintenance checks. Version 2: Course includes lecture, demonstration of engineering (earth-moving and materials-handling) equipment, and practical experience in preventive maintenance and operation of rubber-tired tractors, crawlers, road graders, backhoes, front-end loaders, excavators, and forklifts.
Credit Recommendation: Version 1: In the vocational certificate category, 5 semester hours in earth moving equipment operation and 2 in material handling equipment operation (8/95). Version 2: In the vocational certificate category, 5 semester hours in earth-moving equipment operation and maintenance and 1 in materials-handling equipment operation and maintenance (3/92).

MC-1710-0045

BASIC ENGINEER EQUIPMENT MECHANIC

Course Number: Version 1: 13B. Version 2: 13B.
Location: Engineer School, Cp. Lejeune, NC.
Length: Version 1: 10 weeks (317 hours). Version 2: 10 weeks (317-327 hours).
Exhibit Dates: Version 1: 1/94–Present. Version 2: 3/91–12/93.
Learning Outcomes: Version 1: Upon completion of the course, the student will be able to diagnose service and maintain; diesel engines; construction equipment power trains; electrical, hydraulic, brake and electrical systems. Version 2: Upon completion of the course, the student will be able to understand, test, and repair internal combustion engines and power train systems.
Instruction: Version 1: Instruction will include lecture, demonstration and practical experiences in the diagnosis, service and maintenance of diesel engines, construction equipment power trains, electrical, hydraulic, brake and electrical systems. Version 2: Lectures include laboratory instruction on hand tools and shop safety and maintenance and repair of internal combustion engines and power trains.

Credit Recommendation: Version 1: In the lower-division baccalaureate/associate degree category, 3 semester hours in diesel engine performance, 3 in heavy duty drive trains, 2 in basic hydraulics, and 2 in construction equipment maintenance (8/95). Version 2: In the vocational certificate category, 3 semester hours in engine and power train maintenance and repair (1/94).

MC-1710-0046

VEHICLE RECOVERY

Course Number: Version 1: CAJ. Version 2: CAJ.
Location: Engineer School, Cp. Lejeune, NC.
Length: Version 1: 5-6 weeks (200 hours). Version 2: 4 weeks (200 hours).
Exhibit Dates: Version 1: 4/95–Present. Version 2: 6/91–3/95.
Learning Outcomes: Version 1: Upon completion of the course, the student will be able to operate a five-ton wrecker and recover and tow disabled vehicles safely, perform simple oxyacetylene cutting and welding procedures, and use a wrecker crane. Version 2: Upon completion of the course, the student will have the knowledge and skills required to function effectively as a vehicle recovery man (Tow Truck).
Instruction: Version 1: Instruction includes lectures, demonstrations, and practical exercises in five-ton wrecker operations and maintenance, basic oxyacetylene welding and cutting techniques, safe towing procedures, and use of wrecker crane for loading and unloading operations. Version 2: Lectures and practical exercises in oxyacetylene cutting and welding, crane operations, loading operations, winch operations, tire changing procedures, auxiliary hydraulic operations, and towing procedures. Instruction is concentrated on vehicle recovery and towing procedures. Some minor wrecker maintenance is covered.
Credit Recommendation: Version 1: In the vocational certificate category, 3 semester hours in wrecker and towing procedures (7/96). Version 2: In the vocational certificate category, 4 semester hours in heavy duty vehicle recovery (8/95).

MC-1710-0047

M1A1 TANK SYSTEM MECHANIC

Course Number: A13GBN1.
Location: Armor Center and School, Ft. Knox, KY.
Length: 16-19 weeks (627-714 hours).
Exhibit Dates: 10/93–Present.
Learning Outcomes: Upon completion of the course, the student will be able to maintain the A1/M1A1 Abrams Tank.
Instruction: Instruction includes lectures, demonstrations, and performance exercises in testing, and troubleshooting systems, inspecting, servicing, lubricating, replacing, and adjusting components of the fire control system, main gun, gun turret drive, engine, fuel, exhaust, NBC, cooling, electrical, track, suspension, steering control, hydraulic systems, and engine power trains of the hull of the M1A1 Tank and M88A1 recovery vehicle.
Credit Recommendation: In the lower-division baccalaureate/associate degree category, 3 semester hours in automotive technology (3/97).

MC-1710-0048

UTILITIES OFFICER/CHIEF BY CORRESPONDENCE

Course Number: 11.69.
Location: Marine Corps Institute, Washington, DC.
Length: Maximum, 52 weeks.
Exhibit Dates: 2/92–Present.
Learning Outcomes: Upon completion of the course, the student will be able to supervise the installation of basic utility systems, including electrical power supply, refrigeration and air-conditioning, and water supply systems.
Instruction: Self-paced reading and study cover the areas of electrical power, electrical distribution systems, fundamentals of refrigeration, and general water supply systems. Students will complete a proctored final examination.
Credit Recommendation: In the lower-division baccalaureate/associate degree category, 1 semester hour in basic electrical distribution systems, 1 in basic refrigeration, and 1 in basic water supply systems (6/97).

MC-1710-0049

LIGHT ARMORED VEHICLE BASIC REPAIRMAN

Course Number: 611-2147; A01GBD1.
Location: Marine Corps Detachment, Ordnance Center and School, Aberdeen Proving Ground, MD.
Length: 9 weeks (342 hours).
Exhibit Dates: 3/96–Present.
Learning Outcomes: Upon completion of the course, the student will be able to perform entry level and limited maintenance, repair, and troubleshooting on the engine, transmission, suspension, electrical, pneumatic and hydraulic system of heavy vehicle equipment.
Instruction: Lectures, practical exercises, and performance examinations are used throughout this course.
Credit Recommendation: In the lower-division baccalaureate/associate degree category, 3 semester hours in heavy equipment maintenance (6/97).

MC-1710-0050

LIGHT ARMORED VEHICLE TECHNICIAN

Course Number: 643-2147 (OS); A01GBH1.
Location: Marine Corps Detachment, Ordnance Center and School, Aberdeen Proving Ground, MD.
Length: 10 weeks (379 hours).
Exhibit Dates: 3/96–Present.
Learning Outcomes: Upon completion of the course, the student will be able to perform advanced maintenance, repair, and troubleshooting on heavy vehicle equipment.
Instruction: Lectures, practical exercises, and performance examinations are used throughout this course. Automotive shop management is stressed.
Credit Recommendation: In the lower-division baccalaureate/associate degree category, 3 semester hours in heavy equipment maintenance and 1 in shop management (6/97).

MC-1710-0051

TANK SYSTEMS TECHNICIAN (M1A1)

Course Number: 643-2146.
Location: Marine Corps Detachment, Ordnance Center and School, Aberdeen Proving Ground, MD.
Length: 10-11 weeks (399 hours).
Exhibit Dates: 4/96–Present.
Learning Outcomes: Upon completion of the course, the student will be able to test, troubleshoot, replace, repair, and adjust selected components and systems on various vehicles.
Instruction: Lectures, demonstrations, and practical exercises include the use of tools and test equipment to service suspension, driveline, braking, electrical, and hydraulic systems.
Credit Recommendation: In the lower-division baccalaureate/associate degree category, 1 semester hour in automotive electricity and 2 in heavy equipment repair (6/97).

MC-1712-0001

INSTALLATION, OPERATION, AND MAINTENANCE OF DIESEL ENGINE DRIVEN GENERATOR SETS BY CORRESPONDENCE

Course Number: 11.19c.
Location: Marine Corps Institute, Washington, DC.
Length: Maximum, 52 weeks.
Exhibit Dates: 1/90–1/97.
Objectives: To provide personnel with basic training in the theory of operation and installation of field electrical generator equipment and in first aid for shock victims.
Instruction: This correspondence course contains reading assignments and written lessons with objective questions. Materials include the identification of the type of diesel-driven generators and their installation, operation, and maintenance. The student will demonstrate achievement of course objectives by answering correctly 65 percent of the questions on the proctored final examination.
Credit Recommendation: In the vocational certificate category, credit in first aid based on demonstrated skills (8/83).

MC-1712-0004

COMPRESSION IGNITION ENGINE FUEL SYSTEMS REPAIR

Course Number: 352.
Location: Service Support School, Cp. Lejeune, NC.
Length: 3 weeks (90 hours).
Exhibit Dates: 1/90–3/95.
Learning Outcomes: Upon completion of the course, the student will be able to repair and calibrate Detroit Diesel, Robert Bosch, Stanadyne, and Cummings PT fuel systems.
Instruction: Course covers the inspection, service, repairs, and tests necessary to repair and calibrate the fuel systems of automotive diesel engines. Methodology includes lecture, demonstration, and practical application.
Credit Recommendation: In the lower-division baccalaureate/associate degree category, 3 semester hours in diesel fuel system maintenance and repair (4/89).

MC-1712-0005

MOTOR TRANSPORT MAINTENANCE OFFICER

Course Number: CEB.
Location: Service Support School, Cp. Lejeune, NC.
Length: 2 weeks (64 hours).
Exhibit Dates: 1/90–3/95.
Learning Outcomes: Upon completion of the course, the student will be able to disassemble, inspect components, reassemble, and adjust an automotive generator, alternator, starter, diesel engine, fuel injection pump, front axle assembly with differential, automatic transmission, power steering gear, and air compressor.
Instruction: Course covers intermediate maintenance, organization, facilities and equipment, repair of electrical systems, diesel engine and engine accessory systems, power train systems used in military vehicles, and chassis system components. Methodology includes lectures and laboratory exercises.
Credit Recommendation: In the lower-division baccalaureate/associate degree category, 3 semester hours in automotive and diesel vehicle maintenance (4/89).

MC-1712-0006

FUNDAMENTALS OF DIESEL ENGINES BY CORRESPONDENCE

Course Number: *Version 1:* 13.35c. *Version 2:* 13.35b.
Location: Marine Corps Institute, Washington, DC.
Length: *Version 1:* Maximum, 52 weeks. *Version 2:* Maximum, 52 weeks.
Exhibit Dates: *Version 1:* 10/92–Present. *Version 2:* 1/90–9/92.
Learning Outcomes: *Version 1:* Upon completion of the course, the student will be able to identify and discuss basic internal combustion engine principles including diesel fuel injection and control. *Version 2:* Upon completion of the course, the student will be able to identify and discuss basic internal combustion engine principles including diesel fuel injection and control.
Instruction: *Version 1:* This correspondence course includes the advantages and disadvantages of the diesel engine, engine construction, fuel injection, diesel engine principles, combustion, and control of the engine. *Version 2:* This correspondence course includes the advantages and disadvantages of the diesel engine, engine construction, fuel injection, diesel engine principles, combustion, and control of the engine.
Credit Recommendation: *Version 1:* In the lower-division baccalaureate/associate degree category, 2 semester hours in fundamentals of the diesel engine with emphasis on diesel fuel delivery principles (8/95). *Version 2:* In the lower-division baccalaureate/associate degree category, 2 semester hours in fundamentals of the internal combustion engine with emphasis on diesel fuel delivery principles (6/89).

MC-1712-0007

DIESEL ENGINE MAINTENANCE AND TROUBLESHOOTING BY CORRESPONDENCE

Course Number: 13.43.
Location: Marine Corps Institute, Washington, DC.
Length: Maximum, 52 weeks.
Exhibit Dates: 1/90–Present.
Learning Outcomes: Upon completion of the course, the student will be able to use the multimeter as a diagnostic tool, describe diesel engine maintenance and troubleshooting procedures for different engine systems, and gain knowledge in power, hand tool, and special tool use.
Instruction: The course covers multimeter and special tool care and use. Power and advanced hand tool use will be introduced and

studied. Maintenance and troubleshooting of various diesel engines is studied in-depth.

Credit Recommendation: In the vocational certificate category, 3 semester hours in diesel diagnosis and troubleshooting procedures with special emphasis on multimeter use (6/89).

MC-1712-0008

AUTOMOTIVE INTERMEDIATE MAINTENANCE

Course Number: 35C.
Location: Engineer School, Cp. Lejeune, NC.
Length: 12-15 weeks (491-574 hours).
Exhibit Dates: 1/95–Present.
Learning Outcomes: Upon completion of the course, the students will have the technical knowledge and functional skills that will enable them to effectively perform the repairs to motor vehicle subassemblies and their associated components that are required to be accomplished by intermediate automotive mechanics.
Instruction: Through lectures demonstrations, and extensive practical applications in repair/overhaul of in-line and V-8 compression ignition engines; repair/overhaul of power transmissions and steering and brake system components.
Credit Recommendation: In the lower-division baccalaureate/associate degree category, 1 semester hour in repair of chassis system components, 3 in power transmission components, and 6 in repair of diesel engines (8/95).

MC-1712-0009

SMALL CRAFT MECHANIC

Course Number: EAW.
Location: Engineer School, Cp. Lejeune, NC.
Length: 16 weeks (551 hours).
Exhibit Dates: 2/95–Present.
Learning Outcomes: Upon completion of the course, the student will be able to repair small fiberglass and rubber craft and their subsystems, including engine, instruments, controls, power train, electrical system, and hull.
Instruction: Through lectures, demonstrations, and hands-on applications, the student receives instruction on preventive maintenance associated with small craft, as well as repair of outboard motors, outboard motor installation, small craft metal component repair, fiberglass hull repair, rubber craft repair, and diagnosis and repair of marine diesel engines.
Credit Recommendation: In the lower-division baccalaureate/associate degree category, 3 semester hours in diesel engines and 3 in marine small craft maintenance (7/96).

MC-1714-0011

BASIC ELECTRICIAN

Course Number: *Version 1:* 11B. *Version 2:* 11B.
Location: Engineer School, Cp. Lejeune, NC.
Length: *Version 1:* 7 weeks (219 hours). *Version 2:* 7 weeks (204-239 hours).
Exhibit Dates: *Version 1:* 7/93–Present. *Version 2:* 1/90–6/93.
Learning Outcomes: *Version 1:* Upon completion of the course, the student will be able to apply basic electrical principles, install fixed electrical distribution systems, and operate and maintain power generation units. *Version 2:* Upon completion of the course, the student will be able to operate generators, construct pole lines, and perform interior wiring.
Instruction: *Version 1:* Course includes lectures and laboratories in basic electrical principles; generator installation, operation, and maintenance; pole line construction and wiring; and preventive maintenance procedures for power generators. *Version 2:* Course includes lectures and laboratories in basic electrical principles; generator installation, operation, and maintenance; pole line construction and wiring; and preventive maintenance procedures for power generators.
Credit Recommendation: *Version 1:* In the lower-division baccalaureate/associate degree category, 2 semester hours in basic electricity and 2 in electrical wiring/service installation (7/96). *Version 2:* In the vocational certificate category, 3 semester hours in principles of electricity and 1 in electrical power laboratory (10/89).

MC-1714-0014

ELECTRICAL EQUIPMENT REPAIR SPECIALIST
(Electrical Equipment Repairman)
(Journeyman Electrical Equipment Repairman)

Course Number: *Version 1:* UAA. *Version 2:* UAA.
Location: Engineer School, Cp. Lejeune, NC.
Length: *Version 1:* 15-18 weeks (519-622 hours). *Version 2:* 17 weeks (576 hours).
Exhibit Dates: *Version 1:* 7/93–Present. *Version 2:* 1/90–6/93.
Learning Outcomes: *Version 1:* Upon completion of the course, the student will be able to repair electrical equipment and control devices, including electric motors and motor controls, electronic modules, mobile electrical power systems, electrical circuits, and electrical circuits of utilities equipment and generators. *Version 2:* Upon completion of the course, the student will be able to repair electrical equipment and control devices, including electric motors and motor controls, electronic modules, mobile electrical power systems, electrical circuits, and electrical circuits of utilities equipment other than generators.
Instruction: *Version 1:* Course includes lectures, demonstrations, and practical exercises in maintenance management, AC and DC current, test equipment, electric motors, motor controls, motor repair, diodes, transistors, solid state schematics, module repair, generators, and flood lights. *Version 2:* Course includes lectures, demonstrations, and practical exercises in maintenance management, AC and DC current, test equipment, electric motors, motor controls, motor repair, diodes, transistors, solid state schematics, module repair, generators, and flood lights.
Credit Recommendation: *Version 1:* In the lower-division baccalaureate/associate degree category, 4 semester hours in electrical equipment repair, 3 in AC/DC circuits, and 1 in solid state devices (7/96). *Version 2:* In the vocational certificate category, 12 semester hours in electrical equipment (3/92); in the lower-division baccalaureate/associate degree category, 3 semester hours in AC/DC circuits, 3 in motors/generators, and 1 in solid state devices (3/92).

MC-1714-0016

1. ELECTRICIAN NONCOMMISSIONED OFFICER (NCO)

2. JOURNEYMAN ELECTRICIAN

Course Number: *Version 1:* 11K. *Version 2:* 11K.
Location: Engineer School, Cp. Lejeune, NC.
Length: *Version 1:* 6 weeks (206 hours). *Version 2:* 9 weeks (285-295 hours).
Exhibit Dates: *Version 1:* 1/94–Present. *Version 2:* 1/90–12/93.
Learning Outcomes: *Version 1:* Upon completion of the course, the student will be able to design, plan, install, and inspect interior wiring and external electrical distribution systems; diagnose and repair electrical motors and motor control malfunctions; observe safety principles; and manage an electrical maintenance shop. *Version 2:* Upon completion of the course, the student will be able to perform as journeyman electrician.
Instruction: *Version 1:* Course includes lectures and practical exercises on maintenance management, electrical safety, electrical motors, and electrical distribution systems. *Version 2:* Course includes lectures and practical exercises on equipment and maintenance record management, AC and DC theory, motors and generators, reactance, impedance, inductance and capacitance, interior and exterior wiring, and pole line construction.
Credit Recommendation: *Version 1:* In the lower-division baccalaureate/associate degree category, 2 semester hours in basic electricity and 2 in electrical wiring/service installation (7/96). *Version 2:* In the lower-division baccalaureate/associate degree category, 3 semester hours in DC electrical theory, 1 in AC electrical theory, 2 in motors and generators, and 1 in electrical power laboratory (10/89).

MC-1714-0017

FUEL AND ELECTRICAL COMPONENT REPAIR

Course Number: ACP.
Location: Engineer School, Cp. Lejeune, NC.
Length: 4 weeks (332 hours).
Exhibit Dates: 7/95–Present.
Learning Outcomes: Upon completion of the course, the student will have the knowledge, techniques, and procedures to enable them to perform those inspections, services, repairs, tests and adjustments on components of the fuel, electrical, air induction, and brake systems featured on wheeled and tracked vehicles.
Instruction: Through lectures, demonstrations and extensive practical applications in repairing air induction system components, repairing of electrical system components, and repairing and calibrating of fuel system components.
Credit Recommendation: In the lower-division baccalaureate/associate degree category, 4 semester hours in heavy duty/heavy equipment air induction and fuel system components and 4 in heavy duty/heavy equipment electrical system components (8/95).

MC-1714-0018

FIRE AND ELECTRICAL SYSTEMS COMPONENT REPAIR

Course Number: *Version 1:* ACP. *Version 2:* ACP.
Location: Motor Transport School, Cp. Lejeune, NC.
Length: *Version 1:* 8 weeks (300 hours). *Version 2:* 9 weeks (332 hours).

Exhibit Dates: *Version 1:* 3/97–Present. *Version 2:* 7/95–2/97.

Learning Outcomes: *Version 1:* Upon completion of the course, the student will be able to troubleshoot and repair components of the fuel and electrical systems on wheeled and tracked vehicles. *Version 2:* Upon completion of the course, the student will be able to troubleshoot and repair components of the fuel, electrical, air induction, and brake systems on wheeled and tracked vehicles.

Instruction: *Version 1:* Through lectures, demonstrations, and practical hands-on activities, students learn to maintain and repair fuel injection pumps, fuel injector assemblies, starters, alternators, blowers, turbochargers, and personnel heaters employed on diesel engines. *Version 2:* Through lectures, demonstrations, and practical hands-on activities, students learn to maintain and repair fuel injection pumps, fuel injector assemblies, starters, alternators, blowers, turbochargers, personnel heaters, brake drums, brake rotors and shoes, and air induction systems employed on diesel engines.

Credit Recommendation: *Version 1:* In the lower-division baccalaureate/associate degree category, 3 semester hours in vehicle electrical systems and 3 in diesel fuel systems (6/97). *Version 2:* In the lower-division baccalaureate/associate degree category, 3 semester hours in vehicle electrical systems and 3 in diesel fuel systems (7/96).

MC-1714-0019

ELECTRO-OPTICAL ORDNANCE REPAIR

Course Number: 670-2171.
Location: Marine Corps Detachment, Ordnance Center and School, Aberdeen Proving Ground, MD.
Length: 32 weeks (1289 hours).
Exhibit Dates: 10/96–Present.
Learning Outcomes: Upon completion of the course, the student will be able to operate, maintain, and repair laser infrared observation sets.
Instruction: Course includes maintenance forms and publications; safety procedures; basic AC and DC electronics, including concepts of electricity, safety, voltage, current and resistance, conversion units, Ohm's law, color codes, and analyzing and troubleshooting series, parallel, and series-parallel circuits. Course also covers batteries, multimeters, introduction to oscilloscopes, magnetism, inductance, generation of AC, capacitance, resonance, transistor fundamentals (NPN-PNP), power supplies, and rectifiers. Basic digital circuits are introduced, along with the binary system; Boolean algebra; diode logic; soldering practices; maintaining, troubleshooting, and adjusting infrared observation set; and troubleshooting and adjusting xenon search light fire control system, laser range finder, and ballistics computer.
Credit Recommendation: In the lower-division baccalaureate/associate degree category, 5 semester hours basic electricity and electronics and 2 in computer fundamentals and applications (6/97).

MC-1715-0006

MOBILE COMMUNICATIONS CENTRAL TECHNICIAN (AN/TGC-37(V))
(Communication Central, AN/TGC-37, System Maintenance)
(Mobile Communication Central Technician)

Course Number: 28E.
Location: Communication-Electronics School, Twentynine Palms, CA; Air Ground Combat Center, Twentynine Palms, CA.
Length: 9-12 weeks (302-446 hours).
Exhibit Dates: 1/90–9/90.
Objectives: To train enlisted personnel in the operation, maintenance, and repair of specific types of telegraph systems.
Instruction: Course includes practical experience in the maintenance, operation, repair, and systems analysis of patch panels and switchboards, voice frequency telegraphs, and test equipment.
Credit Recommendation: In the vocational certificate category, 3 semester hours in telephone and telegraph system maintenance (3/91).

MC-1715-0009

GROUND RADIO REPAIR

Course Number: 27M.
Location: Air Ground Combat Center, Twentynine Palms, CA.
Length: 14-19 weeks (467-704 hours).
Exhibit Dates: 1/90–6/95.
Objectives: After 6/95 see MC-1715-0166. To train enlisted personnel to operate, test, and repair ground radio equipment.
Instruction: Course includes lectures and laboratories in FM radio equipment maintenance, including transmitter, receiver, and special circuits operation and various alignment, testing, overhaul, and troubleshooting procedures.
Credit Recommendation: In the lower-division baccalaureate/associate degree category, 3 semester hours in communications equipment maintenance (4/91).

MC-1715-0010

RADIO FUNDAMENTALS

Course Number: 27V.
Location: Air Ground Combat Center, Twentynine Palms, CA.
Length: 6-8 weeks (210-251 hours).
Exhibit Dates: 1/90–3/91.
Objectives: After 3/91 see MC-1715-0160. To prepare enlisted military personnel to maintain, test, adjust, and repair specific military radio sets.
Instruction: Course includes lectures and laboratory exercises in theory of operation and troubleshooting procedures for AM, FM, and SSB transmitters, receivers, and antennas.
Credit Recommendation: In the vocational certificate category, 6 semester hours in radio repair (3/89); in the lower-division baccalaureate/associate degree category, 3 semester hours in electronic communications and 1 in electronic communications laboratory (3/89).

MC-1715-0011

AVIATION RADIO REPAIR

Course Number: 66F.
Location: Basic Electronics School, San Diego, CA; Communication-Electronics School, San Diego, CA; Air Ground Combat Center, Twentynine Palms, CA.
Length: 9-14 weeks (319-462 hours).
Exhibit Dates: 1/90–2/92.
Objectives: After 2/92 see MC-1715-0155. To train enlisted personnel to install, inspect, maintain, and repair aviation and ground communications-electronics equipment.
Instruction: Course includes lectures and practical exercises in the theory, operation, and repair of AM, FM, and SSB radios to the module level. Course includes equipment characteristics, digital logic, circuit analysis, alignment, performance testing, and troubleshooting test equipment.
Credit Recommendation: In the lower-division baccalaureate/associate degree category, 3 semester hours in electronic systems troubleshooting and maintenance (5/90).

MC-1715-0020

TACTICAL AIR COMMAND CENTRAL REPAIRER (Tactical Air Command Central (TACC AN/TYQ-1) Repair)

Course Number: FGV.
Location: Communication-Electronics School, Twentynine Palms, CA; Air Ground Combat Center, Twentynine Palms, CA.
Length: 14-17 weeks (548-597 hours).
Exhibit Dates: 1/90–5/90.
Objectives: After 5/90 see MC-1715-0151. To train enlisted personnel in the operation, testing, maintenance, and repair of the Tactical Air Command Central System. System includes computers, displays, voice communications, and digital data links.
Instruction: Course includes lectures and practical exercises at the introductory level in the operation, testing, and repair of all system components. Course includes wire diagnostics and computer repair to the card level, power supply repair, and system interfaces. Course also includes use of special and general purpose test equipment.
Credit Recommendation: In the lower-division baccalaureate/associate degree category, 2 semester hours in introduction to systems maintenance and 2 in introduction to computer maintenance (5/87).

MC-1715-0024

MICROWAVE EQUIPMENT OPERATOR

Course Number: CGM.
Location: Communication-Electronics School, Twentynine Palms, CA; Air Ground Combat Center, Twentynine Palms, CA.
Length: 3-6 weeks (105-221 hours).
Exhibit Dates: 1/90–4/91.
Objectives: After 4/91 see MC-1715-0161. To train enlisted personnel to operate microwave equipment.
Instruction: Course includes lectures in microwave propagation, introduction to technical manuals, safety precautions, receiver and transmitter analysis, analysis of various radio sets, multiplexer analysis and alignment procedures, profile graph preparation, and preventive maintenance and troubleshooting procedures.
Credit Recommendation: Credit is not recommended because of the limited, specialized nature of the course (5/91).

MC-1715-0025

RESERVE COMMUNICATION OFFICERS, PHASES 1 AND 2

Course Number: CEO.
Location: Communication Officers School, Quantico, VA.
Length: 4 weeks (160 hours).
Exhibit Dates: 1/90–12/94.

Learning Outcomes: After 12/94 see MC-1715-0178. Also see MC-1715-0179. Upon completion of the course, the student will be able to design and install a wire (telephone) communications or a radio communications system and plan and manage communications systems.

Instruction: Lectures, activities, and practical exercises during the four-week resident phase focus on designing and installing communications systems.

Credit Recommendation: In the vocational certificate category, 3 semester hours in installation and maintenance of communications systems for the resident phase (10/90).

MC-1715-0026

TACTICAL DATA COMMUNICATIONS CENTRAL TECHNICIAN

(Tactical Data Communications Central (TDCC AN/TYQ-3) Technician)

Course Number: 0GM.
Location: Communication-Electronics School, Twentynine Palms, CA; Air Ground Combat Center, Twentynine Palms, CA.
Length: 23-29 weeks (846-1140 hours).
Exhibit Dates: 1/90–1/91.
Objectives: After 1/91 see MC-1715-0146. To train enlisted personnel to install, operate, test, and maintain data communications systems.
Instruction: Lectures and practical exercise in the operation and maintenance of a specific tactical data communications system. Topics include systems concepts, programming, data flow, and fault isolation.
Credit Recommendation: In the lower-division baccalaureate/associate degree category, 3 semester hours in computer systems troubleshooting and maintenance and 3 in electronic systems troubleshooting and maintenance (2/89).

MC-1715-0031

AIR SUPPORT OPERATIONS OPERATOR

Course Number: 67L.
Location: Communication-Electronics School, Twentynine Palms, CA; Air Ground Combat Center, Twentynine Palms, CA.
Length: 3-8 weeks (109-267 hours).
Exhibit Dates: 1/90–3/95.
Objectives: After 3/95 see MC-1715-0165. To provide instruction in the operations and tactical employment of a direct air support center, radar directing center, and airborne/mobile direct air support center.
Instruction: Course includes lectures and practical exercises in net operator tasks; installation of power connections; air conditioner, antenna, telephone, and external transceiver connections; operation of plotter and status board keepers; and ASRT radar operating procedures.
Credit Recommendation: Credit is not recommended because of the military-specific nature of the course (11/91).

MC-1715-0040

AVIATION FIRE CONTROL TECHNICIAN

Course Number: 664.
Location: Communication-Electronics School, Twentynine Palms, CA; Air Command and Control School, Twentynine Palms, CA.
Length: 7-11 weeks (267-440 hours).
Exhibit Dates: 1/90–7/91.
Objectives: To provide instruction in the installation, alignment, and advanced troubleshooting of an aviation fire control system.
Instruction: Course includes lectures and practical exercises in the receiving, transmitting, and timing circuitry of the radar system. Course also includes a review of the system computer. Advanced troubleshooting techniques for this system are stressed.
Credit Recommendation: In the lower-division baccalaureate/associate degree category, 3 semester hours in electronic equipment maintenance (5/91).

MC-1715-0046

TACTICAL AIR COMMAND CENTRAL (TACC) TECHNICIAN

Course Number: FGX.
Location: Air Ground Combat Center, Twentynine Palms, CA.
Length: 21-29 weeks (783-1100 hours).
Exhibit Dates: 1/90–7/90.
Objectives: After 7/90 see MC-1715-0142. To train enlisted personnel in the advanced operation, testing, maintenance, and repair of a tactical air command central system. System includes computers, displays, voice communications, and digital data links.
Instruction: Course includes lectures and practical exercises in the operation, testing, and repair of the entire central system. It includes computer diagnostics and repair to the card level, power supply repair, and system interfaces. Use of specialized and general purpose test equipment is also covered.
Credit Recommendation: In the lower-division baccalaureate/associate degree category, 3 semester hours in systems maintenance and 3 in computer maintenance (5/87).

MC-1715-0059

AVIATION RADAR REPAIR (C)

Course Number: 66W.
Location: Communication-Electronics School, San Diego, CA; Air Ground Combat Center, Twentynine Palms, CA.
Length: 13-14 weeks (492-522 hours).
Exhibit Dates: 1/90–12/91.
Objectives: To train enlisted personnel who are graduates of basic electronics and radar courses in the operation, maintenance, and repair of a specific phased array radar system.
Instruction: Course includes lectures and practical exercises in the theory of operation, repair, and maintenance of the computer programmer, synthesizer, transmitter, antenna, receiver, display console, data processor, and other equipment related to the AN/TPS-32 radar system.
Credit Recommendation: In the vocational certificate category, 4 semester hours in electronic equipment troubleshooting (4/91).

MC-1715-0062

GROUND RADAR REPAIR

Course Number: 27E.
Location: Communication-Electronics School, Twentynine Palms, CA.
Length: 5-8 weeks (185-256 hours).
Exhibit Dates: 1/90–Present.
Objectives: After 7/94 see MC-1715-0173. To train enlisted personnel with electronics backgrounds to install, operate, adjust, inspect, and maintain specific ground radar sets.
Instruction: Course includes lectures and practical exercises in the installation, operation, adjustment, inspection, and maintenance of specific ground radar sets, including basic theory of radar sets and auxiliary and test equipment, system circuit analysis, and operation of various subsystems on specific equipment.
Credit Recommendation: In the vocational certificate category, 2 semester hours in systems maintenance (4/91).

MC-1715-0063

RADAR FUNDAMENTALS

Course Number: 27W.
Location: Air Ground Combat Center, Twentynine Palms, CA.
Length: 4 weeks (141-142 hours).
Exhibit Dates: 1/90–3/90.
Objectives: After 3/90 see MC-1715-0154. To train enlisted personnel in the fundamental concepts of radar.
Instruction: Course includes lectures and practical exercises in the theory of electronics for analysis of radar, pulse circuits, and components and includes basic theory of power supplies, timing circuits, blocking oscillators, pulse amplifiers, receiver amplifiers, klystrons, and antenna drive systems. Course covers use of oscilloscopes, signal generators, voltmeters, and special test equipment. Also included is analysis of individual stages in a representative radar training system.
Credit Recommendation: In the lower-division baccalaureate/associate degree category, 3 semester hours in radar systems (2/89).

MC-1715-0068

ADVANCED COMMUNICATION OFFICERS

Course Number: BED.
Location: Communication Officers School, Quantico, VA.
Length: 37-43 weeks (1328-1497 hours).
Exhibit Dates: 1/90–11/91.
Learning Outcomes: After 11/91 see MC-1715-0177. Upon completion of the course, the student will be able to lead and counseling subordinates, develop large scale communications systems, develop an understanding of electronics as theory and as telecommunications, and refine written and oral communications skills.
Instruction: Course includes lectures, exercises, and activities in the design and operation of communications systems to support large scale military activities and lectures and exercises on electricity, principles of electronics, signal multiplexing, computers, communication theory, telecommunications, operation of communications equipment, logistics, planning, organization, and principles of leadership. Course also emphasizes speaking and writing skills for career officers.
Credit Recommendation: In the lower-division baccalaureate/associate degree category, 2 semester hours in principles of leadership, 2 in electronic theory, 2 in written composition, 1 in effective speaking, and 1 in introduction to telecommunications (10/90); in the upper-division baccalaureate category, 2 semester hours in management of communications systems (10/90).

MC-1715-0078

AIR SUPPORT CONTROL OFFICER

Course Number: TOA.
Location: Communication-Electronics School, Twentynine Palms, CA; Air Ground Combat Center, Twentynine Palms, CA.
Length: 6-11 weeks (208-428 hours).
Exhibit Dates: 1/90–2/95.
Objectives: After 2/95 see MC-1715-0176. To train personnel as air support control officers.
Instruction: Course includes lectures and practical exercises in radio net operator duties, plotter/status board techniques, radar operator procedures, and strike controller procedures under simulated field conditions.
Credit Recommendation: Credit is not recommended because of the military-specific nature of the course (8/91).

MC-1715-0089

MICROMINIATURE COMPONENT REPAIR

Course Number: E2D.
Location: Air Ground Combat Center, Twentynine Palms, CA; Communication-Electronics School, Twentynine Palms, CA.
Length: 4-6 weeks (160-212 hours).
Exhibit Dates: 1/90–3/92.
Objectives: After 3/92 see MC-1715-0140. To train selected personnel to conduct repairs on circuits characterized by high density circuitry, highly sensitive components, extremely small size, and/or multilead components.
Instruction: The course encompasses the identification and causes of defective solder connections, the procedures and proper tools required to produce high reliability solder connections, and proper mounting and installation of various components, including such microelectronic devices as integrated circuits, conformal coatings used, and the most effective means of their removal and replacement. Circuit baseboards are identified by composition then rebuilt or repaired. Types of interfacial connections and conductor runs are identified and repaired or replaced. Repairs are made to boards containing multiple or subsurface conductor runs, flexible printed wiring boards and printed wiring harnesses, all without causing further degradation to the circuit under repair. Specifications and techniques required for the various types of terminal and connector pins. In addition to the materials used in the microminiature component repair course, various electroplating chemicals, metals, paints, and etchant materials are provided for plating techniques.
Credit Recommendation: In the lower-division baccalaureate/associate degree category, 3 semester hours in microminiature component repair (9/90).

MC-1715-0091

POLE LINE CONSTRUCTION EQUIPMENT BY CORRESPONDENCE

Course Number: 25.55.
Location: Marine Corps Institute, Washington, DC.
Length: Maximum, 52 weeks.
Exhibit Dates: 1/90–1/97.
Objectives: To teach students to identify pole lines, construction equipment, and hardware and also to teach safety and first aid.
Instruction: This correspondence course consists of study units on safety and first aid, military construction vehicles, and pole installation equipment and hardware. There is a proctored final examination. Course includes no laboratory work.
Credit Recommendation: Credit is not recommended due to the military-specific nature of the course (8/83).

MC-1715-0096

POLE LINE CONSTRUCTION TECHNIQUES BY CORRESPONDENCE

Course Number: 25.56.
Location: Marine Corps Institute, Washington, DC.
Length: Maximum, 52 weeks.
Exhibit Dates: 1/90–1/97.
Objectives: To provide the field wireman with information for the installation of line poles, suspension strands, and aerial cables. In addition, safe operating techniques of pole line construction is stressed.
Instruction: This correspondence course presents study units covering military vehicles used in pole installation and anchor and guy wire installation and suspension. Instruction does not include laboratory exercises.
Credit Recommendation: Credit is not recommended because of the military-specific nature of the course (8/83).

MC-1715-0099

AUTOMATIC TELEPHONE EQUIPMENT BY CORRESPONDENCE

Course Number: 25.54.
Location: Marine Corps Institute, Washington, DC.
Length: Maximum, 52 weeks.
Exhibit Dates: 1/90–1/97.
Objectives: To supply the field wireman with the information necessary for the identification, installation, and operation of military automatic telephone equipment.
Instruction: This correspondence course presents study units covering purpose, use, technical characteristics, and operation of military switchboards and dual-tone multifrequency telephone systems. Instruction does not include any laboratory exercises. A proctored examination concludes the course.
Credit Recommendation: Credit is not recommended due to the military-specific nature of the course (8/83).

MC-1715-0101

COMMUNICATIONS FOR THE COMBAT OPERATIONS CENTER/FIRE SUPPORT COORDINATION CENTER (COC/FSCC) BY CORRESPONDENCE

Course Number: 25.36.
Location: Marine Corps Institute, Washington, DC.
Length: Maximum, 52 weeks.
Exhibit Dates: 1/90–1/97.
Objectives: To provide information and instruction on the communications requirements of a combat operations center and fire support coordination center, including responsibilities, planning, and system description.
Instruction: This correspondence course, without laboratory exercises, provides information and instruction on the purpose, function, and requirements of a combat communications center; the purpose, function, and requirements of a fire support coordination center; radio networks within these centers; radio chief and watch supervisor responsibilities; and other factors concerning the location and placement of these facilities. Some topographical map reading, graph plotting, and simple mathematics are included. A proctored final examination is required for course completion.
Credit Recommendation: Credit is not recommended because of the limited, specialized nature of the course (8/83).

MC-1715-0113

AIMS MAINTENANCE

Course Number: 24X.
Location: Air Ground Combat Center, Twentynine Palms, CA.
Length: 10 weeks (350-399 hours).
Exhibit Dates: 1/90–Present.
Objectives: To train personnel to repair and maintain an air traffic control radar beacon system.
Instruction: Course provides lectures and practical exercises in operation, installation, operational adjustment, and circuitry related to air traffic control radar equipment. Course also includes corrective and preventive maintenance, schematic interpretation to the block diagram level, and the use of technical manuals.
Credit Recommendation: In the vocational certificate category, 3 semester hours in electronic systems maintenance and repair (5/94).

MC-1715-0116

BASIC ELECTRONICS

Course Number: *Version 1:* 272. *Version 2:* 272.
Location: *Version 1:* Air Ground Combat Center, Twentynine Palms, CA. *Version 2:* Communication-Electronics School, Twentynine Palms, CA.
Length: *Version 1:* 10-11 weeks (385 hours). *Version 2:* 11-14 weeks (392-524 hours).
Exhibit Dates: *Version 1:* 2/95–Present. *Version 2:* 1/90–1/95.
Learning Outcomes: *Version 1:* Upon completion of the course, the student will be able to describe the basic concepts of electricity, electronics, digital logic, basic soldering, computer systems, and maintenance management and will be able to perform laboratory measurements using the multimeter and the oscilloscope. *Version 2:* Upon completion of the course, the student will be able to describe the basic concepts of electricity, electronics, digital logic, basic soldering, computer systems, and maintenance management and will be able to perform laboratory measurements using the multimeter and the oscilloscope.
Instruction: *Version 1:* Course includes lectures and practical exercises in basic mathematics, electricity, electronics, fundamentals of digital logic, soldering, and maintenance management. More advanced courses involving repair of specific equipment follow. *Version 2:* Course includes lectures and practical exercises in basic mathematics, electricity, electronics, fundamentals of digital logic, soldering, and maintenance management. More advanced courses involving repair of specific equipment follow.
Credit Recommendation: *Version 1:* In the lower-division baccalaureate/associate degree category, 3 semester hours in basic electronics, 3 in basic electronics laboratory, 1 in computer systems, 2 in AC circuits, and 2 in DC circuits (3/96). *Version 2:* In the vocational certificate

category, 2 semester hours in soldering techniques (9/89); in the lower-division baccalaureate/associate degree category, 2 semester hours in DC circuits, 2 in AC circuits, 3 in solid state electronics, 3 in digital principles, 3 in basic electronics laboratory, and 1 in computer systems and organization (9/89).

MC-1715-0117

GROUND RADAR TECHNICIAN

Course Number: 27F.
Location: Communication-Electronics School, Twentynine Palms, CA.
Length: 8-10 weeks (304-363 hours).
Exhibit Dates: 1/90–11/93.
Learning Outcomes: Upon completion of the course, the student will be able to install, calibrate, and maintain military radar systems; describe the theory of operation of the system; and troubleshoot the system to the card level.
Instruction: Course includes lectures, demonstrations, and practical exercises on topics specific to military radar systems.
Credit Recommendation: In the lower-division baccalaureate/associate degree category, 3 semester hours in electronic systems troubleshooting and maintenance (7/90).

MC-1715-0118

MOBILE DATA COMMUNICATIONS TERMINAL TECHNICIAN

Course Number: BF2.
Location: Communication-Electronics School, Twentynine Palms, CA.
Length: 17 weeks (672-680 hours).
Exhibit Dates: 1/90–5/91.
Learning Outcomes: Upon completion of the course, the student will be able to install, maintain, and troubleshoot mobile data communications equipment including peripherals.
Instruction: Course includes lectures and laboratory exercises on the installation and maintenance of the equipment and its peripherals, including tape punch and reader, printers, interface unit, and modem.
Credit Recommendation: In the lower-division baccalaureate/associate degree category, 3 semester hours in computer systems troubleshooting and maintenance (6/91).

MC-1715-0119

BANCROFT MAINTENANCE

Course Number: DQH.
Location: Air Ground Combat Center, Twentynine Palms, CA; Communication-Electronics School, Twentynine Palms, CA.
Length: 7-9 weeks (259-290 hours).
Exhibit Dates: 1/90 Present.
Learning Outcomes: Upon completion of the course, the student will be able to describe the theory of operation, circuit analysis, and alignment of the Bancroft system and be able to test, troubleshoot, adjust, and perform preventive maintenance on the system.
Instruction: Course includes lectures, demonstrations, and practical exercises on the preventive maintenance of the Bancroft system.
Credit Recommendation: In the lower-division baccalaureate/associate degree category, 3 semester hours in electronic systems troubleshooting and maintenance (1/93).

MC-1715-0120

MICROCOMPUTER REPAIR

(Teletype Repair)

Course Number: *Version 1:* 26D. *Version 2:* 26D. *Version 3:* 26D.
Location: Air Ground Combat Center, Twentynine Palms, CA; Communication-Electronics School, Twentynine Palms, CA.
Length: *Version 1:* 15-16 weeks (552 hours). *Version 2:* 15-16 weeks (559 hours). *Version 3:* 10-11 weeks (384 hours).
Exhibit Dates: *Version 1:* 2/93–Present. *Version 2:* 6/90–1/93. *Version 3:* 1/90–5/90.
Learning Outcomes: *Version 1:* Upon completion of the course, the student will be able to configure a microcomputer system; interface modems, printers, and other communications devices; and use special diagnostics to troubleshoot system components and copiers. *Version 2:* Upon completion of the course, the student will be able to describe the principles of operation, circuit analysis, alignment, interconnection, performance testing, and troubleshooting of teletype equipment and testing and corrective maintenance of a microcomputer-based system. Student will be able to describe site selection and communications security and troubleshoot the system. *Version 3:* Upon completion of the course, the student will be able to describe the principles of operation, circuit analysis, alignment, interconnection, performance testing, and troubleshooting of teletype equipment. Student will be able to describe site selection and communications security and troubleshoot the system.
Instruction: *Version 1:* Instruction includes lectures, demonstrations, and practical exercises on microcomputers. Topics include disk drives, monitors, keyboards, modems, dot matrix and laser printers, copiers, interconnection of communications devices, circuit analysis, microprocessor controlled circuit card tester, and fault isolation and repair using related test equipment. *Version 2:* Lectures, demonstration, and practical exercises on alignment and adjustment of teleprinters. The course also included extensive information regarding operation/maintenance of teleprinters. *Version 3:* Course includes lectures, demonstrations, and practical exercises on alignment and adjustment of teleprinters. Also includes extensive information regarding operation/maintenance of teleprinters.
Credit Recommendation: *Version 1:* In the lower-division baccalaureate/associate degree category, 5 semester hours in microcomputer systems theory and 5 in microcomputer systems laboratory (1/93). *Version 2:* In the vocational certificate category, 3 semester hours in electronic systems troubleshooting and maintenance and 2 in computer systems troubleshooting and maintenance (1/93). *Version 3:* In the vocational certificate category, 3 semester hours in teletype maintenance (2/89); in the lower-division baccalaureate/associate degree category, 3 semester hours in electromechanical systems troubleshooting and repair (2/89).

MC-1715-0121

TELEPHONE/SWITCHBOARD REPAIR

Course Number: *Version 1:* 28W. *Version 2:* 28W.
Location: Communication-Electronics School, Twentynine Palms, CA.
Length: *Version 1:* 13-14 weeks (494 hours). *Version 2:* 15-17 weeks (528-563 hours).
Exhibit Dates: *Version 1:* 2/93–Present. *Version 2:* 1/90–1/93.
Learning Outcomes: *Version 1:* Upon completion of the course, the student will be able to install, troubleshoot, and repair telephone, fiber optic cable, and ancillary cable equipment, and provide technical assistance during installation of the above equipment. *Version 2:* Upon completion of the course, the student will be able to install, operate, troubleshoot, and repair Marine Corps telephone/switchboards and related equipment.
Instruction: *Version 1:* Course includes lectures and practical experience on maintenance of individual telephones, switchboards, radio control sets, and facsimile sets. *Version 2:* Course includes lectures and practical experience on maintenance of individual telephones, switchboards, radio control sets, and facsimile sets.
Credit Recommendation: *Version 1:* In the lower-division baccalaureate/associate degree category, 3 semester hours in electronic systems troubleshooting and maintenance (3/96). *Version 2:* In the vocational certificate category, 3 semester hours in telephone systems (7/91); in the lower-division baccalaureate/associate degree category, 3 semester hours in electronic systems troubleshooting and maintenance (7/91).

MC-1715-0122

AVIATION RADAR TECHNICIAN

Course Number: 66K.
Location: Communication-Electronics School, Twentynine Palms, CA.
Length: 38 weeks (1389 hours).
Exhibit Dates: 1/90–6/90.
Learning Outcomes: Upon completion of the course, the student will be able to troubleshoot and maintain a complex computer-controlled radar system and apply radar circuit principles to the troubleshooting process.
Instruction: Methods include lectures and practical exercises related to the specific radar system.
Credit Recommendation: In the lower-division baccalaureate/associate degree category, 3 semester hours in computer systems troubleshooting and maintenance, 3 in electronic systems troubleshooting and maintenance, and 3 in radar systems (2/89).

MC-1715-0123

MICROWAVE EQUIPMENT MAINTENANCE

Course Number: D3C.
Location: Communication-Electronics School, Twentynine Palms, CA.
Length: 25-28 weeks (882-971 hours).
Exhibit Dates: 1/90–10/92.
Learning Outcomes: Upon completion of the course, the student will be able to apply the principles of a limited number of topics from digital electronics, solid state electronics, and electronic communications and troubleshoot and maintain a military telephone terminal and various radio communications systems.
Instruction: Methods include lectures and practical exercises on topics in electronics. Emphasis is on circuitry, troubleshooting, and maintenance of the specific telephone equipment.

Credit Recommendation: In the vocational certificate category, 3 semester hours in telephone systems (12/90); in the lower-division baccalaureate/associate degree category, 2 semester hours in introduction to electronics and 3 in basic electronics laboratory (12/90).

MC-1715-0124

RADIO TECHNICIAN

Course Number: *Version 1:* 27P. *Version 2:* 27P.

Location: *Version 1:* Air Ground Combat Center, Twentynine Palms, CA. *Version 2:* Communication-Electronics School, Twentynine Palms, CA.

Length: *Version 1:* 8 weeks (315 hours). *Version 2:* 15-16 weeks (308-511 hours).

Exhibit Dates: *Version 1:* 10/95–Present. *Version 2:* 1/90–9/95.

Learning Outcomes: *Version 1:* Upon completion of the course, the student will understand and be able to maintain AM and FM receivers and transmitters, single-sideband receivers and transmitters, digital facsimile sets, and speech security equipment; and troubleshoot, repair, and maintain all of these communication devices. *Version 2:* Upon completion of the course, the student will be able to understand and maintain AM receivers and transmitters, FM receivers and transmitters, and single-sideband receivers and transmitters. Student will also be able to analyze and repair communications security devices.

Instruction: *Version 1:* Lectures, practical exercises and written examination. Emphasis is on the troubleshooting of electronic communication circuits. *Version 2:* Course includes lectures and practical exercises on the characteristics and maintenance of electronic communications equipment.

Credit Recommendation: *Version 1:* In the lower-division baccalaureate/associate degree category, 5 semester hours in communication electronics troubleshooting and maintenance and 1 in communication electronics laboratory (3/96). *Version 2:* In the lower-division baccalaureate/associate degree category, 3 semester hours in electronic systems troubleshooting and maintenance (4/91).

MC-1715-0125

HIGH FREQUENCY MAINTENANCE

Course Number: DPH.

Location: Communication-Electronics School, Twentynine Palms, CA.

Length: 2 weeks (68-80 hours).

Exhibit Dates: 1/90–Present.

Learning Outcomes: Upon completion of the course, the student will be able to maintain a transceiver. The student will be able to describe the components, characteristics, and synchronization of the equipment by tracing the signal flow of the TCS-4B transmitter and the RCS-4B receiver.

Instruction: Course includes lectures and practical exercises in HF propagation, transmitter and receiver characteristics, system synchronization, block diagrams, power amplifiers, and troubleshooting.

Credit Recommendation: In the lower-division baccalaureate/associate degree category, 3 semester hours in electronic systems troubleshooting and maintenance (6/90).

MC-1715-0127

AVIATION RADIO TECHNICIAN

Course Number: 66G.

Location: Communication-Electronics School, Twentynine Palms, CA; Air Ground Combat Center, Twentynine Palms, CA.

Length: 5-10 weeks (175-350 hours).

Exhibit Dates: 1/90–Present.

Learning Outcomes: Upon completion of the course, the student will be able to install, inspect, maintain, and repair aviation radio equipment; use test equipment; and follow corrective maintenance procedures.

Instruction: Course provides technician-level training in selected aviation radio equipment. Lectures include topics on theory of operation and circuit analysis, performance testing, alignment procedures, use of test equipment, and corrective maintenance procedures.

Credit Recommendation: In the lower-division baccalaureate/associate degree category, 3 semester hours in electronic systems troubleshooting and maintenance (1/93).

MC-1715-0128

TECHNICIAN THEORY

Course Number: *Version 1:* TA3. *Version 2:* TA3.

Location: Communication-Electronics School, Twentynine Palms, CA.

Length: *Version 1:* 16 weeks (560 hours). *Version 2:* 16-17 weeks (664 hours).

Exhibit Dates: *Version 1:* 10/90–Present. *Version 2:* 1/90–9/90.

Learning Outcomes: *Version 1:* Upon completion of the course, the student will be able to apply principles of algebra, trigonometry, and complex numbers to analysis of problems in electricity; use common laboratory instruments; understand the principles of DC (does not include network theorems) and AC circuits, semiconductor devices, transistor theory, rectifiers and DC power supplies, amplifiers, operational amplifiers, and analysis using circuit techniques; apply the basic principles of digital electronics, including logic gates, sequential, and combinational logic, counters, adders, and subtractors; apply principles of management and maintenance management; and apply principles of computer organization. *Version 2:* Upon completion of the course, the student will be able to apply principles of algebra, trigonometry, and complex numbers to analysis of problems in electricity and related topics in physics; use common laboratory instruments; understand the principles of DC (does not include network theorems) and AC circuits, semiconductor devices, transistor theory, vacuum tubes, rectifiers and DC power supplies, amplifiers, operational amplifiers, and analysis using circuit techniques; apply the basic principles of digital electronics, including logic gates, sequential and combinational logic, counters, adders, and subtractors; apply principles of management and maintenance management; and apply principles of computer organization.

Instruction: *Version 1:* Course includes lectures and practical exercises in technical mathematics, DC and AC circuits, semiconductors, transistor theory, amplifiers, operational amplifiers, circuit analysis techniques, digital principles, combinational and sequential logic circuits, computer organization, and principles of management including McGregor's Theory X/Theory Y, Maslow's Theory of Motivation, and Herzberg's Motivation-Hygiene Theory. *Version 2:* Course includes lectures and practical exercises in technical mathematics, DC and AC circuits, semiconductors, transistor theory, amplifiers, operational amplifiers, circuit analysis techniques, digital principles, combinational and sequential logic circuits, computer organization, and principles of management, including McGregor's Theory X/Theory Y, Maslow's Theory of Motivation, and Herzberg's Motivation-Hygiene Theory.

Credit Recommendation: *Version 1:* In the vocational certificate category, 1 semester hour in soldering (3/95); in the lower-division baccalaureate/associate degree category, 3 semester hours in basic electronics laboratory, 2 in DC circuits, 2 in AC circuits, 2 in digital principles, 2 in solid state electronics, 3 in computer systems and organization, 2 in technical mathematics, and 2 in maintenance management (3/95); in the upper-division baccalaureate category, 1 semester hour in principles of management (3/95). *Version 2:* In the vocational certificate category, 1 semester hour in soldering (2/89); in the lower-division baccalaureate/associate degree category, 3 semester hours in basic electronics laboratory, 2 in DC circuits, 3 in AC circuits, 3 in digital principles, 5 in solid state electronics, 3 in computer systems and organization, 5 in technical mathematics, and 2 in maintenance management (2/89); in the upper-division baccalaureate category, 1 semester hour in principles of management (2/89).

MC-1715-0129

INTRODUCTION TO TEST EQUIPMENT BY CORRESPONDENCE

Course Number: 28.7.

Location: Marine Corps Institute, Washington, DC.

Length: Maximum, 52 weeks.

Exhibit Dates: 1/90–Present.

Learning Outcomes: Upon completion of the course, the student will be able to recognize measurement instruments used to determine voltage, current, and resistance; recognize analog and digital display, parallax, and series and shunt loading effects, as well as measurement objectives and conditions of proper measurements.

Instruction: Course includes discussion of test equipment, including analog and digital display, electrical characteristics of linear and nonlinear components, measurement methods, parallax errors, and series and shunt-connected loading effects.

Credit Recommendation: In the vocational certificate category, 2 semester hours in electronic test equipment (6/89).

MC-1715-0130

DIRECTION FINDING OPERATIONS BY CORRESPONDENCE

Course Number: 26.1.

Location: Marine Corps Institute, Washington, DC.

Length: Maximum, 52 weeks.

Exhibit Dates: 1/90–Present.

Learning Outcomes: Upon completion of the course, the student will be able to recognize Soviet communications operating characteristics, terms and conditions of electronic warfare, methods and characteristic definitions of direction-finding operations, and the operating procedures of AN/PRO-10 direction finder set and OAR-3054 direction finder set.

Instruction: Course includes description and historical prospective of Soviet communications systems, development of terms and techniques of electronic warfare, basic radio propagation theory, direction-finding operations, and operator and preventive maintenance procedures for the AN/PRO-10 and OAR-3054 direction finder sets.

Credit Recommendation: Credit is not recommended as the subject is limited and specialized as presented and is military-specific in nature (6/89).

MC-1715-0131

FUNDAMENTALS OF DIGITAL LOGIC BY CORRESPONDENCE

Course Number: *Version 1:* 28.6g. *Version 2:* 28.6f.
Location: Marine Corps Institute, Washington, DC.
Length: *Version 1:* Maximum, 52 weeks. *Version 2:* Maximum, 52 weeks.
Exhibit Dates: *Version 1:* 8/92–Present. *Version 2:* 1/90–7/92.
Learning Outcomes: *Version 1:* Upon completion of the course, the student will be able to recognize number system organization and arithmetic operations found in digital systems; fundamental logic elements and functions of digital systems, i.e., AND, OR, NAND, and NOR gates; and electrical representations of logical functions. The student will be aware of monostable and bistable effects of flip flops and the use of flip-flop circuitry in the serial counting function, parallel shift operation, parallel to serial, and serial to parallel conversion. To summarize, upon completion, the student is prepared to learn troubleshooting techniques and schematic reading of digital systems using integrated circuit logic chip gates, flip-flops, latches, and shift registers. *Version 2:* Upon completion of the course, the student will be able to recognize number system organization and arithmetic operations found in digital systems; fundamental logic elements and functions of digital systems, i.e., AND, OR, NAND, and NOR; electrical representation of logic functions; and logical memory units using discrete transistorized designs. Student will be aware of monostable and bistable effects of flip flops; the use of flip-flop circuitry in the serial counting function, parallel shift operation, and parallel-to-serial and serial-to-parallel conversion; the use of discrete transistor circuitry in construction of logical relay elements; the use of magnetic core as a method of memory element storage; and input/output operations of memory storage and removal by usage of shift register transistors. To summarize, upon completion, the student is prepared to learn troubleshooting techniques and schematic reading of computing and control systems designed principally with discrete transistorized circuitry and core memory elements.
Instruction: *Version 1:* Course includes numeric system design, including binary number systems, arithmetic binary operations, and counting process of number systems; logical operator functions and descriptions, including OR, AND, and NOR conditions; development of electrical representation of logical functions (voltage level to logic level relationship); discrete circuit representation of digital storage elements; and timing diagrams. There is a proctored examination at the end of the course. *Version 2:* Course includes numeric system design, including binary number systems, arithmetic binary operations and counting process of number systems; logical operator functions and descriptions including OR, AND, and NOR conditions; development of electrical representation of logical functions (voltage level to logic level relationship); discrete circuit representation of digital storage elements; discrete circuit representation of waveshaping circuits and circuitry used in control system description of magnetic retention of magnetic materials and READ/WRITE operations of core storage; and description of serial counting, parallel register operation, memory storage process, X-OR operation, and SHAFT encoding.

Credit Recommendation: *Version 1:* In the lower-division baccalaureate/associate degree category, 2 semester hours in introduction to digital logic (9/94). *Version 2:* In the lower-division baccalaureate/associate degree category, 2 semester hours in introduction to digital logic (6/89).

MC-1715-0132

SOLID STATE DEVICES BY CORRESPONDENCE

Course Number: 11.42b.
Location: Marine Corps Institute, Washington, DC.
Length: Maximum, 52 weeks.
Exhibit Dates: 1/90–Present.
Learning Outcomes: Upon completion of the course, the student will be able to describe semiconductor atomic theory, semiconductor characteristics, definitions of N-type and P-type material, and dynamic effects of current flow through P-N junctions; the student will be able to recognize NPN & PNP transistors by schematic symbols, wafer diagram, and biasing circuitry required for proper operation; the student will be able to describe distortion conditions, various operating configurations, and proper test, troubleshooting, and repair techniques; the student will be able to identify simple semiconductor devices, including zener diodes, tunnel diodes, varactors, SCRs and TRIACs; the student will be able to identify application circuitry and be aware of the operating functions of dacu devices; the student will be able to identify semiconductor power supply circuits by their schematic representation and identify signal conditioning effects at sequential points of the power supply. The student will be able to to recognize the use of troubleshooting bridge circuitry, filter components, and multiplier circuits for the purpose of troubleshooting; the student will be able to recognize voltage regulation through shunt and series techniques, identify the signal conditions at sequential points of the regulator, and identify, by output effects, component failure of the circuit. The student will be able to recognize power supply multiplier circuits and their signal processing effects.
Instruction: Course includes developmental chronology of solid state growth, topical presentation of devices and their applications, atomic theory and its relation to semiconductor materials, and the PN junction response curve and test methods applied to PN junctions and discussion of transistor types, schematic symbols, biasing techniques, distortion and its effects, amplifier configurations, development of transistor current relationships, troubleshooting and handling procedures. Functional presentations of semiconductor device elements and their applications, zener effects, tunneling, varactor (capacitive) effects, light emitting SCRs, TRIACs, and detecting materials are covered, as well as functional presentation of specialized transistor devices, including the unijunction, field effect, MOSFET elements, their schematic symbols, common packaging configurations, and theory of operation. Course includes discussion of basic power supply concepts, including rectification, filtering, AC/DC conditions, effective value ripple, and bridge operation and loading. The system approach is used with impedance calculations to describe loading effects of filter elements. Discussion includes power supply regulation and doubling operations, including regulation calculations, series and shunt voltage regulation, current regulation, and multiplier configurations. Signal tracing and other troubleshooting techniques are emphasized. All topics are covered at a minimal mathematical level.

Credit Recommendation: In the vocational certificate category, 3 semester hours in introduction to solid state devices (6/89); in the lower-division baccalaureate/associate degree category, 2 semester hours in survey of electronic devices (6/89).

MC-1715-0133

RADIOTELEGRAPH AND VISUAL COMMUNICATION PROCEDURES BY CORRESPONDENCE

Course Number: 25.61.
Location: Marine Corps Institute, Washington, DC.
Length: Maximum, 52 weeks.
Exhibit Dates: 1/90–Present.
Learning Outcomes: Upon completion of the course, the student will be able to properly format a military radio telegraph message and follow proper radiotelegraph procedures in message handling.
Instruction: By using numerous examples the student learns military radiotelegraph network requirements and message protocol, Morse code symbol recognition, message format and preparation procedures, acronyms for condensed messages, techniques to maintain a radio log, and procedures to receive and transmit a message. Alternate signaling protocol using flashing lights and semaphore flags is also presented.
Credit Recommendation: Credit is not recommended because of the limited, specialized nature of the course (6/89).

MC-1715-0134

MILITARY AFFILIATE RADIO SYSTEM (MARS) OPERATOR BY CORRESPONDENCE

Course Number: 25.62.
Location: Marine Corps Institute, Washington, DC.
Length: Maximum, 52 weeks.
Exhibit Dates: 1/90–Present.
Learning Outcomes: Upon completion of the course, the student will be able to describe the history, system network, and message format of the Military Affiliate Radio System.
Instruction: The historical background of the mission, organization, administration, and membership of the Military Affiliate Radio System (MARS) is given. The operator skill requirements for membership within this network are described. The configuration of the networks within the system and their call signals frequencies and power level requirements

are described. Message criteria, types, and formats are given using many examples. Radiotelegraph and radiotelephone operation methods, protocol, and techniques of handling message types are described.

Credit Recommendation: Credit is not recommended because of the limited, specialized nature of the course (6/89).

MC-1715-0135

HIGH FREQUENCY COMMUNICATIONS SYSTEM BY CORRESPONDENCE

Course Number: 25.37.
Location: Marine Corps Institute, Washington, DC.
Length: Maximum, 52 weeks.
Exhibit Dates: 1/90–Present.
Learning Outcomes: Upon completion of the course, the student will be able to describe the purpose and functional operation of major subsystems of the AN/PSC-95 defense communications system, how radio waves propagate in the atmosphere, amplitude modulation, frequency shift keying, and voice frequency carrier telegraph.
Instruction: The physical characteristics of a mobile high frequency, single-sideband, receiver/transmitter radioteletype communications system are described by purpose, front panel controls, operating modes, and data signal paths. The functional description covers the transmitter, receiver, teletypewriter, control panels, antenna couplers, power supplies, heating, lighting, air conditioning systems, and test meters. Safety precautions and general maintenance procedures are presented. High frequency radio wave propagation, amplitude modulation concepts, frequency shift keying, and voice frequency carrier telegraph are described from a nonmathematical viewpoint.
Credit Recommendation: In the lower-division baccalaureate/associate degree category, 1 semester hour in communications technology for nontechnical majors (6/89).

MC-1715-0136

ANTENNA CONSTRUCTION AND PROPAGATION OF RADIO WAVES BY CORRESPONDENCE

Course Number: 25.15g.
Location: Marine Corps Institute, Washington, DC.
Length: 52 weeks.
Exhibit Dates: 1/90–Present.
Learning Outcomes: Upon completion of the course, the student will be able to describe how radio waves propagate through the atmosphere, construct several types of antennas, and describe the characteristics of several types of antenna systems.
Instruction: A discussion of radio wave propagation through the atmosphere covers such subjects as frequency, wavelength, amplitude modulation, and frequency modulation; the effect of various atmospheric layers on wave propagation and reflection; and characteristics of direct waves, ground waves, sky waves, and the causes of skip zones. Definition of usable frequencies due to atmospheric conditions and fading is described. An examination of antenna patterns, polarization, frequency band, gain, impedance, and physical characteristics of log periodic, ground plane, and portable dipole antenna systems are given. The radiation pattern and step-by-step construction methods for half dipole, two element yagi, long wire antenna, half rhombic antenna, half-sloping V, quarter wave, whip and ground plane antennas are described through drawings and tables which provide appropriate wire lengths. The physical characteristics of several types of transmission lines are pictured. Treatment is all nonmathematical.
Credit Recommendation: In the vocational certificate category, 1 semester hour in introduction to antenna theory and construction (6/89).

MC-1715-0137

RADIOTELEPHONE AND ALTERNATE COMMUNICATION PROCEDURES BY CORRESPONDENCE

Course Number: 25.60.
Location: Marine Corps Institute, Washington, DC.
Length: Maximum, 52 weeks.
Exhibit Dates: 1/90–Present.
Learning Outcomes: Upon completion of the course, the student will be able to effectively format military radiotelephone messages, explain administrative rules and procedures of a radiotelephone operator, and utilize alternate visual communication techniques.
Instruction: Students learn proper formulation of a military radiotelephone message in order to be reliable, secure, timely, and of sufficient flexibility to meet requirements. The student learns the kinds of message categories, techniques to rapidly hand print messages, the phonetic alphabet, pronunciation, world time zones, standardized wording and its meaning, operating rules, communications security measures, types of radio interference, and administration rules and procedures in handling various types of messages, including priority levels, cancellations, corrections, and methods of requesting information. Alternate visual communication techniques using lights, semaphore flags, display panels, and pyrotechnics are reviewed with the emphasis placed on the alphanumeric character representation of this visual display. Numerous examples of correct message formats are given.
Credit Recommendation: Credit is not recommended because of the limited, specialized nature of the course (6/89).

MC-1715-0138

INTERROGATOR SET AN/TPX-46 DIRECT SUPPORT/GENERAL SUPPORT REPAIRMAN AND SUPERVISOR

Course Number: None.
Location: Marine Corps Detachment, Redstone Arsenal, AL.
Length: 8 weeks (318 hours).
Exhibit Dates: 1/91–Present.
Learning Outcomes: Upon completion of the course, the student will be able to operate, maintain, and repair the interrogator set using a synchronizer, transmitter, receiver antenna, and signal processor.
Instruction: Repair techniques covered require proficiency in digital electronics, low and high voltage power supplies, and antenna positioning system troubleshooting.
Credit Recommendation: In the lower-division baccalaureate/associate degree category, 3 semester hours in electronic systems troubleshooting and maintenance (10/91).

MC-1715-0139

TEST, MEASUREMENT AND DIAGNOSTIC EQUIPMENT

Course Number: 28T.
Location: Logistics Base, Albany, GA.
Length: 23 weeks (995 hours).
Exhibit Dates: 1/90–1/93.
Learning Outcomes: Upon completion of the course, the student will be able to demonstrate proper soldering techniques on printed circuit board components according to military standard specification; understand and use direct current circuit test procedures using multimeter and meter calibration; understand and use the function generator and oscilloscope as applicable to alternating current and sine and parallel resonant circuits; learn logical troubleshooting techniques on individual circuit functions and parameter testing; draw a simplified schematic of amplifier circuits; use supply and technical publications; troubleshoot and repair power supplies; troubleshoot and repair electrical meters; identify and draw electronic schematic symbols; troubleshoot solid state and integrated circuits and devices; operate and interpret electronic data utilizing oscilloscopes; operate logic probes and current tracers; use common counters in block analysis testing; troubleshoot and repair signal generators; troubleshoot and repair electronic systems on small missile systems; and conduct laboratory experiments using microprocessors.
Instruction: This course covers general and specialized theory, principles of logical troubleshooting, extensive digital logic theory, detailed circuit analysis of representative digital meters, frequency counters, oscilloscopes, signal generators, spectrum analyzers, soldering techniques, microprocessor theory, detailed circuit analysis of microprocessors, logical analyzers, microprocessor testers, and microprocessor-controlled test equipment.
Credit Recommendation: In the lower-division baccalaureate/associate degree category, 3 semester hours in industrial electronics laboratory, 3 in electronic soldering techniques, 3 in microprocessors and controls, 2 in fundamental DC circuits, 2 in fundamental AC circuits, and 3 in solid state electronics (11/91).

MC-1715-0140

CIRCUIT CARD REPAIR
(Microminiature Component Repair)

Course Number: E2D.
Location: Air Ground Combat Center, Twentynine Palms, CA.
Length: 6-8 weeks (238-280 hours).
Exhibit Dates: 4/92–Present.
Learning Outcomes: Before 4/92 see MC-1715-0089. Upon completion of the course, the student will be able to repair circuits characterized by high density circuitry, highly sensitive components, extremely small size, and/or multilead components.
Instruction: The course encompasses the identification and causes of defective solder connections, the procedures and proper tools required to produce high reliability solder connections, and proper mounting and installation of various components, including microelectronic devices, such as integrated circuits, conformal coatings, and the most effective means of removal and replacement. Circuit baseboards are identified by composition, then rebuilt or repaired. Types of interface connections and conductor runs are identified and repaired or

replaced. Repairs are made to boards containing multiple or subsurface conductor runs, flexible printing wiring boards, and printed wiring harnesses, all without causing further degradation to the circuit under repair. Specifications and techniques required for the various types of terminal and connector pins are included. In addition to the materials used in the microminiature component repair course, various electroplating chemicals, metals, paints, and etchant materials are provided for plating.

Credit Recommendation: In the lower-division baccalaureate/associate degree category, 3 semester hours in microminiature component repair (3/95).

MC-1715-0141

TACTICAL DATA COMMUNICATION CENTRAL REPAIR

Course Number: 28Y.
Location: Air Ground Combat Center, Twentynine Palms, CA.
Length: 19 weeks (686 hours).
Exhibit Dates: 1/90–12/93.
Learning Outcomes: Upon completion of this course, the student will be qualified to install, verify operational readiness, configure, perform corrective maintenance, and operate the AN/TYQ-3A tactical data communications central system.
Instruction: Course includes lectures, demonstrations and practical exercises in installation, system verification, configuration, and operation of the AN/TYQ-3A system. Instruction on network operation is also provided.
Credit Recommendation: In the vocational certificate category, 3 semester hours in electronic systems troubleshooting and maintenance and 3 in computer systems troubleshooting and maintenance (1/93).

MC-1715-0142

TACTICAL AIR COMMAND CENTRAL TECHNICIAN

Course Number: FGX.
Location: Air Ground Combat Center, Twentynine Palms, CA.
Length: 28-29 weeks (1008 hours).
Exhibit Dates: 8/90–2/94.
Learning Outcomes: Before 8/90 see MC-1715-0046. Upon completion of the course, the student will be qualified to operate, install, inspect, test, align, maintain, and repair a wide variety of subsystems comprising the AN/TYQ-1B tactical air command control system.
Instruction: Lectures, demonstrations, and practical exercises cover the analysis, alignment, and troubleshooting of power supplies; symptom recognition and fault isolation in display and voice communications equipment; and symptom recognition and fault isolation in the AN/TYQ-1B itself.
Credit Recommendation: In the lower-division baccalaureate/associate degree category, 3 semester hours in electronic systems troubleshooting and maintenance and 3 in computer systems troubleshooting and maintenance (1/93).

MC-1715-0143

HIGH FREQUENCY COMMUNICATION CENTRAL OPERATOR

Course Number: BEZ.
Location: Air Ground Combat Center, Twentynine Palms, CA.
Length: 7 weeks (253 hours).
Exhibit Dates: 1/90–6/94.
Learning Outcomes: Upon completion of the course, the student will be able to operate specific military hardware.
Instruction: Instruction includes lectures, demonstrations, and practical exercises in the operation, installation, maintenance, and administration of communications equipment.
Credit Recommendation: Credit is not recommended because of the military-specific nature of the course (1/93).

MC-1715-0144

AVIATION FIRE CONTROL REPAIR

Course Number: 66L.
Location: Air Ground Combat Center, Twentynine Palms, CA.
Length: 13-14 weeks (469 hours).
Exhibit Dates: 1/90–4/92.
Learning Outcomes: Upon completion of the course, the student will be able to install, operate, adjust, and perform preventive maintenance on the AN/TPB-1D fire control radar system.
Instruction: Instruction provides for the preparation of enlisted personnel in the overall concepts of operation, installation, and adjustment of military aircraft course-directing radar systems. Course includes corrective and preventive maintenance procedures, use of test equipment, and internal testing procedures.
Credit Recommendation: In the vocational certificate category, 3 semester hours in electronic equipment maintenance and 1 in electronic laboratory (1/93).

MC-1715-0145

POSITION LOCATION REPORTING SYSTEM (PLRS) MASTER STATION MAINTENANCE

Course Number: *Version 1:* DQJ. *Version 2:* DQJ.
Location: Air Ground Combat Center, Twentynine Palms, CA.
Length: *Version 1:* 6-10 weeks (342 hours). *Version 2:* 7-8 weeks (272 hours).
Exhibit Dates: *Version 1:* 12/96–Present. *Version 2:* 9/90–11/96.
Learning Outcomes: *Version 1:* This version is pending evaluation. *Version 2:* Upon completion of the course, the student will be able to install, operate, maintain, and troubleshoot the AN/USQ-90 position location reporting system.
Instruction: *Version 1:* This version is pending evaluation. *Version 2:* Lectures, demonstrations and practical exercises cover operation, maintenance, troubleshooting, and fault isolation to the lowest replaceable item.
Credit Recommendation: *Version 1:* Pending evaluation. *Version 2:* In the vocational certificate category, 3 semester hours in electronic systems troubleshooting and maintenance (1/93).

MC-1715-0146

TACTICAL DATA COMMUNICATION CENTRAL TECHNICIAN

Course Number: OGM.
Location: Air Ground Combat Center, Twentynine Palms, CA.
Length: 28 weeks (980 hours).
Exhibit Dates: 2/91–Present.
Learning Outcomes: Before 2/91 see MC-1715-0026. Upon completion of the course, the student will be able to perform corrective maintenance as well as operate and analyze data link operations using the AN/TYQ-3A tactical data communication central system.
Instruction: Lectures, demonstrations and practical exercises cover corrective maintenance to the lowest replaceable unit. There is also thorough coverage of data link operations and the analysis of coded message traffic.
Credit Recommendation: In the lower-division baccalaureate/associate degree category, 3 semester hours in computer systems troubleshooting and maintenance and 3 in electronic systems troubleshooting and maintenance (1/93).

MC-1715-0148

AVIATION RADAR REPAIR (B)

Course Number: 66V.
Location: Air Ground Combat Center, Twentynine Palms, CA.
Length: 8 weeks (280 hours).
Exhibit Dates: 1/90–12/93.
Learning Outcomes: Upon completion of the course, the student will be able to install, operate, adjust, inspect, and maintain computers, electromechanical components, and radar sets.
Instruction: Lectures and practical exercises cover installation, alignment, and maintenance of radar antennas and radar systems; tracing RF energy flow through the system; automatic frequency control; high power pulses; associated computer control circuits; and troubleshooting to the lowest replaceable unit.
Credit Recommendation: In the vocational certificate category, 5 semester hours in communications technology (1/93).

MC-1715-0149

1. POSITION LOCATING REPORTING SYSTEM (PLRS) SUPPORT MAINTENANCE
2. POSITION LOCATION REPORTING SYSTEM (PLRS) SUPPORT MAINTENANCE

Course Number: *Version 1:* DQK. *Version 2:* DQK.
Location: Air Ground Combat Center, Twentynine Palms, CA.
Length: *Version 1:* 3-4 weeks (126 hours). *Version 2:* 3-4 weeks (119-134 hours).
Exhibit Dates: *Version 1:* 10/95–Present. *Version 2:* 1/90–9/95.
Learning Outcomes: *Version 1:* Upon completion of the course, the student will be able to perform circuit analysis, alignment, testing, fault isolation, and corrective/preventive maintenance on the position location reporting system (AN/USQ-90) and its associated subsystems. *Version 2:* Upon completion of the course, the student will be able to perform circuit analysis, alignment, testing, fault isolation, and corrective/preventive maintenance on the position location reporting system (AN/USQ-90) and its associated subsystems (RT-1343 basic user unit and TS-4118 program test set) and describe the operation of the system.
Instruction: *Version 1:* Lectures, demonstrations, practical exercises, and instructional television are used to present the theory of operation, circuit analysis, alignment, performance testing, fault isolation, and corrective/preventive maintenance on the AN/USQ-90

PLRS and its associated subsystems. *Version 2:* Lectures, demonstrations, practical exercises, and instructional television are used to present the theory of operation, circuit analysis, alignment, performance testing, fault isolation, and corrective/preventive maintenance on the AN/USQ-90 PLRS and its associated subsystems.

Credit Recommendation: *Version 1:* In the lower-division baccalaureate/associate degree category, 3 semester hours in electronic systems troubleshooting and maintenance (3/96). *Version 2:* In the lower-division baccalaureate/associate degree category, 3 semester hours in electronic systems troubleshooting and maintenance (1/93).

MC-1715-0150

AIR DEFENSE CONTROL OFFICERS SENIOR

Course Number: 72H.
Location: Air Ground Combat Center, Twentynine Palms, CA.
Length: 2 weeks (81-82 hours).
Exhibit Dates: 8/90–Present.
Learning Outcomes: Upon completion of the course, the student will have a thorough knowledge of the tactical air operations center.
Instruction: The course includes employment of the tactical air operations center and its equipment, data links, electronic warfare, and aircraft tactics and weapons.
Credit Recommendation: Credit is not recommended because of the military-specific nature of the course (1/93).

MC-1715-0151

TACTICAL AIR COMMAND CENTRAL REPAIR

Course Number: FGV.
Location: Air Ground Combat Center, Twentynine Palms, CA.
Length: 16 weeks (581 hours).
Exhibit Dates: 6/90–Present.
Learning Outcomes: Before 6/90 see MC-1715-0020. Upon completion of the course, the student will be able to operate, test, maintain, and repair the tactical air command central system. The system includes computers, displays, voice communications, and digital data links.
Instruction: Lectures and practical exercises at the introductory level cover the operation, testing, and repair of all system components. Course includes the use of simple voltmeters and computerized automated test equipment to locate faults to the card level.
Credit Recommendation: In the lower-division baccalaureate/associate degree category, 2 semester hours in introduction to systems maintenance and 2 in introduction to computer maintenance (1/93).

MC-1715-0152

AVIATION RADAR TECHNICIAN (C)

Course Number: E2W.
Location: Air Ground Combat Center, Twentynine Palms, CA.
Length: 17 weeks (662 hours).
Exhibit Dates: 3/91–8/91.
Learning Outcomes: Upon completion of the course, the student will be able to operate, repair, and maintain the AN/TPS-32 phased array radar system.
Instruction: Lectures, demonstrations and practical exercises cover operational concepts, alignment, self-test facilities, corrective maintenance, test equipment, flow charts, and logic diagrams for the AN/TPS-32 phased array radar system.
Credit Recommendation: In the vocational certificate category, 3 semester hours in electronic systems troubleshooting and maintenance and 3 in computer systems troubleshooting and maintenance (1/93).

MC-1715-0153

AVIATION RADAR REPAIR (A)

Course Number: E2H.
Location: Air Ground Combat Center, Twentynine Palms, CA.
Length: 14 weeks (147 hours).
Exhibit Dates: 1/90–4/93.
Learning Outcomes: Upon completion of the course, the student will be qualified to install, test, adjust, and maintain an aviation radar system.
Instruction: Lectures and practical exercises on installation and maintenance procedures on all parts of the radar system are presented. Also included are practical exercises on the troubleshooting of electronic circuitry.
Credit Recommendation: In the vocational certificate category, 3 semester hours in electronic systems troubleshooting and maintenance (1/93).

MC-1715-0154

RADAR FUNDAMENTALS

Course Number: *Version 1:* 27W. *Version 2:* 27W. *Version 3:* 27W.
Location: Air Ground Combat Center, Twentynine Palms, CA.
Length: *Version 1:* 4 weeks (140 hours). *Version 2:* 4 weeks (140 hours). *Version 3:* 4 weeks (140 hours).
Exhibit Dates: *Version 1:* 9/95–Present. *Version 2:* 10/91–8/95. *Version 3:* 4/90–9/91.
Learning Outcomes: *Version 1:* Upon completion of the course, the student will be able to trace signal flow through a radar system, describe power distribution, waveguides, duplexers, microwave principles, antenna control, receivers, transmitters, indicators, solid-state devices, and operational amplifiers, and able to use basic microwave test equipment to isolate and repair faulty radar systems. *Version 2:* Upon completion of the course, the student will be able to trace signal flow through a radar system and use basic microwave test equipment to isolate and repair a faulty radar system. *Version 3:* Before 4/90 see MC-1715-0063. Upon completion of the course, the student will understand the theory and application of radar principles.
Instruction: *Version 1:* Instruction includes lectures, demonstrations, and practical exercises in radar system, basic microwave theory, radar transmission and modulation, radar receiver and detection, radar indicators, antenna and antenna controls, and microwave test equipment. *Version 2:* Instruction includes lectures, demonstrations, and practical exercises in radar system, basic microwave theory, radar transmission and modulation, radar receiver and detection, radar indicators, antenna and antenna controls, and microwave test equipment. *Version 3:* Instruction includes lectures, demonstrations, and practical exercises in radar system, basic microwave theory, radar transmission and modulation, radar receiver and detection, radar indicators, antenna and antenna controls, and microwave test equipment.

Credit Recommendation: *Version 1:* In the lower-division baccalaureate/associate degree category, 3 semester hours in electronics communication (microwave) and 2 in electronics communication laboratory (3/96). *Version 2:* In the lower-division baccalaureate/associate degree category, 3 semester hours electronic communications (microwave) and 2 in electronic communications laboratory (1/93). *Version 3:* In the lower-division baccalaureate/associate degree category, 3 semester hours in radar systems (1/93).

MC-1715-0155

AVIATION RADIO REPAIR

Course Number: 66F.
Location: Air Ground Combat Center, Twentynine Palms, CA.
Length: 11 weeks (455 hours).
Exhibit Dates: 3/92–Present.
Learning Outcomes: Before 3/92 see MC-1715-0011. Upon completion of the course, the student will be able to install, inspect, maintain, and repair specific aviation and ground communications electronic equipment at the module level.
Instruction: Course includes lectures and practical exercises in the theory, operation, and repair of AM, FM, and SSB radios to the module level. Course also includes equipment characteristics, digital logic, circuit analysis, alignment, performance testing, and troubleshooting test equipment.
Credit Recommendation: In the vocational certificate category, 3 semester hours in electronic systems troubleshooting and maintenance and 3 in electronic systems troubleshooting and maintenance laboratory (1/93); in the lower-division baccalaureate/associate degree category, 1 semester hour in electronic systems troubleshooting and maintenance laboratory (1/93).

MC-1715-0156

AVIATION RADAR TECHNICIAN (A)

Course Number: E2U.
Location: Air Ground Combat Center, Twentynine Palms, CA.
Length: 12 weeks (420 hours).
Exhibit Dates: 3/91–6/95.
Learning Outcomes: Upon completion of the course, students will be able to install, test, adjust, and troubleshoot a specific radar and display system to the module level.
Instruction: The course covers concepts of radar system operation; assembly and disassembly; adjustment; use of built-in test equipment, special tools, and external test equipment; and preventative maintenance.
Credit Recommendation: In the vocational certificate category, 3 semester hours in microwave communications and 3 in microwave communications laboratory (1/93).

MC-1715-0157

TACTICAL AIR COMMAND CENTER OPERATOR

Course Number: 72G.
Location: Air Ground Combat Center, Twentynine Palms, CA.
Length: 1-2 weeks (71-72 hours).
Exhibit Dates: 11/90–12/90.
Learning Outcomes: Upon completion of the course, the student will be able to describe the functions of the tactical air command center.

COURSE EXHIBITS

Instruction: Course includes lectures and practical applications on the function and operation of the tactical air command center.

Credit Recommendation: Credit is not recommended because of the military-specific nature of the course (1/93).

MC-1715-0158

POSITION LOCATION REPORTING SYSTEM (PLRS) MASTER STATION OPERATOR

Course Number: CGN.
Location: Air Ground Combat Center, Twentynine Palms, CA.
Length: 4-5 weeks (167-181 hours).
Exhibit Dates: 9/90–Present.
Learning Outcomes: Upon completion of the course, the student will be able to perform the duties of the position location reporting system master station operator.
Instruction: Lectures, demonstrations, practical exercises, and instructional television are used to illustrate PLRS master station initialization, PLRS data base updates, antijam procedures, PLRS network maintenance, and sending and receiving information using the alternate master station.
Credit Recommendation: Credit is not recommended because military-specific nature of the course (5/96).

MC-1715-0159

AVIATION RADAR TECHNICIAN (B)

Course Number: E2V.
Location: Air Ground Combat Center, Twentynine Palms, CA.
Length: 14-15 weeks (511 hours).
Exhibit Dates: 3/91–6/95.
Learning Outcomes: Upon completion of the course, the student will be qualified to perform radar siting as well as corrective maintenance on the AN/TPS-59 radar.
Instruction: Course includes lectures, practical exercises, and computer-aided instruction in radar site selection, performing software modifications, fault isolation, backplane repair, and fault isolation on support equipment for the AN/TPS-59 radar.
Credit Recommendation: In the vocational certificate category, 3 semester hours in troubleshooting and maintenance and 2 in computer systems troubleshooting and maintenance (8/94).

MC-1715-0160

RADIO FUNDAMENTALS

Course Number: 27V.
Location: Air Ground Combat Center, Twentynine Palms, CA.
Length: 6 weeks (210 hours).
Exhibit Dates: 4/91–Present.
Learning Outcomes: Before 4/91 see MC-1715-0010. Upon completion of the course, the student will apply the theory of operation of RF transmitters and receivers to test and repair them. The student will also understand the theory and use of antennas.
Instruction: Course includes lectures and laboratory exercises in theory of operation and troubleshooting procedures involving AM, FM, and SSB transmitters, receivers, and antennas.
Credit Recommendation: In the vocational certificate category, 6 semester hours in radio repair (6/95); in the lower-division baccalaureate/associate degree category, 3 semester hours in electronic communications and 2 in electronic communications laboratory (6/95).

MC-1715-0161

1. MULTICHANNEL EQUIPMENT OPERATOR
2. MICROWAVE EQUIPMENT OPERATOR

Course Number: *Version 1:* CGM. *Version 2:* CGM.
Location: Air Ground Combat Center, Twentynine Palms, CA.
Length: *Version 1:* 7-8 weeks (288 hours). *Version 2:* 6 weeks (211 hours).
Exhibit Dates: *Version 1:* 2/95–Present. *Version 2:* 5/91–1/95.
Learning Outcomes: *Version 1:* Upon completion of the course, students will demonstrate the ability to install, operate, and maintain multichannel communication systems and microwave equipment. *Version 2:* Before 5/91 see MC-1715-0024. Upon completion of the course, students will apply the fundamental operating characteristics of microwave communications to install, operate, and perform preventative maintenance on a specific microwave system and antenna.
Instruction: *Version 1:* Lecture and practical exercises with "hands-on" operational experience provide the student with multichannel equipment operation and set-up. *Version 2:* The course includes the operating characteristics and limitations of microwave radio systems and equipment. It also includes practical experience with the operation and maintenance of a specific microwave system.
Credit Recommendation: *Version 1:* In the lower-division baccalaureate/associate degree category, 1 semester hour in microwave communication equipment (3/96). *Version 2:* In the lower-division baccalaureate/associate degree category, 1 semester hour in microwave communications (1/93).

MC-1715-0162

1. FIELD WIREMAN
2. FIELD WIRE OPERATORS

Course Number: *Version 1:* 247. *Version 2:* 247.
Location: Air Ground Combat Center, Twentynine Palms, CA.
Length: *Version 1:* 7 weeks (259 hours). *Version 2:* 6 weeks (210 hours).
Exhibit Dates: *Version 1:* 8/95–Present. *Version 2:* 7/93–7/95.
Learning Outcomes: *Version 1:* Upon completion of the course, students will be able to install, maintain, and troubleshoot tactical analog and digital telephony equipment and ancillaries. *Version 2:* Upon completion of the course, students will be able to install, operate, and maintain a field telephone system.
Instruction: *Version 1:* Lecture and practical applications in theory and operation of basic communication terms and definitions, basic job functions related to field wiring of telephony equipment, and also basic instruction supported by performance evaluation of the installation and operation of Digital and Analog telephones. Other topics include networking topics like LAN/WAN technology and basic computer operation in a client/server network. *Version 2:* The course includes the basics of military communications, field wire ties and splices, wire laying and retrieval, terminal equipment, power sources, field telephones, and troubleshooting procedures.

Credit Recommendation: *Version 1:* In the lower-division baccalaureate/associate degree category, 1 in computer network operation and 2 in electrician skill development (3/96). *Version 2:* Credit is not recommended because of the military-specific nature of the course (1/93).

MC-1715-0163

HAWK MISSILE SYSTEM OPERATOR (USMC)

Course Number: 104-7222 (OS).
Location: Air Defense Artillery School, Ft. Bliss, TX.
Length: 9-10 weeks (364 hours).
Exhibit Dates: 2/95–Present.
Learning Outcomes: Upon completion of the course, the student will be able to describe the characteristics, capabilities, and functions of the Hawk missile system.
Instruction: Course includes lectures and practical demonstrations on the Hawk system, safety procedures, emplacement, radar, targeting, and aircraft recognition.
Credit Recommendation: In the vocational certificate category, 1 semester hour in mechanical systems maintenance (6/95).

MC-1715-0165

AIR SUPPORT OPERATIONS OPERATOR

Course Number: 67L.
Location: Air Ground Combat Center, Twentynine Palms, CA.
Length: 7 weeks (267-268 hours).
Exhibit Dates: 4/95–Present.
Learning Outcomes: Before 4/95 see MC-1715-0031. Upon completion of the course, the student will be able to operate the direct air support center communication system.
Instruction: Lecture and practical application are used.
Credit Recommendation: Credit is not recommended because of the limited, specialized nature of the course (3/96).

MC-1715-0166

GROUND RADIO REPAIR

Course Number: 27M.
Location: Air Ground Combat Center, Twentynine Palms, CA.
Length: 14-16 weeks (546 hours).
Exhibit Dates: 7/95–Present.
Learning Outcomes: Before 7/95 see MC-1715-0009. Upon completion of the course, the student will be able to install, operate, troubleshoot, and repair AM, FM, and SSB radio communication equipment; and be able to conduct performance analysis and preventive maintenance on the radio communication equipment.
Instruction: Lecture is supplemented with computer assisted instruction, and practical application on AM, FM, and SSB radio equipment maintenance, operation, and repair.
Credit Recommendation: In the lower-division baccalaureate/associate degree category, 5 semester hours in radio communication equipment repair (3/96).

MC-1715-0167

COMMUNICATION SYSTEMS CHIEF

Course Number: CHK.
Location: Air Ground Combat Center, Twentynine Palms, CA.
Length: 13 weeks (449 hours).
Exhibit Dates: 6/95–Present.

Marine Corps

Learning Outcomes: Upon completion of the course, the student will be able to plan and install communication systems employing radio and telephone equipment; utilize the equipment in performing communication operations using radio and telephone equipment; and develop emergency services of electrical power for both radio and telephone communication systems.

Instruction: Lecture, computer assisted laboratory, and "hands-on" practical experience is used in instruction covering electrical power services, radio systems, telephone and fiber optic systems, and message manipulation methods.

Credit Recommendation: In the lower-division baccalaureate/associate degree category, 3 semester hours in radio or telephone system planning and installation and 1 in DC electricity (3/96).

MC-1715-0168

AN/TSC-120 COMMUNICATION CENTRAL TECHNICIAN

Course Number: DPX.
Location: Air Ground Combat Center, Twentynine Palms, CA.
Length: 5 weeks (180 hours).
Exhibit Dates: 5/95–Present.
Learning Outcomes: Upon completion of the course, the student will be able to operate, troubleshoot, and repair high frequency radio communication equipment.
Instruction: Lecture, computer assisted instruction, and practical application are used in training personnel to operate, troubleshoot, and repair the high frequency (AN/TSC-120) radio communication equipment.
Credit Recommendation: In the lower-division baccalaureate/associate degree category, 2 semester hours in basic electronic/radio communication equipment maintenance and repair (3/96).

MC-1715-0169

MOBILE SECURITY COMMUNICATION TECHNICIAN

Course Number: DRF.
Location: Air Ground Combat Center, Twentynine Palms, CA.
Length: 16-17 weeks (574-610 hours).
Exhibit Dates: 2/93–Present.
Learning Outcomes: Upon completion of the course, the student will be able to maintain the AN/MSC-63A mobile security communication system to include principles of operation, circuit analysis, adjustments, performance testing, and component/system troubleshooting using associated diagnostics and test equipment.
Instruction: Lectures, demonstrations, practical exercises are used to present the physical/electrical characteristics and capabilities of the AN/MSC-63A mobile security system, as well as operation of data terminal equipment, testing, adjusting and troubleshooting, fault isolation, and corrective/preventative maintenance on the AN/MSC-63A mobile security system.
Credit Recommendation: In the lower-division baccalaureate/associate degree category, 3 semester hours in communications technology and 1 in communications technology laboratory (3/97).

MC-1715-0170

AN/MRC-135B REPAIR

Course Number: DRM.
Location: Air Ground Combat Center, Twentynine Palms, CA.
Length: 2 weeks (70 hours).
Exhibit Dates: 9/94–Present.
Learning Outcomes: Upon completion of the course, the student will be able to perform circuit analysis, alignment, performance testing, fault isolation, corrective, and preventive maintenance on a terminal telegraph set.
Instruction: Lectures and practical exercises related to a terminal telegraph set are the two methods of instruction.
Credit Recommendation: In the lower-division baccalaureate/associate degree category, 1 semester hour in maintenance of terminal telegraph equipment (3/96).

MC-1715-0171

AN/TSC-120 COMMUNICATION CENTRAL SYSTEM INSTALLER/MAINTAINER

Course Number: BEZ.
Location: Air Ground Combat Center, Twentynine Palms, CA.
Length: 7-8 weeks (289 hours).
Exhibit Dates: 10/94–Present.
Learning Outcomes: Upon completion of the course, the student will be able to install, operate, and maintain the AN/TSC-120 communication equipment.
Instruction: Lecture, practical application, and performance testing is utilized in learning the operation, maintenance, and installation of the AN/TSC-120 radio communication equipment.
Credit Recommendation: Credit is not recommended because of the limited, specialized nature of the course (3/96).

MC-1715-0173

GROUND RADAR REPAIR

Course Number: 27E.
Location: Air Ground Combat Center, Twentynine Palms, CA.
Length: 8-10 weeks (371 hours).
Exhibit Dates: 8/94–Present.
Learning Outcomes: Before 8/94 see MC-1715-0062. Upon completion of the course, the student will be able to install, operate, adjust, inspect, and perform corrective and preventive maintenance on specific ground radar equipment.
Instruction: Course includes lectures and practical exercises in the installation, operation, adjustment, inspection, and maintenance of specific ground radar sets, including basic theory of radar sets and auxiliary and test equipment, system circuit analysis, and operation of various subsystems on specific equipment.
Credit Recommendation: In the lower-division baccalaureate/associate degree category, 3 semester hours in electrical system maintenance or radar systems maintenance (3/96).

MC-1715-0174

TACTICAL AIR OPERATIONS MODULE REPAIR AN/TYQ-23(V)1

Course Number: E3G.
Location: Air Ground Combat Center, Twentynine Palms, CA.
Length: 7-8 weeks (280 hours).
Exhibit Dates: 2/93–Present.
Learning Outcomes: Upon completion of the course, the student will be able to troubleshoot and replace components in radio, radar display systems, and associated power supplies.
Instruction: Lecture with practical application on specific electronic equipment used for ground and airborne communications, and instruction will focus on radio, radar, fiber optic, and associated power supply maintenance.
Credit Recommendation: In the lower-division baccalaureate/associate degree category, 2 semester hours in basic AC electricity/electronics (3/96).

MC-1715-0175

TACTICAL AIR OPERATIONS MODULE (TAOM), AN/TYQ-23(V)1 TECHNICIAN

Course Number: E3H.
Location: Air Ground Combat Center, Twentynine Palms, CA.
Length: 3 weeks (104 hours).
Exhibit Dates: 2/95–Present.
Learning Outcomes: Upon completion of the course, the student will be able to operate, test, troubleshoot, and repair data processing, digital and voice radio communication, and radar electronic equipment.
Instruction: Lecture, computer aided and practical experience are used to develop intermediate level technician skills. Topics covered are data processing, radio communication using both digital and voice, and radar equipment maintenance. Student demonstrates skills using practical application and written test evaluation methods.
Credit Recommendation: In the lower-division baccalaureate/associate degree category, 4 semester hours in electronic theory and practical application (3/96).

MC-1715-0176

AIR SUPPORT CONTROL OFFICER

Course Number: TOA.
Location: Air Ground Combat Center, Twentynine Palms, CA.
Length: 11-12 weeks (427-428 hours).
Exhibit Dates: 3/95–Present.
Learning Outcomes: Before 3/95 see MC-1715-0078. Upon completion of the course, the student will be able to direct air and assault support during combat using electronic provided data.
Instruction: Course includes lectures and practical exercises in radio network operator duties, plotter/status board techniques, radar operator procedures, and strike controller procedures under simulated field conditions.
Credit Recommendation: Credit is not recommended because of the limited, specialized nature of the course (3/96).

MC-1715-0177

COMMAND AND CONTROL SYSTEMS

Course Number: CHJ.
Location: Command and Control Systems School, Quantico, VA.
Length: 43 weeks (1644 hours).
Exhibit Dates: 12/91–Present.
Learning Outcomes: Before 12/91 see MC-1715-0068. Upon completion of the course, the student will understand those command and control systems, which allow them to plan, direct, and control operations at various levels

within the Marine Corps hierarchy and in the joint arena; students gain an understanding of those activities which have a direct impact on employment of electronic systems supporting the command (executive) decision-making process; students master basic concepts of staff planning, strategic analysis, and interorganizational cooperation; and they gain intermediate level leadership, enhanced research, and english composition skills.

Instruction: The course includes lectures, demonstrations, practical exercises relating to the theory, design, and operation of military electronic communication systems. Lectures and exercises are offered on electricity, principles of electronics, telecommunications, planning, organization, and principles of leadership. The course also provides training in speaking and writing skills.

Credit Recommendation: In the lower-division baccalaureate/associate degree category, 2 semester hours in written composition/technical writing, 1 in effective speaking (speech), and 2 in electronic theory (4/96); in the upper-division baccalaureate category, 3 semester hours in leadership and 6 in telecommunications/computer systems (4/96).

MC-1715-0178

RESERVE BASIC COMMUNICATION OFFICER PHASE 2

Course Number: CGE.
Location: Communication Officers School, Quantico, VA.
Length: 2 weeks (85 hours).
Exhibit Dates: 1/95–Present.
Learning Outcomes: Before 1/95 see MC-1715-0025. Upon completion of the course, the reserve officer will understand the technical requirements, planning, installation, and operation of digital transmission and switching systems, and communication and tactical data networking. The course is intended for graduates of the Basic Communication Officers course or the Data Systems Officers course.
Instruction: Lectures, practical exercises, written projects, and in-class examinations. Topics include communications planning, naval and amphibious communications, communications/electronics maintenance, and large-scale data networks.
Credit Recommendation: In the lower-division baccalaureate/associate degree category, 2 semester hours in communication systems technology and 1 in introduction to computer systems (4/96).

MC-1715-0179

RESERVE BASIC COMMUNICATION OFFICER PHASE 1

Course Number: CEO.
Location: Communication Officers School, Quantico, VA.
Length: 2 weeks (98 hours).
Exhibit Dates: 1/95–Present.
Learning Outcomes: Before 1/95 see MC-1715-0025. Upon completion of the course, the reserve officer will be familiar with communication operating procedures, design and installation of wire, and radio communication systems.
Instruction: Lectures, practical exercises, written projects, and in-class examinations. Topics include introduction to local area networks, radio communications, digital switching systems, wire equipment, digital telephones, and facsimile services.

Credit Recommendation: In the lower-division baccalaureate/associate degree category, 1 semester hour in electronic communication systems technology and 1 in introduction to computer systems (4/96).

MC-1715-0180

SURFACE AIR DEFENSE SYSTEM FIRE CONTROL TECHNICIAN

Course Number: 220; 121-5925 (PIPIII) (OS).
Location: Marine Corps Detachment, Redstone Arsenal, AL.
Length: 46 weeks (1681 hours).
Exhibit Dates: 10/94–5/95.
Learning Outcomes: Upon completion of the course, the student will demonstrate understanding of basic DC/AC circuit theory, diode and transistor operation, digital logic circuits, microcomputer subsystems, and microprocessors. Student will be able to use test equipment to troubleshoot microcomputer systems; perform high reliability soldering; and inspect, test, and repair multilevel printed circuit boards to the component level.
Instruction: Lectures and practical exercises cover series, parallel, and series/parallel DC and AC resistive circuits; AC inductive and capacitive circuits; RL and RC transients; transformer; power supplies; and circuit components. Practical exercises include the use of multimeters, oscilloscopes, and function generators; characteristics of diodes, transistor and basic amplifier circuits, and troubleshooting techniques; practical skills and knowledge of electrical connections; soldering station operation and maintenance; terminal and pin connections; PC board monitoring of components; desoldering coaxial cable connections; cable and multilayer printed wire board repair; and microwave equipment, including transmission lines, waveguides, antennas, and amplifiers (minimum theory). Lecture survey method covers computer arithmetic, logic functions, flip-flops, analog and digital conversion, and basic digital system. Lectures and laboratory exercises cover the 8080 8-bit microprocessor and 80186 16-bit microcprocessor architecture, programming, timing, interrupts, and processing.
Credit Recommendation: In the vocational certificate category, 4 semester hours in precision soldering techniques, 1 in microwave laboratory, and 2 in solid state devices and systems (5/97); in the lower-division baccalaureate/associate degree category, 4 semester hours in introduction to DC/AC circuit analysis, 2 in digital logic systems, 3 in microprocessor operation, and 1 in digital and microcomputer laboratory troubleshooting (5/97).

MC-1715-0181

SURFACE AIR DEFENSE SYSTEM ACQUISITION TECHNICIAN

Course Number: 22P; 121-5924 (OS).
Location: Marine Corps Detachment, Redstone Arsenal, AL.
Length: 49-50 weeks (1774 hours).
Exhibit Dates: 10/94–10/96.
Learning Outcomes: Upon completion of the course, the student will demonstrate understanding of basic DC/AC circuit theory, diode and transistor operation, digital logic circuits, microcomputer subsystems, and microprocessors. Student will be able to use test equipment to troubleshoot microcomputer systems; perform high reliability soldering; and inspect, test, and repair multilevel printed circuit boards to the component level.
Instruction: Lectures and practical exercises cover series, parallel, and series/parallel DC and AC resistive circuits; AC inductive and capacitive circuits; RL and RC transients; transformer; power supplies; and circuit components. Practical exercises include the use of multimeters, oscilloscopes, and function generators; characteristics of diodes, transistor and basic amplifier circuits, and troubleshooting techniques; practical skills and knowledge of electrical connections; soldering station operation and maintenance; terminal and pin connections; PC board monitoring of components; desoldering coaxial cable connections; cable and multilayer printed wire board repair; and microwave equipment, including transmission lines, waveguides, antennas, and amplifiers (minimum theory). Lecture survey method covers computer arithmetic, logic functions, flip-flops, analog and digital conversion, and basic digital system. Lectures and laboratory exercises cover the 8080 8-bit microprocessor and 80186 16-bit microcprocessor architecture, programming, timing, interrupts, and processing.
Credit Recommendation: In the vocational certificate category, 4 semester hours in precision soldering, 1 in microwave laboratory, and 2 in solid state devices and systems (5/97); in the lower-division baccalaureate/associate degree category, 4 semester hours in introduction to AC/DC circuits, 2 in digital logic systems, 3 in microprocessor operation and troubleshooting, and 1 in digital and microprocessor laboratory (5/97).

MC-1717-0004

OPERATIONAL COMMUNICATION CHIEF

Course Number: *Version 1:* 25A. *Version 2:* 25A.
Location: Communication-Electronics School, Twentynine Palms, CA.
Length: *Version 1:* 15-17 weeks (588 hours). *Version 2:* 19 weeks (684 hours).
Exhibit Dates: *Version 1:* 12/93–Present. *Version 2:* 1/90–11/93.
Learning Outcomes: *Version 1:* Upon completion of the course, the student will be able to utilize computer based communication equipment for message development, encoding, and disposal. *Version 2:* Upon completion of the course the student will be able to operate equipment such as receivers, transmitters, teletype equipment, and communications security devices; describe the operation of a communications center; use map reading and communications procedures; and manage and plan communications needs.
Instruction: *Version 1:* Lectures is supplemented with computer aided instruction as a reinforcement for hands-on experience. Topics covered include Lotus Smart Suite, MS-DOS, Windows, and basic computer fundamentals. *Version 2:* Lectures and practical exercises cover radio receiver operation, typing, preventive maintenance of communications security equipment, message verification, communications security, maintenance procedures, electronic warfare, management, and counseling.
Credit Recommendation: *Version 1:* In the lower-division baccalaureate/associate degree category, 1 semester hour in personnel supervision, 1 in maintenance management, and 3 in basic computer skills (3/96). *Version 2:* In the

lower-division baccalaureate/associate degree category, 1 semester hour in personnel supervision, 1 in maintenance management, and 2 in keyboarding (2/89).

MC-1717-0005

BASIC SHOP FUNDAMENTALS FOR THE MECHANIC BY CORRESPONDENCE

Course Number: 13.30.
Location: Marine Corps Institute, Washington, DC.
Length: Maximum, 52 weeks.
Exhibit Dates: 1/90–Present.
Learning Outcomes: Upon completion of the course, the student will be able to identify basic hand tools and their use and identify introductory shop safety procedures and fire prevention.
Instruction: Course includes basic hand tool usage, safety, and care, and introduction to personal and shop safety.
Credit Recommendation: In the vocational certificate category, 1 semester hour in hand tool usage, identification, and safety (6/89).

MC-1717-0006

GROUND ORDNANCE VEHICLE MAINTENANCE CHIEFS

Course Number: 640-2149; A01GBT1.
Location: Marine Corps Detachment, Ordnance Center and School, Aberdeen Proving Ground, MD.
Length: 5 weeks (164 hours).
Exhibit Dates: 3/96–Present.
Learning Outcomes: Upon completion of the course, the student will be able to serve as a vehicle maintenance manager and develop and supervise policies and procedures for repair and maintenance of vehicles.
Instruction: Lectures, demonstrations, and practical experiences cover vehicle maintenance administration in the areas of general shop management, maintenance and supply management, hazardous material storage and disposal, and general maintenance indicators.
Credit Recommendation: In the lower-division baccalaureate/associate degree category, 3 semester hours in vehicle shop maintenance administration (6/97).

MC-1717-0007

GROUND ORDNANCE OFFICER

Course Number: 4E-F15; A01RGZ1.
Location: Marine Corps Detachment, Ordnance Center and School, Aberdeen Proving Ground, MD.
Length: 5 weeks (164 hours).
Exhibit Dates: 3/96–Present.
Learning Outcomes: Upon completion of the course, the student will be able to serve as a vehicle maintenance manager and develop and supervise policies and procedures for repair and maintenance of vehicles.
Instruction: Lectures, demonstrations, and practical experiences cover vehicle maintenance administration in the areas of general shop management, maintenance and supply management, hazardous material storage and disposal, and general maintenance indicators.
Credit Recommendation: In the lower-division baccalaureate/associate degree category, 3 semester hours in vehicle shop maintenance administration (6/97).

MC-1717-0008

GROUND ORDNANCE WEAPONS CHIEF

Course Number: 640-2181; A01GBS1.
Location: Marine Corps Detachment, Ordnance Center and School, Aberdeen Proving Ground, MD.
Length: 5 weeks (164 hours).
Exhibit Dates: 3/96–Present.
Learning Outcomes: Upon completion of the course, the student will be able to serve as a vehicle maintenance manager and develop and supervise policies and procedures for repair and maintenance of vehicles.
Instruction: Lectures, demonstration, and practical experiences cover vehicle maintenance administration in the areas of general shop management, maintenance and supply management, hazardous material storage and disposal, and general maintenance indicators.
Credit Recommendation: In the lower-division baccalaureate/associate degree category, 3 semester hours in vehicle shop maintenance administration (6/97).

MC-1723-0007

JOURNEYMAN METALWORKER

Course Number: ACT.
Location: Service Support School Cp. Lejeune, NC.
Length: 9 weeks (264-344 hours).
Exhibit Dates: 1/90–7/92.
Objectives: To train noncommissioned officers in the duties of a metalworker at the journeyman level to work in the Fleet Marine Force.
Instruction: Lectures, demonstrations and practical exercises cover Marine Corps maintenance management; procedures for requisitioning metalworking tools, equipment, and related material; safety rules and regulations governing welding sites and shops; identification of metals; construction prints and shop drawings; oxyacetylene procedures; sheet metal fabrication; pipe welding and forging; electrical arc welding; and inert gas welding.
Credit Recommendation: In the lower-division baccalaureate/associate degree category, 4 semester hours in advanced welding (4/89).

MC-1723-0008

METAL WORKING AND WELDING OPERATIONS BY CORRESPONDENCE

Course Number: 13.32g.
Location: Marine Corps Institute, Washington, DC.
Length: Maximum, 52 weeks.
Exhibit Dates: 2/90–Present.
Learning Outcomes: Upon completion of the course, the student will be able to describe arc and gas welding and brazing processes, metal treatment processes, and sheet metal forming processes and read welding and fabrication blueprints.
Instruction: Students are required to pass a proctored end-of-course examination in this correspondence course.
Credit Recommendation: In the lower-division baccalaureate/associate degree category, 1 semester hour in arc welding, 1 in gas welding processes, and 1 in sheet metal forming processes (6/90).

MC-1723-0009

BASIC METAL WORKER

Course Number: *Version 1:* 132. *Version 2:* 132.
Location: Engineer School, Cp. Lejeune, NC.
Length: *Version 1:* 15 weeks (511 hours). *Version 2:* 12 weeks (391 hours).
Exhibit Dates: *Version 1:* 1/95–Present. *Version 2:* 1/90–12/94.
Learning Outcomes: *Version 1:* Upon completion of the course, the student will be capable of interpreting welding syllabi and blueprints, performing oxyacetylene and electrical arc welding, and fabricating metal objects. *Version 2:* Upon completion of the course, the student will be capable of performing basic metal welding tasks.
Instruction: *Version 1:* Lectures, demonstrations, and practical exercises cover preparation of sheet metal for welding and welding equipment, tools, and techniques, specifically oxyacetylene, inert gas, and electric arc welding techniques. *Version 2:* Lectures, demonstrations, and practical exercises cover preparation of sheet metal for welding and welding equipment, tools, and techniques, specifically oxyacetylene, inert gas, and electric arc welding techniques.
Credit Recommendation: *Version 1:* In the lower-division baccalaureate/associate degree category, 2 semester hours in oxyacetylene welding, 2 in arc welding, and 2 in welding safety (7/96). *Version 2:* In the vocational certificate category, 3 semester hours in sheet metal welding (3/92).

MC-1723-0010

MACHINIST

Course Number: 702-2161.
Location: Marine Corps Detachment, Ordnance School, Aberdeen Proving Ground.
Length: 12-13 weeks (484 hours).
Exhibit Dates: 1/91–Present.
Learning Outcomes: Upon completion of the course, the student will be able to lay out and measure fabric metal parts by using tools and equipment such as metal lathes, milling machines, shapers, and the metal cutting band saw. Precision measuring tools, print reading, and heat treating are also covered.
Instruction: Lectures, shop work, and performance testing are employed in this course.
Credit Recommendation: In the lower-division baccalaureate/associate degree category, 3 semester hours in milling machine operations, 3 in lathe operations, 3 in precision measurement and bench work, and 3 in machine shop theory (1/98).

MC-1728-0002

MILITARY FUNCTIONS IN CIVIL DISTURBANCES BY CORRESPONDENCE

Course Number: 03.16j.
Location: Marine Corps Institute, Washington, DC.
Length: Maximum, 52 weeks.
Exhibit Dates: 1/90–1/97.
Objectives: To provide the knowledge necessary to control riots or civil disturbances.
Instruction: This correspondence course presents a series of study units covering such topics as crowd and mob behavior, riot control principles, types of disturbances, crowd control

formations, handling explosive devices and bomb threats, weapons, riot control agents, special equipment, and operations planning. There is special emphasis on laws, policies, and legal considerations. A supervised final examination without notes or textbooks is administered.

Credit Recommendation: In the upper-division baccalaureate category, 2 semester hours in civil disturbances and/or riot control (3/85).

MC-1728-0006

MARINE SECURITY GUARD

Course Number: 81H.
Location: Security Guard School, Quantico, VA.
Length: 6-8 weeks (216-281 hours).
Exhibit Dates: 1/90–Present.
Learning Outcomes: Upon completion of the course, the student will be able to perform physical security inspections; demonstrate awareness of external threats to security; recognize/respond to classified material threats; describe techniques used to protect property and reduce threats; identify fire dangers, alarms, locks, and security equipment used; qualify in pistol/shotgun and other self defense procedures; function within environment under State Department and United States Marine Corps rules and regulations; be familiar with potential for terrorism; and be familiar with Embassy conduct expectations.
Instruction: The course covers such topics as emergency planning, knowledge of locks/alarm systems and bombs, anti-terrorist techniques, physical security inspections, and fire protection procedures. Also covered are hostile threats, hostage survival, computer and document safety, and access control measures. Upon reporting to duty assignment, students are required to complete 100 hours of language training.
Credit Recommendation: In the vocational certificate category, 2 semester hours in physical training and weapons (8/87); in the lower-division baccalaureate/associate degree category, 4 semester hours in techniques of physical security (8/87).

MC-1728-0007

CORRECTIONS SUPERVISOR BY CORRESPONDENCE

Course Number: 58.2.
Location: Marine Corps Institute, Washington, DC.
Length: Maximum, 52 weeks.
Exhibit Dates: 1/90–Present.
Learning Outcomes: Upon completion of the course, the student will master techniques of report writing and execution of required forms dealing with confinement, prison release, and general brig operation; supervise and implement emergency plans and training associated with such brig operations as fire, escape, riots, and bomb threats; master administrative procedures connected with transfer of prisoners; supervise day-to-day operations of brig, including staffing, training, and administrative duties; and perform and supervise all functions of custody and control of prisoners.
Instruction: This is a correspondence course which provides fundamentals of corrections supervision and includes exercises designed to provide knowledge of required forms, report writing, implementation of emergency plans, physical security, and general operations of ship's brig and other basic correctional administrative functions.
Credit Recommendation: In the lower-division baccalaureate/associate degree category, 3 semester hours in corrections (9/89).

MC-1728-0008

CORRECTIONS BY CORRESPONDENCE

Course Number: 58.1d.
Location: Marine Corps Institute, Washington, DC.
Length: Maximum, 52 weeks.
Exhibit Dates: 1/90–Present.
Learning Outcomes: Upon completion of the course, the student will be able to describe general correctional policies and procedures relating to control and security of prisoners, emergencies, prisoner discipline, confinement, and general brig operations and maintain various prisoner records and administrative reports.
Instruction: This is a correspondence course designed to provide student with fundamental knowledge of general correctional policies and procedures used in operation of a brig.
Credit Recommendation: In the lower-division baccalaureate/associate degree category, 2 semester hours in corrections (9/89).

MC-1728-0009

SECURITY SUPERVISOR

Course Number: MSS.
Location: Marine Corps Security Force, Chesapeake, VA; Marine Corps Security Force, Mare Island Naval Shipyard, Vallejo, CA.
Length: 2 weeks (97-98 hours).
Exhibit Dates: 4/92–Present.
Learning Outcomes: Upon completion of the course, the student will be proficient in small arms marksmanship, command and control special weapons security teams, and will react to and employ appropriate measures to security breaches.
Instruction: Instruction is provided through lecture, demonstration, and practical exercises covering topics in security, antiterrorism, small arms marksmanship, suspect restraint, and close combat tactics.
Credit Recommendation: In the lower-division baccalaureate/associate degree category, 1 semester hour in military science (leadership) (3/96).

MC-1728-0010

BASIC SECURITY GUARD

Course Number: M4V.
Location: Marine Corps Security Force, Chesapeake, VA; Marine Corps Security Force, Mare Island Naval Shipyard, Vallejo, CA.
Length: 6 weeks (176 hours).
Exhibit Dates: 4/92–Present.
Learning Outcomes: Upon completion of the course, the student will be able to demonstrate proficiency in multiple weaponry and defensive tactics.
Instruction: Instruction in firing range exercises, defensive tactics, and hand to hand combat.
Credit Recommendation: Credit is not recommended because of the limited, specialized nature of the course (3/96).

MC-1729-0006

BAKER NONCOMMISSIONED OFFICER (NCO) (Bakery Noncommissioned Officer (NCO) Leadership)

Course Number: 33K.
Location: Service Support School, Cp. Lejeune, NC.
Length: 10-11 weeks (326-350 hours).
Exhibit Dates: 1/90–3/95.
Objectives: To train food service personnel in leadership behavior essential to effective management and training techniques for a large baking operation.
Instruction: Course covers leadership training for bakery personnel at midmanagement levels and includes communication problem solving and training skills and supervisory techniques. Course also includes laboratory and operational exercises using both stationary and mobile equipment.
Credit Recommendation: In the lower-division baccalaureate/associate degree category, 4 semester hours in quantity food production management (4/88).

MC-1729-0009

BASIC FOOD SERVICE

Course Number: *Version 1:* M0333L6; 33L. *Version 2:* 33L.
Location: *Version 1:* Service Support School, Cp. Lejeune, NC. *Version 2:* Supply School, Cp. Lejeune, NC; Service Support School, Cp. Lejeune, NC.
Length: *Version 1:* 10 weeks (348 hours). *Version 2:* 7-11 weeks (207-365 hours).
Exhibit Dates: *Version 1:* 1/93–Present. *Version 2:* 1/90–12/92.
Learning Outcomes: *Version 1:* Using realistic environmental and definitive performance objectives based on task analysis, the student will be able to perform as a basic cook within a garrison or field messing facility. *Version 2:* Using realistic environmental and definitive performance objectives based on task analysis, the student will be able to perform as a cook within a garrison or field messing facility.
Instruction: *Version 1:* Course includes the introduction and development of basic motor skills as related to food preparation. Management of materials is covered only as it relates to task completion not to the management of people. Basic introduction to preparation techniques is presented utilizing lecture, demonstration, laboratory, and operational settings under constant supervision with frequent critiquing of performance. Specific instruction in operational mathematics, breads, cookies, cakes, pies, egg protein cooking, pasta preparation, salad dressings, meat fish, and poultry preparation, soups, sauces and gravies, vegetables, appetizers, desserts, dining room preparation (cafeteria), and operational procedures specific to military conditions is presented. *Version 2:* Course covers the introduction and development of basic motor skills as related to food preparation. Management of materials is presented only as it relates to task completion and not to the management of people. Basic introduction to preparation techniques utilizing lecture, demonstration, laboratory, and operational settings is presented under constant supervision with frequent critiquing of performance. Specific instruction in operational mathematics, breads, cookies, cakes, pies, egg protein cooking, pasta preparation, salad dressings, meat fish, and poultry preparation, soups, sauces and gravies, vegetables, appetizers, des-

serts, dining room preparation (cafeteria), and operational procedures specific to military conditions is presented.

Credit Recommendation: *Version 1:* In the lower-division baccalaureate/associate degree category, 3 semester hours in basic food preparation (7/95). *Version 2:* In the lower-division baccalaureate/associate degree category, 3 semester hours in basic food preparation (11/87).

MC-1729-0022

SALADS, SANDWICHES, AND DESSERTS BY CORRESPONDENCE

Course Number: 33.20.
Location: Marine Corps Institute, Washington, DC.
Length: Maximum, 52 weeks.
Exhibit Dates: 1/90–1/97.
Objectives: To supply food service personnel with the proper techniques for the preparation of salads, salad dressings, sandwiches, and desserts.
Instruction: This correspondence course consists of reading assignments, written lessons, and objective questions that provide a general background in the fundamentals of preparing salads, salad dressings, sandwiches, and desserts.
Credit Recommendation: In the lower-division baccalaureate/associate degree category, 1 semester hour in food preparation when taken in conjunction with Vegetables, Soups, Sauces, Gravies, and Beverages by Correspondence, 33.19 (MC-1729-0024) (8/83).

MC-1729-0024

VEGETABLES, SOUPS, SAUCES, GRAVIES, AND BEVERAGES BY CORRESPONDENCE

Course Number: 33.19.
Location: Marine Corps Institute, Washington, DC.
Length: Maximum, 52 weeks.
Exhibit Dates: 1/90–7/90.
Objectives: To provide food service personnel with proper techniques to prepare vegetables, soups, sauces, gravies, and beverages.
Instruction: This correspondence course consists of reading assignments, written lessons, and objective questions that provide food service personnel with general background information on the basics of receipt, storage, and preparation of vegetables, soups, sauces, gravies, and beverages in an appealing manner for dining room consumption.
Credit Recommendation: In the lower-division baccalaureate/associate degree category, 1 semester hour in food preparation when taken in conjunction with Salads, Sandwiches, and Desserts by Correspondence, 33.20 (MC-1729-0022) (3/84).

MC-1729-0032

PASTRY BAKING BY CORRESPONDENCE

Course Number: 33.8f.
Location: Marine Corps Institute, Washington, DC.
Length: Maximum, 52 weeks.
Exhibit Dates: 1/90–Present.
Objectives: To describe the sanitation and safety procedures needed for pastries; the types of sweetening, leavening, and flavoring agents; and yeasts and molds inhibitors.
Instruction: This correspondence course covers yeast doughs, danish pastries, pie crusts, pie filling, cakes, cookies, and quick breads. This course has no laboratory component.
Credit Recommendation: Credit is not recommended because of the limited, specialized nature of the course (3/94).

MC-1729-0034

FOOD SERVICE NONCOMMISSIONED OFFICER (NCO)

Course Number: 33N.
Location: Service Support School, Cp. Lejeune, NC.
Length: 10-11 weeks (340-364 hours).
Exhibit Dates: 1/90–Present.
Learning Outcomes: Upon completion of the course, the student will have advanced knowledge of food preparation and baking techniques, while gaining an understanding of food service administrative functions and computerized food management information systems.
Instruction: Course covers computerized food management information system, advanced baking and cooking techniques, and administrative functions. Methodology includes computer-based instruction, lectures, demonstrations, and practical applications.
Credit Recommendation: In the lower-division baccalaureate/associate degree category, 4 semester hours in food production management (7/95).

MC-1729-0035

BASIC BAKER

Course Number: 33A.
Location: Service Support School, Cp. Lejeune, NC.
Length: 9 weeks (302-329 hours).
Exhibit Dates: 1/90–3/95.
Learning Outcomes: Upon completion of the course, the student will be able to apply basic principles of baking to large scale bakery operations, including preparation, product evaluation, sanitation control, and service and produce a variety of bakery products for conventional food service facilities, as well as field operations.
Instruction: Course covers the various techniques of producing bakery products, how to identify their characteristic qualities, the use and maintenance of bakery equipment, the operation of a mobile bakery plant and ice cream plant, and proper sanitation standards. Methodology includes lectures, actual baking in a simulated operation, testing products, and classroom training.
Credit Recommendation: In the lower-division baccalaureate/associate degree category, 3 semester hours in basic baking principles (4/89).

MC-1729-0036

FOOD SERVICE STAFF NONCOMMISSIONED OFFICER (NCO)

Course Number: *Version 1:* M03DA26. *Version 2:* DA2.
Location: Service Support School, Cp. Lejeune, NC.
Length: *Version 1:* 7 weeks (251-252 hours). *Version 2:* 9 weeks (296-315 hours).
Exhibit Dates: *Version 1:* 9/94–Present. *Version 2:* 1/90–8/94.
Learning Outcomes: *Version 1:* Upon completion of the course, the student will be able to manage a food service/bakery operation while utilizing a computerized food management information system. *Version 2:* Upon completion of the course, the student will be able to utilize and further develop technical, conceptual, and human skills in order to perform the administrative and supervisory functions required by middle management in a garrison mess operation or field bakery.
Instruction: *Version 1:* Course covers food service administration, budgeting, and accounting, computerized food management information systems, and bakery/mess supervision and operation. Methodology includes computer-based instruction, lectures, demonstrations, and practical applications. *Version 2:* Course covers service support management requirements; duties and responsibilities related to garrison, mess, field, and bakery operations management, including preparation of requisition forms, inventory control procedures, product evaluation, production planning, yields, food allowances, and cash control; planning for large scale operations, both stationary and mobile; and establishing a field food service facility. Methodology includes lectures, practical exercises, testing, performance, and evaluation.
Credit Recommendation: *Version 1:* In the lower-division baccalaureate/associate degree category, 3 semester hours in food management information systems and 4 semester hours in food production management (7/95). *Version 2:* In the lower-division baccalaureate/associate degree category, 2 semester hours in basic food preparation and 3 in operations management (4/89).

MC-1729-0037

RESERVE FOOD SERVICE REFRESHER

Course Number: FAC.
Location: Service Support School, Cp. Lejeune, NC.
Length: 2 weeks (74 hours).
Exhibit Dates: 1/90–8/94.
Learning Outcomes: Upon completion of the course, the student will be able to assist with the set up of field food service operations; assemble, operate, clean and maintain all required equipment; and serve food in accordance with standards.
Instruction: Course covers operation of immersion water heaters, sanitizers, and field food service equipment; preparation and maintenance of field food service equipment; and preparation and service of basic meals. Methodology includes lectures, demonstrations, and performance in all components.
Credit Recommendation: In the lower-division baccalaureate/associate degree category, 2 semester hours in introduction to basic food preparation (7/95).

MC-1729-0038

FOOD SERVICE FUNDAMENTALS BY CORRESPONDENCE

Course Number: 33.4k.
Location: Marine Corps Institute, Washington, DC.
Length: Maximum, 52 weeks.
Exhibit Dates: 1/90–Present.
Learning Outcomes: Upon completion of the course, the student will be able to identify the organizational structure, sanitation and safety practices, and basic equipment and utensils used in food service operations.

Instruction: In providing an overview of food service fundamentals, this correspondence course emphasizes mission and organization, food microbiology and sanitation, food service equipment, and utensils.

Credit Recommendation: In the lower-division baccalaureate/associate degree category, 1 semester hour in food service operations (6/89).

MC-1729-0039

MESS HALL SANITATION BY CORRESPONDENCE

Course Number: 33.21a.
Location: Marine Corps Institute, Washington, DC.
Length: Maximum, 52 weeks.
Exhibit Dates: 1/90–3/90.
Learning Outcomes: Upon completion of the course, the student will understand and assess basic sanitation and safety practices in food service operations.
Instruction: This correspondence course teaches students to maintain safety and hygienic standards conducive to good health and to control contamination by sanitizing food preparation equipment, dining room equipment, eating and drinking utensils, scullery equipment, and cleaning equipment; receiving and processing foods; sanitary storage of foods; handling and storage of cleaning compounds; safe temperature ranges; and vector control.
Credit Recommendation: In the lower-division baccalaureate/associate degree category, 1 semester hour in food service sanitation when taken taken in conjunction with 33.24, the Mess Hall Subsistence Clerk, MC-1729-0040 (6/89).

MC-1729-0040

MESS HALL SUBSISTENCE CLERK BY CORRESPONDENCE

Course Number: 33.24; 33.24a.
Location: Marine Corps Institute, Washington, DC.
Length: Maximum, 52 weeks.
Exhibit Dates: 1/90–Present.
Learning Outcomes: Upon completion of the course, the student will be able to identify the fundamentals of food forecasting, receiving, storing, issuing, and cost control.
Instruction: Course covers forecasting subsistence requirements, receiving subsistence, issuing subsistence items, administration, stock record and inventory control aid, food cost analysis, and monetary credit. This correspondence course concludes with a proctored comprehensive examination.
Credit Recommendation: In the vocational certificate category, 1 semester hour in food purchasing or inventory control (1/91); in the lower-division baccalaureate/associate degree category, 1 in food service sanitation when taken in conjunction with 33.21a Mess Hall Sanitation by Correspondence (MC-1729-0039) (1/91).

MC-1729-0041

BASIC NUTRITION BY CORRESPONDENCE

Course Number: *Version 1:* 33.16d; 33.16e. *Version 2:* 33.16c.
Location: Marine Corps Institute, Washington, DC.
Length: *Version 1:* Maximum, 52 weeks. *Version 2:* Maximum, 52 weeks.
Exhibit Dates: *Version 1:* 3/90–Present. *Version 2:* 1/90–2/90.
Learning Outcomes: *Version 1:* Upon completion of the course, the student will be able to define nutrition; list factors that influence a person's selection of food; identify the health function of nutrition, minerals, and other food nutrients; and plan nutritional menus. The student will become familiar with the fundamentals of physical fitness training. *Version 2:* Upon completion of the course, the student will be able to define nutrition; list factors that influence a person's selection of food; identify the health function of nutrition, minerals, and other food nutrients; and plan nutritional menus. The student will become familiar with the fundamentals of physical fitness training.
Instruction: *Version 1:* Instruction covers food and health; nutrients which provide fuel and build and repair tissue; vitamins, minerals, water and fiber; food fuel, body weight, weight control factors, and menu planning; physical fitness training. Course makes use of unit exercises and comprehensive review lessons with final exam. *Version 2:* Instruction covers food and health; nutrients which provide fuel and build and repair tissue; vitamins, minerals, water, and fiber; food fuel, body weight, weight control factors, and menu planning; physical fitness training. Course makes use of unit exercises and comprehensive review lessons with final exam.
Credit Recommendation: *Version 1:* In the lower-division baccalaureate/associate degree category, 1 semester hour in nutrition (11/91). *Version 2:* Credit is not recommended because of the limited, specialized nature of the course (6/89).

MC-1729-0042

RESERVE FIELD FOOD SERVICE SUPERVISOR BY CORRESPONDENCE

Course Number: 33.33.
Location: Marine Corps Institute, Washington, DC.
Length: Maximum, 52 weeks.
Exhibit Dates: 1/90–Present.
Learning Outcomes: Upon completion of the course, the student will be able to explain the planning and administrative functions of a food service supervisor, delineate the steps in setting up a food operation, identify food service equipment and operation procedures, describe food ordering and inventory procedures, and recognize accepted standards of sanitation and safety.
Instruction: Course covers planning; setting up mess operations; operational administration; food production work sheets; cost control; sanitation inspections, and using, cleaning, and dismantling food service equipment. Methodology includes unit exercises and a comprehensive review lesson which prepares students for the final exam.
Credit Recommendation: In the lower-division baccalaureate/associate degree category, 1 semester hour in food service supervision (6/89).

MC-1729-0043

FIELD BREAD BAKING BY CORRESPONDENCE

Course Number: 33.10g.
Location: Marine Corps Institute, Washington, DC.
Length: Maximum, 52 weeks.
Exhibit Dates: 1/90–Present.
Learning Outcomes: Upon completion of the course, the student will be able to understand the use of bread baking equipment, bread ingredients and their functions, how to produce field garrison sheet bread, and how to make mathematical computations for determining weight loss, water tempering, and ice calculations. The student will understand bread faults and their remedies and the results of rope and mold disease.
Instruction: The course presents lessons covering a broad spectrum of bread baking activities. The student will be required to understand field bakeries, ice cream production, mobile field baking plant components, bread ingredients and their functions, field bread baking procedures, understand and prevent mold and bacterial invasions, and work with portable field baking equipment. Course covers ice cream production, bread baking, sanitation, operation of machinery, and the theory of fermentation and baking. There is no laboratory included with the course.
Credit Recommendation: In the lower-division baccalaureate/associate degree category, 1 semester hour in theory of baking (5/91).

MC-1729-0044

MEATS AND MEAT COOKERY BY CORRESPONDENCE

Course Number: 33.18b.
Location: Marine Corps Institute, Washington, DC.
Length: Maximum, 52 weeks.
Exhibit Dates: 1/90–Present.
Learning Outcomes: Upon completion of the course, the student will be able to list the grades of beef, veal, lamb, pork, fish, and poultry and, where applicable, state the classification of meats, poultry, and fish used at Marine Corps mess halls; state the inspection process for meats, poultry, and fish and identify the stamps used; describe the freezing and thawing process for meats, poultry, and fish; state the structure and composition of meats, poultry, and fish; and explain carving and boning of meat products.
Instruction: This correspondence course consists of reading assignments, written lessons, and objective questions that provide food service personnel with basic general background information on meats, poultry, and fish.
Credit Recommendation: Credit is not recommended because of the limited, specialized nature of the course (5/91).

MC-1729-0045

CHIEF FOOD SERVICE SPECIALIST BY CORRESPONDENCE

Course Number: 33.25.
Location: Marine Corps Institute, Washington, DC.
Length: Maximum, 52 weeks.
Exhibit Dates: 1/90–Present.
Learning Outcomes: Upon completion of the course, the student will be able to identify areas and items for which the chief food specialist has inspection responsibility; identify food production and service equipment and state cleaning and maintenance procedures associated with this equipment; state administrative responsibilities as they relate to production and service, including the use of work sheets, prebriefing, debriefing, and retrospec-

tive analysis; and list other supervisory duties and areas of responsibility.

Instruction: This correspondence course consists of reading assignments in five study units with study questions at the end of each one. Assignments are submitted to the training officer or mailed to the Marine Corps Institute. Topics include food service inspections, food service administration, food service equipment, and food service training.

Credit Recommendation: In the lower-division baccalaureate/associate degree category, 1 semester hour in introduction to food service supervision (5/91).

MC-1729-0046

CLUB FOOD OPERATIONS BY CORRESPONDENCE

Course Number: 41.6.
Location: Marine Corps Institute, Washington, DC.
Length: Maximum, 52 weeks.
Exhibit Dates: 1/90–Present.
Learning Outcomes: Upon completion of the course, the student will be able to plan and develop menus for a select market while avoiding six major pitfalls; use appropriate menu merchandising techniques; complete the menu price using using food cost percentages and menu food cost; identify effective purchasing, receiving, storage, cost control, and inventory procedures; and state the food preparation methods and/or procedures using standard recipes.

Instruction: This correspondence course consists of reading assignments in four study units, relevant written exercises, and objective questions to reinforce the student's understanding of menu planning, merchandising, and pricing. Food ordering, receiving, storing, and preparation are also included. Topics covered include introduction to menus, the operational cycle, cost control, and food preparation.

Credit Recommendation: In the lower-division baccalaureate/associate degree category, 1 semester hour in menu planning (5/91).

MC-1729-0047

SENIOR FOOD SERVICE

Course Number: FAD.
Location: Service Support School, Cp. Lejeune, NC.
Length: 4 weeks (140 hours).
Exhibit Dates: 9/93–Present.
Learning Outcomes: Upon completion of the course, the student will be able to manage a food service operation, including planning, decision making, communication skills, supply procedures, and computerized food management information systems.

Instruction: Lectures, demonstrations, practical exercises, and computer based instruction.

Credit Recommendation: In the lower-division baccalaureate/associate degree category, 4 semester hours in food service management and 3 semester hours in computerized food management information systems (7/95).

MC-1729-0048

FOOD SERVICE SUBSISTENCE CLERK

Course Number: M033036.
Location: Service Support School, Cp. Lejeune, NC.
Length: 5 weeks (160 hours).
Exhibit Dates: 9/94–Present.

Learning Outcomes: Upon completion of the course, the student will be able to perform as a food storeroom clerk and use a computerized food management information system.

Instruction: Course covers basic skills in food storage and issue procedures and performance of applicable procedures on a computerized food management information system. Methods include lectures, demonstrations, practical applications, and computer-based instruction.

Credit Recommendation: In the lower-division baccalaureate/associate degree category, 3 semester hours in computerized food management information systems and 3 semester hours in food service operations (7/95).

MC-1730-0005

FUNDAMENTALS OF REFRIGERATION BY CORRESPONDENCE

Course Number: 11.61.
Location: Marine Corps Institute, Washington, DC.
Length: Maximum, 52 weeks.
Exhibit Dates: 1/90–1/97.
Learning Outcomes: Upon completion of the course, the student will have a fundamental understanding of refrigeration including the use of psychrometric charts, refrigeration cycle, various refrigerants, the proper storage and handling of refrigerants, compressors, evaporators, motors, and overall system components.

Instruction: Instruction includes the study of fundamentals of refrigeration, various common refrigerants, their toxic effects, complete refrigeration systems, controls, components, and the use of psychrometric charts.

Credit Recommendation: In the vocational certificate category, 3 semester hours in fundamentals of refrigeration (7/87).

MC-1730-0006

BASIC REFRIGERATION MECHANIC

Course Number: *Version 1:* 720-1161 (OS). *Version 2:* 11D. *Version 3:* 11D.
Location: *Version 1:* Marine Corps Detachment, Ordnance Center and School, Aberdeen Proving Ground, MD. *Version 2:* Engineer School, Cp. Lejeune, NC. *Version 3:* Engineer School, Cp. Lejeune, NC.
Length: *Version 1:* 7 weeks (267-268 hours). *Version 2:* 8 weeks (249 hours). *Version 3:* 8 weeks (270 hours).
Exhibit Dates: *Version 1:* 10/96–Present. *Version 2:* 8/94–9/96. *Version 3:* 1/90–7/94.
Learning Outcomes: *Version 1:* Upon completion of the course, the student will be able to install, operate, and maintain refrigeration and air conditioning systems including associated electrical systems and ice making machines. *Version 2:* Upon completion of the course, the student will be able to install, operate, and maintain refrigeration and air conditioning systems and associated electrical systems. Student will apply the theory of evaporative and absorptive refrigeration systems and fundamental circuitry. *Version 3:* Upon completion of the course, the student will be able to install, operate, and maintain refrigeration and air conditioning systems and associated electrical systems. Student will apply the theory of evaporative and absorptive refrigeration systems and fundamental circuitry.

Instruction: *Version 1:* Lectures, demonstrations and hands-on performance are used through-out this course. Emphasis is placed on classroom and laboratory exercises and EPA registration. *Version 2:* Lectures include basic arithmetic, principles of refrigeration, and electrical circuits. Practical exercises focus on installation, operation, maintenance, and troubleshooting of refrigeration and associated electrical systems and components with emphasis on specific refrigeration equipment. Hands-on performance and field maintenance and repair are also included. *Version 3:* Lectures include basic arithmetic, principles of refrigeration, and electrical circuits. Practical exercises focus on installation, operation, maintenance, and troubleshooting of refrigeration and associated electrical systems and components with emphasis on specific refrigeration equipment. Hands-on performance and field maintenance and repair are also included.

Credit Recommendation: *Version 1:* In the lower-division baccalaureate/associate degree category, 4 semester hours in refrigeration, 2 in air conditioning, and 2 in electric motor repair (6/97). *Version 2:* In the lower-division baccalaureate/associate degree category, 5 semester hours in air conditioning/refrigeration (7/96). *Version 3:* In the vocational certificate category, 4 semester hours in refrigeration system service and 2 in air conditioning service and repair (3/92).

MC-1730-0007

JOURNEYMAN REFRIGERATION MECHANIC

Course Number: 11L.
Location: Engineer School, Cp. Lejeune, NC.
Length: 8 weeks (252 hours).
Exhibit Dates: 1/90–1/94.
Learning Outcomes: Upon completion of the course, the student will be able to repair of air conditioning and refrigeration equipment at an advanced level.

Instruction: Student prerequisites include satisfactory completion of the basic refrigeration mechanic course and one year field experience. Lectures, demonstrations, and practical exercises cover maintenance management systems, including routine maintenance and preventive maintenance procedures, electrical concepts for diagnosing and repairing electrical components, electrical systems of refrigeration and air conditioning equipment, and troubleshooting and repair of refrigeration and air conditioning equipment.

Credit Recommendation: In the lower-division baccalaureate/associate degree category, 3 semester hours in refrigeration theory and repair and 3 in air conditioning theory and repair (3/92).

MC-1731-0001

REFRIGERATION SERVICING BY CORRESPONDENCE

Course Number: 11.62.
Location: Marine Corps Institute, Washington, DC.
Length: Maximum, 52 weeks.
Exhibit Dates: 1/90–4/91.
Learning Outcomes: Upon completion of the course, the student will be able to perform preventive and corrective maintenance on refrigeration equipment.

Instruction: This correspondence course contains study units, review exams and final examination and covers installation and preventive maintenance, troubleshooting, servicing refrigeration systems, and repairing major components.

Credit Recommendation: In the vocational certificate category, 1 semester hour in refrigeration theory and 1 in refrigeration maintenance (7/87).

MC-1732-0003

REVERSE OSMOSIS WATER PURIFICATION UNIT BY CORRESPONDENCE

Course Number: 11.22a.
Location: Marine Corps Institute, Washington, DC.
Length: Maximum, 52 weeks.
Exhibit Dates: 1/90–Present.
Learning Outcomes: Upon completion of the course, the student will be able to identify the fundamentals of the reverse osmosis process; state the characteristics, capabilities, and components of the reverse osmosis water purification unit; and operate a reverse osmosis water purification unit, including inspection, start-up, adjustment, shutdown, backwash, and element cleaning.
Instruction: This correspondence course contains reading assignments and study questions with answers to study questions at the end of each unit. Materials cover operating and maintaining a reverse osmosis water purification unit. The student will demonstrate achievement of objectives by answering correctly 70 percent of the total questions on a proctored final examination.
Credit Recommendation: In the vocational certificate category, 3 semester hours in water treatment plant operation (6/89).

MC-2201-0001

TOWED ARTILLERY REPAIRMAN

Course Number: A0108O1; 642-2131.
Location: Marine Corps Detachment, Ordnance Center and School, Aberdeen Proving Ground, MD.
Length: 5 weeks (176 hours).
Exhibit Dates: 3/96–Present.
Learning Outcomes: Upon completion of the course, the student will be able to perform general support maintenance on towed and self-propelled artillery.
Instruction: Instruction includes lectures, demonstrations, and practical exercises in the use and care of hand tools and hardware, basic hydraulic principles, measurement tools, and the maintenance and troubleshooting of artillery weapons systems.
Credit Recommendation: In the lower-division baccalaureate/associate degree category, 1 semester hour in basic hand tools and measurement equipment (6/97).

MC-2202-0002

NAVAL RESERVE OFFICER
(Naval Reserve Officer Bulldog)

Course Number: RMD.
Location: Staff Noncommissioned Officer Academy, Quantico, VA.
Length: 7 weeks (335-523 hours).
Exhibit Dates: 2/90–Present.
Learning Outcomes: Upon completion of the course, the student will be able to demonstrate the leadership, moral, and physical qualities required of a commissioned officer.
Instruction: Lectures, guided discussions, demonstrations, practical exercises, and written examinations cover leadership principles and skills, military courtesy and customs, physical fitness training, and military tactics.
Credit Recommendation: In the lower-division baccalaureate/associate degree category, 2 semester hours in leadership, 1 in physical training, and 5 in military science (5/97).

MC-2202-0003

NAVAL ACADEMY TRAINING

Course Number: RMF.
Location: Staff Noncommissioned Officer Academy, Quantico, VA.
Length: 7 weeks (506 hours).
Exhibit Dates: 2/90–5/97.
Learning Outcomes: Upon completion of the course, the student will be able to demonstrate the leadership, moral, and physical qualities required of a commissioned officer.
Instruction: Lectures, guided discussions, demonstrations, practical exercises, and written examinations cover leadership principles and skills, military courtesy and customs, physical fitness training, and military tactics.
Credit Recommendation: In the lower-division baccalaureate/associate degree category, 2 semester hours in leadership, 1 in physical training, and 5 in military science (3/90).

MC-2204-0014

SEA DUTY INDOCTRINATION
(Sea School)

Course Number: 81N.
Location: Sea School, Portsmouth, VA; Recruit Depot, San Diego, CA.
Length: 3-6 weeks (65-192 hours).
Exhibit Dates: 1/90–8/97.
Objectives: To train enlisted Marines for duty with a ship's Marine detachment.
Instruction: Course includes lectures and practical exercises on duties and shipboard life of a ship's Marine detachment, including drills, honors, and ceremonies, administrative subjects, small arms training, naval orientation, damage control and fire fighting, Marine standards, and gunnery training.
Credit Recommendation: Credit is not recommended because of the military-specific nature of the course (8/87).

MC-2204-0030

AIR DEFENSE CONTROL OFFICER

Course Number: 67P.
Location: Communication-Electronics School, Twentynine Palms, CA; Air Ground Combat Center, Twentynine Palms, CA.
Length: 8-12 weeks (283-479 hours).
Exhibit Dates: 1/90–7/91.
Objectives: After 7/91 see MC-2204-0089. To provide students with knowledge of the functions and operations of a tactical air defense operations center.
Instruction: Lectures and practical exercises cover tactical air operations, air intercept control, weapons system characteristics, flight rules and regulations, ground-controlled intercepts, and related air traffic control procedures. Course contains 5 weeks training in automated systems.
Credit Recommendation: Credit is not recommended because of the military-specific nature of the course (7/90).

MC-2204-0031

AIR CONTROL ELECTRONICS OPERATOR
(Air Control Electronic Operator)

Course Number: 67M.
Location: Air Command and Control School, Twentynine Palms, CA; Air Ground Combat Center, Twentynine Palms, CA.
Length: 4-7 weeks (132-279 hours).
Exhibit Dates: 1/90–7/91.
Objectives: After 7/91 see MC-2204-0090. To train enlisted personnel as operators in tactical air operations centers.
Instruction: Lectures and practical exercises cover principles of an air defense system, tactical air operations center, radar operation, operation of AN/TYQ-2 equipment, basic air traffic control and pertinent flight regulations, electronic warfare fundamentals, and target identification and intercept.
Credit Recommendation: Credit is not recommended because of the military-specific nature of the course (4/90).

MC-2204-0035

WOMAN OFFICER BASIC

Course Number: None.
Location: Marine Corps School, Quantico, VA.
Length: 9 weeks (300-316 hours).
Exhibit Dates: 1/90–4/90.
Objectives: To train newly commissioned women officers in the duties and functions of company and staff officers.
Instruction: Lectures and practical exercises cover military law, leadership, management and administration, military operations, officer assignment and classification, techniques of military instruction, logistics, and communication.
Credit Recommendation: In the upper-division baccalaureate category, credit in advanced military science at schools which normally offer such credit (7/74).

MC-2204-0038

BASIC MILITARY TRAINING
(Recruit Training)

Course Number: 809; 808.
Location: Recruit Depot, Cp. Pendleton, CA; Recruit Depot, San Diego, CA; Recruit Depot, Parris Island, SC.
Length: 10-11 weeks (399-410 hours).
Exhibit Dates: 1/90–5/91.
Objectives: After 5/91 see MC-2204-0088. To provide basic policy guidance and training in the essential subjects required of all marines and to ensure preparedness for follow-on training.
Instruction: Training includes code of conduct, military law, leadership, orientation lectures, history, customs, courtesies, uniform and clothing, mission and organization of the Marine Corps, interior guard, personal health and hygiene, swimming and first aid, close order drill, close combat, observing and reporting, individual movement, camouflage and concealment, field fortifications, helicopter-borne operations, NBC defense, offensive and defensive combat, mines and booby traps, physical conditioning, and parades and ceremonies. Includes both female and male Marine Corps recruits.
Credit Recommendation: In the lower-division baccalaureate/associate degree cate-

gory, 2 semester hours in marksmanship, 1 in personal health/hygiene, 1 in outdoor skills practicum, 3 in personal conditioning/fitness, and 1 in first aid (4/87).

MC-2204-0047

BASIC MORTARMAN
   (Mortarman)

Course Number: 034.
Location: Infantry Training School, Cp. Pendleton, CA; Infantry School, Cp. Lejeune, NC.
Length: 5-7 weeks (173-488 hours).
Exhibit Dates: 1/90–5/97.
Objectives: To provide intensive training in weapons and combat skills.
Instruction: Lectures and practical exercises cover 81mm mortar; 60mm mortar; offensive, defensive, and patrolling operations; land navigation; and offensive and defensive tactics.
Credit Recommendation: Credit is not recommended because of the military-specific nature of the course (5/87).

MC-2204-0048

BASIC MACHINEGUNNER
   (Machinegunner)

Course Number: 033.
Location: Infantry Training School, Cp. Pendleton, CA; Infantry School, Cp. Lejeune, NC.
Length: 5-7 weeks (134-194 hours).
Exhibit Dates: 1/90–5/97.
Objectives: To provide intensive training in weapons and combat skills to produce machine gunners capable of closing with and capturing or destroying the enemy.
Instruction: Course provides instruction in general military subjects pertaining to infantrymen, including tactical employment operations and field exercises in conjunction with riflemen, mortarmen, and antitank assaultmen. Topics include camouflage, patrolling operations, and basic offensive and defensive tactics relating to the 7.62 M-60 machine gun.
Credit Recommendation: Credit is not recommended because of the military-specific nature of the course (5/87).

MC-2204-0051

BASIC RIFLEMAN
   (Rifleman)

Course Number: 031.
Location: Infantry Training School, Cp. Pendleton, CA; Infantry School, Cp. Lejeune, NC.
Length: 5-7 weeks (163-288 hours).
Exhibit Dates: 1/90–5/97.
Objectives: To provide infantrymen with training in weapons and combat skills.
Instruction: Lectures and practical exercises cover land navigation, supporting arms, detection of mines and booby traps, helicopter operations, tracked vehicles, technique of fire and combat firing positions, scouting and patrolling, and offensive and defensive tactics.
Credit Recommendation: Credit is not recommended because of the military-specific nature of the course (5/87).

MC-2204-0054

MINEFIELD MAINTENANCE
   (Minefield Maintenance Man Refresher)
   (Guantanamo Minefield Maintenance Preparation)

Course Number: CE7.
Location: Engineer School, Cp. Lejeune, NC.
Length: 2 weeks (61-70 hours).
Exhibit Dates: 1/90–Present.
Objectives: To familiarize personnel with the tasks, responsibilites, and operating procedures involved with the mines at Guantanamo Bay, Cuba.
Instruction: Lectures and practical experience cover the placement, detection, and disarmament of mines; removal of mines; recording and reporting procedures; maintenance procedures; emergency evacuation procedures; and radio operations and communications.
Credit Recommendation: Credit is not recommended because of the military-specific nature of the course (3/93).

MC-2204-0064

MAINTENANCE OF THE M60 SERIES MACHINE GUN BY CORRESPONDENCE

Course Number: 21.30.
Location: Marine Corps Institute, Washington, DC.
Length: Maximum, 52 weeks.
Exhibit Dates: 1/90–1/97.
Objectives: To provide training in the inspection, assembly/disassembly, and repair of the M60 series (7.62mm) machine gun.
Instruction: This correspondence course consists of six study units, including inspection, assembly/disassembly, functioning, preventive maintenance, and repair of the M60 series (7.62mm) machine gun used by the Marine Corps. This course is followed by a supervised two hour examination without texts or notes.
Credit Recommendation: Credit is not recommended because of the limited, specialized nature of the course (3/85).

MC-2204-0065

INSPECTION AND REPAIR OF THE .50 CALIBER MACHINE GUN BY CORRESPONDENCE

Course Number: 21.31.
Location: Marine Corps Institute, Washington, DC.
Length: Maximum, 52 weeks.
Exhibit Dates: 1/90–1/97.
Objectives: To provide training in the inspection, maintenance, and repair of the .50 caliber machine gun used by the Marine Corps.
Instruction: This correspondence course consists of six study units which include assembly/disassembly, preventive maintenance, functioning, and repair of the .50 caliber machine gun used by the Marine Corps. The course concludes with a two hour supervised examination without texts or notes.
Credit Recommendation: Credit is not recommended because of the limited, specialized nature of the course (3/85).

MC-2204-0066

INSPECTION AND REPAIR OF SHOULDER WEAPONS BY CORRESPONDENCE

Course Number: 21.28c.
Location: Marine Corps Institute, Washington, DC.
Length: Maximum, 52 weeks.
Exhibit Dates: 1/90–1/97.
Objectives: To provide information necessary to inspect and repair infantry shoulder weapons.
Instruction: This course consists of three study units covering the Remington 12 gauge shotgun, 5.56mm rifle and 40mm grenade launcher, followed by a two hour supervised examination without textbooks or notes.
Credit Recommendation: Credit is not recommended because of the limited, specialized nature of the course (3/85).

MC-2204-0069

BASIC LANDING SUPPORT SPECIALIST

Course Number: 13I.
Location: Engineer School, Cp. Lejeune, NC.
Length: 4-6 weeks (140-163 hours).
Exhibit Dates: 1/90–Present.
Learning Outcomes: Upon completion of the course, the student will be able to perform such shore party landing support tasks, as camouflage, field fortifications, and demolitions.
Instruction: Course includes lectures and practical experiences in camouflage, mine warfare, demolitions, amphibious operations, and field fortification construction.
Credit Recommendation: Credit is not recommended because of the military-specific nature of the course (9/89).

MC-2204-0070

AMPHIBIOUS WARFARE

Course Number: RGA.
Location: Amphibious Warfare School, Quantico, VA; Air Ground Training and Education Center, Quantico, VA.
Length: 39 weeks (1183-1303 hours).
Exhibit Dates: 1/90–5/94.
Learning Outcomes: Upon completion of the course, the student will be able to plan, organize, direct, coordinate, and control task force operations, including information gathering, logistics, and human resource management. The officer will also be familiar with aspects of military history and military arts and sciences and will be able to apply management principles to running a small independent organization.
Instruction: Lectures and practical exercises cover management and command and staff operations. Course includes tactics and techniques of amphibious operations, logistics, and human resources and amphibious organization and weapons and equipment with emphasis on command, control, and communications. Course also covers aspects of military history and science and written and oral communications.
Credit Recommendation: In the lower-division baccalaureate/associate degree category, 3 semester hours in written and oral communications (2/90); in the upper-division baccalaureate category, 3 semester hours in principles of management, 3 in logistics management, 1 in military history, 3 in field experience in management, and 2 in small business management (2/90).

MC-2204-0071

1. RESERVE STAFF NONCOMMISSIONED OFFICER CAREER
2. RESIDENT STAFF NONCOMMISSIONED OFFICER (NCO) CAREER RESERVE

Course Number: *Version 1:* CEW. *Version 2:* CEW.
Location: *Version 1:* Staff Noncommissioned Officer Academy, El Toro, CA; Staff Noncommissioned Officer Academy, Quantico, VA. *Version 2:* Staff Noncommissioned Officer Academy, Cp. Lejeune, NC; Staff Noncommissioned Officer Academy, El Toro, CA; Staff Noncommissioned Officer Academy, Quantico, VA.
Length: *Version 1:* 2 weeks (132 hours). *Version 2:* 2 weeks (76-77 hours).
Exhibit Dates: *Version 1:* 4/92–Present. *Version 2:* 1/90–3/92.
Learning Outcomes: *Version 1:* Upon completion of the course, the student will understand the principles of military defensive and offensive tactics and the role of the effective military manager in war and peace. *Version 2:* Upon completion of the course, the student will understand the principles of military defensive and offensive tactics and the role of the effective military manager in war and peace.
Instruction: *Version 1:* Lectures, demonstrations, student presentations, guided discussions, practical exercises, and evaluations cover military management, customs and courtesies of the service, values, and unit level tactics. *Version 2:* Lectures, demonstrations, student presentations, guided discussions, practical exercises, and evaluations cover military management, customs and courtesies of the service, values, and unit level tactics.
Credit Recommendation: *Version 1:* In the lower-division baccalaureate/associate degree category, 4 semester hours in military science (4/96). *Version 2:* In the lower-division baccalaureate/associate degree category, 2 semester hours in military science (2/90).

MC-2204-0072

RESIDENT STAFF NONCOMMISSIONED OFFICER (NCO) ADVANCED RESERVE

Course Number: M4F.
Location: Staff Noncommissioned Officer Academy, Quantico, VA; Staff Noncommissioned Officer Academy, El Toro, CA.
Length: 2 weeks (91 hours).
Exhibit Dates: 1/90–Present.
Learning Outcomes: Upon completion of this course, the student will understand the principles of military defensive and offensive tactics and the role of the effective military manager in peace and in war.
Instruction: Lectures, demonstrations, student presentations, guided discussions, and evaluation cover military management, customs, justice, and unit-level tactics.
Credit Recommendation: In the lower-division baccalaureate/associate degree category, 2 semester hours in military science (3/90).

MC-2204-0074

RESIDENT STAFF NONCOMMISSIONED OFFICER (NCO) CAREER REGULAR
(Marine Leader Training Stage III)

Course Number: *Version 1:* T8A. *Version 2:* T8A.
Location: *Version 1:* Staff Noncommissioned Officer Academy, Cp. Lejeune, NC; Staff Noncommissioned Officer Academy, El Toro, CA; Staff Noncommissioned Officer Academy, Quantico, VA; Staff Noncommissioned Officer Academy, Overseas locations; Staff Noncommissioned Officer Academy, Cp. Butler, Okinawa. *Version 2:* Staff Noncommissioned Officer Academy, El Toro, CA; Staff Noncommissioned Officer Academy, Cp. Butler, Okinawa; Staff Noncommissioned Officer Academy, Cp. Lejeune, NC; Staff Noncommissioned Officer Academy, Quantico, VA.
Length: *Version 1:* 7 weeks (332-333 hours). *Version 2:* 6 weeks (196-228 hours).
Exhibit Dates: *Version 1:* 4/92–Present. *Version 2:* 1/90–3/92.
Learning Outcomes: *Version 1:* Upon completion of the course, the student will be able to apply leadership concepts and theory to contemporary leadership issues in order to stimulate thought and encourage the exchange of ideas; demonstrate the importance of physical fitness and how to establish, maintain, and evaluate a physical fitness training program; and understand the principles of military defensive and offensive tactics and the role of the effective military manager in both peace and war. *Version 2:* Upon completion of the course, the student will be able to apply leadership concepts and theory to contemporary leadership issues in order to stimulate thought and encourage the exchange of ideas; demonstrate the importance of physical fitness and how to establish, maintain, and evaluate a physical fitness training program; and understand the principles of military defensive and offensive tactics and the role of the effective military manager in both peace and war.
Instruction: *Version 1:* Lectures, demonstrations, student presentations, guided discussions, practical exercises, and evaluations cover leadership principles and skills; military management, customs, and unit level tactics; and the design and implementation of physical fitness training. *Version 2:* Lectures, demonstrations, student presentations, guided discussions, practical exercises, and evaluations cover leadership principles and skills; military management, customs, and unit level tactics; and the design and implementation of physical fitness training.
Credit Recommendation: *Version 1:* In the lower-division baccalaureate/associate degree category, 2 semester hours in leadership, 6 in military science, and 1 in physical education (4/96). *Version 2:* In the lower-division baccalaureate/associate degree category, 2 semester hours in leadership, 4 in military science, and 1 in physical education (3/90).

MC-2204-0075

NONCOMMISSIONED OFFICER (NCO) BASIC

Course Number: *Version 1:* 0DD. *Version 2:* 0DD.
Location: Staff Noncommissioned Officer Academies, Individual commands, US and overseas.
Length: *Version 1:* 5 weeks (211 hours). *Version 2:* 3 weeks (106-107 hours).
Exhibit Dates: *Version 1:* 3/90–4/95. *Version 2:* 1/90–2/90.
Learning Outcomes: *Version 1:* After 4/95 see MC-2204-0103. Upon completion of the course, the student will understand the principles of military defensive and offensive tactics and the role of the effective military manager in peace and in war. *Version 2:* Upon completion of the course, the student will be able to apply leadership concepts and theories to contemporary leadership issues in order to stimulate thought and encourage exchange of ideas. The student will understand principles of military offensive and defensive tactics and the role of the effective military manager in peace and in war.
Instruction: *Version 1:* Lectures, demonstrations, student presentations, guided discussions, and evaluations cover military leadership, training, and unit level tactics. *Version 2:* Lectures, demonstrations, student presentation, guided discussion, and practical exercises and evaluations cover leadership principles and skills, military management, customs, courtesies, values, and unit level tactics.
Credit Recommendation: *Version 1:* In the lower-division baccalaureate/associate degree category, 2 semester hours in military science (3/90). *Version 2:* In the lower-division baccalaureate/associate degree category, 1 semester hour in leadership and 2 in military science (3/90).

MC-2204-0076

1. RESIDENT STAFF NONCOMMISSIONED OFFICER (NCO) ADVANCED REGULAR
2. STAFF NONCOMMISSIONED OFFICER (NCO) ADVANCED

Course Number: *Version 1:* T8H. *Version 2:* T8H.
Location: *Version 1:* Staff Noncommissioned Officer Academy, Cp. Butler, Okinawa; Staff Noncommissioned Officer Academy, El Toro, CA; Staff Noncommissioned Officer Academy, Cp. Lejeune, NC; Staff Noncommissioned Officer Academy, Quantico, VA. *Version 2:* Staff Noncommissioned Officer Academy, El Toro, CA; Staff Noncommissioned Officer Academy, Cp. Lejeune, NC; Staff Noncommissioned Officer Academy, Quantico, VA.
Length: *Version 1:* 8-9 weeks (329-369 hours). *Version 2:* 10 weeks (365-427 hours).
Exhibit Dates: *Version 1:* 4/92–Present. *Version 2:* 1/90–3/92.
Learning Outcomes: *Version 1:* Upon completion of the course, the student will be able to apply leadership concepts and theories to contemporary leadership issues in order to stimulate thought and encourage the exchange of ideas; communicate both orally and in writing in a clear and concise manner; and demonstrate an understanding of the importance of physical fitness and know how to establish, maintain, and evaluate a physical fitness training program. The student will understand the principles of military defensive and offensive tactics and will understand the role of the effective military manager in peace and war. *Version 2:* Upon completion of the course, the student will be able to apply leadership concepts and theories to contemporary leadership issues in order to stimulate thought and encourage the exchange of ideas; communicate both orally and in writing in a clear and concise manner; and demonstrate an understanding of the importance of physical fitness and know how to establish, maintain, and evaluate a physical fitness training program. The student will understand the principles of military defensive and offensive tactics and will understand the role of the effective military manager in peace and war.
Instruction: *Version 1:* Lectures, demonstrations, student presentation, guided discussions, practical exercises, and evaluations cover leadership principles and skills; oral and written communication; military management, customs, courtesies, and unit level tactics; and the design and implementation of physical fitness

training. *Version 2:* Lectures, demonstrations, student presentation, guided discussions, practical exercises, and evaluations cover leadership principles and skills; oral and written communication; military management, customs, courtesies, and unit level tactics; and the design and implementation of physical fitness training.

Credit Recommendation: *Version 1:* In the lower-division baccalaureate/associate degree category, 3 semester hours in leadership, 1 in public speaking, 2 in physical education, and 7 in military science (4/96). *Version 2:* In the lower-division baccalaureate/associate degree category, 3 semester hours in leadership, 1 in public speaking, 1 in physical education, and 6 in military science (3/90).

MC-2204-0077

PLATOON LEADERS JUNIOR

Course Number: *Version 1:* RMC. *Version 2:* RMC.
Location: Staff Noncommissioned Officer Academy, Quantico, VA.
Length: *Version 1:* 6-7 weeks (254-519 hours). *Version 2:* 6-7 weeks (254-442 hours).
Exhibit Dates: *Version 1:* 2/90–Present. *Version 2:* 1/90–1/90.
Learning Outcomes: *Version 1:* Upon completion of the course, the student will be able to demonstrate the leadership, moral, and physical qualities required of a commissioned officer. *Version 2:* Upon completion of the course, the student will be able to demonstrate the leadership, moral, and physical qualities required of a commissioned officer.
Instruction: *Version 1:* Lectures, guided discussions, demonstrations, practical exercises, and written examinations cover leadership principles and skills, military courtesy and customs, physical fitness training, and military tactics. *Version 2:* Lectures, guided discussions, demonstrations, practical exercises, and written examinations in leadership principles and skills; military courtesy and customs; physical fitness training; and military tactics.
Credit Recommendation: *Version 1:* In the lower-division baccalaureate/associate degree category, 1 semester hour in leadership, 1 in physical training, and 5 in military science (5/97). *Version 2:* In the lower-division baccalaureate/associate degree category, 1 semester hour in leadership, 1 in physical training, 5 in military science (3/90).

MC-2204-0078

PLATOON LEADERS SENIOR

Course Number: *Version 1:* RMB. *Version 2:* RMB.
Location: Staff Noncommissioned Officer Academy, Quantico, VA.
Length: *Version 1:* 6-7 weeks (276-511 hours). *Version 2:* 6-7 weeks (276-471 hours).
Exhibit Dates: *Version 1:* 2/90–Present. *Version 2:* 1/90–1/90.
Learning Outcomes: *Version 1:* Upon completion of the course, the student will be able to demonstrate the leadership, moral, and physical qualities required of a commissioned officer. *Version 2:* Upon completion of the course, the student will be able to demonstrate the leadership, moral, and physical qualities required of a commissioned officer.
Instruction: *Version 1:* Lectures, guided discussions, demonstrations, practical exercises, and written examinations cover leadership principles and skills, military courtesy and customs, physical fitness training, and military tactics. *Version 2:* Lectures, guided discussions, demonstrations, practical exercises, and written examinations in leadership principles and skills, military courtesy and customs, physical fitness training, and military tactics.
Credit Recommendation: *Version 1:* In the lower-division baccalaureate/associate degree category, 2 semester hours in leadership, 1 in physical training, and 5 in military science (5/97). *Version 2:* In the lower-division baccalaureate/associate degree category, 2 semester hours in leadership, 1 physical training, 5 in military science (3/90).

MC-2204-0079

PLATOON LEADERS COMBINED

Course Number: *Version 1:* RMJ. *Version 2:* RMJ.
Location: Staff Noncommissioned Officer Academy, Quantico, VA.
Length: *Version 1:* 10-12 weeks (405-819 hours). *Version 2:* 10-11 weeks (405-727 hours).
Exhibit Dates: *Version 1:* 2/90–Present. *Version 2:* 1/90–1/90.
Learning Outcomes: *Version 1:* Upon completion of the course, the student will be able to demonstrate the leadership, moral, and physical qualities required of a commissioned officer. *Version 2:* Upon completion of the course, the student will be able to demonstrate the leadership, moral, and physical qualities required of a commissioned officer.
Instruction: *Version 1:* Lectures, guided discussions, demonstrations, practical exercises, and written examinations cover leadership principles and skills, military courtesy and customs, physical fitness training, and military tactics. *Version 2:* Lectures, guided discussions, demonstrations, practical exercises, and written examinations in leadership principles and skills, military courtesy and customs, physical fitness training, and military tactics.
Credit Recommendation: *Version 1:* In the lower-division baccalaureate/associate degree category, 3 semester hours in leadership, 2 in physical training, and 9 in military science (5/97). *Version 2:* In the lower-division baccalaureate/associate degree category, 3 semester hours in leadership, 2 in physical training, and 9 in military science (3/90).

MC-2204-0080

OFFICER CANDIDATE

Course Number: RMA.
Location: Staff Noncommissioned Officer Academy, Quantico, VA.
Length: 10-12 weeks (405-836 hours).
Exhibit Dates: 1/90–Present.
Learning Outcomes: Upon completion of the course, the student will be able to demonstrate the leadership, moral, and physical qualities required of a commissioned officer.
Instruction: Lectures, guided discussions, demonstrations, practical exercises, and written examinations in leadership principles and skills, military courtesy and customs, physical fitness training, and military tactics.
Credit Recommendation: In the lower-division baccalaureate/associate degree category, 3 semester hours in leadership, 2 in physical training, and 9 in military science (5/97).

MC-2204-0081

INFANTRY OFFICER

Course Number: M02RGU4; RGU.
Location: Basic School, Quantico, VA.
Length: 9-10 weeks (532-761 hours).
Exhibit Dates: 1/90–Present.
Learning Outcomes: Upon completion of the course, the student will be able to command and lead an infantry platoon in accomplishing complex missions under conditions of physical and mental stress and to synchronize the applications of supporting organizations and resources as well as build and sustain individual and group esprit.
Instruction: Through lectures, demonstrations and extensive practical application exercises, the course trains students in infantry tactics and in the capabilities of supporting organizations. There is heavy emphasis on the leader as a problem solver and decision maker.
Credit Recommendation: Credit is not recommended because of the military-specific nature of the course (5/97).

MC-2204-0082

RESERVE WARRANT OFFICER BASIC

Course Number: R2T.
Location: Basic School, Quantico, VA.
Length: 2 weeks (95-96 hours).
Exhibit Dates: 1/90–Present.
Learning Outcomes: Upon completion of the course, the student will be able to lead small groups in the completion of complex tasks, effectively manage subordinate personnel, and coordinate the efforts of supporting organizations.
Instruction: Lectures, demonstrations, and discussions present a survey of personnel and resource management techniques, personal fitness and conditioning activities, and basic military skills with emphasis on the role of the leader in small organizations.
Credit Recommendation: Credit is not recommended because of the military-specific nature of the course (7/90).

MC-2204-0083

BASIC OFFICER

Course Number: RMG.
Location: Basic School, Quantico, VA.
Length: 23 weeks (969-1215 hours).
Exhibit Dates: 1/90–Present.
Learning Outcomes: Upon completion of the course, the student will be able to lead small groups in the planning and execution of complex tasks performed under conditions of physical and mental stress; synchronize the application of supporting organizations and resources; develop individual and group proficiency through programs to train, motivate, counsel, evaluate, and discipline subordinates; and build and sustain individual and group esprit and cohesiveness.
Instruction: Lectures, demonstrations and practical application exercises cover leadership and management principles, including personnel training, counseling, motivation, evaluating and disciplining and interpersonal communication skills, information management, and resource management. There is heavy emphasis on problem solving and decision making processes. Course also includes technical instruction in physical education focusing on personal fitness and hygiene, marksmanship, and first

aid/water survival techniques. Through extensive use of case studies and historical examples, the course also provides coverage of defense issues, including total and limited war, guerilla warfare, terrorism, laws of land warfare, and personal conduct during combat.

Credit Recommendation: In the lower-division baccalaureate/associate degree category, 3 semester hours in physical education and 1 in cartography (1/93); in the upper-division baccalaureate category, 3 semester hours in leadership (1/93).

MC-2204-0084

WARRANT OFFICER BASIC

Course Number: RMN.
Location: Basic School, Quantico, VA.
Length: 13 weeks (512-584 hours).
Exhibit Dates: 1/90–Present.
Learning Outcomes: Upon completion of the course, the student will be able to lead small groups in the planning and execution of complex tasks performed under conditions of physical and mental stress; synchronize the application of supporting organizations and resources; develop individual and group proficiency through programs to train, motivate, counsel, evaluate, and discipline subordinates; and build and sustain individual and group esprit and cohesiveness.
Instruction: Lectures, demonstrations and practical application exercises cover leadership and management principles, including personnel training, counseling, motivation, evaluating, and disciplining and interpersonal communication skills, information management, and resource management. There is heavy emphasis on problem solving and decision making processes. The course also includes technical instruction in physical education focusing on personal fitness and hygiene, marksmanship, and first aid/water safety and survival techniques. Through extensive use of case studies and historical examples, the course also provides coverage of defense issues, including total and limited war, guerilla warfare, terrorism, laws of land warfare, and personal conduct during combat.
Credit Recommendation: In the lower-division baccalaureate/associate degree category, 1 semester hour in physical education and 1 in cartography (7/90); in the upper-division baccalaureate category, 3 semester hours in leadership (7/90).

MC-2204-0088

RECRUIT TRAINING
(Basic Training)

Course Number: 809; 808.
Location: Recruit Depot, San Diego, CA; Recruit Depot, Parris Island, SC.
Length: 11-12 weeks (464-511 hours).
Exhibit Dates: 6/91–Present.
Learning Outcomes: Before 6/91 see MC-2204-0038. Upon completion of the course, the male and female graduate recruit will demonstrate knowledge of the code of military conduct, laws of war, history of the U.S. Marine Corps, first aid and field sanitation, and nuclear, biological, and chemical warfare defense; practice military courtesy, good personal health, hygiene, and grooming; will have successfully met prescribed marksmanship standards, physical fitness test requirements, skill level in land navigation, and survival swimming requirements. Male recruit training is 464 hours; female recruit training 510-511 hours.
Instruction: Lectures, demonstrations, performance experiences, tactical problems and repetitive drill cover the above cited topics.
Credit Recommendation: In the lower-division baccalaureate/associate degree category, 1 semester hour in physical fitness and conditioning, 2 in marksmanship, and 1 in orienteering/outdoor skills (3/92).

MC-2204-0089

AIR DEFENSE CONTROL OFFICER (AN/TYQ-23)

Course Number: *Version 1:* M0972M1; 72M. *Version 2:* 72M; 67P.
Location: Air Ground Combat Center, Twentynine Palms, CA.
Length: *Version 1:* 18-19 weeks (751 hours). *Version 2:* 17-18 weeks (656 hours).
Exhibit Dates: *Version 1:* 5/96–Present. *Version 2:* 8/92–4/96.
Learning Outcomes: *Version 1:* This version is pending evaluation. *Version 2:* Before 8/91 see MC-2204-0030. Upon completion of the course, the student will be able to act as a tactical air defense control officer.
Instruction: *Version 1:* This version is pending evaluation. *Version 2:* The course provides a thorough knowledge of tactical air operations, including operation of the AN/TYQ-23 radar, air traffic control, air intercept and missile control, communications, aircraft characteristics, and weapons systems.
Credit Recommendation: *Version 1:* Pending evaluation. *Version 2:* Credit is not recommended because of the military-specific nature of the course (1/93).

MC-2204-0090

AIR CONTROL ELECTRONICS OPERATOR (AN/TYQ-23)

Course Number: 72P; 67M.
Location: Air Ground Combat Center, Twentynine Palms, CA.
Length: 13-14 weeks (382 hours).
Exhibit Dates: 7/92–Present.
Learning Outcomes: Before 8/91 see MC-2204-0031. Upon completion of the course, the student will be able to assist in tactical air defense operations.
Instruction: This course provides a thorough knowledge of tactical air operations, including operation of the AN/TYQ-23 radar, air traffic control, air intercept and missile control, communications, aircraft characteristics, and weapons systems.
Credit Recommendation: Credit is not recommended because of the military-specific nature of the course (1/93).

MC-2204-0091

LANDING FORCE GROUND COMBAT OPERATIONS OFFICER (S-3)

Course Number: 0DY; H-2E-3102.
Location: Mobile training units, US; Landing Force Training Command, Pacific, San Diego, CA.
Length: 2 weeks (110 hours).
Exhibit Dates: 2/89–Present.
Learning Outcomes: Upon completion of the course, the student will be able to plan and initiate amphibious landing operations applying knowledge in landing plan and landing support agencies, operations security, intelligence preparation, communications planning, staff planning and decisions, and supporting arms.
Instruction: Lectures, audiovisuals, demonstrations, and hands-on learning experiences cover amphibious operations management, intelligence gathering and use, communications, logistics, and supporting arms.
Credit Recommendation: Credit is not recommended because of the military-specific nature of the course (11/93).

MC-2204-0092

LANDING FORCE STAFF PLANNING (MARINE EXPEDITIONARY BRIGADE)

Course Number: 0D7; H-2E-3108.
Location: Mobile training units, US; Landing Force Training Command, Pacific, San Diego, CA.
Length: 2 weeks (73 hours).
Exhibit Dates: 7/92–Present.
Learning Outcomes: Upon completion of the course, the student will be able to apply staff planning principles and functions (staff planning sequence); problem solving/decision making techniques; identify background of factors involved in making recommendations for decisions, including concept of operation, medical support planning, amphibious ships, fire support coordination, communications systems, and logistics; and able to organize, direct, coordinate, and control the planning of amphibious operations.
Instruction: Lectures and a practical exercises cover management and staff operations. Course includes tactics of amphibious operations, logistics planning, and coordination of support.
Credit Recommendation: Credit is not recommended because of the limited, specialized nature of the course (11/93).

MC-2204-0093

LANDING FORCE COMBAT SERVICES SUPPORT STAFF PLANNING

Course Number: H-2E-3110.
Location: Landing Force Training Command, Pacific, San Diego, CA.
Length: 2 weeks (64-66 hours).
Exhibit Dates: 7/92–Present.
Learning Outcomes: Upon completion of the course, the student will be able to apply principles and processes of staff planning; describe organization and capabilities of coordinated support operations; identify factors involved in recommendations for decision making, including staff coordination, medical support planning, logistics planning, organization of amphibious forces, and communications planning; and jointly plan and organize the supply and support aspects of an amphibious force.
Instruction: Lectures and a practical exercises cover coordinated staff operations. Course includes organization and tactics of amphibious operations, supply and logistics planning, and coordination of support.
Credit Recommendation: Credit is not recommended because of the limited, specialized nature of the course (11/93).

MC-2204-0094

AVIATION COMBAT ELEMENT STAFF PLANNING

Course Number: OD6; H-2E-3113.

Location: Mobile training units, US; Landing Force Training Command, Pacific, San Diego, CA.

Length: 2 weeks (64-65 hours).

Exhibit Dates: 1/93–Present.

Learning Outcomes: Upon completion of the course, the student will be able to conduct aviation planning in support of amphibious operations.

Instruction: Lectures, practical exercises, and role playing in staff officer positions are methodologies included.

Credit Recommendation: Credit is not recommended because of the military-specific nature of the course (11/93).

MC-2204-0095

FIRE SUPPORT COORDINATION

Course Number: 60C; H-2E-3114.

Location: Mobile training units, US; Landing Force Training Command, Pacific, San Diego, CA.

Length: 2 weeks (50 hours).

Exhibit Dates: 8/91–Present.

Learning Outcomes: Upon completion of the course, the student will be able to describe the capabilities and limitations of weapons systems available to the coordinator as naval guns, air support, and field artillery; describe the coordination system for the delivery of firepower from the above sources; apply principles of fire support coordination, command and control aspects, and fire support scheduling procedures; and advise the commander on capabilities, resources available, and the support plan for firepower.

Instruction: Lectures, conferences, supplemental readings, self-paced texts, and practical exercises cover fire support coordination principles, tactics, procedures, air support systems, field artillery weapons, communications systems available, and military information (intelligence) strategy.

Credit Recommendation: Credit is not recommended because of the military-specific nature of the course (11/93).

MC-2204-0096

BASIC SCOUT SWIMMER

Course Number: H-2E-3742.

Location: Landing Force Training Command, Pacific, San Diego, CA.

Length: 2-3 weeks (94 hours).

Exhibit Dates: 1/93–Present.

Learning Outcomes: Upon completion of the course, the student will be able to plan and conduct scout swimmer missions, including nautical navigation, reconnaissance, and reporting.

Instruction: Course includes planning and conducting practical swimmer missions followed by examinations.

Credit Recommendation: In the lower-division baccalaureate/associate degree category, 1 semester hour in physical education—swimming (11/93).

MC-2204-0097

MARINE COMBAT INSTRUCTOR OF WATER SAFETY

Course Number: 03L; H-010-3921.

Location: Landing Force Training Command, Pacific, San Diego, CA.

Length: 3 weeks (133 hours).

Exhibit Dates: 1/93–Present.

Learning Outcomes: Upon completion of the course, the student will qualify as a Marine Combat Instructor of water survival. Skills are developed in order to conduct water survival training in support of unit training.

Instruction: Lectures, practical exercises, practice, and evaluation in order to certify instructors of water survival. Course includes lifeguarding, swimming, and lifeguarding maneuvers.

Credit Recommendation: In the lower-division baccalaureate/associate degree category, 2 semester hours in physical education—swimming (11/93).

MC-2204-0098

TANK GUNNERY/DIRECT FIRE PROCEDURES (M1A1) BY CORRESPONDENCE

Course Number: 18.46.

Location: Marine Corps Institute, Washington, DC.

Length: Maximum, 52 weeks.

Exhibit Dates: 8/93–Present.

Learning Outcomes: Upon completion of the course, the student will be able to target, fire, operate, and troubleshoot the M1A1 fire control system.

Instruction: Self-instruction and self-examination methods cover target acquisition, range estimation, crew duties, fire commands, fire control system, and machine gun use. A proctored written exam concludes the course.

Credit Recommendation: Credit is not recommended because of the military-specific nature of the course (9/94).

MC-2204-0099

INSPECTION AND REPAIR OF THE MK 19 MACHINEGUN BY CORRESPONDENCE

Course Number: 21.34.

Location: Marine Corps Institute, Washington, DC.

Length: Maximum, 52 weeks.

Exhibit Dates: 1/90–Present.

Learning Outcomes: Upon completion of the course, the student will be able to disassemble, inspect, repair, reassemble, and test the Mk 19 machine gun.

Instruction: Self-instruction and self-quizzes cover the areas of nomenclature, field stripping, disassembly, inspection, repair, reassembly, and troubleshooting the Mk 19 machine gun. A proctored written exam concludes the course.

Credit Recommendation: In the vocational certificate category, 1 semester hour in gunsmithing (9/94).

MC-2204-0100

INFANTRY SQUAD LEADER SQUAD TACTICS BY CORRESPONDENCE

Course Number: 03.83.

Location: Marine Corps Institute, Washington, DC.

Length: Maximum, 52 weeks.

Exhibit Dates: 1/90–Present.

Learning Outcomes: Upon completion of the course, the student will be able to conduct the primary maneuver of the infantry squad or patrol with 6-15 persons, serving as the leader of that squad. As the leader, the student will know the planning considerations for the several formations, provide organized directions for the carrying out of the mission, and serve as the person responsible for its success.

Instruction: Through the use of a text which describes the tactics of defensive and offensive operations, the student focuses on planning for the operation, the offensive tactics, and the defensive operations including considerations of the nuclear or chemical combat environment. Each of the three major themes includes content, situations, and written quizzes for self-evaluation. There is a proctored final examination to evaluate the student's success in the material.

Credit Recommendation: Credit is not recommended because of the military-specific nature of the course (9/94).

MC-2204-0101

INFANTRY SQUAD LEADER: WEAPONS AND FIRE SUPPORT BY CORRESPONDENCE

Course Number: 03.82.

Location: Marine Corps Institute, Washington, DC.

Length: Maximum, 52 weeks.

Exhibit Dates: 1/90–Present.

Learning Outcomes: Upon completion of the course, the student will describe the weapon system's capability and describe the employment of an infantry battalion.

Instruction: Self-instruction and unit quizzes cover machine guns, mortar, support arms, targeting, and naval and air support fire. A proctored written exam concludes the course.

Credit Recommendation: Credit is not recommended because of the military-specific nature of the course (9/94).

MC-2204-0102

RESERVE SERGEANT'S

Course Number: CFF.

Location: Staff Noncommissioned Officer Academy, Quantico, VA; Air Ground Combat Center, Twentynine Palms, CA.

Length: 12 weeks (119-120 hours).

Exhibit Dates: 8/94–Present.

Learning Outcomes: Upon completion of the course, the student will be able to conduct drills, inspections, training, and other aspects of military science as well as demonstrate group leadership and skills.

Instruction: Lectures, demonstrations, student presentations, field work, research, guided discussions, and evaluations cover military leadership, training, unit level tactics, and new materials in the areas of counseling.

Credit Recommendation: In the lower-division baccalaureate/associate degree category, 2 semester hours in military science and 1 in leadership (4/96).

MC-2204-0103

SERGEANT'S COURSE
(Noncommissioned Officer (NCO) Basic)

Course Number: T4M.

Location: Staff Noncommissioned Officer Academy, Quantico, VA; Air Ground Combat Center, Twentynine Palms, CA; Staff Noncommissioned Officer Academy, El Toro, CA; Staff Noncommissioned Officer Academy, Cp. Butler, Okinawa; Air Facility, Kaneohe Bay, HI; Staff Noncommissioned Officer Academy, Cp. Lejeune, NC.

Length: 5 weeks (219 hours).

Exhibit Dates: 5/95–Present.

Learning Outcomes: Before 5/95 see MC-2204-0075. Upon completion of the course, the student will be able to conduct drills, inspections, training, and other aspects of military science as well as demonstrate group leadership skills.

Instruction: Lectures, demonstrations, student presentations, field work, research reference, guided discussions, evaluations cover military leadership, training, unit level tactics, and new materials in the areas of counseling.

Credit Recommendation: In the lower-division baccalaureate/associate degree category, 2 semester hours in leadership and 3 in military science (4/96).

MC-2204-0104

RESERVE STAFF NONCOMMISSIONED OFFICER (NCO) ADVANCED

Course Number: M4L.
Location: Staff Noncommissioned Officer Academy, El Toro, CA; Staff Noncommissioned Officer Academy, Quantico, VA.
Length: 2 weeks (11 days), 152 hours.
Exhibit Dates: 4/92–Present.
Learning Outcomes: Upon completion of the course, the student will be able to apply introductory level leadership concepts and theories to contemporary leadership issues; direct the actions of subordinates in a variety of effective military management techniques; understand the principles of military defensive and offensive tactics; and understand how to establish, implement, and evaluate a physical fitness training program.

Instruction: Lectures, practical application, practical exercises, student presentations, guided discussions, and evaluations cover the principles and skills of leadership; counseling; military management; customs, courtesies, and unit level tactics; and the design and implementation of physical fitness training.

Credit Recommendation: In the lower-division baccalaureate/associate degree category, 1 semester hour in leadership, 3 in military science, and 1 in physical education (4/96).

MC-2204-0105

MARINE COMBAT

Course Number: M92.
Location: Infantry Training School, Cp. Pendleton, CA; Infantry School, Cp. Lejeune, NC.
Length: 4 weeks (185 hours).
Exhibit Dates: 5/94–Present.
Learning Outcomes: Upon completion of the course, the student will possess weapon and infantry skills necessary to function as a member of a rifle squad or machine gun team.

Instruction: Through lecture, demonstration, and practical hands-on training, the student acquires skills in individual weapons, crew-served weapons, patrolling, tactical measures, NBC defense, land navigation, and close combat.

Credit Recommendation: Credit is not recommended because of the military-specific nature of the course (7/96).

MC-2204-0106

BASIC ASSAULTMAN

Course Number: 035.
Location: Infantry School, Cp. Lejeune, NC.
Length: 7 weeks (214 hours).
Exhibit Dates: 1/90–Present.
Learning Outcomes: Upon completion of the course, the student will possess those combat and weapon skills necessary to serve as a basic assaultman.

Instruction: Through lecture, demonstration, and practical hands-on activities, the student receives instruction in threat and NATO vehicles, advanced demolitions, the SMAW and M47 DRAGON, and physical fitness.

Credit Recommendation: Credit is not recommended because of the military-specific nature of the course (7/96).

MC-2204-0107

M1A1 ARMOR CREWMAN

Course Number: 020-1812; A13 TBM.
Location: Armor Center and School, Ft. Knox, KY.
Length: 11 weeks (440 hours).
Exhibit Dates: 9/94–Present.
Learning Outcomes: Upon completion of the course, the student will be able to perform the duties as a tank crewman and operate and maintain the specific equipment as a loader, gunner, and driver.

Instruction: Course includes lectures, demonstrations, conferences, and performance exercises in operating and maintaining the weapons systems, communications systems, NBC defense systems, night vision systems, vehicle PMS, personal weapons, tactical operating systems, and vehicle defenses.

Credit Recommendation: Credit is not recommended because of the military-specific nature of the course (3/97).

MC-2204-0108

RESERVE INFANTRY OFFICER

Course Number: H2C.
Location: Basic School, Quantico, VA.
Length: 3 weeks (175 hours).
Exhibit Dates: 5/95–Present.
Learning Outcomes: Upon completion of the course, the student will be prepared with advanced infantry skills in warfare tactics, fire support, communications, infantry weapons, communications and tactical decision making.

Instruction: Instruction is performance-oriented, covering such topics as close combat, infantry tactics and weapons, communications, and fire support. Methods include field exercises and tactical decision making scenarios. There is emphasis on judgment and timely decision making.

Credit Recommendation: Credit is not recommended because of the military-specific nature of the course (5/97).

MC-2204-0109

AMMUNITION TECHNICIAN, PHASE 3 (USMC)

Course Number: 645-55B10 (OS).
Location: Aviation School, Ft. Rucker, AL; Marine Corps Detachment, Redstone Arsenal, AL.
Length: 3 weeks (110 hours).
Exhibit Dates: 1/94–Present.
Learning Outcomes: Upon completion of the course, the student will be able to conduct automated ammunition inventory and control procedures.

Instruction: Lectures and practical exercises in the computer software the military uses for ammunition stock control and record keeping.

Credit Recommendation: In the lower-division baccalaureate/associate degree category, 1 semester hour in computer based inventory control (5/97).

MC-2204-0110

SMALL ARMS REFRESHER

Course Number: 641-F1; A01GBC1.
Location: Marine Corps Detachment, Ordnance Center and School, Aberdeen Proving Ground, MD.
Length: 2 weeks (70 hours).
Exhibit Dates: 3/96–Present.
Learning Outcomes: Upon completion of the course, the student will have reviewed previous small arms repair training including disassembly, inspection, parts identification, repair, and reassembly of small arms including pistols, rifles, shot guns, light and heavy machine guns, grenade launchers, and mortars. Students will also be able to repair and maintain new weapons and equipment.

Instruction: Lectures, demonstrations, and practical experiences in safety precautions, parts identification, inspection, determination of malfunction, repair, disassembly and reassembly techniques, and test firing of small arms including pistols, rifles, shot guns, light and heavy machine guns, grenade launchers, and mortars.

Credit Recommendation: In the lower-division baccalaureate/associate degree category, 1 semester hour in gunsmithing/small arms repair (6/97).

MC-2204-0111

SMALL ARMS REPAIRER

Course Number: 641-2111.
Location: Marine Corps Detachment, Ordnance Center and School, Aberdeen Proving Ground, MD.
Length: 9 weeks (363 hours).
Exhibit Dates: 3/96–Present.
Learning Outcomes: Upon completion of the course, the student will be able to disassemble, inspect, identify malfunction, repair, and reassemble small arms, including pistols, rifles, shotguns, light and heavy machine guns, grenade launchers, and mortars. Students will also be able to perform preventive maintenance on small arms.

Instruction: Lectures, demonstrations, and practical experiences cover safety precautions, parts identification, inspection, determination of malfunction, repair, disassembly and reassembly techniques, and test firing of small arms including pistols, rifles, shotguns, light and heavy machine guns, grenade launchers, and mortars.

Credit Recommendation: In the lower-division baccalaureate/associate degree category, 3 semester hours gunsmithing/small arms repair (6/97).

MC-2204-0112

WARFIGHTING SKILLS PROGRAM (WARFIGHTING) BY CORRESPONDENCE

Course Number: 7400A.
Location: Marine Corps Institute, Washington, DC.
Length: Maximum, 104 weeks.
Exhibit Dates: 1/91–Present.

Learning Outcomes: Upon completion of the course, the student will understand the theory, nature, and levels of war; development of modern warfare tactics and maneuver warfare concepts and application; planning and writing of orders; techniques of combat leadership; methods of training and qualities needed for successful leadership; and concepts of combined arms operations and fire support.

Instruction: Course consists of assigned reading of provided materials. These include historical analysis and required written exercises. Topics covered include fighting, small unit tactics, combat techniques, Marine Corps leadership, and combined arms.

Credit Recommendation: In the lower-division baccalaureate/associate degree category, 2 semester hours in military studies (9/97).

MC-2204-0113

AMPHIBIOUS WARFARE SCHOOL NONRESIDENT PROGRAM (AWSNP) BY CORRESPONDENCE, PHASE 1

Course Number: 8500, Phase 1.
Location: Marine Corps Institute, Washington, DC.
Length: Maximum, 104 weeks.
Exhibit Dates: 7/95–Present.
Learning Outcomes: Upon completion of the course, the student will understand the theory and nature of war; the practical application of command, control, and communications on the modern battlefield; and organizational command and control concepts at a basic level. Students will be able to apply basic leadership concepts and theories to contemporary leadership issues as a way of stimulating thoughtful and effective interactions with subordinates, formulate organizational plans, and express themselves clearly and forcefully within the administrative structure.

Instruction: Course consists of individual, self-guided study, supplemented with individually implemented practical exercises. Objective examinations are externally administered and graded. Curriculum addresses topics in military history and science, leadership fundamentals, basics of organization and administrative theory, problem solving, decision making, operations planning, and the role of the mid-level military manager in peace and war.

Credit Recommendation: In the lower-division baccalaureate/associate degree category, 3 semester hours in military studies and 2 in introductory management or leadership studies (9/97).

MC-2204-0114

AMPHIBIOUS WARFARE SCHOOL NONRESIDENT PROGRAM (AWSNP) BY CORRESPONDENCE PHASE 2

Course Number: 8600, Phase 2.
Location: Marine Corps Institute, Washington, DC.
Length: Maximum, 104 weeks.
Exhibit Dates: 7/95–Present.
Learning Outcomes: Upon completion of the course, the student will have mastered the fundamentals of offense, defense, amphibious operations, and operations other than war; be able to apply organizational doctrine in a day-to-day operational environment; develop informed and effective plans; make data-based decisions; be capable of communicating directions to subordinates at an intermediate level of organizational responsibility; and will have gained a basic understanding of the organization, capabilities, and employment of Marine Corps assets.

Instruction: Course consists of individual, self guided study, supplemented with individually implemented practical problem solving exercises. Objective examinations are externally administered and graded. Students employ lesson materials and background and supplementary reading to gain skills in developing organizational plans and directives, analyzing the nature of regional security threats as they influence daily operations, conducting task analyses, preparing situational estimates, and developing plans of operation for on-going operations.

Credit Recommendation: In the lower-division baccalaureate/associate degree category, 3 semester hours in introductory military studies (9/97); in the upper-division baccalaureate category, 2 semester hours in introductory management or leadership studies (9/97).

Coast Guard Aviator Exhibits

CGA-C130-001

HC-130 CO-PILOT

Exhibit Dates: 1/90–Present.

Description

Pilots fixed-wing aircraft; demonstrates proficiency in performing standard flight maneuvers; possesses detailed knowledge of FAA regulations, platform-specific aircraft operational characteristics, local area restrictions, and relevant emergency procedures; is designated a military aviator and has the Instrument Flight (IFR) designation; has completed a transition course in a specific aircraft platform including a check flight; is knowledgeable of aircraft systems, communications techniques, aircraft security requirements, and any installed search-and-rescue equipment.

Recommendation

In the lower-division baccalaureate/associate degree category, 18 semester hours in a commercial/instrument rating, 3 in FAA regulations, 3 in air navigation, 3 in flight physiology, 3 in aviation meteorology, 3 in instrument flight operations, 3 in aircraft engines and systems, 3 in aircraft performance, 3 in technical mathematics, 3 aviation safety, 3 in aerodynamics, and 3 in communication skills (oral) (10/96).

CGA-C130-002

HC-130 FIRST PILOT

Exhibit Dates: 1/90–Present.

Description

Progresses to First Pilot designation from the Co-pilot designation; pilots fixed-wing aircraft; commands utility fixed-wing aircraft crews under tactical and non-tactical conditions during all types of meteorological conditions under both day and night operations; plans and conducts search and rescue operations; performs internal/external loads, load dropping and recovery, aerial reconnaissance, drug traffic interdiction, and transportation of personnel and cargo; has knowledge and skills equivalent to an FAA type-rated pilot on fixed-wing aircraft.

Recommendation

In the lower-division baccalaureate/associate degree category, 18 semester hours in commercial/instrument rating, 3 in FAA regulations, 3 in air navigation, 3 in flight physiology, 3 in aviation meteorology, 3 in instrument flight operations, 3 in aircraft engines and systems, 3 in aircraft performance, 3 in technical mathematics, 3 in aviation safety, 3 in aerodynamics, 3 in flight planning, 3 in government rules and regulations, 3 in personnel supervision, and 3 in communication skills (oral). In the upper-division baccalaureate category, 3 semester hours in advanced flight operations, and 3 in interpersonal communication (10/96).

CGA-C130-003

HC-130 AIRCRAFT COMMANDER

Exhibit Dates: 1/90–Present.

Description

Progresses to Aircraft Commander designation from the First Pilot designation; to a more advanced degree, fulfills all requirements for First Pilot; to a high degree, demonstrates the ability to exercise flight discipline and aircrew supervision and to carry out all types of search-and-rescue missions including duty as on-scene commander; completes all reports and documents related to mission; possesses knowledge and skills equivalent to FAA air transport pilot rating.

Recommendation

In the lower-division baccalaureate/associate degree category, 18 semester hours in a commercial/instrument rating, 3 in FAA regulations, 3 in air navigation, 3 in flight physiology, 3 in aviation meteorology, 3 in instrument flight operations, 3 in aircraft engines and systems, 3 in aircraft performance, 3 in technical mathematics, 3 in aviation safety, 3 in aerodynamics, 3 in flight planning, 3 in government rules and regulations, 3 in personnel supervision, 3 in technical writing, and 3 in communication skills (oral). In the upper-division baccalaureate category, 3 semester hours in advanced flight operations, 3 in personnel management, 3 in organizational management, 3 in public relations management, and 3 in interpersonal communication (10/96).

CGA-C130-004

HC-130 INSTRUCTOR PILOT

Exhibit Dates: 1/90–Present.

Description

Progresses to Instructor Pilot designation from First Pilot designation; possesses and demonstrates superior judgment, patience, tact, understanding, and desire to instruct; cultivates leadership skills to foster confidence and respect; has received civilian or military training in methods of instruction; evaluates and documents student proficiency; uses computer-based courses to instruct pilots.

Recommendation

In the lower-division baccalaureate/associate degree category, 18 semester hours in a commercial/instrument rating, 3 in FAA regulations, 3 in air navigation, 3 in flight physiology, 3 in aviation meteorology, 3 in instrument flight operations, 3 in aircraft engines and systems, 3 in aircraft performance, 3 in technical mathematics, 3 in aviation safety, 3 in aerodynamics, 3 in flight planning, 3 in government rules and regulations, 3 in personnel supervision, 3 in technical writing, 3 in computer applications, 1 in needs analysis, and 3 in communication skills (oral). In the upper-division baccalaureate category, 3 in advanced flight operations, 3 in personnel management, 3 in organizational management, 3 in public relations management, 3 in methods of instruction, and 3 in interpersonal communication (10/96).

CGA-HH60-001

HH-60 CO-PILOT

Exhibit Dates: 1/90–Present.

Description

Pilots fixed-wing aircraft; demonstrates proficiency in performing standard flight maneuvers; possesses detailed knowledge of FAA regulations, platform-specific aircraft operational characteristics, local area restrictions, and relevant emergency procedures; is designated a military aviator and has the Instrument Flight (IFR) designation; has completed a transition course in a specific aircraft platform including a check flight; is knowledgeable of aircraft systems, communications techniques, aircraft security requirements, and any installed search-and-rescue equipment.

Recommendation

In the lower-division baccalaureate/associate degree category, 18 semester hours in a commercial/instrument rating, 3 in FAA regulations, 3 in air navigation, 3 in flight physiology, 3 in aviation meteorology, 3 in instrument flight operations, 3 in aircraft engines and systems, 3 in aircraft performance, 3 in technical mathematics, 3 aviation safety, 3 in aerodynamics, and 3 in communication skills (oral) (10/96).

CGA-HH60-002

HH-60 FIRST PILOT

Exhibit Dates: 1/90–Present.

Description

Progresses to First Pilot designation from the Co-pilot designation; pilots fixed-wing aircraft; commands utility fixed-wing aircraft crews under tactical and non-tactical conditions during all types of meteorological conditions under both day and night operations; plans and conducts search and rescue operations; performs internal/external loads, load dropping and recovery, aerial reconnaissance, drug traffic interdiction, and transportation of personnel and cargo; has knowledge and skills equivalent to an FAA type-rated pilot on fixed-wing aircraft.

Recommendation

In the lower-division baccalaureate/associate degree category, 18 semester hours in commercial/instrument rating, 3 in FAA regulations, 3 in air navigation, 3 in flight physiology, 3 in aviation meteorology, 3 in instrument flight operations, 3 in aircraft engines and systems, 3 in aircraft performance, 3 in technical mathematics, 3 in aviation safety, 3 in aerodynamics, 3 in flight planning, 3 in government rules and regulations, 3 in personnel supervision, and 3 in communication skills (oral). In the upper-divi-

CGA-HH60-003

HH-60 AIRCRAFT COMMANDER

Exhibit Dates: 1/90–Present.

Description

Progresses to Aircraft Commander designation from the First Pilot designation; to a more advanced degree, fulfills all requirements for First Pilot; to a high degree, demonstrates the ability to exercise flight discipline and aircrew supervision and to carry out all types of search-and-rescue missions including duty as on-scene commander; completes all reports and documents related to mission; possesses knowledge and skills equivalent to FAA air transport pilot rating.

Recommendation

In the lower-division baccalaureate/associate degree category, 18 semester hours in a commercial/instrument rating, 3 in FAA regulations, 3 in air navigation, 3 in flight physiology, 3 in aviation meteorology, 3 in instrument flight operations, 3 in aircraft engines and systems, 3 in aircraft performance, 3 in technical mathematics, 3 in aviation safety, 3 in aerodynamics, 3 in flight planning, 3 in government rules and regulations, 3 in personnel supervision, 3 in technical writing, and 3 in communication skills (oral). In the upper-division baccalaureate category, 3 semester hours in advanced flight operations, 3 in personnel management, 3 in organizational management, 3 in public relations management, and 3 in interpersonal communication (10/96).

CGA-HH60-004

HH-60 INSTRUCTOR PILOT

Exhibit Dates: 1/90–Present.

Description

Progresses to Instructor Pilot designation from First Pilot designation; possesses and demonstrates superior judgment, patience, tact, understanding, and desire to instruct; cultivates leadership skills to foster confidence and respect; has received civilian or military training in methods of instruction; evaluates and documents student proficiency; uses computer-based courses to instruct pilots.

Recommendation

In the lower-division baccalaureate/associate degree category, 18 semester hours in a commercial/instrument rating, 3 in FAA regulations, 3 in air navigation, 3 in flight physiology, 3 in aviation meteorology, 3 in instrument flight operations, 3 in aircraft engines and systems, 3 in aircraft performance, 3 in technical mathematics, 3 in aviation safety, 3 in aerodynamics, 3 in flight planning, 3 in government rules and regulations, 3 in personnel supervision, 3 in technical writing, 3 in computer applications, 1 in needs analysis, and 3 in communication skills (oral). In the upper-division baccalaureate category, 3 in advanced flight operations, 3 in personnel management, 3 in organizational management, 3 in public relations management, 3 in methods of instruction, and 3 in interpersonal communication (10/96).

CGA-HH65-001

HH-65 CO-PILOT

Exhibit Dates: 1/90–Present.

Description

Pilots fixed-wing aircraft; demonstrates proficiency in performing standard flight maneuvers; possesses detailed knowledge of FAA regulations, platform-specific aircraft operational characteristics, local area restrictions, and relevant emergency procedures; is designated a military aviator and has the Instrument Flight (IFR) designation; has completed a transition course in a specific aircraft platform including a check flight; is knowledgeable of aircraft systems, communications techniques, aircraft security requirements, and any installed search-and-rescue equipment.

Recommendation

In the lower-division baccalaureate/associate degree category, 18 semester hours in a commercial/instrument rating, 3 in FAA regulations, 3 in air navigation, 3 in flight physiology, 3 in aviation meteorology, 3 in instrument flight operations, 3 in aircraft engines and systems, 3 in aircraft performance, 3 in technical mathematics, 3 aviation safety, 3 in aerodynamics, and 3 in communication skills (oral) (10/96).

CGA-HH65-002

HH-65 FIRST PILOT

Exhibit Dates: 1/90–Present.

Description

Progresses to First Pilot designation from the Co-pilot designation; pilots fixed-wing aircraft; commands utility fixed-wing aircraft crews under tactical and non-tactical conditions during all types of meteorological conditions under both day and night operations; plans and conducts search and rescue operations; performs internal/external loads, load dropping and recovery, aerial reconnaissance, drug traffic interdiction, and transportation of personnel and cargo; has knowledge and skills equivalent to an FAA type-rated pilot on fixed-wing aircraft.

Recommendation

In the lower-division baccalaureate/associate degree category, 18 semester hours in commercial/instrument rating, 3 in FAA regulations, 3 in air navigation, 3 in flight physiology, 3 in aviation meteorology, 3 in instrument flight operations, 3 in aircraft engines and systems, 3 in aircraft performance, 3 in technical mathematics, 3 in aviation safety, 3 in aerodynamics, 3 in flight planning, 3 in government rules and regulations, 3 in personnel supervision, and 3 in communication skills (oral). In the upper-division baccalaureate category, 3 semester hours in advanced flight operations, and 3 in interpersonal communication (10/96).

CGA-HH65-003

HH-65 AIRCRAFT COMMANDER

Exhibit Dates: 1/90–Present.

Description

Progresses to Aircraft Commander designation from the First Pilot designation; to a more advanced degree, fulfills all requirements for First Pilot; to a high degree, demonstrates the ability to exercise flight discipline and aircrew supervision and to carry out all types of search-and-rescue missions including duty as on-scene commander; completes all reports and documents related to mission; possesses knowledge and skills equivalent to FAA air transport pilot rating.

Recommendation

In the lower-division baccalaureate/associate degree category, 18 semester hours in a commercial/instrument rating, 3 in FAA regulations, 3 in air navigation, 3 in flight physiology, 3 in aviation meteorology, 3 in instrument flight operations, 3 in aircraft engines and systems, 3 in aircraft performance, 3 in technical mathematics, 3 in aviation safety, 3 in aerodynamics, 3 in flight planning, 3 in government rules and regulations, 3 in personnel supervision, 3 in technical writing, and 3 in communication skills (oral). In the upper-division baccalaureate category, 3 semester hours in advanced flight operations, 3 in personnel management, 3 in organizational management, 3 in public relations management, and 3 in interpersonal communication (10/96).

CGA-HH65-004

HH-65 INSTRUCTOR PILOT

Exhibit Dates: 1/90–Present.

Description

Progresses to Instructor Pilot designation from First Pilot designation; possesses and demonstrates superior judgment, patience, tact, understanding, and desire to instruct; cultivates leadership skills to foster confidence and respect; has received civilian or military training in methods of instruction; evaluates and documents student proficiency; uses computer-based courses to instruct pilots.

Recommendation

In the lower-division baccalaureate/associate degree category, 18 semester hours in a commercial/instrument rating, 3 in FAA regulations, 3 in air navigation, 3 in flight physiology, 3 in aviation meteorology, 3 in instrument flight operations, 3 in aircraft engines and systems, 3 in aircraft performance, 3 in technical mathematics, 3 in aviation safety, 3 in aerodynamics, 3 in flight planning, 3 in government rules and regulations, 3 in personnel supervision, 3 in technical writing, 3 in computer applications, 1 in needs analysis, and 3 in communication skills (oral). In the upper-division baccalaureate category, 3 in advanced flight operations, 3 in personnel management, 3 in organizational management, 3 in public relations management, 3 in methods of instruction, and 3 in interpersonal communication (10/96).

CGA-HU25-001

HU-25 CO-PILOT

Exhibit Dates: 1/90–Present.

Description

Pilots fixed-wing aircraft; demonstrates proficiency in performing standard flight maneuvers; possesses detailed knowledge of FAA regulations, platform-specific aircraft operational characteristics, local area restrictions, and relevant emergency procedures; is designated a military aviator and has the Instrument Flight (IFR) designation; has completed a transition course in a specific aircraft platform including a check flight; is knowledgeable of aircraft systems, communications techniques, aircraft security requirements, and any installed search-and-rescue equipment.

Recommendation

In the lower-division baccalaureate/associate degree category, 18 semester hours in a commercial/instrument rating, 3 in FAA regulations, 3 in air navigation, 3 in flight physiology, 3 in aviation meteorology, 3 in instrument flight operations, 3 in aircraft engines and systems, 3 in aircraft performance, 3 in technical mathematics, 3 aviation safety, 3 in aerodynamics, and 3 in communication skills (oral) (10/96).

CGA-HU25-002

HU-25 FIRST PILOT

Exhibit Dates: 1/90–Present.

Description

Progresses to First Pilot designation from the Co-pilot designation; pilots fixed-wing aircraft; commands utility fixed-wing aircraft crews under tactical and non-tactical conditions during all types of meteorological conditions under both day and night operations; plans and conducts search and rescue operations; performs internal/external loads, load dropping and recovery, aerial reconnaissance, drug traffic interdiction, and transportation of personnel and cargo; has knowledge and skills equivalent to an FAA type-rated pilot on fixed-wing aircraft.

Recommendation

In the lower-division baccalaureate/associate degree category, 18 semester hours in commercial/instrument rating, 3 in FAA regulations, 3 in air navigation, 3 in flight physiology, 3 in aviation meteorology, 3 in instrument flight operations, 3 in aircraft engines and systems, 3 in aircraft performance, 3 in technical mathematics, 3 in aviation safety, 3 in aerodynamics, 3 in flight planning, 3 in government rules and regulations, 3 in personnel supervision, and 3 in communication skills (oral). In the upper-division baccalaureate category, 3 semester hours in advanced flight operations, and 3 in interpersonal communication (10/96).

CGA-HU25-003

HU-25 AIRCRAFT COMMANDER

Exhibit Dates: 1/90–Present.

Description

Progresses to Aircraft Commander designation from the First Pilot designation; to a more advanced degree, fulfills all requirements for First Pilot; to a high degree, demonstrates the ability to exercise flight discipline and aircrew supervision and to carry out all types of search-and-rescue missions including duty as on-scene commander; completes all reports and documents related to mission; possesses knowledge and skills equivalent to FAA air transport pilot rating.

Recommendation

In the lower-division baccalaureate/associate degree category, 18 semester hours in a commercial/instrument rating, 3 in FAA regulations, 3 in air navigation, 3 in flight physiology, 3 in aviation meteorology, 3 in instrument flight operations, 3 in aircraft engines and systems, 3 in aircraft performance, 3 in technical mathematics, 3 in aviation safety, 3 in aerodynamics, 3 in flight planning, 3 in government rules and regulations, 3 in personnel supervision, 3 in technical writing, and 3 in communication skills (oral). In the upper-division baccalaureate category, 3 semester hours in advanced flight operations, 3 in personnel management, 3 in organizational management, 3 in public relations management, and 3 in interpersonal communication (10/96).

CGA-HU25-004

HU-25 INSTRUCTOR PILOT

Exhibit Dates: 1/90–Present.

Description

Progresses to Instructor Pilot designation from First Pilot designation; possesses and demonstrates superior judgment, patience, tact, understanding, and desire to instruct; cultivates leadership skills to foster confidence and respect; has received civilian or military training in methods of instruction; evaluates and documents student proficiency; uses computer-based courses to instruct pilots.

Recommendation

In the lower-division baccalaureate/associate degree category, 18 semester hours in a commercial/instrument rating, 3 in FAA regulations, 3 in air navigation, 3 in flight physiology, 3 in aviation meteorology, 3 in instrument flight operations, 3 in aircraft engines and systems, 3 in aircraft performance, 3 in technical mathematics, 3 in aviation safety, 3 in aerodynamics, 3 in flight planning, 3 in government rules and regulations, 3 in personnel supervision, 3 in technical writing, 3 in computer applications, 1 in needs analysis, and 3 in communication skills (oral). In the upper-division baccalaureate category, 3 in advanced flight operations, 3 in personnel management, 3 in organizational management, 3 in public relations management, 3 in methods of instruction, and 3 in interpersonal communication (10/96).

Coast Guard Enlisted Ratings Exhibits

CGR-AD-001

AVIATION MACHINIST'S MATE
  AD3
  AD2
  AD1
  ADC
  ADCS
  ADCM

Exhibit Dates: 1/90–9/94.

Occupational Group: V (Aviation).

Career Pattern

SN: Seaman (E-3). *AD3:* Aviation Machinist's Mate, Third Class (E-4). *AD2:* Aviation Machinist's Mate, Second Class (E-5). *AD1:* Aviation Machinist's Mate, First Class (E-6). *ADC:* Chief Aviation Machinist's Mate (E-7). *ADCS:* Senior Chief Aviation Machinist's Mate (E-8). *ADCM:* Master Chief Aviation Machinist's Mate (E-9).

Description

Summary: Services and maintains aircraft engines and their related systems, including propellers; performs periodic and special inspections on aircraft, engines, and related systems; field tests, repairs, and adjusts engine and propeller components and accessories; maintains and adjusts helicopter drive trains and flight controls; preserves and depreserves engine propellers and accessories; maintains air conditioning and pressurization systems; supervises aircraft, engine, and propeller shops. *AD3:* Uses drawings, diagrams, blueprints, charts, publications, and work cards to perform inspections, servicing, and maintenance of aircraft and aviation equipment; operates and services gas turbine starting equipment; taxis, tows, and secures aircraft; cleans, inspects, and services installed propellers. *AD2:* Able to perform the duties required for AD3; performs aircraft preflight, through-flight, and post-flight inspections; identifies and locates external leaks to fuel, oil, hydraulic, pneumatic, and oxygen systems; performs aircraft fuel system checks; maintains helicopter rotor heads, gear boxes, and drive shafts; cleans, maintains, removes, installs, and folds helicopter rotor blades; takes engine/gear box oil samples for SOAP analysis; removes, cleans, services, and replaces components; tests cockpits and cabins for pressure tightness, leaks, and proper pressure and temperature controls; replaces air conditioning and pressurization system components; removes and installs ignitor plugs; functionally tests engine power control mechanisms; removes, installs, and makes adjustments to engine fuel controls; rigs and adjusts power controls, fuel selectors, and shut off valve linkages; performs ground run-up of turbine engines; removes, installs, and adjusts propeller governors; checks propellers for blade angle, track, and alignment; removes, cleans, tests, installs, rigs, and adjusts propellers and control mechanisms; performs cleaning and corrosion control on aircraft, propellers, and power plants; services fuel, oil, and hydraulic systems; observes safety precautions while maintaining aircraft. *AD1:* Able to perform the duties required for AD2; supervises aircraft ground handling and servicing; maintains aircraft records, technical publications, directives, and manuals; supervises inspection procedures; performs quality assurance inspection and evaluation of aircraft and equipment; installs and operates calibration test units on turbine engines; inspects compressors and turbines for proper axial and radial clearances; rigs and adjusts propeller control mechanisms; removes, cleans, bench tests, and installs propeller assemblies; adjusts temperature and pressure controls; and rigs and adjusts actuating controls and mechanisms. *ADC:* Able to perform the duties required for AD1; serves as shop supervisor; trains and supervises personnel in all categories of equipment maintenance and repair; prepares preventive maintenance schedules; organizes and maintains technical library; prepares and submits budget requests; plans and supervises on-the-job training. *ADCS:* Able to perform the duties required for ADC; serves as enlisted technical or specialty expert; plans, organizes, and directs the work of personnel maintaining aircraft, engines, propellers, helicopter drive trains, and flight controls. *ADCM:* Able to perform the duties required for ADCS; serves as senior enlisted technical or specialty administrator; manages personnel in the operation, maintenance, and procurement of aircraft; ensures maximum efficiency of the work force and the equipment; prepares general correspondence concerning fiscal, supply, and administrative matters; assists in the formulation of plans, policies, and budget requirements; may supplement the officer corps in the overall supervision and administration of personnel and equipment; may also supervise personnel in other specialty areas.

Recommendation, AD3

In the lower-division baccalaureate/associate degree category, 3 semester hours in aviation maintenance technology (or 1 semester hour in propeller systems, 1 in cleaning and corrosion control, and 1 in ground operation and servicing (5/84).

Recommendation, AD2

In the lower-division baccalaureate/associate degree category, 3 semester hours in aviation maintenance technology (or 1 semester hour in propeller systems, 1 in cleaning and corrosion control, and 1 in ground operation and servicing), and 3 in aircraft fuel systems, 3 in flight control systems, 3 in helicopter rotor systems, 3 in assembly and rigging, and 1 in fluid lines and fittings (5/84).

Recommendation, AD1

In the vocational certificate category, 3 semester hours in aviation maintenance technology (or 1 semester hour in propeller systems, 1 in cleaning and corrosion control, and 1 in ground operation and servicing), and 3 in aircraft fuel systems, 3 in flight control systems, 3 in helicopter rotor systems, 3 in assembly and rigging, 1 in fluid lines and fittings, 3 in turbine engine inspections, and 1 in cabin atmospheric control systems. In the lower-division baccalaureate/associate degree category, 3 semester hours in aviation maintenance technology (or 1 semester hour in propeller systems, 1 in cleaning and corrosion control, and 1 in ground operation and servicing), and 3 in aircraft fuel systems, 3 in flight control systems, 3 in helicopter rotor systems, 3 in assembly and rigging, 1 in fluid lines and fittings, 3 in turbine engine inspections, 1 in cabin atmospheric control systems, and 2 in personnel supervision (5/84).

Recommendation, ADC

In the vocational certificate category, 3 semester hours in aviation maintenance technology (or 1 semester hour in propeller systems, 1 in cleaning and corrosion control, and 1 in ground operation and servicing), and 3 in aircraft fuel systems, 3 in flight control systems, 3 in helicopter rotor systems, 3 in assembly and rigging, 1 in fluid lines and fittings, 3 in turbine engine inspections, 1 in cabin atmospheric control systems, 3 in maintenance management, and 3 in shop management. In the lower-division baccalaureate/associate degree category, 3 semester hours in aviation maintenance technology (or 1 semester hour in propeller systems, 1 in cleaning and corrosion, and 1 in ground operation and servicing), and 3 in aircraft fuel systems, 3 in flight control systems, 3 in helicopter rotor systems, 3 in assembly and rigging, 1 in fluid lines and fittings, 3 in turbine engine inspections, 1 in cabin atmospheric control systems, 3 in personnel supervision, 3 in maintenance management, 3 in shop management, and 3 in principles of management (5/84).

Recommendation, ADCS

In the lower-division baccalaureate/associate degree category, 3 semester hours in aviation maintenance technology (or 1 semester hour in propeller systems, 1 in cleaning and corrosion, and 1 in ground operation and servicing), and 3 in aircraft fuel systems, 3 in flight control systems, 3 in helicopter rotor systems, 3 in assembly and rigging, 1 in fluid lines and fittings, 3 in turbine engine inspections, 1 in cabin atmospheric control systems, 3 in personnel supervision, 3 in maintenance management, 3 in shop management, and 3 in principles of management. In the upper-division baccalaureate category, 3 semester hours for field experience in management and 3 in management problems (5/84).

Recommendation, ADCM

In the lower-division baccalaureate/associate degree category, 3 semester hours in aviation maintenance technology (or 1 semester hour in propeller systems, 1 in cleaning and corrosion, and 1 in ground operation and servicing), and 3 in aircraft fuel systems, 3 in flight control systems, 3 in helicopter rotor systems, 3 in assembly and rigging, 1 in fluid lines and fittings, 3 in turbine engine inspections, 1 in cabin atmospheric control systems, 3 in personnel supervision, 3 in maintenance management, 3 in shop management, and 3 in principles of management. In the upper-division baccalaureate category, 6 semester hours for field experience in management and 3 in management problems (5/84).

CGR-AD-002

AVIATION MACHINIST'S MATE
    AD3
    AD2
    AD1
    ADC
    ADCS
    ADCM

Exhibit Dates: 10/94–Present.

Occupational Group: V (Aviation).

Career Pattern

SN: Seaman (E-3). *AD3:* Aviation Machinist's Mate, Third Class (E-4). *AD2:* Aviation Machinist's Mate, Second Class (E-5). *AD1:* Aviation Machinist's Mate, First Class (E-6). *ADC:* Chief Aviation Machinist's Mate (E-7). *ADCS:* Senior Chief Aviation Machinist's Mate (E-8). *ADCM:* Master Chief Aviation Machinist's Mate (E-9).

Description

Summary: Services and maintains aircraft engines and their related systems, including propellers; performs periodic and special inspections on aircraft, engines, and related systems; field tests, repairs, and adjusts engine and propeller components and accessories; maintains and adjusts helicopter drive trains and flight controls; preserves and depreserves engine propellers and accessories; maintains air conditioning and pressurization systems; supervises aircraft, engine, and propeller shops; uses computers to organize and maintain records. *AD3:* Uses drawings, diagrams, blueprints, charts, publications, and work cards to perform inspections, servicing, and maintenance of aircraft and aviation equipment; operates and services gas turbine starting equipment; taxis, tows, and secures aircraft; cleans, inspects, and services installed propellers. *AD2:* Able to perform the duties required for AD3; performs aircraft preflight, through-flight, and post-flight inspections; identifies and locates external leaks to fuel, oil, hydraulic, pneumatic, and oxygen systems; performs aircraft fuel system checks; maintains helicopter rotor heads, gear boxes, and drive shafts; cleans, maintains, removes, installs, and folds helicopter rotor blades; takes engine/gear box oil samples for SOAP analysis; removes, cleans, services, and replaces components; tests cockpits and cabins for pressure tightness, leaks, and proper pressure and temperature controls; replaces air conditioning and pressurization system components; removes and installs ignitor plugs; functionally tests engine power control mechanisms; removes, installs, and makes adjustments to engine fuel controls; rigs and adjusts power controls, fuel selectors, and shut off valve linkages; performs ground run-up of turbine engines; removes, installs and adjusts propeller governors; checks propellers for blade angle, track, and alignment; removes, cleans, tests, installs, rigs, and adjusts propellers and control mechanisms; performs cleaning and corrosion control on aircraft, propellers, and power plants; services fuel, oil, and hydraulic systems; observes safety precautions while maintaining aircraft. *AD1:* Able to perform the duties required for AD2; supervises aircraft ground handling and servicing; maintains aircraft records, technical publications, directives, and manuals; supervises inspection procedures; performs quality assurance inspection and evaluation of aircraft and equipment; installs and operates calibration test units on turbine engines; inspects compressors and turbines for proper axial and radial clearances; rigs and adjusts propeller control mechanisms; removes, cleans, bench tests, and installs propeller assemblies; adjusts temperature and pressure controls; and rigs and adjusts actuating controls and mechanisms. *ADC:* Able to perform the duties required for AD1; serves as shop supervisor; trains and supervises personnel in all categories of equipment maintenance and repair; prepares preventive maintenance schedules; organizes and maintains technical library; prepares and submits budget requests; plans and supervises on-the-job training. *ADCS:* Able to perform the duties required for ADC; serves as enlisted technical or specialty expert; plans, organizes, and directs the work of personnel maintaining aircraft, engines, propellers, helicopter drive trains, and flight controls. *ADCM:* Able to perform the duties required for ADCS; serves as senior enlisted technical or specialty administrator; manages personnel in the operation, maintenance, and procurement of aircraft; ensures maximum efficiency of the work force and the equipment; prepares general correspondence concerning fiscal, supply, and administrative matters; assists in the formulation of plans, policies, and budget requirements; may supplement the officer corps in the overall supervision and administration of personnel and equipment; may also supervise personnel in other specialty areas.

Recommendation, AD3

In the lower-division baccalaureate/associate degree category, 1 semester hour in propeller systems, 1 in cleaning and corrosion control, 1 in ground operation and servicing, and 3 in introduction to computers (6/95).

Recommendation, AD2

In the lower-division baccalaureate/associate degree category, 1 semester hour in propeller systems, 1 in cleaning and corrosion control, 1 in ground operation and servicing, 3 in introduction to computers, 3 in aircraft fuel systems, 3 in aircraft flight control systems, 3 in helicopter rotor systems, 3 in assembly and rigging, and 1 in fluid lines and fittings (6/95).

Recommendation, AD1

In the lower-division baccalaureate/associate degree category, 1 semester hour in propeller systems, 1 in cleaning and corrosion control, 1 in ground operation and servicing, 3 in introduction to computers, 3 in aircraft fuel systems, 3 in aircraft flight control systems, 3 in helicopter rotor systems, 3 in assembly and rigging, 1 in fluid lines and fittings, 3 in turbine engine inspections, 1 in cabin atmospheric control systems, and 2 in personnel supervision (6/95).

Recommendation, ADC

In the lower-division baccalaureate/associate degree category, 1 semester hour in propeller systems, 1 in cleaning and corrosion, 1 in ground operation and servicing, 3 in introduction to computers, 3 in aircraft fuel systems, 3 in aircraft flight control systems, 3 in helicopter rotor systems, 3 in assembly and rigging, 1 in fluid lines and fittings, 3 in turbine engine inspections, 1 in cabin atmospheric control systems, 3 in personnel supervision, 3 in maintenance management, 3 in shop management, and 3 in organizational management (6/95).

Recommendation, ADCS

In the lower-division baccalaureate/associate degree category, 1 semester hour in propeller systems, 1 in cleaning and corrosion, and 1 in ground operation and servicing, 3 in introduction to computers, 3 in aircraft fuel systems, 3 in aircraft flight control systems, 3 in helicopter rotor systems, 3 in assembly and rigging, 1 in fluid lines and fittings, 3 in turbine engine inspections, 1 in cabin atmospheric control systems, 3 in personnel supervision, 3 in maintenance management, 3 in shop management, and 3 in organizational management. In the upper-division baccalaureate category, 3 semester hours for field experience in management and 3 in management problems (6/95).

Recommendation, ADCM

In the lower-division baccalaureate/associate degree category, 1 semester hour in propeller systems, 1 in cleaning and corrosion, and 1 in ground operation and servicing, 3 in introduction to computers, 3 in aircraft fuel systems, 3 in aircraft flight control systems, 3 in helicopter rotor systems, 3 in assembly and rigging, 1 in fluid lines and fittings, 3 in turbine engine inspections, 1 in cabin atmospheric control systems, 3 in personnel supervision, 3 in maintenance management, 3 in shop management, and 3 in organizational management. In the upper-division baccalaureate category, 6 semester hours for field experience in management and 3 in management problems (6/95).

CGR-AE-001

AVIATION ELECTRICIAN'S MATE
    AE3
    AE2
    AE1
    AEC
    AECS
    AECM

Exhibit Dates: 1/90–9/94.

Occupational Group: V (Aviation).

Career Pattern

SN: Seaman (E-3). *AE3:* Aviation Electrician's Mate, Third Class (E-4). *AE2:* Aviation Electrician's Mate, Second Class (E-5). *AE1:* Aviation Electrician's Mate, First Class (E-6). *AEC:* Chief Aviation Electrician's Mate (E-7). *AECS:* Senior Chief Aviation Electrician's Mate (E-8). *AECM:* Master Chief Aviation Electrician's Mate (E-9).

Description

Summary: Handles and services aircraft; inspects and maintains aircraft electrical and instrument systems, including: power generation, conversion and distribution systems; interior and exterior lighting; electrical components of aircraft controls, including airframe, engine, propeller, and utility control systems; aircraft electrical starting systems, including starters, starting controls and ignition system components; aircraft engine, flight, and flight control instruments; instrument systems; and non-instrument type indicating warning systems; aircraft automatic flight control system, including automatic pilots, flight director systems, aircraft compasses, and altitude reference systems; aircraft batteries and related electrical components. *AE3:* Observes safety precautions while working with energized and de-energized electrical and electronic equipment; checks electrical bonding and shock mountings for condition; replaces and adjusts electrical limit switches; troubleshoots and corrects DC electrical malfunctions; installs instrument indicator range marks; solves simple DC series, parallel and combination circuits; solves voltage divider and bridge circuits; troubleshoots a full-wave bridge PN junction diode vectifier with PL filter; traces signal flow in a common base, emitter, and collector transistor amplifier circuit; plots a frequency response curve for a direct RIC and transformer-coupled amplifier; acts as fireguard during start; ties down and secures aircraft. *AE2:* Able to perform the duties required for AE3; checks and operates ground support equipment and aircraft servicing equipment; supervises aircraft washing; tests, selects, and replaces circuit parts, including electron tubes and solid state devices; adjusts and aligns parts or circuitry; performs pitot-static tests; removes and replaces instrument indicators, transmitters, and sensors; tests fire detection systems; tests and calibrates flight instrument systems; calibrates fuel quantity, engine temperature, air data, and aircraft instrument systems; troubleshoots engine analyzer systems; performs tests on aircraft lighting systems; fabricates and installs aircraft electrical cables; tests power generating systems, aircraft batteries, circuit components, generator armatures, and voltage regulators; adjusts inverter speed control governors; tests hydraulic valve and actuator electrical components, aircraft pressurization and cabin temperature control, and fuel transfer systems; troubleshoots starting control electrical circuits; performs operational checks of aircraft electrical compass systems; swings and calibrates magnetic compasses; operationally checks automatic flight control and stabilization systems. *AE1:* Able to perform the duties required for AE2; supervises aircraft ground handling and servicing; calibrates engine temperature indicating systems, engine performance indicating and warning systems, air data systems, and aircraft instruments; diagnoses engine analyzer system malfunctions; inspects newly installed wiring on aircraft; makes three-phase wye and delta connections of transformers; diagnoses aircraft pressurization and cabin temperature control system malfunctions; maintains, tests, and adjusts electric and/or electronic control systems of aircraft automatic flight control and stabilization systems; troubleshoots automatic flight control and flight director systems. *AEC:* Able to perform the duties required for AE1; inspects work areas, tools, and aviation equipment to detect potentially hazardous and unsafe conditions and takes appropriate corrective action; prepares reports and maintains equipment records; supervises on-the-job training; prepares preventive maintenance schedules. *AECS:* Able to perform the duties required for AEC; serves as enlisted technical or specialty expert; plans, organizes, and directs work of personnel handling and servicing aircraft; plans and administers on-the-job training programs; supervises the preparation of reports. *AECM:* Able to perform the duties required for AECS; serves as senior enlisted technical or specialty administrator; manages personnel in the handling and servicing of aircraft; ensures maximum efficiency of the work force and equipment; prepares general correspondence concerning fiscal, supply, and administrative matters; assists in the formulation of plans, policies, and budget requirements; may supplement the officer corps in the overall supervision and administration of personnel and equipment; may also supervise personnel in other specialty areas.

Recommendation, AE3

In the lower-division baccalaureate/associate degree category, 3 semester hours in aviation maintenance technology (or 1 semester hour in aircraft propeller and electrical systems, 1 in ground operation and servicing, and 1 in cleaning and corrosion control), and 3 in DC electrical fundamentals (5/84).

Recommendation, AE2

In the lower-division baccalaureate/associate degree category, 3 semester hours in aviation maintenance technology (or 1 semester hour in aircraft propeller and electrical systems, 1 in ground operation and servicing, and 1 in cleaning and corrosion control), and 3 in DC electrical fundamentals, 3 in electricity/electronics, 3 in electricity/electronics laboratory, and 12 in aircraft electrical maintenance (5/84).

Recommendation, AE1

In the vocational certificate category, 3 semester hours in aviation maintenance technology (or 1 semester hour in aircraft propeller and electrical systems, 1 in ground operation and servicing, and 1 in cleaning and corrosion control), and 3 in DC electrical fundamentals, 3 in electricity/electronics laboratory, 12 in aircraft electrical maintenance, and 3 in aircraft electrical indicating systems. In the lower-division baccalaureate/associate degree category, 3 semester hours in aviation maintenance technology (or 1 semester hour in aircraft propeller and electrical systems, 1 in ground operation and servicing, and 1 in cleaning and corrosion control), and 3 in DC electrical fundamentals, 3 in electricity/electronics, 3 in electricity/electronics laboratory, 12 in aircraft electrical maintenance, 3 in aircraft electrical indicating systems, and 2 in personnel supervision (5/84).

Recommendation, AEC

In the vocational certificate category, 3 semester hours in aviation maintenance technology (or 1 semester hour in propeller and electrical systems, 1 in ground operation and servicing, and 1 in cleaning and corrosion control), and 3 semester hours in DC electrical fundamentals, 3 in electricity/electronics, 3 in electricity/electronics laboratory, 12 in aircraft electrical maintenance, 3 in aircraft electrical indicating systems, 3 in maintenance management, and 3 in shop management. In the lower-division baccalaureate/associate degree category, 3 semester hours in aviation maintenance technology (or 1 semester hour in aircraft propeller and electrical systems, 1 in ground operation and servicing, and 1 in cleaning and corrosion control), and 3 in DC electrical fundamentals, 3 in electricity/electronics, 3 in electricity/electronics laboratory, 12 in aircraft electrical maintenance, 3 in aircraft electrical indicating systems, 3 in personnel supervision, 3 in maintenance management, 3 in shop management, and 3 in principles of management (5/84).

Recommendation, AECS

In the lower-division baccalaureate/associate degree category, 3 semester hours in aviation maintenance technology (or 1 semester hour in aircraft propeller and electrical systems, 1 in ground operation and servicing, and 1 in cleaning and corrosion control), and 3 in DC electrical fundamentals, 3 in electricity/electronics, 3 in electricity/electronics laboratory, 12 in aircraft electrical maintenance, 3 in aircraft electrical indicating systems, 3 in personnel supervision, 3 in maintenance management, 3 in shop management, and 3 in principles of management. In the upper-division baccalaureate category, 3 semester hours for field experience in management and 3 in management problems (5/84).

Recommendation, AECM

In the lower-division baccalaureate/associate degree category, 3 semester hours in aviation maintenance technology (or 1 semester hour in aircraft propeller and electrical systems, 1 in ground operation and servicing, and 1 in cleaning and corrosion control), and 3 in DC electrical fundamentals, 3 in electricity/electronics, 3 in electricity/electronics laboratory, 12 in aircraft electrical maintenance, 3 in aircraft electrical indicating systems, 3 in personnel supervision, 3 in maintenance management, 3 in shop management, and 3 in principles of management. In the upper-division baccalaureate category, 6 semester hours for field experience in management and 3 in management problems (5/84).

CGR-AE-002

AVIATION ELECTRICIAN'S MATE
AE3
AE2
AE1
AEC
AECS
AECM

Exhibit Dates: 10/94–Present.

Occupational Group: V (Aviation).

Career Pattern

SN: Seaman (E-3). *AE3:* Aviation Electrician's Mate, Third Class (E-4). *AE2:* Aviation Electrician's Mate, Second Class (E-5). *AE1:* Aviation Electrician's Mate, First Class (E-6). *AEC:* Chief Aviation Electrician's Mate (E-7). *AECS:* Senior Chief Aviation Electrician's Mate (E-8). *AECM:* Master Chief Aviation Electrician's Mate (E-9).

Description

Summary: Handles and services aircraft; inspects and maintains aircraft electrical and instrument systems, including: power generation, conversion and distribution systems; interior and exterior lighting; electrical components of aircraft controls, including airframe, engine, propeller, and utility control systems; aircraft electrical starting systems, including starters, starting controls and ignition system components; aircraft engine, flight, and flight control instruments; instrument systems; and noninstrument type indicating warning systems; aircraft automatic flight control system, including automatic pilots, flight director systems, aircraft compasses, and altitude reference systems; aircraft batteries and related electrical components; uses computers to organize and maintain records. *AE3:* Observes safety precautions while working with energized and de-energized electrical and electronic equipment; checks electrical bonding and shock mountings for condition; replaces and adjusts electrical limit switches; troubleshoots and corrects DC electrical malfunctions; installs instrument indicator range marks; solves simple DC series, parallel and combination circuits; solves voltage divider and bridge circuits; troubleshoots a full-wave bridge PN junction diode vectifier with PL filter; traces signal flow in a common base, emitter, and collector transistor amplifier circuit; plots a frequency response curve for a direct RIC and transformer-coupled amplifier; acts as fire guard during start; ties down and secures aircraft. *AE2:* Able to perform the duties required for AE3; checks and operates ground support equipment and aircraft servicing equipment; supervises aircraft washing; tests, selects, and replaces circuit parts, including electron tubes and solid state devices; adjusts and aligns parts or circuitry; performs pitot-static tests; removes and replaces instrument indicators, transmitters, and sensors; tests fire detection systems; tests and calibrates flight instrument systems; calibrates fuel quantity, engine temperature, air data, and aircraft instrument systems; troubleshoots engine analyzer systems; performs tests on aircraft lighting systems; fabricates and installs aircraft electrical cables; tests power generating systems, aircraft batteries, circuit components, generator armatures, and voltage regulators; adjusts inverter speed control governors; tests hydraulic valve and actuator electrical components, aircraft pressurization and cabin temperature control, and fuel transfer systems; troubleshoots starting control electrical circuits; performs operational checks of aircraft electrical compass systems; swings and calibrates magnetic compasses; operationally checks automatic flight control and stabilization systems. *AE1:* Able to perform the duties required for AE2; supervises aircraft ground handling and servicing; calibrates engine temperature indicating systems, engine performance indicating and warning systems, air data systems, and aircraft instruments; diagnoses engine analyzer system malfunctions; inspects newly installed wiring on aircraft; makes three-phase wye and delta connections of transformers; diagnoses aircraft pressurization and cabin temperature control system malfunctions; maintains, tests, and adjusts electric and/or electronic control systems of aircraft automatic flight control and stabilization systems; troubleshoots automatic flight control and flight director systems. *AEC:* Able to perform the duties required for AE1; inspects work areas, tools, and aviation equipment to detect potentially hazardous and unsafe conditions and takes appropriate corrective action; prepares reports and maintains equipment records; supervises on-the-job training; prepares preventive maintenance schedules. *AECS:* Able to perform the duties required for AEC; serves as enlisted technical or specialty expert; plans, organizes, and directs work of personnel handling and servicing aircraft; plans and administers on-the-job training programs; supervises the preparation of reports. *AECM:* Able to perform the duties required for AECS; serves as senior enlisted technical or specialty administrator; manages personnel in the handling and servicing of aircraft; ensures maximum efficiency of the work force and equipment; prepares general correspondence concerning fiscal, supply, and administrative matters; assists in the formulation of plans, policies, and budget requirements; may supplement the officer corps in the overall supervision and administration of personnel and equipment; may also supervise personnel in other specialty areas.

Recommendation, AE3

In the lower-division baccalaureate/associate degree category, 1 semester hour in aircraft propeller and electrical systems, 1 in ground operation and servicing, 1 in cleaning and corrosion control, 3 in DC circuits, and 3 in introduction to computers (5/84).

Recommendation, AE2

In the lower-division baccalaureate/associate degree category, 1 semester hour in aircraft propeller and electrical systems, 1 in ground operation and servicing, 1 in cleaning and corrosion control, 3 in DC circuits, 3 in introduction to computers, 1 in AC circuits, 3 in electricity/electronics laboratory, 3 in electronic systems troubleshooting and maintenance, 1 in solid state electronics, 2 in aircraft instrumentation, and 3 in control systems (5/84).

Recommendation, AE1

In the lower-division baccalaureate/associate degree category, 1 semester hour in aircraft propeller and electrical systems, 1 in ground operation and servicing, 1 in cleaning and corrosion control, 3 in DC circuits, 3 in AC circuits, 3 in introduction to computers, 3 in electronic systems troubleshooting and maintenance, 3 in electricity/electronics laboratory, 1 in solid state electronics, 3 in aircraft instrumentation, 3 in control systems, and 2 in personnel supervision (5/84).

Recommendation, AEC

In the lower-division baccalaureate/associate degree category, 1 semester hour in aircraft propeller and electrical systems, 1 in ground operation and servicing, 1 in cleaning and corrosion control, 3 in DC circuits, 3 in AC circuits, 3 in introduction to computers, 3 electronic systems troubleshooting and maintenance, 1 in solid state electronics, 3 in aircraft instrumentation, 3 in control systems, 3 in electricity/electronics laboratory, 3 in personnel supervision, 3 in maintenance management, 3 in shop management, and 3 in organizational management (5/84).

Recommendation, AECS

In the lower-division baccalaureate/associate degree category, 1 semester hour in aircraft propeller and electrical systems, 1 in ground operation and servicing, 1 in cleaning and corrosion control, 3 in DC circuits, 1 in AC circuits, 3 in introduction to computers, 3 in electronic systems troubleshooting and maintenance, 3 in electricity/electronics laboratory, 1 in solid state electronics, 3 in aircraft instrumentation, 3 in control systems, 3 in personnel supervision, 3 in maintenance management, 3 in shop management, and 3 in organizational management. In the upper-division baccalaureate category, 3 semester hours for field experience in management and 3 in management problems (5/84).

Recommendation, AECM

In the lower-division baccalaureate/associate degree category, 1 semester hour in aircraft propeller and electrical systems, 1 in ground operation and servicing, 1 in cleaning and corrosion control, 3 in DC circuits, 3 in AC circuits, 3 in introduction to computers, 3 in electricity/electronics laboratory, 3 in electronic systems troubleshooting and maintenance, 1 in solid state electronics, 3 in aircraft instrumentation, 3 in control systems, 3 in personnel supervision, 3 in maintenance management, 3 in shop management, and 3 in organizational management. In the upper-division baccalaureate category, 6 semester hours for field experience in management and 3 in management problems (5/84).

CGR-AM-001

AVIATION STRUCTURAL MECHANIC
    AM3
    AM2
    AM1
    AMC
    AMCS
    AMCM

Exhibit Dates: 1/90–9/94.

Occupational Group: V (Aviation).

Career Pattern

SN: Seaman (E-3). *AM3:* Aviation Structural Mechanic, Third Class (E-4). *AM2:* Aviation Structural Mechanic, Second Class (E-5). *AM1:* Aviation Structural Mechanic, First Class (E-6). *AMC:* Chief Aviation Structural Mechanic (E-7). *AMCS:* Senior Chief Aviation Structural Mechanic (E-8). *AMCM:* Master Chief Aviation Structural Mechanic (E-9).

Description

Summary: Maintains aircraft fuselages, wings, fixed and movable control surfaces, and control mechanisms; removes, installs, and rigs flight control surfaces; fabricates and assembles metal parts; makes minor repairs to aircraft skin, performs non-destructive testing procedures using dye penetrant and zyglo methods; maintains hydraulic systems and components, landing gear, wheels and tires, brakes, and related components; maintains fire extinguishing systems plumbing; makes repairs to bladder and integral fuel systems; performs cable swedging; performs electric arc and gas welding operations; performs corrosion control procedures. *AM3:* Services tires and other compressed air systems; operates and maintains paint spray guns; performs corrosion control; applies paints and finishes on aircraft; performs helicopter preflight inspections; lays out patterns and templates for metal work; changes aircraft wheel and tire assemblies. *AM2:* Able to perform the duties required for AM3; maintains and operates ground support equipment; services fuel, oil, and hydraulic systems; performs preflight, throughflight, and postflight inspections; assembles and installs aircraft equipment and parts; replaces airframe pressure seals; performs metal repairs; fabricates and installs metal members of structure of aircraft; inspects and repairs integral and bladder-type fuel cells; services and maintains brake systems; services, maintains, and inspects landing gear controls, struts, locks, doors, and linkages. *AM1:* Able to perform the duties required for AM2; supervises aircraft servicing operations; performs quality assurance inspection of corrosion removal; supervises and interprets maintenance and use of maintenance directives; supervises inspections, maintenance, and repair of aircraft systems; rigs and adjusts landing gear nose wheel steering mechanisms; removes and replaces landing gear assemblies. *AMC:* Able to perform the duties required for AM1; serves as shop supervisor; supervises personnel in all categories of equipment maintenance and repair; prepares preventive maintenance schedules; organizes and maintains technical library; prepares and submits budget requests; plans and supervises on-the-job training. *AMCS:* Able to perform the duties required for AMC; serves as enlisted technical or specialty expert; plans, organizes, and directs the work of personnel maintaining aircraft structures; plans and administers on-the-job training programs; supervises the preparation of reports. *AMCM:* Able to perform the duties required for AMCS; serves as senior enlisted technical or specialty administrator; manages personnel in the maintenance of aircraft structures; ensures maximum efficiency of the work force and the equipment; prepares general correspondence concerning fiscal, supply, and administrative matters; assists in the formulation of plans, policies, and budget requirements; may supplement the officer corps in the overall supervision and administration of personnel and equipment; may also supervise personnel in other specialty areas.

Recommendation, AM3

In the lower-division baccalaureate/associate degree category, 3 semester hours in aviation maintenance technology, (or 1 semester hour in cleaning and corrosion control, 1 in ground operation and servicing, and 1 in flight control systems) (5/84).

Recommendation, AM2

In the lower-division baccalaureate/associate degree category, 3 semester hours in aviation maintenance technology (or 1 semester hour in cleaning and corrosion control, 1 in ground operation and servicing, and 1 in flight control systems), and 3 in aircraft fuel systems, 3 in landing gear systems, 3 in airframe structural repairs, 3 in assembly and rigging, and 3 in aircraft hydraulic systems (5/84).

Recommendation, AM1

In the vocational certificate category, 3 semester hours in aviation maintenance technology (or 1 semester hour in cleaning and corrosion control, 1 in ground operation and servicing, and 1 in flight control systems), and 3 in aircraft fuel systems, 3 in landing gear systems, 3 in airframe structural repairs, 3 in assembly and rigging, 3 in aircraft hydraulic systems, and 3 in airframe inspections. In the lower-division baccalaureate/associate degree category, 3 semester hours in aviation maintenance technology (or 1 semester hour in cleaning and corrosion control, 1 in ground operation and servicing, and 1 in flight control systems), and 3 in aircraft fuel systems, 3 in landing gear systems, 3 in airframe structural repair, 3 in assembly and rigging, 3 in aircraft hydraulic systems, 3 in airframe inspections, and 2 in personnel supervision (5/84).

Recommendation, AMC

In the vocational certificate category, 3 semester hours in aviation maintenance technology (or 1 semester hour in cleaning and corrosion control, 1 in ground operation and servicing, and 1 in flight control systems), and 3 in aircraft fuel systems, 3 in landing gear systems, 3 in airframe structural repair, 3 in assembly and rigging, 3 in aircraft hydraulic systems, 3 in airframe inspections, 3 in maintenance management, and 3 in shop management. In the lower-division baccalaureate/associate degree category, 3 semester hours in aviation maintenance technology (or 1 semester hour in cleaning and corrosion control, 1 in ground operation and servicing, and 1 in flight control systems), and 3 in aircraft fuel systems, 3 in landing gear systems, 3 in airframe structural repair, 3 in assembly and rigging, 3 in aircraft hydraulic systems, 3 in airframe inspections, 3 in person-

nel supervision, 3 in maintenance management, 3 in shop management, and 3 in principles of management (5/84).

Recommendation, AMCS

In the lower-division baccalaureate/associate degree category, 3 semester hours in aviation maintenance technology (or 1 semester hour in cleaning and corrosion control, 1 in ground operation and servicing, and 1 in flight control systems), and 3 in aircraft fuel systems, 3 in landing gear systems, 3 in airframe structural repair, 3 in assembly and rigging, 3 in aircraft hydraulic systems, 3 in airframe inspections, 3 in personnel supervision, 3 in maintenance management, 3 in shop management, and 3 in principles of management. In the upper-division baccalaureate category, 3 semester hours for field experience in management and 3 in management problems (5/84).

Recommendation, AMCM

In the lower-division baccalaureate/associate degree category, 3 semester hours in aviation maintenance technology (or 1 semester hour in cleaning and corrosion control, 1 in ground operation and servicing, and 1 in flight control systems), and 3 in aircraft fuel systems, 3 in landing gear systems, 3 in airframe structural repair, 3 in assembly and rigging, 3 in aircraft hydraulic systems, 3 in airframe inspections, 3 in personnel supervision, 3 in maintenance management, 3 in shop management, and 3 in principles of management. In the upper-division baccalaureate category, 6 semester hours for field experience in management and 3 in management problems (5/84).

CGR-AM-002

AVIATION STRUCTURAL MECHANIC
AM3
AM2
AM1
AMC
AMCS
AMCM

Exhibit Dates: 10/94–Present.

Occupational Group: V (Aviation).

Career Pattern

SN: Seaman (E-3). *AM3:* Aviation Structural Mechanic, Third Class (E-4). *AM2:* Aviation Structural Mechanic, Second Class (E-5). *AM1:* Aviation Structural Mechanic, First Class (E-6). *AMC:* Chief Aviation Structural Mechanic (E-7). *AMCS:* Senior Chief Aviation Structural Mechanic (E-8). *AMCM:* Master Chief Aviation Structural Mechanic (E-9).

Description

Summary: Maintains aircraft fuselages, wings, fixed and movable control surfaces, and control mechanisms; removes, installs, and rigs flight control surfaces; fabricates and assembles metal parts; makes minor repairs to aircraft skin, performs non-destructive testing procedures using dye penetrant and zyglo methods; maintains hydraulic systems and components, landing gear, wheels and tires, brakes, and related components; maintains fire extinguishing systems plumbing; makes repairs to bladder and integral fuel systems; performs cable swaging; performs electric arc and gas welding operations; performs corrosion control procedures; uses computers to organize and maintain records. *AM3:* Services tires and other compressed air systems; operates and maintains paint spray guns; performs corrosion control;

applies paints and finishes on aircraft; performs helicopter preflight inspections; lays out patterns and templates for metal work; changes aircraft wheel and tire assemblies. *AM2:* Able to perform the duties required for AM3; maintains and operates ground support equipment; services fuel, oil, and hydraulic systems; performs preflight, throughflight, and postflight inspections; assembles and installs aircraft equipment and parts; replaces airframe pressure seals; performs metal repairs; fabricates and installs metal members of structure of aircraft; inspects and repairs integral and bladder-type fuel cells; services and maintains brake systems; services, maintains, and inspects landing gear controls, struts, locks, doors, and linkages. *AM1:* Able to perform the duties required for AM2; supervises aircraft servicing operations; performs quality assurance inspection of corrosion removal; supervises and interprets maintenance and use of maintenance directives; supervises inspections, maintenance, and repair of aircraft systems; rigs and adjusts landing gear nose wheel steering mechanisms; removes and replaces landing gear assemblies. *AMC:* Able to perform the duties required for AM1; serves as shop supervisor; supervises personnel in all categories of equipment maintenance and repair; prepares preventive maintenance schedules; organizes and maintains technical library; prepares and submits budget requests; plans and supervises on-the-job training. *AMCS:* Able to perform the duties required for AMC; serves as enlisted technical or specialty expert; plans, organizes, and directs the work of personnel maintaining aircraft structures; plans and administers on-the-job training programs; supervises the preparation of reports. *AMCM:* Able to perform the duties required for AMCS; serves as senior enlisted technical or specialty administrator; manages personnel in the maintenance of aircraft structures; ensures maximum efficiency of the work force and the equipment; prepares general correspondence concerning fiscal, supply, and administrative matters; assists in the formulation of plans, policies, and budget requirements; may supplement the officer corps in the overall supervision and administration of personnel and equipment; may also supervise personnel in other specialty areas.

Recommendation, AM3

In the lower-division baccalaureate/associate degree category, 1 semester hour in cleaning and corrosion control, 1 in ground operation and servicing, 1 in flight control systems, and 3 in introduction to computers (6/95).

Recommendation, AM2

In the lower-division baccalaureate/associate degree category, 1 semester hour in cleaning and corrosion control, 1 in ground operation and servicing, 1 in flight control systems, 3 in introduction to computers, 3 in aircraft fuel systems, 3 in landing gear systems, 3 in airframe structural repairs, 3 in assembly and rigging, and 3 in aircraft hydraulic systems (6/95).

Recommendation, AM1

In the lower-division baccalaureate/associate degree category, 1 semester hour in cleaning and corrosion control, 1 in ground operation and servicing, 1 in flight control systems, 3 in introduction to computers, 3 in aircraft fuel systems, 3 in landing gear systems, 3 in airframe structural repair, 3 in assembly and rigging, 3 in aircraft hydraulic systems, 3 in airframe inspections, and 2 in personnel supervision (6/95).

Recommendation, AMC

In the lower-division baccalaureate/associate degree category, 1 semester hour in cleaning and corrosion control, 1 in ground operation and servicing, 1 in flight control systems, 3 in introduction to computers, 3 in aircraft fuel systems, 3 in landing gear systems, 3 in airframe structural repair, 3 in assembly and rigging, 3 in aircraft hydraulic systems, 3 in airframe inspections, 3 in personnel supervision, 3 in maintenance management, 3 in shop management, and 3 in organizational management (6/95).

Recommendation, AMCS

In the lower-division baccalaureate/associate degree category, 1 semester hour in cleaning and corrosion control, 1 in ground operation and servicing, 1 in flight control systems, 3 in introduction to computers, 3 in aircraft fuel systems, 3 in landing gear systems, 3 in airframe structural repair, 3 in assembly and rigging, 3 in aircraft hydraulic systems, 3 in airframe inspections, 3 in personnel supervision, 3 in maintenance management, 3 in shop management, and 3 in organizational management. In the upper-division baccalaureate category, 3 semester hours for field experience in management and 3 in management problems (6/95).

Recommendation, AMCM

In the lower-division baccalaureate/associate degree category, 1 semester hour in cleaning and corrosion control, 1 in ground operation and servicing, 1 in flight control systems, 3 in introduction to computers, 3 in aircraft fuel systems, 3 in landing gear systems, 3 in airframe structural repair, 3 in assembly and rigging, 3 in aircraft hydraulic systems, 3 in airframe inspections, 3 in personnel supervision, 3 in maintenance management, 3 in shop management, and 3 in organizational management. In the upper-division baccalaureate category, 6 semester hours for field experience in management and 3 in management problems (6/95).

CGR-ASM-001

AVIATION SURVIVALMAN
ASM3
ASM2
ASM1
ASMC
ASMCS
ASMCM

Exhibit Dates: 1/90–9/94.

Occupational Group: V (Aviation).

Career Pattern

SN: Seaman (E-3). ASM3: Aviation Survivalman, Third Class (E-4). ASM2: Aviation Survivalman, Second Class (E-5). ASM1: Aviation Survivalman, First Class (E-6). ASMC: Chief Aviation Survivalman (E-7). ASMCS: Senior Chief Aviation Survivalman (E-8). ASMCM: Master Chief Aviation Survivalman (E-9).

Description

Summary: Inspects, maintains, and repairs parachutes, survival equipment, flight and protective clothing, small arms, oxygen and fire extinguisher systems; instructs aviation personnel in the use and operation of parachutes, survival equipment and techniques, safety practices, and aviation ordnance handling; qualified in aircraft handling and aircraft handling equipment; practices and observes all safety precautions and practices. *ASM3:* Inspects, repairs, packs, and maintains personnel and cargo delivery parachutes; understands parachute loft working procedures; uses hand stitching and sewing machine in making fabric repairs; makes minor repairs and adjustments to sewing machines; inspects survival equipment, salvage pumps, oxygen masks, and ordnance and pyrotechnics; inspects and services inflatable survival equipment; services oxygen systems and components; understands aircraft handling and procedures; inspects and maintains small arms and other aviation ordnance. *ASM2:* Able to perform the duties required for ASM3; instructs aviation personnel in survival equipment, parachutes, oxygen, and fire extinguisher systems; tows, taxis, and secures aircraft; serves as a member of an aircraft handling team; accomplishes preflight and postflight servicing of aircraft systems and equipment; identifies aircraft corrosion factors; prepares survival training programs; uses tools and precision measuring equipment and related electrical testing devices; uses related drawings, blueprints, and manuals. *ASM1:* Able to perform the duties required for ASM2; instructs in procedures for rigging and launching survival equipment; repairs and calibrates special tools; uses and maintains records, forms, publications, and manuals; performs quality assurance; supervises shop procedures; repairs sewing machines, and determines reparability of related equipment; maintains custody of small arms. *ASMC:* Able to perform the duties required for ASM1; trains and supervises personnel in all categories of equipment operation, maintenance, and repair; prepares maintenance schedules; organizes and maintains a technical library; insures safe working conditions; plans and supervises training programs. *ASMCS:* Able to perform the duties required for ASMC; serves as enlisted technical or specialty expert; plans, organizes, and directs the work of personnel operating, servicing, and maintaining parachutes, survival and rescue equipment, related aircraft systems, small arms, aviation ordnance, preflight and post-flight inspections; analyzes system malfunctions; plans and administers on-the-job training programs; supervises the preparation of reports; uses the computerized maintenance system and establishes safety procedures and practices. *ASMCM:* Able to perform the duties required for ASMCS; serves as senior enlisted technical or specialty administrator; manages personnel in the operation, maintenance, and procurement of related equipment; ensures maximum efficiency of the work force and equipment.

Recommendation, ASM3

In the lower-division baccalaureate/associate degree category, 3 semester hours in aviation maintenance technology (or 1 semester hour in cleaning and corrosion control, 1 in ground operation and servicing, and 1 in aircraft systems) (5/84).

Recommendation, ASM2

In the lower-division baccalaureate/associate degree category, 3 semester hours in aviation maintenance technology (or 1 semester hour in cleaning and corrosion control, 1 in ground operation and servicing, and 1 in aircraft systems), and 3 in aircraft system technology and 3 in parachute inspection and rigging (5/84).

Recommendation, ASM1

In the vocational certificate category, 3 semester hours in aviation maintenance technology (or 1 semester hour in cleaning and corrosion control, 1 in ground operation and servicing, and 1 in aircraft systems), and 3 in aircraft system technology, 3 in parachute inspection and rigging, 3 in industrial safety, and 3 in textile machine repair. In the lower-division baccalaureate/associate degree category, 3 semester hours in aviation maintenance technology (or 1 semester hour in cleaning and corrosion control, 1 in ground operation and servicing, and 1 in aircraft systems), and 3 in aircraft system technology, 3 in parachute inspection and rigging, 3 in industrial safety, 3 in textile machine repair, and 2 in personnel supervision (5/84).

Recommendation, ASMC

In the vocational certificate category, 3 semester hours in aviation maintenance technology (or 1 semester hour in cleaning and corrosion control, 1 in ground operation and servicing, and 1 in aircraft systems), and 3 in aircraft system technology, 3 in parachute inspection and rigging, 3 in industrial safety, 3 in textile machine repair, 3 in maintenance management, and 3 in shop management. In the lower-division baccalaureate/associate degree category, 3 semester hours in aviation maintenance technology (or 1 semester hour in cleaning and corrosion control, 1 in ground operation and servicing, and 1 in aircraft systems), and 3 in aircraft system technology, 3 in parachute inspection and rigging, 3 in industrial safety, 3 in textile machine repair, 3 in personnel supervision, 3 in maintenance management, 3 in shop management, and 3 in principles of management (5/84).

Recommendation, ASMCS

In the lower-division baccalaureate/associate degree category, 3 semester hours in aviation maintenance technology (or 1 semester hour in cleaning and corrosion control, 1 in ground operation and servicing, and 1 in aircraft systems), and 3 in aircraft system technology, 3 in parachute inspection and rigging, 3 in industrial safety, 3 in textile machine repair, 3 in personnel supervision, 3 in maintenance management, 3 in shop management, and 3 in principles of management. In the upper-division baccalaureate category, 3 semester hours for field experience in management and 3 in management problems (5/84).

Recommendation, ASMCM

In the lower-division baccalaureate/associate degree category, 3 semester hours in aviation maintenance technology (or 1 semester hour in cleaning and corrosion control, 1 in ground operation and servicing, and 1 in aircraft systems), and 3 in aircraft system technology, 3 in parachute inspection and rigging, 3 in industrial safety, 3 in textile machine repair, 3 in personnel supervision, 3 in maintenance management, 3 in shop management, and 3 in principles of management. In the upper-division baccalaureate category, 6 semester hours for field experience in management and 3 in management problems (5/84).

CGR-ASM-002

AVIATION SURVIVALMAN
    ASM3
    ASM2
    ASM1
    ASMC
    ASMCS
    ASMCM

Exhibit Dates: 10/94–Present.

Occupational Group: V (Aviation).

Career Pattern

SN: Seaman (E-3). *ASM3:* Aviation Survivalman, Third Class (E-4). *ASM2:* Aviation Survivalman, Second Class (E-5). *ASM1:* Aviation Survivalman, First Class (E-6). *ASMC:* Chief Aviation Survivalman (E-7). *ASMCS:* Senior Chief Aviation Survivalman (E-8). *ASMCM:* Master Chief Aviation Survivalman (E-9).

Description

Summary: Inspects, maintains, and repairs parachutes, survival equipment, flight and protective clothing, small arms, oxygen and fire extinguisher systems; instructs aviation personnel in the use and operation of parachutes, survival equipment and techniques, safety practices, and aviation ordnance handling; qualified in aircraft handling and aircraft handling equipment; practices and observes all safety precautions and practices; prepares and delivers formal presentations to a wide variety if civilian and military audiences; practices rescue swimmer techniques; uses computers to organize and maintain records. *ASM3:* Inspects, repairs, packs, and maintains personnel and cargo delivery parachutes; understands parachute loft working procedures; uses hand stitching and sewing machine in making fabric repairs; makes minor repairs and adjustments to sewing machines; inspects survival equipment, salvage pumps, oxygen masks, and ordnance and pyrotechnics; inspects and services inflatable survival equipment; services oxygen systems and components; understands aircraft handling and procedures; inspects and maintains small arms and other aviation ordnance. *ASM2:* Able to perform the duties required for ASM3; instructs aviation personnel in survival equipment, parachutes, oxygen, and fire extinguisher systems; tows, taxis, and secures aircraft; serves as a member of an aircraft handling team; accomplishes preflight and postflight servicing of aircraft systems and equipment; identifies aircraft corrosion factors; prepares survival training programs; uses tools and precision measuring equipment and related electrical testing devices; uses related drawings, blueprints, and manuals. *ASM1:* Able to perform the duties required for ASM2; instructs in procedures for rigging and launching survival equipment; repairs and calibrates special tools; uses and maintains records, forms, publications, and manuals; performs quality assurance; supervises shop procedures; repairs sewing machines, and determines reparability of related equipment; maintains custody of small arms. *ASMC:* Able to perform the duties required for ASM1; trains and supervises personnel in all categories of equipment operation, maintenance, and repair; prepares maintenance schedules; organizes and maintains a technical library; insures safe working conditions; plans and supervises training programs. *ASMCS:* Able to perform the duties required for ASMC; serves as enlisted technical or specialty expert; plans, organizes, and directs the work of personnel operating, servicing, and maintaining parachutes, survival and rescue equipment, related aircraft systems, small arms, aviation ordnance, preflight and post-flight inspections; analyzes system malfunctions; plans and administers on-the-job training programs; supervises the preparation of reports; uses the computerized maintenance system and establishes safety procedures and practices. *ASMCM:* Able to perform the duties required for ASMCS; serves as senior enlisted technical or specialty administrator; manages personnel in the operation, maintenance, and procurement of related equipment; ensures maximum efficiency of the work force and equipment.

Recommendation, ASM3

In the lower-division baccalaureate/associate degree category, 1 semester hour in cleaning and corrosion control, 1 in ground operation and servicing, 1 in aircraft systems, 2 in advanced swimming, 3 in public speaking, and 3 in introduction to computers (6/95).

Recommendation, ASM2

In the lower-division baccalaureate/associate degree category, 1 semester hour in cleaning and corrosion control, 1 in ground operation and servicing, 1 in aircraft systems, 2 in advanced swimming, 3 in public speaking, 3 in introduction to computers, 3 in aircraft life support system maintenance, and 3 in parachute inspection and rigging (6/95).

Recommendation, ASM1

In the lower-division baccalaureate/associate degree category, 1 semester hour in cleaning and corrosion control, 1 in ground operation and servicing, 1 in aircraft oxygen systems, 3 in introduction to computers, 3 in public speaking, 2 in advanced swimming, 3 in aircraft life support system maintenance, 3 in parachute inspection and rigging, 3 in industrial safety, 3 in textile machine repair, and 2 in personnel supervision (6/95).

Recommendation, ASMC

In the lower-division baccalaureate/associate degree category, 1 semester hour in cleaning and corrosion control, 1 in ground operation and servicing, 1 in aircraft oxygen systems, 3 in introduction to computers, 3 in public speaking, 2 in advanced swimming, 3 in aircraft life support system maintenance, 3 in parachute inspection and rigging, 3 in industrial safety, 3 in textile machine repair, 3 in personnel supervision, 3 in maintenance management, 3 in shop management, and 3 in organizational management (6/95).

Recommendation, ASMCS

In the lower-division baccalaureate/associate degree category, 1 semester hour in cleaning and corrosion control, 1 in ground operation and servicing, 1 in aircraft oxygen systems, 3 in introduction to computers, 3 in public speaking, 2 in advanced swimming, 3 in aircraft life support system maintenance, 3 in parachute inspection and rigging, 3 in industrial safety, 3 in textile machine repair, 3 in personnel supervision, 3 in maintenance management, 3 in shop management, and 3 in organizational management. In the upper-division baccalaureate category, 3 semester hours for field experience in management and 3 in management problems (6/95).

Recommendation, ASMCM

In the lower-division baccalaureate/associate degree category, 1 semester hour in cleaning and corrosion control, 1 in ground operation and servicing, 1 in aircraft oxygen systems, 3 in introduction to computers, 3 in public speaking, 2 in advanced swimming, 3 in aircraft life support system maintenance, 3 in parachute inspection and rigging, 3 in industrial safety, 3 in textile machine repair, 3 in personnel supervision, 3 in maintenance management, 3 in shop management, and 3 in organizational management. In the upper-division baccalaureate category, 6 semester hours for field experience in management and 3 in management problems (6/95).

CGR-AT-001

AVIATION ELECTRONICS TECHNICIAN
    AT3
    AT2
    AT1
    ATC
    ATCS
    ATCM

Exhibit Dates: 1/90–9/94.

Occupational Group: V (Aviation).

Career Pattern

SN: Seaman (E-3). *AT3:* Aviation Electronics Technician, Third Class (E-4). *AT2:* Aviation Electronics Technician, Second Class (E-5). *AT1:* Aviation Electronics Technician, First Class (E-6). *ATC:* Chief Aviation Electronics Technician (E-7). *ATCS:* Senior Chief Aviation Electronics Technician (E-8). *ATCM:* Master Chief Aviation Electronics Technician (E-9).

Description

Summary: Handles and services aircraft; performs related aircraft inspections, line maintenance and aircraft changes; operates, maintains, repairs, calibrates, tunes, and aligns avionics systems (communications, navigation, and microwave equipment); makes authorized repairs and adjustments to associated test equipment. *AT3:* Understands electrical and electronic theory, conductors, insulators, resistors, capacitors, and inductors; solves problems using Ohm's Law, basic meter movements including ammeters, voltmeters and ohmmeters; understands AC theory, solid state devices, power supply devices, amplifiers, oscillators, filters, modulation integrated circuits, synchro and servo systems, microphones and head sets; uses binary conversions, Boolean algebra, and truth tables, AND and NAND gates, OR and NOR inverters and memory circuits and multivibrator; sets and adjusts tranceivers; operates communications equipment; performs preventive maintenance, basic troubleshooting, and minor repairs on communication and navigation systems; locates defective circuits and makes electrical connections by soldering and splicing; observes safety procedures and precautions on or around aircraft; eliminates electrical hazards; handles and services aircraft; identifies corrosion problems; operates common aircraft safety devices and uses related drawings, blueprints, and manuals. *AT2:* Able to perform the duties required for AT3; serves as a member of an aircraft handling team; operates ground service equipment; accomplishes normal postflight servicing on aircraft; performs corrosion inspections; demonstrates use of survival and emergency equipment; uses related tools and precision measuring devices; accomplishes equipment changes; fabricates electronic cables; operates, repairs, adjusts, aligns, and calibrates communication and navigation equipment. *AT1:* Able to perform the duties required for AT2; establishes safeguards; supervises aircraft ground handling; supervises aircraft servicing operations; analyzes test equipment for defects and determines corrective action; performs quality control inspections and evaluations; supervises an electronic shop; sets compensation of loop antennas; maintains files, records, publications, reports, and directives. *ATC:* Able to perform the duties required for AT1; inspects work areas, tools, and aviation equipment to detect potentially hazardous and unsafe conditions and takes appropriate corrective action; prepares tool, equipment, and material evaluation reports; maintains aircraft, ground support equipment, and other aviation equipment records; supervises assignment and training programs to ensure that personnel are assigned and trained for maximum utilization; supervises and trains personnel in the maintenance and repair of aeronautical components. *ATCS:* Able to perform the duties required for ATC; serves as enlisted technical or specialty expert; plans, organizes, and directs work of personnel handling and servicing aircraft; plans and administers on-the-job training programs; supervises the preparation of reports. *ATCM:* Able to perform the duties required for ATCS; serves as senior enlisted technical or specialty administrator; manages personnel in the handling and servicing of aircraft; ensures maximum efficiency of the work force and equipment; prepares general correspondence concerning fiscal, supply, and administrative matters; assists in the formulation of plans, policies, and budget requirements; may supplement the officer corps in the overall supervision and administration of personnel and equipment; may also supervise personnel in other specialty areas.

Recommendation, AT3

In the lower-division baccalaureate/associate degree category, 3 semester hours in DC electrical fundamentals, 3 in AC electrical fundamentals, and 3 in navigation and communication systems (5/84).

Recommendation, AT2

In the lower-division baccalaureate/associate degree category, 3 semester hours in aviation maintenance technology (or 1 semester hour in cleaning and corrosion control, 1 in ground operation and servicing, and 1 in aircraft electrical systems), and 3 in DC electrical fundamentals, 3 in AC electrical fundamentals, 3 in navigation and communication systems, 3 in circuit theory, 3 in electrical/electronics laboratory, and 6 in navigation and communication system maintenance (5/84).

Recommendation, AT1

In the vocational certificate category, 3 semester hours in aviation maintenance technology (or 1 semester hour in cleaning and corrosion control, 1 in ground operation and servicing, and 1 in aircraft electrical systems), and 3 in DC electrical fundamentals, 3 in AC electrical fundamentals, 3 in navigation and communication systems, 3 in circuit theory, 3 in electrical/electronics laboratory, 6 in navigation and communication system maintenance, and 6 in electrical systems maintenance. In the lower-division baccalaureate/associate degree category, 3 semester hours in aviation maintenance technology (or 1 semester hour in cleaning and corrosion control, 1 in ground operation and servicing, and 1 in aircraft electrical systems), and 3 in DC electrical fundamentals, 3 in AC electrical fundamentals, 3 in navigation and communication systems, 3 in circuit theory, 3 in electrical/electronics laboratory, 6 in navigation and communication systems maintenance, 6 in electrical systems maintenance, and 2 in personnel supervision (5/84).

Recommendation, ATC

In the vocational certificate category, 3 semester hours in aviation maintenance technology (or 1 semester hour in cleaning and corrosion control, 1 in ground operation and servicing, and 1 in aircraft electrical systems), and 3 in DC electrical fundamentals, 3 in AC electrical fundamentals, 3 in navigation and communication systems, 3 in circuit theory, 3 in electrical/electronics laboratory, 6 in navigation and communication systems maintenance, 6 in electrical systems maintenance, 3 in maintenance management, and 3 in shop management. In the lower-division baccalaureate/associate degree category, 3 semester hours in aviation maintenance technology (or 1 semester hour in cleaning and corrosion control, 1 in ground operation and servicing, and 1 in aircraft electrical systems), and 3 in DC electrical fundamentals, 3 in AC electrical fundamentals, 3 in navigation and communication systems, 3 in circuit theory, 3 in electrical/electronics laboratory, 6 in navigation and communication systems maintenance, 6 in electrical systems maintenance, 3 in personnel supervision, 3 in maintenance management, 3 in shop management, and 3 in principles of management (5/84).

Recommendation, ATCS

In the lower-division baccalaureate/associate degree category, 3 semester hours in aviation maintenance technology (or 1 semester hour in cleaning and corrosion control, 1 in ground operation and servicing, and 1 in aircraft electrical systems), and 3 in DC electrical fundamentals, 3 in AC electrical fundamentals, 3 in navigation and communication systems, 3 in circuit theory, 3 in electrical/electronics laboratory, 6 in navigation and communication systems maintenance, 6 in electrical systems maintenance, 3 in personnel supervision, 3 in maintenance management, and 3 in principles of management. In the upper-division baccalaureate category, 3 semester hours for field experience in management and 3 in management problems (5/84).

Recommendation, ATCM

In the lower-division baccalaureate/associate degree category, 3 semester hours in aviation maintenance technology (or 1 semester hour in cleaning and corrosion control, 1 in ground operation and servicing, and 1 in aircraft electrical systems), and 3 in DC electrical fundamentals, 3 in AC electrical fundamentals, 3 in navigation and communication systems, 3 in circuit theory, 3 in electrical/electronics laboratory, 6 in navigation and communication systems maintenance, 6 in electrical systems maintenance, 3 in personnel supervision, 3 in maintenance management, and 3 in principles of management. In the upper-division baccalaureate category, 6 semester hours for field experience in management and 3 in management problems (5/84).

CGR-AT-002

AVIATION ELECTRONICS TECHNICIAN
AT3
AT2
AT1
ATC
ATCS
ATCM

Exhibit Dates: 10/94–Present.

Occupational Group: V (Aviation).

Career Pattern

SN: Seaman (E-3). *AT3:* Aviation Electronics Technician, Third Class (E-4). *AT2:* Aviation Electronics Technician, Second Class (E-5). *AT1:* Aviation Electronics Technician, First Class (E-6). *ATC:* Chief Aviation Electronics Technician (E-7). *ATCS:* Senior Chief Aviation Electronics Technician (E-8). *ATCM:* Master Chief Aviation Electronics Technician (E-9).

Description

Summary: Handles and services aircraft; performs related aircraft inspections, line maintenance and aircraft changes; operates, maintains, repairs, calibrates, tunes, and aligns avionics systems (communications, navigation, and microwave equipment); makes authorized repairs and adjustments to associated test equipment; uses computers to organize and maintain records. *AT3:* Understands electrical and electronic theory, conductors, insulators, resistors, capacitors, and inductors; solves problems using Ohm's law, basic meter movements including ammeters, voltmeters and ohmmeters; understands AC theory, solid state devices, power supply devices, amplifiers, oscillators, filters, modulation integrated circuits, synchro and servo systems, microphones and head sets; uses binary conversions, Boolean algebra, and truth tables, AND and NAND gates, OR and NOR inverters and memory circuits and multivibrator; sets and adjusts tranceivers; operates communications equipment; performs preventive maintenance, basic troubleshooting, and minor repairs on communication and navigation systems; locates defective circuits and makes electrical connections by soldering and splicing; observes safety procedures and precautions on or around aircraft; eliminates electrical hazards; handles and services aircraft; identifies corrosion problems; operates common aircraft safety devices and uses related drawings, blueprints, and manuals. *AT2:* Able to perform the duties required for AT3; serves as a member of an aircraft handling team; operates ground service equipment; accomplishes normal postflight servicing on aircraft; performs corrosion inspections; demonstrates use of survival and emergency equipment, uses related tools and precision measuring devices; accomplishes equipment changes; fabricates electronic cables; operates, repairs, adjusts, aligns, and calibrates communication and navigation equipment. *AT1:* Able to perform the duties required for AT2; establishes safeguards; supervises aircraft ground handling; supervises aircraft servicing operations; analyzes test equipment for defects and determines corrective action; performs quality control inspections and evaluations; supervises an electronic shop; sets compensation of loop antennas; maintains files, records, publications, reports, and directives. *ATC:* Able to perform the duties required for AT1; inspects work areas, tools, and aviation equipment to detect potentially hazardous and unsafe conditions and takes appropriate corrective action; prepares tool, equipment, and material evaluation reports; maintains aircraft, ground support equipment, and other aviation equipment records; supervises assignment and training programs to ensure that personnel are assigned and trained for maximum utilization; supervises and trains personnel in the maintenance and repair of aeronautical components. *ATCS:* Able to perform the duties required for ATC; serves as enlisted technical or specialty expert; plans, organizes, and directs work of personnel handling and servicing aircraft; plans and administers on-the-job training programs; supervises the preparation of reports. *ATCM:* Able to perform the duties required for ATCS; serves as senior enlisted technical or specialty administrator; manages personnel in the handling and servicing of aircraft; ensures maximum efficiency of the work force and equipment; prepares general correspondence concerning fiscal, supply, and administrative matters; assists in the formulation of plans, policies, and budget requirements; may supplement the officer corps in the overall supervision and administration of personnel and equipment; may also supervise personnel in other specialty areas.

Recommendation, AT3

In the lower-division baccalaureate/associate degree category, 3 semester hours in DC circuits, 3 in AC circuits, 3 in introduction to computers, and 3 in navigation and communication systems (6/95).

Recommendation, AT2

In the lower-division baccalaureate/associate degree category, 1 semester hour in cleaning and corrosion control, 1 in ground operation and servicing, 1 in aircraft electrical systems, 3 in DC circuits, 3 in AC circuits, 3 in introduction to computers, 3 in navigation and communication systems, 3 in electrical/electronics laboratory, and 6 in navigation and communication system maintenance (6/95).

Recommendation, AT1

In the lower-division baccalaureate/associate degree category, 1 semester hour in cleaning and corrosion control, 1 in ground operation and servicing, 1 in aircraft electrical systems, 3 in DC circuits, 3 in AC circuits, 3 in introduction to computers, 3 in navigation and communication systems, 3 in electrical/electronics laboratory, 6 in navigation and communication systems maintenance, 6 in electrical system maintenance, and 2 in personnel supervision (6/95).

Recommendation, ATC

In the lower-division baccalaureate/associate degree category, 1 semester hour in cleaning and corrosion control, 1 in ground operation and servicing, 1 in aircraft electrical systems, 3 in DC circuits, 3 in AC circuits, 3 in introduction to computers, 3 in navigation and communication systems, 3 in electrical/electronics laboratory, 6 in navigation and communication systems maintenance, 6 in electrical systems maintenance, 3 in personnel supervision, 3 in maintenance management, 3 in shop management, and 3 in organizational management (6/95).

Recommendation, ATCS

In the lower-division baccalaureate/associate degree category, 1 semester hour in cleaning and corrosion control, 1 in ground operation and servicing, 1 in aircraft electrical systems, 3 in DC circuits, 3 in AC circuits, 3 in introduction to computers, 3 in navigation and communication systems, 3 in electrical/electronics laboratory, 6 in navigation and communication systems maintenance, 6 in electrical systems maintenance, 3 in personnel supervision, 3 in maintenance management, 3 in shop management, and 3 in organizational management. In the upper-division baccalaureate category, 3 semester hours for field experience in management and 3 in management problems (6/95).

Recommendation, ATCM

In the lower-division baccalaureate/associate degree category, 1 semester hour in cleaning and corrosion control, 1 in ground operation and servicing, 1 in aircraft electrical systems, 3 in DC circuits, 3 in AC circuits, 3 in introduction to computers, 3 in navigation and communication systems, 3 in electrical/electronics laboratory, 6 in navigation and communication systems maintenance, 6 in electrical systems maintenance, 3 in personnel supervision, 3 in maintenance management, 3 in shop management, and 3 in organizational management. In the upper-division baccalaureate category, 6 semester hours for field experience in management and 3 in management problems (6/95).

CGR-BM-001

BOATSWAIN'S MATE
BM3
BM2
BM1
BMC
BMCS
BMCM

Exhibit Dates: 1/90–9/94.

Occupational Group: I (Deck).

Career Pattern

SN: Seaman (E-3). *BM3:* Boatswain's Mate, Third Class (E-4). *BM2:* Boatswain's Mate, Second Class (E-5). *BM1:* Boatswain's Mate, First Class (E-6). *BMC:* Chief Boatswain's Mate (E-7). *BMCS:* Senior Chief Boatswain's Mate (E-8). *BMCM:* Master Chief Boatswain's Mate (E-9).

Description

Summary: Proficient in marlinspike, deck, and boat seamanship; manages, supervises, and administers personnel and facilities; serves as deck watch officer and boat coxswain; trains, directs, and supervises personnel in seamanship, maintenance, rigging, deck equipment and boats; supervises damage control and working parties; maintains discipline; serves as member of gun crews or gun mount captains; operates and maintains heavy equipment used in loading and buoy tending work. *BM3:* Participates as crewman aboard thirty-foot or larger boat during various standard and emergency situations; performing as a coxswain, maneuvers alongside a vessel underway, and docks under average conditions; demonstrates proper procedures for dealing with lawbreakers; interviews suspects and witnesses; collects evidence; demonstrates procedures for search and seizure. *BM2:* Able to perform the duties required for BM3; overhauls blocks; operates sailmaker equipment; performs decorative rope work; makes preparations for refueling and replenishment at sea. *BM1:* Able to perform the duties required for BM2; supervises individual work parties; supervises loading, storage and discharge of cargo and supplies; supervises anchor detail. *BMC:* Able to perform the duties required for BM1; supervises complete replenishment and refueling operations; acts as on-scene commander and/or search and rescue mission coordinator. *BMCS:* Able to perform the duties required for BMC; serves as enlisted technical or specialty expert; plans, organizes, and directs work of personnel; supervises the preparation of reports; plans and administers on-the-job training programs. *BMCM:* Able to perform the duties required for BMCS; serves as senior enlisted technical or specialty administrator; ensures maximum efficiency of the work force; prepares general correspondence concerning fiscal, supply, and administrative matters; assists in the formulation of plans, policies, and budget requirements; may supplement the officer corps in the overall supervision and administration of personnel and equipment; may also supervise personnel in other specialty areas.

Recommendation, BM3

In the vocational certificate category, 1 semester hour in rigging. In the lower-division baccalaureate/associate degree category, 3 semester hours in vessel maintenance and 3 in seamanship; if qualification code was HW (Surfman) or HX (Boat Coxswain), credit in small boat handling on the basis of institutional evaluation (5/84).

Recommendation, BM2

In the vocational certificate category, 4 semester hours in rigging. In the lower-division baccalaureate/associate degree category, 6 semester hours in vessel maintenance, 6 in seamanship, 1 in general office procedures, and 3 in law enforcement; if qualification code was HW (Surfman) or HX (Boat Coxswain), credit in small boat handling on the basis of institutional evaluation (5/84).

Recommendation, BM1

In the vocational certificate category, 4 semester hours in rigging. In the lower-division baccalaureate/associate degree category, 6 semester hours in vessel maintenance, 6 in seamanship, 1 in general office procedures, 3 in law enforcement, and 2 in personnel supervision; if qualification code was HW (Surfman) or HX (Boat Coxswain), credit in small boat handling on the basis of institutional evaluation (5/84).

Recommendation, BMC

In the vocational certificate category, 4 semester hours in rigging. In the lower-division baccalaureate/associate degree category, 6 semester hours in vessel maintenance, 6 in seamanship, 1 in general office procedures, 3 in law enforcement, 3 in personnel supervision, and 3 in principles of management; if qualification code was HW (Surfman) or HX (Boat Coxswain), credit in small boat handling on the basis of institutional evaluation (5/84).

Recommendation, BMCS

In the vocational certificate category, 4 semester hours in rigging. In the lower-division baccalaureate/associate degree category, 6 semester hours in vessel maintenance, 6 in seamanship, 1 in general office procedures, 3 in law enforcement, 3 in personnel supervision, and 3 in principles of management; if qualification code was HW (Surfman) or HX (Boat Coxswain), credit in small boat handling on the basis of institutional evaluation. In the upper-division baccalaureate category, 3 semester hours for field experience in management and 3 in management problems (5/84).

Recommendation, BMCM

In the vocational certificate category, 4 semester hours in rigging. In the lower-division baccalaureate/associate degree category, 6 semester hours in vessel maintenance, 6 in seamanship, 1 in general office procedures, 3 in law enforcement, 3 in personnel supervision, and 3 in principles of management; if qualification code was HW (Surfman) or HX (Boat Coxswain), credit in small boat handling on the basis of institutional evaluation. In the upper-division baccalaureate category, 6 semester hours for field experience in management and 3 in management problems (5/84).

CGR-BM-002

BOATSWAIN'S MATE
BM3
BM2
BM1
BMC
BMCS
BMCM

Exhibit Dates: 10/94–Present.

Occupational Group: Nautical Operations.

Career Pattern

SN: Seaman (E-3). *BM3:* Boatswain's Mate, Third Class (E-4). *BM2:* Boatswain's Mate, Second Class (E-5). *BM1:* Boatswain's Mate, First Class (E-6). *BMC:* Chief Boatswain's Mate (E-7). *BMCS:* Senior Chief Boatswain's Mate (E-8). *BMCM:* Master Chief Boatswain's Mate (E-9).

Description

Summary: Proficient in marlinspike, deck, and boat seamanship; manages, supervises, and administers personnel and facilities; serves as deck watch officer and boat coxswain; trains, directs, and supervises personnel in seamanship, maintenance, rigging, deck equipment and boats; supervises damage control and working parties; maintains discipline; serves as member of gun crews or gun mount captains; operates and maintains heavy equipment used in loading and buoy tending work. *BM3:* Participates as crewman aboard thirty-foot or larger boat during various standard and emergency situations; performing as a coxswain, maneuvers alongside a vessel underway, and docks under average conditions; demonstrates proper procedures for dealing with lawbreakers; interviews suspects and witnesses; collects evidence; demonstrates procedures for search and seizure; uses computer software for communications and data base. *BM2:* Able to perform the duties required for BM3; overhauls blocks; operates sailmaker equipment; performs decorative rope work; makes preparations for refueling and replenishment at sea. *BM1:* Able to perform the duties required for BM2; supervises individual work parties; supervises loading, storage and discharge of cargo and supplies; supervises anchor detail. *BMC:* Able to perform the duties required for BM1; supervises complete replenishment and refueling operations; acts as on-scene commander and/or search and rescue mission coordinator. *BMCS:* Able to perform the duties required for BMC; serves as enlisted technical or specialty expert; plans, organizes, and directs work of personnel; supervises the preparation of reports; plans and administers on-the-job training programs. *BMCM:* Able to perform the duties required for BMCS; serves as senior enlisted technical or specialty administrator; ensures maximum efficiency of the work force; prepares general correspondence concerning fiscal, supply, and administrative matters; assists in the formulation of plans, policies, and budget requirements; may supplement the officer corps in the overall supervision and administration of personnel and equipment; may also supervise personnel in other specialty areas.

Recommendation, BM3

In the vocational certificate category, 1 semester hour in rigging. In the lower-division baccalaureate/associate degree category, 3 semester hours in vessel maintenance, 3 in coastal piloting, and 3 in seamanship; if qualification code was HW (Surfman) or HX (Boat Coxswain), credit in small boat handling on the basis of institutional evaluation (8/95).

Recommendation, BM2

In the vocational certificate category, 3 semester hours in rigging. In the lower-division baccalaureate/associate degree category, 3 semester hours in vessel maintenance, 3 in seamanship, 1 in general office procedures, 3 in coastal piloting, 3 in computer science, 3 in navigation rules, and 3 in law enforcement; if qualification code was HW (Surfman) or HX (Boat Coxswain), credit in small boat handling on the basis of institutional evaluation (8/95).

Recommendation, BM1

In the vocational certificate category, 3 semester hours in rigging. In the lower-division baccalaureate/associate degree category, 3 semester hours in vessel maintenance, 3 in seamanship, 1 in general office procedures, 3 in coastal piloting, 3 in computer science, 3 in navigation rules, 3 in law enforcement, 3 in technical communications, and 2 in personnel supervision; if qualification code was HW (Surfman) or HX (Boat Coxswain), credit in small boat handling on the basis of institutional evaluation (8/95).

Recommendation, BMC

In the vocational certificate category, 3 semester hours in rigging. In the lower-division baccalaureate/associate degree category, 3 semester hours in vessel maintenance, 3 in seamanship, 1 in general office procedures, 3 in coastal piloting, 3 in computer science, 3 in navigation rules, 3 in law enforcement, 3 in technical communications, and 3 in personnel supervision; if qualification code was HW (Surfman) or HX (Boat Coxswain), credit in small boat handling on the basis of institutional evaluation (8/95).

Recommendation, BMCS

In the vocational certificate category, 3 semester hours in rigging. In the lower-division baccalaureate/associate degree category, 3 semester hours in vessel maintenance, 3 in seamanship, 1 in general office procedures, 3 in coastal piloting, 3 in computer science, 3 in navigation rules, 3 in law enforcement, 3 in technical communications, and 3 in personnel supervision; if qualification code was HW (Surfman) or HX (Boat Coxswain), credit in small boat handling on the basis of institutional evaluation. In the upper-division baccalaureate category, 3 semester hours for field experience in management and 3 in management problems (8/95).

Recommendation, BMCM

In the vocational certificate category, 3 semester hours in rigging. In the lower-division baccalaureate/associate degree category, 3 semester hours in vessel maintenance, 3 in seamanship, 1 in general office procedures, 3 in law enforcement, and 3 in personnel supervision; if qualification code was HW (Surfman) or HX (Boat Coxswain), credit in small boat handling on the basis of institutional evaluation. In the upper-division baccalaureate category, 6 semester hours for field experience in management and 3 in management problems (8/95).

CGR-DC-001

DAMAGE CONTROLMAN
DC3
DC2
DC1
DCC
DCCS
DCCM

Exhibit Dates: 1/90–7/93.

Occupational Group: III (Engineering and Hull).

Career Pattern

FN: Fireman (E-3). *DC3:* Damage Controlman, Third Class (E-4). *DC2:* Damage Controlman, Second Class (E-5). *DC1:* Damage Controlman, First Class (E-6). *DCC:* Chief Damage Controlman (E-7). *DCCS:* Senior Chief Damage Controlman (E-8). *DCCM:* Master Chief Damage Controlman (E-9).

Description

Summary: Repairs and performs maintenance on damage control equipment for preserving water tight integrity; utilizes techniques and skills in carpentry, firefighting, pipefitting, anchoring, welding, laying out, assembling, measuring devices, and nuclear, biological, and chemical (NBC) warfare and defense; instructs and coordinates damage control parties; instructs and supervises personnel in technique of NBC warfare defense, including the use of personnel decontamination stations and protective shelters. *DC3:* Operates, stores, and performs maintenance on firefighting equipment and damage control equipment; inspects and tests installed damage control equipment and systems; performs brazing, electrical arc welding, and oxyacetylene welding and cutting; repairs ship plumbing and piping systems; performs carpentry tasks; controls discharging and pumping of waste in restricted water; completes maintenance data forms; orders repair parts and tools. *DC2:* Able to perform the duties required for DC3; performs gas-free testing of voids and compartments to assure safe entry; inspects fire main, sprinklers, and ventilation systems; performs repairs to ship structure using heavy gauge metal; completes maintenance reports; inventories installed equipment. *DC1:* Able to perform the duties required for DC2; organizes and supervises damage control and firefighting parties; prepares training records; reviews completed maintenance data forms; prepares weekly maintenance schedules. *DCC:* Able to perform the duties required for DC1; plans and conducts damage control exercises; maintains shop records and prepares progress reports; supervises training programs; estimates time, personnel, and material requirements; prepares quarterly maintenance schedules; supervises damage control maintenance technician shop. *DCCS:* Able to perform the duties required for

DCC; prepares directives and instruction for attaining organizational objectives; establishes and implements a program for interviewing, evaluating, and assigning personnel to assure maximum utilization; prepares correspondence; organizes and evaluates training programs; administers a long-range planned maintenance program. *DCCM:* Able to perform the duties required for DCCS; reviews personnel, equipment, and material requirements and forecasts future requirements; plans, organizes, implements, and controls activities; develops operating budgets and monitors expenditures.

Recommendation, DC3

In the lower-division baccalaureate/associate degree category, 2 semester hours in welding, 2 in blueprint interpretation, 2 in carpentry, and 2 in plumbing and pipefitting (5/84).

Recommendation, DC2

In the lower-division baccalaureate/associate degree category, 3 semester hours in welding, 3 in blueprint interpretation, 3 in carpentry, 3 in plumbing and pipefitting, and 3 in introduction to hydraulic systems (5/84).

Recommendation, DC1

In the lower-division baccalaureate/associate degree category, 3 semester hours in welding, 3 in blueprint interpretation, 3 in carpentry, 3 in plumbing and pipefitting, 3 in introduction to hydraulic systems, 3 in construction technology, and 2 in personnel supervision (5/84).

Recommendation, DCC

In the lower-division baccalaureate/associate degree category, 3 semester hours in welding, 3 in blueprint interpretation, 3 in carpentry, 3 in plumbing and pipefitting, 3 in introduction to hydraulic systems, 3 in construction technology, 3 in personnel supervision, and 3 in principles of management (5/84).

Recommendation, DCCS

In the lower-division baccalaureate/associate degree category, 3 semester hours in welding, 3 in blueprint interpretation, 3 in carpentry, 3 in plumbing and pipefitting, 3 in introduction to hydraulic systems, 3 in construction technology, 3 in personnel supervision, and 3 in principles of management. In the upper-division baccalaureate category, 3 semester hours for field experience in management and 3 in management problems (5/84).

Recommendation, DCCM

In the lower-division baccalaureate/associate degree category, 3 semester hours in welding, 3 in blueprint interpretation, 3 in carpentry, 3 in plumbing and pipefitting, 3 in introduction to hydraulic systems, 3 in construction technology, 3 in personnel supervision, and 3 in principles of management. In the upper-division baccalaureate category, 6 semester hours for field experience in management and 3 in management problems (5/84).

CGR-DC-002

DAMAGE CONTROLMAN
DC3
DC2
DC1
DCC
DCCS
DCCM

Exhibit Dates: 8/93–Present.

Occupational Group: Nautical Operations.

Career Pattern

FN: Fireman (E-3). *DC3:* Damage Controlman, Third Class (E-4). *DC2:* Damage Controlman, Second Class (E-5). *DC1:* Damage Controlman, First Class (E-6). *DCC:* Chief Damage Controlman (E-7). *DCCS:* Senior Chief Damage Controlman (E-8). *DCCM:* Master Chief Damage Controlman (E-9).

Description

Summary: Skilled in the use and maintenance of equipment necessary for carpentry, fire and contamination control, plumbing, pipe fitting, welding, assembly, planning and layout, and general damage control; responsible for the maintenance and repair of damage control equipment and chemical/biological/radiological defense. *DC3:* Operates, stores, and performs maintenance on firefighting equipment and damage control equipment; inspects and tests installed damage control equipment and systems; performs brazing, electrical arc welding, and oxyacetylene welding and cutting; repairs ship plumbing and piping systems; performs carpentry tasks; controls discharging and pumping of waste in restricted water; completes maintenance data forms; orders repair parts and tools. *DC2:* Able to perform the duties required for DC3; performs gas-free testing of voids and compartments to assure safe entry; inspects fire main, sprinklers, and ventilation systems; performs repairs to ship structure using heavy gauge metal; completes maintenance reports; inventories installed equipment. *DC1:* Able to perform the duties required for DC2; organizes and supervises damage control and firefighting parties; prepares training records; reviews completed maintenance data forms; prepares weekly maintenance schedules. *DCC:* Able to perform the duties required for DC1; plans and conducts damage control exercises; maintains shop records and prepares progress reports; supervises training programs; estimates time, personnel, and material requirements; prepares quarterly maintenance schedules; supervises damage control maintenance technician shop. *DCCS:* Able to perform the duties required for DCC; prepares directives and instruction for attaining organizational objectives; establishes and implements a program for interviewing, evaluating, and assigning personnel to assure maximum utilization; prepares correspondence; organizes and evaluates training programs; administers a long-range planned maintenance program. *DCCM:* Able to perform the duties required for DCCS; reviews personnel, equipment, and material requirements and forecasts future requirements; plans, organizes, implements, and controls activities; develops operating budgets and monitors expenditures.

Recommendation, DC3

In the lower-division baccalaureate/associate degree category, 2 semester hours in welding (MIG, TIG, ARC, and Oxyacetylene), 1 in blueprint reading, 2 in carpentry, 2 in plumbing and pipefitting, 3 in construction technology, and 2 in safety and hazardous materials (8/95).

Recommendation, DC2

In the lower-division baccalaureate/associate degree category, 3 semester hours in welding (MIG, TIG, ARC, and Oxyacetylene), 2 in blueprint reading, 2 in carpentry, 3 in plumbing and pipefitting, 4 in construction technology, and 3 in safety and hazardous materials (8/95).

Recommendation, DC1

In the lower-division baccalaureate/associate degree category, 6 semester hours in welding (MIG, TIG, ARC, and Oxyacetylene), 3 in blueprint reading, 3 in carpentry, 3 in plumbing and pipefitting, 6 in construction technology, 3 in safety and hazardous materials, and 2 in personnel supervision (8/95).

Recommendation, DCC

In the lower-division baccalaureate/associate degree category, 6 semester hours in welding (MIG, TIG, ARC, and Oxyacetelene), 3 in blueprint reading, 3 in carpentry, 3 in plumbing and pipefitting, 6 in construction technology, 3 in safety and hazardous materials, and 3 in personnel supervision (8/95).

Recommendation, DCCS

In the lower-division baccalaureate/associate degree category, 6 semester hours in welding (MIG, TIG, ARC, and Oxyacetelene), 3 in blueprint reading, 3 in carpentry, 3 in plumbing and pipefitting, 6 in construction technology, 3 in safety and hazardous materials, and 3 in personnel supervision. In the upper-division baccalaureate category, 3 semester hours for field experience in management and 3 in management problems (8/95).

Recommendation, DCCM

In the lower-division baccalaureate/associate degree category, 6 semester hours in welding (MIG, TIG, ARC, and Oxyacetelene), 3 in blueprint reading, 3 in carpentry, 3 in plumbing and pipefitting, 6 in construction technology, 3 in safety and hazardous materials, and 3 in personnel supervision. In the upper-division baccalaureate category, 6 semester hours for field experience in management and 3 in management problems (8/95).

CGR-DP-001

DATA PROCESSING TECHNICIAN
DP3
DP2
DP1
DPC
DPCS
DPCM

Exhibit Dates: 1/90–Present.

Career Pattern

SN: Seaman (E-3). *DP3:* Data Processing Technician, Third Class (E-4). *DP2:* Data Processing Technician, Second Class (E-5). *DP1:* Data Processing Technician, First Class (E-6). *DPC:* Chief Data Processing Technician (E-7). *DPCS:* Senior Chief Data Processing Technician (E-8). *DPCM:* Master Chief Data Processing Technician (E-9).

Description

Summary: Manages and maintains networked computer systems; installs hardware and software; troubleshoots problems related to hardware, software, and the network; maintains databases; establishes and provides customized training; writes user guides; and establishes system security and network integrity. *DP3:* Writes computer programs and routines; installs and maintains proprietary operating and application software; installs and maintains work stations and server hardware. *DP2:* Able to perform the duties required for DP3; installs and maintains network communication software and hardware; assists in maintaining record entry and file management routines. *DP1:* Able to perform the duties required for DP2; installs and maintains proprietary software for client-server data management system; conducts program and network troubleshooting and software and hardware evaluation. *DPC:* Able to perform the duties required for DP1; supervises installation and maintenance of proprietary software for local-area network (LAN) and wide-area network (WAN) operations; supervises and advises file management systems operation and maintenance; trains users on software applications. *DPCS:* Able to perform the duties required for DPC; supervises installation and maintenance of proprietary software for network systems, file management, and relational databases; provides advice and assistance for automation, policy, and procedures; trains users on systems management; establishes customized training programs; evaluates staff. *DPCM:* Able to perform the duties required for DPCS; manages technical support; analyzes system requirements, and makes recommendations for installation and maintenance of proprietary software for local area and wide area networks, client-server information systems, and network operation, security, and integrity; writes user guides; reviews system security, policy, and procedures.

Recommendation, DP3

In the lower-division baccalaureate/associate degree category, 3 semester hours in computer programming (3/96).

Recommendation, DP2

In the lower-division baccalaureate/associate degree category, 3 semester hours in computer programming and 3 in data communications (3/96).

Recommendation, DP1

In the lower-division baccalaureate/associate degree category, 3 semester hours in computer programming, 3 in data communications, and 3 in basic management information systems (3/96).

Recommendation, DPC

In the lower-division baccalaureate/associate degree category, 3 semester hours in computer programming, 3 in data communications, and 3 in basic management information systems. In the upper-division baccalaureate category, 3 semester hours in advanced management information systems (3/96).

Recommendation, DPCS

In the lower-division baccalaureate/associate degree category, 3 semester hours in computer programming, 3 in data communications, and 3 in basic management information systems. In the upper-division baccalaureate category, 3 semester hours in advanced management information systems and 3 in network systems management (3/96).

Recommendation, DPCM

In the lower-division baccalaureate/associate degree category, 3 semester hours in computer programming, 3 in data communications, and 3 in basic management information systems. In the upper-division baccalaureate category, 3 semester hours in advanced management information systems, 3 in network systems management, and 3 in systems administration (3/96).

CGR-EM-001

ELECTRICIAN'S MATE
EM3
EM2
EM1
EMC
EMCS
EMCM

Exhibit Dates: 1/90–7/93.

Occupational Group: III (Engineering and Hull).

Career Pattern

FN: Fireman (E-3). *EM3:* Electrician's Mate, Third Class (E-4). *EM2:* Electrician's Mate, Second Class (E-5). *EM1:* Electrician's Mate, First Class (E-6). *EMC:* Chief Electrician's Mate (E-7). *EMCS:* Senior Chief Electrician's Mate (E-8). *EMCM:* Master Chief Electrician's Mate (E-9).

Description

Summary: Operates, installs, maintains, and repairs motors, generators, switchboards, and control equipment; installs, maintains, and repairs power and lighting circuits and electrical fixtures; performs tests for short circuits, grounds, and other casualties and repairs or rebuilds electrical equipment in an electric shop. *EM3:* Draws and interprets schematic diagrams of electrical circuits; operates test equipment used in servicing electrical and electronic equipment; has proficiency in DC and AC circuitry, including resonance; understands repair and servicing procedures for vacuum tubes and solid-state devices as well as the operating principles and construction of AC and DC motors and generators, gyroscopes, circuit breakers and storage batteries; knows procedures for starting and paralleling generators and switchboards. *EM2:* Able to perform the duties required for EM3; operates and constructs AC/DC voltage regulators and magnetic amplifiers; services circuit breakers, degaussing systems, armature commutators, solenoids, servo-mechanisms, electric galley equipment, and normal, alternate, and emergency power distribution systems for shipboard lighting and power; stands watch on main propulsion machinery and controls; locates, prepares, and maintains records, reports, and publications; maintains inventory. *EM1:* Able to perform the duties required for EM2; makes authorized repairs and calibration of test equipment; services AC and DC motor and generator controllers; operates and maintains ship propulsion equipment along with the control and auxiliary control systems; understands the principles of gyro-compass and dead-reckoning equipment. *EMC:* Able to perform the duties required for EM1; takes complete charge of the engine room on a large vessel or the engineering department on a small vessel or shore installation; instructs classes; prepares power failure reports; makes time and material repair estimates and prepares shipyard availability work requests and schedules; may also serve as an inspector in a shipyard; prepares preventive maintenance schedules; organizes and maintains technical library; prepares and submits budget requests; plans and supervises on-the-job training. *EMCS:* Able to perform the duties required for EMC; serves as enlisted technical or specialty expert; plans, organizes, and directs the work of personnel operating and maintaining electrical systems; plans and administers on-the-job training programs; supervises the preparation of reports. *EMCM:* Able to perform the duties required for EMCS; serves as a senior enlisted technical or specialty administrator; manages personnel in the operation, maintenance, procurement, and survey of electrical equipment; ensures maximum efficiency of the work force and the equipment; prepares general correspondence concerning fiscal, supply, and administrative matters; assists in the formulation of plans, policies, and budget requirements; may supplement the officer corps in the overall supervision and administration of personnel and equipment; may also supervise personnel in other specialty areas.

Recommendation, EM3

In the vocational certificate category, 6 semester hours in electrical maintenance procedures and 9 in basic electricity. In the lower-division baccalaureate/associate degree category, 3 semester hours in basic AC/DC theory (3/84).

Recommendation, EM2

In the vocational certificate category, 6 semester hours in electrical maintenance procedures and 9 in basic electricity. In the lower-division baccalaureate/associate degree category, 3 semester hours in basic AC/DC theory and 3 in recordkeeping (3/84).

Recommendation, EM1

In the vocational certificate category, 6 semester hours in electrical maintenance procedures and 9 in basic electricity. In the lower-division baccalaureate/associate degree category, 3 semester hours in basic AC/DC theory, 3 in recordkeeping, 3 in basic logic circuits, 3 in maintenance management, and 2 in personnel supervision (3/84).

Recommendation, EMC

In the vocational certificate category, 6 semester hours in electrical maintenance procedures and 9 in basic electricity. In the lower-division baccalaureate/associate degree category, 3 semester hours in basic AC/DC theory, 3 in recordkeeping, 3 in basic logic circuits, 3 in maintenance management, 3 in personnel supervision, and 3 in principles of management (3/84).

Recommendation, EMCS

In the vocational certificate category, 6 semester hours in electrical maintenance procedures and 9 in basic electricity. In the lower-division baccalaureate/associate degree category, 3 semester hours in basic AC/DC theory, 3 in recordkeeping, 3 in basic logic circuits, 3 in maintenance management, 3 in personnel supervision, and 3 in principles of management. In the upper-division baccalaureate category, 3 semester hours for field experience in management and 3 in management problems (3/84).

Recommendation, EMCM

In the vocational certificate category, 6 semester hours in electrical maintenance procedures and 9 in basic electricity. In the lower-division baccalaureate/associate degree category, 3 semester hours in basic AC/DC theory, 3 in recordkeeping, 3 in basic logic circuits, 3 in maintenance management, 3 in personnel supervision, and 3 in principles of management. In the upper-division baccalaureate category, 6 semester hours for field experience in management and 3 in management problems (3/84).

CGR-EM-002

ELECTRICIAN'S MATE
- EM3
- EM2
- EM1
- EMC
- EMCS
- EMCM

Exhibit Dates: 8/93–Present.

Occupational Group: Nautical and Shore Electrical Engineering.

Career Pattern

FN: Fireman (E-3). *EM3:* Electrician's Mate, Third Class (E-4). *EM2:* Electrician's Mate, Second Class (E-5). *EM1:* Electrician's Mate, First Class (E-6). *EMC:* Chief Electrician's Mate (E-7). *EMCS:* Senior Chief Electrician's Mate (E-8). *EMCM:* Master Chief Electrician's Mate (E-9).

Description

Summary: Operates, installs, maintains, and repairs motors, generators, switchboards, and solid state control equipment; installs, maintains, and repairs power and lighting circuits and electrical fixtures; performs tests for short circuits, grounds, and other casualties and repairs or rebuilds electrical equipment in an electric shop. *EM3:* Draws and interprets schematic diagrams of electrical circuits; operates test equipment used in servicing electrical and electronic equipment; has proficiency in DC and AC circuitry, including resonance; understands repair and servicing procedures for solid-state devices and integrated circuits as well as the operating principles and construction of AC and DC motors and generators, gyroscopes, circuit breakers and storage batteries; knows procedures for starting and paralleling generators and switchboards. *EM2:* Able to perform the duties required for EM3; operates and constructs AC/DC voltage regulators and magnetic amplifiers; services circuit breakers, degaussing systems, armature commutators, solenoids, servo-mechanisms, electric galley equipment, and normal, alternate, and emergency power distribution systems for shipboard lighting and power; stands watch on main propulsion machinery and controls; locates, prepares, and maintains records, reports, and publications; maintains inventory. *EM1:* Able to perform the duties required for EM2; makes authorized repairs and calibration of test equipment; services AC and DC motor and generator controllers; operates and maintains ship propulsion equipment along with the control and auxiliary control systems; understands the principles of gyro-compass and dead-reckoning equipment. *EMC:* Able to perform the duties required for EM1; takes complete charge of the engine room on a large vessel or the engineering department on a small vessel or shore installation; instructs classes; prepares power failure reports; makes time and material repair estimates and prepares shipyard availability work requests and schedules; may also serve as an inspector in a shipyard; prepares preventive maintenance schedules; organizes and maintains technical library; prepares and submits budget requests; plans and supervises on-the-job training. *EMCS:* Able to perform the duties required for EMC; serves as enlisted technical or specialty expert; plans, organizes, and directs the work of personnel operating and maintaining electrical systems; plans and administers on-the-job training programs; supervises the preparation of reports. *EMCM:* Able to perform the duties required for EMCS; serves as a senior enlisted technical or specialty administrator; manages personnel in the operation, maintenance, procurement, and survey of electrical equipment; ensures maximum efficiency of the work force and the equipment; prepares general correspondence concerning fiscal, supply, and administrative matters; assists in the formulation of plans, policies, and budget requirements; may supplement the officer corps in the overall supervision and administration of personnel and equipment; may also supervise personnel in other specialty areas.

Recommendation, EM3

In the lower-division baccalaureate/associate degree category, 3 semester hours in basic AC/DC theory and 6 in electrical systems troubleshooting and maintenance (8/95).

Recommendation, EM2

In the lower-division baccalaureate/associate degree category, 3 semester hours in basic AC/DC theory, 6 in electrical systems troubleshooting and maintenance, and 3 in recordkeeping (8/95).

Recommendation, EM1

In the lower-division baccalaureate/associate degree category, 3 semester hours in basic AC/DC theory, 6 in electrical systems troubleshooting and maintenance, 3 in recordkeeping, 3 in industrial electronics, 3 in maintenance management, 2 in technical writing, and 2 in personnel supervision (8/95).

Recommendation, EMC

In the lower-division baccalaureate/associate degree category, 3 semester hours in basic AC/DC theory, 6 in electrical systems troubleshooting and maintenance, 3 in recordkeeping, 3 in industrial electronics, 3 in maintenance management, 3 in personnel supervision, and 2 in technical writing (8/95).

Recommendation, EMCS

In the lower-division baccalaureate/associate degree category, 3 semester hours in basic AC/DC theory, 6 in electrical systems troubleshooting and maintenance, 3 in recordkeeping, 3 in industrial electronics, 3 in maintenance management, 3 in personnel supervision, and 2 in technical writing. In the upper-division baccalaureate category, 3 semester hours for field experience in management and 3 in management problems (8/95).

Recommendation, EMCM

In the lower-division baccalaureate/associate degree category, 3 semester hours in basic AC/DC theory, 6 in electrical systems troubleshooting and maintenance, 3 in recordkeeping, 3 in industrial electronics, 3 in maintenance management, 3 in personnel supervision, and 2 in technical writing. In the upper-division baccalaureate category, 6 semester hours for field experience in management and 3 in management problems (8/95).

CGR-ET-001

ELECTRONICS TECHNICIAN
- ET3
- ET2
- ET1
- ETC
- ETCS
- ETCM

Exhibit Dates: 1/90–9/94.

Occupational Group: III (Engineering and Hull).

Career Pattern

SN: Seaman (E-3). *ET3:* Electronics Technician, Third Class (E-4). *ET2:* Electronics Technician, Second Class (E-5). *ET1:* Electronics Technician, First Class (E-6). *ETC:* Chief Electronics Technician (E-7). *ETCS:* Senior Chief Electronics Technician (E-8). *ETCM:* Master Chief Electronics Technician (E-9).

Description

Summary: Operates, maintains, repairs, calibrates, tunes, and adjusts electronic equipment used for communication, cryptography, detection, recognition, identification, navigation, and electronic surveillance. *ET3:* Has knowledge of electricity/electronics; reads and interprets schematic diagrams; uses electronic test equipment to localize and replace faulty components. *ET2:* Able to perform the duties required for ET3; provides technical guidance to subordinate personnel; performs preventive maintenance on electronic equipment; keeps equipment maintenance logs. *ET1:* Able to perform the duties required for ET2; prepares preventive maintenance schedules and work requests; maintains equipment status reports; supervises shipping and handling procedures. *ETC:* Able to perform the duties required for ET1; supervises the operation of the electronic shop; supervises the use, filing, and maintenance of publications, logs, and records; plans, organizes, and administers the maintenance program for the repair of electronic equipment; generates work requests; plans and supervises on-the-job training programs; prepares and submits budget requests; organizes and maintains technical library. *ETCS:* Able to perform the duties required for ETC; serves as enlisted technical or specialty expert; plans, organizes, and directs personnel in the operation and maintenance of electronics equipment; plans and administers on-the-job training programs; supervises the preparation of reports. *ETCM:* Able to perform the duties required for ETCS; serves as senior enlisted technical or specialty administrator; ensures maximum efficiency of the work force and the equipment; manages the operation and maintenance procedures of electronic equipment; prepares general correspondence concerning fiscal, supply, and administrative matters; assists in the formulation of plans, policies, and budget requirements; may supplement the officer corps in the overall supervision and administration of personnel and equipment; may also supervise personnel in other specialty areas.

Recommendation, ET3

In the lower-division baccalaureate/associate degree category, 3 semester hours in basic AC/DC theory, 3 in diagnostic procedures, 3 in principles of instrumentation, 3 in circuit theory, and 3 in basic computer principles (3/84).

Recommendation, ET2

In the lower-division baccalaureate/associate degree category, 3 semester hours in basic AC/DC theory, 3 in diagnostic procedures, 3 in principles of instrumentation, 3 in circuit theory, and 3 in basic computer principles (3/84).

Recommendation, ET1

In the lower-division baccalaureate/associate degree category, 3 semester hours in basic AC/DC theory, 3 in diagnostic procedures, 3 in principles of instrumentation, 3 in circuit theory, 3 in basic computer principles, 3 in maintenance management, and 2 in personnel supervision (3/84).

Recommendation, ETC
In the lower-division baccalaureate/associate degree category, 3 semester hours in basic AC/DC theory, 3 in diagnostic procedures, 3 in principles of instrumentation, 3 in circuit theory, 3 in basic computer principles, 3 in maintenance management, 3 in personnel supervision, and 3 in principles of management (3/84).

Recommendation, ETCS
In the lower-division baccalaureate/associate degree category, 3 semester hours in basic AC/DC theory, 3 in diagnostic procedures, 3 in principles of instrumentation, 3 in circuit theory, 3 in basic computer principles, 3 in maintenance management, 3 in personnel supervision, and 3 in principles of management. In the upper-division baccalaureate category, 3 semester hours for field experience in management and 3 in management problems (3/84).

Recommendation, ETCM
In the lower-division baccalaureate/associate degree category, 3 semester hours in basic AC/DC theory, 3 in diagnostic procedures, 3 in principles of instrumentation, 3 in circuit theory, 3 in basic computer principles, 3 in maintenance management, 3 in personnel supervision, and 3 in principles of management. In the upper-division baccalaureate category, 6 semester hours for field experience in management and 3 in management problems (3/84).

CGR-ET-002

ELECTRONICS TECHNICIAN
    ET3
    ET2
    ET1
    ETC
    ETCS
    ETCM

Exhibit Dates: 10/94–Present.

Occupational Group: Nautical and Shore Electronic Engineering.

Career Pattern
SN: Seaman (E-3). *ET3:* Electronics Technician, Third Class (E-4). *ET2:* Electronics Technician, Second Class (E-5). *ET1:* Electronics Technician, First Class (E-6). *ETC:* Chief Electronics Technician (E-7). *ETCS:* Senior Chief Electronics Technician (E-8). *ETCM:* Master Chief Electronics Technician (E-9).

Description
Summary: Operates, maintains, repairs, calibrates, tunes, and adjusts electronic equipment used for communication, cryptography, detection, recognition, identification, navigation, and electronic surveillance. *ET3:* Has knowledge of electricity/electronics; reads and interprets schematic diagrams; uses electronic test equipment to localize and replace faulty components. *ET2:* Able to perform the duties required for ET3; provides technical guidance to subordinate personnel; performs preventive maintenance on electronic equipment; keeps equipment maintenance logs. *ET1:* Able to perform the duties required for ET2; prepares preventive maintenance schedules and work requests; maintains equipment status reports; supervises shipping and handling procedures. *ETC:* Able to perform the duties required for ET1; supervises the operation of the electronic shop; supervises the use, filing, and maintenance of publications, logs, and records; plans, organizes, and administers the maintenance program for the repair of electronic equipment; generates work requests; plans and supervises on-the-job training programs; prepares and submits budget requests; organizes and maintains technical library. *ETCS:* Able to perform the duties required for ETC; serves as enlisted technical or specialty expert; plans, organizes, and directs personnel in the operation and maintenance of electronics equipment; plans and administers on-the-job training programs; supervises the preparation of reports. *ETCM:* Able to perform the duties required for ETCS; serves as senior enlisted technical or specialty administrator; ensures maximum efficiency of the work force and the equipment; manages the operation and maintenance procedures of electronic equipment; prepares general correspondence concerning fiscal, supply, and administrative matters; assists in the formulation of plans, policies, and budget requirements; may supplement the officer corps in the overall supervision and administration of personnel and equipment; may also supervise personnel in other specialty areas.

Recommendation, ET3
In the lower-division baccalaureate/associate degree category, 3 semester hours in basic AC/DC theory, 3 in electronic systems troubleshooting and maintenance, 3 in principles of instrumentation, 3 in basic circuit theory, and 3 in basic digital principles (8/95).

Recommendation, ET2
In the lower-division baccalaureate/associate degree category, 3 semester hours in basic AC/DC theory, 3 in electronic systems troubleshooting and maintenance, 3 in principles of instrumentation, 3 in basic circuit theory, 3 in basic digital principles, and 2 in technical writing (8/95).

Recommendation, ET1
In the lower-division baccalaureate/associate degree category, 3 semester hours in basic AC/DC theory, 3 in electronic systems troubleshooting and maintenance, 3 in principles of instrumentation, 3 in basic circuit theory, 3 in basic digital principles, 3 in maintenance management, 2 in personnel supervision, and 2 in technical writing (8/95).

Recommendation, ETC
In the lower-division baccalaureate/associate degree category, 3 semester hours in basic AC/DC theory, 3 in electronic systems troubleshooting and maintenance, 3 in principles of instrumentation, 3 in basic circuit theory, 3 in basic digital principles, 3 in maintenance management, 3 in personnel supervision, and 2 in technical writing (8/95).

Recommendation, ETCS
In the lower-division baccalaureate/associate degree category, 3 semester hours in basic AC/DC theory, 3 in electronic systems troubleshooting and maintenance, 3 in principles of instrumentation, 3 in basic circuit theory, 3 in basic digital principles, 3 in maintenance management, 3 in personnel supervision, and 3 in technical writing. In the upper-division baccalaureate category, 3 semester hours for field experience in management and 3 in management problems (8/95).

Recommendation, ETCM
In the lower-division baccalaureate/associate degree category, 3 semester hours in basic AC/DC theory, 3 in electronic systems troubleshooting and maintenance, 3 in principles of instrumentation, 3 in basic circuit theory, 3 in basic digital principles, 3 in maintenance management, 3 in personnel supervision, and 3 in technical writing. In the upper-division baccalaureate category, 6 semester hours for field experience in management and 3 in management problems (8/95).

CGR-FN-001

FIREMAN
    FN

Exhibit Dates: 1/90–Present.

Occupational Group: Technical and Administrative Assistance.

Career Pattern
Fireman is a general rate (apprenticeship) for persons at paygrade E-1 (recruit), E-2 (Fireman apprentice), and E-3 (Fireman). At paygrade E-4 (petty officer third class), the person may enter any one of the following ratings: Damage Controlman (DC), Electrician's Mate (EM), or Machinery Technician (MK).

Description
Performs all basic fireman apprenticeship functions in engineering areas aboard ship, involving cleanliness, operation, maintenance, and preservation of main propulsion, auxiliary steam or diesel machinery, steam or diesel generators, various pumps, motors, and associated equipment; identifies basic types and components of boilers, steam turbines, reduction gears, propellers and shafting, shipboard electric systems, and internal combustion engines; performs routine maintenance procedures.

Recommendation
In the lower-division baccalaureate/associate degree category, 3 semester hours in introduction to fire science technology and 1 in blueprint reading; credit in swimming on the basis of institutional evaluation. Credit for Fireman should be granted only after paygrade E-3 has been achieved (3/96).

CGR-FT-001

FIRE CONTROL TECHNICIAN
    FT3
    FT2
    FT1
    FTC
    FTCS
    FTCM

Exhibit Dates: 1/90–9/94.

Occupational Group: II (Ordnance).

Career Pattern
SN: Seaman (E-3). *FT3:* Fire Control Technician, Third Class (E-4). *FT2:* Fire Control Technician, Second Class (E-5). *FT1:* Fire Control Technician, First Class (E-6). *FTC:* Chief Fire Control Technician (E-7). *FTCS:* Senior Chief Fire Control Technician (E-8). *FTCM:* Master Chief Fire Control Technician (E-9).

Description
Summary: Operates and maintains shipboard weapons control systems; performs preventive and corrective maintenance, including troubleshooting, testing, analyzing, adjusting, and repairing. *FT3:* Operates all weapons control

systems; performs equipment maintenance, including troubleshooting, testing, analyzing, adjusting, and repairing of power drives, fire control radar equipment, target designation equipment, radar signal processing equipment, and analog and digital computer (comprised of electromechanical assemblies, optical devices, electron tube circuitry and solid state circuitry); performs optical and electrical alignment of weapons control systems and gun mount installations. *FT2:* Able to perform the duties required for FT3; calculates conventional weapons ballistic corrections; maintains records and logs; provides technical assistance to subordinate personnel. *FT1:* Able to perform the duties required for FT2; supervises and trains personnel in all phases of weapons control systems maintenance; prepares reports and work lists. *FTC:* Able to perform the duties required for FT1; supervises and trains personnel in all phases of weapons control systems operation and employment; implements weapons department administrative procedures; prepares preventive maintenance schedules; organizes and maintains technical library; prepares and submits budget request; plans and supervises on-the-job training. *FTCS:* Able to perform the duties required for FTC; serves as enlisted technical or specialty expert; plans, organizes, and directs work of personnel operating and maintaining weapons control systems; plans and administers on-the-job training programs; supervises the preparation of reports. *FTCM:* Able to perform the duties required for FTCS; serves as senior enlisted technical or specialty administrator; manages personnel in the operation, maintenance, and procurement of weapons control systems equipment; ensures maximum efficiency of the work force and the equipment; prepares general correspondence concerning fiscal, supply, and administrative matters; assists in the formulation of plans, policies, and budget requirements; may supplement the officer corps in overall supervision and administration of personnel and equipment; may also supervise personnel in other specialty area.

Recommendation, FT3

In the lower-division baccalaureate/associate degree category, 3 semester hours in basic AC/DC theory, 3 in diagnostic principles, 3 in circuit theory, 3 in radar principles, and 3 in introduction to analog computers (3/84).

Recommendation, FT2

In the lower-division baccalaureate/associate degree category, 3 semester hours in basic AC/DC theory, 3 in diagnostic principles, 3 in circuit theory, 3 in radar principles, 3 in introduction to analog computers, and 3 in recordkeeping (3/84).

Recommendation, FT1

In the lower-division baccalaureate/associate degree category, 3 semester hours in basic AC/DC theory, 3 in diagnostic principles, 3 in circuit theory, 3 in radar principles, 3 in introduction to analog computers, 3 in recordkeeping, 2 in personnel supervision, and 3 in maintenance management (3/84).

Recommendation, FTC

In the lower-division baccalaureate/associate degree category, 3 semester hours in basic AC/DC theory, 3 in diagnostic principles, 3 in circuit theory, 3 in radar principles, 3 in introduction to analog computers, 3 in recordkeeping, 3 in personnel supervision, 3 in maintenance management, and 3 in principles of management (3/84).

Recommendation, FTCS

In the lower-division baccalaureate/associate degree category, 3 semester hours in basic AC/DC theory, 3 in diagnostic principles, 3 in circuit theory, 3 in radar principles, 3 in introduction to analog computers, 3 in recordkeeping, 3 in personnel supervision, 3 in maintenance management, and 3 in principles of management. In the upper-division baccalaureate category, 3 semester hours for field experience in management and 3 in management problems (3/84).

Recommendation, FTCM

In the lower-division baccalaureate/associate degree category, 3 semester hours in basic AC/DC theory, 3 in diagnostic principles, 3 in circuit theory, 3 in radar principles, 3 in introduction to analog computers, 3 in recordkeeping, 3 in personnel supervision, 3 in maintenance management, and 3 in principles of management. In the upper-division baccalaureate category, 6 semester hours for field experience in management and 3 in management problems (3/84).

CGR-FT-002

FIRE CONTROL TECHNICIAN
FT3
FT2
FT1
FTC
FTCS
FTCM

Exhibit Dates: 10/94–Present.

Career Pattern

SN: Seaman (E-3). *FT3:* Fire Control Technician, Third Class (E-4). *FT2:* Fire Control Technician, Second Class (E-5). *FT1:* Fire Control Technician, First Class (E-6). *FTC:* Chief Fire Control Technician (E-7). *FTCS:* Senior Chief Fire Control Technician (E-8). *FTCM:* Master Chief Fire Control Technician (E-9).

Description

Summary: Operates and maintains shipboard weapons control systems; performs preventive and corrective maintenance, including troubleshooting, testing, analyzing, adjusting, and repairing. *FT3:* Operates all weapons control systems; performs equipment maintenance, including troubleshooting, testing, analyzing, adjusting, and repairing of power drives, fire control radar equipment, target designation equipment, radar signal processing equipment, and digital computer (comprised of electromechanical assemblies, optical devices, electron tube circuitry and solid state circuitry); performs optical and electrical alignment of weapons control systems and gun mount installations. *FT2:* Able to perform the duties required for FT3; calculates conventional weapons ballistic corrections; maintains records and logs; provides technical assistance to subordinates. *FT1:* Able to perform the duties required for FT2; supervises and trains staff in all phases of weapons control systems maintenance; prepares reports and work lists. *FTC:* Able to perform the duties required for FT1; supervises and trains staff in all phases of weapons control systems operation and employment; implements weapons department administrative procedures; prepares preventive maintenance schedules; organizes and maintains technical library; prepares and submits budget request; plans and supervises on-the-job training. *FTCS:* Able to perform the duties required for FTC; serves as enlisted technical or specialty expert; plans, organizes, and directs work of personnel operating and maintaining weapons control systems; plans and administers on-the-job training programs; supervises the preparation of reports. *FTCM:* Able to perform the duties required for FTCS; serves as senior enlisted technical or specialty administrator; manages the operation, maintenance, and procurement of weapons control systems equipment; ensures maximum efficiency of the work force and the equipment; prepares general correspondence concerning fiscal, supply, and administrative matters; assists in the formulation of plans, policies, and budget requirements; may supplement the officer corps in overall supervision and administration of personnel and equipment; may also supervise personnel in other specialty area.

Recommendation, FT3

In the lower-division baccalaureate/associate degree category, 3 semester hours in basic electricity, 3 in diagnostic principles, 3 in basic electronics, 3 in radar principles, and 3 in introduction to digital circuits. (3/96).

Recommendation, FT2

In the lower-division baccalaureate/associate degree category, 3 semester hours in basic electricity, 3 in diagnostic principles, 3 in basic electronics, 3 in radar principles, 3 in introduction to digital circuits, and 3 in record keeping (3/96).

Recommendation, FT1

In the lower-division baccalaureate/associate degree category, 3 semester hours in basic electricity, 3 in diagnostic principles, 3 in basic electronics, 3 in radar principles, 3 in introduction to digital circuits, 3 in recordkeeping, 2 in personnel supervision, 3 in maintenance management, and 2 in technical communications (3/96).

Recommendation, FTC

In the lower-division baccalaureate/associate degree category, 3 semester hours in basic electricity, 3 in diagnostic principles, 3 in basic electronics, 3 in radar principles, 3 in introduction to digital circuits, 3 in recordkeeping, 2 in personnel supervision, 3 in maintenance management, and 2 in technical communications. In the upper-division baccalaureate category, 3 semester hours in personnel management (3/96).

Recommendation, FTCS

In the lower-division baccalaureate/associate degree category, 3 semester hours in basic electricity, 3 in diagnostic principles, 3 in basic electronics, 3 in radar principles, 3 in introduction to digital circuits, 3 in recordkeeping, 2 in personnel supervision, 3 in maintenance management, and 2 in technical communications. In the upper-division baccalaureate category, 3 semester hours for field experience in management and 3 in personnel management (3/96).

Recommendation, FTCM

In the lower-division baccalaureate/associate degree category, 3 semester hours in basic electricity, 3 in diagnostic principles, 3 in basic electronics, 3 in radar principles, 3 in introduction to digital circuits, 3 in recordkeeping, 2 in personnel supervision, 3 in maintenance management, and 2 in technical communications. In the upper-division baccalaureate category, 3 semester hours for field experience in management, 3 in personnel management, and 3 in curriculum development (3/96).

CGR-GM-001

GUNNER'S MATE
  GM3
  GM2
  GM1
  GMC
  GMCS
  GMCM

Exhibit Dates: 1/90–9/94.

Occupational Group: II (Ordnance).

Career Pattern
  SN: Seaman (E-3). *GM3:* Gunner's Mate, Third Class (E-4). *GM2:* Gunner's Mate, Second Class (E-5). *GM1:* Gunner's Mate, First Class (E-6). *GMC:* Chief Gunner's Mate (E-7). *GMCS:* Senior Chief Gunner's Mate (E-8). *GMCM:* Master Chief Gunner's Mate (E-9).

Description
  Summary: Operates, maintains, and repairs small arms, torpedo tubes, guns and mounts, launchers, and associated handling equipment. *GM3:* Operates guns, mounts, launchers, ammunition hoists, projectile hoists, and associated equipment; operates, maintains, stows, and issues small arms and associated equipment; constructs basic electronic circuits; measures current and voltage in electronic circuits; uses schematic diagrams; uses and cares for hand tools; maintains ordnance logs, and ammunition stock records. *GM2:* Able to perform the duties required for GM3; diagnoses and repairs electronic control systems; conducts inventories and prepares requisitions; organizes and administers a program on safety instruction. *GM1:* Able to perform the duties required for GM2; instructs in the handling and storage of ammunition and maintenance of records; develops, implements, and supervises a weapons departmental training program. *GMC:* Able to perform the duties required for GM1; supervises maintenance of ordnance electrical and electronic systems; supervises handling and storage of explosive ordnance components; serves as shop supervisor; prepares preventive maintenance schedules; organizes and maintains technical library; prepares and submits budget requests; plans and supervises on-the-job training. *GMCS:* Able to perform the duties required for GMC; serves as enlisted technical or specialty expert; plans, organizes, and directs the work of personnel in operation, maintenance, and repair procedures; plans and administers on-the-job training programs; supervises the preparation of reports. *GMCM:* Able to perform the duties required for GMCS; serves as senior enlisted technical or specialty administrator; manages personnel in operation, maintenance, and repair procedures; ensures maximum efficiency of the work force and the equipment; prepares general correspondence concerning fiscal, supply, and administrative matters; assists in the formulation of plans, policies, and budget requirements; may supplement the officer corps in the overall supervision and administration of personnel and equipment; may also supervise personnel in other specialty areas.

Recommendation, GM3
  In the lower-division baccalaureate/associate degree category, 2 semester hours in introduction to electricity and electronics, 2 in hydraulic systems maintenance, 2 in electronics systems maintenance, 2 in gunsmithing, 1 in industrial safety, 1 in use and care of hand tools, and 1 in blueprint reading (5/84).

Recommendation, GM2
  In the lower-division baccalaureate/associate degree category, 3 semester hours in introduction to electricity and electronics, 3 in hydraulic systems maintenance, 3 in electronics systems maintenance, 3 in gunsmithing, 1 in industrial safety, 1 in use and care of hand tools, and 1 in blueprint reading (5/84).

Recommendation, GM1
  In the lower-division baccalaureate/associate degree category, 4 semester hours in introduction to electricity and electronics, 4 in hydraulic systems maintenance, 4 in electronics systems maintenance, 4 in gunsmithing, 1 in industrial safety, 1 in use and care of hand tools, 1 in blueprint reading, and 2 in personnel supervision (5/84).

Recommendation, GMC
  In the lower-division baccalaureate/associate degree category, 4 semester hours in introduction to electricity and electronics, 4 in hydraulic systems maintenance, 4 in electronics systems maintenance, 4 in gunsmithing, 1 in industrial safety, 1 in use and care of hand tools, 1 in blueprint reading, 3 in personnel supervision, and 3 in principles of management (5/84).

Recommendation, GMCS
  In the lower-division baccalaureate/associate degree category, 4 semester hours in introduction to electricity and electronics, 4 in hydraulic systems maintenance, 4 in electronics systems maintenance, 4 in gunsmithing, 1 in industrial safety, 1 in use and care of hand tools, 1 in blueprint reading, 3 in personnel supervision, and 3 in principles of management. In the upper-division baccalaureate category, 3 semester hours for field experience in management and 3 in management problems (5/84).

Recommendation, GMCM
  In the lower-division baccalaureate/associate degree category, 4 semester hours in introduction to electricity and electronics, 4 in hydraulic systems maintenance, 4 in electronics systems maintenance, 4 in gunsmithing, 1 in industrial safety, 1 in use and care of hand tools, 1 in blueprint reading, 3 in personnel supervision, and 3 in principles of management. In the upper-division baccalaureate category, 6 semester hours for field experience in management and 3 in management problems (5/84).

CGR-GM-002

GUNNER'S MATE
  GM3
  GM2
  GM1
  GMC
  GMCS
  GMCM

Exhibit Dates: 10/94–Present.

Occupational Group: Nautical Operations-Ordnance.

Career Pattern
  SN: Seaman (E-3). *GM3:* Gunner's Mate, Third Class (E-4). *GM2:* Gunner's Mate, Second Class (E-5). *GM1:* Gunner's Mate, First Class (E-6). *GMC:* Chief Gunner's Mate (E-7). *GMCS:* Senior Chief Gunner's Mate (E-8). *GMCM:* Master Chief Gunner's Mate (E-9).

Description
  Summary: Operates, maintains, and repairs small arms, torpedo tubes, guns and mounts, launchers, and associated handling equipment. *GM3:* Operates guns, mounts, launchers, ammunition hoists, projectile hoists, and associated equipment; operates, maintains, stows, and issues small arms and associated equipment; measures current and voltage; uses blueprint diagrams; uses and cares for hand tools; maintains ordnance logs, and ammunition stock records. *GM2:* Able to perform the duties required for GM3; conducts inventories and prepares requisitions; organizes and administers a program on safety instruction. *GM1:* Able to perform the duties required for GM2; instructs in the handling and storage of ammunition and maintenance of records; develops, implements, and supervises a weapons departmental training program. *GMC:* Able to perform the duties required for GM1; supervises maintenance of ordnance electrical and electronic systems; supervises handling and storage of explosive ordnance components; serves as shop supervisor; prepares preventive maintenance schedules; organizes and maintains technical library; prepares and submits budget requests; plans and supervises on-the-job training. *GMCS:* Able to perform the duties required for GMC; serves as enlisted technical or specialty expert; plans, organizes, and directs the work of personnel in operation, maintenance, and repair procedures; plans and administers on-the-job training programs; supervises the preparation of reports. *GMCM:* Able to perform the duties required for GMCS; serves as senior enlisted technical or specialty administrator; manages personnel in operation, maintenance, and repair procedures; ensures maximum efficiency of the work force and the equipment; prepares general correspondence concerning fiscal, supply, and administrative matters; assists in the formulation of plans, policies, and budget requirements; may supplement the officer corps in the overall supervision and administration of personnel and equipment; may also supervise personnel in other specialty areas.

Recommendation, GM3
  In the lower-division baccalaureate/associate degree category, 1 semester hour in fire and hazardous material safety (8/95).

Recommendation, GM2
  In the lower-division baccalaureate/associate degree category, 1 semester hour in fire and hazardous material safety (8/95).

Recommendation, GM1
  In the lower-division baccalaureate/associate degree category, 1 semester hour in fire and hazardous material safety, 1 in hydraulics systems troubleshooting and maintenance, 1 in blueprint reading, and 2 in personnel supervision (8/95).

Recommendation, GMC
  In the lower-division baccalaureate/associate degree category, 1 semester hour in fire and hazardous material, 1 in blueprint reading, and 3 in personnel supervision (8/95).

Recommendation, GMCS
  In the lower-division baccalaureate/associate degree category, 1 semester hour in fire and hazardous material safety, 1 in blueprint reading, and 3 in personnel supervision. In the

COAST GUARD ENLISTED RATINGS EXHIBITS

upper-division baccalaureate category, 3 semester hours for field experience in management (8/95).

Recommendation, GMCM

In the lower-division baccalaureate/associate degree category, 1 semester hour in fire and hazardous material safety, 1 in blueprint reading, and 3 in personnel supervision. In the upper-division baccalaureate category, 6 semester hours for field experience in management (8/95).

CGR-HS-001

HEALTH SERVICES TECHNICIAN
  HS3
  HS2
  HS1
  HSC
  HSCS
  HSCM

Exhibit Dates: 1/90–2/93.

Occupational Group: VIII (Medical).

Career Pattern

SN: Seaman (E-3). *HS3:* Health Services Technician, Third Class (E-4). *HS2:* Health Services Technician, Second Class (E-5). *HS1:* Health Services Technician, First Class (E-6). *HSC:* Chief Health Services Technician (E-7). *HSCS:* Senior Chief Health Services Technician (E-8). *HSCM:* Master Chief Health Services Technician (E-9).

Description

Summary: Performs duties as assistant in the prevention and treatment of disease and injuries, as assistant to the dental officer in treatment of patients, and in the administration of medical departments ashore and afloat. *HS3:* Performs basic first aid; assists with physical examinations; provides nursing care to patients; assists in the procurement, storage, and issue of medical supplies; instructs in personal hygiene, first aid, and self aid; assists in the maintenance of sanitary conditions; is prepared to assist in the prevention and treatment of nuclear, biological, and chemical warfare casualties; composes health care and dental records; prepares patients for dental treatment; assists dental officers in all phases of dentistry; provides dental health prevention education and performs basic x-ray procedures. *HS2:* Able to perform the duties required for HS3; performs advanced first aid procedures; performs the duties required for HS3; performs advanced first aid procedures; performs cardiopulmonary resuscitation; assists in the transportation of the sick and injured; administers medicines and parental solutions; dispenses commonly used pharmaceutical preparations; prepares and maintains medical reports; prepares basic radiologic reports; performs basic radiologic procedures; processes x-rays; performs basic laboratory procedures including urinalysis and blood counts; supervises central sterilization procedures; performs basic preventive maintenance on instruments; acts as oral surgery assistant. *HS1:* Able to perform the duties required for HS2; performs advanced laboratory procedures including microbiology and parasitology; performs supervisory duties; performs minor surgical techniques; performs special radiographic techniques; coordinates a radiation health program; mixes compounds; supervises both medical and dental offices and accounting systems, including supplies. *HSC:* Able to perform the duties required for HS1; supervises assigned personnel; maintains a medical and dental records department. *HSCS:* Able to perform the duties required for HSC; serves as enlisted technical or specialty expert; administers medical/dental training programs; reviews all medical reports and presents them to appropriate commands; provides direction and supervision to enlisted personnel in the performance of their functions; supervises procurement, storage, accounting, preservation, and allocation of medical and dental materials.) *HSCM:* Able to perform the duties required for HSCS; serves as senior enlisted technical or specialty administrator; provides senior level supervision and administration to all enlisted personnel; ensures maximum efficiency of the work force and the equipment; prepares general correspondence concerning fiscal, supply, and administrative matters; assists in the formulation of plans, policies, and budget requirements; may supplement the officer corps in the overall supervision and administration of personnel and equipment.

Recommendation, HS3

In the lower-division baccalaureate/associate degree category, 2-3 semester hours in clinical experience (in areas related to nursing techniques, laboratory techniques, dental office procedures, and preventive medicine) based on institutional evaluation of the student's work experience; add credit for the Health Services Technician A School in exhibit CG-0709-0009 (10/87).

Recommendation, HS2

In the lower-division baccalaureate/associate degree category, 7 semester hours in clinical experience (in areas related to nursing techniques, laboratory techniques, dental office procedures, and preventive medicine); add credit for the Health Services Technician A School in exhibit CG-0709-0009 (10/87).

Recommendation, HS1

In the lower-division baccalaureate/associate degree category, 11 semester hours in clinical experience (in areas related to nursing techniques, laboratory techniques, dental office procedures, and preventive medicine), and 1 in personnel supervision; add credit for the Health Services Technician A School in exhibit CG-0709-0009 (10/87).

Recommendation, HSC

In the lower-division baccalaureate/associate degree category, 11 semester hours in clinical experience (in areas related to nursing techniques, laboratory techniques, radiation prevention, dental office procedures, and preventive medicine), and 6 in personnel supervision; add credit for the Health Services Technician A School in exhibit CG-0709-0009 (10/87).

Recommendation, HSCS

In the lower-division baccalaureate/associate degree category, 11 semester hours in clinical experience (in areas related to nursing techniques, laboratory techniques, radiation prevention, dental office procedures, and preventive medicine), and 6 in personnel supervision; add credit for the Health Services Technician A School in exhibit CG-0709-0009. In the upper-division baccalaureate category, 3 semester hours for field experience in management and 3 in management problems (10/87).

Recommendation, HSCM

In the lower-division baccalaureate/associate degree category, 11 semester hours in clinical experience (in areas related to nursing techniques, laboratory techniques, radiation prevention, dental office procedures, and preventive medicine), and 6 in personnel supervision; add credit for the Health Services Technician A School in exhibit CG-0709-0009. In the upper-division baccalaureate category, 3 semester hours for field experience in management, 3 in management problems, 3 for field experience in health/dental care management, and 2 in introduction to health care/dental administration (10/87).

CGR-HS-002

HEALTH SERVICES TECHNICIAN
  HS3
  HS2
  HS1
  HSC
  HSCS
  HSCM

Exhibit Dates: 3/93–Present.

Occupational Group: Medical Care.

Career Pattern

SN: Seaman (E-3). *HS3:* Health Services Technician, Third Class (E-4). *HS2:* Health Services Technician, Second Class (E-5). *HS1:* Health Services Technician, First Class (E-6). *HSC:* Chief Health Services Technician (E-7). *HSCS:* Senior Chief Health Services Technician (E-8). *HSCM:* Master Chief Health Services Technician (E-9).

Description

Summary: Performs duties as assistant in the prevention and treatment of disease and injuries, as assistant to the dental officer in treatment of patients, and in the administration of medical departments ashore and afloat. *HS3:* Performs emergency intervention procedures consistent with Basic EMT skills, triage, extraction and transportation; assesses clinic patients and performs basic treatments including infections; performs basic laboratory procedures including blood counts; prepares and dispenses pharmaceuticals under supervision, maintains medical history and examination records; assists in all phases of dentistry and performs basic dental X-ray procedures. *HS2:* Able to perform the duties required for HS3; performs mass casualty triage; performs minor surgical procedures; under supervision assesses and treats all clinic patients; instructs patients on personal health; performs limited radiography; performs environmental inspections and testing; reviews and maintains medical records; inventories and orders supplies; acts as an oral surgery assistant; performs quality assurance procedures on medical and dental equipment. *HS1:* Able to perform the duties required for HS2; acts as a health benefit advisor; maintains and procures clinic equipment and supplies; reviews medical bills; monitors advanced medical/dental quality assurance; supervises medical and dental clinics and their accounting systems. *HSC:* Able to perform the duties required for HS1; supervises assigned personnel; maintains a medical and dental records department; performs general clinic management; monitors budgets. *HSCS:* Able to perform the duties required for HSC; serves as enlisted technical or specialty expert; administers medical/dental training programs; reviews all medical reports and presents them to appropriate commands; provides direction and supervision to enlisted personnel in the performance of their functions;

supervises procurement, storage, accounting, preservation, and allocation of medical and dental materials; ensures regulation compliance. *HSCM:* Able to perform the duties required for HSCS; serves as senior enlisted technical or specialty administrator; provides senior level supervision and administration to all enlisted personnel; ensures maximum efficiency of the work force and the equipment; prepares general correspondence concerning fiscal, supply, and administrative matters; assists in the formulation of plans, policies, and budget requirements; may supplement the officer corps in the overall supervision and administration of personnel and equipment.

Recommendation, HS3
In the lower-division baccalaureate/associate degree category, 6 semester hours in patient care clinical experience (3/96).

Recommendation, HS2
In the lower-division baccalaureate/associate degree category, 12 semester hours in patient care clinical experience (3/96).

Recommendation, HS1
In the lower-division baccalaureate/associate degree category, 18 semester hours in patient care clinical experience and 3 in personnel supervision (3/96).

Recommendation, HSC
In the lower-division baccalaureate/associate degree category, 18 semester hours in patient care clinical experience and 3 in personnel supervision. In the upper-division baccalaureate category, 3 semester hours for field experience in health services administration and 3 in management problems (3/96).

Recommendation, HSCS
In the lower-division baccalaureate/associate degree category, 18 semester hours in patient care clinical experience and 6 in personnel supervision. In the upper-division baccalaureate category, 6 semester hours for field experience in management and 6 in management problems (3/96).

Recommendation, HSCM
In the lower-division baccalaureate/associate degree category, 18 semester hours in clinical experience and 6 in personnel supervision. In the upper-division baccalaureate category, 6 semester hours for field experience in management, 6 in management problems, 3 in health service policy administration, 3 in health care organization and management, and 3 in report writing (3/96).

CGR-HS-003

HEALTH SERVICES TECHNICIAN
HS3
HS2
HS1
HSC
HSCS
HSCM

Exhibit Dates: 3/98–Present.

Occupational Group: Medical Care.

Career Pattern
SN: Seaman (E-3). *HS3:* Health Services Technician, Third Class (E-4). *HS2:* Health Services Technician, Second Class (E-5). *HS1:* Health Services Technician, First Class (E-6). *HSC:* Chief Health Services Technician (E-7). *HSCS:* Senior Chief Health Services Technician (E-8). *HSCM:* Master Chief Health Services Technician (E-9).

Description
Summary: Performs duties as assistant in the prevention and treatment of disease and injuries and in the administration of medical departments ashore and afloat. *HS3:* Performs emergency intervention procedures consistent with Basic EMT skills, triage, extraction and transportation; assesses clinic patients and performs basic treatments including infections; performs basic laboratory procedures including blood counts; prepares and dispenses pharmaceuticals under supervision, maintains medical history and examination records. *HS2:* Able to perform the duties required for HS3; performs mass casualty triage; performs minor surgical procedures; under supervision assesses and treats all clinic patients; instructs patients on personal health; performs limited radiography; performs environmental inspections and testing; reviews and maintains medical records; inventories and orders supplies; performs quality assurance procedures on medical equipment. *HS1:* Able to perform the duties required for HS2; acts as a health benefit advisor; maintains and procures clinic equipment and supplies; reviews medical bills; monitors advanced medical quality assurance; supervises medical clinics and their accounting systems. *HSC:* Able to perform the duties required for HS1; supervises assigned personnel; maintains a medical records department; performs general clinic management; monitors budgets. *HSCS:* Able to perform the duties required for HSC; serves as enlisted technical or specialty expert; administers medical training programs; reviews all medical reports and presents them to appropriate commands; provides direction and supervision to enlisted personnel in the performance of their functions; supervises procurement, storage, accounting, preservation, and allocation of medical materials; ensures regulation compliance. *HSCM:* Able to perform the duties required for HSCS; serves as senior enlisted technical or specialty administrator; provides senior level supervision and administration to all enlisted personnel; ensures maximum efficiency of the work force and the equipment; prepares general correspondence concerning fiscal, supply, and administrative matters; assists in the formulation of plans, policies, and budget requirements; may supplement the officer corps in the overall supervision and administration of personnel and equipment.

Recommendation, HS3
In the lower-division baccalaureate/associate degree category, 6 semester hours in patient care clinical experience (3/96).

Recommendation, HS2
In the lower-division baccalaureate/associate degree category, 12 semester hours in patient care clinical experience (3/96).

Recommendation, HS1
In the lower-division baccalaureate/associate degree category, 18 semester hours in patient care clinical experience and 3 in personnel supervision (3/96).

Recommendation, HSC
In the lower-division baccalaureate/associate degree category, 18 semester hours in patient care clinical experience and 3 in personnel supervision. In the upper-division baccalaureate category, 3 semester hours for field experience in health services administration and 3 in management problems (3/96).

Recommendation, HSCS
In the lower-division baccalaureate/associate degree category, 18 semester hours in patient care clinical experience and 6 in personnel supervision. In the upper-division baccalaureate category, 6 semester hours for field experience in management and 6 in management problems (3/96).

Recommendation, HSCM
In the lower-division baccalaureate/associate degree category, 18 semester hours in clinical experience and 6 in personnel supervision. In the upper-division baccalaureate category, 6 semester hours for field experience in management, 6 in management problems, 3 in health service policy administration, 3 in health care organization and management, and 3 in report writing (3/96).

CGR-HSD-001

HEALTH SERVICES TECHNICIAN, DENTAL
HSD3

Exhibit Dates: 3/93–Present.

Career Pattern
SN: Seaman (E-3). *HSD3:* Health Services Technician, Dental, Third Class (E-4).

Description
Summary: Performs duties as assistant in the prevention and treatment of disease and injuries and as assistant to the dental officer in treatment of patients. Performs emergency intervention procedures consistent with basic EMT skills, triage, extraction, and transportation; assists in all phases of dentistry; performs basic dental X-ray procedures; prepares dental tray set-ups; performs as an oral surgery assistant; performs cleaning and routine maintenance on dental equipment; provides instruction to patients on oral hygiene; maintains dental records.

Recommendation
In the lower-division baccalaureate/associate degree category, 6 semester hours in dental clinical experience (3/98).

CGR-IV-001

INVESTIGATOR
IV3
IV2
IV1
IVC
IVCS
IVCM

Exhibit Dates: 1/90–2/98.

Occupational Group: Police/Detective Operations.

Career Pattern
SN: Seaman (E-3). *IV3:* Investigator Third Class (E-4). *IV2:* Investigator Second Class (E-5). *IV1:* Investigator First Class (E-6). *IVC:* Chief Investigator (E-7). *IVCS:* Senior Chief Investigator (E-8). *IVCM:* Master Chief Investigator (E-9).

Description
Summary: Provides criminal and internal investigation services; investigates crimes against persons and property through the use of standard investigative tools and techniques; composes and processes necessary incident investigative and intelligence reports. *IV3:* Conducts criminal and background investiga-

tions; processes crime scenes; interviews witnesses; conducts felony stops; serves search warrants; arrests and conducts interrogations; processes prisoners; proficient in the use of investigative equipment and tools including fingerprint and moulage kits, narcotics field test, UV lights, and a variety of cameras. *IV2:* Able to perform the duties required for IV3; sketches crime scenes; conducts crime scene search; conducts photographic and physical lineups; plots coordinates and symbols on maps; collects intelligence information; collects data for use in comprehensive intelligence plans. *IV1:* Able to perform the duties required for IV2; prepares sworn affidavits for search warrants; prepares quarterly case load reports; supervises execution of search warrants, tactical arrest operations, and initial bomb and hostage investigations; supervises protective details and collection of intelligence data. *IVC:* Able to perform the duties required for IV1; coordinates compound crime scenes; provides liaison with other law enforcement agencies; reviews and evaluates investigative procedures; supervises an attack scenario for protective service operations; analyzes multisource intelligence information; conducts risk assessments and determine appropriate countermeasures. *IVCS:* Able to perform the duties required for IVC; trains command on regulations and procedures for investigations, intelligence, and access to Special Agents; establishes and maintains training program for Special Agents and Investigators; develops intelligence production plans; coordinates the dissemination of intelligence information. *IVCM:* Able to perform the duties required for IVCS; evaluates large protective service operations; evaluates intelligence estimates; develops criteria for dissemination of intelligence information; evaluates intelligence production room plans; conducts evaluations of a district or unit's intelligence methods; evaluates training and emergency operation plans.

Recommendation, IV3

In the lower-division baccalaureate/associate degree category, 3 semester hours in criminal investigations and 3 in criminalistics (3/96).

Recommendation, IV2

In the lower-division baccalaureate/associate degree category, 3 semester hours in criminal investigations and 3 in criminalistics (3/96).

Recommendation, IV1

In the lower-division baccalaureate/associate degree category, 3 semester hours in criminal investigations and 3 in criminalistics (3/96).

Recommendation, IVC

In the lower-division baccalaureate/associate degree category, 3 semester hours in criminal investigations, 3 in criminalistics, and 3 in personnel supervision (3/96).

Recommendation, IVCS

In the lower-division baccalaureate/associate degree category, 3 semester hours in criminal investigations, 3 in criminalistics, and 3 in personnel supervision. In the upper-division baccalaureate category, 3 semester hours in curriculum development and 3 in personnel management (3/96).

Recommendation, IVCM

In the lower-division baccalaureate/associate degree category, 3 semester hours in criminal investigations, 3 in criminalistics, and 3 in personnel supervision. In the upper-division baccalaureate category, 3 semester hours in curriculum development, 3 in personnel management, and 3 for field experience in management (3/96).

CGR-IV-002

INVESTIGATOR
(Special Agent)
IV3
IV2
IV1
IVC
IVCS
IVCM

Exhibit Dates: 3/98–Present.

Occupational Group: Police/Detective Operations.

Career Pattern

SN: Seaman (E-3). *IV3:* Investigator Third Class (E-4). *IV2:* Investigator Second Class (E-5). *IV1:* Investigator First Class (E-6). *IVC:* Chief Investigator (E-7). *IVCS:* Senior Chief Investigator (E-8). *IVCM:* Master Chief Investigator (E-9).

Description

Summary: Provides criminal internal and external investigation services; investigates crimes against persons and property through the use of standard investigative tools and techniques; prepares sworn affidavits for search warrants; prepares quarterly case load reports; composes and processes necessary incident investigative and intelligence reports. *IV3:* Conducts criminal and background investigations; processes crime scenes; interviews witnesses; conducts felony stops; serves search and arrest warrants; conducts interrogations; processes prisoners; proficient in the use of investigative equipment and tools including fingerprint and narcotics field test, UV lights, and a variety of cameras. *IV2:* Able to perform the duties required for IV3; sketches crime scenes; conducts crime scene search; conducts photographic and physical lineups; plots coordinates and symbols on maps; collects data for use in comprehensive intelligence plans. *IV1:* Able to perform the duties required for IV2; prepares sworn affidavits for search warrants; prepares quarterly case load reports. *IVC:* Able to perform the duties required for IV1; supervises execution of search warrants, tactical arrest operations, and initial bomb and hostage investigations; supervises protective details and collection of intelligence data; coordinates compound crime scenes; provides liaison with other law enforcement agencies; reviews and evaluates investigative procedures; supervises an attack scenario for protective service operations. *IVCS:* Able to perform the duties required for IVC; trains command on regulations and procedures for investigations, intelligence, and access to Special Agents; establishes and maintains training program for Special Agents and Investigators; coordinates the dissemination of intelligence information. *IVCM:* Able to perform the duties required for IVCS; evaluates large protective service operations; evaluates intelligence estimates; develops criteria for dissemination of intelligence information; evaluates intelligence production room plans; conducts evaluations of a district or unit's intelligence methods; evaluates training and emergency operation plans.

Recommendation, IV3

In the lower-division baccalaureate/associate degree category, 3 semester hours in criminal investigations and 3 in criminal evidence and procedures (3/98).

Recommendation, IV2

In the lower-division baccalaureate/associate degree category, 3 semester hours in criminal investigations and 3 in criminal evidence and procedures (3/98).

Recommendation, IV1

In the lower-division baccalaureate/associate degree category, 3 semester hours in criminal investigations and 3 in criminal evidence and procedures (3/98).

Recommendation, IVC

In the lower-division baccalaureate/associate degree category, 3 semester hours in criminal investigations, 3 in criminal evidence and procedures, and 3 in personnel supervision (3/98).

Recommendation, IVCS

In the lower-division baccalaureate/associate degree category, 3 semester hours in criminal investigations, 3 in criminal evidence and procedures, and 3 in personnel supervision. In the upper-division baccalaureate category, 3 semester hours in personnel management (3/98).

Recommendation, IVCM

In the lower-division baccalaureate/associate degree category, 3 semester hours in criminal investigations, 3 in criminal evidence and procedures, and 3 in personnel supervision. In the upper-division baccalaureate category, 3 semester hours in personnel management and 3 for field experience in management (3/98).

CGR-MK-001

MACHINERY TECHNICIAN
MK3
MK2
MK1
MKC
MKCS
MKCM

Exhibit Dates: 1/90–9/94.

Occupational Group: III (Engineering and Hull).

Career Pattern

FN: Fireman (E-3). *MK3:* Machinery Technician, Third Class (E-4). *MK2:* Machinery Technician, Second Class (E-5). *MK1:* Machinery Technician, First Class (E-6). *MKC:* Chief Machinery Technician (E-7). *MKCS:* Senior Chief Machinery Technician (E-8). *MKCM:* Master Chief Machinery Technician (E-9).

Description

Summary: Operates, maintains, and repairs internal combustion engines, boilers, steam turbines, and main propulsion power transmission equipment; operates, maintains, and repairs auxiliary fireroom, refrigeration, air conditioning, electrical, and machine shop equipment; organizes, leads, and participates in damage control repair parties, and performs maintenance-related administrative functions. *MK3:* Starts, operates, and checks diesel, gas turbine and gasoline engines; lines up fuel systems for receiving and transferring fuel; operates fuel oil centrifuges; performs maintenance or separator/coalescer filter units; operates lubricating oil and cooling systems; conducts tests on engine oil and cooling water; operates distilling plants;

operates refrigeration and air conditioning systems; using hot gas method, defrosts refrigeration units; tests for refrigerant leaks; repacks and adjusts pump stuffing boxes; repacks pressure valves; grinds valve seats and discs; operates auxiliary equipment such as air compressors, laundry, galley, and boat handling equipment; operates hydraulic equipment; performs minor maintenance on hydraulic equipment; uses and cares for basic machine shop handtools and measuring instruments; interprets simple diagrams and blueprints; use oxyacetylene and electric-arc equipment to perform minor brazing, cutting and welding operations; performs specific gravity tests on storage batteries; performs operational maintenance on small boats. *MK2:* Able to perform the duties required for MK3; inspects gas turbine compressors and turbines for axial and radial clearances; inspects test-runs and adjusts diesel and gasoline engines; removes, inspects, and repairs liners, pistons, cylinder heads, valves, piston rings and pins, bearings, pumps, gears, and shafting on gasoline and diesel engines; performs functional tests on gas turbine starting systems; tests unit injectors and other types of fuel nozzles on diesel and gas turbine engines; purges diesel injection systems; chemically treats internal combustion engine cooling system water; services cooling system head exchangers; repairs and maintains distilling plants; performs required maintenance and vacuum pumps and sealing systems; changes and adds lubricating oil in refrigerant compressors; adjusts temperature and pressure controls; evacuates, dehydrates, tests, and recharges refrigerant systems; overhauls, checks and aligns pumps; adjusts valves; repairs auxiliary equipment; fabricates and installs hydraulic system hoses, tubing and fittings; uses dial indicators, micrometers, burdge gages, and depth gages; operates engine lathe for cutting threads, turning tapers, and plain turning; checks electrical distribution systems for grounds; parallels two AC generators; performs megger tests on motors and generators; uses appropriate manuals, prints and other materials to obtain technical repair information. *MK1:* Able to perform the duties required for MK2; overhauls and/or repairs internal combustion engines; inspects and adjusts mechanical and hydraulic governors; coordinates, monitors and controls operation of gas turbines and gas turbine generators; maintains and adjusts fuel system of gasoline, diesel, and gas turbine engines; performs procedures for refueling helicopter; takes reduction gear and thrust bearing clearances; makes minor and emergency repairs to reduction gears; plugs and/or replaces heat exchanger tubes on distilling plants; tests and renews oil seals in refrigerant compressors; performs major repairs on refrigeration and air conditioning compressors; repairs or replaces temperature control valves; troubleshoots and corrects hydraulic system malfunctions; supervises all phases of lathe operations; troubleshoots electrical control circuits; starts and secures gyro compasses; prepares and maintains engineering maintenance reports and records. *MKC:* Able to perform the duties required for MK1; conducts operational tests and makes required adjustments upon completion of an engine overhaul; analyzes reports of discrepancies and malfunctions, and determines corrective action; performs inspections on gas turbine engines, and boiler watersides and firesides; disassembles, cleans, repairs and assembles fuel oil heaters, and conducts required hydrostatic tests; checks main reduction gears for backlash and alignment; disassembles, inspects, repairs, reassembles, and tests constant pressure pump governors; estimates time, labor, and materials required for repair of machinery, structures, equipment, or systems; supervises engineering department on ship; prepares work requests and schedules. *MKCS:* Able to perform the duties required for MKC; serves as enlisted technical or specialty expert; provides technical information concerning maintenance, operation, capabilities, and limitation of engineering equipment and machinery; trains personnel in the principles of operation, and supervises them on all engineering equipment; trains personnel in casualty control procedures; prepares technical and nontechnical lesson plans for instruction of enlisted personnel in maintenance, operating, training, and administration; plans, organizes, and directs work of personnel; supervises the preparation of reports. *MKCM:* Able to perform the duties required for MKCS; serves as senior enlisted technical or specialty administrator; organizes and directs subordinate personnel in the operation, repair, overhaul, and procurement of ship propulsion and auxiliary equipment and supplies; prepares general correspondence concerning fiscal, supply, and administrative matters; assists in the formulation of plans, policies, and budget requirements; may supplement the officer corps in the overall supervision and administration of personnel and equipment; may also supervise personnel in other specialty areas.

Recommendation, MK3

In the lower-division baccalaureate/associate degree category, 1 semester hour in internal combustion engines and 2 in air conditioning and refrigeration (8/95).

Recommendation, MK2

In the lower-division baccalaureate/associate degree category, 3 semester hours in internal combustion engines, 1 in hydraulics, 3 in air conditioning and refrigeration, 2 in basic electricity, 2 in lathe operations, and 2 in precision measuring instruments (8/95).

Recommendation, MK1

In the lower-division baccalaureate/associate degree category, 4 semester hours in internal combustion engines, 2 in hydraulics, 6 in air conditioning and refrigeration, 2 in basic electricity, 2 in lathe operations, 2 in precision measuring instruments, and 2 in personnel supervision (8/95).

Recommendation, MKC

In the lower-division baccalaureate/associate degree category, 6 semester hours in internal combustion engines, 6 in air conditioning and refrigeration, 2 in hydraulics, 2 in lathe operations, 2 in basic electricity, 2 in precision measuring instruments, 3 in personnel supervision, and 3 in principles of management (8/95).

Recommendation, MKCS

In the lower-division baccalaureate/associate degree category, 6 semester hours in internal combustion engines, 6 in air conditioning and refrigeration, 2 in hydraulics, 2 in lathe operations, 2 in basic electricity, 2 in precision measuring instruments, 3 in personnel supervision, and 3 in principles of management. In the upper-division baccalaureate category, 3 semester hours for field experience in management and 3 in management problems (8/95).

Recommendation, MKCM

In the lower-division baccalaureate/associate degree category, 6 semester hours in internal combustion engines, 6 in air conditioning and refrigeration, 2 in hydraulics, 2 in lathe operations, 2 in basic electricity, 2 in precision measuring instruments, 3 in personnel supervision, and 3 in principles of management. In the upper-division baccalaureate category, 6 semester hours for field experience in management and 3 in management problems (8/95).

CGR-MK-002

MACHINERY TECHNICIAN
MK3
MK2
MK1
MKC
MKCS
MKCM

Exhibit Dates: 10/94–Present.

Occupational Group: Nautical Operations/Mechanical Engineering.

Career Pattern

FN: Fireman (E-3). *MK3:* Machinery Technician, Third Class (E-4). *MK2:* Machinery Technician, Second Class (E-5). *MK1:* Machinery Technician, First Class (E-6). *MKC:* Chief Machinery Technician (E-7). *MKCS:* Senior Chief Machinery Technician (E-8). *MKCM:* Master Chief Machinery Technician (E-9).

Description

Summary: Operates, maintains, and repairs internal combustion engines, boilers, steam turbines, and main propulsion power transmission equipment; operates, maintains, and repairs auxiliary fireroom, refrigeration, air conditioning, electrical, and machine shop equipment; organizes, leads, and participates in damage control repair parties, and performs maintenance-related administrative functions. *MK3:* Starts, operates, and checks diesel, gas turbine and gasoline engines; lines up fuel systems for receiving and transferring fuel; operates fuel oil centrifuges; performs maintenance or separator/coalescer filter units; operates lubricating oil and cooling systems; conducts tests on engine oil and cooling water; operates distilling plants; operates refrigeration and air conditioning systems; using hot gas method, defrosts refrigeration units; tests for refrigerant leaks; repacks and adjusts pump stuffing boxes; repacks pressure valves; grinds valve seats and discs; operates auxiliary equipment such as air compressors, laundry, galley, and boat handling equipment; operates hydraulic equipment; performs minor maintenance on hydraulic equipment; uses and cares for basic machine shop handtools and measuring instruments; interprets simple diagrams and blueprints; use oxyacetylene and electric-arc equipment to perform minor brazing, cutting and welding operations; performs specific gravity tests on storage batteries; performs operational maintenance on small boats. *MK2:* Able to perform the duties required for MK3; inspects gas turbine compressors and turbines for axial and radial clearances; inspects test-runs and adjusts diesel and gasoline engines; removes, inspects, and repairs liners, pistons, cylinder heads, valves, piston rings and pins, bearings, pumps, gears, and shafting on gasoline and diesel engines; performs functional tests on gas turbine starting systems; tests unit injectors and other types of fuel nozzles on diesel and gas turbine engines; purges diesel

COAST GUARD ENLISTED RATINGS EXHIBITS 233

injection systems; chemically treats internal combustion engine cooling system water; services cooling system head exchangers; repairs and maintains distilling plants; performs required maintenance and vacuum pumps and sealing systems; changes and adds lubricating oil in refrigerant compressors; adjusts temperature and pressure controls; evacuates, dehydrates, tests, and recharges refrigerant systems; overhauls, checks and aligns pumps; adjusts valves; repairs auxiliary equipment; fabricates and installs hydraulic system hoses, tubing and fittings; uses dial indicators, micrometers, burdge gages, and depth gages; operates engine lathe for cutting threads, turning tapers, and plain turning; checks electrical distribution systems for grounds; parallels two AC generators; performs megger tests on motors and generators; uses appropriate manuals, prints and other materials to obtain technical repair information. *MK1:* Able to perform the duties required for MK2; overhauls and/or repairs internal combustion engines; inspects and adjusts mechanical and hydraulic governors; coordinates, monitors and controls operation of gas turbines and gas turbine generators; maintains and adjusts fuel system of gasoline, diesel, and gas turbine engines; performs procedures for refueling helicopter; takes reduction gear and thrust bearing clearances; makes minor and emergency repairs to reduction gears; plugs and/or replaces heat exchanger tubes on distilling plants; tests and renews oil seals in refrigerant compressors; performs major repairs on refrigeration and air conditioning compressors; repairs or replaces temperature control valves; troubleshoots and corrects hydraulic system malfunctions; supervises all phases of lathe operations; troubleshoots electrical control circuits; starts and secures gyro compasses; prepares and maintains engineering maintenance reports and records. *MKC:* Able to perform the duties required for MK1; conducts operational tests and makes required adjustments upon completion of an engine overhaul; analyzes reports of discrepancies and malfunctions, and determines corrective action; performs inspections on gas turbine engines, and boiler watersides and firesides; disassembles, cleans, repairs, and assembles fuel oil heaters, and conducts required hydrostatic tests; checks main reduction gears for backlash and alignment; disassembles, inspects, repairs, reassembles, and tests constant pressure pump governors; estimates time, labor, and materials required for repair of machinery, structures, equipment, or systems; supervises engineering department on ship; prepares work requests and schedules. *MKCS:* Able to perform the duties required for MKC; serves as enlisted technical or specialty expert; provides technical information concerning maintenance, operation, capabilities, and limitation of engineering equipment and machinery; trains personnel in the principles of operation, and supervises them on all engineering equipment; trains personnel in casualty control procedures; prepares technical and nontechnical lesson plans for instruction of enlisted personnel in maintenance, operating, training, and administration; plans, organizes, and directs work of personnel; supervises the preparation of reports. *MKCM:* Able to perform the duties required for MKCS; serves as senior enlisted technical or specialty administrator; organizes and directs subordinate personnel in the operation, repair, overhaul, and procurement of ship propulsion and auxiliary equipment and supplies; prepares general correspondence concerning fiscal, supply, and administrative matters; assists in the formulation of plans, policies, and budget requirements; may supplement the officer corps in the overall supervision and administration of personnel and equipment; may also supervise personnel in other specialty areas.

Recommendation, MK3

In the lower-division baccalaureate/associate degree category, 1 semester hour in internal combustion engines, 2 in air conditioning and refrigeration, 2 in machine tool technology, 1 in low pressure boiler operations, and 1 in blueprint reading (8/95).

Recommendation, MK2

In the lower-division baccalaureate/associate degree category, 3 semester hours in internal combustion engines, 1 in hydraulics, 3 in air conditioning and refrigeration, 2 in machine tool technology, 2 in low pressure boiler operations, 1 in blueprint reading, and 2 in computer applications (8/95).

Recommendation, MK1

In the lower-division baccalaureate/associate degree category, 4 semester hours in internal combustion engines, 2 in hydraulics, 6 in air conditioning and refrigeration, 3 in low pressure boiler operations, 3 in machine tool technology, 1 in blueprint reading, 2 in computer applications, and 2 in personnel supervision (8/95).

Recommendation, MKC

In the lower-division baccalaureate/associate degree category, 6 semester hours in internal combustion engines, 6 in air conditioning and refrigeration, 2 in hydraulics, 3 in computer applications, 3 in low pressure boiler operations, 3 in machine tool technology,1 in blueprint reading, and 3 in personnel supervision (8/95).

Recommendation, MKCS

In the lower-division baccalaureate/associate degree category, 6 semester hours in internal combustion engines, 6 in air conditioning and refrigeration, 2 in hydraulics, 3 in computer applications, 3 in low pressure boiler operations, 3 in machine tool technology, 1 in blueprint reading, and 3 in personnel supervision. In the upper-division baccalaureate category, 3 semester hours for field experience in management and 3 in management problems (8/95).

Recommendation, MKCM

In the lower-division baccalaureate/associate degree category, 6 semester hours in internal combustion engines, 6 in air conditioning and refrigeration, 2 in hydraulics, 3 in computer applications, 3 in machine tool technology, 3 in low pressure boiler operations, and 3 in personnel supervision. In the upper-division baccalaureate category, 6 semester hours for field experience in management and 3 in management problems (8/95).

CGR-MST-001

MARINE SCIENCE TECHNICIAN
- MST3
- MST2
- MST1
- MSTC
- MSTCS
- MSTCM

Exhibit Dates: 1/90–8/92.

Occupational Group: I (Deck).

Career Pattern

SN: Seaman (E-3). MST3: Marine Science Technician, Third Class (E-4). MST2: Marine Science Technician, Second Class (E-5). MST1: Marine Science Technician, First Class (E-6). MSTC: Chief Marine Science Technician (E-7). MSTCS: Senior Chief Marine Science Technician (E-8). MSTCM: Master Chief Marine Science Technician (E-9).

Description

Summary: Observes, records, and analyzes environmental and scientific data; conducts field monitoring and laboratory analysis for chemical and oil identification; operates and maintains Coast Guard data processing systems; retrieves and assimilates data from remote sensing and local observations; employs numerical models and provides environmental information for conducting Coast Guard operations. *MST3:* Prepares and maintains user documentation of basic data processing procedures; operates small programmable computers; uses standard meteorological equipment; conducts meteorological observations; works with and maintains standard oceanographic instrumentation; demonstrates safety precautions in operation of a chemistry laboratory in accordance with Coast Guard regulations. *MST2:* Able to perform the duties required for MST3; maintains a tape or disk library; decodes and plots standard meteorological observations; analyzes surface and upper air charts; inspects areas for hazardous waste materials and suggests corrective measures; forecasts and plots ice, tide, and current data. *MST1:* Able to perform the duties required for MST2; schedules program runs; troubleshoots terminal, peripheral, and program failures and corrects as necessary; performs operator checks and preventive maintenance on small computers; develops oil sampling programs; processes results of pollution investigations; interprets results of chemical analysis; supervises personnel in making standard meteorological observations; prepares written instructions; organizes and evaluates training programs; develops and implements a program for evaluating personnel. *MSTC:* Able to perform the duties required for MST1; plans, implements, and monitors safety programs; prepares and submits budget requests; instructs and supervises personnel in the use and maintenance of scientific data collection. *MSTCS:* Able to perform the duties required for MSTC; serves as enlisted technical or specialty expert; plans and organizes activities to achieve objectives; develops budgets and monitors expenditures. *MSTCM:* Able to perform the duties required for MSTCS; serves as senior enlisted technical or specialty administrator; performs operational administrative functions; establishes objectives and sets priorities; prepares general correspondence concerning fiscal, supply, and administrative matters; assists in the formulation of plans, policies, and budget requirements; may supplement the officer corps in the overall supervision and administration of personnel and equipment; may also supervise personnel in other specialty areas.

Recommendation, MST3

In the lower-division baccalaureate/associate degree category, 3 semester hours in marine instrumentation, 3 in basic computer science, and 3 in basic meteorology (7/84).

Recommendation, MST2

In the lower-division baccalaureate/associate degree category, 3 semester hours in marine instrumentation, 3 in basic computer science, 3 in basic meteorology, 3 in physical science, and 2 in meteorology (7/84).

Recommendation, MST1

In the lower-division baccalaureate/associate degree category, 3 semester hours in marine instrumentation, 3 in basic computer science, 3 in basic meteorology, 3 in physical science, 2 in meteorology, and 2 in personnel supervision (7/84).

Recommendation, MSTC

In the lower-division baccalaureate/associate degree category, 3 semester hours in marine instrumentation, 3 in basic computer science, 3 in basic meteorology, 3 in physical science, 2 in meteorology, 3 in personnel supervision, and 3 in principles of management (7/84).

Recommendation, MSTCS

In the lower-division baccalaureate/associate degree category, 3 semester hours in marine instrumentation, 3 in basic computer science, 3 in basic meteorology, 3 in physical science, 2 in meteorology, 3 in personnel supervision, and 3 in principles of management. In the upper-division baccalaureate category, 3 semester hours for field experience in management and 3 in management problems (7/84).

Recommendation, MSTCM

In the lower-division baccalaureate/associate degree category, 3 semester hours in marine instrumentation, 3 in basic computer science, 3 in basic meteorology, 3 in physical science, 2 in meteorology, 3 in personnel supervision, and 3 in principles of management. In the upper-division baccalaureate category, 6 semester hours for field experience in management and 3 in management problems (7/84).

CGR-MST-002

MARINE SCIENCE TECHNICIAN
  MST3
  MST2
  MST1
  MSTC
  MSTCS
  MSTCM

Exhibit Dates: 9/92–Present.

Occupational Group: Environmental Science.

Career Pattern

SN: Seaman (E-3). *MST3:* Marine Science Technician, Third Class (E-4). *MST2:* Marine Science Technician, Second Class (E-5). *MST1:* Marine Science Technician, First Class (E-6). *MSTC:* Chief Marine Science Technician (E-7). *MSTCS:* Senior Chief Marine Science Technician (E-8). *MSTCM:* Master Chief Marine Science Technician (E-9).

Description

Summary: Observes, records, and analyzes environmental and scientific data; conducts field monitoring and laboratory analysis for chemical and oil identification; operates and maintains Coast Guard data processing systems; retrieves and assimilates data from remote sensing and local observations; employs numerical models and provides environmental information for conducting Coast Guard operations. *MST3:* Prepares and maintains user documentation of basic data processing procedures; operates small programmable computers; uses standard meteorological equipment; conducts meteorological observations; works with and maintains standard oceanographic instrumentation; demonstrates safety precautions in operation of a chemistry laboratory in accordance with Coast Guard regulations. *MST2:* Able to perform the duties required for MST3; maintains a tape or disk library; decodes and plots standard meteorological observations; analyzes surface and upper air charts; inspects areas for hazardous waste materials and suggests corrective measures; forecasts and plots ice, tide, and current data. *MST1:* Able to perform the duties required for MST2; schedules program runs; troubleshoots terminal, peripheral, and program failures and corrects as necessary; performs operator checks and preventive maintenance on small computers; develops oil sampling programs; processes results of pollution investigations; interprets results of chemical analysis; supervises staff in making standard meteorological observations; prepares written instructions; organizes and evaluates training programs; develops and implements a program for evaluating staff. *MSTC:* Able to perform the duties required for MST1; plans, implements, and monitors safety programs; prepares and submits budget requests; instructs and supervises staff in the use and maintenance of scientific data collection. *MSTCS:* Able to perform the duties required for MSTC; serves as enlisted technical or specialty expert; plans and organizes activities to achieve objectives; develops budgets and monitors expenditures. *MSTCM:* Able to perform the duties required for MSTCS; serves as senior enlisted technical or specialty administrator; performs operational administrative functions; establishes objectives and sets priorities; prepares general correspondence concerning fiscal, supply, and administrative matters; assists in the formulation of plans, policies, and budget requirements; develop training materials to be used by Marine Science Technicians for rate advancement; may supplement the officer corps in the overall supervision and administration of personnel and equipment; may also supervise personnel in other specialty areas.

Recommendation, MST3

In the lower-division baccalaureate/associate degree category, 3 semester hours in marine instrumentation, 3 in records and information management, and 3 in basic meteorology (3/96).

Recommendation, MST2

In the lower-division baccalaureate/associate degree category, 3 semester hours in marine instrumentation, 3 in records and information management, 3 in basic meteorology, 3 in physical science, 3 in environmental safety, and 1 in hazardous material investigation (3/96).

Recommendation, MST1

In the lower-division baccalaureate/associate degree category, 3 semester hours in marine instrumentation, 3 in records and information management, 3 in basic meteorology, 3 in physical science, 3 in environmental safety, 1 in hazardous material investigation, and 2 in personnel supervision (3/96).

Recommendation, MSTC

In the lower-division baccalaureate/associate degree category, 3 semester hours in marine instrumentation, 3 in records and information management, 3 in basic meteorology, 3 in physical science, 3 in environmental safety, 1 in hazardous material investigation, and 2 in personnel supervision. In the upper-division baccalaureate category, 3 semester hours in personnel management (3/96).

Recommendation, MSTCS

In the lower-division baccalaureate/associate degree category, 3 semester hours in marine instrumentation, 3 in records and information management, 3 in basic meteorology, 3 in physical science, 3 in environmental safety, 1 in hazardous material investigation, and 2 in personnel supervision. In the upper-division baccalaureate category, 3 semester hours for field experience in management and 3 in personnel management (3/96).

Recommendation, MSTCM

In the lower-division baccalaureate/associate degree category, 3 semester hours in marine instrumentation, 3 in records and information management, 3 in basic meteorology, 3 in physical science, 3 in environmental safety, 1 in hazardous material investigation, and 2 in personnel supervision.. In the upper-division baccalaureate category, 3 semester hours for field experience in management, 3 in personnel management, and 3 in curriculum development (3/96).

CGR-PA-001

PUBLIC AFFAIRS SPECIALIST
  (Photojournalist)
  PA3
  PA2
  PA1
  PAC
  PACS
  PACM

Exhibit Dates: 1/90–5/91.

Occupational Group: IV (Administrative and Clerical).

Career Pattern

SN: Seaman (E-3). *PA3:* Photojournalist Third Class (E-4). *PA2:* Photojournalist Second Class (E-5). *PA1:* Photojournalist First Class (E-6). *PAC:* Chief Photojournalist (E-7). *PACS:* Senior Chief Photojournalist (E-8). *PACM:* Master Chief Photojournalist (E-9).

Description

Summary: Supervises or participates in the administration of public affairs activities in the area of media relations, community relations, and internal information programs; arranges and conducts interviews; prepares news releases, fact sheets, and feature stories; writes and edits news copy; operates photographic equipment; prepares art work and designs layout for various publications. *PA3:* Researches and writes news releases and feature articles; covers news events; processes news photographs and writes captions; prepares material for media release; arranges and conducts interviews; edits news copy; prepares art work and designs layout for various Coast Guard publications. *PA2:* Able to perform the duties required for PA3; supervises inventories of office equipment; evaluates photographic equipment and makes acquisition recommendations; has increased practical experience and knowledge of visual/photographic theory. *PA1:* Able to perform the duties required for PA2; supervises operation of a district office; has increased knowledge of how public affairs programs interact with the media; acts as representative of the Coast Guard to the community; operates 16 millimeter equipment and uses 35 millimeter

COAST GUARD ENLISTED RATINGS EXHIBITS

for special photographic techniques. *PAC:* Able to perform the duties required for PA1; administers public affairs office; supervises work assignments; prepares budgets; coordinates interagency efforts. *PACS:* Able to perform the duties required for PAC; serves as enlisted specialty expert; provides leadership, supervision and administration to the entire rating; organizes, directs and coordinates instructional and training programs. *PACM:* Able to perform the duties required for PACS; serves as senior enlisted specialty administrator; prepares general correspondence concerning fiscal, supply, and administrative matters; assists in the formulation of plans, policies, and budget requirements; may supplement the officer corps in the overall supervision and administration of personnel.

Recommendation, PA3

In the lower-division baccalaureate/associate degree category, 3 semester hours in interviewing, 3 in news writing, 1 in feature writing, 3 in news reporting, 2 in news editing, 2 in technical writing, 3 in basic audio-visual materials, 3 in still photography, 1 in technical photography, 3 in layout/design, and 1 for field experience in public affairs; if served as Cinematographer (identified by qualification code 41), add 3 semester hours in film production; if served as electronic Media Specialist (identified by qualification code 42), add 3 semester hours in video production (6/84).

Recommendation, PA2

In the lower-division baccalaureate/associate degree category, 3 semester hours in interviewing, 3 in news writing, 2 in feature writing, 3 in news reporting, 3 in news editing, 3 in technical writing, 3 in basic audio-visual materials, 3 in still photography, 1 in technical photography, 3 in layout/design, 2 for field experience in public affairs, 2 in art direction, 2 in office management, 1 in principles of accounting, and 2 for field experience in film and video; if served as Cinematographer (identified by qualification code 41), add 3 semester hours in film production; if served as Electronic Media Specialist (identified by qualification code 42), add 3 semester hours in video production (6/84).

Recommendation, PA1

In the lower-division baccalaureate/associate degree category, 3 semester hours in interviewing, 3 in newswriting, 3 in feature writing, 3 in news reporting, 3 in news editing, 3 in technical writing, 3 in basic audio-visual materials, 3 in still photography, 1 in technical photography, 3 in layout/design, 3 for field experience in public affairs, 2 in art direction, 3 in office management, 1 in principles of accounting, 2 for field experience in film and video, and 2 in personnel supervision; if served as Cinematographer (identified by qualification code 41), add 3 semester hours in film production; if served as Electronic Media Specialist (identified by qualification code 42), add 3 semester hours in video production (6/84).

Recommendation, PAC

In the lower-division baccalaureate/associate degree category, 3 semester hours in interviewing, 3 in news writing, 3 in feature writing, 3 in news reporting, 3 in news editing, 3 in technical writing, 3 in basic audio-visual materials, 3 in still photography, 1 in technical photography, 3 in layout/design, 3 for field experience in public affairs, 2 in art direction, 3 in office management, 1 in principles of accounting, 2 for field experience in film and video, 3 in personnel supervision, and 3 in principles of management; if served as Cinematographer (identified by qualification code 41), add 3 semester hours in film production; if served as Electronic Media Specialist (identified by qualification code 42), add 3 semester hours in video production. In the upper-division baccalaureate category, 2 semester hours in organizational communication and 3 for field experience in management (6/84).

Recommendation, PACS

In the lower-division baccalaureate/associate degree category, 3 semester hours in interviewing, 3 in news writing, 3 in feature writing, 3 in news reporting, 3 in news editing, 3 in technical writing, 3 in basic audio-visual materials, 3 in still photography, 1 in technical photography, 3 in layout/design, 3 for field experience in public affairs, 2 in art direction, 3 in office management, 1 in principles of accounting, 2 for field experience in film and video, 3 in personnel supervision, and 3 in principles of management; if served as Cinematographer (identified by qualification code 41), add 3 semester hours in film production; if served as Electronic Media Specialist (identified by qualification code 42), add 3 semester hours in video production. In the upper-division baccalaureate category, 3 semester hours in organizational communication and 3 for field experience in management (6/84).

Recommendation, PACM

In the lower-division baccalaureate/associate degree category, 3 semester hours in interviewing, 3 in news writing, 3 in feature writing, 3 in news reporting, 3 in news editing, 3 in technical writing, 3 in basic audio-visual materials, 3 in still photography, 1 in technical photography, 3 in layout/design, 3 for field experience in public affairs, 2 in art direction, 3 in office management, 1 in principles of accounting, 2 for field experience in film and video, 3 in personnel supervision, and 3 in principles of management; if served as Cinematographer (identified by qualification code 41), add 3 semester hours in film production; if served as Electronic Media Specialist (identified by qualification code 42), add 3 semester hours in video production. In the upper-division baccalaureate category, 3 semester hours in organizational communication and 6 for field experience in management (6/84).

CGR-PA-002

PUBLIC AFFAIRS SPECIALIST
   PA3
   PA2
   PA1
   PAC
   PACS
   PACM

Exhibit Dates: 6/91–Present.

Occupational Group: IV (Administrative and Clerical).

Career Pattern

SN: Seaman (E-3). *PA3:* Public Affairs Specialist Third Class (E-4). *PA2:* Public Affairs Specialist Second Class (E-5). *PA1:* Public Affairs Specialist First Class (E-6). *PAC:* Chief Public Affairs Specialist (E-7). *PACS:* Senior Chief Public Affairs Specialist (E-8). *PACM:* Master Chief Public Affairs Specialist (E-9).

Description

Summary: Participates in or supervises the administration of public affairs activities in the area of media relations, community relations, and internal information programs; arranges and conducts interviews; prepares news releases, fact sheets, and feature stories; writes and edits news copy; operates photographic equipment; prepares art work and designs layout for various publications. *PA3:* Researches and writes news releases and feature articles; covers news events; processes news photographs and writes captions; prepares material for media releases; arranges and conducts interviews; edits news copy. *PA2:* Able to perform the duties required for PA3; supervises inventories of office equipment; evaluates photographic equipment and makes acquisition recommendations; has increased practical experience and knowledge of visual/photographic theory. *PA1:* Able to perform the duties required for PA2; supervises operation of a district office; has increased knowledge of how public affairs programs interact with the media; acts as representative of the Coast Guard to the community. *PAC:* Able to perform the duties required for PA1; administers public affairs office; supervises work assignments; prepares budgets; coordinates interagency efforts. *PACS:* Able to perform the duties required for PAC; serves as enlisted specialty expert; provides leadership, supervision, and administration to the entire rating; organizes, directs, and coordinates instructional and training programs. *PACM:* Able to perform the duties required for PACS; coordinates public affairs; serves as senior enlisted specialty administrator; prepares general correspondence concerning fiscal, supply, and office management and administrative matters; assists in the formulation of plans, policies, and budget requirements; may supplement the officer corps in the overall supervision and administration of personnel.

Recommendation, PA3

In the lower-division baccalaureate/associate degree category, 3 semester hours in basic news reporting, 3 in basic photography, 2 in copy editing, and 1 in layout and design (6/91).

Recommendation, PA2

In the lower-division baccalaureate/associate degree category, 3 semester hours in basic news reporting, 3 in basic photography, 3 in copy editing, 3 in layout and design, 3 in feature writing, and 3 in public relations writing (6/91).

Recommendation, PA1

In the lower-division baccalaureate/associate degree category, 3 semester hours in basic news reporting, 3 in basic photography, 3 in copy editing, 3 in layout and design, 3 in feature writing, and 3 in public relations writing. In the upper-division baccalaureate category, 3 semester hours in photojournalism, 3 in newspaper production, 3 in advanced news reporting, and 3 in media relations (6/91).

Recommendation, PAC

In the lower-division baccalaureate/associate degree category, 3 semester hours in basic news reporting, 3 in basic photography, 3 in copy editing, 3 in layout and design, 3 in feature writing, and 3 in public relations writing. In the upper-division baccalaureate category, 3 semester hours in photojournalism, 3 in newspaper production, 3 in advanced news reporting, 3 in media relations, 3 in technical writing, 3 in publication management, 3 in personnel management, and 3 in advanced media relations (6/91).

Recommendation, PACS

In the lower-division baccalaureate/associate degree category, 3 semester hours in basic news reporting, 3 in basic photography, 3 in copy editing, 3 in layout and design, 3 in feature writing, and 3 in public relations writing. In the upper-division baccalaureate category, 3 semester hours in photojournalism, 3 in newspaper production, 3 in advanced news reporting, 3 in media relations, 3 in technical writing, 3 in publication management, 3 in personnel management, 3 in advanced media relations, 3 in business administration, 3 in personnel management, and 3 in interpersonal communications (6/91).

Recommendation, PACM

In the lower-division baccalaureate/associate degree category, 3 semester hours in basic news reporting, 3 in basic photography, 3 in copy editing, 3 in layout and design, 3 in feature writing, and 3 in public relations writing. In the upper-division baccalaureate category, 3 semester hours in photojournalism, 3 in newspaper production, 3 in advanced news reporting, 3 in media relations, 3 in technical writing, 3 in publication management, 3 in personnel management, 3 in advanced media relations, 3 in business administration, 3 in personnel management, 3 in interpersonal communications, 3 in organizational management, 3 in public affairs planning, and 3 in public relations policy management (6/91).

CGR-PS-001

PORT SECURITYMAN
- PS3
- PS2
- PS1
- PSC
- PSCS
- PSCM

Exhibit Dates: 1/90–1/93.

Occupational Group: VI (Port Security).

Career Pattern

SN: Seaman (E-3). *PS3:* Port Securityman, Third Class (E-4). *PS2:* Port Securityman, Second class (E-5). *PS1:* Port Securityman, First Class (E-6). *PSC:* Chief Port Securityman (E-7). *PSCS:* Senior Chief Port Securityman (E-8). *PSCM:* Master Chief Port Securityman (E-9).

Description

Summary: Enforces the regulations and orders relative to the protection and security of vessels, harbors, ports, and waterfront facilities; supervises and controls the safe handling, transportation, stowage, and storage of explosives and other dangerous cargoes; in restricted areas, prevents unauthorized persons from entering upon the vessel or waterfront facility, and carefully examines all authorized persons; assists in combating fires. *PS3:* Performs general police patrol duty; arrests persons and collects evidence; controls access to restricted areas; conducts surveillance of vessels and identifies offenses; inspects vessels, cargo, and waterfront facilities; inspects documentation; reports on pollution and monitors cleanup; supervises cargo loading; inspects and operates fire equipment. *PS2:* Able to perform the duties required for PS3; conducts investigations and interviews; processes applications; takes fingerprints; conducts vessel searches, using electronic survey equipment; identifies various types of oil. *PS1:* Able to perform the duties required for PS2; writes investigative reports; instructs subordinate personnel in fingerprinting techniques, procedures for manning a security zone, use of service pistol, and safety requirements for explosives; serves as liaison with other law enforcement agencies; completes permit applications for explosives and other dangerous cargo; conducts investigations of pollution incidents; analyzes pollution reports; supervises cleanup of pollution incidents; knows fire rescue techniques; supervises use of loading equipment to determine safe operation; supervises layout of firefighting equipment. *PSC:* Able to perform the duties required for PS1; plans and conducts training programs; supervises cargo transfers and inspection procedures; prepares local pollution contingency plan; cooperates with private cleanup contractors; evaluates fire equipment performance; plans, implements, and monitors security of vessels, harbors, and ports. *PSCS:* Able to perform the duties required for PSC; serves as safety officer; develops budgets and monitors expenditures; supervises and directs law enforcement procedures, explosives-loading detail, and the loading, stowing, security, and offloading of dangerous cargo. *PSCM:* Able to perform the duties required for PSCS; serves as senior enlisted technical or specialty administrator; manages and supervises operations and procedures involving security and law enforcement, dangerous cargo, and safety and fire prevention; develops training and safety programs; establishes objectives and sets priorities; performs operational administrative functions; prepares correspondence concerning fiscal, supply, and administrative matters; assists in the formulation of plans, policies, and budget requirements; may supplement the officer corps in the overall supervision and administration of personnel and equipment; may also supervise personnel in other specialty areas.

Recommendation, PS3

In the lower-division baccalaureate/associate degree category, 6 semester hours in criminal procedures, 6 in marine cargo handling, 2 in basic marine pollution analysis, and 3 in basic fire safety (7/84).

Recommendation, PS2

In the lower-division baccalaureate/associate degree category, 6 semester hours in criminal procedures, 6 in marine cargo handling, 2 in basic marine pollution analysis, and 3 in basic fire safety. In the upper-division baccalaureate category, 3 semester hours in advanced criminal procedures, 3 in advanced marine cargo handling, and 2 in advanced marine pollution analysis (7/84).

Recommendation, PS1

In the lower-division baccalaureate/associate degree category, 6 semester hours in criminal procedures, 6 in cargo handling, 2 in basic marine pollution analysis, 3 in basic fire safety, and 2 in personnel supervision. In the upper-division baccalaureate category, 3 semester hours in advanced criminal procedures, 3 in advanced marine cargo handling, 2 in advanced marine pollution analysis, 3 in interagency law enforcement liaison, and 6 in advanced fire safety technology (7/84).

Recommendation, PSC

In the lower-division baccalaureate/associate degree category, 6 semester hours in criminal procedures, 6 in marine cargo handling, 2 in basic marine pollution analysis, 3 in basic fire safety, 3 in personnel supervision, and 3 in principles of management. In the upper-division baccalaureate category, 3 semester hours in advanced criminal procedures, 3 in advanced marine cargo handling, 2 in advanced marine pollution analysis, 3 in interagency law enforcement liaison, 6 in advanced fire safety technology, 3 in advanced hazardous cargo handling, and 3 in advanced marine pollution control and cleanup (7/84).

Recommendation, PSCS

In the lower-division baccalaureate/associate degree category, 6 semester hours in criminal procedures, 6 in marine cargo handling, 2 in basic marine pollution analysis, 3 in basic fire safety, 3 in personnel supervision, and 3 in principles of management. In the upper-division baccalaureate category, 3 semester hours in advanced criminal procedures, 3 in advanced marine cargo handling, 2 in advanced marine pollution analysis, 3 in interagency law enforcement liaison, 6 in advanced fire safety technology, 3 in advanced hazardous cargo handling, 3 in advanced marine pollution control and cleanup, 3 for field experience in management, and 3 in management problems (7/84).

Recommendation, PSCM

In the lower-division baccalaureate/associate degree category, 6 semester hours in criminal procedures, 6 in marine cargo handling, 2 in basic marine pollution analysis, 3 in basic fire safety, 3 in personnel supervision, and 3 in principles of management. In the upper-division baccalaureate category, 3 semester hours in advanced criminal procedures, 3 in advanced marine cargo handling, 2 in advanced marine pollution analysis, 3 in interagency law enforcement liaison, 6 in advanced fire safety technology, 3 in advanced hazardous cargo handling, 3 in advanced marine pollution control and cleanup, 6 for field experience in management, and 3 in management problems (7/84).

CGR-PS-002

PORT SECURITYMAN
- PS3
- PS2
- PS1
- PSC
- PSCS
- PSCM

Exhibit Dates: 2/93–Present.

Occupational Group: Police/Security Operations.

Career Pattern

SN: Seaman (E-3). *PS3:* Port Securityman, Third Class (E-4). *PS2:* Port Securityman, Second class (E-5). *PS1:* Port Securityman, First Class (E-6). *PSC:* Chief Port Securityman (E-7). *PSCS:* Senior Chief Port Securityman (E-8). *PSCM:* Master Chief Port Securityman (E-9).

Description

Summary: Enforces the regulations and orders relative to the protection and security of vessels, harbors, ports, and waterfront facilities; supervises and controls the safe handling, transportation, stowage, and storage of explosives and other dangerous cargoes; in restricted areas, prevents unauthorized persons from entering upon the vessel or waterfront facility, and carefully examines all authorized persons; assists in combating fires. *PS3:* Performs general police patrol duty; arrests persons and collects evidence; controls access to restricted areas; conducts surveillance of vessels and identifies offenses; inspects vessels, cargo, and waterfront facilities; inspects documentation; reports on pollution and monitors cleanup; supervises cargo loading; inspects and operates fire equipment. *PS2:* Able to perform the duties required for PS3; conducts investigations and interviews; processes applications; takes fingerprints; conducts vessel searches, using electronic survey equipment; identifies various types of oil. *PS1:* Able to perform the duties required for PS2; writes investigative reports; instructs subordinate personnel in fingerprinting techniques, procedures for manning a security zone, use of service pistol, and safety requirements for explosives; serves as liaison with other law enforcement agencies; completes permit applications for explosives and other dangerous cargo; conducts investigations of pollution incidents; analyzes pollution reports; supervises cleanup of pollution incidents; knows fire rescue techniques; supervises use of loading equipment to determine safe operation; supervises layout of fire fighting equipment. *PSC:* Able to perform the duties required for PS1; plans and conducts training programs; supervises cargo transfers and inspection procedures; prepares local pollution contingency plan; cooperates with private cleanup contractors; evaluates fire equipment performance; plans, implements, and monitors security of vessels, harbors, and ports. *PSCS:* Able to perform the duties required for PSC; serves as safety officer; develops budgets and monitors expenditures; supervises and directs law enforcement procedures, explosives-loading detail, and the loading, stowing, security, and offloading of dangerous cargo. *PSCM:* Able to perform the duties required for PSCS; serves as senior enlisted technical or specialty administrator; manages and supervises operations and procedures involving security and law enforcement, dangerous cargo, and safety and fire prevention; develops training and safety programs; establishes objectives and sets priorities; performs operational administrative functions; prepares correspondence concerning fiscal, supply, and administrative matters; assists in the formulation of plans, policies, and budget requirements; may supplement the officer corps in the overall supervision and administration of personnel and equipment; may also supervise personnel in other specialty areas.

Recommendation, PS3

In the lower-division baccalaureate/associate degree category, 3 semester hours in maritime regulations, 3 in law enforcement, 3 in fire suppression, 3 in warehousing, and 2 in environmental safety (3/96).

Recommendation, PS2

In the lower-division baccalaureate/associate degree category, 3 semester hours in maritime regulations, 3 in law enforcement, 3 in fire suppression, 3 in warehousing, and 2 in environmental safety (3/96).

Recommendation, PS1

In the lower-division baccalaureate/associate degree category, 3 semester hours in maritime regulations, 3 in law enforcement, 3 in fire suppression, 3 in warehousing, and 2 in environmental safety (3/96).

Recommendation, PSC

In the lower-division baccalaureate/associate degree category, 3 semester hours in maritime regulations, 3 in law enforcement, 3 in fire suppression, 3 in warehousing, 2 in environmental safety, 3 in personnel supervision, and 3 in introduction to OSHA regulations. In the upper-division baccalaureate category, 3 semester hours for field experience in management, 3 in personnel management, and 3 in curriculum development (3/96).

Recommendation, PSCS

In the lower-division baccalaureate/associate degree category, 3 semester hours in maritime regulations, 3 in law enforcement, 3 in fire suppression, 3 in warehousing, 2 in environmental safety, 3 in personnel supervision, and 3 in introduction to OSHA regulations. In the upper-division baccalaureate category, 3 semester hours for field experience in management, 3 in personnel management, and 3 in curriculum development (3/96).

Recommendation, PSCM

In the lower-division baccalaureate/associate degree category, 3 semester hours in maritime regulations, 3 in law enforcement, 3 in fire suppression, 3 in warehousing, 2 in environmental safety, 3 in personnel supervision, and 3 in introduction to OSHA regulations. In the upper-division baccalaureate category, 3 semester hours for field experience in management, 3 in personnel management, and 3 in curriculum development (3/96).

CGR-QM-001

QUARTERMASTER
QM3
QM2
QM1
QMC
QMCS
QMCM

Exhibit Dates: 1/90–1/94.

Occupational Group: 1 (Deck).

Career Pattern

SN: Seaman (E-3). *QM3:* Quartermaster Third Class (E-4). *QM2:* Quartermaster Second Class (E-5). *QM1:* Quartermaster First Class (E-6). *QMC:* Chief Quartermaster (E-7). *QMCS:* Senior Chief Quartermaster (E-8). *QMCM:* Master Chief Quartermaster (E-9).

Description

Summary: Stands watch as assistant to the officers of the deck and to the navigator; serves as helmsman during time when precise ship control is required; performs communications, navigation, and bridgewatch duties; procures, corrects, uses, and stows navigational charts and publications; maintains navigational instruments and keeps navigational time; participates in ceremonies conducted in accordance with national and foreign observance and customs; sends and receives visual messages; serves as executive petty officer. *QM3:* Inventories, procures, uses and corrects nautical charts and publications; uses basic course plotting instruments, lead line, depth sounder, and compass; indentifies aids to navigation; handles plain language radio communications and visual communication; maintains compass record book and weather observation sheet. *QM2:* Able to perform the duties required for QM3; uses Coast Guard publications concerning Nautical Rules of the Road, manuevering board and relative bearings; selects charts for voyage planning; determines danger angles and danger bearings; determines ship's position by celestial observations; conducts basic weather observations and understands their significance; supervises bridge personnel; inventories installed equipment and spare parts; orders repair parts and tools. *QM1:* Able to perform the duties required for QM2; adjusts compasses and prepares deviation tables; adjusts and aligns sextants and stadimeters; interprets weather charts; plots probable path and location of storm centers; prepares a great circle track using various methods; maintains the ship's equipment configuration accounting system; prepares weekly schedules of preventive maintenance; supervises preventive maintenance procedures; prepares oceanographic reports such as wave observation log, ship's ice log, adjusted track log, and sounding journal; supervises and trains personnel in navigation, watchstanding, and use and care of navigational charts, tables, and publications. *QMC:* Able to perform the duties required for QM1; understands the use and time requirements on various files and logs; understands various types of messages; prepares quarterly schedules of preventive maintenance; requisitions equipment; interviews, selects, and evaluates personnel for the navigation department; plans emergency drills; conducts briefings; understands the law of cyclonic storms and storm avoidance measures; understands the limitations of radar and loran aids to navigation; understands the use of radio direction finding (RDF) equipment; organizes and maintains technical library; prepares and submits budget requests; plans and supervises on-the-job training. *QMCS:* Able to perform the duties required for QMC; serves as enlisted technical or specialty expert; plans, organizes, and directs the work of subordinate personnel; plans and administers on-the-job training programs; supervises the preparation of reports. *QMCM:* Able to perform the duties required for QMCS; serves as senior enlisted technical or specialty administrator; ensures maximum efficiency of the work force and the equipment; prepares general correspondence concerning fiscal, supply, and administrative matters; assists in the formulation of plans, policies, and budget requirements; may supplement the officer corps in overall supervision and administration of personnel and equipment; may also supervise personnel in other specialty areas.

Recommendation, QM3

In the lower-division baccalaureate/associate degree category, 3 semester hours in seamanship, 3 in technical mathematics, 3 in introduction to meteorology, 3 in coastwise navigation and piloting, and 1 in office procedures, and credit in celestial navigation on the basis of institutional evaluation (5/84).

Recommendation, QM2

In the lower-division baccalaureate/associate degree category, 3 semester hours in seamanship, 3 in technical mathematics, 1 in introduction to meteorology, 3 in coastwise navigation and piloting, and 3 in office procedures, and credit in celestial navigation on the basis of institutional evaluation (5/84).

Recommendation, QM1

In the lower-division baccalaureate/associate degree category, 3 semester hours in seamanship, 3 in technical mathematics, 1 in introduction to meteorology, 3 in coastwise navigation and piloting, 3 in office procedures, and 2 in personnel supervision, and credit in celestial navigation on the basis of institutional evaluation (5/84).

Recommendation, QMC

In the lower-division baccalaureate/associate degree category, 3 semester hours in seamanship, 3 in technical mathematics, 1 in introduction to meteorology, 3 in coastwise navigation and piloting, 3 in office procedures, 3 in personnel supervision, and 3 in principles of management, and credit in celestial navigation on the basis of institutional evaluation (5/84).

Recommendation, QMCS

In the lower-division baccalaureate/associate degree category, 3 semester hours in seamanship, 3 in technical mathematics, 1 in introduction to meteorology, 3 in coastwise navigation and piloting, 3 in office procedures, 3 in personnel supervision, and 3 in principles of management, and credit in celestial navigation on the basis of institutional evaluation. In the upper-division baccalaureate category, 3 semester hours for field experience in management and 3 in management problems (5/84).

Recommendation, QMCM

In the lower-division baccalaureate/associate degree category, 3 semester hours in seamanship, 3 in technical mathematics, 1 in introduction to meteorology, 3 in coastwise navigation and piloting, 3 in office procedures, 3 in personnel supervision, and 3 in principles of management, and credit in celestial navigation on the basis of institutional evaluation. In the upper-division baccalaureate category, 6 semester hours for field experience in management and 3 in management problems (5/84).

CGR-QM-002

QUARTERMASTER
    QM3
    QM2
    QM1
    QMC
    QMCS
    QMCM

Exhibit Dates: 2/93–Present.

Occupational Group: Nautical Operations-Navigation.

Career Pattern

SN: Seaman (E-3). *QM3:* Quartermaster Third Class (E-4). *QM2:* Quartermaster Second Class (E-5). *QM1:* Quartermaster First Class (E-6). *QMC:* Chief Quartermaster (E-7). *QMCS:* Senior Chief Quartermaster (E-8). *QMCM:* Master Chief Quartermaster (E-9).

Description

Summary: Stands watch as assistant to the officers of the deck and to the navigator; serves as helmsman during time when precise ship control is required; performs communications, navigation, and bridgewatch duties; procures, corrects, uses, and stows navigational charts and publications; maintains navigational instruments and keeps navigational time; participates in ceremonies conducted in accordance with national and foreign observance and customs; sends and receives visual messages; serves as executive petty officer. *QM3:* Inventories, procures, uses and corrects nautical charts and publications; uses basic course plotting instruments, lead line, depth sounder, and compass; identifies aids to navigation; handles plain language radio communications and visual communication; maintains compass record book and weather observation sheet. *QM2:* Able to perform the duties required for QM3; uses Coast Guard publications concerning Nautical Rules of the Road, maneuvering board and relative bearings; selects charts for voyage planning; determines danger angles and danger bearings; determines ship's position by celestial observations; conducts basic weather observations and understands their significance; supervises bridge personnel; inventories installed equipment and spare parts; orders repair parts and tools. *QM1:* Able to perform the duties required for QM2; adjusts compasses and prepares deviation tables; adjusts and aligns sextants and stadimeters; interprets weather charts; plots probable path and location of storm centers; prepares a great circle track using various methods; maintains the ship's equipment configuration accounting system; prepares weekly schedules of preventive maintenance; supervises preventive maintenance procedures; prepares oceanographic reports such as wave observation log, ship's ice log, adjusted track log, and sounding journal; supervises and trains personnel in navigation, watchstanding, and use and care of navigational charts, tables, and publications. *QMC:* Able to perform the duties required for QM1; understands the use and time requirements on various files and logs; understands various types of messages; prepares quarterly schedules of preventive maintenance; requisitions equipment; interviews, selects, and evaluates personnel for the navigation department; plans emergency drills; conducts briefings; uses computer to track cyclonic storms and plan storm avoidance measures; understands the limitations of radar and loran aids to navigation; understands the use of radio direction finding (RDF) equipment; organizes and maintains technical library; using the computer, prepares and submits budget requests; plans and supervises on-the-job training. *QMCS:* Able to perform the duties required for QMC; serves as enlisted technical or specialty expert; plans, organizes, and directs the work of subordinate personnel; plans and administers on-the-job training programs; supervises the preparation of reports. *QMCM:* Able to perform the duties required for QMCS; serves as senior enlisted technical or specialty administrator; ensures maximum efficiency of the work force and the equipment; prepares general correspondence concerning fiscal, supply, and administrative matters; assists in the formulation of plans, policies, and budget requirements; may supplement the officer corps in overall supervision and administration of personnel and equipment; may also supervise personnel in other specialty areas.

Recommendation, QM3

In the lower-division baccalaureate/associate degree category, 3 semester hours in seamanship, 3 in technical mathematics, 1 in introduction to meteorology, 3 in coastwise navigation and piloting, 1 in computer science, 1 in technical communications, and 1 in office procedures, and credit in celestial navigation on the basis of institutional evaluation (8/95).

Recommendation, QM2

In the lower-division baccalaureate/associate degree category, 3 semester hours in seamanship, 3 in technical mathematics, 2 in introduction to meteorology, 3 in coastwise navigation and piloting, 2 in computer science, 2 in technical communications, and 2 in office procedures, and credit in celestial navigation on the basis of institutional evaluation (8/95).

Recommendation, QM1

In the lower-division baccalaureate/associate degree category, 3 semester hours in seamanship, 3 in technical mathematics, 3 in introduction to meteorology, 3 in coastwise navigation and piloting, 3 in computer science, 3 in technical communications, 3 in office procedures, and 2 in personnel supervision, and credit in celestial navigation on the basis of institutional evaluation (8/95).

Recommendation, QMC

In the lower-division baccalaureate/associate degree category, 3 semester hours in seamanship, 3 in technical mathematics, 3 in introduction to meteorology, 3 in coastwise navigation and piloting, 3 in computer science, 3 in technical communications, 3 in office procedures, and 3 in personnel supervision, and credit in celestial navigation on the basis of institutional evaluation (8/95).

Recommendation, QMCS

In the lower-division baccalaureate/associate degree category, 3 semester hours in seamanship, 3 in technical mathematics, 3 in introduction to meteorology, 3 in coastwise navigation and piloting, 3 in computer science, 3 in technical communications, 3 in office procedures, and 3 in personnel supervision, and credit in celestial navigation on the basis of institutional evaluation. In the upper-division baccalaureate category, 3 semester hours for field experience in management and 3 in management problems (8/95).

Recommendation, QMCM

In the lower-division baccalaureate/associate degree category, 3 semester hours in seamanship, 3 in technical mathematics, 3 in introduction to meteorology, 3 in coastwise navigation and piloting, 3 in computer science, 3 in technical communications, 3 in office procedures, and 3 in personnel supervision, and credit in celestial navigation on the basis of institutional evaluation. In the upper-division baccalaureate category, 6 semester hours for field experience in management and 3 in management problems (8/95).

CGR-RD-001

RADARMAN
    RD3
    RD2
    RD1
    RDC
    RDCS
    RDCM

Exhibit Dates: 1/90–7/93.

Occupational Group: 1 (Deck).

Career Pattern

SN: Seaman (E-3). *RD3:* Radarman Third Class (E-4). *RD2:* Radarman Second Class (E-5). *RD1:* Radarman First Class (E-6). *RDC:* Chief Radarman (E-7). *RDCS:* Senior Chief Radarman (E-8). *RDCM:* Master Chief Radarman (E-9).

Description

Summary: Operates radar and associated equipment in order to collect, process, display, evaluate, and disseminate information related to the movement of ships, aircraft, missiles, and natural objects; performs duties related to navigation and piloting by plotting own ship's track by radar fixes and radio direction finders; is conversant with Nautical Rules of the Road as used in collision avoidance. *RD3:* Prepares and maintains records and logs for Combat Information Center operations and operating equipment; has knowledge of interior communications and equipment; plots ranges, bearings, and fixes on nautical charts, including dead reckoning; operates fathometers and radio direction finders; knows Nautical Rules of the Road as used in collision avoidance and use of radar in fog; has knowledge of basic search procedures, and solution of basic maneuvering board problems; operates military-specific equipment. *RD2:* Able to perform the duties required for RD3; prepares requisitions for supplies; instructs personnel in principles of equipment operation; sets up controlled radio equipment in accordance with plans; has knowledge of whistle signals, basic lights, distress signals, and the use of bridge-to-bridge radio telephone; knows search and rescue procedures; performs routine preventive maintenance on electronic equipment; solves advanced maneuvering board problems; knows procedures in advisory air control in situations requiring control of aircraft. *RD1:* Able to perform the duties required for RD2; computes statistics necessary for operational reports; organizes communication equipment and personnel to implement communications plans; plans, organizes and schedules preventive maintenance programs for electronic equipment; provides command with technical advice concerning capabilities, limitations, reliability, and operation of radar equipment. *RDC:* Able to perform the duties required for RD1; prepares training reports; understands concepts of electronic warfare, including early warning and utilization of radar under conditions of radar jamming and interference; has knowledge of current naval publications concerning functions and procedures of radar and related equipment; prepares preventive maintenance schedules; organizes and maintains technical library; prepares and submits budget requests; plans and supervises on-the-job training programs; supervises the preparation of reports. *RDCS:* Able to perform the duties required for RDC; plans, organizes, and directs work of personnel operating and maintaining radar and associated equipment; conducts training programs; supervises the preparation of reports. *RDCM:* Able to perform the duties required for RDCS; serves as senior enlisted technical or specialty administrator; manages personnel in the operation, maintenance, and procurement of radar and associated equipment; ensures maximum efficiency of the work force and the equipment; prepares general correspondence concerning fiscal, supply, and administrative matters; assists in the formulation of plans, policies, and budget requirements; may supplement the officer corps in the overall supervision and administration of personnel and equipment; may also supervise personnel in other specialty areas.

Recommendation, RD3

In the lower-division baccalaureate/associate degree category, 3 semester hours in basic radar use and operation and 3 in coastwise navigation and piloting (radar navigation) (3/84).

Recommendation, RD2

In the lower-division baccalaureate/associate degree category, 3 semester hours in basic radar use and operation, 3 in coastwise navigation and piloting (radar navigation), and 3 in record keeping (3/84).

Recommendation, RD1

In the lower-division baccalaureate/associate degree category, 3 semester hours in basic radar use and operation, 3 in coastwise navigation and piloting (radar navigation), 3 in record keeping, 3 in maintenance management, and 2 in personnel supervision (3/84).

Recommendation, RDC

In the lower-division baccalaureate/associate degree category, 3 semester hours in basic radar use and operation, 3 in coastwise navigation and piloting (radar navigation), 3 in record keeping, 3 in maintenance management, 3 in personnel supervision, and 3 in principles of management (3/84).

Recommendation, RDCS

In the lower-division baccalaureate/associate degree category, 3 semester hours in basic radar use and operation, 3 in coastwise navigation and piloting (radar navigation), 3 in record keeping, 3 in maintenance management, 3 in personnel supervision, and 3 in principles of management. In the upper-division baccalaureate category, 3 semester hours for field experience in management and 3 in management problems (3/84).

Recommendation, RDCM

In the lower-division baccalaureate/associate degree category, 3 semester hours in basic radar use and operation, 3 in coastwise navigation and piloting (radar navigation), 3 in record keeping, 3 in maintenance management, 3 in personnel supervision, and 3 in principles of management. In the upper-division baccalaureate category, 6 semester hours for field experience in management and 3 in management problems (3/84).

CGR-RD-002

RADARMAN
  RD3
  RD2
  RD1
  RDC
  RDCS
  RDCM

Exhibit Dates: 8/93–Present.

Occupational Group: Nautical Operations-Navigation.

Career Pattern

SN: Seaman (E-3). *RD3:* Radarman Third Class (E-4). *RD2:* Radarman Second Class (E-5). *RD1:* Radarman First Class (E-6). *RDC:* Chief Radarman (E-7). *RDCS:* Senior Chief Radarman (E-8). *RDCM:* Master Chief Radarman (E-9).

Description

Summary: Operates radar and associated equipment in order to collect, process, display, evaluate, and disseminate information related to the movement of ships, aircraft, missiles, and natural objects; performs duties related to navigation and piloting by plotting own ship's track by radar fixes and radio direction finders; is conversant with Nautical Rules of the Road as used in collision avoidance. *RD3:* Prepares and maintains records and logs for Combat Information Center operations and operating equipment; has knowledge of interior communications and equipment; plots ranges, bearings, and fixes on nautical charts, including dead reckoning; operates fathometers and radio direction finders; knows Nautical Rules of the Road as used in collision avoidance and use of radar in fog; has knowledge of basic search procedures, and solution of basic maneuvering board problems; operates military-specific equipment. *RD2:* Able to perform the duties required for RD3; prepares requisitions for supplies; instructs personnel in principles of equipment operation; sets up controlled radio equipment in accordance with plans; has knowledge of whistle signals, basic lights, distress signals, and the use of bridge-to-bridge radio telephone; knows search and rescue procedures; performs routine preventive maintenance on electronic equipment; solves advanced maneuvering board problems; knows procedures in advisory air control in situations requiring control of aircraft; is trained in law enforcement procedures. *RD1:* Able to perform the duties required for RD2; computes statistics necessary for operational reports; organizes communication equipment and personnel to implement communications plans; plans, organizes and schedules preventive maintenance programs for electronic equipment; provides command with technical advice concerning capabilities, limitations, reliability, and operation of radar equipment. *RDC:* Able to perform the duties required for RD1; prepares training reports; understands concepts of electronic warfare, including early warning and utilization of radar under conditions of radar jamming and interference; has knowledge of current naval publications concerning functions and procedures of radar and related equipment; prepares preventive maintenance schedules; organizes and maintains technical library; prepares and submits budget requests; plans and supervises on-the-job training programs; supervises the preparation of reports. *RDCS:* Able to perform the duties required for RDC; plans, organizes, and directs work of personnel operating and maintaining radar and associated equipment; conducts training programs; supervises the preparation of reports. *RDCM:* Able to perform the duties required for RDCS; serves as senior enlisted technical or specialty administrator; manages personnel in the operation, maintenance, and procurement of radar and associated equipment; ensures maximum efficiency of the work force and the equipment; prepares general correspondence concerning fiscal, supply, and administrative matters; assists in the formulation of plans, policies, and budget requirements; may supplement the officer corps in the overall supervision and administration of personnel and equipment; may also supervise personnel in other specialty areas.

Recommendation, RD3

In the lower-division baccalaureate/associate degree category, 3 semester hours in basic radar use and operation and 3 in coastwise navigation and piloting (radar navigation) (3/96).

Recommendation, RD2

In the lower-division baccalaureate/associate degree category, 3 semester hours in basic radar use and operation, 3 in coastwise navigation and piloting (radar navigation), 3 in law enforcement, and 3 in record keeping (3/96).

Recommendation, RD1

In the lower-division baccalaureate/associate degree category, 3 semester hours in basic radar use and operation, 3 in coastwise navigation and piloting (radar navigation), 3 in law enforcement, 3 in record keeping, 3 in maintenance management, and 2 in personnel supervision (3/96).

Recommendation, RDC

In the lower-division baccalaureate/associate degree category, 3 semester hours in basic radar use and operation, 3 in coastwise navigation and piloting (radar navigation), 3 in law enforcement, 3 in record keeping, 3 in maintenance management, and 3 in personnel supervision. In the upper-division baccalaureate category, 3 semester hours in personnel management (3/96).

Recommendation, RDCS

In the lower-division baccalaureate/associate degree category, 3 semester hours in basic radar use and operation, 3 in coastwise navigation and piloting (radar navigation), 3 in law enforcement, 3 in record keeping, 3 in maintenance management, and 3 in personnel supervision. In the upper-division baccalaureate category, 3 semester hours for field experience in management, 3 in personnel management, and 3 in management problems (3/96).

Recommendation, RDCM

In the lower-division baccalaureate/associate degree category, 3 semester hours in basic radar use and operation, 3 in coastwise navigation and piloting (radar navigation), 3 in law enforcement, 3 in record keeping, 3 in maintenance management, 3 in personnel supervision, and 3 in principles of management. In the upper-division baccalaureate category, 3 semester hours for field experience in management, 3 in personnel management, 3 in curriculum development, and 3 in management problems (3/96).

CGR-RM-001

RADIOMAN
  RM3
  RM2
  RM1
  RMC
  RMCS
  RMCM

Exhibit Dates: 1/90–Present.

Occupational Group: IV (Administrative and Clerical).

Career Pattern

SN: Seaman (E-3). *RM3:* Radioman Third Class (E-4). *RM2:* Radioman Second Class (E-5). *RM1:* Radioman First Class (E-6). *RMC:* Chief Radioman (E-7). *RMCS:* Senior Chief Radioman (E-8). *RMCM:* Master Chief Radioman (E-9).

Description

Summary: Transmits, receives, and processes all forms of telecommunications through various transmission media; operates, monitors, and controls telecommunication transmission reception, terminal, and processing equipment. *RM3:* Prepares messages in correct format for transmission via military and commercial circuits; prepares telegrams in international and domestic form; transmits and receives on radiotelephone and radio-teletype circuits using standard procedures and keeping the required logs; completely processes, routes, and files traffic within own unit; has knowledge of basic communication systems; types 45 words per minute. *RM2:* Able to perform the duties required for RM3; provides technical guidance to subordinate personnel. *RM1:* Able to perform the duties required for RM2; instructs and supervises subordinate personnel in operation and procedures of communication system; organizes duties and assigns personnel; requisitions supplies and parts. *RMC:* Able to perform the duties required for RM1; coordinates maintenance efforts to optimize equipment operating condition; instructs classes; prepares and submits budget requests; plans and supervises on-the-job training; organizes and maintains technical library. *RMCS:* Able to perform the duties required for RMC; serves as enlisted technical or specialty expert; plans, organizes, and supervises communication activities; prepares equipment reports; plans and administers on-the-job training programs. *RMCM:* Able to perform the duties required for RMCS; serves as senior enlisted technical or specialty administrator; manages personnel in the operation of telecommunications equipment; ensures maximum efficiency of the work force and the equipment; prepares general correspondence concerning fiscal, supply, and administrative matters; assists in the formulation of plans, policies, and budget requirements; may supplement the officer corps in the overall supervision and administration of personnel and equipment; may also supervise personnel in other specialty areas.

Recommendation, RM3

In the lower-division baccalaureate/associate degree category, 3 semester hours in typing, 3 in office practices, 3 in radio communications, and 3 in word processing (3/84).

Recommendation, RM2

In the lower-division baccalaureate/associate degree category, 3 semester hours in typing, 3 in office practices, 3 in radio communications, and 3 in word processing (3/84).

Recommendation, RM1

In the lower-division baccalaureate/associate degree category, 3 semester hours in typing, 3 in office practices, 3 in radio communications, 3 in word processing, 3 in maintenance management, and 3 in personnel supervision (3/84).

Recommendation, RMC

In the lower-division baccalaureate/associate degree category, 3 semester hours in typing, 3 in office practices, 3 in radio communications, 3 in word processing, 3 in maintenance management, 3 in personnel supervision, and 3 in principles of management (3/84).

Recommendation, RMCS

In the lower-division baccalaureate/associate degree category, 3 semester hours in typing, 3 in office practices, 3 in radio communications, 3 in word processing, 3 in maintenance management, 3 in personnel supervision, and 3 in principles of management. In the upper-division baccalaureate category, 3 semester hours for field experience in management and 3 in management problems (3/84).

Recommendation, RMCM

In the lower-division baccalaureate/associate degree category, 3 semester hours in typing, 3 in office practices, 3 in radio communications, 3 in word processing, 3 in maintenance management, 3 in personnel supervision, and 3 in principles of management. In the upper-division baccalaureate category, 6 semester hours for field experience in management and 3 in management problems (3/84).

CGR-SK-001

STOREKEEPER
  SK3
  SK2
  SK1
  SKC
  SKCS
  SKCM

Exhibit Dates: 1/90–7/92.

Occupational Group: IV (Administrative and Clerical).

Career Pattern

SN: Seaman (E-3). *SK3:* Storekeeper Third Class (E-4). *SK2:* Storekeeper Second Class (E-5). *SK1:* Storekeeper First Class (E-6). *SKC:* Chief Storekeeper (E-7). *SKCS:* Senior Chief Storekeeper (E-8). *SKCM:* Master Chief Storekeeper (E-9).

Description

Summary: Opens, maintains and closes military pay records; prepares payroll certification sheets, travel transportation requests and pay records; accounts for property equipage, supplies and materials; operates data processing equipment; prepares and maintains all required forms, records, correspondence, reports and files; provides counseling on travel/transportation entitlements and procedures. *SK3:* Operates office machines including adding machines, duplicating equipment, copying machines, microfiche reading equipment, and microfiche copying equipment; files correspondence; prepares and maintains payroll records; provides supply support including preparation of requisitions for units; familiar with major components of automated data processing (ADP) equipment and knows the common terminology. *SK2:* Able to perform the duties required for SK3; initiates routine correspondence and establishes correspondence files; initiates and supervises the preparation of required reports; controls inventory; prepares bills of lading and processes shipments in accordance with regulations; operates office machines including graphotypes and addressographs. *SK1:* Able to perform the duties required for SK2; operates an office typewriter at 40 words per minute; applies automated supply procedures; assigns, supervises and trains personnel in the supply department; trains subordinates in the proper use of supply publications; audits required inventory store reports for all categories of materials; identifies sources and uses of funds and supervises budgeting procedures. *SKC:* Able to perform the duties required for SK1; computes the cubic dimensions of weight of material and equipment; lays out physical areas of offices, issue rooms, and storerooms plans, and supervises on-the-job training; organizes and supervises procedures for inventory management; audits year-end financial reports; prepares store replenishment data and forecasts unit requirements for various lengths of time and climatic conditions. *SKCS:* Able to perform the duties required for SKC; serves as enlisted specialty expert; maintains and directs supply functions, including those related to security, fiscal control procedures, and transportation; analyzes supply reports, identifies problem areas, and recommends corrective strategies; plans and administers on-the-job training programs; supervises the preparation of reports. *SKCM:* Able to perform the duties required for SKCS; serves as senior enlisted specialty administrator; plans and organizes controls in compliance with policy statements; establishes goals and priorities; reviews and evaluates personnel, equipment and material requirements; establishes and directs training programs; prepares general correspondence concerning fiscal, supply, and administrative matters; assists in the formulation of plans, policies, and budget requirements; may supplement the officer corp in the overall supervision and administration of personnel and equipment; may also supervise personnel in other specialty areas.

Recommendation, SK3

In the vocational certificate category, 1 semester hour in filing and records management, 1 in office machines, and 1 in clerical procedures. In the lower-division baccalaureate/associate degree category, 1 semester hour in filing and records management, 1 in office machines, 1 in clerical procedures, and 2 in data processing concepts (6/84).

Recommendation, SK2

In the vocational certificate category, 2 semester hours in filing and records management, 2 in office machines, 1 in clerical procedures, and 2 in record keeping. In the lower-division baccalaureate/associate degree category, 2 semester hours in filing and records management, 2 in office machines, 1 in clerical procedures, 2 in record keeping, 3 in inventory control, and 2 in data processing concepts (6/84).

Recommendation, SK1

In the vocational certificate category, 2 semester hours in filing and records management, 2 in office machines, 2 in clerical procedures, 2 in record keeping, 1 in automated record keeping, 2 in typing, and 2 in business mathematics. In the lower-division baccalaureate/associate degree category, 2 semester hours in filing and records management, 2 in office machines, 2 in clerical procedures, 2 in record keeping, 1 in automated record keeping, 3 in inventory control, 2 in data processing concepts, 2 in typing, 2 in business mathematics, and 2 in personnel supervision (6/84).

Recommendation, SKC

In the vocational certificate category, 2 semester hours in filing and records management, 2 in office machines, 2 in clerical procedures, 2 in record keeping, 1 in automated record keeping, 2 in typing, 2 in business mathematics, and 3 in communication skills. In the lower-division baccalaureate/associate degree category, 2 semester hours in filing and records management, 2 in office machines, 2 in clerical procedures, 2 in record keeping, 1 in automated record keeping, 3 in inventory control, 2 in data processing concepts, 2 in typing, 2 in business mathematics, 3 in personnel supervision, 3 in communication skills, and 3 in principles of management (6/84).

Recommendation, SKCS

In the lower-division baccalaureate/associate degree category, 2 semester hours in filing and records management, 2 in office machines, 2 in clerical procedures, 2 in record keeping, 1 in automated record keeping, 3 in inventory control, 2 in data processing concepts, 2 in typing, 2 in business mathematics, 3 in personnel supervision, 3 in communication skills, and 3 in principles of management. In the upper-division baccalaureate category, 3 semester hours for field experience in management and 3 in management problems (6/84).

Recommendation, SKCM

In the lower-division baccalaureate/associate degree category, 2 semester hours in filing and records management, 2 in office machines, 2 in clerical procedures, 2 in record keeping, 1 in automated record keeping, 3 in inventory control, 2 in data processing concepts, 2 in typing, 2 in business mathematics, 3 in personnel supervision, 3 in communication skills, and 3 in principles of management. In the upper-division baccalaureate category, 6 semester hours for field experience in management, 3 in management problems, and 3 in supply management (6/84).

CGR-SK-002

STOREKEEPER
SK3
SK2
SK1
SKC
SKCS
SKCM

Exhibit Dates: 8/92–Present.

Occupational Group: Administrative/Supply Management.

Career Pattern

SN: Seaman (E-3). *SK3:* Storekeeper Third Class (E-4). *SK2:* Storekeeper Second Class (E-5). *SK1:* Storekeeper First Class (E-6). *SKC:* Chief Storekeeper (E-7). *SKCS:* Senior Chief Storekeeper (E-8). *SKCM:* Master Chief Storekeeper (E-9).

Description

Summary: Purchases, receives, stores, issues, and maintains inventory records of property, supplies, and materials; operates computer and accounting software; provides technical support on software and the preparation of accounting records and reports; prepares and maintains all required forms, records, correspondence, reports and files; provides counsel on transportation procedures; maintains retail operations of a base exchange. *SK3:* Operates computer and other office equipment such as fax, copier, and microfiche; files correspondence and records; prepares requisitions; prepares bills of lading and processes shipments in accordance with regulation; familiar with basic concepts and application of accounting software. *SK2:* Able to perform the duties required for SK3; initiates and supervises the preparation of required reports; controls inventory; serves as liaison with finance office; performs shipping and receiving tasks for warehousing operations. *SK1:* Able to perform the duties required for SK2; applies automated supply procedures; assigns, supervises and trains personnel in the supply department; trains subordinates in the proper use of supply publications; audits required inventory store reports; identifies sources and uses of funds and supervises budgeting procedures; provides technical support on accounting software; maintains retail operations of a base exchange. *SKC:* Able to perform the duties required for SK1; supervises on-the-job training; organizes and supervises procedures for inventory management; audits year-end financial reports; prepares store replenishment data and forecasts unit requirements; performs system administration duties; evaluates and directs staff. *SKCS:* Able to perform the duties required for SKC; serves as enlisted specialty expert; maintains and directs supply functions, including those related to security, fiscal control procedures, and transportation; analyzes supply reports, identifies problem areas, and recommends corrective strategies; plans and administers on-the-job training programs; supervises the preparation of reports. *SKCM:* Able to perform the duties required for SKCS; serves as senior enlisted specialty administrator; plans and organizes controls in compliance with policy statements; establishes goals and priorities; reviews and evaluates personnel, equipment and material requirements; establishes and directs training programs; prepares general correspondence concerning fiscal, supply, and administrative matters; assists in the formulation of plans, policies, and budget requirements; may supplement the officer corps in the overall supervision and administration of staff and equipment; may also supervise staff in other specialty areas.

Recommendation, SK3

In the lower-division baccalaureate/associate degree category, 1 semester hour in records management, 2 in clerical procedures, 1 in business software applications, and 1 in keyboarding (3/96).

Recommendation, SK2

In the lower-division baccalaureate/associate degree category, 2 semester hours in records management, 3 in clerical procedures, 2 in business software applications, 2 in accounting, 1 in keyboarding, and 2 in supply management (3/96).

Recommendation, SK1

In the lower-division baccalaureate/associate degree category, 3 semester hours in records management, 3 in clerical procedures, 3 in business software applications, 3 in accounting, 2 in keyboarding, 2 in supply management, and 2 in personnel supervision. In the upper-division baccalaureate category, 2 semester hours in principles of retailing (3/96).

Recommendation, SKC

In the lower-division baccalaureate/associate degree category, 3 semester hours in records management, 3 in clerical procedures, 3 in business software applications, 3 in accounting, 2 in keyboarding, 3 in supply management, and 3 in personnel supervision. In the upper-division baccalaureate category, 3 semester hours in principles of retailing and 2 in organizational management (3/96).

Recommendation, SKCS

In the lower-division baccalaureate/associate degree category, 3 semester hours in records management, 3 in clerical procedures, 3 in business software applications, 3 in accounting, 2 in keyboarding, 3 in supply management, and 3 in personnel supervision. In the upper-division baccalaureate category, 3 semester hours for field experience in management, 3 in organizational management, and 3 in principles of retailing (3/96).

Recommendation, SKCM

In the lower-division baccalaureate/associate degree category, 3 semester hours in records management, 3 in clerical procedures, 3 in business software applications, 3 in accounting, 2 in keyboarding, 3 in supply management, and 3 in personnel supervision. In the upper-division baccalaureate category, 3 semester hours for field experience in management, 3 in manage-

CGR-SN-001

SEAMAN
    SN

Exhibit Dates: 1/90–Present.

Occupational Group: Shipboard and Administrative Assistance.

Career Pattern

Seaman (SN) is a general rate (apprenticeship) for persons at paygrades E-1 (recruit), E-2 (apprentice), and E-3 (Seaman). At paygrade E-4 (petty officer third class), the person may enter any one of the following ratings: Aviation Machinist's Mate (AD), Aviation Electrician's Mate (AE), Aviation Structural Mechanic (AM), Aviation Survivalman (ASM), Aviation Electronics Technician (AT), Boatswain's Mate (BM), Data Processing Technician (DP), Electronics Technician (ET), Fire Control Technician (FC), Gunner's Mate (GM), Health Services Technician (HS), Investigator (IV), Marine Science Technician (MST), Musician (MU), Port Securityman (PS), Quartermaster (QM), Radarman (RD), Radioman (RM), Storekeeper (SK), Subsistence Specialist (SS), Telecommunications Specialist (TC), Telephone Technician (TT), or Yeoman (YN).

Description

Maintains vessels, boats, shore facility structures, deck machinery and equipment, lines, and rigging; stands underway watches on board ships and boats as helmsmen, lookouts, and messengers; stands anchor, communications, and other special watches in port or at shore facilities; assists in the maintenance of aids-to-navigation; operates boats, booms, cranes, and winches; acts as member of gun crews and damage control parties.

Recommendation

In the lower-division baccalaureate/associate degree category, 1 semester hour in deck seamanship. NOTE: Credit for Seaman (SN) should be granted only after paygrade E-3 has been achieved (8/95).

CGR-SS-001

SUBSISTENCE SPECIALIST
    SS3
    SS2
    SS1
    SSC
    SSCS
    SSCM

Exhibit Dates: 1/90–9/94.

Occupational Group: VI (Administrative and Clerical).

Career Pattern

SN: Seaman (E-3). *SS3:* Subsistence Specialist, Third Class (E-4). *SS2:* Subsistence Specialist, Second Class (E-5). *SS1:* Subsistence Specialist, First Class (E-6). *SSC:* Chief Subsistence Specialist (E-7). *SSCS:* Senior Chief Subsistence Specialist (E-8). *SSCM:* Master Chief Subsistence Specialist (E-9).

Description

Summary: Prepares foods using standardized recipes; operates food equipment; maintains sanitary food service, preparation, and storage areas; receives, stores, and maintains stock levels of subsistence items. *SS3:* Measures, weighs, blends and mixes foods; prepares foods progressively in accordance with standardized recipes; under supervision, prepares fruits, vegetables, salads, meats, baked products, beverages, and soups; portions and serves food to consumers; carves meat and poultry for serving; operates specialized food service equipment; knows transmission methods of food borne diseases. *SS2:* Able to perform the duties required for SS3; establishes schedules and supervises serving line; assists in maintaining cost control, ordering, maintenance, and rotation of stock; prepares food in accordance with stated menu. *SS1:* Able to perform the duties required for SS2; decorates cakes; plans special meals; applies culinary techniques; assists in maintaining cost control procedures; supervises sanitation; modifies menus and methods of preparation; maintains inventory control to insure operating safety levels; supervises food service facility; assigns food service duties; maintains sales analysis records and ensures that proper costs to portion control are maintained. *SSC:* Able to perform the duties required for SS1; serves as administrator of small food service facility or as a supervisor in a large facility; plans and conducts training; prepares schedules; estimates, maintains, and interprets financial records. *SSCS:* Able to perform the duties required for SSC; develops and implements policies and procedures to attain organizational goals and objectives; selects, assigns and evaluates personnel; reviews personnel, equipment and material requirements; organizes, schedules and evaluates training programs; determines future requirements and space utilization; develops operating budget and monitors expenditures. *SSCM:* Able to perform the duties required for SSCS; provides guidance to subordinates.

Recommendation, SS3

In the lower-division baccalaureate/associate degree category, 1 semester hour in quantity food preparation, 1 in sanitation, and 3 for a food service internship (9/84).

Recommendation, SS2

In the lower-division baccalaureate/associate degree category, 2 semester hours in quantity food preparation, 2 in sanitation, 2 in food service operations, and 5 for a food service internship (9/84).

Recommendation, SS1

In the lower-division baccalaureate/associate degree category, 3 semester hours in quantity food preparation, 2 in sanitation, 3 in food service operations, 3 in food service management, and 6 for a food service internship (9/84).

Recommendation, SSC

In the lower-division baccalaureate/associate degree category, 3 semester hours in quantity food preparation, 3 in sanitation, 3 in food service operations, 4 in food service management, 6 for a food service internship, and 3 in personnel supervision. In the upper-division baccalaureate category, 2 semester hours for a food service internship (9/84).

Recommendation, SSCS

In the lower-division baccalaureate/associate degree category, 3 semester hours in quantity food preparation, 3 in sanitation, 3 in food service operations, 5 in food service management, 6 for a food service internship, and 3 in personnel supervision. In the upper-division baccalaureate category, 3 semester hours for a food service internship (9/84).

Recommendation, SSCM

In the lower-division baccalaureate/associate degree category, 3 semester hours in quantity food preparation, 3 in sanitation, 3 in food service operations, 5 in food service management, 6 for a food service internship, and 3 in personnel supervision. In the upper-division baccalaureate category, 3 semester hours for a food service internship (9/84).

CGR-SS-002

SUBSISTENCE SPECIALIST
    SS3
    SS2
    SS1
    SSC
    SSCS
    SSCM

Exhibit Dates: 10/94–Present.

Occupational Group: Culinary Specialists.

Career Pattern

SN: Seaman (E-3). *SS3:* Subsistence Specialist, Third Class (E-4). *SS2:* Subsistence Specialist, Second Class (E-5). *SS1:* Subsistence Specialist, First Class (E-6). *SSC:* Chief Subsistence Specialist (E-7). *SSCS:* Senior Chief Subsistence Specialist (E-8). *SSCM:* Master Chief Subsistence Specialist (E-9).

Description

Summary: Prepares foods using standardized recipes; operates food equipment; maintains sanitary food service, preparation, and storage areas; receives, stores, and maintains stock levels of subsistence items. *SS3:* Measures, weighs, blends and mixes foods; prepares foods progressively in accordance with standardized recipes; under supervision, prepares fruits, vegetables, salads, meats, baked products, beverages, and soups; portions and serves food to consumers; carves meat and poultry for serving; operates specialized food service equipment; knows transmission methods of food-borne diseases. *SS2:* Able to perform the duties required for SS3; establishes schedules and supervises serving line; assists in maintaining cost control, ordering, maintenance, and rotation of stock; prepares food in accordance with stated menu. *SS1:* Able to perform the duties required for SS2; decorates cakes; plans special meals; applies culinary techniques; assists in maintaining cost control procedures; supervises sanitation; modifies menus and methods of preparation; maintains inventory control to insure operating safety levels; supervises food service facility; assigns food service duties; maintains sales analysis records and ensures that proper costs to portion control are maintained. *SSC:* Able to perform the duties required for SS1; serves as administrator of small food service facility or as a supervisor in a large facility; plans and conducts training; prepares schedules; estimates, maintains, and interprets financial records. *SSCS:* Able to perform the duties required for SSC; develops and implements policies and procedures to attain organizational goals and objectives; selects, assigns and evaluates personnel; reviews personnel, equipment and material requirements; organizes, schedules and evaluates training programs; determines future requirements and space utilization; develops operating budget and monitors expenditures. *SSCM:* Able to perform the duties required for SSCS; provides guidance to subordinates.

Recommendation, SS3

In the lower-division baccalaureate/associate degree category, 1 semester hour in quantity food preparation, 1 in sanitation, and 3 for a food service internship (3/96).

Recommendation, SS2

In the lower-division baccalaureate/associate degree category, 2 semester hours in quantity food preparation, 2 in sanitation, 2 in food service operations, and 5 for a food service internship (3/96).

Recommendation, SS1

In the lower-division baccalaureate/associate degree category, 3 semester hours in quantity food preparation, 2 in sanitation, 3 in food service operations, 3 in food service management, and 6 for a food service internship (3/96).

Recommendation, SSC

In the lower-division baccalaureate/associate degree category, 3 semester hours in quantity food preparation, 3 in sanitation, 3 in food service operations, 4 in food service management, 6 for a food service internship, and 3 in personnel supervision. In the upper-division baccalaureate category, 2 semester hours for a food service internship (3/96).

Recommendation, SSCS

In the lower-division baccalaureate/associate degree category, 3 semester hours in quantity food preparation, 3 in sanitation, 3 in food service operations, 5 in food service management, 6 for a food service internship, and 3 in personnel supervision. In the upper-division baccalaureate category, 3 semester hours for a food service internship (3/96).

Recommendation, SSCM

In the lower-division baccalaureate/associate degree category, 3 semester hours in quantity food preparation, 3 in sanitation, 3 in food service operations, 5 in food service management, 6 for a food service internship, and 3 in personnel supervision. In the upper-division baccalaureate category, 3 semester hours for a food service internship and 3 in management problems (3/96).

CGR-ST-001

SONAR TECHNICIAN
ST3
ST2
ST1
STC
STCS
STCM

Exhibit Dates: 1/90–Present.

Occupational Group: II (Ordnance).

Career Pattern

SN: Seaman (E-3). *ST3:* Sonar Technician, Third Class (E-4). *ST2:* Sonar Technician, Second Class (E-5). *ST1:* Sonar Technician, First Class (E-6). *STC:* Chief Sonar Technician (E-7). *STCS:* Senior Chief Sonar Technician (E-8). *STCM:* Master Chief Sonar Technician (E-9).

Description

Summary: Operates, troubleshoots, repairs, and performs maintenance on shipboard sonar, oceanographic, and underwater fire control equipment, and associated equipment for the solution of anti-submarine warfare problems. *ST3:* Operates, locates, and analyzes equipment casualties; makes repairs, adjustments, alignments, and performs organizational and intermediate maintenance in surface sonar and allied equipment; performs corrective and preventive maintenance on all anti-submarine warfare sensors, weapons, and countermeasure systems and test equipment, including those utilizing solid state microminiature technology. *ST2:* Able to perform the duties required for ST3; provides technical assistance to subordinate personnel. *ST1:* Able to perform the duties required for ST2; organizes anti-submarine attack teams; supervises the use and upkeep of surface ship sonar and underwater equipment; evaluates equipment operation and determines sonar performance data. *STC:* Able to perform the duties required for ST1; trains and supervises personnel in all categories of equipment maintenance and repair; prepares preventive maintenance schedules; organizes and maintains technical library; prepares and submits budget requests; plans and supervises on-the-job training. *STCS:* Able to perform the duties required for STC; serves as enlisted technical or specialty expert; plans, organizes, and directs the work of personnel operating and maintaining sonar oceanographic and underwater fire control equipment; plans and administers on-the-job training programs; supervises the preparation of reports. *STCM:* Able to perform the duties required for STCS; serves as senior enlisted technical or specialty administrator; manages personnel in the operation, maintenance, and procurement of sonar, oceanographic, and underwater fire control equipment; ensure maximum efficiency of the work force and the equipment; prepares general correspondence concerning fiscal, supply, and administrative matters; assists in the formulation of plans, policies, and budget requirements; may supplement the officer corps in the overall supervision and administration of personnel and equipment; may also supervise personnel in other specialty areas.

Recommendation, ST3

In the lower-division baccalaureate/associate degree category, 3 semester hours in basic AC/DC theory, 3 in diagnostic procedures, 3 in circuit theory, 3 in microminiature technology, 3 in introduction to analog computers, and credit in oceanography on the basis of institutional evaluation (3/84).

Recommendation, ST2

In the lower-division baccalaureate/associate degree category, 3 semester hours in basic AC/DC theory, 3 in diagnostic procedures, 3 in circuit theory, 3 in microminiature technology, 3 in introduction to analog computers, and credit in oceanography on the basis of institutional evaluation (3/84).

Recommendation, ST1

In the vocational certificate category, 3 semester hours in basic AC/DC theory, 3 in diagnostic procedures, 3 in circuit theory, 3 in microminiature technology, 3 in introduction to analog computers, 3 in maintenance management, and credit in oceanography on the basis of institutional evaluation. In the lower-division baccalaureate/associate degree category, 3 semester hours in basic AC/DC theory, 3 in diagnostic procedures, 3 in circuit theory, 3 in microminiature technology, 3 in introduction to analog computers, 3 in maintenance management, 2 in personnel supervision, and credit in oceanography on the basis of institutional evaluation (3/84).

Recommendation, STC

In the vocational certificate category, 3 semester hours in basic AC/DC theory, 3 in diagnostic procedures, 3 in circuit theory, 3 in microminiature technology, 3 in introduction to analog computers, 3 in maintenance management, and credit in oceanography on the basis of institutional evaluation. In the lower-division baccalaureate/associate degree category, 3 semester hours in basic AC/DC theory, 3 in diagnostic procedures, 3 in circuit theory, 3 in microminiature technology, 3 in introduction to analog computers, 3 in maintenance management, 3 in personnel supervision, 3 in principles of management, and credit in oceanography on the basis of institutional evaluation (3/84).

Recommendation, STCS

In the vocational certificate category, 3 semester hours in basic AC/DC theory, 3 in diagnostic procedures, 3 in circuit theory, 3 in microminiature technology, 3 in introduction to analog computers, 3 in maintenance management, and credit in oceanography on the basis of institutional evaluation. In the lower-division baccalaureate/associate degree category, 3 semester hours in basic AC/DC theory, 3 in diagnostic procedures, 3 in circuit theory, 3 in microminiature technology, 3 in introduction to analog computers, 3 in maintenance management, 3 in personnel supervision, 3 in principles of management, and credit in oceanography on the basis of institutional evaluation. In the upper-division baccalaureate category, 3 semester hours for field experience in management and 3 in management problems (3/84).

Recommendation, STCM

In the vocational certificate category, 3 semester hours in basic AC/DC theory, 3 in diagnostic procedures, 3 in circuit theory, 3 in microminiature technology, 3 in introduction to analog computers, 3 in maintenance management, and credit in oceanography on the basis of institutional evaluation. In the lower-division baccalaureate/associate degree category, 3 semester hours in basic AC/DC theory, 3 in diagnostic procedures, 3 in circuit theory, 3 in microminiature technology, 3 in introduction to analog computers, 3 in maintenance management, 3 in personnel supervision, 3 in principles of management, and credit in oceanography on the basis of institutional evaluation. In the upper-division baccalaureate category, 6 semester hours for field experience in management and 3 in management problems (3/84).

CGR-TC-001

TELECOMMUNICATIONS SPECIALIST
TC3
TC2
TC1
TCC
TCCS
TCCM

Exhibit Dates: 10/94–Present.

Occupational Group: Telecommunications Operations.

Career Pattern

SN: Seaman (E-3). *TC3:* Telecommunications Specialist Third Class (E-4). *TC2:* Telecommunications Specialist Second Class (E-5). *TC1:* Telecommunications Specialist First Class (E-6). *TCC:* Chief Telecommunications Specialist (E-7). *TCCS:* Senior Chief Telecommunications Specialist (E-8). *TCCM:* Master Chief Telecommunications Specialist (E-9).

Description

Summary: Transmits, receives, and processes all forms of telecommunications through various transmission media; operates, monitors, and controls telecommunication transmission reception, terminal, and processing equipment. *TC3:* Prepares messages in correct format for transmission via military and commercial circuits; prepares telegrams in international and domestic form; transmits and receives on radiotelephone and radio-teletype circuits using standard procedures and keeping the required logs; completely processes, routes, and files traffic within own unit; has knowledge of basic commmunications systems; types 45 words per minute. *TC2:* Able to perform the duties required for TC3; provides technical guidance to subordinates. *TC1:* Able to perform the duties required for TC2; instructs and supervises subordinate personnel in operation and procedures of communication system; organizes duties and assignments; requisitions supplies and parts. *TCC:* Able to perform the duties required for TC1; coordinates maintenance efforts to optimize equipment operating condition; instructs classes; prepares and submits budget requests; plans and supervises on-the-job training programs; organizes and maintains technical library. *TCCS:* Able to perform the duties required for TCC; serves as enlisted technical or specialty expert; plans, organizes, and supervises communications activities; prepares equipment reports; plans and administers on-the-job training. *TCCM:* Able to perform the duties required for TCCS; serves as senior enlisted technical or specialty administrator; manages telecommunications equipment operations facilities; ensures maximum efficiency of the work force and the equipment; prepares general correspondence concerning fiscal, supply, and administrative matters; assists in the formulation of plans, policies, and budget requirements; may supplement the officer corps in overall supervision and administration; may also supervise subordinates in other specialty areas.

Recommendation, TC3

In the lower-division baccalaureate/associate degree category, 3 semester hours in keyboarding, 3 in office practices, 3 in communication systems operations, 3 in computer applications, and 3 in word processing (8/95).

Recommendation, TC2

In the lower-division baccalaureate/associate degree category, 3 semester hours in keyboarding, 3 in office practices, 3 in communication systems operations, 3 in computer applications, and 3 in word processing (8/95).

Recommendation, TC1

In the lower-division baccalaureate/associate degree category, 3 semester hours in keyboarding, 3 in office practices, 3 in communication systems operations, 3 in computer applications, 3 in word processing, and 3 in personnel supervision (8/95).

Recommendation, TCC

In the lower-division baccalaureate/associate degree category, 3 semester hours in keyboarding, 3 in office practices, 3 in communication systems operations, 3 in computer applications, 3 in word processing, 3 in personnel supervision, and 2 in technical writing (8/95).

Recommendation, TCCS

In the lower-division baccalaureate/associate degree category, 3 semester hours in keyboarding, 3 in office practices, 3 in communication systems operations, 3 in computer applications, 3 in word processing, 3 in personnel supervision, and 3 in technical writing. In the upper-division baccalaureate category, 3 semester hours for field experience in management and 3 in management problems (8/95).

Recommendation, TCCM

In the lower-division baccalaureate/associate degree category, 3 semester hours in keyboarding, 3 in office practices, 3 in communication systems operations, 3 in computer applications, 3 in word processing, 3 in technical writing, and 3 in personnel supervision. In the upper-division baccalaureate category, 6 semester hours for field experience in management and 3 in management problems (8/95).

CGR-TT-001

TELEPHONE TECHNICIAN
 TT3
 TT2
 TT1
 TTC
 TTCS
 TTCM

Exhibit Dates: 1/90–9/94.

Occupational Group: III (Engineering and Hull).

Career Pattern

SN: Seaman (E-3). *TT3:* Telephone Technician, Third Class (E-4). *TT2:* Telephone Technician, Second Class (E-5). *TT1:* Telephone Technician, First Class (E-6). *TTC:* Chief Telephone Technician (E-7). *TTCS:* Senior Chief Telephone Technician (E-8). *TTCM:* Master Chief Telephone Technician (E-9).

Description

Summary: Installs, operates, maintains, and repairs all types of communications and terminal equipment including: telephone, telegraph, teletype equipment, data communications and terminal equipment, key telephone, private branch exchange, subscriber, and transmission carrier systems; microwave radio; inter-office and intraship communications; public address systems. *TT3:* Has knowledge of electricity/electronics; reads and interprets schematics; uses electronic test equipment to localize and replace faulty components; places and splices aerial, underground and submarine cables; surveys locations and constructs pole line and cable plant for communications, signaling, and power distribution systems; contacts the public relative to acquisition of right of way; plans and estimates needs; arranges for the procurement of materials; installs antennas and antenna ground system. *TT2:* Able to perform the duties required for TT3; provides technical guidance to subordinate personnel. *TT1:* Able to perform the duties required for TT2; supervises the installation of telecommunications systems and equipment; estimates equipment and material needs; prepares work requests; reviews completed work logs and checklists. *TTC:* Able to perform the duties required for TT1; supervises the operation of the electronics shop; plans, organizes, and administers the maintenance and repair program; prepares preventive maintenance schedules; prepares and submits budget requests; plans and supervises on-the-job training; organizes and maintains technical library. *TTCS:* Able to perform the duties required for TTC; serves as enlisted technical or specialty expert; plans, organizes, and directs personnel in the operation and maintenance of communications equipment; plans and administers on-the-job training programs; supervises the preparation of reports. *TTCM:* Able to perform the duties required for TTCS; serves as senior enlisted technical or specialty administrator; ensures maximum efficiency of the work force and the equipment; manages the operation and maintenance procedures of communications equipment; prepares general correspondence concerning fiscal, supply, and administrative matters; assists in the formulation of plans, policies, and budget requirements; may supplement the officer corps in the overall supervision and administration of personnel and equipment; may also supervise personnel in other specialty areas.

Recommendation, TT3

In the lower-division baccalaureate/associate degree category, 3 semester hours in basic AC/DC theory, 3 in diagnostic procedures, 3 in telecommunication systems, 3 in circuit theory, and 3 in basic computer principles (3/84).

Recommendation, TT2

In the lower-division baccalaureate/associate degree category, 3 semester hours in basic AC/DC theory, 3 in diagnostic procedures, 3 in telecommunication systems, 3 in circuit theory, and 3 in basic computer principles (3/84).

Recommendation, TT1

In the lower-division baccalaureate/associate degree category, 3 semester hours in basic AC/DC theory, 3 in diagnostic procedures, 3 in telecommunication systems, 3 in circuit theory, 3 in basic computer principles, and 3 in personnel supervision (3/84).

Recommendation, TTC

In the lower-division baccalaureate/associate degree category, 3 semester hours in basic AC/DC theory, 3 in diagnostic procedures, 3 in telecommunication systems, 3 in circuit theory, 3 in basic computer principles, 3 in personnel supervision, 3 in applied psychology, and 3 in principles of management (3/84).

Recommendation, TTCS

In the lower-division baccalaureate/associate degree category, 3 semester hours in basic AC/DC theory, 3 in diagnostic procedures, 3 in telecommunication systems, 3 in circuit theory, 3 in basic computer principles, 3 in personnel supervision, 3 in applied psychology, and 3 in principles of management. In the upper-division baccalaureate category, 3 semester hours for field experience in management and 3 in management problems (3/84).

Recommendation, TTCM

In the lower-division baccalaureate/associate degree category, 3 semester hours in basic AC/DC theory, 3 in diagnostic procedures, 3 in telecommunication systems, 3 in circuit theory, 3 in basic computer principles, 3 in personnel supervision, 3 in applied psychology, and 3 in principles of management. In the upper-division baccalaureate category, 6 semester hours for field experience in management and 3 in management problems (3/84).

COAST GUARD ENLISTED RATINGS EXHIBITS 245

CGR-TT-002

TELEPHONE TECHNICIAN
- TT3
- TT2
- TT1
- TTC
- TTCS
- TTCM

Exhibit Dates: 10/94–Present.

Occupational Group: Communication Systems Installation and Maintenance.

Career Pattern

SN: Seaman (E-3). *TT3:* Telephone Technician, Third Class (E-4). *TT2:* Telephone Technician, Second Class (E-5). *TT1:* Telephone Technician, First Class (E-6). *TTC:* Chief Telephone Technician (E-7). *TTCS:* Senior Chief Telephone Technician (E-8). *TTCM:* Master Chief Telephone Technician (E-9).

Description

Summary: Installs, operates, maintains, and repairs telecommunications systems including: voice, data, and video networks; intercom, public address and antenna systems; switching and routing equipment; terminal equipment; telecommunication links; and interior and exterior distribution systems. *TT3:* Has knowledge of electricity/electronics; reads and interprets schematics; uses electronic test equipment to localize and replace faulty components; places and splices aerial, underground and submarine cables; surveys locations and constructs pole line and cable plant for communications, signaling, and power distribution systems; contacts the public relative to acquisition of right of way; plans and estimates needs; arranges for the procurement of materials; installs antennas and antenna ground system. *TT2:* Able to perform the duties required for TT3; provides technical guidance to subordinate personnel. *TT1:* Able to perform the duties required for TT2; supervises the installation of telecommunications systems and equipment; estimates equipment and material needs; prepares work requests; reviews completed work logs and checklists. *TTC:* Able to perform the duties required for TT1; supervises the operation of the electronics shop; plans, organizes, and administers the maintenance and repair program; prepares preventive maintenance schedules; prepares and submits budget requests; plans and supervises on-the-job training; organizes and maintains technical library. *TTCS:* Able to perform the duties required for TTC; serves as enlisted technical or specialty expert; plans, organizes, and directs personnel in the operation and maintenance of communications equipment; plans and administers on-the-job training programs; supervises the preparation of reports. *TTCM:* Able to perform the duties required for TTCS; serves as senior enlisted technical or specialty administrator; ensures maximum efficiency of the work force and the equipment; manages the operation and maintenance procedures of communications equipment; prepares general correspondence concerning fiscal, supply, and administrative matters; assists in the formulation of plans, policies, and budget requirements; may supplement the officer corps in the overall supervision and administration of personnel and equipment; may also supervise personnel in other specialty areas.

Recommendation, TT3

In the lower-division baccalaureate/associate degree category, 3 semester hours in basic AC/DC theory, 3 in electronic systems troubleshooting and maintenance, 3 in telecommunication systems, and 3 in basic circuit theory (8/95).

Recommendation, TT2

In the lower-division baccalaureate/associate degree category, 3 semester hours in basic AC/DC theory, 3 in electronic systems troubleshooting and maintenance, 3 in telecommunication systems, and 3 in basic circuit theory (8/95).

Recommendation, TT1

In the lower-division baccalaureate/associate degree category, 3 semester hours in basic AC/DC theory, 3 in electronic systems troubleshooting and maintenance, 3 in telecommunication systems, 3 in basic circuit theory, and 1 in personnel supervision (8/95).

Recommendation, TTC

In the lower-division baccalaureate/associate degree category, 3 semester hours in basic AC/DC theory, 3 in electronic systems troubleshooting and maintenance, 3 in telecommunication systems, 3 in basic circuit theory, 2 in personnel supervision, and 2 in maintenance management (8/95).

Recommendation, TTCS

In the lower-division baccalaureate/associate degree category, 3 semester hours in basic AC/DC theory, 3 in electronic systems troubleshooting and maintenance, 3 in telecommunication systems, 3 in basic circuit theory, 3 in personnel supervision, and 3 in maintenance management. In the upper-division baccalaureate category, 3 semester hours for field experience in management and 3 in management problems (8/95).

Recommendation, TTCM

In the lower-division baccalaureate/associate degree category, 3 semester hours in basic AC/DC theory, 3 in electronic systems troubleshooting and maintenance, 3 in telecommunication systems, 3 in basic circuit theory, 3 in personnel supervision, and 3 in maintenance management. In the upper-division baccalaureate category, 6 semester hours for field experience in management and 3 in management problems (8/95).

CGR-YN-001

YEOMAN
- YN3
- YN2
- YN1
- YNC
- YNCS
- YNCM

Exhibit Dates: 1/90–8/92.

Occupational Group: IV (Administrative and Clerical).

Career Pattern

SN: Seaman (E-3). *YN3:* Yeoman Third Class (E-4). *YN2:* Yeoman Second Class (E-5). *YN1:* Yeoman First Class (E-6). *YNC:* Chief Yeoman (E-7). *YNCS:* Senior Chief Yeoman (E-8). *YNCM:* Master Chief Yeoman (E-9).

Description

Summary: Performs general clerical administrative and secretarial duties including typewriting, filing, records management, office publications maintenance, and office equipment operation; maintains personnel service and accounting records; serves as reporter and fact finding body, and acts as office manager. *YN3:* Prepares, routes, and forwards office correspondence; establishes, maintains and disposes of office files; prepares, interprets, and transcribes messages (a straight-copy typing rate of at least 30 words per minute); prepares and distributes mail; performs duties as receptionist; prepares applications, statistical reports, and personnel management information entries; prepares security and legal correspondence, benefits and entitlements forms; familiar with current word processing concepts. *YN2:* Able to perform the duties required for YN3; orders publications and forms; performs personnel transfer procedures; types personnel papers for changes in rating (a straight-copy ending rate of at least 40 words per minute); prepares personnel papers relating to enlistment, discharge, financial records, and courts martial; counsels personnel regarding rights, benefits, and pay systems; able to perform basic functions on word processing equipment; monitors filing and record maintenance, retention and disposal. *YN1:* Able to perform the duties required for YN2; coordinates records disposal program; composes and types office correspondence (a straight-copy typing rate of at least 50 words per minute); transcribes recorded dictation at a minimum rate of 6 words per minute; demonstrates knowledge of forms management; prepares preliminary papers for general courts martial; assists in supervising and training office personnel in military entitlements and procedures; assists in the supervision of maintenance of records; performs advanced functions of word processing equipment. *YNC:* Able to perform the duties required for YN1; takes recorded dictation at 60 words per minute; transcribes recorded dictation material at 8 words per minute; composes correspondence directives; demonstrates proper office management concepts and techniques; prepares recommendations for administrative discharges and handling classified matter; demonstrates knowledge of military justice system; administers training programs for clerical personnel; supervises and trains office personnel in military pay entitlements and procedures. *YNCS:* Able to perform the duties required for YNC; serves as enlisted specialty supervisor; provides direction and supervision to enlisted personnel; plans and administers on-the-job training programs; serves as senior enlisted advisor in matters concerning enlisted personnel; demonstrates technical and specialty expertise; functions as office supervisor; interviews, assigns, supervises, and evaluates office personnel; devises, implements and evaluates word processing systems. *YNCM:* Able to perform the duties required for YNCS; functions as senior enlisted technical and specialty administrator; assumes management, supervisory, and administrative responsibility; responsible for organizing, directing and coordinating programs of instruction for subordinates; supplements officer corps in overall supervision and administration; establishes goals, objectives, and priorities in administration of office functions; evaluates and makes recommendations regarding office personnel, equipment, supplies, and budgets.

Recommendation, YN3

In the lower-division baccalaureate/associate degree category, 2 semester hours in typing, 1 in filing and records management, 1 in clerical procedures, 1 in office machines, 1 in communication skills (written), and 2 in word processing concepts; if qualification code was 02 (Verbatim Reporter) or 03 (Legal Clerk), add 3 semester hours in legal practices and procedures (6/84).

Recommendation, YN2

In the lower-division baccalaureate/associate degree category, 3 semester hours in typing, 2 in filing and records management, 2 in clerical procedures, 1 in office machines, 1 in communication skills (written), 3 in word processing concepts, and 2 in word processing applications; if qualification code was 02 (Verbatim Reporter) or 03 (Legal Clerk), add 3 semester hours in legal practices and procedures (6/84).

Recommendation, YN1

In the lower-division baccalaureate/associate degree category, 4 semester hours in typing, 2 in filing and records management, 2 in clerical procedures, 1 in office machines, 2 in communication skills (written), 3 in word processing concepts, 3 in word processing applications, 2 in machine transcription, and 2 in personnel supervision; if qualification code was 02 (Verbatim Reporter) or 03 (Legal Clerk), add 3 semester hours in legal practices and procedures (6/84).

Recommendation, YNC

In the lower-division baccalaureate/associate degree category, 4 semester hours in typing, 2 in filing and records management, 3 in clerical procedures, 1 in office machines, 3 in communication skills (written), 3 in word processing concepts, 3 in word processing applications, 2 in machine transcription, 3 in personnel supervision, and 1 in office management; if qualification code was 02 (Verbatim Reporter) or 03 (Legal Clerk), add 3 semester hours in legal practices and procedures (6/84).

Recommendation, YNCS

In the lower-division baccalaureate/associate degree category, 4 semester hours in typing, 2 in filing and records management, 3 in clerical procedures, 1 in office machines, 3 in communication skills (written), 3 in word processing concepts, 3 in word processing applications, 2 in machine transcription, 3 in personnel supervision, 3 in office management, 3 in principles of management, and 3 in word processing management; if qualification code was 02 (Verbatim Reporter) or 03 (Legal Clerk), add 3 semester hours in legal practices and procedures. In the upper-division baccalaureate category, 3 semester hours for field experience in management (6/84).

Recommendation, YNCM

In the lower-division baccalaureate/associate degree category, 4 semester hours in typing, 2 in filing and records management, 3 in clerical procedures, 1 in office machines, 3 in communication skills (written), 3 in word processing concepts, 3 in word processing applications, 2 in machine transcription, 3 in personnel supervision, 3 in office management, 3 in principles of management, and 3 in word processing management; if qualification code was 02 (Verbatim Reporter) or 03 (Legal Clerk), add 3 semester hours in legal practices and procedures. In the upper-division baccalaureate category, 6 semester hours for field experience in management and 3 in management problems (6/84).

CGR-YN-002

YEOMAN
    YN3
    YN2
    YN1
    YNC
    YNCS
    YNCM

Exhibit Dates: 9/92–Present.

Occupational Group: Administrative.

Career Pattern

SN: Seaman (E-3). *YN3:* Yeoman Third Class (E-4). *YN2:* Yeoman Second Class (E-5). *YN1:* Yeoman First Class (E-6). *YNC:* Chief Yeoman (E-7). *YNCS:* Senior Chief Yeoman (E-8). *YNCM:* Master Chief Yeoman (E-9).

Description

Summary: Performs general clerical administrative and secretarial duties including typing, filing, records management, office publications maintenance, and office equipment operation; maintains personnel service and accounting records; serves as reporter and fact finding body; serves as office manager.) *YN3:* Prepares, routes, and forwards office correspondence; establishes, maintains, and disposes of office files; prepares, interprets, and transcribes messages; prepares and distributes mail; performs receptionist duties; prepares applications, statistical reports, and personnel management information entries; prepares benefits and entitlements forms; familiar with basic word processing concepts. *YN2:* Able to perform the duties required for YN3; orders publications and forms; originates correspondence; prepares personnel papers relating to enlistment, discharge, transfer, rating change, financial records, and courts martial; provides counsel regarding rights, benefits, and pay systems; able to perform basic functions on word processing equipment; monitors filing and record maintenance, retention, and disposal. *YN1:* Able to perform the duties required for YN2; coordinates records disposal program; composes and types office correspondence; transcribes recorded dictation; demonstrates knowledge of forms management; prepares preliminary papers for general courts martial; prepares security and legal correspondence; assists in supervising and training office staff in military entitlements and procedures; assists in the supervision of records maintenance; performs advanced functions of word processing equipment. *YNC:* Able to perform the duties required for YN1; takes recorded dictation; composes correspondence directives; demonstrates proper office management concepts and techniques; prepares recommendations for administrative discharges and handling classified matter; demonstrates knowledge of military justice system; administers training programs for clerical staff; supervises and trains office staff. *YNCS:* Able to perform the duties required for YNC; serves as enlisted specialty supervisor; provides direction and supervision to enlisted personnel; plans and administers on-the-job training programs; serves as senior enlisted advisor in matters concerning enlisted personnel; demonstrates technical and specialty expertise; functions as office supervisor; interviews, assigns, supervises, and evaluates office personnel. *YNCM:* Able to perform the duties required for YNCS; functions as senior enlisted technical and specialty administrator; assumes management, supervisory, and administrative responsibility; responsible for organizing, directing and coordinating programs of instruction for subordinates; supplements officer corps in overall supervision and administration; establishes goals, objectives, and priorities in administration of office functions; evaluates and makes recommendations regarding office personnel, equipment, supplies, and budgets.

Recommendation, YN3

In the lower-division baccalaureate/associate degree category, 2 semester hours in keyboarding, 1 in records and information management, 2 in clerical procedures, 1 in business communication, and 2 in business software applications (3/96).

Recommendation, YN2

In the lower-division baccalaureate/associate degree category, 3 semester hours in keyboarding, 2 in records and information management, 3 in clerical procedures, 1 in business communications, 3 in business software applications, and 1 in office procedures (3/96).

Recommendation, YN1

In the lower-division baccalaureate/associate degree category, 4 semester hours in keyboarding, 2 in records and information management, 3 in clerical procedures, 3 in business communications, 3 in business software applications, 2 in office procedures, and 2 in personnel supervision; if qualification code was 02 (Verbatim Reporter) or 03 (Legal Clerk), add 3 semester hours in legal practices and procedures (3/96).

Recommendation, YNC

In the lower-division baccalaureate/associate degree category, 4 semester hours in keyboarding, 2 in records and information management, 3 in clerical procedures, 3 in business communications skills, 3 in business software applications, 3 in office procedures, 3 in personnel supervision, and 2 in office administration; if qualification code was 02 (Verbatim Reporter) or 03 (Legal Clerk), add 3 semester hours in legal practices and procedures (3/96).

Recommendation, YNCS

In the lower-division baccalaureate/associate degree category, 4 semester hours in keyboarding, 2 in records and information management, 3 in clerical procedures, 3 in business communication, 3 in business software applications, 3 in office procedures, 3 in personnel supervision, and 3 in office administration; if qualification code was 02 (Verbatim Reporter) or 03 (Legal Clerk), add 3 semester hours in legal practices and procedures. In the upper-division baccalaureate category, 3 semester hours for field experience in management and 3 in organizational management (3/96).

Recommendation, YNCM

In the lower-division baccalaureate/associate degree category, 4 semester hours in keyboarding, 2 in records and information management, 3 in clerical procedures, 3 in business communications, 3 in software applications, 3 in office procedures, 3 in personnel supervision, and 3 in office administration; if qualification code was 02 (Verbatim Reporter) or 03 (Legal Clerk), add 3 semester hours in legal practices and procedures. In the upper-division baccalaureate category, 3 semester hours for field experience in management, 3 in management problems, and 3 in organizational management (3/96).

Coast Guard Warrant Officer Exhibits

CGW-AVI-001

AVIATION ENGINEERING

Exhibit Dates: 1/90–9/94.

Career Pattern

May have progressed to Aviation Engineering Warrant Officer from Aviation Machinist's Mate (AD), Aviation Electrician's Mate (AE), or Aviation Structural Mechanic (AM).

Description

Serves as an officer technical specialist in aircraft maintenance; provides technical advice and information concerning capabilities, limitations, and reliability of aircraft powerplants, accessories, airframes, and equipment; directs and supervises practices and procedures for service, maintenance, overhaul, repair, inspection, alteration, modification, adjustment, preservation, and depreservation of aircraft powerplants, accessories, airframes, and equipment; formulates and supervises training programs; prepares, maintains, and submits personnel and material records, logs, reports, and accounts.

Recommendation

In the lower-division baccalaureate/associate degree category, 3 semester hours in technical communications, 3 in personnel supervision, and 3 for field experience in aviation maintenance systems operations. In the upper-division baccalaureate category, 3 semester hours in principles of management, 3 in management problems, 3 in budget administration, and 3 for field experience in management (3/86).

CGW-AVI-002

AVIATION ENGINEERING

Exhibit Dates: 10/94–Present.

Career Pattern

May have progressed to Aviation Engineering Warrant Officer from Aviation Machinist's Mate (AD), Aviation Electrician's Mate (AE), or Aviation Structural Mechanic (AM).

Description

Serves as an officer technical specialist in aircraft maintenance; provides technical advice and information concerning capabilities, limitations, and reliability of aircraft powerplants, accessories, airframes, and equipment; directs and supervises practices and procedures for service, maintenance, overhaul, repair, inspection, alteration, modification, adjustment, preservation, and depreservation of aircraft powerplants, accessories, airframes, and equipment; formulates and supervises training programs; prepares, maintains, and submits personnel and material records, logs, reports, and accounts.

Recommendation

In the lower-division baccalaureate/associate degree category, 3 semester hours in technical communications, 3 in personnel supervision, and 3 for field experience in aviation maintenance systems operations. In the upper-division baccalaureate category, 3 semester hours in organizational management, 3 in management problems, 3 in budget administration, and 3 for field experience in management (6/95).

CGW-BOSN-001

BOATSWAIN

Exhibit Dates: 1/90–9/94.

Career Pattern

May have progressed to Boatswain Warrant Officer from Boatswain's Mate (BM), Marine Science Technician (MST), Quartermaster (QM), or Radarman (RD).

Description

Supervises and directs personnel in all aspects of deck seamanship, small craft operations, emergency skills, navigation, search operations, gunnery operations, and cargo and buoy handling operations; serves as specialist on inspection duties; develops and supervises training programs; prepares, maintains, and submits personnel, material, and operational records, reports, and accounts; prepares correspondence and administers budgets.

Recommendation

In the lower-division baccalaureate/associate degree category, 3 semester hours in technical communications, 3 in personnel supervision, and 3 for field experience in general ship operations In the upper-division baccalaureate category, 3 semester hours in principles of management, 3 in management problems, 3 for field experience in management, and 3 in budget administration (12/85).

CGW-BOSN-002

BOATSWAIN

Exhibit Dates: 10/94–Present.

Career Pattern

May have progressed to Boatswain Warrant Officer from Boatswain's Mate (BM), Marine Science Technician (MST), Quartermaster (QM), or Radarman (RD).

Description

Supervises and directs all aspects of deck seamanship, small craft operations, emergency skills, navigation, search operations, gunnery operations, and cargo and buoy handling operations; serves as specialist on inspection duties; develops and supervises training programs; prepares, maintains, and submits personnel, material, and operational records, reports, and accounts; prepares correspondence and administers budgets; serves as officer in charge of small vessels; plans and supervises hull maintenance; orders supplies and equipment.

Recommendation

In the lower-division baccalaureate/associate degree category, 3 semester hours in technical writing and communications, 3 in safety and hazardous materials, 3 in personnel supervision, and 3 for field experience in general ship operations. In the upper-division baccalaureate category, 3 semester hours in principles of management, 6 for field experience in management, 3 in budget planning and administration, and 3 in law enforcement (8/95).

CGW-COMM-001

COMMUNICATIONS

Exhibit Dates: 1/90–9/94.

Career Pattern

May have progressed to Communications Warrant Officer from Radioman (RM).

Description

Serves as an officer technical specialist in the field of communications; supervises and directs communication personnel in conduct of radio and wire communications; plans and supervises instructional programs; prepares, maintains, and submits personnel, material, operational records, and accounts; prepares correspondence and administrative budgets; serves as technical liaison with other civilian and military communication facilities.

Recommendation

In the lower-division baccalaureate/associate degree category, 3 semester hours in technical communications, 3 in personnel supervision, and 3 for field experience in communications systems operations. In the upper-division baccalaureate category, 3 semester hours in principles of management, 3 in management problems, 3 in budget administration, and 3 for field experience in management (2/86).

CGW-COMM-002

COMMUNICATIONS

Exhibit Dates: 10/94–Present.

Career Pattern

May have progressed to Communications Warrant Officer from Radioman (RM) or Telecommunications Specialist (TC).

Description

Serves as an officer technical specialist in the field of communications; supervises and directs communication personnel in conduct of computer, satellite, and telecommunications; plans and supervises instructional programs; prepares, maintains, and submits personnel, material, operational records, and accounts; prepares correspondence and administrative budgets; manages projects; reviews policy and procedures; designs communications systems; analyzes equipment needs and recommends new equipment; serves as technical liaison with other civilian and military communication facilities.

Recommendation

In the lower-division baccalaureate/associate degree category, 3 semester hours in technical writing, 3 in personnel supervision, and 3 for field experience in telecommunications systems operations. In the upper-division baccalaureate category, 3 semester hours in principles of management, 3 in management problems, 3 in strategic planning/forecasting, and 3 for field experience in project management (8/95).

CGW-ELC-001

ELECTRONICS

Exhibit Dates: 1/90–9/94.

Career Pattern

May have progressed to Electronics Warrant Officer from Aviation Electronics Technician (AT), Electronics Technician (ET), Sonar Technician (ST), or Telephone Technician (TT).

Description

Serves as an officer technical specialist in the field of electronics; supervises and directs personnel in all aspects of electronic repair of equipment, operation of electronic repair facilities, and maintenance of operational equipment; plans and supervises instructional programs; handles personnel duties, budget matters, records, and requisition of supplies; serves as technical liaison with other services and commercial organizations; prepares technical and administrative reports.

Recommendation

In the lower-division baccalaureate/associate degree category, 3 semester hours in technical communications, 3 in personnel supervision, and 3 for field experience in electronics systems operations. In the upper-division baccalaureate category, 3 semester hours in principles of management, 3 in management problems, 3 in budget administration, and 3 for field experience in management (2/86).

CGW-ELC-002

ELECTRONICS

Exhibit Dates: 10/94–Present.

Career Pattern

May have progressed to Electronics Warrant Officer from Aviation Electronics Technician (AT), Electronics Technician (ET), or Telephone Technician (TT).

Description

Serves as an officer technical specialist in the field of electronics; supervises and directs personnel in all aspects of electronic repair of equipment, operation of electronic repair facilities, and maintenance of operational equipment; plans and supervises instructional programs; handles personnel duties, budget matters, records, and requisition of supplies; serves as technical liaison with other services and commercial organizations; prepares technical and administrative reports.

Recommendation

In the lower-division baccalaureate/associate degree category, 3 semester hours in technical communications, 3 in personnel supervision, and 3 for field experience in electronics systems operations. In the upper-division baccalaureate category, 3 semester hours in principles of management, 3 in management problems, 3 in budget administration, and 3 in project management (3/96).

CGW-ENG-001

NAVAL ENGINEERING

Exhibit Dates: 1/90–9/94.

Career Pattern

May have progressed to Naval Engineering Warrant Officer from Electrician's Mate (EM) or Machinery Technician (MK).)

Description

Serves as an officer technical specialist in the field of engineering and in machinery repair on uses, capabilities, limitattions, and reliability of engineering equipment; directs and supervises handling, storage, installation, operation, testing, maintenance, and repair of engineering equipment; develops and supervises training programs; prepares, maintains, and submits personnel and material records and reports; supervises procurement, preservation, and accounting practices for supplies and repair parts.

Recommendation

In the lower-division baccalaureate/associate degree category, 3 semester hours in technical communications, 3 in personnel supervision, and 3 for field experience in marine propulsion systems operations. In the upper-division baccalaureate category, 3 semester hours in principles of management, 3 in management problems, 3 in budget administration, and 3 for field experience in management (3/86).

CGW-ENG-002

NAVAL ENGINEERING

Exhibit Dates: 10/94–Present.

Career Pattern

May have progressed to Naval Engineering Warrant Officer from Electrician's Mate (EM) or Machinery Technician (MK).

Description

Serves as an officer technical specialist in the field of engineering and in machinery repair on uses, capabilities, limitations, and reliability of engineering equipment; directs and supervises handling, storage, installation, operation, testing, maintenance, and repair of engineering equipment; develops and supervises training programs; prepares, maintains, and submits personnel and material records and reports; supervises procurement, preservation, and accounting practices for supplies and repair parts.

Recommendation

In the lower-division baccalaureate/associate degree category, 3 semester hours in technical writing and communications and 3 for field experience in marine propulsion systems operations. In the upper-division baccalaureate category, 3 semester hours in strategic planning/forecasting and 6 for field experience in management (8/95).

CGW-FS-001

FINANCE AND SUPPLY

Exhibit Dates: 1/90–9/94.

Career Pattern

May have progressed to Finance and Supply Warrant Officer from Storekeeper (SK) or Subsistence Specialist (SS).

Description

Serves as a technical specialist in the field of finance and supply; organizes, plans, and supervises the work of 5-30 persons engaged in procuring, storing, inventorying, and issuing supplies, including persons who work in disbursing offices; supervises and directs people in the preparation of estimates of requirements, inventories of supplies and equipment, audit of records and budget, computation of pay, preparation of vouchers, and allocation of materials; disburses public funds as agent cashier and as assistant disbursing officer; supervises computing of pay, posting of payments, the balancing of cash and check disbursements, and the auditing of accounts; supervises the preparation and the maintaining of supply department records and inventories; supervises preparation of correspondence and the maintaining of filing systems for publications and directives.

Recommendation

In the lower-division baccalaureate/associate degree category, 3 semester hours in technical communications, 3 in accounting systems, 3 in personnel supervision, and 3 for field experience in budgeting. In the upper-division baccalaureate category, 3 in principles of management, 3 in management problems, 3 in budget administration, and 3 for field experience in management (3/86).

CGW-FS-002

FINANCE AND SUPPLY

Exhibit Dates: 10/94–Present.

Career Pattern

May have progressed to Finance and Supply Warrant Officer from Storekeeper (SK) or Subsistence Specialist (SS).

Description

Serves as a technical specialist in the field of finance and supply; organizes, plans, and supervises the work of 5-30 persons engaged in procuring, storing, inventorying, and issuing supplies, including persons who work in disbursing offices; serves as an authorized contracting agent responsible for contract preparation, award and supervision; supervises and directs people in the preparation of estimates of requirements, inventories of supplies and equipment, audit of records and budget, computation of pay, preparation of vouchers, and allocation of materials; disburses public funds as agent cashier and as assistant disbursing officer; supervises the balancing of cash and check disbursements, and the auditing of accounts; supervises the preparation and the maintaining of supply department records and inventories; supervises the preparation of correspondence and the maintenance of filing systems for publications and directives.

Recommendation

In the lower-division baccalaureate/associate degree category, 3 semester hours in report writing, 3 in management information systems, 3 in accounting systems, and 3 in personnel supervision. In the upper-division baccalaureate category, 3 semester hours in organizational management, 3 in management problems, 3 in budget administration, 3 in logistics management, 3 in purchasing, and 3 for field experience in management (3/96).

CGW-INF-001

PUBLIC INFORMATION

Exhibit Dates: 1/90–9/94.

Career Pattern

Progressed to Public Information Warrant Officer from Public Affairs Specialist (PA).

Description

Serves as an officer technical specialist in public information; provides information on photography, the news media, community relations, and internal relations; organizes, directs, and coordinates operations and upkeep of a public information office; prepares and monitors budgets for the operation of a public information office; writes and edits major stories, speeches, and television and motion picture scripts; establishes and supervises comprehensive training programs.

Recommendation

In the lower-division baccalaureate/associate degree category, 3 semester hours in writing for mass media, 3 in curriculum design and organization, 3 in business communications, and 3 in audio/visual-photographic technology In the upper-division baccalaureate category, 3 semester hours in public relations, 3 in budget administration, and 6 for field experience in public affairs (3/86).

CGW-INF-002

PUBLIC INFORMATION

Exhibit Dates: 10/94–Present.

Career Pattern

Progressed to Public Information Warrant Officer from Public Affairs Specialist (PA).

Description

Serves as an officer technical specialist in public information; provides information on photography, the news media, community relations, and internal relations; organizes, directs, and coordinates operations and upkeep of a public information office; prepares and monitors budgets for the operation of a public information office; writes and edits major stories, speeches, and television and motion picture scripts; establishes and supervises training programs for unit public affairs personnel; maintains public affairs files, media contacts, clipping service and contingency plans; conducts press conferences, prepares press releases, issues press passes, and advises the organization on public affairs policy and media relations.

Recommendation

In the lower-division baccalaureate/associate degree category, 3 semester hours in writing for electronic media, 3 in basic news writing, 3 in principles of public relations, 3 in business communications, and 3 in electronic media production. In the upper-division baccalaureate category, 3 semester hours in advanced public relations, 3 in organizational management, and 6 for field experience in public affairs planning and management (3/96).

CGW-MAT-001

MATERIEL MAINTENANCE

Exhibit Dates: 1/90–9/94.

Career Pattern

May have progressed to Materiel Maintenance Warrant Officer from Damage Controlman (DC) or Aviation Survivalman (ASM).

Description

Serves as an operational and technical specialist in the fields of repair, maintenance, damage control, and fire fighting aboard ship and at shore units; serves as assistant to engineering and repair officers, as shop superintendent, and as technical advisor concerning uses, capabilities, limitations, and reliability of ship repair, fire fighting, and damage control equipment and as officer in charge of maintenance and repair detachments ashore; organizes and supervises personnel in ship repair and maintenance; provides technical advice and information concerning use, characteristics, and limitations of building and construction materials; organizes and supervises maintenance and repair forces on work involving repairs to buildings, towers, docks, bulkheads, street paving, and pipelines, including water and sewer lines; develops and supervises training programs; supervises preparation, maintenance, and submission of personnel and materiel records and reports; supervises procurement, stowage, preservation, and utilization practices for repair parts, building materials, and equipment.

Recommendation

In the lower-division baccalaureate/associate degree category, 3 semester hours in technical communications, 3 in personnel supervision, and 3 for field experience in maintenance activities and programs. In the upper-division baccalaureate category, 3 semester hours in principles of management, 3 in management problems, 3 in budget administration, and 3 for field experience in management (3/86).

CGW-MAT-002

MATERIEL MAINTENANCE

Exhibit Dates: 10/94–Present.

Career Pattern

May have progressed to Materiel Maintenance Warrant Officer from Damage Controlman (DC) or Aviation Survivalman (ASM).

Description

Serves as an operational and technical specialist in the fields of repair, maintenance, damage control, and fire fighting aboard ship and at shore units; serves as assistant to engineering and repair officers, as shop superintendent, and as technical advisor concerning uses, capabilities, limitations, and reliability of ship repair, fire fighting, and damage control equipment and as officer in charge of maintenance and repair detachments ashore; organizes and supervises personnel in ship repair and maintenance; provides technical advice and information concerning use, characteristics, and limitations of building and construction materials; organizes and supervises maintenance and repair forces on work involving repairs to buildings, towers, docks, bulkheads, street paving, and pipelines, including water and sewer lines; develops and supervises training programs; supervises preparation, maintenance, and submission of personnel and materiel records and reports; supervises procurement, stowage, preservation, and utilization practices for repair parts, building materials, and equipment.

Recommendation

In the lower-division baccalaureate/associate degree category, 3 semester hours in technical communications, 3 in personnel supervision, and 3 for field experience in maintenance activities and programs. In the upper-division baccalaureate category, 3 semester hours in principles of management, 3 in management problems, 3 in budget administration, and 3 in project management (3/96).

CGW-MED-001

MEDICAL ADMINISTRATION

Exhibit Dates: 1/90–9/94.

Career Pattern

Progressed to Medical Administration Warrant Officer from Health Services Technician (HS).

Description

Serves as an officer technical specialist in the field of medical administration; administers nonprofessional aspects of medical and dental facilities; manages administrative functions such as fiscal, supply, personnel records, and other related medical matters; promotes and manages environmental sanitation programs; administers and serves as instructor in medical training programs; performs medical service planning and logistics duties; serves as assistant to inspectors in reviewing administrative organization and operations of medical and dental facilities.

Recommendation

In the upper-division baccalaureate category, 3 semester hours in organizational management, 3 in personnel management, 3 in accounting and control, 3 in administrative policy, and 3 in data collection and record management (2/86).

CGW-MED-002

MEDICAL ADMINISTRATION

Exhibit Dates: 10/94–Present.

Career Pattern

Progressed to Medical Administration Warrant Officer from Health Services Technician (HS).

Description

Serves as an officer and mid level manager in the field of medical and health services administration; administers non-professional aspects of medical and dental facilities; manages administrative functions including fiscal, supply, personnel records, and other related matters; promotes and manages environmental safety and sanitation programs; supervises individual training of health service personnel and general health services training for the organization; performs health and medical services planning and logistics duties; provides contract support for civilian medical services needed to supplement on-base facilities or for off-base referrals. Tasks include: record keeping and files management; automated information systems management; report writing; statistical analysis; quality assurance; health insurance benefits administration; health and medical services regulation compliance review and revision.

Recommendation

In the lower-division baccalaureate/associate degree category, 3 semester hours in records and ADP information management, 3 in report writing, and 3 in office administration. In the upper-division baccalaureate category, 3 semester hours in organizational management, 3 in management problems in health services, 3 for field experience in management, 3 in human resources management, 3 in personnel management, and 3 in health services managed care administration (3/96).

CGW-PERS-001

PERSONNEL ADMINISTRATION

Exhibit Dates: 1/90–9/94.

Career Pattern

May have progressed to Personnel Administration Warrant Officer from Data Processing Technician (Emergency Rating) or Yeoman (YN).

Description

Serves as a technical specialist in general and personnel administration; provides technical advice and information concerning officer and enlisted personnel regulations and administration; organizes and supervises personnel in an administrative or personnel office of up to 20 persons; supervises personnel in preparing and processing correspondence, reports, personnel records, and accounts; uses computers in records administration; administers funds for travel and procurement, and the allocation of officer materials and equipment; prepares office organization, work flow, and layout charts; prescribes procedures and techniques for work improvement; prepares official correspondence, administrative directives, regulations, and orders.

Recommendation

In the lower-division baccalaureate/associate degree category, 3 semester hours in technical communications, 3 in business communications, 3 in personnel supervision, and 3 for field experience in office management. In the upper-division baccalaureate category, 3 semester hours in principles of management, 3 in management problems, 3 in budget administration, and 3 for field experience in management (3/86).

CGW-PERS-002

PERSONNEL ADMINISTRATION

Exhibit Dates: 10/94–Present.

Career Pattern

Progressed to Personnel Administration Warrant Officer from Yeoman (YN).

Description

Serves as a technical specialist in general and personnel administration; provides technical advice and information concerning officer and enlisted personnel regulations and administration; organizes and supervises an administrative or personnel office of up to 20 persons; supervises staff in preparing and processing correspondence, reports, personnel records, and accounts; uses work station in records administration; administers funds for travel and procurement, and the allocation of billeting materials and equipment; prepares office organization, work flow, and layout charts; conducts performance analysis and prescribes procedures and techniques for work improvement; prepares official correspondence, administrative directives, regulations, and orders.

Recommendation

In the lower-division baccalaureate/associate degree category, 3 semester hours in report writing, 3 in business communications, 3 in management information systems, 3 in personnel supervision, and 3 for field experience in office management. In the upper-division baccalaureate category, 3 semester hours in organizational management, 3 in personnel management, 3 in management problems, and 6 for field experience in management (3/96).

CGW-PSS-001

PORT SAFETY AND SECURITY

Exhibit Dates: 1/90–2/98.

Career Pattern

May have progressed to Port Safety and Security Warrant Officer from Port Securityman (PS).

Description

Serves as a port safety and security specialist; provides information and advice concerning physical security procedures, Coast Guard law enforcement procedures, and jurisdiction; supervises port safety and security personnel in port emergency response operations, including fire fighting and hazardous material spills; prepares and reviews reports; prescribes methods and procedures for law enforcement and security operations; supervises dissemination of regulations to subordinate personnel.

Recommendation

In the lower-division baccalaureate/associate degree category, 3 semester hours in technical communications and 3 in personnel supervision (12/85).

CGW-PSS-002

PORT SAFETY AND SECURITY

Exhibit Dates: 3/98–Present.

Career Pattern

May have progressed to Port Safety and Security Warrant Officer from Port Securityman (PS) or Investigator (IV).

Description

If warrant officer serves as a port safety and security specialist, performs the following duties: provides information and advice concerning physical security procedures, Coast Guard law enforcement procedures, and jurisdiction; supervises port safety and security personnel in port emergency response operations, including fire fighting and hazardous material spills; prepares and reviews reports; prescribes methods and procedures for law enforcement and security operations; supervises dissemination of regulations to subordinate personnel.

If warrant officer serves as an investigator, performs the following duties: supervises execution of search warrants, tactical arrest operations, and initial bomb and hostage investigations; supervises protective details and collection of intelligence data; coordinates compound crime scenes; provides liaison with other law enforcement agencies; reviews and evaluates investigative procedures; supervises an attack scenario for protective service operations; trains command on regulations and procedures for investigations, intelligence, and access to Special Agents; establishes and maintains training program for Special Agents and Investigators; coordinates the dissemination of intelligence information; evaluates large protective service operations; evaluates intelligence estimates; develops criteria for dissemination of intelligence information; evaluates intelligence production room plans; conducts evaluations of a district or unit's intelligence methods; evaluates training and emergency operation plans.

Recommendation

In the lower-division baccalaureate/associate degree category, 3 semester hours in technical communications and 3 in personnel supervision. In the upper-division baccalaureate category, 3 semester hours for field experience in management (3/98).

CGW-PYA-001

PHYSICIAN'S ASSISTANT

Exhibit Dates: 1/90–9/94.

Career Pattern

May have progressed to Physician's Assistant from Hospital Corpsman (HM) or Health Services Technician (HS).

Description

Serves as an officer technical specialist, qualified by academic and practical training to provide patient services under the supervision and direction of a licensed physician; performs diagnostic and therapeutic tasks in support of a primary care physician; engages in continuing medical education, health maintenance, and community health; has practical training in principles of education and applied psychology as it relates to day-to-day clinical activities; may serve in Naval hospitals, clinics, and branch clinics. NOTE: Appointment to Physician's Assistant Warrant Officer is based upon graduation from an accredited Physician's Assistant program and satisfactory completion of the appropriate licensing examination.

Recommendation

In the upper-division baccalaureate category, 6 semester hours in management, 6 in methods of instruction, and 2 in applied psychology In the graduate degree category, 3 semester hours in educational principles (2/86).

CGW-PYA-002

PHYSICIAN'S ASSISTANT

Exhibit Dates: 10/94–Present.

Career Pattern

Progressed to Physician's Assistant from Health Services Technician (HS).

Description

Serves as an officer technical specialist, qualified by academic and practical training to provide patient services under the supervision and direction of a licensed physician; provides diagnostic and therapeutic services in support of a primary care physician; engages in continuing medical education, health maintenance, and community health; has practical training and experience in principles of education and applied psychology as it relates to day-to-day clinical activities; may serve in hospitals, clinics, and branch clinics; may serve as health services staff officer in a command structure. NOTE: Appointment to Physician's Assistant Warrant Officer or Commissioned Officer is based upon graduation from an accredited Physician's Assistant program and satisfactory completion of the appropriate licensing examination.

Recommendation

In the upper-division baccalaureate category, 3 semester hours for field experience in management and 3 in health services administration. In the graduate degree category, 3 semester hours in advanced clinical practicum,

3 in patient counseling, and 3 in primary patient care (3/96).

CGW-WEPS-001

WEAPONS

Exhibit Dates: 1/90–9/94.

Career Pattern

May have progressed to Weapons Warrant Officer from Fire Control Technician (FT) or Gunner's Mate (GM).

Description

Serves as an operational and technical specialist in gunnery and ordnance; serves as assistant gunnery and ordnance repair officer; directs and supervises assembly, installation, operation, testing, maintenance, and repair of ordnance equipment; supervises testing, handling, stowage, preservation, requisitioning, issuing, and accounting practices and procedures for all ammunition and ammunition components; supervises stowage, preservation, security, requisitioning, and accounting practices and procedures for all ordnance equipment and repair parts; develops and supervises training programs; prepares, maintains, and submits ordnance, personnel, material, and operational record, reports, and accounts.

Recommendation

In the lower-division baccalaureate/associate degree category, 3 semester hours in technical communications, 3 in personnel supervision, and 3 for field experience in inventory systems management. In the upper-division baccalaureate category, 3 semester hours in principles of management, 3 in management problems, 3 in budget administration, and 3 for field experience in management (3/86).

CGW-WEPS-002

WEAPONS

Exhibit Dates: 10/94–Present.

Career Pattern

May have progressed to Weapons Warrant Officer from Fire Control Technician (FT) or Gunner's Mate (GM).

Description

Serves as an operational and technical specialist in gunnery and ordnance; serves as assistant gunnery and ordnance repair officer; directs and supervises assembly, installation, operation, testing, maintenance, and repair of ordnance equipment; supervises testing, handling, stowage, preservation, requisitioning, issuing, and accounting practices and procedures for all ammunition and ammunition components; supervises stowage, preservation, security, requisitioning, and accounting practices and procedures for all ordnance equipment and repair parts; develops and supervises training programs; prepares, maintains, and submits ordnance, personnel, material, and operational records, reports, and accounts.

Recommendation

In the lower-division baccalaureate/associate degree category, 3 semester hours in technical communications, 3 in personnel supervision, and 3 for field experience in inventory systems management. In the upper-division baccalaureate category, 3 semester hours in principles of management, 3 in management problems, 3 in budget administration, and 3 in project management (3/96).

Marine Corps Enlisted MOS Exhibits

MCE-6012-001

AIRCRAFT MECHANIC, A-4/TA-4/QA-4
6012

Exhibit Dates: 1/90–1/98.

Occupational Field: 60 (Aircraft Maintenance).

Career Pattern

PVT: Private (E-1). PFC: Private First Class (E-2). LCP: Lance Corporal (E-3). CPL: Corporal (E-4). SGT: Sergeant (E-5). SSGT: Staff Sergeant (E-6). GYSGT: Gunnery Sergeant (E-7). NOTE: MOS duties are not defined in terms of rank or paygrade, but by levels. Level I identifies tasks that are taught in the classroom; the individual is able to perform simple parts of tasks. Level II identifies those tasks that are part of on-the-job training; the trainee can do most tasks pertaining to a particular system and needs supervision on more difficult parts of tasks. Level III indicates that the individual can perform all essential tasks without supervision. Level IV indicates a high level of proficiency in job performance; individual can perform advanced technical functions, instruction, inspection, and supervision. The MOS designator 6011 (Aircraft Mechanic - Trainee) is used to identify individuals in Levels I and II. Upon completion of Level II, MOS 6012 is awarded. Credit is recommended for Levels III and IV.

Description

Summary: Inspects and maintains aircraft airframes and airframe components; performs duties related to flight-line operations. *Levels I and II:* Operates and maintains applicable aircraft maintenance ground support equipment; obtains license as required; directs aircraft taxiing; performs daily preflight and postflight inspections and assists the pilot and crew in the performance of those inspections; assists during periodic airframe inspection; practices applicable safety procedures and precautions on the flight line and in hangar deck areas; performs hot and cold refueling of aircraft; performs oil analysis sampling and uses various types of fuels and lubricants; removes and installs aircraft power plants, components, and parts as required; recognizes and assists in corrective treatment and/or preventive maintenance for corrosion; uses precision measuring equipment required in the performance of duties; assists in the performance of calendar inspections; operates assigned fire fighting equipment; follows weight and balance limitations for applicable aircraft; performs starting and ground turn-up functions when authorized; uses technical publications in skill area. *Level III:* Able to perform the duties required for Levels I and II; performs duties as flight-line section leader or standardization team chief; conducts informal on-the-job technical training. *Level IV:* Able to perform the duties required for Level III; assists in the supervision and administration of flight-line hangar deck area and related operations ashore and afloat; plans, schedules, and directs work center assignments; performs duties as maintenance department noncommissioned officer-in-charge.

Recommendation, Level III

In the lower-division baccalaureate/associate degree category, 9 semester hours in aircraft airframe maintenance and inspection, 3 in aviation maintenance technology, and 3 in industrial safety (8/89).

Recommendation, Level IV

In the lower-division baccalaureate/associate degree category, 12 semester hours in aircraft airframe maintenance and inspection, 6 in aviation maintenance technology, 3 in industrial safety, 3 in personnel supervision, and 3 in maintenance management. In the upper-division baccalaureate category, 2 semester hours for field experience in management, if rank was Staff Sergeant (SSGT) and 3 semester hours for field experience in management, if rank was Gunnery Sergeant (GYSGT) (8/89).

MCE-6013-001

AIRCRAFT MECHANIC, A-6/EA-6
6013

Exhibit Dates: 1/90–1/98.

Occupational Field: 60 (Aircraft Maintenance).

Career Pattern

PVT: Private (E-1). PFC: Private First Class (E-2). LCP: Lance Corporal (E-3). CPL: Corporal (E-4). SGT: Sergeant (E-5). SSGT: Staff Sergeant (E-6). GYSGT: Gunnery Sergeant (E-7). NOTE: MOS duties are not defined in terms of rank or paygrade, but by levels. Level I identifies tasks that are taught in the classroom; the individual is able to perform simple parts of tasks. Level II identifies those tasks that are part of on-the-job training; the trainee can do most tasks pertaining to a particular system and needs supervision on more difficult parts of tasks. Level III indicates that the individual can perform all essential tasks without supervision. Level IV indicates a high level of proficiency in job performance; the individual can perform advanced technical functions, instruction, inspection, and supervision. The MOS designator 6011 (Aircraft Mechanic—Trainee) is used to identify individuals in Levels I and II. Upon completion of Level II, MOS 6013 is awarded. Credit is recommended for Levels III and IV.

Description

Summary: Inspects and maintains aircraft airframes and airframe components; performs duties related to flight-line operations. *Levels I and II:* Operates and maintains applicable aircraft maintenance ground support equipment; obtains license as required; directs aircraft taxiing; performs daily preflight and postflight inspections and assists the pilot and crew in the performance of those inspections; assists during periodic airframe inspection; practices applicable safety procedures and precautions on the flight line and in hangar deck areas; performs hot and cold refueling on aircraft; performs oil analysis sampling and uses various types of fuels and lubricants; removes and installs aircraft power plants, components, and parts as required; recognizes and assists in corrective treatment and/or preventive maintenance of corrosion; uses precision measuring equipment required in the performance of duties; assists in the performance of calendar inspections; operates assigned fire fighting equipment; follows weight and balance limitations for applicable aircraft; performs starting and ground turn-up functions when authorized; uses technical publications in skill area. *Level III:* Able to perform the duties required for Levels I and II; performs duties as flight-line section leader or standardization team chief; conducts informal on-the-job technical training. *Level IV:* Able to perform the duties required for Level III; assists in the supervision and administration of flight-line hangar deck area and related operations ashore and afloat; plans, schedules, and directs work center assignments; performs duties as maintenance department noncommissioned officer-in-charge.

Recommendation, Level III

In the lower-division baccalaureate/associate degree category, 9 semester hours in aircraft airframe maintenance and inspection, 3 in aviation maintenance technology, and 3 in industrial safety (8/89).

Recommendation, Level IV

In the lower-division baccalaureate/associate degree category, 12 semester hours in aircraft airframe maintenance and inspection, 6 in aviation maintenance technology, 3 in industrial safety, 3 in personnel supervision, and 3 in maintenance management. In the upper-division baccalaureate category, 2 semester hours for field experience in management, if rank was Staff Sergeant (SSGT) and 3 semester hours for field experience in management, if rank was Gunnery Sergeant (GYSGT) (8/89).

MCE-6014-001

AIRCRAFT MECHANIC, F-4/RF-4
6014

Exhibit Dates: 1/90–1/98.

Occupational Field: 60 (Aircraft Maintenance).

Career Pattern

PVT: Private (E-1). PFC: Private First Class (E-2). LCP: Lance Corporal (E-3). CPL: Corporal (E-4). SGT: Sergeant (E-5). SSGT: Staff Sergeant (E-6). GYSGT: Gunncry Scrgeant (E-7). NOTE: MOS duties are not defined in terms of rank or paygrade, but by levels. Level I identifies tasks that are taught in the classroom; the individual is able to perform simple parts of tasks. Level II identifies those tasks that are part of on-the-job training; the trainee can do most tasks pertaining to a particular system and needs supervision on more difficult parts of tasks. Level III indicates that the individual can perform all essential tasks without supervision. Level IV indicates a high level of proficiency in job performance; the individual can perform advanced technical functions, instruction, inspection, and supervision. The MOS designator 6011 (Aircraft Mechanic—Trainee) is used to identify individuals in Levels I and II. Upon completion of Level II, MOS 6014 is awarded. Credit is recommended for Levels III and IV.

Description

Summary: Inspects and maintains aircraft airframes and airframe components; performs duties related to flight-line operations. *Levels I and II:* Operates and maintains applicable aircraft maintenance ground support equipment; obtains license as required; directs aircraft taxiing; performs daily preflight and postflight inspections and assists the pilot and crew in the performance of those inspections; assists during periodic airframe inspection; practices applicable safety procedures and precautions on the flight line and in hangar deck areas; performs hot and cold refueling of aircraft; performs oil analysis sampling and uses various types of fuels and lubricants; removes and installs aircraft power plants, components, and parts as required; recognizes and assists in corrective treatment and/or preventive maintenance for corrosion; uses precision measuring equipment required in the performance of duties; assists in the performance of calendar inspections; operates assigned fire fighting equipment; follows weight and balance limitations for applicable aircraft; performs starting and ground turn-up functions when authorized; uses technical publications in skill area. *Level III:* Able to perform the duties required for Levels I and II; performs duties as flight-line section leader or standardization team chief; conducts informal on-the-job technical training. *Level IV:* Able to perform the duties required for Level III; assists in the supervision and administration of flight-line hangar deck area and related operations ashore and afloat; plans, schedules, and directs work center assignments; performs duties as maintenance department noncommissioned officer-in-charge.

Recommendation, Level III

In the lower-division baccalaureate/associate degree category, 9 semester hours in aircraft airframe maintenance and inspection, 3 in aviation maintenance technology, and 3 in industrial safety (8/89).

Recommendation, Level IV

In the lower-division baccalaureate/associate degree category, 12 semester hours in aircraft airframe maintenance and inspection, 6 in aviation maintenance technology, 3 in industrial safety, 3 in personnel supervision, and 3 in maintenance management. In the upper-division baccalaureate category, 2 semester hours for field experience in management, if rank was Staff Sergeant (SSGT) and 3 semester hours for field experience in management, if rank was Gunnery Sergeant (GYSGT) (8/89).

MCE-6015-001

AIRCRAFT MECHANIC, AV-8A/TAV-8
6015

Exhibit Dates: 1/90–1/98.

Occupational Field: 60 (Aircraft Maintenance).

Career Pattern

PVT: Private (E-1). PFC: Private First Class (E-2). LCP: Lance Corporal (E-3). CPL: Corporal (E-4). SGT: Sergeant (E-5). SSGT: Staff Sergeant (E-6). GYSGT: Gunnery Sergeant (E-7). NOTE: MOS duties are not defined in terms of rank or paygrade, but by levels. Level I identifies tasks that are taught in the classroom; the individual is able to perform simple parts of tasks. Level II identifies those tasks that are part of on-the-job training; the trainee can do most tasks pertaining to a particular system and needs supervision on more difficult parts of tasks. Level III indicates that the individual can perform all essential tasks without supervision. Level IV indicates a high level of proficiency in job performance; the individual can perform advanced technical functions, instruction, inspection, and supervision. The MOS designator 6011 (Aircraft Mechanic—Trainee) is used to identify individuals in Levels I and II. Upon completion of Level II, MOS 6015 is awarded. Credit is recommended for Levels III and IV.

Description

Summary: Inspects and maintains aircraft airframes and airframe components; performs duties related to flight-line operations. *Levels I and II:* Operates and maintains applicable aircraft maintenance ground support equipment; obtains license as required; directs aircraft taxiing; performs daily preflight and postflight inspections and assists the pilot and crew in the performance of those inspections; assists during periodic airframe inspection; practices applicable safety procedures and precautions on the flight line and in hangar deck areas; performs hot and cold refueling of aircraft; performs oil analysis sampling and uses various types of fuels and lubricants; removes and installs aircraft power plants, components, and parts as required; recognizes and assists in corrective treatment and/or preventive maintenance for corrosion; uses precision measuring equipment required in the performance of duties; assists in the performance of calendar inspections; operates assigned fire fighting equipment; follows weight and balance limitations for applicable aircraft; performs starting and ground turn-up functions when authorized; uses technical publications in skill area. *Level III:* Able to perform the duties required for Levels I and II; performs duties as flight-line section leader or standardization team chief; conducts informal on-the-job technical training. *Level IV:* Able to perform the duties required for Level III; assists in the supervision and administration of flight-line hangar deck area and related operations ashore and afloat; plans, schedules, and directs work center assignments; performs duties as maintenance department noncommissioned officer-in-charge.

Recommendation, Level III

In the lower-division baccalaureate/associate degree category, 9 semester hours in aircraft airframe maintenance and inspection, 3 in aviation maintenance technology, and 3 in industrial safety (8/89).

Recommendation, Level IV

In the lower-division baccalaureate/associate degree category, 12 semester hours in aircraft airframe maintenance and inspection, 6 in aviation maintenance technology, 3 in industrial safety, 3 in personnel supervision, and 3 in maintenance management. In the upper-division baccalaureate category, 2 semester hours for field experience in management, if rank was Staff Sergeant (SSGT) and 3 semester hours for field experience in management, if rank was Gunnery Sergeant (GYSGT) (8/89).

MCE-6016-001

AIRCRAFT MECHANIC, KC-130
6016

Exhibit Dates: 1/90–1/98.

Occupational Field: 60 (Aircraft Maintenance).

Career Pattern

PVT: Private (E-1). PFC: Private First Class (E-2). LCP: Lance Corporal (E-3). CPL: Corporal (E-4). SGT: Sergeant (E-5). SSGT: Staff Sergeant (E-6). GYSGT: Gunnery Sergeant (E-7). NOTE: MOS duties are not defined in terms of rank or paygrade, but by levels. Level I identifies tasks that are taught in the classroom; the individual is able to perform simple parts of tasks. Level II identifies those tasks that are part of on-the-job training; the trainee can do most tasks pertaining to a particular system and needs supervision on more difficult parts of tasks. Level III indicates that the individual can perform all essential tasks without supervision. Level IV indicates a high level of proficiency in job performance; the individual can perform advanced technical functions, instruction, inspection, and supervision. The MOS designator 6011 (Aircraft Mechanic—Trainee) is used to identify individuals in Levels I and II. Upon completion of Level II, MOS 6016 is awarded. Credit is recommended for Levels III and IV.

Description

Summary: Inspects and maintains aircraft airframes and airframe components; performs duties related to flight-line operations. *Levels I and II:* Operates and maintains applicable aircraft maintenance ground support equipment; obtains license as required; directs aircraft taxiing; performs daily preflight and postflight inspections and assists the pilot and crew in the performance of those inspections; assists during periodic airframe inspection; practices applicable safety procedures and precautions on the flight line and in hangar deck areas; performs hot and cold refueling of aircraft; performs oil analysis sampling and uses various types of fuels and lubricants; removes and installs aircraft power plants, components, and parts as required; recognizes and assists in corrective treatment and/or preventive maintenance for corrosion; uses precision measuring equipment required in the performance of duties; assists in the performance of calendar inspections; operates assigned fire fighting equipment; follows weight and balance limitations for applicable aircraft; performs starting and ground turn-up functions when authorized; uses technical publications in skill area. *Level III:* Able to perform the duties required for Levels I and II; performs duties as flight-line section leader or standardization team chief; conducts informal on-the-job technical training. *Level IV:* Able to perform the duties required for Level III; assists in the supervision and administration of flight-line hangar deck area and related operations ashore and afloat; plans, schedules, and directs work center assignments; performs duties as maintenance department noncommissioned officer-in-charge.

Recommendation, Level III

In the lower-division baccalaureate/associate degree category, 9 semester hours in aircraft airframe maintenance and inspection, 3 in aviation maintenance technology, and 3 in industrial safety (8/89).

Recommendation, Level IV

In the lower-division baccalaureate/associate degree category, 12 semester hours in aircraft airframe maintenance and inspection, 6 in aviation maintenance technology, 3 in industrial safety, 3 in personnel supervision, and 3 in maintenance management. In the upper-division baccalaureate category, 2 semester hours for field experience in management, if rank was Staff Sergeant (SSGT) and 3 semester hours for field experience in management, if rank was Gunnery Sergeant (GYSGT) (8/89).

MCE-6017-001

AIRCRAFT MECHANIC, F/A-18
6017

Exhibit Dates: 1/90–1/98.

Occupational Field: 60 (Aircraft Maintenance).

Career Pattern

PVT: Private (E-1). PFC: Private First Class (E-2). LCP: Lance Corporal (E-3). CPL: Corporal (E-4). SGT: Sergeant (E-5). SSGT: Staff Sergeant (E-6). GYSGT: Gunnery Sergeant (E-7). NOTE: MOS duties are not defined in terms of rank or paygrade, but by levels. Level I identifies tasks that are taught in the classroom; the individual is able to perform simple parts of tasks. Level II identifies those tasks that are part of on-the-job training; the trainee can do most tasks pertaining to a particular system and needs supervision on more difficult parts of tasks. Level III indicates that the individual can perform all essential tasks without supervision. Level IV indicates a high level of proficiency in job performance; the individual can perform advanced technical functions, instruction, inspection, and supervision. The MOS designator 6011 (Aircraft Mechanic—Trainee) is used to identify individuals in Levels I and II. Upon completion of Level II, MOS 6017 is awarded. Credit is recommended for Levels III and IV.

Description

Summary: Inspects and maintains aircraft airframes and airframe components; performs duties related to flight-line operations. *Levels I and II:* Operates and maintains applicable aircraft maintenance ground support equipment; obtains license as required; directs aircraft taxiing; performs daily preflight and postflight inspections and assists the pilot and crew in the performance of those inspections; assists during periodic airframe inspection; practices applicable safety procedures and precautions on the flight line and in hangar deck areas; performs hot and cold refueling of aircraft; performs oil analysis sampling and uses various types of fuels and lubricants; removes and installs aircraft power plants, components, and parts as required; recognizes and assists in corrective treatment and/or preventive maintenance for corrosion; uses precision measuring equipment required in the performance of duties; assists in the performance of calendar inspections; operates assigned fire fighting equipment; follows weight and balance limitations for applicable aircraft; performs starting and ground turn-up functions when authorized; uses technical publications in skill area. *Level III:* Able to perform the duties required for Levels I and II; performs duties as flight-line section leader or standardization team chief; conducts informal on-the-job technical training. *Level IV:* Able to perform the duties required for Level III; assists in the supervision and administration of flight-line hangar deck area and related operations ashore and afloat; plans, schedules, and directs work center assignments; performs duties as maintenance department noncommissioned officer-in-charge.

Recommendation, Level III

In the lower-division baccalaureate/associate degree category, 9 semester hours in aircraft airframe maintenance and inspection, 3 in aviation maintenance technology, and 3 in industrial safety (8/89).

Recommendation, Level IV

In the lower-division baccalaureate/associate degree category, 12 semester hours in aircraft airframe maintenance and inspection, 6 in aviation maintenance technology, 3 in industrial safety, 3 in personnel supervision, and 3 in maintenance management. In the upper-division baccalaureate category, 2 semester hours for field experience in management, if rank was Staff Sergeant (SSGT) and 3 semester hours for field experience in management, if rank was Gunnery Sergeant (GYSGT) (8/89).

MCE-6018-001

AIRCRAFT MECHANIC, OV-10
6018

Exhibit Dates: 1/90–1/98.

Occupational Field: 60 (Aircraft Maintenance).

Career Pattern

PVT: Private (E-1). PFC: Private First Class (E-2). LCP: Lance Corporal (E-4). CPL: Corporal (E-4). SGT: Sergeant (E-5). SSGT: Staff Sergeant (E-6). GYSGT: Gunnery Sergeant (E-7). NOTE: MOS duties are not defined in terms of rank or paygrade, but by levels. Level I identifies tasks that are taught in the classroom; the individual is able to perform simple parts of tasks. Level II identifies those tasks that are part of on-the-job training; the trainee can do most tasks pertaining to a particular system and needs supervision on more difficult parts of tasks. Level III indicates that the individual can perform all essential tasks without supervision. Level IV indicates a high level of proficiency in job performance; the individual can perform advanced technical functions, instruction, inspection, and supervision. The MOS designator 6011 (Aircraft Mechanic—Trainee) is used to identify individuals in Levels I and II. Upon completion of Level II, MOS 6018 is awarded. Credit is recommended for Levels III and IV.

Description

Summary: Inspects and maintains aircraft airframes and airframe components; performs duties related to flight-line operations. *Levels I and II:* Operates and maintains applicable aircraft maintenance ground support equipment; obtains license as required; directs aircraft taxiing; performs daily preflight and postflight inspections and assists the pilot and crew in the performance of those inspections; assists during periodic airframe inspection; practices applicable safety procedures and precautions on the flight line and in hangar deck areas; performs hot and cold refueling of aircraft; performs oil analysis sampling and uses various types of fuels and lubricants; removes and installs aircraft power plants, components, and parts as required; recognizes and assists in corrective treatment and/or prevention maintenance for corrosion; uses precision measurement equipment required in the performance of duties; assists in the performance of calendar inspections; operates assigned fire fighting equipment; follows weight and balance limitations for applicable aircraft; performs starting and ground turn-up functions when authorized; uses technical publications in skill area. *Level III:* Able to perform the duties required for Levels I and II; performs duties as flight-line section leader or standardization team chief; conducts informal on-the-job technical training. *Level IV:* Able to perform the duties required for Level III; assists in the supervision and administration of flight-line hangar deck area and related operations ashore and afloat; plans, schedules, and directs work center assignments; performs duties as maintenance department noncommissioned officer-in-charge.

Recommendation, Level III

In the lower-division baccalaureate/associate degree category, 9 semester hours in aircraft airframe maintenance and inspection, 3 in aviation maintenance technology, and 3 in industrial safety (8/89).

Recommendation, Level IV

In the lower-division baccalaureate/associate degree category, 12 semester hours in aircraft airframe maintenance and inspection, 6 in aviation maintenance technology, 3 in industrial safety, 3 in personnel supervision, and 3 in maintenance management. In the upper-division baccalaureate category, 2 semester hours for field experience in management, if rank was Staff Sergeant (SSGT) and 3 semester hours for field experience in management, if rank was Gunnery Sergeant (GYSGT) (8/89).

MCE-6019-001

AIRCRAFT MAINTENANCE CHIEF
6019

Exhibit Dates: 5/92–Present.

Occupational Field: 60 (Aircraft Maintenance).

Career Pattern

May progress to MOS 6019 from MOS's 6012-18, 6022-27, 6031, 6032, 6044, 6052-58, 6060, 6072, 6075-78, 6082-88, or 6092-98 (Private through Gunnery Sergeant). MSGT: Master Sergeant (E-8). MGYSGT: Master Gunnery Sergeant (E-9). NOTE: Chiefs must be qualified in Level IV of previously-held MOS. Descriptions and credit recommendations for chiefs are by rank. Check the individual's fitness report or brief sheet for the rank and for proficiency report.

Description

Summary: Supervises aircraft maintenance and repair facility. *Master Sergeant:* Supervises the establishment and functioning of aircraft repair and maintenance facilities; assists in directing, supervising, and coordinating maintenance activities; prepares reports, schedules, and rosters regarding aircraft maintenance and repair; requisitions spare parts, replacement parts, supplies, and equipment; maintains maintenance, inspection, and technical training records. *Master Gunnery Sergeant:* Able to perform the duties required for Master Sergeant;

provides staff support in planning and implementing maintenance activities; has a thorough knowledge of Marine Corps administrative procedures and aviation staff organization and functioning; prepares and presents reports on logistics and readiness to higher command levels.

Recommendation, Master Sergeant

In the lower-division baccalaureate/associate degree category, 3 semester hours in personnel supervision, 3 in technical writing, and 3 in records and information management. In the upper-division baccalaureate category, 3 semester hours in organizational management, 3 in management problems, and 3 for field experience in management. NOTE: Credit should also be granted for the prerequisite MOS (5/92).

Recommendation, Master Gunnery Sergeant

In the lower-division baccalaureate/associate degree category, 3 semester hours in personnel supervision, 3 in technical writing, and 3 in records and information management. In the upper-division baccalaureate category, 3 semester hours in organizational management, 6 in management problems, and 6 for field experience in management. NOTE: Credit should also be granted for the prerequisite MOS (5/92).

MCE-6022-001

AIRCRAFT POWER PLANTS MECHANIC, J-52
6022

Exhibit Dates: 1/90–1/98.

Occupational Field: 60 (Aircraft Maintenance).

Career Pattern

PVT: Private (E-1). PFC: Private First Class (E-2). LCP: Lance Corporal (E-3). CPL: Corporal (E-4). SGT: Sergeant (E-5). SSGT: Staff Sergeant (E-6). GYSGT: Gunnery Sergeant (E-7). NOTE: MOS duties are not defined in terms of rank or paygrade, but by levels. Level I identifies tasks that are taught in the classroom; the individual is able to perform simple parts of tasks. Level II identifies those tasks that are part of on-the-job training; the trainee can do most tasks pertaining to a particular system and needs supervision on more difficult parts of tasks. Level III indicates that the individual can perform all essential tasks without supervision. Level IV indicates a high level of proficiency in job performance; the individual can perform advanced technical functions, instruction, inspection, and supervision. The MOS designator 6011 (Aircraft Mechanic—Trainee) is used to identify individuals in Levels I and II. Upon completion of Level II, MOS 6022 is awarded. Credit is recommended for Levels III and IV.

Description

Summary: Inspects, maintains, and repairs aircraft power plants and performs duties related to flight-line operations. *Levels I and II:* Interprets shop sketches, drawings, schematics, and blueprints; uses technical publications including inspection work cards and troubleshooting logic trees; under direct and indirect supervision, disassembles, repairs, and assembles power plants and related systems; in performance of maintenance functions, uses assorted hand and power tools such as jewelers' files, micrometer, and borescope; assists in minor maintenance of ground handling equipment and power plant test cells; knows types and designations of fuels and lubricants and uses color-coded charts for identification of lines and tubing; assists with corrosion prevention procedures and practices ground safety precautions in the maintenance area. *Level III:* Able to perform the duties required for Levels I and II; performs complete repair of power plants; performs ground testing and engine run-up; prepares power plants for storage or shipment; changes engines; prepares and submits work center reports; conducts technical training; plans, schedules, and supervises maintenance center operations. *Level IV:* Able to perform the duties required for Level III; performs and supervises complete repair of power plants and systems; functions as a maintenance supervisor and quality assurance inspector.

Recommendation, Level III

In the lower-division baccalaureate/associate degree category, 9 semester hours in jet turbine maintenance and repair, 3 in aviation maintenance technology, and 3 in industrial safety (8/89).

Recommendation, Level IV

In the lower-division baccalaureate/associate degree category, 12 semester hours in jet turbine maintenance and repair, 6 in aviation maintenance technology, 3 in industrial safety, 3 in personnel supervision, and 3 in maintenance management. In the upper-division baccalaureate category, 2 semester hours for field experience in management, if rank was Staff Sergeant (SSGT) and 3 semester hours for field experience in management, if rank was Gunnery Sergeant (GYSGT) (8/89).

MCE-6023-001

AIRCRAFT POWER PLANTS MECHANIC, T-76
6023

Exhibit Dates: 1/90–1/98.

Occupational Field: 60 (Aircraft Maintenance).

Career Pattern

PVT: Private (E-1). PFC: Private First Class (E-2). LCP: Lance Corporal (E-3). CPL: Corporal (E-4). SGT: Sergeant (E-5). SSGT: Staff Sergeant (E-6). GYSGT: Gunnery Sergeant (E-7). NOTE: MOS duties are not defined in terms of rank or paygrade, but by levels. Level I identifies tasks that are taught in the classroom; the individual is able to perform simple parts of tasks. Level II identifies those tasks that are part of on-the-job training; the trainee can do most tasks pertaining to a particular system and needs supervision on more difficult parts of tasks. Level III indicates that the individual can perform all essential tasks without supervision. Level IV indicates a high level of proficiency in job performance; the individual can perform advanced technical functions, instruction, inspection, and supervision. The MOS designator 6011 (Aircraft Mechanic—Trainee), is used to identify individuals in Levels I and II. Upon completion of Level II, MOS 6023 is awarded. Credit is recommended for Levels III and IV.

Description

Summary: Inspects, maintains, and repairs aircraft power plants, and performs duties related to flight-line operations. *Levels I and II:* Interprets shop sketches, drawings, schematics, and blueprints; uses technical publications including inspection work cards and troubleshooting logic trees; under direct and indirect supervision, disassembles, repairs, and assembles power plants and related systems; in performance of maintenance functions, uses assorted hand and power tools such as jewelers' files, micrometer, and borescope; assists in minor maintenance of ground handling equipment and power plant test cells; knows types and designations of fuels and lubricants and uses color-coded charts for identification of lines and tubing; assists with corrosion prevention procedures and practices ground safety precautions in the maintenance area. *Level III:* Able to perform the duties required for Levels I and II; performs complete repair of power plants; performs ground testing and engine run-up; prepares power plants for storage or shipment; changes engines; prepares and submits work center reports; conducts technical training; plans, schedules, and supervises maintenance center operations. *Level IV:* Able to perform the duties required for Level III; performs and supervises complete repair of power plants and systems; functions as a maintenance supervisor and quality assurance inspector.

Recommendation, Level III

In the lower-division baccalaureate/associate degree category, 9 semester hours in jet turbine maintenance and repair, 3 in aviation maintenance technology, and 3 in industrial safety (8/89).

Recommendation, Level IV

In the lower-division baccalaureate/associate degree category, 12 semester hours in jet turbine maintenance and repair, 6 in aviation maintenance technology, 3 in industrial safety, 3 in personnel supervision, and 3 in maintenance management. In the upper-division baccalaureate category, 2 semester hours for field experience in management, if rank was Staff Sergeant (SSGT) and 3 semester hours for field experience in management, if rank was Gunnery Sergeant (GYSGT) (8/89).

MCE-6024-001

AIRCRAFT POWER PLANTS MECHANIC, J-79
6024

Exhibit Dates: 1/90–1/98.

Occupational Field: 60 (Aircraft Maintenance).

Career Pattern

PVT: Private (E-1). PFC: Private First Class (E-2). LCP: Lance Corporal (E-3). CPL: Corporal (E-4). SGT: Sergeant (E-5). SSGT: Staff Sergeant (E-6). GYSGT: Gunnery Sergeant (E-7). NOTE: MOS duties are not defined in terms of rank or paygrade, but by levels. Level I identifies tasks that are taught in the classroom; the individual is able to perform simple parts of tasks. Level II identifies those tasks that are part of on-the-job training; the trainee can do most tasks pertaining to a particular system and needs supervision on more difficult parts of tasks. Level III indicates that the individual can perform all essential tasks without supervision. Level IV indicates a high level of proficiency in job performance; the individual can perform advanced technical functions, instruction, inspection, and supervision. The MOS designator 6011 (Aircraft Mechanic—Trainee), is used to identify individuals in Levels I and II. Upon completion of Level II, MOS 6024 is awarded. Credit is recommended for Levels III and IV.

Description

Summary: Inspects, maintains, and repairs aircraft power plants, and performs duties related to flight-line operations. *Levels I and II:* Interprets shop sketches, drawings, schematics, and blueprints; uses technical publications including inspection work cards and troubleshooting logic trees; under direct and indirect supervision, disassembles, repairs, and assembles power plants and related systems; in performance of maintenance functions, uses assorted hand and power tools such as jewelers' files, micrometer, and borescope; assists in minor maintenance of ground handling equipment and power plant test cells; knows types and designations of fuels and lubricants and uses color-coded charts for identification of lines and tubing; assists with corrosion prevention procedures and practices ground safety precautions in the maintenance area. *Level III:* Able to perform the duties required for Levels I and II; performs complete repair of power plants; performs ground testing and engine run-up; prepares power plants for storage or shipment; changes engines; prepares and submits work center reports; conducts technical training; plans, schedules, and supervises maintenance center operations. *Level IV:* Able to perform the duties required for Level III; performs and supervises complete repair of power plants and systems; functions as a maintenance supervisor and quality assurance inspector.

Recommendation, Level III

In the lower-division baccalaureate/associate degree category, 9 semester hours in jet turbine maintenance and repair, 3 in aviation maintenance technology, and 3 in industrial safety (8/89).

Recommendation, Level IV

In the lower-division baccalaureate/associate degree category, 12 semester hours in jet turbine maintenance and repair, 6 in aviation maintenance technology, 3 in industrial safety, 3 in personnel supervision, and 3 in maintenance management. In the upper-division baccalaureate category, 2 semester hours for field experience in management, if rank was Staff Sergeant (SSGT) and 3 semester hours for field experience in management, if rank was Gunnery Sergeant (GYSGT) (8/89).

MCE-6025-001

AIRCRAFT POWER PLANTS MECHANIC, ROLLS ROYCE PEGASUS
6025

Exhibit Dates: 1/90–1/98.

Occupational Field: 60 (Aircraft Maintenance).

Career Pattern

PVT: Private (E-1). PFC: Private First Class (E-2). LCP: Lance Corporal (E-3). CPL: Corporal (E-4). SGT: Sergeant (E-5). SSGT: Staff Sergeant (E-6). GYSGT: Gunnery Sergeant (E-7). NOTE: MOS duties are not defined in terms of rank or paygrade, but by levels. Level I identifies tasks that are taught in the classroom; the individual is able to perform simple parts of tasks. Level II identifies those tasks that are part of on-the-job training; the trainee can do most tasks pertaining to a particular system and needs supervision on more difficult parts of tasks. Level III indicates that the individual can perform all essential tasks without supervision. Level IV indicates a high level of proficiency in job performance; the individual can perform advanced technical functions, instruction, inspection, and supervision. The MOS designator 6011 (Aircraft Mechanic—Trainee), is used to identify individuals in Levels I and II. Upon completion of Level II, MOS 6025 is awarded. Credit is recommended for Levels III and IV.

Description

Summary: Inspects, maintains, and repairs aircraft power plants, and performs duties related to flight-line operations. *Levels I and II:* Interprets shop sketches, drawings, schematics, and blueprints; uses technical publications including inspection work cards and troubleshooting logic trees; under direct and indirect supervision, disassembles, repairs, and assembles power plants and related systems; in performance of maintenance functions, uses assorted hand and power tools such as jewelers' files, micrometer, and borescope; assists in minor maintenance of ground handling equipment and power plant test cells; knows types and designations of fuels and lubricants and uses color-coded charts for identification of lines and tubing; assists with corrosion prevention procedures and practices ground safety precautions in the maintenance area. *Level III:* Able to perform the duties required for Levels I and II; performs complete repair of power plants; performs ground testing and engine run-up; prepares power plants for storage or shipment; changes engines; prepares and submits work center reports; conducts technical training; plans, schedules, and supervises maintenance center operations. *Level IV:* Able to perform the duties required for Level III; performs and supervises complete repair of power plants and systems; functions as a maintenance supervisor and quality assurance inspector.

Recommendation, Level III

In the lower-division baccalaureate/associate degree category, 9 semester hours in jet turbine maintenance and repair, 3 in aviation maintenance technology, and 3 in industrial safety (8/89).

Recommendation, Level IV

In the lower-division baccalaureate/associate degree category, 12 semester hours in jet turbine maintenance and repair, 6 in aviation maintenance technology, 3 in industrial safety, 3 in personnel supervision, and 3 in maintenance management. In the upper-division baccalaureate category, 2 semester hours for field experience in management, if rank was Staff Sergeant (SSGT) and 3 semester hours for field experience in management, if rank was Gunnery Sergeant (GYSGT) (8/89).

MCE-6026-001

AIRCRAFT POWER PLANTS MECHANIC, T-56
6026

Exhibit Dates: 1/90–1/98.

Occupational Field: 60 (Aircraft Maintenance).

Career Pattern

PVT: Private (E-1). PFC: Private First Class (E-2). LCP: Lance Corporal (E-3). CPL: Corporal (E-4). SGT: Sergeant (E-5). SSGT: Staff Sergeant (E-6). GYSGT: Gunnery Sergeant (E-7). NOTE: MOS duties are not defined in terms of rank or paygrade, but by levels. Level I identifies tasks that are taught in the classroom; the individual is able to perform simple parts of tasks. Level II identifies those tasks that are part of on-the-job training; the trainee can do most tasks pertaining to a particular system and needs supervision on more difficult parts of tasks. Level III indicates that the individual can perform all essential tasks without supervision. Level IV indicates a high level of proficiency in job performance; the individual can perform advanced technical functions, instruction, inspection, and supervision. The MOS designator 6011 (Aircraft Mechanic—Trainee), is used to identify individuals in Levels I and II. Upon completion of Level II, MOS 6026 is awarded. Credit is recommended for Levels III and IV.

Description

Summary: Inspects, maintains, and repairs aircraft power plants, and performs duties related to flight-line operations. *Levels I and II:* Interprets shop sketches, drawings, schematics, and blueprints; uses technical publications including inspection work cards and troubleshooting logic trees; under direct and indirect supervision, disassembles, repairs, and assembles power plants and related systems; in performance of maintenance functions, uses assorted hand and power tools such as jewelers' files, micrometer, and borescope; assists in minor maintenance on ground handling equipment and power plant test cells; knows types and designations of fuels and lubricants and uses color-coded charts for identification of lines and tubing; assists with corrosion prevention procedures and practices ground safety precautions in the maintenance area. *Level III:* Able to perform the duties required for Levels I and II; performs complete repair of power plants; performs ground testing and engine run-up; prepares power plants for storage or shipment; changes engine; prepares and submits work center reports; conducts technical training; plans, schedules, and supervises maintenance center operations. *Level IV:* Able to perform the duties required for Level III; performs and supervises complete repair of power plants and systems; functions as a maintenance supervisor and quality assurance inspector.

Recommendation, Level III

In the lower-division baccalaureate/associate degree category, 9 semester hours in jet turbine maintenance and repair, 3 in aviation maintenance technology, and 3 in industrial safety (8/89).

Recommendation, Level IV

In the lower-division baccalaureate/associate degree category, 12 semester hours in jet turbine maintenance and repair, 6 in aviation maintenance technology, 3 in industrial safety, 3 in personnel supervision, and 3 in maintenance management. In the upper-division baccalaureate category, 2 semester hours for field experience in management, if rank was Staff Sergeant (SSGT) and 3 semester hours for field experience in management, if rank was Gunnery Sergeant (GYSGT) (8/89).

MCE-6027-001

AIRCRAFT POWER PLANTS MECHANIC, F-404
6027

Exhibit Dates: 1/90–1/98.

Occupational Field: 60 (Aircraft Maintenance).

Career Pattern

PVT: Private (E-1). PFC: Private First Class (E-2). LCP: Lance Corporal (E-3). CPL: Corporal (E-4). SGT: Sergeant (E-5). SSGT: Staff Sergeant (E-6). GYSGT: Gunnery Sergeant (E-7). NOTE: MOS duties are not defined in terms of rank or paygrade, but by levels. Level I identifies tasks that are taught in the classroom; the individual is able to perform simple parts of tasks. Level II identifies those tasks that are part of on-the-job training; the trainee can do most tasks pertaining to a particular system and needs supervision on more difficult parts of tasks. Level III indicates that the individual can perform all essential tasks without supervision. Level IV indicates a high level of proficiency in job performance; the individual can perform advanced technical functions, instruction, inspection, and supervision. The MOS designator 6011 (Aircraft Mechanic—Trainee), is used to identify individuals in Levels I and II. Upon completion of Level II, MOS 6027 is awarded. Credit is recommended for Levels III and IV.

Description

Summary: Inspects, maintains, and repairs aircraft power plants, and performs duties related to flight line operations. *Levels I and II:* Interprets shop sketches, drawings, schematics, and blueprints; uses technical publications including inspection work cards and troubleshooting logic trees; under direct and indirect supervision, disassembles, repairs, and assembles power plants and related systems; in performance of maintenance functions, uses assorted hand and power tools such as jewelers' files, micrometer, and borescope; assists in minor maintenance on ground handling equipment and power plant test cells; knows types and designations of fuels and lubricants and uses color-coded charts for identification of lines and tubing; assists with corrosion prevention procedures and practices ground safety precautions in the maintenance area. *Level III:* Able to perform the duties required for Levels I and II; performs complete repair of power plants; performs ground testing and engine run-up; prepares power plants for storage or shipment; changes engine; prepares and submits work center reports; conducts technical training; plans, schedules, and supervises maintenance center operations. *Level IV:* Able to perform the duties required for Level III; performs and supervises complete repair of power plants and systems; functions as a maintenance supervisor and quality assurance inspector.

Recommendation, Level III

In the lower-division baccalaureate/associate degree category, 9 semester hours in jet turbine maintenance and repair, 3 in aviation maintenance technology, and 3 in industrial safety (8/89).

Recommendation, Level IV

In the lower-division baccalaureate/associate degree category, 12 semester hours in jet turbine maintenance and repair, 6 in aviation maintenance technology, 3 in industrial safety, 3 in personnel supervision, and 3 in maintenance management. In the upper-division baccalaureate category, 2 semester hours for field experience in management, if rank was Staff Sergeant (SSGT) and 3 semester hours for field experience in management, if rank was Gunnery Sergeant (GYSGT) (8/89).

MCE-6035-001

AIRCRAFT POWER PLANTS TEST CELL OPERATOR, FIXED WING
6035

Exhibit Dates: 5/92–Present.

Occupational Field: 60 (Aircraft Maintenance).

Career Pattern

Must be qualified in one of the aircraft power plants MOS's (6022, 6023, 6024, 6025, 6026, or 6027). PVT: Private (E-1). PFC: Private First Class (E-2). LCP: Lance Corporal (E-3). CPL: Corporal (E-4). SGT: Sergeant (E-5). SSGT: Staff Sergeant (E-6). GYSGT: Gunnery Sergeant (E-7). MSGT: Master Sergeant (E-8). MGYSGT: Master Gunnery Sergeant (E-9). NOTE: MOS duties are not defined in terms of rank or paygrade, but by levels. Level I identifies tasks that are taught in the classroom; the individual is able to perform simple parts of tasks. Level II identifies those tasks that are part of on-the-job training; the trainee can do most tasks pertaining to a particular system and needs supervision on more difficult parts of tasks. Level III indicates that the individual can perform all essential tasks without supervision. Level IV indicates a high level of proficiency in job performance; the individual can perform advanced technical functions, instruction, inspection, and supervision. Credit is recommended for Levels III and IV.

Description

Summary: Inspects, tests, and performs corrective maintenance on fixed-wing gas turbine engines and engine systems; performs test cell operations and minor repairs to test cells. *Levels I and II:* Interprets shop sketches, drawings, schematics, and blueprints; performs minor maintenance on gas turbine engines and test cells; uses precision measuring equipment; observes safety procedures; completes maintenance data forms; performs test cell engines run-ups. *Level III:* Able to perform the duties required for Levels I and II; removes and replaces power plant components and accessories; conducts informal on-the-job training. *Level IV:* Able to perform the duties required for Level III; plans, schedules, and directs work center assignments; supervises and performs complete power plant repair in work center; supervises, directs, and administers fixed-wing gas turbine engine test cell operations.

Recommendation, Level III

NOTE: Credit is recommended for the prerequisite MOS only. See the appropriate exhibit for previously-held MOS (5/92).

Recommendation, Level IV

In the upper-division baccalaureate category, if rank was Staff Sergeant (SSGT), 2 semester hours for field experience in management; if rank was Gunnery Sergeant (GYSGT), 3 semester hours for field experience in management; if rank was Master Sergeant (MSGT), 3 semester hours for field experience in management and 3 in organizational management; if rank was Master Gunnery Sergeant (MGYSGT), 6 semester hours for field experience in management, 3 in organizational management, and 3 in management problems. NOTE: Credit should also be granted for the prerequisite MOS (5/92).

MCE-6044-001

AIRCRAFT NON-DESTRUCTIVE INSPECTION TECHNICIAN
6044

Exhibit Dates: 5/92–Present.

Occupational Field: 60 (Aircraft Maintenance).

Career Pattern

Must be qualified in one of the Aircraft Structures Mechanic MOS's (6092, 6093, 6094, 6095, 6096, 6097, or 6098). CPL: Corporal (E-4). SGT: Sergeant (E-5). SSGT: Staff Sergeant (E-6). GYSGT: Gunnery Sergeant (E-7). MSGT: Master Sergeant (E-8), MGYSGT: Master Gunnery Sergeant (E-9). NOTE: MOS duties are not defined in terms of rank or paygrade, but by levels. Level I identifies tasks that are taught in the classroom; the individual is able to perform simple parts of tasks. Level II identifies those tasks that are part of on-the-job training; the trainee can do most tasks pertaining to a particular system and needs supervision on more difficult parts of tasks. Level III indicates that the individual can perform all essential tasks without supervision. Level IV indicates a high level of proficiency in job performance; the individual can perform advanced technical functions, instruction, inspection, and supervision. Credit is recommended for Levels III and IV.

Description

Summary: Inspects and tests aircraft component parts using nondestructive testing procedures on both aircraft structures and engine components. *Levels I and II:* Performs nondestructive tests on aircraft components; reads schematic diagrams and blueprints; operates ground support equipment; practices safety procedures; sets up and operates laboratory equipment; uses technical publications. *Level III:* Able to perform the duties required for Levels I and II; collects maintenance data and practices material control procedures; inspects test results to verify component facilities. *Level IV:* Able to perform the duties required for Level III; conducts informal on-the-job technical training; plans, schedules, and directs work center assignments; supervises maintenance-level nondestructive testing work center.

Recommendation, Level III

In the lower-division baccalaureate/associate degree category, 3 semester hours in materials testing and 1 in technical writing. NOTE: Credit should also be granted for the prerequisite MOS (5/92).

Recommendation, Level IV

In the lower-division baccalaureate/associate degree category, 3 semester hours in materials testing, 1 in technical writing, 3 in personnel supervision, and 3 in maintenance management. In the upper-division baccalaureate category, if rank was Staff Sergeant (SSGT), 2 semester hours for field experience in management; if rank was Gunnery Sergeant (GYSGT), 3 semester hours for field experience in management; if rank was Master Sergeant (MSGT), 3 semester hours for field experience in management and 3 in organizational management; if rank was Master Gunnery Sergeant (MGYSGT), 6 semester hours for field experience in management, 3 in organizational management, and 3 in management problems. NOTE: Credit should also be granted for the prerequisite MOS (5/92).

MCE-6046-001

AIRCRAFT MAINTENANCE ADMINISTRATION CLERK
6046

Exhibit Dates: 1/90–1/98.

Occupational Field: 60 (Aircraft Maintenance).

Career Pattern

PVT: Private (E-1). PFC: Private First Class (E-2). LCP: Lance Corporal (E-3). CPL: Corporal (E-4). SGT: Sergeant (E-5). SSGT: Staff Sergeant (E-6). GYSGT: Gunnery Sergeant (E-7). NOTE: MOS duties are not defined in terms of rank or paygrade, but by levels. Level I identifies tasks that are taught in the classroom; the individual is able to perform simple parts of tasks. Level II identifies those tasks that are part of on-the-job training; the trainee can do most tasks pertaining to a particular system and needs supervision on more difficult parts of tasks. Level III indicates that the individual can perform all essential tasks without supervision. Level IV indicates a high level of proficiency in job performance; the individual can perform advanced technical functions, instruction, inspection, and supervision. Credit is recommended for Levels III and IV.

Description

Summary: Prepares and maintains various types of aircraft maintenance records, including log books, directives, and files; maintains status records of aircraft maintenance and repair; types 25-30 words per minute and/or uses word processor for correspondence; uses aircraft accounting and inventory systems. *Levels I and II:* Prepares reports, records, directives and correspondence; maintains aircraft and engine status boards; maintains files of repair publications, correspondence, and records; assists in inventory of aircraft; conducts informal technical training within assigned skill area. *Level III:* Able to perform the duties required for Levels I and II; plans, coordinates, and performs work and administrative assignments; organizes and supervises the administrative functions in maintenance; may have responsibility for organizing, supervising, and serving as the head of an intermediate administrative section. *Level IV:* Able to perform the duties required for Level III; performs management responsibilities at a facility for the administration of aircraft maintenance; instructs and supervises training of personnel; may have assignment as inspector of administrative facilities for aircraft maintenance; may have responsibility for the mathematical calculations and the supervision of aircraft loading to ensure proper distribution of cargo weights within large aircraft.

Recommendation, Level III

In the lower-division baccalaureate/associate degree category, 3 semester hours in record keeping (12/89).

Recommendation, Level IV

In the lower-division baccalaureate/associate degree category, 3 semester hours in record keeping and 2 in personnel supervision. In the upper-division baccalaureate category, 3 semester hours in management problems. If rank was Staff Sergeant (SSGT), 2 semester hours for field experience in management; if rank was Gunnery Sergeant (GYSGT), 3 semester hours for field experience in management (12/89).

MCE-6047-001

AIRCRAFT MAINTENANCE DATA ANALYSIS TECHNICIAN
6047

Exhibit Dates: 1/90–1/98.

Occupational Field: 60 (Aircraft Maintenance).

Career Pattern

PVT: Private (E-1). PFC: Private First Class (E-2). LCP: Lance Corporal (E-3). CPL: Corporal (E-4). SGT: Sergeant (E-5). SSGT: Staff Sergeant (E-6). GYSGT: Gunnery Sergeant (E-7). NOTE: MOS duties are not defined in terms of rank or paygrade, but by levels. Level I identifies tasks that are taught in the classroom; the individual is able to perform simple parts of tasks. Level II identifies those tasks that are part of on-the-job training; the trainee can do most tasks pertaining to a particular system and needs supervision on more difficult parts of tasks. Level III indicates that the individual can perform all essential tasks without supervision. Level IV indicates a high level of proficiency in job performance; the individual can perform advanced technical functions, instruction, inspection, and supervision. Credit is recommended for Levels III and IV.

Description

Summary: Provides information and recommendations to aid the maintenance manager and logistician in the performance of their tasks by extracting, analyzing, and collating maintenance data from detailed reports; develops and analyzes maintenance summaries; develops charts, tables, and graphs; isolates maintenance trends and determines effectiveness and efficiency of the maintenance effort; presents summaries and recommendations. *Levels I and II:* Operates data entry equipment such as interpreter, sorter, collator, reproducer, calculating punch, alphabetic accounting machines, and personal computers; operates standard attachments for data entry equipment and optical scanners; wires control panels from wiring diagrams for electronic accounting machines; operates data entry and data verifying equipment; maintains equipment utilization records; extracts, analyzes, collates, and prepares maintenance summaries; analyzes and identifies material deficiencies; develops charts, tables, and graphs. *Level III:* Able to perform the duties required for Levels I and II; assists personnel in the interpretation and use of reports of maintenance actions and man-hour accounting systems; identifies areas of deficiencies in training and readiness; develops uniform factors for determining maintenance capability; analyzes maintenance data for maintenance planning and estimation; evaluates exceptions to standards to determine cause and effect. *Level IV:* Able to perform the duties required for Level III; isolates trends and determines standards for evaluation of the maintenance effort; coordinates analysis functions with other branches; maintains personnel job performance requirements; conducts training of subordinate personnel; develops and applies uniform methods for improving the maintenance effort; evaluates limitations to determine the effect of the maintenance effort; performs management responsibilities at a facility conducting aircraft maintenance data analysis.

Recommendation, Level III

In the lower-division baccalaureate/associate degree category, 3 semester hours in introduction to computers, 3 in records and information management, and 2 in college algebra (12/89).

Recommendation, Level IV

In the lower-division baccalaureate/associate degree category, 3 semester hours in introduction to computers, 3 in records and information management, 2 in college algebra, and 2 in personnel supervision. In the upper-division baccalaureate category, 3 semester hours in management problems. If rank was Staff Sergeant (SSGT), 2 semester hours for field experience in management; if rank was Gunnery Sergeant (GYSGT), 3 semester hours for field experience in management (12/89).

MCE-6052-001

AIRCRAFT HYDRAULIC/PNEUMATIC MECHANIC, A-4/TA-4/OA-4
6052

Exhibit Dates: 1/90–1/98.

Occupational Field: 60 (Aircraft Maintenance).

Career Pattern

PVT: Private (E-1). PFC: Private First Class (E-2). LCP: Lance Corporal (E-3). CPL: Corporal (E-4). SGT: Sergeant (E-5). SSGT: Staff Sergeant (E-6). GYSGT: Gunnery Sergeant (E-7). NOTE: MOS duties are not defined in terms of rank or paygrade, but by levels. Level I identifies tasks that are taught in the classroom; the individual is able to perform simple parts of tasks. Level II identifies those tasks that are part of on-the-job training; the trainee can do most tasks pertaining to a particular system and needs supervision on more difficult parts of tasks. Level III indicates that the individual can perform all essential tasks without supervision. Level IV indicates a high level of proficiency in job performance; the individual can perform advanced technical functions, instruction, inspection, and supervision. The MOS designator 6051 (Aircraft Hydraulic/Pneumatic Mechanic—Trainee), is used to identify individuals in Levels I and II. Upon completion of Level II, MOS 6052 is awarded. Credit is recommended for Levels III and IV.

Description

Summary: Inspects, maintains, and repairs aircraft hydraulic/pneumatic systems and system components. *Levels I and II:* Uses color-coded charts of aircraft piping and lines; reads schematic diagrams, shop sketches, and blueprints and interprets applicable technical data; under supervision, performs component repairs; uses fluids and gases in applicable systems; uses precision measuring equipment; assists in or performs required inspections; fabricates replacement lines of tubing and hoses. *Level III:* Able to perform the duties required for Levels I and II; performs component repair of hydraulic/pneumatic components as required; uses and maintains hydraulic/pneumatic test equipment; observes applicable material control procedures; conducts informal on-the-job training; plans, schedules, and directs work center assignments. *Level IV:* Able to perform the duties required for Level III; performs and supervises the maintenance/repair requirements for aircraft hydraulic/pneumatic systems components; supervises an aircraft hydraulic/pneumatic check crew or maintenance unit; performs duties as quality assurance inspector; supervises and administers an organizational maintenance level aircraft hydraulic/pneumatic work center.

Recommendation, Level III

In the lower-division baccalaureate/associate degree category, 6 semester hours in hydraulic/pneumatic systems repair, 3 in aviation maintenance technology, and 3 in industrial safety (8/89).

Recommendation, Level IV

In the lower-division baccalaureate/associate degree category, 9 semester hours in hydraulic/pneumatic systems repair, 6 in aviation maintenance technology, 3 in industrial safety, 3 in personnel supervision, and 3 in maintenance management. In the upper-division baccalaureate category, 2 semester hours for field experience in management, if rank was Staff Sergeant (SSGT) and 3 semester hours for field experience in management, if rank was Gunnery Sergeant (GYSGT) (8/89).

MCE-6053-001

AIRCRAFT HYDRAULIC/PNEUMATIC MECHANIC, A-6/EA-6
6053

Exhibit Dates: 1/90–1/98.

Occupational Field: 60 (Aircraft Maintenance).

Career Pattern

PVT: Private (E-1). PFC: Private First Class (E-2). LCP: Lance Corporal (E-3). CPL: Corporal (E-4). SGT: Sergeant (E-5). SSGT: Staff Sergeant (E-6). GYSGT: Gunnery Sergeant (E-7). NOTE: MOS duties are not defined in terms of rank or paygrade, but by levels. Level I identifies tasks that are taught in the classroom; the individual is able to perform simple parts of tasks. Level II identifies those tasks that are part of on-the-job training; the trainee can do most tasks pertaining to a particular system and needs supervision on more difficult parts of tasks. Level III indicates that the individual can perform all essential tasks without supervision. Level IV indicates a high level of proficiency in job performance; the individual can perform advanced technical functions, instruction, inspection, and supervision. The MOS designator 6051 (Aircraft Hydraulic/Pneumatic Mechanic—Trainee), is used to identify individuals in Levels I and II. Upon completion of Level II, MOS 6053 is awarded. Credit is recommended for Levels III and IV.

Description

Summary: Inspects, maintains, and repairs aircraft hydraulic/pneumatic systems and system components. *Levels I and II:* Uses color-coded charts of aircraft piping and lines; reads schematic diagrams, shop sketches, and blueprints and interprets applicable technical data; under supervision, performs component repairs; uses fluids and gases in applicable systems; uses precision measuring equipment; assists in or performs required inspections; fabricates replacement lines of tubing and hoses. *Level III:* Able to perform the duties required for Levels I and II; performs component repair of hydraulic/pneumatic components as required; uses and maintains hydraulic/pneumatic test equipment; observes applicable material control procedures; conducts informal on-the-job training; plans, schedules, and directs work center assignments. *Level IV:* Able to perform the duties required for Level III; performs and supervises the maintenance/repair requirements for aircraft hydraulic/pneumatic system components; supervises an aircraft hydraulic/pneumatic check crew or maintenance unit; performs duties as quality assurance inspector; supervises and administers an organizational maintenance level aircraft hydraulic/pneumatic work center.

Recommendation, Level III

In the lower-division baccalaureate/associate degree category, 6 semester hours in hydraulic/pneumatic systems repair, 3 in aviation maintenance technology, and 3 in industrial safety (8/89).

Recommendation, Level IV

In the lower-division baccalaureate/associate degree category, 9 semester hours in hydraulic/pneumatic systems repair, 6 in aviation maintenance technology, 3 in industrial safety, 3 in personnel supervision, and 3 in maintenance management. In the upper-division baccalaureate category, 2 semester hours for field experience in management, if rank was Staff Sergeant (SSGT) and 3 semester hours for field experience in management, if rank was Gunnery Sergeant (GYSGT) (8/89).

MCE-6054-001

AIRCRAFT HYDRAULIC/PNEUMATIC MECHANIC, F-4/RF-4
6054

Exhibit Dates: 1/90–1/98.

Occupational Field: 60 (Aircraft Maintenance).

Career Pattern

PVT: Private (E-1). PFC: Private First Class (E-2). LCP: Lance Corporal (E-3). CPL: Corporal (E-4). SGT: Sergeant (E-5). SSGT: Staff Sergeant (E-6). GYSGT: Gunnery Sergeant (E-7). NOTE: MOS duties are not defined in terms of rank or paygrade, but by levels. Level I identifies tasks that are taught in the classroom; the individual is able to perform simple parts of tasks. Level II identifies those tasks that are part of on-the-job training; the trainee can do most tasks pertaining to a particular system and needs supervision on more difficult parts of tasks. Level III indicates that the individual can perform all essential tasks without supervision. Level IV indicates a high level of proficiency in job performance; individual can perform advanced technical functions, instruction, inspection, and supervision. The MOS designator 6051 (Aircraft Hydraulic/Pneumatic Mechanic—Trainee), is used to identify individuals in Levels I and II. Upon completion of Level II, MOS 6054 is awarded. Credit is recommended for Levels III and IV.

Description

Summary: Inspects, maintains, and repairs aircraft hydraulic/pneumatic systems and system components. *Levels I and II:* Uses color-coded charts of aircraft piping and lines; reads schematic diagrams, shop sketches, and blueprints and interprets applicable technical data; under supervision, performs component repairs; uses fluids and gases in applicable systems; uses precision measuring equipment; assists in or performs required inspections; fabricates replacement lines of tubing and hoses. *Level III:* Able to perform the duties required for Levels I and II; performs component repair of hydraulic/pneumatic components as required; uses and maintains hydraulic/pneumatic test equipment; observes applicable material control procedures; conducts informal on-the-job training; plans, schedules, and directs work center assignments. *Level IV:* Able to perform the duties required for Level III; performs and supervises the maintenance/repair requirements for aircraft hydraulic/pneumatic systems components; supervises an aircraft hydraulic/pneumatic check crew or maintenance unit; performs duties as quality assurance inspector; supervises and administers an organizational maintenance level aircraft hydraulic/pneumatic work center.

Recommendation, Level III

In the lower-division baccalaureate/associate degree category, 6 semester hours in hydraulic/pneumatic systems repair, 3 in aviation maintenance technology, and 3 in industrial safety (8/89).

Recommendation, Level IV

In the lower-division baccalaureate/associate degree category, 9 semester hours in hydraulic/pneumatic systems repair, 6 in aviation maintenance technology, 3 in industrial safety, 3 in personnel supervision, and 3 in maintenance management. In the upper-division baccalaureate category, 2 semester hours for field experience in management, if rank was Staff Sergeant (SSGT) and 3 semester hours for field experience in management, if rank was Gunnery Sergeant (GYSGT) (8/89).

MCE-6055-001

AIRCRAFT HYDRAULIC/PNEUMATIC MECHANIC, AV-8/TAV-8
6055

Exhibit Dates: 1/90–1/98.

Occupational Field: 60 (Aircraft Maintenance).

Career Pattern

PVT: Private (E-1). PFC: Private First Class (E-2). LCP: Lance Corporal (E-3). CPL: Corporal (E-4). SGT: Sergeant (E-5). SSGT: Staff Sergeant (E-6). GYSGT: Gunnery Sergeant (E-7). NOTE: MOS duties are not defined in terms of rank or paygrade, but by levels. Level I identifies tasks that are taught in the classroom; the individual is able to perform simple parts of tasks. Level II identifies those tasks that are part of on-the-job training; the trainee can do most tasks pertaining to a particular system and needs supervision on more difficult parts of tasks. Level III indicates that the individual can perform all essential tasks without supervision. Level IV indicates a high level of proficiency in job performance; the individual can perform advanced technical functions, instruction, inspection, and supervision. The MOS designator 6051 (Aircraft Hydraulic/Pneumatic Mechanic—Trainee), is used to identify individuals in Levels I and II. Upon completion of Level II, MOS 6055 is awarded. Credit is recommended for Levels III and IV.

Description

Summary: Inspects, maintains, and repairs aircraft hydraulic/pneumatic systems and system components. *Levels I and II:* Uses color-coded charts of aircraft piping and lines; reads schematic diagrams, shop sketches, and blueprints and interprets applicable technical data; under supervision, performs component repairs; uses fluids and gases in applicable systems; uses precision measuring equipment; assists in or performs required inspections; fabricates replacement lines of tubing and hoses. *Level III:* Able to perform the duties required for Levels I and II; performs component repair of hydraulic/pneumatic components as required; uses and maintains hydraulic/pneumatic test equipment; observes applicable material control procedures; conducts informal on-the-job training; plans, schedules, and directs work center assignments. *Level IV:* Able to perform the duties required for Level III; performs and supervises the maintenance/repair requirements for aircraft hydraulic/pneumatic systems components; supervises an aircraft hydraulic/pneumatic check crew or maintenance unit; performs duties as quality assurance inspector; supervises and administers an organizational maintenance level aircraft hydraulic/pneumatic work center.

Recommendation, Level III

In the lower-division baccalaureate/associate degree category, 6 semester hours in hydraulic/pneumatic systems repair, 3 in aviation maintenance technology, and 3 in industrial safety (8/89).

Recommendation, Level IV

In the lower-division baccalaureate/associate degree category, 9 semester hours in hydraulic/pneumatic systems repair, 6 in aviation maintenance technology, 3 in industrial safety, 3 in personnel supervision, and 3 in maintenance management. In the upper-division baccalaureate category, 2 semester hours for field experience in management, if rank was Staff Sergeant (SSGT) and 3 semester hours for field experience in management, if rank was Gunnery Sergeant (GYSGT) (8/89).

MCE-6056-001

AIRCRAFT HYDRAULIC/PNEUMATIC MECHANIC, KC-130
6056

Exhibit Dates: 1/90–1/98.

Occupational Field: 60 (Aircraft Maintenance).

Career Pattern

PVT: Private (E-1). PFC: Private First Class (E-2). LCP: Lance Corporal (E-3). CPL: Corporal (E-4). SGT: Sergeant (E-5). SSGT: Staff Sergeant (E-6). GYSGT: Gunnery Sergeant (E-7). NOTE: MOS duties are not defined in terms of rank or paygrade, but by levels. Level I identifies tasks that are taught in the classroom; the individual is able to perform simple parts of tasks. Level II identifies those tasks that are part of on-the-job training; the trainee can do most tasks pertaining to a particular system and needs supervision on more difficult parts of tasks. Level III indicates that the individual can perform all essential tasks without supervision. Level IV indicates a high level of proficiency in job performance; the individual can perform advanced technical functions, instruction, inspection, and supervision. The MOS designator 6051 (Aircraft Hydraulic/Pneumatic Mechanic—Trainee), is used to identify individuals in Levels I and II. Upon completion of Level II, MOS 6056 is awarded. Credit is recommended for Levels III and IV.

Description

Summary: Inspects, maintains, and repairs aircraft hydraulic/pneumatic systems and system components. *Levels I and II:* Uses color-coded charts of aircraft piping and lines; reads schematic diagrams, shop sketches, and blueprints and interprets applicable technical data; under supervision, performs component repairs; uses fluids and gases in applicable systems; uses precision measuring equipment; assists in or performs required inspections; fabricates replacement lines of tubing and hoses. *Level III:* Able to perform the duties required for Levels I and II; performs component repair of hydraulic/pneumatic components as required; uses and maintains hydraulic/pneumatic test equipment; observes applicable material control procedures; conducts informal on-the-job training; plans, schedules, and directs work center assignments. *Level IV:* Able to perform the duties required for Level III; performs and supervises the maintenance/repair requirements for aircraft hydraulic/pneumatic systems components; supervises an aircraft hydraulic/pneumatic check crew or maintenance unit; performs duties as quality assurance inspector; supervises and administers an organizational maintenance level aircraft hydraulic/pneumatic work center.

Recommendation, Level III

In the lower-division baccalaureate/associate degree category, 6 semester hours in hydraulic/pneumatic systems repair, 3 in aviation maintenance technology, and 3 in industrial safety (8/89).

Recommendation, Level IV

In the lower-division baccalaureate/associate degree category, 9 semester hours in hydraulic/pneumatic systems repair, 6 in aviation maintenance technology, 3 in industrial safety, 3 in personnel supervision, and 3 in maintenance management. In the upper-division baccalaureate category, 2 semester hours for field experience in management, if rank was Staff Sergeant (SSGT) and 3 semester hours for field experience in management, if rank was Gunnery Sergeant (GYSGT) (8/89).

MCE-6057-001

AIRCRAFT HYDRAULIC/PNEUMATIC MECHANIC, F/A-18
6057

Exhibit Dates: 1/90–1/98.

Occupational Field: 60 (Aircraft Maintenance).

Career Pattern

PVT: Private (E-1). PFC: Private First Class (E-2). LCP: Lance Corporal (E-3). CPL: Corporal (E-4). SGT: Sergeant (E-5). SSGT: Staff Sergeant (E-6). GYSGT: Gunnery Sergeant (E-7). NOTE: MOS duties are not defined in terms of rank or paygrade, but by levels. Level I identifies tasks that are taught in the classroom; the individual is able to perform simple parts of tasks. Level II identifies those tasks that are part of on-the-job training; the trainee can do most tasks pertaining to a particular system and needs supervision on more difficult parts of tasks. Level III indicates that the individual can perform all essential tasks without supervision. Level IV indicates a high level of proficiency in job performance; the individual can perform advanced technical functions, instruction, inspection, and supervision. The MOS designator 6051 (Aircraft Hydraulic/Pneumatic Mechanic—Trainee), is used to identify individuals in Levels I and II. Upon completion of Level II, MOS 6057 is awarded. Credit is recommended for Levels III and IV.

Description

Summary: Inspects, maintains, and repairs aircraft hydraulic/pneumatic systems and system components. *Levels I and II:* Uses color-coded charts of aircraft piping and lines; reads schematic diagrams, shop sketches, and blueprints and interprets applicable technical data; under supervision, performs component repairs; uses fluids and gases in applicable systems; uses precision measuring equipment; assists in or performs required inspections; fabricates replacement lines of tubing and hoses. *Level III:* Able to perform the duties required for Levels I and II; performs component repair of hydraulic/pneumatic components as required; uses and maintains hydraulic/pneumatic test equipment; observes applicable material control procedures; conducts informal on-the-job training; plans, schedules, and directs work center assignments. *Level IV:* Able to perform the duties required for Level III; performs and supervises the maintenance/repair requirements for aircraft hydraulic/pneumatic systems components; supervises an aircraft hydraulic/pneumatic check crew or maintenance unit; performs duties as quality assurance inspector; supervises and administers an organizational maintenance level aircraft hydraulic/pneumatic work center.

Recommendation, Level III

In the lower-division baccalaureate/associate degree category, 6 semester hours in hydraulic/pneumatic systems repair, 3 in aviation maintenance technology, and 3 in industrial safety (8/89).

Recommendation, Level IV

In the lower-division baccalaureate/associate degree category, 9 semester hours in hydraulic/pneumatic systems repair, 6 in aviation maintenance technology, 3 in industrial safety, 3 in personnel supervision, and 3 in maintenance management. In the upper-division baccalaureate category, 2 semester hours for field experience in management, if rank was Staff Sergeant (SSGT) and 3 semester hours for field experience in management, if rank was Gunnery Sergeant (GYSGT) (8/89).

MCE-6058-001

AIRCRAFT HYDRAULIC/PNEUMATIC MECHANIC, OV-10
6058

Exhibit Dates: 1/90–1/98.

Occupational Field: 60 (Aircraft Maintenance).

Career Pattern

PVT: Private (E-1). PFC: Private First Class (E-2). LCP: Lance Corporal (E-3). CPL: Corporal (E-4). SGT: Sergeant (E-5). SSGT: Staff Sergeant (E-6). GYSGT: Gunnery Sergeant (E-7). NOTE: MOS duties are not defined in terms of rank or paygrade, but by levels. Level I identifies tasks that are taught in the classroom; the individual is able to perform simple parts of tasks. Level II identifies those tasks that are part of on-the-job training; the trainee can do most tasks pertaining to a particular system and needs supervision on more difficult parts of tasks. Level III indicates that the individual can perform all essential tasks without supervision. Level IV indicates a high level of proficiency in job performance; the individual can perform advanced technical functions, instruction, inspection, and supervision. The MOS designator 6051 (Aircraft Hydraulic/Pneumatic Mechanic—Trainee), is used to identify individuals in Levels I and II. Upon completion of Level II, MOS 6058 is awarded. Credit is recommended for Levels III and IV.

Description

Summary: Inspects, maintains, and repairs aircraft hydraulic/pneumatic systems and system components. *Levels I and II:* Uses color-coded charts of aircraft piping and lines; reads schematic diagrams, shop sketches, and blueprints and interprets applicable technical data; under supervision, performs component repairs; uses fluids and gases in applicable systems; uses precision measuring equipment; assists in or performs required inspections; fabricates replacement lines of tubing and hoses. *Level III:* Able to perform the duties required for Levels I and II; performs component repair of hydraulic/pneumatic components as required; uses and maintains hydraulic/pneumatic test equipment; observes applicable material control procedures; conducts informal on-the-job training; plans, schedules, and directs work center assignments. *Level IV:* Able to perform the duties required for Level III; performs and supervises the maintenance/repair requirements for aircraft hydraulic/pneumatic systems components; supervises an aircraft hydraulic/pneumatic check crew or maintenance unit; performs duties as quality assurance inspector; supervises and administers an organizational maintenance level aircraft hydraulic/pneumatic work center.

Recommendation, Level III

In the lower-division baccalaureate/associate degree category, 6 semester hours in hydraulic/pneumatic systems repair, 3 in aviation maintenance technology, and 3 in industrial safety (8/89).

Recommendation, Level IV

In the lower-division baccalaureate/associate degree category, 9 semester hours in hydraulic/pneumatic systems repair, 6 in aviation maintenance technology, 3 in industrial safety, 3 in personnel supervision, and 3 in maintenance management. In the upper-division baccalaureate category, 2 semester hours for field experience in management, if rank was Staff Sergeant (SSGT) and 3 semester hours for field experience in management, if rank was Gunnery Sergeant (GYSGT) (8/89).

MCE-6060-001

FLIGHT EQUIPMENT MARINE
6060

Exhibit Dates: 5/92–Present.

Occupational Field: 60 (Aircraft Maintenance).

Career Pattern

PVT: Private (E-1). PFC: Private First Class (E-2). LCP: Lance Corporal (E-3). CPL: Corporal (E-3). SGT: Sergeant (E-5). SSGT: Staff Sergeant (E-6). GYSGT: Gunnery Sergeant (E-7). NOTE: MOS duties are not defined in terms of rank or paygrade, but by levels. Level I identifies tasks that are taught in the classroom; the individual is able to perform simple parts of tasks. Level II identifies those tasks that are part of on-the-job training; the trainee can do most tasks pertaining to a particular system and needs supervision on more difficult parts of tasks. Level III indicates that the individual can perform all essential tasks without supervision. Level IV indicates a high level of proficiency in job performance; the individual can perform advanced technical functions, instruction, inspection, and supervision. The MOS designator 6081 (Aircraft Safety Equipment Mechanic—Trainee) is used to identify individuals in Levels I and II. Upon completion of Level II, MOS 6060 is awarded. Credit is recommended for Levels III and IV.

Description

Summary: Inspects, tests, and repairs parachutes, flight survival equipment, flight equipment, carbon dioxide cylinders and systems, and gaseous and liquid oxygen equipment. *Levels I and II:* Inspects and replaces aircraft flight equipment and flight survival equipment; packs and rigs parachutes; uses technical and maintenance publications; maintains and repairs aircraft-related safety equipment, fire extinguishing systems, and oxygen systems; follows approved procedures related to special explosives used in fire extinguishing systems. *Level III:* Able to perform the duties required for Levels I and II; tests, adjusts, and repairs oxygen masks and fire extinguishing systems; conducts informal on-the-job training; tests and adjusts automatic parachute actuators and measures temperature and humidity in parachute dry lockers and storage rooms. *Level IV:* Able to perform the duties required for Level III; plans, schedules, and directs work assignments; performs duties as quality assurance inspector; prepares and submits required reports and records.

Recommendation, Level III

In the lower-division baccalaureate/associate degree category, 3 semester hours in occupational safety and 1 in technical writing (5/92).

Recommendation, Level IV

In the lower-division baccalaureate/associate degree category, 3 semester hours in occupational safety, 1 in technical writing, 3 in personnel supervision, and 3 in maintenance management. In the upper-division baccalaureate category, 2 semester hours for field experience in management, if rank was Staff Sergeant (SSGT) and 3 semester hours for field experience in management, if rank was Gunnery Sergeant (GYSGT) (5/92).

MCE-6072-001

AIRCRAFT MAINTENANCE GROUND SUPPORT
EQUIPMENT MECHANIC HYDRAULIC/
PNEUMATIC/STRUCTURE MECHANIC
6072

Exhibit Dates: 5/92–Present.

Occupational Field: 60 (Aircraft Maintenance).

Career Pattern

PVT: Private (E-1). PFC: Private First Class (E-2). LCP: Lance Corporal (E-3). CPL: Corporal (E-3). SGT: Sergeant (E-5). SSGT: Staff Sergeant (E-6). GYSGT: Gunnery Sergeant (E-7). NOTE: MOS duties are not defined in terms of rank or paygrade, but by levels. Level I identifies tasks that are taught in the classroom; the individual is able to perform simple parts of tasks. Level II identifies those tasks that are part of on-the-job training; the trainee can do most tasks pertaining to a particular system and needs supervision on more difficult parts of tasks. Level III indicates that the individual can perform all essential tasks without supervision. Level IV indicates a high level of proficiency in job performance; the individual can perform advanced technical functions, instruction, inspection, and supervision. The MOS designator 6071 (Aircraft Maintenance Ground Support Equipment Mechanic—Trainee) is used to identify individuals in Levels I and II. Upon completion of Level II, MOS 6072 is awarded. Credit is recommended for Levels III and IV.

Description

Summary: Inspects, maintains, repairs, and tests aircraft maintenance ground support equipment, hydraulic/pneumatic/structure systems, and system components; performs duties related to the operation of support equipment and the licensing of aircraft maintenance personnel in the operation of ground support/special support equipment. *Levels I and II:* Practices work center safety procedures and regulations and applicable emergency first aid procedures; uses technical publications; assists or performs periodic inspections as required by maintenance requirement cards and completes maintenance and repairs as required; incorporates bulletins and/or changes to supported equipment; operates and maintains applicable support equipment and is licensed as required; detects corrosion and performs corrosion control procedures; completes applicable maintenance data collection forms; inspects, tests and repairs hydraulic and pneumatic systems and components; maintains and repairs all applicable ground support equipment and components; inspects, maintains, and repairs gasoline diesel and gas turbine power plants. *Level III:* Able to perform the duties required for Levels I and II; follows material control procedures; uses precision measuring equipment; conducts informal on-the-job technical training; prepares and maintains reports and records pertaining to the fabrication, installation, operation, maintenance, and repair of ground support equipment; performs the duties of a quality assurance inspector; prepares, monitors, and submits required work center reports and records; ensures that all equipment requiring calibration is within its calibration cycle; requisitions all necessary repair parts and supplies; is qualified and capable of instructing personnel in operator care, use, and maintenance of aircraft ground support equipment. *Level IV:* Able to perform the duties required for Level III; plans, schedules, and directs work center assignments; supervises, directs, and administers personnel and facilities for the maintenance of aircraft ground support hydraulic/pneumatic systems ashore and afloat.

Recommendation, Level III

In the lower-division baccalaureate/associate degree category, 3 semester hours in fluid power principles, 3 in power plant operation and maintenance, 2 in introduction to DC electricity, 3 in occupational safety, 2 in records and information management, and 1 in technical writing (5/92).

Recommendation, Level IV

In the lower-division baccalaureate/associate degree category, 3 semester hours in fluid power principles, 3 in power plant operation and maintenance, 2 in introduction to DC electricity, 3 in occupational safety, 2 in records and information management, 1 in technical writing, 3 in personnel supervision, and 3 in maintenance management. In the upper-division baccalaureate category, 2 semester hours for field experience in management, if rank was Staff Sergeant (SSGT) and 3 semester hours for field experience in management, if rank was Gunnery Sergeant (GYSGT) (5/92).

MCE-6073-001

AIRCRAFT MAINTENANCE GROUND SUPPORT EQUIPMENT ELECTRICIAN/REFRIGERATION MECHANIC
6073

Exhibit Dates: 5/92–Present.

Occupational Field: 60 (Aircraft Maintenance).

Career Pattern

PVT: Private (E-1). PFC: Private First Class (E-2). LCP: Lance Corporal (E-3). CPL: Corporal (E-3). SGT: Sergeant (E-5). SSGT: Staff Sergeant (E-6). GYSGT: Gunnery Sergeant (E-7). NOTE: MOS duties are not defined in terms of rank or paygrade, but by levels. Level I identifies tasks that are taught in the classroom; the individual is able to perform simple parts of tasks. Level II identifies those tasks that are part of on-the-job training; the trainee can do most tasks pertaining to a particular system and needs supervision on more difficult parts of tasks. Level III indicates that the individual can perform all essential tasks without supervision. Level IV indicates a high level of proficiency in job performance; the individual can perform advanced technical functions, instruction, inspection, and supervision. The MOS designator 6071 (Aircraft Safety Equipment Mechanic—Trainee) is used to identify individuals in Levels I and II. Upon completion of Level II, MOS 6073 is awarded. Credit is recommended for Levels III and IV.

Description

Summary: Installs, inspects, tests, maintains, and repairs aircraft support equipment, electrical and instrument systems, and refrigeration and air conditioning equipment, systems, and accessories; performs duties related to the operation of support equipment and to the licensing of aircraft maintenance personnel. *Levels I and II:* Follows work center safety procedures and regulations and applicable emergency first aid procedures; uses technical publications; knows the theory of operation of transformers, motor generators, and associated circuit components as related to electrical and instrument systems; installs, operates, maintains, and repairs aviation ground support equipment, electrical, air conditioning, and refrigeration systems; is licensed as appropriate for aviation ground support equipment operation; knows the operating principles and standard fittings, devices, equipment, materials, and accessories for refrigeration and air conditioning systems; completes applicable maintenance data forms; inspects, maintains, and repairs gasoline, diesel, and gas turbine power plants. *Level III:* Able to perform the duties required for Levels I and II; follows material control procedures; conducts formal and informal on-the-job technical training; prepares and maintains logs, records, and reports pertaining to ground service, electrical, and air conditioning equipment; supervises work center personnel; performs the duties of quality assurance inspector; instructs personnel in the operation, care, use, and maintenance of appropriate ground service equipment. *Level IV:* Able to perform the duties required for Level III; plans, schedules, and directs work center assignments; supervises, directs, and administers maintenance and repair personnel and facilities.

Recommendation, Level III

In the lower-division baccalaureate/associate degree category, 2 semester hours in introduction to DC electricity, 3 in AC and DC circuits, 3 in principles of refrigeration and air conditioning, 3 in power plant operation and maintenance, 3 in occupational safety, 3 in records and information management, and 1 in technical writing (5/92).

Recommendation, Level IV

In the lower-division baccalaureate/associate degree category, 2 semester hours in introduction to DC electricity, 3 in AC and DC circuits, 3 in principles of refrigeration and air conditioning, 3 in power plant operation and maintenance, 3 in occupational safety, 3 in records and information management, 1 in technical writing, 3 in personnel supervision, and 3 in maintenance management. In the upper-division baccalaureate category, 2 semester hours for field experience in management, if rank was Staff Sergeant (SSGT) and 3 semester hours for field experience in management, if rank was Gunnery Sergeant (GYSGT) (5/92).

MCE-6075-001

CRYOGENICS EQUIPMENT OPERATOR
6075

Exhibit Dates: 5/92–Present.

Occupational Field: 60 (Aircraft Maintenance).

Career Pattern

PVT: Private (E-1). PFC: Private First Class (E-2). LCP: Lance Corporal (E-3). CPL: Corporal (E-4). SGT: Sergeant (E-5). SSGT: Staff Sergeant (E-6). GYSGT: Gunnery Sergeant (E-7). NOTE: MOS duties are not defined in terms of rank or paygrade, but by levels. Level I identifies tasks that are taught in the classroom; the individual is able to perform simple parts of tasks. Level II identifies those tasks that are part of on-the-job training; the trainee can do most tasks pertaining to a particular system and needs supervision on more difficult parts of tasks. Level III indicates that the individual can perform all essential tasks without supervision. Level IV indicates a high level of proficiency in job performance; the individual can perform advanced technical functions, instruction, inspection, and supervision. The MOS designator 6071 (Aircraft Maintenance Ground Support Equipment Mechanic—Trainee) is used to identify individuals in Levels I and II. Upon completion of Level II, MOS 6075 is awarded. Credit is recommended for Levels III and IV.

Description

Summary: Assembles, operates, and maintains liquid oxygen/nitrogen generating plants, storage, and aircraft servicing equipment, vaporizing equipment, vacuum pumps, and liquid oxygen (LOX) tank purging units; operates and maintains purity analysis test equipment to ensure product acceptability. *Levels I and II:* Operates and maintains appropriate shop equipment; monitors product purity; operates aircraft servicing equipment; practices safety procedures; uses technical publications, schematic diagrams, sketches, and blueprints; recharges gaseous cylinders; prepares chemical solutions; activities and uses LOX generating plant. *Level III:* Able to perform the duties required for Levels I and II; conducts informal on-the-job training; operates and maintains purity analysis equipment. *Level IV:* Able to perform the duties required for Level III; supervises shop work loads; supervises and evaluates shop personnel.

Recommendation, Level III

In the lower-division baccalaureate/associate degree category, 2 semester hours in refrigeration principles and 2 in occupational safety (5/92).

Recommendation, Level IV

In the lower-division baccalaureate/associate degree category, 2 semester hours in refrigeration principles, 2 in occupational safety, 3 in personnel supervision, and 3 in maintenance management. In the upper-division baccalaureate category, 2 semester hours for field experience in management, if rank was Staff Sergeant (SSGT) and 3 semester hours for field experience in management, if rank was Gunnery Sergeant (GYSGT) (5/92).

MCE-6082-001

AIRCRAFT SAFETY EQUIPMENT MECHANIC, A-4/TA-4/OA-4
6082

Exhibit Dates: 1/90–1/98.

Occupational Field: 60 (Aircraft Maintenance).

Career Pattern

PVT: Private (E-1). PFC: Private First Class (E-2). LCP: Lance Corporal (E-3). CPL: Corporal (E-4). SGT: Sergeant (E-5). SSGT: Staff Sergeant (E-6). GYSGT: Gunnery Sergeant (E-7). NOTE: MOS duties are not defined in terms of rank or paygrade, but by levels. Level I identifies tasks that are taught in the classroom; the individual is able to perform simple parts of tasks. Level II identifies those tasks that are part of on-the-job training; the trainee can do most tasks pertaining to a particular system and needs supervision on more difficult parts of tasks. Level III indicates that the individual can perform all essential tasks without supervision. Level IV indicates a high level of proficiency in job performance; the individual can perform advanced technical functions, instruction, inspection, and supervision. The MOS designator 6081 (Aircraft Safety Equipment Mechanic—Trainee), is used to identify individuals in Levels I and II. Upon completion of Level II, MOS 6082 is awarded. Credit is recommended for Levels III and IV.

Description

Summary: Inspects and maintains aircraft safety equipment. *Levels I and II:* Performs, under close supervision and training, routine duties incident to inspection or replacement of aircraft safety systems and components; uses technical and maintenance publications; disassembles, inspects, and assembles aircraft-related safety equipment, air conditioning and pressurization systems, fire extinguishing systems, oxygen systems; follows approved procedures related to special explosive cartridges used in escape systems. *Level III:* Able to perform the duties required for Levels I and II; tests aircraft for pressure tightness; removes, checks, rigs, and adjusts ejection seats, components, and/or seat and canopy components; analyzes aircraft oxygen system malfunctions; inspects and adjusts firing mechanisms for emerging escape systems; conducts informal on-the-job training within assigned skill designator; tests for carbon monoxide contamination; inspects and adjusts pressure control units; replaces defective indicators and controls. *Level*

IV: Able to perform the duties required for Level III; plans, schedules, and directs work assignments; performs duties as quality assurance inspector; prepares and submits required reports and records.

Recommendation, Level III

In the lower-division baccalaureate/associate degree category, 3 semester hours in industrial safety (8/89).

Recommendation, Level IV

In the lower-division baccalaureate/associate degree category, 3 semester hours in industrial safety, 3 in personnel supervision, and 3 in maintenance management. In the upper-division baccalaureate category, 2 semester hours for field experience in management, if rank was Staff Sergeant (SSGT) and 3 semester hours for field experience in management, if rank was Gunnery Sergeant (GYSGT) (8/89).

MCE-6083-001

AIRCRAFT SAFETY EQUIPMENT MECHANIC, A-6/ EA-6
6083

Exhibit Dates: 1/90–1/98.

Occupational Field: 60 (Aircraft Maintenance).

Career Pattern

PVT: Private (E-1). PFC: Private First Class (E-2). LCP: Lance Corporal (E-3). CPL: Corporal (E-4). SGT: Sergeant (E-5). SSGT: Staff Sergeant (E-6). GYSGT: Gunnery Sergeant (E-7). NOTE: MOS duties are not defined in terms of rank or paygrade, but by levels. Level I identifies tasks that are taught in the classroom; the individual is able to perform simple parts of tasks. Level II identifies those tasks that are part of on-the-job training; the trainee can do most tasks pertaining to a particular system and needs supervision on more difficult parts of tasks. Level III indicates that the individual can perform all essential tasks without supervision. Level IV indicates a high level of proficiency in job performance; the individual can perform advanced technical functions, instruction, inspection, and supervision. The MOS designator 6081 (Aircraft Safety Equipment Mechanic—Trainee), is used to identify individuals in Levels I and II. Upon completion of Level II, MOS 6083 is awarded. Credit is recommended for Levels III and IV.

Description

Summary: Inspects and maintains aircraft safety equipment.) *Levels I and II:* Performs, under close supervision and training, routine duties incident to inspection or replacement of aircraft safety systems and components; uses technical and maintenance publications; disassembles, inspects, and assembles aircraft-related safety equipment, air conditioning and pressurization systems, fire extinguishing systems, and oxygen systems; follows approved procedures related to special explosive cartridges used in escape systems. *Level III:* Able to perform the duties required for Levels I and II; tests aircraft for pressure tightness; removes, checks, rigs, and adjusts ejection seats, components, and/or seat and canopy components; analyzes aircraft oxygen system malfunctions; inspects and adjusts firing mechanisms for emerging escape systems; conducts informal on-the-job training within assigned skill designator; tests for carbon monoxide contamination; inspects and adjusts pressure control units; replaces defective indicators and controls. *Level IV:* Able to perform the duties required for Level III; plans, schedules, and directs work assignments; performs duties as quality assurance inspector; prepares and submits required reports and records.

Recommendation, Level III

In the lower-division baccalaureate/associate degree category, 3 semester hours in industrial safety (8/89).

Recommendation, Level IV

In the lower-division baccalaureate/associate degree category, 3 semester hours in industrial safety, 3 in personnel supervision, and 3 in maintenance management. In the upper-division baccalaureate category, 2 semester hours for field experience in management, if rank was Staff Sergeant (SSGT) and 3 semester hours for field experience in management, if rank was Gunnery Sergeant (GYSGT) (8/89).

MCE-6084-001

AIRCRAFT SAFETY EQUIPMENT MECHANIC, F-4/RF-4
6084

Exhibit Dates: 1/90–1/98.

Occupational Field: 60 (Aircraft Maintenance).

Career Pattern

PVT: Private (E-1). PFC: Private First Class (E-2). LCP: Lance Corporal (E-3). CPL: Corporal (E-4). SGT: Sergeant (E-5). SSGT: Staff Sergeant (E-6). GYSGT: Gunnery Sergeant (E-7). NOTE: MOS duties are not defined in terms of rank or paygrade, but by levels. Level I identifies tasks that are taught in the classroom; the individual is able to perform simple parts of tasks. Level II identifies those tasks that are part of on-the-job training; the trainee can do most tasks pertaining to a particular system and needs supervision on more difficult parts of tasks. Level III indicates that the individual can perform all essential tasks without supervision. Level IV indicates a high level of proficiency in job performance; the individual can perform advanced technical functions, instruction, inspection, and supervision. The MOS designator 6081 (Aircraft Safety Equipment Mechanic—Trainee), is used to identify individuals in Levels I and II. Upon completion of Level II, MOS 6084 is awarded. Credit is recommended for Levels III and IV.

Description

Summary: Inspects and maintains aircraft safety equipment.) *Levels I and II:* Performs, under close supervision and training, routine duties incident to inspection or replacement of aircraft safety systems and components; uses technical and maintenance publications; disassembles, inspects, and assembles aircraft-related safety equipment, air conditioning and pressurization systems, fire extinguishing systems, and oxygen systems; follows approved procedures related to special explosive cartridges used in escape systems. *Level III:* Able to perform the duties required for Levels I and II; tests aircraft for pressure tightness; removes, checks, rigs, and adjusts ejection seats, components, and/or seat and canopy components; analyzes aircraft oxygen system malfunctions; inspects and adjusts firing mechanisms for emerging escape systems; conducts informal on-the-job training within assigned skill designator; tests for carbon monoxide contamination; inspects and adjusts pressure control units; replaces defective indicators and controls. *Level IV:* Able to perform the duties required for Level III; plans, schedules, and directs work assignments; performs duties as quality assurance inspector; prepares and submits required reports and records.

Recommendation, Level III

In the lower-division baccalaureate/associate degree category, 3 semester hours in industrial safety (8/89).

Recommendation, Level IV

In the lower-division baccalaureate/associate degree category, 3 semester hours in industrial safety, 3 in personnel supervision, and 3 in maintenance management. In the upper-division baccalaureate category, 2 semester hours for field experience in management, if rank was Staff Sergeant (SSGT) and 3 semester hours for field experience in management, if rank was Gunnery Sergeant (GYSGT) (8/89).

MCE-6085-001

AIRCRAFT SAFETY EQUIPMENT MECHANIC, AV-8/ TAV-8
6085

Exhibit Dates: 1/90–1/98.

Occupational Field: 60 (Aircraft Maintenance).

Career Pattern

PVT: Private (E-1). PFC: Private First Class (E-2). LCP: Lance Corporal (E-3). CPL: Corporal (E-4). SGT: Sergeant (E-5). SSGT: Staff Sergeant (E-6). GYSGT: Gunnery Sergeant (E-7). NOTE: MOS duties are not defined in terms of rank or paygrade, but by levels. Level I identifies tasks that are taught in the classroom; the individual is able to perform simple parts of tasks. Level II identifies those tasks that are part of on-the-job training; the trainee can do most tasks pertaining to a particular system and needs supervision on more difficult parts of tasks. Level III indicates that the individual can perform all essential tasks without supervision. Level IV indicates a high level of proficiency in job performance; the individual can perform advanced technical functions, instruction, inspection, and supervision. The MOS designator 6081 (Aircraft Safety Equipment Mechanic—Trainee), is used to identify individuals in Levels I and II. Upon completion of Level II, MOS 6085 is awarded. Credit is recommended for Levels III and IV.

Description

Summary: Inspects and maintains aircraft safety equipment.) *Levels I and II:* Performs, under close supervision and training, routine duties incident to inspection or replacement of aircraft safety systems and components; uses technical and maintenance publications; disassembles, inspects, and assembles aircraft-related safety equipment, air conditioning and pressurization systems, fire extinguishing systems, and oxygen systems; follows approved procedures related to special explosive cartridges used in escape systems. *Level III:* Able to perform the duties required for Levels I and II; tests aircraft for pressure tightness; removes, checks, rigs, and adjusts ejection seats, components, and/or seat and canopy components; analyzes aircraft oxygen system malfunctions; inspects and adjusts firing mechanisms for emerging escape systems; conducts informal on-the-job training within assigned skill designator; tests for carbon monoxide contamination; inspects and adjusts pressure control units; replaces defective indicators and controls. *Level IV:* Able to perform the duties required for Level III; plans, schedules, and directs work assignments; performs duties as quality assurance inspector; prepares and submits required reports and records.

Recommendation, Level III

In the lower-division baccalaureate/associate degree category, 3 semester hours in industrial safety (8/89).

Recommendation, Level IV

In the lower-division baccalaureate/associate degree category, 3 semester hours in industrial safety, 3 in personnel supervision, and 3 in maintenance management. In the upper-division baccalaureate category, 2 semester hours for field experience in management, if rank was Staff Sergeant (SSGT) and 3 semester hours for field experience in management, if rank was Gunnery Sergeant (GYSGT) (8/89).

MCE-6086-001

AIRCRAFT SAFETY EQUIPMENT MECHANIC, KC-130
6086

Exhibit Dates: 1/90–1/98.

Occupational Field: 60 (Aircraft Maintenance).

Career Pattern

PVT: Private (E-1). PFC: Private First Class (E-2). LCP: Lance Corporal (E-3). CPL: Corporal (E-4). SGT: Sergeant (E-5). SSGT: Staff Sergeant (E-6). GYSGT: Gunnery Sergeant (E-7). NOTE: MOS duties are not defined in terms of rank or paygrade, but by levels. Level I identifies tasks that are taught in the classroom; the individual is able to perform simple parts of tasks. Level II identifies those tasks that are part of on-the-job training; the trainee can do most tasks pertaining to a particular system and needs supervision on more difficult parts of tasks. Level III indicates that the individual can perform all essential tasks without supervision. Level IV indicates a high level of proficiency in job performance; the individual can perform advanced technical functions, instruction, inspection, and supervision. The MOS designator 6081 (Aircraft Safety Equipment Mechanic—Trainee), is used to identify individuals in Levels I and II. Upon completion of Level II, MOS 6086 is awarded. Credit is recommended for Levels III and IV.

Description

Summary: Inspects and maintains aircraft safety equipment.) *Levels I and II:* Performs, under close supervision and training, routine duties incident to inspection or replacement of aircraft safety systems and components; uses technical and maintenance publications; disassembles, inspects, and assembles aircraft-related safety equipment, air conditioning and pressurization systems, fire extinguishing systems, and oxygen systems; follows approved procedures related to special explosive cartridges used in escape systems. *Level III:* Able to perform the duties required for Levels I and II; tests aircraft for pressure tightness; removes, checks, rigs, and adjusts ejection seats, components, and/or seat and canopy components; analyzes aircraft oxygen system malfunctions; inspects and adjusts firing mechanisms for emerging escape systems; conducts informal on-the-job training within assigned skill desig-

nator; tests for carbon monoxide contamination; inspects and adjusts pressure control units; replaces defective indicators and controls. *Level IV:* Able to perform the duties required for Level III; plans, schedules, and directs work assignments; performs duties as quality assurance inspector; prepares and submits required reports and records.

Recommendation, Level III

In the lower-division baccalaureate/associate degree category, 3 semester hours in industrial safety (8/89).

Recommendation, Level IV

In the lower-division baccalaureate/associate degree category, 3 semester hours in industrial safety, 3 in personnel supervision, and 3 in maintenance management. In the upper-division baccalaureate category, 2 semester hours for field experience in management, if rank was Staff Sergeant (SSGT); and 3 semester hours for field experience in management, if rank was Gunnery Sergeant (GYSGT) (8/89).

MCE-6087-001

AIRCRAFT SAFETY EQUIPMENT MECHANIC, F/A-18
6087

Exhibit Dates: 1/90–1/98.

Occupational Field: 60 (Aircraft Maintenance).

Career Pattern

PVT: Private (E-1). PFC: Private First Class (E-2). LCP: Lance Corporal (E-3). CPL: Corporal (E-4). SGT: Sergeant (E-5). SSGT: Staff Sergeant (E-6). GYSGT: Gunnery Sergeant (E-7). NOTE: MOS duties are not defined in terms of rank or paygrade, but by levels. Level I identifies tasks that are taught in the classroom; the individual is able to perform simple parts of tasks. Level II identifies those tasks that are part of on-the-job training; the trainee can do most tasks pertaining to a particular system and needs supervision on more difficult parts of tasks. Level III indicates that the individual can perform all essential tasks without supervision. Level IV indicates a high level of proficiency in job performance; the individual can perform advanced technical functions, instruction, inspection, and supervision. The MOS designator 6081 (Aircraft Safety Equipment Mechanic—Trainee), is used to identify individuals in Levels I and II. Upon completion of Level II, MOS 6087 is awarded. Credit is recommended for Levels III and IV.

Description

Summary: Inspects and maintains aircraft safety equipment.) *Levels I and II:* Performs, under close supervision and training, routine duties incident to inspection or replacement of aircraft safety systems and components; uses technical and maintenance publications; disassembles, inspects, and assembles aircraft-related safety equipment, air conditioning and pressurization systems, fire extinguishing systems, and oxygen systems; follows approved procedures related to special explosive cartridges used in escape systems. *Level III:* Able to perform the duties required for Levels I and II; tests aircraft for pressure tightness; removes, checks, rigs, and adjusts ejection seats, components, and/or seat and canopy components; analyzes aircraft oxygen system malfunctions; inspects and adjusts firing mechanisms for emerging escape systems; conducts informal on-the-job training within assigned skill desig-

nator; tests for carbon monoxide contamination; inspects and adjusts pressure control units; replaces defective indicators and controls. *Level IV:* Able to perform the duties required for Level III; plans, schedules, and directs work assignments; performs duties as quality assurance inspector; prepares and submits required reports and records.

Recommendation, Level III

In the lower-division baccalaureate/associate degree category, 3 semester hours in industrial safety (8/89).

Recommendation, Level IV

In the lower-division baccalaureate/associate degree category, 3 semester hours in industrial safety, 3 in personnel supervision, and 3 in maintenance management. In the upper-division baccalaureate category, 2 semester hours for field experience in management, if rank was Staff Sergeant (SSGT) and 3 semester hours for field experience in management, if rank was Gunnery Sergeant (GYSGT) (8/89).

MCE-6088-001

AIRCRAFT SAFETY EQUIPMENT MECHANIC, OV-10
6088

Exhibit Dates: 1/90–1/98.

Occupational Field: 60 (Aircraft Maintenance).

Career Pattern

PVT: Private (E-1). PFC: Private First Class (E-2). LCP: Lance Corporal (E-3). CPL: Corporal (E-4). SGT: Sergeant (E-5). SSGT: Staff Sergeant (E-6). GYSGT: Gunnery Sergeant (E-7). NOTE: MOS duties are not defined in terms of rank or paygrade, but by levels. Level I identifies tasks that are taught in the classroom; the individual is able to perform simple parts of tasks. Level II identifies those tasks that are part of on-the-job training; the trainee can do most tasks pertaining to a particular system and needs supervision on more difficult parts of tasks. Level III indicates that the individual can perform all essential tasks without supervision. Level IV indicates a high level of proficiency in job performance; the individual can perform advanced technical functions, instruction, inspection, and supervision. The MOS designator 6081 (Aircraft Safety Equipment Mechanic—Trainee), is used to identify individuals in Levels I and II. Upon completion of Level II, MOS 6088 is awarded. Credit is recommended for Levels III and IV.

Description

Summary: Inspects and maintains aircraft safety equipment.) *Levels I and II:* Performs, under close supervision and training, routine duties incident to inspection or replacement of aircraft safety systems and components; uses technical and maintenance publications; disassembles, inspects, and assembles aircraft-related safety equipment, air conditioning and pressurization systems, fire extinguishing systems, and oxygen systems; follows approved procedures related to special explosive cartridges used in escape systems. *Level III:* Able to perform the duties required for Levels I and II; tests aircraft for pressure tightness; removes, checks, rigs, and adjusts ejection seats, components, and/or seat and canopy components; analyzes aircraft oxygen system malfunctions; inspects and adjusts firing mechanisms for emerging escape systems; conducts informal on-the-job training within assigned skill designator; tests for carbon monoxide contamination; inspects and adjusts pressure control units; replaces defective indicators and controls. *Level IV:* Able to perform the duties required for Level III; plans, schedules, and directs work assignments; performs duties as quality assurance inspector; prepares and submits required reports and records.

Recommendation, Level III

In the lower-division baccalaureate/associate degree category, 3 semester hours in industrial safety (8/89).

Recommendation, Level IV

In the lower-division baccalaureate/associate degree category, 3 semester hours in industrial safety, 3 in personnel supervision, and 3 in maintenance management. In the upper-division baccalaureate category, 2 semester hours for field experience in management, if rank was Staff Sergeant (SSGT) and 3 semester hours for field experience in management, if rank was Gunnery Sergeant (GYSGT) (8/89).

MCE-6092-001

AIRCRAFT STRUCTURES MECHANIC, A-4/TA-4/ OA-4
6092

Exhibit Dates: 1/90–1/98.

Occupational Field: 60 (Aircraft Maintenance).

Career Pattern

PVT: Private (E-1). PFC: Private First Class (E-2). LCP: Lance Corporal (E-3). CPL: Corporal (E-4). SGT: Sergeant (E-5). SSGT: Staff Sergeant (E-6). GYSGT: Gunnery Sergeant (E-7). NOTE: MOS duties are not defined in terms of rank or paygrade, but by levels. Level I identifies tasks that are taught in the classroom; the individual is able to perform simple parts of tasks. Level II identifies those tasks that are part of on-the-job training; the trainee can do most tasks pertaining to a particular system and needs supervision on more difficult parts of tasks. Level III indicates that the individual can perform all essential tasks without supervision. Level IV indicates a high level of proficiency in job performance; the individual can perform advanced technical functions, instruction, inspection, and supervision. The MOS designator 6091 (Aircraft Structures Mechanic—Trainee), is used to identify individuals in Levels I and II. Upon completion of Level II, MOS 6092 is awarded. Credit is recommended for Levels III and IV.

Description

Summary: Inspects, maintains, and repairs aircraft structures and structural components. *Levels I and II:* Uses metalworking tools in the maintenance and repair of aircraft structures and structural components; uses shears, brakes, grinders, and presses; refers to and complies with technical publications such as color-code charts, schematic diagrams, sketches, blueprints and instructions; inspects and maintains powered and nonpowered hand tools and related support equipment; performs periodic inspections of system components and assists in removal and replacement of aircraft component parts such as wings, control surfaces, landing gear, arresting gear, and tanks; inspects and repairs aircraft metal (aluminum alloy), composites, fiberglass, rubber, and plastic; makes simple dies, templates, and jigs to be used in metalworking and fabrication of aircraft parts; does riveting of all types; uses established procedures to control and prevent aircraft corrosion; observes acceptable industrial safety practices; documents work performed. *Level III:* Able to perform the duties required for Levels I and II; organizes work assignments; heat-treats metal alloys used in aircraft; classifies parts for use in the repair of aircraft; repairs, maintains, and rigs wings, airframe, and cockpit control mechanisms; performs major structural inspections and repairs; repairs aircraft de-icing systems, cockpit enclosures, radomes, and surfaces; conducts on-the-job and technical training. *Level IV:* Able to perform the duties required for Level III; designs cradle structures for moving, handling, and securing aircraft or equipment ashore/afloat; plans, schedules, and directs work center assignments; performs and supervises the complete repair of aircraft structures and structural components; performs aircraft weighing; supervises, directs, and administers an organizational maintenance level aircraft structural repair work center ashore/afloat.

Recommendation, Level III

In the lower-division baccalaureate/associate degree category, 9 semester hours in airframe structures repair, 3 in aviation maintenance technology, and 3 in industrial safety (8/89).

Recommendation, Level IV

In the lower-division baccalaureate/associate degree category, 12 semester hours in airframe structures repair, 6 in aviation maintenance technology, 3 in personnel supervision, 3 in maintenance management, and 3 in industrial safety. In the upper-division baccalaureate category, 2 semester hours for field experience in management, if rank was Staff Sergeant (SSGT) and 3 semester hours for field experience in management, if rank was Gunnery Sergeant (GYSGT) (8/89).

MCE-6093-001

AIRCRAFT STRUCTURES MECHANIC, A-6/EA-6
6093

Exhibit Dates: 1/90–1/98.

Occupational Field: 60 (Aircraft Maintenance).

Career Pattern

PVT: Private (E-1). PFC: Private First Class (E-2). LCP: Lance Corporal (E-3). CPL: Corporal (E-4). SGT: Sergeant (E-5). SSGT: Staff Sergeant (E-6). GYSGT: Gunnery Sergeant (E-7). NOTE: MOS duties are not defined in terms of rank or paygrade, but by levels. Level I identifies tasks that are taught in the classroom; the individual is able to perform simple parts of tasks. Level II identifies those tasks that are part of on-the-job training; the trainee can do most tasks pertaining to a particular system and needs supervision on more difficult parts of tasks. Level III indicates that the individual can perform all essential tasks without supervision. Level IV indicates a high level of proficiency in job performance; the individual can perform advanced technical functions, instruction, inspection, and supervision. The MOS designator 6091 (Aircraft Structures Mechanic—Trainee), is used to identify individuals in Levels I and II. Upon completion of Level II, MOS 6093 is awarded. Credit is recommended for Levels III and IV.

Description

Summary: Inspects, maintains, and repairs aircraft structures and structural components. *Levels I and II:* Uses metalworking tools in the maintenance and repair of aircraft structures and structural components; uses shears, brakes, grinders, and presses; refers to and complies with technical publications such as color-code charts, schematic diagrams, sketches, blueprints, and instructions; inspects and maintains powered and nonpowered hand tools and related support equipment; performs periodic inspections of system components and assists in removal and replacement of aircraft component parts such as wings, control surfaces, landing gear, arresting gear, and tanks; inspects and repairs aircraft metal (aluminum alloy), composites, fiberglass, rubber, and plastic; makes simple dies, templates, and jigs to be used in metalworking and fabrication of aircraft parts; does riveting of all types; uses established procedures to control and prevent aircraft corrosion; observes acceptable industrial safety practices; documents work performed. *Level III:* Able to perform the duties required for Levels I and II; organizes work assignments; heat-treats metal alloys used in aircraft; classifies parts for use in the repair of aircraft; repairs, maintains, and rigs wings, airframe, and cockpit control mechanisms; performs major structural inspections and repairs; repairs aircraft de-icing systems, cockpit enclosures, radomes, and surfaces; conducts on-the-job and technical training. *Level IV:* Able to perform the duties required for Level III; designs cradle structures for moving, handling, and securing aircraft or equipment ashore/afloat; plans, schedules, and directs work center assignments; performs and supervises the complete repair of aircraft structures and structural components; performs aircraft weighing; supervises, directs, and administers an organizational maintenance level aircraft structural repair work center ashore/afloat.

Recommendation, Level III

In the lower-division baccalaureate/associate degree category, 9 semester hours in airframe structures repair, 3 in aviation maintenance technology, and 3 in industrial safety (8/89).

Recommendation, Level IV

In the lower-division baccalaureate/associate degree category, 12 semester hours in airframe structures repair, 6 in aviation maintenance technology, 3 in personnel supervision, 3 in maintenance management, and 3 in industrial safety. In the upper-division baccalaureate category, 2 semester hours for field experience in management, if rank was Staff Sergeant (SSGT) and 3 semester hours for field experience in management, if rank was Gunnery Sergeant (GYSGT) (8/89).

MCE-6094-001

AIRCRAFT STRUCTURES MECHANIC, F-4/RF-4
6094

Exhibit Dates: 1/90–1/98.

Occupational Field: 60 (Aircraft Maintenance).

Career Pattern

PVT: Private (E-1). PFC: Private First Class (E-2). LCP: Lance Corporal (E-3). CPL: Corporal (E-4). SGT: Sergeant (E-5). SSGT: Staff Sergeant (E-6). GYSGT: Gunnery Sergeant (E-7). NOTE: MOS duties are not defined in terms of rank or paygrade, but by levels. Level I identifies tasks that are taught in the classroom; the individual is able to perform simple parts of tasks. Level II identifies those tasks that are part of on-the-job training; the trainee can do most tasks pertaining to a particular system and needs supervision on more difficult parts of tasks. Level III indicates that the individual can perform all essential tasks without supervision. Level IV indicates a high level of proficiency in job performance; the individual can perform advanced technical functions, instruction, inspection, and supervision. The MOS designator 6091 (Aircraft Structures Mechanic—Trainee), is used to identify individuals in Levels I and II. Upon completion of Level II, MOS 6094 is awarded. Credit is recommended for Levels III and IV.

Description

Summary: Inspects, maintains, and repairs aircraft structures and structural components. *Levels I and II:* Uses metalworking tools in the maintenance and repair of aircraft structures and structural components; uses shears, brakes, grinders, and presses; refers to and complies with technical publications such as color-code charts, schematic diagrams, sketches, blueprints, and instructions; inspects and maintains powered and nonpowered hand tools and related support equipment; performs periodic inspections of system components and assists in removal and replacement of aircraft component parts such as wings, control surfaces, landing gear, arresting gear, and tanks; inspects and repairs aircraft metal (aluminum alloy), composites, fiberglass, rubber, and plastic; makes simple dies, templates, and jigs to be used in metalworking and fabrication of aircraft parts; does riveting of all types; uses established procedures to control and prevent aircraft corrosion; observes acceptable industrial safety practices; documents work performed. *Level III:* Able to perform the duties required for Levels I and II; organizes work assignments; heat-treats metal alloys used in aircraft; classifies parts for use in the repair of aircraft; repairs, maintains, and rigs wings, airframe, and cockpit control mechanisms; performs major structural inspections and repairs; repairs aircraft de-icing systems, cockpit enclosures, radomes, and surfaces; conducts on-the-job and technical training. *Level IV:* Able to perform the duties required for Level III; designs cradle structures for moving, handling, and securing aircraft or equipment ashore/afloat; plans, schedules, and directs work center assignments; performs and supervises the complete repair of aircraft structures and structural components; performs aircraft weighing; supervises, directs, and administers an organizational maintenance level aircraft structural repair work center ashore/afloat.

Recommendation, Level III

In the lower-division baccalaureate/associate degree category, 9 semester hours in airframe structures repair, 3 in aviation maintenance technology, and 3 in industrial safety (8/89).

Recommendation, Level IV

In the lower-division baccalaureate/associate degree category, 12 semester hours in airframe structures repair, 6 in aviation maintenance technology, 3 in personnel supervision, 3 in maintenance management, and 3 in industrial safety. In the upper-division baccalaureate category, 2 semester hours for field experience in management, if rank was Staff Sergeant (SSGT) and 3 semester hours for field experience in management, if rank was Gunnery Sergeant (GYSGT) (8/89).

MCE-6095-001

AIRCRAFT STRUCTURES MECHANIC, AV-8/TAV-8
6095

Exhibit Dates: 1/90–1/98.

Occupational Field: 60 (Aircraft Maintenance).

Career Pattern

PVT: Private (E-1). PFC: Private First Class (E-2). LCP: Lance Corporal (E-3). CPL: Corporal (E-4). SGT: Sergeant (E-5). SSGT: Staff Sergeant (E-6). GYSGT: Gunnery Sergeant (E-7). NOTE: MOS duties are not defined in terms of rank or paygrade, but by levels. Level I identifies tasks that are taught in the classroom; the individual is able to perform simple parts of tasks. Level II identifies those tasks that are part of on-the-job training; the trainee can do most tasks pertaining to a particular system and needs supervision on more difficult parts of tasks. Level III indicates that the individual can perform all essential tasks without supervision. Level IV indicates a high level of proficiency in job performance; the individual can perform advanced technical functions, instruction, inspection, and supervision. The MOS designator 6091 (Aircraft Structures Mechanic—Trainee), is used to identify individuals in Levels I and II. Upon completion of Level II, MOS 6095 is awarded. Credit is recommended for Levels III and IV.

Description

Summary: Inspects, maintains, and repairs aircraft structures and structural components. *Levels I and II:* Uses metalworking tools in the maintenance and repair of aircraft structures and structural components; uses shears, brakes, grinders, and presses; refers to and complies with technical publications such as color-code charts, schematic diagrams, sketches, blueprints, and instructions; inspects and maintains powered and nonpowered hand tools and related support equipment; performs periodic inspections of system components and assists in removal and replacement of aircraft component parts such as wings, control surfaces, landing gear, arresting gear, and tanks; inspects and repairs aircraft metal (aluminum alloy), composites, fiberglass, rubber, and plastic; makes simple dies, templates, and jigs to be used in metalworking and fabrication of aircraft parts; does riveting of all types; uses established procedures to control and prevent aircraft corrosion; observes acceptable industrial safety practices; documents work performed. *Level III:* Able to perform the duties required for Levels I and II; organizes work assignments; heat-treats metal alloys used in aircraft; classifies parts for use in the repair of aircraft; repairs, maintains, and rigs wings, airframe, and cockpit control mechanisms; performs major structural inspections and repairs; repairs aircraft de-icing systems, cockpit enclosures, radomes, and surfaces; conducts on-the-job and technical training. *Level IV:* Able to perform the duties required for Level III; designs cradle structures for moving, handling, and securing aircraft or equipment ashore/afloat; plans, schedules, and directs work center assignments; performs and supervises the complete repair of aircraft structures and structural components; performs aircraft weighing; supervises, directs, and administers an organizational maintenance level aircraft structural repair work center ashore/afloat.

Recommendation, Level III

In the lower-division baccalaureate/associate degree category, 9 semester hours in airframe structures repair, 3 in aviation maintenance technology, and 3 in industrial safety (8/89).

Recommendation, Level IV

In the lower-division baccalaureate/associate degree category, 12 semester hours in airframe structures repair, 6 in aviation maintenance technology, 3 in personnel supervision, 3 in maintenance management, and 3 in industrial safety. In the upper-division baccalaureate category, 2 semester hours for field experience in management, if rank was Staff Sergeant (SSGT) and 3 semester hours for field experience in management, if rank was Gunnery Sergeant (GYSGT) (8/89).

MCE-6096-001

AIRCRAFT STRUCTURES MECHANIC, KC-130
6096

Exhibit Dates: 1/90–1/98.

Occupational Field: 60 (Aircraft Maintenance).

Career Pattern

PVT: Private (E-1). PFC: Private First Class (E-2). LCP: Lance Corporal (E-3). CPL: Corporal (E-4). SGT: Sergeant (E-5). SSGT: Staff Sergeant (E-6). GYSGT: Gunnery Sergeant (E-7). NOTE: MOS duties are not defined in terms of rank or paygrade, but by levels. Level I identifies tasks that are taught in the classroom; the individual is able to perform simple parts of tasks. Level II identifies those tasks that are part of on-the-job training; the trainee can do most tasks pertaining to a particular system and needs supervision on more difficult parts of tasks. Level III indicates that the individual can perform all essential tasks without supervision. Level IV indicates a high level of proficiency in job performance; the individual can perform advanced technical functions, instruction, inspection, and supervision. The MOS designator 6091 (Aircraft Structures Mechanic—Trainee), is used to identify individuals in Levels I and II. Upon completion of Level II, MOS 6096 is awarded. Credit is recommended for Levels III and IV.

Description

Summary: Inspects, maintains, and repairs aircraft structures and structural components. *Levels I and II:* Uses metalworking tools in the maintenance and repair of aircraft structures and structural components; uses shears, brakes, grinders, and presses; refers to and complies with technical publications such as color-code charts, schematic diagrams, sketches, blueprints, and instructions; inspects and maintains powered and nonpowered hand tools and related support equipment; performs periodic inspections of system components and assists in removal and replacement of aircraft component parts such as wings, control surfaces, landing gear, arresting gear, and tanks; inspects and repairs aircraft metal (aluminum alloy), composites, fiberglass, rubber, and plastic; makes simple dies, templates, and jigs to be used in metalworking and fabrication of aircraft parts; does riveting of all types; uses established procedures to control and prevent aircraft corrosion; observes acceptable industrial safety practices; documents work performed. *Level III:* Able to perform the duties required for Levels I and II; organizes work assignments; heat-treats metal alloys used in aircraft; classifies parts for use in the repair of aircraft; repairs, maintains, and rigs wings, airframe, and cockpit control mechanisms; performs major structural inspections and repairs; repairs aircraft de-icing systems, cockpit enclosures, radomes, and surfaces; conducts on-the-job and technical training. *Level IV:* Able to perform the duties required for Level III; designs cradle structures for moving, handling, and securing aircraft or equipment ashore/afloat; plans, schedules, and directs work center assignments; performs and supervises the complete repair of aircraft structures and structural components; performs aircraft weighing; supervises, directs, and administers an organizational maintenance level aircraft structural repair work center ashore/afloat.

Recommendation, Level III

In the lower-division baccalaureate/associate degree category, 9 semester hours in airframe structures repair, 3 in aviation maintenance technology, and 3 in industrial safety (8/89).

Recommendation, Level IV

In the lower-division baccalaureate/associate degree category, 12 semester hours in airframe structures repair, 6 in aviation maintenance technology, 3 in personnel supervision, 3 in maintenance management, and 3 in industrial safety. In the upper-division baccalaureate category, 2 semester hours for field experience in management, if rank was Staff Sergeant (SSGT) and 3 semester hours for field experience in management, if rank was Gunnery Sergeant (GYSGT) (8/89).

MCE-6097-001

AIRCRAFT STRUCTURES MECHANIC, F/A-18
6097

Exhibit Dates: 1/90–1/98.

Occupational Field: 60 (Aircraft Maintenance).

Career Pattern

PVT: Private (E-1). PFC: Private First Class (E-2). LCP: Lance Corporal (E-3). CPL: Corporal (E-4). SGT: Sergeant (E-5). SSGT: Staff Sergeant (E-6). GYSGT: Gunnery Sergeant (E-7). NOTE: MOS duties are not defined in terms of rank or paygrade, but by levels. Level I identifies tasks that are taught in the classroom; the individual is able to perform simple parts of tasks. Level II identifies those tasks that are part of on-the-job training; the trainee can do most tasks pertaining to a particular system and needs supervision on more difficult parts of tasks. Level III indicates that the individual can perform all essential tasks without supervision. Level IV indicates a high level of proficiency in job performance; the individual can perform advanced technical functions, instruction, inspection, and supervision. The MOS designator 6091 (Aircraft Structures Mechanic—Trainee), is used to identify individuals in Levels I and II. Upon completion of Level II, MOS 6097 is awarded. Credit is recommended for Levels III and IV.

Description

Summary: Inspects, maintains, and repairs aircraft structures and structural components. *Levels I and II:* Uses metalworking tools in the maintenance and repair of aircraft structures and structural components; uses shears, brakes, grinders, and presses; refers to and complies with technical publications such as color-code charts, schematic diagrams, sketches, blueprints, and instructions; inspects and maintains powered and nonpowered hand tools and related support equipment; performs periodic inspections of system components and assists in removal and replacement of aircraft component parts such as wings, control surfaces, landing gear, arresting gear, and tanks; inspects and repairs aircraft metal (aluminum alloy), composites, fiberglass, rubber, and plastic; makes simple dies, templates, and jigs to be used in metalworking and fabrication of aircraft parts; does riveting of all types; uses established procedures to control and prevent aircraft corrosion; observes acceptable industrial safety practices; documents work performed. *Level III:* Able to perform the duties required for Levels I and II; organizes work assignments; heat-treats metal alloys used in aircraft; classifies parts for use in the repair of aircraft; repairs, maintains, and rigs wings, airframe, and cockpit control mechanisms; performs major structural inspections and repairs; repairs aircraft de-icing systems, cockpit enclosures, radomes, and surfaces; conducts on-the-job and technical training. *Level IV:* Able to perform the duties required for Level III; designs cradle structures for moving, handling, and securing aircraft or equipment ashore/afloat; plans, schedules, and directs work center assignments; performs and supervises the complete repair of aircraft structures and structural components; performs aircraft weighing; supervises, directs, and administers an organizational maintenance level aircraft structural repair work center ashore/afloat.

Recommendation, Level III

In the lower-division baccalaureate/associate degree category, 9 semester hours in airframe structures repair, 3 in aviation maintenance technology, and 3 in industrial safety (8/89).

Recommendation, Level IV

In the lower-division baccalaureate/associate degree category, 12 semester hours in airframe structures repair, 6 in aviation maintenance technology, 3 in personnel supervision, 3 in maintenance management, and 3 in industrial safety. In the upper-division baccalaureate category, 2 semester hours for field experience in management, if rank was Staff Sergeant (SSGT) and 3 semester hours for field experience in management, if rank was Gunnery Sergeant (GYSGT) (8/89).

MCE-6098-001

AIRCRAFT STRUCTURES MECHANIC, OV-10
6098

Exhibit Dates: 1/90–1/98.

Occupational Field: 60 (Aircraft Maintenance).

Career Pattern

PVT: Private (E-1). PFC: Private First Class (E-2). LCP: Lance Corporal (E-3). CPL: Corporal (E-4). SGT: Sergeant (E-5). SSGT: Staff Sergeant (E-6). GYSGT: Gunnery Sergeant (E-7). NOTE: MOS duties are not defined in terms of rank or paygrade, but by levels. Level I identifies tasks that are taught in the classroom; the individual is able to perform simple parts of tasks. Level II identifies those tasks that are part of on-the-job training; the trainee can do most tasks pertaining to a particular system and needs supervision on more difficult parts of tasks. Level III indicates that the individual can perform all essential tasks without supervision. Level IV indicates a high level of proficiency in job performance; the individual can perform advanced technical functions, instruction, inspection, and supervision. The MOS designator 6091 (Aircraft Structures Mechanic—Trainee), is used to identify individuals in Levels I and II. Upon completion of Level II, MOS 6098 is awarded. Credit is recommended for Levels III and IV.

Description

Summary: Inspects, maintains, and repairs aircraft structures and structural components. *Levels I and II:* Uses metalworking tools in the maintenance and repair of aircraft structures and structural components; uses shears, brakes, grinders, and presses; refers to and complies with technical publications such as color-code charts, schematic diagrams, sketches, blueprints, and instructions; inspects and maintains powered and nonpowered hand tools and related support equipment; performs periodic inspections of system components and assists in removal and replacement of aircraft component parts such as wings, control surfaces, landing gear, arresting gear, and tanks; inspects and repairs aircraft metal (aluminum alloy), composites, fiberglass, rubber, and plastic; makes simple dies, templates, and jigs to be used in metalworking and fabrication of aircraft parts; does riveting of all types; uses established procedures to control and prevent aircraft corrosion; observes acceptable industrial safety practices; documents work performed. *Level III:* Able to perform the duties required for Levels I and II; organizes work assignments; heat-treats metal alloys used in aircraft; classifies parts for use in the repair of aircraft; repairs, maintains, and rigs wings, airframe, and cockpit control mechanisms; performs major structural inspections and repairs; repairs aircraft de-icing systems, cockpit enclosures, radomes, and surfaces; conducts on-the-job and technical training. *Level IV:* Able to perform the duties required for Level III; designs cradle structures for moving, handling, and securing aircraft or equipment ashore/afloat; plans, schedules, and directs work center assignments; performs and supervises the complete repair of aircraft structures and structural components; performs aircraft weighing; supervises, directs, and administers an organizational maintenance level aircraft structural repair work center ashore/afloat.

Recommendation, Level III

In the lower-division baccalaureate/associate degree category, 9 semester hours in airframe structures repair, 3 in aviation maintenance technology, and 3 in industrial safety (8/89).

Recommendation, Level IV

In the lower-division baccalaureate/associate degree category, 12 semester hours in airframe structures repair, 6 in aviation maintenance technology, 3 in personnel supervision, 3 in maintenance management, and 3 in industrial safety. In the upper-division baccalaureate category, 2 semester hours for field experience in management, if rank was Staff Sergeant (SSGT); and 3 semester hours for field experience in management, if rank was Gunnery Sergeant (GYSGT) (8/89).

MARINE CORPS ENLISTED MOS EXHIBITS

MCE-6112-001

HELICOPTER MECHANIC (CH-46)
6112

Exhibit Dates: 1/90–1/98.

Occupational Field: 61 (Aircraft Maintenance).

Career Pattern

PVT: Private (E-1). PFC: Private First Class (E-2). LCP: Lance Corporal (E-3). CPL: Corporal (E-4). SGT: Sergeant (E-5). SSGT: Staff Sergeant (E-6). GYSGT: Gunnery Sergeant (E-7). NOTE: MOS duties are not defined in terms of rank or paygrade, but by levels. Level I identifies tasks that are taught in the classroom; the individual is able to perform simple parts of tasks. Level II identifies those tasks that are part of on-the-job training; the trainee can do most tasks pertaining to a particular system and needs supervision on more difficult parts of tasks. Level III indicates that the individual can perform all essential tasks without supervision. Level IV indicates a high level of proficiency in job performance; the individual can perform advanced technical functions, instruction, inspection, and supervision. The MOS designator 6111 (Aircraft Mechanic—Trainee) is used to identify individuals in Levels I and II. Upon completion of Level II, MOS 6112 is awarded. Credit is recommended for Levels III and IV.

Description

Summary: Inspects and maintains airframes, power plants, and aircraft components; performs duties related to flight-line operations. *Levels I and II:* Operates and maintains special shop and ground support equipment; assists in performing preflight and postflight inspections; practices ground safety procedures in flight-line and hangar areas; performs hot and cold refueling of aircraft and understands the use of airframe and power plant fluids and lubricants including the color codes for lines and tubing; under supervision, removes and installs aircraft components and parts as required; uses technical publications, bulletins, diagrams, and material control procedures; observes weight and balance limitations; uses precision measuring tools. *Level III:* Able to perform the duties required for Levels I and II; performs inspection and maintenance duties on the various systems of the aircraft including the fuel, flight control, rotor, utility, and power plant systems; checks cockpit controls, switches, and safety devices; performs engine starting, tune-up procedures (when authorized), and taxiing of aircraft; removes and installs equipment, components, and parts as required; recognizes corrosion and applies corrective treatment or preventive maintenance; performs duties as section or crew leader or quality assurance inspector; conducts informal technical training programs; assists in the supervision and administration of aircraft maintenance operations. *Level IV:* Able to perform the duties required for Level III; plans, schedules, and directs aircraft maintenance work center activities; supervises flight-line and hangar operations; administers formal and informal technical training.

Recommendation, Level III

In the lower-division baccalaureate/associate degree category, 9 semester hours in airframe structures repair, 3 in aviation maintenance technology, and 3 in industrial safety (8/89).

Recommendation, Level IV

In the lower-division baccalaureate/associate degree category, 12 semester hours in airframe structures repair, 6 in aviation maintenance technology, 3 in personnel supervision, 3 in maintenance management, and 3 in industrial safety. In the upper-division baccalaureate category, 2 semester hours for field experience in management, if rank was Staff Sergeant (SSGT) and 3 semester hours for field experience in management, if rank was Gunnery Sergeant (GYSGT) (8/89).

MCE-6113-001

HELICOPTER MECHANIC (CH-53)
6113

Exhibit Dates: 1/90–1/98.

Occupational Field: 61 (Aircraft Maintenance).

Career Pattern

PVT: Private (E-1). PFC: Private First Class (E-2). LCP: Lance Corporal (E-3). CPL: Corporal (E-4). SGT: Sergeant (E-5). SSGT: Staff Sergeant (E-6). GYSGT: Gunnery Sergeant (E-7). NOTE: MOS duties are not defined in terms of rank or paygrade, but by levels. Level I identifies tasks that are taught in the classroom; the individual is able to perform simple parts of tasks. Level II identifies those tasks that are part of on-the-job training; the trainee can do most tasks pertaining to a particular system and needs supervision on more difficult parts of tasks. Level III indicates that the individual can perform all essential tasks without supervision. Level IV indicates a high level of proficiency in job performance; the individual can perform advanced technical functions, instruction, inspection, and supervision. The MOS designator 6111 (Aircraft Mechanic—Trainee) is used to identify individuals in Levels I and II. Upon completion of Level II, MOS 6113 is awarded. Credit is recommended for Levels III and IV.

Description

Summary: Inspects and maintains airframes, power plants, and aircraft components; performs duties related to flight-line operations. *Levels I and II:* Operates and maintains special shop and ground support equipment; assists in performing preflight and postflight inspections; practices ground safety procedures in flight-line and hangar areas; performs hot and cold refueling of aircraft and understands the use of airframe and power plant fluids and lubricants including the color codes for lines and tubing; under supervision, removes and installs aircraft components and parts as required; uses technical publications, bulletins, diagrams, and material control procedures; observes weight and balance limitations; uses precision measuring tools. *Level III:* Able to perform the duties required for Levels I and II; performs inspection and maintenance duties on the various systems of the aircraft, including the fuel, flight control, rotor, utility, and power plant systems; checks cockpit controls, switches, and safety devices; performs engine starting, tune-up procedures (when authorized), and taxiing of aircraft; removes and installs equipment, components, and parts as required; recognizes corrosion and applies corrective treatment or preventive maintenance; performs duties as section or crew leader or quality assurance inspector; conducts informal technical training programs, assists in the supervision and administration of aircraft maintenance operations. *Level IV:* Able to perform the duties required for Level III; plans, schedules, and directs aircraft maintenance work center activities; supervises flight-line and hangar operations; administers formal and informal technical training.

Recommendation, Level III

In the lower-division baccalaureate/associate degree category, 9 semester hours in airframe structures repair, 3 in aviation maintenance technology, and 3 in industrial safety (8/89).

Recommendation, Level IV

In the lower-division baccalaureate/associate degree category, 12 semester hours in airframe structures repair, 6 in aviation maintenance technology, 3 in personnel supervision, 3 in maintenance management, and 3 in industrial safety. In the upper-division baccalaureate category, 2 semester hours for field experience in management, if rank was Staff Sergeant (SSGT) and 3 semester hours for field experience in management, if rank was Gunnery Sergeant (GYSGT) (8/89).

MCE-6114-001

HELICOPTER MECHANIC (U/AH-1)
6114

Exhibit Dates: 1/90–1/98.

Occupational Field: 61 (Aircraft Maintenance).

Career Pattern

PVT: Private (E-1). PFC: Private First Class (E-2). LCP: Lance Corporal (E-3). CPL: Corporal (E-4). SGT: Sergeant (E-5). SSGT: Staff Sergeant (E-6). GYSGT: Gunnery Sergeant (E-7). NOTE: MOS duties are not defined in terms of rank or paygrade, but by levels. Level I identifies tasks that are taught in the classroom; the individual is able to perform simple parts of tasks. Level II identifies those tasks that are part of on-the-job training; the trainee can do most tasks pertaining to a particular system and needs supervision on more difficult parts of tasks. Level III indicates that the individual can perform all essential tasks without supervision. Level IV indicates a high level of proficiency in job performance; the individual can perform advanced technical functions, instruction, inspection, and supervision. The MOS designator 6111 (Aircraft Mechanic—Trainee) is used to identify individuals in Levels I and II. Upon completion of Level II, MOS 6114 is awarded. Credit is recommended for Levels III and IV.

Description

Summary: Inspects and maintains airframes, power plants, and aircraft components; performs duties related to flight-line operations. *Levels I and II:* Operates and maintains special shop and ground support equipment; assists in performing preflight and postflight inspections; practices ground safety procedures in flight-line and hangar areas; performs hot and cold refueling of aircraft and understands the use of airframe and power plant fluids and lubricants including the color codes for lines and tubing; under supervision, removes and installs aircraft components and parts as required; uses technical publications, bulletins, diagrams, and material control procedures; observes weight and balance limitations; uses precision measuring tools. *Level III:* Able to perform the duties required for Levels I and II; performs inspection and maintenance duties on the various systems of the aircraft including the fuel, flight control, rotor, utility, and power plant systems; checks cockpit controls, switches, and safety devices; performs engine starting, tune-up procedures (when authorized), and taxiing of aircraft; removes and installs equipment, components, and parts as required; recognizes corrosion and applies corrective treatment or preventive maintenance; performs duties as section or crew leader or quality assurance inspector; conducts informal technical training programs; assists in the supervision and administration of aircraft maintenance operations. *Level IV:* Able to perform the duties required for Level III; plans, schedules, and directs aircraft maintenance work center activities; supervises flight-line and hangar operations; administers formal and informal technical training.

Recommendation, Level III

In the lower-division baccalaureate/associate degree category, 9 semester hours in airframe structures repair, 3 in aviation maintenance technology, and 3 in industrial safety (8/89).

Recommendation, Level IV

In the lower-division baccalaureate/associate degree category, 12 semester hours in airframe structures repair, 6 in aviation maintenance technology, 3 in personnel supervision, 3 in maintenance management, and 3 in industrial safety. In the upper-division baccalaureate category, 2 semester hours for field experience in management, if rank was Staff Sergeant (SSGT) and 3 semester hours for field experience in management, if rank was Gunnery Sergeant (GYSGT) (8/89).

MCE-6115-001

HELICOPTER MECHANIC (CH-53E)
6115

Exhibit Dates: 1/90–1/98.

Occupational Field: 61 (Aircraft Maintenance).

Career Pattern

PVT: Private (E-1). PFC: Private First Class (E-2). LCP: Lance Corporal (E-3). CPL: Corporal (E-4). SGT: Sergeant (E-5). SSGT: Staff Sergeant (E-6). GYSGT: Gunnery Sergeant (E-7). NOTE: MOS duties are not defined in terms of rank or paygrade, but by levels. Level I identifies tasks that are taught in the classroom; the individual is able to perform simple parts of tasks. Level II identifies those tasks that are part of on-the-job training; the trainee can do most tasks pertaining to a particular system and needs supervision on more difficult parts of tasks. Level III indicates that the individual can perform all essential tasks without supervision. Level IV indicates a high level of proficiency in job performance; the individual can perform advanced technical functions, instruction, inspection, and supervision. The MOS designator 6111 (Aircraft Mechanic—Trainee) is used to identify individuals in Levels I and II. Upon completion of Level II, MOS 6115 is awarded. Credit is recommended for Levels III and IV.

Description

Summary: Inspects and maintains airframes, power plants, and aircraft components; performs duties related to flight-line operations. *Levels I and II:* Operates and maintains special shop and ground support equipment; assists in performing preflight and postflight inspections; practices ground safety procedures in flight-line and hangar areas; performs hot and cold refueling of aircraft and understands the use of airframe and power plant fluids and lubricants including the color codes for lines and tubing; under supervision, removes and installs aircraft components and parts as required; uses technical publications, bulletins, diagrams, and material control procedures; observes weight and balance limitations; uses precision measuring tools. *Level III:* Able to perform the duties required for Levels I and II; performs inspection and maintenance on the various systems of the aircraft including the fuel, flight control, rotor, utility, and power plant systems; checks cockpit controls, switches, and safety devices; performs engine starting, tune-up procedures (when authorized), and taxiing of aircraft; removes and installs equipment, components, and parts as required; recognizes corrosion and applies corrective treatment or preventive maintenance; performs duties as section or crew leader or quality assurance inspector; conducts informal technical training programs; assists in the supervision and administration of aircraft maintenance operations. *Level IV:* Able to perform the duties required for Level III; plans, schedules, and directs aircraft maintenance work center activities; supervises flight-line and hangar operations; administers formal and informal technical training.

Recommendation, Level III

In the lower-division baccalaureate/associate degree category, 9 semester hours in airframe structures repair, 3 in aviation maintenance technology, and 3 in industrial safety (8/89).

Recommendation, Level IV

In the lower-division baccalaureate/associate degree category, 12 semester hours in airframe structures repair, 6 in aviation maintenance technology, 3 in personnel supervision, 3 in maintenance management, and 3 in industrial safety. In the upper-division baccalaureate category, 2 semester hours for field experience in management, if rank was Staff Sergeant (SSGT) and 3 semester hours for field experience in management, if rank was Gunnery Sergeant (GYSGT) (8/89).

MCE-6119-001

HELICOPTER MAINTENANCE CHIEF
6119

Exhibit Dates: 5/92–Present.

Occupational Field: 61 (Aircraft Maintenance).

Career Pattern

May progress to MOS 6119 from: MOS's 6112-15, 6122-25, 6132, 6135, 6142-44, or 6152-55 (Private through Gunnery Sergeant). MSGT: Master Sergeant (E-8). MGYSGT: Master Gunnery Sergeant (E-9). NOTE: Chiefs must be qualified in Level IV of previously-held MOS. Descriptions and credit recommendations for chiefs are by rank. Check the individual's fitness report or brief sheet for the rank and for proficiency report.

Description

Summary: Supervises a helicopter maintenance and repair facility. *Master Sergeant:* Supervises the establishment and functioning of helicopter repair and maintenance facilities; assists in directing, supervising, and coordinating maintenance activities; prepares reports, schedules, and rosters regarding helicopter maintenance and repair; requisitions spare parts, replacement parts, supplies, and equipment; maintains maintenance, inspection, and technical training records. *Master Gunnery Sergeant:* Able to perform the duties required for Master Sergeant; provides staff support in planning and implementing maintenance activities; has a thorough knowledge of Marine Corps administrative procedures and aviation staff organization and functioning; prepares and presents reports on logistics and readiness to higher command levels.

Recommendation, Master Sergeant

In the lower-division baccalaureate/associate degree category, 3 semester hours in personnel supervision, 3 in technical writing, and 3 in records and information management. In the upper-division baccalaureate category, 3 semester hours in organizational management, 3 in management problems, and 3 for field experience in management. NOTE: Credit should also be granted for the prerequisite MOS (5/92).

Recommendation, Master Gunnery Sergeant

In the lower-division baccalaureate/associate degree category, 3 semester hours in personnel supervision, 3 in technical writing, and 3 in records and information management. In the upper-division baccalaureate category, 3 semester hours in organizational management, 6 in management problems, and 6 for field experience in management. NOTE: Credit should also be granted for the prerequisite MOS (5/92).

MCE-6122-001

HELICOPTER POWER PLANTS MECHANIC (T-58)
6122

Exhibit Dates: 1/90–1/98.

Occupational Field: 61 (Aircraft Maintenance).

Career Pattern

PVT: Private (E-1). PFC: Private First Class (E-2). LCP: Lance Corporal (E-3). CPL: Corporal (E-4). SGT: Sergeant (E-5). SSGT: Staff Sergeant (E-6). GYSGT: Gunnery Sergeant (E-7). NOTE: MOS duties are not defined in terms of rank or paygrade, but by levels. Level I identifies tasks that are taught in the classroom; the individual is able to perform simple parts of tasks. Level II identifies those tasks that are part of on-the-job training; the trainee can do most tasks pertaining to a particular system and needs supervision on more difficult parts of tasks. Level III indicates that the individual can perform all essential tasks without supervision. Level IV indicates a high level of proficiency in job performance; the individual can perform advanced technical functions, instruction, inspection, and supervision. The MOS designator 6111 (Aircraft Mechanic—Trainee) is used to identify individuals in Levels I and II. Upon completion of Level II, MOS 6122 is awarded. Credit is recommended for Levels III and IV.

Description

Summary: Inspects, maintains, and repairs aircraft power plants and performs duties related to flight-line operations. *Levels I and II:* Interprets shop sketches, drawings, schematics, and blueprints; uses technical publications and other maintenance data; under supervision, assembles, disassembles, and repairs power plants and power plant systems; uses precision measuring instruments and special power tools in performing maintenance functions; assists in minor maintenance on ground handling equipment and power plant test cells; knows the types and designation of fuels and lubricants and utilizes color-coded charts for lines and tubing; assists in performing periodic inspections and removing and replacing power plants and power plant components; assists in corrosion prevention procedures; practices ground safety precautions in the power plant maintenance area. *Level III:* Able to perform the duties required for Levels I and II; performs the complete repair of power plants; performs ground testing and engine run-up; prepares power plants for storage or shipment; changes engines; prepares and submits work center reports; conducts technical training; plans, schedules, and supervises work center operations. *Level IV:* Able to perform the duties required for Level III; performs and supervises the complete repair of power plant and power plant systems; serves as a power plant maintenance supervisor and performs the duties of a quality assurance inspector.

Recommendation, Level III

In the lower-division baccalaureate/associate degree category, 9 semester hours in gas turbine maintenance and repair, 3 in aviation maintenance technology, and 3 in industrial safety (12/88).

Recommendation, Level IV

In the lower-division baccalaureate/associate degree category, 12 semester hours in gas turbine maintenance and repair, 6 in aviation maintenance technology, 3 in industrial safety, 3 in personnel supervision, and 3 in maintenance management. In the upper-division baccalaureate category, 2 semester hours for field experience in management, if rank was Staff Sergeant (SSGT) and 3 semester hours for field experience in management, if rank was Gunnery Sergeant (GYSGT) (12/88).

MCE-6123-001

HELICOPTER POWER PLANTS MECHANIC (T-64)
6123

Exhibit Dates: 1/90–1/98.

Occupational Field: 61 (Aircraft Maintenance).

Career Pattern

PVT: Private (E-1). PFC: Private First Class (E-2). LCP: Lance Corporal (E-3). CPL: Corporal (E-4). SGT: Sergeant (E-5). SSGT: Staff Sergeant (E-6). GYSGT: Gunnery Sergeant (E-7). NOTE: MOS duties are not defined in terms of rank or paygrade, but by levels. Level I identifies tasks that are taught in the classroom; the individual is able to perform simple parts of tasks. Level II identifies those tasks that are part of on-the-job training; the trainee can do most tasks pertaining to a particular system and needs supervision on more difficult parts of tasks. Level III indicates that the individual can perform all essential tasks without supervision. Level IV indicates a high level of proficiency in job performance; the individual can perform advanced technical functions, instruction, inspection, and supervision. The MOS designator 6111 (Aircraft Mechanic—Trainee) is used to identify individuals in Levels I and II. Upon completion of Level II, MOS 6123 is awarded. Credit is recommended for Levels III and IV.

Description

Summary: Inspects, maintains, and repairs aircraft power plants and performs duties related to flight-line operations. *Levels I and II:* Interprets shop sketches, drawings, schematics, and blueprints; uses technical publications and other maintenance data; under supervision, assembles, disassembles, and repairs power plants and power plant systems; uses precision measuring instruments and special power tools in performing maintenance functions; assists in minor maintenance on ground handling equipment and power plant test cells; knows the types and designation of fuels and lubricants and utilizes color-coded charts for lines and tubing; assists in performing periodic inspections and removing and replacing power plants and power plant components; assists in corrosion prevention procedures; practices ground safety precautions in the power plant maintenance area. *Level III:* Able to perform the duties required for Levels I and II; performs the complete repair of power plants; performs ground testing and engine run-up; prepares power plants for storage or shipment; changes engines; prepares and submits work center reports; conducts technical training; plans, schedules, and supervises work center operations. *Level IV:* Able to perform the duties required for Level III; performs and supervises the complete repair of power plant and power plant systems; serves as a power plant maintenance supervisor and performs the duties of a quality assurance inspector.

Recommendation, Level III

In the lower-division baccalaureate/associate degree category, 9 semester hours in gas turbine maintenance and repair, 3 in aviation maintenance technology, and 3 in industrial safety (12/88).

Recommendation, Level IV

In the lower-division baccalaureate/associate degree category, 12 semester hours in gas turbine maintenance and repair, 6 in aviation maintenance technology, 3 in industrial safety, 3 in personnel supervision, and 3 in maintenance management. In the upper-division baccalaureate category, 2 semester hours for field experience in management, if rank was Staff Sergeant (SSGT) and 3 semester hours for field experience in management, if rank was Gunnery Sergeant (GYSGT) (12/88).

MCE-6125-001

HELICOPTER POWER PLANTS MECHANIC (T-400)
6125

Exhibit Dates: 1/90–1/98.

Occupational Field: 61 (Aircraft Maintenance).

Career Pattern

PVT: Private (E-1). PFC: Private First Class (E-2). LCP: Lance Corporal (E-3). CPL: Corporal (E-4). SGT: Sergeant (E-5). SSGT: Staff Sergeant (E-6). GYSGT: Gunnery Sergeant (E-7). NOTE: MOS duties are not defined in terms of rank or paygrade, but by levels. Level I identifies tasks that are taught in the classroom; the individual is able to perform simple parts of tasks. Level II identifies those tasks that are part of on-the-job training; the trainee can do most tasks pertaining to a particular system and needs supervision on more difficult parts of tasks. Level III indicates that the individual can perform all essential tasks without supervision. Level IV indicates a high level of proficiency in job performance; the individual can perform advanced technical functions, instruction, inspection, and supervision. The MOS designator 6111 (Aircraft Mechanic—Trainee) is used to identify individuals in Levels I and II. Upon completion of Level II, MOS 6125 is awarded. Credit is recommended for Levels III and IV.

Description

Summary: Inspects, maintains, and repairs aircraft power plants and performs duties related to flight-line operations. *Levels I and II:* Interprets shop sketches, drawings, schematics, and blueprints; uses technical publications and other maintenance data; under supervision, assembles, disassembles, and repairs power plants and power plant systems; uses precision measuring instruments and special power tools in performing maintenance functions; assists in minor maintenance on ground handling equipment and power plant test cells; knows the types and designation of fuels and lubricants and utilizes color-coded charts for lines and tubing; assists in performing periodic inspections, removing and replacing power plants and power plant components; assists in corrosion prevention procedures; practices ground safety precautions in the power plant maintenance area. *Level III:* Able to perform the duties required for Levels I and II; performs the complete repair of power plants; performs ground testing and engine run-up; prepares power plants for storage or shipment; changes engines; prepares and submits work center reports; conducts technical training; plans, schedules, and supervises work center operations. *Level IV:* Able to perform the duties required for Level III; performs and supervises the complete repair of power plant and power plant systems; serves as a power plant maintenance supervisor and performs the duties of a quality assurance inspector.

Recommendation, Level III

In the lower-division baccalaureate/associate degree category, 9 semester hours in gas turbine maintenance and repair, 3 in aviation maintenance technology, and 3 in industrial safety (12/88).

Recommendation, Level IV

In the lower-division baccalaureate/associate degree category, 12 semester hours in gas turbine maintenance and repair, 6 in aviation maintenance technology, 3 in industrial safety, 3 in personnel supervision, and 3 in maintenance management. In the upper-division baccalaureate category, 2 semester hours for field experience in management, if rank was Staff Sergeant (SSGT) and 3 semester hours for field experience in management, if rank was Gunnery Sergeant (GYSGT) (12/88).

MCE-6135-001

AIRCRAFT POWER PLANT TEST CELL OPERATOR, ROTARY WING
6135

Exhibit Dates: 5/92–Present.

Occupational Field: 61 (Aircraft Maintenance).

Career Pattern

Must be qualified in one of the helicopter power plants MOS's (6122, 6123, or 6125). PVT: Private (E-1). PFC: Private First Class (E-2). LCP: Lance Corporal (E-3). CPL: Corporal (E-3). SGT: Sergeant (E-5). SSGT: Staff Sergeant (E-6). GYSGT: Gunnery Sergeant (E-7). NOTE: MOS duties are not defined in terms of rank or paygrade, but by levels. Level I identifies tasks that are taught in the classroom; the individual is able to perform simple parts of tasks. Level II identifies those tasks that are part of on-the-job training; the trainee can do most tasks pertaining to a particular system and needs supervision on more difficult parts of tasks. Level III indicates that the individual can perform all essential tasks without supervision. Level IV indicates a high level of proficiency in job performance; the individual can perform advanced technical functions, instruction, inspection, and supervision. Credit is recommended for Levels III and IV.

Description

Summary: Inspects, tests, and performs corrective maintenance on rotary-wing gas turbine engines and engine systems; performs minor repairs to test cells. *Levels I and II:* Interprets shop sketches, drawings, schematics, and blueprints; performs minor maintenance on gas turbine engines and test cells; uses precision measuring equipment; observes safety procedures; completes maintenance data forms; performs test cell engine run-ups. *Level III:* Able to perform the duties required for Levels I and II; removes and replaces power plant components and accessories; conducts informal on-the-job training. *Level IV:* Able to perform the duties required for Level III; plans, schedules, and directs work center assignments; supervises and performs complete power plant repair in work center; supervises, directs, and administers helicopter gas turbine engine test cell operations.

Recommendation, Level III

In the lower-division baccalaureate/associate degree category, credit is recommended for the prerequisite MOS only; see exhibit MCE-6122-001, MCE-6123-001, or MCE-6125-001 (5/92).

Recommendation, Level IV

In the lower-division baccalaureate/associate degree category, credit is recommended for the prerequisite MOS only; see exhibits MCE-6122-001, MCE-6123-001, or MCE-6125-001. In the upper-division baccalaureate category, 2 semester hours for field experience in management, if rank was Staff Sergeant (SSGT) and 3 semester hours for field experience in management, if rank was Gunnery Sergeant (GYSGT) (5/92).

MCE-6142-001

HELICOPTER STRUCTURES MECHANIC (CH-46)
6142

Exhibit Dates: 1/90–1/98.

Occupational Field: 61 (Aircraft Maintenance).

Career Pattern

PVT: Private (E-1). PFC: Private First Class (E-2). LCP: Lance Corporal (E-3). CPL: Corporal (E-4). SGT: Sergeant (E-5). SSGT: Staff Sergeant (E-6). GYSGT: Gunnery Sergeant (E-7). NOTE: MOS duties are not defined in terms of rank or paygrade, but by levels. Level I identifies tasks that are taught in the classroom; the individual is able to perform simple parts of tasks. Level II identifies those tasks that are part of on-the-job training; the trainee can do most tasks pertaining to a particular system and needs supervision on more difficult parts of tasks. Level III indicates that the individual can perform all essential tasks without supervision. Level IV indicates a high level of proficiency in job performance; the individual can perform advanced technical functions, instruction, inspection, and supervision. The MOS designator 6111 (Aircraft Mechanic—Trainee) is used to identify individuals in Levels I and II. Upon completion of Level II, MOS 6142 is awarded. Credit is recommended for Levels III and IV.

Description

Summary: Performs inspections, maintenance, and repair of helicopter structures and components; work is applicable to airplanes and related systems. *Levels I and II:* Uses metalworking tools in the maintenance and repair of helicopter components; uses shears, brakes, grinders, and presses; refers to and complies with technical publications such as color-coded charts, schematic diagrams, sketches, blueprints, and instructions; inspects and maintains powered and nonpowered hand tools and related support equipment; performs periodic inspections of system components and assists in removal and replacement of blades, control surfaces, landing gear, and tanks; inspects and repairs aircraft metal (aluminum), composites, fiber glass, rubber, and plastic; makes simple dies, templates and jigs to be used in metalworking and fabrication of aircraft parts; does riveting of all types; uses established procedures to control and prevent aircraft corrosion; organizes work assignments; observes acceptable industrial safety practices; documents work performed. *Level III:* Able to perform the duties required for Levels I and II; performs major structural inspections and repairs; heat-treats metal alloys used in aircraft; repairs, maintains, and rigs blades, airframe, and control mechanisms; conducts on-the-job and technical training. *Level IV:* Able to perform the duties required for Level III; designs cradle structures for moving, handling, and securing aircraft and related equipment; performs aircraft weight and balance checks; supervises complete repair of helicopter structures and components; administers a helicopter structural repair center.

Recommendation, Level III

In the lower-division baccalaureate/associate degree category, 12 semester hours in airframe structures repair and 3 in aviation maintenance technology (12/88).

Recommendation, Level IV

In the lower-division baccalaureate/associate degree category, 15 semester hours in airframe structures repair, 6 in aviation maintenance technology, 3 in personnel supervision, and 3 in maintenance management. In the upper-division baccalaureate category, 2 semester hours for field experience in management, if rank was Staff Sergeant (SSGT) and 3 semester hours for field experience in management, if rank was Gunnery Sergeant (GYSGT) (12/88).

MCE-6143-001

HELICOPTER STRUCTURES MECHANIC (CH-53)
6143

Exhibit Dates: 1/90–1/98.

Occupational Field: 61 (Aircraft Maintenance).

Career Pattern

PVT: Private (E-1). PFC: Private First Class (E-2). LCP: Lance Corporal (E-3). CPL: Corporal (E-4). SGT: Sergeant (E-5). SSGT: Staff Sergeant (E-6). GYSGT: Gunnery Sergeant (E-7). NOTE: MOS duties are not defined in terms of rank or paygrade, but by levels. Level I identifies tasks that are taught in the classroom; the individual is able to perform simple parts of tasks. Level II identifies those tasks that are part of on-the-job training; the trainee can do most tasks pertaining to a particular system and needs supervision on more difficult parts of tasks. Level III indicates that the individual can perform all essential tasks without supervision. Level IV indicates a high level of proficiency in job performance; the individual can perform advanced technical functions, instruction, inspection, and supervision. The MOS designator 6111 (Aircraft Mechanic—Trainee) is used to identify individuals in Levels I and II. Upon completion of Level II, MOS 6143 is awarded. Credit is recommended for Levels III and IV.

Description

Summary: Performs inspections, maintenance, and repair of helicopter structures and components; work is applicable to airplanes and related systems. *Levels I and II:* Uses metalworking tools in the maintenance and repair of helicopter components; uses shears, brakes, grinders, and presses; refers to and complies with technical publications such as color-coded charts, schematic diagrams, sketches, blueprints, and instructions; inspects and maintains powered and nonpowered hand tools and related support equipment; performs periodic inspections of system components and assists in removal and replacement of blades, control surfaces, landing gear, and tanks; inspects and repairs aircraft metal (aluminum), composites, fiber glass, rubber, and plastic; makes simple dies, templates and jigs to be used in metalworking and fabrication of aircraft parts; does riveting of all types; uses established procedures to control and prevent aircraft corrosion; organizes work assignments; observes acceptable industrial safety practices; documents work performed. *Level III:* Able to perform the duties required for Levels I and II; performs major structural inspections and repairs; heat-treats metal alloys used in aircraft; repairs, maintains, and rigs blades, airframe and control mechanisms; conducts on-the-job and technical training. *Level IV:* Able to perform the duties required for Level III; designs cradle structures for moving, handling, and securing aircraft and related equipment; performs aircraft weight and balance checks; supervises complete repair of helicopter structures and components; administers a helicopter structural repair center.

Recommendation, Level III

In the lower-division baccalaureate/associate degree category, 12 semester hours in airframe structures repair and 3 in aviation maintenance technology (12/88).

Recommendation, Level IV

In the lower-division baccalaureate/associate degree category, 15 semester hours in airframe structures repair, 6 in aviation maintenance technology, 3 in personnel supervision, and 3 in maintenance management. In the upper-division baccalaureate category, 2 semester hours for field experience in management, if rank was Staff Sergeant (SSGT) and 3 semester hours for field experience in management, if rank was Gunnery Sergeant (GYSGT) (12/88).

MCE-6144-001

HELICOPTER STRUCTURES MECHANIC (U/AH-1)
6144

Exhibit Dates: 1/90–1/98.

Occupational Field: 61 (Aircraft Maintenance).

Career Pattern

PVT: Private (E-1). PFC: Private First Class (E-2). LCP: Lance Corporal (E-3). CPL: Corporal (E-4). SGT: Sergeant (E-5). SSGT: Staff Sergeant (E-6). GYSGT: Gunnery Sergeant (E-7). NOTE: MOS duties are not defined in terms of rank or paygrade, but by levels. Level I identifies tasks that are taught in the classroom; the individual is able to perform simple parts of tasks. Level II identifies those tasks that are part of on-the-job training; the trainee can do most tasks pertaining to a particular system and needs supervision on more difficult parts of tasks. Level III indicates that the individual can perform all essential tasks without supervision. Level IV indicates a high level of proficiency in job performance; the individual can perform advanced technical functions, instruction, inspection, and supervision. The MOS designator 6111 (Aircraft Mechanic—Trainee) is used to identify individuals in Levels I and II. Upon completion of Level II, MOS 6144 is awarded. Credit is recommended for Levels III and IV.

Description

Summary: Performs inspections, maintenance, and repair of helicopter structures and components; work is applicable to airplanes and related systems. *Levels I and II:* Uses metalworking tools in the maintenance and repair of helicopter components; uses shears, brakes, grinders, and presses; refers to and complies with technical publications such as color-coded charts, schematic diagrams, sketches, blueprints, and instructions; inspects and maintains powered and nonpowered hand tools and related support equipment; performs periodic inspections of system components and assists in removal and replacement of blades, control surfaces, landing gear, and tanks; inspects and repairs aircraft metal (aluminum), composites, fiber glass, rubber, and plastic; makes simple dies, templates and jigs to be used in metalworking and fabrication of aircraft parts; does riveting of all types; uses established procedures to control and prevent aircraft corrosion; organizes work assignments; observes acceptable industrial safety practices; documents work performed. *Level III:* Able to perform the duties required for Levels I and II; performs major structural inspections and repairs; heat-treats metal alloys used in aircraft; repairs, maintains, and rigs blades, airframe, and control mechanisms; conducts on-the-job and technical training. *Level IV:* Able to perform the duties required for Level III; designs cradle structures for moving, handling, and securing aircraft and related equipment; performs aircraft weight and balance checks; supervises complete repair of helicopter structures and components; administers a helicopter structural repair center.

Recommendation, Level III

In the lower-division baccalaureate/associate degree category, 12 semester hours in airframe structures repair and 3 in aviation maintenance technology (12/88).

Recommendation, Level IV

In the lower-division baccalaureate/associate degree category, 15 semester hours in airframe structures repair, 6 in aviation maintenance technology, 3 in personnel supervision, and 3 in maintenance management. In the upper-division baccalaureate category, 2 semester hours for field experience in management, if rank was Staff Sergeant (SSGT) and 3 semester hours for field experience in management, if rank was Gunnery Sergeant (GYSGT) (12/88).

MCE-6152-001

HELICOPTER HYDRAULIC/PNEUMATIC MECHANIC (CH-46)
6152

Exhibit Dates: 1/90–1/98.

Occupational Field: 61 (Aircraft Maintenance).

Career Pattern

PVT: Private (E-1). PFC: Private First Class (E-2). LCP: Lance Corporal (E-3). CPL: Corporal (E-4). SGT: Sergeant (E-5). SSGT: Staff Sergeant (E-6). GYSGT: Gunnery Sergeant (E-7). NOTE: MOS duties are not defined in terms of rank or paygrade, but by levels. Level I identifies tasks that are taught in the classroom; the individual is able to perform simple parts of tasks. Level II identifies those tasks that are part of on-the-job training; the trainee can do most tasks pertaining to a particular system and needs supervision on more difficult parts of tasks. Level III indicates that the individual can perform all essential tasks without supervision. Level IV indicates a high level of proficiency in job performance; the individual can perform advanced technical functions, instruction, inspection, and supervision. The MOS designator 6111 (Aircraft Mechanic—Trainee) is used to identify individuals in Levels I and II. Upon completion of Level II, MOS 6152 is awarded. Credit is recommended for Levels III and IV.

Description

Summary: Maintains, inspects, and repairs helicopter hydraulic/pneumatic systems and system components. *Levels I and II:* Uses color-coded charts of helicopter piping and lines; reads schematic diagrams, shop sketches, and blueprints and interprets applicable technical data; under supervision, performs component repairs; uses fluids and gases in applicable systems; uses precision measuring equipment; assists in or performs inspections required by maintenance requirement cards and completes maintenance/repairs as required. *Level III:* Able to perform the duties required for Levels I and II; performs complete repair of helicopter braking systems; removes and replaces hydraulic/pneumatic systems and/or components; uses and maintains hydraulic/pneumatic test equipment; observes applicable material control procedures; conducts informal on-the-job training; plans, schedules, and directs work center assignments. *Level IV:* Able to perform the duties required for Level III; performs and supervises the maintenance/repair requirements of helicopter hydraulic/pneumatic systems components; supervises a helicopter hydraulic/pneumatic check crew or maintenance unit; performs duties as quality assurance inspector; supervises and administers an organizational maintenance level helicopter hydraulic/pneumatic work center.

Recommendation, Level III

In the lower-division baccalaureate/associate degree category, 12 semester hours in hydraulic systems repair and 3 in aviation maintenance technology (12/88).

Recommendation, Level IV

In the lower-division baccalaureate/associate degree category, 15 semester hours in hydraulic systems repair, 6 in aviation maintenance technology, 3 in personnel supervision, and 3 in maintenance management. In the upper-division baccalaureate category, 2 semester hours for field experience in management, if rank was Staff Sergeant (SSGT) and 3 semester hours for field experience in management, if rank was Gunnery Sergeant (GYSGT) (12/88).

MCE-6153-001

HELICOPTER HYDRAULIC/PNEUMATIC MECHANIC (CH-53)
6153

Exhibit Dates: 1/90–1/98.

Occupational Field: 61 (Aircraft Maintenance).

Career Pattern

PVT: Private (E-1). PFC: Private First Class (E-2). LCP: Lance Corporal (E-3). CPL: Corporal (E-4). SGT: Sergeant (E-5). SSGT: Staff Sergeant (E-6). GYSGT: Gunnery Sergeant (E-7). NOTE: MOS duties are not defined in terms of rank or paygrade, but by levels. Level I identifies tasks that are taught in the classroom; the individual is able to perform simple parts of tasks. Level II identifies those tasks that are part of on-the-job training; the trainee can do most tasks pertaining to a particular system and needs supervision on more difficult parts of tasks. Level III indicates that the individual can perform all essential tasks without supervision. Level IV indicates a high level of proficiency in job performance; the individual can perform advanced technical functions, instruction, inspection, and supervision. The MOS designator 6111 (Aircraft Mechanic—Trainee) is used to identify individuals in Levels I and II. Upon completion of Level II, MOS 6153 is awarded. Credit is recommended for Levels III and IV.

Description

Summary: Maintains, inspects, and repairs helicopter hydraulic/pneumatic systems and system components. *Levels I and II:* Uses color-code d charts of helicopter piping and lines; reads schematic diagrams, shop sketches, and blueprints and interprets applicable technical data; under supervision, performs component repairs; uses fluids and gases in applicable systems; uses precision measuring equipment; assists in or performs inspections required by maintenance requirement cards and completes maintenance/repairs as required. *Level III:* Able to perform the duties required for Levels I and II; performs complete repair of helicopter braking systems; removes and replaces hydraulic/pneumatic systems and/or components; uses and maintains hydraulic/pneumatic test equipment; observes applicable material control procedures; conducts informal on-the-job training; plans, schedules, and directs work center assignments. *Level IV:* Able to perform the duties required for Level III; performs and supervises the maintenance/repair requirements of helicopter hydraulic/pneumatic systems components; supervises a helicopter hydraulic/pneumatic check crew or maintenance unit; performs duties as quality assurance inspector; supervises and administers an organizational maintenance level helicopter hydraulic/pneumatic work center.

Recommendation, Level III

In the lower-division baccalaureate/associate degree category, 12 semester hours in hydraulic systems repair and 3 in aviation maintenance technology (12/88).

Recommendation, Level IV

In the lower-division baccalaureate/associate degree category, 15 semester hours in hydraulic systems repair, 6 in aviation maintenance technology, 3 in personnel supervision, and 3 in maintenance management. In the upper-division baccalaureate category, 2 semester hours for field experience in management, if rank was Staff Sergeant (SSGT) and 3 semester hours for field experience in management, if rank was Gunnery Sergeant (GYSGT) (12/88).

MCE-6154-001

HELICOPTER HYDRAULIC/PNEUMATIC MECHANIC (U/AH-1)
6154

Exhibit Dates: 1/90–1/98.

Occupational Field: 61 (Aircraft Maintenance).

Career Pattern

PVT: Private (E-1). PFC: Private First Class (E-2). LCP: Lance Corporal (E-3). CPL: Corporal (E-4). SGT: Sergeant (E-5). SSGT: Staff Sergeant (E-6). GYSGT: Gunnery Sergeant (E-7). NOTE: MOS duties are not defined in terms of rank or paygrade, but by levels. Level I identifies tasks that are taught in the classroom; the individual is able to perform simple parts of tasks. Level II identifies those tasks that are part of on-the-job training; the trainee can do most tasks pertaining to a particular system and needs supervision on more difficult parts of tasks. Level III indicates that the individual can perform all essential tasks without supervision. Level IV indicates a high level of proficiency in job performance; the individual can perform advanced technical functions, instruction, inspection, and supervision. The MOS designator 6111 (Aircraft Mechanic—Trainee) is used to identify individuals in Levels I and II. Upon completion of Level II, MOS 6154 is awarded. Credit is recommended for Levels III and IV.

Description

Summary: Maintains, inspects, and repairs helicopter hydraulic/pneumatic systems and system components. *Levels I and II:* Uses color-coded charts of helicopter piping and lines; reads schematic diagrams, shop sketches, and blueprints and interprets applicable technical data; under supervision, performs component repairs; uses fluids and gases in applicable systems; uses precision measuring equipment; assists in or performs inspections required by maintenance requirement cards and completes maintenance/repairs as required. *Level III:* Able to perform the duties required for Levels I and II; performs complete repair of helicopter braking systems; removes and replaces hydraulic/pneumatic systems and/or components; uses and maintains hydraulic/pneumatic test equipment; observes applicable material control procedures; conducts informal on-the-job training; plans, schedules, and directs work center assignments. *Level IV:* Able to perform the duties required for Level III; performs and supervises the maintenance/repair requirements of helicopter hydraulic/pneumatic system components; supervises a helicopter hydraulic/pneumatic check crew or maintenance unit; performs duties as quality assurance inspector; supervises and administers an organizational maintenance level helicopter hydraulic/pneumatic work center.

Recommendation, Level III

In the lower-division baccalaureate/associate degree category, 12 semester hours in hydraulic systems repair and 3 in aviation maintenance technology (12/88).

Recommendation, Level IV

In the lower-division baccalaureate/associate degree category, 15 semester hours in hydraulic systems repair, 6 in aviation maintenance technology, 3 in personnel supervision, and 3 in maintenance management. In the upper-division baccalaureate category, 2 semester hours for field experience in management, if rank was Staff Sergeant (SSGT) and 3 semester hours for field experience in management, if rank was Gunnery Sergeant (GYSGT) (12/88).

MCE-6155-001

HELICOPTER HYDRAULIC/PNEUMATIC MECHANIC (CH-53E)
6155

Exhibit Dates: 1/90–1/98.

Occupational Field: 61 (Aircraft Maintenance).

Career Pattern

PVT: Private (E-1). PFC: Private First Class (E-2). LCP: Lance Corporal (E-3). CPL: Corporal (E-4). SGT: Sergeant (E-5). SSGT: Staff Sergeant (E-6). GYSGT: Gunnery Sergeant (E-7). NOTE: MOS duties are not defined in terms of rank or paygrade, but by levels. Level I identifies tasks that are taught in the classroom; the individual is able to perform simple parts of tasks. Level II identifies those tasks that are part of on-the-job training; the trainee can do most tasks pertaining to a particular system and needs supervision on more difficult parts of tasks. Level III indicates that the individual can perform all essential tasks without supervision. Level IV indicates a high level of proficiency in job performance; the individual can perform advanced technical functions, instruction, inspection, and supervision. The MOS designator 6111 (Aircraft Mechanic—Trainee) is used to identify individuals in Levels I and II. Upon completion of Level II, MOS 6155 is awarded. Credit is recommended for Levels III and IV.

Description

Summary: Maintains, inspects, and repairs helicopter hydraulic/pneumatic systems and system components. *Levels I and II:* Uses color-coded charts of helicopter piping and lines; reads schematic diagrams, shop sketches, and blueprints and interprets applicable technical data; under supervision, performs component repairs; uses fluids and gases in applicable systems; uses precision measuring equipment; assists in or performs inspections required by maintenance requirement cards and completes maintenance/repairs as required. *Level III:* Able to perform the duties required for Levels I and II; performs complete repair of helicopter braking systems; removes and replaces hydraulic/pneumatic systems and/or components; uses and maintains hydraulic/pneumatic test equipment; observes applicable material control procedures; conducts informal on-the-job training; plans, schedules, and directs work center assignments. *Level IV:* Able to perform the duties required for Level III; performs and supervises the maintenance/repair requirements of helicopter hydraulic/pneumatic system components; supervises a helicopter hydraulic/pneumatic check crew or maintenance unit; performs duties as quality assurance inspector; supervises and administers an organizational maintenance level helicopter hydraulic/pneumatic work center.

Recommendation, Level III

In the lower-division baccalaureate/associate degree category, 12 semester hours in hydraulic systems repair and 3 in aviation maintenance technology (12/88).

Recommendation, Level IV

In the lower-division baccalaureate/associate degree category, 15 semester hours in hydraulic systems repair, 6 in aviation maintenance technology, 3 in personnel supervision, and 3 in maintenance management. In the upper-division baccalaureate category, 2 semester hours for field experience in management, if rank was Staff Sergeant (SSGT) and 3 semester hours for field experience in management, if rank was Gunnery Sergeant (GYSGT) (12/88).

MCE-6172-001

HELICOPTER CREW CHIEF, CH-46
6172

Exhibit Dates: 5/92–Present.

Occupational Field: 61 (Aircraft Maintenance).

Career Pattern

May progress to MOS 6172 from any MOS in the Aircraft Maintenance (60/61) or Avionics (63) occupational field. LCP: Lance Corporal (E-3). CPL: Corporal (E-4). SGT: Sergeant (E-5). NOTE: MOS duties are not defined in terms of rank or paygrade, but by levels. Level I identifies tasks that are taught in the classroom; the individual is able to perform simple parts of tasks. Level II identifies those tasks that are part of on-the-job training; the trainee can do most tasks pertaining to a particular system and needs supervision on more difficult parts of tasks. Level III indicates that the individual can perform all essential tasks without supervision. Level IV indicates a high level of proficiency in job performance; the individual can perform advanced technical functions, instruction, inspection, and supervision. Crew chiefs must be qualified at Level III in their prerequisite MOS.

Description

Summary: Serves as flight crewmember, operating and performing maintenance on the CH-46 helicopter. *Levels I and II:* Performs daily inspections on assigned aircraft; assists in preflight inspections; monitors aircraft performance during flight; acts as a lookout and advises pilot of obstacles and other aircraft; is responsible for passengers and cargo; loads cargo; serves as a mechanic. *Level III:* Able to perform the duties required for Levels I and II; trains subordinates; keeps service records; orders supplies; manages maintenance information. *Level IV:* Able to perform the duties required for Level III; supervises and qualifies subordinates; prepares technical and advisory reports on maintenance efforts.

Recommendation, Level III

In the lower-division baccalaureate/associate degree category, 3 semester hours in records and information management and 1 in technical writing. NOTE: Credit should be added for previously-held MOS (5/92).

Recommendation, Level IV

In the lower-division baccalaureate/associate degree category, 3 semester hours in records and information management and 1 in technical writing. NOTE: Credit should be added for previously-held MOS (5/92).

MCE-6173-001

HELICOPTER CREW CHIEF, CH-53 A/D
6173

Exhibit Dates: 5/92–Present.

Occupational Field: 61 (Aircraft Maintenance).

Career Pattern

May progress to MOS 6173 from any MOS in the Aircraft Maintenance (60/61) or Avionics (63) occupational field. LCP: Lance Corporal (E-3). CPL: Corporal (E-4). SGT: Sergeant (E-5). NOTE: MOS duties are not defined in terms of rank or paygrade, but by levels. Level I identifies tasks that are taught in the classroom; the individual is able to perform simple parts of tasks. Level II identifies those tasks that are part of on-the-job training; the trainee can do most tasks pertaining to a particular system and needs supervision on more difficult parts of tasks. Level III indicates that the individual can perform all essential tasks without supervision. Level IV indicates a high level of proficiency in job performance; the individual can perform advanced technical functions, instruction, inspection, and supervision. Crew chiefs must be qualified at Level III in their prerequisite MOS.

Description

Summary: Serves as flight crewmember, operating and performing maintenance on the CH-53 A/D helicopter. *Levels I and II:* Performs daily inspections on assigned aircraft; assists in preflight inspections; monitors aircraft performance during flight; acts as a lookout and advises pilot of obstacles and other aircraft; is responsible for passengers and cargo; loads cargo; serves as a mechanic. *Level III:* Able to perform the duties required for Levels I and II; trains subordinates; keeps service records; orders supplies; manages maintenance information. *Level IV:* Able to perform the duties required for Level III; supervises and qualifies subordinates; prepares technical and advisory reports on maintenance efforts.

Recommendation, Level III

In the lower-division baccalaureate/associate degree category, 3 semester hours in records and information management and 1 in technical writing. NOTE: Credit should be added for previously-held MOS (5/92).

Recommendation, Level IV

In the lower-division baccalaureate/associate degree category, 3 semester hours in records and information management and 1 in technical writing. NOTE: Credit should be added for previously-held MOS (5/92).

MCE-6174-001

HELICOPTER CREW CHIEF, CH-53E
6174

Exhibit Dates: 5/92–Present.

Occupational Field: 61 (Aircraft Maintenance).

Career Pattern

May progress to MOS 6174 from any MOS in the Aircraft Maintenance (60/61) or Avionics (63) occupational field. LCP: Lance Corporal (E-3). CPL: Corporal (E-4). SGT: Sergeant (E-5). NOTE: MOS duties are not defined in terms of rank or paygrade, but by levels. Level I identifies tasks that are taught in the classroom; the individual is able to perform simple parts of tasks. Level II identifies those tasks that are part of on-the-job training; the trainee can do most tasks pertaining to a particular system and needs supervision on more difficult parts of tasks. Level III indicates that the individual can perform all essential tasks without supervision. Level IV indicates a high level of proficiency in job performance; the individual can perform advanced technical functions, instruction, inspection, and supervision. Crew chiefs must be qualified at Level III in their prerequisite MOS.

Description

Summary: Serves as flight crewmember, operating and performing maintenance on the CH-53E helicopter. *Levels I and II:* Performs daily inspections on assigned aircraft; assists in preflight inspections; monitors aircraft performance during flight; acts as a lookout and advises pilot of obstacles and other aircraft; is responsible for passengers and cargo; loads cargo; serves as a mechanic. *Level III:* Able to perform the duties required for Levels I and II; trains subordinates; keeps service records; orders supplies; manages maintenance information. *Level IV:* Able to perform the duties required for Level III; supervises and qualifies subordinates; prepares technical and advisory reports on maintenance efforts.

Recommendation, Level III

In the lower-division baccalaureate/associate degree category, 3 semester hours in records and information management and 1 in technical writing. NOTE: Credit should be added for previously-held MOS (5/92).

Recommendation, Level IV

In the lower-division baccalaureate/associate degree category, 3 semester hours in records and information management and 1 in technical writing. NOTE: Credit should be added for previously-held MOS (5/92).

MCE-6175-001

HELICOPTER CREW CHIEF, UH-N
6175

Exhibit Dates: 5/92–Present.

Occupational Field: 61 (Aircraft Maintenance).

Career Pattern

May progress to MOS 6175 from any MOS in the Aircraft Maintenance (60/61) or Avionics (63) occupational field. LCP: Lance Corporal (E-3). CPL: Corporal (E-4). SGT: Sergeant (E-5). NOTE: MOS duties are not defined in terms of rank or paygrade, but by levels. Level I identifies tasks that are taught in the classroom; the individual is able to perform simple parts of tasks. Level II identifies those tasks that are part of on-the-job training; the trainee can do most tasks pertaining to a particular system and needs supervision on more difficult parts of tasks. Level III indicates that the individual can perform all essential tasks without supervision. Level IV indicates a high level of proficiency in job performance; the individual can perform advanced technical functions, instruction, inspection, and supervision. Crew chiefs must be qualified at Level III in their prerequisite MOS.

Description

Summary: Serves as flight crewmember, operating and performing maintenance on the UN-1N helicopter. *Levels I and II:* Performs daily inspections on assigned aircraft; assists in preflight inspections; monitors aircraft performance during flight; acts as a lookout and advises pilot of obstacles and other aircraft; is responsible for passengers and cargo; loads cargo; serves as a mechanic. *Level III:* Able to perform the duties required for Levels I and II; trains subordinates; keeps service records; orders supplies; manages maintenance information. *Level IV:* Able to perform the duties required for Level III; supervises and qualifies subordinates; prepares technical and advisory reports on maintenance efforts.

Recommendation, Level III

In the lower-division baccalaureate/associate degree category, 3 semester hours in records and information management and 1 in technical writing. NOTE: Credit should be added for previously-held MOS (5/92).

Recommendation, Level IV

In the lower-division baccalaureate/associate degree category, 3 semester hours in records and information management and 1 in technical writing. NOTE: Credit should be added for previously-held MOS (5/92).

MCE-6176-001

HELICOPTER CREW CHIEF, V-22
6176

Exhibit Dates: 5/92–Present.

Occupational Field: 61 (Aircraft Maintenance).

Career Pattern

May progress to MOS 6176 from any MOS in the Aircraft Maintenance (60/61) or Avionics (63) occupational field. LCP: Lance Corporal (E-3). CPL: Corporal (E-4). SGT: Sergeant (E-5). NOTE: MOS duties are not defined in terms of rank or paygrade, but by levels. Level I identifies tasks that are taught in the classroom; the individual is able to perform simple parts of tasks. Level II identifies those tasks that are part of on-the-job training; the trainee can do most tasks pertaining to a particular system and needs supervision on more difficult parts of tasks. Level III indicates that the individual can perform all essential tasks without supervision. Level IV indicates a high level of proficiency in job performance; the individual can perform advanced technical functions, instruction, inspection, and supervision. Crew chiefs must be qualified at Level III in their prerequisite MOS.

Description

Summary: Serves as flight crewmember, operating and performing maintenance on the V-22 helicopter. *Levels I and II:* Performs daily inspections on assigned aircraft; assists in preflight inspections; monitors aircraft performance during flight; acts as a lookout and advises pilot of obstacles and other aircraft; is responsible for passengers and cargo; loads cargo; serves as a mechanic. *Level III:* Able to perform the duties required for Levels I and II; trains subordinates; keeps service records;

orders supplies; manages maintenance information. *Level IV:* Able to perform the duties required for Level III; supervises and qualifies subordinates; prepares technical and advisory reports on maintenance efforts.

Recommendation, Level III

In the lower-division baccalaureate/associate degree category, 3 semester hours in records and information management and 1 in technical writing. NOTE: Credit should be added for previously-held MOS (5/92).

Recommendation, Level IV

In the lower-division baccalaureate/associate degree category, 3 semester hours in records and information management and 1 in technical writing. NOTE: Credit should be added for previously-held MOS (5/92).

MCE-6312-001

AIRCRAFT COMMUNICATIONS/NAVIGATION SYSTEMS TECHNICIAN, A-4/TA-4/QA-4
6312

Exhibit Dates: 1/90–1/98.

Occupational Field: 63 (Avionics).

Career Pattern

PVT: Private (E-1). PFC: Private First Class (E-2). LCP: Lance Corporal (E-3). CPL: Corporal (E-4). SGT: Sergeant (E-5). SSGT: Staff Sergeant (E-6). GYSGT: Gunnery Sergeant (E-7). NOTE: MOS duties are not defined in terms of rank or paygrade, but by levels. Level I identifies tasks that are taught in the classroom; the individual is able to perform simple parts of tasks. Level II identifies those tasks that are part of on-the-job training; the trainee can do most tasks pertaining to a particular system and needs supervision on more difficult parts of tasks. Level III indicates that the individual can perform all essential tasks without supervision. Level IV indicates a high level of proficiency in job performance; the individual can perform advanced technical functions, instruction, inspection, and supervision. The MOS designator 6311 (Aircraft Communications/Navigation Systems Technician—Trainee) is used to identify individuals in Levels I and II. Upon completion of Level II, MOS 6312 is awarded. Credit is recommended for Levels III and IV.

Description

Summary: Installs, removes, inspects, tests, maintains, and repairs components, subsystems, and ancillary equipment of installed aircraft communications, navigation, and deceptive electronic countermeasures systems. *Levels I and II:* Applies theory of operation of vacuum tubes, transistors, solid state devices, integrated circuit components, transformers, motors, generators, and electronic control circuitry, including resistance, capacitance, inductance, as applicable; uses electronic schematics, wiring diagrams, and technical data contained in publications pertaining to avionics systems, test equipment, and aircraft; observes safety precautions while working on and around aircraft; diagnoses and isolates system failures on installed equipment using appropriate test equipment and publications; performs corrective actions for corrosion of electronic components; completes maintenance test forms; operates ground support equipment to perform maintenance of installed systems; performs avionic systems maintenance in accordance with acceptable procedures; performs periodic inspections and completes maintenance repair actions in accordance with appropriate technical documents; performs operational tests of installed systems. *Level III:* Able to perform the duties required for Levels I and II; performs systems maintenance functions, including removal, installation, adjustment, and alignment of system components and repair or replacement of associated aircraft wiring and interconnecting devices; performs inspections of completed maintenance actions; conducts technical training within assigned area; monitors source data collection and assists in the preparation of reports and records; assists in the planning and scheduling of work assignments; supervises maintenance performed on installed systems. *Level IV:* Able to perform the duties required for Level III; possesses a through understanding and working knowledge of sound management principles, workshop supervision, proper training methods, and approved maintenance procedures; coordinates maintenance actions and ensures appropriate maintenance documentation is performed.

Recommendation, Level III

In the lower-division baccalaureate/associate degree category, 3 semester hours in electrical/electronic laboratory, 9 in aircraft electronic maintenance, and 3 in communications/navigation maintenance (8/89).

Recommendation, Level IV

In the lower-division baccalaureate/associate degree category, 3 semester hours in electrical/electronic laboratory, 9 in aircraft electronic maintenance, 3 in communications/navigation maintenance, 3 in personnel supervision, and 3 in maintenance management. In the upper-division baccalaureate category, 2 semester hours for field experience in management, if rank was Staff Sergeant (SSGT) and 3 semester hours for field experience in management, if rank was Gunnery Sergeant (GYSGT) (8/89).

MCE-6313-001

AIRCRAFT COMMUNICATIONS/NAVIGATION SYSTEMS TECHNICIAN, A-6/TC-4C/EA-6A
6313

Exhibit Dates: 1/90–1/98.

Occupational Field: 63 (Avionics).

Career Pattern

PVT: Private (E-1). PFC: Private First Class (E-2). LCP: Lance Corporal (E-3). CPL: Corporal (E-4). SGT: Sergeant (E-5). SSGT: Staff Sergeant (E-6). GYSGT: Gunnery Sergeant (E-7). NOTE: MOS duties are not defined in terms of rank or paygrade, but by levels. Level I identifies tasks that are taught in the classroom; the individual is able to perform simple parts of tasks. Level II identifies those tasks that are part of on-the-job training; the trainee can do most tasks pertaining to a particular system and needs supervision on more difficult parts of tasks. Level III indicates that the individual can perform all essential tasks without supervision. Level IV indicates a high level of proficiency in job performance; the individual can perform advanced technical functions, instruction, inspection, and supervision. The MOS designator 6311 (Aircraft Communications/Navigation Systems Technician—Trainee) is used to identify individuals in Levels I and II. Upon completion of Level II, MOS 6313 is awarded. Credit is recommended for Levels III and IV.

Description

Summary: Installs, removes, inspects, tests, maintains, and repairs components, subsystems, and ancillary equipment on installed aircraft communications, navigation, and deceptive electronic countermeasures systems. *Levels I and II:* Applies theory of operation of vacuum tubes, transistors, solid state devices, integrated circuit components, transformers, motors, generators, and electronic control circuitry, including resistance, capacitance, inductance, as applicable; uses electronic schematics, wiring diagrams, and technical data contained in publications pertaining to avionic systems, test equipment, and aircraft; observes safety precautions while working on and around aircraft; diagnoses and isolates system failures on installed equipment using appropriate test equipment and publications; performs corrective actions for corrosion of electronic components; completes maintenance test forms; operates ground support equipment to perform maintenance of installed systems; performs avionic systems maintenance in accordance with acceptable procedures; performs operational tests of installed systems. *Level III:* Able to perform the duties required for Levels I and II; performs maintenance functions for systems, including removal, installation, adjustment, and alignment of system components and repair or replacement of associated aircraft wiring and interconnecting devices; performs inspections of completed maintenance actions; conducts technical training within assigned area; monitors source data collection and assists in the preparation of reports and records; assists in the planning and scheduling of work assignments; supervises maintenance performed on installed systems. *Level IV:* Able to perform the duties required for Level III; possesses a thorough understanding and working knowledge of sound management principles, workshop supervision, proper training methods, and approved maintenance procedures; coordinates maintenance actions and ensures appropriate maintenance documentation is performed.

Recommendation, Level III

In the lower-division baccalaureate/associate degree category, 3 semester hours in electrical/electronic laboratory, 9 in aircraft electronic maintenance, and 3 in communications/navigation maintenance (8/89).

Recommendation, Level IV

In the lower-division baccalaureate/associate degree category, 3 semester hours in electrical/electronic laboratory, 9 in aircraft electronic maintenance, 3 in communications/navigation maintenance, 3 in personnel supervision, and 3 in maintenance management. In the upper-division baccalaureate category, 2 semester hours for field experience in management, if rank was Staff Sergeant (SSGT); and 3 semester hours for field experience in management, if rank was Gunnery Sergeant (GYSGT) (8/89).

MCE-6314-001

AIRCRAFT COMMUNICATIONS/NAVIGATION SYSTEMS TECHNICIAN, RF-4/F-4
6314

Exhibit Dates: 1/90–1/98.

Occupational Field: 63 (Avionics).

Career Pattern

PVT: Private (E-1). PFC: Private First Class (E-2). LCP: Lance Corporal (E-3). CPL: Corporal (E-4). SGT: Sergeant (E-5). SSGT: Staff Sergeant (E-6). GYSGT: Gunnery Sergeant (E-7). NOTE: MOS duties are not defined in terms of rank or paygrade, but by levels. Level I identifies tasks that are taught in the classroom; the individual is able to perform simple parts of tasks. Level II identifies those tasks that are part of on-the-job training; the trainee can do most tasks pertaining to a particular system and needs supervision on more difficult parts of tasks. Level III indicates that the individual can perform all essential tasks without supervision. Level IV indicates a high level of proficiency in job performance; the individual can perform advanced technical functions, instruction, inspection, and supervision. The MOS designator 6311 (Aircraft Communications/Navigation Systems Technician—Trainee) is used to identify individuals in Levels I and II. Upon completion of Level II, MOS 6314 is awarded. Credit is recommended for Levels III and IV.

Description

Summary: Installs, removes, inspects, tests, maintains, and repairs components, subsystems, and ancillary equipment of installed aircraft communications, navigation, and deceptive electronic countermeasures systems. *Levels I and II:* Applies theory of operation of vacuum tubes, transistors, solid state devices, integrated circuit components, transformers, motors, generators, and electronic control circuitry, including resistance, capacitance, inductance, as applicable; uses electronic schematics, wiring diagrams, and technical data contained in publications pertaining to avionics systems, test equipment, and aircraft; observes safety precautions while working on and around aircraft; diagnoses and isolates systems failure of installed equipment using appropriate test equipment and publications; performs corrective action for corrosion of electronic components; completes maintenance test forms; operates ground support equipment to perform maintenance of installed systems; performs avionics systems maintenance in accordance with acceptable procedures; performs operational tests of installed systems. *Level III:* Able to perform the duties required for Levels I and II; performs maintenance functions for systems, including removal, installation, adjustment, and alignment of system components and repair or replacement of associated aircraft wiring and interconnecting devices; performs inspections of completed maintenance actions; conducts technical training within assigned area; monitors source data collection and assists in the preparation of reports and records; assists in the planning and scheduling of work assignments; supervises maintenance performed on installed systems. *Level IV:* Able to perform the duties required for Level III; possesses a thorough understanding and working knowledge of sound management principles, workshop supervision, proper training methods, and approved maintenance procedures; coordinates maintenance actions and ensures appropriate maintenance documentation is performed.

Recommendation, Level III

In the lower-division baccalaureate/associate degree category, 3 semester hours in electrical/electronic laboratory, 9 in aircraft electronic maintenance, and 3 in communications/navigation maintenance (8/89).

Recommendation, Level IV

In the lower-division baccalaureate/associate degree category, 3 semester hours in electrical/electronic laboratory, 9 in aircraft electronic maintenance, 3 in communications/navigation maintenance, 3 in personnel supervision, and 3 in maintenance management. In the upper-division baccalaureate category, 2 semester hours for field experience in management, if rank was Staff Sergeant (SSGT) and 3 semester hours for field experience in management, if rank was Gunnery Sergeant (GYSGT) (8/89).

MCE-6315-001

AIRCRAFT COMMUNICATIONS/NAVIGATION SYSTEMS TECHNICIAN, AV-8
6315

Exhibit Dates: 1/90–1/98.

Occupational Field: 63 (Avionics).

Career Pattern

PVT: Private (E-1). PFC: Private First Class (E-2). LCP: Lance Corporal (E-3). CPL: Corporal (E-4). SGT: Sergeant (E-5). SSGT: Staff Sergeant (E-6). GYSGT: Gunnery Sergeant (E-7). NOTE: MOS duties are not defined in terms of rank or paygrade, but by levels. Level I identifies tasks that are taught in the classroom; the individual is able to perform simple parts of tasks. Level II identifies those tasks that are part of on-the-job training; the trainee can do most tasks pertaining to a particular system and needs supervision on more difficult parts of tasks. Level III indicates that the individual can perform all essential tasks without supervision. Level IV indicates a high level of proficiency in job performance; the individual can perform advanced technical functions, instruction, inspection, and supervision. The MOS designator 6311 (Aircraft Communications/Navigation Systems Technician—Trainee) is used to identify individuals in Levels I and II. Upon completion of Level II, MOS 6315 is awarded. Credit is recommended for Levels III and IV.

Description

Summary: Installs, removes, inspects, tests, maintains, and repairs components, subsystems, and ancillary equipment of installed aircraft communications, navigation, and deceptive electronic countermeasures systems. *Levels I and II:* Applies theory of operation of vacuum tubes, transistors, solid state devices, integrated circuit components, transformers, motors, generators, and electronic control circuitry, including resistance, capacitance, inductance, as applicable; uses electronic schematics, wiring diagrams, and technical data contained in publications pertaining to avionics systems, test equipment, and aircraft; observes safety precautions while working on and around aircraft; diagnoses and isolates system failures of installed equipment using appropriate test equipment and publications; performs corrective action for corrosion of electronic components; completes maintenance test forms; operates ground support equipment to perform maintenance of installed systems; performs avionics systems maintenance in accordance with acceptable procedures; performs operational tests of installed systems. *Level III:* Able to perform the duties required for Levels I and II; performs maintenance functions for systems, including removal, installation, adjustment, and alignment of system components and repair or replacement of associated aircraft wiring and interconnecting devices; performs inspections of completed maintenance actions; conducts technical training within assigned area; monitors source data collection and assists in the preparation of reports and records; assists in the planning and scheduling of work assignments; supervises maintenance performed on installed systems. *Level IV:* Able to perform the duties required for Level III; possesses a thorough understanding and working knowledge of sound management principles, workshop supervision, proper training methods, and approved maintenance procedures; coordinates maintenance actions and ensures appropriate maintenance documentation is performed.

Recommendation, Level III

In the lower-division baccalaureate/associate degree category, 3 semester hours in electrical/electronic laboratory, 9 in aircraft electronic maintenance, and 3 in communications/navigation maintenance (8/89).

Recommendation, Level IV

In the lower-division baccalaureate/associate degree category, 3 semester hours in electrical/electronic laboratory, 9 in aircraft electronic maintenance, 3 in communications/navigation maintenance, 3 in personnel supervision, and 3 in maintenance management. In the upper-division baccalaureate category, 2 semester hours for field experience in management, if rank was Staff Sergeant (SSGT) and 3 semester hours for field experience in management, if rank was Gunnery Sergeant (GYSGT) (8/89).

MCE-6316-001

AIRCRAFT COMMUNICATIONS/NAVIGATION SYSTEMS TECHNICIAN, KC-130/OV-10
6316

Exhibit Dates: 1/90–1/98.

Occupational Field: 63 (Avionics).

Career Pattern

PVT: Private (E-1). PFC: Private First Class (E-2). LCP: Lance Corporal (E-3). CPL: Corporal (E-4). SGT: Sergeant (E-5). SSGT: Staff Sergeant (E-6). GYSGT: Gunnery Sergeant (E-7). NOTE: MOS duties are not defined in terms of rank or paygrade, but by levels. Level I identifies tasks that are taught in the classroom; the individual is able to perform simple parts of tasks. Level II identifies those tasks that are part of on-the-job training; the trainee can do most tasks pertaining to a particular system and needs supervision on more difficult parts of tasks. Level III indicates that the individual can perform all essential tasks without supervision. Level IV indicates a high level of proficiency in job performance; the individual can perform advanced technical functions, instruction, inspection, and supervision. The MOS designator 6311 (Aircraft Communications/Navigation Systems Technician—Trainee) is used to identify individuals in Levels I and II. Upon completion of Level II, MOS 6316 is awarded. Credit is recommended for Levels III and IV.

Description

Summary: Installs, removes, inspects, tests, maintains, and repairs components, subsystems, and ancillary equipment of installed aircraft communications, navigation, and deceptive electronic countermeasures systems. *Levels I and II:* Applies theory of operation of vacuum tubes, transistors, solid state devices, integrated circuit components, transformers, motors, generators, and electronic control circuitry, including resistance, capacitance, inductance, as applicable; uses electronic schematics, wiring diagrams, and technical data contained in publications pertaining to avionics systems, test equipment, and aircraft; observes safety precautions while working on and around aircraft; diagnoses and isolates system failures on installed equipment using appropriate test equipment and publications; performs corrective action for corrosion of electronic components; completes maintenance test forms; operates ground support equipment to perform maintenance of installed systems; performs avionics systems maintenance in accordance with acceptable procedures; performs operational tests of installed systems. *Level III:* Able to perform the duties required for Levels I and II; performs maintenance functions for systems, including removal, installation, adjustment, and alignment of systems components and repair or replacement of associated aircraft wiring and interconnecting devices; performs inspections of completed maintenance actions; conducts technical training within assigned area; monitors source data collection and assists in the preparation of reports and records; assists in the planning and scheduling of work assignments; supervises maintenance performed on installed systems. *Level IV:* Able to perform the duties required for Level III; possesses a thorough understanding and working knowledge of sound management principles, workshop supervision, proper training methods, and approved maintenance procedures; coordinates maintenance actions and ensures appropriate maintenance documentation is performed.

Recommendation, Level III

In the lower-division baccalaureate/associate degree category, 3 semester hours in electrical/electronic laboratory, 9 in aircraft electronic maintenance, and 3 in communications/navigation maintenance (8/89).

Recommendation, Level IV

In the lower-division baccalaureate/associate degree category, 3 semester hours in electrical/electronic laboratory, 9 in aircraft electronic maintenance, 3 in communications/navigation maintenance, 3 in personnel supervision, and 3 in maintenance management. In the upper-division baccalaureate category, 2 semester hours for field experience in management, if rank was Staff Sergeant (SSGT) and 3 semester hours for field experience in management, if rank was Gunnery Sergeant (GYSGT) (8/89).

MCE-6317-001

AIRCRAFT COMMUNICATIONS/NAVIGATION SYSTEMS TECHNICIAN, F/A-18
6317

Exhibit Dates: 1/90–1/98.

Occupational Field: 63 (Avionics).

Career Pattern

PVT: Private (E-1). PFC: Private First Class (E-2). LCP: Lance Corporal (E-3). CPL: Corporal (E-4). SGT: Sergeant (E-5). SSGT: Staff Sergeant (E-6). GYSGT: Gunnery Sergeant (E-7). NOTE: MOS duties are not defined in terms of rank or paygrade, but by levels. Level I identifies tasks that are taught in the classroom; the individual is able to perform simple parts of tasks. Level II identifies those tasks that are part of on-the-job training; the trainee can do most tasks pertaining to a particular system and needs supervision on more difficult parts of tasks. Level III indicates that the individual can perform all essential tasks without supervision. Level IV indicates a high level of proficiency in job performance; the individual can perform advanced technical functions, instruction, inspection, and supervision. The MOS designator 6311 (Aircraft Communications/Navigation Systems Technician—Trainee) is used to identify individuals in Levels I and II. Upon completion of Level II, MOS 6317 is awarded. Credit is recommended for Levels III and IV.

Description

Summary: Installs, removes, inspects, tests, maintains, and repairs components, subsystems, and ancillary equipment on installed aircraft communications, navigation, and deceptive electronic countermeasures systems. *Levels I and II:* Applies theory of operation of vacuum tubes, transistors, solid state devices, integrated circuit components, transformers, motors, generators, and electronic control circuitry, including resistance, capacitance, inductance, as applicable; uses electronic schematics, wiring diagrams, and technical data contained in publications pertaining to avionics systems, test equipment, and aircraft; observes safety precautions while working on and around aircraft; diagnoses and isolates system failures on installed equipment using appropriate test equipment and publications; performs corrective action for corrosion of electronic components; completes maintenance test forms; operates ground support equipment to perform maintenance of installed systems; performs avionic systems maintenance in accordance with acceptable procedures; performs operational tests of installed systems. *Level III:* Able to perform the duties required for Levels I and II; performs maintenance functions for systems, including removal, installation, adjustment, and alignment of systems components and repair or replacement of associated aircraft wiring and interconnecting devices; performs inspections of completed maintenance actions; conducts technical training within assigned area; monitors source data collection and assists in the preparation of reports and records; assists in the planning and scheduling of work assignments; supervises maintenance performed on installed systems. *Level IV:* Able to perform the duties required for Level III; possesses a thorough understanding and working knowledge of sound management principles, workshop supervision, proper training methods, and approved maintenance procedures; coordinates maintenance actions and ensures appropriate maintenance documentation is performed.

Recommendation, Level III

In the lower-division baccalaureate/associate degree category, 3 semester hours in electrical/electronic laboratory, 9 in aircraft electronic maintenance, and 3 in communications/navigation maintenance (8/89).

Recommendation, Level IV

In the lower-division baccalaureate/associate degree category, 3 semester hours in electrical/electronic laboratory, 9 in aircraft electronic maintenance, 3 in communications/navigation maintenance, 3 in personnel supervision, and 3 in maintenance management. In the upper-division baccalaureate category, 2 semester hours for field experience in management, if rank was Staff Sergeant (SSGT) and 3 semester hours for field experience in management, if rank was Gunnery Sergeant (GYSGT) (8/89).

MCE-6318-001

AIRCRAFT COMMUNICATIONS/NAVIGATION SYSTEMS TECHNICIAN, OV-10
6318

Exhibit Dates: 1/90–1/98.

Occupational Field: 63 (Avionics).

Career Pattern

PVT: Private (E-1). PFC: Private First Class (E-2). LCP: Lance Corporal (E-3). CPL: Corporal (E-4). SGT: Sergeant (E-5). SSGT: Staff Sergeant (E-6). GYSGT: Gunnery Sergeant (E-7). NOTE: MOS duties are not defined in terms of rank or paygrade, but by levels. Level I identifies tasks that are taught in the classroom; the individual is able to perform simple parts of tasks. Level II identifies those tasks that are part of on-the-job training; the trainee can do most tasks pertaining to a particular system and needs supervision on more difficult parts of tasks. Level III indicates that the individual can perform all essential tasks without supervision. Level IV indicates a high level of proficiency in job performance; the individual can perform advanced technical functions, instruction, inspection, and supervision. The MOS designator 6311 (Aircraft Communications/Navigation Systems Technician—Trainee) is used to identify individuals in Levels I and II. Upon completion of Level II, MOS 6318 is awarded. Credit is recommended for Levels III and IV.

Description

Summary: Installs, removes, inspects, tests, maintains, and repairs components, subsystems, and ancillary equipment of installed aircraft communications, navigation, and deceptive electronic countermeasures systems. *Levels I and II:* Applies theory of operation of vacuum tubes, transistors, solid state devices, integrated circuit components, transformers, motors, generators, and electronic control circuitry, including resistance, capacitance, inductance, as applicable; uses electronic schematics, wiring diagrams, and technical data contained in publications pertaining to avionics systems, test equipment, and aircraft; observes safety precautions while working on and around aircraft; diagnoses and isolates system failures on installed equipment using appropriate test equipment and publications; performs corrective action for corrosion of electronic components; completes maintenance test forms; operates ground support equipment to perform maintenance of installed systems; performs avionic systems maintenance in accordance with acceptable procedures; performs periodic inspections and completes maintenance repair actions in accordance with appropriate technical documents; performs operational tests of installed systems. *Level III:* Able to perform the duties required for Levels I and II; performs maintenance functions for systems, including removal, installation, adjustment, and alignment of system components and repair or replacement of associated aircraft wiring and interconnecting devices; performs inspections of completed maintenance actions; conducts technical training within assigned area; monitors source data collection and assists in the preparation of reports and records; assists in the planning and scheduling of work assignments; supervises maintenance performed on installed systems. *Level IV:* Able to perform the duties required for Level III; possesses a through understanding and working knowledge of sound management principles, workshop supervision, proper training methods, and approved maintenance procedures; coordinates maintenance actions and ensures appropriate maintenance documentation is performed.

Recommendation, Level III

In the lower-division baccalaureate/associate degree category, 3 semester hours in electrical/electronic laboratory, 9 in aircraft electronic maintenance, and 3 in communications/navigation maintenance (8/89).

Recommendation, Level IV

In the lower-division baccalaureate/associate degree category, 3 semester hours in electrical/electronic laboratory, 9 in aircraft electronic maintenance, 3 in communications/navigation maintenance, 3 in personnel supervision, and 3 in maintenance management. In the upper-division baccalaureate category, 2 semester hours for field experience in management, if rank was Staff Sergeant (SSGT) and 3 semester hours for field experience in management, if rank was Gunnery Sergeant (GYSGT) (8/89).

MCE-6322-001

AIRCRAFT COMMUNICATIONS/NAVIGATION SYSTEMS TECHNICIAN (CH-46)
6322

Exhibit Dates: 1/90–1/98.

Occupational Field: 63 (Avionics).

Career Pattern

PVT: Private (E-1). PFC: Private First Class (E-2). LCP: Lance Corporal (E-3). CPL: Corporal (E-4). SGT: Sergeant (E-5). SSGT: Staff Sergeant (E-6). GYSGT: Gunnery Sergeant (E-7). NOTE: MOS duties are not defined in terms of rank or paygrade, but by levels. Level I identifies tasks that are taught in the classroom; the individual is able to perform simple parts of tasks. Level II identifies those tasks that are part of on-the-job training; the trainee can do most tasks pertaining to a particular system and needs supervision on more difficult parts of tasks. Level III indicates that the individual can perform all essential tasks without supervision. Level IV indicates a high level of proficiency in job performance; the individual can perform advanced technical functions, instruction, inspection, and supervision. The MOS designator 6311 (Aircraft Communications/Navigation Systems Technician—Trainee) is used to identify individuals in Levels I and II. Upon completion of Level II, MOS 6322 is awarded. Credit is recommended for Levels III and IV.

Description

Summary: Installs, removes, inspects, tests, maintains, and repairs components, subsystems, and ancillary equipment of installed aircraft communications, navigation, and deceptive electronic countermeasures systems. *Levels I and II:* Applies theory of operation of vacuum tubes, transistors, solid state devices, integrated circuit components, transformers, motors, generators, and electronic control circuitry, including resistance, capacitance, and inductance, as applicable; uses electronic schematics, wiring diagrams, and technical data contained in publications pertaining to avionics systems, test equipment, and aircraft; observes safety precautions while working on and around aircraft; diagnoses and isolates system failures on installed equipment using appropriate test equipment and publications; performs corrective action for corrosion of electronic components; completes maintenance test forms; operates ground support equipment to perform maintenance of installed systems; performs avionic systems maintenance in accordance with acceptable procedures; performs periodic inspections and completes maintenance repair actions in accordance with acceptable procedures; performs periodic inspections and completes maintenance repair actions in accordance with appropriate technical documents; performs operational tests of installed systems. *Level III:* Able to perform the duties required for Levels I and II; performs maintenance functions for systems, including removal, installation, adjustment, and alignment of system components and repair or replacement of associated aircraft wiring and interconnecting devices; performs inspections of completed maintenance actions; conducts technical training within assigned area; monitors source data collection and assists in the preparation of reports and records; assists in the planning and scheduling of work assignments; supervises maintenance performed on installed systems. *Level IV:* Able to perform the duties required for Level III; possesses a thorough understanding and working knowledge of sound management principles, workshop supervision, proper training methods, and approved maintenance procedures; coordinates maintenance actions and ensures appropriate maintenance documentation is performed.

Recommendation, Level III

In the lower-division baccalaureate/associate degree category, 3 semester hours in electrical/electronic laboratory, 9 in aircraft electronic maintenance, and 3 in communications/navigation maintenance (12/88).

Recommendation, Level IV

In the lower-division baccalaureate/associate degree category, 3 semester hours in electrical/electronic laboratory, 9 in aircraft electronic maintenance, 3 in communications/navigation maintenance, 3 in personnel supervision, and 3 in maintenance management. In the upper-division baccalaureate category, 2 semester hours for field experience in management, if rank was Staff Sergeant (SSGT) and 3 semester hours for field experience in management, if rank was Gunnery Sergeant (GYSGT) (12/88).

MCE-6323-001

AIRCRAFT COMMUNICATIONS/NAVIGATION SYSTEMS TECHNICIAN (CH-53)
6323

Exhibit Dates: 1/90–1/98.

Occupational Field: 63 (Avionics).

Career Pattern

PVT: Private (E-1). PFC: Private First Class (E-2). LCP: Lance Corporal (E-3). CPL: Corporal (E-4). SGT: Sergeant (E-5). SSGT: Staff Sergeant (E-6). GYSGT: Gunnery Sergeant (E-7). NOTE: MOS duties are not defined in terms of rank or paygrade, but by levels. Level I identifies tasks that are taught in the classroom; the individual is able to perform simple parts of tasks. Level II identifies those tasks that are part of on-the-job training; the trainee can do most tasks pertaining to a particular system and needs supervision on more difficult parts of tasks. Level III indicates that the individual can perform all essential tasks without supervision. Level IV indicates a high level of proficiency in job performance; the individual can perform advanced technical functions, instruction, inspection, and supervision. The MOS designator 6311 (Aircraft Communications/Navigation Systems Technician—Trainee) is used to identify individuals in Levels I and II. Upon completion of Level II, MOS 6323 is awarded. Credit is recommended for Levels III and IV.

Description

Summary: Installs, removes, inspects, tests, maintains, and repairs components, subsystems, and ancillary equipment of installed aircraft communications, navigation, and deceptive electronic countermeasures systems. *Levels I and II:* Applies theory of operation of vacuum tubes, transistors, solid state devices, integrated circuit components, transformers, motors, generators, and electronic control circuitry, including resistance, capacitance, inductance, as applicable; uses electronic schematics, wiring diagrams, and technical data contained in publications pertaining to avionics systems, test equipment, and aircraft; observes safety precautions while working on and around aircraft; diagnoses and isolates system failures on installed equipment using appropriate test equipment and publications; performs corrective action for corrosion of electronic components; completes maintenance test forms; operates ground support equipment to perform maintenance on installed systems; performs avionic systems maintenance in accordance with acceptable procedures; performs operational tests on installed systems. *Level III:* Able to perform the duties required for Levels I and II; performs maintenance functions for systems, including removal, installation, adjustment, and alignment of system components and repair or replacement of associated aircraft wiring and interconnecting devices; performs inspections of completed maintenance actions; conducts technical training within assigned area; monitors source data collection and assists in the preparation of reports and records; assists in the planning and scheduling of work assignments; supervises maintenance performed on installed systems. *Level IV:* Able to perform the duties required for Level III; possesses a thorough understanding and working knowledge of sound management principles, workshop supervision, proper training methods, and approved maintenance procedures; coordinates maintenance actions and ensures appropriate maintenance documentation is performed.

Recommendation, Level III

In the lower-division baccalaureate/associate degree category, 3 semester hours in electrical/electronic laboratory, 9 in aircraft electronic maintenance, and 3 in communications/navigation maintenance (8/89).

Recommendation, Level IV

In the lower-division baccalaureate/associate degree category, 3 semester hours in electrical/electronic laboratory, 9 in aircraft electronic maintenance, 3 in communications/navigation maintenance, 3 in personnel supervision, and 3 in maintenance management. In the upper-division baccalaureate category, 2 semester hours for field experience in management, if rank was Staff Sergeant (SSGT) and 3 semester hours for field experience in management, if rank was Gunnery Sergeant (GYSGT) (8/89).

MCE-6324-001

AIRCRAFT COMMUNICATIONS/NAVIGATION SYSTEMS TECHNICIAN, (U/AH-1)
6324

Exhibit Dates: 1/90–1/98.

Occupational Field: 63 (Avionics).

Career Pattern

PVT: Private (E-1). PFC: Private First Class (E-2). LCP: Lance Corporal (E-3). CPL: Corporal (E-4). SGT: Sergeant (E-5). SSGT: Staff Sergeant (E-6). GYSGT: Gunnery Sergeant (E-7). NOTE: MOS duties are not defined in terms of rank or paygrade, but by levels. Level I identifies tasks that are taught in the classroom; the individual is able to perform simple parts of tasks. Level II identifies those tasks that are part of on-the-job training; the trainee can do most tasks pertaining to a particular system and needs supervision on more difficult parts of tasks. Level III indicates that the individual can perform all essential tasks without supervision. Level IV indicates a high level of proficiency in job performance; the individual can perform advanced technical functions, instruction, inspection, and supervision. The MOS designator 6311 (Aircraft Communications/Navigation Systems Technician—Trainee) is used to identify individuals in Levels I and II. Upon completion of Level II, MOS 6324 is awarded. Credit is recommended for Levels III and IV.

Description

Summary: Installs, removes, inspects, tests, maintains, and repairs components, subsystems, and ancillary equipment of installed aircraft communications, navigation, and deceptive electronic countermeasures systems. *Levels I and II:* Applies theory of operation of vacuum tubes, transistors, solid state devices, integrated circuit components, transformers, motors, generators, and electronic control circuitry, including resistance, capacitance, inductance, as applicable; uses electronic schematics, wiring diagrams, and technical data contained in publications pertaining to avionic systems, test equipment, and aircraft; observes safety precautions while working on and around aircraft; diagnoses and isolates system failures on installed equipment using appropriate test equipment and publications; performs corrective action for corrosion of electronic components; completes maintenance test forms; operates ground support equipment to perform maintenance of installed systems; performs avionic systems maintenance in accordance with acceptable procedures; performs operational tests on installed systems. *Level III:* Able to perform the duties required for Levels I and II; performs maintenance functions for systems, including removal, installation, adjustment, and alignment of system components and repair or replacement of associated aircraft wiring and interconnecting devices; performs inspections of completed maintenance actions; conducts technical training within assigned area; monitors source data collection and assists in the preparation of reports and records; assists in the planning and scheduling of work assignments; supervises maintenance performed on installed systems. *Level IV:* Able to perform the duties required for Level III; possesses a thorough understanding and working knowledge of sound management principles, workshop supervision, proper training methods, and approved maintenance procedures; coordinates maintenance actions and ensures appropriate maintenance documentation is performed.

Recommendation, Level III

In the lower-division baccalaureate/associate degree category, 3 semester hours in electrical/electronic laboratory, 9 in aircraft electronic maintenance, and 3 in communications/navigation maintenance (8/89).

Recommendation, Level IV

In the lower-division baccalaureate/associate degree category, 3 semester hours in electrical/electronic laboratory, 9 in aircraft electronic maintenance, 3 in communications/navigation maintenance, 3 in personnel supervision, and 3 in maintenance management. In the upper-division baccalaureate category, 2 semester hours for field experience in management, if rank was Staff Sergeant (SSGT) and 3 semester hours for field experience in management, if rank was Gunnery Sergeant (GYSGT) (8/89).

MCE-6325-001

AIRCRAFT COMMUNICATIONS/NAVIGATION SYSTEMS TECHNICIAN V-22
6325

Exhibit Dates: 1/90–1/98.

Occupational Field: 63 (Avionics).

Career Pattern

PVT: Private (E-1). PFC: Private First Class (E-2). LCP: Lance Corporal (E-3). CPL: Corporal (E-4). SGT: Sergeant (E-5). SSGT: Staff Sergeant (E-6). GYSGT: Gunnery Sergeant (E-7). NOTE: MOS duties are not defined in terms of rank or paygrade, but by levels. Level I identifies tasks that are taught in the classroom; the individual is able to perform simple parts of tasks. Level II identifies those tasks that are part of on-the-job training; the trainee can do most tasks pertaining to a particular system and needs supervision on more difficult parts of tasks. Level III indicates that the individual can perform all essential tasks without supervision. Level IV indicates a high level of proficiency in job performance; the individual can perform advanced technical functions, instruction, inspection, and supervision. The MOS designator 6311 (Aircraft Communications/Navigation Systems Technician—Trainee) is used to identify individuals in Levels I and II. Upon completion of Level II, MOS 6325 is awarded. Credit is recommended for Levels III and IV.

Description

Summary: Installs, removes, inspects, tests, maintains, and repairs components, subsystems, and ancillary equipment of installed aircraft communications, navigation, and deceptive electronic countermeasures systems. *Levels I and II:* Applies theory of operation of vacuum tubes, transistors, solid state devices, integrated circuit components, transformers, motors, generators, and electronic control circuitry, including resistance, capacitance, and inductance, as applicable; uses electronic schematics, wiring diagrams and technical data contained in publications pertaining to avionic systems, test equipment, and aircraft; observes safety precautions while working on and around aircraft; diagnoses and isolates system failures of installed equipment using appropriate test equipment and publications; performs corrective action for corrosion of electronic components; completes maintenance test forms; operates ground support equipment to perform maintenance of installed systems; performs avionic systems maintenance in accordance with acceptable procedures; performs periodic inspections and completes maintenance repair actions in accordance with acceptable procedures and technical documents; performs operational tests of installed systems. *Level III:* Able to perform the duties required for Levels I and II; performs maintenance functions for systems, including removal, installation, adjustment, and alignment of systems components and repair on replacement of associated aircraft wiring and interconnecting devices; performs inspections of completed maintenance actions; conducts technical training within assigned area; monitors source data collection and assists in the preparation of reports and records; assists in the planning and scheduling of work assignments; supervises maintenance performed on installed systems. *Level IV:* Able to perform the duties required for Level III; possesses a thorough understanding and working knowledge of sound management principles, workshop supervision, proper training methods, and approved maintenance procedures; coordinates maintenance actions and ensures appropriate maintenance documentation is performed.

Recommendation, Level III

In the lower-division baccalaureate/associate degree category, 3 semester hours in electrical/electronic laboratory, 9 in aircraft electronic maintenance, and 3 in communications/navigation maintenance (12/88).

Recommendation, Level IV

In the lower-division baccalaureate/associate degree category, 3 semester hours in electrical/electronic laboratory, 9 in aircraft electronic maintenance, 3 in communications/navigation maintenance, 3 in personnel supervision, and 3 in maintenance management. In the upper-division baccalaureate category, 2 semester hours for field experience in management, if rank was Staff Sergeant (SSGT) and 3 semester hours for field experience in management, if rank was Gunnery Sergeant (GYSGT) (12/88).

MCE-6333-001

AIRCRAFT ELECTRICAL SYSTEMS TECHNICIAN, A-6/EA-6/TC-4C
6333

Exhibit Dates: 1/90–1/98.

Occupational Field: 63 (Avionics).

Career Pattern

PVT: Private (E-1). PFC: Private First Class (E-2). LCP: Lance Corporal (E-3). CPL: Corporal (E-4). SGT: Sergeant (E-5). SSGT: Staff Sergeant (E-6). GYSGT: Gunnery Sergeant (E-7). NOTE: MOS duties are not defined in terms of rank or paygrade, but by levels. Level I identifies tasks that are taught in the classroom; the individual is able to perform simple parts of tasks. Level II identifies those tasks that are part of on-the-job training; the trainee can do most tasks pertaining to a particular system and needs supervision on more difficult parts of tasks. Level III indicates that the individual can perform all essential tasks without supervision. Level IV indicates a high level of proficiency in job performance; the individual can perform advanced technical functions, instruction, inspection, and supervision. The MOS designator 6331 (Aircraft Electrical Systems Technician—Trainee) is used to identify individuals in Levels I and II. Upon completion of Level II, MOS 6333 is awarded. Credit is recommended for Levels III and IV.

Description

Summary: Installs, removes, inspects, tests, maintains, and repairs aircraft electrical components, systems, and auxiliary equipment. *Levels*

I and II: Operates and maintains applicable shop and ground support equipment; practices ground safety procedures and precautions in flight-line or hangar operations or in test areas where dangerous voltages or other conditions may exist; demonstrates and applies the knowledge of wire repair; assists in the inspection and testing of electrical components by applying the theory of operation of transistors, diodes, solid state components, integrated circuits, motors, servos, power transmissions, and other electrical devices; demonstrates the use of electrical schematics, wiring diagrams, color codes, publications, and other technical data; under supervision and by using appropriate test equipment, diagnoses failures on aircraft electrical systems, including the power supply system, lighting and control systems, power plant system, and automatic flight control system. *Level III:* Able to perform the duties required for Levels I and II; performs corrective action on corroded or deteriorated electronic components; performs the removal, installation, adjustment, alignment, and replacement of components and wiring in interconnecting devices; conducts preflight and postflight operational tests on the aircraft electrical system. *Level IV:* Able to perform the duties required for Level III; possesses a thorough understanding of sound management principles, work center supervision, proper methods of training personnel, and approved electrical systems maintenance; organizes and administers facilities for the maintenance and repair of aircraft electrical systems; performs quality assurance inspections for maintenance and repair of electrical equipment; assists in the planning and supervision of work center assignments.

Recommendation, Level III

In the lower-division baccalaureate/associate degree category, 3 semester hours in electrical/electronic laboratory, 9 in aircraft electrical maintenance, and 3 in electromechanical laboratory (8/89).

Recommendation, Level IV

In the lower-division baccalaureate/associate degree category, 3 semester hours in electrical/electronic laboratory, 9 in aircraft electrical maintenance, 3 in electromechanical laboratory, 3 in personnel supervision, and 3 in maintenance management. In the upper-division baccalaureate category, 2 semester hours for field experience in management, if rank was Staff Sergeant (SSGT) and 3 semester hours for field experience in management, if rank was Gunnery Sergeant (GYSGT) (8/89).

MCE-6335-001

AIRCRAFT ELECTRICAL SYSTEMS TECHNICIAN, AV-8

6335

Exhibit Dates: 1/90–1/98.

Occupational Field: 63 (Avionics).

Career Pattern

PVT: Private (E-1). PFC: Private First Class (E-2). LCP: Lance Corporal (E-3). CPL: Corporal (E-4). SGT: Sergeant (E-5). SSGT: Staff Sergeant (E-6). GYSGT: Gunnery Sergeant (E-7). NOTE: MOS duties are not defined in terms of rank or paygrade, but by levels. Level I identifies tasks that are taught in the classroom; the individual is able to perform simple parts of tasks. Level II identifies those tasks that are part of on-the-job training; the trainee can do most tasks pertaining to a particular system and needs supervision on more difficult parts of tasks. Level III indicates that the individual can perform all essential tasks without supervision. Level IV indicates a high level of proficiency in job performance; the individual can perform advanced technical functions, instruction, inspection, and supervision. The MOS designator 6331 (Aircraft Electrical Systems Technician—Trainee) is used to identify individuals in Levels I and II. Upon completion of Level II, MOS 6335 is awarded. Credit is recommended for Levels III and IV.

Description

Summary: Installs, removes, inspects, tests, maintains, and repairs aircraft electrical components, systems, and auxiliary equipment. *Levels I and II:* Operates and maintains applicable shop and ground support equipment; practices ground safety procedures and precautions in flight-line or hangar operations or in test areas where dangerous voltages or other conditions may exist; demonstrates and applies the knowledge of wire repair; assists in the inspection and testing of electrical components by applying the theory of operation of transistors, diodes, solid state components, integrated circuits, motors, servos, power transmissions, and other electrical devices; demonstrates the use of electrical schematics, wiring diagrams, color codes, publications, and other technical data; under supervision and by using appropriate test equipment, diagnoses failures on aircraft electrical systems, including the power supply system, lighting and control systems, power plant system, and automatic flight control system. *Level III:* Able to perform the duties required for Levels I and II; performs corrective action on corroded or deteriorated electronic components; performs the removal, installation, adjustment, alignment, and replacement of components and wiring in interconnecting devices; conducts preflight and postflight operational tests on the aircraft electrical system. *Level IV:* Able to perform the duties required for Level III; possesses a thorough understanding of sound management principles, work center supervision, proper methods of training personnel, and approved electrical systems maintenance; organizes and administers facilities for the maintenance and repair of aircraft electrical systems; performs quality assurance inspections for maintenance and repair of electrical equipment; assists in the planning and supervision of work center assignments.

Recommendation, Level III

In the lower-division baccalaureate/associate degree category, 3 semester hours in electrical/electronic laboratory, 9 in aircraft electrical maintenance, and 3 in electromechanical laboratory (8/89).

Recommendation, Level IV

In the lower-division baccalaureate/associate degree category, 3 semester hours in electrical/electronic laboratory, 9 in aircraft electrical maintenance, 3 in electromechanical laboratory, 3 in personnel supervision, and 3 in maintenance management. In the upper-division baccalaureate category, 2 semester hours for field experience in management, if rank was Staff Sergeant (SSGT) and 3 semester hours for field experience in management, if rank was Gunnery Sergeant (GYSGT) (8/89).

MCE-6336-001

AIRCRAFT ELECTRICAL SYSTEMS TECHNICIAN, KC-130/OV-10

6336

Exhibit Dates: 1/90–1/98.

Occupational Field: 63 (Avionics).

Career Pattern

PVT: Private (E-1). PFC: Private First Class (E-2). LCP: Lance Corporal (E-3). CPL: Corporal (E-4). SGT: Sergeant (E-5). SSGT: Staff Sergeant (E-6). GYSGT: Gunnery Sergeant (E-7). NOTE: MOS duties are not defined in terms of rank or paygrade, but by levels. Level I identifies tasks that are taught in the classroom; the individual is able to perform simple parts of tasks. Level II identifies those tasks that are part of on-the-job training; the trainee can do most tasks pertaining to a particular system and needs supervision on more difficult parts of tasks. Level III indicates that the individual can perform all essential tasks without supervision. Level IV indicates a high level of proficiency in job performance; the individual can perform advanced technical functions, instruction, inspection, and supervision. The MOS designator 6331 (Aircraft Electrical Systems Technician—Trainee) is used to identify individuals in Levels I and II. Upon completion of Level II, MOS 6336 is awarded. Credit is recommended for Levels III and IV.

Description

Summary: Installs, removes, inspects, tests, maintains, and repairs aircraft electrical components, systems, and auxiliary equipment. *Levels I and II:* Operates and maintains applicable shop and ground support equipment; practices ground safety procedures and precautions in flight-line or hangar operations or in test areas where dangerous voltages or other conditions may exist; demonstrates and applies knowledge of wire repair; assists in the inspection and testing of electrical components by applying the theory of operation of transistors, diodes, solid state components, integrated circuits, motors, servos, power transmissions, and other electrical devices; demonstrates the use of electrical schematics, wiring diagrams, color codes, publications, and other technical data; under supervision and by using appropriate test equipment, diagnoses failures on aircraft electrical systems, including the power supply system, lighting and control systems, power plant system, and automatic flight control system. *Level III:* Able to perform the duties required for Levels I and II; performs corrective action on corroded or deteriorated electronic components; performs the removal, installation, adjustment, alignment, and replacement of components and wiring in interconnecting devices; conducts preflight and postflight operational tests on the aircraft electrical system. *Level IV:* Able to perform the duties required for Level III; possesses a thorough understanding of sound management principles, work center supervision, proper methods of training personnel, and approved electrical systems maintenance; organizes and administers facilities for the maintenance and repair of aircraft electrical systems; performs quality assurance inspections for maintenance and repair of electrical equipment; assists in the planning and supervision of work center assignments.

Recommendation, Level III

In the lower-division baccalaureate/associate degree category, 3 semester hours in electrical/electronic laboratory, 9 in aircraft electrical maintenance, and 3 in electromechanical laboratory (8/89).

Recommendation, Level IV

In the lower-division baccalaureate/associate degree category, 3 semester hours in electrical/electronic laboratory, 9 in aircraft electrical maintenance, 3 in electromechanical laboratory, 3 in personnel supervision, and 3 in maintenance management. In the upper-division baccalaureate category, 2 semester hours for field experience in management, if rank was Staff Sergeant (SSGT) and 3 semester hours for field experience in management, if rank was Gunnery Sergeant (GYSGT) (8/89).

MCE-6337-001

AIRCRAFT ELECTRICAL SYSTEMS TECHNICIAN, F/A-18
6337

Exhibit Dates: 1/90–1/98.

Occupational Field: 63 (Avionics).

Career Pattern

PVT: Private (E-1). PFC: Private First Class (E-2). LCP: Lance Corporal (E-3). CPL: Corporal (E-4). SGT: Sergeant (E-5). SSGT: Staff Sergeant (E-6). GYSGT: Gunnery Sergeant (E-7). NOTE: MOS duties are not defined in terms of rank or paygrade, but by levels. Level I identifies tasks that are taught in the classroom; the individual is able to perform simple parts of tasks. Level II identifies those tasks that are part of on-the-job training; the trainee can do most tasks pertaining to a particular system and needs supervision on more difficult parts of tasks. Level III indicates that the individual can perform all essential tasks without supervision. Level IV indicates a high level of proficiency in job performance; the individual can perform advanced technical functions, instruction, inspection, and supervision. The MOS designator 6331 (Aircraft Electrical Systems Technician—Trainee) is used to identify individuals in Levels I and II. Upon completion of Level II, MOS 6337 is awarded. Credit is recommended for Levels III and IV.

Description

Summary: Installs, removes, inspects, tests, maintains, and repairs aircraft electrical components, systems, and auxiliary equipment. *Levels I and II:* Operates and maintains applicable shop and ground support equipment; practices ground safety procedures and precautions in flight-line or hangar operations or in test areas where dangerous voltages or other conditions may exist; demonstrates and applies knowledge of wire repair; assists in the inspection and testing of electrical components by applying the theory of operation of transistors, diodes, solid state components, integrated circuits, motors, servos, power transmissions, and other electrical devices; demonstrates the use of electrical schematics, wiring diagrams, color codes, publications, and other technical data; under supervision and by using appropriate test equipment, diagnoses failures on aircraft electrical systems, including the power supply system, lighting and control systems, power plant system, and automatic flight control system. *Level III:* Able to perform the duties required for Levels I and II; performs corrective action on corroded or deteriorated electronic components; performs the removal, installation, adjustment, alignment, and replacement of components and wiring in interconnecting devices; conducts preflight and postflight operational tests on the aircraft electrical system. *Level IV:* Able to perform the duties required for Level III; possesses a thorough understanding of sound management principles, work center supervision, proper methods of training personnel, and approved electrical systems maintenance; organizes and administers facilities for the maintenance and repair of aircraft electrical systems; performs quality assurance inspections for maintenance and repair of electrical equipment; assists in the planning and supervision of work center assignments.

Recommendation, Level III

In the lower-division baccalaureate/associate degree category, 3 semester hours in electrical/electronic laboratory, 9 in aircraft electrical maintenance, and 3 in electromechanical laboratory (8/89).

Recommendation, Level IV

In the lower-division baccalaureate/associate degree category, 3 semester hours in electrical/electronic laboratory, 9 in aircraft electrical maintenance, 3 in electromechanical laboratory, 3 in personnel supervision, and 3 in maintenance management. In the upper-division baccalaureate category, 2 semester hours for field experience in management, if rank was Staff Sergeant (SSGT) and 3 semester hours for field experience in management, if rank was Gunnery Sergeant (GYSGT) (8/89).

MCE-6353-001

AIRCRAFT WEAPONS SYSTEMS SPECIALIST, A-6/TC-4C
6353

Exhibit Dates: 1/90–1/98.

Occupational Field: 63 (Avionics).

Career Pattern

PVT: Private (E-1). PFC: Private First Class (E-2). LCP: Lance Corporal (E-3). CPL: Corporal (E-4). SGT: Sergeant (E-5). SSGT: Staff Sergeant (E-6). GYSGT: Gunnery Sergeant (E-7). NOTE: MOS duties are not defined in terms of rank or paygrade, but by levels. Level I identifies tasks that are taught in the classroom; the individual is able to perform simple parts of tasks. Level II identifies those tasks that are part of on-the-job training; the trainee can do most tasks pertaining to a particular system and needs supervision on more difficult parts of tasks. Level III indicates that the individual can perform all essential tasks without supervision. Level IV indicates a high level of proficiency in job performance; the individual can perform advanced technical functions, instruction, inspection, and supervision. The MOS designator 6351 (Advanced Avionics Technician—Trainee), is used to identify individuals in Levels I and II. Upon completion of Level II, MOS 6353 is awarded. Credit is recommended for Levels III and IV.

Description

Summary: Inspects, removes, tests, repairs, installs, and maintains components, subsystems, and related equipment of aircraft attack systems. *Levels I and II:* Applies operating theory of vacuum tubes, transistors, generators, synchros, servos, electromechanical drive mechanisms, transmission lines, and antenna and theory of impedance, resistance, capacitance, and inductance; uses electronic schematics, block diagrams, wiring diagrams, color codes, and other technical data in appropriate technical publications; uses technical publications to obtain data for supply purposes; observes safety precautions; demonstrates working knowledge of first aid; isolates and diagnoses system failures; operates ground support equipment used in maintenance; possesses necessary operating licenses and performs maintenance and inspections on ground support equipment; performs maintenance action including inspection and repair according to applicable technical publications. *Level III:* Able to perform the duties required for Levels I and II; performs operational tests of installed systems; removes, adjusts, aligns, and installs components and repairs or replaces associated wiring and interconnecting devices; performs necessary corrective action on corroded electrical components. *Level IV:* Able to perform the duties required for Level III; performs quality assurance inspection for maintenance actions; plans and schedules workshop assignments; supervises maintenance actions; conducts technical training; monitors collection of source data and completes reports and records; possesses working knowledge of sound management principles, workshop supervision, proper personnel training methods, and accepted avionics maintenance procedures; organizes and administers maintenance and repair facilities.

Recommendation, Level III

In the lower-division baccalaureate/associate degree category, 3 semester hours in electronic systems troubleshooting, 3 in electromechanical systems troubleshooting, and 3 in industrial safety (8/89).

Recommendation, Level IV

In the lower-division baccalaureate/associate degree category, 3 semester hours in electronic systems troubleshooting, 3 in electromechanical systems troubleshooting, 3 in industrial safety, 3 in maintenance management, and 3 in personnel supervision. In the upper-division baccalaureate category, 2 semester hours for field experience in management, if rank was Staff Sergeant (SSGT) and 3 semester hours for field experience in management, if rank was Gunnery Sergeant (GYSGT) (8/89).

MCE-6354-001

AIRCRAFT WEAPONS SYSTEMS SPECIALIST, F-4J/S
6354

Exhibit Dates: 1/90–1/98.

Occupational Field: 63 (Avionics).

Career Pattern

PVT: Private (E-1). PFC: Private First Class (E-2). LCP: Lance Corporal (E-3). CPL: Corporal (E-4). SGT: Sergeant (E-5). SSGT: Staff Sergeant (E-6). GYSGT: Gunnery Sergeant (E-7). NOTE: MOS duties are not defined in terms of rank or paygrade, but by levels. Level I identifies tasks that are taught in the classroom; the individual is able to perform simple parts of tasks. Level II identifies those tasks that are part of on-the-job training; the trainee can do most tasks pertaining to a particular system and needs supervision on more difficult parts of tasks. Level III indicates that the individual can perform all essential tasks without supervision. Level IV indicates a high level of proficiency in job performance; the individual can perform advanced technical functions, instruction, inspection, and supervision. The MOS designator 6351 (Advanced Avionics Technician—Trainee), is used to identify individuals in Levels I and II. Upon completion of Level II, MOS 6354 is awarded. Credit is recommended for Levels III and IV.

Description

Summary: Inspects, removes, tests, repairs, installs, and maintains components, subsystems, and related equipment of aircraft attack systems. *Levels I and II:* Applies operating theory of vacuum tubes, transistors, generators, synchros, servos, electromechanical drive mechanisms, transmission lines, and antennas and the theory of impedance, resistance, capacitance, and inductance; uses electronic schematics, block diagrams, wiring diagrams, color codes, and other technical data in appropriate technical publications; uses technical publications to obtain data for supply purposes; observes safety precautions; demonstrates working knowledge of first aid; isolates and diagnoses system failures; operates ground support equipment used in maintenance; possesses necessary operating licenses and performs maintenance and inspections on ground support equipment; performs maintenance action including inspection and repair according to applicable technical publications. *Level III:* Able to perform the duties required for Levels I and II; performs operational tests of installed systems; removes, adjusts, aligns, and installs components and repairs or replaces associated wiring and interconnecting devices; performs necessary corrective action on corroded electrical components. *Level IV:* Able to perform the duties required for Level III; performs quality assurance inspection for maintenance actions; plans and schedules workshop assignments; supervises maintenance actions; conducts technical training; monitors collection of source data and completes reports and records; possesses working knowledge of sound management principles, workshop supervision, proper personnel training methods, and accepted avionics maintenance procedures; organizes and administers maintenance and repair facilities.

Recommendation, Level III

In the lower-division baccalaureate/associate degree category, 3 semester hours in electronic systems troubleshooting, 3 in electromechanical systems troubleshooting, and 3 in industrial safety (8/89).

Recommendation, Level IV

In the lower-division baccalaureate/associate degree category, 3 semester hours in electronic systems troubleshooting, 3 in electromechanical systems troubleshooting, 3 in industrial safety, 3 in maintenance management, and 3 in personnel supervision. In the upper-division baccalaureate category, 2 semester hours for field experience in management, if rank was Staff Sergeant (SSGT) and 3 semester hours for field experience in management, if rank was Gunnery Sergeant (GYSGT) (8/89).

MCE-6386-001

AIRCRAFT ELECTRONIC COUNTERMEASURES SYSTEMS TECHNICIAN, EA-6B, ORGANIZATIONAL MAINTENANCE ACTIVITY (OMA)
6386

Exhibit Dates: 1/90–Present.

Occupational Field: 63 (Avionics).

Career Pattern

PVT: Private (E-1). PFC: Private First Class (E-2). LCP: Lance Corporal (E-3). CPL: Corporal (E-4). SGT: Sergeant (E-5). SSGT: Staff Sergeant (E-6). GYSGT: Gunnery Sergeant (E-7). NOTE: MOS duties are not defined in terms of rank or paygrade, but by levels. Level I identifies tasks that are taught in the classroom; the individual is able to perform simple parts of tasks. Level II identifies those tasks that are part of on-the-job training; the trainee can do most tasks pertaining to a particular system and needs supervision on more difficult parts of tasks. Level III indicates that the individual can perform all essential tasks without supervision. Level IV indicates a high level of proficiency in job performance; the individual can perform advanced technical functions, instruction, inspection, and supervision. The MOS designator 6311 (Aircraft Communications/Navigation/Electrical/Weapon Systems Technician—Trainee) is used to identify individuals in Levels I and II. Upon completion of Level II, MOS 6386 is awarded. Credit is recommended for Levels III and IV.

Description

Summary: Installs, removes, inspects, tests, maintains, and repairs systems, components, and ancillary equipment of installed aircraft electronic countermeasures systems. *Levels I and II:* Applies operating theory of vacuum tubes, transistors, diodes, solid state devices, integrated circuit components, transmission lines, and antennas and theory of impedance, resistance, and inductance as applicable to aircraft electronic countermeasures systems; uses electronic schematics, block diagrams, wiring diagrams, color codes, and other technical data contained in appropriate technical publications; uses technical and supply publications and lists to obtain complete data for supply purposes; observes safety precautions while working on and around aircraft, shops, and hangar areas; demonstrates a working knowledge of first aid procedures; isolates and diagnoses aircraft electronic countermeasures system failures, using appropriate test equipment and publications; recognizes corrosion of aircraft systems and electronic components and performs necessary corrective action; completes maintenance data forms; performs avionic systems maintenance actions; performs periodic inspections and completes maintenance repair actions as required by periodic maintenance requirement cards. *Level III:* Able to perform the duties required for Levels I and II; performs maintenance functions for aircraft electronic countermeasures systems, including removal, installation, adjustment, and alignment of system components and the repair or replacement of associated aircraft wiring and interconnecting devices; conducts technical on-the-job training. *Level IV:* Able to perform the duties required for Level III; performs quality assurance inspections on all maintenance actions; supervises maintenance activities; plans and schedules daily work load; monitors source data collection and assists in the preparation of work center reports and records. If rank of Gunnery Sergeant has been attained, coordinates maintenance actions of all assigned work centers; possesses a through understanding and working knowledge of sound management principles, work center supervision, proper methods of training personnel, and accepted and approved avionic maintenance procedures.

Recommendation, Level III

In the lower-division baccalaureate/associate degree category, 3 semester hours in basic electronics laboratory, 2 in principles of DC circuits, and 2 in principles of AC circuits (10/91).

Recommendation, Level IV

In the lower-division baccalaureate/associate degree category, 3 semester hours in basic electronics laboratory, 2 in principles of DC circuits, 2 in principles of AC circuits, 3 in personnel supervision, and 3 in maintenance management; if rank of Staff Sergeant has been attained, 2 semester hours in records and information management. In the upper-division baccalaureate category, if rank of Gunnery Sergeant has been attained, 3 semester hours for field experience in management (10/91).

MCE-6391-001

AVIONICS MAINTENANCE CHIEF
6391

Exhibit Dates: 5/92–Present.

Occupational Field: 63/64 (Avionics).

Career Pattern

May progress to MOS 6391 from any MOS in the Avionics Occupational Field (63/64) (Private through Gunnery Sergeant). MSGT: Master Sergeant (E-8). MGYSGT: Master Gunnery Sergeant (E-9). NOTE: Chiefs must be qualified in Level IV of previously-held MOS. Descriptions and credit recommendations for chiefs are by rank. Check the individual's fitness report or brief sheet for the rank and for proficiency report.

Description

Summary: Supervises maintenance and repair of aircraft avionic systems, equipment, and components. *Master Sergeant:* Supervises the establishment and functioning of aviation repair and maintenance facilities; assists in directing, supervising, and coordinating maintenance activities; prepares reports, schedules, and rosters regarding aircraft maintenance and repair; requisitions spare parts, replacement parts, supplies, and equipment; maintains maintenance, inspection, and technical training records. *Master Gunnery Sergeant:* Able to perform the duties required for Master Sergeant; provides staff support in planning and implementing maintenance activities; has a thorough knowledge of Marine Corps administrative procedures and aviation staff organization and functioning; prepares and presents reports on logistics and readiness to higher command levels.

Recommendation, Master Sergeant

In the lower-division baccalaureate/associate degree category, 3 semester hours in personnel supervision, 3 in technical writing, and 3 in records and information management. In the upper-division baccalaureate category, 3 semester hours in organizational management, 3 in management problems, and 3 for field experience in management. NOTE: Credit should also be granted for the prerequisite MOS (5/92).

Recommendation, Master Gunnery Sergeant

In the lower-division baccalaureate/associate degree category, 3 semester hours in personnel supervision, 3 in technical writing, and 3 in records and information management. In the upper-division baccalaureate category, 3 semester hours in organizational management, 6 in management problems, and 6 for field experience in management. NOTE: Credit should also be granted for the prerequisite MOS (5/92).

MCE-6412-001

AIRCRAFT COMMUNICATIONS SYSTEMS TECHNICIAN, INTERMEDIATE MAINTENANCE ACTIVITY (IMA)
6412

Exhibit Dates: 1/90–Present.

Occupational Field: 64 (Avionics).

Career Pattern

PVT: Private (E-1). PFC: Private First Class (E-2). LCP: Lance Corporal (E-3). CPL: Corporal (E-4). SGT: Sergeant (E-5). NOTE: MOS duties are not defined in terms of rank or paygrade, but by levels. Level I identifies tasks that are taught in the classroom; the individual is able to perform simple parts of tasks. Level II identifies those tasks that are part of on-the-job training; the trainee can do most tasks pertaining to a particular system and needs supervision on more difficult parts of tasks. Level III indicates that the individual can perform all essential tasks without supervision. Level IV indicates a high level of proficiency in job performance; the individual can perform advanced technical functions, instruction, inspection, and supervision. The MOS designator 6411 (Aircraft Communications/Navigation Systems Technician—Trainee) is used to identify individuals in Levels I and II. Upon completion of level II, MOS 6412 is awarded. Credit is recommended for Levels III and IV.

Description

Summary: Inspects, tests, maintains, and repairs components, assemblies, subassemblies, modules, cards, printed circuit boards, and ancillary equipment whose aggregate constitutes a complete aircraft communications system or subsystem. *Levels I and II:* Applies theory of operation of vacuum tubes, transistors, diodes, solid state devices, integrated circuit components, transformers, motors, generators, synchros, servos, electromechanical drive mechanisms, transmission lines, and antennas and the theory of impedance, resistance, capacitance, and inductance to aircraft communications/navigation systems; uses electronic schematics, block diagrams, wiring diagrams, color codes, and other technical publications; uses technical and supply publications to obtain accurate data for supply purposes; observes shop safety precautions; demonstrates a working knowledge of first aid; isolates and diagnoses aircraft communications systems/subsystems failures; performs the necessary corrective actions prescribed in applicable maintenance instructions for corroded and deteriorated electronic components. *Level III:* Able to perform the duties required for Levels I and II; submits accurate and timely maintenance and supply data information on applicable maintenance action forms; ensures that all "in use" test and measuring equipment is within its calibration/qualification cycle; requisitions replacement parts and necessary supplies; performs fault isolation and repair on components of aircraft communications systems; incorporates appropriate avionics changes in aircraft communications systems as directed; conducts technical and on-the-job training; works with the aviation supply office and the electronic supply office supply systems. *Level IV:* Able to perform the duties required for Level III; supervises maintenance activities; performs quality assurance inspections on all maintenance actions completed on system components; plans and schedules daily work load; prepares quality deficiency reports.

Recommendation, Level III

In the lower-division baccalaureate/associate degree category, 3 semester hours in basic electronics laboratory, 2 in principles of DC circuits, 2 in principles of AC circuits, 3 in solid state electronics, and 3 in electronic systems troubleshooting and maintenance (10/91).

Recommendation, Level IV

In the lower-division baccalaureate/associate degree category, 3 semester hours in basic electronics laboratory, 2 in principles of DC circuits, 2 in principles of AC circuits, 3 in solid state electronics, 3 in electronic systems troubleshooting and maintenance, 3 in personnel supervision, and 3 in maintenance management (10/91).

MCE-6413-001

AIRCRAFT NAVIGATION SYSTEMS TECHNICIAN, IFF/RADAR/TACAN, INTERMEDIATE MAINTENANCE ACTIVITY (IMA)
6413

Exhibit Dates: 1/90–Present.

Occupational Field: 64 (Avionics).

Career Pattern

PVT: Private (E-1). PFC: Private First Class (E-2). LCP: Lance Corporal (E-3). CPL: Corporal (E-4). SGT: Sergeant (E-5). NOTE: MOS duties are not defined in terms of rank or paygrade, but by levels. Level I identifies tasks that are taught in the classroom; the individual is able to perform simple parts of tasks. Level II identifies those tasks that are part of on-the-job training; the trainee can do most tasks pertaining to a particular system and needs supervision on more difficult parts of tasks. Level III indicates that the individual can perform all essential tasks without supervision. Level IV indicates a high level of proficiency in job performance; the individual can perform advanced technical functions, instruction, inspection, and supervision. The MOS designator 6411 (Aircraft Communications/Navigation Systems Technician—Trainee) is used to identify individuals in Levels I and II. Upon completion of level II, MOS 6413 is awarded. Credit is recommended for Levels III and IV.

Description

Summary: Inspects, tests, maintains, and repairs components, assemblies, subassemblies, modules, cards, printed circuits boards, and ancillary equipment whose aggregate constitutes a complete aircraft communications system or subsystem. *Levels I and II:* Applies theory of operation of vacuum tubes, transistors, diodes, solid state devices, integrated circuit components, transformers, motors, generators, synchros, servos, electromechanical drive mechanisms, transmission lines, and antennas and the theory of impedance, resistance, capacitance, and inductance to aircraft communications/navigation systems; uses electronic schematics, block diagrams, wiring diagrams, color codes, and other technical publications; uses technical and supply publications to obtain accurate data for supply purposes; observes shop safety precautions; demonstrates a working knowledge of first aid; isolates and diagnoses aircraft communications systems/subsystems failures; performs the necessary corrective action prescribed in applicable maintenance instructions for corroded and deteriorated electronic components. *Level III:* Able to perform the duties required for Levels I and II; submits accurate and timely maintenance and supply data information on applicable maintenance action forms; ensures that all "in use" test and measuring equipment is within its calibration/qualification cycle; requisitions replacement parts and necessary supplies; performs fault isolation and repair on components of aircraft communications systems; incorporates appropriate avionics changes in aircraft communications systems as directed; conducts technical and on-the-job training; uses the aviation supply office and the electronic supply office supply systems. *Level IV:* Able to perform the duties required for Level III; supervises maintenance activities; performs quality assurance inspections on all maintenance actions completed on system components; plans and schedules daily work load; prepares quality deficiency reports.

Recommendation, Level III

In the lower-division baccalaureate/associate degree category, 3 semester hours in basic electronics laboratory, 2 in principles of DC circuits, 2 in principles of AC circuits, 3 in solid state electronics, and 3 in electronic systems troubleshooting and maintenance (10/91).

Recommendation, Level IV

In the lower-division baccalaureate/associate degree category, 3 semester hours in basic electronics laboratory, 2 in principles of DC circuits, 2 in principles of AC circuits, 3 in solid state electronics, 3 in electronic systems troubleshooting and maintenance, 3 in personnel supervision, and 3 in maintenance management (10/91).

MCE-6414-001

ADVANCED AIRCRAFT COMMUNICATIONS/NAVIGATION SYSTEMS TECHNICIAN, INTERMEDIATE MAINTENANCE ACTIVITY (IMA)
6414

Exhibit Dates: 1/90–Present.

Occupational Field: 64 (Avionics).

Career Pattern

May progress to MOS 6414 from MOS 6412, 6413, or 6483 (Private through Sergeant). SSGT: Staff Sergeant (E-6). GYSGT: Gunnery Sergeant (E-7). NOTE: MOS duties are not defined in terms of rank or paygrade, but by levels. Level I identifies tasks that are taught in the classroom; the individual is able to perform simple parts of tasks. Level II identifies those tasks that are part of on-the-job training; the trainee can do most tasks pertaining to a particular system and needs supervision on more difficult parts of tasks. Level III indicates that the individual can perform all essential tasks without supervision. Level IV indicates a high level of proficiency in job performance; the individual can perform advanced technical functions, instruction, inspection, and supervision. Credit is recommended for Levels III and IV. An individual in MOS 6414 must be qualified at Level IV.

Description

Summary: Tests, maintains, and repairs aircraft replaceable assemblies, shop replaceable assemblies, and ancillary equipment; able to perform the level IV duties of MOS 6412,

6413, or 6483. Possesses a thorough understanding and working knowledge of sound management principles, work center supervision, proper methods of training personnel, and accepted and approved naval aviation maintenance procedures; performs quality assurance inspections for maintenance actions performed on equipment and components; assists in the planning and scheduling of work center assignments; monitors source data collection and prepares work center reports and records; performs expanded troubleshooting using concise fault isolation procedures; organizes and administers facilities for the maintenance and repair of aircraft communications/navigation systems.

Recommendation

In the lower-division baccalaureate/associate degree category, 2 semester hours in records and information management. In the upper-division baccalaureate category, if the rank of Gunnery Sergeant (GYSGT) has been attained, 3 semester hours for field experience in management. NOTE: Add credit for Level IV of previously-held MOS (10/91).

MCE-6422-001

AIRCRAFT CRYPTOGRAPHIC SYSTEMS TECHNICIAN,
INTERMEDIATE MAINTENANCE ACTIVITY
(IMA)
6422

Exhibit Dates: 1/90–Present.

Occupational Field: 64 (Avionics).

Career Pattern

PVT: Private (E-1). PFC: Private First Class (E-2). LCP: Lance Corporal (E-3). CPL: Corporal (E-4). SGT: Sergeant (E-5). SSGT: Staff Sergeant (E-6). GYSGT: Gunnery Sergeant (E-7). NOTE: MOS duties are not defined in terms of rank or paygrade, but by levels. Level I identifies tasks that are taught in the classroom; the individual is able to perform simple parts of tasks. Level II identifies those tasks that are part of on-the-job training; the trainee can do most tasks pertaining to a particular system and needs supervision on more difficult parts of tasks. Level III indicates that the individual can perform all essential tasks without supervision. Level IV indicates a high level of proficiency in job performance; the individual can perform advanced technical functions, instruction, inspection, and supervision. The MOS designator 6411 (Aircraft Communications/Navigation Systems Technician—Trainee) is used to identify individuals in Levels I and II. Upon completion of Level II, MOS 6422 is awarded. Credit is recommended for Levels III and IV.

Description

Summary: Inspects, tests, maintains, and repairs equipment, assemblies, subassemblies, modules, cards, and other components of airborne cryptographic systems. *Levels I and II:* Applies theory of operation vacuum tubes, transistors, diodes, solid state devices, integrated circuit components, transformers, motors, generators, synchros, servos, electromechanical drive mechanisms, transmission lines, and antennas and theory of impedance, resistance, capacitance, and inductance to aircraft cryptographic systems; uses electronic schematics, block diagrams, wiring diagrams, color codes, and other technical data contained in appropriate technical publications; uses available technical and supply publications to obtain accurate data for supply purposes; observes shop safety precautions; uses first aid procedures; diagnoses failures in assemblies, subassemblies, modules, cards, boards, or other discrete components of an item under test using specified general, special, or peculiar test equipment and following the appropriate maintenance instruction manuals; certifies the installation of aviation cryptographic systems in aircraft to provide reasonable assurance that no compromise of cryptographics could occur from an improper installation. *Level III:* Able to perform the duties required for Levels I and II; submits maintenance and supply information on applicable forms using existing directives; ensures that all "in use" test and measuring equipment assigned to a given test bench is within its calibration/qualification cycle per applicable instructions and directives; requisitions replacement parts and supplies; conducts technical on-the-job training; uses maintenance directives and the publications systems; uses the aviation supply office and the electronics supply office supply systems. *Level IV:* Able to perform the duties required for Level III; supervises maintenance activities; performs quality assurance inspections on all maintenance actions; plans and schedules daily work load; possesses a thorough understanding and working knowledge of sound management principles, work center supervision, proper methods of training personnel, and accepted and approved avionics maintenance procedures; monitors source data collection and work center reports and records; performs expanded troubleshooting using fault isolation techniques. If rank of Gunnery Sergeant has been attained, organizes and administers facilities for the repair of cryptographic equipment.

Recommendation, Level III

In the lower-division baccalaureate/associate degree category, 3 semester hours in basic electronics laboratory, 2 in principles of DC circuits, 2 in principles of AC circuits, 3 in solid state electronics, and 3 in electronic systems troubleshooting and maintenance (10/91).

Recommendation, Level IV

In the lower-division baccalaureate/associate degree category, 3 semester hours in basic electronics laboratory, 2 in principles of DC circuits, 2 in principles of AC circuits, 3 in solid state electronics, 3 in electronic systems troubleshooting and maintenance, 3 in personnel supervision, and 3 in maintenance management; if rank of Staff Sergeant was attained, 2 semester hours in records and information management. In the upper-division baccalaureate category, if rank of Gunnery Sergeant was attained, 3 semester hours for field experience in management (10/91).

MCE-6423-001

AVIATION ELECTRONIC MICRO-MINIATURE/
INSTRUMENT AND CABLE REPAIR
TECHNICIAN
6423

Exhibit Dates: 1/90–Present.

Occupational Field: 64 (Avionics).

Career Pattern

PVT: Private (E-1). PFC: Private First Class (E-2). LCP: Lance Corporal (E-3). CPL: Corporal (E-4). SGT: Sergeant (E-5). NOTE: MOS duties are not defined in terms of rank or paygrade, but by levels. Level I identifies tasks that are taught in the classroom; the individual is able to perform simple parts of tasks. Level II identifies those tasks that are part of on-the-job training; the trainee can do most tasks pertaining to a particular system and needs supervision on more difficult parts of tasks. Level III indicates that the individual can perform all essential tasks without supervision. Level IV indicates a high level of proficiency in job performance; the individual can perform advanced technical functions, instruction, inspection, and supervision. The MOS designator 6411 (Aircraft Communications/Navigation Systems Technician—Trainee) is used to identify individuals in Levels I and II. Upon completion of level II, MOS 6423 is awarded. Credit is recommended for Levels III and IV.

Description

Summary: Inspects, tests, maintains, and repairs modules, cards, printed circuit boards, cables, instruments, and miniature and microminiature components. *Levels I and II:* Applies theory of operation and use of vacuum tubes, transistors, diodes, solid state devices, integrated circuit components, transformers, motors, generators, synchros, servos, electromechanical drive mechanisms, transmission lines, and antennas and the theory of impedance, resistance, capacitance, and inductance to aircraft avionics systems; uses electronic schematics, block diagrams, wiring diagrams, color codes, and other technical data contained in appropriate technical publications; uses special tools and equipment to repair modules, cards, printed circuit boards, multilayer boards, instruments, cables, and other solid state circuit components which have transistors, diodes, integrated circuit components, and other miniature and microminiature components; uses technical and supply publications and lists to obtain accurate data for supply purposes; observes shop safety precautions; demonstrates a working knowledge of first aid; uses available technical publications and equipment to perform corrosion control on avionics equipment. *Level III:* Able to perform the duties required for Levels I and II; submits accurate and timely maintenance/supply data information on applicable maintenance action forms; ensures that all "in use" test and measure equipment is within its calibration/qualification cycle per applicable instructions and directives; requisitions replacement parts and necessary supplies using applicable publications and supply documents; conducts technical on-the-job training; uses the aviation supply office and the electronics supply office supply systems. *Level IV:* Able to perform the duties required for Level III; performs quality assurance inspections on all maintenance actions completed on system components; plans and schedules daily work load; supervises maintenance activities.

Recommendation, Level III

In the lower-division baccalaureate/associate degree category, 3 semester hours in basic electronics laboratory, 2 in principles of DC circuits, 2 in principles of AC circuits, 3 in solid state electronics, 3 in electronic systems troubleshooting and maintenance, and 3 in industrial safety (10/91).

Recommendation, Level IV

In the lower-division baccalaureate/associate degree category, 3 semester hours in basic electronics laboratory, 2 in principles of DC circuits, 2 in principles of AC circuits, 3 solid state electronics, 3 in electronic systems troubleshooting and maintenance, 3 in industrial safety, 3 in personnel supervision, and 3 in maintenance management (10/91).

MCE-6432-001

AIRCRAFT ELECTRICAL/INSTRUMENT/FLIGHT CONTROL SYSTEMS TECHNICIAN, INTERMEDIATE MAINTENANCE ACTIVITY (IMA), FIXED TECHNICIAN, INTERMEDIATE MAINTENANCE ACTIVITY (IMA), FIXED WING
6432

Exhibit Dates: 1/90–Present.

Occupational Field: 64 (Avionics).

Career Pattern
PVT: Private (E-1). PFC: Private First Class (E-2). LCP: Lance Corporal (E-3). CPL: Corporal (E-4). SGT: Sergeant (E-5). NOTE: MOS duties are not defined in terms of rank or paygrade, but by levels. Level I identifies tasks that are taught in the classroom; the individual is able to perform simple parts of tasks. Level II identifies those tasks that are part of on-the-job training; the trainee can do most tasks pertaining to a particular system and needs supervision on more difficult parts of tasks. Level III indicates that the individual can perform all essential tasks without supervision. Level IV indicates a high level of proficiency in job performance; the individual can perform advanced technical functions, instruction, inspection, and supervision. The MOS designator 6431 (Aircraft Electrical Systems Technician—Trainee) is used to identify individuals in Levels I and II. Upon completion of Level II, MOS 6432 is awarded. Credit is recommended for Levels III and IV.

Description
Summary: Tests, repairs, and maintains components and related equipment for aircraft electrical/flight control systems. *Levels I and II:* Applies theory of operation and use of vacuum tubes, transistors, diodes, solid state devices, integrated circuit components, transformers, motors, generators, synchros, servos, electro-mechanical drive mechanisms, transmission lines, and antennas and the theory of impedance, resistance, capacitance, and inductance to aircraft avionic systems; uses electronic schematics, block diagrams, wiring diagrams, color codes, and other technical data contained in appropriate technical publications; uses special tools and equipment to repair modules, cards, printed circuit boards, multilayer boards, instruments, cables, and other solid state circuit components which have transistors, diodes, integrated circuit components, and other miniature and microminiature components; uses technical and supply publications to obtain accurate data for supply purposes; observes shop safety precautions; demonstrates a working knowledge of first aid; uses available technical publications and equipment to perform corrosion control on avionics equipment. *Level III:* Able to perform the duties required for Levels I and II; submits accurate and timely maintenance and supply information on applicable maintenance action forms; ensures that all "in use" test and measuring equipment is within its calibration/qualification cycle; requisitions replacement parts and necessary supplies; conducts technical on-the-job training; uses the aviation supply office and the electronics supply office supply systems. *Level IV:* Able to perform the duties required for Level III; supervises maintenance activities; performs quality assurance inspections on all maintenance actions; plans and schedules daily work loads.

Recommendation, Level III
In the lower-division baccalaureate/associate degree category, 3 semester hours in basic electronics laboratory, 2 in principles of DC circuits, 2 in principles of AC circuits, 3 in solid state electronics, and 3 in electronic systems troubleshooting and maintenance (10/91).

Recommendation, Level IV
In the lower-division baccalaureate/associate degree category, 3 semester hours in basic electronics laboratory, 2 in principles of DC circuits, 2 in principles of AC circuits, 3 in solid state electronics, 3 in electronic systems troubleshooting and maintenance, 3 in personnel supervision, and 3 in maintenance management (10/91).

MCE-6433-001

AIRCRAFT ELECTRICAL/INSTRUMENT/FLIGHT CONTROL SYSTEMS TECHNICIAN, HELICOPTER/OV-10, INTERMEDIATE MAINTENANCE ACTIVITY (IMA)
6433

Exhibit Dates: 1/90–Present.

Occupational Field: 64 (Avionics).

Career Pattern
PVT: Private (E-1). PFC: Private First Class (E-2). LCP: Lance Corporal (E-3). CPL: Corporal (E-4). SGT: Sergeant (E-5). NOTE: MOS duties are not defined in terms of rank or paygrade, but by levels. Level I identifies tasks that are taught in the classroom; the individual is able to perform simple parts of tasks. Level II identifies those tasks that are part of on-the-job training; the trainee can do most tasks pertaining to a particular system and needs supervision on more difficult parts of tasks. Level III indicates that the individual can perform all essential tasks without supervision. Level IV indicates a high level of proficiency in job performance; the individual can perform advanced technical functions, instruction, inspection, and supervision. The MOS designator 6431 (Aircraft Electrical Systems Technician—Trainee) is used to identify individuals in Levels I and II. Upon completion of level II, MOS 6433 is awarded. Credit is recommended for Levels III and IV.

Description
Summary: Tests, repairs and maintains components and related equipment for aircraft electrical/flight control systems. *Levels I and II:* Applies theory of operation and use of vacuum tubes, transistors, diodes, solid state devices, integrated circuit components, transformers, motors, generators, synchros, servos, electro-mechanical drive mechanisms, transmission lines, and antennas and the theory of impedance, resistance, capacitance, and inductance to aircraft avionic systems; uses electronic schematics, block diagrams, wiring diagrams, color codes, and other technical data contained in appropriate technical publications; uses special tools and equipment to repair modules, cards, printed circuit boards, multilayer boards, instruments, cables, and other solid state circuit components which have transistors, diodes, integrated circuit components, and other miniature and microminiature components; uses technical and supply publications to obtain accurate data for supply purposes; observes shop safety precautions; demonstrates a working knowledge of first aid; uses available technical publications and equipment to perform corrosion control on avionics equipment. *Level III:* Able to perform the duties required for Levels I and II; submits accurate and timely maintenance and supply information on applicable maintenance action forms; ensures that all "in use" test and measuring equipment is within its calibration/qualification cycle; requisitions replacement parts and necessary supplies; conducts technical on-the-job training; uses the aviation supply office and the electronics supply office supply systems. *Level IV:* Able to perform the duties required for Level III; supervises maintenance activities; performs quality assurance inspections on all maintenance actions; plans and schedules daily work load.

Recommendation, Level III
In the lower-division baccalaureate/associate degree category, 3 semester hours in basic electronics laboratory, 2 in principles of DC circuits, 2 in principles of AC circuits, 3 in solid state electronics, and 3 in electronic systems troubleshooting and maintenance (10/91).

Recommendation, Level IV
In the lower-division baccalaureate/associate degree category, 3 semester hours in basic electronics laboratory, 2 in principles of DC circuits, 2 in principles of AC circuits, 3 in solid state electronics, 3 in electronic systems troubleshooting and maintenance, 3 in personnel supervision, and 3 in maintenance management (10/91).

MCE-6434-001

ADVANCED AIRCRAFT ELECTRICAL INSTRUMENT/ FLIGHT CONTROL SYSTEMS TECHNICIAN, INTERMEDIATE MAINTENANCE ACTIVITY (IMA)
6434

Exhibit Dates: 1/90–Present.

Occupational Field: 64 (Avionics).

Career Pattern
May progress to MOS 6434 from MOS 6432, 6433, or 6423 (Private through Sergeant). SSGT: Staff Sergeant (E-6). GYSGT: Gunnery Sergeant (E-7). NOTE: MOS duties are not defined in terms of rank or paygrade, but by levels. Level I identifies tasks that are taught in the classroom; the individual is able to perform simple parts of tasks. Level II identifies those tasks that are part of on-the-job training; the trainee can do most tasks pertaining to a particular system and needs supervision on more difficult parts of tasks. Level III indicates that the individual can perform all essential tasks without supervision. Level IV indicates a high level of proficiency in job performance; the individual can perform advanced technical functions, instruction, inspection, and supervision. Credit is recommended for Levels III and IV. An Individual in MOS 6434 must be qualified at Level IV.

Description
Summary: Tests, repairs, and maintains assemblies, components, and related equipment for aircraft electrical and flight control systems beyond normal fault isolation procedures. Able to perform the Level IV duties of MOS's 6432, 6433, and 6423; possesses a thorough understanding and working knowledge of sound management principles, work center supervision, proper methods of training personnel, and

accepted and approved naval aviation maintenance procedures; performs quality assurance inspections for maintenance actions performed on equipment and components; assists in the planning and scheduling of work center assignments; monitors source data collection and prepares work center reports and records; performs expanded troubleshooting using concise fault isolation procedures. If rank of Gunnery Sergeant has been attained, organizes and administers facilities for the maintenance and repair of aircraft electrical/flight control systems.

Recommendation

In the lower-division baccalaureate/associate degree category, 2 semester hours in records and information management. In the upper-division baccalaureate category, if rank of Gunnery Sergeant (GYSGT) has been attained, 3 semester hours for field experience in management. NOTE: Add credit for Level IV of previously-held MOS (10/91).

MCE-6462-001

AVIONICS TEST SET TECHNICIAN, INTERMEDIATE MAINTENANCE ACTIVITY (IMA)
6462

Exhibit Dates: 1/90–Present.

Occupational Field: 64 (Avionics).

Career Pattern

PVT: Private (E-1). PFC: Private First Class (E-2). LCP: Lance Corporal (E-3). CPL: Corporal (E-4). SGT: Sergeant (E-5). NOTE: MOS duties are not defined in terms of rank or paygrade, but by levels. Level I identifies tasks that are taught in the classroom; the individual is able to perform simple parts of tasks. Level II identifies those tasks that are part of on-the-job training; the trainee can do most tasks pertaining to a particular system and needs supervision on more difficult parts of tasks. Level III indicates that the individual can perform all essential tasks without supervision. Level IV indicates a high level of proficiency in job performance; the individual can perform advanced technical functions, instruction, inspection, and supervision. The MOS designator 6411 (Aircraft Communications/Navigation Systems Technician—Trainee) is used to identify individuals in Levels I and II. Upon completion of level II, MOS 6462 is awarded. Credit is recommended for Levels III and IV.

Description

Summary: Inspects and maintains automatic electronic test equipment at the intermediate level. *Levels I and II:* Applies theory of operation and use of vacuum tubes and solid state devices, such as diodes, transistors, integrated circuits, transformers, motors, generators, servomechanisms, electromechanical drives, transmission lines, waveguides, and similar components as applicable to automatic test systems; uses applicable automatic test systems; uses applicable service manuals and technical publications; practices proper safety procedures. *Level III:* Able to perform the duties required for Levels I and II; demonstrates knowledge of first aid practices; isolates and diagnoses failures in replaceable assemblies and ancillary equipment; performs necessary corrective actions prescribed in applicable maintenance directives. *Level IV:* Able to perform the duties required for Level III; submits accurate maintenance and supply information; ensures that all test and measuring equipment is within its calibration tolerances prescribed by directives; plans work schedules; incorporates appropriate avionics changes as directed; conducts technical on-the-job training; prepares quality deficiency reports.

Recommendation, Level III

In the lower-division baccalaureate/associate degree category, 3 semester hours in basic electronics laboratory, 2 in principles of DC circuits, 2 in principles of AC circuits, 3 in solid state electronics, and 3 in electronic systems troubleshooting and maintenance (10/91).

Recommendation, Level IV

In the lower-division baccalaureate/associate degree category, 3 semester hours in basic electronics laboratory, 2 in principles of DC circuits, 2 in principles of AC circuits, 3 in solid state electronics, 3 in electronic systems troubleshooting and maintenance, 3 in personnel supervision, and 3 in maintenance management (10/91).

MCE-6463-001

RADAR TEST STATION/RADAR SYSTEMS TEST STATION TECHNICIAN, INTERMEDIATE MAINTENANCE ACTIVITY (IMA)
6463

Exhibit Dates: 1/90–Present.

Occupational Field: 64 (Avionics).

Career Pattern

PVT: Private (E-1). PFC: Private First Class (E-2). LCP: Lance Corporal (E-3). CPL: Corporal (E-4). SGT: Sergeant (E-5). NOTE: MOS duties are not defined in terms of rank or paygrade, but by levels. Level I identifies tasks that are taught in the classroom; the individual is able to perform simple parts of tasks. Level II identifies those tasks that are part of on-the-job training; the trainee can do most tasks pertaining to a particular system and needs supervision on more difficult parts of tasks. Level III indicates that the individual can perform all essential tasks without supervision. Level IV indicates a high level of proficiency in job performance; the individual can perform advanced technical functions, instruction, inspection, and supervision. The MOS designator 6411 (Aircraft Communications/Navigation Systems Technician—Trainee) is used to identify individuals in Levels I and II. Upon completion of level II, MOS 6463 is awarded. Credit is recommended for Levels III and IV.

Description

Summary: Inspects and maintains automatic electronic test equipment at the intermediate level. *Levels I and II:* Applies theory of operation and use of vacuum tubes and solid state devices, such as diodes, transistors, integrated circuits, transformers, motors, generators, servomechanisms, electromechanical drives, transmission lines, waveguides, and similar components according to service manuals and technical publications; practices proper safety procedures. *Level III:* Able to perform the duties required for Levels I and II; demonstrates knowledge of first aid practices; isolates and diagnoses failures in replaceable assemblies and ancillary equipment; performs necessary corrective actions prescribed in applicable maintenance directives. *Level IV:* Able to perform the duties required for Level III; submits accurate maintenance and supply information; ensures that all test and measuring equipment is within its calibration tolerances prescribed by directives; plans work schedules; incorporates appropriate avionics changes as directed; conducts technical on-the-job training; prepares quality deficiency reports.

Recommendation, Level III

In the lower-division baccalaureate/associate degree category, 3 semester hours in basic electronics laboratory, 2 in principles of DC circuits, 2 in principles of AC circuits, 3 in solid state electronics, and 3 in electronic systems troubleshooting and maintenance (10/91).

Recommendation, Level IV

In the lower-division baccalaureate/associate degree category, 3 semester hours in basic electronics laboratory, 2 in principles of DC circuits, 2 in principles of AC circuits, 3 in solid state electronics, 3 in electronic systems troubleshooting and maintenance, 3 in personnel supervision, and 3 in maintenance management (10/91).

MCE-6464-001

AIRCRAFT INERTIAL NAVIGATION SYSTEM TECHNICIAN, INTERMEDIATE MAINTENANCE ACTIVITY (IMA)
6464

Exhibit Dates: 1/90–Present.

Occupational Field: 64 (Avionics).

Career Pattern

PVT: Private (E-1). PFC: Private First Class (E-2). LCP: Lance Corporal (E-3). CPL: Corporal (E-4). SGT: Sergeant (E-5). NOTE: MOS duties are not defined in terms of rank or paygrade, but by levels. Level I identifies tasks that are taught in the classroom; the individual is able to perform simple parts of tasks. Level II identifies those tasks that are part of on-the-job training; the trainee can do most tasks pertaining to a particular system and needs supervision on more difficult parts of tasks. Level III indicates that the individual can perform all essential tasks without supervision. Level IV indicates a high level of proficiency in job performance; the individual can perform advanced technical functions, instruction, inspection, and supervision. The MOS designator 6411 (Aircraft Communications/Navigation Systems Technician—Trainee) is used to identify individuals in Levels I and II. Upon completion of level II, MOS 6464 is awarded. Credit is recommended for Levels III and IV.

Description

Summary: Inspects and maintains automatic electronic test equipment at the intermediate level. *Levels I and II:* Applies theory of operation and use of vacuum tubes and solid state devices, such as diodes, transistors, integrated circuits, transformers, motors, generators, servomechanisms, electromechanical drives, transmission lines, waveguides, and similar components as applicable to automatic test systems; uses applicable service manuals and technical publications; practices proper safety procedures. *Level III:* Able to perform the duties required for Levels I and II; demonstrates knowledge of first aid practices; isolates and diagnoses failures in replaceable assemblies and ancillary equipment; performs necessary corrective actions prescribed in maintenance directives. *Level IV:* Able to perform the duties required for Level III; submits accurate maintenance and supply information; ensures that all test and measuring equipment is within its calibration tolerances prescribed by directives;

plans work schedules; incorporates appropriate avionics changes as directed; conducts technical on-the-job training; prepares quality deficiency reports.

Recommendation, Level III

In the lower-division baccalaureate/associate degree category, 3 semester hours in basic electronics laboratory, 2 in principles of DC circuits, 2 in principles of AC circuits, 3 in solid state electronics, and 3 in electronic systems troubleshooting and maintenance (10/91).

Recommendation, Level IV

In the lower-division baccalaureate/associate degree category, 3 semester hours in basic electronics laboratory, 2 in principles of DC circuits, 2 in principles of AC circuits, 3 in solid state electronics, 3 in electronic systems troubleshooting and maintenance, 3 in personnel supervision, and 3 in maintenance management (10/91).

MCE-6465-001

AIRCRAFT INERTIAL NAVIGATION SYSTEM TECHNICIAN INTERMEDIATE MAINTENANCE ACTIVITY (IMA)
6465

Exhibit Dates: 1/90–Present.

Occupational Field: 64 (Avionics).

Career Pattern

PVT: Private (E-1). PFC: Private First Class (E-2). LCP: Lance Corporal (E-3). CPL: Corporal (E-4). SGT: Sergeant (E-5). NOTE: MOS duties are not defined in terms of rank or paygrade, but by levels. Level I identifies tasks that are taught in the classroom; the individual is able to perform simple parts of tasks. Level II identifies those tasks that are part of on-the-job training; the trainee can do most tasks pertaining to a particular system and needs supervision on more difficult parts of tasks. Level III indicates that the individual can perform all essential tasks without supervision. Level IV indicates a high level of proficiency in job performance; the individual can perform advanced technical functions, instruction, inspection, and supervision. The MOS designator 6411 (Aircraft Communications/Navigation Systems Technician—Trainee) is used to identify individuals in Levels I and II. Upon completion of level II, MOS 6465 is awarded. Credit is recommended for Levels III and IV.

Description

Summary: Inspects and maintains automatic electronic test equipment at the intermediate level. *Levels I and II:* Applies theory of operation and use of vacuum tubes and solid state devices, such as diodes, transistors, integrated circuits, transformers, motors, generators, servomechanisms, electromechanical drives, transmission lines, waveguides, and similar components as applicable to automatic test systems; uses applicable service manuals and technical publications; practices proper safety procedures. *Level III:* Able to perform the duties required for Levels I and II; demonstrates knowledge of first aid practices; isolates and diagnoses failures in replaceable assemblies and ancillary equipment; performs necessary corrective actions prescribed in maintenance directives. *Level IV:* Able to perform the duties required for Level III; submits accurate maintenance and supply information; ensures that all test and measuring equipment is within its calibration tolerances prescribed by directives;

plans work schedules; incorporates appropriate avionics changes as directed; conducts technical on-the-job training; prepares quality deficiency reports.

Recommendation, Level III

In the lower-division baccalaureate/associate degree category, 3 semester hours in basic electronics laboratory, 2 in principles of DC circuits, 2 in principles of AC circuits, 3 in solid state electronics, and 3 in electronic systems troubleshooting and maintenance (10/91).

Recommendation, Level IV

In the lower-division baccalaureate/associate degree category, 3 semester hours in basic electronics laboratory, 2 in principles of DC circuits, 2 in principles of AC circuits, 3 in solid state electronics, 3 in electronic systems troubleshooting and maintenance, 3 in personnel supervision, and 3 in maintenance management (10/91).

MCE-6466-001

AIRCRAFT FORWARD LOOKING INFRARED/ELECTRO-OPTICAL TECHNICIAN, INTERMEDIATE MAINTENANCE ACTIVITY (IMA)
6466

Exhibit Dates: 1/90–Present.

Occupational Field: 64 (Avionics).

Career Pattern

PVT: Private (E-1). PFC: Private First Class (E-2). LCP: Lance Corporal (E-3). CPL: Corporal (E-4). SGT: Sergeant (E-5). NOTE: MOS duties are not defined in terms of rank or paygrade, but by levels. Level I identifies tasks that are taught in the classroom; the individual is able to perform simple parts of tasks. Level II identifies those tasks that are part of on-the-job training; the trainee can do most tasks pertaining to a particular system and needs supervision on more difficult parts of tasks. Level III indicates that the individual can perform all essential tasks without supervision. Level IV indicates a high level of proficiency in job performance; the individual can perform advanced technical functions, instruction, inspection, and supervision. The MOS designator 6411 (Aircraft Communications/Navigation Systems Technician—Trainee) is used to identify individuals in Levels I and II. Upon completion of level II, MOS 6466 is awarded. Credit is recommended for Levels III and IV.

Description

Summary: Inspects and maintains automatic electronic test equipment at the intermediate level. *Levels I and II:* Applies theory of operation and use of vacuum tubes and solid state devices, such as diodes, transistors, integrated circuits, transformers, motors, generators, servomechanisms, electromechanical drives, transmission lines, waveguides, and similar components as applicable to automatic test systems; uses applicable service manuals and technical publications; practices proper safety procedures. *Level III:* Able to perform the duties required for Levels I and II; demonstrates knowledge of first aid practices; isolates and diagnoses failures in replaceable assemblies and ancillary equipment; performs necessary corrective actions prescribed in applicable maintenance directives. *Level IV:* Able to perform the duties required for Level III; submits accurate maintenance and supply information; ensures that all test and measuring equipment is within its calibration tolerances prescribed by

directives; plans work schedules; incorporates appropriate avionics changes as directed; conducts technical on-the-job training; prepares quality deficiency reports.

Recommendation, Level III

In the lower-division baccalaureate/associate degree category, 3 semester hours in basic electronics laboratory, 2 in principles of DC circuits, 2 in principles of AC circuits, 3 in solid state electronics, and 3 in electronic systems troubleshooting and maintenance (10/91).

Recommendation, Level IV

In the lower-division baccalaureate/associate degree category, 3 semester hours in basic electronics laboratory, 2 in principles of DC circuits, 2 in principles of AC circuits, 3 in solid state electronics, 3 in electronic systems troubleshooting and maintenance, 3 in personnel supervision, and 3 in maintenance management (10/91).

MCE-6467-001

AIRCRAFT RADCOM/CAT IIID TECHNICIAN, INTERMEDIATE MAINTENANCE ACTIVITY (IMA)
6467

Exhibit Dates: 1/90–Present.

Occupational Field: 64 (Avionics).

Career Pattern

PVT: Private (E-1). PFC: Private First Class (E-2). LCP: Lance Corporal (E-3). CPL: Corporal (E-4). SGT: Sergeant (E-5). NOTE: MOS duties are not defined in terms of rank or paygrade, but by levels. Level I identifies tasks that are taught in the classroom; the individual is able to perform simple parts of tasks. Level II identifies those tasks that are part of on-the-job training; the trainee can do most tasks pertaining to a particular system and needs supervision on more difficult parts of tasks. Level III indicates that the individual can perform all essential tasks without supervision. Level IV indicates a high level of proficiency in job performance; the individual can perform advanced technical functions, instruction, inspection, and supervision. The MOS designator 6411 (Aircraft Communications/Navigation Systems Technician—Trainee) is used to identify individuals in Levels I and II. Upon completion of level II, MOS 6467 is awarded. Credit is recommended for Levels III and IV.

Description

Summary: Inspects and maintains automatic electronic test equipment at the intermediate level. *Levels I and II:* Applies theory of operation and use of vacuum tubes and solid state devices, such as diodes, transistors, integrated circuits, transformers, motors, generators, servomechanisms, electromechanical drives, transmission lines, waveguides, and similar components as applicable to automatic test systems; uses applicable service manuals and technical publications; practices proper safety procedures. *Level III:* Able to perform the duties required for Levels I and II; demonstrates knowledge of first aid practices; isolates and diagnoses failures in replaceable assemblies and ancillary equipment; performs necessary corrective actions prescribed in applicable maintenance directives. *Level IV:* Able to perform the duties required for Level III; submits accurate maintenance and supply information; ensures that all test and measuring equipment is within its calibration tolerances prescribed by directives; plans work schedules; incorporates appropriate avionics changes as directed; conducts technical on-the-job training; prepares quality deficiency reports.

Recommendation, Level III

In the lower-division baccalaureate/associate degree category, 3 semester hours in basic electronics laboratory, 2 in principles of DC circuits, 2 in principles of AC circuits, 3 in solid state electronics, and 3 in electronic systems troubleshooting and maintenance (10/91).

Recommendation, Level IV

In the lower-division baccalaureate/associate degree category, 3 semester hours in basic electronics laboratory, 2 in principles of DC circuits, 2 in principles of AC circuits, 3 in solid state electronics, 3 in electronic systems troubleshooting and maintenance, 3 in personnel supervision, and 3 in maintenance management (10/91).

MCE-6468-001

AIRCRAFT ELECTRICAL EQUIPMENT TEST SET/
MOBILE ELECTRONIC TEST SET
TECHNICIAN, INTERMEDIATE
MAINTENANCE ACTIVITY (IMA)
6468

Exhibit Dates: 1/90–Present.

Occupational Field: 64 (Avionics).

Career Pattern

PVT: Private (E-1). PFC: Private First Class (E-2). LCP: Lance Corporal (E-3). CPL: Corporal (E-4). SGT: Sergeant (E-5). NOTE: MOS duties are not defined in terms of rank or paygrade, but by levels. Level I identifies tasks that are taught in the classroom; the individual is able to perform simple parts of tasks. Level II identifies those tasks that are part of on-the-job training; the trainee can do most tasks pertaining to a particular system and needs supervision on more difficult parts of tasks. Level III indicates that the individual can perform all essential tasks without supervision. Level IV indicates a high level of proficiency in job performance; the individual can perform advanced technical functions, instruction, inspection, and supervision. The MOS designator 6411 (Aircraft Communications/Navigation Systems Technician—Trainee) is used to identify individuals in Levels I and II. Upon completion of level II, MOS 6468 is awarded. Credit is recommended for Levels III and IV.

Description

Summary: Inspects and maintains automatic electronic test equipment at the intermediate maintenance level. *Levels I and II:* Applies theory of operation and use of vacuum tubes and solid state devices, such as diodes, transistors, integrated circuits, transformers, motors, generators, servomechanisms, electromechanical drives, transmission lines, waveguides, and similar components as applicable to automatic test systems; uses applicable service manuals and technical publications; practices proper safety procedures. *Level III:* Able to perform the duties required for Levels I and II; demonstrates knowledge of first aid practices; isolates and diagnoses failures in replaceable assemblies and ancillary equipment; performs necessary corrective actions prescribed in applicable maintenance directives. *Level IV:* Able to perform the duties required for Level III; submits accurate maintenance and supply information; ensures that all test and measuring equipment is within its calibration tolerances prescribed by directives; plans work schedules; incorporates appropriate avionics changes as directed; conducts technical on-the-job training; prepares quality deficiency reports.

Recommendation, Level III

In the lower-division baccalaureate/associate degree category, 3 semester hours in basic electronics laboratory, 2 in principles of DC circuits, 2 in principles of AC circuits, 3 in solid state electronics, and 3 in electronic systems troubleshooting and maintenance (10/91).

Recommendation, Level IV

In the lower-division baccalaureate/associate degree category, 3 semester hours in basic electronics laboratory, 2 in principles of DC circuits, 2 in principles of AC circuits, 3 in solid state electronics, 3 in electronic systems troubleshooting and maintenance, 3 in personnel supervision, and 3 in maintenance management (10/91).

MCE-6469-001

ADVANCED AUTOMATIC TEST EQUIPMENT
TECHNICIAN, INTERMEDIATE
MAINTENANCE ACTIVITY (IMA)
6469

Exhibit Dates: 1/90–Present.

Occupational Field: 64 (Avionics).

Career Pattern

May progress to MOS 6469 from MOS 6462, 6463, 6464, 6465, 6466, 6467, or 6468 (Private through Sergeant). SSGT: Staff Sergeant (E-6). GYSGT: Gunnery Sergeant (E-7). NOTE: MOS duties are not defined in terms of rank or paygrade, but by levels. Level I identifies tasks that are taught in the classroom; the individual is able to perform simple parts of tasks. Level II identifies those tasks that are part of on-the-job training; the trainee can do most tasks pertaining to a particular system and needs supervision on more difficult parts of tasks. Level III indicates that the individual can perform all essential tasks without supervision. Level IV indicates a high level of proficiency in job performance; the individual can perform advanced technical functions, instruction, inspection, and supervision. Credit is recommended for Levels III and IV. An individual in MOS 6469 must be qualified at Level IV.

Description

Summary: Tests, maintains, repairs, and analyzes airborne weapon replaceable assemblies, shop replaceable assemblies, automatic test equipment, and ancillary equipment beyond normal fault isolation procedures. Able to perform the Level IV duties of MOS's 6462, 6463, 6464, 6465, 6466, 6467, and 6468; possesses thorough understanding and working knowledge of sound management principles, work center supervision, and accepted maintenance procedures; performs quality assurance inspections; monitors data sources and prepares work center reports and records; performs expanded test procedures that use concise fault isolation procedures; possesses a fundamental knowledge of computer language at the programmer level; if rank of Gunnery Sergeant has been attained, organizes and administers facilities for the maintenance and repair of airborne weapon replaceable assemblies.

Recommendation

In the lower-division baccalaureate/associate degree category, 2 semester hours in records and information management and 3 in introduction to computing. In the upper-division baccalaureate category, if rank of Gunnery Sergeant (GYSGT) has been attained, 3 semester hours for field experience in management. NOTE: Add credit for Level IV of previously-held MOS (10/91).

MCE-6482-001

AIRCRAFT ELECTRICAL COUNTERMEASURES
SYSTEMS TECHNICIAN, FIXED WING,
INTERMEDIATE MAINTENANCE ACTIVITY
(IMA)
6482

Exhibit Dates: 1/90–Present.

Occupational Field: 64 (Avionics).

Career Pattern

PVT: Private (E-1). PFC: Private First Class (E-2). LCP: Lance Corporal (E-3). CPL: Corporal (E-4). SGT: Sergeant (E-5). NOTE: MOS duties are not defined in terms of rank or paygrade, but by levels. Level I identifies tasks that are taught in the classroom; the individual is able to perform simple parts of tasks. Level II identifies those tasks that are part of on-the-job training; the trainee can do most tasks pertaining to a particular system and needs supervision on more difficult parts of tasks. Level III indicates that the individual can perform all essential tasks without supervision. Level IV indicates a high level of proficiency in job performance; the individual can perform advanced technical functions, instruction, inspection, and supervision. The MOS designator 6411 (Aircraft Communications/Navigation Systems Technician—Trainee) is used to identify individuals in Levels I and II. Upon completion of level II, MOS 6482 is awarded. Credit is recommended for Levels III and IV.

Description

Summary: Maintains, tests, and repairs replaceable components and related equipment for aircraft electronic countermeasures. *Levels I and II:* Applies theory of operation and use of vacuum tubes, transistors, diodes, solid state devices, integrated circuit components, transformers, motors, generators, synchros, servos, electromechanical drive mechanisms, transmission lines, and antennas and the theory of impedance, resistance, capacitance, and inductance to aircraft electronic countermeasure systems/subsystems; uses electronic schematics, block diagrams, wiring diagrams, and other technical data contained in appropriate technical publications; uses technical and supply publications to obtain accurate data for requisitioning replacement parts; observes shop safety precautions; demonstrates a working knowledge of first aid; diagnoses replaceable assemblies and ancillary equipment, using general and specialized test equipment and following appropriate maintenance instruction manuals; recognizes corrosion/deterioration of electronic components and performs the necessary corrective action. *Level III:* Able to perform the duties required for Levels I and II; submits accurate and timely maintenance/supply data information on applicable maintenance action forms; ensures that all "in use" test and measuring equipment assigned are within their calibration/qualification cycle per applicable instructions and directives; incorporates appropriate avionics changes in aircraft electronic countermeasures systems equipment as directed; conducts technical and on-the-job training; prepares complete and accurate quality deficiency reports. *Level IV:* Able to perform the duties required for Level III; supervises maintenance activities; performs quality assurance inspections on all maintenance actions; plans and schedules daily work load.

Recommendation, Level III

In the lower-division baccalaureate/associate degree category, 3 semester hours in basic electronics laboratory, 2 in principles of DC circuits, 2 in principles of AC circuits, 3 in solid state electronics, 3 in electronic systems troubleshooting and maintenance, and 3 in digital principles (10/91).

Recommendation, Level IV

In the lower-division baccalaureate/associate degree category, 3 semester hours in basic electronics laboratory, 2 in principles of DC circuits, 2 in principles of AC circuits, 3 in solid state electronics, 3 in electronic systems troubleshooting and maintenance, 3 in digital principles, 3 in personnel supervision, and 3 in maintenance management (10/91).

MCE-6483-001

AIRCRAFT ELECTRONIC COUNTERMEASURES
  SYSTEMS TECHNICIAN, HELICOPTER,
  INTERMEDIATE MAINTENANCE ACTIVITY
  (IMA)
  6483

Exhibit Dates: 1/90–Present.

Occupational Field: 64 (Avionics).

Career Pattern

PVT: Private (E-1). PFC: Private First Class (E-2). LCP: Lance Corporal (E-3). CPL: Corporal (E-4). SGT: Sergeant (E-5). NOTE: MOS duties are not defined in terms of rank or paygrade, but by levels. Level I identifies tasks that are taught in the classroom; the individual is able to perform simple parts of tasks. Level II identifies those tasks that are part of on-the-job training; the trainee can do most tasks pertaining to a particular system and needs supervision on more difficult parts of tasks. Level III indicates that the individual can perform all essential tasks without supervision. Level IV indicates a high level of proficiency in job performance; the individual can perform advanced technical functions, instruction, inspection, and supervision. The MOS designator 6411 (Aircraft Communications/Navigation Systems Technician—Trainee) is used to identify individuals in Levels I and II. Upon completion of the level II, MOS 6483 is awarded. Credit is recommended for Levels III and IV.

Description

Summary: Maintains, tests, and repairs replaceable components and related equipment for aircraft electronic countermeasures. *Levels I and II:* Applies theory of operation and use of vacuum tubes, transistors, diodes, solid state devices, integrated circuit components, transformers, motors, generators, synchros, servos, electromechanical drive mechanism, transmission lines, and antennas and the theory of impedance, resistance, capacitance, and inductance as applicable to aircraft electronic countermeasure systems/subsystems; uses electronic schematics, block diagrams, wiring diagrams, and other technical data contained in appropriate technical publications; uses technical and supply publications to obtain accurate data for requisitioning replacement parts; observes shop safety precautions; demonstrates a working knowledge of first aid; diagnoses replaceable assemblies and ancillary equipment, using general and specialized test equipment and following appropriate maintenance instruction manuals; recognizes corrosion/deterioration of electronic components and performs the necessary corrective action. *Level III:* Able to perform the duties required for Levels I and II; submits accurate and timely maintenance/supply data information on applicable maintenance action forms; ensures that all "in use" test and measuring equipment assigned are within their calibration/qualification cycle per applicable instructions and directives; incorporates appropriate avionics changes in aircraft electronic countermeasure systems equipment as directed; conducts technical and on-the-job training; prepares complete and accurate quality deficiency reports. *Level IV:* Able to perform the duties required for Level III; supervises maintenance activities; performs quality assurance inspections on all maintenance actions; plans and schedules daily work load.

Recommendation, Level III

In the lower-division baccalaureate/associate degree category, 3 semester hours in basic electronics laboratory, 2 in principles of DC circuits, 2 in principles of AC circuits, 3 in solid state electronics, 3 in electronic systems troubleshooting and maintenance, and 3 in digital principles (10/91).

Recommendation, Level IV

In the lower-division baccalaureate/associate degree category, 3 semester hours in basic electronics laboratory, 2 in principles of DC circuits, 2 in principles of AC circuits, 3 in solid state electronics, 3 in electronic systems troubleshooting and maintenance, 3 in digital principles, 3 in personnel supervision, and 3 in maintenance management (10/91).

MCE-6484-001

AIRCRAFT ELECTRONIC COUNTERMEASURES
  SYSTEMS TECHNICIAN, EA-6,
  INTERMEDIATE MAINTENANCE ACTIVITY
  (IMA)
  6484

Exhibit Dates: 1/90–Present.

Occupational Field: 64 (Avionics).

Career Pattern

PVT: Private (E-1). PFC: Private First Class (E-2). LCP: Lance Corporal (E-3). CPL: Corporal (E-4). SGT: Sergeant (E-5). NOTE: MOS duties are not defined in terms of rank or pay grade, but by levels. Level I identifies tasks that are taught in the classroom; the individual is able to perform simple parts of tasks. Level II identifies those tasks that are part of on-the-job training; the trainee can do most tasks pertaining to a particular system and needs supervision on more difficult parts of tasks. Level III indicates that the individual can perform all essential tasks without supervision. Level IV indicates a high level of proficiency in job performance; the individual can perform advanced technical functions, instruction, inspection, and supervision. The MOS designator 6411 (Aircraft Communications/Navigation Systems Technician—Trainee) is used to identify individuals in Levels I and II. Upon completion of level II, MOS 6484 is awarded. Credit is recommended for Levels III and IV.

Description

Summary: Maintains, tests, and repairs replaceable components and related equipment for aircraft electronic countermeasures. *Levels I and II:* Applies theory of operation and use of vacuum tubes, transistors, diodes, solid state devices, integrated circuit components, transformers, motors, generators, synchros, servos, electromechanical drive mechanisms, transmission lines, and antennas and the theory of impedance, resistance, capacitance, and inductance to aircraft electronic countermeasure systems/subsystems; uses electronic schematics, block diagrams, wiring diagrams, and other technical data contained in appropriate technical publications; uses technical and supply publications to obtain accurate data for requisitioning replacement parts; observes shop safety precautions; demonstrates a working knowledge of first aid; diagnoses replaceable assemblies and ancillary equipment, using general and specialized test equipment and following appropriate maintenance instruction manuals; recognizes corrosion/deterioration of electronic components and performs the necessary corrective action. *Level III:* Able to perform the duties required for Levels I and II; submits accurate and timely maintenance/supply data on applicable maintenance action forms; ensures that all "in use" test and measuring equipment assigned are within their calibration/qualification cycle per applicable instructions and directives; incorporates avionics changes in aircraft electronic countermeasure systems equipment as directed; conducts technical and on-the-job training; prepares complete and accurate quality deficiency reports. *Level IV:* Able to perform the duties required for Level III; supervises maintenance activities; performs quality assurance inspections on all maintenance actions; plans and schedules daily work load.

Recommendation, Level III

In the lower-division baccalaureate/associate degree category, 3 semester hours in basic electronics laboratory, 2 in principles of DC circuits, 2 in principles of AC circuits, 3 in solid state electronics, 3 in electronic systems troubleshooting and maintenance, and 3 in digital principles (10/91).

Recommendation, Level IV

In the lower-division baccalaureate/associate degree category, 3 semester hours in basic electronic laboratory, 2 in principles of DC circuits, 2 in principles of AC circuits, 3 in solid state electronics, 3 in electronic systems troubleshooting and maintenance, 3 in digital principles, 3 in personnel supervision, and 3 in maintenance management (10/91).

MCE-6485-001

ADVANCED AIRCRAFT ELECTRONIC
  COUNTERMEASURES TECHNICIAN,
  INTERMEDIATE MAINTENANCE ACTIVITY
  (IMA)
  6485

Exhibit Dates: 1/90–Present.

Occupational Field: 64 (Avionics).

Career Pattern

May progress to MOS 6485 from MOS 6482 or 6484 (Private through Sergeant). SSGT: Staff Sergeant (E-6). GYSGT: Gunnery Sergeant (E-

7). NOTE: MOS duties are not defined in terms of rank or paygrade, but by levels. Level I identifies tasks that are taught in the classroom; the individual is able to perform simple parts of tasks. Level II identifies those tasks that are part of on-the-job training; the trainee can do most tasks pertaining to a particular system and needs supervision on more difficult parts of tasks. Level III indicates that the individual can perform all essential tasks without supervision. Level IV indicates a high level of proficiency in job performance; the individual can perform advanced technical functions, instruction, inspection, and supervision. Credit is recommended for Levels III and IV. An individual in MOS 6485 must be qualified at Level IV.

Description

Summary: Tests, maintains, repairs, and analyzes replaceable assemblies, test equipment, and ancillary equipment beyond normal fault isolation procedures. Able to perform the Level IV duties of MOS's 6482 and 6484; possesses a thorough understanding and working knowledge of sound management principles, work center supervision, proper methods of training personnel, and naval aircraft maintenance procedures; performs quality assurance inspections for maintenance actions performed on equipment and components; assists in the planning and scheduling of work center assignments; monitors source data collection and assists in the preparation of reports and records; performs troubleshooting using concise fault isolation procedures; has a fundamental understanding of computer language at the programmer level; if rank of Gunnery Sergeant has been attained, organizes and administers facilities for the maintenance and repair of airborne weapon replaceable assemblies and shop replaceable assemblies.

Recommendation

In the lower-division baccalaureate/associate degree category, 2 semester hours in records and information management and 3 in introduction to computing. In the upper-division baccalaureate category, if rank of Gunnery Sergeant (GYSGT) has been attained, 3 semester hours for field experience in management. NOTE: Add credit for Level IV of previously-held MOS (10/91).

MCE-6492-001

AVIATION PRECISION MEASUREMENT EQUIPMENT/
AUTOMATIC TEST EQUIPMENT
CALIBRATION AND REPAIR TECHNICIAN,
INTERMEDIATE MAINTENANCE ACTIVITY
(IMA)
6492

Exhibit Dates: 1/90–Present.

Occupational Field: 64 (Avionics).

Career Pattern

PVT: Private (E-1). PFC: Private First Class (E-2). LCP: Lance Corporal (E-3). CPL: Corporal (E-4). SGT: Sergeant (E-5). SSGT: Staff Sergeant (E-6). GYSGT: Gunnery Sergeant (E-7). NOTE: MOS duties are not defined in terms of rank or paygrade, but by levels. Level I identifies tasks that are taught in the classroom; the individual is able to perform simple parts of tasks. Level II identifies those tasks that are part of on-the-job training; the trainee can do most tasks pertaining to a particular system and needs supervision on more difficult parts of tasks. Level III indicates that the individual can perform all essential tasks without supervision. Level IV indicates a high level of proficiency in job performance; the individual can perform advanced technical functions, instruction, inspection, and supervision. The MOS designator 6411 (Aircraft Communications/Navigation Systems Technician—Trainee) is used to identify individuals in Levels I and II. Upon completion of level II, MOS 6492 is awarded. Credit is recommended for Levels III and IV.

Description

Summary: Tests, maintains, repairs, and calibrates aviation precision measurement and automatic test equipment. *Levels I and II:* Applies theory of operation and use of vacuum tubes, solid state devices, integrated circuit, mechanical mechanisms, servomechanism systems, electronic and mechanical components to test and measuring equipment; uses electronic and mechanical documentation to determine precision and sensitivity of test equipment; observes shop safety practices for electrical, chemical, microwave, and mechanical systems; demonstrates normal first aid procedures; disassembles precision measuring equipment for calibration and/or repair, reassembles, and performs recalibration; submits documentation relative to repair and calibration of measurement and test equipment; uses mechanical, optical, and electrical standards to document accuracy of measurement and automatic test equipment. *Level III:* Able to perform the duties required for Levels I and II; requisitions supplies; installs support equipment changes and completes necessary documentation; ensures that all "in use" standards are within calibration cycle; instructs subordinates. *Level IV:* Able to perform the duties required for Level III; performs quality assurance inspections on all maintenance actions; supervises maintenance activities; coordinates requirements with aviation and electronic supply systems; prepares quality deficiency reports; plans and schedules daily work schedules; maintains appropriate records of all calibration/repair of subordinates; coordinates activities with user facilities; if rank of of Gunnery Sergeant was attained, organizes and administers maintenance facilities.

Recommendation, Level III

In the lower-division baccalaureate/associate degree category, 3 semester hours in basic electronics laboratory, 2 in principles of DC circuits, 2 in principles of AC circuits, 3 in solid state electronics, 3 in electronic systems troubleshooting and maintenance, and 2 in dimensional metrology (10/91).

Recommendation, Level IV

In the lower-division baccalaureate/associate degree category, 3 semester hours in basic electronics laboratory, 2 in principles of DC circuits, 2 in principles of AC circuits, 3 in solid state electronics, 3 in electronic systems troubleshooting and maintenance, 2 in dimensional metrology, 3 in personnel supervision, and 3 in maintenance management; if rank of Staff Sergeant was attained, 2 semester hours in records and information management. In the upper-division baccalaureate category, if rank of Gunnery Sergeant was attained, 3 semester hours for field experience in management (10/91).

MCE-6521-001

AVIATION ORDNANCE MUNITIONS TECHNICIAN
6521

Exhibit Dates: 5/92–Present.

Occupational Field: 65 (Aviation Ordnance).

Career Pattern

PVT: Private (E-1). PFC: Private First Class (E-2). LCP: Lance Corporal (E-3). CPL: Corporal (E-4). SGT: Sergeant (E-5). SSGT: Staff Sergeant (E-6). GYSGT: Gunnery Sergeant (E-7). NOTE: MOS duties are not defined in terms of rank or paygrade, but by levels. Level I identifies tasks that are taught in the classroom; the individual is able to perform simple parts of tasks. Level II identifies those tasks that are part of on-the-job training; the trainee can do most tasks pertaining to a particular system and needs supervision on more difficult parts of tasks. Level III indicates that the individual can perform all essential tasks without supervision. Level IV indicates a high level of proficiency in job performance; individual can perform advanced technical functions, instruction, inspection, and supervision. The MOS designator 6511 (Aviation Ordnance—Trainee) is used to identify individuals in Levels I and II. Upon completion of Level II, MOS 6521 is awarded. Credit is recommended for Levels III and IV.

Description

Summary: Performs duties incident to ordering, receiving, storage, assembly, issuing, transporting, and accountability of air-delivered and certain ground ammunition; builds, operates, and maintains ready storage of air-delivered ammunition/missiles. *Levels I and II:* Provides safety, security, and record keeping for ordnance; uses and maintains aviation ammunition-handling and transportable equipment; identifies, stores and inventories ammunition by types, nomenclature, and explosive hazard; assembles and transports ammunition using appropriate equipment. *Level III:* Able to perform the duties required for Levels I and II; operates and maintains a magazine service storage area; ensures that all applicable shipping regulations and inspection procedures applicable to ammunition handling are observed; requisitions ammunition; maintains files; prepares messages, correspondence, and reports; determines the serviceability of ammunition; trains aircraft ordnance technicians in all phases of aircraft and ground ammunition procurement, handling, and transportation; trains technicians in weapon assembly equipment and procedures. *Level IV:* Able to perform the duties required for Level III; establishes a ready service storage missile facility; operates ammunition stock recording system; trains ordnance personnel in all phases of aviation ammunition-handling procedures; coordinates ammunition support at other activities and organizations; ensures that administrative procedures necessary to operate an aviation ammunition-handling and assembly facility and ready service storage area are followed; conducts inspections; prepares naval messages, orders, instructions, and safety bulletins; performs the duties of quality assurance inspector; establishes training programs for all aspects of ammunition technician duties, tasks, and responsibilities.

Recommendation, Level III

In the lower-division baccalaureate/associate degree category, 3 semester hours in occupational safety, 2 in records and information management, and 1 in technical writing (5/92).

Recommendation, Level IV

In the lower-division baccalaureate/associate degree category, 3 semester hours in occupational safety, 2 in records and information management, 1 in technical writing, 3 in personnel supervision, and 3 in maintenance management. In the upper-division baccalaureate category, 2 semester hours for field experience in management, if rank was Staff Sergeant (SSGT) and 3 semester hours for field experience in management, if rank was Gunnery Sergeant (GYSGT) (5/92).

MCE-6531-001

AIRCRAFT ORDNANCE TECHNICIAN
6531

Exhibit Dates: 5/92–Present.

Occupational Field: 65 (Aviation Ordnance).

Career Pattern

PVT: Private (E-1). PFC: Private First Class (E-2). LCP: Lance Corporal (E-3). CPL: Corporal (E-4). SGT: Sergeant (E-5). SSGT: Staff Sergeant (E-6). GYSGT: Gunnery Sergeant (E-7). NOTE: MOS duties are not defined in terms of rank or paygrade, but by levels. Level I identifies tasks that are taught in the classroom; the individual is able to perform simple parts of tasks. Level II identifies those tasks that are part of on-the-job training; the trainee can do most tasks pertaining to a particular system and needs supervision on more difficult parts of tasks. Level III indicates that the individual can perform all essential tasks without supervision. Level IV indicates a high level of proficiency in job performance; the individual can perform advanced technical functions, instruction, inspection, and supervision. The MOS designator 6511 (Aviation Ordnance—Trainee) is used to identify individuals in Levels I and II. Upon completion of Level II, MOS 6531 is awarded. Credit is recommended for Levels III and IV.

Description

Summary: Inspects, maintains, and repairs armament equipment; loads aviation ordnance an aircraft. *Levels I and II:* Employs safety, security, and record keeping techniques for ordnance; operates and maintains armament weapons support equipment; loads and unloads weapons and stores them on aircraft; tests aircraft weapons and release and control systems of missiles and gun systems; arms and de-arms aircraft; tests and maintains racks, launchers, adapters, and electrical components of aircraft armament circuits; installs and maintains aerial target towing equipment and its associated support equipment; when directed, loads nuclear weapons; conducts functional tests of aircraft armament electrical fusing, firing, and release circuits and maintains them in an operational manner; applies boresight procedures and techniques to aircraft sight and weapons systems. *Level III:* Able to perform the duties required for Levels I and II; procures supplies; maintains records, prepares reports, and utilizes publications pertinent to aviation ordnance and aircraft armament equipment; performs quality control on all ordnance functions; reports unsatisfactory or defective equipment and material; trains aircraft ordnance technicians in all phases of squadron-level aviation ordnance. *Level IV:* Able to perform the duties required for Level III; establishes a training program; applies administrative procedures necessary to operate a squadron ordnance section; conducts administrative and material inspections; performs the duties of quality assurance inspector; prepares naval messages, orders, instructions, and safety notices.

Recommendation, Level III

In the lower-division baccalaureate/associate degree category, 2 semester hours in introduction to DC electricity, 3 in occupational safety, 2 in records and information management, and 1 in technical writing (5/92).

Recommendation, Level IV

In the lower-division baccalaureate/associate degree category, 2 semester hours in introduction to DC electricity, 3 in occupational safety, 2 in records and information management, 1 in technical writing, 3 in personnel supervision, and 3 in maintenance management. In the upper-division baccalaureate category, 2 semester hours for field experience in management, if rank was Staff Sergeant (SSGT) and 3 semester hour for field experience, if rank was Gunnery Sergeant (GYSGT) (5/92).

MCE-6541-001

AVIATION ORDNANCE EQUIPMENT REPAIR TECHNICIAN
6541

Exhibit Dates: 5/92–Present.

Occupational Field: 65 (Aviation Ordnance).

Career Pattern

PVT: Private (E-1). PFC: Private First Class (E-2). LCP: Lance Corporal (E-3). CPL: Corporal (E-4). SGT: Sergeant (E-5). SSGT: Staff Sergeant (E-6). GYSGT: Gunnery Sergeant (E-7). NOTE: MOS duties are not defined in terms of rank or paygrade, but by levels. Level I identifies tasks that are taught in the classroom; the individual is able to perform simple parts of tasks. Level II identifies those tasks that are part of on-the-job training; the trainee can do most tasks pertaining to a particular system and needs supervision on more difficult parts of tasks. Level III indicates that the individual can perform all essential tasks without supervision. Level IV indicates a high level of proficiency in job performance; the individual can perform advanced technical functions, instruction, inspection, and supervision. The MOS designator 6511 (Aircraft Ordnance—Trainee) is used to identify individuals in Levels I and II. Upon completion of Level II, MOS 6541 is awarded. Credit is recommended for Levels III and IV.

Description

Summary: Repairs, inspects, tests, adjusts, stows, and accounts for airborne equipment, missiles, and weapons; maintains wide variety of highly technical aircraft armament weapon systems; performs quality assurance, safety, record keeping, and maintenance management duties. *Levels I and II:* Performs inspections, tests, checks, adjustments, preventive maintenance, and repair on support equipment, missile launching equipment, multiple ejection/bomb racks, aircraft guns, turrets, aerial targets, and associated equipment; performs operator maintenance on armament weapon systems support/transporting equipment; assembles and performs maintenance on air-launched guided missiles; performs preventive maintenance and minor repair on testing, handling, and missile launching equipment. *Level III:* Able to perform the duties required for Levels I and II; performs the duties of collateral duty inspector for all ordnance equipment functions; conducts training programs; maintains files; records modifications to missiles and equipment; prepares technical reports; performs corrosion control actions. *Level IV:* Able to perform the duties required for Level III; performs duties of quality assurance inspector; establishes a rigid training program for aviation ordnance equipment repair technicians in all phases of aviation ordnance/missile functions; uses aviation ordnance administrative procedures necessary to establish and operate an intermediate maintenance activity; conducts administrative and material inspections; prepares naval messages, orders, instructions, and safety notices.

Recommendation, Level III

In the lower-division baccalaureate/associate degree category, 2 semester hours in introduction to DC electricity, 3 in occupational safety, 2 in records and information management, and 1 in technical writing (5/92).

Recommendation, Level IV

In the lower-division baccalaureate/associate degree category, 2 semester hours in introduction to DC electricity, 3 in occupational safety, 2 in records and information management, 1 in technical writing, 3 in personnel supervision, and 3 in maintenance management. In the upper-division baccalaureate category, 2 semester hours for field experience in management, if rank was Staff Sergeant (SSGT) and 3 semester hours for field experience in management, if rank was Gunnery Sergeant (GYSGT) (5/92).

MCE-6591-001

AVIATION ORDNANCE CHIEF
6591

Exhibit Dates: 5/92–Present.

Occupational Field: 63/64 (Avionics).

Career Pattern

May progress to MOS 6591 from MOS's 6521, 6531, 6541, or 6561 (Private through Gunnery Sergeant). MSGT: Master Sergeant (E-8). MGYSGT: Master Gunnery Sergeant (E-9). NOTE: Chiefs must be qualified in Level IV of previously-held MOS. Descriptions and credit recommendations for chiefs are by rank. Check the individual's fitness report or brief sheet for the rank and for proficiency report.

Description

Summary: Supervises an aviation ordnance facility. *Master Sergeant:* Supervises the establishment and functioning of aviation ordnance activities; writes reports on quality control readiness analysis, personnel documentation, manpower documentation, and equipment and safety performance; makes arrangements for equipment and materiel transportation. *Master Gunnery Sergeant:* Able to perform the duties required for Master Sergeant; provides authoritative staff support in planning and implementing aviation ordnance, as a member of a major staff; prepares and presents reports on logistics and readiness to highest command levels; has a thorough knowledge of Marine Corps administrative procedures and aviation ordnance staff organization and functioning.

Recommendation, Master Sergeant

In the lower-division baccalaureate/associate degree category, 3 semester hours in personnel supervision, 3 in technical writing, and 3 in records and information management. In the upper-division baccalaureate category, 3 semester hours in logistics management, 3 in management problems, and 3 for field experience in management. NOTE: Credit should also be granted for the prerequisite MOS (5/92).

Recommendation, Master Gunnery Sergeant

In the lower-division baccalaureate/associate degree category, 3 semester hours in personnel supervision, 3 in technical writing, and 3 in records and information management. In the upper-division baccalaureate category, 3 semester hours in logistics management, 6 in management problems, and 6 for field experience in management. NOTE: Credit should also be granted for the prerequisite MOS (5/92).

Appendix A

The Evaluation Systems

BACKGROUND

Early editions of the *Guide to the Evaluation of Educational Experiences in the Armed Services* were prepared in response to specific needs. Immediately after World War II, the consensus in the educational community was that the practice of granting blanket credit to World War I veterans as a reward for length of service was educationally unsound. Educators concluded that military learning experiences applicable to civilian curricula should be assessed by faculty for potential credit. Therefore, in December 1945, at the request of civilian educational institutions and the regional accrediting associations, the American Council on Education (ACE), established the Commission on Accreditation of Service Experiences, renamed the Commission on Educational Credit and Credentials in 1979, to evaluate military educational programs and to assist institutions in granting credit for such experiences. The first edition of the *Guide* was published in 1946.

The extension of the World War II G.I. Bill to include veterans of the Korean conflict, and the subsequent enrollment of many veterans in colleges and universities, created a need for the second edition, published in 1954.

The 1968 edition was prepared in anticipation of the increased enrollment of veterans resulting from the educational assistance provided under the Veterans Readjustment Benefits Act of 1966, and with the expectation that many would apply for educational credit for their learning experiences in the armed services. In addition, technological advances had necessitated major changes in service training, with a resulting need for new or revised educational credit recommendations.

The 1974 edition was prepared primarily to respond to three emerging considerations. First, because of the growth in vocational and technical programs and the emergence of the concept of postsecondary education, there was a need to evaluate courses for possible credit in the vocational and technical categories in addition to the baccalaureate and graduate categories of previous editions. Second, active-duty servicemembers were enrolling in increasing numbers in civilian educational programs and were seeking credit for military formal courses soon after completing their service school training. Third, credit recommendations were needed for the many courses initiated or revised by the military since 1968.

The 1974 edition marked the beginning of a new approach to reporting evaluations of formal military training. At its fall 1973 meeting, the Commission approved the concept of an ongoing *Guide* system. Elements of that system included the publication of biennial editions of the *Guide* through computerized composition, continual staff review of courses, and the computerized storage of course information for a more rapid updating of credit recommendations. In 1994, the computerized *Guide* system came in-house, and all data are managed by the Military Evaluations Program staff.

Over the years the recommendations contained in the *Guide* have assisted education institutions in granting credit to hundreds of thousands of servicemembers. Surveys showed that most of the nation's colleges and universities use the formal course recommendations in awarding credit to veterans and active-duty service personnel. The recommendations have been widely accepted because military formal courses share certain key elements with traditional postsecondary programs. They are formally approved and administered, are designed for the purpose of achieving learning outcomes, are conducted by qualified persons with specific subject-matter expertise, and are structured to provide for the reliable and valid assessment of student learning.

The recommendations reflect the Commission's belief that it is sound educational practice to give recognition for learning, no matter how or where that learning has been attained, provided that the learning is at the appropriate level, is in the appropriate area, and is applicable to an individual's postsecondary program of study.

Until 1975, however, no mechanism existed for providing recognition for the learning a servicemember attained through such learning experiences as self-instruction, on-the-job training, and work experience. In 1975, the Commission implemented a program for the evaluation of learning represented by demonstrated proficiency in Army enlisted military occupational specialties (MOS's). The MOS evaluation procedures were developed, tested, and refined during a feasibility study conducted by ACE and sponsored by the Department of the Army. Evaluators made recommendations for educational credit and advanced standing in apprentice training programs. Subsequently, the occupational assessment program of the Commission was expanded to to include Navy general rates, ratings, warrant officers and limited duty officers, Army warrant officer MOS's, Navy warrant officer and limited duty officer specialties, Coast Guard enlisted ratings and warrant officers, and selected Marine Corps MOS's. A small number of Naval Enlisted Classifications (NECs) have also been evaluated.

In 1994, ACE published the *1954–1989 Guide to the Evaluation of Educational Experiences in the Armed Services*. It contains all courses and occupations with exhibit dates of 1954 to December 1989. **Please retain the *1954–1989 Guide* as a permanent reference and use it with the current *Guide to the Evaluation of Educational Experiences in the Armed Services.***

APPENDIX A

Beginning with the 1994 edition, the *Guide* contains all course and occupation exhibits with start dates of 1/90 and later. This is also true for the 1996 *Guide*.

THE COURSE EVALUATION SYSTEM

Courses listed in the *Guide* are service school courses conducted on a formal basis, i.e., approved by a central authority within each service and listed by the service in its catalog. These courses are conducted for a specified period of time with a prescribed course of instruction, in a structured learning situation, and with qualified instructors.

Most courses are given on a full-time basis. After 1981, ACE began evaluation of courses that are 45 academic hours in length. Prior to that time courses evaluated were at least of two weeks duration, or, if less than two weeks in length, the course had to include a minimum of 60 contact hours of instruction. Before 1973, the minimum length requirement was three weeks or 90 contact hours.

In the fall of 1973, the Commission approved the following procedures and guidelines for the evaluation of military formal courses.

The Evaluation Process

Courses are evaluated by teams of at least three subject-matter specialists (college and university professors, deans, and other academicians). Through discussion and the application of evaluation procedures and guidelines, team members reach a consensus on the amount and category of credit to be recommended.

Evaluation materials include the course syllabus, training materials, tests, textbooks, technical manuals, and examinations. Additional information may be obtained from discussions with instructors and program administrators, classroom observations, and examination of instructional equipment and laboratory facilities.

Evaluators have two major tasks for each course: the formulation of a credit recommendation and the preparation of the course's description. The credit recommendation consists of the category of credit, the number of semester hours recommended, and the appropriate subject area. Evaluators phrase the course description (which appears in the *Guide* exhibits under the headings Learning Outcomes or Objectives and Instruction) in terms meaningful to civilian educators. The course description supplements the credit recommendations by summarizing the nature of a given course.

Selection of Evaluators

Nominations for course evaluators are requested from postsecondary institutions, professional and disciplinary societies, education associations, other evaluators, and regional accrediting associations.

The criteria for the selection of formal course evaluators are as follows:

1. The area of an evaluator's competence will closely approximate the area of the training to be evaluated.

2. Preference will be given to candidates with five or more years of postsecondary teaching or administrative experience, including curriculum development.

3. Preference will be given to candidates who are generally receptive to the recognition of learning that occurs in a variety of settings.

An evaluator candidate is interviewed by a staff member to determine whether the individual meets the selection criteria.

An effort is also made to obtain a diverse geographic representation on the team. Subject-matter specialists represent a variety of postsecondary institutional types.

THE COAST GUARD ENLISTED OCCUPATION CLASSIFICATION SYSTEM

Enlisted Rating Structure

The Coast Guard Enlisted Rating Structure is used for classifying enlisted personnel, identifying personnel qualifications, and reporting personnel requirements and resources. It also provides the framework for enlisted career development through paths of advancement from paygrades E–1 (recruit) through E–9 (master chief petty officer). For ACE purposes, there are two main types of occupational classifications in the Enlisted Rating Structure.

1. General Rates (Apprenticeships)—Identifications assigned to personnel at paygrades E–1, E–2, and E–3. There are two general rates: Seaman and Fireman. They each involve the performance of entry-level tasks and lead to the ratings.

2. Ratings—Broad occupational fields that encompass similar duties and functions and that, in most instances, provide paths of advancement and career development for personnel from paygrades E–4 (petty officer third class) to E–9 (master chief petty officer).

Ratings require performance of routine tasks at the lower paygrades and more difficult tasks at progressively higher paygrades. The relationship between petty officer designations and paygrades is shown below.

Petty Officer Classification	Pay Grade
Petty Officer Third Class (PO3)	E–4
Petty Officer Second Class (PO2)	E–5
Petty Officer First Class (PO1)	E–6
Chief Petty Officer (CPO)	E–7
Senior Chief Petty Officer (SCPO)	E–8
Master Chief Petty Officer (MCPO)	E–9

General rates and ratings are organized in career patterns. A career pattern provides the normal path of advancement from recruit (paygrade E–1) to master chief petty officer (paygrade E–9). An example of a career pattern using the Seaman general rate and the Quartermaster rating is shown here.

Title	Pay Grade
Seaman Recruit	E–1
Seaman Apprentice	E–2
Quartermaster Third Class	E–4
Quartermaster Second Class	E–5
Quartermaster First Class	E–6
Chief Quartermaster	E–7
Senior Chief Quartermaster	E–8
Master Chief Quartermaster	E–9

Standards for Advancement

The Coast Guard's requirements for enlisted minimum skills are defined and contained in two types of minimum standards: military standards and occupational standards. Military standards consist of qualifications (knowledge and practical factors) that specify the skills and knowledge required as a minimum for advancement to specific paygrades. These include military requirements and professional development standards. Occupational standards specify the skills and knowledge that apply to enlisted personnel as a minimum for advancement in a specific general rate or rating in addition to the military standards. They are divided into practical and knowledge factors and are presented as individual qualification items. For both military standards and occupational standards, each higher paygrade represents more complex duties, increased skills, and greater responsibility.

Special Qualifications

The Coast Guard's special qualifications, identified by qualification codes, supplement the enlisted rating structure by identifying special skills and knowledge that require a more refined or specific identification by the assignment of qualification codes to enlisted personnel who meet the stated eligibility requirements. Qualification codes are normally attained by the completion of a technical school or, in some cases, on-the-job training. Because of the specialized nature and limited scope of the skills and qualifications required, qualification codes have not been evaluated.

The Coast Guard Enlisted Evaluation and Advancement System

The Coast Guard regularly evaluates the occupational proficiency of its men and women. In fact, the demonstration of occupational proficiency is directly linked to the advancement system. Individuals are allowed to take the advancement examination after meeting these four criteria: (1) demonstrate that they can perform the tasks required for the next higher paygrade, (2) complete the appropriate correspondence and/or residence courses for the next higher paygrade, (3) serve a minimum length of time in their paygrade and in the service, and (4) be recommended by their commanding officer.

The primary evaluation technique is the written examination for each paygrade of each rating. It contains 175–200 multiple-choice questions that are based on the occupational standards and tasks for each paygrade of a given rating. It is given more weight than any other factor in the evaluation process. Persons are not advanced in their rating until they have demonstrated that they are proficient in the next higher paygrade of the rating. A final multiple score is computed for each individual. It is composed of the following factors:

Factor	Maximum Credit
Servicewide Exam Score	80
Performance Marks	50
Time in Service	20
Time in Paygrade	20
Medals and Awards	10
Bonus Points	2
Total Possible	182

THE COAST GUARD WARRANT OFFICER CLASSIFICATION SYSTEM

Warrant officers are technical officer specialists who perform duties that are technically oriented and acquired through experience and training and that are limited in scope in relation to other officer categories.

There are four grades of warrant officers: Warrant Officer, W–1 (WO1) (discontinued in 1975); Chief Warrant Officer, W–2 (CWO2); Chief Warrant Officer, W–3 (CWO3); and Chief Warrant Officer, W–4 (CWO4). The grades reflect salary increases and are normally indicative of the length of time a person has served as a warrant officer. *The grades do not signify differences in job duties.* A position requiring a warrant officer may be filled by any qualified warrant from WO1 to CWO4.

Selection and Evaluation

The warrant officer program provides an opportunity for appointment to commissioned warrant officer status for selected senior enlisted personnel. Competition is keen and individuals generally start preparing early in their careers. Potential candidates normally improve their skills through on-the-job training and specialized training through schools and correspondence courses.

Selection boards, composed of senior officers of experience, maturity, and varied backgrounds, are convened annually. Each candidate's qualifications are evaluated, and the board recommends those deemed best qualified.

Selection boards also meet annually to recommend those qualified for promotion. Performance is the major key to promotion. The selection procedure for promotion has always recognized the outstanding performer; mediocre performers are judged not to be competitive enough. Evaluations are recorded on Officer Evaluation Reports, an objective appraisal of performance by the senior reporting officer.

Time-in-grade is another criterion for promotion. Until May 1967, the time-in-grade requirements were:

Promotion to W–2: 2 years in grade W–1
Promotion to W–3: 4 years in grade W–2
Promotion to W–4: 4 years in grade W–3.

After May 1, 1967, grade W–1 was discontinued; time-in-grade requirements are now:

Promotion to W–3: 4 years in grade W–2
Promotion to W-4: 4 years in grade W-3.

THE MARINE CORPS ENLISTED OCCUPATION CLASSIFICATION SYSTEM

The Marine Corps MOS system is composed of occupational fields and MOS's within each field. The occupational system is constructed on the concept that similar skill and knowledge requirements are grouped in functional areas (occupational fields), which provide for the most efficient and effective classification, assignment, promotion, and utilization of personnel.

An MOS code has four digits and a descriptive title. The MOS describes a group of related duties and tasks that extend over one or more grades. The first two digits designate the occupational field and the last two digits identify the promotional channel and the specialty within the occupational field. The fourth digit may be a skill designator, which identifies additional skill or knowledge requirements usually acquired through experience or advanced schooling. A "0" in the fourth digit indicates a *basic* MOS (see next paragraph), and a "1" indicates trainee status. In some occupational fields, primarily the aviation fields, the skill designator may refer to the particular type of aircraft on which the Marine is qualified.

Each occupational field contains a *basic* MOS, which represents the learner level (e.g., MOS 6000 is a Basic Aircraft Maintenance Marine). It is assigned as a specific MOS in that field when assigned for training. When in retraining status, a Marine's primary MOS will be changed to the basic MOS.

There are two main categories of MOS's. Category "A" enlisted MOS's (occupational fields 01 through 73) are assigned as primary MOS's. Enlisted personnel are promoted on the basis of holding a category A primary MOS. A category A MOS identifies the primary skill and knowledge of a Marine.

Category "B" MOS's (8000–9599 series) are used to identify billets. They may be assigned only as additional MOS's. These MOS's designate a particular skill or training that is in addition to a Marine's primary MOS.

The relationship between rank and paygrade is shown here:

Private (PVT)	E–1
Private First Class (PFC)	E–2
Lance Corporal (LCP)	E–3
Corporal (CPL)	E–4
Sergeant (SGT)	E–5
Staff Sergeant (SSGT)	E–6
Gunnery Sergeant (GYSGT)	E–7
First Sergeant (1STSGT)/Master Sergeant (MSGT)	E–8
Sergeant Major (SGT/MAJ)/ Master Gunnery Sergeant (MGYSGT)	E–9

Performance Evaluation System

At this time only the *aircraft maintenance* and *avionics* occupational fields have established a performance evaluation system that meets ACE's criteria. The evaluation program, known as *Individual Training Standards System (ITSS), Maintenance Training Management and Evaluation Program (MATMEP)*, provides for the evaluation of a Marine's job performance. It is a standardized, documentable, level-progressive, technical skills management and evaluation program for enlisted aviation technical maintenance training.

The *ITSS (MATMEP)* is organized as a two-way grid. The row headings identify the specific duties and tasks that define an MOS. The column headings identify the skill levels at which each MOS duty and task are performed. When a Marine can perform a particular task at a specified level of skill, the Marine's supervisor evaluates the performance. If the Marine performs satisfactorily, the supervisor dates and signs the *ITSS (MATMEP)*. When all duties and tasks defining a particular skill level can be performed satisfactorily by a Marine, the supervisor dates and signs the *ITSS (MATMEP)* summary sheet in the location designated "Completed Level _____." When a Marine's *ITSS (MATMEP)* summary sheet indicates that the Marine has completed Level III, this indicates that *all* the MOS-requisite tasks have been certified at Level III and that the ACE credit recommendations may be applied.

For aircraft maintenance and avionics MOS's. Marines typically attend one of the formal training schools. This training is then followed by formal and informal on-the-job training at a fleet duty post. The *ITSS (MATMEP)* Levels I and II reflect the initial training or apprentice levels of an MOS. Upon completion of the initial training school, a Marine is typically certifiable at Level I. After one or two years of post school on-the-job experience, the Marine may be certified as a Level II performer.

Level III MOS attainment represents an advanced or skilled level of performance of an assigned job and indicates that a Marine can perform the technical aspects of the task or job with little or no supervision. It ordinarily takes three or four years of working in an MOS for a Marine to attain Level III certification.

Level IV indicates that the Marine is a highly skilled performer and is able to supervise lower-skill-level-MOS holders on a regular basis and can perform shop management duties related to the MOS.

A Marine's level of certification is not directly related to the Marine's military rank. Both military rank and skill level are associated with the length of time a Marine has served—the longer the time in service the more likely is a Marine to hold a higher rank and the more likely is a Marine to have acquired a higher level of MOS skill. However, rank and MOS skill proceed on independent tracks in many instances. In certain instances (e.g., when a longer-serving Marine switches from one MOS to another), a corporal may be certified at a higher skill level in a specific MOS than a sergeant in the same MOS.

ACE Credit Recommendations

Credit has been recommended for Levels III and IV in aircraft maintenance and avionics MOS's.

To document proficiency, institutions should verify that Level III or Level IV has been attained by checking the *ITSS (MATMEP)* summary sheet. This sheet is available to Marines from their duty station supervisors. The form must be signed by a certifying officer.

THE OCCUPATION EVALUATION SYSTEM

The ACE evaluation system for occupations has three major components: the selection of evaluators, the materials required for evaluation, and the procedures and guidelines evaluators use in reaching decisions and making recommendations.

Selection of Evaluators

Nominations for evaluators are requested from postsecondary institutions; professional and disciplinary societies; education associations; and regional accrediting associations.

The criteria for selection of evaluators are as follows:
1. The area of an evaluator's competence will closely approximate the area of the training to be evaluated.
2. Preference will be given to candidates with five or more years of postsecondary teaching or administrative experience, including curriculum development.
3. Preference will be given to candidates who are generally receptive to the recognition of learning that occurs in a variety of settings.

An evaluator candidate is interviewed by a staff member to determine whether the individual meets the selection criteria.

An effort is also made to obtain a diverse geographic representation on the team. Subject-matter specialists represent a variety of postsecondary institutional types.

Materials Required for Evaluation

In order to make a recommendation, evaluators first identify the skills, competencies, and knowledge associated with a given occupational specialty. The materials relevant to the evaluation are made available to staff members and evaluators by the military services. Materials include the official Coast Guard manuals that describe the duties and qualifications for each occupation; rate training manuals and other publications used by Coast Guard enlisted men and women in the day-to-day performance of their duties and to prepare for their advancement examinations; and the advancement examination. Marine Corps materials include the official MOS manual that describes the duties and qualifications for each MOS; individual training standards manuals; and the Maintenance Training Management and Evaluation Program (MATMEP) task list. Additional information is obtained by observing and interviewing Coast Guard servicemembers and Marines during visits to installations.

The Evaluation Process

Evaluators identify the skills, competencies, and knowledge required of servicemembers who are qualified in a given occupational specialty and relate that demonstrated learning to the same attributes acquired by students who have completed a comparable postsecondary course or curriculum. Because the evaluations are based on a comparison of learning outcomes, the amount of time a given enlisted man or woman may have spent acquiring occupational proficiency is not taken into consideration. The emphasis is on translating the learning demonstrated through occupational proficiency into terms used in formal civilian postsecondary education systems to recognize the same learning. This reflects the belief of the Commission that the value of learning is not dependent on where or how the learning occurs.

Evaluation teams are assigned three tasks in the evaluation process: to identify the learning represented by occupational proficiency by reviewing the pertinent written materials and by observing men and women performing their occupation and interviewing them and their supervisors; to prepare a description of the duties, skills, competencies, and knowledge required for each specialty; and to make recommendations for each specialty based on discussion and consensus.

Throughout the evaluation process, evaluators exercise professional judgment in applying the evaluative criteria and procedures. This position reflects the Commission's belief that sound educational evaluation is more dependent on professional judgment and expertise than on rigid application of criteria.

The Commission continually reviews its criteria and procedures. Evaluators are encouraged to provide feedback and recommendations for consideration by the Commission.

CREDIT CATEGORIES

Educational credit is a concept used by postsecondary institutions to quantify and record a student's successful completion of a unit of study. Postsecondary education consists of courses and programs of instruction for persons who are high school graduates or the equivalent, or who are beyond compulsory school age. ACE evaluators utilize the following categories of educational credit when formulating credit recommendations:

Vocational Certificate. This category describes course work of the type normally found in certificate or diploma (nondegree) programs that are usually a year or less in length and designed to provide students with occupational skills. Course content is specialized, and the accompanying shop, laboratory, or similar practical components emphasize procedural more than analytical skills.

Lower-Division Baccalaureate/Associate Degree. This category describes course work of the type normally found in the first two years of a baccalaureate program and in programs leading to the associate degree. The instruction stresses development of analytical abilities at the introductory level. Verbal, mathematical, and scientific concepts associated with an academic discipline are introduced, as are basic principles. Occupationally oriented courses in this category are normally designed to prepare a student to function as a technician in a particular field.

Upper-Division Baccalaureate. This category describes courses of the type found in the last two years of a baccalaureate program. The courses involve specialization of a theoretical or analytical nature beyond the introductory level. Successful performance by students normally requires prior study in the area.

Graduate Degree. This category describes courses with content of the type found in graduate programs. These courses often require independent study, original research, critical analysis, and the scholarly and professional application of the specialized knowledge or discipline. Students enrolled in such courses normally have completed a baccalaureate program.

Semester Hours

Credit recommendations for courses are not derived by simple arithmetic conversion. Evaluators exercise professional judgment and consider only those competencies that can be equated with civilian postsecondary curricula. Intensive courses offered by the military do not necessarily require as much outside preparation as many regular college courses. Evaluators consider the factors of pre- and post-course assignments, prior work-related experience, the concentrated nature of the learning experience, and the reinforcement of the course material gained in the subsequent work setting.

The occupation recommendations are based on the skills, competencies, and knowledge gained, as demonstrated through proficiency in the rating, without reference to how much time elapsed during the learning process. The semester hour is used as a standard to express how many semester hours of appropriate course work a student would normally complete to attain the same learning outcomes or attest to the same level of competency.

Credit recommendations are expressed in semester credit hours. In determining semester hour recommendations, evaluators will be guided by, but not restricted to, the following standard definitions:

1. One semester credit hour for the equivalent of 15 hours of classroom contact plus 30 hours of outside preparation; or
2. One semester credit hour for the equivalent of 30 hours of laboratory work plus necessary outside preparation, normally expected to be 15 hours; or
3. One semester credit hour for the equivalent of not less than 45 hours (contact hours) of shop instruction.

Other Resources

The Defense Activity for Non-Traditional Education Support (DANTES) maintains the educational records of the servicemembers who have completed DANTES Subject Standardized Tests (DSST's), CLEP examinations, USAFI (United States Armed Forces Institute), and GED tests.

Before July 1, 1974, the results of courses and tests taken under the auspices of USAFI (United States Armed Forces Institute, disestablished 1974) are available from the DANTES Program:

DANTES Program
The Chauncey Group International
P.O. Box 6605
Princeton, NJ 08541-6605

There is a $10.00 fee charged for *each* transcript requested. There is no charge for transcripts sent to military Test Control Officers (TCOs) for counseling purposes.

For GED tests taken overseas after July 1, 1974 write to:

GED Testing Service
One Dupont Circle
Washington, DC 2036-1193

For GED tests taken within the United States after July 1, 1974, write to the Department of Education in the state where the test was taken.

The results of DANTES Subject Standardized Tests (DSSTs) and CLEP tests taken under the auspices of the DANTES Program after July 1, 1974, are available from:

DANTES Program
The Chauncey Group International
P.O. Box 6604
Princeton, NJ 08541-6604
Telephone: 609-720-6740

A fee of $8.00 is charged for *each* transcript requested. A transcript may include any or all DSST and CLEP examinations taken while in the military. There is no charge for transcripts sent to military Test Control Officers (TCOs) for counseling purposes.

You may call the DANTES Progra at 609-720-6740 to order a transcript request form. Or you may request a transcript if you include all of the following information along with the appropriate fee:

Name (include all names tests will be registered under)
Your current address and phone number
Your Social Security Number
The address to which you would like your transcript mailed
Which test you would like included on the transcript; i.e. passing scores only, certain tests, etc.
Your signature is required -- this authorizes the DANTES Program to release the information

Transcript request fees may be paid by check or money order payable to the DANTES program. Payment may also be made with a Master Card or Visa (please include the expiration date). If paying by credit card, you may fax your request to 609-720-6800.

Records of individuals tested by the Veteran's Administration after October 1, 1989, may be obtained from:

Manager
Military Testing
GED Testing Service
American Council on Education
One Dupont Circle, Suite 250
Washington, DC 20036-1163

Appendix B

Sample Military Records

- DD Form 295 (1986) Application for the Evaluation of Learning Experiences During Military Service, page B–2
- DD Form 214 (1979) Certificate of Release or Discharge from Active Duty, page B–6
- DD Form 214 (1988) Certificate of Release or Discharge from Active Duty, page B–7
- Sample Course Completion Certificate, page B–8
- CG–3303 (1959) (Coast Guard enlisted) Achievement Sheet, page B–9
- CG–5311 (1988) Officer Evaluation Report (OER), page B–11
- ITSS (MATMEP) Individual Duty Area Qualification Summary, page B–15
- Request Pertaining to Military Records (Standard Form 180), page B–17

APPLICATION FOR THE EVALUATION OF
LEARNING EXPERIENCES DURING MILITARY SERVICE

_____
(Date)

TO: (Name and address of educational institution, agency, or employer)

EVALUATION REQUEST FOR:

_____
(Name of Applicant)

_____
(Social Security Number)

ATTENTION:

Dear Official:

    The applicant named above has requested that the attached summary of educational achievements, accomplished while in the Armed Forces of the United States, be forwarded to you for review and evaluation.

    The American Council on Education publishes the *Guide to the Evaluation of Educational Experiences in the Armed Services* which includes postsecondary credit evaluations of military learning experiences. The 1954 edition of the *Guide* contains recommendations for formal courses offered by the Armed Services during the period 1941 to 1954. The current edition contains credit recommendations for (1) military training courses offered after 1954; (2) Army military occupational specialties (MOS's) for enlisted personnel and warrant officers; (3) ratings held by Navy and Coast Guard enlisted personnel; and (4) occupational designators held by Navy and Coast Guard warrant officers and Navy limited duty officers. In addition to recommendations for semester hour credits, some Army enlisted MOS's and Navy ratings also have recommendations for advanced standing in apprentice training programs.

    The American Council on Education maintains an advisory service to provide credit recommendations for courses and tests, MOS's, ratings, and other occupations evaluated after the publication date of the current *Guide*. Credit recommendations are provided to officials of schools, state departments of education or other educational institutions, employers, apprenticeship training directors, labor union and trade association officials, military education officers and applicants. *Credit recommendations are not provided to officials at the applicant's request.* Authorized persons may write directly to the Military Evaluations Program Office, American Council on Education, One Dupont Circle, N.W., Washington, D.C. 20036-1193.

    The evaluation of this applicant's learning experiences, as well as any guidance which you may provide, should be sent directly to the applicant at the address shown in block 6 on page 3. Your interest is genuinely appreciated.

Sincerely,

_____
(Education Officer)

DD Form 295, NOV 86   Previous editions are obsolete.

Privacy Act Statement

AUTHORITY: 5 USC 301 and EO 9397, November 1943 (SSN).

PRINCIPAL PURPOSE: To permit authorized agencies to evaluate military experience for academic placement and/or employment.

ROUTINE USES: Used at the request of the individual for the evaluation of military training.

DISCLOSURE: Voluntary; however, failure to provide requested information impedes the evaluation process by educational institutions or potential employers.

INSTRUCTIONS TO APPLICANT

DD Form 295 is for your convenience in applying for evaluation of your educational experiences during military service. Give as much detailed information as possible. Include additional information on separate sheets, if necessary.

You are encouraged to write a preliminary letter to the school or agency concerned, explaining your interest in its evaluation of your records for the continuance of your education. Training, correspondence study, or special experiences not described on this form, which you believe would be of interest to those reviewing your case, should be included in this letter.

The applicant should:

a. Complete items 1 through 15.

b. If you have attended college or completed any college correspondence courses, ask that college to send a transcript to the Registrar of the evaluating agency that this form is addressed to. DO NOT LIST ANY COLLEGE OR UNIVERSITY COURSES ON THIS FORM.

c. If you have completed any college-level standardized examinations for credit, such as USAFI or DANTES Subject Standardized Tests, or CLEP, ask the appropriate agency to send a score report to the Registrar of the evaluating agency that this form is addressed to. DO NOT LIST ANY EXAMINATIONS ON THIS FORM.

d. After completion, submit this DD Form 295 to the Certifying Officer.

INSTRUCTIONS TO CERTIFYING OFFICER
(Custodian of Personnel Records)

DD Form 295 is intended to provide factual information that schools and other evaluating agencies require for evaluation of the applicant's educational achievement. CERTIFYING OFFICERS WILL NOT MAKE RECOMMENDATIONS REGARDING CREDIT TO BE AWARDED.

The certifying officer should:

a. Complete items 16 through 18.

b. Insure that the information provided in Section II is documented in the applicant's Service Record. Names of schools or courses should not be abbreviated.

c. Send this DD Form 295 to the Education Officer.

INSTRUCTIONS TO EDUCATION OFFICER

The education officer should:

a. Complete item 19.

b. Counsel the service member.

c. Complete page 1. The name and address of the evaluating agency should be the same as that listed at the top of page 3 of this form.

PAGE 1 IS IN ADDITION TO, AND NOT A SUBSTITUTE FOR, THE LETTER TO BE WRITTEN TO THE EVALUATING AGENCY BY THE APPLICANT.

d. Mail DD Form 295 directly to the designated evaluating agency.

APPLICATION FOR THE EVALUATION OF LEARNING EXPERIENCES DURING MILITARY SERVICE

TO (Name and address of educational institution, agency, or employer)

SECTION I - TO BE COMPLETED BY APPLICANT

1. NAME (Last, First, Middle Initial)	2. GRADE/RANK OR RATING	3. SOCIAL SECURITY NO.	4. PREVIOUS SERVICE NUMBER(S)

5. PRESENT BRANCH OF SERVICE (Includes National Guard and Reserve components)
☐ a. ARMY   ☐ b. NAVY   ☐ c. AIR FORCE   ☐ d. MARINE CORPS   ☐ e. COAST GUARD

6. APPLICANT'S MAILING ADDRESS FOR REPLY FROM EDUCATIONAL INSTITUTION

7. DATE OF BIRTH	8. PERMANENT HOME ADDRESS

CIVILIAN EDUCATION

9. HIGHEST GRADE OF SCHOOL COMPLETED (X one)
☐ 6   ☐ 7   ☐ 8   ☐ 9   ☐ 10   ☐ 11   ☐ 12

10. HIGHEST YEAR OF COLLEGE COMPLETED (X one)
☐ a. NONE   ☐ b. FRESHMAN   ☐ c. SOPHOMORE   ☐ d. JUNIOR   ☐ e. SENIOR

11. COLLEGE DEGREE EARNED (X if applicable)
☐ a. ASSOCIATE   ☐ b. BACHELOR

12. EDUCATIONAL INSTITUTION LAST ATTENDED

a. NAME	b. MAILING ADDRESS

13. USAFI COURSES COMPLETED IN SERVICE (Prior to 1974)
(The applicant should request a transcript for all courses to be forwarded directly to the evaluating agency.)

a. CATALOG NUMBER AND TITLE OF COURSE (If no courses were taken, print NONE)	b. METHOD OF STUDY (Correspondence, self-teaching, locally conducted classes, etc.)	c. LOCATION WHERE COMPLETED	d. DATE COURSE COMPLETED
(1)			
(2)			
(3)			
(4)			
(5)			
(6)			
(7)			
(8)			

14. MILITARY CORRESPONDENCE COURSE COMPLETED
(The applicant should attach a copy of the course completion letter or certificate.)

a. COURSE NAME (If no courses were taken, print NONE)	b. COURSE SPONSOR (AIPD, MCI, ECI, CGI)	c. DATE COURSE COMPLETED
(1)		
(2)		
(3)		
(4)		
(5)		
(6)		
(7)		
(8)		
(9)		

15. APPLICANT CERTIFICATION: I have read the Privacy Act Statement on Page 2.

a. SIGNATURE	b. DATE SIGNED

DD Form 295, NOV 86

SECTION II - TO BE COMPLETED BY CERTIFYING OFFICER
(Read Instructions on Page 2 before completing this page)

16. FORMAL SERVICE SCHOOLS ATTENDED *(If longer than one week) (If none, print NONE)*

	a. COURSE TITLE	b. MILITARY COURSE NUMBER	c. NAME OF SCHOOL, CITY, STATE	d. DATE ENTERED	e. LENGTH[1] (In weeks)	f. DATE COMPLETED	g. FINAL MARK AND/OR CLASS STANDING[2]	19. ACE GUIDE COURSE OR OCCUPATION IDENTIFICATION NO. *(To be filled out in Education Center)*
(1)								
(2)								
(3)								
(4)								
(5)								
(6)								
(7)								
(8)								
(9)								
(10)								

17. MILITARY OCCUPATIONAL HISTORY

	a. MILITARY SPEC. CODE (MOS, AFSC, Rate, etc.)[3]	b. MILITARY OCCUPATIONAL TITLE (Do Not Abbreviate)	c. DATES HELD From (Mo/yr)	To (Mo/yr)	d. MOS/SQT SCORE (For Army Enlisted Personnel[4])
(1)					
(2)					
(3)					

NOTES:
[1] Print SP if course length was self paced.
[2] If information is available, give grade received. If class standing is shown, give number in class, e.g., 10 in 241.
[3] List most recent skill levels or grade.
[4] MOS/SQT Evaluation Score and Date of evaluation.

THIS APPLICATION MUST BE SIGNED BY AN OFFICER OR A DULY AUTHORIZED NONCOMMISSIONED OFFICER.
I certify that the information contained herein has been compared with official records, and that this information is correct.

18. CERTIFYING OFFICER

a. NAME *(Print or Type)*	b. GRADE/RANK	c. MILITARY ADDRESS *(Include ZIP Code)*
d. SIGNATURE	e. DATE SIGNED	

DD Form 295, NOV 86

B-6 DD Form 214 (July 1979)

CAUTION: NOT TO BE USED FOR IDENTIFICATION PURPOSES
THIS IS AN IMPORTANT RECORD SAFEGUARD IT
ANY ALTERATIONS IN SHADED AREAS RENDER FORM VOID

DD FORM 214 (1 JUL 79)
PREVIOUS EDITIONS OF THIS FORM ARE OBSOLETE.
CERTIFICATE OF RELEASE OR DISCHARGE FROM ACTIVE DUTY

1. NAME (Last, first, middle)	
2. DEPARTMENT, COMPONENT AND BRANCH	NAVY - USN
3. SOCIAL SECURITY NO.	189 32 1767
4a. GRADE, RATE OR RANK	EWCM
4b. PAY GRADE	E9
5. DATE OF BIRTH	14 AUG 42
6. PLACE OF ENTRY INTO ACTIVE DUTY	San Diego, California
7. LAST DUTY ASSIGNMENT AND MAJOR COMMAND	USS ALBANY (CG-10)
8. STATION WHERE SEPARATED	USS ALBANY (CG-10) at Gaeta, Italy
9. COMMAND TO WHICH TRANSFERRED	Not applicable
10. SGLI COVERAGE AMOUNT $	20,000

11. PRIMARY SPECIALTY NUMBER, TITLE AND YEARS AND MONTHS IN SPECIALTY (Additional specialty numbers and titles involving periods of one or more years)
EW-1774 Electronics Warfare Systems Technician (SLQ-22/24)

12. RECORD OF SERVICE

	YEAR(s)	MON(s)	DAY(s)
a. Date Entered AD This Period	74	01	04
b. Separation Date This Period	80	05	13
c. Net Active Service This Period	06	04	10
d. Total Prior Active Service	13	03	22
e. Total Prior Inactive Service	00	00	00
f. Foreign Service	00	00	00
g. Sea Service	03	01	07
h. Effective Date of Pay Grade	79	12	16
i. Reserve Oblig. Term. Date	NA	NA	NA

13. DECORATIONS, MEDALS, BADGES, CITATIONS AND CAMPAIGN RIBBONS AWARDED OR AUTHORIZED (All periods of service)
Good Conduct Award (FIFTH)
Navy Expeditionary Medal
Armed Forces Expeditionary Medal
National Defense Service Medal
Republic of Viet-Nam Campaign Medal w/device
Viet-Nam Service Medal (7 awards)
Navy Unit Commendation Medal (2 awards)
Joint Service Commendation Medal

14. MILITARY EDUCATION (Course Title, number weeks, and month and year completed)
Instructor Basic - 4 weeks - Mar 74
OHS Advanced - 2 weeks - Apr 77
RD "B" School - 32 weeks - Apr 71
Electronics Warfare Operator "C" - 6 weeks - May 71
CET WLR-1 Series (7107) - 7 weeks - Jul 71
CET AN/ULQ-6 - 8 weeks - Sep 71
ASW Tactical Course - 1 week - Jan 63
RD "A" - 24 weeks - Jan 62

15. MEMBER CONTRIBUTED TO POST-VIETNAM ERA VETERANS' EDUCATIONAL ASSISTANCE PROGRAM: [X] NO
16. HIGH SCHOOL GRADUATE OR EQUIVALENT: [X] YES
17. DAYS ACCRUED LEAVE PAID: NONE

18. REMARKS
Immediate reenlistment, 14 May 1980. This form was administratively issued 14 May 1980. Block 13 continued: Combat Action Ribbon - Navy "E" Ribbon (2 awards).

19. MAILING ADDRESS AFTER SEPARATION
404 Twin Oaks Drive, Havertown, PA 19083

20. MEMBER REQUESTS COPY 6 BE SENT TO DIR. OF VET AFFAIRS: [X] NO

21. SIGNATURE OF MEMBER BEING SEPARATED

22. TYPED NAME, GRADE, TITLE AND SIGNATURE OF OFFICIAL AUTHORIZED TO SIGN
Assistant Personnel Officer

S/N 0102-LF-000-2140
MEMBER - 1

DD Form 214 (Nov 1988)   B-7

CAUTION: NOT TO BE USED FOR IDENTIFICATION PURPOSES   **THIS IS AN IMPORTANT RECORD. SAFEGUARD IT.**   **ANY ALTERATIONS IN SHADED AREAS RENDER FORM VOID**

CERTIFICATE OF RELEASE OR DISCHARGE FROM ACTIVE DUTY

1. NAME (Last, First, Middle)	2. DEPARTMENT, COMPONENT AND BRANCH	3. SOCIAL SECURITY NO.

4.a. GRADE, RATE OR RANK	4.b. PAY GRADE	5. DATE OF BIRTH (YYMMDD)	6. RESERVE OBLIG. TERM. DATE		
			Year	Month	Day

7.a. PLACE OF ENTRY INTO ACTIVE DUTY	7.b. HOME OF RECORD AT TIME OF ENTRY (City and state, or complete address if known)

8.a. LAST DUTY ASSIGNMENT AND MAJOR COMMAND	8.b. STATION WHERE SEPARATED

9. COMMAND TO WHICH TRANSFERRED	10. SGLI COVERAGE Amount: $	None

11. PRIMARY SPECIALTY (List number, title and years and months in specialty. List additional specialty numbers and titles involving periods of one or more years.)	12. RECORD OF SERVICE	Year(s)	Month(s)	Day(s)
	a. Date Entered AD This Period			
	b. Separation Date This Period			
	c. Net Active Service This Period			
	d. Total Prior Active Service			
	e. Total Prior Inactive Service			
	f. Foreign Service			
	g. Sea Service			
	h. Effective Date of Pay Grade			

13. DECORATIONS, MEDALS, BADGES, CITATIONS AND CAMPAIGN RIBBONS AWARDED OR AUTHORIZED (All periods of service)

14. MILITARY EDUCATION (Course title, number of weeks, and month and year completed)

15.a. MEMBER CONTRIBUTED TO POST-VIETNAM ERA VETERANS' EDUCATIONAL ASSISTANCE PROGRAM	Yes	No	15.b. HIGH SCHOOL GRADUATE OR EQUIVALENT	Yes	No	16. DAYS ACCRUED LEAVE PAID

17. MEMBER WAS PROVIDED COMPLETE DENTAL EXAMINATION AND ALL APPROPRIATE DENTAL SERVICES AND TREATMENT WITHIN 90 DAYS PRIOR TO SEPARATION		Yes	

18. REMARKS

19.a. MAILING ADDRESS AFTER SEPARATION (Include Zip Code)	19.b. NEAREST RELATIVE (Name and address - include Zip Code)

20. MEMBER REQUESTS COPY 6 BE SENT TO	DIR. OF VET AFFAIRS	Yes	No	22. OFFICIAL AUTHORIZED TO SIGN (Typed name, grade, title and signature)
21. SIGNATURE OF MEMBER BEING SEPARATED				

DD Form 214, NOV 88   S/N 0102-LF-006-5500   Previous editions are obsolete.   MEMBER

B-8   Sample Course Completion Certificate

Department Of The Navy

Graduate Certificate

Awarded To

_____

In recognition for successfully completing a ___160___ course
                                                     Clock Hours
entitled __Group Paced Instructor Course__ and was graduated
this __22nd__ day of __July__ 19__88__

Commanding Officer

Accredited By
Southern Association
Of Colleges & Schools

DEPARTMENT OF TRANSPORTATION U. S. COAST GUARD CG-3303 (Rev. 9-71)	ACHIEVEMENT SHEET (See instructions on reverse)

1. RATE RECORD

DATE	RATE	DESIG-NATOR	QUALIFICATION CODE	AUTHORITY	SIGNATURE (Grade and Title)	UNIT

2. DECORATIONS, MEDALS, BADGES, COMMENDATIONS, CITATIONS, AND CAMPAIGN RIBBONS AWARDED OR AUTHORIZED (Use "Remarks" block on back if necessary)

4. EDUCATION AND TEST RECORD

TEST	GCT	ARI	MECH	CLER	ETST	RCT	SPMT
DATE							
FORM NO.							
SCORE							
SIGNATURE							
DATE OF RETEST							
FORM NO.							
SCORE							
SIGNATURE							

3. RECORD OF SERVICE SCHOOLS ATTENDED

NAME AND LOCATION

DATE ENROLLED	CLASS NO.	COURSE LENGTH	RATE ON GRADUATION
DATE COMPLETED	CLASS STANDING	FINAL MARK (Std Score)	
	NUMBER IN CLASS		

GRADUATED (If, no, give reason) ☐ YES ☐ NO

SIGNATURE

HIGHEST SCHOOL LEVEL ATTAINED

LANGUAGE QUALIFICATIONS

NAME AND LOCATION

DATE ENROLLED	CLASS NO.	COURSE LENGTH	RATE ON GRADUATION
DATE COMPLETED	CLASS STANDING	FINAL MARK (Std Score)	
	NUMBER IN CLASS		

GRADUATED (If no, give reason) ☐ YES ☐ NO

SIGNATURE

5. RECORD OF COAST GUARD INSTITUTE COURSES COMPLETED

DATE COMPLETED	NAME OF COURSE	MARK	SIGNATURE

NAME AND LOCATION

DATE ENROLLED	CLASS NO.	COURSE LENGTH	RATE ON GRADUATION
DATE COMPLETED	CLASS STANDING	FINAL MARK (Std Score)	
	NUMBER IN CLASS		

GRADUATED (If no, give reason) ☐ YES ☐ NO

SIGNATURE

6. RECORD OF OTHER OFF-DUTY STUDY (USAFI, College Courses, etc.)

DATE COMPLETED	NAME OF COURSE	MARK	SIGNATURE

NAME (Last)	(First)	(Middle)	SERVICE NUMBER/SOCIAL SECURITY NO.	RATE

ACHIEVEMENT SHEET — PREVIOUS EDITIONS MAY BE USED — PAGE 3

Reverse of CG-3303 (Rev. 9-71)

INSTRUCTIONS

1. Prepare in duplicate. Retain original and duplicate in service record until separated from the Coast Guard. In case of immediate reenlistment, transfer original to new service record and forward duplicate with closed out service record to Commandant (PE). If individual does not immediately reenlist, the original copy shall be securely stapled to the individual's copy of Report of Separation from the Armed Forces of the United States (DD 214) and the individual notified that he should retain this page and present it with his discharge certificate should he desire to reenlist at any time either in the Regular Coast Guard or in the Coast Guard Reserve.

2. ITEM 1. Enter, as occurring, any changes in rate and acquirement of a designator and/or mechanical skill, such as completion of propeller (aviation), carburetor (aviation), refrigeration training.

3. ITEM 2. Enter date of receipt of all medals and awards received during current and/or prior period of service.

4. ITEM 3. School commands will complete upon graduation or disenrollment from a service school.

5. ITEM 4. Complete all entries as soon as practicable after enlistment. A man who has reached E-4 and has no test scores in his records, and no apparent future need for them exists, need not be examined solely for the purpose of recording such scores.

6. ITEM 5. To be completed when an individual completes Coast Guard Institute Courses.

7. ITEM 6. To be completed when an individual completes an USAFI, Navy, or Civilian Institution correspondence course, or an off-duty residence course.

REMARKS

DEPARTMENT OF TRANSPORTATION
U. S. COAST GUARD
CG-5311 (Page 1) (Rev. 12-88)

OFFICER EVALUATION REPORT (OER)
LEVEL I

1. ADMINISTRATIVE DATA

a. NAME (Last, First, Middle Initial)
b. SSN
c. STATUS INDICATOR/SPECIALTY
d. GRADE
e. DATE OF RANK — YR MO DAY

f. UNIT
g. DIST - OPFAC
h. OBC
i. DATE REPORTED

j. OCCASION FOR REGULAR REPORT
- Annual/Semiannual
- Detachment/Change of Reporting Officer
- Detachment of Officer
- Promotion of Officer

k. EXCEPTION REPORT
- Special
- Concurrent

l. PERIOD OF REPORT — TO

m. REPORTED-ON OFFICER SIGNATURE

n. DAYS NOT OBSERVED — TAD LV OTHER

o. DATE

2. DESCRIPTION OF DUTIES:

DOCUMENTS ATTACHED:

3. PERFORMANCE OF DUTIES: Measures an officer's ability to get things done.

Category	1	3	5	7	N/O
a. BEING PREPARED: Demonstrated ability to anticipate, to identify what must be done, to set priorities, and to prepare for accomplishing unit and organizational missions under both predictable and uncertain conditions.	Got caught by the unexpected. Appeared to be controlled by events/crises. Set vague or unrealistic goals, if any. Set wrong priorities. Tended not to follow existing operating procedures, plans, or systems. Not always prepared to meet operational or administrative responsibilities.	Anticipated well. Rarely caught unprepared. Set high but realistic goals. Took prompt positive action to meet changing or unexpected situations. Skillfully used existing operating procedures, plans, or systems and "did homework" to stay well prepared for responsibilities and missions.	Always ready. Never caught unprepared. Always looked beyond the immediate events/problems. Set the "right" priorities and controlled events. Achieved highest possible state of preparation for accomplishing responsibilities and missions. Turned potential adversity into opportunity.		
b. USING RESOURCES: Demonstrated ability to delegate, to provide follow-up control, and to utilize people, money, material, and time effectively.	Overlooked/underused available resources. Wasted materials, or improperly utilized publications and equipment. Did not always provide subordinates adequate resources and direction. Over/undersupervised; did not delegate wisely. Assigned wrong personnel to a given job. Failed to follow-up.	Successfully used available resources, publications, and equipment to complete assigned tasks. Budgeted own/others time productively. Found ways to cut waste. Delegates; made logical work assignments. Ensured subordinates had adequate tools, materials, time, and direction. Followed-up.	Got the most out of people. Used all available resources to the best advantage. Consistently came up with ways to save own and subordinates time, eliminate waste, and "do more with less" in producing high-quality work. Always followed up and knew what was going on.		
c. GETTING RESULTS: The quality/quantity of the officer's work accomplishments. The effectiveness or impact the results had on the officer's unit and/or the Coast Guard.	Usually met specified goals in routine situations. Occasionally produced or accepted work that needed upgrading or redoing. Results maintained the status quo.	Got the job done in all routine situations and in many unusual ones. Fulfilled identified goals and requirements even when resources were scarce. Produced finished quality work and required same from subordinates. Results had a positive impact on department and/or unit.	Got results which far surpassed your expectations in all situations. Always found ways to do more and do it better in spite of resource constraints. Own work and that of subordinates was consistently of high quality; never needed redoing. Results had significant positive impact on department and/or unit.		
d. RESPONSIVENESS: The degree to which the officer responded, replied, or met deadlines in a timely manner.	Needed reminding; did not report back. Tended to miss due dates or deadlines without justification. Slow or late responding to requests, memos, letters or calls. Resisted changes in policy, direction, or responsibilities.	Reported back; kept you informed. Dependably completed projects and met deadlines. Made timely responses to requests, memos, letters and calls. Took changes in policy, direction, or responsibilities in stride.	Highly conscientious; kept superiors well informed. Always completed projects early. Was unusually prompt in responding to all requests, memos, letters, and calls. Extremely flexible; responded enthusiastically to changes in policy, direction, or responsibilities.		
e. OPERATIONAL/SPECIALTY EXPERTISE: The acquisition of both knowledge and skills and the demonstration of both technical competency and proficiency in an operational/specialty billet. (Includes seamanship, airmanship, engineering, commercial vessel safety, SAR, law, etc., as appropriate.)	Failed to meet acceptable standards or demonstrate satisfactory progress in operational or specialty qualification. Required excessive guidance or supervision. Experienced difficulty grasping concepts or demonstrating proficiency. Failed to maintain qualifications. Recommendations were occasionally unreliable. Avoided opportunities to further develop or demonstrate operational or specialty expertise.	Competent authority on specialty or operational issues. Excellent acquisition and application of operational or specialty expertise (knowledge and skills) for assigned duties. Needed minimal supervision. Sought increased responsibility. Recommendations were reliable. Showed steady professional growth through education, training, and professional reading.	Superior operational or specialty expertise (knowledge and skills). Remarkable grasp of complex issues, concepts, and situations. Rarely needed guidance or supervision. Attitude reflected a "follow my lead!" approach. Rapid professional growth. An achiever. Advice typically flawless. Professional development beyond requirements. Significant achievements beyond performance of duties. Noteworthy examples.		
f. COLLATERAL DUTY/ADMINISTRATIVE EXPERTISE: The level of service knowledge, technical and managerial skills the officer demonstrated in collateral duties or in administrative responsibilities. (Includes CMCO, morale, civil rights, committees, etc., as appropriate.)	Required excessive guidance or supervision in routine activities. Slow to develop or "come up to speed."	Rapidly acquired necessary knowledge. Very competent dealing with complex issues, problems, or situations. Adept at determining, and then applying, correct procedures to manage the department or unit efficiently and accomplish command objectives. Rarely needed guidance or supervision.	Significant efficiency or organizational contributions to the unit or Coast Guard, or improvements to existing methods in areas of professional responsibility. Accomplishments had wide-ranging impacts. Noteworthy examples.		
g. WARFARE EXPERTISE: The acquisition of both knowledge and skills and the demonstration of both technical competency and proficiency. The officer's interest in the Coast Guard's warfare role as demonstrated by involvement in warfare-related education, training, and experience, regardless of billet.	Failed to meet acceptable standards or demonstrate satisfactory progress in acquisition of warfare expertise in a readiness billet. Lacked either motivation, interest, adaptability, or aptitude. Tunnel vision. Avoided opportunities to develop expertise including acquisition of essential knowledge or enhancement of skills when not in a readiness billet. Lacking basic vocabulary.	Utilized available opportunities to participate in operational readiness exercises or sought knowledge and experience in a designated crisis action position. Conversant on military readiness and warfare issues with both knowledge and appreciation for Coast Guard's role. Steady growth in warfare expertise through education, training, experience, or professional reading, regardless of billet.	Superior in-depth plan format knowledge and content development skills. Authority on maritime naval strategy, overall joint military strategy, and national political strategy. Participated in planning, execution and/or after-action exercise analysis in CPX, fleet or field Command Post environment. Professional development or contribution beyond billet requirements. Excellent candidate for demanding readiness billet.		

CG-5311 (Page 2) (Rev. 12-88)

h. COMMENTS:

4. INTERPERSONAL RELATIONS: Measures how an officer affects or is affected by others.

	1		3		5		7	N/O
a. WORKING WITH OTHERS: Demonstrated ability to promote a team effort, to cooperate, and to work with other people or units to achieve common goals.		Sometimes disregarded the ideas and feelings of others, or caused hostility because of failure to inform or consult. Impatient or impolite; talked too much or listened too little. Was inflexible, lost temper or control. Was slow to resolve conflicts. Not a team player. ○		Encouraged open expression of ideas and respected the views/ideas of others. Worked comfortably with others of all ranks/positions. Kept others informed; consulted others. Got different people and organizations to work together without mandates. Carried share of load. Helped others resolve conflicts and stay focused on team goals. ○		Excelled at getting all ranks/positions to work together. Skillfully used knowledge of group dynamics. Inspired cooperation among diverse individuals or groups. Stimulated open expression of ideas. Channeled group conflict into creative energy; achieved goals not otherwise obtainable. ○	○	○
b. HUMAN RELATIONS: The degree to which this officer fulfilled the letter and spirit of the Commandant's Human Relations Policy in personal relationships and official actions.		Exhibited discriminatory tendencies toward others due to their religion, age, sex, race, or ethnic background. Allowed bias to influence appraisals or the treatment of others. Used position to harass others; was disrespectful; made slurring remarks. Did not hold subordinates accountable for their human relations responsibilities. ○		Treated others fairly and with dignity regardless of religion, age, sex, race, or ethnic background. Carried out work, training, and appraisal responsibilities without bias. Held subordinates accountable for living up to the spirit of the Commandant's Human Relations Policy. ○		Through leadership and demonstrated strong personal commitment, promoted fair and equal treatment of others in all situations, regardless of religion, age, sex, race, or ethnic background. Actively campaigned against prejudicial actions or behavior by others. Made clearly noteworthy contributions to this end. ○	○	○

c. COMMENTS:

5. LEADERSHIP SKILLS: Measures an officer's ability to guide, direct, develop, influence, and support others in their performance of work.

	1		3		5		7	N/O
a. LOOKING OUT FOR OTHERS: The officer's sensitivity and responsiveness to the needs, problems, goals, and achievements of others.		Showed little concern for the safety, problems, needs, or goals of others. Overlooked or tolerated unfair, insensitive, or abusive treatment of others. May have been accessible to others, but unresponsive to their personal needs. Seldom acknowledged or recognized subordinates' achievements. ○		Cared about people. Recognized and responded to their needs. Concerned for their safety/well-being. Was accessible. Listened and helped with personal or job-related problems, needs, and goals. When unable to assist, suggested or provided other resources. "Went to bat" for people. Rewarded deserving subordinates in a timely fashion. ○		Demonstrated a commitment to develop and nurture a caring community in others. Personally ensured resources were available to meet people's needs and that limits of endurance were not exceeded. Was always accessible to others and their problems. Extremely conscientious in ensuring subordinates received appropriate and timely recognition. ○	○	○
b. DEVELOPING SUBORDINATES: The extent to which an officer used coaching, counseling, and training and provided opportunities for growth to increase the skills, knowledge, and proficiency of subordinates.		Showed little interest in training or development of subordinates. May have unnecessarily withheld authority or over-supervised. Did not challenge subordinates' abilities. Tolerated marginal performance, or criticized excessively. Did not keep subordinates informed; provided little constructive feedback. ○		Provided opportunities and encouraged subordinates to expand their roles, handle important tasks, and learn by doing. Held subordinates accountable; provided timely praise and constructive criticism. Provided opportunities for training which supported professional growth. ○		Created challenging situations which prompted an unusually high level development of people. Unit or work group always ran like "clockwork." People always knew what was going on and routinely handled the unexpected. Developed comprehensive and creative training programs; promoted a commitment to learning and personal development. ○	○	○
c. DIRECTING OTHERS: The officer's effectiveness in influencing or directing others in the accomplishment of tasks or missions.		An officer who had difficulty controlling and influencing others effectively. Did not instill confidence or enhance cooperation among subordinates and others. Set work standards which were vague or misunderstood. Tolerated late or marginal performance. Faltered in difficult situations. ○		A leader who earned the support and commitment of others. Set high work standards and expectations which were clearly understood and required subordinates to meet them. Evenhanded. Kept others motivated and on track even when "the going got tough." ○		A strong leader who commanded respect and inspired others to achieve results not normally attainable. People wanted to serve under his/her leadership. Communicated high work standards and expectations which were clearly understood. Got superior results even in time-critical and difficult situations. Won people over rather than imposing will. ○	○	○

CG-5311 (Page 3) (Rev. 12-88)

e. COMMENTS:

6. COMMUNICATIONS SKILLS: Measures an officer's ability to communicate in a positive, clear, and convincing manner.

a. SPEAKING AND LISTENING: How well an officer spoke and listened in individual exchanges, large or small groups, briefings or public situations; demonstrated ability to express verbal thoughts clearly, coherently, logically and extemporaneously.	1	Weak speaking or listening skills. Utilized inappropriate language or mannerisms. Expressed thoughts lacked preparation, confidence, common sense, or logic. Rambled or lost the audience. Failed to listen carefully. Argumentative. Identify specific situations that required better skills. ○	3	Accomplished speaker; comfortable in both public and private situations. Spoke in an articulate, confident, and credible manner with appropriate gestures and without distracting mannerisms. Not visibly uncomfortable in extemporaneous presentations. Listened attentively to others and the audience. ○	5	Displayed a remarkable ability to identify and discuss key issues, and to express thoughts clearly, coherently, and extemporaneously with credibility. Captivated and persuaded audiences. Chosen by superiors to make presentations on complex or sensitive issues, or when audience had unusual significance. ○	7	N/O
								○	○
b. WRITING: How well an officer communicated through written material and proofread before submission; demonstrated ability to prepare or review communication for superiors, self or subordinates and to express written thoughts clearly, coherently, logically and persuasively.		Written material frequently required revision for clarity, lack of proofreading, or requirements of the Coast Guard Correspondence or Style Manuals. ○		Written material set example for brevity, clarity, logic, persuasion, and tact. Correspondence grammatically correct and appropriate for the audience. Conscientious proofreader. Material from subordinates reflected the same high standards. ○		Expressed complex and controversial material in such a lucid and persuasive way that achievement of stated objectives was materially aided. Meticulous proofreader. Written material responsible for unit achievement or mission accomplishment, or published material brought credit upon CG. Provide noteworthy examples. ○		○ ○

c. COMMENTS:

7. SUPERVISOR AUTHENTICATION

a. NAME AND SIGNATURE	b. GRADE	c. SSN	d. TITLE OF POSITION	e. DATE

8. REPORTING OFFICER COMMENTS:

9. PERSONAL QUALITIES: Measures selected qualities which illustrate the character of the individual.

a. INITIATIVE: Demonstrated ability to move forward, make changes, and seek responsibility without guidance and supervision.	1	Postponed needed action. Implemented change only when confronted by necessity or directed to do so. Often overtaken by events. May have suppressed initiative of subordinates. Was unsupportive of changes directed by higher authority. ○	3	Strove to do the job better. Developed new ideas, methods, and practices. Got things done. Made improvements; "worked smarter, not harder." Self-starter; not afraid of making mistakes. Supported new ideas/methods/practices and efforts of others to bring about constructive change. Anticipated problems and took timely action to avoid/resolve them. ○	5	Aggressively sought additional responsibility. Was extremely innovative. Originated, nurtured, promoted, or brought about new ideas, methods, or practices which resulted in significant improvements to unit and/or Coast Guard. Did not promote change for sake of change. Made worthwhile ideas/practices work when others may have given up. ○	7	N/O
								○	○
b. JUDGMENT: Demonstrated ability to arrive at sound decisions and make sound recommendations by using experience, common sense, and analytical thought in the decision process.		Sometimes indecisive or showed uncertainty when making decisions. May have acted too quickly or too late. Did not take advantage of good sources of information. Needed watching; repeated mistakes. Made too many wrong decisions/recommendations. ○		Demonstrated analytical thought and common sense in making proper decisions or recommendations. Recognized developing problems and considered facts and alternatives. Asked for help when needed. Results demonstrated sound judgment in most cases. ○		Always did the "right" thing at the "right" time. Combined keen analytical thought and insight to make timely and successful decisions. Focused on the key issues and the most relevant information, even in complex situations. ○		○ ○
c. RESPONSIBILITY: Demonstrated commitment to getting the job done and to hold one's self accountable for own and subordinates' actions; courage of convictions; ability to accept decisions contrary to own views and make them work.		Usually could be depended upon to do the right thing. Normally accountable for own work. May have accepted less than satisfactory work or tolerated indifference. Tended not to get involved or speak up. Provided minimal support for decisions counter to own ideas. ○		Placed goals of Coast Guard above personal ambitions and gains. Possesses high standard of honor and integrity. Held self and subordinates accountable. Kept commitments even when uncomfortable or difficult to do so. Spoke up when necessary, even when position was unpopular. Supported organizational policies/decisions which may have been counter to own ideas. ○		Uncompromising honor and integrity. "Went the extra mile, and more." Always held self and subordinates accountable for production and actions. Had the courage to stand up and be counted. Succeeded in making even unpopular policies/decisions work. ○		○ ○
d. STAMINA: The officer's ability to think and act effectively under conditions that were stressful and/or mentally or physically fatiguing.		Performance became marginal under stress or during periods of extended work. Made poor decisions, overlooked key factors, focused on wrong priorities, or lost sight of safety considerations. Balked at putting in necessary overtime. Became rattled in stressful situations. ○		Performance was sustained at a high level when under stress or during periods of extended work without loss of productivity or safety. Stayed cool when the pressure was on. Willingly worked extra hours when necessary to get the job done. ○		Thrived under stressful situations. Performance reached an unusually high level when under stress or during periods of extended work. Productivity remained at an extremely high level with no increased risk to personnel and/or equipment. ○		○ ○
e. HEALTH AND WELL-BEING: The extent to which an officer exercised moderation in the use of alcohol. The degree to which an officer maintained weight standards. The measure of an officer's effort to invest in the Coast Guard's future by caring for his or her health.		Failed to meet minimum standards of weight control or sobriety. ○		Maintained weight standards. Used alcohol only discriminately or not at all; job performance and social behavior was never affected. Encouraged similar behavior in others and held subordinates accountable. Intemperate alcohol use by subordinates not tolerated. ○		Remarkable vitality, enthusiasm, alertness, and energy level. Consistently contributed at high standards. Demonstrated a significant commitment, beyond setting an example, to the well-being of self and subordinates. Contributed a leadership role in the civilian/military community outside normal duties. Noteworthy examples. ○		○ ○

CG-5311 (Page 4) (Rev. 12-88)

f. COMMENTS:

10. REPRESENTING THE COAST GUARD: Measures how an officer's ability to bring credit to the Coast Guard through looks and actions.

	1	3	5	7	N/O
a. MILITARY BEARING: The extent to which an officer appeared neat, smart and well groomed in uniform or civilian attire; conformed to military traditions, customs, and courtesies; and set standards for subordinates' performance.	Occasionally failed to conform to military traditions, or customs and courtesies. Unable or unwilling to consistently appear neat, smart, and well-groomed in uniform and civilian attire. Standards set in Uniform Regulations not maintained. Performance of subordinates was marginal or unacceptable.	The typically excellent officer. Demonstrated great care in maintaining and wearing uniforms. Meticulous grooming. Immaculate civilian attire. Precise in rendering military courtesies. Maintained military formality, precedence, etiquette, and deference to both rank and privilege. Required same of subordinates.	The typically distinguished officer. Clearly set standards for CG uniform and grooming excellence. Set or inspired similar standards in others. Performance of subordinates was exceptional. Exemplified the finest traditions of military customs, etiquette and protocol in very visible situations. Significant contributions or public recognition. Noteworthy examples.		
b. PROFESSIONALISM: How an officer applied knowledge and skills in providing service to the public. The manner in which the officer represented the Coast Guard.	Misinformed/unaware of Coast Guard policies and objectives and how they relate to own areas of responsibility. Bluffed rather than admit ignorance. Did little to enhance self-image or image of Coast Guard. Was ineffective when working with others. Led a personal life which infringed on Coast Guard responsibilities or image.	Well-versed in how Coast Guard objectives, policies, procedures serve the public; considered an expert in some areas. Was straightforward, cooperative, and evenhanded in dealing with the public and government. Aware of impact actions/ impressions may cause on others. Supported CG ideals. Personal life reinforced CG image.	The ideal officer to represent the Coast Guard. Inspired confidence and trust; clearly conveyed dedication to CG ideals in both public and private life. Worked creatively and confidently with representatives of public and government. Left everyone with a very positive image of self and Coast Guard.		
c. DEALING WITH THE PUBLIC: How an officer acted when dealing with other services, agencies, businesses, the media, or the public.	Appeared ill-at-ease with the public or media. Inconsistent in application of CG programs to public sector. Faltered under pressure. Took antagonistic or condescending approach. Made inappropriate statements. Embarrassed Coast Guard in a social situation.	Dealt fairly and honestly with the public, media and others at all levels. Responded promptly. Showed no favoritism. Didn't falter when faced with difficult situations. Was comfortable in social situations. Sensitive to concerns.	Always self-assured and in control when dealing with public, media and others at all levels. Straightforward, impartial, and diplomatic. Applied CG rules/programs fairly and uniformly. Showed unusual social grace. Responded with great poise to provocative actions of others.		

d. COMMENTS:

11. LEADERSHIP AND POTENTIAL. (Describe demonstrated leadership ability and overall potential for greater responsibility, promotion, special assignment, and command.)

12. COMPARISON SCALE AND DISTRIBUTION. (Compare this officer with others of the same grade whom you have known in your career.)

UNSATISFACTORY	A QUALIFIED OFFICER		ONE OF THE MANY COMPETENT PROFESSIONALS WHO FORM THE MAJORITY OF THIS GRADE			AN EXCEPTIONAL OFFICER	A DISTINGUISHED OFFICER
○	○	○	○	○	○	○	○

13. REPORTING OFFICER AUTHENTICATION

a. NAME AND SIGNATURE	b. GRADE	c. SSN	d. TITLE OF POSITION	e. DATE

14. REVIEWER AUTHENTICATION — COMMENTS ATTACHED (Required when the Reporting Officer is not a Coast Guard Officer.)

a. NAME AND SIGNATURE	b. GRADE	c. SSN	d. TITLE OF POSITION	e. DATE

15. RETURN ADDRESS. (Name and address to which a copy will be sent when the original is filed in the officer's record.)

16. HEADQUARTERS VALIDATION

PRIVACY ACT STATEMENT
This information is requested under the authority of 14 U.S.C. 633 to determine an officer's suitability for promotion or job assignment. Submission of this information is mandatory. Failure to provide it could adversely affect promotion opportunities and job assignments or lead to disciplinary action.

ITSS (MATMEP) AV-8B

INDIVIDUAL DUTY AREA QUALIFICATION SUMMARY
AIRCRAFT COMMUNICATIONS/NAVIGATION SYSTEMS TECHNICIAN (MOS 6315)

NAME/SSN: _____ Granted MOS 6311 ___/___ LEVEL II COMPLETED ___/___
                           Granted MOS 6315 ___/___ LEVEL III COMPLETED ___/___

DUTY #	DUTY DESCRIPTION	LEVEL I DATE / SIGN	LEVEL II DATE / SIGN	LEVEL III DATE / SIGN	LEVEL IV DATE / SIGN
A.	GENERAL, OPERATIONAL AND SAFETY DUTIES	___/___	xxxxxxxxxxxxxxxxxx	xxxxxxxxxxxxxxxxxx	xxxxxxxxxxxxxxxxxx
A.1	SUPPORT/SPECIAL EQUIPMENT	xxxxxxxxxxxxxxxxxx	___/___	xxxxxxxxxxxxxxxxxx	xxxxxxxxxxxxxxxxxx
A.2	SAFETY PRECAUTIONS/PROCEDURES	___/___	xxxxxxxxxxxxxxxxxx	xxxxxxxxxxxxxxxxxx	xxxxxxxxxxxxxxxxxx
A.3	AIRCRAFT PUBS, DIAGRAMS, SKETCHES & DRAWINGS	___/___	xxxxxxxxxxxxxxxxxx	xxxxxxxxxxxxxxxxxx	xxxxxxxxxxxxxxxxxx
A.4	PRECISION MEASURING EQUIPMENT	xxxxxxxxxxxxxxxxxx	___/___	xxxxxxxxxxxxxxxxxx	xxxxxxxxxxxxxxxxxx
A.5	PRINCIPLES OF ESD & EMC	___/___	xxxxxxxxxxxxxxxxxx	xxxxxxxxxxxxxxxxxx	xxxxxxxxxxxxxxxxxx
B.	SCHEDULED AND UNSCHEDULED MAINTENANCE DUTIES	xxxxxxxxxxxxxxxxxx	xxxxxxxxxxxxxxxxxx	xxxxxxxxxxxxxxxxxx	___/___
B.1	REQUIRED SCHEDULED/UNSCHEDULED INSPECTIONS	xxxxxxxxxxxxxxxxxx	___/___	___/___	xxxxxxxxxxxxxxxxxx
B.2	HIGH TIME/SPECIAL/CONDITIONAL INSPECTIONS	xxxxxxxxxxxxxxxxxx	xxxxxxxxxxxxxxxxxx	___/___	xxxxxxxxxxxxxxxxxx
B.3	TECHNICAL DIRECTIVES	xxxxxxxxxxxxxxxxxx	xxxxxxxxxxxxxxxxxx	___/___	xxxxxxxxxxxxxxxxxx
B.4	CORROSION CONTROL	___/___	xxxxxxxxxxxxxxxxxx	xxxxxxxxxxxxxxxxxx	xxxxxxxxxxxxxxxxxx
B.5	COMMUNICATION SYSTEM	___/___	xxxxxxxxxxxxxxxxxx	xxxxxxxxxxxxxxxxxx	xxxxxxxxxxxxxxxxxx
B.6	IFF SYSTEM	___/___	xxxxxxxxxxxxxxxxxx	xxxxxxxxxxxxxxxxxx	xxxxxxxxxxxxxxxxxx
B.7	TACAN SYSTEM	___/___	xxxxxxxxxxxxxxxxxx	xxxxxxxxxxxxxxxxxx	xxxxxxxxxxxxxxxxxx
B.8	ALL WEATHER LANDING SYSTEM	___/___	xxxxxxxxxxxxxxxxxx	xxxxxxxxxxxxxxxxxx	xxxxxxxxxxxxxxxxxx
B.9	RADAR BEACON SYSTEM	___/___	xxxxxxxxxxxxxxxxxx	xxxxxxxxxxxxxxxxxx	xxxxxxxxxxxxxxxxxx
B.10	ELECTRIC ALTIMETER SYSTEM	___/___	xxxxxxxxxxxxxxxxxx	xxxxxxxxxxxxxxxxxx	xxxxxxxxxxxxxxxxxx
B.11	COUNTERMEASURES DISPENSING SYSTEM	___/___	xxxxxxxxxxxxxxxxxx	xxxxxxxxxxxxxxxxxx	xxxxxxxxxxxxxxxxxx
B.12	RADAR WARNING RECEIVER SYSTEM	___/___	xxxxxxxxxxxxxxxxxx	xxxxxxxxxxxxxxxxxx	xxxxxxxxxxxxxxxxxx
B.13	DEFENSE ELECTRONIC COUNTERMEASURE SYSTEM	___/___	xxxxxxxxxxxxxxxxxx	xxxxxxxxxxxxxxxxxx	xxxxxxxxxxxxxxxxxx
B.14	COMNAV & RELATED SYSTEMS	xxxxxxxxxxxxxxxxxx	___/___	xxxxxxxxxxxxxxxxxx	xxxxxxxxxxxxxxxxxx
B.15	WIRE REPAIR	___/___	xxxxxxxxxxxxxxxxxx	xxxxxxxxxxxxxxxxxx	xxxxxxxxxxxxxxxxxx
C.	MAINTENANCE ADMINISTRATION DUTIES	xxxxxxxxxxxxxxxxxx	xxxxxxxxxxxxxxxxxx	xxxxxxxxxxxxxxxxxx	___/___
C.1	DATA COLLECTION FORMS	xxxxxxxxxxxxxxxxxx	___/___	xxxxxxxxxxxxxxxxxx	xxxxxxxxxxxxxxxxxx
C.2	MATERIAL CONTROL PROCEDURES	xxxxxxxxxxxxxxxxxx	___/___	xxxxxxxxxxxxxxxxxx	xxxxxxxxxxxxxxxxxx

DATE: JANUARY 1988

ITSS (MATMEP) AV-8B

IQS MOS 6315 (Continued)

DUTY #	DUTY DESCRIPTION	LEVEL I DATE / SIGN	LEVEL II DATE / SIGN	LEVEL III DATE / SIGN	LEVEL IV DATE / SIGN
D.	PRODUCTIVE INDIRECT WORK CENTER DUTIES	xxxxxxxxxxxxxxxxx	xxxxxxxxxxxxxxxxx	xxxxxxxxxxxxxxxxx	xxxxxxxxxxxxxxxxx
D.1	COLLATERAL DUTY INSPECTION	xxxxxxxxxxxxxxxxx	xxxxxxxxxxxxxxxxx	/	/
D.2	WORK CENTER SUPERVISOR DUTIES	xxxxxxxxxxxxxxxxx	xxxxxxxxxxxxxxxxx	/	/
D.3	FORMAL/INFORMAL MGT/TECHNICAL TRAINING	xxxxxxxxxxxxxxxxx	xxxxxxxxxxxxxxxxx	/	/

DATE: JANUARY 1988

REQUEST PERTAINING TO MILITARY RECORDS

Please read instructions on the reverse. If more space is needed, use plain paper.

PRIVACY ACT OF 1974 COMPLIANCE INFORMATION. The following information is provided in accordance with 5 U.S.C. 552a(e)(3) and applies to this form. Authority for collection of the information is 44 U.S.C. 2907, 3101, and 3103, and E.O. 9397 of November 22, 1943. Disclosure of the information is voluntary. The principal purpose of the information is to assist the facility servicing the records in locating and verifying the correctness of the requested records or information to answer your inquiry. Routine uses of the information as established and published in accordance with 5 U.S.C.a(e)(4)(D) include the transfer of relevant information to appropriate Federal, State, local, or foreign agencies for use in civil, criminal, or regulatory investigations or prosecution. In addition, this form will be filed with the appropriate military records and may be transferred along with the record to another agency in accordance with the routine uses established by the agency which maintains the record. If the requested information is not provided it may not be possible to service your inquiry.

SECTION I—INFORMATION NEEDED TO LOCATE RECORDS (Furnish as much as possible)

1. NAME USED DURING SERVICE *(Last, first, and middle)*
2. SOCIAL SECURITY NO.
3. DATE OF BIRTH
4. PLACE OF BIRTH

5. ACTIVE SERVICE, PAST AND PRESENT   (For an effective records search, it is important that ALL service be shown below)

BRANCH OF SERVICE *(Also, show last organization, if known)*	DATES OF ACTIVE SERVICE		Check one		SERVICE NUMBER DURING THIS PERIOD
	DATE ENTERED	DATE RELEASED	OFFICER	ENLISTED	

6. RESERVE SERVICE, PAST OR PRESENT   *If "none," check here* ▶ ☐

a. BRANCH OF SERVICE	b. DATES OF MEMBERSHIP		c. Check one		d. SERVICE NUMBER DURING THIS PERIOD
	FROM	TO	OFFICER ☐	ENLISTED ☐	

7. NATIONAL GUARD MEMBERSHIP   *(Check one):*   a. ARMY ☐   b. AIR FORCE ☐   c. NONE ☐

d. STATE	e. ORGANIZATION	f. DATES OF MEMBERSHIP		g. Check one		h. SERVICE NUMBER DURING THIS PERIOD
		FROM	TO	OFFICER ☐	ENLISTED ☐	

8. IS SERVICE PERSON DECEASED   ☐ YES   ☐ NO   *If "yes," enter date of death.*

9. IS (WAS) INDIVIDUAL A MILITARY RETIREE OR FLEET RESERVIST   ☐ YES   ☐ NO

SECTION II—REQUEST

1. EXPLAIN WHAT INFORMATION OR DOCUMENTS YOU NEED; OR, CHECK ITEM 2; OR, COMPLETE ITEM 3

2. IF YOU ONLY NEED A STATEMENT OF SERVICE   check here ☐

3. LOST SEPARATION DOCUMENT REPLACEMENT REQUEST *(Complete a or b. and c.)*

		YEAR ISSUED	
☐	a. REPORT OF SEPARATION *(DD Form 214 or equivalent)*		This contains information normally needed to determine eligibility for benefits. It may be furnished only to the veteran, the surviving next of kin, or to a representative with veteran's signed release (item 5 of this form).
☐	b. DISCHARGE CERTIFICATE		This shows only the date and character at discharge. It is of little value in determining eligibility for benefits. It may be issued only to veterans discharged honorably or under honorable conditions; or, if deceased, to the surviving spouse.

c. EXPLAIN HOW SEPARATION DOCUMENT WAS LOST

4. EXPLAIN PURPOSE FOR WHICH INFORMATION OR DOCUMENTS ARE NEEDED

6. REQUESTER

a. IDENTIFICATION *(check appropriate box)*
☐ Same person identified in Section I   ☐ Surviving spouse
☐ Next of kin (relationship) _____
☐ Other (specify)

b. SIGNATURE *(see instruction 3 on reverse side)*   DATE OF REQUEST

5. RELEASE AUTHORIZATION, IF REQUIRED
(Read instruction 3 on reverse side)
I hereby authorize release of the requested information/documents to the person indicated at right (item 7).

VETERAN SIGN HERE ▶ _____
(If signed by other than veteran show relationship to veteran.)

7. Please type or print clearly —   COMPLETE RETURN ADDRESS
Name, number and street, city, State and ZIP code

TELEPHONE NO. *(include area code)* ▶

INSTRUCTIONS

1. Information needed to locate records. Certain identifying information is necessary to determine the location of an individual's record of military service. Please give careful consideration to and answer each item on this form. If you do not have and cannot obtain the information for an item, show "NA," meaning the information is "not available." Include as much of the requested information as you can. This will help us to give you the best possible service.

2. Charges for service. A nominal fee is charged for certain types of service. In most instances service fees cannot be determined in advance. If your request involves a service fee you will be notified as soon as that determination is made.

3. Restrictions on release of information. Information from records of military personnel is released subject to restrictions imposed by the military departments consistent with the provisions of the Freedom of Information Act of 1967 (as amended in 1974) and the Privacy Act of 1974. A service person has access to almost any information contained in his own record. The next of kin, if the veteran is deceased, and Federal officers for official purposes, are authorized to receive information from a military service or medical record only as specified in the above cited Acts. Other requesters must have the release authorization, in item 5 of the form, signed by the veteran or, if deceased, by the next of kin. Employers and others needing proof of military service are expected to accept the information shown on documents issued by the Armed Forces at the time a service person is separated.

4. Location of military personnel records. The various categories of military personnel records are described in the chart below. For each category there is a code number which indicates the address at the bottom of the page to which this request should be sent. For each military service there is a note explaining approximately how long the records are held by the military service before they are transferred to the National Personnel Records Center, St. Louis. Please read these notes carefully and make sure you send your inquiry to the right address. Please note especially that the record is not sent to the National Personnel Records Center as long as the person retains any sort of reserve obligation, whether drilling or non-drilling.

(If the person has two or more periods of service within the same branch, send your request to the office having the record for the last period of service.)

5. Definitions for abbreviations used below:
NPRC—National Personnel Records Center PERS—Personnel Records
TDRL—Temporary Disability Retirement List MED—Medical Records

SERVICE	NOTE: (See paragraph 4 above.)	CATEGORY OF RECORDS	WHERE TO WRITE ADDRESS CODE	
AIR FORCE (USAF)	Except for TDRL and general officers retired with pay, Air Force records are transferred to NPRC from Code 1, 90 days after separation and from Code 2, 150 days after separation.	Active members (includes National Guard on active duty in the Air Force), TDRL, and general officers retired with pay.		1
		Reserve, retired reservist in nonpay status, current National Guard officers not on active duty in Air Force, and National Guard released from active duty in Air Force.		2
		Current National Guard enlisted not on active duty in Air Force.		13
		Discharged, deceased, and retired with pay.		14
COAST GUARD (USCG)	Coast Guard officer and enlisted records are transferred to NPRC 7 months after separation.	Active, reserve, and TDRL members.		3
		Discharged, deceased, and retired members *(see next item)*.		14
		Officers separated before 1/1/29 and enlisted personnel separated before 1/1/15.		6
MARINE CORPS (USMC)	Marine Corps records are transferred to NPRC between 6 and 9 months after separation.	Active, TDRL, and Selected Marine Corps Reserve members.		4
		Individual Ready Reserve and Fleet Marine Corps Reserve members.		5
		Discharged, deceased, and retired members *(see next item)*.		14
		Members separated before 1/1/1905.		6
ARMY (USA)	Army records are transferred to NPRC as follows: Active Army and Individual Ready Reserve Control Groups: About 60 days after separation. U.S. Army Reserve Troop Unit personnel: About 120 to 180 days after separation.	Reserve, living retired members, retired general officers, and active duty records of current National Guard members who performed service in the U.S. Army before 7/1/72.*		7
		Active officers (including National Guard on active duty in the U.S. Army).		8
		Active enlisted (including National Guard on active duty in the U.S. Army) and enlisted TDRL.		9
		Current National Guard officers not on active duty in the U.S. Army.		12
		Current National Guard enlisted not on active duty in the U.S. Army.		13
		Discharged and deceased members *(see next item)*.		14
		Officers separated before 7/1/17 and enlisted separated before 11/1/12.		6
		Officers and warrant officers TDRL.		8
NAVY (USN)	Navy records are transferred to NPRC 6 months after retirement or complete separation.	Active members (including reservists on duty)—PERS and MED		10
		Discharged, deceased, retired (with and without pay) less than six months, TDRL, drilling and nondrilling reservists	PERS ONLY	10
			MED ONLY	11
		Discharged, deceased, retired (with and without pay) more than six months *(see next item)*—PERS & MED		14
		Officers separated before 1/1/03 and enlisted separated before 1/1/1886—PERS and MED		6

*Code 12 applies to active duty records of current National Guard officers who performed service in the U.S. Army after 6/30/72.
Code 13 applies to active duty records of current National Guard enlisted members who performed service in the U.S. Army after 6/30/72.

ADDRESS LIST OF CUSTODIANS (BY CODE NUMBERS SHOWN ABOVE)—Where to write / send this form for each category of records

1	Air Force Manpower and Personnel Center Military Personnel Records Division Randolph AFB, TX 78150-6001	**5**	Marine Corps Reserve Support Center 10950 El Monte Overland Park, KS 66211-1408	**8**	USA MILPERCEN ATTN: DAPC-MSR 200 Stovall Street Alexandria, VA 22332-0400	**12**	Army National Guard Personnel Center Columbia Pike Office Building 5600 Columbia Pike Falls Church, VA 22041
2	Air Reserve Personnel Center Denver, CO 80280-5000	**6**	Military Archives Division National Archives and Records Administration Washington, DC 20408	**9**	Commander U.S. Army Enlisted Records and Evaluation Center Ft. Benjamin Harrison, IN 46249-5301	**13**	The Adjutant General (of the appropriate State, DC, or Puerto Rico)
3	Commandant U.S. Coast Guard Washington, DC 20593-0001	**7**	Commander U.S. Army Reserve Personnel Center ATTN: DARP-PAS 9700 Page Boulevard St. Louis, MO 63132-5200	**10**	Commander Naval Military Personnel Command ATTN: NMPC-036 Washington, DC 20370-5036	**14**	National Personnel Records Center (Military Personnel Records) 9700 Page Boulevard St. Louis, MO 63132
4	Commandant of the Marine Corps (Code MMRB-10) Headquarters, U.S. Marine Corps Washington, DC 20380-0001			**11**	Naval Reserve Personnel Center New Orleans, LA 70146-5000		

Occupation Title Index

This index is designed to provide access to the occupation exhibits in this volume. The titles are listed in alphabetical order. When the occupational title is found, note the exhibit ID number to its right. Locate that number in the proper occupation exhibit section.

Occupations are grouped by military service, using the following prefixes: **MOS:** Army; **CGR** and **CGW:** Coast Guard; **MCE:** Marine Corps; and **NER, LDO,** and **NWO:** Navy.

Advanced Aircraft Communications/Navigation Systems Technician, Intermediate Maintenance Activity (IMA)	MCE-6414-001
Advanced Aircraft Electrical Instrument/Flight Control Systems Technician, Intermediate Maintenance Activity (IMA)	MCE-6434-001
Advanced Aircraft Electronic Countermeasures Technician, Intermediate Maintenance Activity (IMA)	MCE-6485-001
Advanced Automatic Test Equipment Technician, Intermediate Maintenance Activity (IMA)	MCE-6469-001
Aircraft Communications Systems Technician, Intermediate Maintenance Activity (IMA)	MCE-6412-001
Aircraft Communications/Navigation Systems Technician (CH-46)	MCE-6322-001
Aircraft Communications/Navigation Systems Technician (CH-53)	MCE-6323-001
Aircraft Communications/Navigation Systems Technician V-22	MCE-6325-001
Aircraft Communications/Navigation Systems Technician, (U/AH-1)	MCE-6324-001
Aircraft Communications/Navigation Systems Technician, A-4/TA-4/QA-4	MCE-6312-001
Aircraft Communications/Navigation Systems Technician, A-6/TC-4C/EA-6A	MCE-6313-001
Aircraft Communications/Navigation Systems Technician, AV-8	MCE-6315-001
Aircraft Communications/Navigation Systems Technician, F/A-18	MCE-6317-001
Aircraft Communications/Navigation Systems Technician, KC-130/OV-10	MCE-6316-001
Aircraft Communications/Navigation Systems Technician, OV-10	MCE-6318-001
Aircraft Communications/Navigation Systems Technician, RF-4/F-4	MCE-6314-001
Aircraft Cryptographic Systems Technician, Intermediate Maintenance Activity (IMA)	MCE-6422-001
Aircraft Electrical Countermeasures Systems Technician, Fixed Wing, Intermediate Maintenance Activity (IMA)	MCE-6482-001
Aircraft Electrical Equipment Test Set/Mobile Electronic Test Set Technician, Intermediate Maintenance Activity (IMA)	MCE-6468-001
Aircraft Electrical Systems Technician, A-6/EA-6/TC-4C	MCE-6333-001
Aircraft Electrical Systems Technician, AV-8	MCE-6335-001
Aircraft Electrical Systems Technician, F/A-18	MCE-6337-001
Aircraft Electrical Systems Technician, KC-130/OV-10	MCE-6336-001
Aircraft Electrical/Instrument/Flight Control Systems Technician, Helicopter/OV-10, Intermediate Maintenance Activity (IMA)	MCE-6433-001
Aircraft Electrical/Instrument/Flight Control Systems Technician, Intermediate Maintenance Activity (IMA), Fixed Technician, Intermediate Maintenance Activity (IMA), Fixed Wing	MCE-6432-001
Aircraft Electronic Countermeasures Systems Technician, EA-6, Intermediate Maintenance Activity (IMA)	MCE-6484-001
Aircraft Electronic Countermeasures Systems Technician, EA-6B, Organizational Maintenance Activity (OMA)	MCE-6386-001
Aircraft Electronic Countermeasures Systems Technician, Helicopter, Intermediate Maintenance Activity (IMA)	MCE-6483-001
Aircraft Forward Looking Infrared/Electro-optical Technician, Intermediate Maintenance Activity (IMA)	MCE-6466-001
Aircraft Hydraulic/Pneumatic Mechanic, A-4/TA-4/OA-4	MCE-6052-001
Aircraft Hydraulic/Pneumatic Mechanic, A-6/EA-6	MCE-6053-001
Aircraft Hydraulic/Pneumatic Mechanic, AV-8/TAV-8	MCE-6055-001
Aircraft Hydraulic/Pneumatic Mechanic, F-4/RF-4	MCE-6054-001
Aircraft Hydraulic/Pneumatic Mechanic, F/A-18	MCE-6057-001
Aircraft Hydraulic/Pneumatic Mechanic, KC-130	MCE-6056-001
Aircraft Hydraulic/Pneumatic Mechanic, OV-10	MCE-6058-001
Aircraft Inertial Navigation System Technician Intermediate Maintenance Activity (IMA)	MCE-6465-001
Aircraft Inertial Navigation System Technician, Intermediate Maintenance Activity (IMA)	MCE-6464-001
Aircraft Maintenance Administration Clerk	MCE-6046-001
Aircraft Maintenance Chief	MCE-6019-001
Aircraft Maintenance Data Analysis Technician	MCE-6047-001
Aircraft Maintenance Ground Support Equipment Electrician/Refrigeration Mechanic	MCE-6073-001
Aircraft Maintenance Ground Support Equipment Mechanic Hydraulic/Pneumatic/Structure Mechanic	MCE-6072-001
Aircraft Mechanic, A-4/TA-4/QA-4	MCE-6012-001
Aircraft Mechanic, A-6/EA-6	MCE-6013-001
Aircraft Mechanic, AV-8A/TAV-8	MCE-6015-001
Aircraft Mechanic, F-4/RF-4	MCE-6014-001
Aircraft Mechanic, F/A-18	MCE-6017-001
Aircraft Mechanic, KC-130	MCE-6016-001
Aircraft Mechanic, OV-10	MCE-6018-001
Aircraft Navigation Systems Technician, IFF/Radar/TACAN, Intermediate Maintenance Activity (IMA)	MCE-6413-001
Aircraft Non-Destructive Inspection Technician	MCE-6044-001
Aircraft Ordnance Technician	MCE-6531-001

OCCUPATION TITLE INDEX

Aircraft Power Plant Test Cell Operator, Rotary Wing	MCE-6135-001
Aircraft Power Plants Mechanic, F-404	MCE-6027-001
Aircraft Power Plants Mechanic, J-52	MCE-6022-001
Aircraft Power Plants Mechanic, J-79	MCE-6024-001
Aircraft Power Plants Mechanic, Rolls Royce Pegasus	MCE-6025-001
Aircraft Power Plants Mechanic, T-56	MCE-6026-001
Aircraft Power Plants Mechanic, T-76	MCE-6023-001
Aircraft Power Plants Test Cell Operator, Fixed Wing	MCE-6035-001
Aircraft RADCOM/CAT IIID Technician, Intermediate Maintenance Activity (IMA)	MCE-6467-001
Aircraft Safety Equipment Mechanic, A-4/TA-4/OA-4	MCE-6082-001
Aircraft Safety Equipment Mechanic, A-6/EA-6	MCE-6083-001
Aircraft Safety Equipment Mechanic, AV-8/TAV-8	MCE-6085-001
Aircraft Safety Equipment Mechanic, F-4/RF-4	MCE-6084-001
Aircraft Safety Equipment Mechanic, F/A-18	MCE-6087-001
Aircraft Safety Equipment Mechanic, KC-130	MCE-6086-001
Aircraft Safety Equipment Mechanic, OV-10	MCE-6088-001
Aircraft Structures Mechanic, A-4/TA-4/OA-4	MCE-6092-001
Aircraft Structures Mechanic, A-6/EA-6	MCE-6093-001
Aircraft Structures Mechanic, AV-8/TAV-8	MCE-6095-001
Aircraft Structures Mechanic, F-4/RF-4	MCE-6094-001
Aircraft Structures Mechanic, F/A-18	MCE-6097-001
Aircraft Structures Mechanic, KC-130	MCE-6096-001
Aircraft Structures Mechanic, OV-10	MCE-6098-001
Aircraft Weapons Systems Specialist, A-6/TC-4C	MCE-6353-001
Aircraft Weapons Systems Specialist, F-4J/S	MCE-6354-001
Aviation Electrician's Mate	CGR-AE-001 CGR-AE-002
Aviation Electronic Micro-Miniature/Instrument and Cable Repair Technician	MCE-6423-001
Aviation Electronics Technician	CGR-AT-001 CGR-AT-002
Aviation Engineering	CGW-AVI-001 CGW-AVI-002
Aviation Machinist's Mate	CGR-AD-001 CGR-AD-002
Aviation Ordnance Chief	MCE-6591-001
Aviation Ordnance Equipment Repair Technician	MCE-6541-001
Aviation Ordnance Munitions Technician	MCE-6521-001
Aviation Precision Measurement Equipment/Automatic Test Equipment Calibration and Repair Technician, Intermediate Maintenance Activity (IMA)	MCE-6492-001
Aviation Structural Mechanic	CGR-AM-001 CGR-AM-002
Aviation Survivalman	CGR-ASM-001 CGR-ASM-002
Avionics Maintenance Chief	MCE-6391-001
Avionics Test Set Technician, Intermediate Maintenance Activity (IMA)	MCE-6462-001
Boatswain	CGW-BOSN-001 CGW-BOSN-002
Boatswain's Mate	CGR-BM-001 CGR-BM-002
Communications	CGW-COMM-001 CGW-COMM-002
Cryogenics Equipment Operator	MCE-6075-001
Damage Controlman	CGR-DC-001 CGR-DC-002
Data Processing Technician	CGR-DP-001
Electrician's Mate	CGR-EM-001 CGR-EM-002
Electronics	CGW-ELC-001 CGW-ELC-002
Electronics Technician	CGR-ET-001 CGR-ET-002
Finance and Supply	CGW-FS-001 CGW-FS-002
Fire Control Technician	CGR-FT-001 CGR-FT-002
Fireman	CGR-FN-001
Flight Equipment Marine	MCE-6060-001
Gunner's Mate	CGR-GM-001 CGR-GM-002
HC-130 Aircraft Commander	CGA-C130-003
HC-130 Co-Pilot	CGA-C130-001
HC-130 First Pilot	CGA-C130-002
HC-130 Instructor Pilot	CGA-C130-004
Health Services Technician	CGR-HS-001 CGR-HS-002 CGR-HS-003
Health Services Technician, Dental	CGR-HSD-001
Helicopter Crew Chief, CH-46	MCE-6172-001
Helicopter Crew Chief, CH-53 A/D	MCE-6173-001
Helicopter Crew Chief, CH-53E	MCE-6174-001
Helicopter Crew Chief, UH-N	MCE-6175-001
Helicopter Crew Chief, V-22	MCE-6176-001
Helicopter Hydraulic/Pneumatic Mechanic (CH-46)	MCE-6152-001
Helicopter Hydraulic/Pneumatic Mechanic (CH-53)	MCE-6153-001
Helicopter Hydraulic/Pneumatic Mechanic (CH-53E)	MCE-6155-001
Helicopter Hydraulic/Pneumatic Mechanic (U/AH-1)	MCE-6154-001
Helicopter Maintenance Chief	MCE-6119-001
Helicopter Mechanic (CH-46)	MCE-6112-001
Helicopter Mechanic (CH-53)	MCE-6113-001
Helicopter Mechanic (CH-53E)	MCE-6115-001
Helicopter Mechanic (U/AH-1)	MCE-6114-001
Helicopter Power Plants Mechanic (T-400)	MCE-6125-001
Helicopter Power Plants Mechanic (T-58)	MCE-6122-001
Helicopter Power Plants Mechanic (T-64)	MCE-6123-001
Helicopter Structures Mechanic (CH-46)	MCE-6142-001
Helicopter Structures Mechanic (CH-53)	MCE-6143-001
Helicopter Structures Mechanic (U/AH-1)	MCE-6144-001
HH-60 Aircraft Commander	CGA-HH60-003
HH-60 Co-Pilot	CGA-HH60-001
HH-60 First Pilot	CGA-HH60-002
HH-60 Instructor Pilot	CGA-HH60-004
HH-65 Aircraft Commander	CGA-HH65-003
HH-65 Co-Pilot	CGA-HH65-001
HH-65 First Pilot	CGA-HH65-002
HH-65 Instructor Pilot	CGA-HH65-004
HU-25 Aircraft Commander	CGA-HU25-003
HU-25 Co-Pilot	CGA-HU25-001
HU-25 First Pilot	CGA-HU25-002

OCCUPATION TITLE INDEX

HU-25 Instructor Pilot	CGA-HU25-004
Investigator	CGR-IV-001
	CGR-IV-002
Machinery Technician	CGR-MK-001
	CGR-MK-002
Marine Science Technician	CGR-MST-001
	CGR-MST-002
Materiel Maintenance	CGW-MAT-001
	CGW-MAT-002
Medical Administration	CGW-MED-001
	CGW-MED-002
Naval Engineering	CGW-ENG-001
	CGW-ENG-002
Personnel Administration	CGW-PERS-001
	CGW-PERS-002
Photojournalist	CGR-PA-001
Physician's Assistant	CGW-PYA-001
	CGW-PYA-002
Port Safety and Security	CGW-PSS-001
	CGW-PSS-002
Port Securityman	CGR-PS-001
	CGR-PS-002
Public Affairs Specialist	CGR-PA-001
	CGR-PA-002
Public Information	CGW-INF-001
	CGW-INF-002
Quartermaster	CGR-QM-001
	CGR-QM-002
Radar Test Station/Radar Systems Test Station Technician, Intermediate Maintenance Activity (IMA)	MCE-6463-001
Radarman	CGR-RD-001
	CGR-RD-002
Radioman	CGR-RM-001
Seaman	CGR-SN-001
Sonar Technician	CGR-ST-001
Special Agent	CGR-IV-002
Storekeeper	CGR-SK-001
	CGR-SK-002
Subsistence Specialist	CGR-SS-001
	CGR-SS-002
Telecommunications Specialist	CGR-TC-001
Telephone Technician	CGR-TT-001
	CGR-TT-002
Weapons	CGW-WEPS-001
	CGW-WEPS-002
Yeoman	CGR-YN-001
	CGR-YN-002

Keyword Index

This index is designed to provide access to the courses listed in the course exhibit sections of this volume. Course titles are arranged alphabetically under keywords extracted directly from the titles. For example, the keyword *Dental* is followed by all course titles containing that word.

To use this index:
- Identify a word (or group of words) that appears to be unique or descriptive.
- Locate the keyword(s) in this index.
- If the keyword or the course title cannot be found, identify another descriptive word in the title and try again.
- When the course title is found, note the exhibit ID number to the right of the title. Locate that number in the course exhibit section.

Course exhibits are grouped by military service, using the following prefixes: **AF:** Air Force; **AR:** Army; **CG:** Coast Guard; **DD:** Department of Defense; **MC:** Marine Corps; and **NV:** Navy.

3"/50
3"/50 Gun Mount Mk 22 Operation and Maintenance, Class C
CG-2204-0002
Gunner's Mate, Class A, Phase 2 (Advanced Electricity, 3"/50 Caliber Gun, 5"/38 Caliber Gun)
CG-1714-0013

5"/38
Gunner's Mate, Class A, Phase 2 (Advanced Electricity, 3"/50 Caliber Gun, 5"/38 Caliber Gun)
CG-1714-0013

618M-3
Aviation Electronics Technician 618M-3, Class C
CG-1715-0079
CG-1715-0142

A-10
Fighter Weapons Instructor A-10
AF-1406-0061
Weapon Control Systems Mechanic (A-7D, A-10, AC-130, F-5) by Correspondence
AF-1715-0050
Weapons Control Systems Mechanic (A-7D, A-10, AC-130, F-5) by Correspondence
AF-1715-0800

A-7D
A-7D Avionics Aerospace Ground Equipment Specialist by Correspondence
AF-1715-0078
Weapon Control Systems Mechanic (A-7D, A-10, AC-130, F-5) by Correspondence
AF-1715-0050
Weapon Control Systems Mechanic (A-7D: AN/APQ-126) by Correspondence
AF-1715-0054
Weapons Control Systems Mechanic (A-7D, A-10, AC-130, F-5) by Correspondence
AF-1715-0800

A/OA-10
Basic Operational Training A/OA-10
AF-1704-0281
Transition/Requalification Training A/OA-10
AF-1704-0283

AB/SPS-64(V)
AB/SPS-64(V) Small Cutter Radar Maintenance
CG-1715-0140

AC-130
F-4 and AC-130 Weapons Control Systems Technician by Correspondence
AF-1715-0805
Weapon Control Systems Mechanic (A-7D, A-10, AC-130, F-5) by Correspondence
AF-1715-0050
Weapons Control Systems Mechanic (A-7D, A-10, AC-130, F-5) by Correspondence
AF-1715-0800

AC-130E
Special Operations Training, AC-130E Pilot
AF-1606-0020

Academy
Air National Guard Academy of Military Science
AF-2203-0051
Chief Petty Officer Academy
CG-1511-0002
Naval Academy Training
MC-2202-0003

Accounting
Accounting and Finance Officer
AF-1408-0101
Accounting for Plant Property by Correspondence
MC-1405-0025
Commercial Services and Automated Travel Record Accounting System by Correspondence
AF-1408-0074
Fiscal Accounting
MC-1402-0055
Fiscal Accounting for Supply Clerks by Correspondence
MC-1401-0010
Introduction to Marine Corps Accounting by Correspondence
MC-1401-0004

Acquisition
Advanced Information Systems Acquisition
DD-0326-0005
Communication-Electronics Systems Acquisition and Management by Correspondence
AF-1715-0753
Defense Acquisition Engineering, Manufacturing, and Quality Control
DD-1408-0010
Executive Acquisition Logistics Management
DD-0326-0004
Fundamentals of Systems Acquisition Management
DD-1408-0012
Intermediate Information Systems Acquisition
DD-0326-0006
Intermediate Software Acquisition Management
DD-1408-0013
Intermediate Systems Acquisition
DD-1408-0011
DD-1408-0020
Introduction to Acquisition Management by Correspondence
AF-1408-0106
Software Acquisition Management
DD-1402-0004
Surface Air Defense System Acquisition Technician
MC-1715-0181
Systems Acquisition for Contracting Personnel (Executive)
DD-1408-0009

Actions
Air Reserve Forces Social Actions Technician (Drug/Alcohol) by Correspondence
AF-0708-0004
Social Actions Technician (Drugs/Alcohol) by Correspondence
AF-1406-0054
Social Actions Technician (Equal Opportunity/Human Relations) by Correspondence
AF-1406-0053

Adjutant
Adjutant
MC-1405-0031

ADL-81
ADL-81 Loran-C Receiver, Class C
CG-1715-0082

Administration
Administration Specialist by Correspondence
AF-1406-0045
Administration Technician by Correspondence
AF-1406-0046
Advanced Personnel Administration
MC-1405-0032
Aviation Administration by Correspondence
CG-1704-0038
Aviation Engineering Administration
CG-1405-0005

KEYWORD INDEX

Aviation Maintenance Administration by Correspondence
    CG-1408-0003
Basic Typing and Personnel Administration
    MC-1406-0021
Contract Administration by Correspondence
    AF-1408-0087
Engineering Administration
    CG-1408-0035
Food Service Specialist Administration and Management
    CG-1729-0010
General Personnel Administration for Reserve by Correspondence
    MC-1406-0025
Health Services Administration
    AF-0709-0034
Independent Duty Administration
    MC-1403-0011
Intermediate Contract Administration
    DD-1405-0005
Legal Administration Clerk by Correspondence
    MC-1406-0024
Legal Administration for the Reporting Unit by Correspondence
    MC-1406-0024
Personnel Administration for the Reporting Unit by Correspondence
    MC-1406-0026
Reserve Administration
    MC-1406-0022
Subsistence Specialist Administration and Management
    CG-1729-0010
Weapons Administration by Correspondence
    CG-1408-0007

Administrative
Administrative Clerk
    MC-1403-0014
Medical Administrative Specialist by Correspondence
    AF-0709-0027

Advanced
Advanced Analog Electronics Technology, Class C
    CG-1715-0149
Advanced Digital Electronics Technology, Class C
    CG-1715-0148
Advanced Electricity, Electronics, and Hydraulics
    CG-1715-0147
Chinese Advanced
    DD-0602-0234
Russian Advanced
    DD-0602-0128
    DD-0602-0180
Staff Noncommissioned Officers Advanced Nonresident Program (SNCOANP) by Correspondence
    MC-1408-0028

Advisory
Career Advisory Technician by Correspondence
    AF-1406-0051

Aerial
Aerial Navigation
    MC-1606-0002
Aerial Port Operation and Management
    AF-0419-0036

Aeromedical
Medical Service Specialist, Aeromedical by Correspondence
    AF-0709-0033

Aerospace
A-7D Avionics Aerospace Ground Equipment Specialist by Correspondence
    AF-1715-0078
Aerospace Control and Warning Systems Operator by Correspondence
    AF-1715-0754
Aerospace Ground Equipment Mechanic by Correspondence
    AF-1704-0187
    AF-1710-0038
Aerospace Ground Equipment Technician by Correspondence
    AF-1704-0192
Aerospace Photographic Systems Specialist by Correspondence
    AF-1715-0139
Aerospace Physiology Specialist by Correspondence
    AF-0709-0032
Aerospace Propulsion Specialist (Jet Engine) by Correspondence
    AF-1704-0256
Aerospace Propulsion Specialist (Turboprop) by Correspondence
    AF-1704-0257
Aerospace Propulsion Technician (Jet Engine) by Correspondence
    AF-1704-0229
Aerospace Propulsion Technician (Turboprop) by Correspondence
    AF-1710-0037
Avionics Aerospace Ground Equipment Specialist (F/RF-4 Peculiar Avionics AGE) by Correspondence
    AF-1715-0077
Avionics Aerospace Ground Equipment Specialist by Correspondence
    AF-1715-0076

Aftercare
Aftercare Program Management
    MC-0801-0002

Aids
Advanced Minor Aids to Navigation Maintenance
    CG-1715-0131
Aids to Navigation Officer Advanced, Class C
    CG-2205-0008
Aids to Navigation Operations Management
    CG-1408-0042
Aids to Navigation Positioning
    CG-2205-0025
Automated Aids to Navigation Lighthouse Technician
    CG-1715-0128
Enemy Defense Penetration Aids
    AF-1715-0744
Integrated Avionic Communication, Navigation and Penetration Aids Systems Specialist by Correspondence
    AF-1714-0039
Integrated Avionic Communications, Navigation, and Penetration Aids Systems Specialist (F/FB-111) by Correspondence
    AF-1715-0779
Integrated Avionics Communication, Navigation, and Penetration Aids Systems Specialist (F-16) by Correspondence
    AF-1704-0220
Integrated Avionics Communication, Navigation, and Penetration Aids Systems Specialist (F/FB-111) by Correspondence
    AF-1715-0101
Integrated Avionics Communication, Navigation, and Penetration Aids Systems Specialist by Correspondence
    AF-1715-0100
Integrated Avionics Communications, Navigation, and Penetration Aids Systems Specialist (F-15) by Correspondence
    AF-1704-0225
Integrated Avionics Communications, Navigation, and Penetration Aids Systems Specialist (F/FB-111) by Correspondence
    AF-1704-0263
Navigational Aids Equipment Specialist by Correspondence
    AF-1715-0012
Navigational Aids Equipment Technician by Correspondence
    AF-1715-0017
Officer In Charge, Aids to Navigation Team
    CG-1406-0009
Officers Advanced Aids to Navigation
    CG-2205-0008

Aiming
Aiming Circle Operator by Correspondence
    MC-1601-0035

AIMS
AIMS Maintenance
    MC-1715-0113

Air
Air Cargo Specialist by Correspondence
    AF-0419-0004
Air Command and Staff College
    AF-1511-0001
Air Command and Staff College Resident Program
    AF-1511-0010
Air Command and Staff Correspondence Associate Program
    AF-1511-0013
Air Command and Staff Correspondence Program
    AF-1511-0013
Air Command and Staff Nonresident Seminar Associate Program
    AF-1511-0009
Air Command and Staff Nonresident Seminar Program
    AF-1511-0009
Air Conditioning and Refrigeration, Class C
    CG-1701-0002
Air Control Electronic Operator
    MC-2204-0031
Air Control Electronics Operator
    MC-2204-0031
Air Control Electronics Operator (AN/TYQ-23)
    MC-2204-0090
Air Launched Missile Systems Specialist by Correspondence
    AF-1714-0041
    AF-1715-0814
Air Movement Planning
    MC-0419-0007
Air National Guard Academy of Military Science
    AF-2203-0051
Air National Guard Fighter Weapons Instructor
    AF-1606-0151
Air Navigation
    MC-1606-0002

Air Passenger Specialist by Correspondence
AF-0419-0021
Air Support Control Officer
MC-1715-0078
Air Support Operations Operator
MC-1715-0031
Air Transportation Journeyman by Correspondence
AF-0419-0038
Air War College
AF-1511-0011
Air War College Associate Programs Nonresident and Correspondence
AF-1511-0012
Air War College Correspondence Program
AF-1511-0012
Air War College Nonresident Seminar Program
AF-1511-0012
Apprentice Air Cargo Specialist by Correspondence
AF-0419-0029
Civil Air Patrol Mission Observer Level II by Correspondence
AF-1704-0195
Civil Air Patrol Public Affairs Officer by Correspondence
AF-0505-0004
Civil Air Patrol Scanner by Correspondence
AF-1704-0194
Civil Air Patrol-Safety Officer (Level II Technical Rating) by Correspondence
AF-0801-0005
Fighter Weapons Instructor Air Weapons Controller
AF-1406-0062
Heating, Ventilation, Air Conditioning, and Refrigeration Journeyman by Correspondence
AF-1701-0012
AF-1701-0013
Heating, Ventilation, Air Conditioning, and Refrigeration Specialist by Correspondence
AF-1701-0012
AF-1701-0013
Intelligence for the Marine Air-Ground Task Force by Correspondence
MC-1606-0005
Introduction to Civil Air Patrol Emergency Services by Correspondence
AF-1704-0196
Marine Air-Ground Task Force Deployment Support System/Computer-Aided Embarkation Management System
MC-1402-0059
Refrigeration and Air Conditioning
CG-1730-0001
Refrigeration/Air Conditioning Operation and Maintenance, Class C
CG-1701-0002
Tactical Air Command and Control Specialist by Correspondence
AF-1406-0068
Tactical Air Command Center Operator
MC-1715-0157
Tactical Air Command Central (TACC) Technician
MC-1715-0046
Tactical Air Command Central Repair
MC-1715-0151
Tactical Air Command Central Technician
MC-1715-0142
Tactical Air Control Party
MC-1606-0013

Air Defense
Air Defense Control Officer
MC-2204-0030
Air Defense Control Officer (AN/TYQ-23)
MC-2204-0089
Air Defense Control Officers Senior
MC-1715-0150
Tactical Air Defense Controller
MC-1704-0004
Tactical Air Defense Controller (AN/TYQ-23)
MC-1704-0009
Tactical Air Defense Controller Enlisted
MC-1704-0004

Air Traffic Control
Air Traffic Control Radar Repairman by Correspondence
AF-1715-0003
Air Traffic Control Radar Specialist by Correspondence
AF-1715-0003
Air Traffic Control Radar Technician by Correspondence
AF-1715-0009

Airborne
Airborne Command and Control Communications Equipment Journeyman by Correspondence
AF-1715-0811
AF-1715-0816
Airborne Command and Control Communications Equipment Specialist by Correspondence
AF-1715-0816
Airborne Command and Control Equipment Specialist by Correspondence
AF-1715-0811
Airborne Command Post Communications Equipment Specialist by Correspondence
AF-1715-0122
Airborne Communications Systems Operator by Correspondence
AF-1704-0198
Airborne Warning and Control Radar Specialist by Correspondence
AF-1715-0123
AF-1715-0801

Aircraft
Aircraft Armament Systems Specialist by Correspondence
AF-1715-0035
Aircraft Communication/Navigation Systems Journeyman by Correspondence
AF-1715-0817
Aircraft Control and Warning Radar Specialist by Correspondence
AF-1715-0005
Aircraft Electrical and Environmental Systems Journeyman by Correspondence
AF-1714-0045
Aircraft Electrical Systems Specialist by Correspondence
AF-1714-0003
Aircraft Electro-Environmental System Technician by Correspondence
AF-1704-0221
Aircraft Electro-Environmental Systems Technician by Correspondence
AF-1704-0222
AF-1704-0223
Aircraft Engine Familiarization by Correspondence
CG-1704-0034

KEYWORD INDEX D-3

Aircraft Environmental Systems Mechanic by Correspondence
AF-1704-0200
Aircraft Fuel Systems Mechanic by Correspondence
AF-1704-0191
AF-1704-0258
Aircraft Fuel Systems Technician by Correspondence
AF-1704-0202
Aircraft Guidance and Control Systems Technician by Correspondence
AF-1704-0261
Aircraft Logs and Records
CG-1405-0005
Aircraft Maintenance Noncommissioned Officer (NCO) by Correspondence
MC-1704-0008
Aircraft Maintenance Officer
AF-1704-0251
Aircraft Maintenance Officer (Accelerated)
AF-1704-0253
Aircraft Maintenance Officer (Accelerated/Air Reserve Forces)
AF-1704-0252
AF-1704-0253
Aircraft Maintenance Officer (Bridge)
AF-1704-0286
Aircraft Maintenance Officer, Air Reserve
AF-1704-0252
Aircraft Maintenance Specialist and Bombardment Aircraft by Correspondence
AF-1704-0184
Aircraft Maintenance Specialist, Airlift and Bombardment Aircraft by Correspondence
AF-1704-0181
Aircraft Maintenance Specialist, Tactical Aircraft by Correspondence
AF-1704-0102
Aircraft Maintenance/Munitions Officer
AF-1704-0251
Aircraft Metals Technology (Machinist) by Correspondence
AF-1717-0027
Aircraft Metals Technology (Welding) by Correspondence
AF-1723-0011
AF-1724-0007
Aircraft Pneudraulic Systems Mechanic by Correspondence
AF-1704-0201
Aircraft Pneudraulic Systems Technician by Correspondence
AF-1704-0230
AF-1704-0231
Aircraft Structural Maintenance Technician (Airframe Repair) by Correspondence
AF-1704-0250
AF-1717-0028
Airlift Aircraft Maintenance Specialist by Correspondence
AF-1704-0184
Airlift Aircraft Maintenance Technician by Correspondence
AF-1704-0249
HC-131A Aircraft Maintenance, Class C
CG-1704-0006
Strategic Aircraft Maintenance Specialist by Correspondence
AF-1704-0181
AF-1704-0188
AF-1704-0209
AF-1704-0210
AF-1704-0211
AF-1704-0247

KEYWORD INDEX

Aircrew (continued)
Tactical Aircraft Maintenance Technician by Correspondence
　　AF-1704-0228

Aircrew
Aircrew Basic by Correspondence
　　CG-1704-0040
Aircrew Egress Systems Mechanic by Correspondence
　　AF-1717-0005
Aircrew Life Support Specialist by Correspondence
　　AF-1704-0199
Search and Rescue HU-25A Basic Aircrew by Correspondence
　　CG-0802-0008

Aircrewman
Basic Search and Rescue Aircrewman by Correspondence
　　CG-0802-0012

Airfield
Airfield Management Specialist by Correspondence
　　AF-1704-0215
Apprentice Airfield Management Specialist by Correspondence
　　AF-1704-0214

Airframe
Aircraft Structural Maintenance Technician (Airframe Repair) by Correspondence
　　AF-1704-0250
　　AF-1717-0028
Airframe Repair Specialist by Correspondence
　　AF-1723-0010
HH-60J Airframe and Power Train Maintenance, Class C
　　CG-1704-0048
HH-65A Airframe/Power Train Maintenance, Class C
　　CG-1704-0050
HU-25A Airframe, Class C
　　CG-1704-0044

Airlift
Aircraft Maintenance Specialist, Airlift and Bombardment Aircraft by Correspondence
　　AF-1704-0181
Airlift Aircraft Maintenance Specialist by Correspondence
　　AF-1704-0184
Airlift Aircraft Maintenance Technician by Correspondence
　　AF-1704-0249

Albanian
Albanian Basic
　　DD-0602-0018

Alcohol
Air Reserve Forces Social Actions Technician (Drug/Alcohol) by Correspondence
　　AF-0708-0004
Social Actions Technician (Drugs/Alcohol) by Correspondence
　　AF-1406-0054

Algebra
Elementary Algebra by Correspondence
　　CG-1104-0001

Allied
Allied Visual Communications, Class C
　　CG-2205-0039

Aluminum
Aluminum Welding, Class C
　　CG-1710-0016

Ammunition
Ammunition Technician, Phase 3 (USMC)
　　MC-2204-0109

Amphibious
Amphibious Vehicle Unit Leader
　　MC-1710-0010
Amphibious Vehicle Unit Leaders (Enlisted)
　　MC-1710-0010
Amphibious Warfare
　　MC-2204-0070
Amphibious Warfare School Nonresident Program (AWSNP) by Correspondence Phase 2
　　MC-2204-0114
Amphibious Warfare School Nonresident Program (AWSNP) by Correspondence, Phase 1
　　MC-2204-0113
Assault Amphibious Vehicle Unit Leader
　　MC-1710-0010

AN/APQ-126
Weapon Control Systems Mechanic (A-7D: AN/APQ-126) by Correspondence
　　AF-1715-0054

AN/APS-127
Aviation Electronics Technician AN/APS-127
　　CG-1715-0080
Aviation Electronics Technician AN/APS-127 Radar, Class C
　　CG-1715-0080

AN/ARN-118(V)
Aviation Electronics Technician AN/ARN-118(V), Class C
　　CG-1715-0084

AN/ARN-133V2
Aviation Electronics Technician AN/ARN-133V2, Class C
　　CG-1715-0083

AN/FPN-39
AN/FPN-39 Loran Transmitter, Class C
　　CG-1715-0118
AN/FPN-39 Transmitter Maintenance, Class C
　　CG-1715-0118

AN/FPN-42
AN/FPN-42 Loran Transmitter, Class C
　　CG-1715-0117

AN/FPN-44A
AN/FPN-44A Loran Transmitter, Class C
　　CG-1715-0110
AN/FPN-44A Transmitter Maintenance, Class C
　　CG-1715-0110

AN/FPN-64(V)
AN/FPN-64(V) Loran Solid State Transmitter, Class C
　　CG-1715-0123
AN/FPN-64(V) Transmitter Maintenance
　　CG-1715-0123

AN/SPS-29
AN/SPS-29 Maintenance and Repair, Class C
　　CG-1715-0049

AN/SPS-29D
AN/SPS-29D Radar Systems, Class C
　　CG-1715-0049

AN/SPS-64(V)
AN/SPS-64(V) Large Cutter Radar Systems Maintenance
　　CG-1715-0139

AN/SPS-64(V)1,2,3
AN/SPS-64(V)1,2,3 Radars
　　CG-1715-0119

AN/SPS-64(V)4
AN/SPS-64(V)4 Radar
　　CG-1715-0070
　　CG-1715-0120

AN/SPS-66
AN/SPS-66 and AN/SPS-66A Radars, Class C
　　CG-1715-0122

AN/SPS-66A
AN/SPS-66 and AN/SPS-66A Radars, Class C
　　CG-1715-0122

AN/TGC-37
Communication Central, AN/TGC-37, System Maintenance
　　MC-1715-0006
Mobile Communications Central Technician (AN/TGC-37(V))
　　MC-1715-0006

AN/TPX-46
Interrogator Set AN/TPX-46 Direct Support/General Support Repairman and Supervisor
　　MC-1715-0138

AN/TYQ-1
Tactical Air Command Central (TACC AN/TYQ-1) Repair
　　MC-1715-0020

AN/TYQ-23
Air Control Electronics Operator (AN/TYQ-23)
　　MC-2204-0090
Air Defense Control Officer (AN/TYQ-23)
　　MC-2204-0089
Tactical Air Defense Controller (AN/TYQ-23)
　　MC-1704-0009

AN/TYQ-3
Tactical Data Communications Central (TDCC AN/TYQ-3) Technician
　　MC-1715-0026

AN/URC-114(V)
AN/URC-114(V) 1 KW Transceiver System Maintenance
　　CG-1715-0124
AN/URC-114(V) Low Power (1KW) Communications System, Class C
　　CG-1715-0124

AN/URC-9
AN/URC-9 Radio Set, Class C
　　CG-1715-0121

AN/URT-41(V)2
AN/URT-41(V)2 10 KW Transmitter System Maintenance
　　CG-1715-0112

AN/UYK-7
AN/UYK-7 Computer and Data Auxiliary Console Course Development
　　CG-1402-0002
AN/UYK-7 Computer and OJ-172 Data Exchange Auxiliary Console
　　CG-1402-0002

Analysis
Advanced Terrain Analysis
　　DD-1601-0028
Basic Terrain Analysis
　　DD-1601-0020
Cost Analysis Officer
　　AF-1401-0019

KEYWORD INDEX D-5

Cost Analysis Specialist by Correspondence
AF-1408-0109
Cost and Management Analysis Specialist by Correspondence
AF-1408-0083
Cost and Management Analysis Technician by Correspondence
AF-1408-0084
Financial Analysis Journeyman by Correspondence
AF-1408-0111
Financial Analysis Specialist by Correspondence
AF-1408-0109
Financial Management Officer (Financial Analysis)
AF-1401-0002
Fundamentals of Cost Analysis
DD-1115-0001
Intermediate ELINT Collection and Analysis
DD-1715-0016
Maintenance Data Systems Analysis Specialist by Correspondence
AF-1107-0002
Maintenance Systems Analysis Specialist by Correspondence
AF-1402-0069
Supply Systems Analysis Supervisor by Correspondence
AF-1405-0066
Terrain Analysis Warrant Officer Certification
DD-1601-0029
Vehicle Maintenance Control and Analysis Specialist by Correspondence
AF-1710-0036

Analyst
Supply Systems Analyst Journeyman by Correspondence
AF-1405-0072

Analytical
Analytical Photogrammetric Positioning System
DD-1601-0026

Anesthetist
Nurse Anesthetist
AF-0703-0013

Antenna
Antenna and Cable Systems Installation/Maintenance Specialist by Correspondence
AF-1715-0802
Antenna and Cable Systems Projects/Maintenance Specialist by Correspondence
AF-1715-0802
Antenna Construction and Propagation of Radio Waves by Correspondence
MC-1715-0136

APA-165
Weapon Control Systems Mechanic (F-4C/D: APQ-109/APA-165) by Correspondence
AF-1715-0052
Weapon Control Systems Technician (F-4C/D: APQ-109/APA-165) by Correspondence
AF-1715-0056

APQ-109
Weapon Control Systems Mechanic (F-4C/D: APQ-109/APA-165) by Correspondence
AF-1715-0052
Weapon Control Systems Technician (F-4C/D: APQ-109/APA-165) by Correspondence
AF-1715-0056

APQ-120
Weapon Control Systems Mechanic (F-4E: APQ-120) by Correspondence
AF-1715-0053
Weapon Control Systems Technician (F-4E: APQ-120) by Correspondence
AF-1715-0063

Arabic
Arabic Basic (Modern Standard Arabic)
DD-0602-0232
Arabic Intermediate
DD-0602-0220
Arabic-Egyptian Extended
DD-0602-0211
Arabic-Egyptian Special
DD-0602-0208
Arabic-Iraqi Extended
DD-0602-0209
Arabic-Iraqi Special
DD-0602-0207
Arabic-Syrian Extended
DD-0602-0210
Arabic-Syrian Special
DD-0602-0206

Armament
Aircraft Armament Systems Specialist by Correspondence
AF-1715-0035

Armed
Armed Forces Staff College
DD-0326-0001

Armor
M1A1 Armor Crewman
MC-2204-0107

Armored
Light Armored Vehicle Basic Repairman
MC-1710-0049
Light Armored Vehicle Technician
MC-1710-0050

Arms
Combat Arms Training and Maintenance Specialist/Technician by Correspondence
AF-2203-0054
Small Arms Instructor
CG-1408-0041
Small Arms Instructor, Class C
CG-1408-0041
Small Arms Refresher
MC-2204-0110
Small Arms Repairer
MC-2204-0111

Artillery
Artillery Survey Instruments by Correspondence
MC-1601-0032
Introduction to Field Artillery Survey by Correspondence
MC-1601-0039
Towed Artillery Repairman
MC-2201-0001

ASG-15
Defensive Fire Control Systems Mechanic (B-52D/G: MD-9, ASG-15 Turrets) by Correspondence
AF-1715-0047

Ashore
Firefighting Ashore by Correspondence
CG-1728-0005

Asia
Intelligence Brief: Southwest Asia by Correspondence
MC-1511-0001

ASQ-151
Bomb Navigation Systems Specialist (B-52G/H: ASQ-176, ASQ-151 Systems) by Correspondence
AF-1715-0046

ASQ-176
Bomb Navigation Systems Specialist (B-52G/H: ASQ-176, ASQ-151 Systems) by Correspondence
AF-1715-0046

Assault
Assault Amphibious Vehicle Unit Leader
MC-1710-0010

Assaultman
Basic Assaultman
MC-2204-0106

Assembler
Assembler Language Code (ALC) Programming (Combination)
MC-1402-0051
IBM System 360 Assembler Language Coding (Entry)
MC-1402-0051

Associate
Air Command and Staff Correspondence Associate Program
AF-1511-0013
Air Command and Staff Nonresident Seminar Associate Program
AF-1511-0009
Air War College Associate Programs Nonresident and Correspondence
AF-1511-0012

Attack
F-15 Integrated Avionic Attack Control Systems Specialist by Correspondence
AF-1715-0767
Integrated Avionic Attack Control Systems Specialist (F-16) by Correspondence
AF-1715-0096
Integrated Avionic Attack Control Systems Specialist by Correspondence
AF-1715-0759
Integrated Avionics Attack Control Systems Specialist (F-15) by Correspondence
AF-1704-0218
AF-1715-0097
Integrated Avionics Attack Control Systems Specialist (F/FB-111) by Correspondence
AF-1715-0090
AF-1715-0760
Integrated Avionics Attack Control Systems Specialist by Correspondence
AF-1715-0091

Audiovisual
Audiovisual Equipment Repairer
AF-1715-0741
Audiovisual Media Specialist by Correspondence
AF-1406-0073
Precision Imagery and Audiovisual Media Maintenance Specialist by Correspondence
AF-1715-0136

Auditor
Auditor Retrieval Systems
AF-1402-0074

Aural Comprehension
German Aural Comprehension
DD-0602-0016

KEYWORD INDEX

Automated
Automated Aids to Navigation Lighthouse Technician
  CG-1715-0128
Automated Data Processing Equipment Fleet Marine Force Programmer
  MC-1402-0050
Automated Data Processing Equipment for the Fleet Marine
  MC-1402-0050
Automated Information Systems Contracting
  DD-1402-0006
Commercial Services and Automated Travel Record Accounting System by Correspondence
  AF-1408-0074

Automatic
Automatic Flight Control Systems Specialist by Correspondence
  AF-1715-0075
Automatic Telephone Equipment by Correspondence
  MC-1715-0099
Automatic Tracking Radar Journeyman by Correspondence
  AF-1715-0812
Automatic Tracking Radar Specialist by Correspondence
  AF-1715-0007
  AF-1715-0776
  AF-1715-0794
  AF-1715-0795
  AF-1715-0812
Automatic Tracking Radar Technician by Correspondence
  AF-1715-0011
CEJT-SX-100/200 Mitel Electronic Private Automatic Branch Exchange Generic 217, Class C
  CG-1715-0113
CEJT-SX-200 Mitel Electronic Private Automatic Branch Exchange Generic 1000 or 1001 Telephone System
  CG-1715-0116
F-15 Avionics Automatic Test Station and Component Specialist by Correspondence
  AF-1715-0772
F/FB-111 Automatic Test Station and Component Specialist by Correspondence
  AF-1715-0773
HH-3F Automatic Flight Control System and Selected Electrical Maintenance, Class C
  CG-1714-0015
HH-60J Electrical/Automatic Flight Control Systems Maintenance, Class C
  CG-1704-0051
HH-65A Electrical/Automatic Flight Control Systems Maintenance, Class C
  CG-1704-0049
TEL 14 CDXC-SG-1/1A Pulse 120 Electronic Private Automatic Branch Exchange Telephone System, Class C
  CG-1715-0115

Automotive
Automotive Brake System by Correspondence
  MC-1703-0033
Automotive Cooling and Lubricating Systems by Correspondence
  MC-1703-0032
Automotive Engine Maintenance and Repair by Correspondence
  MC-1703-0024
Automotive Fuel and Exhaust Systems by Correspondence
  MC-1703-0031
Automotive Intermediate Maintenance
  MC-1712-0008
Automotive Organizational Maintenance
  MC-1703-0028
Automotive Power Trains by Correspondence
  MC-1703-0022
Reserve Automotive Mechanic
  MC-1408-0020

Auxiliary
AN/UYK-7 Computer and Data Auxiliary Console Course Development
  CG-1402-0002
AN/UYK-7 Computer and OJ-172 Data Exchange Auxiliary Console
  CG-1402-0002

Aviation
Aviation Administration by Correspondence
  CG-1704-0038
Aviation Combat Element Staff Planning
  MC-2204-0094
Aviation Electrician's Mate First Class by Correspondence
  CG-1714-0022
Aviation Electrician's Mate Second Class by Correspondence
  CG-1714-0025
Aviation Electrician's Mate, Class A Rotary Wing Training
  CG-1704-0025
Aviation Electronics Technician 618M-3, Class C
  CG-1715-0079
  CG-1715-0142
Aviation Electronics Technician AN/APS-127
  CG-1715-0080
Aviation Electronics Technician AN/APS-127 Radar, Class C
  CG-1715-0080
Aviation Electronics Technician AN/ARC-160, Class C
  CG-1715-0078
Aviation Electronics Technician AN/ARN-118(V), Class C
  CG-1715-0084
Aviation Electronics Technician AN/ARN-133V2, Class C
  CG-1715-0083
Aviation Electronics Technician First Class by Correspondence
  CG-1715-0132
Aviation Electronics Technician Second Class by Correspondence
  CG-1715-0031
Aviation Electronics Technician, Class A
  CG-1715-0001
Aviation Engineering Administration
  CG-1405-0005
Aviation Fire Control Repair
  MC-1715-0144
Aviation Fire Control Technician
  MC-1715-0040
Aviation Machinist Mate First Class by Correspondence
  CG-1704-0008
Aviation Machinist's Mate
  CG-1704-0001
Aviation Machinist's Mate (AD), Class A
  CG-1704-0001
Aviation Machinist's Mate First Class by Correspondence
  CG-1704-0008
  CG-1704-0052
Aviation Machinist's Mate Fixed Wing, Class A
  CG-1704-0022
Aviation Machinist's Mate Rotary Wing, Class A
  CG-1704-0023
Aviation Machinist's Mate Second Class by Correspondence
  CG-1704-0007
Aviation Machinist's Mate, Class A
  CG-1704-0001
Aviation Maintenance Administration by Correspondence
  CG-1408-0003
Aviation Quality Assurance Supervisor by Correspondence
  MC-1405-0047
Aviation Radar Repair (A)
  MC-1715-0153
Aviation Radar Repair (B)
  MC-1715-0148
Aviation Radar Repair (C)
  MC-1715-0059
Aviation Radar Technician
  MC-1715-0122
Aviation Radar Technician (A)
  MC-1715-0156
Aviation Radar Technician (B)
  MC-1715-0159
Aviation Radar Technician (C)
  MC-1715-0152
Aviation Radio Repair
  MC-1715-0011
  MC-1715-0155
Aviation Radio Technician
  MC-1715-0127
Aviation Structural Mechanic (AM), Class A
  CG-1704-0027
Aviation Structural Mechanic First Class by Correspondence
  CG-1704-0033
Aviation Structural Mechanic Second Class by Correspondence
  CG-1704-0039
Aviation Structural Mechanic, Class A
  CG-1704-0047
Aviation Survivalman First Class by Correspondence
  CG-1704-0041
Aviation Survivalman Second Class by Correspondence
  CG-1704-0018
  CG-1704-0043

Avionic
Avionic Guidance and Control Systems Technician by Correspondence
  AF-1704-0232
Avionic Sensor Systems Specialist (Electro-Optical Sensors) by Correspondence
  AF-1715-0071
Avionic Sensor Systems Specialist (Reconnaissance Electronic Sensors) by Correspondence
  AF-1715-0068
Avionic Sensor Systems Specialist (Tactical/Real Time Display Electronic Sensors) by Correspondence
  AF-1715-0070
F-15 Integrated Avionic Attack Control Systems Specialist by Correspondence
  AF-1715-0767
F-15/F-111 Avionic Systems Journeyman, Instrument, by Correspondence
  AF-1715-0815

KEYWORD INDEX   D-7

Integrated Avionic Attack Control Systems Specialist (F-16) by Correspondence
AF-1715-0096

Integrated Avionic Attack Control Systems Specialist by Correspondence
AF-1715-0759

Integrated Avionic Communication, Navigation and Penetration Aids Systems Specialist by Correspondence
AF-1714-0039

Integrated Avionic Communications, Navigation, and Penetration Aids Systems Specialist (F/FB-111) by Correspondence
AF-1715-0779

Integrated Avionic Instrument and Flight Control Systems by Correspondence
AF-1704-0255

Integrated Avionic Instrument and Flight Control Systems Specialist by Correspondence
AF-1704-0226
AF-1704-0227
AF-1715-0095

Integrated Organizational Avionic System Specialist by Correspondence
AF-1715-0758

Integrated Organizational Avionic Systems Specialist by Correspondence
AF-1405-0069

Avionics

A-7D Avionics Aerospace Ground Equipment Specialist by Correspondence
AF-1715-0078

Avionics Aerospace Ground Equipment Specialist (F/RF-4 Peculiar Avionics AGE) by Correspondence
AF-1715-0077

Avionics Aerospace Ground Equipment Specialist by Correspondence
AF-1715-0076

Avionics Guidance and Control Systems Technician by Correspondence
AF-1704-0234
AF-1704-0235
AF-1704-0236
AF-1704-0237
AF-1704-0238
AF-1704-0239
AF-1704-0240
AF-1704-0241
AF-1704-0259
AF-1715-0761

Avionics Instrument Systems Specialist by Correspondence
AF-1715-0080

Avionics Sensors Maintenance Journeyman by Correspondence
AF-1715-0818

Avionics Test Station and Component Specialist (F-16/A-10) by Correspondence
AF-1715-0756

Avionics Test Station and Component Specialist by Correspondence
AF-1715-0791

B-1B Avionics Test Station and Component Specialist by Correspondence
AF-1408-0097

B-1B Avionics Test Station and Component Technician by Correspondence
AF-1715-0793

Defensive Avionics/Communications/Navigation Systems by Correspondence
AF-1715-0789

F-15 Avionics Automatic Test Station and Component Specialist by Correspondence
AF-1715-0772

F-15 Avionics Manual/ECM Test Station and Component Specialist by Correspondence
AF-1715-0771

F-15 Integrated Organizational Avionics Systems Specialist by Correspondence
AF-1704-0254

F-16 A/B Avionics Test Station and Component Specialist by Correspondence
AF-1715-0780

F-16 C/D Avionics Test Station and Component Specialist by Correspondence
AF-1715-0785

F-16 Integrated Organizational Avionics Systems Specialist by Correspondence
AF-1715-0808

F/FB-111 Avionics Manual/Electronic Countermeasures (ECM) Test Station and Component Specialist by Correspondence
AF-1704-0224

General Subjects for F-111 Avionics System Specialist by Correspondence
AF-1715-0804

HC-130 Avionics by Correspondence
CG-1715-0085

HH-60J Avionics Maintenance, Class C
CG-1715-0145

HH-65A Avionics Maintenance, Class C
CG-1715-0144

HU-25A Avionics Maintenance, Class C
CG-1704-0045

HU-25A Avionics, Class C
CG-1704-0031

Integrated Avionics Attack Control Systems Specialist (F-15) by Correspondence
AF-1704-0218
AF-1715-0097

Integrated Avionics Attack Control Systems Specialist (F/FB-111) by Correspondence
AF-1715-0090
AF-1715-0760

Integrated Avionics Attack Control Systems Specialist by Correspondence
AF-1715-0091

Integrated Avionics Communication, Navigation, and Penetration Aids Systems Specialist (F-16) by Correspondence
AF-1704-0220

Integrated Avionics Communication, Navigation, and Penetration Aids Systems Specialist (F/FB-111) by Correspondence
AF-1715-0101

Integrated Avionics Communication, Navigation, and Penetration Aids Systems Specialist by Correspondence
AF-1715-0100

Integrated Avionics Communications, Navigation, and Penetration Aids Systems Specialist (F-15) by Correspondence
AF-1704-0225
AF-1715-0106

Integrated Avionics Communications, Navigation, and Penetration Aids Systems Specialist (F/FB-111) by Correspondence
AF-1704-0263

Integrated Avionics Computerized Test Station and Component Specialist (F-15) by Correspondence
AF-1715-0086

Integrated Avionics Computerized Test Station and Component Specialist (F/FB-111) by Correspondence
AF-1715-0087

Integrated Avionics Electronic Warfare Equipment and Component Specialist (F-15) by Correspondence
AF-1715-0082

Integrated Avionics Electronic Warfare Equipment and Component Specialist (F/FB-111) by Correspondence
AF-1715-0081

Integrated Avionics Instrument and Flight Control Systems Specialist (F-15) by Correspondence
AF-1715-0098

Integrated Avionics Instrument and Flight Control Systems Specialist (F-16) by Correspondence
AF-1715-0099

Integrated Avionics Instrument and Flight Control Systems Specialist (F/FB/EF-111) by Correspondence
AF-1715-0094

Integrated Avionics Manual Test Station and Component Specialist (F-15) by Correspondence
AF-1715-0092

Integrated Avionics Manual Test Station and Component Specialist (F/FB-111) by Correspondence
AF-1715-0093

Offensive Avionics Systems Specialist (B-1B) by Correspondence
AF-1704-0248

Avionicsman

HC-130 Avionicsman by Correspondence
CG-1715-0085

HH-3F Avionicsman by Correspondence
CG-1715-0086

HU-25A Avionicsman by Correspondence
CG-0802-0007

B-1B

B-1B Avionics Test Station and Component Specialist by Correspondence
AF-1408-0097

B-1B Avionics Test Station and Component Technician by Correspondence
AF-1715-0793

Instruments/Flight Control Systems Specialist (B-1B) by Correspondence
AF-1715-0770

Offensive Avionics Systems Specialist (B-1B) by Correspondence
AF-1704-0248

B-52

B-52 Bomber Weapons Instructor
AF-1704-0273

Bomber Weapons Instructor B-52
AF-1704-0273

B-52D/G

Defensive Fire Control Systems Mechanic (B-52D/G: MD-9, ASG-15 Turrets) by Correspondence
AF-1715-0047

B-52G/H

Bomb Navigation Systems Specialist (B-52G/H: ASQ-176, ASQ-151 Systems) by Correspondence
AF-1715-0046

B1

Bomber Weapons Instructor B1
AF-1704-0274

Baker

Baker Noncommissioned Officer (NCO)
MC-1729-0006

Basic Baker
MC-1729-0035

KEYWORD INDEX

Bakery
  Bakery Noncommissioned Officer (NCO) Leadership
    MC-1729-0006

Baking
  Field Bread Baking by Correspondence
    MC-1729-0043
  Pastry Baking by Correspondence
    MC-1729-0032

Bancroft
  Bancroft Maintenance
    MC-1715-0119

Base
  Base Transportation Officer
    AF-0419-0033

Basic
  Albanian Basic
    DD-0602-0018
  Arabic Basic (Modern Standard Arabic)
    DD-0602-0232
  Basic Military Training
    MC-2204-0038
  Basic Officer
    MC-2204-0083
  Basic Operational Training F-16
    AF-1704-0280
  Basic Operational Training F-16 A/B
    AF-1704-0280
  Basic Training
    CG-2205-0035
    MC-2204-0088
  Chinese Cantonese Basic
    DD-0602-0021
  Chinese Mandarin Basic
    DD-0602-0116
  Czech Basic
    DD-0602-0229
  Dutch Basic
    DD-0602-0223
  French Basic
    DD-0602-0213
  German Basic
    DD-0602-0215
  Greek Basic
    DD-0602-0029
  Hebrew Basic
    DD-0602-0235
  Hungarian Basic
    DD-0602-0132
  Indonesian Basic
    DD-0602-0032
  Indonesian-Malay Basic
    DD-0602-0012
  Italian Basic
    DD-0602-0108
    DD-0602-0226
  Japanese Basic
    DD-0602-0117
  Korean Basic
    DD-0602-0118
  Light Armored Vehicle Basic Repairman
    MC-1710-0049
  Malay Basic
    DD-0602-0012
  Noncommissioned Officer (NCO) Basic
    MC-2204-0103
  Norwegian Basic
    DD-0602-0224
  Persian Basic
    DD-0602-0230
  Pilipino/Tagalog Basic
    DD-0602-0237
  Polish Basic
    DD-0602-0112
  Portuguese Basic
    DD-0602-0212
  Romanian Basic
    DD-0602-0225
  Russian Basic
    DD-0602-0227
  Serbo-Croatian Basic
    DD-0602-0114
  Short Basic Turkish
    DD-0602-0042
  Spanish Basic
    DD-0602-0222
  Swedish Basic
    DD-0602-0131
  Tagalog Basic
    DD-0602-0237
  Thai Basic
    DD-0602-0119
  Turkish Basic
    DD-0602-0231
  Ukrainian Basic
    DD-0602-0105
  Vietnamese Basic
    DD-0602-0238

Beverages
  Vegetables, Soups, Sauces, Gravies, and Beverages by Correspondence
    MC-1729-0024

Bioenvironmental
  Bioenvironmental Engineering Specialist by Correspondence
    AF-0707-0010

Biological
  Basic Nuclear, Biological, and Chemical (NBC) Defense
    MC-0801-0014
  Intelligence Brief: The Opposing Forces Nuclear, Biological, Chemical (NBC) Threat by Correspondence
    MC-1606-0011

Biomedical
  Biomedical Equipment Maintenance Specialist by Correspondence
    AF-1715-0745

Blood
  Medical Laboratory Craftsman (Hematology, Serology, Blood Banking and Immunohematology) by Correspondence
    AF-0702-0010
  Medical Laboratory Technician (Hematology, Serology, Blood Banking and Immunohematology) by Correspondence
    AF-0702-0010
    AF-0709-0031

Boarding
  Maritime Law Enforcement Boarding Officer, Class C
    CG-1728-0021
  Maritime Law Enforcement Boarding Team Member, Class C
    CG-1728-0045

Boat
  Reserve Small Boat Crewmember
    CG-2205-0027
  Small Boat Crewmember
    CG-2205-0027
  Small Boat Engineer, Class C
    CG-1714-0019
  Small Boat Engineering
    CG-1714-0019
  Small Boat Magnetic Compass Calibration by Correspondence
    CG-1722-0006

Boating
  National Boating Safety
    CG-0802-0014

Boatswain's
  Boatswain's Mate First Class by Correspondence
    CG-1708-0004
  Boatswain's Mate Second Class By Correspondence
    CG-1708-0021
  Boatswain's Mate Second Class by Correspondence
    CG-1708-0021
  Boatswain's Mate Third Class by Correspondence
    CG-1708-0022
  Boatswain's Mate, Class A
    CG-1708-0016

Body
  General Purpose Vehicle and Body Maintenance by Correspondence
    AF-1703-0022
  General Purpose Vehicle and Body Maintenance Supervisor by Correspondence
    AF-1703-0022
  Vehicle Body Mechanic by Correspondence
    AF-1710-0035

Boiler
  Fire Tube and Boiler/Flash Type Evaporator, Class C
    CG-1710-0008

Boilers
  Electrical/Electronic Control for Fire Tube Boilers/Oily Water Separator and Electrical Generator
    CG-1715-0107
  Electrical/Electronic Control for Fire Tube Boilers/Oily Water Separator/Electrical Generators Operation and Maintenance, Class C
    CG-1715-0107

Bomb
  Bomb Navigation Systems Specialist (B-52G/H: ASQ-176, ASQ-151 Systems) by Correspondence
    AF-1715-0046
  Bomb Navigation Systems Specialist by Correspondence
    AF-1715-0774

Bombardment
  Aircraft Maintenance Specialist and Bombardment Aircraft by Correspondence
    AF-1704-0184
  Aircraft Maintenance Specialist, Airlift and Bombardment Aircraft by Correspondence
    AF-1704-0181

Brake
  Automotive Brake System by Correspondence
    MC-1703-0033

Bread
  Field Bread Baking by Correspondence
    MC-1729-0043

Bridge
  Aircraft Maintenance Officer (Bridge)
    AF-1704-0286

Broadcast
  Armed Forces Radio Television System Broadcast Manager
    DD-0505-0005

Broadcaster
  Basic Broadcaster
    DD-0505-0004

KEYWORD INDEX D-9

Broadcasting
Introduction to Broadcasting Reserve, Phase 2
DD-0504-0016
Radio and Television Broadcasting Specialist by Correspondence
AF-0505-0002

Budget
Budget by Correspondence
AF-1408-0092
Budget Officer
AF-1408-0100
Budget, Cost Estimating, and Financial Management Workshop
DD-1408-0017
Field Budget Formulation by Correspondence
MC-1401-0011
Financial Management Specialist (Budget) by Correspondence
AF-1408-0077
Fiscal/Budget Technician
MC-1402-0055

Bulgarian
Bulgarian Intermediate
DD-0602-0022

C-141
C-141 Flight Engineer Technician
AF-1704-0031

C-5
C-5 Flight Engineer Technician
AF-1704-0033
C-5 Pilot
AF-1606-0023
Flight Engineer School, C-5
AF-1704-0033

Cable
Antenna and Cable Systems Installation/Maintenance Specialist by Correspondence
AF-1715-0802
Antenna and Cable Systems Projects/Maintenance Specialist by Correspondence
AF-1715-0802
Cable Splicing Installation and Maintenance Specialist by Correspondence
AF-1715-0126
Communications Cable Systems Installation/Maintenance Specialist by Correspondence
AF-1714-0036
AF-1714-0037

Cadre
Marine Cadre Trainer
MC-1406-0035

CAEMS
Introduction to MDSS II/CAEMS
MC-1402-0059

Caliber
Gunner's Mate, Class A, Phase 2 (Advanced Electricity, 3"/50 Caliber Gun, 5"/38 Caliber Gun)
CG-1714-0013

Calibration
DCLF Reference Measurement and Calibration
AF-1715-0493
Microwave Measurement and Calibration
AF-1715-0740
Physical Measurement and Calibration
AF-1715-0743
Small Boat Magnetic Compass Calibration by Correspondence
CG-1722-0006

Candidate
Officer Candidate
CG-2202-0005
MC-2204-0080
Reserve Officer Candidate Indoctrination
CG-2202-0004

Card
Circuit Card Repair
MC-1715-0140

Career
Career Advisory Technician by Correspondence
AF-1406-0051
Resident Staff Noncommissioned Officer (NCO) Career Regular
MC-2204-0074
Resident Staff Noncommissioned Officer (NCO) Career Reserve
MC-2204-0071

Cargo
Air Cargo Specialist by Correspondence
AF-0419-0004
Apprentice Air Cargo Specialist by Correspondence
AF-0419-0029

Cartographic
Cartographic/Geodetic Officer
DD-1713-0006

Cartography
Advanced Cartography
DD-1713-0007
Basic Cartography
DD-1601-0025

Celestial
Celestial Navigation by Correspondence
CG-1304-0021

Central
Communication Central, AN/TGC-37, System Maintenance
MC-1715-0006
Mobile Communication Central Technician
MC-1715-0006
Mobile Communications Central Technician (AN/TGC-37(V))
MC-1715-0006
Tactical Air Command Central (TACC AN/TYQ-1) Repair
MC-1715-0020
Tactical Air Command Central (TACC) Technician
MC-1715-0046
Tactical Air Command Central Repairer
MC-1715-0020
Tactical Data Communications Central (TDCC AN/TYQ-3) Technician
MC-1715-0026
Tactical Data Communications Central Technician
MC-1715-0026
Telephone Central Office Switching Equipment Specialist, Electronic/Electromechanical by Correspondence
AF-1715-0127

Ceremonies
Honors and Ceremonies by Correspondence
CG-2205-0032

Certification
Terrain Analysis Warrant Officer Certification
DD-1601-0029

Chapel
Chapel Management Specialist by Correspondence
AF-1408-0085
AF-1408-0108
Chapel Management Technician by Correspondence
AF-1408-0086
Chapel Service Support Journeyman by Correspondence
AF-1408-0113

Chaplain
Chaplain Service Support Craftsman by Correspondence
AF-1408-0112

Charge
Officer In Charge/Executive Petty Officer
CG-2205-0020

Charting
Mapping, Charting, and Geodesy Officer
DD-1601-0021
Remotely Sensed Imagery for Mapping, Charting, and Geodesy
DD-1601-0030

Chemical
Basic Nuclear, Biological, and Chemical (NBC) Defense
MC-0801-0014
Chemical Warfare Defense by Correspondence
MC-0801-0013
Intelligence Brief: The Opposing Forces Nuclear, Biological, Chemical (NBC) Threat by Correspondence
MC-1606-0011

Chemistry
Medical Laboratory Craftsman (Chemistry and Urinalysis) by Correspondence
AF-0702-0008
Medical Laboratory Technician (Chemistry and Urinalysis) by Correspondence
AF-0702-0008

Chief
Chief Food Service Specialist by Correspondence
MC-1729-0045
Chief Petty Officer Academy
CG-1511-0002
Chief, Inspection Department
CG-1406-0010
Communication Systems Chief
MC-1408-0023
Damage Controlman Chief by Correspondence
CG-1408-0037
Electronics Technician Chief by Correspondence
CG-1402-0001
Engineer Equipment Chief
MC-1710-0043
Engineer Equipment Chief by Correspondence
MC-1601-0016
Engineer Operations Chief
MC-1601-0027
Military Requirements for Chief Petty Officer by Correspondence
CG-2205-0036
Operational Communication Chief
MC-1717-0004
Reserve Operational Communication Chief Refresher
MC-1402-0067

KEYWORD INDEX

Sonar Technician Chief by Correspondence
CG-1408-0039
Utilities Chief
MC-1710-0035
Utilities Officer/Chief by Correspondence
MC-1710-0048

Chinese
Chinese Advanced
DD-0602-0234
Chinese Intermediate
DD-0602-0233

Chinese Cantonese
Chinese Cantonese Basic
DD-0602-0021

Chinese Mandarin
Chinese Mandarin Basic
DD-0602-0116

Circle
Aiming Circle Operator by Correspondence
MC-1601-0035

Circuit
Circuit Card Repair
MC-1715-0140

Circuits
Communications Electronics Equipment, Circuits and Systems by Correspondence
AF-1715-0792

Civil
Civil Air Patrol Mission Observer Level II by Correspondence
AF-1704-0195
Civil Air Patrol Public Affairs Officer by Correspondence
AF-0505-0004
Civil Air Patrol Scanner by Correspondence
AF-1704-0194
Civil Air Patrol-Safety Officer (Level II Technical Rating) by Correspondence
AF-0801-0005
Civil Engineering Control System Specialist by Correspondence
AF-1710-0039
Civil Engineering Control Systems Specialist by Correspondence
AF-1710-0040
Introduction to Civil Air Patrol Emergency Services by Correspondence
AF-1704-0196
Military Civil Rights by Correspondence
CG-1512-0001
Military Functions in Civil Disturbances by Correspondence
MC-1728-0002

Clerk
Additional Demands Listing Clerk by Correspondence
MC-1408-0022
Administrative Clerk
MC-1403-0014
Basic Disbursing Clerk
MC-1401-0013
Basic Travel Clerk
MC-1403-0009
Personal Financial Records Clerk
MC-1403-0010
Personnel Clerk
MC-1405-0034
Senior Clerk
MC-1403-0015
Unit Diary Clerk
MC-1403-0013

Validation and Reconciliation Clerk by Correspondence
MC-1403-0017

Club
Club Food Operations by Correspondence
MC-1729-0046
Club Management Supervisor by Correspondence
AF-1406-0059

Coastal
Coastal Defense Command and Staff, Class C
CG-0419-0003
Coastal Defense Exercise Planner, Port Level
CG-1722-0013
Coastal Defense Planner, Port Level
CG-1722-0012
Coastal Search Planning, Class C
CG-1722-0010

COBOL
COBOL Programming
MC-1402-0013
IBM System 360 (OS) COBOL Programming
MC-1402-0013
IBM System/360 (OS) COBOL Programming Entry-level
MC-1402-0013

Code
Assembler Language Code (ALC) Programming (Combination)
MC-1402-0051

Coding
IBM System 360 Assembler Language Coding (Entry)
MC-1402-0051

COL-URG-II
COL-URG-II HF Transmitting System
CG-1715-0072
COL-URG-II High Frequency Transmitting System
CG-1715-0072

Cold
Cold Weather Medicine
MC-0707-0001
Cold Weather Survival
MC-0804-0002

Collection
Intermediate ELINT Collection and Analysis
DD-1715-0016

College
Industrial College of the Armed Forces
DD-1511-0010
Industrial College of the Armed Forces (Resident Program)
DD-1511-0003
National War College
DD-1511-0009

Combat
Aviation Combat Element Staff Planning
MC-2204-0094
Basic Combat Engineer
MC-1601-0024
Combat Arms Training and Maintenance Specialist/Technician by Correspondence
AF-2203-0054
Combat Control Operator by Correspondence
AF-1704-0216
Combat Engineer
MC-1601-0026

Combat Engineer Chief, Construction Support by Correspondence
MC-1601-0012
Combat Engineer Noncommissioned Officer (NCO)
MC-1601-0028
Combat Engineer Noncommissioned Officer (NCO) by Correspondence
MC-1601-0036
Combat Engineer Officer
MC-1601-0026
Communications for the Combat Operations Center/Fire Support Coordination Center (COC/FSCC) by Correspondence
MC-1715-0101
Introduction to Combat Intelligence by Correspondence
MC-1606-0009
Journeyman Combat Engineer
MC-1601-0028
Landing Force Combat Services Support Staff Planning
MC-2204-0093
Landing Force Ground Combat Communication Staff Planning
MC-1601-0046
Landing Force Ground Combat Operations Officer (S-3)
MC-2204-0091
Marine Combat
MC-2204-0105
Marine Combat Instructor of Water Safety
MC-2204-0097
Reserve Combat Engineer Noncommissioned Officer (NCO)
MC-1601-0030
Reserve Combat Engineer Officer
MC-1601-0029
Reserve Combat Engineer, Phase 2
MC-1601-0042

Command
Air Command and Staff College
AF-1511-0001
Air Command and Staff College Resident Program
AF-1511-0010
Air Command and Staff Correspondence Associate Program
AF-1511-0013
Air Command and Staff Correspondence Program
AF-1511-0013
Air Command and Staff Nonresident Seminar Associate Program
AF-1511-0009
Air Command and Staff Nonresident Seminar Program
AF-1511-0009
Airborne Command and Control Communications Equipment Journeyman by Correspondence
AF-1715-0811
AF-1715-0816
Airborne Command and Control Communications Equipment Specialist by Correspondence
AF-1715-0816
Airborne Command and Control Equipment Specialist by Correspondence
AF-1715-0811
Airborne Command Post Communications Equipment Specialist by Correspondence
AF-1715-0122
Coastal Defense Command and Staff, Class C
CG-0419-0003

KEYWORD INDEX   D-11

Command and Control Specialist by Correspondence
    AF-1704-0217
Command and Control Systems
    MC-1715-0177
Command and Staff College Reserve Phase 1, Phase 2
    MC-1408-0015
Command and Staff Nonresident Program by Correspondence
    MC-1408-0025
Marine Corps Command and Staff College
    MC-1408-0017
Tactical Air Command and Control Specialist by Correspondence
    AF-1406-0068
Tactical Air Command Center Operator
    MC-1715-0157
Tactical Air Command Central (TACC AN/TYQ-1) Repair
    MC-1715-0020
Tactical Air Command Central Repair
    MC-1715-0151
Tactical Air Command Central Repairer
    MC-1715-0020
Tactical Air Command Central Technician
    MC-1715-0142

Commercial
Commercial Services and Automated Travel Record Accounting System by Correspondence
    AF-1408-0074
Financial Management Specialist (Commercial Services) by Correspondence
    AF-1408-0074

Commission
Direct Commission Officer
    CG-2202-0003

Communication
Advanced Communication Officers
    MC-1715-0068
Aircraft Communication/Navigation Systems Journeyman by Correspondence
    AF-1715-0817
AN/TSC-120 Communication Central System Installer/Maintainer
    MC-1715-0171
AN/TSC-120 Communication Central Technician
    MC-1715-0168
Basic Communication Officer
    MC-1405-0045
Basic Communication Officers
    MC-1405-0045
Communication Center Operator
    MC-1409-0005
Communication Central, AN/TGC-37, System Maintenance
    MC-1715-0006
Communication Noncommissioned Officer (NCO)
    MC-1405-0048
Communication Systems Chief
    MC-1408-0023
    MC-1715-0167
Communication-Electronics Systems Acquisition and Management by Correspondence
    AF-1715-0753
Communication/Navigation Systems Technician (Doppler Systems) by Correspondence
    AF-1704-0260
Communication/Navigation Systems Technician by Correspondence
    AF-1704-0242
    AF-1704-0243
    AF-1704-0244
    AF-1704-0245
    AF-1704-0246
High Frequency Communication Central Operator
    MC-1715-0143
Integrated Avionic Communication, Navigation and Penetration Aids Systems Specialist by Correspondence
    AF-1714-0039
Integrated Avionics Communication, Navigation, and Penetration Aids Systems Specialist (F-16) by Correspondence
    AF-1704-0220
Integrated Avionics Communication, Navigation, and Penetration Aids Systems Specialist (F/FB-111) by Correspondence
    AF-1715-0101
Integrated Avionics Communication, Navigation, and Penetration Aids Systems Specialist by Correspondence
    AF-1715-0100
Landing Force Ground Combat Communication Staff Planning
    MC-1601-0046
Mobile Communication Central Technician
    MC-1715-0006
Mobile Security Communication Technician
    MC-1715-0169
Operational Communication Chief
    MC-1717-0004
Radiotelegraph and Visual Communication Procedures by Correspondence
    MC-1715-0133
Radiotelephone and Alternate Communication Procedures by Correspondence
    MC-1715-0137
Reserve Basic Communication Officer Phase 1
    MC-1715-0179
Reserve Basic Communication Officer Phase 2
    MC-1715-0178
Reserve Communication Officers, Phases 1 and 2
    MC-1715-0025
Secure Communication Systems Maintenance Specialist by Correspondence
    AF-1715-0786
Tactical Data Communication Central Repair
    MC-1715-0141
Tactical Data Communication Central Technician
    MC-1715-0146

Communications
Airborne Command and Control Communications Equipment Journeyman by Correspondence
    AF-1715-0811
    AF-1715-0816
Airborne Command and Control Communications Equipment Specialist by Correspondence
    AF-1715-0816
Airborne Command Post Communications Equipment Specialist by Correspondence
    AF-1715-0122
Airborne Communications Systems Operator by Correspondence
    AF-1704-0198
Allied Visual Communications, Class C
    CG-2205-0039
AN/URC-114(V) Low Power (1KW) Communications System, Class C
    CG-1715-0124
Communications Cable Systems Installation/Maintenance Specialist by Correspondence
    AF-1714-0036
    AF-1714-0037
Communications Computer Systems Control Specialist by Correspondence
    AF-1715-0778
Communications Computers Systems Control Specialist by Correspondence
    AF-1715-0788
Communications Electronics Equipment, Circuits and Systems by Correspondence
    AF-1715-0792
Communications for the Combat Operations Center/Fire Support Coordination Center (COC/FSCC) by Correspondence
    MC-1715-0101
Communications Officer by Correspondence
    CG-1404-0004
Communications Systems Radio Operator by Correspondence
    AF-1715-0781
Communications-Computer Systems Control Technician by Correspondence
    AF-1715-0790
Communications-Computer Systems Program Management Specialist by Correspondence
    AF-1715-0787
Communications-Electronics Employment by Correspondence
    AF-1715-0748
Communications-Electronics Systems Technology by Correspondence
    AF-1715-0747
Communications/Navigation Systems Technician by Correspondence
    AF-1715-0116
Defensive Avionics/Communications/Navigation Systems by Correspondence
    AF-1715-0789
Electronic Cryptographic Communications Equipment Specialist by Correspondence
    AF-1715-0024
Electronics Fundamentals Communications Track
    CG-1715-0090
Ground Radio Communications Specialist (Unit) by Correspondence
    AF-1715-0782
Ground Radio Communications Specialist by Correspondence
    AF-1715-0227
Ground Radio Communications Technician by Correspondence
    AF-1715-0465
High Frequency Communications System by Correspondence
    MC-1715-0135
Integrated Avionic Communications, Navigation, and Penetration Aids Systems Specialist (F/FB-111) by Correspondence
    AF-1715-0779
Integrated Avionics Communications, Navigation, and Penetration Aids Systems Specialist (F-15) by Correspondence
    AF-1704-0225
    AF-1715-0106
Integrated Avionics Communications, Navigation, and Penetration Aids Systems Specialist (F/FB-111) by Correspondence
    AF-1704-0263

KEYWORD INDEX

Missile Control Communications Systems Repairman by Correspondence
AF-1715-0130
Missile Control Communications Systems Specialist by Correspondence
AF-1714-0038
Mobile Data Communications Terminal Technician
MC-1715-0118
National Communications Security (COMSEC)
DD-1404-0004
Satellite and Wideband Communications Equipment Specialist by Correspondence
AF-1715-0797
AF-1715-0799
Satellite Communications System Equipment Specialist by Correspondence
AF-1715-0014
Secure Communications Systems Maintenance by Correspondence
AF-1715-0813
AF-1715-0820
Space Communications Systems Equipment by Correspondence
AF-1715-0014
Space Communications Systems Equipment Operator/Technician by Correspondence
AF-1715-0021
Tactical Data Communications Central (TDCC AN/TYQ-3) Technician
MC-1715-0026
Tactical Data Communications Central Technician
MC-1715-0026
Wideband Communications Equipment Specialist by Correspondence
AF-1715-0208
Wideband Communications Equipment Technician by Correspondence
AF-1715-0015

Compass
Compass Systems by Correspondence
CG-1708-0008
Small Boat Magnetic Compass Calibration by Correspondence
CG-1722-0006

Component
Avionics Test Station and Component Specialist (F-16/A-10) by Correspondence
AF-1715-0756
Avionics Test Station and Component Specialist by Correspondence
AF-1715-0791
B-1B Avionics Test Station and Component Specialist by Correspondence
AF-1408-0097
B-1B Avionics Test Station and Component Technician by Correspondence
AF-1715-0793
F-15 Avionics Automatic Test Station and Component Specialist by Correspondence
AF-1715-0772
F-15 Avionics Manual/ECM Test Station and Component Specialist by Correspondence
AF-1715-0771
F-16 A/B Avionics Test Station and Component Specialist by Correspondence
AF-1715-0780
F-16 C/D Avionics Test Station and Component Specialist by Correspondence
AF-1715-0785
F/FB-111 Automatic Test Station and Component Specialist by Correspondence
AF-1715-0773

F/FB-111 Avionics Manual/Electronic Countermeasures (ECM) Test Station and Component Specialist by Correspondence
AF-1704-0224
Fire and Electrical Systems Component Repair
MC-1714-0018
Integrated Avionics Computerized Test Station and Component Specialist (F-15) by Correspondence
AF-1715-0086
Integrated Avionics Computerized Test Station and Component Specialist (F/FB-111) by Correspondence
AF-1715-0087
Integrated Avionics Electronic Warfare Equipment and Component Specialist (F-15) by Correspondence
AF-1715-0082
Integrated Avionics Electronic Warfare Equipment and Component Specialist (F/FB-111) by Correspondence
AF-1715-0081
Integrated Avionics Manual Test Station and Component Specialist (F-15) by Correspondence
AF-1715-0092
Integrated Avionics Manual Test Station and Component Specialist (F/FB-111) by Correspondence
AF-1715-0093
Microminiature Component Repair
MC-1715-0089
MC-1715-0140

Compression
Compression Ignition Engine Fuel Systems Repair
MC-1712-0004

Comptroller
Comptroller Staff Officer
AF-1408-0104
Comptroller Staff Officer, Air Reserve Forces
AF-1401-0018
Professional Military Comptroller
AF-1408-0004
Professional Military Comptroller (PMCC)
AF-1408-0099

Computer
AN/UYK-7 Computer and Data Auxiliary Console Course Development
CG-1402-0002
AN/UYK-7 Computer and OJ-172 Data Exchange Auxiliary Console
CG-1402-0002
Communications Computer Systems Control Specialist by Correspondence
AF-1715-0778
Communications-Computer Systems Control Technician by Correspondence
AF-1715-0790
Communications-Computer Systems Program Management Specialist by Correspondence
AF-1715-0787
Computer Aided Search Planning by Correspondence
CG-0802-0017
Computer Operator
MC-1402-0023
Computer Orientation for Intermediate Executives
DD-1402-0003
Computer Technician
MC-1402-0056

Electronic Computer and Switching Systems Specialist by Correspondence
AF-1715-0765
Ground Computer Technician
MC-1402-0056
IBM S/360 OS Computer Operator
MC-1402-0023
System Computer Technician
MC-1402-0065
Tactical General Purpose Computer Technician
MC-1402-0056

Computer-Aided
Marine Air-Ground Task Force Deployment Support System/Computer-Aided Embarkation Management System
MC-1402-0059

Computerized
Integrated Avionics Computerized Test Station and Component Specialist (F-15) by Correspondence
AF-1715-0086
Integrated Avionics Computerized Test Station and Component Specialist (F/FB-111) by Correspondence
AF-1715-0087

Computers
Communications Computers Systems Control Specialist by Correspondence
AF-1715-0788

COMSEC
National Communications Security (COMSEC)
DD-1404-0004

Conditioning
Air Conditioning and Refrigeration, Class C
CG-1701-0002
Heating, Ventilation, Air Conditioning, and Refrigeration Journeyman by Correspondence
AF-1701-0012
AF-1701-0013
Heating, Ventilation, Air Conditioning, and Refrigeration Specialist by Correspondence
AF-1701-0012
AF-1701-0013
Refrigeration and Air Conditioning
CG-1730-0001
Refrigeration/Air Conditioning Operation and Maintenance, Class C
CG-1701-0002

Connectors
Soldering and Electrical Connectors by Correspondence
AF-1714-0033
AF-1723-0012

Conning
Watchstanding: The Conning Officer by Correspondence
CG-1708-0012

Console
AN/UYK-7 Computer and Data Auxiliary Console Course Development
CG-1402-0002
AN/UYK-7 Computer and OJ-172 Data Exchange Auxiliary Console
CG-1402-0002

Construction
Antenna Construction and Propagation of Radio Waves by Correspondence
MC-1715-0136

KEYWORD INDEX　　D-13

Apprentice Construction Equipment Operator by Correspondence
　　　　　　　　　　AF-1710-0029
Combat Engineer Chief, Construction Support by Correspondence
　　　　　　　　　　MC-1601-0012
Construction Equipment Operator by Correspondence
　　　　　　　　　　AF-1710-0028
Construction Print Reading by Correspondence
　　　　　　　　　　MC-1601-0041
Pavement and Construction Equipment Journeyman by Correspondence
　　　　　　　　　　AF-1601-0051
　　　　　　　　　　AF-1601-0052
Pavement and Construction Equipment Specialist by Correspondence
　　　　　　　　　　AF-1601-0051
　　　　　　　　　　AF-1601-0052
Pole Line Construction Equipment by Correspondence
　　　　　　　　　　MC-1715-0091
Pole Line Construction Techniques by Correspondence
　　　　　　　　　　MC-1715-0096
Theory and Construction of Turbine Engines by Correspondence
　　　　　　　　　　MC-1710-0004

Contingency
General Contingency Responsibilities by Correspondence
　　　　　　　　　　AF-2203-0057

Contract
Contract Administration by Correspondence
　　　　　　　　　　AF-1408-0087
Contract Performance Management Fundamentals
　　　　　　　　　　DD-1408-0014
Contract Pricing
　　　　　　　　　　DD-1405-0002
　　　　　　　　　　DD-1405-0004
Government Contract Law
　　　　　　　　　　DD-0326-0007
Intermediate Contract Administration
　　　　　　　　　　DD-1405-0005
Intermediate Contract Performance Management
　　　　　　　　　　DD-1408-0015
Intermediate Contract Pricing
　　　　　　　　　　DD-1405-0006
Operational Level Contract Pricing
　　　　　　　　　　DD-1405-0001
Principles of Contract Pricing by Correspondence
　　　　　　　　　　AF-1408-0091

Contracting
Automated Information Systems Contracting
　　　　　　　　　　DD-1402-0006
Contracting Specialist by Correspondence
　　　　　　　　　　AF-1408-0071
Contracting Supervisor by Correspondence
　　　　　　　　　　AF-1408-0072
Facilities Contracting Fundamentals
　　　　　　　　　　DD-1402-0009
Intermediate Facilities Contracting
　　　　　　　　　　DD-1402-0007
Operational Level Contracting Fundamentals
　　　　　　　　　　DD-1405-0003
Systems Acquisition for Contracting Personnel (Executive)
　　　　　　　　　　DD-1408-0009

Contracts
Facilities Contracts Pricing
　　　　　　　　　　DD-1402-0008

Control
270' WMEC Machinery Plant Control and Monitoring System
　　　　　　　　　　CG-1710-0013
270' WMEC Machinery Plant Control and Monitoring System, Class C
　　　　　　　　　　CG-1710-0013
270' WMEC Main Propulsion Control and Monitoring System (Electrical) Operation and Maintenance, Class C
　　　　　　　　　　CG-1710-0013
378 Class WHEC Control Systems Operation and Maintenance, Class C
　　　　　　　　　　CG-1714-0021
Aerospace Control and Warning Systems Operator by Correspondence
　　　　　　　　　　AF-1715-0754
Air Control Electronics Operator (AN/TYQ-23)
　　　　　　　　　　MC-2204-0090
Air Defense Control Officer
　　　　　　　　　　MC-2204-0030
Air Defense Control Officer (AN/TYQ-23)
　　　　　　　　　　MC-2204-0089
Air Defense Control Officers Senior
　　　　　　　　　　MC-1715-0150
Air Support Control Officer
　　　　　　　　　　MC-1715-0078
　　　　　　　　　　MC-1715-0176
Airborne Command and Control Communications Equipment Journeyman by Correspondence
　　　　　　　　　　AF-1715-0811
　　　　　　　　　　AF-1715-0816
Airborne Command and Control Communications Equipment Specialist by Correspondence
　　　　　　　　　　AF-1715-0816
Airborne Command and Control Equipment Specialist by Correspondence
　　　　　　　　　　AF-1715-0811
Airborne Warning and Control Radar Specialist by Correspondence
　　　　　　　　　　AF-1715-0123
　　　　　　　　　　AF-1715-0801
Aircraft Control and Warning Radar Specialist by Correspondence
　　　　　　　　　　AF-1715-0005
Aircraft Guidance and Control Systems Technician by Correspondence
　　　　　　　　　　AF-1704-0261
Avionic Guidance and Control Systems Technician by Correspondence
　　　　　　　　　　AF-1704-0232
Avionics Guidance and Control Systems Technician by Correspondence
　　　　　　　　　　AF-1704-0234
　　　　　　　　　　AF-1704-0235
　　　　　　　　　　AF-1704-0236
　　　　　　　　　　AF-1704-0237
　　　　　　　　　　AF-1704-0238
　　　　　　　　　　AF-1704-0239
　　　　　　　　　　AF-1704-0240
　　　　　　　　　　AF-1704-0241
　　　　　　　　　　AF-1704-0259
　　　　　　　　　　AF-1715-0761
Basic Infection Control Surveillance (Infection Control and Epidemiology)
　　　　　　　　　　AF-0707-0011
Basic Supply Stock Control
　　　　　　　　　　MC-1405-0035

Civil Engineering Control System Specialist by Correspondence
　　　　　　　　　　AF-1710-0039
Civil Engineering Control Systems Specialist by Correspondence
　　　　　　　　　　AF-1710-0040
Combat Control Operator by Correspondence
　　　　　　　　　　AF-1704-0216
Command and Control Specialist by Correspondence
　　　　　　　　　　AF-1704-0217
Communications Computers Systems Control Specialist by Correspondence
　　　　　　　　　　AF-1715-0788
Communications-Computer Systems Control Technician by Correspondence
　　　　　　　　　　AF-1715-0790
Corrosion Control Specialist by Correspondence
　　　　　　　　　　AF-1717-0013
Damage Control and Stability by Correspondence
　　　　　　　　　　CG-2205-0024
Data Control Techniques
　　　　　　　　　　MC-1402-0042
F-15 Integrated Avionic Attack Control Systems Specialist by Correspondence
　　　　　　　　　　AF-1715-0767
F-4 and AC-130 Weapons Control Systems Technician by Correspondence
　　　　　　　　　　AF-1715-0805
Guidance and Control Systems Technician by Correspondence
　　　　　　　　　　AF-1704-0233
IBM System 360 (OS) Data Control Techniques
　　　　　　　　　　MC-1402-0042
Integrated Avionic Attack Control Systems Specialist (F-16) by Correspondence
　　　　　　　　　　AF-1715-0096
Integrated Avionic Attack Control Systems Specialist by Correspondence
　　　　　　　　　　AF-1715-0759
Integrated Avionics Attack Control Systems Specialist (F-15) by Correspondence
　　　　　　　　　　AF-1704-0218
　　　　　　　　　　AF-1715-0097
Integrated Avionics Attack Control Systems Specialist (F/FB-111) by Correspondence
　　　　　　　　　　AF-1715-0090
　　　　　　　　　　AF-1715-0760
Integrated Avionics Attack Control Systems Specialist by Correspondence
　　　　　　　　　　AF-1715-0091
Loran-C Control Station Operations
　　　　　　　　　　CG-1715-0114
Loran-C Control Station Operations, Class C
　　　　　　　　　　CG-1715-0114
Loran-C Timing and Control Operations and Maintenance, Class C
　　　　　　　　　　CG-1715-0111
Missile Control Communications Systems Repairman by Correspondence
　　　　　　　　　　AF-1715-0130
Missile Control Communications Systems Specialist by Correspondence
　　　　　　　　　　AF-1714-0038
OS/VS2 MVS Data Control Techniques
　　　　　　　　　　MC-1402-0042
Production Control Specialist by Correspondence
　　　　　　　　　　AF-1405-0060
Small Cutter Damage Control
　　　　　　　　　　CG-0801-0001

KEYWORD INDEX

Tactical Air Command and Control Specialist by Correspondence
AF-1406-0068
Tactical Air Control Party
MC-1606-0013
Telecommunications Systems Control Technician by Correspondence
AF-1715-0034
Timing and Control Equipment, Class C
CG-1715-0111
Vehicle Maintenance Control and Analysis Specialist by Correspondence
AF-1710-0036
Weapons Control Systems Mechanic (A-7D, A-10, AC-130, F-5) by Correspondence
AF-1715-0800

Controller
Fighter Weapons Instructor Air Weapons Controller
AF-1406-0062
Maritime Reserve Coordination Center Controller
CG-1722-0011
Tactical Air Defense Controller
MC-1704-0004
Tactical Air Defense Controller Enlisted
MC-1704-0004

Controlman
Damage Controlman Advanced Reserve Active Duty for Training, Class C
CG-1710-0015
Damage Controlman First Class By Correspondence
CG-1710-0020
Damage Controlman Refresher Reserve Active Duty for Training, Class C
CG-1710-0015
Damage Controlman Second Class by Correspondence
CG-1710-0021
Damage Controlman Third Class by Correspondence
CG-1710-0019
Damage Controlman, Class A
CG-1710-0014

Conversion
Conversion and Lantirn Training F16 C/D Block 40/42
AF-1704-0278
Conversion Training F-15 Track 1A
AF-1606-0159
Conversion Training F-16C/D
AF-1704-0267
AF-1704-0268

Cookery
Meats and Meat Cookery by Correspondence
MC-1729-0044

Cooling
Automotive Cooling and Lubricating Systems by Correspondence
MC-1703-0032
Cooling and Lubrication System Maintenance by Correspondence
MC-1703-0032

Coordination
Communications for the Combat Operations Center/Fire Support Coordination Center (COC/FSCC) by Correspondence
MC-1715-0101
Maritime Reserve Coordination Center Controller
CG-1722-0011

Corrections
Corrections by Correspondence
MC-1728-0008
Corrections Supervisor by Correspondence
MC-1728-0007

Correspondence
A-7D Avionics Aerospace Ground Equipment Specialist by Correspondence
AF-1715-0078
Accounting for Plant Property by Correspondence
MC-1405-0025
Additional Demands Listing Clerk by Correspondence
MC-1408-0022
Administration Specialist by Correspondence
AF-1406-0045
Administration Technician by Correspondence
AF-1406-0046
Aerospace Control and Warning Systems Operator by Correspondence
AF-1715-0754
Aerospace Ground Equipment Mechanic by Correspondence
AF-1704-0187
AF-1710-0038
Aerospace Ground Equipment Technician by Correspondence
AF-1704-0192
Aerospace Photographic Systems Specialist by Correspondence
AF-1715-0139
Aerospace Physiology Specialist by Correspondence
AF-0709-0032
Aerospace Propulsion Specialist (Jet Engine) by Correspondence
AF-1704-0256
Aerospace Propulsion Specialist (Turboprop) by Correspondence
AF-1704-0257
Aerospace Propulsion Technician (Jet Engine) by Correspondence
AF-1704-0229
Aerospace Propulsion Technician (Turboprop) by Correspondence
AF-1710-0037
Aiming Circle Operator by Correspondence
MC-1601-0035
Air Cargo Specialist by Correspondence
AF-0419-0004
Air Command and Staff Correspondence Associate Program
AF-1511-0013
Air Command and Staff Correspondence Program
AF-1511-0013
Air Force Joint Service Supervisor Safety by Correspondence
AF-0801-0001
Air Force Logistics Command Directorate of Materiel Management by Correspondence
AF-1408-0089
Air Force Technical Order System by Correspondence
AF-2203-0053
Air Launched Missile Systems Specialist by Correspondence
AF-1714-0041
AF-1715-0814
Air Passenger Specialist by Correspondence
AF-0419-0021

Air Reserve Forces Social Actions Technician (Drug/Alcohol) by Correspondence
AF-0708-0004
Air Traffic Control Radar Repairman by Correspondence
AF-1715-0003
Air Traffic Control Radar Specialist by Correspondence
AF-1715-0003
Air Traffic Control Radar Technician by Correspondence
AF-1715-0009
Air Transportation Journeyman by Correspondence
AF-0419-0038
Air War College Associate Programs Nonresident and Correspondence
AF-1511-0012
Air War College Correspondence Program
AF-1511-0012
Airborne Command and Control Communications Equipment Journeyman by Correspondence
AF-1715-0811
AF-1715-0816
Airborne Command and Control Communications Equipment Specialist by Correspondence
AF-1715-0816
Airborne Command and Control Equipment Specialist by Correspondence
AF-1715-0811
Airborne Command Post Communications Equipment Specialist by Correspondence
AF-1715-0122
Airborne Communications Systems Operator by Correspondence
AF-1704-0198
Airborne Warning and Control Radar Specialist by Correspondence
AF-1715-0123
AF-1715-0801
Aircraft Armament Systems Specialist by Correspondence
AF-1715-0035
Aircraft Communication/Navigation Systems Journeyman by Correspondence
AF-1715-0817
Aircraft Control and Warning Radar Specialist by Correspondence
AF-1715-0005
Aircraft Electrical and Environmental Systems Journeyman by Correspondence
AF-1714-0045
Aircraft Electrical Systems Specialist by Correspondence
AF-1714-0003
Aircraft Electro-Environmental System Technician by Correspondence
AF-1704-0221
Aircraft Electro-Environmental Systems Technician by Correspondence
AF-1704-0222
AF-1704-0223
Aircraft Engine Familiarization by Correspondence
CG-1704-0034
Aircraft Environmental Systems Mechanic by Correspondence
AF-1704-0200
Aircraft Fuel Systems Mechanic by Correspondence
AF-1704-0191
AF-1704-0258
Aircraft Fuel Systems Technician by Correspondence
AF-1704-0202

KEYWORD INDEX D-15

Aircraft Guidance and Control Systems Technician by Correspondence
AF-1704-0261

Aircraft Maintenance Noncommissioned Officer (NCO) by Correspondence
MC-1704-0008

Aircraft Maintenance Specialist and Bombardment Aircraft by Correspondence
AF-1704-0184

Aircraft Maintenance Specialist, Airlift and Bombardment Aircraft by Correspondence
AF-1704-0181

Aircraft Maintenance Specialist, Tactical Aircraft by Correspondence
AF-1704-0102

Aircraft Metals Technology (Machinist) by Correspondence
AF-1717-0027

Aircraft Metals Technology (Welding) by Correspondence
AF-1723-0011
AF-1724-0007

Aircraft Pneudraulic Systems Mechanic by Correspondence
AF-1704-0201

Aircraft Pneudraulic Systems Technician by Correspondence
AF-1704-0230
AF-1704-0231

Aircraft Structural Maintenance Technician (Airframe Repair) by Correspondence
AF-1704-0250
AF-1717-0028

Aircrew Basic by Correspondence
CG-1704-0040

Aircrew Egress Systems Mechanic by Correspondence
AF-1717-0005

Aircrew Life Support Specialist by Correspondence
AF-1704-0199

Airfield Management Specialist by Correspondence
AF-1704-0215

Airframe Repair Specialist by Correspondence
AF-1723-0010

Airlift Aircraft Maintenance Specialist by Correspondence
AF-1704-0184

Airlift Aircraft Maintenance Technician by Correspondence
AF-1704-0249

Amphibious Warfare School Nonresident Program (AWSNP) by Correspondence Phase 2
MC-2204-0114

Amphibious Warfare School Nonresident Program (AWSNP) by Correspondence, Phase 1
MC-2204-0113

Antenna and Cable Systems Installation/Maintenance Specialist by Correspondence
AF-1715-0802

Antenna and Cable Systems Projects/Maintenance Specialist by Correspondence
AF-1715-0802

Antenna Construction and Propagation of Radio Waves by Correspondence
MC-1715-0136

Applied Sciences Technician by Correspondence
AF-1715-0775

Apprentice Air Cargo Specialist by Correspondence
AF-0419-0029

Apprentice Airfield Management Specialist by Correspondence
AF-1704-0214

Apprentice Construction Equipment Operator by Correspondence
AF-1710-0029

Apprentice Electrician by Correspondence
AF-1714-0023

Apprentice Fitness and Recreation Specialist by Correspondence
AF-1406-0055

Apprentice Food Service Specialist by Correspondence
AF-1729-0010

Apprentice Graphics Specialized by Correspondence
AF-1719-0009

Apprentice Mason by Correspondence
AF-1710-0030

Apprentice Materiel Facilities Specialist by Correspondence
AF-1405-0018

Apprentice Nondestructive Inspection Specialist by Correspondence
AF-1724-0006

Apprentice Operations Resource Management Specialist by Correspondence
AF-1402-0073

Apprentice Pavements Maintenance Specialist by Correspondence
AF-1601-0048

Apprentice Plumber by Correspondence
AF-1710-0027

Apprentice Reprographic Specialist by Correspondence
AF-1719-0008

Apprentice Services Specialist by Correspondence
AF-1729-0013

Apprentice Still Photographic Specialist by Correspondence
AF-1709-0029

Apprentice Subsistence Operations Specialist by Correspondence
AF-1729-0015

Apprentice Vehicle Operator/Dispatcher by Correspondence
AF-0419-0026

Army Joint Service Supervisor Safety by Correspondence
AF-0801-0002

Artillery Survey Instruments by Correspondence
MC-1601-0032

Audiovisual Media Specialist by Correspondence
AF-1406-0073

Automatic Flight Control Systems Specialist by Correspondence
AF-1715-0075

Automatic Telephone Equipment by Correspondence
MC-1715-0099

Automatic Tracking Radar Journeyman by Correspondence
AF-1715-0812

Automatic Tracking Radar Specialist by Correspondence
AF-1715-0007
AF-1715-0776
AF-1715-0794
AF-1715-0795
AF-1715-0812

Automatic Tracking Radar Technician by Correspondence
AF-1715-0011

Automotive Brake System by Correspondence
MC-1703-0033

Automotive Cooling and Lubricating Systems by Correspondence
MC-1703-0032

Automotive Engine Maintenance and Repair by Correspondence
MC-1703-0024

Automotive Fuel and Exhaust Systems by Correspondence
MC-1703-0031

Automotive Power Trains by Correspondence
MC-1703-0022

Aviation Administration by Correspondence
CG-1704-0038

Aviation Electrician's Mate First Class by Correspondence
CG-1714-0022

Aviation Electrician's Mate Second Class by Correspondence
CG-1714-0025

Aviation Electronics Technician First Class by Correspondence
CG-1715-0132

Aviation Electronics Technician Second Class by Correspondence
CG-1715-0031

Aviation Machinist Mate First Class by Correspondence
CG-1704-0008

Aviation Machinist's Mate First Class by Correspondence
CG-1704-0008
CG-1704-0052

Aviation Machinist's Mate Second Class by Correspondence
CG-1704-0007

Aviation Maintenance Administration by Correspondence
CG-1408-0003

Aviation Quality Assurance Supervisor by Correspondence
MC-1405-0047

Aviation Structural Mechanic First Class by Correspondence
CG-1704-0033

Aviation Structural Mechanic Second Class by Correspondence
CG-1704-0039

Aviation Survivalman First Class by Correspondence
CG-1704-0041

Aviation Survivalman Second Class by Correspondence
CG-1704-0018
CG-1704-0043

Avionic Guidance and Control Systems Technician by Correspondence
AF-1704-0232

Avionic Sensor Systems Specialist (Electro-Optical Sensors) by Correspondence
AF-1715-0071

Avionic Sensor Systems Specialist (Reconnaissance Electronic Sensors) by Correspondence
AF-1715-0068

Avionic Sensor Systems Specialist (Tactical/Real Time Display Electronic Sensors) by Correspondence
AF-1715-0070

Avionics Aerospace Ground Equipment Specialist (F/RF-4 Peculiar Avionics AGE) by Correspondence
AF-1715-0077

KEYWORD INDEX

Avionics Aerospace Ground Equipment Specialist by Correspondence
AF-1715-0076
Avionics Guidance and Control Systems Technician by Correspondence
AF-1704-0234
AF-1704-0235
AF-1704-0236
AF-1704-0237
AF-1704-0238
AF-1704-0239
AF-1704-0240
AF-1704-0241
AF-1704-0259
AF-1715-0761
Avionics Instrument Systems Specialist by Correspondence
AF-1715-0080
Avionics Sensors Maintenance Journeyman by Correspondence
AF-1715-0818
Avionics Test Station and Component Specialist (F-16/A-10) by Correspondence
AF-1715-0756
Avionics Test Station and Component Specialist by Correspondence
AF-1715-0791
B-1B Avionics Test Station and Component Specialist by Correspondence
AF-1408-0097
B-1B Avionics Test Station and Component Technician by Correspondence
AF-1715-0793
Basic Mathematics by Correspondence
CG-1107-0001
Basic Nutrition by Correspondence
MC-1729-0041
Basic Pay Entitlement by Correspondence
MC-1401-0008
Basic Radar Use and Operation by Correspondence
CG-1715-0076
Basic Search and Rescue Aircrewman by Correspondence
CG-0802-0012
Basic Search and Rescue by Correspondence
CG-0802-0009
Basic Shop Fundamentals for the Mechanic by Correspondence
MC-1717-0005
Basic Techniques of Wave Form Measurement Using an Oscilloscope by Correspondence
AF-1715-0784
Basic Warehousing by Correspondence
MC-1405-0046
Bioenvironmental Engineering Specialist by Correspondence
AF-0707-0010
Biomedical Equipment Maintenance Specialist by Correspondence
AF-1715-0745
Boatswain's Mate First Class by Correspondence
CG-1708-0004
Boatswain's Mate Second Class By Correspondence
CG-1708-0021
Boatswain's Mate Second Class by Correspondence
CG-1708-0021
Boatswain's Mate Third Class by Correspondence
CG-1708-0022

Bomb Navigation Systems Specialist (B-52G/H: ASQ-176, ASQ-151 Systems) by Correspondence
AF-1715-0046
Bomb Navigation Systems Specialist by Correspondence
AF-1715-0774
Budget by Correspondence
AF-1408-0092
Cable Splicing Installation and Maintenance Specialist by Correspondence
AF-1715-0126
Career Advisory Technician by Correspondence
AF-1406-0051
Celestial Navigation by Correspondence
CG-1304-0021
Chapel Management Specialist by Correspondence
AF-1408-0085
AF-1408-0108
Chapel Management Technician by Correspondence
AF-1408-0086
Chapel Service Support Journeyman by Correspondence
AF-1408-0113
Chaplain Service Support Craftsman by Correspondence
AF-1408-0112
Chemical Warfare Defense by Correspondence
MC-0801-0013
Chief Food Service Specialist by Correspondence
MC-1729-0045
Civil Air Patrol Mission Observer Level II by Correspondence
AF-1704-0195
Civil Air Patrol Public Affairs Officer by Correspondence
AF-0505-0004
Civil Air Patrol Scanner by Correspondence
AF-1704-0194
Civil Air Patrol-Safety Officer (Level II Technical Rating) by Correspondence
AF-0801-0005
Civil Engineering Control System Specialist by Correspondence
AF-1710-0039
Civil Engineering Control Systems Specialist by Correspondence
AF-1710-0040
Civil Law by Correspondence
AF-1511-0007
Club Food Operations by Correspondence
MC-1729-0046
Club Management Supervisor by Correspondence
AF-1406-0059
Coast Guard Division Officer by Correspondence
CG-2205-0033
Coast Guard Joint Service Supervisor Safety by Correspondence
AF-0801-0004
Coast Guard Orientation by Correspondence
CG-2205-0013
Coast Guard Orientation Officers by Correspondence
CG-2205-0019
Combat Arms Training and Maintenance Specialist/Technician by Correspondence
AF-2203-0054
Combat Control Operator by Correspondence
AF-1704-0216

Combat Engineer Chief, Construction Support by Correspondence
MC-1601-0012
Combat Engineer Noncommissioned Officer (NCO) by Correspondence
MC-1601-0036
Command and Control Specialist by Correspondence
AF-1704-0217
Commercial Services and Automated Travel Record Accounting System by Correspondence
AF-1408-0074
Communication-Electronics Systems Acquisition and Management by Correspondence
AF-1715-0753
Communication/Navigation Systems Technician (Doppler Systems) by Correspondence
AF-1704-0260
Communication/Navigation Systems Technician by Correspondence
AF-1704-0242
AF-1704-0243
AF-1704-0244
AF-1704-0245
AF-1704-0246
Communications Cable Systems Installation/Maintenance Specialist by Correspondence
AF-1714-0036
AF-1714-0037
Communications Computer Systems Control Specialist by Correspondence
AF-1715-0778
Communications Computers Systems Control Specialist by Correspondence
AF-1715-0788
Communications Electronics Equipment, Circuits and Systems by Correspondence
AF-1715-0792
Communications for the Combat Operations Center/Fire Support Coordination Center (COC/FSCC) by Correspondence
MC-1715-0101
Communications Officer by Correspondence
CG-1404-0004
Communications Systems Radio Operator by Correspondence
AF-1715-0781
Communications-Computer Systems Control Technician by Correspondence
AF-1715-0790
Communications-Computer Systems Program Management Specialist by Correspondence
AF-1715-0787
Communications-Electronics Employment by Correspondence
AF-1715-0748
Communications-Electronics Systems Technology by Correspondence
AF-1715-0747
Communications/Navigation Systems Technician by Correspondence
AF-1715-0116
Compass Systems by Correspondence
CG-1708-0008
Computer Aided Search Planning by Correspondence
CG-0802-0017
Construction Equipment Operator by Correspondence
AF-1710-0028
Construction Print Reading by Correspondence
MC-1601-0041
Contract Administration by Correspondence
AF-1408-0087

KEYWORD INDEX D-17

Contracting Specialist by Correspondence
AF-1408-0071
Contracting Supervisor by Correspondence
AF-1408-0072
Cooling and Lubrication System Maintenance by Correspondence
MC-1703-0032
Corrections by Correspondence
MC-1728-0008
Corrections Supervisor by Correspondence
MC-1728-0007
Correspondence Course of the Industrial College of the Armed Forces
DD-1511-0001
Corrosion Control Specialist by Correspondence
AF-1717-0013
Cost Analysis Specialist by Correspondence
AF-1408-0109
Cost and Management Analysis Specialist by Correspondence
AF-1408-0083
Cost and Management Analysis Technician by Correspondence
AF-1408-0084
Counseling for Marines by Correspondence
MC-1406-0032
Crane Operator by Correspondence
MC-1710-0039
Crime Prevention by Correspondence
AF-1728-0042
Damage Control and Stability by Correspondence
CG-2205-0024
Damage Control Deck Group Ratings by Correspondence
CG-1710-0011
Damage Controlman Chief by Correspondence
CG-1408-0037
Damage Controlman First Class By Correspondence
CG-1710-0020
Damage Controlman Second Class by Correspondence
CG-1710-0021
Damage Controlman Third Class by Correspondence
CG-1710-0019
Deck Seamanship by Correspondence
CG-1708-0010
Deck Watch Officer Navigation Rules by Correspondence
CG-1708-0014
Defensive Avionics/Communications/Navigation Systems by Correspondence
AF-1715-0789
Defensive Fire Control Systems Mechanic (B-52D/G: MD-9, ASG-15 Turrets) by Correspondence
AF-1715-0047
Dental Assistant Specialist by Correspondence
AF-0701-0003
Dental Laboratory Specialist by Correspondence
AF-0701-0018
Desert Operations by Correspondence
MC-0803-0007
Diesel Engine Maintenance and Troubleshooting by Correspondence
MC-1712-0007
Diet Therapy Specialist by Correspondence
AF-0104-0003
Diet Therapy Supervisor by Correspondence
AF-0104-0002

Direction Finding Operations by Correspondence
MC-1715-0130
Disaster Preparedness Specialist by Correspondence
AF-0802-0026
Education and Training Manager by Correspondence
AF-1406-0078
Education and Training Officer by Correspondence
AF-1406-0079
Education Specialist by Correspondence
AF-1406-0069
Electrical Power Production Journeyman by Correspondence
AF-1714-0040
AF-1714-0043
Electrical Power Production Specialist by Correspondence
AF-1714-0030
AF-1714-0043
Electrical Power Production Technician by Correspondence
AF-1408-0096
AF-1714-0024
AF-1714-0031
Electrical Systems Journeyman by Correspondence
AF-1714-0042
AF-1714-0044
Electrical Systems Technician by Correspondence
AF-1714-0042
Electrician by Correspondence
AF-1714-0028
Electrician's Mate First Class by Correspondence
CG-1714-0011
Electrician's Mate Second Class by Correspondence
CG-1714-0023
Electrician's Mate Third Class by Correspondence
CG-1714-0005
CG-1714-0024
Electronic Computer and Switching Systems Specialist by Correspondence
AF-1715-0765
Electronic Cryptographic Communications Equipment Specialist by Correspondence
AF-1715-0024
Electronic Tubes and Special Purpose Tubes by Correspondence
AF-1715-0749
Electronic Warfare (EW) System Journeyman by Correspondence
AF-1715-0810
Electronic Warfare Systems Journeyman by Correspondence
AF-1715-0819
Electronic Warfare Systems Specialist by Correspondence
AF-1715-0466
AF-1715-0769
Electronic Warfare Systems Technician (Unit) by Correspondence
AF-1715-0783
Electronics Fundamentals by Correspondence
AF-1715-0746
Electronics Technician Chief by Correspondence
CG-1402-0001
Electronics Technician First Class by Correspondence
CG-1715-0038

Electronics Technician Second Class by Correspondence
CG-1715-0037
Electronics Technician Second Class by Correspondence (ET2)
CG-1715-0146
Elementary Algebra by Correspondence
CG-1104-0001
Engineer Equipment Chief by Correspondence
MC-1601-0016
Engineer Equipment Operator by Correspondence
MC-1601-0034
Engineer Forms and Records by Correspondence
MC-1601-0015
MC-1601-0037
Engineering Assistant Specialist by Correspondence
AF-1601-0046
Environmental Medicine Specialist by Correspondence
AF-0707-0009
Environmental Support Specialist by Correspondence
AF-1710-0041
F-100 Jet Engine Mechanic by Correspondence
AF-1704-0205
F-15 Avionics Automatic Test Station and Component Specialist by Correspondence
AF-1715-0772
F-15 Avionics Manual/ECM Test Station and Component Specialist by Correspondence
AF-1715-0771
F-15 Integrated Avionic Attack Control Systems Specialist by Correspondence
AF-1715-0767
F-15 Integrated Organizational Avionics Systems Specialist by Correspondence
AF-1704-0254
F-15/F-111 Avionic Systems Journeyman, Instrument, by Correspondence
AF-1715-0815
F-16 A/B Avionics Test Station and Component Specialist by Correspondence
AF-1715-0780
F-16 C/D Avionics Test Station and Component Specialist by Correspondence
AF-1715-0785
F-16 Integrated Organizational Avionics Systems Specialist by Correspondence
AF-1715-0808
F-4 and AC-130 Weapons Control Systems Technician by Correspondence
AF-1715-0805
F/FB-111 Automatic Test Station and Component Specialist by Correspondence
AF-1715-0773
F/FB-111 Avionics Manual/Electronic Countermeasures (ECM) Test Station and Component Specialist by Correspondence
AF-1704-0224
Fabrication and Parachute Specialist by Correspondence
AF-1733-0002
Field Bread Baking by Correspondence
MC-1729-0043
Field Budget Formulation by Correspondence
MC-1401-0011
Financial Analysis Journeyman by Correspondence
AF-1408-0111

KEYWORD INDEX

Financial Analysis Specialist by Correspondence
AF-1408-0109
Financial Management by Correspondence
MC-1401-0012
Financial Management Specialist (Budget) by Correspondence
AF-1408-0077
Financial Management Specialist (Commercial Services) by Correspondence
AF-1408-0074
Financial Management Specialist (Military Pay) by Correspondence
AF-1408-0074
Financial Management Specialist by Correspondence
AF-1408-0073
AF-1408-0074
AF-1408-0075
AF-1408-0076
Financial Management Supervisor by Correspondence
AF-1408-0081
Financial Management/Services Supervisor (Functions and Responsibilities) by Correspondence
AF-1408-0082
Financial Management/Services Supervisor by Correspondence
AF-1408-0081
Financial Services Specialist (Introduction) by Correspondence
AF-1408-0078
Financial Services Specialist (Military Pay) by Correspondence
AF-1408-0079
Financial Services Specialist (Travel) by Correspondence
AF-1408-0080
Financial Services Specialist by Correspondence
AF-1408-0078
Fire and Safety Technician Second Class by Correspondence
CG-1728-0034
Fire Control Technician First Class by Correspondence
CG-1715-0135
Fire Control Technician Second Class by Correspondence
CG-1715-0137
Fire Protection Specialist by Correspondence
AF-1728-0038
Fire Protection Supervisor by Correspondence
AF-1728-0046
Firefighting Ashore by Correspondence
CG-1728-0005
Firefighting on Vessels by Correspondence
CG-1728-0007
Fireman by Correspondence
CG-1722-0016
First Sergeant by Correspondence
AF-2203-0052
Fiscal Accounting for Supply Clerks by Correspondence
MC-1401-0010
Fitness and Recreation Specialist by Correspondence
AF-1406-0056
Fitness and Recreation Supervisor by Correspondence
AF-1406-0057
Flight Engineer Specialist (Helicopter Qualified) by Correspondence
AF-1704-0197

Flight Engineer Specialist by Correspondence
AF-1704-0094
Food Service Fundamentals by Correspondence
MC-1729-0038
Food Service Specialist First Class by Correspondence
CG-1729-0005
Food Service Specialist Second Class by Correspondence
CG-1729-0004
Food Service Specialist Third Class by Correspondence
CG-1729-0006
Food Service Supervisor by Correspondence
AF-1729-0012
Freight Traffic Specialist by Correspondence
AF-0419-0032
Fuel Specialist by Correspondence
AF-1703-0019
Fundamental Principles of Electronic Data Processing Equipment by Correspondence
AF-1402-0072
Fundamentals of Diesel Engines by Correspondence
MC-1712-0006
Fundamentals of Digital Logic by Correspondence
MC-1715-0131
Fundamentals of Electricity by Correspondence
AF-1714-0029
Fundamentals of Marine Corps Leadership by Correspondence
MC-1406-0023
Fundamentals of Refrigeration by Correspondence
MC-1730-0005
Fundamentals of Solid State Devices by Correspondence
AF-1715-0750
General Contingency Responsibilities by Correspondence
AF-2203-0057
General Personnel Administration for Reserve by Correspondence
MC-1406-0025
General Purpose Vehicle and Body Maintenance by Correspondence
AF-1703-0022
General Purpose Vehicle and Body Maintenance Supervisor by Correspondence
AF-1703-0022
General Purpose Vehicle Mechanic by Correspondence
AF-1703-0018
General Subjects for F-111 Avionics System Specialist by Correspondence
AF-1715-0804
Graphics Specialist by Correspondence
AF-1719-0010
Ground Radio Communications Specialist (Unit) by Correspondence
AF-1715-0782
Ground Radio Communications Specialist by Correspondence
AF-1715-0227
Ground Radio Communications Technician by Correspondence
AF-1715-0465
Guidance and Control Systems Technician by Correspondence
AF-1704-0233
Gunner's Mate First Class by Correspondence
CG-1714-0027

Gunner's Mate Second Class by Correspondence
CG-1714-0009
CG-1714-0026
HC-130 Avionics by Correspondence
CG-1715-0085
HC-130 Avionicsman by Correspondence
CG-1715-0085
HC-130 Flight Engineer by Correspondence
CG-1704-0009
HC-130 Loadmaster (AC130L) by Correspondence
CG-0419-0002
HC-130 Loadmaster by Correspondence
CG-0419-0002
Health Services Management Journeyman by Correspondence
AF-0799-0008
AF-0799-0009
AF-0799-0010
Health Services Technician First Class by Correspondence
CG-0709-0006
Health Services Technician Second Class by Correspondence
CG-0709-0005
Heating Systems Specialist by Correspondence
AF-1701-0010
AF-1701-0011
Heating Systems Technician by Correspondence
AF-1701-0009
Heating, Ventilation, Air Conditioning, and Refrigeration Journeyman by Correspondence
AF-1701-0012
AF-1701-0013
Heating, Ventilation, Air Conditioning, and Refrigeration Specialist by Correspondence
AF-1701-0012
AF-1701-0013
Helicopter Mechanic by Correspondence
AF-1704-0206
HH-3F Avionicsman by Correspondence
CG-1715-0086
HH-3F Flight Mechanic by Correspondence
CG-1704-0011
HH-52A Flight Mechanic by Correspondence
CG-1704-0036
HH-60 Flight Mechanic by Correspondence
CG-1704-0042
HH-65A Flight Mechanic by Correspondence
CG-1704-0035
High Frequency Communications System by Correspondence
MC-1715-0135
Honors and Ceremonies by Correspondence
CG-2205-0032
HU-25A Avionicsman by Correspondence
CG-0802-0007
HU-25A Dropmaster by Correspondence
CG-1704-0037
Hydraulic Principles and Troubleshooting by Correspondence
MC-1704-0007
Ice Observer by Correspondence
CG-1304-0006
Imagery Interpreter Specialist by Correspondence
AF-1709-0027
Imagery Production Specialist by Correspondence
AF-1709-0030

KEYWORD INDEX D-19

Imagery Systems Maintenance Specialist by Correspondence
AF-1715-0806
AF-1715-0807

Individual Personnel Records by Correspondence
MC-1403-0016

Infantry Squad Leader Squad Tactics by Correspondence
MC-2204-0100

Infantry Squad Leader: Land Navigation by Correspondence
MC-1601-0048

Infantry Squad Leader: Weapons and Fire Support by Correspondence
MC-2204-0101

Infantry Staff Noncommissioned Officer (NCO): Military Symbols and Overlays by Correspondence
MC-1601-0040

Information Management Specialist by Correspondence
AF-1406-0045

Information Systems Operator by Correspondence
AF-1402-0070

Information Systems Programming Specialist by Correspondence
AF-1402-0071

Inspection and Repair of Shoulder Weapons by Correspondence
MC-2204-0066

Inspection and Repair of the .50 Caliber Machine Gun by Correspondence
MC-2204-0065

Inspection and Repair of the Mk 19 Machinegun by Correspondence
MC-2204-0099

Installation, Operation, and Maintenance of Diesel Engine Driven Generator Sets by Correspondence
MC-1712-0001

Instructional System Development for Training Managers by Correspondence
AF-1406-0080

Instrumentation Mechanic by Correspondence
AF-1715-0045

Instruments/Flight Control Systems Specialist (B-1B) by Correspondence
AF-1715-0770

Integrated Avionic Attack Control Systems Specialist (F-16) by Correspondence
AF-1715-0096

Integrated Avionic Attack Control Systems Specialist by Correspondence
AF-1715-0759

Integrated Avionic Communication, Navigation and Penetration Aids Systems Specialist by Correspondence
AF-1714-0039

Integrated Avionic Communications, Navigation, and Penetration Aids Systems Specialist (F/FB-111) by Correspondence
AF-1715-0779

Integrated Avionic Instrument and Flight Control Systems by Correspondence
AF-1704-0255

Integrated Avionic Instrument and Flight Control Systems Specialist by Correspondence
AF-1704-0226
AF-1704-0227
AF-1715-0095

Integrated Avionics Attack Control Systems Specialist (F-15) by Correspondence
AF-1704-0218
AF-1715-0097

Integrated Avionics Attack Control Systems Specialist (F/FB-111) by Correspondence
AF-1715-0090
AF-1715-0760

Integrated Avionics Attack Control Systems Specialist by Correspondence
AF-1715-0091

Integrated Avionics Communication, Navigation, and Penetration Aids Systems Specialist (F-16) by Correspondence
AF-1704-0220

Integrated Avionics Communication, Navigation, and Penetration Aids Systems Specialist (F/FB-111) by Correspondence
AF-1715-0101

Integrated Avionics Communication, Navigation, and Penetration Aids Systems Specialist by Correspondence
AF-1715-0100

Integrated Avionics Communications, Navigation, and Penetration Aids Systems Specialist (F-15) by Correspondence
AF-1704-0225
AF-1715-0106

Integrated Avionics Communications, Navigation, and Penetration Aids Systems Specialist (F/FB-111) by Correspondence
AF-1704-0263

Integrated Avionics Computerized Test Station and Component Specialist (F-15) by Correspondence
AF-1715-0086

Integrated Avionics Computerized Test Station and Component Specialist (F/FB-111) by Correspondence
AF-1715-0087

Integrated Avionics Electronic Warfare Equipment and Component Specialist (F-15) by Correspondence
AF-1715-0082

Integrated Avionics Electronic Warfare Equipment and Component Specialist (F/FB-111) by Correspondence
AF-1715-0081

Integrated Avionics Instrument and Flight Control Systems Specialist (F-15) by Correspondence
AF-1715-0098

Integrated Avionics Instrument and Flight Control Systems Specialist (F-16) by Correspondence
AF-1715-0099

Integrated Avionics Instrument and Flight Control Systems Specialist (F/FB/EF-111) by Correspondence
AF-1715-0094

Integrated Avionics Manual Test Station and Component Specialist (F-15) by Correspondence
AF-1715-0092

Integrated Avionics Manual Test Station and Component Specialist (F/FB-111) by Correspondence
AF-1715-0093

Integrated Organizational Avionic System Specialist by Correspondence
AF-1715-0758

Integrated Organizational Avionic Systems Specialist by Correspondence
AF-1405-0069

Intelligence Brief: Southwest Asia by Correspondence
MC-1511-0001

Intelligence Brief: The Military Threat by Correspondence
MC-1606-0012

Intelligence Brief: The Opposing Forces Nuclear, Biological, Chemical (NBC) Threat by Correspondence
MC-1606-0011

Intelligence for the Marine Air-Ground Task Force by Correspondence
MC-1606-0005

Intelligence Fundamentals by Correspondence
AF-1606-0119

Intelligence Officer by Correspondence
AF-1606-0120

Intelligence Operations Specialist by Correspondence
AF-1606-0121

Intelligence Operations Technician by Correspondence
AF-1606-0123

Interior Wiring by Correspondence
MC-1601-0033

Introduction to Acquisition Management by Correspondence
AF-1408-0106

Introduction to Air Force Initial Provisioning by Correspondence
AF-1408-0090

Introduction to Civil Air Patrol Emergency Services by Correspondence
AF-1704-0196

Introduction to Combat Intelligence by Correspondence
MC-1606-0009

Introduction to Field Artillery Survey by Correspondence
MC-1601-0039

Introduction to Labor Relations for Air Force Supervisors by Correspondence
AF-1408-0088

Introduction to Logistics by Correspondence
AF-1408-0110

Introduction to Marine Corps Accounting by Correspondence
MC-1401-0004

Introduction to Retailing by Correspondence
MC-0406-0001

Introduction to Test Equipment by Correspondence
MC-1715-0129

Introduction to the Quality Function by Correspondence
AF-1405-0067

Inventory Management Specialist (Munitions) by Correspondence
AF-1405-0064

Inventory Management Specialist by Correspondence
AF-1405-0051

Inventory Management Supervisor by Correspondence
AF-1405-0059
AF-1405-0071

Investigator 1 by Correspondence
CG-1728-0014

Investigator Second Class by Correspondence
CG-1728-0042

Jet Engine Mechanic by Correspondence
AF-1704-0137

Land Navigation by Correspondence
MC-1601-0048

Law Enforcement by Correspondence
CG-1728-0013

KEYWORD INDEX

Law Enforcement Specialist by Correspondence
AF-1728-0044
Legal Administration Clerk by Correspondence
MC-1406-0024
Legal Administration for the Reporting Unit by Correspondence
MC-1406-0024
Legal Services Specialist by Correspondence
AF-1406-0047
Liquid Fuel Systems Maintenance Specialist by Correspondence
AF-1601-0030
Machinery Technician First Class by Correspondence
CG-1723-0006
Machinist by Correspondence
AF-1723-0008
Maintenance Data Systems Analysis Specialist by Correspondence
AF-1107-0002
Maintenance of Bulk Fuel Equipment by Correspondence
MC-1601-0038
Maintenance of the M60 Series Machine Gun by Correspondence
MC-2204-0064
Maintenance Systems Analysis Specialist by Correspondence
AF-1402-0069
Management of Value Engineering by Correspondence
AF-1405-0068
Maneuvering Boards by Correspondence
CG-1708-0015
Manpower Management Technician by Correspondence
AF-1406-0052
Marine Corps Technical Publications System by Correspondence
MC-1408-0021
Marine Environmental Protection (MEP) by Correspondence
CG-1303-0001
Marine Safety Initial Indoctrination by Correspondence
CG-1708-0019
Marine Safety Initial Indoctrination Lesson Plan Series by Correspondence
CG-1708-0019
Marine Safety Initial Indoctrination Marine Inspection by Correspondence
CG-1708-0024
Marine Safety Initial Indoctrination Port Operations by Correspondence
CG-1708-0025
Marine Science Technician First Class by Correspondence
CG-1304-0018
CG-1304-0022
Marine Science Technician Second Class by Correspondence
CG-1304-0016
Materials Handling Equipment by Correspondence
MC-1710-0040
Materiel Facilities Specialist by Correspondence
AF-1405-0013
Materiel Facilities Supervisor by Correspondence
AF-1405-0033
Materiel Storage and Distribution Specialist by Correspondence
AF-1405-0070

Mathematics for Marines by Correspondence
MC-1107-0001
Meats and Meat Cookery by Correspondence
MC-1729-0044
Medical Administrative Specialist by Correspondence
AF-0709-0027
Medical Laboratory Craftsman (Chemistry and Urinalysis) by Correspondence
AF-0702-0008
Medical Laboratory Craftsman (Hematology, Serology, Blood Banking and Immunohematology) by Correspondence
AF-0702-0010
Medical Laboratory Craftsman (Microbiology) by Correspondence
AF-0702-0009
Medical Laboratory Technician (Chemistry and Urinalysis) by Correspondence
AF-0702-0008
Medical Laboratory Technician (Hematology, Serology, Blood Banking and Immunohematology) by Correspondence
AF-0702-0010
AF-0709-0031
Medical Laboratory Technician (Microbiology) by Correspondence
AF-0702-0009
AF-0709-0030
Medical Materiel Specialist by Correspondence
AF-1406-0094
Medical Service Specialist by Correspondence
AF-0703-0016
AF-0703-0018
Medical Service Specialist, Aeromedical by Correspondence
AF-0709-0033
Medical Service Technician by Correspondence
AF-0703-0011
Mental Health Service Specialist by Correspondence
AF-0703-0017
Mental Health Unit Specialist by Correspondence
AF-0708-0002
Mess Hall Sanitation by Correspondence
MC-1729-0039
Mess Hall Subsistence Clerk by Correspondence
MC-1729-0040
Metal Fabricating Specialist by Correspondence
AF-1723-0014
AF-1723-0015
Metal Fabrication Specialist by Correspondence
AF-1723-0013
Metals Processing Specialist by Correspondence
AF-1723-0007
AF-1723-0009
Meteorological and Navigation Systems Journeyman by Correspondence
AF-1304-0017
AF-1715-0821
Meteorological and Navigation Systems Specialist by Correspondence
AF-1304-0017
AF-1715-0821
Meteorological Technician by Correspondence (Specialized)
AF-1304-0015

Military Affiliate Radio System (MARS) Operator by Correspondence
MC-1715-0134
Military Civil Rights by Correspondence
CG-1512-0001
Military Functions in Civil Disturbances by Correspondence
MC-1728-0002
Military Justice by Correspondence
AF-1728-0043
Military Operations on Urban Terrain by Correspondence
MC-1606-0006
Military Requirements for Chief Petty Officer by Correspondence
CG-2205-0036
Military Requirements for E-3 by Correspondence
CG-2205-0028
Military Requirements for E-4 by Correspondence
CG-2205-0029
Military Requirements for E-5 by Correspondence
CG-2205-0030
Military Requirements for E-6 by Correspondence
CG-2205-0031
Military Requirements for Second Class Petty Officer by Correspondence
CG-2205-0038
Missile and Space Systems Electronic Maintenance Journeyman by Correspondence
AF-1714-0041
AF-1715-0814
Missile and Space Systems Maintenance Journeyman by Correspondence
AF-1704-0265
Missile Control Communications Systems Repairman by Correspondence
AF-1715-0130
Missile Control Communications Systems Specialist by Correspondence
AF-1714-0038
Missile Facilities Specialist by Correspondence
AF-1704-0264
AF-1717-0026
Missile Maintenance Specialist (WS-133) by Correspondence
AF-1710-0031
Missile Maintenance Specialist by Correspondence
AF-1704-0265
Missile Systems Maintenance Specialist by Correspondence
AF-1715-0038
AF-1715-0043
Morale, Welfare, Recreation, and Services by Correspondence
AF-0804-0002
Morale, Welfare, Recreation, and Services Journeyman by Correspondence
AF-0804-0001
National Security Management (Correspondence/Nonresident/Seminar Course of the National Defense University)
DD-1511-0001
National Security Management by Correspondence
DD-1511-0008
Navigation Rules by Correspondence
CG-1708-0020
Navigation Systems by Correspondence
CG-1708-0011

KEYWORD INDEX D-21

Navigational Aids Equipment Specialist by Correspondence
    AF-1715-0012
Navigational Aids Equipment Technician by Correspondence
    AF-1715-0017
Navy Joint Service Supervisor Safety by Correspondence
    AF-0801-0003
Noncommissioned Officer (NCO) by Correspondence
    MC-1406-0023
Nondestructive Inspection Specialist by Correspondence
    AF-1723-0005
Nuclear Warfare Defense by Correspondence
    MC-0705-0001
Nursing Service Management by Correspondence
    AF-0703-0014
Offensive Avionics Systems Specialist (B-1B) by Correspondence
    AF-1704-0248
Open Mess Management Specialist by Correspondence
    AF-1406-0058
Operations Resource Management by Correspondence
    AF-1408-0095
Optometry Specialist by Correspondence
    AF-0706-0003
Paralegal Journeyman by Correspondence
    AF-1407-0003
Passenger and Household Goods Specialist by Correspondence
    AF-0419-0031
Pastry Baking by Correspondence
    MC-1729-0032
Pavement and Construction Equipment Journeyman by Correspondence
    AF-1601-0051
    AF-1601-0052
Pavement and Construction Equipment Specialist by Correspondence
    AF-1601-0051
    AF-1601-0052
Pavements Maintenance Specialist by Correspondence
    AF-1601-0047
Personal Affairs Specialist by Correspondence
    AF-1406-0049
Personal Finance by Correspondence
    MC-1401-0009
Personal Financial Management by Correspondence
    MC-1401-0009
Personnel Administration for the Reporting Unit by Correspondence
    MC-1406-0026
Personnel Reporting for REMMPS by Correspondence
    MC-1406-0027
Personnel Specialist by Correspondence
    AF-1406-0076
Personnel Systems Management Journeyman by Correspondence
    AF-1402-0075
    AF-1402-0076
    AF-1402-0077
    AF-1402-0078
Personnel Systems Management Specialist by Correspondence
    AF-1406-0077

Pest Management Specialist by Correspondence
    AF-0101-0005
Pharmacy Specialist by Correspondence
    AF-0799-0003
Photo-Sensors Maintenance Specialist (Tactical/Reconnaissance Electronic Sensors) by Correspondence
    AF-1715-0777
Photo-Sensors Maintenance Specialist by Correspondence
    AF-1715-0764
Physical Therapy Specialist by Correspondence
    AF-0704-0004
Piloting Navigation by Correspondence
    CG-1708-0006
Plumbing and Sewage Disposal by Correspondence
    MC-1710-0005
Plumbing Specialist by Correspondence
    AF-1710-0021
Plumbing Technician by Correspondence
    AF-1710-0023
Pole Line Construction Equipment by Correspondence
    MC-1715-0091
Pole Line Construction Techniques by Correspondence
    MC-1715-0096
Port Securityman First Class by Correspondence
    CG-1728-0038
Port Securityman Second Class by Correspondence
    CG-1728-0032
Position Location Reporting System (PLRS) Capabilities and Employment by Correspondence
    MC-1606-0008
Power Measurements by Correspondence
    AF-1115-0008
    AF-1714-0034
Power Supplies by Correspondence
    AF-1715-0752
Precision Imagery and Audiovisual Media Maintenance Specialist by Correspondence
    AF-1715-0136
Precision Measuring Equipment Specialist by Correspondence
    AF-1715-0073
Precision Measuring Equipment Technician by Correspondence
    AF-1715-0074
Preventive Maintenance and Operating Techniques for Heavy Vehicles by Correspondence
    MC-1703-0034
Principles of Contract Pricing by Correspondence
    AF-1408-0091
Principles of Instruction for the Marine Noncommissioned Officer (NCO) by Correspondence
    MC-1406-0028
Production Control Specialist by Correspondence
    AF-1405-0060
Public Affairs Officer by Correspondence
    AF-0401-0002
Public Affairs Specialist by Correspondence
    AF-0401-0001
Public Affairs Specialist First Class by Correspondence
    CG-1709-0002

Public Affairs Specialist Second Class by Correspondence
    CG-1709-0001
Quality Management by Correspondence
    AF-1406-0081
Quartermaster First Class by Correspondence
    CG-2205-0026
Quartermaster Second Class by Correspondence
    CG-1722-0015
Quartermaster Third Class by Correspondence
    CG-1722-0014
Quartermaster Third Class, by Correspondence
    CG-1722-0014
Radar Operator by Correspondence
    CG-1715-0074
Radarman First Class by Correspondence
    CG-1408-0036
Radarman Second Class by Correspondence
    CG-2205-0023
Radarman Third Class by Correspondence
    CG-2205-0017
Radio and Television Broadcasting Specialist by Correspondence
    AF-0505-0002
Radiology Technician by Correspondence
    AF-0705-0003
Radioman First Class by Correspondence
    CG-1715-0136
Radioman Second Class by Correspondence
    CG-1404-0002
Radiotelegraph and Visual Communication Procedures by Correspondence
    MC-1715-0133
Radiotelephone and Alternate Communication Procedures by Correspondence
    MC-1715-0137
Refrigeration and Cryogenics Specialist by Correspondence
    AF-1730-0016
Refrigeration Servicing by Correspondence
    MC-1731-0001
Relays, Generators, Motors, and Electromechanical Devices by Correspondence
    AF-1714-0032
    AF-1715-0768
Reprographic Specialist by Correspondence
    AF-1408-0098
Reserve Field Food Service Supervisor by Correspondence
    MC-1729-0042
Resource Advisor by Correspondence
    AF-1408-0093
Reverse Osmosis Water Purification Unit by Correspondence
    MC-1732-0003
Safety and Security of the Port by Correspondence
    CG-0802-0011
Safety and Security of the Port: Marine Environmental Protection by Correspondence
    CG-1728-0036
Safety and Security of the Port: Military Explosives by Correspondence
    CG-1728-0033
Salads, Sandwiches, and Desserts by Correspondence
    MC-1729-0022
SASSY: Organic Procedures by Correspondence
    MC-1405-0044

KEYWORD INDEX

Satellite and Wideband Communications Equipment Specialist by Correspondence
AF-1715-0797
AF-1715-0799
Satellite Communications System Equipment Specialist by Correspondence
AF-1715-0014
Scientific Measurements Technician by Correspondence
AF-1715-0762
Seaman by Correspondence
CG-1708-0009
Search and Rescue by Correspondence
CG-0802-0018
Search and Rescue Fundamentals by Correspondence
CG-0802-0016
Search and Rescue HU-25A Basic Aircrew by Correspondence
CG-0802-0008
Secure Communication Systems Maintenance Specialist by Correspondence
AF-1715-0786
Secure Communications Systems Maintenance by Correspondence
AF-1715-0813
AF-1715-0820
Security of Classified Information by Correspondence
CG-1728-0035
Security Specialist by Correspondence
AF-1728-0045
Senior Petty Officers by Correspondence
CG-2205-0034
Sergeants Nonresident Program by Correspondence
MC-1408-0027
Services Specialist by Correspondence
AF-1729-0014
Services Supervisor by Correspondence
AF-0419-0028
AF-1408-0107
Shiphandling by Correspondence
CG-1708-0013
Sinewave Oscillators-Modulation/Demodulation by Correspondence
AF-1715-0751
Small Boat Magnetic Compass Calibration by Correspondence
CG-1722-0006
Small Unit Paperwork by Correspondence
CG-1403-0001
Social Actions Technician (Drugs/Alcohol) by Correspondence
AF-1406-0054
Social Actions Technician (Equal Opportunity/Human Relations) by Correspondence
AF-1406-0053
Soldering and Electrical Connectors by Correspondence
AF-1714-0033
AF-1723-0012
Solid State Devices by Correspondence
MC-1715-0132
Sonar Technician Chief by Correspondence
CG-1408-0039
Sonar Technician Second Class by Correspondence
CG-1715-0133
Sonobuoy by Correspondence
CG-2205-0037
Space Communications Systems Equipment by Correspondence
AF-1715-0014
Space Communications Systems Equipment Operator/Technician by Correspondence
AF-1715-0021

Space Systems Equipment Maintenance Specialist by Correspondence
AF-1715-0036
Space Systems Operations Specialist by Correspondence
AF-1715-0755
Special Investigations and Counterintelligence Technician by Correspondence
AF-1728-0039
Special Purpose Vehicle and Equipment Craftsman by Correspondence
AF-1703-0021
Special Purpose Vehicle and Equipment Maintenance Apprentice by Correspondence
AF-1703-0020
Special Purpose Vehicle and Equipment Mechanic by Correspondence
AF-1703-0020
AF-1710-0032
Special Purpose Vehicle and Equipment Supervisor by Correspondence
AF-1703-0021
Special Vehicle Mechanic (Firetrucks) by Correspondence
AF-1710-0033
Special Vehicle Mechanic (Refueling Vehicles) by Correspondence
AF-1710-0034
Staff Noncommissioned Officers Advanced Nonresident Program (SNCOANP) by Correspondence
MC-1408-0028
Still Photographic Specialist by Correspondence
AF-1709-0028
Storekeeper First Class by Correspondence
CG-1405-0008
Storekeeper Second Class by Correspondence
CG-1405-0009
Storekeeper Third Class by Correspondence
CG-1405-0002
Strategic Aircraft Maintenance Specialist by Correspondence
AF-1704-0181
AF-1704-0188
AF-1704-0209
AF-1704-0210
AF-1704-0211
AF-1704-0247
Structural Journeyman by Correspondence
AF-1732-0015
AF-1732-0016
Structural Maintenance Technician by Correspondence
AF-1704-0250
Structural Technician by Correspondence
AF-1601-0049
Subsistence Operations Specialist by Correspondence
AF-1729-0016
Subsistence Operations Technician by Correspondence
AF-1729-0017
Subsistence Specialist First Class by Correspondence
CG-1729-0005
Subsistence Specialist Second Class by Correspondence
CG-1729-0004
Subsistence Specialist Third Class by Correspondence
CG-1729-0006
Supply Management by Correspondence
MC-1405-0027

Supply Systems Analysis Supervisor by Correspondence
AF-1405-0066
Supply Systems Analyst Journeyman by Correspondence
AF-1405-0072
Surgical Service Specialist by Correspondence
AF-0703-0015
Systems Repair Technician by Correspondence
AF-1715-0763
T58-GE-8B and T58-GE-5 Turboshaft Engine Familiarization by Correspondence
CG-1704-0021
Tactical Air Command and Control Specialist by Correspondence
AF-1406-0068
Tactical Aircraft Maintenance Technician by Correspondence
AF-1704-0228
Tank Gunnery/Direct Fire Procedures (M1A1) by Correspondence
MC-2204-0098
Target Intelligence Specialist by Correspondence
AF-1606-0122
Telecommunications Operations Specialist by Correspondence
AF-0504-0003
Telecommunications Specialist First Class by Correspondence
CG-1403-0002
Telecommunications Specialist Second Class by Correspondence
CG-1402-0004
Telecommunications System Control Specialist by Correspondence
AF-1715-0390
Telecommunications System Maintenance Specialist by Correspondence
AF-1715-0031
Telecommunications Systems Control Technician by Correspondence
AF-1715-0034
Telephone Central Office Switching Equipment Specialist, Electronic/Electromechanical by Correspondence
AF-1715-0127
Telephone Equipment Installation and Repair Specialist by Correspondence
AF-1715-0131
Telephone Switching Specialist by Correspondence
AF-1715-0803
Telephone Technician First Class by Correspondence
CG-1715-0134
Telephone Technician Second Class by Correspondence
CG-1715-0063
Television Equipment Repairman by Correspondence
AF-1715-0013
Television Systems Specialist by Correspondence
AF-1715-0796
AF-1715-0798
Terrorism Counteraction for Marines by Correspondence
MC-1606-0010
The Noncommissioned Officer (NCO) by Correspondence
MC-1406-0023
Theory and Construction of Turbine Engines by Correspondence
MC-1710-0004

KEYWORD INDEX D-23

Tractor-Trailer Operator by Correspondence
MC-1710-0041
Traffic Management Journeyman by Correspondence
AF-0419-0037
Training Systems Technician by Correspondence
AF-1406-0072
Training Technician by Correspondence
AF-1406-0070
Turboprop Propulsion Mechanic by Correspondence
AF-1704-0204
Utilities Officer/Chief by Correspondence
MC-1710-0048
Utilities System Journeyman by Correspondence
AF-1732-0004
AF-1732-0005
Validation and Reconciliation Clerk by Correspondence
MC-1403-0017
Vegetables, Soups, Sauces, Gravies, and Beverages by Correspondence
MC-1729-0024
Vehicle Body Mechanic by Correspondence
AF-1710-0035
Vehicle Maintenance Control and Analysis Specialist by Correspondence
AF-1710-0036
Vehicle Operations Supervisor by Correspondence
AF-0419-0023
Vehicle Operator/Dispatcher Journeyman by Correspondence
AF-1703-0023
Visual Information Production-Documentation Journeyman by Correspondence
AF-0505-0005
Visual Information Production-Documentation Specialist by Correspondence
AF-0505-0003
AF-0505-0005
Warehousing Operations by Correspondence
MC-1405-0021
Warfighting Skills Program (Warfighting) by Correspondence
MC-2204-0112
Watchstanding: The Conning Officer by Correspondence
CG-1708-0012
Waterfront Protection by Correspondence
CG-1728-0012
Weapon Control Systems Mechanic (A-7D, A-10, AC-130, F-5) by Correspondence
AF-1715-0050
Weapon Control Systems Mechanic (A-7D: AN/APQ-126) by Correspondence
AF-1715-0054
Weapon Control Systems Mechanic (F-4C/D: APQ-109/APA-165) by Correspondence
AF-1715-0052
Weapon Control Systems Mechanic (F-4E: APQ-120) by Correspondence
AF-1715-0053
Weapon Control Systems Technician (F-4C/D: APQ-109/APA-165) by Correspondence
AF-1715-0056
Weapon Control Systems Technician (F-4D/E) by Correspondence
AF-1715-0766
Weapon Control Systems Technician (F-4E: APQ-120) by Correspondence
AF-1715-0063
Weapons Administration by Correspondence
CG-1408-0007
Weapons Control Systems Mechanic (A-7D, A-10, AC-130, F-5) by Correspondence
AF-1715-0800
Weapons Officer by Correspondence
CG-2205-0018
Weather Equipment Specialist by Correspondence
AF-1304-0014
Weather Forecasting and Flight Briefing by Correspondence
CG-1304-0014
Weather Specialist by Correspondence
AF-1304-0016
Wideband Communications Equipment Specialist by Correspondence
AF-1715-0208
Wideband Communications Equipment Technician by Correspondence
AF-1715-0015
Yeoman First Class by Correspondence
CG-1408-0006
Yeoman Second Class by Correspondence
CG-1408-0004
Yeoman Third Class by Correspondence
CG-1408-0005

Corrosion
Corrosion Control Specialist by Correspondence
AF-1717-0013

Cost
Budget, Cost Estimating, and Financial Management Workshop
DD-1408-0017
Cost Analysis Officer
AF-1401-0019
Cost Analysis Specialist by Correspondence
AF-1408-0109
Cost and Management Analysis Specialist by Correspondence
AF-1408-0083
Cost and Management Analysis Technician by Correspondence
AF-1408-0084
Fundamentals of Cost Analysis
DD-1115-0001
Software Cost Estimating
DD-1402-0005

Counseling
Counseling for Marines by Correspondence
MC-1406-0032

Counteraction
Terrorism Counteraction for Marines by Correspondence
MC-1606-0010

Counterintelligence
Special Investigations and Counterintelligence Technician by Correspondence
AF-1728-0039

Countermeasures
F/FB-111 Avionics Manual/Electronic Countermeasures (ECM) Test Station and Component Specialist by Correspondence
AF-1704-0224

Coxswain
Coxswain 41
CG-1712-0008
Coxswain Skills
MC-1708-0003
Coxswain, Class C
CG-1708-0017

Craft
Small Craft Mechanic
MC-1712-0009

Crane
Crane Operator by Correspondence
MC-1710-0039

Crew
Master Crew Chief
AF-1704-0070

Crewman
Crewman 41, Class C
CG-2205-0040

Crewmember
Reserve Small Boat Crewmember
CG-2205-0027
Small Boat Crewmember
CG-2205-0027

Crime
Crime Prevention by Correspondence
AF-1728-0042

Cryogenics
Refrigeration and Cryogenics Specialist by Correspondence
AF-1730-0016

Cryptographic
Electronic Cryptographic Communications Equipment Specialist by Correspondence
AF-1715-0024

Cryptologic
Senior Military Cryptologic Supervisors
DD-1404-0001

Cutter
AB/SPS-64(V) Small Cutter Radar Maintenance
CG-1715-0140
AN/SPS-64(V) Large Cutter Radar Systems Maintenance
CG-1715-0139
Small Cutter Damage Control
CG-0801-0001

Czech
Czech Basic
DD-0602-0229
Czech Intermediate
DD-0602-0024

Damage
Damage Control and Stability by Correspondence
CG-2205-0024
Damage Control Deck Group Ratings by Correspondence
CG-1710-0011
Damage Controlman Advanced Reserve Active Duty for Training, Class C
CG-1710-0015
Damage Controlman Chief by Correspondence
CG-1408-0037
Damage Controlman First Class By Correspondence
CG-1710-0020
Damage Controlman Refresher Reserve Active Duty for Training, Class C
CG-1710-0015
Damage Controlman Second Class by Correspondence
CG-1710-0021
Damage Controlman Third Class by Correspondence
CG-1710-0019
Damage Controlman, Class A
CG-1710-0014
Small Cutter Damage Control
CG-0801-0001

KEYWORD INDEX

Data
AN/UYK-7 Computer and Data Auxiliary Console Course Development
    CG-1402-0002
AN/UYK-7 Computer and OJ-172 Data Exchange Auxiliary Console
    CG-1402-0002
Data Control Techniques
    MC-1402-0042
IBM System 360 (OS) Data Control Techniques
    MC-1402-0042
Maintenance Data Systems Analysis Specialist by Correspondence
    AF-1107-0002
Mobile Data Communications Terminal Technician
    MC-1715-0118
OS/VS2 MVS Data Control Techniques
    MC-1402-0042
Tactical Data Communication Central Repair
    MC-1715-0141
Tactical Data Communication Central Technician
    MC-1715-0146
Tactical Data Communications Central (TDCC AN/TYQ-3) Technician
    MC-1715-0026
Tactical Data Communications Central Technician
    MC-1715-0026

Data Processing
Automated Data Processing Equipment Fleet Marine Force Programmer
    MC-1402-0050
Automated Data Processing Equipment for the Fleet Marine
    MC-1402-0050
Fundamental Principles of Electronic Data Processing Equipment by Correspondence
    AF-1402-0072

DCLF
DCLF Reference Measurement and Calibration
    AF-1715-0493

Deck
Damage Control Deck Group Ratings by Correspondence
    CG-1710-0011
Deck Seamanship by Correspondence
    CG-1708-0010
Deck Watch Officer Navigation Rules by Correspondence
    CG-1708-0014

Defense
Chemical Warfare Defense by Correspondence
    MC-0801-0013
Coastal Defense Command and Staff, Class C
    CG-0419-0003
Coastal Defense Exercise Planner, Port Level
    CG-1722-0013
Coastal Defense Planner, Port Level
    CG-1722-0012
Defense Acquisition Engineering, Manufacturing, and Quality Control
    DD-1408-0010
Defense Basic Preservation and Packing
    DD-0419-0004
Defense Equal Opportunity Management Institute
    DD-1512-0003
Defense Management Systems
    DD-1408-0006
Defense Packaging Design
    DD-0419-0006
Defense Packaging Management Training
    DD-0419-0005
Defense Packaging of Hazardous Materials for Transportation
    DD-0419-0003
Defense Packing and Unitization
    DD-0419-0002
Defense Preservation and Intermediate Protection
    DD-0419-0001
Defense Resources Management
    DD-1408-0006
Enemy Defense Penetration Aids
    AF-1715-0744
Inter-American Defense College
    DD-1511-0011
National Security Management (Correspondence/Nonresident/Seminar Course of the National Defense University)
    DD-1511-0001
Nuclear Warfare Defense by Correspondence
    MC-0705-0001
Surface Air Defense System Acquisition Technician
    MC-1715-0181
Surface Air Defense System Fire Control Technician
    MC-1715-0180

Defensive
Defensive Avionics/Communications/Navigation Systems by Correspondence
    AF-1715-0789
Defensive Fire Control Systems Mechanic (B-52D/G: MD-9, ASG-15 Turrets) by Correspondence
    AF-1715-0047

Demodulation
Sinewave Oscillators-Modulation/Demodulation by Correspondence
    AF-1715-0751

Dental
Dental Assistant Specialist by Correspondence
    AF-0701-0003
Dental Laboratory Specialist by Correspondence
    AF-0701-0018
Dental Technician, Class C
    CG-0701-0002

Deployment
Marine Air-Ground Task Force Deployment Support System/Computer-Aided Embarkation Management System
    MC-1402-0059

Desert
Desert Operations by Correspondence
    MC-0803-0007

Design
Defense Packaging Design
    DD-0419-0006

Designer
Course Designer
    CG-1406-0006

Desserts
Salads, Sandwiches, and Desserts by Correspondence
    MC-1729-0022

Detector
Videograph B Fog Detector Maintenance
    CG-1715-0127

Developer
Instructor/Course Developer, Class C
    CG-1406-0004

Development
Advanced Systems Planning, Research, Development, and Engineering
    DD-1408-0016
Instructional System Development for Training Managers by Correspondence
    AF-1406-0080
Nursing Executive Development
    AF-0709-0037
Nursing Staff Development Officer Basic
    AF-0703-0020

Devices
Fundamentals of Solid State Devices by Correspondence
    AF-1715-0750
Relays, Generators, Motors, and Electromechanical Devices by Correspondence
    AF-1715-0768
Solid State Devices by Correspondence
    MC-1715-0132

Diagnostic
Test, Measurement and Diagnostic Equipment
    MC-1715-0139

Diagnostics
Multiple Virtual Storage (MVS) Diagnostics
    MC-1402-0053

Diary
Unit Diary Clerk
    MC-1403-0013

Diesel
Caterpillar Diesel Engine Maintenance, Class C
    CG-1712-0004
Diesel Engine Maintenance and Troubleshooting by Correspondence
    MC-1712-0007
Fundamentals of Diesel Engines by Correspondence
    MC-1712-0006
Installation, Operation, and Maintenance of Diesel Engine Driven Generator Sets by Correspondence
    MC-1712-0001
Lister Diesel Overhaul and Engine Power System Equipment Maintenance
    CG-1712-0007
Lister Diesel Overhaul and Power System Equipment Maintenance
    CG-1712-0007
VT903M Cummins Diesel Engine Disassembly-Assembly and Tune-up
    CG-1712-0006

Diet
Diet Therapy Specialist by Correspondence
    AF-0104-0003
Diet Therapy Supervisor by Correspondence
    AF-0104-0002

Digital
Advanced Digital Electronics Technology, Class C
    CG-1715-0148
Digital Microprocessor, Class C
    CG-1715-0081
Fundamentals of Digital Logic by Correspondence
    MC-1715-0131

KEYWORD INDEX D-25

Direction
Direction Finding Operations by Correspondence
MC-1715-0130

Disaster
Disaster Preparedness Specialist by Correspondence
AF-0802-0026

Disbursing
Advanced Disbursing
MC-1401-0006
Basic Disbursing Clerk
MC-1401-0013
Disbursing Officer
MC-1408-0016

Disk
RP11/RP03 Disk Pack Maintenance
DD-1715-0008

Dispatcher
Apprentice Vehicle Operator/Dispatcher by Correspondence
AF-0419-0026
Vehicle Operator/Dispatcher Journeyman by Correspondence
AF-1703-0023

Display
Avionic Sensor Systems Specialist (Tactical/Real Time Display Electronic Sensors) by Correspondence
AF-1715-0070

Disposal
Plumbing and Sewage Disposal by Correspondence
MC-1710-0005

Distribution
Materiel Storage and Distribution Specialist by Correspondence
AF-1405-0070

Disturbances
Military Functions in Civil Disturbances by Correspondence
MC-1728-0002

Documentation
Visual Information Production-Documentation Journeyman by Correspondence
AF-0505-0005
Visual Information Production-Documentation Specialist by Correspondence
AF-0505-0003
AF-0505-0005

Doppler
Communication/Navigation Systems Technician (Doppler Systems) by Correspondence
AF-1704-0260

Dropmaster
HU-25A Dropmaster by Correspondence
CG-1704-0037

Drug
Air Reserve Forces Social Actions Technician (Drug/Alcohol) by Correspondence
AF-0708-0004

Drugs
Social Actions Technician (Drugs/Alcohol) by Correspondence
AF-1406-0054

Dutch
Dutch Basic
DD-0602-0223

Duty
Enlisted Supply Independent Duty
MC-1405-0040

ECM
F-15 Avionics Manual/ECM Test Station and Component Specialist by Correspondence
AF-1715-0771

Editors
Editors
DD-0504-0015

Education
Education and Training Manager by Correspondence
AF-1406-0078
Education and Training Officer by Correspondence
AF-1406-0079
Education Specialist by Correspondence
AF-1406-0069

Egress
Aircrew Egress Systems Mechanic by Correspondence
AF-1717-0005

Egyptian
Arabic Egyptian/Syrian Intermediate
DD-0602-0023
Arabic-Egyptian Extended
DD-0602-0211
Arabic-Egyptian Special
DD-0602-0208

Electrical
270' WMEC Main Propulsion Control and Monitoring System (Electrical) Operation and Maintenance, Class C
CG-1710-0013
Advanced Electrical/Electronics
CG-1715-0105
Aircraft Electrical and Environmental Systems Journeyman by Correspondence
AF-1714-0045
Aircraft Electrical Systems Specialist by Correspondence
AF-1714-0003
Electrical Equipment Repair Specialist
MC-1714-0014
Electrical Equipment Repairman
MC-1714-0014
Electrical Power Production Journeyman by Correspondence
AF-1714-0040
AF-1714-0043
Electrical Power Production Specialist
AF-1714-0040
Electrical Power Production Specialist by Correspondence
AF-1714-0030
AF-1714-0043
Electrical Power Production Technician by Correspondence
AF-1408-0096
AF-1714-0024
AF-1714-0031
Electrical Systems Journeyman by Correspondence
AF-1714-0042
AF-1714-0044
Electrical Systems Technician by Correspondence
AF-1714-0042
Electrical/Electronic Control for Fire Tube Boilers/Oily Water Separator and Electrical Generator
CG-1715-0107

Electrical/Electronic Control for Fire Tube Boilers/Oily Water Separator/Electrical Generators Operation and Maintenance, Class C
CG-1715-0107
Fire and Electrical Systems Component Repair
MC-1714-0018
Fuel and Electrical Component Repair
MC-1714-0017
HH-3F Automatic Flight Control System and Selected Electrical Maintenance, Class C
CG-1714-0015
HH-60J Electrical/Automatic Flight Control Systems Maintenance, Class C
CG-1704-0051
HH-65A Electrical/Automatic Flight Control Systems Maintenance, Class C
CG-1704-0049
HU-25A Electrical, Class C
CG-1704-0046
Journeyman Electrical Equipment Repairman
MC-1714-0014
Soldering and Electrical Connectors by Correspondence
AF-1714-0033
AF-1723-0012

Electrician
Apprentice Electrician by Correspondence
AF-1714-0023
Basic Electrician
MC-1714-0011
Electrician by Correspondence
AF-1714-0028
Electrician Mate Advanced Reserve Active Duty for Training
CG-1714-0020
Electrician Mate, Class A
CG-1714-0018
Electrician Noncommissioned Officer (NCO)
MC-1714-0016
Journeyman Electrician
MC-1714-0016

Electrician's
Aviation Electrician's Mate, Class A Rotary Wing Training
CG-1704-0025

Electrician's Mate
Aviation Electrician's Mate First Class by Correspondence
CG-1714-0022
Aviation Electrician's Mate Second Class by Correspondence
CG-1714-0025
Electrician's Mate First Class by Correspondence
CG-1714-0011
Electrician's Mate Second Class by Correspondence
CG-1714-0023
Electrician's Mate Third Class by Correspondence
CG-1714-0005
CG-1714-0024
Electrician's Mate, Class A
CG-1714-0018

Electricity
Advanced Electricity, Electronics, and Hydraulics
CG-1715-0147
Fundamentals of Electricity by Correspondence
AF-1714-0029

KEYWORD INDEX

Gunner's Mate, Class A, Phase 2 (Advanced Electricity, 3"/50 Caliber Gun, 5"/38 Caliber Gun)
    CG-1714-0013

Electro-Environmental
Aircraft Electro-Environmental System Technician by Correspondence
    AF-1704-0221
Aircraft Electro-Environmental Systems Technician by Correspondence
    AF-1704-0222
    AF-1704-0223

Electro-Optical
Avionic Sensor Systems Specialist (Electro-Optical Sensors) by Correspondence
    AF-1715-0071

Electromechanical
Relays, Generators, Motors, and Electromechanical Devices by Correspondence
    AF-1714-0032
    AF-1715-0768
Telephone Central Office Switching Equipment Specialist, Electronic/Electromechanical by Correspondence
    AF-1715-0127

Electronic
Advanced Electronic Journalism
    DD-0505-0006
Air Control Electronic Operator
    MC-2204-0031
Avionic Sensor Systems Specialist (Reconnaissance Electronic Sensors) by Correspondence
    AF-1715-0068
Avionic Sensor Systems Specialist (Tactical/Real Time Display Electronic Sensors) by Correspondence
    AF-1715-0070
CEJT-SX-100/200 Mitel Electronic Private Automatic Branch Exchange Generic 217, Class C
    CG-1715-0113
CEJT-SX-200 Mitel Electronic Private Automatic Branch Exchange Generic 1000 or 1001 Telephone System
    CG-1715-0116
Electrical/Electronic Control for Fire Tube Boilers/Oily Water Separator and Electrical Generator
    CG-1715-0107
Electrical/Electronic Control for Fire Tube Boilers/Oily Water Separator/Electrical Generators Operation and Maintenance, Class C
    CG-1715-0107
Electronic Computer and Switching Systems Specialist by Correspondence
    AF-1715-0765
Electronic Cryptographic Communications Equipment Specialist by Correspondence
    AF-1715-0024
Electronic Journalism
    DD-0505-0003
    DD-0505-0007
Electronic Tubes and Special Purpose Tubes by Correspondence
    AF-1715-0749
F/FB-111 Avionics Manual/Electronic Countermeasures (ECM) Test Station and Component Specialist by Correspondence
    AF-1704-0224
Fundamental Principles of Electronic Data Processing Equipment by Correspondence
    AF-1402-0072

Missile and Space Systems Electronic Maintenance Journeyman by Correspondence
    AF-1714-0041
    AF-1715-0814
Photo-Sensors Maintenance Specialist (Tactical/Reconnaissance Electronic Sensors) by Correspondence
    AF-1715-0777
TEL-14 CDXC-SG-1/1A Pulse 120 Electronic Private Automatic Branch Exchange Telephone System, Class C
    CG-1715-0115
Telephone Central Office Switching Equipment Specialist, Electronic/Electromechanical by Correspondence
    AF-1715-0127

Electronic Warfare
Electronic Warfare (EW) System Journeyman by Correspondence
    AF-1715-0810
Electronic Warfare Systems Journeyman by Correspondence
    AF-1715-0819
Electronic Warfare Systems Specialist by Correspondence
    AF-1715-0466
    AF-1715-0769
Electronic Warfare Systems Technician (Unit) by Correspondence
    AF-1715-0783
Integrated Avionics Electronic Warfare Equipment and Component Specialist (F-15) by Correspondence
    AF-1715-0082
Integrated Avionics Electronic Warfare Equipment and Component Specialist (F/FB-111) by Correspondence
    AF-1715-0081

Electronics
Advanced Analog Electronics Technology, Class C
    CG-1715-0149
Advanced Electrical/Electronics
    CG-1715-0105
Air Control Electronics Operator
    MC-2204-0031
Air Control Electronics Operator (AN/TYQ-23)
    MC-2204-0090
Aviation Electronics Technician 618M-3, Class C
    CG-1715-0142
Basic Electronics
    MC-1715-0116
Communication-Electronics Systems Acquisition and Management by Correspondence
    AF-1715-0753
Communications Electronics Equipment, Circuits and Systems by Correspondence
    AF-1715-0792
Communications-Electronics Employment by Correspondence
    AF-1715-0748
Communications-Electronics Systems Technology by Correspondence
    AF-1715-0747
Electronics Fundamentals by Correspondence
    AF-1715-0746
Electronics Fundamentals Communications Track
    CG-1715-0090
Electronics Fundamentals V/S Track #1
    CG-1715-0091
Electronics Fundamentals V/S Track #2
    CG-1715-0093

Electronics Technician
Aviation Electronics Technician 618M-3, Class C
    CG-1715-0079
Aviation Electronics Technician AN/APS-127
    CG-1715-0080
Aviation Electronics Technician AN/APS-127 Radar, Class C
    CG-1715-0080
Aviation Electronics Technician AN/ARC-160, Class C
    CG-1715-0078
Aviation Electronics Technician AN/ARN-118(V), Class C
    CG-1715-0084
Aviation Electronics Technician AN/ARN-133V2, Class C
    CG-1715-0083
Aviation Electronics Technician First Class by Correspondence
    CG-1715-0132
Aviation Electronics Technician Second Class by Correspondence
    CG-1715-0031
Aviation Electronics Technician, Class A
    CG-1715-0001
Electronics Technician Chief by Correspondence
    CG-1402-0001
Electronics Technician First Class by Correspondence
    CG-1715-0038
Electronics Technician Second Class by Correspondence
    CG-1715-0037
Electronics Technician Second Class by Correspondence (ET2)
    CG-1715-0146
Electronics Technician, Class A
    CG-1715-0108

ELINT
Intermediate ELINT Collection and Analysis
    DD-1715-0016

Embarkation
Basic Logistics/Embarkation Specialist
    MC-1402-0058
Marine Air-Ground Task Force Deployment Support System/Computer-Aided Embarkation Management System
    MC-1402-0059
Reserve Embarkation
    MC-1402-0061
Team Embarkation Officer/Assistant
    MC-1402-0060
    MC-1402-0061

Emergency
Emergency Medical Technician Basic
    CG-0709-0004
Emergency Medical Technician, Class C
    CG-0709-0004
Introduction to Civil Air Patrol Emergency Services by Correspondence
    AF-1704-0196

Enemy
Enemy Defense Penetration Aids
    AF-1715-0744

Enforcement
Law Enforcement by Correspondence
    CG-1728-0013
Law Enforcement Specialist by Correspondence
    AF-1728-0044

KEYWORD INDEX D-27

Maritime Law Enforcement Boarding Officer
CG-1728-0021
Maritime Law Enforcement Boarding Officer, Class C
CG-1728-0021
Maritime Law Enforcement Boarding Team Member, Class C
CG-1728-0045
Maritime Law Enforcement Instructor
CG-1728-0024

Engine
Aerospace Propulsion Technician (Jet Engine) by Correspondence
AF-1704-0229
Aircraft Engine Familiarization by Correspondence
CG-1704-0034
Automotive Engine Maintenance and Repair by Correspondence
MC-1703-0024
Caterpillar Diesel Engine Maintenance, Class C
CG-1712-0004
Compression Ignition Engine Fuel Systems Repair
MC-1712-0004
Diesel Engine Maintenance and Troubleshooting by Correspondence
MC-1712-0007
F-100 Jet Engine Mechanic by Correspondence
AF-1704-0205
Installation, Operation, and Maintenance of Diesel Engine Driven Generator Sets by Correspondence
MC-1712-0001
Jet Engine Mechanic by Correspondence
AF-1704-0137
Lister Diesel Overhaul and Engine Power System Equipment Maintenance
CG-1712-0007
LTS-101 Engine Maintenance, Class C
CG-1710-0022
T58-5 and T62 Engine Maintenance, Class C
CG-1704-0026
T58-GE-8B and T58-GE-5 Turboshaft Engine Familiarization by Correspondence
CG-1704-0021
T58-GE-8B Engine Maintenance, Class C
CG-1704-0004
VT903M Cummins Diesel Engine Disassembly-Assembly and Tune-up
CG-1712-0006

Engineer
Basic Combat Engineer
MC-1601-0024
Basic Engineer Equipment Mechanic
MC-1710-0001
MC-1710-0045
Basic Engineer Equipment Operator
MC-1710-0044
C-141 Flight Engineer Technician
AF-1704-0031
C-5 Flight Engineer Technician
AF-1704-0033
Combat Engineer
MC-1601-0026
Combat Engineer Chief, Construction Support by Correspondence
MC-1601-0012
Combat Engineer Noncommissioned Officer (NCO)
MC-1601-0028
Combat Engineer Noncommissioned Officer (NCO) by Correspondence
MC-1601-0036
Combat Engineer Officer
MC-1601-0026
Engineer Equipment Chief
MC-1710-0043
Engineer Equipment Chief by Correspondence
MC-1601-0016
Engineer Equipment Mechanic Noncommissioned Officer (NCO)
MC-1710-0033
Engineer Equipment Officer
MC-1601-0043
Engineer Equipment Operator by Correspondence
MC-1601-0034
Engineer Equipment Operator Noncommissioned Officer (NCO)
MC-1710-0034
Engineer Forms and Records by Correspondence
MC-1601-0015
MC-1601-0037
Engineer Operations Chief
MC-1601-0027
Flight Engineer School, C-5
AF-1704-0033
Flight Engineer Specialist (Helicopter Qualified) by Correspondence
AF-1704-0197
Flight Engineer Specialist by Correspondence
AF-1704-0094
Flight Test Engineer/Navigator
AF-1606-0153
HC-130 Flight Engineer by Correspondence
CG-1704-0009
Journeyman Combat Engineer
MC-1601-0028
Journeyman Engineer Equipment Mechanic
MC-1710-0033
Journeyman Engineer Equipment Operator
MC-1710-0034
Reserve Combat Engineer Noncommissioned Officer (NCO)
MC-1601-0030
Reserve Combat Engineer Officer
MC-1601-0029
Reserve Combat Engineer, Phase 2
MC-1601-0042
Reserve Engineer Equipment Supervisor
MC-1601-0031
Reserve Engineer Equipment Supervisor Phase 2
MC-1601-0031
Small Boat Engineer, Class C
CG-1714-0019

Engineering
Advanced Systems Planning, Research, Development, and Engineering
DD-1408-0016
Aviation Engineering Administration
CG-1405-0005
Bioenvironmental Engineering Specialist by Correspondence
AF-0707-0010
Civil Engineering Control System Specialist by Correspondence
AF-1710-0039
Civil Engineering Control Systems Specialist by Correspondence
AF-1710-0040
Defense Acquisition Engineering, Manufacturing, and Quality Control
DD-1408-0010
Engineering Administration
CG-1408-0035
Engineering Assistant Specialist by Correspondence
AF-1601-0046
Engineering Petty Officer Indoctrination
CG-1408-0035
Intermediate Systems Planning, Research Development, and Engineering
DD-1402-0011
Management of Value Engineering by Correspondence
AF-1405-0068
Small Boat Engineering
CG-1714-0019

Engines
Fundamentals of Diesel Engines by Correspondence
MC-1712-0006
Theory and Construction of Turbine Engines by Correspondence
MC-1710-0004

Enlisted
Enlisted Supply Advanced
MC-1405-0037
Enlisted Supply Basic
MC-1405-0035
Enlisted Supply Independent Duty
MC-1405-0040
Enlisted Supply Intermediate
MC-1405-0039
Enlisted Supply Reorientation
MC-1405-0037
Enlisted Warehousing Intermediate
MC-1405-0036

Entertainment
Shipboard Information, Training, and Entertainment (SITE) System
DD-0504-0012

Entitlement
Basic Pay Entitlement by Correspondence
MC-1401-0008

Environmental
Aircraft Electrical and Environmental Systems Journeyman by Correspondence
AF-1714-0045
Aircraft Environmental Systems Mechanic by Correspondence
AF-1704-0200
Environmental Medicine Specialist by Correspondence
AF-0707-0009
Environmental Support Specialist by Correspondence
AF-1710-0041
Marine Environmental Protection (MEP) by Correspondence
CG-1303-0001
Safety and Security of the Port: Marine Environmental Protection by Correspondence
CG-1728-0036

Equal
Defense Equal Opportunity Management Institute
DD-1512-0003
Social Actions Technician (Equal Opportunity/Human Relations) by Correspondence
AF-1406-0053

KEYWORD INDEX

Equipment
Apprentice Construction Equipment Operator by Correspondence
AF-1710-0029
Audiovisual Equipment Repairer
AF-1715-0741
Basic Engineer Equipment Mechanic
MC-1710-0001
MC-1710-0045
Basic Engineer Equipment Operator
MC-1710-0044
Basic Hygiene Equipment Operator
MC-1601-0025
Biomedical Equipment Maintenance Specialist by Correspondence
AF-1715-0745
Construction Equipment Operator by Correspondence
AF-1710-0028
Electrical Equipment Repair Specialist
MC-1714-0014
Electrical Equipment Repairman
MC-1714-0014
Engineer Equipment Chief by Correspondence
MC-1601-0016
Engineer Equipment Mechanic Noncommissioned Officer (NCO)
MC-1710-0033
Engineer Equipment Officer
MC-1601-0043
Engineer Equipment Operator Noncommissioned Officer (NCO)
MC-1710-0034
Hydraulic Systems and Equipment Operation and Maintenance
CG-1704-0032
Hydraulic Systems and Equipment, Class C
CG-1704-0032
Hygiene Equipment Operator Noncommissioned Officer (NCO)
MC-1601-0044
Introduction to Test Equipment by Correspondence
MC-1715-0129
Journeyman Electrical Equipment Repairman
MC-1714-0014
Journeyman Hygiene Equipment Operator
MC-1601-0044
Lister Diesel Overhaul and Engine Power System Equipment Maintenance
CG-1712-0007
Lister Diesel Overhaul and Power System Equipment Maintenance
CG-1712-0007
Microwave Equipment Maintenance
MC-1715-0123
Microwave Equipment Operator
MC-1715-0161
Multichannel Equipment Operator
MC-1715-0161
Navigational Aids Equipment Specialist by Correspondence
AF-1715-0012
Reserve Engineer Equipment Supervisor
MC-1601-0031
Reserve Engineer Equipment Supervisor Phase 2
MC-1601-0031
Space Systems Equipment Maintenance Specialist by Correspondence
AF-1715-0036
Special Purpose Vehicle and Equipment Mechanic by Correspondence
AF-1710-0032
Telephone Equipment Installation and Repair Specialist by Correspondence
AF-1715-0131
Television Equipment Repairman by Correspondence
AF-1715-0013
Test, Measurement and Diagnostic Equipment
MC-1715-0139
Wideband Communications Equipment Technician by Correspondence
AF-1715-0015

Estimating
Budget, Cost Estimating, and Financial Management Workshop
DD-1408-0017
Software Cost Estimating
DD-1402-0005

Evaluation
Intermediate Test and Evaluation (T & E)
DD-1408-0022

Evaporator
270' WMEC S/S Generator, Waste Heat Recovery System and Evaporator Operation and Maintenance, Class C
CG-1712-0005
Fire Tube and Boiler/Flash Type Evaporator, Class C
CG-1710-0008

EW
Electronic Warfare (EW) System Journeyman by Correspondence
AF-1715-0810

Exchange
CEJT-SX-100/200 Mitel Electronic Private Automatic Branch Exchange Generic 217, Class C
CG-1715-0113
CEJT-SX-200 Mitel Electronic Private Automatic Branch Exchange Generic 1000 or 1001 Telephone System
CG-1715-0116
TEL-14 CDXC-SG-1/1A Pulse 120 Electronic Private Automatic Branch Exchange Telephone System, Class C
CG-1715-0115

Executive
Executive Acquisition Logistics Management
DD-0326-0004
Executive Program Manager
DD-1408-0019
Nursing Executive Development
AF-0709-0037
Officer In Charge and Executive Petty Officer
CG-1715-0020
Officer In Charge/Executive Petty Officer
CG-2205-0020
Systems Acquisition for Contracting Personnel (Executive)
DD-1408-0009

Executives
Automated Information Systems Management for Intermediate Executives
DD-1402-0003
Computer Orientation for Intermediate Executives
DD-1402-0003

Exhaust
Automotive Fuel and Exhaust Systems by Correspondence
MC-1703-0031

Expeditionary
Landing Force Staff Planning (Marine Expeditionary Brigade)
MC-2204-0092

Explosive
Explosive Handling Supervisor
CG-0802-0013
Marine Safety Explosive Handling Supervisor
CG-0802-0015

Explosives
Safety and Security of the Port: Military Explosives by Correspondence
CG-1728-0033

Extended
Arabic-Egyptian Extended
DD-0602-0211
Arabic-Iraqi Extended
DD-0602-0209
Arabic-Syrian Extended
DD-0602-0210
German Extended
DD-0602-0217
Russian Extended
DD-0602-0115

F-100
F-100 Jet Engine Mechanic by Correspondence
AF-1704-0205

F-111
Fighter Weapons Instructor F-111
AF-1406-0063
General Subjects for F-111 Avionics System Specialist by Correspondence
AF-1715-0804

F-15
Basic Operational Training F-15
AF-1606-0158
Conversion Training F-15 Track 1A
AF-1606-0159
F-15 Avionics Automatic Test Station and Component Specialist by Correspondence
AF-1715-0772
F-15 Avionics Manual/ECM Test Station and Component Specialist by Correspondence
AF-1715-0771
F-15 Basic Qualification Training
AF-1606-0158
F-15 Integrated Avionic Attack Control Systems Specialist by Correspondence
AF-1715-0767
F-15 Integrated Organizational Avionics Systems Specialist by Correspondence
AF-1704-0254
Fighter Weapons Instructor F-15
AF-1406-0066
Instructor Pilot Training (F-15)
AF-1606-0157
Integrated Avionics Attack Control Systems Specialist (F-15) by Correspondence
AF-1704-0218
AF-1715-0097
Integrated Avionics Communications, Navigation, and Penetration Aids Systems Specialist (F-15) by Correspondence
AF-1704-0225
AF-1715-0106
Integrated Avionics Computerized Test Station and Component Specialist (F-15) by Correspondence
AF-1715-0086

KEYWORD INDEX D-29

Integrated Avionics Electronic Warfare Equipment and Component Specialist (F-15) by Correspondence
AF-1715-0082

Integrated Avionics Instrument and Flight Control Systems Specialist (F-15) by Correspondence
AF-1715-0098

Integrated Avionics Manual Test Station and Component Specialist (F-15) by Correspondence
AF-1715-0092

Transition/Requalification F-15
AF-1606-0159

F-15/F-111

F-15/F-111 Avionic Systems Journeyman, Instrument, by Correspondence
AF-1715-0815

F-15E

Basic Operational Training F-15E
AF-1704-0269

Basic Qualification Training F-15E
AF-1704-0269

Fighter Weapons Instructor F-15E
AF-1406-0095

Instructor Qualification Training F-15E
AF-1406-0097

Transition/Requalification Training F-15E
AF-1704-0266

F-16

F-16 A/B Avionics Test Station and Component Specialist by Correspondence
AF-1715-0780

F-16 C/D Avionics Test Station and Component Specialist by Correspondence
AF-1715-0785

F-16 Integrated Organizational Avionics Systems Specialist by Correspondence
AF-1715-0808

Fighter Weapons Instructor F-16
AF-1406-0064

Integrated Avionic Attack Control Systems Specialist (F-16) by Correspondence
AF-1715-0096

Integrated Avionics Communication, Navigation, and Penetration Aids Systems Specialist (F-16) by Correspondence
AF-1704-0220

Integrated Avionics Instrument and Flight Control Systems Specialist (F-16) by Correspondence
AF-1715-0099

Transition/Requalification Training F-16 A/B
AF-1704-0275

Transition/Requalification Training F-16 C/D (Block 40/42)
AF-1704-0276

F-16C/D

Basic Operational Training F-16C/D
AF-1704-0270

Conversion Training F-16C/D
AF-1704-0267
AF-1704-0268

Instructor Pilot Upgrade Training F-16C/D
AF-1704-0271

Transition/Requalification Training F-16C/D
AF-1704-0272

F-4

F-4 and AC-130 Weapons Control Systems Technician by Correspondence
AF-1715-0805

F-4C/D

Weapon Control Systems Mechanic (F-4C/D: APQ-109/APA-165) by Correspondence
AF-1715-0052

F-4D/E

Weapon Control Systems Technician (F-4D/E) by Correspondence
AF-1715-0766

F-4E

Weapon Control Systems Mechanic (F-4E: APQ-120) by Correspondence
AF-1715-0053

Weapon Control Systems Technician (F-4E: APQ-120) by Correspondence
AF-1715-0063

F-5

Weapon Control Systems Mechanic (A-7D, A-10, AC-130, F-5) by Correspondence
AF-1715-0050

Weapons Control Systems Mechanic (A-7D, A-10, AC-130, F-5) by Correspondence
AF-1715-0800

F/FB-111

F/FB-111 Automatic Test Station and Component Specialist by Correspondence
AF-1715-0773

F/FB-111 Avionics Manual/Electronic Countermeasures (ECM) Test Station and Component Specialist by Correspondence
AF-1704-0224

Integrated Avionic Communications, Navigation, and Penetration Aids Systems Specialist (F/FB-111) by Correspondence
AF-1715-0779

Integrated Avionics Attack Control Systems Specialist (F/FB-111) by Correspondence
AF-1715-0090
AF-1715-0760

Integrated Avionics Communication, Navigation, and Penetration Aids Systems Specialist (F/FB-111) by Correspondence
AF-1715-0101

Integrated Avionics Communications, Navigation, and Penetration Aids Systems Specialist (F/FB-111) by Correspondence
AF-1704-0263

Integrated Avionics Computerized Test Station and Component Specialist (F/FB-111) by Correspondence
AF-1715-0087

Integrated Avionics Electronic Warfare Equipment and Component Specialist (F/FB-111) by Correspondence
AF-1715-0081

Integrated Avionics Manual Test Station and Component Specialist (F/FB-111) by Correspondence
AF-1715-0093

F/FB/EF-111

Integrated Avionics Instrument and Flight Control Systems Specialist (F/FB/EF-111) by Correspondence
AF-1715-0094

Fabricating

Metal Fabricating Specialist by Correspondence
AF-1723-0014
AF-1723-0015

Fabrication

Fabrication and Parachute Specialist by Correspondence
AF-1733-0002

Metal Fabrication Specialist by Correspondence
AF-1723-0013

Facilities

Apprentice Materiel Facilities Specialist by Correspondence
AF-1405-0018

Facilities Contracting Fundamentals
DD-1402-0009

Facilities Contracts Pricing
DD-1402-0008

Intermediate Facilities Contracting
DD-1402-0007

Materiel Facilities Specialist by Correspondence
AF-1405-0013

Materiel Facilities Supervisor by Correspondence
AF-1405-0033

Missile Facilities Specialist by Correspondence
AF-1704-0264
AF-1717-0026

Field

Field Medical Service Technic Enlisted
MC-0709-0002

Field Radio Operator
MC-1403-0018

Field Service Medical Service Technic
MC-0709-0002

Field Wire Operators
MC-1715-0162

Field Wireman
MC-1715-0162

Introduction to Field Artillery Survey by Correspondence
MC-1601-0039

Fighter

Air National Guard Fighter Weapons Instructor
AF-1606-0151

Fighter Weapons Instructor
AF-1704-0193

Fighter Weapons Instructor A-10
AF-1406-0061

Fighter Weapons Instructor Air Weapons Controller
AF-1406-0062

Fighter Weapons Instructor F-111
AF-1406-0063

Fighter Weapons Instructor F-15
AF-1406-0066

Fighter Weapons Instructor F-15E
AF-1406-0095

Fighter Weapons Instructor F-16
AF-1406-0064

Fighter Weapons Instructor Intelligence
AF-1406-0096

RF-4 Fighter Weapons Instructor
AF-1606-0152

Finance

Accounting and Finance Officer
AF-1408-0101

Personal Finance by Correspondence
MC-1401-0009

Financial

Budget, Cost Estimating, and Financial Management Workshop
DD-1408-0017

Financial Analysis Journeyman by Correspondence
AF-1408-0111

KEYWORD INDEX

Financial Analysis Specialist by Correspondence
   AF-1408-0109
Financial Management by Correspondence
   MC-1401-0012
Financial Management Officer
   MC-1408-0024
Financial Management Officer (Financial Analysis)
   AF-1401-0002
Financial Management Officer (Financial Services)
   AF-1408-0102
Financial Management Specialist (Budget) by Correspondence
   AF-1408-0077
Financial Management Specialist (Commercial Services) by Correspondence
   AF-1408-0074
Financial Management Specialist (Military Pay) by Correspondence
   AF-1408-0074
Financial Management Specialist by Correspondence
   AF-1408-0073
   AF-1408-0074
   AF-1408-0075
   AF-1408-0076
Financial Management Staff Officer
   AF-1408-0103
Financial Management Staff Officer, Air Reserve Forces
   AF-1408-0105
Financial Management Supervisor by Correspondence
   AF-1408-0081
Financial Management/Services Supervisor (Functions and Responsibilities) by Correspondence
   AF-1408-0082
Financial Management/Services Supervisor by Correspondence
   AF-1408-0081
Financial Services Specialist (Introduction) by Correspondence
   AF-1408-0078
Financial Services Specialist (Military Pay) by Correspondence
   AF-1408-0079
Financial Services Specialist (Travel) by Correspondence
   AF-1408-0080
Financial Services Specialist by Correspondence
   AF-1408-0078
Personal Financial Management by Correspondence
   MC-1401-0009
Personal Financial Records Clerk
   MC-1403-0010

Fire
Communications for the Combat Operations Center/Fire Support Coordination Center (COC/FSCC) by Correspondence
   MC-1715-0101
Electrical/Electronic Control for Fire Tube Boilers/Oily Water Separator and Electrical Generator
   CG-1715-0107
Electrical/Electronic Control for Fire Tube Boilers/Oily Water Separator/Electrical Generators Operation and Maintenance, Class C
   CG-1715-0107
Fire and Electrical Systems Component Repair
   MC-1714-0018
Fire and Safety Technician Second Class by Correspondence
   CG-1728-0034
Fire Protection Specialist by Correspondence
   AF-1728-0038
Fire Protection Supervisor by Correspondence
   AF-1728-0046
Fire Support Coordination
   MC-2204-0095
Fire Tube and Boiler/Flash Type Evaporator, Class C
   CG-1710-0008
Infantry Squad Leader: Weapons and Fire Support by Correspondence
   MC-2204-0101
Surface Air Defense System Fire Control Technician
   MC-1715-0180
Tank Gunnery/Direct Fire Procedures (M1A1) by Correspondence
   MC-2204-0098

Fire Control
Aviation Fire Control Repair
   MC-1715-0144
Aviation Fire Control Technician
   MC-1715-0040
Defensive Fire Control Systems Mechanic (B-52D/G: MD-9, ASG-15 Turrets) by Correspondence
   AF-1715-0047
Fire Control Mk 56 System, Class C
   CG-1715-0016
Fire Control System Mk 92 Mod 1 Operation and Maintenance
   CG-1715-0103
Fire Control Technician First Class by Correspondence
   CG-1715-0135
Fire Control Technician Second Class by Correspondence
   CG-1715-0137
Fire Control Technician, Class C
   CG-1715-0016
Gun Fire Control Systems Mk 52 and Mk 56
   CG-1715-0016
Mk 92 Mod 1 Fire Control System
   CG-1715-0103

Firefighting
Firefighting Ashore by Correspondence
   CG-1728-0005
Firefighting on Vessels by Correspondence
   CG-1728-0007

Fireman
Fireman by Correspondence
   CG-1722-0016

Firetrucks
Special Vehicle Mechanic (Firetrucks) by Correspondence
   AF-1710-0033

Fiscal
Fiscal Accounting
   MC-1402-0055
Fiscal Accounting for Supply Clerks by Correspondence
   MC-1401-0010
Fiscal/Budget Technician
   MC-1402-0055

Fitness
Apprentice Fitness and Recreation Specialist by Correspondence
   AF-1406-0055
Fitness and Recreation Specialist by Correspondence
   AF-1406-0056
Fitness and Recreation Supervisor by Correspondence
   AF-1406-0057

Fixed
Aviation Machinist's Mate Fixed Wing, Class A
   CG-1704-0022

Fleet
Automated Data Processing Equipment Fleet Marine Force Programmer
   MC-1402-0050
Automated Data Processing Equipment for the Fleet Marine
   MC-1402-0050

Flight
C-141 Flight Engineer Technician
   AF-1704-0031
C-5 Flight Engineer Technician
   AF-1704-0033
Flight Engineer School, C-5
   AF-1704-0033
Flight Engineer Specialist (Helicopter Qualified) by Correspondence
   AF-1704-0197
Flight Engineer Specialist by Correspondence
   AF-1704-0094
Flight Test Engineer/Navigator
   AF-1606-0153
HC-130 Flight Engineer by Correspondence
   CG-1704-0009
HH-3F Flight Mechanic by Correspondence
   CG-1704-0011
HH-52A Flight Mechanic by Correspondence
   CG-1704-0036
HH-60 Flight Mechanic by Correspondence
   CG-1704-0042
HH-65A Flight Mechanic by Correspondence
   CG-1704-0035
Weather Forecasting and Flight Briefing by Correspondence
   CG-1304-0014

Flight Control
Automatic Flight Control Systems Specialist by Correspondence
   AF-1715-0075
HH-3F Automatic Flight Control System and Selected Electrical Maintenance, Class C
   CG-1714-0015
HH-60J Electrical/Automatic Flight Control Systems Maintenance, Class C
   CG-1704-0051
HH-65A Electrical/Automatic Flight Control Systems Maintenance, Class C
   CG-1704-0049
Instruments/Flight Control Systems Specialist (B-1B) by Correspondence
   AF-1715-0770
Integrated Avionic Instrument and Flight Control Systems by Correspondence
   AF-1704-0255
Integrated Avionic Instrument and Flight Control Systems Specialist by Correspondence
   AF-1704-0226
   AF-1704-0227
   AF-1715-0095

KEYWORD INDEX D-31

Integrated Avionics Instrument and Flight Control Systems Specialist (F-15) by Correspondence
AF-1715-0098

Integrated Avionics Instrument and Flight Control Systems Specialist (F-16) by Correspondence
AF-1715-0099

Integrated Avionics Instrument and Flight Control Systems Specialist (F/FB/EF-111) by Correspondence
AF-1715-0094

Fog
Videograph B Fog Detector Maintenance
CG-1715-0127

Food
Apprentice Food Service Specialist by Correspondence
AF-1729-0010
Basic Food Service
MC-1729-0009
Chief Food Service Specialist by Correspondence
MC-1729-0045
Club Food Operations by Correspondence
MC-1729-0046
Food Service Fundamentals by Correspondence
MC-1729-0038
Food Service Noncommissioned Officer (NCO)
MC-1729-0034
Food Service Staff Noncommissioned Officer (NCO)
MC-1729-0036
Food Service Subsistence Clerk
MC-1729-0048
Food Service Supervisor by Correspondence
AF-1729-0012
Reserve Field Food Service Supervisor by Correspondence
MC-1729-0042
Reserve Food Service Refresher
MC-1729-0037
Senior Food Service
MC-1729-0047

Force
Intelligence for the Marine Air-Ground Task Force by Correspondence
MC-1606-0005
Landing Force Combat Services Support Staff Planning
MC-2204-0093
Landing Force Ground Combat Communication Staff Planning
MC-1601-0046
Landing Force Ground Combat Operations Officer (S-3)
MC-2204-0091
Landing Force Staff Planning (Marine Expeditionary Brigade)
MC-2204-0092

Forces
Armed Forces Staff College
DD-0326-0001
Intelligence Brief: The Opposing Forces Nuclear, Biological, Chemical (NBC) Threat by Correspondence
MC-1606-0011

Forecasting
Weather Forecasting and Flight Briefing by Correspondence
CG-1304-0014

FORTRAN
FORTRAN Program Specialist
MC-1402-0034
IBM System 360 OS FORTRAN Programming
MC-1402-0034

Freight
Basic Freight Operation
MC-0419-0001
Freight Traffic Specialist by Correspondence
AF-0419-0032

French
French Basic
DD-0602-0213
French Intermediate
DD-0602-0221

Frequency
COL-URG-II High Frequency Transmitting System
CG-1715-0072
High Frequency Communication Central Operator
MC-1715-0143
High Frequency Communications System by Correspondence
MC-1715-0135
High Frequency Maintenance
MC-1715-0125

Fuel
Aircraft Fuel Systems Mechanic by Correspondence
AF-1704-0191
AF-1704-0258
Aircraft Fuel Systems Technician by Correspondence
AF-1704-0202
Automotive Fuel and Exhaust Systems by Correspondence
MC-1703-0031
Compression Ignition Engine Fuel Systems Repair
MC-1712-0004
Fuel and Electrical Component Repair
MC-1714-0017
Fuel Specialist by Correspondence
AF-1703-0019
Liquid Fuel Systems Maintenance Specialist by Correspondence
AF-1601-0030
Maintenance of Bulk Fuel Equipment by Correspondence
MC-1601-0038

Fuels
Fuels Management Officer
AF-1601-0050

Fundamentals
Electronics Fundamentals by Correspondence
AF-1715-0746
Electronics Fundamentals Communications Track
CG-1715-0090
Electronics Fundamentals V/S Track #1
CG-1715-0091
Facilities Contracting Fundamentals
DD-1402-0009
Fundamentals of Cost Analysis
DD-1115-0001
Fundamentals of Electricity by Correspondence
AF-1714-0029
Fundamentals of Solid State Devices by Correspondence
AF-1715-0750
Intermediate Production and Quality Management Fundamentals
DD-1408-0024
Multiple Virtual Storage (MVS) Fundamentals and Logic
MC-1402-0052
Operational Level Contracting Fundamentals
DD-1405-0003
Production and Quality Management Fundamentals
DD-1408-0023
Radar Fundamentals
MC-1715-0154
Radio Fundamentals
MC-1715-0160

Gas
Pratt and Whitney FT4A Gas Turbine
CG-1710-0018

Gateway
German Gateway
DD-0602-0205
Korean Gateway
DD-0602-0129
Turkish Gateway
DD-0602-0216

General
General Purpose Vehicle and Body Maintenance by Correspondence
AF-1703-0022
General Purpose Vehicle and Body Maintenance Supervisor by Correspondence
AF-1703-0022

Generator
Installation, Operation, and Maintenance of Diesel Engine Driven Generator Sets by Correspondence
MC-1712-0001

Generators
Electrical/Electronic Control for Fire Tube Boilers/Oily Water Separator/Electrical Generators Operation and Maintenance, Class C
CG-1715-0107
Relays, Generators, Motors, and Electromechanical Devices by Correspondence
AF-1714-0032
AF-1715-0768

Geodesy
Mapping, Charting, and Geodesy Officer
DD-1601-0021
Remotely Sensed Imagery for Mapping, Charting, and Geodesy
DD-1601-0030

Geodetic
Advanced Geodetic Survey
DD-1601-0022
Basic Geodetic Survey
DD-1601-0023
Cartographic/Geodetic Officer
DD-1713-0006

Geographic
Geographic Information Systems (GIS)
DD-1601-0031

German
German Aural Comprehension
DD-0602-0016
German Basic
DD-0602-0215
German Extended
DD-0602-0217
German Gateway
DD-0602-0205

KEYWORD INDEX

German Intermediate
    DD-0602-0121
    DD-0602-0219
German Short
    DD-0602-0214

Graphics
Apprentice Graphics Specialized by Correspondence
    AF-1719-0009
Graphics Specialist by Correspondence
    AF-1719-0010

Gravies
Vegetables, Soups, Sauces, Gravies, and Beverages by Correspondence
    MC-1729-0024

Greek
Greek Basic
    DD-0602-0029

Ground
A-7D Avionics Aerospace Ground Equipment Specialist by Correspondence
    AF-1715-0078
Aerospace Ground Equipment Mechanic by Correspondence
    AF-1704-0187
    AF-1710-0038
Aerospace Ground Equipment Technician by Correspondence
    AF-1704-0192
Avionics Aerospace Ground Equipment Specialist (F/RF-4 Peculiar Avionics AGE) by Correspondence
    AF-1715-0077
Avionics Aerospace Ground Equipment Specialist by Correspondence
    AF-1715-0076
Ground Computer Technician
    MC-1402-0056
Ground Operations Specialist
    MC-1601-0047
Ground Ordnance Officer
    MC-1717-0007
Ground Ordnance Vehicle Maintenance Chiefs
    MC-1717-0006
Ground Ordnance Weapons Chief
    MC-1717-0008
Ground Radar Repair
    MC-1715-0062
    MC-1715-0173
Ground Radar Technician
    MC-1715-0117
Ground Radio Communications Specialist (Unit) by Correspondence
    AF-1715-0782
Ground Radio Communications Specialist by Correspondence
    AF-1715-0227
Ground Radio Communications Technician by Correspondence
    AF-1715-0465
Ground Radio Repair
    MC-1715-0009
    MC-1715-0166
Ground Supply Officer
    MC-1405-0038
Landing Force Ground Combat Communication Staff Planning
    MC-1601-0046
Landing Force Ground Combat Operations Officer (S-3)
    MC-2204-0091
Marine Air-Ground Task Force Deployment Support System/Computer-Aided Embarkation Management System
    MC-1402-0059

Guard
Air National Guard Academy of Military Science
    AF-2203-0051
Air National Guard Fighter Weapons Instructor
    AF-1606-0151
Basic Security Guard
    MC-1728-0010
Marine Security Guard
    MC-1728-0006

Guidance
Aircraft Guidance and Control Systems Technician by Correspondence
    AF-1704-0261
Avionic Guidance and Control Systems Technician by Correspondence
    AF-1704-0232
Avionics Guidance and Control Systems Technician by Correspondence
    AF-1704-0234
    AF-1704-0235
    AF-1704-0236
    AF-1704-0237
    AF-1704-0238
    AF-1704-0239
    AF-1704-0240
    AF-1704-0241
    AF-1704-0259
    AF-1715-0761
Guidance and Control Systems Technician by Correspondence
    AF-1704-0233

Gun
20mm Mk 16 Mod 5 Machine Gun and Magazine Sprinklers Operation and Maintenance, Class C
    CG-2204-0001
3"/50 Gun Mount Mk 22 Operation and Maintenance, Class C
    CG-2204-0002
Gun Fire Control Systems Mk 52 and Mk 56
    CG-1715-0016
Gunner's Mate, Class A, Phase 2 (Advanced Electricity, 3"/50 Caliber Gun, 5"/38 Caliber Gun)
    CG-1714-0013
Inspection and Repair of the .50 Caliber Machine Gun by Correspondence
    MC-2204-0065
Maintenance of the M60 Series Machine Gun by Correspondence
    MC-2204-0064

Gunner's
Gunner's Mate First Class by Correspondence
    CG-1714-0027
Gunner's Mate Second Class by Correspondence
    CG-1714-0009
    CG-1714-0026
Gunner's Mate, Class A
    CG-1714-0017
Gunner's Mate, Class A, Phase 2 (Advanced Electricity, 3"/50 Caliber Gun, 5"/38 Caliber Gun)
    CG-1714-0013

Gunnery
Tank Gunnery/Direct Fire Procedures (M1A1) by Correspondence
    MC-2204-0098

Gyrocompass
Mk 27 Gyrocompass System
    CG-1715-0104
Mk 27 Gyrocompass System Operation and Maintenance
    CG-1715-0104
Mk 29 Mod 1 Gyrocompass Operation and Maintenance
    CG-1715-0106

Handling
Materials Handling Equipment by Correspondence
    MC-1710-0040

Hawk
Hawk Missile System Operator (USMC)
    MC-1715-0163

Hazardous
Defense Packaging of Hazardous Materials for Transportation
    DD-0419-0003

HC-130
HC-130 Avionics by Correspondence
    CG-1715-0085
HC-130 Avionicsman by Correspondence
    CG-1715-0085
HC-130 Flight Engineer by Correspondence
    CG-1704-0009
HC-130 Loadmaster (AC130L) by Correspondence
    CG-0419-0002
HC-130 Loadmaster by Correspondence
    CG-0419-0002

HC-131A
HC-131A Aircraft Maintenance, Class C
    CG-1704-0006

Health
Health Services Administration
    AF-0709-0034
Health Services Management Journeyman by Correspondence
    AF-0799-0008
    AF-0799-0009
    AF-0799-0010
Health Services Technician First Class by Correspondence
    CG-0709-0006
Health Services Technician Second Class by Correspondence
    CG-0709-0005
Health Services Technician, Class A
    CG-0709-0009
Mental Health Service Specialist by Correspondence
    AF-0703-0017
Mental Health Unit Specialist by Correspondence
    AF-0708-0002

Heat
270' WMEC S/S Generator, Waste Heat Recovery System and Evaporator Operation and Maintenance, Class C
    CG-1712-0005

Heating
Heating Systems Specialist by Correspondence
    AF-1701-0010
    AF-1701-0011

KEYWORD INDEX D-33

Heating Systems Technician by Correspondence
AF-1701-0009

Heating, Ventilation, Air Conditioning, and Refrigeration Journeyman by Correspondence
AF-1701-0012
AF-1701-0013

Heating, Ventilation, Air Conditioning, and Refrigeration Specialist by Correspondence
AF-1701-0012
AF-1701-0013

Heavy
Preventive Maintenance and Operating Techniques for Heavy Vehicles by Correspondence
MC-1703-0034

Hebrew
Hebrew Basic
DD-0602-0235

Helicopter
Flight Engineer Specialist (Helicopter Qualified) by Correspondence
AF-1704-0197
Helicopter Mechanic by Correspondence
AF-1704-0206

Hematology
Medical Laboratory Craftsman (Hematology, Serology, Blood Banking and Immunohematology) by Correspondence
AF-0702-0010
Medical Laboratory Technician (Hematology, Serology, Blood Banking and Immunohematology) by Correspondence
AF-0702-0010
AF-0709-0031

HH-3F
HH-3F Automatic Flight Control System and Selected Electrical Maintenance, Class C
CG-1714-0015
HH-3F Avionicsman by Correspondence
CG-1715-0086
HH-3F Flight Mechanic by Correspondence
CG-1704-0011

HH-52A
HH-52A Flight Mechanic by Correspondence
CG-1704-0036

HH-60
HH-60 Flight Mechanic by Correspondence
CG-1704-0042

HH-60J
HH-60J Avionics Maintenance, Class C
CG-1715-0145

HH-65A
HH-65A Airframe/Power Train Maintenance, Class C
CG-1704-0050
HH-65A Avionics Maintenance, Class C
CG-1715-0144
HH-65A Electrical/Automatic Flight Control Systems Maintenance, Class C
CG-1704-0049
HH-65A Flight Mechanic by Correspondence
CG-1704-0035

High
COL-URG-II High Frequency Transmitting System
CG-1715-0072

High Frequency Communication Central Operator
MC-1715-0143
High Frequency Communications System by Correspondence
MC-1715-0135
High Frequency Maintenance
MC-1715-0125
High Reliability Soldering, Class C
CG-1710-0012

Honors
Honors and Ceremonies by Correspondence
CG-2205-0032

Household
Passenger and Household Goods Specialist by Correspondence
AF-0419-0031

HU-25A
HU-25A Airframe, Class C
CG-1704-0044
HU-25A Avionics Maintenance, Class C
CG-1704-0045
HU-25A Avionics, Class C
CG-1704-0031
HU-25A Avionicsman by Correspondence
CG-0802-0007
HU-25A Dropmaster by Correspondence
CG-1704-0037
HU-25A Electrical, Class C
CG-1704-0046
Search and Rescue HU-25A Basic Aircrew by Correspondence
CG-0802-0008

Human
Human Resource Management (HRMC)
AF-1406-0075
Social Actions Technician (Equal Opportunity/Human Relations) by Correspondence
AF-1406-0053

Hungarian
Hungarian Basic
DD-0602-0132

Hydraulic
Hydraulic Principles and Troubleshooting by Correspondence
MC-1704-0007
Hydraulic Systems and Equipment Operation and Maintenance
CG-1704-0032
Hydraulic Systems and Equipment, Class C
CG-1704-0032

Hydrographic
Hydrographic Survey
DD-1722-0001

Hygiene
Basic Hygiene Equipment Operator
MC-1601-0025
Hygiene Equipment Operator Noncommissioned Officer (NCO)
MC-1601-0044
Journeyman Hygiene Equipment Operator
MC-1601-0044

IBM
IBM S/360 OS Computer Operator
MC-1402-0023
IBM System 360 (OS) COBOL Programming
MC-1402-0013
IBM System 360 (OS) Data Control Techniques
MC-1402-0042

IBM System 360 Assembler Language Coding (Entry)
MC-1402-0051
IBM System 360 Operating System (OS) Operations
MC-1402-0023
IBM System 360 OS FORTRAN Programming
MC-1402-0034
IBM System/360 (OS) COBOL Programming Entry-level
MC-1402-0013

Ice
Ice Observer by Correspondence
CG-1304-0006

Ignition
Compression Ignition Engine Fuel Systems Repair
MC-1712-0004

Imagery
Imagery Interpreter Specialist by Correspondence
AF-1709-0027
Imagery Production Specialist by Correspondence
AF-1709-0030
Imagery Systems Maintenance Specialist by Correspondence
AF-1715-0806
AF-1715-0807
Precision Imagery and Audiovisual Media Maintenance Specialist by Correspondence
AF-1715-0136
Remotely Sensed Imagery for Mapping, Charting, and Geodesy
DD-1601-0030
Remotely Sensed Imagery for MC&G
DD-1601-0030

Immunohematology
Medical Laboratory Craftsman (Hematology, Serology, Blood Banking and Immunohematology) by Correspondence
AF-0702-0010
Medical Laboratory Technician (Hematology, Serology, Blood Banking and Immunohematology) by Correspondence
AF-0702-0010
AF-0709-0031

Independent
Enlisted Supply Independent Duty
MC-1405-0040
Food Service Specialist, Independent Duty
CG-1729-0009
Independent Duty Administration
MC-1403-0011
Subsistence Specialist, Independent Duty
CG-1729-0009

Indoctrination
Engineering Petty Officer Indoctrination
CG-1408-0035
Marine Safety Initial Indoctrination by Correspondence
CG-1708-0019
Marine Safety Initial Indoctrination Lesson Plan Series by Correspondence
CG-1708-0019
Marine Safety Initial Indoctrination Marine Inspection by Correspondence
CG-1708-0024
Marine Safety Initial Indoctrination Port Operations by Correspondence
CG-1708-0025
Reserve Officer Candidate Indoctrination
CG-2202-0004

KEYWORD INDEX

Sea Duty Indoctrination
MC-2204-0014

Indonesian
Indonesian Basic
DD-0602-0032

Indonesian-Malay
Indonesian-Malay Basic
DD-0602-0012

Industrial
Correspondence Course of the Industrial College of the Armed Forces
DD-1511-0001
Industrial College of the Armed Forces
DD-1511-0010
Industrial College of the Armed Forces (Resident Program)
DD-1511-0003

Infantry
Infantry Officer
MC-2204-0081
Infantry Squad Leader Squad Tactics by Correspondence
MC-2204-0100
Infantry Squad Leader: Land Navigation by Correspondence
MC-1601-0048
Infantry Squad Leader: Weapons and Fire Support by Correspondence
MC-2204-0101
Infantry Staff Noncommissioned Officer (NCO): Military Symbols and Overlays by Correspondence
MC-1601-0040
Reserve Infantry Officer
MC-2204-0108

Infection
Basic Infection Control Surveillance (Infection Control and Epidemiology)
AF-0707-0011

Information
Advanced Information Systems Acquisition
DD-0326-0005
Automated Information Systems Contracting
DD-1402-0006
Automated Information Systems Management for Intermediate Executives
DD-1402-0003
Automated Information Systems Management for Intermediate Managers
DD-1402-0003
Geographic Information Systems (GIS)
DD-1601-0031
Information Management Specialist by Correspondence
AF-1406-0045
Information Specialist (Journalist)
DD-0504-0001
Information Systems Operator by Correspondence
AF-1402-0070
Information Systems Programming Specialist by Correspondence
AF-1402-0071
Intermediate Information Systems Acquisition
DD-0326-0006
Security of Classified Information by Correspondence
CG-1728-0035
Shipboard Information, Training, and Entertainment (SITE) System
DD-0504-0012

Visual Information Production-Documentation Journeyman by Correspondence
AF-0505-0005
Visual Information Production-Documentation Specialist by Correspondence
AF-0505-0003
AF-0505-0005

Initial
Marine Safety Initial Indoctrination Marine Inspection by Correspondence
CG-1708-0024
Marine Safety Initial Indoctrination Port Operations by Correspondence
CG-1708-0025

Inspection
Apprentice Nondestructive Inspection Specialist by Correspondence
AF-1724-0006
Chief, Inspection Department
CG-1406-0010
Inspection and Repair of Shoulder Weapons by Correspondence
MC-2204-0066
Inspection and Repair of the .50 Caliber Machine Gun by Correspondence
MC-2204-0065
Marine Safety Initial Indoctrination Marine Inspection by Correspondence
CG-1708-0024
Marine Safety Inspection
CG-1728-0039
Nondestructive Inspection Specialist by Correspondence
AF-1723-0005

Inspector
Marine Inspector
CG-1728-0039

Installation
Antenna and Cable Systems Installation/Maintenance Specialist by Correspondence
AF-1715-0802
Cable Splicing Installation and Maintenance Specialist by Correspondence
AF-1715-0126
Communications Cable Systems Installation/Maintenance Specialist by Correspondence
AF-1714-0036
AF-1714-0037
Installation, Operation, and Maintenance of Diesel Engine Driven Generator Sets by Correspondence
MC-1712-0001
Mitel SX-200D Installation and Maintenance, Class C
CG-1715-0116
Telephone Equipment Installation and Repair Specialist by Correspondence
AF-1715-0131

Installer
AN/TSC-120 Communication Central System Installer/Maintainer
MC-1715-0171

Instruction
PDP-11 Instruction Set
DD-1715-0011
Principles of Instruction for the Marine Noncommissioned Officer (NCO) by Correspondence
MC-1406-0028

Instructional
Instructional Management
MC-1406-0031

Instructional System Development for Training Managers by Correspondence
AF-1406-0080

Instructor
Adversary Tactics Instructor (AFSC K1115H)
AF-1406-0065
Adversary Tactics Instructor (AFSC K1745)
AF-1406-0067
Air National Guard Fighter Weapons Instructor
AF-1606-0151
B-52 Bomber Weapons Instructor
AF-1704-0273
Basic Instructor
CG-1406-0005
Bomber Weapons Instructor B-52
AF-1704-0273
Bomber Weapons Instructor B1
AF-1704-0274
ENJPT Pilot Instructor Training Shepherd (T-37/T-38)
AF-1406-0101
Fighter Weapons Instructor
AF-1704-0193
Fighter Weapons Instructor A-10
AF-1406-0061
Fighter Weapons Instructor Air Weapons Controller
AF-1406-0062
Fighter Weapons Instructor F-111
AF-1406-0063
Fighter Weapons Instructor F-15
AF-1406-0066
Fighter Weapons Instructor F-15E
AF-1406-0095
Fighter Weapons Instructor F-16
AF-1406-0064
Fighter Weapons Instructor Intelligence
AF-1406-0096
Formal Instructor
MC-1406-0029
Formal School Instructor
MC-1406-0029
IFF Instructor Training (AT-38B)
AF-1406-0100
Instructor Orientation
MC-1406-0030
Instructor Pilot Training (F-15)
AF-1606-0157
Instructor Pilot Upgrade
AF-1704-0284
Instructor Pilot Upgrade Training F-16 C
AF-1406-0102
Instructor Pilot Upgrade Training F-16 C/D
AF-1406-0102
Instructor Pilot Upgrade Training F-16C/D
AF-1704-0271
Instructor Qualification Training F-15E
AF-1406-0097
AF-1406-0098
Instructor Training
MC-1406-0030
Instructor/Course Developer, Class C
CG-1406-0004
Intelligence Weapons Instructor
AF-1406-0096
Marine Combat Instructor of Water Safety
MC-2204-0097
Maritime Law Enforcement Instructor
CG-1728-0024
Pilot Instructor Training
AF-1406-0074
Pilot Instructor Training (T-37)
AF-1406-0074

KEYWORD INDEX   D-35

Pilot Instructor Training (T-38)
AF-1406-0074
RF-4 Fighter Weapons Instructor
AF-1606-0152
T-1A Pilot Instructor Training
AF-1406-0099
Weapons Instructor (AWC)
AF-1406-0062

Instrument
Avionics Instrument Systems Specialist by Correspondence
AF-1715-0080
F-15/F-111 Avionic Systems Journeyman, Instrument, by Correspondence
AF-1715-0815
Integrated Avionic Instrument and Flight Control Systems by Correspondence
AF-1704-0255
Integrated Avionic Instrument and Flight Control Systems Specialist by Correspondence
AF-1704-0226
AF-1704-0227
AF-1715-0095
Integrated Avionics Instrument and Flight Control Systems Specialist (F-15) by Correspondence
AF-1715-0098
Integrated Avionics Instrument and Flight Control Systems Specialist (F-16) by Correspondence
AF-1715-0099
Integrated Avionics Instrument and Flight Control Systems Specialist (F/FB/EF-111) by Correspondence
AF-1715-0094
Survey Instrument Maintenance
DD-1721-0003

Instrumentation
Instrumentation Mechanic by Correspondence
AF-1715-0045

Instruments
Artillery Survey Instruments by Correspondence
MC-1601-0032
Instruments/Flight Control Systems Specialist (B-1B) by Correspondence
AF-1715-0770

Integrated
F-15 Integrated Avionic Attack Control Systems Specialist by Correspondence
AF-1715-0767
F-15 Integrated Organizational Avionics Systems Specialist by Correspondence
AF-1704-0254
F-16 Integrated Organizational Avionics Systems Specialist by Correspondence
AF-1715-0808
Integrated Avionic Attack Control Systems Specialist (F-16) by Correspondence
AF-1715-0096
Integrated Avionic Attack Control Systems Specialist by Correspondence
AF-1715-0759
Integrated Avionic Communication, Navigation and Penetration Aids Systems Specialist by Correspondence
AF-1714-0039
Integrated Avionic Communications, Navigation, and Penetration Aids Systems Specialist (F/FB-111) by Correspondence
AF-1715-0779
Integrated Avionic Instrument and Flight Control Systems by Correspondence
AF-1704-0255
Integrated Avionic Instrument and Flight Control Systems Specialist by Correspondence
AF-1715-0095
Integrated Avionics Attack Control Systems Specialist (F-15) by Correspondence
AF-1715-0097
Integrated Avionics Attack Control Systems Specialist (F/FB-111) by Correspondence
AF-1715-0090
AF-1715-0760
Integrated Avionics Attack Control Systems Specialist by Correspondence
AF-1715-0091
Integrated Avionics Communication, Navigation, and Penetration Aids Systems Specialist (F/FB-111) by Correspondence
AF-1715-0101
Integrated Avionics Communication, Navigation, and Penetration Aids Systems Specialist by Correspondence
AF-1715-0100
Integrated Avionics Communications, Navigation, and Penetration Aids Systems Specialist (F-15) by Correspondence
AF-1715-0106
Integrated Avionics Communications, Navigation, and Penetration Aids Systems Specialist (F/FB-111) by Correspondence
AF-1704-0263
Integrated Avionics Computerized Test Station and Component Specialist (F-15) by Correspondence
AF-1715-0086
Integrated Avionics Computerized Test Station and Component Specialist (F/FB-111) by Correspondence
AF-1715-0087
Integrated Avionics Electronic Warfare Equipment and Component Specialist (F-15) by Correspondence
AF-1715-0082
Integrated Avionics Electronic Warfare Equipment and Component Specialist (F/FB-111) by Correspondence
AF-1715-0081
Integrated Avionics Instrument and Flight Control Systems Specialist (F-15) by Correspondence
AF-1715-0098
Integrated Avionics Instrument and Flight Control Systems Specialist (F-16) by Correspondence
AF-1715-0099
Integrated Avionics Instrument and Flight Control Systems Specialist (F/FB/EF-111) by Correspondence
AF-1715-0094
Integrated Avionics Manual Test Station and Component Specialist (F-15) by Correspondence
AF-1715-0092
Integrated Avionics Manual Test Station and Component Specialist (F/FB-111) by Correspondence
AF-1715-0093
Integrated Organizational Avionic System Specialist by Correspondence
AF-1715-0758
Marine Corps Integrated Maintenance Management Systems
MC-1402-0057

Intelligence
Fighter Weapons Instructor Intelligence
AF-1406-0096
Intelligence Brief: Southwest Asia by Correspondence
MC-1511-0001
Intelligence Brief: The Military Threat by Correspondence
MC-1606-0012
Intelligence Brief: The Opposing Forces Nuclear, Biological, Chemical (NBC) Threat by Correspondence
MC-1606-0011
Intelligence for the Marine Air-Ground Task Force by Correspondence
MC-1606-0005
Intelligence Fundamentals by Correspondence
AF-1606-0119
Intelligence Officer by Correspondence
AF-1606-0120
Intelligence Operations Specialist by Correspondence
AF-1606-0121
Intelligence Operations Technician by Correspondence
AF-1606-0123
Intelligence Weapons Instructor
AF-1406-0096
Introduction to Combat Intelligence by Correspondence
MC-1606-0009
Target Intelligence Specialist by Correspondence
AF-1606-0122

Inter-American
Inter-American Defense College
DD-1511-0011

Interior
Interior Wiring by Correspondence
MC-1601-0033

Intermediate
Arabic Egyptian/Syrian Intermediate
DD-0602-0023
Arabic Intermediate
DD-0602-0220
Bulgarian Intermediate
DD-0602-0022
Chinese Intermediate
DD-0602-0233
Czech Intermediate
DD-0602-0024
Enlisted Supply Intermediate
MC-1405-0039
Enlisted Warehousing Intermediate
MC-1405-0036
French Intermediate
DD-0602-0221
German Intermediate
DD-0602-0121
DD-0602-0219
Intermediate Systems Acquisition
DD-1408-0011
Intermediate Systems Planning, Research Development, and Engineering
DD-1402-0011
Korean Intermediate
DD-0602-0236
Polish Intermediate
DD-0602-0122
Russian Intermediate
DD-0602-0123
Spanish Intermediate
DD-0602-0218

KEYWORD INDEX

Interpreter (continued)
Thai Intermediate
DD-0602-0239
Vietnamese Intermediate
DD-0602-0125

Interpreter
Imagery Interpreter Specialist by Correspondence
AF-1709-0027

Interrogator
Interrogator Set AN/TPX-46 Direct Support/General Support Repairman and Supervisor
MC-1715-0138

Inventory
Inventory Management Specialist (Munitions) by Correspondence
AF-1405-0064
Inventory Management Specialist by Correspondence
AF-1405-0051
Inventory Management Supervisor by Correspondence
AF-1405-0059
AF-1405-0071

Investigation
Investigation Department
CG-1728-0017

Investigations
Special Investigations and Counterintelligence Technician by Correspondence
AF-1728-0039

Investigator
Investigator 1 by Correspondence
CG-1728-0014
Investigator Second Class by Correspondence
CG-1728-0042

Iraqi
Arabic-Iraqi Extended
DD-0602-0209
Arabic-Iraqi Special
DD-0602-0207

Italian
Italian Basic
DD-0602-0108
DD-0602-0226

Japanese
Japanese Basic
DD-0602-0117

Jet
Aerospace Propulsion Specialist (Jet Engine) by Correspondence
AF-1704-0256
Aerospace Propulsion Technician (Jet Engine) by Correspondence
AF-1704-0229
F-100 Jet Engine Mechanic by Correspondence
AF-1704-0205
Jet Engine Mechanic by Correspondence
AF-1704-0137

Joint
Coast Guard Joint Service Supervisor Safety by Correspondence
AF-0801-0004
Joint and Combined Staff Officer
DD-0326-0002
Joint and Combined Warfighting
DD-0326-0003
Navy Joint Service Supervisor Safety by Correspondence
AF-0801-0003

Journalism
Advanced Electronic Journalism
DD-0505-0006
Electronic Journalism
DD-0505-0003
DD-0505-0007
Introduction to Journalism Reserve, Phase 2
DD-0504-0017

Journalist
Basic Journalist
DD-0504-0001
Basic Military Journalist
DD-0504-0001
Information Specialist (Journalist)
DD-0504-0001

Journeyman
Air Transportation Journeyman by Correspondence
AF-0419-0038
Airborne Command and Control Communications Equipment Journeyman by Correspondence
AF-1715-0811
AF-1715-0816
Aircraft Communication/Navigation Systems Journeyman by Correspondence
AF-1715-0817
Aircraft Electrical and Environmental Systems Journeyman by Correspondence
AF-1714-0045
Automatic Tracking Radar Journeyman by Correspondence
AF-1715-0812
Avionics Sensors Maintenance Journeyman by Correspondence
AF-1715-0818
Chapel Service Support Journeyman by Correspondence
AF-1408-0113
Electrical Power Production Journeyman by Correspondence
AF-1714-0040
AF-1714-0043
Electrical Systems Journeyman by Correspondence
AF-1714-0042
AF-1714-0044
Electronic Warfare (EW) System Journeyman by Correspondence
AF-1715-0810
Electronic Warfare Systems Journeyman by Correspondence
AF-1715-0819
F-15/F-111 Avionic Systems Journeyman, Instrument, by Correspondence
AF-1715-0815
Financial Analysis Journeyman by Correspondence
AF-1408-0111
Health Services Management Journeyman by Correspondence
AF-0799-0008
AF-0799-0009
AF-0799-0010
Heating, Ventilation, Air Conditioning, and Refrigeration Journeyman by Correspondence
AF-1701-0012
AF-1701-0013
Journeyman Combat Engineer
MC-1601-0028
Journeyman Electrician
MC-1714-0016
Journeyman Engineer Equipment Mechanic
MC-1710-0033
Journeyman Engineer Equipment Operator
MC-1710-0034
Journeyman Hygiene Equipment Operator
MC-1601-0044
Journeyman Metalworker
MC-1723-0007
Journeyman Refrigeration Mechanic
MC-1730-0007
Meteorological and Navigation Systems Journeyman by Correspondence
AF-1304-0017
AF-1715-0821
Missile and Space Systems Electronic Maintenance Journeyman by Correspondence
AF-1714-0041
AF-1715-0814
Missile and Space Systems Maintenance Journeyman by Correspondence
AF-1704-0265
Morale, Welfare, Recreation, and Services Journeyman by Correspondence
AF-0804-0001
Paralegal Journeyman by Correspondence
AF-1407-0003
Pavement and Construction Equipment Journeyman by Correspondence
AF-1601-0051
AF-1601-0052
Personnel Systems Management Journeyman by Correspondence
AF-1402-0075
AF-1402-0076
AF-1402-0077
AF-1402-0078
Structural Journeyman by Correspondence
AF-1732-0015
AF-1732-0016
Supply Systems Analyst Journeyman by Correspondence
AF-1405-0072
Traffic Management Journeyman by Correspondence
AF-0419-0037
Utilities System Journeyman by Correspondence
AF-1732-0004
AF-1732-0005
Vehicle Operator/Dispatcher Journeyman by Correspondence
AF-1703-0023
Visual Information Production-Documentation Journeyman by Correspondence
AF-0505-0005

Junior
Platoon Leaders Junior
MC-2204-0077

Justice
Military Justice by Correspondence
AF-1728-0043

Korean
Korean Basic
DD-0602-0118
Korean Gateway
DD-0602-0129
Korean Intermediate
DD-0602-0236

Labor
Introduction to Labor Relations for Air Force Supervisors by Correspondence
AF-1408-0088

Laboratory
Dental Laboratory Specialist by Correspondence
AF-0701-0018

KEYWORD INDEX   D-37

Medical Laboratory Craftsman (Chemistry and Urinalysis) by Correspondence
AF-0702-0008
Medical Laboratory Craftsman (Hematology, Serology, Blood Banking and Immunohematology) by Correspondence
AF-0702-0010
Medical Laboratory Craftsman (Microbiology) by Correspondence
AF-0702-0009
Medical Laboratory Technician (Chemistry and Urinalysis) by Correspondence
AF-0702-0008
Medical Laboratory Technician (Hematology, Serology, Blood Banking and Immunohematology) by Correspondence
AF-0702-0010
AF-0709-0031
Medical Laboratory Technician (Microbiology) by Correspondence
AF-0702-0009
AF-0709-0030

Land
Infantry Squad Leader: Land Navigation by Correspondence
MC-1601-0048
Land Navigation by Correspondence
MC-1601-0048

Landing
Basic Landing Support Specialist
MC-2204-0069
Landing Force Combat Services Support Staff Planning
MC-2204-0093
Landing Force Ground Combat Communication Staff Planning
MC-1601-0046
Landing Force Ground Combat Operations Officer (S-3)
MC-2204-0091
Landing Force Staff Planning (Marine Expeditionary Brigade)
MC-2204-0092

Language
Assembler Language Code (ALC) Programming (Combination)
MC-1402-0051
IBM System 360 Assembler Language Coding (Entry)
MC-1402-0051

Launched
Air Launched Missile Systems Specialist by Correspondence
AF-1714-0041
AF-1715-0814

Law
Civil Law by Correspondence
AF-1511-0007
Government Contract Law
DD-0326-0007
Law Enforcement by Correspondence
CG-1728-0013
Law Enforcement Specialist by Correspondence
AF-1728-0044
Maritime Law Enforcement Boarding Officer
CG-1728-0021
Maritime Law Enforcement Boarding Team Member, Class C
CG-1728-0045
Maritime Law Enforcement Instructor
CG-1728-0024

Leader
Amphibious Vehicle Unit Leader
MC-1710-0010
Assault Amphibious Vehicle Unit Leader
MC-1710-0010
Infantry Squad Leader Squad Tactics by Correspondence
MC-2204-0100
Infantry Squad Leader: Land Navigation by Correspondence
MC-1601-0048
Infantry Squad Leader: Weapons and Fire Support by Correspondence
MC-2204-0101
Marine Leader Training Stage III
MC-2204-0074

Leaders
Amphibious Vehicle Unit Leaders (Enlisted)
MC-1710-0010
Platoon Leaders Combined
MC-2204-0079
Platoon Leaders Junior
MC-2204-0077
Platoon Leaders Senior
MC-2204-0078
Winter Mountain Leaders (A)
MC-0804-0003
Winter Mountain Leaders (B)
MC-0804-0004

Leadership
Bakery Noncommissioned Officer (NCO) Leadership
MC-1729-0006
Fundamentals of Marine Corps Leadership by Correspondence
MC-1406-0023
Leadership and Management
CG-1406-0008
Officer Leadership and Management
CG-1717-0009
Reserve Officer Leadership and Management
CG-1717-0008
Reserve Senior Petty Officer Leadership and Management
CG-1717-0007
Senior Petty Officer Leadership and Management
CG-1717-0010

Legal
Advanced Legal Services
MC-1406-0033
Legal Administration Clerk by Correspondence
MC-1406-0024
Legal Administration for the Reporting Unit by Correspondence
MC-1406-0024
Legal Services
MC-1406-0034
Legal Services Specialist by Correspondence
AF-1406-0047

Life
Aircrew Life Support Specialist by Correspondence
AF-1704-0199

Lighthouse
Automated Aids to Navigation Lighthouse Technician
CG-1715-0128
Lighthouse Technician
CG-1715-0128
Lighthouse Technician, Class C
CG-1715-0128

Line
Pole Line Construction Equipment by Correspondence
MC-1715-0091
Pole Line Construction Techniques by Correspondence
MC-1715-0096

Liquid
Liquid Fuel Systems Maintenance Specialist by Correspondence
AF-1601-0030

Listing
Additional Demands Listing Clerk by Correspondence
MC-1408-0022

Lithographer
Navy/Air Force Basic Lithographer
DD-1719-0007

Lithography
Advanced Lithography
DD-1719-0011

Loadmaster
HC-130 Loadmaster (AC130L) by Correspondence
CG-0419-0002
HC-130 Loadmaster by Correspondence
CG-0419-0002

Location
Position Location Reporting System (PLRS) Capabilities and Employment by Correspondence
MC-1606-0008
Position Location Reporting System (PLRS) Master Station Maintenance
MC-1715-0145
Position Location Reporting System (PLRS) Master Station Operator
MC-1715-0158
Position Location Reporting System (PLRS) Support Maintenance
MC-1715-0149

Logic
Fundamentals of Digital Logic by Correspondence
MC-1715-0131
Multiple Virtual Storage (MVS) Fundamentals and Logic
MC-1402-0052

Logistics
Basic Logistics/Embarkation Specialist
MC-1402-0058
Executive Acquisition Logistics Management
DD-0326-0004
Introduction to Logistics by Correspondence
AF-1408-0110
Logistics Vehicle System (LVS) Maintenance
MC-1408-0019
Logistics Vehicle System Maintenance
MC-1408-0019
Logistics Vehicle System Operator
MC-1703-0030

Loran
ADL-81 Loran-C Receiver, Class C
CG-1715-0082
AN/FPN-39 Loran Transmitter, Class C
CG-1715-0118
AN/FPN-42 Loran Transmitter, Class C
CG-1715-0117
AN/FPN-44A Loran Transmitter, Class C
CG-1715-0110

KEYWORD INDEX

AN/FPN-64(V) Loran Solid State Transmitter, Class C
  CG-1715-0123
Loran Track #1 for Lorstas, Malone, Senca
  CG-1715-0092
Loran Track #10 Lorsta, Shetland Island
  CG-1715-0102
Loran Track #2 for Lorstas, Kodiac, Yakota
  CG-1715-0095
Loran Track #3 for Lorsta, Honolulu
  CG-1715-0097
Loran Track #4 for Lorstas, Grangeville, Raymondville
  CG-1715-0096
Loran Track #5 for Lorstas, Attu, Baudette, Caribou, Carolina Beach, Estartit, Jupiter, Kargabarun, Kure, Nantucket, Port Clarence, Shoal Cove, Sylt, Upolu Point
  CG-1715-0094
Loran Track #6 For Lorstas, Dana, Fallon, George, Gesashi, Hokkaido, Iwo Jima, Lampeduska, Marcus, Narrowcape Searchlight, Tok, Yap
  CG-1715-0098
Loran Track #7 for Lorsta, Middletown
  CG-1715-0099
Loran Track #8 for Lorsta, Sellia Marina
  CG-1715-0100
Loran Track #9 for Lorsta, Johnston Island, St. Paul Island
  CG-1715-0101

Loran-C
ADL-81/82 Loran-C Receiver Maintenance, Class C
  CG-1715-0141
Loran-C Control Station Operations
  CG-1715-0114
Loran-C Control Station Operations, Class C
  CG-1715-0114
Loran-C Timing and Control Operations and Maintenance, Class C
  CG-1715-0111

Lubricating
Automotive Cooling and Lubricating Systems by Correspondence
  MC-1703-0032

M1A1
M1A1 Armor Crewman
  MC-2204-0107
M1A1 Tank System Mechanic
  MC-1710-0047
Tank Gunnery/Direct Fire Procedures (M1A1) by Correspondence
  MC-2204-0098

M60
Maintenance of the M60 Series Machine Gun by Correspondence
  MC-2204-0064

Machine
20mm Mk 16 Mod 5 Machine Gun and Magazine Sprinklers Operation and Maintenance, Class C
  CG-2204-0001

Machinegun
Inspection and Repair of the Mk 19 Machinegun by Correspondence
  MC-2204-0099

Machinegunner
Basic Machinegunner
  MC-2204-0048
Machinegunner
  MC-2204-0048

Machinery
270' WMEC Machinery Plant Control and Monitoring System
  CG-1710-0013
270' WMEC Machinery Plant Control and Monitoring System, Class C
  CG-1710-0013
Machinery Technician First Class by Correspondence
  CG-1723-0006
Machinery Technician, Class A
  CG-1723-0005

Machinist
Aircraft Metals Technology (Machinist) by Correspondence
  AF-1717-0027
Machinist
  MC-1723-0010
Machinist by Correspondence
  AF-1723-0008

Machinist's Mate
Aviation Machinist's Mate
  CG-1704-0001
Aviation Machinist's Mate (AD), Class A
  CG-1704-0001
Aviation Machinist's Mate First Class by Correspondence
  CG-1704-0008
  CG-1704-0052
Aviation Machinist's Mate Fixed Wing, Class A
  CG-1704-0022
Aviation Machinist's Mate Rotary Wing, Class A
  CG-1704-0023
Aviation Machinist's Mate Second Class by Correspondence
  CG-1704-0007
Aviation Machinist's Mate, Class A
  CG-1704-0001

Magazine
20mm Mk 16 Mod 5 Machine Gun and Magazine Sprinklers Operation and Maintenance, Class C
  CG-2204-0001

Magnetic
Small Boat Magnetic Compass Calibration by Correspondence
  CG-1722-0006

Maintenance
ADL-81/82 Loran-C Receiver Maintenance, Class C
  CG-1715-0141
Aircraft Maintenance Officer
  AF-1704-0251
Aircraft Maintenance Officer (Accelerated/Air Reserve Forces)
  AF-1704-0252
  AF-1704-0253
Aircraft Maintenance Officer (Bridge)
  AF-1704-0286
Aircraft Maintenance Officer, Air Reserve
  AF-1704-0252
Aircraft Maintenance Specialist and Bombardment Aircraft by Correspondence
  AF-1704-0184
Aircraft Maintenance Specialist, Airlift and Bombardment Aircraft by Correspondence
  AF-1704-0181
Aircraft Maintenance/Munitions Officer
  AF-1704-0251
Airlift Aircraft Maintenance Specialist by Correspondence
  AF-1704-0184

APS-127 Radar System Maintenance, Class C
  CG-1715-0143
Automotive Intermediate Maintenance
  MC-1712-0008
Aviation Maintenance Administration by Correspondence
  CG-1408-0003
Fire Control System Mk 92 Mod 1 Operation and Maintenance
  CG-1715-0103
HH-60J Airframe and Power Train Maintenance, Class C
  CG-1704-0048
HH-60J Avionics Maintenance, Class C
  CG-1715-0145
HH-65A Avionics Maintenance, Class C
  CG-1715-0144
HH-65A Electrical/Automatic Flight Control Systems Maintenance, Class C
  CG-1704-0049
HU-25A Avionics Maintenance, Class C
  CG-1704-0045
Logistics Vehicle System Maintenance
  MC-1408-0019
LTS-101 Engine Maintenance, Class C
  CG-1710-0022
Maintenance Systems Analysis Specialist by Correspondence
  AF-1402-0069
Marine Corps Integrated Maintenance Management Systems
  MC-1402-0057
Strategic Aircraft Maintenance Specialist by Correspondence
  AF-1704-0181
  AF-1704-0188
T58-5 and T62 Engine Maintenance, Class C
  CG-1704-0026
T58-GE-8B Engine Maintenance, Class C
  CG-1704-0004
Telecommunications System Maintenance Specialist by Correspondence
  AF-1715-0031

Malay
Indonesian-Malay Basic
  DD-0602-0012
Malay Basic
  DD-0602-0012

Management
Advanced Management Program
  DD-1408-0008
Advanced Production and Quality Management
  DD-1408-0010
Advanced Program Management
  DD-1408-0018
Aerial Port Operation and Management
  AF-0419-0036
Aftercare Program Management
  MC-0801-0002
Aids to Navigation Operations Management
  CG-1408-0042
Air Force Logistics Command Directorate of Materiel Management by Correspondence
  AF-1408-0089
Airfield Management Specialist by Correspondence
  AF-1704-0215
Apprentice Airfield Management Specialist by Correspondence
  AF-1704-0214
Apprentice Operations Resource Management Specialist by Correspondence
  AF-1402-0073

KEYWORD INDEX D-39

Automated Information Systems Management for Intermediate Executives
DD-1402-0003
Automated Information Systems Management for Intermediate Managers
DD-1402-0003
Budget, Cost Estimating, and Financial Management Workshop
DD-1408-0017
Chapel Management Specialist by Correspondence
AF-1408-0085
AF-1408-0108
Chapel Management Technician by Correspondence
AF-1408-0086
Club Management Supervisor by Correspondence
AF-1406-0059
Communication-Electronics Systems Acquisition and Management by Correspondence
AF-1715-0753
Communications-Computer Systems Program Management Specialist by Correspondence
AF-1715-0787
Contract Performance Management Fundamentals
DD-1408-0014
Cost and Management Analysis Specialist by Correspondence
AF-1408-0083
Cost and Management Analysis Technician by Correspondence
AF-1408-0084
Defense Equal Opportunity Management Institute
DD-1512-0003
Defense Management Systems
DD-1408-0006
Defense Packaging Management Training
DD-0419-0005
Defense Resources Management
DD-1408-0006
Executive Acquisition Logistics Management
DD-0326-0004
Financial Management by Correspondence
MC-1401-0012
Financial Management Officer
MC-1408-0024
Financial Management Officer (Financial Analysis)
AF-1401-0002
Financial Management Officer (Financial Services)
AF-1408-0102
Financial Management Specialist (Budget) by Correspondence
AF-1408-0077
Financial Management Specialist (Commercial Services) by Correspondence
AF-1408-0074
Financial Management Specialist (Military Pay) by Correspondence
AF-1408-0074
Financial Management Specialist by Correspondence
AF-1408-0073
AF-1408-0074
AF-1408-0075
AF-1408-0076
Financial Management Staff Officer
AF-1408-0103
Financial Management Staff Officer, Air Reserve Forces
AF-1408-0105

Financial Management Supervisor by Correspondence
AF-1408-0081
Financial Management/Services Supervisor (Functions and Responsibilities) by Correspondence
AF-1408-0082
Financial Management/Services Supervisor by Correspondence
AF-1408-0081
Food Service Specialist Administration and Management
CG-1729-0010
Food Service Specialist, Large Mess Management
CG-1729-0008
Fuels Management Officer
AF-1601-0050
Fundamentals of Systems Acquisition Management
DD-1408-0012
Health Services Management Journeyman by Correspondence
AF-0799-0008
AF-0799-0009
AF-0799-0010
Human Resource Management (HRMC)
AF-1406-0075
Information Management Specialist by Correspondence
AF-1406-0045
Instructional Management
MC-1406-0031
Intermediate Contract Performance Management
DD-1408-0015
Intermediate Production and Quality Management Fundamentals
DD-1408-0024
Intermediate Software Acquisition Management
DD-1408-0013
Introduction to Acquisition Management by Correspondence
AF-1408-0106
Inventory Management Specialist (Munitions) by Correspondence
AF-1405-0064
Inventory Management Specialist by Correspondence
AF-1405-0051
Inventory Management Supervisor by Correspondence
AF-1405-0059
AF-1405-0071
Leadership and Management
CG-1406-0008
Management of Value Engineering by Correspondence
AF-1405-0068
Manpower Management Technician by Correspondence
AF-1406-0052
Marine Air-Ground Task Force Deployment Support System/Computer-Aided Embarkation Management System
MC-1402-0059
Marine Corps Integrated Maintenance Management Systems
MC-1402-0057
National Security Management (Correspondence/Nonresident/Seminar Course of the National Defense University)
DD-1511-0001
National Security Management by Correspondence
DD-1511-0008

Navy Management Systems
DD-1408-0006
Nursing Service Management
AF-0709-0036
Nursing Service Management by Correspondence
AF-0703-0014
Nursing Service Management for Air Reserve Components
AF-0709-0035
Officer Leadership and Management
CG-1717-0009
Open Mess Management Specialist by Correspondence
AF-1406-0058
Operating Room Management
AF-0703-0021
Operations Resource Management by Correspondence
AF-1408-0095
Personal Financial Management by Correspondence
MC-1401-0009
Personnel Systems Management Journeyman by Correspondence
AF-1402-0075
AF-1402-0076
AF-1402-0077
AF-1402-0078
Personnel Systems Management Specialist by Correspondence
AF-1406-0077
Pest Management Specialist by Correspondence
AF-0101-0005
Port Physical Security Management
CG-1728-0026
Production and Quality Management Fundamentals
DD-1408-0023
Professional Personnel Management
AF-1406-0033
Program Management
DD-1408-0002
DD-1408-0007
Quality Management by Correspondence
AF-1406-0081
Reserve Officer Leadership and Management
CG-1717-0008
Reserve Senior Petty Officer Leadership and Management
CG-1717-0007
Senior Petty Officer Leadership and Management
CG-1717-0010
Software Acquisition Management
DD-1402-0004
Subsistence Specialist Administration and Management
CG-1729-0010
Subsistence Specialist, Large Mess Management
CG-1729-0008
Supply Management by Correspondence
MC-1405-0027
Traffic Management Journeyman by Correspondence
AF-0419-0037

Manager
Armed Forces Radio Television System Broadcast Manager
DD-0505-0005
Education and Training Manager by Correspondence
AF-1406-0078

KEYWORD INDEX

Executive Program Manager
DD-1408-0019
Standard Workstation Regional System Manager
CG-1402-0003
System Manager
CG-1402-0003

Manager's
Program Manager's Survival
DD-1408-0021

Managers
Instructional System Development for Training Managers by Correspondence
AF-1406-0080

Maneuvering
Maneuvering Boards by Correspondence
CG-1708-0015

Manpower
Manpower Management Technician by Correspondence
AF-1406-0052
Professional Manpower (PMPMC)
AF-1406-0075

Manual
F-15 Avionics Manual/ECM Test Station and Component Specialist by Correspondence
AF-1715-0771
F/FB-111 Avionics Manual/Electronic Countermeasures (ECM) Test Station and Component Specialist by Correspondence
AF-1704-0224
Integrated Avionics Manual Test Station and Component Specialist (F-15) by Correspondence
AF-1715-0092
Integrated Avionics Manual Test Station and Component Specialist (F/FB-111) by Correspondence
AF-1715-0093

Manufacturing
Defense Acquisition Engineering, Manufacturing, and Quality Control
DD-1408-0010

Mapping
Mapping, Charting, and Geodesy Officer
DD-1601-0021
Remotely Sensed Imagery for Mapping, Charting, and Geodesy
DD-1601-0030

Marine
Marine Cadre Trainer
MC-1406-0035
Marine Combat
MC-2204-0105
Marine Corps War College
MC-1511-0002
Marine Environmental Protection (MEP) by Correspondence
CG-1303-0001
Marine Inspector
CG-1728-0039
Marine Safety Initial Indoctrination by Correspondence
CG-1708-0019
Marine Safety Initial Indoctrination Lesson Plan Series by Correspondence
CG-1708-0019
Marine Safety Initial Indoctrination Marine Inspection by Correspondence
CG-1708-0024
Marine Safety Initial Indoctrination Port Operations by Correspondence
CG-1708-0025
Marine Safety Inspection
CG-1728-0039
Marine Safety Officer ADT
CG-1728-0029
Marine Safety Petty Officer
CG-1728-0030
Marine Safety Port Operations
CG-1728-0041
Marine Science Technician First Class by Correspondence
CG-1304-0018
CG-1304-0022
Marine Science Technician Second Class by Correspondence
CG-1304-0016
Marine Science Technician, Class A
CG-1304-0017
Safety and Security of the Port: Marine Environmental Protection by Correspondence
CG-1728-0036

Maritime
Long Range (OTH) Maritime Navigation
MC-1708-0002
Maritime Law Enforcement Boarding Officer
CG-1728-0021
Maritime Law Enforcement Boarding Officer, Class C
CG-1728-0021
Maritime Law Enforcement Boarding Team Member, Class C
CG-1728-0045
Maritime Law Enforcement Instructor
CG-1728-0024
Maritime Reserve Coordination Center Controller
CG-1722-0011
Maritime Search and Rescue Planning, Class C
CG-1728-0043

Mason
Apprentice Mason by Correspondence
AF-1710-0030

Master
Master Crew Chief
AF-1704-0070
Position Location Reporting System (PLRS) Master Station Maintenance
MC-1715-0145
Position Location Reporting System (PLRS) Master Station Operator
MC-1715-0158

Materials
Materials Handling Equipment by Correspondence
MC-1710-0040

Materiel
Air Force Logistics Command Directorate of Materiel Management by Correspondence
AF-1408-0089
Apprentice Materiel Facilities Specialist by Correspondence
AF-1405-0018
Materiel Facilities Specialist by Correspondence
AF-1405-0013
Materiel Facilities Supervisor by Correspondence
AF-1405-0033
Materiel Storage and Distribution Specialist by Correspondence
AF-1405-0070
Medical Materiel Specialist by Correspondence
AF-1406-0094

Mathematics
Basic Mathematics by Correspondence
CG-1107-0001
Mathematics for Marines by Correspondence
MC-1107-0001

MD-9
Defensive Fire Control Systems Mechanic (B-52D/G: MD-9, ASG-15 Turrets) by Correspondence
AF-1715-0047

Measurement
Basic Techniques of Wave Form Measurement Using an Oscilloscope by Correspondence
AF-1715-0784
DCLF Reference Measurement and Calibration
AF-1715-0493
Microwave Measurement and Calibration
AF-1715-0740
Physical Measurement and Calibration
AF-1715-0743
Test, Measurement and Diagnostic Equipment
MC-1715-0139

Measurements
Power Measurements by Correspondence
AF-1115-0008
AF-1714-0034
Scientific Measurements Technician by Correspondence
AF-1715-0762

Measuring
Precision Measuring Equipment Specialist by Correspondence
AF-1715-0073
Precision Measuring Equipment Technician by Correspondence
AF-1715-0074

Meat
Meats and Meat Cookery by Correspondence
MC-1729-0044

Meats
Meats and Meat Cookery by Correspondence
MC-1729-0044

Mechanic
Aviation Structural Mechanic, Class A
CG-1704-0047
Basic Engineer Equipment Mechanic
MC-1710-0045
Basic Refrigeration Mechanic
MC-1730-0006
Basic Shop Fundamentals for the Mechanic by Correspondence
MC-1717-0005
M1A1 Tank System Mechanic
MC-1710-0047
Small Craft Mechanic
MC-1712-0009

Media
Audiovisual Media Specialist by Correspondence
AF-1406-0073
Precision Imagery and Audiovisual Media Maintenance Specialist by Correspondence
AF-1715-0136

KEYWORD INDEX D-41

Medical
Emergency Medical Technician Basic
　　CG-0709-0004
Emergency Medical Technician, Class C
　　CG-0709-0004
Field Medical Service Technic Enlisted
　　MC-0709-0002
Field Service Medical Service Technic
　　MC-0709-0002
Medical Administrative Specialist by Correspondence
　　AF-0709-0027
Medical Department Officers Orientation
　　MC-0707-0002
Medical Laboratory Craftsman (Chemistry and Urinalysis) by Correspondence
　　AF-0702-0008
Medical Laboratory Craftsman (Hematology, Serology, Blood Banking and Immunohematology) by Correspondence
　　AF-0702-0010
Medical Laboratory Craftsman (Microbiology) by Correspondence
　　AF-0702-0009
Medical Laboratory Technician (Chemistry and Urinalysis) by Correspondence
　　AF-0702-0008
Medical Laboratory Technician (Hematology, Serology, Blood Banking and Immunohematology) by Correspondence
　　AF-0702-0010
　　AF-0709-0031
Medical Laboratory Technician (Microbiology) by Correspondence
　　AF-0702-0009
　　AF-0709-0030
Medical Materiel Specialist by Correspondence
　　AF-1406-0094
Medical Personnel Augmentation System/ Mobile Medical Augmentation Readiness Team
　　MC-0801-0012
Medical Service Specialist by Correspondence
　　AF-0703-0012
　　AF-0703-0016
　　AF-0703-0018
Medical Service Specialist, Aeromedical by Correspondence
　　AF-0709-0033
Medical Service Technician by Correspondence
　　AF-0703-0011
Mobile Medical Augmentation Readiness Team
　　MC-0801-0012

Medicine
Cold Weather Medicine
　　MC-0707-0001
Environmental Medicine Specialist by Correspondence
　　AF-0707-0009

Mental
Mental Health Service Specialist by Correspondence
　　AF-0703-0017
Mental Health Unit Specialist by Correspondence
　　AF-0708-0002

Mess
Food Service Specialist, Large Mess Management
　　CG-1729-0008

Mess Hall Sanitation by Correspondence
　　MC-1729-0039
Mess Hall Subsistence Clerk by Correspondence
　　MC-1729-0040
Open Mess Management Specialist by Correspondence
　　AF-1406-0058
Subsistence Specialist, Large Mess Management
　　CG-1729-0008

Metal
Basic Metal Worker
　　MC-1723-0009
Metal Fabricating Specialist by Correspondence
　　AF-1723-0014
　　AF-1723-0015
Metal Fabrication Specialist by Correspondence
　　AF-1723-0013
Metal Working and Welding Operations by Correspondence
　　MC-1723-0008

Metals
Aircraft Metals Technology (Machinist) by Correspondence
　　AF-1717-0027
Aircraft Metals Technology (Welding) by Correspondence
　　AF-1723-0011
　　AF-1724-0007
Metals Processing Specialist by Correspondence
　　AF-1723-0007
　　AF-1723-0009

Metalworker
Journeyman Metalworker
　　MC-1723-0007
Metalworker
　　MC-1703-0035

Meteorological
Meteorological and Navigation Systems Journeyman by Correspondence
　　AF-1304-0017
　　AF-1715-0821
Meteorological and Navigation Systems Specialist by Correspondence
　　AF-1304-0017
　　AF-1715-0821
Meteorological Technician by Correspondence (Specialized)
　　AF-1304-0015

Microbiology
Medical Laboratory Craftsman (Microbiology) by Correspondence
　　AF-0702-0009
Medical Laboratory Technician (Microbiology) by Correspondence
　　AF-0702-0009
　　AF-0709-0030

Microcomputer
Microcomputer Repair
　　MC-1715-0120

Microminiature
Microminiature Component Repair
　　MC-1715-0089
　　MC-1715-0140

Microprocessor
Digital Microprocessor, Class C
　　CG-1715-0081

Microwave
Microwave Equipment Maintenance
　　MC-1715-0123
Microwave Equipment Operator
　　MC-1715-0024
　　MC-1715-0161
Microwave Measurement and Calibration
　　AF-1715-0740

Military
Basic Military Journalist
　　DD-0504-0001
Financial Services Specialist (Military Pay) by Correspondence
　　AF-1408-0079
Infantry Staff Noncommissioned Officer (NCO): Military Symbols and Overlays by Correspondence
　　MC-1601-0040
Intelligence Brief: The Military Threat by Correspondence
　　MC-1606-0012
Military Affiliate Radio System (MARS) Operator by Correspondence
　　MC-1715-0134
Military Civil Rights by Correspondence
　　CG-1512-0001
Military Justice by Correspondence
　　AF-1728-0043
Military Operations on Urban Terrain by Correspondence
　　MC-1606-0006
Military Requirements for Chief Petty Officer by Correspondence
　　CG-2205-0036
Military Requirements for E-3 by Correspondence
　　CG-2205-0028
Military Requirements for E-4 by Correspondence
　　CG-2205-0029
Military Requirements for E-5 by Correspondence
　　CG-2205-0030
Military Requirements for E-6 by Correspondence
　　CG-2205-0031
Military Requirements for Second Class Petty Officer by Correspondence
　　CG-2205-0038
Professional Military Comptroller (PMCC)
　　AF-1408-0099
Safety and Security of the Port: Military Explosives by Correspondence
　　CG-1728-0033

Minefield
Guantanamo Minefield Maintenance Preparation
　　MC-2204-0054
Minefield Maintenance
　　MC-2204-0054
Minefield Maintenance Man Refresher
　　MC-2204-0054

Minor
Advanced Minor Aids to Navigation Maintenance
　　CG-1715-0131

Missile
Air Launched Missile Systems Specialist by Correspondence
　　AF-1714-0041
　　AF-1715-0814
Hawk Missile System Operator (USMC)
　　MC-1715-0163

KEYWORD INDEX

Missile and Space Systems Electronic Maintenance Journeyman by Correspondence
AF-1714-0041
AF-1715-0814
Missile and Space Systems Maintenance Journeyman by Correspondence
AF-1704-0265
Missile Control Communications Systems Repairman by Correspondence
AF-1715-0130
Missile Control Communications Systems Specialist by Correspondence
AF-1714-0038
Missile Facilities Specialist by Correspondence
AF-1704-0264
AF-1717-0026
Missile Maintenance Specialist (WS-133) by Correspondence
AF-1710-0031
Missile Maintenance Specialist by Correspondence
AF-1704-0265
Missile Systems Maintenance Specialist by Correspondence
AF-1715-0038
AF-1715-0043

Mission
Civil Air Patrol Mission Observer Level II by Correspondence
AF-1704-0195

Mitel
CEJT-SX-100/200 Mitel Electronic Private Automatic Branch Exchange Generic 217, Class C
CG-1715-0113
CEJT-SX-200 Mitel Electronic Private Automatic Branch Exchange Generic 1000 or 1001 Telephone System
CG-1715-0116
Mitel SX-200D Installation and Maintenance, Class C
CG-1715-0116

Mk 16
20mm Mk 16 Mod 5 Machine Gun and Magazine Sprinklers Operation and Maintenance, Class C
CG-2204-0001

Mk 19
Inspection and Repair of the Mk 19 Machinegun by Correspondence
MC-2204-0099

Mk 22
3"/50 Gun Mount Mk 22 Operation and Maintenance, Class C
CG-2204-0002

Mk 27
Mk 27 Gyrocompass System
CG-1715-0104
Mk 27 Gyrocompass System Operation and Maintenance
CG-1715-0104

Mk 29
Mk 29 Mod 1 Gyrocompass Operation and Maintenance
CG-1715-0106

Mk 52
Gun Fire Control Systems Mk 52 and Mk 56
CG-1715-0016

Mk 56
Fire Control Mk 56 System, Class C
CG-1715-0016
Gun Fire Control Systems Mk 52 and Mk 56
CG-1715-0016

Mk 92
Fire Control System Mk 92 Mod 1 Operation and Maintenance
CG-1715-0103
Mk 92 Mod 1 Fire Control System
CG-1715-0103

Mobile
Medical Personnel Augmentation System/ Mobile Medical Augmentation Readiness Team
MC-0801-0012
Mobile Communication Central Technician
MC-1715-0006
Mobile Communications Central Technician (AN/TGC-37(V))
MC-1715-0006
Mobile Data Communications Terminal Technician
MC-1715-0118
Mobile Medical Augmentation Readiness Team
MC-0801-0012
Mobile Security Communication Technician
MC-1715-0169
Mobile Technical Control
MC-1402-0063

Modulation
Sinewave Oscillators-Modulation/Demodulation by Correspondence
AF-1715-0751

Module
Tactical Air Operations Module (TAOM), AN/TYQ-23(V)1 Technician
MC-1715-0175
Tactical Air Operations Module Repair AN/TYQ-23(V)1
MC-1715-0174

Monitor
Primary Chain Monitor Set Maintenance, Class C
CG-1715-0125

Monitoring
270' WMEC Machinery Plant Control and Monitoring System
CG-1710-0013
270' WMEC Machinery Plant Control and Monitoring System, Class C
CG-1710-0013
270' WMEC Main Propulsion Control and Monitoring System (Electrical) Operation and Maintenance, Class C
CG-1710-0013

Morale
Morale, Welfare, Recreation, and Services by Correspondence
AF-0804-0002
Morale, Welfare, Recreation, and Services Journeyman by Correspondence
AF-0804-0001

Mortarman
Basic Mortarman
MC-2204-0047
Mortarman
MC-2204-0047

Motor
Motor Transport Maintenance Officer
MC-1712-0005
Motor Transport Officer
MC-0419-0005
Motor Transport Operations Noncommissioned Officer (NCO)
MC-1408-0018
Motor Transport Staff Noncommissioned Officer (NCO)
MC-1703-0025
Motor Vehicle Operator
MC-1703-0027
Reserve Motor Transport Officer Refresher
MC-1703-0029

Motors
Relays, Generators, Motors, and Electromechanical Devices by Correspondence
AF-1714-0032
AF-1715-0768

Mountain
Mountain Survival
MC-0804-0005
Winter Mountain Leaders (A)
MC-0804-0003
Winter Mountain Leaders (B)
MC-0804-0004

Movement
Air Movement Planning
MC-0419-0007

Multichannel
Multichannel Equipment Operator
MC-1715-0161

Multiple
Multiple Virtual Storage (MVS) Diagnostics
MC-1402-0053
Multiple Virtual Storage (MVS) Fundamentals and Logic
MC-1402-0052
Multiple Virtual Storage (MVS) Performance and Training
MC-1402-0054

Munitions
Aircraft Maintenance/Munitions Officer
AF-1704-0251
Inventory Management Specialist (Munitions) by Correspondence
AF-1405-0064

National
Air National Guard Academy of Military Science
AF-2203-0051
Air National Guard Fighter Weapons Instructor
AF-1606-0151
National Boating Safety
CG-0802-0014
National Communications Security (COMSEC)
DD-1404-0004
National Security Management (Correspondence/Nonresident/Seminar Course of the National Defense University)
DD-1511-0001
National Security Management by Correspondence
DD-1511-0008
National War College
DD-1511-0002
DD-1511-0009
Reserve Components National Security Seminar
DD-1511-0006

Nautel
Nautel Radiobeacon Maintenance
CG-1715-0127

KEYWORD INDEX   **D-43**

Naval
Naval Academy Training
    MC-2202-0003
Naval Reserve Officer
    MC-2202-0002
Naval Reserve Officer Bulldog
    MC-2202-0002

Navigation
Advanced Minor Aids to Navigation Maintenance
    CG-1715-0131
Aerial Navigation
    MC-1606-0002
Aids to Navigation Officer Advanced, Class C
    CG-2205-0008
Aids to Navigation Operations Management
    CG-1408-0042
Aids to Navigation Positioning
    CG-2205-0025
Air Navigation
    MC-1606-0002
Aircraft Communication/Navigation Systems Journeyman by Correspondence
    AF-1715-0817
Automated Aids to Navigation Lighthouse Technician
    CG-1715-0128
Bomb Navigation Systems Specialist (B-52G/H: ASQ-176, ASQ-151 Systems) by Correspondence
    AF-1715-0046
Bomb Navigation Systems Specialist by Correspondence
    AF-1715-0774
Celestial Navigation by Correspondence
    CG-1304-0021
Communication/Navigation Systems Technician (Doppler Systems) by Correspondence
    AF-1704-0260
Communication/Navigation Systems Technician by Correspondence
    AF-1704-0242
    AF-1704-0243
    AF-1704-0244
    AF-1704-0245
    AF-1704-0246
Communications/Navigation Systems Technician by Correspondence
    AF-1715-0116
Deck Watch Officer Navigation Rules by Correspondence
    CG-1708-0014
Defensive Avionics/Communications/Navigation Systems by Correspondence
    AF-1715-0789
Infantry Squad Leader: Land Navigation by Correspondence
    MC-1601-0048
Integrated Avionic Communication, Navigation and Penetration Aids Systems Specialist by Correspondence
    AF-1714-0039
Integrated Avionic Communications, Navigation, and Penetration Aids Systems Specialist (F/FB-111) by Correspondence
    AF-1715-0779
Integrated Avionics Communication, Navigation, and Penetration Aids Systems Specialist (F-16) by Correspondence
    AF-1704-0220
Integrated Avionics Communication, Navigation, and Penetration Aids Systems Specialist (F/FB-111) by Correspondence
    AF-1715-0101
Integrated Avionics Communication, Navigation, and Penetration Aids Systems Specialist by Correspondence
    AF-1715-0100
Integrated Avionics Communications, Navigation, and Penetration Aids Systems Specialist (F-15) by Correspondence
    AF-1704-0225
    AF-1715-0106
Integrated Avionics Communications, Navigation, and Penetration Aids Systems Specialist (F/FB-111) by Correspondence
    AF-1704-0263
Land Navigation by Correspondence
    MC-1601-0048
Long Range (OTH) Maritime Navigation
    MC-1708-0002
Meteorological and Navigation Systems Journeyman by Correspondence
    AF-1304-0017
    AF-1715-0821
Meteorological and Navigation Systems Specialist by Correspondence
    AF-1304-0017
    AF-1715-0821
Navigation Rules by Correspondence
    CG-1708-0005
    CG-1708-0020
Navigation Systems by Correspondence
    CG-1708-0011
Officer In Charge, Aids to Navigation Team
    CG-1406-0009
Officers Advanced Aids to Navigation
    CG-2205-0008
Piloting Navigation by Correspondence
    CG-1708-0006

Navigational
Navigational Aids Equipment Specialist by Correspondence
    AF-1715-0012
Navigational Aids Equipment Technician by Correspondence
    AF-1715-0017

Navigator
Flight Test Engineer/Navigator
    AF-1606-0153
HC-130H Navigator Basic, Class C
    CG-1708-0023

NAVMACS
NAVMACS/SATCOM Systems Maintenance
    CG-1715-0138

NBC
Intelligence Brief: The Opposing Forces Nuclear, Biological, Chemical (NBC) Threat by Correspondence
    MC-1606-0011

NCO
Aircraft Maintenance Noncommissioned Officer (NCO) by Correspondence
    MC-1704-0008
Baker Noncommissioned Officer (NCO)
    MC-1729-0006
Bakery Noncommissioned Officer (NCO) Leadership
    MC-1729-0006
Combat Engineer Noncommissioned Officer (NCO)
    MC-1601-0028
Combat Engineer Noncommissioned Officer (NCO) by Correspondence
    MC-1601-0036
Communication Noncommissioned Officer (NCO)
    MC-1405-0048
Electrician Noncommissioned Officer (NCO)
    MC-1714-0016
Engineer Equipment Mechanic Noncommissioned Officer (NCO)
    MC-1710-0033
Engineer Equipment Operator Noncommissioned Officer (NCO)
    MC-1710-0034
Food Service Noncommissioned Officer (NCO)
    MC-1729-0034
Food Service Staff Noncommissioned Officer (NCO)
    MC-1729-0036
Hygiene Equipment Operator Noncommissioned Officer (NCO)
    MC-1601-0044
Infantry Staff Noncommissioned Officer (NCO): Military Symbols and Overlays by Correspondence
    MC-1601-0040
Motor Transport Operations Noncommissioned Officer (NCO)
    MC-1408-0018
Motor Transport Staff Noncommissioned Officer (NCO)
    MC-1703-0025
Noncommissioned Officer (NCO) Basic
    MC-2204-0075
    MC-2204-0103
Noncommissioned Officer (NCO) by Correspondence
    MC-1406-0023
Principles of Instruction for the Marine Noncommissioned Officer (NCO) by Correspondence
    MC-1406-0028
Reserve Combat Engineer Noncommissioned Officer (NCO)
    MC-1601-0030
Reserve Staff Noncommissioned Officer (NCO) Advanced
    MC-2204-0104
Resident Staff Noncommissioned Officer (NCO) Advanced Regular
    MC-2204-0076
Resident Staff Noncommissioned Officer (NCO) Advanced Reserve
    MC-2204-0072
Resident Staff Noncommissioned Officer (NCO) Career Regular
    MC-2204-0074
Resident Staff Noncommissioned Officer (NCO) Career Reserve
    MC-2204-0071
Staff Noncommissioned Officer (NCO) Advanced
    MC-2204-0076
The Noncommissioned Officer (NCO) by Correspondence
    MC-1406-0023

Nondestructive
Apprentice Nondestructive Inspection Specialist by Correspondence
    AF-1724-0006
Nondestructive Inspection Specialist by Correspondence
    AF-1723-0005

Nonresident
Air Command and Staff Nonresident Seminar Associate Program
    AF-1511-0009

KEYWORD INDEX

Air Command and Staff Nonresident Seminar Program
AF-1511-0009
Air War College Associate Programs Nonresident and Correspondence
AF-1511-0012
Air War College Nonresident Seminar Program
AF-1511-0012
Amphibious Warfare School Nonresident Program (AWSNP) by Correspondence Phase 2
MC-2204-0114
Amphibious Warfare School Nonresident Program (AWSNP) by Correspondence, Phase 1
MC-2204-0113
Command and Staff Nonresident Program by Correspondence
MC-1408-0025
National Security Management (Correspondence/Nonresident/Seminar Course of the National Defense University)
DD-1511-0001
Sergeants Nonresident Program by Correspondence
MC-1408-0027
Squadron Officers School Nonresident
AF-2203-0056
Staff Noncommissioned Officers Advanced Nonresident Program (SNCOANP) by Correspondence
MC-1408-0028

Norwegian
Norwegian Basic
DD-0602-0224

Nuclear
Basic Nuclear, Biological, and Chemical (NBC) Defense
MC-0801-0014
Intelligence Brief: The Opposing Forces Nuclear, Biological, Chemical (NBC) Threat by Correspondence
MC-1606-0011
Nuclear Warfare Defense by Correspondence
MC-0705-0001

Nurse
Nurse Anesthetist
AF-0703-0013
OB/GYN Nurse Practitioner
AF-0703-0022

Nursing
Accelerated Basic Obstetrical Nursing
AF-0703-0023
Nursing Executive Development
AF-0709-0037
Nursing Service Management
AF-0709-0036
Nursing Service Management by Correspondence
AF-0703-0014
Nursing Service Management for Air Reserve Components
AF-0709-0035
Nursing Staff Development Officer Basic
AF-0703-0020
Operating Room Nursing
AF-0703-0019

Nutrition
Basic Nutrition by Correspondence
MC-1729-0041

OB/GYN
OB/GYN Nurse Practitioner
AF-0703-0022

Observer
Civil Air Patrol Mission Observer Level II by Correspondence
AF-1704-0195
Ice Observer by Correspondence
CG-1304-0006

Obstetrical
Accelerated Basic Obstetrical Nursing
AF-0703-0023

Offensive
Offensive Avionics Systems Specialist (B-1B) by Correspondence
AF-1704-0248

Office
Telephone Central Office Switching Equipment Specialist, Electronic/Electromechanical by Correspondence
AF-1715-0127

Officer
Basic Officer
MC-2204-0083
Coast Guard Division Officer by Correspondence
CG-2205-0033
Naval Reserve Officer
MC-2202-0002
Naval Reserve Officer Bulldog
MC-2202-0002
Officer Candidate
CG-2202-0005
MC-2204-0080
Woman Officer Basic
MC-2204-0035

Offset
Basic Offset Printing
DD-1719-0006

Open
Open Mess Management Specialist by Correspondence
AF-1406-0058

Operating
IBM System 360 Operating System (OS) Operations
MC-1402-0023
Operating Room Management
AF-0703-0021
Operating Room Nursing
AF-0703-0019

Operational
Basic Operational and Transition/Requalification Training A-10
AF-1704-0285
Basic Operational Training
AF-1704-0277
Basic Operational Training A/OA-10
AF-1704-0281
Basic Operational Training F-15
AF-1606-0158
Basic Operational Training F-15E
AF-1704-0269
Basic Operational Training F-16
AF-1704-0280
Basic Operational Training F-16 A/B
AF-1704-0280
Basic Operational Training F-16C/D
AF-1704-0270
Basic Operational Training F16 C/D Block 40/42
AF-1704-0279

Reserve Operational Communication Chief Refresher
MC-1402-0067

Operations
Air Support Operations Operator
MC-1715-0031
MC-1715-0165
Ground Operations Specialist
MC-1601-0047
Motor Transport Operations Noncommissioned Officer (NCO)
MC-1408-0018
Operations Resource Management by Correspondence
AF-1408-0095
Port Operations Department
CG-1728-0041
Space Systems Operations Specialist by Correspondence
AF-1715-0755
Tactical Air Operations Module (TAOM), AN/TYQ-23(V)1 Technician
MC-1715-0175
Tactical Air Operations Module Repair AN/TYQ-23(V)1
MC-1715-0174

Operator
Basic Equipment Operator
MC-1710-0044
Engineer Equipment Operator by Correspondence
MC-1601-0034
Motor Vehicle Operator
MC-1703-0027
Reserve Field Radio Operator Refresher
MC-1402-0062

Opportunity
Defense Equal Opportunity Management Institute
DD-1512-0003
Social Actions Technician (Equal Opportunity/Human Relations) by Correspondence
AF-1406-0053

Optical
Electro-Optical Ordnance Repair
MC-1714-0019

Optometry
Optometry Specialist by Correspondence
AF-0706-0003

Ordnance
Electro-Optical Ordnance Repair
MC-1714-0019
Ground Ordnance Officer
MC-1717-0007
Ground Ordnance Vehicle Maintenance Chiefs
MC-1717-0006
Ground Ordnance Weapons Chief
MC-1717-0008

Organic
SASSY: Organic Procedures by Correspondence
MC-1405-0044

Organizational
Automotive Organizational Maintenance
MC-1703-0028
F-15 Integrated Organizational Avionics Systems Specialist by Correspondence
AF-1704-0254
F-16 Integrated Organizational Avionics Systems Specialist by Correspondence
AF-1715-0808

KEYWORD INDEX D-45

Integrated Organizational Avionic System Specialist by Correspondence
AF-1715-0758
Integrated Organizational Avionic Systems Specialist by Correspondence
AF-1405-0069

Orientation
Coast Guard Orientation by Correspondence
CG-2205-0013
Coast Guard Orientation Officers by Correspondence
CG-2205-0019
Medical Department Officers Orientation
MC-0707-0002

OS
IBM S/360 OS Computer Operator
MC-1402-0023
IBM System 360 (OS) COBOL Programming
MC-1402-0013
IBM System 360 Operating System (OS) Operations
MC-1402-0023
IBM System 360 OS FORTRAN Programming
MC-1402-0034
IBM System/360 (OS) COBOL Programming Entry-level
MC-1402-0013

Oscillators
Sinewave Oscillators-Modulation/Demodulation by Correspondence
AF-1715-0751

Oscilloscope
Basic Techniques of Wave Form Measurement Using an Oscilloscope by Correspondence
AF-1715-0784

Osmosis
Reverse Osmosis Water Purification Unit by Correspondence
MC-1732-0003

Packaging
Defense Packaging Design
DD-0419-0006
Defense Packaging Management Training
DD-0419-0005
Defense Packaging of Hazardous Materials for Transportation
DD-0419-0003

Packing
Defense Basic Preservation and Packing
DD-0419-0004
Defense Packing and Unitization
DD-0419-0002

Paperwork
Small Unit Paperwork by Correspondence
CG-1403-0001

Parachute
Fabrication and Parachute Specialist by Correspondence
AF-1733-0002

Paralegal
Paralegal Journeyman by Correspondence
AF-1407-0003

Passenger
Air Passenger Specialist by Correspondence
AF-0419-0021
Passenger and Household Goods Specialist by Correspondence
AF-0419-0031

Pastry
Pastry Baking by Correspondence
MC-1729-0032

Patrol
Civil Air Patrol Mission Observer Level II by Correspondence
AF-1704-0195
Civil Air Patrol Public Affairs Officer by Correspondence
AF-0505-0004
Civil Air Patrol Scanner by Correspondence
AF-1704-0194
Civil Air Patrol-Safety Officer (Level II Technical Rating) by Correspondence
AF-0801-0005
Introduction to Civil Air Patrol Emergency Services by Correspondence
AF-1704-0196

Pavement
Pavement and Construction Equipment Journeyman by Correspondence
AF-1601-0051
AF-1601-0052
Pavement and Construction Equipment Specialist by Correspondence
AF-1601-0051
AF-1601-0052

Pavements
Apprentice Pavements Maintenance Specialist by Correspondence
AF-1601-0048
Pavements Maintenance Specialist by Correspondence
AF-1601-0047

Pay
Basic Pay Entitlement by Correspondence
MC-1401-0008
Financial Management Specialist (Military Pay) by Correspondence
AF-1408-0074
Financial Services Specialist (Military Pay) by Correspondence
AF-1408-0079

PDP-11
PDP-11 Instruction Set
DD-1715-0011

Penetration
Enemy Defense Penetration Aids
AF-1715-0744
Integrated Avionic Communication, Navigation and Penetration Aids Systems Specialist by Correspondence
AF-1714-0039
Integrated Avionic Communications, Navigation, and Penetration Aids Systems Specialist (F/FB-111) by Correspondence
AF-1715-0779
Integrated Avionics Communication, Navigation, and Penetration Aids Systems Specialist (F/FB-111) by Correspondence
AF-1715-0101
Integrated Avionics Communication, Navigation, and Penetration Aids Systems Specialist by Correspondence
AF-1715-0100
Integrated Avionics Communications, Navigation, and Penetration Aids Systems Specialist (F-15) by Correspondence
AF-1704-0225
AF-1715-0106
Integrated Avionics Communications, Navigation, and Penetration Aids Systems Specialist (F/FB-111) by Correspondence
AF-1704-0263

Persian
Persian Basic
DD-0602-0230

Personal
Personal Affairs Specialist by Correspondence
AF-1406-0049
Personal Finance by Correspondence
MC-1401-0009
Personal Financial Management by Correspondence
MC-1401-0009
Personal Financial Records Clerk
MC-1403-0010

Personnel
Advanced Personnel Administration
MC-1405-0032
Basic Typing and Personnel Administration
MC-1406-0021
General Personnel Administration for Reserve by Correspondence
MC-1406-0025
Individual Personnel Records by Correspondence
MC-1403-0016
Medical Personnel Augmentation System/Mobile Medical Augmentation Readiness Team
MC-0801-0012
Personnel Administration for the Reporting Unit by Correspondence
MC-1406-0026
Personnel Clerk
MC-1405-0034
Personnel Officer
MC-1405-0033
Personnel Reporting for REMMPS by Correspondence
MC-1406-0027
Personnel Specialist by Correspondence
AF-1406-0076
Personnel Systems Management Journeyman by Correspondence
AF-1402-0075
AF-1402-0076
AF-1402-0077
AF-1402-0078
Personnel Systems Management Specialist by Correspondence
AF-1406-0077
Professional Personnel Management
AF-1406-0033
Systems Acquisition for Contracting Personnel (Executive)
DD-1408-0009

Pest
Pest Management Specialist by Correspondence
AF-0101-0005

Petty
Chief Petty Officer Academy
CG-1511-0002
Engineering Petty Officer Indoctrination
CG-1408-0035
Marine Safety Petty Officer
CG-1728-0030
Military Requirements for Chief Petty Officer by Correspondence
CG-2205-0036
Officer In Charge and Executive Petty Officer
CG-1715-0020
Officer In Charge/Executive Petty Officer
CG-2205-0020

KEYWORD INDEX

Reserve Senior Petty Officer Leadership and Management
CG-1717-0007
Senior Petty Officer Leadership and Management
CG-1717-0010
Senior Petty Officers by Correspondence
CG-2205-0034

Pharmacy
Pharmacy Specialist by Correspondence
AF-0799-0003

Photo-Sensors
Photo-Sensors Maintenance Specialist (Tactical/Reconnaissance Electronic Sensors) by Correspondence
AF-1715-0777
Photo-Sensors Maintenance Specialist by Correspondence
AF-1715-0764

Photogrammetric
Analytical Photogrammetric Positioning System
DD-1601-0026

Photographic
Aerospace Photographic Systems Specialist by Correspondence
AF-1715-0139
Apprentice Still Photographic Specialist by Correspondence
AF-1709-0029
Still Photographic Specialist by Correspondence
AF-1709-0028

Photojournalism
Intermediate Photojournalism
DD-1709-0002

Photolithographic
Basic Photolithographic Processes
DD-1601-0019

Physical
Physical Therapy Specialist by Correspondence
AF-0704-0004
Port Physical Security Management
CG-1728-0026
Port Physical Security Practical
CG-1728-0025

Physiology
Aerospace Physiology Specialist by Correspondence
AF-0709-0032

Pilipino
Pilipino/Tagalog Basic
DD-0602-0237

Pilot
C-5 Pilot
AF-1606-0023
ENJPT Pilot Instructor Training Shepherd (T-37/T-38)
AF-1406-0101
Experimental Test Pilot
AF-1606-0150
Instructor Pilot Training (F-15)
AF-1606-0157
Instructor Pilot Upgrade
AF-1704-0284
Instructor Pilot Upgrade Training F-16 C
AF-1406-0102
Instructor Pilot Upgrade Training F-16 C/D
AF-1406-0102
Instructor Pilot Upgrade Training F-16C/D
AF-1704-0271
Pilot Instructor Training
AF-1406-0074
Pilot Instructor Training (T-37)
AF-1406-0074
Pilot Instructor Training (T-38)
AF-1406-0074
Special Operations Training, AC-130E Pilot
AF-1606-0020
T-1A Pilot Instructor Training
AF-1406-0099
Undergraduate Pilot Training
AF-1606-0154

Piloting
Piloting Navigation by Correspondence
CG-1708-0006

Planner
Coastal Defense Exercise Planner, Port Level
CG-1722-0013
Coastal Defense Planner, Port Level
CG-1722-0012

Planning
Advanced Systems Planning, Research, Development, and Engineering
DD-1408-0016
Air Movement Planning
MC-0419-0007
Aviation Combat Element Staff Planning
MC-2204-0094
Coastal Search Planning, Class C
CG-1722-0010
Comprehensive Search and Rescue Planning, Class C
CG-1722-0011
Computer Aided Search Planning by Correspondence
CG-0802-0017
Intermediate Systems Planning, Research Development, and Engineering
DD-1402-0011
Landing Force Combat Services Support Staff Planning
MC-2204-0093
Landing Force Ground Combat Communication Staff Planning
MC-1601-0046
Landing Force Staff Planning (Marine Expeditionary Brigade)
MC-2204-0092
Seabee Tactical Shipboard Planning
MC-0419-0008

Plant
270' WMEC Machinery Plant Control and Monitoring System
CG-1710-0013
270' WMEC Machinery Plant Control and Monitoring System, Class C
CG-1710-0013
Accounting for Plant Property by Correspondence
MC-1405-0025

Platoon
Platoon Leaders Combined
MC-2204-0079
Platoon Leaders Junior
MC-2204-0077
Platoon Leaders Senior
MC-2204-0078

PLRS
Position Locating Reporting System (PLRS) Support Maintenance
MC-1715-0149
Position Location Reporting System (PLRS) Master Station Maintenance
MC-1715-0145
Position Location Reporting System (PLRS) Master Station Operator
MC-1715-0158
Position Location Reporting System (PLRS) Support Maintenance
MC-1715-0149

Plumber
Apprentice Plumber by Correspondence
AF-1710-0027

Plumbing
Plumbing and Sewage Disposal by Correspondence
MC-1710-0005
Plumbing Specialist by Correspondence
AF-1710-0021
Plumbing Technician by Correspondence
AF-1710-0023

Pneudraulic
Aircraft Pneudraulic Systems Mechanic by Correspondence
AF-1704-0201
Aircraft Pneudraulic Systems Technician by Correspondence
AF-1704-0230
AF-1704-0231

Point
Point Positioning Systems
DD-1601-0026

Polish
Polish Basic
DD-0602-0112
Polish Intermediate
DD-0602-0122

Polygraph
Polygraph Examiner Training
DD-1728-0003

Port
Aerial Port Operation and Management
AF-0419-0036
Coastal Defense Exercise Planner, Port Level
CG-1722-0013
Coastal Defense Planner, Port Level
CG-1722-0012
Marine Safety Initial Indoctrination Port Operations by Correspondence
CG-1708-0025
Marine Safety Port Operations
CG-1728-0041
Port Operations Department
CG-1728-0041
Port Physical Security Management
CG 1728 0026
Port Physical Security Practical
CG-1728-0025
Port Security and Safety Officer
CG-1728-0029
Port Security Safety Enlisted
CG-1728-0027
Port Security, Class A
CG-1728-0040
Port Securityman Direct Entry
CG-1728-0044
Port Securityman First Class by Correspondence
CG-1728-0038
Port Securityman Second Class by Correspondence
CG-1728-0032

KEYWORD INDEX D-47

Port Securityman Third Class by Correspondence
    CG-1728-0009
Port Securityman, Class A
    CG-1728-0040
Safety and Security of the Port by Correspondence
    CG-0802-0011
Safety and Security of the Port: Marine Environmental Protection by Correspondence
    CG-1728-0036
Safety and Security of the Port: Military Explosives by Correspondence
    CG-1728-0033

Portuguese
Portuguese Basic
    DD-0602-0212

Position
Position Locating Reporting System (PLRS) Support Maintenance
    MC-1715-0149
Position Location Reporting System (PLRS) Capabilities and Employment by Correspondence
    MC-1606-0008
Position Location Reporting System (PLRS) Master Station Maintenance
    MC-1715-0145
Position Location Reporting System (PLRS) Master Station Operator
    MC-1715-0158
Position Location Reporting System (PLRS) Support Maintenance
    MC-1715-0149

Positioning
Aids to Navigation Positioning
    CG-2205-0025
Analytical Photogrammetric Positioning System
    DD-1601-0026
Point Positioning Systems
    DD-1601-0026

Power
AN/URC-114(V) Low Power (1KW) Communications System, Class C
    CG-1715-0124
Automotive Power Trains by Correspondence
    MC-1703-0022
Electrical Power Production Journeyman by Correspondence
    AF-1714-0040
    AF-1714-0043
Electrical Power Production Specialist
    AF-1714-0040
Electrical Power Production Specialist by Correspondence
    AF-1714-0030
    AF-1714-0043
Electrical Power Production Technician by Correspondence
    AF-1408-0096
    AF-1714-0024
    AF-1714-0031
HH-60J Airframe and Power Train Maintenance, Class C
    CG-1704-0048
HH-65A Airframe/Power Train Maintenance, Class C
    CG-1704-0050
Lister Diesel Overhaul and Engine Power System Equipment Maintenance
    CG-1712-0007
Lister Diesel Overhaul and Power System Equipment Maintenance
    CG-1712-0007
Power Measurements by Correspondence
    AF-1115-0008
    AF-1714-0034
Power Supplies by Correspondence
    AF-1715-0752

Practitioner
OB/GYN Nurse Practitioner
    AF-0703-0022

Precision
Precision Imagery and Audiovisual Media Maintenance Specialist by Correspondence
    AF-1715-0136
Precision Measuring Equipment Specialist by Correspondence
    AF-1715-0073
Precision Measuring Equipment Technician by Correspondence
    AF-1715-0074

Preparedness
Disaster Preparedness Specialist by Correspondence
    AF-0802-0026

Preservation
Defense Basic Preservation and Packing
    DD-0419-0004
Defense Preservation and Intermediate Protection
    DD-0419-0001

Prevention
Crime Prevention by Correspondence
    AF-1728-0042

Preventive
Preventive Maintenance and Operating Techniques for Heavy Vehicles by Correspondence
    MC-1703-0034

Pricing
Contract Pricing
    DD-1405-0002
    DD-1405-0004
Facilities Contracts Pricing
    DD-1402-0008
Intermediate Contract Pricing
    DD-1405-0006
Operational Level Contract Pricing
    DD-1405-0001
Principles of Contract Pricing by Correspondence
    AF-1408-0091

Primary
Primary Chain Monitor Set Maintenance, Class C
    CG-1715-0125

Principles
Principles of Contract Pricing by Correspondence
    AF-1408-0091

Print
Construction Print Reading by Correspondence
    MC-1601-0041

Printing
Basic Offset Printing
    DD-1719-0006

Processing
Metals Processing Specialist by Correspondence
    AF-1723-0007
    AF-1723-0009

Production
Advanced Production and Quality Management
    DD-1408-0010
Electrical Power Production Journeyman by Correspondence
    AF-1714-0040
    AF-1714-0043
Electrical Power Production Specialist
    AF-1714-0040
Electrical Power Production Specialist by Correspondence
    AF-1714-0043
Electrical Power Production Technician by Correspondence
    AF-1408-0096
    AF-1714-0024
Imagery Production Specialist by Correspondence
    AF-1709-0030
Intermediate Production and Quality Management Fundamentals
    DD-1408-0024
Production and Quality Management Fundamentals
    DD-1408-0023
Production Control Specialist by Correspondence
    AF-1405-0060
Visual Information Production-Documentation Journeyman by Correspondence
    AF-0505-0005
Visual Information Production-Documentation Specialist by Correspondence
    AF-0505-0003
    AF-0505-0005

Professional
Professional Manpower (PMPMC)
    AF-1406-0075
Professional Military Comptroller
    AF-1408-0004
Professional Military Comptroller (PMCC)
    AF-1408-0099
Professional Personnel Management
    AF-1406-0033

Program
Advanced Management Program
    DD-1408-0008
Advanced Program Management
    DD-1408-0018
Command and Staff Nonresident Program by Correspondence
    MC-1408-0025
Executive Program Manager
    DD-1408-0019
Program Management
    DD-1408-0002
    DD-1408-0007
Program Manager's Survival
    DD-1408-0021
Warfighting Skills Program (Warfighting) by Correspondence
    MC-2204-0112

Programmer
Automated Data Processing Equipment Fleet Marine Force Programmer
    MC-1402-0050

KEYWORD INDEX

Programming
Assembler Language Code (ALC) Programming (Combination)
MC-1402-0051
COBOL Programming
MC-1402-0013
IBM System 360 (OS) COBOL Programming
MC-1402-0013
IBM System 360 OS FORTRAN Programming
MC-1402-0034
IBM System/360 (OS) COBOL Programming Entry-level
MC-1402-0013
Information Systems Programming Specialist by Correspondence
AF-1402-0071

Property
Accounting for Plant Property by Correspondence
MC-1405-0025

Propulsion
270' WMEC Main Propulsion Control and Monitoring System (Electrical) Operation and Maintenance, Class C
CG-1710-0013
Aerospace Propulsion Specialist (Jet Engine) by Correspondence
AF-1704-0256
Aerospace Propulsion Specialist (Turboprop) by Correspondence
AF-1704-0257
Aerospace Propulsion Technician (Jet Engine) by Correspondence
AF-1704-0229
Aerospace Propulsion Technician (Turboprop) by Correspondence
AF-1710-0037
Turboprop Propulsion Mechanic by Correspondence
AF-1704-0204

Protection
Defense Preservation and Intermediate Protection
DD-0419-0001
Fire Protection Specialist by Correspondence
AF-1728-0038
Fire Protection Supervisor by Correspondence
AF-1728-0046
Marine Environmental Protection (MEP) by Correspondence
CG-1303-0001
Waterfront Protection by Correspondence
CG-1728-0012

Provisioning
Introduction to Air Force Initial Provisioning by Correspondence
AF-1408-0090

Public
Advanced Public Affairs Supervisor
DD-1709-0003
Civil Air Patrol Public Affairs Officer by Correspondence
AF-0505-0004
Public Affairs Officer
DD-0504-0009
DD-0504-0018
Public Affairs Officer by Correspondence
AF-0401-0002

Public Affairs Officer Course Reserve Component
DD-0504-0013
Public Affairs Officer Reserve Component, Phase 2
DD-0504-0013
Public Affairs Specialist by Correspondence
AF-0401-0001
Public Affairs Specialist First Class by Correspondence
CG-1709-0002
Public Affairs Specialist Second Class by Correspondence
CG-1709-0001
Public Affairs Supervisor
DD-0504-0014

Publications
Marine Corps Technical Publications System by Correspondence
MC-1408-0021

Pulse
TEL-14 CDXC-SG-1/1A Pulse 120 Electronic Private Automatic Branch Exchange Telephone System, Class C
CG-1715-0115

Punctuation
Punctuation by Correspondence
MC-0501-0002

Purification
Reverse Osmosis Water Purification Unit by Correspondence
MC-1732-0003

Qualification
Basic Qualification Training F-15E
AF-1704-0269
F-15 Basic Qualification Training
AF-1606-0158
Instructor Qualification Training F-15E
AF-1406-0097
AF-1406-0098

Quality
Advanced Production and Quality Management
DD-1408-0010
Aviation Quality Assurance Supervisor by Correspondence
MC-1405-0047
Defense Acquisition Engineering, Manufacturing, and Quality Control
DD-1408-0010
Intermediate Production and Quality Management Fundamentals
DD-1408-0024
Introduction to the Quality Function by Correspondence
AF-1405-0067
Production and Quality Management Fundamentals
DD-1408-0023
Quality Management by Correspondence
AF-1406-0081

Quartermaster
Quartermaster First Class by Correspondence
CG-2205-0026
Quartermaster Second Class by Correspondence
CG-1722-0015
Quartermaster Third Class by Correspondence
CG-1722-0014
Quartermaster Third Class, by Correspondence
CG-1722-0014

Quartermaster, Class A
CG-1708-0018

Radar
AB/SPS-64(V) Small Cutter Radar Maintenance
CG-1715-0140
Air Traffic Control Radar Repairman by Correspondence
AF-1715-0003
Air Traffic Control Radar Specialist by Correspondence
AF-1715-0003
Air Traffic Control Radar Technician by Correspondence
AF-1715-0009
Airborne Warning and Control Radar Specialist by Correspondence
AF-1715-0123
AF-1715-0801
Aircraft Control and Warning Radar Specialist by Correspondence
AF-1715-0005
AN/SPS-29D Radar Systems, Class C
CG-1715-0049
AN/SPS-64(V) Large Cutter Radar Systems Maintenance
CG-1715-0139
AN/SPS-64(V)4 Radar
CG-1715-0070
CG-1715-0120
APS-127 Radar System Maintenance, Class C
CG-1715-0143
Automatic Tracking Radar Journeyman by Correspondence
AF-1715-0812
Automatic Tracking Radar Specialist by Correspondence
AF-1715-0007
AF-1715-0776
AF-1715-0794
AF-1715-0795
AF-1715-0812
Automatic Tracking Radar Technician by Correspondence
AF-1715-0011
Aviation Electronics Technician AN/APS-127 Radar, Class C
CG-1715-0080
Aviation Radar Repair (A)
MC-1715-0153
Aviation Radar Repair (B)
MC-1715-0148
Aviation Radar Repair (C)
MC-1715-0059
Aviation Radar Technician
MC-1715-0122
Aviation Radar Technician (A)
MC-1715-0156
Aviation Radar Technician (B)
MC-1715-0159
Aviation Radar Technician (C)
MC-1715-0152
Basic Radar Use and Operation by Correspondence
CG-1715-0076
Ground Radar Repair
MC-1715-0062
MC-1715-0173
Ground Radar Technician
MC-1715-0117
Radar Fundamentals
MC-1715-0063
MC-1715-0154
Radar Operator by Correspondence
CG-1715-0074

KEYWORD INDEX D-49

Radarman
   Radarman First Class by Correspondence
      CG-1408-0036
   Radarman Second Class by Correspondence
      CG-2205-0023
   Radarman Third Class by Correspondence
      CG-2205-0017
   Radarman, Class A
      CG-1715-0130

Radars
   AN/SPS-64(V)1,2,3 Radars
      CG-1715-0119
   AN/SPS-66 and AN/SPS-66A Radars, Class C
      CG-1715-0122

Radio
   AN/URC-9 Radio Set, Class C
      CG-1715-0121
   Antenna Construction and Propagation of Radio Waves by Correspondence
      MC-1715-0136
   Armed Forces Radio Television System Broadcast Manager
      DD-0505-0005
   Aviation Radio Repair
      MC-1715-0011
      MC-1715-0155
   Aviation Radio Technician
      MC-1715-0127
   Communications Systems Radio Operator by Correspondence
      AF-1715-0781
   Field Radio Operator
      MC-1403-0018
   Ground Radio Communications Specialist (Unit) by Correspondence
      AF-1715-0782
   Ground Radio Communications Specialist by Correspondence
      AF-1715-0227
   Ground Radio Communications Technician by Correspondence
      AF-1715-0465
   Ground Radio Repair
      MC-1715-0009
      MC-1715-0166
   Military Affiliate Radio System (MARS) Operator by Correspondence
      MC-1715-0134
   Radio and Television Broadcasting Specialist by Correspondence
      AF-0505-0002
   Radio Fundamentals
      MC-1715-0010
      MC-1715-0160
   Radio Technician
      MC-1715-0124
   Reserve Field Radio Operator Refresher
      MC-1402-0062
   Reserve Radio Chief Refresher
      MC-1402-0064

Radiobeacon
   Nautel Radiobeacon Maintenance
      CG-1715-0127

Radiology
   Radiology Technician by Correspondence
      AF-0705-0003

Radioman
   Radioman First Class by Correspondence
      CG-1715-0136
   Radioman Second Class by Correspondence
      CG-1404-0002
   Radioman, Class A
      CG-1404-0005

Radiotelegraph
   Radiotelegraph and Visual Communication Procedures by Correspondence
      MC-1715-0133

Radiotelephone
   Radiotelephone and Alternate Communication Procedures by Correspondence
      MC-1715-0137

Range
   Long Range (OTH) Maritime Navigation
      MC-1708-0002

Readiness
   Medical Personnel Augmentation System/ Mobile Medical Augmentation Readiness Team
      MC-0801-0012
   Mobile Medical Augmentation Readiness Team
      MC-0801-0012

Reading
   Construction Print Reading by Correspondence
      MC-1601-0041

Receiver
   ADL-81/82 Loran-C Receiver Maintenance, Class C
      CG-1715-0141

Reconciliation
   Validation and Reconciliation Clerk by Correspondence
      MC-1403-0017

Reconnaissance
   Avionic Sensor Systems Specialist (Reconnaissance Electronic Sensors) by Correspondence
      AF-1715-0068
   Basic Reconnaissance
      MC-1601-0045
   Photo-Sensors Maintenance Specialist (Tactical/Reconnaissance Electronic Sensors) by Correspondence
      AF-1715-0777

Records
   Aircraft Logs and Records
      CG-1405-0005
   Engineer Forms and Records by Correspondence
      MC-1601-0015
      MC-1601-0037
   Individual Personnel Records by Correspondence
      MC-1403-0016
   Personal Financial Records Clerk
      MC-1403-0010

Recovery
   270' WMEC S/S Generator, Waste Heat Recovery System and Evaporator Operation and Maintenance, Class C
      CG-1712-0005
   Vehicle Recovery
      MC-1710-0046

Recreation
   Apprentice Fitness and Recreation Specialist by Correspondence
      AF-1406-0055
   Fitness and Recreation Specialist by Correspondence
      AF-1406-0056
   Fitness and Recreation Supervisor by Correspondence
      AF-1406-0057
   Morale, Welfare, Recreation, and Services by Correspondence
      AF-0804-0002
   Morale, Welfare, Recreation, and Services Journeyman by Correspondence
      AF-0804-0001

Recruit
   Recruit Training
      CG-2205-0035
      MC-2204-0038
      MC-2204-0088

Recruiter
   Basic Recruiter
      MC-0327-0001
   Marine Corps Recruiter
      MC-0327-0001
   Recruiter Trainer
      CG-1406-0007

Reference
   DCLF Reference Measurement and Calibration
      AF-1715-0493

Refrigeration
   Air Conditioning and Refrigeration, Class C
      CG-1701-0002
   Basic Refrigeration Mechanic
      MC-1730-0006
   Fundamentals of Refrigeration by Correspondence
      MC-1730-0005
   Heating, Ventilation, Air Conditioning, and Refrigeration Journeyman by Correspondence
      AF-1701-0012
      AF-1701-0013
   Heating, Ventilation, Air Conditioning, and Refrigeration Specialist by Correspondence
      AF-1701-0012
      AF-1701-0013
   Journeyman Refrigeration Mechanic
      MC-1730-0007
   Refrigeration and Air Conditioning
      CG-1730-0001
   Refrigeration and Cryogenics Specialist by Correspondence
      AF-1730-0016
   Refrigeration Servicing by Correspondence
      MC-1731-0001
   Refrigeration/Air Conditioning Operation and Maintenance, Class C
      CG-1701-0002

Refueler
   Semitrailer Refueler Operator
      MC-1710-0038

Refueling
   Special Vehicle Mechanic (Refueling Vehicles) by Correspondence
      AF-1710-0034

Regional
   Standard Workstation Regional System Manager
      CG-1402-0003

Relations
   Social Actions Technician (Equal Opportunity/Human Relations) by Correspondence
      AF-1406-0053

Relays
   Relays, Generators, Motors, and Electromechanical Devices by Correspondence
      AF-1714-0032
      AF-1715-0768

KEYWORD INDEX

Reliability
High Reliability Soldering, Class C
    CG-1710-0012

Remotely
Remotely Sensed Imagery for Mapping, Charting, and Geodesy
    DD-1601-0030
Remotely Sensed Imagery for MC&G
    DD-1601-0030

Repair
AN/MRC-135B Repair
    MC-1715-0170
Fuel and Electrical Component Repair
    MC-1714-0017

Reporting
Legal Administration for the Reporting Unit by Correspondence
    MC-1406-0024
Personnel Administration for the Reporting Unit by Correspondence
    MC-1406-0026
Personnel Reporting for REMMPS by Correspondence
    MC-1406-0027
Position Locating Reporting System (PLRS) Support Maintenance
    MC-1715-0149
Position Location Reporting System (PLRS) Capabilities and Employment by Correspondence
    MC-1606-0008
Position Location Reporting System (PLRS) Master Station Maintenance
    MC-1715-0145
Position Location Reporting System (PLRS) Master Station Operator
    MC-1715-0158
Position Location Reporting System (PLRS) Support Maintenance
    MC-1715-0149

Reproduction
Reproduction Equipment Repair
    DD-1706-0003

Reprographic
Apprentice Reprographic Specialist by Correspondence
    AF-1719-0008
Reprographic Specialist by Correspondence
    AF-1408-0098

Requalification
Basic Operational and Transition/Requalification Training A-10
    AF-1704-0285
Transition/Requalification F-15
    AF-1606-0159
Transition/Requalification Training A/OA-10
    AF-1704-0283
Transition/Requalification Training F-15E
    AF-1704-0266
Transition/Requalification Training F-16C/D
    AF-1704-0272

Requirements
Military Requirements for Chief Petty Officer by Correspondence
    CG-2205-0036
Military Requirements for E-3 by Correspondence
    CG-2205-0028
Military Requirements for E-4 by Correspondence
    CG-2205-0029
Military Requirements for E-5 by Correspondence
    CG-2205-0030
Military Requirements for E-6 by Correspondence
    CG-2205-0031
Military Requirements for Second Class Petty Officer by Correspondence
    CG-2205-0038

Rescue
Basic Search and Rescue Aircrewman by Correspondence
    CG-0802-0012
Basic Search and Rescue by Correspondence
    CG-0802-0009
Comprehensive Search and Rescue Planning, Class C
    CG-1722-0011
Maritime Search and Rescue Planning, Class C
    CG-1728-0043
Search and Rescue by Correspondence
    CG-0802-0018
Search and Rescue Fundamentals by Correspondence
    CG-0802-0016
Search and Rescue HU-25A Basic Aircrew by Correspondence
    CG-0802-0008

Research
Advanced Systems Planning, Research, Development, and Engineering
    DD-1408-0016

Reserve
Air Reserve Forces Social Actions Technician (Drug/Alcohol) by Correspondence
    AF-0708-0004
Aircraft Maintenance Officer (Accelerated/Air Reserve Forces)
    AF-1704-0252
    AF-1704-0253
Aircraft Maintenance Officer, Air Reserve
    AF-1704-0252
Command and Staff College Reserve Phase 1, Phase 2
    MC-1408-0015
Comptroller Staff Officer, Air Reserve Forces
    AF-1401-0018
Damage Controlman Advanced Reserve Active Duty for Training, Class C
    CG-1710-0015
Damage Controlman Refresher Reserve Active Duty for Training, Class C
    CG-1710-0015
Electrician Mate Advanced Reserve Active Duty for Training
    CG-1714-0020
Financial Management Staff Officer, Air Reserve Forces
    AF-1408-0105
General Personnel Administration for Reserve by Correspondence
    MC-1406-0025
Introduction to Broadcasting Reserve, Phase 2
    DD-0504-0016
Introduction to Journalism Reserve, Phase 2
    DD-0504-0017
Maritime Reserve Coordination Center Controller
    CG-1722-0011
Naval Reserve Officer
    MC-2202-0002
Naval Reserve Officer Bulldog
    MC-2202-0002
Nursing Service Management for Air Reserve Components
    AF-0709-0035
Public Affairs Officer Course Reserve Component
    DD-0504-0013
Public Affairs Officer Reserve Component, Phase 2
    DD-0504-0013
Reserve Administration
    MC-1406-0022
Reserve Automotive Mechanic
    MC-1408-0020
Reserve Basic Communication Officer Phase 1
    MC-1715-0179
Reserve Basic Communication Officer Phase 2
    MC-1715-0178
Reserve Combat Engineer Noncommissioned Officer (NCO)
    MC-1601-0030
Reserve Combat Engineer Officer
    MC-1601-0029
Reserve Combat Engineer, Phase 2
    MC-1601-0042
Reserve Communication Officers, Phases 1 and 2
    MC-1715-0025
Reserve Components National Security Seminar
    DD-1511-0006
Reserve Embarkation
    MC-1402-0061
Reserve Engineer Equipment Supervisor
    MC-1601-0031
Reserve Engineer Equipment Supervisor Phase 2
    MC-1601-0031
Reserve Field Food Service Supervisor by Correspondence
    MC-1729-0042
Reserve Field Radio Operator Refresher
    MC-1402-0062
Reserve Food Service Refresher
    MC-1729-0037
Reserve Infantry Officer
    MC-2204-0108
Reserve Motor Transport Officer Refresher
    MC-1703-0029
Reserve Officer Candidate Indoctrination
    CG-2202-0004
Reserve Officer Leadership and Management
    CG-1717-0008
Reserve Operational Communication Chief Refresher
    MC-1402-0067
Reserve Radio Chief Refresher
    MC-1402-0064
Reserve Senior Petty Officer Leadership and Management
    CG-1717-0007
Reserve Sergeant's
    MC-2204-0102
Reserve Small Boat Crewmember
    CG-2205-0027
Reserve Staff Noncommissioned Officer (NCO) Advanced
    MC-2204-0104
Reserve Staff Noncommissioned Officer Career
    MC-2204-0071
Reserve Unit Supply Refresher
    MC-1405-0043

KEYWORD INDEX D-51

Reserve
Reserve Warrant Officer Basic
MC-2204-0082
Reserve Yeoman Basic
CG-1409-0004
Resident Staff Noncommissioned Officer (NCO) Advanced Reserve
MC-2204-0072
Resident Staff Noncommissioned Officer (NCO) Career Reserve
MC-2204-0071
Storekeeper Basic Reserve
CG-1405-0010
Yeoman Reserve
CG-1409-0004

Resident
Industrial College of the Armed Forces (Resident Program)
DD-1511-0003
Resident Staff Noncommissioned Officer (NCO) Advanced Regular
MC-2204-0076
Resident Staff Noncommissioned Officer (NCO) Advanced Reserve
MC-2204-0072
Resident Staff Noncommissioned Officer (NCO) Career Regular
MC-2204-0074
Resident Staff Noncommissioned Officer (NCO) Career Reserve
MC-2204-0071
Squadron Officers School Resident
AF-2203-0055

Resource
Apprentice Operations Resource Management Specialist by Correspondence
AF-1402-0073
Human Resource Management (HRMC)
AF-1406-0075
Operations Resource Management by Correspondence
AF-1408-0095
Resource Advisor by Correspondence
AF-1408-0093

Resources
Defense Resources Management
DD-1408-0006

Retailing
Introduction to Retailing by Correspondence
MC-0406-0001

Retrieval
Auditor Retrieval Systems
AF-1402-0074

RF-4
RF-4 Fighter Weapons Instructor
AF-1606-0152

Rifleman
Basic Rifleman
MC-2204-0051
Rifleman
MC-2204-0051

Romanian
Romanian Basic
DD-0602-0225

Rotary
Aviation Electrician's Mate, Class A Rotary Wing Training
CG-1704-0025
Aviation Machinist's Mate Rotary Wing, Class A
CG-1704-0023

Rules
Navigation Rules by Correspondence
CG-1708-0020

Russian
Russian Advanced
DD-0602-0128
DD-0602-0180
Russian Basic
DD-0602-0227
Russian Extended
DD-0602-0115
Russian Intermediate
DD-0602-0123

Safety
Air Force Joint Service Supervisor Safety by Correspondence
AF-0801-0001
Army Joint Service Supervisor Safety by Correspondence
AF-0801-0002
Civil Air Patrol-Safety Officer (Level II Technical Rating) by Correspondence
AF-0801-0005
Coast Guard Joint Service Supervisor Safety by Correspondence
AF-0801-0004
Fire and Safety Technician Second Class by Correspondence
CG-1728-0034
Marine Combat Instructor of Water Safety
MC-2204-0097
Marine Safety Explosive Handling Supervisor
CG-0802-0015
Marine Safety Initial Indoctrination by Correspondence
CG-1708-0019
Marine Safety Initial Indoctrination Lesson Plan Series by Correspondence
CG-1708-0019
Marine Safety Initial Indoctrination Marine Inspection by Correspondence
CG-1708-0024
Marine Safety Initial Indoctrination Port Operations by Correspondence
CG-1708-0025
Marine Safety Inspection
CG-1728-0039
Marine Safety Officer ADT
CG-1728-0029
Marine Safety Petty Officer
CG-1728-0030
Marine Safety Port Operations
CG-1728-0041
National Boating Safety
CG-0802-0014
Navy Joint Service Supervisor Safety by Correspondence
AF-0801-0003
Port Security and Safety Officer
CG-1728-0029
Port Security Safety Enlisted
CG-1728-0027
Safety and Security of the Port by Correspondence
CG-0802-0011
Safety and Security of the Port: Marine Environmental Protection by Correspondence
CG-1728-0036
Safety and Security of the Port: Military Explosives by Correspondence
CG-1728-0033

Salads
Salads, Sandwiches, and Desserts by Correspondence
MC-1729-0022

Sandwiches
Salads, Sandwiches, and Desserts by Correspondence
MC-1729-0022

Sanitation
Mess Hall Sanitation by Correspondence
MC-1729-0039

SASSY
SASSY: Organic Procedures by Correspondence
MC-1405-0044

SATCOM
NAVMACS/SATCOM Systems Maintenance
CG-1715-0138

Satellite
Satellite and Wideband Communications Equipment Specialist by Correspondence
AF-1715-0797
AF-1715-0799
Satellite Communications System Equipment Specialist by Correspondence
AF-1715-0014

Sauces
Vegetables, Soups, Sauces, Gravies, and Beverages by Correspondence
MC-1729-0024

Scanner
Civil Air Patrol Scanner by Correspondence
AF-1704-0194

Science
Air National Guard Academy of Military Science
AF-2203-0051
Marine Science Technician First Class by Correspondence
CG-1304-0018
CG-1304-0022
Marine Science Technician Second Class by Correspondence
CG-1304-0016
Marine Science Technician, Class A
CG-1304-0017

Sciences
Applied Sciences Technician by Correspondence
AF-1715-0775

Scientific
Scientific Measurements Technician by Correspondence
AF-1715-0762

Scout
Basic Scout Swimmer
MC-2204-0096

Sea
Sea Duty Indoctrination
MC-2204-0014
Sea School
MC-2204-0014

Seabee
Seabee Tactical Shipboard Planning
MC-0419-0008

Seaman
Seaman by Correspondence
CG-1708-0009

KEYWORD INDEX

Seamanship
Deck Seamanship by Correspondence
    CG-1708-0010

Search
Basic Search and Rescue Aircrewman by Correspondence
    CG-0802-0012
Basic Search and Rescue by Correspondence
    CG-0802-0009
Coastal Search Planning, Class C
    CG-1722-0010
Comprehensive Search and Rescue Planning, Class C
    CG-1722-0011
Computer Aided Search Planning by Correspondence
    CG-0802-0017
Search and Rescue by Correspondence
    CG-0802-0018
Search and Rescue Fundamentals by Correspondence
    CG-0802-0016
Search and Rescue HU-25A Basic Aircrew by Correspondence
    CG-0802-0008

Secure
Secure Communication Systems Maintenance Specialist by Correspondence
    AF-1715-0786
Secure Communications Systems Maintenance by Correspondence
    AF-1715-0813
    AF-1715-0820

Security
Basic Security Guard
    MC-1728-0010
Marine Security Guard
    MC-1728-0006
Mobile Security Communication Technician
    MC-1715-0169
National Communications Security (COMSEC)
    DD-1404-0004
National Security Management (Correspondence/Nonresident/Seminar Course of the National Defense University)
    DD-1511-0001
National Security Management by Correspondence
    DD-1511-0008
Port Physical Security Management
    CG-1728-0026
Port Physical Security Practical
    CG-1728-0025
Port Security and Safety Officer
    CG-1728-0029
Port Security Safety Enlisted
    CG-1728-0027
Port Security, Class A
    CG-1728-0040
Reserve Components National Security Seminar
    DD-1511-0006
Safety and Security of the Port by Correspondence
    CG-0802-0011
Safety and Security of the Port: Marine Environmental Protection by Correspondence
    CG-1728-0036
Safety and Security of the Port: Military Explosives by Correspondence
    CG-1728-0033
Security of Classified Information by Correspondence
    CG-1728-0035
Security Specialist by Correspondence
    AF-1728-0045
Security Supervisor
    MC-1728-0009

Securityman
Port Securityman Direct Entry
    CG-1728-0044
Port Securityman First Class by Correspondence
    CG-1728-0038
Port Securityman Second Class by Correspondence
    CG-1728-0032
Port Securityman Third Class by Correspondence
    CG-1728-0009

Seminar
Air War College Nonresident Seminar Program
    AF-1511-0012
National Security Management (Correspondence/Nonresident/Seminar Course of the National Defense University)
    DD-1511-0001

Semitrailer
Semitrailer Refueler Operator
    MC-1710-0038

Senior
Air Defense Control Officers Senior
    MC-1715-0150
Platoon Leaders Senior
    MC-2204-0078
Reserve Senior Petty Officer Leadership and Management
    CG-1717-0007
Senior Clerk
    MC-1403-0015
Senior Food Service
    MC-1729-0047
Senior Military Cryptologic Supervisors
    DD-1404-0001
Senior Petty Officer Leadership and Management
    CG-1717-0010
Senior Petty Officers by Correspondence
    CG-2205-0034

Sensed
Remotely Sensed Imagery for Mapping, Charting, and Geodesy
    DD-1601-0030
Remotely Sensed Imagery for MC&G
    DD-1601-0030

Sensor
Avionic Sensor Systems Specialist (Electro-Optical Sensors) by Correspondence
    AF-1715-0071
Avionic Sensor Systems Specialist (Reconnaissance Electronic Sensors) by Correspondence
    AF-1715-0068
Avionic Sensor Systems Specialist (Tactical/Real Time Display Electronic Sensors) by Correspondence
    AF-1715-0070

Sensors
Avionics Sensors Maintenance Journeyman by Correspondence
    AF-1715-0818
Photo-Sensors Maintenance Specialist (Tactical/Reconnaissance Electronic Sensors) by Correspondence
    AF-1715-0777

Serbo-Croatian
Serbo-Croatian Basic
    DD-0602-0114

Sergeant
First Sergeant by Correspondence
    AF-2203-0052

Sergeant's
Reserve Sergeant's
    MC-2204-0102
Sergeant's Course
    MC-2204-0103

Sergeants
Sergeants Nonresident Program by Correspondence
    MC-1408-0027

Serology
Medical Laboratory Craftsman (Hematology, Serology, Blood Banking and Immunohematology) by Correspondence
    AF-0702-0010
Medical Laboratory Technician (Hematology, Serology, Blood Banking and Immunohematology) by Correspondence
    AF-0702-0010
    AF-0709-0031

Service
Air Force Joint Service Supervisor Safety by Correspondence
    AF-0801-0001
Army Joint Service Supervisor Safety by Correspondence
    AF-0801-0002
Basic Food Service
    MC-1729-0009
Chapel Service Support Journeyman by Correspondence
    AF-1408-0113
Chaplain Service Support Craftsman by Correspondence
    AF-1408-0112
Coast Guard Joint Service Supervisor Safety by Correspondence
    AF-0801-0004
Food Service Fundamentals by Correspondence
    MC-1729-0038
Food Service Noncommissioned Officer (NCO)
    MC-1729-0034
Food Service Staff Noncommissioned Officer (NCO)
    MC-1729-0036
Medical Service Specialist by Correspondence
    AF-0703-0016
    AF-0703-0018
Medical Service Specialist, Aeromedical by Correspondence
    AF-0709-0033
Medical Service Technician by Correspondence
    AF-0703-0011
Mental Health Service Specialist by Correspondence
    AF-0703-0017
Navy Joint Service Supervisor Safety by Correspondence
    AF-0801-0003
Nursing Service Management
    AF-0709-0036

KEYWORD INDEX D-53

Nursing Service Management by Correspondence
  AF-0703-0014
Reserve Field Food Service Supervisor by Correspondence
  MC-1729-0042
Reserve Food Service Refresher
  MC-1729-0037
Surgical Service Specialist by Correspondence
  AF-0703-0015

Services
Apprentice Services Specialist by Correspondence
  AF-1729-0013
Commercial Services and Automated Travel Record Accounting System by Correspondence
  AF-1408-0074
Financial Management Specialist (Commercial Services) by Correspondence
  AF-1408-0074
Financial Management/Services Supervisor (Functions and Responsibilities) by Correspondence
  AF-1408-0082
Financial Management/Services Supervisor by Correspondence
  AF-1408-0081
Financial Services Specialist (Introduction) by Correspondence
  AF-1408-0078
Financial Services Specialist (Travel) by Correspondence
  AF-1408-0080
Financial Services Specialist by Correspondence
  AF-1408-0078
Health Services Technician First Class by Correspondence
  CG-0709-0006
Health Services Technician Second Class by Correspondence
  CG-0709-0005
Legal Services Specialist by Correspondence
  AF-1406-0047
Morale, Welfare, Recreation, and Services by Correspondence
  AF-0804-0002
Morale, Welfare, Recreation, and Services Journeyman by Correspondence
  AF-0804-0001
Services Specialist by Correspondence
  AF-1729-0014
Services Supervisor by Correspondence
  AF-0419-0028
  AF-1408-0107

Sewage
Plumbing and Sewage Disposal by Correspondence
  MC-1710-0005

Shipboard
Seabee Tactical Shipboard Planning
  MC-0419-0008
Shipboard Information, Training, and Entertainment (SITE) System
  DD-0504-0012

Shiphandling
Shiphandling by Correspondence
  CG-1708-0013

Shop
Basic Shop Fundamentals for the Mechanic by Correspondence
  MC-1717-0005

Short Basic
Short Basic Turkish
  DD-0602-0042

Sinewave
Sinewave Oscillators-Modulation/Demodulation by Correspondence
  AF-1715-0751

Small
Reserve Small Boat Crewmember
  CG-2205-0027
Small Arms Instructor
  CG-1408-0041
Small Arms Instructor, Class C
  CG-1408-0041
Small Arms Refresher
  MC-2204-0110
Small Arms Repairer
  MC-2204-0111
Small Boat Crewmember
  CG-2205-0027
Small Boat Engineer, Class C
  CG-1714-0019
Small Craft Mechanic
  MC-1712-0009
Small Cutter Damage Control
  CG-0801-0001

Social
Air Reserve Forces Social Actions Technician (Drug/Alcohol) by Correspondence
  AF-0708-0004
Social Actions Technician (Drugs/Alcohol) by Correspondence
  AF-1406-0054
Social Actions Technician (Equal Opportunity/Human Relations) by Correspondence
  AF-1406-0053

Software
Intermediate Software Acquisition Management
  DD-1408-0013
Software Acquisition Management
  DD-1402-0004
Software Cost Estimating
  DD-1402-0005

Soldering
Soldering and Electrical Connectors by Correspondence
  AF-1714-0033
  AF-1723-0012

Solid State
AN/FPN-64(V) Loran Solid State Transmitter, Class C
  CG-1715-0123
Fundamentals of Solid State Devices by Correspondence
  AF-1715-0750
Solid State Devices by Correspondence
  MC-1715-0132

Sonar
Sonar Technician Chief by Correspondence
  CG-1408-0039
Sonar Technician First Class by Correspondence
  CG-1715-0034
Sonar Technician Second Class by Correspondence
  CG-1715-0133

Sonobuoy
Sonobuoy by Correspondence
  CG-2205-0037

Soups
Vegetables, Soups, Sauces, Gravies, and Beverages by Correspondence
  MC-1729-0024

Space
Missile and Space Systems Electronic Maintenance Journeyman by Correspondence
  AF-1714-0041
  AF-1715-0814
Missile and Space Systems Maintenance Journeyman by Correspondence
  AF-1704-0265
Space Communications Systems Equipment by Correspondence
  AF-1715-0014
Space Communications Systems Equipment Operator/Technician by Correspondence
  AF-1715-0021
Space Systems Equipment Maintenance Specialist by Correspondence
  AF-1715-0036
Space Systems Operations Specialist by Correspondence
  AF-1715-0755

Spanish
Spanish Basic
  DD-0602-0222
Spanish Intermediate
  DD-0602-0218

Special
Arabic-Egyptian Special
  DD-0602-0208
Arabic-Iraqi Special
  DD-0602-0207
Arabic-Syrian Special
  DD-0602-0206
Electronic Tubes and Special Purpose Tubes by Correspondence
  AF-1715-0749
Special Purpose Vehicle and Equipment Craftsman by Correspondence
  AF-1703-0021
Special Purpose Vehicle and Equipment Maintenance Apprentice by Correspondence
  AF-1703-0020
Special Purpose Vehicle and Equipment Mechanic by Correspondence
  AF-1703-0020
  AF-1710-0032
Special Purpose Vehicle and Equipment Supervisor by Correspondence
  AF-1703-0021
Special Vehicle Mechanic (Refueling Vehicles) by Correspondence
  AF-1710-0034

Special Operations
Special Operations Training, AC-130E Pilot
  AF-1606-0020

Spelling
Spelling by Correspondence
  MC-0501-0001

Splicing
Cable Splicing Installation and Maintenance Specialist by Correspondence
  AF-1715-0126

Sprinklers
20mm Mk 16 Mod 5 Machine Gun and Magazine Sprinklers Operation and Maintenance, Class C
  CG-2204-0001

Squadron
Squadron Officers School Nonresident
  AF-2203-0056

KEYWORD INDEX

Squadron Officers School Resident
       AF-2203-0055

Stability
Damage Control and Stability by Correspondence
       CG-2205-0024

Staff
Air Command and Staff College
       AF-1511-0001
Air Command and Staff College Resident Program
       AF-1511-0010
Air Command and Staff Correspondence Associate Program
       AF-1511-0013
Air Command and Staff Correspondence Program
       AF-1511-0013
Air Command and Staff Nonresident Seminar Associate Program
       AF-1511-0009
Air Command and Staff Nonresident Seminar Program
       AF-1511-0009
Armed Forces Staff College
       DD-0326-0001
Aviation Combat Element Staff Planning
       MC-2204-0094
Coastal Defense Command and Staff, Class C
       CG-0419-0003
Command and Staff College Reserve Phase 1, Phase 2
       MC-1408-0015
Comptroller Staff Officer
       AF-1408-0104
Comptroller Staff Officer, Air Reserve Forces
       AF-1401-0018
Financial Management Staff Officer
       AF-1408-0103
Financial Management Staff Officer, Air Reserve Forces
       AF-1408-0105
Food Service Staff Noncommissioned Officer (NCO)
       MC-1729-0036
Infantry Staff Noncommissioned Officer (NCO): Military Symbols and Overlays by Correspondence
       MC-1601-0040
Joint and Combined Staff Officer
       DD-0326-0002
Landing Force Combat Services Support Staff Planning
       MC-2204-0093
Landing Force Ground Combat Communication Staff Planning
       MC-1601-0046
Landing Force Staff Planning (Marine Expeditionary Brigade)
       MC-2204-0092
Marine Corps Command and Staff College
       MC-1408-0017
Motor Transport Staff Noncommissioned Officer (NCO)
       MC-1703-0025
Nursing Staff Development Officer Basic
       AF-0703-0020
Reserve Staff Noncommissioned Officer Career
       MC-2204-0071
Resident Staff Noncommissioned Officer (NCO) Advanced Regular
       MC-2204-0076
Resident Staff Noncommissioned Officer (NCO) Advanced Reserve
       MC-2204-0072
Resident Staff Noncommissioned Officer (NCO) Career Regular
       MC-2204-0074
Resident Staff Noncommissioned Officer (NCO) Career Reserve
       MC-2204-0071
Staff Noncommissioned Officer (NCO) Advanced
       MC-2204-0076
Staff Noncommissioned Officers Advanced Nonresident Program (SNCOANP) by Correspondence
       MC-1408-0028
Transportation Staff Officer
       AF-0419-0035

Station
Avionics Test Station and Component Specialist (F-16/A-10) by Correspondence
       AF-1715-0756
Avionics Test Station and Component Specialist by Correspondence
       AF-1715-0791
B-1B Avionics Test Station and Component Technician by Correspondence
       AF-1715-0793
F-15 Avionics Automatic Test Station and Component Specialist by Correspondence
       AF-1715-0772
F-15 Avionics Manual/ECM Test Station and Component Specialist by Correspondence
       AF-1715-0771
F-16 A/B Avionics Test Station and Component Specialist by Correspondence
       AF-1715-0780
F-16 C/D Avionics Test Station and Component Specialist by Correspondence
       AF-1715-0785
F/FB-111 Automatic Test Station and Component Specialist by Correspondence
       AF-1715-0773
Integrated Avionics Computerized Test Station and Component Specialist (F-15) by Correspondence
       AF-1715-0086
Integrated Avionics Computerized Test Station and Component Specialist (F/FB-111) by Correspondence
       AF-1715-0087
Integrated Avionics Manual Test Station and Component Specialist (F-15) by Correspondence
       AF-1715-0092
Integrated Avionics Manual Test Station and Component Specialist (F/FB-111) by Correspondence
       AF-1715-0093
Position Location Reporting System (PLRS) Master Station Maintenance
       MC-1715-0145
Position Location Reporting System (PLRS) Master Station Operator
       MC-1715-0158

Steel
Steel Welding, Class C
       CG-1710-0017

Still
Apprentice Still Photographic Specialist by Correspondence
       AF-1709-0029
Still Photographic Specialist by Correspondence
       AF-1709-0028

Stock
Basic Supply Stock Control
       MC-1405-0035

Storage
Materiel Storage and Distribution Specialist by Correspondence
       AF-1405-0070
Multiple Virtual Storage (MVS) Diagnostics
       MC-1402-0053
Multiple Virtual Storage (MVS) Fundamentals and Logic
       MC-1402-0052
Multiple Virtual Storage (MVS) Performance and Training
       MC-1402-0054

Storekeeper
Storekeeper Basic Reserve
       CG-1405-0010
Storekeeper First Class by Correspondence
       CG-1405-0008
Storekeeper Second Class by Correspondence
       CG-1405-0009
Storekeeper Third Class by Correspondence
       CG-1405-0002
Storekeeper, Class A
       CG-1405-0006

Strategic
Strategic Aircraft Maintenance Specialist by Correspondence
       AF-1704-0181
       AF-1704-0188
       AF-1704-0209
       AF-1704-0210
       AF-1704-0211
       AF-1704-0247

Structural
Aircraft Structural Maintenance Technician (Airframe Repair) by Correspondence
       AF-1704-0250
       AF-1717-0028
Aviation Structural Mechanic (AM), Class A
       CG-1704-0027
Aviation Structural Mechanic First Class by Correspondence
       CG-1704-0033
Aviation Structural Mechanic Second Class by Correspondence
       CG-1704-0039
Aviation Structural Mechanic, Class A
       CG-1704-0047
Structural Journeyman by Correspondence
       AF-1732-0015
       AF-1732-0016
Structural Maintenance Technician by Correspondence
       AF-1704-0250
Structural Technician by Correspondence
       AF-1601-0049

Subsistence
Apprentice Subsistence Operations Specialist by Correspondence
       AF-1729-0015
Food Service Subsistence Clerk
       MC-1729-0048
Mess Hall Subsistence Clerk by Correspondence
       MC-1729-0040
Subsistence Operations Specialist by Correspondence
       AF-1729-0016
Subsistence Operations Technician by Correspondence
       AF-1729-0017

KEYWORD INDEX D-55

Subsistence Specialist Administration and Management
 CG-1729-0010
Subsistence Specialist First Class by Correspondence
 CG-1729-0005
Subsistence Specialist Second Class by Correspondence
 CG-1729-0004
Subsistence Specialist Third Class by Correspondence
 CG-1729-0006
Subsistence Specialist, Class A
 CG-1729-0007
Subsistence Specialist, Independent Duty
 CG-1729-0009
Subsistence Specialist, Large Mess Management
 CG-1729-0008
Subsistence Supply
 MC-1401-0007

Supervisor
Advanced Public Affairs Supervisor
 DD-1709-0003
Aviation Quality Assurance Supervisor by Correspondence
 MC-1405-0047
Public Affairs Supervisor
 DD-0504-0014
Services Supervisor by Correspondence
 AF-0419-0028

Supplies
Power Supplies by Correspondence
 AF-1715-0752

Supply
Basic Supply Stock Control
 MC-1405-0035
Enlisted Supply Advanced
 MC-1405-0037
Enlisted Supply Basic
 MC-1405-0035
Enlisted Supply Independent Duty
 MC-1405-0040
Enlisted Supply Intermediate
 MC-1405-0039
Enlisted Supply Reorientation
 MC-1405-0037
Fiscal Accounting for Supply Clerks by Correspondence
 MC-1401-0010
Ground Supply Officer
 MC-1405-0038
Reserve Unit Supply Refresher
 MC-1405-0043
Subsistence Supply
 MC-1401-0007
Supply Management by Correspondence
 MC-1405-0027
Supply Systems Analysis Supervisor by Correspondence
 AF-1405-0066
Supply Systems Analyst Journeyman by Correspondence
 AF-1405-0072

Support
Air Support Control Officer
 MC-1715-0078
 MC-1715-0176
Aircrew Life Support Specialist by Correspondence
 AF-1704-0199
Basic Landing Support Specialist
 MC-2204-0069
Chapel Service Support Journeyman by Correspondence
 AF-1408-0113
Chaplain Service Support Craftsman by Correspondence
 AF-1408-0112
Communications for the Combat Operations Center/Fire Support Coordination Center (COC/FSCC) by Correspondence
 MC-1715-0101
Environmental Support Specialist by Correspondence
 AF-1710-0041
Fire Support Coordination
 MC-2204-0095
Infantry Squad Leader: Weapons and Fire Support by Correspondence
 MC-2204-0101
Interrogator Set AN/TPX-46 Direct Support/General Support Repairman and Supervisor
 MC-1715-0138
Landing Force Combat Services Support Staff Planning
 MC-2204-0093
Position Locating Reporting System (PLRS) Support Maintenance
 MC-1715-0149
Position Location Reporting System (PLRS) Support Maintenance
 MC-1715-0149

Surface
Surface Air Defense System Acquisition Technician
 MC-1715-0181
Surface Air Defense System Fire Control Technician
 MC-1715-0180

Surgical
Surgical Service Specialist by Correspondence
 AF-0703-0015

Surveillance
Basic Infection Control Surveillance (Infection Control and Epidemiology)
 AF-0707-0011

Survey
Advanced Geodetic Survey
 DD-1601-0022
Artillery Survey Instruments by Correspondence
 MC-1601-0032
Basic Geodetic Survey
 DD-1601-0023
Hydrographic Survey
 DD-1722-0001
Introduction to Field Artillery Survey by Correspondence
 MC-1601-0039
Survey Instrument Maintenance
 DD-1721-0003

Survival
Cold Weather Survival
 MC-0804-0002
Mountain Survival
 MC-0804-0005
Program Manager's Survival
 DD-1408-0021

Survivalman
Aviation Survivalman First Class by Correspondence
 CG-1704-0041
Aviation Survivalman Second Class by Correspondence
 CG-1704-0018
 CG-1704-0043

Swedish
Swedish Basic
 DD-0602-0131

Swimmer
Basic Scout Swimmer
 MC-2204-0096

Switchboard
Telephone/Switchboard Repair
 MC-1715-0121

Switching
Electronic Computer and Switching Systems Specialist by Correspondence
 AF-1715-0765
Telephone Central Office Switching Equipment Specialist, Electronic/Electromechanical by Correspondence
 AF-1715-0127
Telephone Switching Specialist by Correspondence
 AF-1715-0803

SX-200D
Mitel SX-200D Installation and Maintenance, Class C
 CG-1715-0116

Syrian
Arabic Egyptian/Syrian Intermediate
 DD-0602-0023
Arabic-Syrian Extended
 DD-0602-0210
Arabic-Syrian Special
 DD-0602-0206

System
AN/MSC-63A System Manager
 MC-1402-0068
AN/TSC-120 Communication Central System Installer/Maintainer
 MC-1715-0171
Logistics Vehicle System (LVS) Maintenance
 MC-1408-0019

System 360
IBM System 360 (OS) COBOL Programming
 MC-1402-0013
IBM System 360 (OS) Data Control Techniques
 MC-1402-0042
IBM System 360 Assembler Language Coding (Entry)
 MC-1402-0051
IBM System 360 Operating System (OS) Operations
 MC-1402-0023
IBM System 360 OS FORTRAN Programming
 MC-1402-0034

Systems
Advanced Information Systems Acquisition
 DD-0326-0005
Automated Information Systems Contracting
 DD-1402-0006
Communication Systems Chief
 MC-1408-0023
 MC-1715-0167
Fundamentals of Systems Acquisition Management
 DD-1408-0012

KEYWORD INDEX

HH-60J Electrical/Automatic Flight Control Systems Maintenance, Class C
   CG-1704-0051
HH-65A Electrical/Automatic Flight Control Systems Maintenance, Class C
   CG-1704-0049
Intermediate Information Systems Acquisition
   DD-0326-0006
Intermediate Systems Acquisition
   DD-1408-0011
   DD-1408-0020
Systems Acquisition for Contracting Personnel (Executive)
   DD-1408-0009
Systems Repair Technician by Correspondence
   AF-1715-0763

T58-5
T58-5 and T62 Engine Maintenance, Class C
   CG-1704-0026

T58-GE-5
T58-GE-8B and T58-GE-5 Turboshaft Engine Familiarization by Correspondence
   CG-1704-0021

T58-GE-8B
T58-GE-8B and T58-GE-5 Turboshaft Engine Familiarization by Correspondence
   CG-1704-0021
T58-GE-8B Engine Maintenance, Class C
   CG-1704-0004

T62
T58-5 and T62 Engine Maintenance, Class C
   CG-1704-0026

TACC
Tactical Air Command Central (TACC AN/TYQ-1) Repair
   MC-1715-0020
Tactical Air Command Central (TACC) Technician
   MC-1715-0046

Tactical
Aircraft Maintenance Specialist, Tactical Aircraft by Correspondence
   AF-1704-0102
Avionic Sensor Systems Specialist (Tactical/Real Time Display Electronic Sensors) by Correspondence
   AF-1715-0070
Photo-Sensors Maintenance Specialist (Tactical/Reconnaissance Electronic Sensors) by Correspondence
   AF-1715-0777
Seabee Tactical Shipboard Planning
   MC-0419-0008
Tactical Air Command and Control Specialist by Correspondence
   AF-1406-0068
Tactical Air Command Center Operator
   MC-1715-0157
Tactical Air Command Central (TACC AN/TYQ-1) Repair
   MC-1715-0020
Tactical Air Command Central (TACC) Technician
   MC-1715-0046
Tactical Air Command Central Repair
   MC-1715-0151
Tactical Air Command Central Repairer
   MC-1715-0020
Tactical Air Command Central Technician
   MC-1715-0142
Tactical Air Control Party
   MC-1606-0013
Tactical Air Defense Controller
   MC-1704-0004
Tactical Air Defense Controller (AN/TYQ-23)
   MC-1704-0009
Tactical Air Defense Controller Enlisted
   MC-1704-0004
Tactical Air Operations Module (TAOM), AN/TYQ-23(V)1 Technician
   MC-1715-0175
Tactical Air Operations Module Repair AN/TYQ-23(V)1
   MC-1715-0174
Tactical Aircraft Maintenance Technician by Correspondence
   AF-1704-0228
Tactical Data Communication Central Repair
   MC-1715-0141
Tactical Data Communication Central Technician
   MC-1715-0146
Tactical Data Communications Central (TDCC AN/TYQ-3) Technician
   MC-1715-0026
Tactical Data Communications Central Technician
   MC-1715-0026
Tactical General Purpose Computer Technician
   MC-1402-0056

Tactics
Adversary Tactics Instructor (AFSC K1115H)
   AF-1406-0065
Adversary Tactics Instructor (AFSC K1745)
   AF-1406-0067
Infantry Squad Leader Squad Tactics by Correspondence
   MC-2204-0100

Tagalog
Pilipino/Tagalog Basic
   DD-0602-0237
Tagalog Basic
   DD-0602-0237

Tank
M1A1 Tank System Mechanic
   MC-1710-0047
Tank Gunnery/Direct Fire Procedures (M1A1) by Correspondence
   MC-2204-0098
Tank Systems Technician (M1A1)
   MC-1710-0051

Target
Target Intelligence Specialist by Correspondence
   AF-1606-0122

Team
Team Embarkation Officer/Assistant
   MC-1402-0060
   MC-1402-0061

Technical
Air Force Technical Order System by Correspondence
   AF-2203-0053
Marine Corps Technical Publications System by Correspondence
   MC-1408-0021
Mobile Technical Control
   MC-1402-0063

Technician
AN/TSC-120 Communication Central Technician
   MC-1715-0168
System Computer Technician
   MC-1402-0065
Tank Systems Technician (M1A1)
   MC-1710-0051
Technician Theory
   MC-1715-0128

Technology
Advanced Digital Electronics Technology, Class C
   CG-1715-0148

Telecommunications
Telecommunications Operations Specialist by Correspondence
   AF-0504-0003
Telecommunications Specialist First Class by Correspondence
   CG-1403-0002
Telecommunications Specialist Second Class by Correspondence
   CG-1402-0004
Telecommunications Specialist, Class A
   CG-1404-0005
Telecommunications System Control Specialist by Correspondence
   AF-1715-0390
Telecommunications System Maintenance Specialist by Correspondence
   AF-1715-0031
Telecommunications Systems Control Technician by Correspondence
   AF-1715-0034

Telephone
Automatic Telephone Equipment by Correspondence
   MC-1715-0099
CEJT-SX-200 Mitel Electronic Private Automatic Branch Exchange Generic 1000 or 1001 Telephone System
   CG-1715-0116
TEL-14 CDXC-SG-1/1A Pulse 120 Electronic Private Automatic Branch Exchange Telephone System, Class C
   CG-1715-0115
Telephone Central Office Switching Equipment Specialist, Electronic/Electromechanical by Correspondence
   AF-1715-0127
Telephone Equipment Installation and Repair Specialist by Correspondence
   AF-1715-0131
Telephone Switching Specialist by Correspondence
   AF-1715-0803
Telephone Technician First Class by Correspondence
   CG-1715-0134
Telephone Technician Second Class by Correspondence
   CG-1715-0063
Telephone Technician, Class A
   CG-1715-0109
Telephone/Switchboard Repair
   MC-1715-0121

Teletype
Teletype Repair
   MC-1715-0120
Tempest Model 40 Teletype, Class C
   CG-1715-0068

Television
Armed Forces Radio Television System Broadcast Manager
   DD-0505-0005

KEYWORD INDEX D-57

Radio and Television Broadcasting Specialist by Correspondence
  AF-0505-0002
Television Equipment Repairman by Correspondence
  AF-1715-0013
Television Systems Specialist by Correspondence
  AF-1715-0796
  AF-1715-0798

Tempest
Tempest Model 40 Teletype, Class C
  CG-1715-0068

Terminal
Mobile Data Communications Terminal Technician
  MC-1715-0118

Terrain
Advanced Terrain Analysis
  DD-1601-0028
Basic Terrain Analysis
  DD-1601-0020
Military Operations on Urban Terrain by Correspondence
  MC-1606-0006
Terrain Analysis Warrant Officer Certification
  DD-1601-0029

Terrorism
Terrorism Counteraction for Marines by Correspondence
  MC-1606-0010

Test
Avionics Test Station and Component Specialist (F-16/A-10) by Correspondence
  AF-1715-0756
Avionics Test Station and Component Specialist by Correspondence
  AF-1715-0791
B-1B Avionics Test Station and Component Specialist by Correspondence
  AF-1408-0097
B-1B Avionics Test Station and Component Technician by Correspondence
  AF-1715-0793
Experimental Test Pilot
  AF-1606-0150
F-15 Avionics Automatic Test Station and Component Specialist by Correspondence
  AF-1715-0772
F-15 Avionics Manual/ECM Test Station and Component Specialist by Correspondence
  AF-1715-0771
F-16 A/B Avionics Test Station and Component Specialist by Correspondence
  AF-1715-0780
F-16 C/D Avionics Test Station and Component Specialist by Correspondence
  AF-1715-0785
F/FB-111 Automatic Test Station and Component Specialist by Correspondence
  AF-1715-0773
F/FB-111 Avionics Manual/Electronic Countermeasures (ECM) Test Station and Component Specialist by Correspondence
  AF-1704-0224
Flight Test Engineer/Navigator
  AF-1606-0153
Integrated Avionics Computerized Test Station and Component Specialist (F-15) by Correspondence
  AF-1715-0086
Integrated Avionics Computerized Test Station and Component Specialist (F/FB-111) by Correspondence
  AF-1715-0087
Integrated Avionics Manual Test Station and Component Specialist (F-15) by Correspondence
  AF-1715-0092
Integrated Avionics Manual Test Station and Component Specialist (F/FB-111) by Correspondence
  AF-1715-0093
Intermediate Test and Evaluation (T & E)
  DD-1408-0022
Introduction to Test Equipment by Correspondence
  MC-1715-0129
Test, Measurement and Diagnostic Equipment
  MC-1715-0139

Thai
Thai Basic
  DD-0602-0119
Thai Intermediate
  DD-0602-0239

Theory
Technician Theory
  MC-1715-0128

Therapy
Diet Therapy Specialist by Correspondence
  AF-0104-0003
Diet Therapy Supervisor by Correspondence
  AF-0104-0002
Physical Therapy Specialist by Correspondence
  AF-0704-0004

Threat
Intelligence Brief: The Military Threat by Correspondence
  MC-1606-0012
Intelligence Brief: The Opposing Forces Nuclear, Biological, Chemical (NBC) Threat by Correspondence
  MC-1606-0011

Timing
Loran-C Timing and Control Operations and Maintenance, Class C
  CG-1715-0111
Timing and Control Equipment, Class C
  CG-1715-0111

Towed
Towed Artillery Repairman
  MC-2201-0001

Tracking
Automatic Tracking Radar Journeyman by Correspondence
  AF-1715-0812
Automatic Tracking Radar Specialist by Correspondence
  AF-1715-0007
  AF-1715-0776
  AF-1715-0794
  AF-1715-0795
  AF-1715-0812
Automatic Tracking Radar Technician by Correspondence
  AF-1715-0011

Tractor
Tractor-Trailer Operator by Correspondence
  MC-1710-0041

Traffic
Freight Traffic Specialist by Correspondence
  AF-0419-0032
Traffic Management Journeyman by Correspondence
  AF-0419-0037

Trailer
Tractor-Trailer Operator by Correspondence
  MC-1710-0041

Trainer
Recruiter Trainer
  CG-1406-0007

Training
Basic Operational Training
  AF-1704-0277
Conversion and Lantirn Training F16 C/D Block 40/42
  AF-1704-0278
Education and Training Officer by Correspondence
  AF-1406-0079
ENJPT Pilot Instructor Training Shepherd (T-37/T-38)
  AF-1406-0101
IFF Instructor Training (AT-38B)
  AF-1406-0100
Instructional System Development for Training Managers by Correspondence
  AF-1406-0080
Instructor Pilot Upgrade Training F-16 C
  AF-1406-0102
Instructor Pilot Upgrade Training F-16 C/D
  AF-1406-0102
Lantirn Training F-16C/D Block 40/42
  AF-1704-0278
Training Specialist by Correspondence
  AF-1406-0071
Training Systems Technician by Correspondence
  AF-1406-0072
Training Technician by Correspondence
  AF-1406-0070
Transition/Requalification Training F-16 A/B
  AF-1704-0275
Transition/Requalification Training F-16 C/D (Block 40/42)
  AF-1704-0276

Trains
Automotive Power Trains by Correspondence
  MC-1703-0022

Transceiver
AN/URC-114(V) 1 KW Transceiver System Maintenance
  CG-1715-0124

Transition
Basic Operational and Transition/Requalification Training A-10
  AF-1704-0285
Transition/Requalification F-15
  AF-1606-0159
Transition/Requalification Training A/OA-10
  AF-1704-0283

Transmitter
AN/FPN-39 Loran Transmitter, Class C
  CG-1715-0118
AN/FPN-39 Transmitter Maintenance, Class C
  CG-1715-0118
AN/FPN-42 Loran Transmitter, Class C
  CG-1715-0117
AN/FPN-44A Loran Transmitter, Class C
  CG-1715-0110

KEYWORD INDEX

AN/FPN-44A Transmitter Maintenance, Class C
 CG-1715-0110
AN/FPN-64(V) Loran Solid State Transmitter, Class C
 CG-1715-0123
AN/FPN-64(V) Transmitter Maintenance
 CG-1715-0123
AN/URT-41(V)2 10 KW Transmitter System Maintenance
 CG-1715-0112

Transmitting
AN/URT-41(V) Transmitting System, Class C
 CG-1715-0112
COL-URG-II HF Transmitting System
 CG-1715-0072
COL-URG-II High Frequency Transmitting System
 CG-1715-0072

Transport
Motor Transport Maintenance Officer
 MC-1712-0005
Motor Transport Officer
 MC-0419-0005
Motor Transport Operations Noncommissioned Officer (NCO)
 MC-1408-0018
Motor Transport Staff Noncommissioned Officer (NCO)
 MC-1703-0025
Reserve Motor Transport Officer Refresher
 MC-1703-0029

Transportation
Air Transportation Journeyman by Correspondence
 AF-0419-0038
Base Transportation Officer
 AF-0419-0033
Defense Packaging of Hazardous Materials for Transportation
 DD-0419-0003
Transportation Officer
 AF-0419-0034
Transportation Staff Officer
 AF-0419-0035

Travel
Basic Travel Clerk
 MC-1403-0009
Commercial Services and Automated Travel Record Accounting System by Correspondence
 AF-1408-0074
Financial Services Specialist (Travel) by Correspondence
 AF-1408-0080

Turbine
Pratt and Whitney FT4A Gas Turbine
 CG-1710-0018
Theory and Construction of Turbine Engines by Correspondence
 MC-1710-0004

Turboprop
Aerospace Propulsion Specialist (Turboprop) by Correspondence
 AF-1704-0257
Aerospace Propulsion Technician (Turboprop) by Correspondence
 AF-1710-0037
Turboprop Propulsion Mechanic by Correspondence
 AF-1704-0204

Turboshaft
T58-GE-8B and T58-GE-5 Turboshaft Engine Familiarization by Correspondence
 CG-1704-0021

Turkish
Short Basic Turkish
 DD-0602-0042
Turkish Basic
 DD-0602-0231
Turkish Gateway
 DD-0602-0216

Turrets
Defensive Fire Control Systems Mechanic (B-52D/G: MD-9, ASG-15 Turrets) by Correspondence
 AF-1715-0047

Typing
Basic Typing
 MC-1406-0021
Basic Typing and Personnel Administration
 MC-1406-0021

Ukrainian
Ukrainian Basic
 DD-0602-0105

Undergraduate
Undergraduate Pilot Training
 AF-1606-0154

Unit
Electronic Warfare Systems Technician (Unit) by Correspondence
 AF-1715-0783
Ground Radio Communications Specialist (Unit) by Correspondence
 AF-1715-0782
Legal Administration for the Reporting Unit by Correspondence
 MC-1406-0024
Personnel Administration for the Reporting Unit by Correspondence
 MC-1406-0026
Reserve Unit Supply Refresher
 MC-1405-0043
Reverse Osmosis Water Purification Unit by Correspondence
 MC-1732-0003
Small Unit Paperwork by Correspondence
 CG-1403-0001
Unit Diary Clerk
 MC-1403-0013

Unitization
Defense Packing and Unitization
 DD-0419-0002

Upgrade
Upgrade Training A-10
 AF-1704-0282

Urban
Military Operations on Urban Terrain by Correspondence
 MC-1606-0006

Urinalysis
Medical Laboratory Craftsman (Chemistry and Urinalysis) by Correspondence
 AF-0702-0008
Medical Laboratory Technician (Chemistry and Urinalysis) by Correspondence
 AF-0702-0008

Utilities
Utilities Chief
 MC-1710-0035
Utilities Officer
 MC-1710-0036
Utilities Officer/Chief by Correspondence
 MC-1710-0048
Utilities System Journeyman by Correspondence
 AF-1732-0004
 AF-1732-0005

V/S
Electronics Fundamentals V/S Track #1
 CG-1715-0091
Electronics Fundamentals V/S Track #2
 CG-1715-0093

Validation
Validation and Reconciliation Clerk by Correspondence
 MC-1403-0017

Vegetables
Vegetables, Soups, Sauces, Gravies, and Beverages by Correspondence
 MC-1729-0024

Vehicle
Amphibious Vehicle Unit Leader
 MC-1710-0010
Amphibious Vehicle Unit Leaders (Enlisted)
 MC-1710-0010
Apprentice Vehicle Operator/Dispatcher by Correspondence
 AF-0419-0026
Assault Amphibious Vehicle Unit Leader
 MC-1710-0010
General Purpose Vehicle and Body Maintenance by Correspondence
 AF-1703-0022
General Purpose Vehicle and Body Maintenance Supervisor by Correspondence
 AF-1703-0022
General Purpose Vehicle Mechanic by Correspondence
 AF-1703-0018
Ground Ordnance Vehicle Maintenance Chiefs
 MC-1717-0006
Light Armored Vehicle Basic Repairman
 MC-1710-0049
Light Armored Vehicle Technician
 MC-1710-0050
Logistics Vehicle System Maintenance
 MC-1408-0019
Logistics Vehicle System Operator
 MC-1703-0030
Motor Vehicle Operator
 MC-1703-0027
Special Purpose Vehicle and Equipment Craftsman by Correspondence
 AF-1703-0021
Special Purpose Vehicle and Equipment Maintenance Apprentice by Correspondence
 AF-1703-0020
Special Purpose Vehicle and Equipment Mechanic by Correspondence
 AF-1703-0020
 AF-1710-0032
Special Purpose Vehicle and Equipment Supervisor by Correspondence
 AF-1703-0021
Special Vehicle Mechanic (Firetrucks) by Correspondence
 AF-1710-0033
Special Vehicle Mechanic (Refueling Vehicles) by Correspondence
 AF-1710-0034
Vehicle Body Mechanic by Correspondence
 AF-1710-0035
Vehicle Maintenance Control and Analysis Specialist by Correspondence
 AF-1710-0036

KEYWORD INDEX D-59

Vehicle Operations Supervisor by Correspondence
AF-0419-0023

Vehicle Operator/Dispatcher Journeyman by Correspondence
AF-1703-0023

Vehicle Recovery
MC-1710-0046

Vehicles
Preventive Maintenance and Operating Techniques for Heavy Vehicles by Correspondence
MC-1703-0034

Ventilation
Heating, Ventilation, Air Conditioning, and Refrigeration Journeyman by Correspondence
AF-1701-0012
AF-1701-0013
Heating, Ventilation, Air Conditioning, and Refrigeration Specialist by Correspondence
AF-1701-0012
AF-1701-0013

Videograph
Videograph B Fog Detector Maintenance
CG-1715-0127

Vietnamese
Vietnamese Basic
DD-0602-0238
Vietnamese Intermediate
DD-0602-0125

Virtual
Multiple Virtual Storage (MVS) Diagnostics
MC-1402-0053
Multiple Virtual Storage (MVS) Fundamentals and Logic
MC-1402-0052
Multiple Virtual Storage (MVS) Performance and Training
MC-1402-0054

Visual
Allied Visual Communications, Class C
CG-2205-0039
Radiotelegraph and Visual Communication Procedures by Correspondence
MC-1715-0133
Visual Information Production-Documentation Journeyman by Correspondence
AF-0505-0005
Visual Information Production-Documentation Specialist by Correspondence
AF-0505-0003
AF-0505-0005

War
Air War College
AF-1511-0011
Air War College Associate Programs Nonresident and Correspondence
AF-1511-0012
Air War College Correspondence Program
AF-1511-0012
Air War College Nonresident Seminar Program
AF-1511-0012
Marine Corps War College
MC-1511-0002
National War College
DD-1511-0002
DD-1511-0009

Warehousing
Basic Warehousing by Correspondence
MC-1405-0046
Enlisted Warehousing Intermediate
MC-1405-0036
Warehousing Operations by Correspondence
MC-1405-0021

Warfare
Amphibious Warfare
MC-2204-0070
Amphibious Warfare School Nonresident Program (AWSNP) by Correspondence Phase 2
MC-2204-0114
Amphibious Warfare School Nonresident Program (AWSNP) by Correspondence, Phase 1
MC-2204-0113
Chemical Warfare Defense by Correspondence
MC-0801-0013
Nuclear Warfare Defense by Correspondence
MC-0705-0001

Warfighting
Joint and Combined Warfighting
DD-0326-0003
Warfighting Skills Program (Warfighting) by Correspondence
MC-2204-0112

Warning
Aerospace Control and Warning Systems Operator by Correspondence
AF-1715-0754
Airborne Warning and Control Radar Specialist by Correspondence
AF-1715-0123
AF-1715-0801
Aircraft Control and Warning Radar Specialist by Correspondence
AF-1715-0005

Warrant
Reserve Warrant Officer Basic
MC-2204-0082
Terrain Analysis Warrant Officer Certification
DD-1601-0029
Warrant Officer Basic
MC-2204-0084

Waste
270' WMEC S/S Generator, Waste Heat Recovery System and Evaporator Operation and Maintenance, Class C
CG-1712-0005

Watch
Deck Watch Officer Navigation Rules by Correspondence
CG-1708-0014

Watchstanding
Watchstanding: The Conning Officer by Correspondence
CG-1708-0012

Water
Electrical/Electronic Control for Fire Tube Boilers/Oily Water Separator and Electrical Generator
CG-1715-0107
Electrical/Electronic Control for Fire Tube Boilers/Oily Water Separator/Electrical Generators Operation and Maintenance, Class C
CG-1715-0107
Marine Combat Instructor of Water Safety
MC-2204-0097
Reverse Osmosis Water Purification Unit by Correspondence
MC-1732-0003

Waterfront
Waterfront Protection by Correspondence
CG-1728-0012

Wave
Basic Techniques of Wave Form Measurement Using an Oscilloscope by Correspondence
AF-1715-0784

Waves
Antenna Construction and Propagation of Radio Waves by Correspondence
MC-1715-0136

Weapon Control
Weapon Control Systems Mechanic (A-7D, A-10, AC-130, F-5) by Correspondence
AF-1715-0050
Weapon Control Systems Mechanic (A-7D: AN/APQ-126) by Correspondence
AF-1715-0054
Weapon Control Systems Mechanic (F-4C/D: APQ-109/APA-165) by Correspondence
AF-1715-0052
Weapon Control Systems Mechanic (F-4E: APQ-120) by Correspondence
AF-1715-0053
Weapon Control Systems Technician (F-4C/D: APQ-109/APA-165) by Correspondence
AF-1715-0056
Weapon Control Systems Technician (F-4D/E) by Correspondence
AF-1715-0766
Weapon Control Systems Technician (F-4E: APQ-120) by Correspondence
AF-1715-0063

Weapons
Air National Guard Fighter Weapons Instructor
AF-1606-0151
B-52 Bomber Weapons Instructor
AF-1704-0273
Bomber Weapons Instructor B-52
AF-1704-0273
Bomber Weapons Instructor B1
AF-1704-0274
F-4 and AC-130 Weapons Control Systems Technician by Correspondence
AF-1715-0805
Fighter Weapons Instructor
AF-1704-0193
Fighter Weapons Instructor A-10
AF-1406-0061
Fighter Weapons Instructor Air Weapons Controller
AF-1406-0062
Fighter Weapons Instructor F-111
AF-1406-0063
Fighter Weapons Instructor F-15
AF-1406-0066
Fighter Weapons Instructor F-15E
AF-1406-0095
Fighter Weapons Instructor F-16
AF-1406-0064
Fighter Weapons Instructor Intelligence
AF-1406-0096
Ground Ordnance Weapons Chief
MC-1717-0008
Infantry Squad Leader: Weapons and Fire Support by Correspondence
MC-2204-0101
Inspection and Repair of Shoulder Weapons by Correspondence
MC-2204-0066
Intelligence Weapons Instructor
AF-1406-0096

KEYWORD INDEX

RF-4 Fighter Weapons Instructor
  AF-1606-0152
Weapons Administration by Correspondence
  CG-1408-0007
Weapons Control Systems Mechanic (A-7D, A-10, AC-130, F-5) by Correspondence
  AF-1715-0800
Weapons Instructor (AWC)
  AF-1406-0062
Weapons Officer by Correspondence
  CG-2205-0018

Weather
Cold Weather Medicine
  MC-0707-0001
Cold Weather Survival
  MC-0804-0002
Weather Briefer Analyst, Class C
  CG-1304-0020
Weather Briefer, Class C
  CG-1304-0020
Weather Equipment Specialist by Correspondence
  AF-1304-0014
Weather Forecasting and Flight Briefing by Correspondence
  CG-1304-0014
Weather Specialist by Correspondence
  AF-1304-0016

Welding
Aircraft Metals Technology (Welding) by Correspondence
  AF-1723-0011
  AF-1724-0007
Aluminum Welding, Class C
  CG-1710-0016
Metal Working and Welding Operations by Correspondence
  MC-1723-0008
Steel Welding, Class C
  CG-1710-0017

Welfare
Morale, Welfare, Recreation, and Services by Correspondence
  AF-0804-0002
Morale, Welfare, Recreation, and Services Journeyman by Correspondence
  AF-0804-0001

Wideband
Satellite and Wideband Communications Equipment Specialist by Correspondence
  AF-1715-0797
  AF-1715-0799
Wideband Communications Equipment Specialist by Correspondence
  AF-1715-0208
Wideband Communications Equipment Technician by Correspondence
  AF-1715-0015

Winter
Winter Mountain Leaders (A)
  MC-0804-0003
Winter Mountain Leaders (B)
  MC-0804-0004

Wire
Field Wire Operators
  MC-1715-0162

Wireman
Field Wireman
  MC-1715-0162

Wiring
Interior Wiring by Correspondence
  MC-1601-0033

Woman
Woman Officer Basic
  MC-2204-0035

Workstation
Standard Workstation Regional System Manager
  CG-1402-0003

WS-133
Missile Maintenance Specialist (WS-133) by Correspondence
  AF-1710-0031

Yeoman
Reserve Yeoman Basic
  CG-1409-0004
Yeoman First Class by Correspondence
  CG-1408-0006
Yeoman Reserve
  CG-1409-0004
Yeoman School, Class A
  CG-1409-0003
Yeoman Second Class by Correspondence
  CG-1408-0004
Yeoman Third Class by Correspondence
  CG-1408-0005

Course Number Index

This index is designed to provide access to courses listed in the course exhibit section of this volume. Military course numbers are listed in alphanumeric order. When the official military course number is found, note the exhibit ID number to its right. Locate that number in the proper course exhibit section.

Course exhibits are grouped by military service, using the following prefixes: **AF:** Air Force; **AR:** Army; **CG:** Coast Guard; **DD:** Department of Defense; **MC:** Marine Corps; and **NV:** Navy.

Course #	Exhibit ID
00.1a	MC-1406-0028
00.4	MC-1606-0008
0015-1	CG-1408-0037
0021-1	CG-1402-0001
0043-1	CG-1408-0039
01.12	MC-1406-0032
01.18j	MC-0501-0001
01.19f	MC-0501-0002
01.28	MC-1406-0025
01.28a	MC-1406-0025
01.35g	MC-1403-0016
01.36e	MC-1406-0026
01.36f	MC-1406-0026
01.39	MC-1406-0027
01.43	MC-1406-0024
01.43a	MC-1406-0024
0101-5	CG-1714-0022
0105-7	CG-1704-0052
0105-8	CG-1704-0052
0107-7	CG-1704-0033
0107-8	CG-1704-0033
0108-4	CG-1704-0041
0109-7	CG-1708-0004
0109-8	CG-1708-0004
011	MC-1405-0033
0110	CG-0709-0006
0115-5	CG-1710-0020
0115-6	CG-1710-0020
0119-6	CG-1714-0011
0121-7	CG-1715-0038
01230A	AF-1704-0194
0127-8	CG-1715-0135
0129-7	CG-1714-0027
0130-1	CG-0709-0006
0132-3	CG-1723-0006
0134-2	CG-1304-0018
0134-3	CG-1304-0022
0134-4	CG-1304-0022
0136-2	CG-1709-0002
0137-5	CG-2205-0026
0139-6	CG-1408-0036
0141-7	CG-1715-0136
0141-8	CG-1403-0002
	CG-1715-0136
0149-8	CG-1405-0008
0150-1	CG-1405-0008
0151-4	CG-1729-0005
0151-5	CG-1729-0005
0155-1	CG-1408-0006
0165-1	CG-1728-0038
0166-1	CG-1728-0014
0167-1	CG-1728-0038
0175-1	CG-1408-0006
017A47	DD-0602-0237
01AD	DD-0602-0232
01AD47	DD-0602-0232
01C	MC-1405-0034
01CX	DD-0602-0229
01CX47	DD-0602-0229
01DU	DD-0602-0223
01DU25	DD-0602-0223
01FR	DD-0602-0213
01FR25	DD-0602-0213
01GM	DD-0602-0215
01GM34	DD-0602-0215
01HE47	DD-0602-0235
01J	MC-1406-0022
01JT	DD-0602-0226
01JT25	DD-0602-0226
01LA	DD-0602-0222
01LA25	DD-0602-0222
01NR	DD-0602-0224
01NR25	DD-0602-0224
01PF	DD-0602-0230
01PF47	DD-0602-0230
01PL47	DD-0602-0112
01PQ	DD-0602-0212
01PQ25	DD-0602-0212
01RQ	DD-0602-0225
01RQ34	DD-0602-0225
01RU	DD-0602-0227
01RU47	DD-0602-0227
01S	MC-1403-0013
01T	MC-1403-0014
01TU	DD-0602-0231
01TU47	DD-0602-0231
01VN	DD-0602-0238
01VN47	DD-0602-0238
02.01	MC-1511-0001
02.10	MC-1606-0010
02.12	MC-1606-0012
02.13	MC-1606-0011
02.8a	MC-1606-0009
02.9a	MC-1606-0005
020-1812	MC-2204-0107
02010	AF-0505-0004
0209-9	CG-1708-0021
021-9	CG-1402-0001
02130B	AF-1704-0195
02130D	AF-1704-0196
0215-7	CG-1710-0021
0215-8	CG-1710-0021
02170	AF-0801-0005
0221-7	CG-1715-0037
	CG-1715-0146
0225-3	CG-1728-0034
0227-1	CG-1715-0137
0229-5	CG-1714-0026
0229-6	CG-1714-0026
0229-7	CG-1714-0026
0229-8	CG-1714-0026
0230	CG-0709-0005
0230-1	CG-0709-0005
0230-2	CG-0709-0005
0234-3	CG-1304-0016
0234-4	CG-1304-0016
0234-5	CG-1304-0016
0237-6	CG-1722-0015
0237-7	CG-1722-0015
0237-8	CG-1722-0015
0239-2	CG-2205-0023
0241-9	CG-1402-0004
0245-7	CG-1715-0063
0249-8	CG-1405-0009
0251-4	CG-1729-0004
0255-1	CG-1408-0004
0265-2	CG-1728-0032
0266-1	CG-1728-0042
0267-1	CG-1728-0032
0275-2	CG-1408-0004
03.16j	MC-1728-0002
03.3K	MC-1406-0023
03.3M	MC-1406-0023
03.54	MC-0803-0007
03.66a	MC-1606-0006
03.81	MC-1601-0048
03.81a	MC-1601-0048
03.82	MC-2204-0101
03.83	MC-2204-0100
03.97	MC-1601-0040
03035	AF-1715-0768
03036	AF-1723-0012
03037	AF-1714-0034
03039	AF-1715-0784
0309-2	CG-1708-0022
0309-3	CG-1708-0022
0309-4	CG-1708-0022
031	MC-2204-0051
	CG-1710-0019
0315-8	CG-1710-0019
033	MC-2204-0048
0337-1	CG-1722-0014
0337-7	CG-1722-0014
0337-8	CG-1722-0014
0337-9	CG-1722-0014
034	MC-2204-0047
035	MC-2204-0106
0350-1	CG-1405-0002
0351-4	CG-1729-0006
0375-1	CG-1408-0005
03GM	DD-0602-0205
03GM06	DD-0602-0205
03L	MC-2204-0097

COURSE NUMBER INDEX

03Q	MC-0804-0004
03TU	DD-0602-0216
03TU12	DD-0602-0216
04.15-1	MC-1403-0017
04.15-2	MC-1403-0017
0401-6	CG-1404-0004
0401-7	CG-1404-0004
0402-2	CG-1728-0035
0402-3	CG-1728-0035
0402-4	CG-1728-0035
0407-1	CG-2205-0019
0409-5	CG-0802-0018
0414-1	CG-0802-0017
0419-1	CG-2205-0037
0425-1	CG-1728-0033
043-9	CG-1408-0039
0431-1	CG-0802-0016
0440-3	CG-1704-0040
0440-4	CG-1704-0040
0441-3	CG-1704-0036
0442-2	CG-1715-0086
0443-2	CG-1704-0011
0444-3	CG-1715-0085
0444-4	CG-1715-0085
0445-2	CG-1704-0009
0445-3	CG-1704-0009
0447-2	CG-0419-0002
0448-3	CG-1704-0038
0448-4	CG-1704-0038
0450-1	CG-1722-0016
0452-3	CG-2205-0029
0452-4	CG-2205-0029
0453-3	CG-2205-0030
0453-4	CG-2205-0030
0453-5	CG-2205-0038
0454-2	CG-2205-0031
0454-3	CG-2205-0031
0455-2	CG-2205-0028
0455-3	CG-2205-0028
0456-1	CG-2205-0036
0456-2	CG-2205-0036
0456-3	CG-2205-0036
0463-5	CG-1304-0021
0464-2	CG-1708-0015
0467-1	CG-2205-0033
0469-3	CG-1708-0020
0469-4	CG-1708-0020
0480-2	CG-2205-0024
0483-1	CG-1704-0042
0483-2	CG-1704-0042
0485-1	CG-1107-0001
0486-1	CG-1104-0001
04GM	DD-0602-0214
04GM12	DD-0602-0214
04II	MC-1402-0058
0512-1	CG-0802-0007
0512-2	CG-1704-0035
0513-1	CG-0802-0008
0514-1	CG-1704-0037
0515-1	CG-1704-0035
0551-1	CG-1704-0021
0552-1	CG-1704-0034
0552-2	CG-1704-0034
0575	CG-1708-0019
0575-1	CG-1708-0019
0580-1	CG-1708-0024
0585-1	CG-1708-0025
05AE	DD-0602-0211
05AE16	DD-0602-0211
05AP	DD-0602-0210
05AP16	DD-0602-0210
05DG	DD-0602-0209
05DG16	DD-0602-0209
0641	CG-1722-0006
06613	AF-1406-0081
06AD	DD-0602-0220
06AD32	DD-0602-0220
06CM37	DD-0602-0233
06FR	DD-0602-0221
06FR24	DD-0602-0221
06GM24	DD-0602-0219
06KP	DD-0602-0236
06KP47	DD-0602-0236
06LA	DD-0602-0218
06LA24	DD-0602-0218
06TH	DD-0602-0239
06TH37	DD-0602-0239
0700-1	CG-2205-0034
07340	AF-0708-0004
0766-1	CG-1728-0036
07920	AF-0401-0002
07CM	DD-0602-0234
07CM37	DD-0602-0234
08.11	MC-1601-0035
08.12	MC-1601-0032
08.9	MC-1601-0039
0800-1	CG-1512-0001
09AE	DD-0602-0208
09AE24	DD-0602-0208
09AP	DD-0602-0206
09AP24	DD-0602-0206
09DG	DD-0602-0207
09DG24	DD-0602-0207
0D5	MC-1601-0046
0D7	MC-2204-0092
0DD	MC-2204-0075
0DY	MC-2204-0091
0GM	MC-1715-0026
0IC	MC-1405-0034
0ICC47	DD-0602-0021
101-4	CG-1714-0022
101-5	CG-1714-0022
103-7	CG-1715-0132
104-7222 (OS)	MC-1715-0163
104140Z	AF-1606-0020
105-6	CG-1704-0008
109-6	CG-1708-0004
1090	AF-2203-0052
10GM	DD-0602-0217
10GM24	DD-0602-0217
11.19c	MC-1712-0001
11.21b	MC-1710-0005
11.22a	MC-1732-0003
11.42b	MC-1715-0132
11.43	MC-1601-0033
11.61	MC-1730-0005
11.62	MC-1731-0001
11.69	MC-1710-0048
110	MC-1601-0025
11350B	AF-1704-0197
11350C	AF-1704-0094
11650	AF-1704-0198
11851A	AF-1715-0816
11851B	AF-1715-0811
119-3	CG-1714-0011
119-4	CG-1714-0011
119-5	CG-1714-0011
11B	MC-1714-0011
11D	MC-1730-0006
11E	MC-1710-0035
11K	MC-1714-0016
11L	MC-1730-0007
1200	AF-2203-0053
121-5924 (OS)	MC-1715-0181
121-5925 (PIPIII) (OS)	MC-1715-0180
121-6	CG-1715-0038
121-7	CG-1715-0038
12250	AF-1704-0199
127-6	CG-1715-0135
127-7	CG-1715-0135
129-6	CG-1714-0027
13.19h	MC-1601-0036
13.19i	MC-1601-0036
13.28d	MC-1601-0016
13.30	MC-1717-0005
13.31i	MC-1601-0034
13.32g	MC-1723-0008
13.34g	MC-1107-0001
13.34h	MC-1107-0001
13.35b	MC-1712-0006
13.35c	MC-1712-0006
13.38c	MC-1601-0012
13.39	MC-1601-0038
13.40	MC-1710-0040
13.42c	MC-1601-0037
13.43	MC-1712-0007
13.44b	MC-1601-0041
13.44c	MC-1601-0041
13.45	MC-1704-0007
13.52a	MC-1710-0039
130	MC-1601-0024
130-2	CG-0709-0006
132	MC-1723-0009
139-5	CG-1408-0036
13B	MC-1710-0001
	MC-1710-0045
13E	MC-1710-0043
13F	MC-1710-0044
13G	MC-1601-0027
13I	MC-2204-0069
14.42b	MC-1601-0015
143-2	CG-1715-0034
145-6	CG-1715-0134
145-7	CG-1715-0134
151-3	CG-1729-0005
155-8	CG-1408-0006
174501AA	AF-1406-0067
1745IDWN	AF-1406-0062
18.46	MC-2204-0098
1900	AF-0801-0001
1901	AF-0801-0002
1902	AF-0801-0003
1903	AF-0801-0004
19AF syllabus B/F-V5A-M	AF-1406-0100

COURSE NUMBER INDEX E-3

201-5	CG-1714-0025
201-6	CG-1714-0025
20150	AF-1606-0121
20151	AF-1606-0122
20170	AF-1606-0123
203-5	CG-1715-0031
203-6	CG-1715-0031
205-1	CG-1704-0007
205-2	CG-1704-0007
205-9	CG-1704-0007
20650	AF-1709-0027
207-7	CG-1704-0039
208-2	CG-1704-0018
208-4	CG-1704-0018
208-5	CG-1704-0018
	CG-1704-0043
209-8	CG-1708-0021
21.28c	MC-2204-0066
21.30	MC-2204-0064
21.31	MC-2204-0065
21.34	MC-2204-0099
219-1	CG-1714-0023
220	MC-1715-0180
225-2	CG-1728-0034
227-8	CG-1715-0137
227-9	CG-1715-0137
229-4	CG-1714-0009
22P	MC-1715-0181
23131	AF-1719-0009
23132	AF-1709-0029
23150	AF-1406-0073
23151	AF-1719-0010
23152	AF-1709-0028
23153	AF-0505-0005
23250	AF-0505-0003
23350	AF-1709-0030
236-4	CG-1709-0001
239-9	CG-2205-0023
241-6	CG-1404-0002
241-7	CG-1404-0002
241-8	CG-1404-0002
24250	AF-0802-0026
243-2	CG-1715-0133
245-6	CG-1715-0063
247	MC-1715-0162
24X	MC-1715-0113
25.15g	MC-1715-0136
25.36	MC-1715-0101
25.37	MC-1715-0135
25.54	MC-1715-0099
25.55	MC-1715-0091
25.56	MC-1715-0096
25.60	MC-1715-0137
25.61	MC-1715-0133
25.62	MC-1715-0134
250-1	CG-1405-0009
251-3	CG-1729-0004
25150	AF-1304-0016
254	MC-1409-0005
255-7	CG-1408-0004
255-8	CG-1408-0004
2570	AF-1304-0015
25A	MC-1717-0004
25U	MC-1403-0018
26.1	MC-1715-0130
26D	MC-1715-0120
27131	AF-1704-0214
27132	AF-1402-0073
27151	AF-1704-0215
27152	AF-1408-0095
272	MC-1715-0116
27350	AF-1704-0216
27450	AF-1704-0217
27550	AF-1406-0068
27650	AF-1715-0754
27750	AF-1715-0755
27E	MC-1715-0062
	MC-1715-0173
27F	MC-1715-0117
27M	MC-1715-0009
	MC-1715-0166
27P	MC-1715-0124
27V	MC-1715-0010
	MC-1715-0160
27W	MC-1715-0063
	MC-1715-0154
28-R-701.1	DD-0504-0001
28.6f	MC-1715-0131
28.6g	MC-1715-0131
28.7	MC-1715-0129
28E	MC-1715-0006
28T	MC-1715-0139
28V	MC-1405-0045
28W	MC-1715-0121
28Y	MC-1715-0141
29150	AF-0504-0003
2A252B	AF-1715-0810
2A351B	AF-1715-0815
2EK	MC-1402-0062
2S052	AF-1405-0072
30.1m	MC-1405-0046
30.1n	MC-1405-0046
30.20	MC-1405-0027
30.23a	MC-1408-0021
30.24	MC-1408-0022
30.3h	MC-1405-0021
30.9g-1	MC-1405-0044
3025	AF-1715-0746
30250	AF-1304-0014
3026	AF-1715-0747
3027	AF-1715-0753
3028	AF-1715-0748
303	MC-1401-0007
3030	AF-1714-0029
3031	AF-1715-0749
3032	AF-1715-0750
3034	AF-1715-0751
3035	AF-1714-0032
30351	AF-1715-0003
30352	AF-1715-0005
30353	AF-1715-0007
30353A	AF-1715-0776
30353B	AF-1715-0794
30353C	AF-1715-0795
	AF-1715-0812
3036	AF-1714-0033
3037	AF-1115-0008
30371	AF-1715-0009
30373	AF-1715-0011
3038	AF-1715-0752
30450	AF-1715-0208
30451	AF-1715-0012
30452A	AF-1715-0821
30452B	AF-1304-0017
30454	AF-1715-0227
30454A	AF-1715-0782
30454B	AF-1715-0792
30455	AF-1715-0013
30455A	AF-1715-0796
30455B	AF-1715-0798
30456	AF-1715-0014
30457A	AF-1715-0797
30457B	AF-1715-0799
30470	AF-1715-0015
30471	AF-1715-0017
30474	AF-1715-0465
30476	AF-1715-0021
30554	AF-1715-0765
30650	AF-1715-0024
30653	AF-1715-0031
30656	AF-1715-0786
30656A	AF-1715-0820
30656B	AF-1715-0813
30750	AF-1715-0390
30770	AF-1715-0034
30950	AF-1715-0036
30A	MC-1405-0039
30G	MC-1405-0037
30H	MC-1405-0036
30V	MC-1405-0035
315-7	CG-1710-0019
31650G	AF-1715-0038
31650T	AF-1715-0043
31653	AF-1715-0045
319-7	CG-1714-0005
319-8	CG-1714-0024
32150	AF-1715-0046
32151G	AF-1715-0047
32152	AF-1715-0050
	AF-1715-0800
32152P	AF-1715-0052
32152Q	AF-1715-0053
32152S	AF-1715-0054
32172P	AF-1715-0056
32172Q	AF-1715-0063
32252A	AF-1715-0068
32252B	AF-1715-0070
32252C	AF-1715-0071
32450	AF-1715-0073
32470	AF-1715-0074
32550	AF-1715-0075
32551	AF-1715-0080
32650	AF-1715-0076
32650C	AF-1715-0077
32650D	AF-1715-0078
32653A	AF-1715-0081
32653B	AF-1715-0082
32654A	AF-1715-0087
32654B	AF-1715-0086
32655A	AF-1715-0093
32655B	AF-1715-0092
32656	AF-1715-0091
32656A	AF-1715-0090
32656B	AF-1704-0218
	AF-1715-0097

COURSE NUMBER INDEX

32656C	AF-1715-0096
32657	AF-1715-0095
32657A	AF-1715-0094
32657B	AF-1715-0098
32657C	AF-1715-0099
32658	AF-1715-0100
32658A	AF-1715-0101
32658B	AF-1715-0106
32658C	AF-1704-0220
32852	AF-1715-0123
	AF-1715-0801
32853	AF-1715-0466
32855	AF-1715-0122
32870	AF-1715-0116
33.10g	MC-1729-0043
33.16c	MC-1729-0041
33.16d	MC-1729-0041
33.16e	MC-1729-0041
33.18b	MC-1729-0044
33.19	MC-1729-0024
33.20	MC-1729-0022
33.21a	MC-1729-0039
33.24	MC-1729-0040
33.24a	MC-1729-0040
33.25	MC-1729-0045
33.33	MC-1729-0042
33.4k	MC-1729-0038
33.8f	MC-1729-0032
339-7	CG-2205-0017
33A	MC-1729-0035
33K	MC-1729-0006
33L	MC-1729-0009
33N	MC-1729-0034
34.10a	MC-1401-0004
34.11	MC-1401-0010
34.12	MC-1401-0011
34.14	MC-1401-0012
34.20c	MC-1401-0009
34.20d	MC-1401-0009
34.21a	MC-1401-0008
34.5c	MC-1405-0025
349-9	CG-1405-0002
34A	MC-1403-0009
34B	MC-1401-0006
34D	MC-1401-0013
34M	MC-1403-0010
34N	MC-1402-0055
35.13a	MC-1703-0032
35.13b	MC-1703-0032
35.15b	MC-1703-0033
35.24a	MC-1703-0034
35.25b	MC-1703-0031
35.33	MC-1710-0041
35.8b	MC-1703-0024
35.8c	MC-1703-0024
35.9f	MC-1703-0022
351-3	CG-1729-0006
352	MC-1712-0004
355-1	CG-1408-0005
355-2	CG-1408-0005
355-3	CG-1408-0005
35C	MC-1712-0008
35F	MC-1703-0025
35H	MC-1703-0028
35X	MC-1703-0027
35Z	MC-1703-0030
36150	AF-1715-0802
36151	AF-1715-0126
36151A	AF-1714-0036
36151B	AF-1714-0037
36251	AF-1715-0127
	AF-1715-0803
36253	AF-1714-0038
36254	AF-1715-0131
3625B	AF-1715-0130
365-1	CG-1728-0009
38R	MC-1402-0051
38U	MC-1402-0050
39150	AF-1402-0069
39170	AF-1107-0002
395	MC-1402-0013
39G	MC-1402-0042
3A/24-710	DD-1601-0030
3E050	AF-2203-0057
3S200	AF-1406-0078
403-3	CG-1728-0013
404-1	CG-2205-0013
40450	AF-1715-0136
40450A	AF-1715-0806
40450B	AF-1715-0807
40451	AF-1715-0139
406-3	CG-2205-0018
409-701/4M-701	DD-1601-0021
40L	MC-1402-0023
41.6	MC-1729-0046
41.7	MC-0406-0001
410-1	CG-1408-0007
411-200	DD-1601-0025
411-200/411-81C10	DD-1601-0025
411-207	DD-1601-0026
411-208	DD-1713-0007
411-81C10	DD-1601-0025
411-81C30	DD-1713-0007
411-APPS	DD-1601-0026
41151A	AF-1704-0265
41152A	AF-1704-0264
412-101	DD-1601-0023
412-103	DD-1722-0001
412-104	DD-1601-0022
412-82D10	DD-1601-0023
416-4	CG-1708-0006
418-1	CG-1708-0008
420-1	CG-1304-0014
421-1	CG-2205-0032
42153	AF-1704-0187
42173	AF-1704-0192
422-1	CG-0802-0009
423-1	CG-1728-0012
42350	AF-1714-0003
42351	AF-1704-0200
42352	AF-1717-0005
42353	AF-1704-0191
42354	AF-1704-0201
42373	AF-1704-0202
424-1	CG-1708-0011
426-1	CG-1708-0013
42652	AF-1704-0137
42653	AF-1704-0204
42654	AF-1704-0205
427	CG-1728-0007
427-3	CG-1728-0007
42750	AF-1723-0008
42751	AF-1717-0013
42752	AF-1723-0005
42753	AF-1733-0002
42754	AF-1723-0009
42755	AF-1723-0010
429-1	CG-0802-0011
430-1	CG-1708-0012
4313	DD-0505-0003
	DD-0505-0007
43150	AF-1704-0206
43151	AF-1704-0102
43152A	AF-1704-0209
43152C	AF-1704-0181
43152E	AF-1704-0188
43152G	AF-1704-0184
43152J	AF-1704-0210
43152Z	AF-1704-0211
432-1	CG-1715-0074
4321	DD-0505-0003
	DD-0505-0007
433-1	CG-1708-0010
4391	DD-0505-0003
	DD-0505-0007
43E	DD-1709-0002
43F	MC-1601-0030
440-2	CG-0802-0012
442	MC-1406-0033
44350G	AF-1710-0031
445-1	CG-1704-0009
445-2	CG-1704-0009
44550G	AF-1717-0026
447-1	CG-0419-0002
448-1	CG-1408-0003
448-2	CG-1408-0003
45070	AF-1704-0221
45070A	AF-1704-0222
45070B	AF-1704-0223
451-9	CG-1708-0009
45154A	AF-1715-0772
45154B	AF-1715-0771
45155	AF-1715-0756
45155A	AF-1715-0780
45155B	AF-1715-0791
45155C	AF-1715-0785
45156A	AF-1715-0773
45156B	AF-1704-0224
45157	AF-1408-0097
45177	AF-1715-0793
45243C	AF-1715-0779
45250A	AF-1405-0069
45250B	AF-1715-0758
45251	AF-1704-0254
45251A	AF-1715-0767
45251C	AF-1704-0225
45252	AF-1715-0808
45252A	AF-1715-0759
45252B	AF-1704-0226
	AF-1704-0255
45252C	AF-1714-0039
45253	AF-1715-0804
45253A	AF-1715-0760
45253B	AF-1704-0227
45253C	AF-1704-0263
45274	AF-1704-0228
45352	AF-1715-0817
45371	AF-1704-0261

COURSE NUMBER INDEX E-5

45450A	AF-1704-0256
45450B	AF-1704-0257
45451	AF-1710-0038
45453	AF-1704-0258
45455	AF-1714-0045
45470A	AF-1704-0229
45470B	AF-1710-0037
45474	AF-1704-0230
45474A	AF-1704-0231
45517E	AF-1704-0236
45550	AF-1715-0818
45550A	AF-1715-0777
45550B	AF-1715-0764
45571	AF-1715-0761
45571A	AF-1704-0232
45571B	AF-1704-0233
45571C	AF-1704-0234
45571D	AF-1704-0235
	AF-1704-0259
45571F	AF-1704-0237
45571G	AF-1704-0238
45571H	AF-1704-0239
45571J	AF-1704-0240
45571K	AF-1704-0241
45571X	AF-1715-0761
45572A	AF-1704-0242
45572B	AF-1704-0243
45572C	AF-1704-0244
45572D	AF-1704-0245
45572E	AF-1704-0246
	AF-1704-0260
45573A	AF-1715-0805
45573C	AF-1715-0766
45650	AF-1715-0774
45651	AF-1715-0769
45651A	AF-1715-0819
45671	AF-1715-0783
45750	AF-1704-0247
45753A	AF-1704-0248
45753B	AF-1715-0770
45753C	AF-1715-0789
45772	AF-1704-0249
45851	AF-1724-0006
45870A	AF-1717-0027
45870B	AF-1723-0011
	AF-1724-0007
45872A	AF-1717-0028
45872B	AF-1704-0250
46250	AF-1715-0035
464-1	CG-1708-0015
46650A	AF-1714-0041
46650B	AF-1715-0814
469-2	CG-1708-0005
470-1	CG-1403-0001
47230	AF-1703-0020
47250	AF-1710-0032
47251A	AF-1710-0033
47251B	AF-1710-0034
47252	AF-1703-0018
47253	AF-1710-0035
47254	AF-1710-0036
47271	AF-1703-0021
47275	AF-1703-0022
476-1	CG-1304-0006
476-2	CG-1304-0006
477-1	CG-1710-0011
478-2	CG-1728-0005
491-400	DD-1601-0020
491-400/491-81Q10	DD-1601-0020
491-402	DD-1601-0028
491-81Q30	DD-1601-0028
49151	AF-1402-0070
49152	AF-1402-0071
49251	AF-1715-0781
49350A	AF-1715-0778
49350B	AF-1715-0788
49370	AF-1715-0790
49650	AF-1715-0787
4E-F15	MC-1717-0007
4M-21C	DD-1601-0021
4M-701	DD-1601-0021
4M-701AF	DD-1713-0006
4M-841A	DD-1601-0029
4M/41-708	DD-1601-0031
4N-215D	DD-1601-0029
50AY9746	AF-0703-0013
5123	AF-1402-0072
53151	AF-1723-0007
54230	AF-1714-0023
54250	AF-1714-0028
54250A	AF-1714-0042
54250B	AF-1714-0044
54252A	AF-1714-0030
	AF-1714-0043
54252B	AF-1714-0040
54272A	AF-1408-0096
54272B	AF-1714-0031
54370	AF-1714-0024
54533A	AF-1710-0039
54533B	AF-1710-0040
54550	AF-1730-0016
54550A	AF-1701-0012
54550B	AF-1701-0013
54551	AF-1601-0030
54552A	AF-1701-0010
54552B	AF-1701-0011
54572	AF-1701-0009
55130	AF-1601-0048
55131	AF-1710-0029
55150	AF-1601-0047
55151	AF-1710-0028
55151A	AF-1601-0051
55151B	AF-1601-0052
55231	AF-1710-0030
55235	AF-1710-0027
55250A	AF-1732-0015
55250B	AF-1732-0016
55252	AF-1723-0013
55252A	AF-1723-0014
55252B	AF-1723-0015
55255	AF-1710-0021
55271	AF-1601-0049
55275	AF-1710-0023
55350	AF-1601-0046
55530	AF-1405-0060
56650	AF-0101-0005
56651	AF-1710-0041
56651A	AF-1732-0004
56651B	AF-1732-0005
57.6d	MC-0801-0013
57.7i	MC-0705-0001
570	DD-0505-0004
570-46Q/R30	DD-0504-0014
570-46Q/R40	DD-1709-0003
570-46R10	DD-0505-0004
570-71Q10	DD-0504-0001
570-71Q20	DD-0504-0001
570-ASIJ8	DD-1709-0002
570-F3	DD-0505-0003
	DD-0505-0007
57150	AF-1728-0038
57170	AF-1728-0046
58.1d	MC-1728-0008
58.2	MC-1728-0007
58X	MC-1406-0034
5OB7921 002	DD-0504-0018
5OBA7921 002 (USAF)	DD-0504-0009
60.1g	MC-1704-0008
60.2	MC-1710-0004
60.6a	MC-1405-0047
60250	AF-0419-0031
60251	AF-0419-0032
60253	AF-0419-0037
60330	AF-0419-0026
60350	AF-1703-0023
60370	AF-0419-0023
60531	AF-0419-0029
60550	AF-0419-0021
60551	AF-0419-0004
60555	AF-0419-0038
60C	MC-2204-0095
611-1	CG-1303-0001
611-2147	MC-1710-0049
61130	AF-1729-0013
61150	AF-1729-0014
61170	AF-0419-0028
61231	AF-1729-0015
61251	AF-1729-0016
61271	AF-1729-0017
62230	AF-1729-0010
62270	AF-1729-0012
62370	AF-1408-0107
631-1	CG-1708-0014
63150	AF-1703-0019
640-2149	MC-1717-0006
640-2181	MC-1717-0008
641-2111	MC-2204-0111
641-F1	MC-2204-0110
642-2131	MC-2201-0001
643-2146	MC-1710-0051
643-2147 (OS)	MC-1710-0050
645-55B10 (OS)	MC-2204-0109
64531	AF-1405-0018
64550	AF-1405-0051
64550A	AF-1405-0064
64551	AF-1405-0013
	AF-1405-0070
64570	AF-1405-0059
	AF-1405-0071
64571	AF-1405-0033
64572	AF-1405-0066
651-1	CG-1715-0076
65150	AF-1408-0071
65170	AF-1408-0072
6601	AF-1405-0067
6603	AF-1405-0068
6604	AF-1408-0088
6605	AF-1408-0089
6606	AF-1408-0087
6608	AF-1408-0090
6610	AF-1408-0091
6611	AF-1408-0106
6612	AF-1408-0110
664	MC-1715-0040

COURSE NUMBER INDEX

66F	MC-1715-0011
	MC-1715-0155
66G	MC-1715-0127
66K	MC-1715-0122
66L	MC-1715-0144
66V	MC-1715-0148
66W	MC-1715-0059
670-2171	MC-1714-0019
670-41B10	DD-1721-0003
670-601	DD-1721-0003
670-602	DD-1721-0003
6701	AF-1408-0092
6702	AF-1408-0093
67251A	AF-1408-0073
67251B	AF-1408-0074
67251C	AF-1408-0075
67251D	AF-1408-0076
67251E	AF-1408-0077
67252A	AF-1408-0078
67252B	AF-1408-0079
67252C	AF-1408-0080
67273A	AF-1408-0081
67273B	AF-1408-0082
674	MC-1606-0013
67450	AF-1408-0109
	AF-1408-0111
67L	MC-1715-0031
	MC-1715-0165
67M	MC-2204-0031
	MC-2204-0090
67P	MC-2204-0030
	MC-2204-0089
690-620	DD-1706-0003
690-ASIJ6	DD-1706-0003
69150	AF-1408-0083
69170	AF-1408-0084
70150	AF-1408-0085
70170	AF-1408-0086
702-2161	MC-1723-0010
70250	AF-1406-0045
70270	AF-1406-0046
70330	AF-1719-0008
70350	AF-1408-0098
704-1316	MC-1703-0035
70550	AF-1406-0047
720-1161 (OS)	MC-1730-0006
7200	MC-1408-0028
72G	MC-1715-0157
72H	MC-1715-0150
72M	MC-2204-0089
72N	MC-1704-0009
72P	MC-2204-0090
73150	AF-1406-0077
73150A	AF-1402-0075
73150B	AF-1402-0076
73150C	AF-1402-0077
73150D	AF-1402-0078
73250	AF-1406-0076
73251	AF-1406-0049
73274	AF-1406-0051
73371	AF-1406-0052
73470A	AF-1406-0053
73470B	AF-1406-0054
740-303	DD-1719-0006
740-306	DD-1601-0019
740-309	DD-1719-0007
740-310	DD-1719-0011
740-310 (83F30)	DD-1719-0011
740-83E10	DD-1601-0019
740-83F10	DD-1719-0006
7400A	MC-2204-0112
74131	AF-1406-0055
74151	AF-1406-0056
74171	AF-1406-0057
74250	AF-1406-0058
74270	AF-1406-0059
7504	AF-1406-0079
75132	AF-1406-0071
7515	AF-1406-0080
75150	AF-1406-0069
75171	AF-1406-0072
75172	AF-1406-0070
75350	AF-2203-0054
7802A	AF-1511-0007
78150A	AF-0804-0001
78150B	AF-0804-0002
79150	AF-0401-0001
79151	AF-0505-0002
7G	DD-0504-0018
7G (USMC)	DD-0504-0009
7G-46A	DD-0504-0009
	DD-0504-0018
7G-46A (USCG)	DD-0504-0009
7G-F11/570-F2	DD-0504-0015
7G-F3	DD-0504-0013
7H-F11	DD-1728-0003
8000	AF-1606-0119
	MC-1408-0027
8001	AF-1606-0120
807501DOZN	AF-1406-0096
808	MC-2204-0038
	MC-2204-0088
809	MC-2204-0038
	MC-2204-0088
8100	AF-1728-0042
81150	AF-1728-0045
81152	AF-1728-0044
81C	MC-0327-0001
81H	MC-1728-0006
81K	MC-1708-0003
81N	MC-2204-0014
82170	AF-1728-0039
822-F1 (JT)	DD-0419-0001
822 F13 (JT)	DD-0419-0004
822-F2 (JT)	DD-0419-0002
822-F7 (JT)	DD-0419-0003
8500	MC-2204-0113
8600	MC-2204-0114
8700	MC-1408-0025
8800	AF-1728-0043
88150	AF-1407-0003
89350	AF-1408-0108
	AF-1408-0113
89370	AF-1408-0112
8B-F1 (JT)	DD-0419-0001
8B-F16 (JT)	DD-0419-0006
8B-F2 (JT)	DD-0419-0002
8B-F26 (JT)	DD-0419-0005
8B-F7 (JT)	DD-0419-0003
90150	AF-0709-0033
90250	AF-0703-0012
90250A	AF-0703-0018
90250B	AF-0703-0016
90252	AF-0703-0015
90270	AF-0703-0011
90370	AF-0705-0003
90412	AF-0709-0030
90413	AF-0709-0031
90550	AF-0799-0003
90650	AF-0709-0027
90650A	AF-0799-0008
90650B	AF-0799-0009
90650C	AF-0799-0010
90750	AF-0707-0010
90850	AF-0707-0009
91150	AF-0709-0032
91255	AF-0706-0003
91350	AF-0704-0004
91450	AF-0703-0017
91451	AF-0708-0002
91550	AF-1406-0094
91850	AF-1715-0745
92470A	AF-0702-0008
92470B	AF-0702-0009
92470C	AF-0702-0010
92650	AF-0104-0003
92670	AF-0104-0002
9711	AF-0703-0014
98150	AF-0701-0003
98250	AF-0701-0018
99104-5	AF-1715-0763
99105-5	AF-1715-0762
99106-5	AF-1715-0775
A-532-0015	MC-1402-0013
A-532-0030	MC-1402-0051
A-532-0031	MC-1402-0034
A-570-0010	DD-0504-0012
	DD-0505-0004
A-570-0011 (USN)	DD-0504-0001
A-570-0013	DD-0504-0015
A-580-0017	DD-1709-0002
A-740-0025	DD-1719-0007
A-7G-0010	DD-0504-0018
A-7G-0010 (USN)	DD-0504-0009
A-7G-0011/CDP 0313	DD-0505-0005
A-7G0013	DD-0504-0013
A0108O1	MC-2201-0001
A01GBC1	MC-2204-0110
A01GBD1	MC-1710-0049
A01GBH1	MC-1710-0050
A01GBS1	MC-1717-0008
A01GBT1	MC-1717-0006
A01RGZ1	MC-1717-0007
A0214A1	DD-1601-0023
A0214R1	DD-1601-0025
A021A1	DD-1601-0023
A024621	DD-1601-0019
A0247B1	DD-1601-0026
A02PAA1	DD-1719-0006
A10001A0PD	AF-1704-0284
A10001DOPN	AF-1406-0061
A1000B	AF-1704-0282
A1000B/TXA/TXB/TXC	AF-1704-0282

COURSE NUMBER INDEX E-7

A1000BOAPD	AF-1704-0281
A1000BOOAD	AF-1704-0285
A1000IOAPD	AF-1704-0284
A1000TRAAD	AF-1704-0285
A1000TXA	AF-1704-0282
A1000TXAPD	AF-1704-0283
A1000TXB	AF-1704-0282
A1000TXC	AF-1704-0282
A1143D1 (43D)	DD-0504-0015
A13 TBM	MC-2204-0107
A13GBN1	MC-1710-0047
A43570C-4	AF-1704-0033
A435X0C-1	AF-1704-0031
A7000FW	AF-1704-0193
AAB	MC-1406-0021
AAC	MC-1403-0015
AAD	MC-1405-0032
AAE	MC-1403-0011
AAF	MC-1405-0031
ABA79130-1 (USAF)	DD-0504-0001
ACC	MC-1601-0026
ACE	MC-1710-0036
ACN	MC-1601-0043
ACP	MC-1714-0017
	MC-1714-0018
ACQ 101	DD-1408-0012
ACQ 201	DD-1408-0020
ACS	MC-1601-0028
ACT	MC-1723-0007
ACU	MC-1710-0033
ACX	MC-1710-0034
AFIS-AEJC	DD-0505-0006
AFIS-BBC	DD-0505-0004
AFIS-IB-RC	DD-0504-0016
AFIS-IJ-RC	DD-0504-0017
AFIS-IPC	DD-1709-0002
AFIS-PAOC-RC	DD-0504-0013
AHK	MC-1601-0045
ANC-5	CG-2205-0008
ANC-AM	CG-1715-0131
ANC-ANT	CG-1406-0009
ANC-AP	CG-2205-0025
ANC-FD	CG-1715-0127
ANC-LT	CG-1715-0128
ANC-M	CG-1712-0007
ANC-OPS	CG-1408-0042
ANC-RB	CG-1715-0127
AO214J1	DD-1601-0020
AO214S1	DD-1601-0021
AO215F1	DD-1706-0003
AO2DC1	DD-1722-0001
AVCC	CG-2205-0039
B-300-0013	MC-0709-0002
B-300-0053	MC-0709-0002
B1000IDOAE/WE	AF-1704-0274
B1BWIC	AF-1704-0274
B52001DOAB/JB/EB	AF-1704-0273
B5200IDOAE/WE/EE	AF-1704-0273
B52BWIC	AF-1704-0273
BCE 101	DD-1115-0001
BCE 208	DD-1402-0005
BCF 101	DD-1115-0001
BCF 208	DD-1402-0005
BCF 301	DD-1408-0017
BED	MC-1715-0068
BEZ	MC-1715-0143
	MC-1715-0171
BF2	MC-1715-0118
BFM 102	DD-1408-0014
BFM 203	DD-1408-0015
BMA	CG-1708-0016
C30BR4021 004	AF-1704-0251
C3OBR4001 001/002	AF-1704-0251
C3OBR4021 003	AF-1704-0252
C3OBR4021 004	AF-1704-0253
C3OZR6421 000	AF-1601-0050
CAJ	MC-1710-0046
CDA	MC-1408-0016
	MC-1408-0024
CDG	MC-1408-0018
CE7	MC-2204-0054
CEB	MC-1712-0005
CEJ	MC-0419-0005
CEL	MC-1703-0029
CEO	MC-1715-0025
	MC-1715-0179
CEU	MC-1408-0020
CEW	MC-2204-0071
CEX	MC-1405-0043
CEY	MC-1601-0031
CEZ	MC-1601-0029
CFF	MC-2204-0102
CGE	MC-1715-0178
CGM	MC-1715-0024
	MC-1715-0161
CGN	MC-1715-0158
CHJ	MC-1715-0177
CHK	MC-1408-0023
	MC-1715-0167
CO9	MC-1710-0038
COG	MC-1405-0038
COM-02	CG-1715-0112
COM-03	CG-1715-0124
COM-04	CG-1715-0121
COM-05	CG-1715-0138
COM-12	CG-1715-0072
CON 102	DD-1405-0003
CON 103	DD-1402-0009
CON 104	DD-1405-0004
CON 105	DD-1405-0001
CON 106	DD-1402-0008
CON 201	DD-0326-0007
CON 221	DD-1405-0005
CON 223	DD-1402-0007
CON 231	DD-1405-0006
CON 241	DD-1402-0006
COX C	CG-1708-0017
CUMM (R)	CG-1712-0006
CY-200	DD-1404-0001
CY-300	DD-1404-0004
D2C	MC-1402-0052
D2H	MC-1402-0053
D2J	MC-1402-0054
D3C	MC-1715-0123
DA2	MC-1729-0036
DA9	MC-1402-0056
DC ADV(R)	CG-1710-0015
DC-1	CG-1710-0017
DC-2	CG-1710-0016
DC-3	CG-0801-0001
DCA	CG-1710-0014
DCJ	DD-0504-0013
	DD-0505-0005
DCO	CG-2202-0003
DPH	MC-1715-0125
DPX	MC-1715-0168
DQH	MC-1715-0119
DQJ	MC-1715-0145
DQK	MC-1715-0149
DRF	MC-1715-0169
DRG	MC-1402-0065
DRJ	MC-1402-0063
DRM	MC-1715-0170
E2D	MC-1715-0089
	MC-1715-0140
E2H	MC-1715-0153
E2U	MC-1715-0156
E2V	MC-1715-0159
E2W	MC-1715-0152
E3G	MC-1715-0174
E3H	MC-1715-0175
E5ABD70330 000	DD-1719-0007
E5OXZ4924 002	MC-1402-0051
E5OZX4924 000/RRM	MC-1402-0034
EA-280	DD-1715-0016
EAT	MC-1601-0042
EAW	MC-1712-0009
EM-1	CG-1715-0149
EM-17	CG-1715-0105
EM-18	CG-1714-0021
EM-2	CG-1715-0148
EM-20	CG-1715-0104
EM-21	CG-1715-0107
EM-25	CG-1710-0013
EM-26	CG-1715-0106
EM-ADV-(R)	CG-1714-0020
EMA	CG-1714-0018
ES-403	DD-1715-0008
ES-411	DD-1715-0011
ESABD70330 000	DD-1601-0019
F-16 COIOOPL/M	AF-1704-0271
F-16COIOOPL/M	AF-1704-0271
F-16COTXOPL	AF-1704-0272
F-16COTXOPL/M	AF-1704-0272
F-400 FWS	AF-1606-0151
F-V5A-A(T-37)	AF-1406-0074
F-V5A-B(T-38)	AF-1406-0074
F-V5A-E	AF-1406-0099
F-V5A-Y	AF-1406-0074
F-V5A-Z	AF-1406-0074
F-V5N-A/B	AF-1406-0101
F11101DOA1	AF-1406-0063
F11101DOW1	AF-1406-0063
F1110IDOAC	AF-1406-0063

COURSE NUMBER INDEX

Course	Number
F1110IDOWC	AF-1406-0063
F1500100AL/T	AF-1606-0157
F1500B00 AL/T	AF-1606-0158
F1500IDOPN	AF-1406-0066
F1500TX0AL/T	AF-1606-0159
F15AC100AT	AF-1606-0157
F15ACB00AT	AF-1606-0158
F15ACTX00AT	AF-1606-0159
F15ACTX0AT	AF-1606-0159
F15E01DOAN/WN	AF-1406-0095
F15E0B00AL/WL	AF-1704-0269
F15EOB00AL/WL	AF-1704-0269
F15EOIOAL/WL	AF-1406-0098
F15EOIOOAL/WL	AF-1406-0097
F15EOTX0AL/WL	AF-1704-0266
F16001DOPN	AF-1406-0064
F16A0B00AL/M	AF-1704-0280
F16AOTX0PL/M	AF-1704-0275
F16C0CX0PL	AF-1704-0267
F16CGB00OPL	AF-1704-0279
F16CGCLOPL	AF-1704-0278
F16CGCX0PL	AF-1704-0268
F16CGLOOPL	AF-1704-0278
F16CGTX0PL	AF-1704-0276
F16COB00AL	AF-1704-0277
F16COB0OPL/M	AF-1704-0270
F16COCX0PL/M	AF-1704-0267
F16COIOOAL	AF-1406-0102
F50000A1AN	AF-1406-0065
FAC	MC-1729-0037
FAD	MC-1729-0047
FG-46B/570-F4	DD-0505-0005
FG5	MC-1704-0004
	MC-1704-0009
FGV	MC-1715-0020
	MC-1715-0151
FGX	MC-1715-0046
	MC-1715-0142
FT-2	CG-1715-0103
FT-3	CG-1402-0002
FT-5	CG-1715-0103
G-2G-4318	MC-1606-0013
G-551-4407	MC-1402-0058
G3ABR40430 005	AF-1715-0741
G3AZR32470 017	AF-1715-0493
G3AZR32470 018	AF-1715-0740
G3AZR32470 023	AF-1715-0743
G502A7921 001	DD-0504-0013
G502A7921 004	DD-0505-0005
G50BD5731 000	DD-1713-0006
G5ABA79131 000	DD-0505-0004
G5ABD22230 000	DD-1601-0023
G5AZA79151	DD-0505-0007
G5AZD22150 000	DD-1601-0026
GA2A79150	DD-0504-0015
GM-1	CG-2204-0001
GM-10	CG-1715-0147
GM-3	CG-2204-0002
GMA	CG-1714-0017
GSAZA79150 002	DD-1709-0002
H-010-3921	MC-2204-0097
H-010-3923	MC-1601-0045
H-201-3334	MC-1405-0048
H-250-3166	MC-1601-0047
H-2E-3102	MC-2204-0091
H-2E-3108	MC-2204-0092
H-2E-3110	MC-2204-0093
H-2E-3113	MC-2204-0094
H-2E-3114	MC-2204-0095
H-2E-3741	MC-1708-0003
H-2E-3742	MC-2204-0096
H-2E-3743	MC-1708-0002
H-2G-3615	MC-1606-0013
H-4C-3331	MC-1601-0046
H-551-3553	MC-0419-0008
H-551-3554	MC-1402-0059
H-551-3555	MC-1402-0058
H-8A-3550	MC-1402-0061
H-8A-3551	MC-1402-0060
H-8A-3558	MC-0419-0007
H2C	MC-2204-0108
INTELIDOZN	AF-1406-0096
IRM 201	DD-0326-0006
IRM 303	DD-0326-0005
J30AR6011 000	AF-0419-0035
J30AR6711 001	AF-1408-0103
J30AR6751 000	AF-1408-0104
J30BR21A1 008	AF-1704-0253
J30BR21A1 009	AF-1704-0252
J30BR4021	AF-1704-0251
J30BR41A1 001	AF-0709-0034
J30BR6051 000	AF-0419-0034
J30BR6721 001	AF-1408-0101
J30BR6721 002	AF-1408-0102
J30BR6721 003	AF-1401-0002
J30BR6731 000	AF-1408-0100
J30BR6741 001	AF-1401-0019
J30BR9021 001	AF-0709-0034
J30LR21A1 008	AF-1704-0286
J30RR46A3004	AF-0709-0035
J30RR6011 000	AF-0419-0036
J30RR6051 002	AF-0419-0033
J30ZR4000 011	AF-0707-0011
J30ZR46A3005	AF-0709-0036
J30ZR46N3D002	AF-0703-0020
J30ZR6751 000	AF-1401-0018
J30ZR6784 002	AF-1402-0074
J30ZR9756A	AF-0703-0022
J30ZR9756D002	AF-0703-0020
J3OBR21A1 006	AF-1704-0251
J3OBR4021 001	AF-1704-0253
J3OBR65F1 003	AF-1401-0002
J3ORR9711 004	AF-0709-0035
J3OZP46S3 002	AF-0703-0021
J3OZP9731 003	AF-0703-0019
J3OZP9736 002	AF-0703-0021
	AF-0703-0023
J3OZR6421 000	AF-1601-0050
J3OZR6711 000	AF-1408-0105
J3OZR9576 001	AF-0707-0011
J3OZR9711 005	AF-0709-0036
J5Z046N3000	AF-0703-0023
J5Z046S1000	AF-0703-0019
J5Z046S3000	AF-0703-0021
J5OZO9716 002	AF-0709-0037
L302R64P3 016	DD-1405-0002
L3Z	MC-1402-0068
L40ST64P3 016	DD-1405-0002
L5OZN6916 000	DD-1408-0006
LMDC 400	AF-1408-0004
LMDC 501	AF-1406-0033
LOG 304	DD-0326-0004
LOR-02	CG-1715-0111
LOR-03	CG-1715-0118
LOR-04	CG-1715-0117
LOR-05	CG-1715-0110
LOR-06	CG-1715-0123
LOR-07	CG-1715-0125
LOR-08	CG-1715-0114
M02RGU4	MC-2204-0081
M033036	MC-1729-0048
M0333L6	MC-1729-0009
M03DA26	MC-1729-0036
M0972M1	MC-2204-0089
M4F	MC-2204-0072
M4L	MC-2204-0104
M4T	MC-1406-0035
M4V	MC-1728-0010
M59	MC-1511-0002
M5C	MC-0804-0005
M6D	MC-0709-0002
M6E	MC-0801-0012
M6F	MC-0707-0002
M7B	MC-0804-0003
M92	MC-2204-0105
MBC	MC-1408-0019
MK-1	CG-1408-0035
MK-1(R)	CG-1408-0035
MK-22	CG-1701-0002
MK-24	CG-1712-0004
MK-27	CG-1712-0005
MK-4	CG-1710-0018
MK-5	CG-1710-0008
MK-6	CG-1704-0032
MKA	CG-1723-0005
MLE IC	CG-1728-0024
MLMDC 400	AF-1406-0075
MLMDC 501	AF-1408-0099
MS 400R	CG-1728-0030
MS 402R	CG-1728-0029
MS 420R	CG-1728-0040
MS 421R	CG-1728-0027
	CG-1728-0044
MS 422R	CG-1728-0041
MS 423R	CG-1728-0025
MS 425R	CG-1728-0026
MS 426R	CG-0802-0013
MS 452R	CG-1728-0039
MS 457R	CG-1406-0010
MS 472R	CG-1728-0017
MS 496R	CG-0802-0013
	CG-0802-0015
MS 722R	CG-1304-0020
MS 729R	CG-1304-0017
MS 732	CG-1722-0012
MS 733R	CG-0419-0003
MS 735R	CG 1722 0013
MSS	MC-1728-0009
NAV-01	CG-1715-0122
NAV-02	CG-1715-0119
NAV-03	CG-1715-0120
NAV-05	CG-1715-0049
NAV-06	CG-1715-0139
NAV-07	CG-1715-0140
NAV-14	CG-1715-0070
NBSC	CG-0802-0014
OCS	CG-2202-0005
OD1	MC-1402-0060
OD6	MC-2204-0094
OGM	MC-1715-0146

COURSE NUMBER INDEX E-9

P-00-3306	DD-1408-0006	RMB	MC-2204-0078	T3B	MC-0801-0014
P-V4A-B	AF-1606-0154	RMC	MC-2204-0077	T4B	MC-1406-0029
PMT 201	DD-1408-0011	RMD	MC-2202-0002	T4M	MC-2204-0103
PMT 302	DD-1408-0018	RMF	MC-2202-0003	T8A	MC-2204-0074
PMT 303	DD-1408-0019	RMG	MC-2204-0083	T8H	MC-2204-0076
PMT 305	DD-1408-0021	RMJ	MC-2204-0079	TA3	MC-1715-0128
PMT 341	DD-1408-0009	RMN	MC-2204-0084	TEL-10	CG-1715-0113
PQM 101	DD-1408-0023	ROCI	CG-2202-0004	TEL-13	CG-1715-0116
PQM 201	DD-1408-0024	S-00-3306	DD-1408-0006	TEL-14	CG-1715-0115
PQM 301	DD-1408-0010	S-501-0001	MC-0801-0002	TOA	MC-1715-0078
PRD 301	DD-1408-0010	SAF	MC-1405-0040		MC-1715-0176
R2D	MC-1402-0067	SAI	CG-1408-0041	TPS 2865	AF-1606-0150
R2E	MC-1402-0064	SAM 201	DD-1408-0013	TPS 2875	AF-1606-0153
R2T	MC-2204-0082	SAM 301	DD-1402-0004	TPS 28XX	AF-1606-0153
RAC-ADV	CG-1730-0001	SBC	CG-2205-0027	TST 202	DD-1408-0022
RCC	MC-1601-0047	SBE	CG-1714-0019	TT A	CG-1715-0109
RD	CG-1715-0130	SBE-(R)	CG-1714-0019	UAA	MC-1714-0014
RF-400 RWS	AF-1606-0152	SCA	MC-1405-0043	UAC	MC-1601-0044
RF9	MC-1406-0031	SYS 201	DD-1402-0011	WAB	MC-0804-0002
RGA	MC-2204-0070	SYS 301	DD-1408-0016	WAC	MC-0707-0001
RGU	MC-2204-0081	T1115005	AF-1715-0744	X502D8016 011	DD-1601-0030
RHA	MC-1408-0017	T2H	MC-1402-0034	XRG	MC-1406-0030
RMA	MC-2204-0080			YAMS-000	AF-2203-0051

F-1

REQUEST FOR COURSE RECOMMENDATION

The applicant for credit must fill out one form for *each* service school course completed. The institutional official is responsible for verifying from official military records that the student completed the entire course, and for submitting the form to The Center for Adult Learning and Educational Credentials, American Council on Education, One Dupont Circle, Washington, DC 20036-1193. ATTN: Military Evaluations. *Please Print.*

1. *Exact* course title *(do not abbreviate)* _____

_____

2. Service branch offering the course:
   - ☐ Air Force
   - ☐ Army
   - ☐ Coast Guard
   - ☐ Department of Defense
   - ☐ Marine Corps
   - ☐ Navy

3. Name of service school attended: _____

4. Location (installation, state): _____

5. Length of course *(in weeks):* _____

6. Dates of attendance: From:_____ To:_____
                                       day/month/year   day/month/year

7. Official military course number: _____

8. MOS/AFSC/Rating: _____

9. Course was designed for:
   - ☐ Warrant Officers
   - ☐ Officer Candidates
   - ☐ Commissioned Officers
   - ☐ Enlisted Personnel
   - ☐ Aviation Cadets
   - ☐ Noncommissioned Officers

10. Rank or rating upon completion of the course: _____

11. Please give some indication of subjects studied in course:

_____

_____

NAME OF STUDENT

STATUS (FRESHMAN, SOPHOMORE, ETC.)

**DO NOT WRITE IN THIS SPACE
STAFF USE**

SIGNATURE OF COLLEGE OFFICIAL

NAME OF COLLEGE OFFICIAL

TITLE

INSTITUTION

STREET

CITY   STATE   ZIP CODE

AREA CODE   NUMBER   EXT.

REQUEST FOR COAST GUARD RATING AND WARRANT OFFICER EXHIBITS

Officials should use this form only for requesting exhibits that contain the phrase, "Pending evaluation." As occupations are evaluated, they will be listed in *The Center Update*. When you want to obtain the recommendation for a newly-evaluated occupation, identify the exhibit you are requesting by using the complete *exhibit I.D.* (e.g., CGR-AM-001), and the title of the occupation. Include the applicant's name if you would like the name mentioned in the reply. Submit the form to The Center for Adult Learning and Educational Credentials, American Council on Education, One Dupont Circle, Washington, DC 20036-1193, ATTN: Military Evaluations.

Exhibit I.D. Number	Title of Occupation Please print; do not abbreviate.	Name of Applicant

DO NOT WRITE IN THIS SPACE — STAFF USE

SIGNATURE OF OFFICIAL

NAME OF OFFICIAL

TITLE

INSTITUTION OR ORGANIZATION

STREET

CITY   STATE   ZIP CODE

AREA CODE   NUMBER   EXT.

Please retain file copies of any occupation recommendations received from the Advisory Service.

REQUEST FOR MARINE CORPS ENLISTED MOS EXHIBITS

Officials should use this form only for requesting exhibits that are not in the *Handbook*. As occupations are evaluated, they will be listed in *The Center Update*. When you want to obtain the recommendation for a newly-evaluated occupation, identify the exhibit you are requesting by using the Marine Corps MOS designator and the title of the occupation. Include the applicant's name if you would like the name mentioned in the reply. Submit the form to The Center for Adult Learning and Educational Credentials, American Council on Education, One Dupont Circle, Washington, DC 20036-1193, ATTN: Military Evaluations.

Exhibit I.D. Number	MOS Title Please print; do not abbreviate.	Name of Applicant

DO NOT WRITE IN THIS SPACE — STAFF USE

SIGNATURE OF OFFICIAL

NAME OF OFFICIAL

TITLE

INSTITUTION OR ORGANIZATION

STREET

CITY   STATE   ZIP CODE

AREA CODE   NUMBER   EXT.

Please retain file copies of any occupation recommendations received from the Advisory Service.